Engineering Noise Control

This classic and authoritative textbook contains material that is not over-simplified and can be used to solve real-world noise control engineering problems. *Engineering Noise Control,* 6th edition covers theoretical concepts, and practical application of current noise control technology.

Topics extensively covered or revised from the 5th edition include: beating; addition and subtraction of noise levels; combining multi-path noise level reductions; hearing damage assessment and protection; speech intelligibility; noise weighting curves; instrumentation, including MEMS, IEPE and TEDS sensors; noise source types, including transportation noise and equipment noise estimations; outdoor sound propagation, including noise barriers, meteorological effects and sloping ground effects; sound in rooms, muffling devices, including 4-pole analysis, self noise and pressure drop calculations; sound transmission through single, double and triple partitions; vibration measurement and control, finite element analysis; boundary element methods; and statistical energy analysis.

- Discusses all aspects of industrial and environmental noise control
- An ideal textbook for advanced undergraduate and graduate courses in noise control
- An excellent reference text for acoustic consultants and engineers
- Practical applications are used to demonstrate theoretical concepts
- Includes material not available in other books

A wide range of example problems and solutions that are linked to noise control practice are available for download from **www.causalsystems.com**.

Colin H. Hansen is an Emeritus Professor in Mechanical Engineering at the University of Adelaide, Australia and past President and Honorary Fellow of the International Institute of Acoustics and Vibration and Honorary Fellow of the Australian Acoustical Society. He was awarded the 2009 Rayleigh Medal by the UK Institute of Acoustics, the 2013 Michell Medal by Engineers Australia and the 2014 Rossing Prize in Acoustics Education by the Acoustical Society of America.

Carl Q. Howard is a Professor in Mechanical Engineering at the University of Adelaide, Australia.
Kristy L. Hansen is a Senior Lecturer in Mechanical Engineering at Flinders University, Australia.
The late **David A. Bies** was formerly reader at the University of Adelaide, Australia.

Engineering Noise Control
Sixth Edition

David A. Bies
Colin H. Hansen
Carl Q. Howard
Kristy L. Hansen

CRC Press
Taylor & Francis Group
Boca Raton London New York

CRC Press is an imprint of the
Taylor & Francis Group, an **informa** business

Cover image: Shutterstock ©

Sixth edition published 2024
by CRC Press
6000 Broken Sound Parkway NW, Suite 300, Boca Raton, FL 33487-2742

and by CRC Press
4 Park Square, Milton Park, Abingdon, Oxon, OX14 4RN

CRC Press is an imprint of Taylor & Francis Group, LLC

Fifth edition published by CRC Press 2018

Library of Congress Cataloging-in-Publication Data

Names: Bies, David A., 1925- author. | Hansen, Colin H., 1951- author. |
Howard, Carl Q., 1970- author. | Hansen, Kristy L., author.
Title: Engineering noise control / David A. Bies, Colin H. Hansen, Carl Q.
Howard, Kristy L. Hansen.
Description: Sixth edition. | Boca Raton : CRC Press, 2024. | Includes
bibliographical references and index.
Identifiers: LCCN 2023001141 | ISBN 9780367414788 (paperback) | ISBN
9780367414795 (hardback) | ISBN 9780367814908 (ebook)
Subjects: LCSH: Noise control. | Soundproofing. | Machinery--Noise.
Classification: LCC TD892 .B54 2024 | DDC 620.2/3--dc23/eng/20230118
LC record available at https://lccn.loc.gov/2023001141

ISBN: 978-0-367-41479-5 (hbk)
ISBN: 978-0-367-41478-8 (pbk)
ISBN: 978-0-367-81490-8 (ebk)

DOI: 10.1201/9780367814908

Typeset in Latin Modern font
by KnowledgeWorks Global Ltd.

Publisher's note: This book has been prepared from camera-ready copy provided by the authors.

Dedication

This book is dedicated to our families.

Contents

Preface to the First Edition

This book grew out of a perceived need for a textbook on engineering noise control suitable for use in an undergraduate or postgraduate course in Engineering or Applied Physics and which, at the same time, would be useful as a practical yet rigorous and up-to-date reference for practising engineers, architects and acoustical consultants.

Preliminary versions of this volume have been used for the past eight years as a basis of an engineering noise control course for final-year Mechanical Engineering students at the University of Adelaide and also for short courses on industrial noise control for practising engineers and industrial design personnel.

The book begins with an introduction to terminology and basic concepts of acoustic wave description and propagation. Included is a discussion of acoustical flow resistance, a simple but fundamental property of porous materials which is basic to the understanding of a variety of acoustical phenomena. Use is made of previously unpublished algorithms and flow resistance information to calculate accurately properties of porous sound-absorbing materials.

A discussion of the ear, the subject of Chapter 2, is used as a basis for the understanding of such matters as acoustical masking and hearing damage due to excessive noise exposure. A practical guide to instrumentation follows in Chapter 3. Microphone calibration is considered in some detail and similarities between the microphone and the ear, in the perception of the aural field, are pointed out. Criteria which always form the basis for a noise control solution are considered in Chapter 4. Particular care has been given to the discussion of ambient level specification for spaces where a number of different methods of specification are commonly encountered in practice. A review of long-accepted data is used to show that hearing loss due to noise exposure is related to the integral of pressure, which is expressed in this book in terms of a hearing deterioration index (HDI), and not the integral of energy (pressure squared), as universally assumed and implemented in legislation throughout Europe and Australia.

In Chapter 5 a variety of commonly used idealised sources are briefly but comprehensively considered. This information provides the basis for source identification; it also provides the basis for understanding active (that is, noise-cancelling) control techniques. Also included is a comprehensive discussion of outdoor sound propagation. The concept of flow resistance as applied to ground reflection has only recently been recognised in the literature and is incorporated in the discussion.

The importance of the internal impedance of sound sources is emphasised and discussed in the first part of Chapter 6. This is followed by a consideration of the radiation field of sources, its significance, and criteria for identification of the field. In particular, source directivity as a far-field phenomenon is emphasised and carefully defined. The remainder of this chapter is concerned with the many methods of laboratory and field measurement of source sound power and the uses of such information.

In Chapter 7 the elements of sound fields in enclosed spaces are considered. Some effort is expended in establishing that rooms of any shape will exhibit modal response; use of splayed walls or odd-shaped rooms will not change the basic room behaviour. The point is also made that the traditional description of the sound field in terms of sound absorption, being a property of the wall treatment, is a gross simplification which makes tractable an otherwise very difficult problem, but at the expense of precision. Within the limitations of the analysis the optimisation of reverberation control is discussed.

In the discussion of barriers and enclosures of Chapter 8, new procedures are provided for estimating sound transmission loss of single and double-panel wall constructions. Similarly, new procedures are provided for investigating diffraction around barriers located either out-of-doors or indoors, where wall and ceiling reflections must be taken into account. While these discussions are based upon published literature, such sources are generally not understandable except to the specialist, so that the authors have made a special effort to put the information into simple and usable textbook form.

Muffling devices are considered in Chapter 9. Simple new procedures for estimating expected attenuation of lined ducts of both rectangular and circular cross-sections are presented. These procedures are based upon exact analyses and make use of the algorithms introduced in Chapter 1 and presented in an appendix for describing accurately the acoustical properties of porous materials in terms of flow resistance. The analysis of insertion loss for the commonly used side branch resonator, low pass filter and small engine exhaust muffler has been prepared especially for this book. Results are given in terms of insertion loss for various types of source and termination, thus providing an estimate of the expected difference in noise level obtained before and after insertion of the muffler. This analysis is much more useful than the commonly used transmission loss analysis, which gives no indication of insertion loss. Consideration is also given to both flow noise generation and pressure drop. Exhaust duct directivity information is new and is based upon previously unreported model studies.

Chapter 10 is concerned with vibration control from the point of view of noise suppression. Vibration isolation and the use of vibration absorbers are discussed in detail. The concepts of resonant and non-resonant vibration modes, and how they contribute to sound radiation, are discussed with reference to the appropriate use of vibration-damping material. Semi-empirical prediction schemes for estimating the sound power radiated by a number of commonly encountered industrial noise sources are presented in Chapter 11.

The book concludes with Chapter 12, in which several recently developed analytical techniques for the estimation of radiated sound power of complex structures are reviewed.

Problems suitable for students using the text are included in Appendix A. Appendix B lists important properties of a number of materials. The final three appendices are concerned with the measurement and prediction of the statistical absorption coefficients and normal acoustic impedances of constructions commonly used for sound absorption.

Where possible, references are given within the text to books, reports and technical papers, which may provide the reader with a more detailed treatment of their subject matter than is possible here. The reference list at the end of the book is intended as a first source for further reading and is by no means claimed to be comprehensive; thus, omission of a reference is not intended as a reflection on its value.

We have spent considerable effort in the elimination of errors in the text, but in the event that any more are found, we will be grateful for notification from our readers.

David A. Bies
Colin H. Hansen

Preface to the Fourth Edition

Although this fourth edition follows the same basic style and format as the first, second and third editions, the content has been considerably updated and expanded, yet again. This is partly in response to significant advances in the practice of acoustics and in the associated technology during the six years since the third edition and partly in response to improvements, corrections, suggestions and queries raised by various practitioners and students. The major additions are outlined below. However, there are many other minor additions and corrections that have been made to the text but which are not specifically identified here.

The emphasis of this edition is purely on passive means of noise control, and the chapter on active noise control that appeared in the second and third editions has been replaced with a chapter on practical numerical acoustics, where it is shown how free, open-source software can be used to solve some difficult acoustics problems, which are too complex for theoretical analysis. The removal of Chapter 12 on active noise control is partly due to lack of space and partly because a more comprehensive and a more useful treatment is available in the book, Understanding Active Noise Cancellation by Colin H. Hansen.

Chapter 1 includes updated material on the speed of sound in compliant ducts, and the entire section on speed of sound has been rewritten with a more unified treatment of solids, liquids and gases.

Chapter 2 has been updated to include some recent discoveries regarding the mechanism of hearing damage.

Chapter 3 has been considerably updated and expanded to include a discussion of expected measurement precision and errors using the various forms of instrumentation, as well as a discussion of more advanced instrumentation for noise source localisation using near-field acoustic holography and beamforming. The discussion on spectrum analysers and recording equipment has been completely rewritten to reflect more modern instrumentation.

In Chapter 4, the section on evaluation of environmental noise has been updated and rewritten.

Additions in Chapter 5 include a better definition of incoming solar radiation for enabling the excess attenuation due to meteorological influences to be determined. Many parts of Section 5.11 on outdoor sound propagation have been rewritten in an attempt to clarify some ambiguities in the third edition. The treatment of a vibrating sphere dipole source has also been considerably expanded.

In Chapter 7, the section on speech intelligibility in auditoria has been considerably expanded and includes some guidance on the design of sound reinforcement systems. In the low-frequency analysis of sound fields, cylindrical rooms are now included in addition to rectangular rooms. The section on the measurement of the room constant has been expanded and explained more clearly. In the section on auditoria, a discussion of the optimum reverberation time in classrooms has now been included.

In Chapter 8, the discussion on STC and weighted sound reduction index has been revised. The prediction scheme for estimating the transmission loss of single isotropic panels has been extended to low frequencies in the resonance and stiffness-controlled ranges, and the Davy method for estimating the Transmission Loss of double panel walls has been completely revised and corrected. The discussion now explains how to calculate the TL of multi-leaf and composite panels. Multi-leaf panels are described as those made up of different layers (or leaves) of the same material connected together in various ways whereas composite panels are described as those made up of two leaves of different materials bonded rigidly together. A procedure to calculate the transmission loss of very narrow slits such as found around doors with weather seals has also been added. A section on the calculation of flanking transmission has now been included with details provided for the calculation of flanking transmission via suspended ceilings. The section on calculating the Insertion Loss of barriers according to ISO9613-2 has been rewritten to more clearly reflect the intention of the standard. In addition, expressions are now provided for calculating the path lengths for sound diffracted around the ends of a barrier.

Chapter 9 has had a number of additions: Transmission Loss calculations (in addition to Insertion Loss calculations) for side branch resonators and expansion chambers; a much more detailed and accurate analysis of Helmholtz resonators, including better estimates for the effective length of the neck; an expanded discussion of higher-order mode propagation, with expressions for modal cut-on frequencies of circular section ducts; a number of new models for calculating the Transmission Loss of plenum chambers; and a more detailed treatment of directivity of exhaust stacks.

In Chapter 10, the effect of the mass of the spring on the resonance frequency of isolated systems has been included in addition to the inclusion of a discussion of the surge phenomenon in coil springs. The treatment of vibration absorbers has been revised and expanded to include a discussion of vibration neutralisers, and plots of performance of various configurations are provided. The treatment of two-stage vibration isolation has been expanded and non-dimensional plots provided to allow estimation of the effect of various parameters on the isolation performance.

Chapter 11 remains unchanged and Chapter 12 has been replaced with Chapter 13, where the previous content of Chapter 13 now serves as an introduction to a much-expanded chapter on practical numerical acoustics written by Dr Carl Howard. This chapter covers the analysis of complex acoustics problems using boundary element analysis, finite element analysis and MATLAB®. Emphasis is not on the theoretical aspects of these analyses but rather on the practical application of various software packages including a free open-source boundary element package.

Appendix A, which in the first edition contained example problems, has been replaced with a simple derivation of the wave equation. A comprehensive selection of example problems tailored especially for the book are now available on the internet for no charge at: http://www.causalsystems.com.

Appendix B has been updated and considerably expanded with many more materials and their properties covered.

In Appendix C, the discussion of flow resistance measurement using an impedance tube has been expanded and clarified. Expressions for the acoustic impedance of porous fibreglass and rockwool materials have been extended to include polyester fibrous materials and plastic foams. The impedance expressions towards the end of Appendix C now include a discussion of multi-layered materials.

Preface to the Fifth Edition

The fifth edition of the book has been thoroughly updated and reorganised. It also contains a considerable amount of new material. We have tried to keep the book as a suitable text for later undergraduate and graduate students, while at the same time extending content that will make the book more useful to acoustical consultants and noise control engineers.

Chapter 1 now has a section on the Doppler shift caused by moving sources and receivers as well as a section on amplitude modulation and amplitude variation to complement the section on beating.

Chapter 2 is now a combination of Chapters 2 and 4 of the earlier editions. It made sense to us to combine the description of the ear, hearing response and loudness with criteria. New measures for quantifying noise have been included as well as a section on low-frequency loudness and the response of the hearing mechanism to infrasound. The discussion on weighting networks has been extended to the G-weighting and Z-weighting networks, and the frequency range covered by weighting networks has been extended at the low end to 0.25 Hz.

Chapter 3 is now an updated version of what was in Chapter 3 in earlier editions. It has been updated to reflect current digital instrumentation and current methods of data acquisition and recording.

Chapter 4 has been rearranged so that it only includes the sound source descriptions that were in Chapter 5 of earlier editions, and it now includes sound power estimation schemes for various sources that were previously considered in a separate Chapter 6 of the previous editions.

The new Chapter 5 is entirely devoted to outdoor sound propagation. The chapter has been rearranged to first discuss principles underlying outdoor sound propagation calculations (including infrasound propagation) and second to discuss the various sound propagation models that are currently in use, with particular emphasis on the new model to be used by the European Union and the detailed Harmonoise model. The required input data for each model and the limitations of each model are also discussed. Finally, a section is dedicated to uncertainty analysis as we believe that it is very important that any noise level predictions are presented with an associated uncertainty.

Chapter 6 on room acoustics and sound absorption is similar in content to Chapter 7 in previous editions except that auditorium acoustics has been excluded, as there was insufficient space available to do justice to such a complex topic that has been the sole subject of a number of other excellent books.

Chapter 7 (Chapter 8 in previous editions) on sound transmission loss calculations and enclosure design has been updated to reflect recent advances in Davy's prediction model for double panel walls and a more accurate model on the inclusion of the flexibility of connecting studs. In addition, the European model (EN12354-1) for sound transmission loss calculations has been included as well as a section on the effects of various stud spacings on the sound transmission loss of double panel walls.

Chapter 8, which was Chapter 9 in previous editions, has been extended to include an expanded treatment of the calculation of the insertion loss for splitter silencers in both circular and rectangular section ducts, as well as an expanded treatment of pressure loss due to flow through lined ducts and splitter silencers. The self-noise produced by silencers is now discussed in more detail, and procedures are now provided for its estimation. Noise generated by flow through silencers is now discussed in more detail, including a discussion of pin noise in gas turbine exhausts. Calculation procedures are provided for the estimation of noise generated by flow through silencers and past pins holding on heat-insulating material. The treatment of exhaust stack directivity in previous editions did not mention the effects of exhaust stack temperature.

In this edition, a qualitative description is provided of how exhaust stack temperature may be taken into account. Finally a section, with a number of examples, has been added on 4-pole analysis, which is an excellent technique for the analysis of sound propagation through duct and muffler systems.

In Chapter 9, which was Chapter 10 in previous editions, the treatment of vibration absorbers has been expanded; otherwise, not much else has changed from the fourth edition.

The Chapter 10 (Chapter 11 in previous editions) sections on fan noise and transportation noise have been updated. In particular, the sections on the calculation of train noise and road traffic noise have been extended, and a number of models in current use have been described in detail.

Chapter 11, which was Chapter 12 in the fourth edition, has been updated, but the content has not changed much from the fourth edition.

Chapter 12 is a new chapter concerned entirely with frequency analysis. It is a very much expanded version of Appendix D in previous editions. This material, which was partly covered in an Appendix in previous editions, is an important part of understanding noise and vibration problems and, as such, it deserves a chapter of its own.

Appendix A is new and has been included to provide the background material on matrix algebra needed to fully appreciate the 4-pole material in Chapter 8 and the acoustic material property measurement procedures in Appendix D.

Appendices A and B of the fourth edition are now Appendices B and C in this edition. Appendix D (which was Appendix C in the fourth edition) has been updated with a description of the two-microphone method for measuring the normal incidence absorption coefficient and the normal impedance over a wide frequency range in a very short measurement time.

Appendix E is new and includes background material needed to appreciate the material on the Harmonoise propagation model in Chapter 5. This appendix includes explanations of how the Fresnel weighting coefficient for ground-reflected waves, as well as the effects of ground reflections and barriers on sound propagation, may be calculated.

Colin H. Hansen
Carl Q. Howard

May, 2017

Preface to the Sixth Edition

The sixth edition of the book has been reviewed and updated, including the addition of a substantial amount of new material and the correction of errata that plagued the 5th edition. However, the organisation of material remains the same as in the 5th edition. As the book was beginning to become too large to be manageable, we decided to move all the example problems and solutions to the web site, www.causalsystems.com (menu item, "Textbook"). We believe that, along with the many example problems available on our web sites, the book remains a suitable text for later undergraduate and graduate students, while at the same time being useful for acoustical consultants and noise control engineers. Any errata that we or our readers find will also be posted on causalsystems.com (menu item, "Textbook").

In Chapter 1, the section on sound intensity has been reviewed and updated, and subsections have been added to introduce complex number formulations for pressure and particle velocity variables and to explain the difference between logarithmic and arithmetic addition and subtraction of decibels and situations in which each method is used. In Chapter 2, the section on relative loudness and the sone has been revised using the 2017 ISO standards as a basis and peak, maximum and impulsive descriptors for sound pressure levels are now explained. Also in Chapter 2, the section on hearing damage risk assessment has been updated according to the most recent ISO standard and the approach adopted in the USA has now been included. In addition, a section on speech transmissibility index and speech intelligibility index and their calculation has been added, and acceptable ambient noise specifications for various occupied spaces have been updated according to the most recent standards. In the discussion of overall A-weighted noise criteria, there are now sections discussing the assessment of low-frequency noise, transportation noise, tonality, intermittency and impulsiveness. Also included is a section on the effects of noise on sleep. As uncertainty is gradually gaining more attention and coverage in international standards, a section on this topic applied to measured data has also been added.

Parts of Chapter 3 have been updated to include more recent instrumentation, including PC-based spectrum analysers, data logging instrumentation, personal sound exposure meters, digital microphones, IEPE/IPC sensors and data acquisition and recording instrumentation.

In Chapter 4, sound power measurement procedures for anechoic rooms, semi-free field conditions and near field conditions have been updated. A section has been added that describes techniques for the measurement of the sound power radiated by wind turbines, as this is an important input into the modelling and assessment of noise levels in surrounding communities. References to international standards describing the sound power measurement of a range of other special noise sources are also provided. Uncertainty in sound power measurement data is also discussed at length.

Chapter 5 on outdoor sound propagation has been substantially modified and updated. This includes the spherical wave reflection analysis, which has been updated according to the new ANSI standard as well as a revision of how atmospheric turbulence effects are included. The assessment of meteorological and ground effects using ray tracing contains a considerable amount of new material. The section on the ISO sound propagation model has been revised, and the section on the Harmonoise propagation model has been deleted as it is a model that is not used very often, and its complexity does not justify any perceived increase in prediction accuracy. The section on finite-length outdoor barriers has been thoroughly revised and extended, including calculations for sloping ground and more detailed calculations for sound diffracting around the ends of barriers. The section on propagation model prediction uncertainty has also been updated and clarified, particularly the part involving the conversion of uncertainties to dB levels.

Chapter 6 on room acoustics and sound absorption remains essentially unchanged. However, sections have been added on micro-perforated panels, thin film micro-perforated sheeting and meta-materials. Chapter 7 has been revised and updated with the most recent models for sound transmission loss calculations included. Sections on the measurement of sound transmission loss and impact isolation have been considerably expanded and clarified, including a discussion of measurement uncertainty and acceptable single-number criteria for various buildings. The calculation of the noise reduction of enclosures has been revised and expanded, including a discussion of how to take the directivity of enclosure wall radiation into account when predicting noise levels and enclosure noise reductions at various locations and in various directions outdoors.

Chapter 8 has expanded discussions on the use of resonator mufflers for broadband noise control as well as tonal control, the impedance of a perforated plate, the effect of duct end reflections on exhaust stack sound radiation and the effect of exhaust gas temperature and cross flow on exhaust stack directivity.

Chapters 9, 10 and 12 have been revised and updated, and in Chapter 11, the boundary element and finite element analyses have been revised, while the statistical energy analysis section has been considerably expanded. The appendices remain essentially unchanged from the 5th edition, apart from some added clarifications and corrections.

Finally, we apologise that we have removed example problems due to lack of space. The good news is that these problems and their solutions can be found on www.causalsystems.com <textbook menu>.

Colin H. Hansen
Carl Q. Howard
Kristy L. Hansen

December, 2022

Acknowledgements

We would like to thank all of those who took the time to offer constructive criticisms of the first, second, third, fourth and fifth editions, our graduate students and the many final-year mechanical engineering students at the University of Adelaide, Australia who have used the first, second, third, fourth and fifth editions as texts in their engineering acoustics course. In particular, we extend our thanks to Tom Bohdan, who found quite a few errata in the 5th edition and took the time to write to us.

We are also grateful for our families for their unwavering support in allowing us to complete this 6th edition, which addresses the deficiencies of the 5th edition as well as adding a considerable amount of new material.

1

Fundamentals and Basic Terminology

LEARNING OBJECTIVES

In this chapter, the reader is introduced to:

- fundamentals and basic terminology of noise control;
- noise-control strategies for new and existing facilities;
- the most effective noise-control solutions;
- the wave equation;
- plane and spherical waves;
- sound intensity;
- units of measurement;
- concepts of sound pressure, intensity and power level;
- frequency analysis and sound spectra;
- adding and subtracting sound levels;
- three kinds of impedance; and
- flow resistance.

1.1 Introduction

The recognition of noise as a source of annoyance began in antiquity, but the relationship, sometimes subtle, that may exist between noise and money seems to be a development of more recent times. For example, the manager of a large wind tunnel once told one of the authors that in the evening he liked to hear, from the back porch of his home, the steady hum of his machine approximately 2 km away, for to him, the hum meant money. However, to his neighbours it meant only annoyance, and he eventually had to do without his evening pleasure.

The conflicts of interest associated with noise that arise from the staging of rock concerts and motor races, or from the operation of airports, are well-known. In such cases, the relationship between noise and money is not at all subtle. Clearly, as noise may be the desired end or an inconsequential by-product of the desired end for one group, and the bane of another, a need for its control exists. Each group can have what it wants only to the extent that noise control is possible.

The recognition of noise as a serious health hazard is a development of modern times. The World Health Organisation (World Health Organisation, 2011) has stated that "There is sufficient evidence from large-scale epidemiological studies linking the population's exposure to

DOI: 10.1201/9780367814908-1

environmental noise with adverse health effects. Therefore, environmental noise should be considered not only as a cause of nuisance but also a concern for public health and environmental health."

With modern industry has come noise-induced deafness; amplified music also takes its toll. While amplified music may give pleasure to many, the excessive noise of much modern industry probably gives pleasure to very few or none at all. However, the relationship between noise and money still exists and cannot be ignored. If financially compensating people who become deaf is little more expensive than implementing industrial noise control, then the incentive definitely exists to do nothing, and hope that such a decision is not questioned.

A common method of noise control is a barrier or enclosure, and in some cases, this may be the only practical solution. However, experience has shown that noise control at the design stage is generally accomplished at about one-tenth of the cost of adding a barrier or an enclosure to an existing installation. At the design stage, the noise-producing mechanism may be selected for least noise and again experience suggests that the quieter process often results in a better machine overall. These unexpected advantages then provide the economic incentive for implementation, and noise control becomes an incidental benefit. Unfortunately, in most industries, engineers are seldom in the position of being able to make fundamental design changes to noisy equipment. They must often make do with what they are supplied and learn to apply effective "add-on" noise-control technology. Such "add-on" measures often prove cumbersome in use, and experience has shown that quite often, "add-on" controls are quietly sabotaged by employees who experience little immediate benefit and find them an impediment to their work.

In the following text, the chapters have been arranged to follow a natural progression, leading the reader from the basic fundamentals of acoustics through to advanced methods of noise control. However, each chapter has been written to stand alone, so that those with some training in noise control or acoustics can use the text as a ready reference. The emphasis is on sufficient precision of noise-control design to provide effectiveness at minimum cost and means of anticipating and avoiding possible noise problems in new facilities.

Simplification has been avoided so as not to obscure the basic physics of a problem and possibly mislead the reader. Where simplifications are necessary, their consequences are brought to the reader's attention. Discussion of complex problems has also not been avoided for the sake of simplicity of presentation. Where the discussion is complex, as with diffraction around buildings or with ground-plane reflection, results of calculations, which are sufficient for engineering estimates, are provided. In many cases, procedures also are provided to enable serious readers to carry out the calculations for themselves. For those who wish to avoid tedious calculations, there is a software package, ENC, available that follows this text very closely. See www.causalsystems.com.

In writing the equations that appear throughout the text, a consistent set of symbols is used: these symbols are mostly defined following their first use in each chapter. Where convenient, the equations are expressed in dimensionless form; otherwise, SI units are implied unless explicitly stated otherwise.

To apply noise-control technology successfully, it is necessary to have a basic understanding of the physical principles of acoustics and how these may be applied to the reduction of excessive noise. Chapter 1 has been written with the aim of providing these basic principles in sufficient detail to enable the reader to understand the applications in the rest of the book. Chapter 2 is concerned with the subjective response to sound and includes a description of the human hearing mechanism as well as measures used to quantify noise, occupational hearing damage risk and community noise annoyance. Chapter 3 describes instrumentation and techniques for quantifying noise. In summary, Chapters 1 to 3 have been written with the aim of providing the reader with the means to assess and quantify a noise problem.

Chapter 4 has been written with the aim of providing the reader with the basis for identifying noise sources and estimating sound pressure levels in the surrounding environment, as well as providing the means for rank ordering sources in terms of their emitted sound power. Chapter 5 is about quantifying noise propagation from a noise source to a receiver outdoors and includes a description of commonly used noise propagation models for predicting sound pressure levels at locations remote from the sources of noise, using the source sound power levels as a basis. The contents of Chapters 4 and 5 may be used in either a predictive mode for new proposed facilities or products or an analytical mode for the analysis of existing facilities or products to identify and rank order noise sources.

Chapter 6 describes the analysis of sound in enclosed spaces. Means are also provided for designing acoustic treatments and for determining their effectiveness. Chapter 7 includes methods for calculating the sound transmission loss of partitions and the design of enclosures and pipe lagging systems. Chapter 8 is concerned with the design and analysis of dissipative and reactive mufflers as well as plenum chambers and exhaust stacks.

Chapter 9 is about vibration isolation and control, and also gives attention to the problem of determining when vibration damping will be effective in the control of emitted noise and when it will be ineffective. Chapter 10 provides means for the prediction of noise radiated by many common noise sources and is largely empirical, but is generally guided by considerations such as those of Chapter 4. Chapter 11 is concerned with numerical acoustics and its application to the solution of complex sound radiation problems and interior noise problems.

Chapter 12 is focussed entirely on acoustical signal processing. With the development of more complex instrumentation and computer analysis tools, it is important that practitioners understand the fundamentals of analysis techniques so that they are familiar with the limitations associated with the analysis as well as the potential of these techniques to properly quantify and clarify a noise problem.

Appendix A provides background material on matrix algebra, which is useful for following parts of Chapter 8 and Appendix D. Appendix B contains a derivation of the linear wave equation, which is useful for appreciating its limitations of applicability. Appendix C contains properties of a range of acoustical materials that are relevant to acoustics calculations. Appendix D is devoted to porous materials and their characterisation for acoustic analyses.

1.2 Noise-Control Strategies

Possible strategies for noise control are always more numerous for new facilities and products than for existing facilities and products. Consequently, it is always more cost-effective to implement noise control at the design stage than to wait for complaints about a finished facility or product.

In existing facilities, controls may be required in response to specific complaints from within the workplace or from the surrounding community, and excessive noise levels may be quantified by suitable measurements. In proposed new facilities, possible complaints must be anticipated, and expected excessive noise levels must be estimated by some procedure. Often it is not possible to eliminate unwanted noise entirely and to do so is often prohibitively expensive; thus minimum acceptable levels of noise must be formulated, and these levels constitute the criteria for acceptability.

Criteria for acceptability are generally established with reference to appropriate regulations for the workplace and community. In addition, for community noise, it is advisable that at worst, any facility should not increase background (or ambient) sound pressure levels in a community by more than 5 dB(A) over levels that existed prior to construction of the facility, irrespective of what local regulations may allow. Note that this 5 dB(A) increase applies to broadband noise, and that clearly distinguishable tones (single frequencies) are less acceptable.

When dealing with community complaints (predicted or observed), it is wise to be conservative; that is, to aim for adequate control for the worst case, noting that community sound pressure levels may vary greatly (± 10 dB) about the mean as a result of atmospheric conditions (wind and temperature gradients and turbulence). It is noteworthy that complainants tend to be more conscious of noise after making a complaint and thus subconsciously tend to listen for it. Thus, even after considerable noise reduction may have been achieved and regulations satisfied, complaints may continue. Clearly, it is better to avoid complaints in the first place, which is another argument supporting the assertion of cost-effectiveness in the design stage.

For both existing and proposed new facilities and products, an important part of the process is to identify noise sources and rank order them in terms of contributions to excessive noise. When the requirements for noise control have been quantified and sources identified and ranked, it is possible to consider various options for control and finally to determine the cost-effectiveness of the various options. As was mentioned earlier, the cost of enclosing a noise source is generally much greater than modifying the source or process producing the noise. Thus, an argument based on cost-effectiveness is provided for extending the process of noise source identification to specific sources on a particular item of equipment and rank ordering these contributions within the limits of practicality.

Community sound pressure level predictions and calculations of the effects of noise control are generally carried out in octave frequency bands. Current models for prediction are not sufficiently accurate to allow finer frequency resolution and less fine frequency resolution does not allow a proper account of frequency-dependent effects. Generally, octave band analysis provides a satisfactory compromise between too much and too little detail. Where greater spectrum detail is required, 1/3-octave band analysis is often sufficient, although narrower band analysis (1 Hz bandwidth for example) is useful for identifying tones and associated noise sources.

If complaints arise from the workplace, then the issues should be addressed so that regulations are satisfied, and to minimise hearing damage compensation claims, the goal of any noise-control program should be to reach a workplace sound pressure level of no more than 85 dB(A). Criteria for other situations in the workplace are discussed in Chapter 2. Measurements and calculations are generally carried out in standardised octave or 1/3-octave bands, but particular care must be given to the identification of any tones that may be present, as these must be treated separately. More details on noise-control measures can be found in the remainder of this text and also in ISO 11690-2 (2020).

Any noise problem may be described in terms of a sound source, a transmission path and a receiver, and noise control may take the form of altering any one or all of these elements.

When faced with an industrial noise problem, reducing its hazard can be achieved in a number of ways, and these are listed below in order of effectiveness.

1. Eliminate the hazard, which means physically removing it (modification of the source).
2. Substitute the noisy process with a quieter one (modification of the source).
3. Reduce the hazard by good design (modification of the source).
4. Isolate personnel from the hazard via physical barriers or mufflers (modification of the transmission path).
5. Change the way people work by moving them out of noisy areas or by introducing quieter ways of doing things (modification of the receiver).
6. Provide earplugs and earmuffs (modification of the receiver).

When considered in terms of cost-effectiveness and acceptability, modification of the source is well ahead of either modification of the transmission path or the receiver. On the other hand, in existing facilities, the last two may be the only feasible options.

1.2.1 Sound Source Modification

Modification of the energy source to reduce the generated noise often provides the best means of noise control. For example, where impacts are involved, as in punch presses, any reduction of the peak impact force (even at the expense of the force acting over a longer time period) will dramatically reduce the noise generated. Generally, when a choice between various mechanical processes is possible to accomplish a given task, the best choice, from the point of view of minimum noise, will be the process that minimises the time rate of change of force, or jerk (time rate of change of acceleration). Alternatively, when the process is aerodynamic, a similar principle applies; that is, the process that minimises pressure gradients will produce minimum noise. In general, whether a process is mechanical or aerodynamic, minimum rate of change of force is associated with minimum noise.

Mechanical shock between solids should be minimised; for example, impact noise may be generated by parts falling into metal bins, and the height that the parts fall could be reduced by using an adjustable height collector (see Figure 1.1(a)) or the collector could be lined with conveyor belt material. Alternatively, the collector could have rubber flaps installed to break the fall of the parts (see Figure 1.1(b)).

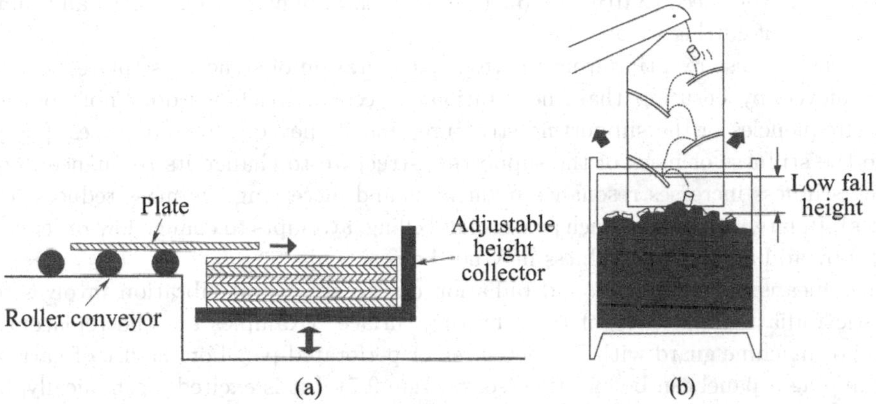

FIGURE 1.1 Impact noise reduction: (a) variable height collector; (b) interrupted fall.

The control of noise at its source may require maintenance, substitution of materials, substitution of equipment or parts of equipment, specification of quiet equipment, substitution of processes, substitution of mechanical power generation and transmission equipment, change of work methods, reduction of vibration of large structures such as plates, beams, etc. or reduction of noise resulting from fluid flow.

Maintenance includes balancing moving parts, replacement or adjustment of worn or loose parts, modifying parts to prevent rattles and ringing, lubrication of moving parts and use of properly shaped and sharpened cutting tools.

Substitution of materials includes replacing metal with plastic; an example being the replacement of steel sprockets in chain drives with sprockets made from flexible polyamide plastics.

Substitution of equipment includes use of electric tools rather than pneumatic tools (e.g. hand tools), use of stepped dies rather than single-operation dies, use of rotating shears rather than square shears, use of hydraulic rather than mechanical presses, use of presses rather than hammers and use of belt conveyors rather than roller conveyors.

Substitution of parts of equipment includes modification of gear teeth, by replacing spur gears with helical gears – generally resulting in 10 dB of noise reduction, replacement of straight-edged cutters with spiral cutters (for example, in woodworking machines a 10 dB(A) reduction may be achieved), replacement of gear drives with belt drives, replacement of metal gears with plastic gears (beware of additional maintenance problems) and replacement of steel or solid wheels with pneumatic tyres.

Substitution of processes includes using mechanical ejectors rather than pneumatic ejectors, hot rather than cold working, pressing rather than rolling or forging, welding or squeeze riveting rather than impact riveting, use of cutting fluid in machining processes, changing from impact action (e.g. hammering a metal bar) to progressive pressure action (e.g. bending a metal bar with pliers), replacement of circular saw blades with damped blades and replacement of mechanical limit stops with micro-switches.

Substitution of mechanical power generation and transmission equipment includes use of electric motors rather than internal combustion engines or gas turbines, or the use of belts or hydraulic power transmissions rather than gear boxes.

Change of work methods includes replacing ball machines with selective demolition in building demolition, replacing pneumatic tools by changing manufacturing methods, such as moulding holes in concrete rather than cutting after production of the concrete component, use of remote control of noisy equipment such as pneumatic tools, separating noisy workers, but keeping noisy operations in the same area, separating noisy operations from non-noisy processes. Changing work methods may also involve selecting the slowest machine speed appropriate for a job (selecting large, slow machines rather than smaller, faster ones), minimising the width of tools in contact with the workpiece (2 dB(A) reduction for each halving of tool width) and minimising protruding parts of cutting tools.

Reductions of noise resulting from the resonant vibration of structures (plates, beams, etc.) may be achieved by ensuring that the rotational speed of machines does not coincide with resonance frequencies of the supporting structure, and if they do, in some cases it is possible to change the stiffness or mass of the supporting structure to change its resonance frequencies (increasing stiffness increases resonance frequencies and increasing the mass reduces resonance frequencies). In large structures, such as a roof or ceiling, attempts to change low-order resonance frequencies by adding mass or stiffness may not be practical.

Another means for reducing sound radiation due to structural vibration involves reducing the acoustic radiation efficiency of the vibrating surface. Examples are the replacement of a solid panel or machine guard with a woven mesh or perforated panel or the use of narrower belt drives. Damping a panel can be effective (see Section 9.7) if it is excited mechanically, but note that if the panel is excited by an acoustic field, damping will have little or no effect on its sound radiation. Blocking vibration transmission along a noise-radiating structure by the placement of a heavy mass on the structure close to the original source of the noise can also be effective.

Reduction of noise resulting from fluid flow may involve providing machines with adequate cooling fins so that noisy fans are no longer needed, using centrifugal rather than propeller fans, locating fans in smooth, undisturbed air flow, using fan blades designed using computational fluid dynamics software to minimise turbulence, using large low-speed fans rather than smaller faster ones, minimising the velocity of fluid flow and maximising the cross-section of fluid streams. Fluid flow noise reduction may also involve reducing the pressure drop across any one component in a fluid flow system (for example, by using multi-stage control valves), minimising fluid turbulence where possible (e.g. avoiding obstructions in the flow), choosing quiet pumps in hydraulic systems, choosing quiet nozzles for compressed air systems (see Figure 10.4), isolating pipes carrying the fluid from support structures, using flexible connectors in pipe systems to control energy travelling in the fluid as well as the pipe wall and using flexible fabric sections in low-pressure air ducts (near the noise source such as a fan).

In hydraulic systems, the choice of pumps is important and in compressed air systems the choice of nozzles is important. Other alternatives include minimising the number of noisy machines running at any one time, relocating noisy equipment to less sensitive areas, or if noise is adversely affecting a community, avoiding running noisy machines at night.

Guidelines for the design of low-noise machinery and equipment are provided in ISO/TR 11688-1 (1995) and ISO/TR 11688-2 (1998).

1.2.2 Control of the Transmission Path

In considering control of the noise path from the source to the receiver, some or all of the following treatments need to be considered: barriers (walls), partial enclosures or full equipment enclosures, local enclosures for noisy components on a machine, reactive or dissipative mufflers (the former for low-frequency noise or small exhausts, the latter for high frequencies or large diameter exhaust outlets), lined ducts or lined plenum chambers for air-handling systems, vibration isolation of machines from noise-radiating structures, vibration absorbers and dampers, active noise control and the addition of sound-absorbing material to reverberant spaces to reduce reflected noise fields.

1.2.3 Modification of the Receiver

In some cases, when all else fails, it may be necessary to apply noise control to the receiver of the excessive noise. This type of control may involve use of earmuffs, earplugs or other forms of hearing protection; the enclosure of personnel, if this is practical; moving personnel further from the noise sources; switching the locations of personnel to reduce individual noise exposure times; and education and emphasis on public relations to address excessive noise generated by the facility in the community as well as on-site. Clearly, in the context of treatment of the noise receiver, the latter action is all that would be effective for noise affecting a community, although sometimes it may be less expensive to purchase the houses of residents affected by noise, even at prices well above market value.

1.2.4 Existing Facilities

In existing facilities or products, quantification of the noise problem requires identification of noise sources, determination of the transmission paths from the sources to the receivers, rank ordering of the various contributors to the problems and, finally, determination of acceptable solutions. To begin, sound pressure levels must be determined at potentially sensitive locations or at locations from which the complaints arise. For noise that affects communities, these measurements may not be straightforward; for example, noise propagation may be strongly affected by variable weather conditions and measurements over a representative time period may be required. This is usually done using remote data logging equipment in addition to periodic manual measurements to check the validity of the logged data. Guidelines for taking community noise measurements are provided in ASTM E1780-12 (2021) and ASTM E1503-22 (2022).

The next step is to apply acceptable noise level criteria to each location and thus determine the required noise reductions, generally as a function of octave or 1/3-octave frequency bands (see Section 1.14). Noise level criteria are usually set by regulations and appropriate standards.

Next, the transmission paths by which the noise reaches the place of issue are determined. For some cases, this step is often obvious. However, cases may occasionally arise when this step may present some difficulty, but it may be very important in helping to identify the source of the issue.

Having identified the possible transmission paths, the next step is to identify (understand) the noise generation mechanism or mechanisms, as noise control at the source always gives the best solution. Where the problem is one of occupational noise, this task is often straightforward. However, where noise problems are caused by consumer products or where industrial noise is adversely affecting communities, this task may prove difficult and/or expensive. However, altering the noise-generating mechanisms should always be considered as a means for possible control.

Often noise sources are either vibrating surfaces or unsteady fluid flow (air, gas or steam). The latter aerodynamic sources are often associated with exhausts. In most cases, it is worthwhile determining the source of the energy that is causing the structure or the aerodynamic source to radiate sound, as control may best start there. For a product, considerable ingenuity may be required to determine the nature and solution to the problem. One such example is a CPAP machine, which acts to pressurise the lungs of sleepers to counteract sleep apnoea. Such a machine contains a blower, but it must be so quiet that it does not disturb a sleeping person or prevent sleep initialisation. Clearly, quieter models of such a device will have a much greater market advantage.

For airborne noise sources, it is important to determine the sound power and directivity of each, to determine their relative contributions to the noise problem. The radiated sound power and directivity of sources can be determined by reference to the equipment manufacturer's data, reference to Chapter 10, or by measurement using methods outlined in Chapter 4. The sound power should be characterised in octave or 1/3-octave frequency bands (see Section 1.14), and dominant single frequencies should be identified. Any background noise contaminating the sound power measurements must be taken into account (see Section 1.11.5).

Having identified the noise sources and determined their radiated sound power levels, the next task is to determine the relative contribution of each noise source to the sound pressure level at each location where the measured sound pressure levels are considered to be excessive. For a facility involving just a few noise sources, this is a relatively straightforward task. However, for a facility involving tens or hundreds of noise sources, the task of rank ordering can be intimidating, especially when the receiver locations are in the surrounding community. In the latter case, the effect of the ground terrain and surface, absorption due to propagation through the air and the influence of atmospheric conditions must also be taken into account, as well as the decrease in sound pressure level with distance due to the "spreading out" of the sound waves.

Commercial computer software is available to assist with the calculation of the contributions of noise sources to sound pressure levels at sensitive locations in the community or in the workplace. Alternatively, one may write one's own software (see Chapter 4). In either case, for an existing facility, measured sound pressure levels can be compared with predicted levels to validate the calculations. Once the computer model is validated, it is then a simple matter to investigate various options for control and their cost-effectiveness.

In summary, a noise-control program for an existing facility includes:

- undertaking an assessment of the current environment where there appears to be a problem, including the preparation of both worst-case and average sound pressure level contours where required;
- establishment of the noise-control objectives or criteria to be met;
- identification of noise transmission paths and generation mechanisms;
- rank ordering noise sources contributing to any excessive levels;
- formulating a noise-control program and implementation schedule;
- carrying out the program; and
- verifying the achievement of the objectives of the program.

More detail on noise-control strategies for existing facilities can be found in ISO 11690-1 (2020).

1.2.5 Facilities in the Design Stage

In new facilities and products, quantification of potential excessive noise issues at the design stage may range from simple to difficult. At the design stage, the problems are the same as for existing facilities and products; they are, identification of the source or sources, determination of the transmission paths of the noise from the sources to the receivers, rank ordering of the various contributors to the problem and finally determination of acceptable solutions. Most importantly, at the design stage, the options for noise control are generally many and may include rejection of the proposed design.

Consideration of the possible need for noise control in the design stage has the very great advantage that an opportunity is provided to choose a process or processes that may avoid or greatly reduce the need for noise control. Experience suggests that processes chosen because they make less noise often have the additional advantage of being more efficient.

The first step for new facilities is to determine the noise criteria (see Section 2.12) for sensitive locations, which may typically include areas of the surrounding residential community that will be closest to the planned facility, locations along the boundary of the land owned by the industrial company responsible for the new facility, and within the facility at locations of operators of noisy machinery. Again, care must be taken to be conservative where surrounding communities are concerned so that noise problems are avoided.

To prevent the creation of noise issues affecting communities, the next step is to estimate expected sound pressure levels (in octave frequency bands) at the sensitive locations, based on machinery sound power level and directivity information (the latter may not always be available), and outdoor sound propagation prediction procedures as described in Chapter 5 and ISO 9613-2 (1996). Previous experience or the local weather bureau can provide expected ranges in atmospheric weather conditions (wind and temperature gradients and turbulence levels) so that a likely range and worst-case sound pressure levels can be predicted for each community location. When directivity information is not available, it is generally assumed that the source radiates uniformly in all directions.

If the estimated sound pressure levels at any sensitive location exceed the established criteria, then the equipment contributing most to the excess levels should be targeted for noise control, which could take the form of:

- specifying lower equipment sound power levels, or sound pressure levels at the operator position, to the equipment manufacturer;
- including noise-control fixtures (mufflers, barriers, enclosures, or factory walls with a higher sound transmission loss) in the factory design; or
- rearrangement and careful planning of buildings and equipment within them.

Sufficient noise control should be specified to leave no doubt that the noise criteria will be met at every sensitive location. Saving money at this stage is not cost-effective. If predicting equipment sound power levels with sufficient accuracy proves difficult, it may be helpful to make measurements on a similar existing facility or product.

More detail on noise-control strategies and noise prediction in workplaces for facilities at the design stage can be found in ISO 11690-1 (2020), ISO 11690-2 (2020) and ISO 11690-3 (1997).

1.2.6 Airborne versus Structure-Borne Noise

Very often in existing facilities, it is relatively straightforward to track down the original source(s) of the noise, but it can sometimes be difficult to determine how the noise propagates from its source to a receiver. A classic example of this type of problem is associated with noise onboard ships. When excessive noise (usually associated with the ship's engines) is experienced in a cabin close to the engine room (or in some cases, far from the engine room), or on the deck above the engine room, it is necessary to determine how the noise propagates from the engine. If the problem arises from airborne noise passing through the deck or bulkheads, then a solution may include one or more of the following: enclosing the engine, adding sound-absorbing material to the engine room, increasing the sound transmission loss of the deck or bulkhead by using double wall constructions or replacing the engine exhaust muffler. On the other hand, if the noise problem is caused by the engine exciting the hull into vibration through its mounts or through other rigid connections between the engine and the hull (for example, bolting the muffler to the engine and hull), then an entirely different approach would be required. In this latter case, it would be the mechanically excited deck, hull and bulkhead vibrations that would be responsible for the unwanted noise. The solution would be to vibration isolate the engine (perhaps through a well-constructed floating platform) or any items such as mufflers from the surrounding structure. In some cases, standard engine vibration isolation mounts designed especially for a marine environment can be used.

As both types of control are expensive, it is important to determine conclusively and in advance, the sound transmission path. The simplest way to do this is to measure the sound pressure levels in octave frequency bands at a number of locations in the engine room with the engine running and also at locations in the ship where the noise is excessive. Then the engine should be shut down and a loudspeaker placed in the engine room and driven so that it produces sound pressure levels in the engine room sufficiently high for them to be readily detected at the locations where noise reduction is required. Usually, an octave band filter is used with the speaker so that only noise in the octave band of interest at any one time is generated. This aids both in generating sufficient level and in detection.

The sound pressure level data measured throughout the ship with just the loudspeaker operating should be increased by the difference between the engine room levels with the engine as source and with the speaker as source, to give corrected levels for comparison with levels measured with the engine running. In many cases, it will be necessary for the loudspeaker system to produce noise of a similar level to that produced by the engine to ensure that measurements made elsewhere on the ship are above the background noise. In some cases, this may be difficult to achieve in practice with loudspeakers. The most suitable noise input to the speaker is a recording of the engine noise, but in some cases, a white noise generator may be acceptable. If the corrected sound pressure levels in the spaces of concern with the speaker excited are substantially less than those with the engine running, then it is clear that engine isolation is the first noise control that should be implemented. In this case, the best control that could be expected from engine isolation would be the difference in corrected sound pressure level with the speaker excited and sound pressure level with the engine running. If the corrected sound pressure levels in the spaces of concern with the speaker excited are similar to those measured with the engine running, then acoustic noise transmission is the likely path, although structure-borne noise may also be important, but at a slightly lower level. In this case, treatment to minimise airborne noise should be undertaken and following this treatment, the speaker test should be repeated to determine if the treatment has been effective and to determine if structure-borne noise has subsequently become the problem.

Another example of the importance of determining the noise transmission path is demonstrated in the solution to an intense tonal noise problem in the cockpit of a fighter aircraft, which was thought to be due to a pump, as the frequency of the tone corresponded to a multiple of the pump rotational speed. Much fruitless effort was expended to determine the sound transmission path until it was shown that the source was the broadband aerodynamic noise at the air-conditioning outlet into the cockpit, and the reason for the tonal quality was because the cockpit responded modally (see Chapter 7). The frequency of a strong cockpit resonance coincided with a multiple of the rotational speed of the pump but was unrelated. In this case, the obvious lack of any reasonable transmission path led to an alternative hypothesis and a solution.

1.3 Acoustical Standards and Software

Acoustical standards are available that describe standardised methods for undertaking various calculations in acoustics and noise control. International standards are published by the International Standards Organisation (ISO) and the International Electrotechnical Commission (IEC). Relevant American standards are published by the American National Standards Institute (ANSI) and the American Society of Mechanical Engineers (ASME). Relevant standards are referenced throughout this textbook, but readers should also search the relevant standards organisation database for a possible standard to describe any measurement or calculation that is to be undertaken. Where guidelines in American standards (ANSI and ASTM) differ from those in international standards (ISO and IEC), most countries (except USA and Canada) base local standards on international guidelines.

Software that is able to do the calculations outlined in this book and according to the various standards is also available for purchase. Examples of packages for outdoor sound propagation and prediction of sound pressure levels at various locations due to various sound sources include SoundPlan and CadnaA. An example of a package to perform room acoustics calculations is Odeon, and a package that can perform all of the calculations in this book (and more) is ENC (causalsystems.com). These packages are widely used by consultants in acoustics and noise control, but the packages that provide sound pressure level contours are usually quite expensive. However, the purpose of this book is to assist readers to gain an understanding of the fundamental principles that underpin these software packages, such that the input data required by the various programs will be better understood in terms of which data needs to be accurate and which can be estimated. In addition, the knowledge gained from this book and example problems available from www.causalsystems.com will be of assistance in understanding and interpreting the results obtained from the various software packages.

1.4 Acoustic Field Variables

1.4.1 Variables

Sound is the sensation produced at the ear by very small pressure fluctuations in the air. The fluctuations in the surrounding air constitute a sound field. These pressure fluctuations are usually caused by a solid vibrating surface but may be generated in other ways; for example, by the turbulent mixing of air masses in a jet exhaust. Saw teeth in high-speed motion (60 ms^{-1}) produce a very loud broadband noise of aerodynamic origin, which has nothing to do with vibration of the blade. As the disturbance that produces the sensation of sound may propagate from the source to the ear through any elastic medium, the concept of a sound field will be extended to include structure-borne as well as airborne vibrations. A sound field is described as a perturbation of steady-state variables, which describe a medium through which sound is transmitted.

For a fluid, expressions for the pressure, P_{tot}, velocity, U_{tot}, temperature, T_{tot}, and density, ρ_{tot}, may be written in terms of the steady-state (mean values), shown as P_s, $\boldsymbol{U}$, T and ρ and the variable (perturbation) values, p, $\boldsymbol{u}$, τ and σ, as follows:

Pressure :	$P_{\text{tot}} = P_s + p(\boldsymbol{r}, t)$	(Pa)
Velocity :	$\boldsymbol{U}_{\text{tot}} = \boldsymbol{U} + \boldsymbol{u}(\boldsymbol{r}, t)$	(m/s)
Temperature :	$T_{\text{tot}} = T + \tau(\boldsymbol{r}, t)$	(°C)
Density :	$\rho_{\text{tot}} = \rho + \sigma(\boldsymbol{r}, t)$	(kg/m^3)

where $\boldsymbol{r}$ is the position vector, t is time and the variables in bold font are vector quantities.

Pressure, temperature and density are familiar scalar quantities that do not require discussion. However, an explanation is required for the particle velocity $\boldsymbol{u}(\boldsymbol{r}, t)$ and the vector equation above that involves it. The notion of particle velocity is based on the assumption of a continuous rather than a molecular medium. The term, "particle", refers to a small part of the assumed continuous medium and not to the molecules of the medium. Thus, even though the actual motion associated with the passage of an acoustic disturbance through the conducting medium, such as air at high frequencies, may be of the order of the molecular motion, the particle velocity describes a macroscopic average motion superimposed upon the inherent Brownian motion of the medium. In the case of a convected medium moving with a mean velocity, $\boldsymbol{U}$, which itself may be a function of the position vector, $\boldsymbol{r}$, and time, t, the perturbating particle velocity, $\boldsymbol{u}(\boldsymbol{r}, t)$, associated with the passage of an acoustic disturbance may be thought of as adding to the mean velocity to give the total velocity.

Any variable could be chosen for the description of a sound field, but it is easiest to measure pressure in a fluid and strain, or more generally acceleration, in a solid. Consequently, these are the variables usually considered. These choices have the additional advantage of providing a scalar description of the sound field from which all other variables may be derived. For example, the particle velocity is important for the determination of sound intensity, but it is a vector quantity and requires three measurements as opposed to one for pressure. Nevertheless, instrumentation (Microflown) is available that allows the instantaneous measurement of particle velocity along all three Cartesian coordinate axes at the same time. In solids, it is generally easiest to measure acceleration, especially in thin panels, although strain might be preferred as the measured variable in some special cases. If non-contact measurement is necessary, then instrumentation known as laser vibrometers are available that can measure vibration velocity along all three Cartesian coordinate axes at the same time and also allow scanning of the surface being measured so that a complete picture of the surface vibration response can be obtained for any frequency of interest.

1.4.2 Acoustic Field

In Section 1.4.1, the concept of an acoustic field was introduced and extended to include structure-borne as well as airborne disturbances, with the implicit assumption that a disturbance initiated at a source will propagate with finite speed to a receiver.

It is of interest to consider the nature of an acoustic disturbance and the speed with which it propagates. To begin, it should be understood that small perturbations of the acoustic field may always be described as the sum of cyclic disturbances of appropriate frequencies, amplitudes and relative phases. In a fluid, a sound field will be manifested by variations in local pressure of generally very small amplitude with associated variations in density, displacement, particle velocity and temperature. Thus in a fluid, a small compression, followed by a compensating rarefaction, may propagate away from a source as a sound wave. The associated particle velocity lies parallel to the direction of propagation of the disturbance, the local particle displacement being first in the direction of propagation, then reversing to return the particle to its initial position after passage of the disturbance. This is a description of a compressional or longitudinal wave.

The viscosity of the fluids of interest in this text is sufficiently small for shear forces to play a very small part in the propagation of acoustic disturbances. A solid surface, vibrating in its plane without any normal component of motion, will produce shear waves in the adjacent fluid in which the local particle displacement is parallel to the exciting surface, but normal to the direction of propagation of the disturbance. However, such motion is always confined to a very narrow region near to the vibrating surface and does not result in energy transport away from the near field region. Alternatively, a compressional wave propagating in a fluid parallel to a solid bounding surface will give rise to a similar type of disturbance at the fixed boundary, but again the shear wave will be confined to a very thin viscous boundary layer in the fluid. Temperature variations associated with the passage of an acoustic disturbance through a gas next to a solid boundary, which is characterised by a very much greater thermal capacity, will likewise give rise to a thermal wave propagating into the boundary; but again, as with the shear wave, the thermal wave will be confined to a very thin thermal boundary layer in the fluid of the same order of size as the viscous boundary layer. Such viscous and thermal effects, generally referred to as the acoustic boundary layer, are usually negligible for energy transport and are generally neglected, except in the analysis of sound propagation in tubes and porous media, where they provide the energy dissipation mechanisms. Viscous and thermal effects are also important in thermoacoustics, where high-amplitude sound waves are intentionally used to transport heat to create an acoustic heat pump.

It has been mentioned that sound propagates in liquids and gases predominantly as longitudinal compressional waves; shear and thermal waves play no significant part. In solids, however, the situation is much more complicated, as shear stresses are readily supported. Not only are longitudinal waves possible, but so are transverse, shear and torsional waves. In addition, the types of waves that propagate in solids depend strongly on boundary conditions. For example, in thin plates bending waves, which are really a mixture of longitudinal and shear waves, predominate, with important consequences for acoustics and noise control. Bending waves are of importance in the consideration of sound radiation from extended surfaces and the transmission of sound from one space to another through an intervening partition.

1.4.3 Magnitudes

The minimum acoustic pressure audible to the young human ear judged to be in good health, and unsullied by too much exposure to excessively loud music, is approximately 20×10^{-6} Pa, or 2×10^{-10} atmospheres (since one atmosphere equals 101.3×10^{3} Pa). The minimum audible level occurs between 3000 and 4000 Hz and is a physical limit; lower sound pressure levels would be swamped by thermal noise due to molecular motion in air. For the normal human ear, pain is experienced at sound pressures of the order of 60 Pa or 6×10^{-4} atmospheres. Evidently, acoustic pressures ordinarily are quite small fluctuations about the mean atmospheric pressure.

1.4.4 Speed of Sound

Sound is conducted to the ear through the surrounding medium, which in general will be air and sometimes water, but sound may be conducted by any fluid or solid. In fluids, which readily support compression, sound is transmitted as longitudinal waves and the associated particle motion in the transmitting medium is parallel to the direction of wave propagation. However, as fluids support shear very weakly, waves dependent on shear are weakly transmitted, but usually they may be neglected. Consequently, longitudinal waves are often called sound waves. For example, the speed of sound waves travelling in plasma has provided information about the interior of the sun. In solids, which can support both compression and shear, energy may be transmitted by all types of waves, but only longitudinal wave propagation is referred to as "sound".

The concept of an "unbounded medium" will be introduced as a convenient and often-used idealisation. In practice, the term "unbounded medium" has the meaning that longitudinal wave propagation may be considered sufficiently remote from the influence of any boundaries that such influence may be neglected. The concept of an unbounded medium is generally referred to as "free field", and this alternative expression will also be used where appropriate in this text. The propagation speed of sound waves, called the phase speed, in any conducting medium (solid or fluid) is dependent on the stiffness, D, and the density, ρ, of the medium. The stiffness, D, however, may be complicated by the boundary conditions of the medium, and in some cases, it may also be frequency-dependent. These matters will be discussed in the following text. In this format, the phase speed, c, takes the following simple form:

$$c = \sqrt{D/\rho} \quad \text{(m/s)} \tag{1.1}$$

The effect of boundaries on the longitudinal wave speed will now be considered but with an important omission for the purpose of simplification. The discussion will not include boundaries between solids, which generally is a seismic wave propagation problem not ordinarily encountered in noise-control problems. Only propagation at boundaries between solids and fluids and between fluids will be considered, as they affect longitudinal wave propagation. At boundaries between solids and gases, the characteristic impedance mismatch (see Section 1.16) is generally so great that the effect of the gas on wave propagation in the solid may be neglected; in such cases, the gas may be considered to be a simple vacuum in terms of its effect on wave propagation in the solid. However, sound waves impinging on a solid can excite it into vibration, and this is discussed in Chapter 7 for the case of acoustic excitation of a panel.

In solids, the effect of boundaries is to relieve stresses in the medium at the unsupported boundary faces, as a result of expansion at the boundaries normal to the direction of compression. At locations in the solid that are not near any boundaries, such expansion is not possible. Thus, in a solid medium, the free field (parts of the medium not near any boundaries) is very stiff. On the other hand, for the case of boundaries being very close together, wave propagation may not take place at all, and in this case, the field within such space, known as evanescent, is commonly assumed to be uniform. It may be noted that the latter conclusion follows from an argument generally applied to an acoustic field in a fluid within rigid walls. Here the latter argument has been applied to an acoustic field within a rigid medium with unconstrained walls.

For longitudinal wave propagation in solids, the stiffness, D, depends on the ratio of the dimensions of the solid to the wavelength of a propagating longitudinal wave. Let the solid be characterised by three orthogonal dimensions $h_i, i = 1, 2, 3$, which determine its overall size. Let h be the greatest of the three dimensions of the solid, where E denotes Young's modulus and f denotes the frequency of a longitudinal wave propagating in the solid. Then the criterion proposed for determining D is that the ratio of the dimension, h, to the half wavelength of the propagating longitudinal wave in the solid is greater than or equal to one. For example, wave propagation may take place along dimension h when the half wavelength of the propagating wave is less than or just equal to the dimension, h. This observation suggests that the following inequality must be satisfied for wave propagation to take place:

$$2hf \geq \sqrt{D/\rho} \tag{1.2}$$

For the case that only one dimension, h, satisfies the inequality and two dimensions do not, then the solid must be treated as a wire or thin rod along dimension, h, on which waves may travel. In this case, the stiffness, D, is that of a rod, D_r, and takes the following form:

$$D_r = E \qquad \text{(Pa)} \tag{1.3}$$

The latter result constitutes the definition of Young's modulus of elasticity, E. In the case that two dimensions satisfy the inequality and one dimension does not, the solid must be treated as a plate over which waves may travel. In this case, where ν is Poisson's ratio (ν is approximately 0.3 for steel), the stiffness, $D = D_p$, takes the following form:

$$D_p = E/(1 - \nu^2) \qquad \text{(Pa)} \tag{1.4}$$

For a material for which Poisson's ratio is equal to 0.3, $D = 1.099E$. If all three dimensions, h_i, satisfy the criterion, then wave travel may take place in all directions in the solid. In this case, the stiffness, $D = D_s$, takes the following form:

$$D_s = \frac{E(1 - \nu)}{(1 + \nu)(1 - 2\nu)} \qquad \text{(Pa)} \tag{1.5}$$

For fluids, the stiffness, D_F, is the bulk modulus or the reciprocal of the more familiar compressibility, given by:

$$D_F = -V(\partial V/\partial P_s)^{-1} = \rho(\partial P_s/\partial \rho) \qquad \text{(Pa)} \tag{1.6}$$

where V is a unit volume and $\partial V/\partial P_s$ is the incremental change in volume associated with an incremental change in absolute pressure, P_s.

The effect of boundaries on the longitudinal wave speed in fluids will now be considered. For fluids (gases and liquids) in pipes at frequencies below the first higher order mode cut-on frequency (see Section 8.10.3.3), where only plane waves propagate, the close proximity of the wall of the pipe to the fluid within may have a very strong effect in decreasing the medium stiffness. The stiffness of a fluid in a pipe, tube, or more generally, a conduit will be written as D_C. The difference between D_F and D_C represents the effect of the pipe wall on the stiffness of the contained fluid. This effect will depend on the ratio of the mean pipe radius, R, to wall thickness, t, the ratio of the density, ρ_w, of the pipe wall to the density, ρ, of the fluid within it, Poisson's ratio, ν, of the pipe wall material, as well as the ratio of the fluid stiffness, D_F, to Young's modulus, E, of the pipe wall. The expression for the stiffness, D_C, of a fluid in a conduit is (Pavic, 2006):

$$D_C = \frac{D_F}{1 + \dfrac{D_F}{E}\left(\dfrac{2R}{t} + \dfrac{\rho_w}{\rho}\nu^2\right)} \tag{1.7}$$

The compliance of a pipe wall will tend to increase the effective compressibility of the fluid and thus decrease the speed of longitudinal wave propagation in pipes. Generally, the effect will be small for gases, but for water in plastic pipes, the effect may be large. In liquids, the effect may range from negligible in heavy-walled, small-diameter pipes to large in large-diameter conduits.

For fluids (gases and liquids), thermal conductivity and viscosity are two other mechanisms, besides chemical processes, by which fluids may interact with boundaries. Generally, thermal conductivity and viscosity in fluids are very small, and such acoustical effects as may arise from them are only of importance very close to boundaries and in consideration of damping mechanisms. Where a boundary is at the interface between fluids or between fluids and a solid, the effects may be large, but as such effects are often confined to a very thin layer at the boundary, they are commonly neglected.

Variations in pressure are associated with variations in temperature as well as density; thus, heat conduction during the passage of an acoustic wave is important. In gases, for acoustic waves at frequencies ranging from infrasonic up to well into the ultrasonic frequency range, the associated gradients are so small that pressure fluctuations may be considered to be essentially adiabatic; that is, no sensible heat transfer takes place between adjacent gas particles and, to a very good approximation, the process is reversible. However, at very high frequencies and in porous media at low frequencies, the compression process tends to be isothermal. In the latter cases, heat transfer tends to be complete and in phase with the compression.

For gases, use of Equation (1.1), the equation for adiabatic compression (which gives $D = \gamma P_s$), and the equation of state for gases gives the following for the adiabatic speed of sound:

$$c = \sqrt{\gamma P_s/\rho} = \sqrt{\gamma RT/M} \quad \text{(m/s)} \tag{1.8}$$

and the isothermal speed of sound is:

$$c = \sqrt{P_s/\rho} = \sqrt{RT/M} \quad \text{(m/s)} \tag{1.9}$$

where γ is the ratio of specific heats (1.40 for air), T is the temperature in Kelvin (K), R is the universal gas constant which has the value 8.314 J mol^{-1} K^{-1} and M is the molecular weight, which for air is 0.029 kg/mol. For sound propagating in free space, P_s is the atmospheric pressure, P_0. Equations (1.1) and (1.8) are derived in many standard texts: for example, Morse (1948); Pierce (1981); and Kinsler et al. (1999). The isothermal speed of sound in air is about 18% less than the adiabatic speed of sound. We use the adiabatic speed of sound for general problems involving sound propagation in air, but for problems where sound is propagating through a porous acoustic material, we use the isothermal speed of sound.

For gases, the speed of sound depends on the temperature of the gas through which the acoustic wave propagates. For sound propagating in air at audio frequencies, the process is adiabatic. In this case, for temperature, T, in degrees Celsius (and not greatly different from 20°C), the speed of sound may be calculated to three significant figures using the following approximation:

$$c = 331 + 0.6T \qquad \text{(m/s)} \tag{1.10}$$

For calculations in example problems associated with this text and available on www.causalsystems.com, unless otherwise stated, a temperature of 20°C for air is assumed, resulting in a speed of sound of 343 m/s and an air density of 1.206 kg/m^3 at sea level, thus giving $\rho c = 414$. Some representative speeds of sound are given in Appendix C.

1.4.5 Dispersion

The speed of sound wave propagation as given by Equation (1.1) is quite general and permits the possibility that the stiffness, D, may either be constant or a function of frequency. For the cases considered thus far, it has been assumed that the stiffness, D, is independent of the frequency of the sound wave, with the consequence that all associated wave components of whatever frequency will travel at the same speed and thus the wave will propagate without dispersion, meaning wave travel takes place without changing the wave shape.

On the other hand, there are many cases where the stiffness, D, is a function of frequency and in such cases, the associated wave speed will also be a function of frequency. A familiar example is that of an ocean wave, the speed of which is dependent on the ocean depth. As a wave advances into shallow water, its higher frequency components travel faster than the lower frequency components, as the speed of each component is proportional to the depth of water relative to its wavelength. The greater the depth of water relative to the component wavelength, the greater the component speed. In deep water, the relative difference in the ratio of water depth to wavelength between low- and high-frequency components is small. However, as the water becomes shallow near the shore, this difference becomes larger and eventually causes the wave to break. A dramatic example is that of a relatively small and fast-moving ocean swell, produced by an earthquake deep beneath the ocean far out to sea, which may lead to a tsunami on the beach.

In Chapter 7, bending waves that occur in panels, which are a combination of longitudinal and shear waves, are introduced, as they play an important role in sound transmission through and from panels. Bending wave speed is dependent on the frequency of the disturbance and thus bending waves are dispersive. A dispersive wave means that it will propagate at a *phase velocity* that depends on its wavelength. Particle motion associated with bending waves is normal to the direction of propagation, in contrast with longitudinal waves, for which it is in the same direction.

In liquids and gases, dispersive propagation is observed above the audio frequency range at ultrasonic frequencies where relaxation effects are encountered. Such effects make longitudinal wave propagation frequency-dependent and consequently dispersive. Although not strictly dispersive, the speed of propagation of longitudinal waves associated with higher order modes in ducts is an example of a case where the effective wave speed along the duct axis is frequency-dependent. However, this is because the number of reflections of the wave from the duct wall per unit axial length is frequency-dependent, rather than the speed of propagation of the wave through the medium in the duct.

When an acoustic disturbance is produced, some work must be done on the conducting medium to produce the disturbance. Furthermore, as the disturbance propagates, energy stored in the field is convected with the advancing disturbance. When the wave propagation is non-dispersive, the energy of the disturbance propagates with the speed of sound; that is, with the phase speed of the longitudinal compressional waves. On the other hand, when propagation is dispersive, the frequency components of the disturbance all propagate with different phase speeds; the energy of the disturbance, however, propagates with a speed that is referred to as the group speed. Thus in the case of dispersive propagation, one might imagine a disturbance that changes its shape as it advances, while at the same time maintaining a group identity, and travelling at a group speed different from that of any of its frequency components. The group speed is quantified later in Equation (1.34). For non-dispersive wave propagation, the group speed, c_g, is equal to the phase speed, c.

1.4.6 Acoustic Potential Function

The hydrodynamic equations, from which the equations governing acoustic phenomena derive, are generally complex and well beyond solution in closed form. Fortunately, acoustic phenomena are generally associated with very small perturbations. Thus, in such cases it is possible to greatly simplify the governing equations to obtain the relatively simple linear equations of acoustics. Phenomena, which may be described by relatively simple linear equations, are referred to as linear acoustics and the equations are referred to as linearised.

However, situations may arise in which the simplifications of linear acoustics are inappropriate; the associated phenomena are then referred to as nonlinear. For example, a sound wave incident on a perforated plate may incur large energy dissipation due to nonlinear effects under special circumstances. Convection of sound through or across a small hole, due either to a superimposed steady flow or to relatively large amplitudes associated with the sound field, may convert the cyclic flow of the sound field into local fluid streaming. Such nonlinear effects take energy from the sound field, thus reducing the sound pressure level to produce local streaming of the fluid medium, which produces no sound. Similar nonlinear effects also may be associated with acoustic energy dissipation at high sound pressure levels, in excess of 130 dB re 20 µPa.

In general, except for special cases, such as those mentioned, which may be dealt with separately, the losses associated with an acoustic field are quite small, and consequently, the acoustic field may be treated as conservative. This means that for the purposes of the wave equation, energy dissipation may be considered insignificant and thus neglected, although for long distance sound propagation and in reverberation rooms, the absorption of sound due to propagation through the air is significant and accounted for separately.

For a conservative acoustic field, it is possible to define a potential function, ϕ, which, as will be shown in Section 1.4.7, is a solution to the wave equation (Pierce, 1981) with two very important advantages. The potential function may be either real or complex and most importantly it provides a means for determining both the acoustic pressure and the particle velocity by simple differentiation. The acoustic potential function, ϕ, is defined so that its negative gradient provides the particle velocity, $\boldsymbol{u}$, as:

$$\boldsymbol{u} = -\nabla\phi \tag{1.11}$$

Alternatively, differentiation of the acoustic potential function with respect to time, t, provides the acoustic pressure, which for negligible convection velocity, $\boldsymbol{U}$, is given by:

$$p = \rho \partial\phi/\partial t \tag{1.12}$$

At high sound pressure levels, or in cases where the particle velocity is large (as in the case when intense sound induces streaming through a small hole or many small holes in parallel), Equation (1.12) takes the form (Morse and Ingard, 1968):

$$p = \rho\left[\partial\phi/\partial t - \frac{1}{2}(\partial\phi/\partial x)^2\right] \qquad \text{(Pa)} \tag{1.13}$$

where the coordinate, x, is along the centre line (axis) of a hole. In writing Equation (1.13) a third term on the right side of the equation given in the reference has been omitted as it is inversely proportional to the square of the phase speed and thus for the cases considered here, it is negligible. Alternatively, if a convection velocity, $\boldsymbol{U}$, is present and large and the particle velocity, $\boldsymbol{u}$, is small, Equation (1.12) takes the form:

$$p = \rho\left[\partial\phi/\partial t - \boldsymbol{U}\partial\phi/\partial x\right] \qquad \text{(Pa)} \tag{1.14}$$

Taking the gradient of Equation (1.12), interchanging the order of differentiation on the right-hand side of the equation and introducing Equation (1.11) gives Euler's famous equation of motion for a unit volume of fluid acted on by a pressure gradient:

$$\rho\frac{\partial \boldsymbol{u}}{\partial t} = -\nabla p \tag{1.15}$$

1.4.7 Wave Equation

In Section 1.4.6, it was postulated that an acoustic potential function, ϕ, may be defined which by simple differentiation provides solutions to the wave equation for the particle velocity, $\boldsymbol{u}$, and acoustic pressure, p. The acoustic potential function satisfies the well-known linearised wave equation as (Kinsler et al., 1999):

$$\nabla^2\phi = (1/c^2)\partial^2\phi/\partial t^2 \tag{1.16}$$

Equation (1.16) is the general three-dimensional form of the acoustic wave equation in which the Laplacian operator, ∇^2, is determined by the choice of curvilinear coordinates (Morse and Ingard (1968), p307–8). For the present purpose, it will be sufficient to restrict attention to rectangular, cylindrical and spherical coordinates. Equation (1.16) also applies if the acoustic pressure variable, p, is used to replace ϕ in Equation (1.16). However, the wave equation for the acoustic particle velocity is more complicated. Derivations of the wave equation in terms of acoustic particle velocity with and without the presence of a mean flow are given by Hansen (2018, p. 21). Other useful books containing derivations of the wave equation are Fahy (2001) and Fahy and Thompson (2015). A brief derivation of the wave equation is given in this text in Appendix B.

1.4.8 Complex Number Formulations

In the remainder of the book, sound waves will often be represented using complex numbers with real and imaginary parts, so that, for example, the time varying sound pressure, p, at a particular location of an acoustic wave of frequency, ω (radians/s), may be represented as, $p = (a + \mathrm{j}b)\mathrm{e}^{\mathrm{j}\omega t} = \hat{p}\mathrm{e}^{\mathrm{j}\omega t} = |\hat{p}|\mathrm{e}^{\mathrm{j}(\omega t+\beta)}$, where $\mathrm{j} = \sqrt{-1}$, $\hat{p}$ is the complex pressure amplitude of the sound wave and $|\hat{p}| = \sqrt{a^2 + b^2}$. The angle, β (radians), is included in $\hat{p}$ and is the phase of p relative to some reference which could be its particle velocity, the sound pressure at another location or the sound pressure of another sound wave of the same frequency. The purpose in using complex numbers to represent a relative phase is to enable representation of the phase and amplitude of a sound wave in a form that is easily amenable to mathematical manipulation. In the remainder of this book, for any sinusoidally-time-varying complex quantity, x, its complex amplitude will be denoted, $\hat{x}$, and its scalar amplitude will be denoted $|\hat{x}|$.

1.5 Plane, Cylindrical and Spherical Waves

In general, sound wave propagation is quite complicated and not amenable to simple analysis. However, sound wave propagation can often be described in terms of the propagation properties of plane, cylindrical and spherical waves. Plane and spherical waves, in turn, have the convenient property that they can be described in terms of one dimension. Thus, the investigation of plane and spherical waves, although by no means exhaustive, is useful as a means of greatly simplifying and rendering tractable what, in general, may be a very complicated problem.

1.5.1 Plane Wave Propagation

For the case of plane wave propagation, only one spatial dimension, x, the direction of propagation, is required to describe the acoustic field. An example of plane wave propagation is sound propagating along the centre line of a tube with rigid walls. In this case, Equation (1.16) reduces to:

$$\partial^2\phi/\partial x^2 = (1/c^2)\partial^2\phi/\partial t^2 \tag{1.17}$$

A solution of Equation (1.17), which may be verified by direct substitution, is:

$$\phi = \mathrm{F}(ct \pm x) \tag{1.18}$$

The function, F, in Equation (1.18) describes a distribution along the x-axis at any fixed time, t, as well as the variation with time at any fixed place, x, along the direction of propagation. If the argument, $(ct \pm x)$, is fixed and the positive sign is chosen, then with increasing time, t, x must decrease with speed, c. Alternatively, if the argument $(ct \pm x)$ is fixed and the negative sign is chosen, then with increasing time, t, x must increase with speed, c. Consequently, a wave travelling in the positive x-direction is represented by taking the negative sign and a wave travelling in the negative x-direction is represented by taking the positive sign in the argument of Equation (1.18). A very important relationship between acoustic pressure and particle velocity will now be determined. A prime sign, $'$, will indicate differentiation of a function by its argument; that is, $\mathrm{dF}(w)/\mathrm{d}w = \mathrm{F}'(w)$. Substitution of Equation (1.18) in Equation (1.11) gives Equation (1.19) and substitution in Equation (1.12) gives Equation (1.20) as:

$$u = \mp \mathrm{F}'(ct \pm x) \tag{1.19}$$

$$p = \rho c \mathrm{F}'(ct \pm x) \tag{1.20}$$

Division of Equation (1.20) by Equation (1.19) gives:

$$p/u = \pm \rho c \tag{1.21}$$

which is a very important result – the characteristic impedance, ρc, of a plane wave.

In Equation (1.21), the positive sign is taken for waves travelling in the positive x-direction, while the negative sign is taken for waves travelling in the negative x-direction. The characteristic impedance is one of three kinds of impedance used in acoustics. It provides a very useful relationship between acoustic pressure and particle velocity in a plane wave. It also has the property that a duct terminated in its characteristic impedance will respond as an infinite duct, as no wave will be reflected at its termination. Fourier analysis enables the representation of any function, $\mathrm{F}(ct \pm x)$, as a sum or integral of harmonic functions. Thus, it will be useful for consideration of the wave equation to investigate the special properties of harmonic solutions. Consideration will begin with the following harmonic solution for the acoustic potential function:

$$\phi = A\cos(k(ct \pm x) + \beta) \tag{1.22}$$

where k is a constant, which will be investigated, and β is an arbitrary constant representing an arbitrary relative phase. As β is arbitrary in Equation (1.22), for fixed time, t, β may be chosen so that:

$$kct + \beta = 0 \tag{1.23}$$

In this case, Equation (1.18) reduces to the following representation of the spatial distribution:

$$\phi = A\cos(kx) = A\cos(2\pi x/\lambda) \tag{1.24a,b}$$

From Equations (1.24a&b) it may be concluded that the unit of length, λ, defined as the wavelength of the propagating wave and the constant, k, defined as the wavenumber are related as:

$$2\pi/\lambda = k \tag{1.25}$$

An example of harmonic (single frequency) plane wave propagation in a tube is illustrated in Figure 1.2. The type of wave generated is longitudinal, as shown in Figure 1.2(a) and the corresponding pressure fluctuations as a function of time are shown in Figure 1.2(b).

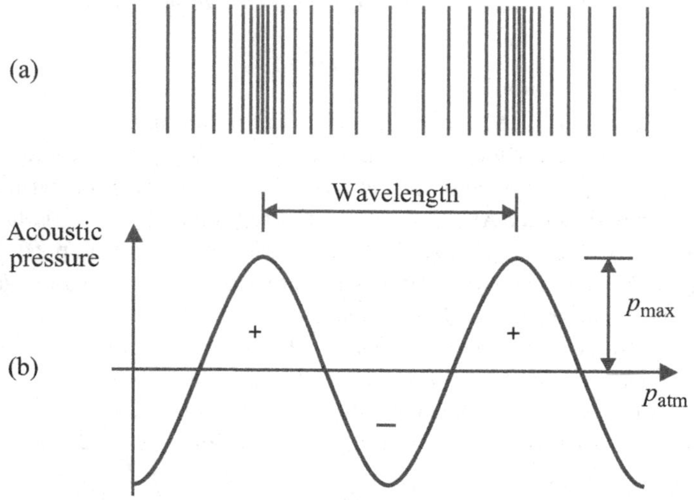

FIGURE 1.2 Representation of a sound wave: (a) compressions and rarefactions of a sound wave in space at a fixed instance in time; (b) graphical representation of the sound pressure variation.

The distribution in space has been considered and now the distribution in time for a fixed point in space will be considered. The arbitrary phase constant, β, of Equation (1.22) will be chosen so that, for fixed position, x:

$$\beta \pm kx = 0 \tag{1.26}$$

Equation (1.22) then reduces to the following representation for the temporal distribution:

$$\phi = A\cos(kct) = A\cos\frac{2\pi}{T_p}t \tag{1.27}$$

The period, T_p, of the propagating wave is given by:

$$2\pi/kc = T_p \qquad \text{(s)} \tag{1.28}$$

Its reciprocal is the more familiar frequency, f (Hz). Since the angular frequency, $\omega = 2\pi f$ (radians/s), is quite often used as well, the following relation should be noted:

$$2\pi/T_p = 2\pi f = \omega \qquad (\text{s}^{-1}) \tag{1.29}$$

and from Equations (1.28) and (1.29):

$$k = \omega/c = 2\pi f/c = 2\pi/\lambda \qquad (\text{m}^{-1}) \tag{1.30}$$

and from Equations (1.25), (1.29), and (1.30):

$$f\lambda = c \qquad \text{(m/s)} \tag{1.31}$$

The wavenumber, k, may be thought of as a spatial frequency, where k is the analogue of frequency, f, and wavelength, λ, is the analogue of the period, T_p.

The relationship between wavelength and frequency, for sound propagating in air, is illustrated in Figure 1.3.

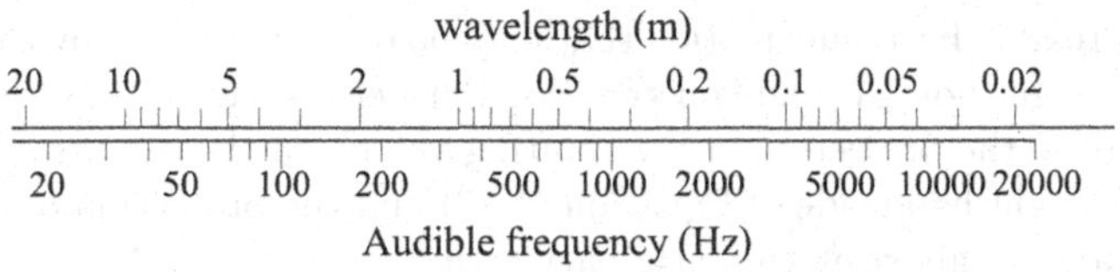

FIGURE 1.3 Wavelength in air versus frequency under normal conditions.

The wavelength of generally audible sound varies by a factor of about one thousand. The shortest audible wavelength is 17 mm (corresponding to 20000 Hz) and the longest is 17 m (corresponding to 20 Hz), although humans can detect sound via their vestibular system (which the ear is part of) at much lower frequencies if it is sufficiently loud. Letting $A = B/\rho\omega$ in Equation (1.22) and use of Equation (1.30) and either (1.11) or (1.12) gives the following useful expressions for the acoustic pressure and particle velocity respectively, for a plane wave:

$$p = B\sin(\omega t \mp kx + \beta) \qquad \text{(Pa)} \tag{1.32}$$

$$u = \pm\frac{B}{\rho c}\sin(\omega t \mp kx + \beta) \qquad \text{(m/s)} \tag{1.33}$$

where the upper sign represents wave propagation in the positive x-direction.

It may be mentioned in passing that the group speed, briefly introduced in Section 1.4.5, has the following form:

$$c_g = \mathrm{d}\omega/\mathrm{d}k \qquad \text{(m/s)} \tag{1.34}$$

By differentiating Equation (1.30) with respect to wavenumber, k, it may be concluded that for non-dispersive wave propagation, where the wave speed is independent of frequency, as for longitudinal compressional waves in unbounded media, the phase and group speeds are equal. Thus, in the case of longitudinal waves propagating in unbounded media, the rate of acoustic energy transport is the same as the speed of sound, as earlier stated.

Acoustic velocity potential and acoustic pressure harmonic solutions to the wave equation are complex and may be written as follows in terms of either exponential or trigonometric functions:

$$\phi = A\mathrm{e}^{\mathrm{j}(\omega t \mp kx + \beta)} = A\cos(\omega t \mp kx + \beta) + \mathrm{j}A\sin(\omega t \mp kx + \beta) \tag{1.35}$$

$$p = B\mathrm{e}^{\mathrm{j}(\omega t \mp kx + \beta)} = B\cos(\omega t \mp kx + \beta) + \mathrm{j}B\sin(\omega t \mp kx + \beta) \tag{1.36}$$

where $\mathrm{j} = \sqrt{-1}$. and $B = A/(\mathrm{j}\rho\omega) = A\mathrm{e}^{-\mathrm{j}\pi/2}/(\rho\omega)$, indicating that the pressure phase lags the velocity potential phase by 90°. In Equations (1.35) and (1.36), the negative sign represents a wave travelling in the positive x-direction, while the positive sign represents a wave travelling in the negative x-direction. The real parts of Equation (1.35) are just the solutions given by Equation (1.22). The imaginary parts of Equation (1.35) are also solutions, but in quadrature (90° out of phase) with the real solutions. It can also be seen that the 90° phase difference between the real and imaginary solutions represents a fixed time shift so the two solutions may be considered the same if the start time of one is adjusted by the time equivalent to a 90° phase shift.

By convention, the complex notation is defined so that what is measured with an instrument corresponds to the real part; the imaginary part is then inferred from the real part. The complex exponential form of the harmonic solution to the wave equation is used as a mathematical convenience, as it greatly simplifies mathematical manipulations, allows waves with different phases to be added together easily and allows graphical representation of the solution as a rotating vector in the complex plane. Setting $\beta = 0$ and $x = 0$, allows Equation (1.35) to be rewritten as:

$$Ae^{j\omega t} = A\exp(j\omega t) = A(\cos\omega t + j\sin\omega t) \tag{1.37}$$

Equation (1.37) represents harmonic motion that may be represented at any time, t, as a rotating vector of constant magnitude, A, and constant angular velocity, ω, as illustrated in Figure 1.4. Referring to the figure, the projection of the rotating vector on the abscissa, x-axis, is given by the real term on the right hand side of Equation (1.37) and the projection of the rotating vector on the ordinate, y-axis, is given by the imaginary term.

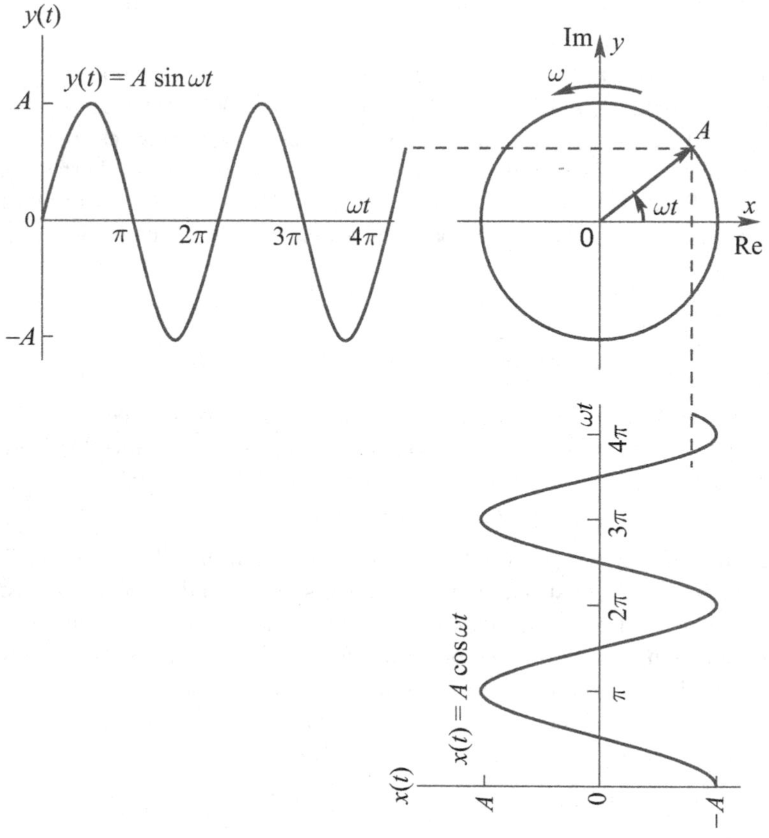

FIGURE 1.4 Harmonic motions represented as a rotating vector.

Use of the complex form of the solution makes integration and differentiation particularly simple. Also, impedances are conveniently described using this notation. For these reasons, the complex notation will be used throughout this book. However, care must be taken in the use of the complex notation when multiplying one function by another. In the calculation of products of quantities expressed in complex notation, it is important to remember that the product implies that in general, the real parts of the quantities are multiplied. This is important, for example, in the calculation of sound intensity associated with single frequency sound fields, expressed in complex notation.

1.5.2 Cylindrical Wave Propagation

A second important case is that of cylindrical wave propagation; an example is the propagation of sound waves from a pipeline at low frequencies such that the pipeline wall is vibrating in phase along the length of the pipeline (which is many wavelengths long) and the pipe radius is very small compared to a wavelength. In this case, sound radiates from the pipe in the form of cylindrical waves which satisfy Equation (1.38).

$$\frac{\partial^2(\phi)}{\partial r^2} + \frac{1}{r}\frac{\partial \phi}{\partial r} = \frac{1}{c^2}\frac{\partial^2(\phi)}{\partial t^2} \tag{1.38}$$

The solution to Equation (1.38), which is the wave equation in cylindrical coordinates is complicated, involving Bessel and Neumann functions, so will not be discussed further here. More detail is provided in Skudrzyk (1971).

An important property of cylindrical waves is that at distances greater than a wavelength from the source, the sound pressure level decays at a rate of 3 dB for each doubling of distance from the sound source whereas for spherical waves (discussed in the next section), the decay rate is 6 dB for each doubling of distance from the source. Note that plane waves (such as found in a duct at low frequencies) do not decay as the distance from the source increases.

1.5.3 Spherical Wave Propagation

A third important case is that of spherical wave propagation; an example is the propagation of sound waves from a small source in free space with no boundaries nearby. In this case, the wave Equation (1.16) may be written in spherical coordinates in terms of a radial term only, since no angular dependence is implied. Thus, Equation (1.16) becomes (Morse and Ingard (1968), p. 309):

$$\frac{1}{r^2}\frac{\partial}{\partial r}\left[r^2\frac{\partial \phi}{\partial r}\right] = \frac{1}{c^2}\frac{\partial^2 \phi}{\partial t^2} \tag{1.39}$$

The LHS can be rewritten as:

$$\frac{1}{r^2}\frac{\partial}{\partial r}\left[r^2\frac{\partial \phi}{\partial r}\right] = \frac{2}{r}\frac{\partial \phi}{\partial r} + \frac{\partial^2 \phi}{\partial r^2} = \frac{1}{r}\frac{\partial}{\partial r}\left[\phi + r\frac{\partial \phi}{\partial r}\right] = \frac{1}{r}\frac{\partial^2(r\phi)}{\partial r^2} \tag{1.40}$$

Thus, the wave equation may be rewritten as:

$$\frac{\partial^2(r\phi)}{\partial r^2} = \frac{1}{c^2}\frac{\partial^2(r\phi)}{\partial t^2} \tag{1.41}$$

The difference between, and similarity of, Equations (1.17) and (1.41) should be noted. Evidently, $r\phi = \mathrm{F}(ct \mp r)$ is a solution of Equation (1.41) where the source is located at the origin. Thus:

$$\phi = \frac{\mathrm{F}(ct \mp r)}{r} \tag{1.42}$$

The implications of the latter solution will now be investigated. To proceed, Equations (1.11) and (1.12) are used to write expressions for the acoustic pressure and particle velocity in terms of the potential function given by Equation (1.42). The expression for the acoustic pressure is:

$$p = \rho c\frac{\mathrm{F}'(ct \mp r)}{r} \quad \text{(Pa)} \tag{1.43}$$

and the expression for the acoustic particle velocity is:

$$u = \frac{\mathrm{F}(ct \mp r)}{r^2} \pm \frac{\mathrm{F}'(ct \mp r)}{r} \quad \text{(m/s)} \tag{1.44}$$

In Equations (1.42), (1.43) and (1.44) the upper sign describes a spherical wave that decreases in amplitude as it diverges outward from the origin, where the source is located. Alternatively, the lower sign describes a converging spherical wave, which increases in amplitude as it converges towards the origin. The characteristic impedance of the spherical wave may be computed, as was done earlier for the plane wave, by dividing Equation (1.43) by Equation (1.44) to obtain the expression:

$$\frac{p}{u} = \rho c \frac{r\mathrm{F}'(ct \mp r)}{\mathrm{F}(ct \mp r) \pm r\mathrm{F}'(ct \mp r)} \tag{1.45}$$

If the distance, r, from the origin is very large, the quantity, $r\mathrm{F}'$, will be sufficiently large compared to the quantity, F, for the latter to be neglected; in this case, for outward-going waves the characteristic impedance becomes ρc, while for inward-going waves it becomes $-\rho c$. In summary, at large enough distance from the origin of a spherical wave, the curvature of any part of the wave finally becomes negligible, and the characteristic impedance becomes that of a plane wave, as given by Equation (1.30). See the discussion following Equation (1.18) in Section 1.5.1 for a definition of the use of the prime, $'$. A moment's reflection, however, immediately raises the question: how large is a large distance? The answer concerns the curvature of the wavefront; a large distance must be where the curvature or radius of the wavefront as measured in wavelengths is large. For example, referring to Equation (1.25), a large distance, r, must be where:

$$kr \gg 1 \tag{1.46}$$

For harmonic waves, the solution given by Equation (1.42) can also be written as:

$$\phi = \frac{\mathrm{F}(k(ct \pm r))}{r} = \frac{\mathrm{F}(\omega t \pm kr)}{r} = \frac{A}{r}\mathrm{e}^{\mathrm{j}(\omega t \pm kr)} \tag{1.47a,b,c}$$

which is the same as the solution for acoustic pressure (but not particle velocity), except that the coefficient, A, would be different (but related to the coefficient, A, in Equation (1.47) by a factor of $\mathrm{j}\rho\omega$). This factor is obtained by substituting Equation (1.47) into Equation (1.12), resulting in the following expression for the acoustic pressure for outwardly travelling waves (corresponding to the negative sign in Equation (1.47)).

$$p = \frac{\mathrm{j}\omega A \rho}{r}\mathrm{e}^{\mathrm{j}(\omega t - kr)} = \frac{\mathrm{j}k\rho c A}{r}\mathrm{e}^{\mathrm{j}(\omega t - kr)} \tag{1.48a,b}$$

Substitution of Equation (1.47) into Equation (1.11) gives an expression for the acoustic particle velocity, as:

$$u = \frac{A}{r^2}\mathrm{e}^{\mathrm{j}(\omega t - kr)} + \frac{\mathrm{j}kA}{r}\mathrm{e}^{\mathrm{j}(\omega t - kr)} \tag{1.49}$$

Dividing Equation (1.48) by Equation (1.49) gives:

$$\frac{p}{u} = \rho c \frac{\mathrm{j}kr}{1 + \mathrm{j}kr} \tag{1.50}$$

which holds for a harmonic wave characterised by a wavenumber k, and also for a narrow band of noise characterised by a narrow range of wavenumbers around k. For inward-travelling waves, the sign of k is negative.

Consideration of Equation (1.50) now gives explicit meaning to large distance, as according to Equations (1.25) and (1.46), large distance means that the distance measured in wavelengths is large; for example, $r > 3\lambda$. Note that when Equation (1.46) is satisfied, Equation (1.50) reduces to the positive, outward-travelling form of Equation (1.21), which is a plane wave. For the case of a narrow band of noise, for example, an octave band, the wavelength is conveniently taken as the wavelength associated with the centre frequency of the band.

1.5.4 Wave Summation

It will be shown that any number of harmonic waves, of the same frequency travelling in one particular direction, combine to produce one wave travelling in the same direction. For example, a wave that is reflected back and forth between two terminations many times may be treated as a single wave travelling in each direction.

Assume that many waves, all of the same frequency, travel together in the same direction. The waves may each be represented by a rotating vector, as shown in Figure 1.5. The wave vectors in the figure will all rotate together with the passage of time and thus they will add vectorially as illustrated in the figure for the simple case of two waves separated in phase by β.

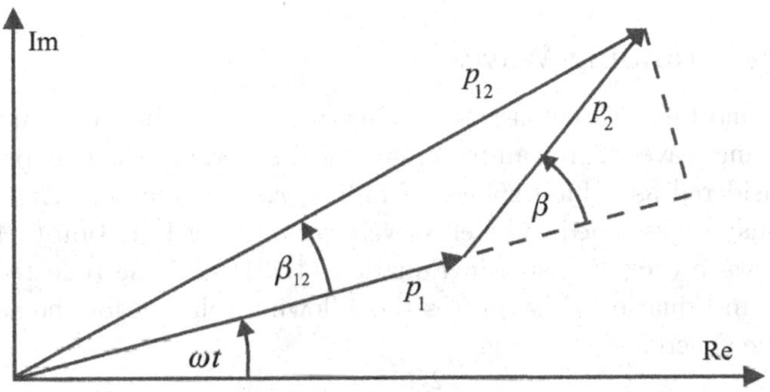

FIGURE 1.5 Graphical representation of the addition of two vectors in the complex plane, which represents the sum of two complex-valued harmonic pressure waves.

Consider any two waves travelling in one direction, which may be described as $p_1 = B_1 \mathrm{e}^{\mathrm{j}\omega t}$ and $p_2 = B_2 \mathrm{e}^{\mathrm{j}(\omega t + \beta)}$, where β is the phase difference between the two waves. The cosine rule gives for the magnitude, B_{12}, and the relative phase, β_{12}, of the combined wave:

$$B_{12}^2 = B_1^2 + B_2^2 + 2B_1 B_2 \cos\beta \tag{1.51}$$

$$\beta_{12} = \tan^{-1} \frac{B_2 \sin\beta}{B_1 + B_2 \cos\beta} \tag{1.52}$$

Equations (1.51) and (1.52) define the vector sum of the two complex-valued harmonic pressure waves, as:

$$p_{12} = B_{12} \mathrm{e}^{\mathrm{j}(\omega t + \beta_{12})} \tag{1.53}$$

The process is then repeated, each time adding to the cumulative sum of the previous waves, a new wave not already considered, until the sum of all waves travelling in the same direction has been obtained. It can be observed that the sum will always be like any of its constituent parts; thus it may be concluded that the cumulative sum may be represented by a single wave travelling in the same direction as all of its constituent parts.

1.5.5 Plane Standing Waves

If a loudspeaker emitting a tone is placed at one end of a closed tube, there will be multiple reflections of waves from each end of the tube. As has been shown, all of the reflections in the tube result in two waves, one propagating in each direction. These two waves will also combine, and form a "standing wave". This effect can be illustrated by writing the following expression for sound pressure at any location in the tube as a result of the two waves of amplitudes A and B, respectively, travelling in the two opposite directions, where $A \geq B$:

$$p = A\mathrm{e}^{\mathrm{j}(\omega t + kx)} + B\mathrm{e}^{\mathrm{j}(\omega t - kx + \beta)} \tag{1.54}$$

Equation (1.54) can be rewritten making use of the identity:

$$0 = -Be^{\mathrm{j}(kx+\beta)} + Be^{\mathrm{j}(kx+\beta)} \tag{1.55}$$

Thus:

$$p = (A - Be^{\mathrm{j}\beta})e^{\mathrm{j}(\omega t+kx)} + 2Be^{\mathrm{j}(\omega t+\beta)}\cos kx \tag{1.56}$$

Equation (1.56) consists of two terms. The first term on the right-hand side is a left travelling wave of amplitude $(A - Be^{\mathrm{j}\beta})$ and the second term on the right-hand side is a standing wave of amplitude $2Be^{\mathrm{j}\beta}$. In the latter case, the wave amplitude is described by a cosine, which varies with spatial location.

1.5.6 Spherical Standing Waves

Standing waves are most easily demonstrated using plane waves, but any harmonic wave motion may produce standing waves. An example of spherical standing waves is provided by the sun, which may be considered as a fluid sphere of radius, r, in a vacuum. At the outer edge, the acoustic pressure may be assumed to be effectively zero. Using Equation (1.48), the sum of the outward travelling wave (negative sign in Equation (1.48)) and the reflected inward travelling wave (positive sign in Equation (1.48)) gives the following relation for the acoustic pressure, p, at the surface of the sphere:

$$p = \mathrm{j}k\rho c\frac{2Be^{\mathrm{j}\omega t}}{r}\cos kr = 0 \tag{1.57}$$

where the identity, $e^{-\mathrm{j}kr} + e^{\mathrm{j}kr} = 2\cos kr$ has been used. Evidently, the simplest solution for Equation (1.57) is $kr = (2N-1)/2$ where N is an integer. If it is assumed that there are no losses experienced by the wave travelling through the media making up the sun, the first half of the equation is valid everywhere except at the centre, where $r = 0$ and the solution is singular. Inspection of Equation (1.57) shows that it describes a standing wave. Note that the largest difference between maximum and minimum pressures occurs in the standing wave when $p = 0$ at the boundary. However, standing waves (with smaller differences between the maximum and minimum pressures) will also be generated for conditions where the pressure at the outer boundary is not equal to 0.

1.6 Mean Square Quantities

In Section 1.4.1 the variables of acoustics were listed and discussed. For the case of fluids, they were shown to be small perturbations in steady-state quantities, such as pressure, density, velocity and temperature. Alternatively, in solids they are small perturbations in displacement, stress and strain variables. In all cases, acoustic fields are concerned with time-varying quantities with mean values of zero; thus, the variables of acoustics are commonly determined by measurement as mean square or as root-mean-square (RMS) quantities.

In some cases, however, we are concerned with the product of two time-varying quantities. For example, sound intensity will be discussed in Section 1.8, where it will be shown that the quantity of interest is the product of the two time-varying quantities, acoustic pressure and acoustic particle velocity averaged over time. The time-average of the product of two time-dependent variables, $\mathrm{F}(t)$ and $\mathrm{G}(t)$, will be referred to frequently in the following text and will be indicated by the following notation: $\langle \mathrm{F}(t)\mathrm{G}(t)\rangle$. Sometimes the time dependence indicated by (t) will be suppressed to simplify the notation. The time-average of the product of $\mathrm{F}(t)$ and $\mathrm{G}(t)$, averaged over time, T_A, is defined as:

$$\langle \mathrm{F}(t)\mathrm{G}(t)\rangle = \langle \mathrm{FG}\rangle = \lim_{T_A\to\infty}\frac{1}{T_A}\int_0^{T_A}\mathrm{F}(t)\mathrm{G}(t)\,\mathrm{d}t \tag{1.58}$$

When F(t) = G(t), the mean square of the variable is obtained. Thus, the mean square sound pressure, $\langle p^2(\boldsymbol{r},t)\rangle$, and mean square particle velocity, $\langle p^2(\boldsymbol{r},t)\rangle$, at position $\boldsymbol{r}$ are:

$$\langle p^2(\boldsymbol{r},t)\rangle = \lim_{T_A\to\infty}\frac{1}{T_A}\int_0^{T_A} p(\boldsymbol{r},t)p(\boldsymbol{r},t)\,\mathrm{d}t; \;\; \text{and} \;\; \langle u^2(\boldsymbol{r},t)\rangle = \lim_{T_A\to\infty}\frac{1}{T_A}\int_0^{T_A} u(\boldsymbol{r},t)u(\boldsymbol{r},t)\,\mathrm{d}t \quad (1.59)$$

The root-mean-square (RMS) sound pressure at location, $\boldsymbol{r}$, which will be shown later is used to evaluate the sound pressure level, is calculated as

$$p_{\mathrm{RMS}}(\boldsymbol{r}) = \sqrt{\langle p^2(\boldsymbol{r},t)\rangle} \quad (1.60)$$

The angled brackets, $\langle\;\rangle$, were used in the previous equation to indicate the time-average of the function within the brackets. They are sometimes used to indicate other types of averages of the function within the brackets; for example, the space-average of the function. Where there may be a possibility of confusion, the averaging variable is added as a subscript; for example, the mean square sound pressure averaged over space and time may also be written as $\langle p^2(\boldsymbol{r},t)\rangle_{S,t}$.

The hat symbol above a variable is used to represent the complex amplitude of a sinusoidally time-varying quantity, whose modulus is related to the RMS quantity by a factor of $\sqrt{2}$. For example, the acoustic pressure amplitude, $|\hat{p}| = \sqrt{2}\,p_{\mathrm{RMS}}$.

1.7 Energy Density

Any propagating sound wave has both potential and kinetic energy associated with it. The total energy (kinetic + potential) present in a unit volume of fluid is referred to as the energy density. Energy density is of interest because it is used as the quantity that is minimised in active noise cancellation systems for reducing noise in enclosed spaces. The kinetic energy per unit volume is given by the standard expression for the kinetic energy of a moving mass divided by the volume occupied by the mass. Thus:

$$\psi_k(t) = \frac{1}{2}\rho u^2(t) \quad (1.61)$$

The derivation of the potential energy per unit volume is a little more complex and may be found in Fahy (2001) or Fahy and Thompson (2015). The resulting expression is:

$$\psi_p(t) = \frac{p^2(t)}{2\rho c^2} \quad (1.62)$$

The total instantaneous energy density at time, t, is the sum of the instantaneous kinetic and potential energies and can be written as:

$$\psi_t(t) = \psi_k(t) + \psi_p(t) = \frac{\rho}{2}\left[u^2(t) + \frac{p^2(t)}{(\rho c)^2}\right] \quad (1.63)$$

Note that for a plane wave, the pressure and particle velocity are related by $u(t) = p(t)/\rho c$, and the total time-averaged energy density is then:

$$\langle\psi_t(t)\rangle = \frac{\langle p^2(t)\rangle}{\rho c^2} \quad (1.64)$$

where the brackets, $\langle\;\rangle$, in the equation indicate the time-average (mean square value).

1.8 Sound Intensity

Sound waves propagating through a fluid result in a transmission of energy. The time-averaged rate at which the energy is transmitted is the sound intensity. This is a vector quantity, as it is associated with the direction in which the energy is being transmitted. This property makes sound intensity particularly useful in many acoustical applications. The measurement of sound intensity is discussed in Section 3.10, and its use for the determination of sound power is discussed in Section 4.13. Other uses include identifying noise sources on items of equipment, measuring the sound transmission loss of building partitions, measuring impedance and sound-absorbing properties of materials and evaluating flanking sound transmission in buildings. Here, discussion is restricted to general principles and definitions, and the introduction of the concepts of instantaneous intensity and time-averaged intensity. The concept of time-averaged intensity applies to all kinds of noise and for simplicity, where the context allows, will be referred to in this text as simply the intensity.

For the special case of sounds characterised by a single frequency or a very narrow frequency band of noise, where either a unique or at least an approximate phase can be assigned to the particle velocity relative to the pressure fluctuations, the concept of instantaneous intensity allows extension and identification of an active component and a reactive component, which can be defined and given physical meaning. Reactive intensity is observed in the near field of sources (see Section 4.12), near reflecting surfaces and in standing wave fields. The time-average of the reactive component of instantaneous intensity is zero, as the reactive component is a measure of the instantaneous stored energy in the field, which does not propagate. However, this extension is not possible for other than the cases stated. For the case of freely propagating sound; for example, in the far field of a source (see Section 4.12), the acoustic pressure and particle velocity are always in phase and the reactive intensity is identically zero in all cases.

1.8.1 Definitions

Sound intensity is a vector quantity determined as the product of sound pressure and the component of particle velocity in the direction of the intensity vector. It is a measure of the rate at which work is done on a conducting medium by an advancing sound wave and thus the rate of power transmission through a surface normal to the intensity vector. As the process of sound propagation is cyclic, so is the power transmission; consequently an instantaneous and a time-average intensity may be defined. However, in either case, the intensity is the product of pressure and particle velocity. For the case of single frequency sound, represented in complex notation, this has the meaning that intensity is computed as the product of like quantities; for example, both pressure and particle velocity must be real quantities, defined by an amplitude and a phase. Thus, only the cosine term of the exponential function is used, where:

$$\mathrm{e}^{\mathrm{j}(\omega t+kr+\beta)} = \cos(\omega t + kr + \beta) + \mathrm{j}\sin(\omega t + kr + \beta) \tag{1.65}$$

The instantaneous sound intensity, $\boldsymbol{I}_i(\boldsymbol{r},t)$, in an acoustic field at a location given by the field vector, $\boldsymbol{r}$, is a vector quantity describing the instantaneous acoustic power transmission per unit area in the direction of the vector particle velocity, $\boldsymbol{u}(\boldsymbol{r},t)$. The general expression for the instantaneous sound intensity is:

$$\boldsymbol{I}_i(\boldsymbol{r},t) = p(\boldsymbol{r},t)\boldsymbol{u}(\boldsymbol{r},t) \qquad (\mathrm{W/m^2}) \tag{1.66}$$

A general expression for the sound intensity, $\boldsymbol{I}(\boldsymbol{r})$, is the time-average of the instantaneous intensity given by Equation (1.66). Referring to Equation (1.58), let $\mathrm{F}(t)$ be replaced with $p(\boldsymbol{r},t)$ and $\mathrm{G}(t)$ be replaced with $\boldsymbol{u}(\boldsymbol{r},t)$, then the time-averaged sound intensity may be written as:

$$\boldsymbol{I}(\boldsymbol{r}) = \langle p(\boldsymbol{r},t)\boldsymbol{u}(\boldsymbol{r},t)\rangle = \lim_{T_A\to\infty}\frac{1}{T_A}\int_0^{T_A} p(\boldsymbol{r},t)\boldsymbol{u}(\boldsymbol{r},t)\,\mathrm{d}t \qquad (\mathrm{W/m^2}) \tag{1.67}$$

For the special case of single frequency sound, the acoustic pressure of complex amplitude, $\hat{p}$, and the particle velocity of complex amplitude, $\hat{\boldsymbol{u}}$, at any location in three-dimensional space may be written in the following general form:

$$p(\boldsymbol{r},t) = \hat{p}(\boldsymbol{r})\mathrm{e}^{\mathrm{j}\omega t} = |\hat{p}(\boldsymbol{r})|\mathrm{e}^{\mathrm{j}(\omega t + \beta_p(\boldsymbol{r}))} \tag{1.68}$$

and:

$$\boldsymbol{u}(\boldsymbol{r},t) = \hat{\boldsymbol{u}}(\boldsymbol{r})\mathrm{e}^{\mathrm{j}\omega t} = |\hat{\boldsymbol{u}}(\boldsymbol{r})|\mathrm{e}^{\mathrm{j}(\omega t + \beta_u(\boldsymbol{r}))} \tag{1.69}$$

where the amplitudes, $|\hat{p}(\boldsymbol{r})|$ and $|\hat{\boldsymbol{u}}(\boldsymbol{r})|$, and the phases, $\beta_p(\boldsymbol{r})$ and $\beta_p(\boldsymbol{r})$, are real, space dependent quantities, and both $\hat{p}(\boldsymbol{r})$ and $\hat{\boldsymbol{u}}(\boldsymbol{r})$ are complex quantities with real and imaginary parts. The phase term, $\beta_p(\boldsymbol{r})$, includes the term, $-k_a\boldsymbol{r}$. The particle velocity, $\boldsymbol{u}$, is shown as bold as it is a vector in the direction of wave propagation.

The instantaneous intensity cannot be determined simply by multiplying Equations (1.68) and (1.69) together. This is because the complex notation formulation, $\mathrm{e}^{\mathrm{j}(\omega t+\beta_p(\boldsymbol{r}))}$, can only be used for linear quantities. Thus, the product of two quantities represented in complex notation is given by the product of their real components only (Skudrzyk, 1971).

Integration with respect to time of Equation (1.15), introducing the unit vector $\boldsymbol{n} = \boldsymbol{r}/r$, taking the gradient in the direction $\boldsymbol{n}$ and using Equation (1.68), gives:

$$\boldsymbol{u}(\boldsymbol{r},t) = \frac{\boldsymbol{n}\mathrm{j}}{\omega\rho}\frac{\partial p(\boldsymbol{r},t)}{\partial r} = \frac{\boldsymbol{n}}{\omega\rho}\left[-|\hat{p}(\boldsymbol{r})|\frac{\partial\beta_p(\boldsymbol{r})}{\partial r} + \mathrm{j}\frac{\partial|\hat{p}(\boldsymbol{r})|}{\partial r}\right]\mathrm{e}^{\mathrm{j}(\omega t+\beta_p(\boldsymbol{r}))} \quad \text{(m/s)} \tag{1.70}$$

Substitution of the real part of Equation (1.68) into Equation (1.66) gives the following result for the instantaneous sound intensity, $\boldsymbol{I}(\boldsymbol{r},t)$, in direction, $\boldsymbol{n}$, as:

$$\begin{aligned}\boldsymbol{I}_i(\boldsymbol{r},t) = -\frac{\boldsymbol{n}}{\omega\rho}\Bigg[&|\hat{p}(\boldsymbol{r})|^2\frac{\partial\beta_p(\boldsymbol{r})}{\partial r}\cos^2(\omega t+\beta_p(\boldsymbol{r})) \\ &+ |\hat{p}(\boldsymbol{r})|\frac{\partial|\hat{p}(\boldsymbol{r})|}{\partial r}\cos(\omega t+\beta_p(\boldsymbol{r}))\sin(\omega t+\beta_p(\boldsymbol{r}))\Bigg] \quad (\mathrm{W/m^2})\end{aligned} \tag{1.71}$$

The first term in brackets on the right-hand side of Equation (1.71) is the product of the real part of the acoustic pressure and the in-phase component of the real part of the particle velocity and is defined as the active intensity. The second term on the right-hand side of the equation is the product of the real part of the acoustic pressure and the in-quadrature component of the real part of the particle velocity and is defined as the reactive intensity. The reactive intensity is a measure of the energy stored in the field during each cycle but is not transmitted. Using well-known trigonometric identities (Abramowitz and Stegun, 1965), Equation (1.71) may be rewritten as:

$$\begin{aligned}\boldsymbol{I}_i(\boldsymbol{r},t) = -\frac{\boldsymbol{n}}{2\omega\rho}\Bigg\{&|\hat{p}(\boldsymbol{r})|^2\frac{\partial\beta_p(\boldsymbol{r})}{\partial r}\left[1+\cos2(\omega t+\beta_p(\boldsymbol{r}))\right] \\ &+|\hat{p}(\boldsymbol{r})|\frac{\partial|\hat{p}(\boldsymbol{r})|}{\partial r}\sin2(\omega t+\beta_p(\boldsymbol{r}))\Bigg\} \quad (\mathrm{W/m^2})\end{aligned} \tag{1.72}$$

Equation (1.72) shows that both the active and the reactive components of the instantaneous intensity vary sinusoidally but the active component has a constant part.

Taking the time-average of Equation (1.72) gives the expression for the active intensity as:

$$\boldsymbol{I}(\boldsymbol{r}) = -\frac{\boldsymbol{n}}{2\omega\rho}|\hat{p}(\boldsymbol{r})|^2\frac{\partial\beta_p(\boldsymbol{r})}{\partial r} \quad (\mathrm{W/m^2}) \tag{1.73}$$

Equation (1.73) is a measure of the acoustic power transmission in the direction of the intensity vector.

Alternatively, substitution of the real parts of Equations (1.68) and (1.69) into Equation (1.66) gives the instantaneous intensity in direction, $\boldsymbol{n}$ (unit vector), as:

$$\boldsymbol{I}_i(\boldsymbol{r},t) = \boldsymbol{n}|\hat{p}(\boldsymbol{r})||\hat{\boldsymbol{u}}(\boldsymbol{r})|\cos(\omega t + \beta_p(\boldsymbol{r}))\cos(\omega t + \beta_u(\boldsymbol{r})) \qquad (\mathrm{W/m^2}) \tag{1.74}$$

Using well-known trigonometric identities (Abramowitz and Stegun, 1965), Equation (1.74) may be rewritten as:

$$\begin{aligned}\boldsymbol{I}_i(\boldsymbol{r},t) = \frac{\boldsymbol{n}|\hat{p}(\boldsymbol{r})||\hat{\boldsymbol{u}}(\boldsymbol{r})|}{2}\Big\{&\big[1+\cos 2(\omega t + \beta_p(\boldsymbol{r}))\big]\cos(\beta_p(\boldsymbol{r}) - \beta_u(\boldsymbol{r}))\big] \\ &+ \sin 2(\omega t + \beta_p(\boldsymbol{r}))\sin(\beta_p(\boldsymbol{r}) - \beta_u(\boldsymbol{r}))\Big\} \ (\mathrm{W/m^2})\end{aligned} \tag{1.75}$$

Equation (1.75) is an alternative form of Equation (1.72). The first two terms in curly brackets on the right-hand side of Equation (1.75) (the 1 and the cosine term) is the active intensity, which has a mean value given by:

$$\boldsymbol{I}(\boldsymbol{r}) = \frac{\boldsymbol{n}|\hat{p}(\boldsymbol{r})||\hat{\boldsymbol{u}}(\boldsymbol{r})|}{2}\cos(\beta_p(\boldsymbol{r}) - \beta_u(\boldsymbol{r})) = \frac{\boldsymbol{n}}{2}\mathrm{Re}\{\hat{p}(\boldsymbol{r})\hat{\boldsymbol{u}}^*(\boldsymbol{r})\} \qquad (\mathrm{W/m^2}) \tag{1.76}$$

where the * indicates the complex conjugate. The amplitude of the active intensity is twice the mean value given by Equation (1.76).

Although the time-averaged reactive intensity is zero (as it is a quantity that oscillates back and forth in direction, $\boldsymbol{n}$), its amplitude, $\boldsymbol{I}_r(\boldsymbol{r})$, is given by the amplitude of the sine term in Equation (1.75) as:

$$\boldsymbol{I}_r(\boldsymbol{r}) = \frac{\boldsymbol{n}|\hat{p}(\boldsymbol{r})||\hat{\boldsymbol{u}}(\boldsymbol{r})|}{2}\sin(\beta_p(\boldsymbol{r}) - \beta_u(\boldsymbol{r})) = \frac{\boldsymbol{n}}{2}\mathrm{Im}\{\hat{p}(\boldsymbol{r})\hat{\boldsymbol{u}}^*(\boldsymbol{r})\} \qquad (\mathrm{W/m^2}) \tag{1.77}$$

1.8.2 Plane Wave and Far-Field Intensity

Waves radiating outward, away from any source, tend to become planar. Consequently, the equations derived in this section also apply in the far field of any source. For this purpose, the radius of curvature of an acoustic wave should be greater than about ten times the radiated wavelength. For a point source, this would imply a distance of 10 times the acoustic wavelength.

For a propagating plane wave, the characteristic impedance ρc is a real quantity and thus, according to Equation (1.21), the acoustic pressure and particle velocity are in phase and consequently, acoustic power is transmitted. The intensity is a vector quantity but, where direction is understood, the magnitude is of greater interest and will frequently find use throughout the rest of this book. Consequently, the intensity will be written in scalar form as a magnitude. If Equation (1.21) is used to replace $\boldsymbol{u}$ in Equation (1.76) the expression for the plane wave sound intensity at location, $\boldsymbol{r}$, becomes:

$$I(\boldsymbol{r}) = |\hat{p}(\boldsymbol{r})|^2/(2\rho c) = \langle p^2(\boldsymbol{r},t)\rangle/(\rho c) \qquad (\mathrm{W/m^2}) \tag{1.78}$$

In Equation (1.78) the intensity has been written in terms of the mean square pressure. If Equation (1.21) is used to replace p in the expression for intensity, the following alternative form of the expression for the harmonic plane wave sound intensity is obtained:

$$I(\boldsymbol{r}) = \rho c|\hat{u}(\boldsymbol{r})|^2/2 = \rho c\langle u^2(\boldsymbol{r},t)\rangle \qquad (\mathrm{W/m^2}) \tag{1.79}$$

where again the vector intensity has been written in scalar form as a magnitude in terms of the scalar particle velocity, u. The mean square particle velocity is defined in a similar way as the mean square sound pressure.

1.8.3 Spherical Wave Intensity

If Equations (1.43) and (1.44) are substituted into Equation (1.67) and use is made of Equation (4.2) (see Section 4.2.1) then Equation (1.78) is obtained, showing that the latter equation also holds for a spherical wave at any distance, r, from a harmonic point source or for a spherical wave at a sufficient distance from a finite-size source to be in the far field. Alternatively, similar reasoning shows that Equation (1.79) is only true of a spherical wave at large distances from a point source (see Section 1.5.3). To simplify the notation in the following discussion, the location dependence, $\boldsymbol{r}$, and time dependence, t, of the quantities p and u will be assumed, and specific reference to these dependencies will be omitted. It is convenient to rewrite Equation (1.50) in terms of its magnitude and phase. Carrying out the indicated algebra gives:

$$\frac{p}{u} = \rho c \, \mathrm{e}^{\mathrm{j}\beta} \cos\beta \qquad (\mathrm{W/m^2}) \tag{1.80}$$

where $\beta = (\beta_p - \beta_u)$ is the phase angle by which the acoustic pressure leads the particle velocity and is defined as:

$$\beta = \tan^{-1}[1/(kr)] \tag{1.81}$$

Solving Equation (1.80) for the particle velocity in terms of the pressure and replacing instantaneous values of pressure and particle velocity with scalar amplitudes at location $\boldsymbol{r}$ gives:

$$|\hat{u}(\boldsymbol{r})| = |\hat{p}(\boldsymbol{r})|/(\rho c \cos\beta) \tag{1.82}$$

Equation (1.75) gives the instantaneous intensity for the case considered here in terms of the scalar pressure amplitude, $|\hat{p}(\boldsymbol{r})|$, and scalar particle velocity amplitude, $|\hat{u}(\boldsymbol{r})|$, at location $\boldsymbol{r}$. Substitution of Equations (1.82) and (1.81) into Equation (1.75) gives the following expression for the scalar instantaneous intensity of a spherical wave:

$$I_{si}(\boldsymbol{r},t) = \frac{|\hat{p}(\boldsymbol{r})|^2}{2\rho c}\left\{\left[1 + \cos 2(\omega t + \beta_p(\boldsymbol{r}))\right] + \frac{1}{kr}\sin 2(\omega t + \beta_p(\boldsymbol{r}))\right\} \qquad (\mathrm{W/m^2}) \tag{1.83}$$

Consideration of Equation (1.83) shows that the time-average of the term in square brackets on the right-hand side is non-zero (active intensity) and is the same as that of a plane wave given by Equation (1.78), while the time-average of the sine term is zero and thus the sine term is associated with the non-propagating reactive intensity. The sine term (reactive intensity) tends to zero as the distance r from the source to observation point becomes large; that is, the sine term is negligible in the far field of the source. On the other hand, the reactive intensity becomes quite large close to the source, which is a near field effect. Integration over time of Equation (1.83), taking note that the integral of the sine term is zero, gives the same expression for the intensity of a spherical wave as was obtained previously for a plane wave (see Equation (1.78)).

1.9 Sound Power

As mentioned in Section 1.8, when sound propagates, transmission of acoustic power is implied. The sound intensity, as a measure of the energy passing through a unit area of the acoustic medium per unit time, was defined for plane and spherical waves and found to be the same. It will be assumed that the expression given by Equation (1.78) holds in general for sources that radiate more complicated acoustic waves, at least at sufficient distance from the source so that, in general, the power, W, measured in units of watts (W) radiated by any acoustic source is:

$$W = \int_S \boldsymbol{I} \cdot \boldsymbol{n} \, \mathrm{d}S \qquad (\mathrm{W}) \tag{1.84}$$

where $\boldsymbol{n}$ is the unit vector normal to the surface of area S.

For the cases of the plane wave and spherical wave, the mean square pressure, $\langle p^2 \rangle$, is a function of a single spatial variable in the direction of propagation. The meaning is now extended to include, for example, variations with angular direction, as is the case for sources that radiate more power in some directions than in others. A loudspeaker that radiates most power on-axis to the front would be such a source.

According to Equation (1.84), the sound power, W, radiated by a source is defined as the integral of the sound intensity over a surface surrounding the source. Most often, a convenient surface is an encompassing sphere or spherical section, but sometimes other surface shapes are chosen, as dictated by the circumstances of the particular case considered. For a sound source producing uniformly spherical waves (or radiating equally in all directions), a spherical surface is most convenient, and in this case Equation (1.84) leads to the following expression:

$$W = 4\pi r^2 I \quad \text{(W)} \tag{1.85}$$

where the magnitude of the sound intensity, I (W/m^2), is measured at a distance r from the source. In this case, the source has been treated as though it radiates uniformly in all directions. Consideration is given to sources that do not radiate uniformly in all directions in Section 4.8.

In practice, the accuracy of sound intensity measurement is affected by the magnitude of the reactive intensity, which is why it is best not to take sound intensity measurements too close to the sound source or in highly reverberant environments.

1.10 Decibels

Pressure is an engineering unit, which is measured relatively easily; however, the ear responds approximately logarithmically to energy input, which is proportional to the square of the sound pressure. The minimum sound pressure that the ear may detect is less than 20 µPa, while the greatest sound pressure before pain is experienced is 60 Pa. A linear scale based on the square of the sound pressure would require 10^{13} unit divisions to cover the range of human experience; however, the human brain is not organised to encompass such an enormous range in a linear way. The remarkable dynamic range of the ear suggests that some kind of compressed scale should be used. A scale suitable for expressing the square of the sound pressure in units best matched to subjective response is logarithmic rather than linear (see Sections 2.2.3 and 2.2.5).

The logarithmic scale provides a convenient way of comparing the sound pressure of one sound with another. To avoid a scale that is too compressed, a factor of 10 is introduced, giving rise to the decibel. The level of sound pressure, p, is then said to be L_p decibels (dB) greater than or less than a reference sound pressure, p_{ref}, according to the following equation:

$$L_p = 10\log_{10}\frac{\langle p^2 \rangle}{p_{\text{ref}}^2} = 10\log_{10}\langle p^2 \rangle - 10\log_{10} p_{\text{ref}}^2 \quad \text{(dB re 20 µPa)} \tag{1.86}$$

For the purpose of absolute level determination, the sound pressure is expressed in terms of a datum pressure corresponding approximately to the lowest sound pressure that the young normal ear can detect. The result is called the sound pressure level, L_p (or SPL), which has the units of decibels (dB) and should be written as "dB re 20 µPa" when referring to measurements conducted in air. When it can be assumed that a discussion concerns sound pressure level measurements in air, the reference value "re 20 µPa" is dropped and the sound pressure level is simply written with units of "dB". This is the quantity that is measured with a sound level meter.

The sound pressure is a measured root-mean-square (RMS) value and the reference pressure $p_{\text{ref}} = 2 \times 10^{-5}$ N/m^2 or 20 µPa. When this value for the reference pressure is substituted into Equation (1.86), the following convenient alternative form is obtained:

$$L_p = 10\log_{10}\langle p^2 \rangle + 94 \quad \text{(dB re 20 µPa)}eq \tag{1.87}$$

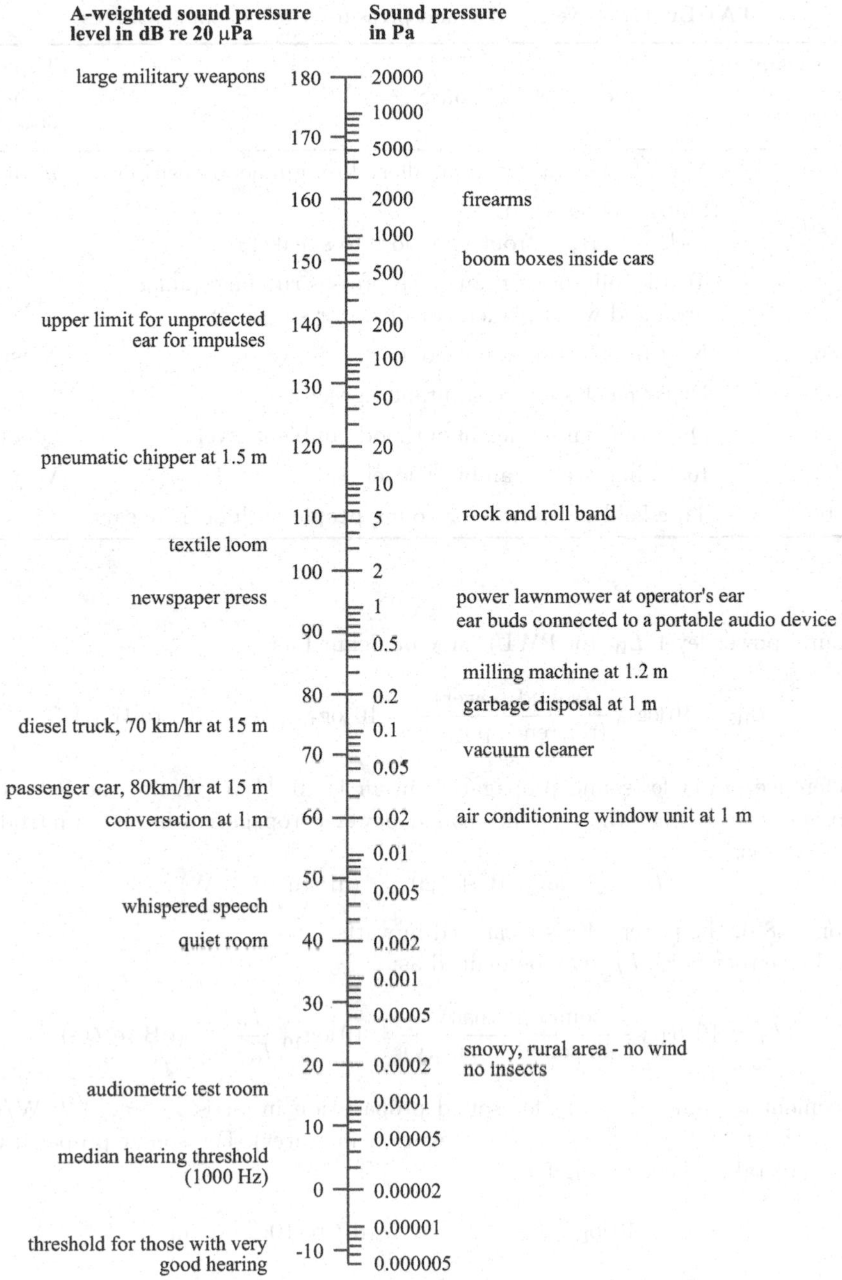

FIGURE 1.6 Sound pressure levels of some sources.

In Equation (1.87), the acoustic pressure, p, is measured in Pascals. Some feeling for the relation between subjective loudness and sound pressure level may be gained by reference to Figure 1.6 and Table 1.1, which show A-weighted sound pressure levels produced by a range of noise sources. A-weighting is discussed in more detail in Section 2.3.

For wave propagation in water (and any other liquids), Equation (1.86) also applies, but the reference pressure used is 1 μPa, so the "94" in Equation (1.87) becomes "120".

TABLE 1.1 A-weighted sound pressure levels of some sources

Sound pressure level (dB re 20 µPa)	Description of sound source	Typical subjective description
140	Moon launch at 100 m; artillery fire, gunner's position	Intolerable
120	Ship's engine room; rock concert, in front and close to speakers	
100	Textile mill; press room with presses running; punch press and wood planers, at operator's position	Very noisy
80	Next to busy highway, shouting	Noisy
60	Department store, restaurant, speech levels	
40	Quiet residential neighbourhood, ambient level	Quiet
20	Recording studio, ambient level	Very quiet
0	Threshold of hearing for young people with healthy ears	

The sound power level, L_W (or PWL), may be defined as:

$$L_W = 10\log_{10}\frac{\text{(sound power)}}{\text{(reference power)}} = 10\log_{10}\frac{W}{W_{\text{ref}}} \quad (\text{dB re } W_{\text{ref}}) \tag{1.88}$$

The reference power for sound propagation in air is 10^{-12} W. Again, the following convenient form is obtained when the reference sound power (propagation in air) is introduced into Equation (1.88) as:

$$L_W = 10\log_{10} W + 120 \qquad (\text{dB re } 10^{-12}\,\text{W}) \tag{1.89}$$

In Equation (1.89), the power, W, is measured in watts.

A sound intensity level, L_I, may be defined as:

$$L_I = 10\log_{10}\frac{\text{(sound intensity)}}{\text{(ref. sound intensity)}} = 10\log_{10}\frac{I}{I_{\text{ref}}} \quad (\text{dB re } I_{\text{ref}}) \tag{1.90}$$

A convenient reference intensity for sound propagation in air is $I_{\text{ref}} = 10^{-12}\,\text{W/m}^2$ (compared to 6.7×10^{-19} W/m^2 for sound propagation in water). For sound propagation in air, Equation (1.90) takes the following form:

$$L_I = 10\log_{10} I + 120 \qquad (\text{dB re } 10^{-12}\,\text{W/m}^2) \tag{1.91}$$

The introduction of Equation (1.78) into Equation (1.91) and use of Equation (1.87) gives the following useful result for plane waves:

$$L_I = L_p - 10\log_{10}(\rho c/400) = L_p + 26 - 10\log_{10}(\rho c) \approx L_p - 0.15 \qquad (\text{dB re } 10^{-12}\,\text{W/m}^2) \tag{1.92}$$

The constant "0.15" in Equation (1.92) comes from the evaluation of the term $10\log_{10}(\rho c/400)$. The properties of air at sea level and 20°C (see Appendix C) are: $\rho = 1.206$ kg/m^3 and $c = 343$ m/s; thus, the characteristic impedance is $\rho c = 414$ kg m^{-2} s^{-1} and $10\log_{10}(414/400) = 0.15$.

1.11 Combining Sound Pressures

1.11.1 Coherent and Incoherent Sounds

Thus far, the sounds that have been considered have been harmonic, being characterised by single frequencies. Sounds of the same frequency bear fixed phase relationships with each other and as observed in Section 1.5.4, their summation is strongly dependent on their phase relationship. Such sounds are known as coherent sounds. Coherent sounds are quite rare, although sound radiated from different parts of a large tonal source such as an electrical transformer in a substation is an example of coherent sound. Coherent sounds can also be easily generated electronically. When coherent sounds combine, they sum vectorially and their relative phase will determine the sum (see Section 1.5.4). It is more common to encounter sounds that are characterised by varying relative phases. For example, in an orchestra, the musical instruments of a section may all play in pitch, but in general, their relative phases will be random. The violin section may play beautifully but the phases of the sounds of the individual violins will vary randomly, one from another. Thus, the sounds of the violins will be incoherent with one another, and their contributions at an observer will sum incoherently. Incoherent sounds are sounds of random relative phase and they sum as scalar quantities on an energy basis. The mathematical expressions describing the combining of incoherent sounds may be considered as special limiting cases of those describing the combining of coherent sound.

1.11.2 Addition of Coherent Sound Pressures

When coherent sounds (which must be tonal and of the same frequency) are to be combined, the phase between the sounds must be included in the calculation.

Let the instantaneous total acoustic pressure, $p_t = p_1 + p_2$ and $p_i = |\hat{p}_i|\cos(\omega t + \beta_i), i = 1, 2,$ then:

$$p_t^2 = |\hat{p}_1^2|\cos^2(\omega t + \beta_1) + |\hat{p}_2^2|\cos^2(\omega t + \beta_2) + 2|\hat{p}_1|\,|\hat{p}_2|\cos(\omega t + \beta_1)\cos(\omega t + \beta_2) \tag{1.93}$$

where $|\hat{p}|$ denotes a scalar amplitude. Use of well-known trigonometric identities (Abramowitz and Stegun, 1965) allows Equation (1.93) to be rewritten as:

$$p_t^2 = \frac{1}{2}|\hat{p}_1^2|\,[1 + \cos2(\omega t + \beta_1)] + \frac{1}{2}|\hat{p}_2^2|\,[1 + \cos2(\omega t + \beta_2)] + |\hat{p}_1|\,|\hat{p}_2|\,[\cos(2\omega t + \beta_1 + \beta_2) + \cos(\beta_1 - \beta_2)] \tag{1.94}$$

Substitution of Equation (1.94) into Equation (1.59) and carrying out the indicated operations gives the time-averaged total mean square sound pressure, $\langle p_t^2\rangle$. Thus, for two sounds of the same frequency, characterised by mean square sound pressures $\langle p_1^2\rangle$ and $\langle p_2^2\rangle$ and phase difference $\beta_1 - \beta_2$, the total mean square sound pressure is given by:

$$\langle p_t^2\rangle = \langle p_1^2\rangle + \langle p_2^2\rangle + 2\langle p_1 p_2\rangle\cos(\beta_1 - \beta_2) \tag{1.95}$$

The mean square sound pressure resulting from the addition of N tones of the same frequency but different phases can be obtained using a similar procedure as was used for two tones, with the following result.

$$\langle p_t^2\rangle = \sum_{i=1}^{N}\sum_{k=1}^{N}\langle p_i p_k\rangle\cos(\beta_i - \beta_k) \tag{1.96}$$

The total sound pressure level, L_{pt} is calculated using Equation (1.87).

1.11.3 Addition of Incoherent Sounds (Logarithmic Addition)

When bands of noise are added and the phases are random, the limiting form of Equation (1.95) reduces to the case of addition of two incoherent sounds as:

$$\langle p_t^2 \rangle = \langle p_1^2 \rangle + \langle p_2^2 \rangle \tag{1.97}$$

which, by use of Equation (1.86), may be written in a general form for the logarithmic addition of N incoherent sounds, represented as decibels, as:

$$L_{pt} = 10 \log_{10} \left(10^{L_1/10} + 10^{L_2/10} + \ldots + 10^{L_N/10} \right) \quad \text{(dB re 20 µPa)} \tag{1.98}$$

Incoherent sounds add together on a linear energy (pressure squared) basis. The simple procedure embodied in Equation (1.98) may easily be performed on a standard calculator. The procedure accounts for the addition of sounds on a linear energy basis and their representation on a logarithmic basis. Note that the division by 10, rather than 20 in the exponent, is because the process involves the addition of squared pressures.

1.11.4 Logarithmic versus Arithmetic Addition and Averaging

Logarithmic addition and averaging are used when it is desired to add or average quantities expressed as decibels. When a series of sound pressure levels (usually L_{Aeq} or L_{eq} levels) are to be averaged (often needed when measuring sound power of an item of equipment - see Ch. 4), the sound pressure levels must be converted to sound pressures squared, which are then averaged and converted back to average sound pressure level. The average, $L_{p,av}$, of N sound pressure level measurements, $L_i, (i = 1, N)$, may thus be written using the following equation. As the reference sound pressure, p_{ref}, is used in both conversions, it cancels out and does not appear in the equation.

$$L_{p,av} = 10 \log_{10} \frac{1}{N} \left(10^{L_1/10} + 10^{L_2/10} + \ldots + 10^{L_N/10} \right) \quad \text{(dB re 20 µPa)} \tag{1.99}$$

Non-logarithmic quantities such as sound absorption coefficient are averaged arithmetically, so the octave band absorption coefficient (see Section 6.7) is obtained by arithmetically averaging the values, α_1, α_2, α_3, for the three 1/3-octave bands included in the octave band (see Section 1.14.1). Thus:

$$\alpha_{\text{oct}} = (\alpha_1 + \alpha_2 + \alpha_3)/3 \tag{1.100}$$

An exception is the averaging of reverberation times (see Section 6.6.2), for which reciprocals are arithmetically averaged to obtain the average reciprocal, which is then inverted to obtain the average reverberation time. To find the octave band average reverberation time from measurements of the reverberation times for each of the 1/3-octave bands contained in the octave band, we use:

$$T_{60,\text{oct}} = \left[\left(1/T_{60,1} + 1/T_{60,1} + 1/T_{60,1} \right) /3 \right]^{-1} \quad \text{(s)} \tag{1.101}$$

1.11.5 Subtraction of Sound Pressure Levels

Sometimes it is necessary to logarithmically subtract one noise level in decibels from another to obtain a resulting decibel level; for example, when background noise must be subtracted from the total noise to obtain the sound produced by a machine alone. The method used is similar to that described in the logarithmic addition of levels. The sound pressure level, L_p, resulting from subtracting one sound pressure level, L_{p2} from another, L_{p1}, is:

$$L_p = 10 \log_{10} \left(10^{L_{p1}/10} - 10^{L_{p2}/10} \right) \quad \text{(dB re 20 µPa)} \tag{1.102}$$

For example, if the sound pressure level measured at a particular location in a factory with a noisy machine operating nearby is 92.0 dB and the sound pressure level at the same location without the machine operating is 88.0 dB, then the sound pressure level, L_{pm}, produced by the machine alone is, $L_{pm} = 10\log_{10}\left(10^{92/10} - 10^{88/10}\right) = 89.8$ (dB re 20 µPa).

For noise-testing purposes, this procedure should be used only when the total sound pressure level exceeds the background noise by 3 dB or more. If the difference is less than 3 dB a valid sound test probably cannot be made. Note that here subtraction is between squared pressures.

1.11.6 Combining Level Reductions

Sometimes it is necessary to determine the effect of the placement or removal of constructions such as barriers and reflectors on the sound pressure level at an observation point. The difference between levels before and after an alteration (placement or removal of a construction) is called the noise reduction, NR. If the level decreases after the alteration, the NR is positive; if the level increases, the NR is negative. The problem of assessing the effect of an alteration is complex because the number of possible paths along which sound may travel from the source to the observer may increase or decrease as a result of the alteration. In assessing the overall effect of any alteration, the combined effect of all possible propagation paths must be considered. Initially, it is supposed that a reference level L_{pR} may be defined at the point of observation, as a level that would or does exist due only to straight-line propagation from source to receiver (excluding any ground reflections). Noise reduction due to propagation over any other path is then assessed in terms of this reference level. Calculated noise reductions would include spreading due to travel over a longer path, losses due to barriers, reflection losses at reflectors and losses due to source directivity effects (see Section 5.3). For octave band analysis, it will be assumed that the noise arriving at the point of observation by different paths combines incoherently. Thus, the total observed sound level may be determined by adding together logarithmically the contributing levels due to each propagation path. The problem that will now be addressed is how to combine noise reductions to obtain an overall noise reduction due to an alteration. Either before alteration or after alteration, the sound pressure level at the point of observation due to the ith path may be written in terms of the ith path noise reduction, NR_i, as:

$$L_{pi} = L_{pR} - \mathrm{NR}_i \qquad \text{(dB re 20 µPa)} \tag{1.103}$$

In either case, the observed overall sound pressure level due to contributions over N paths, including the direct path, is:

$$L_p = L_{pR} + 10\log_{10}\sum_{i=1}^{N} 10^{-(\mathrm{NR}_i/10)} \qquad \text{(dB re 20 µPa)} \tag{1.104}$$

If the direct path exists, it must be included in the sum in Equation (1.104) as one of the i paths with a corresponding NR_i value of 0 dB.

The effect of an alteration will now be considered, where note is taken that, after alteration, the propagation paths, associated noise reductions and number of paths may differ from those before alteration. In both before and after cases, the direct path (characterised by 0 dB NR) must be included if it exists as a possible propagation path. Of course, if the line-of-sight between the source and receiver does not exist (perhaps due to an obstruction such as a barrier), then the direct path would not exist. Introducing subscripts to indicate cases, A (before alteration) and B (after alteration), the overall noise reduction ($\mathrm{NR} = L_{pA} - L_{pB}$) due to the alteration is:

$$\begin{aligned}\mathrm{NR} &= L_{pR} + 10\log_{10}\sum_{i=1}^{N_A} 10^{-(\mathrm{NR}_{A,i}/10)} - L_{pR} - 10\log_{10}\sum_{i=1}^{N_B} 10^{-(\mathrm{NR}_{B,i}/10)} \qquad \text{(dB)}\\ &= 10\log_{10}\sum_{i=1}^{N_A} 10^{-(\mathrm{NR}_{A,i}/10)} - 10\log_{10}\sum_{i=1}^{N_B} 10^{-(\mathrm{NR}_{B,i}/10)} \qquad \text{(dB)}\end{aligned} \tag{1.105}$$

1.12 Beating

When two tones of very small frequency difference are presented to the ear, one tone, which varies in amplitude with a frequency modulation equal to the difference in frequency of the two tones, will be heard. When the two tones have exactly the same frequency, the frequency modulation will cease. When the tones are separated by a frequency difference greater than what is called the "critical bandwidth", two tones are heard. When the tones are separated by less than the critical bandwidth, one tone of modulated amplitude is heard where the frequency of modulation is equal to the difference in frequency of the two tones. The latter phenomenon is known as beating.

Let two tonal sounds of amplitudes A_1 and A_2 and of slightly different frequencies, ω and $\omega+\Delta\omega$, be added together. It will be shown that a third amplitude modulated sound is obtained. The total pressure due to the two tones may be written as:

$$p_1 + p_2 = A_1 \cos \omega t + A_2 \cos(\omega + \Delta\omega)t \tag{1.106}$$

where one tone is described by the first term and the other tone is described by the second term in Equation (1.106). Assuming that $A_1 \geq A_2$, defining $A = A_1 + A_2$ and $B = A_1 - A_2$, using well-known trigonometric identities, and replacing ω in the first term with $\omega + \Delta\omega - \Delta\omega$, Equation (1.106) may be rewritten as:

$$\begin{aligned} p_1 + p_2 = {} & A\cos[(\Delta\omega/2)t]\cos[(\omega + \Delta\omega/2)t] \\ & + B\cos[(\Delta\omega/2 - \pi/2)t]\cos[(\omega + \Delta\omega/2 - \pi/2)t] \end{aligned} \tag{1.107}$$

When $A_1 = A_2, B = 0$ and the second term in Equation (1.107) is zero. The first term is a cosine wave of frequency $(\omega + \Delta\omega)$ modulated by a frequency $\Delta\omega/2$. At certain values of time, t, the amplitude of the wave is zero; thus, the wave is described as fully modulated. If B is non-zero as a result of the two waves having different amplitudes, a modulated wave is still obtained, but the depth of the modulation decreases with increasing B and the wave is described as partially modulated. If $\Delta\omega$ is small, the familiar beating phenomenon is obtained (see Figure 1.7). The figure shows a beating phenomenon where the two waves are slightly different in amplitude resulting in partial modulation and incomplete cancellation at the null points.

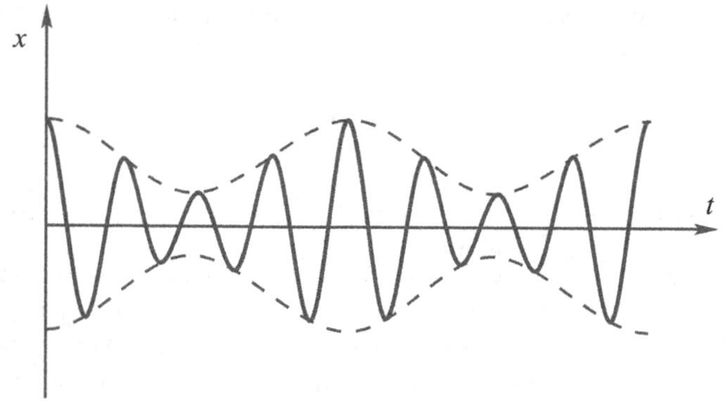

FIGURE 1.7 Illustration of beating.

It is interesting to note that if a signal containing two tones closely spaced in frequency were analysed on a very fine resolution spectrum analyser, only two peaks would be seen; one at each of the two interacting frequencies. There would be no peak seen at the beat frequency as there is no energy at that frequency, even though humans apparently "hear" that frequency.

1.13 Amplitude Modulation and Amplitude Variation

Amplitude modulation (AM) is defined as the periodic variation in amplitude of a noise or vibration signal. The extent of amplitude modulation is often characterised in terms of a modulation depth. This concept can be quite difficult to define for a realistic noise spectrum so we will begin here by defining it for a modulated single-frequency tone.

The sound pressure as a function of time that will be experienced when a tonal sound of frequency, f, is modulated with a frequency, f_m, is given by:

$$p(t) = A(1 + \mu m(t)) \cos(2\pi f t + \phi) \tag{1.108}$$

where μ is defined as the modulation index (Oerlemans, 2013) and ϕ is an arbitrary phase angle, which may be set equal to zero for our purposes here. The modulation function, $m(t)$, can be represented as a simple cosine function as:

$$m(t) = \cos(2\pi f_m t + \phi_m) \tag{1.109}$$

where ϕ_m is an arbitrary phase angle that may be set equal to zero for our discussion here. The quantity, A, is the time-averaged value of the amplitude of the signal being modulated. The modulation signal, $m(t)$, of Equation (1.109) can be derived directly from $p(t)$ of Equation (1.108) by using a Hilbert transform (see Section 12.3.11) to find an expression for the envelope signal, $e(t)$, which is equal to $A(1 + \mu m(t))$.

The maximum and minimum amplitudes, A_{max} and A_{min}, respectively, of the modulated waveform are given by:

$$A_{\text{max}} = A(1 + \mu) \quad \text{and} \quad A_{\text{min}} = A(1 - \mu) \tag{1.110}$$

and

$$\mu = \frac{A_{\text{max}} - A_{\text{min}}}{A_{\text{max}} + A_{\text{min}}} \tag{1.111}$$

where A_{max} and A_{min} are defined in Figure 1.8(b). As mentioned previously, A is the mean amplitude of the signal being modulated (the 10 Hz signal in Figure 1.8(b)).

The modulation depth, R_{mm}, in dB is defined as (Oerlemans, 2013):

$$R_{mm} = 20 \log_{10} \left(\frac{A_{\text{max}}}{A_{\text{min}}} \right) = 20 \log_{10} \left(\frac{1 + \mu}{1 - \mu} \right) = (L_p)_{\text{max}} - (L_p)_{\text{min}} \tag{1.112}$$

where $(L_p)_{\text{max}}$ is the maximum sound pressure level if the y-axis scale is in dB and $(L_p)_{\text{min}}$ is the minimum sound pressure level.

Amplitude variation of a signal includes both periodic amplitude modulation as well as random amplitude variations. These latter variations are generally a result of completely different mechanisms to those that are responsible for amplitude modulation, and may include amplitude variation as a result of beating. Other causes of amplitude variation in wind farm noise are variations in the interference between noise coming from different turbines as a result of varying atmospheric conditions causing the relative phase of the sounds arriving at a residence to vary over time.

Beating, amplitude modulation and random amplitude variation are all illustrated in Figure 1.8. In Figure 1.8(a) for the beating case, the two signals are 10 Hz and 11 Hz, resulting in a difference frequency (modulation frequency) of 1 Hz. The amplitude of the 11 Hz signal is 0.8 times the amplitude of the 10 Hz signal. In Figure 1.8(b), a signal of frequency, f = 10 Hz, is amplitude modulated at a frequency of f_m = 1 Hz, with a modulation index of 0.8. Figure 1.8(c) represents a time series measurement at a distance of 3 km from a wind farm. Figure 1.8(d)

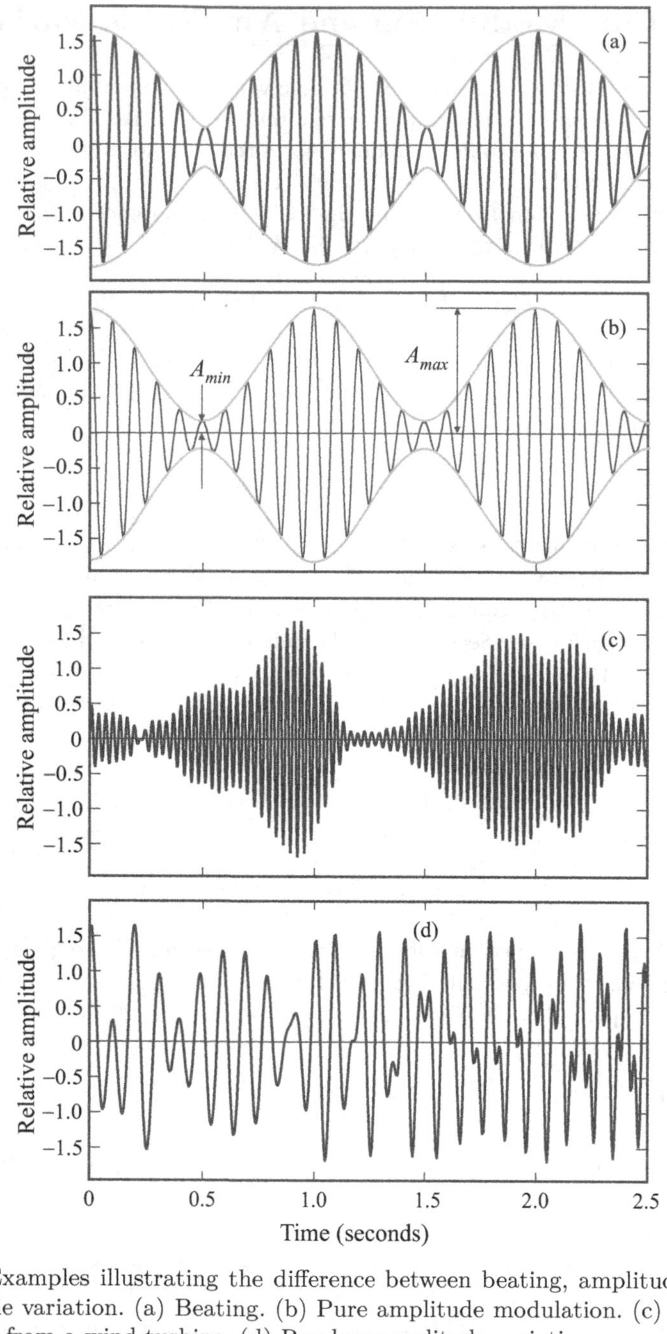

FIGURE 1.8 Examples illustrating the difference between beating, amplitude modulation and random amplitude variation. (a) Beating. (b) Pure amplitude modulation. (c) Typical amplitude modulated signal from a wind turbine. (d) Random amplitude variation.

shows a typical time varying signal representing wind farm noise but which is not amplitude modulated in any regular way. Thus, it is referred to as an amplitude varying noise.

Although the time domain representations of beating and amplitude modulation look to be very similar in the time domain representations shown in Figures 1.8a and b, respectively, they are very different in the frequency domain. The "beating" signal will show only two peaks in the frequency spectrum, one at each of the two frequencies that are combining to form the beating waveform, 10 Hz and 11 Hz in this case. The "amplitude modulated" signal will show three peaks, one at the frequency of the tone being modulated and one at each side of this frequency and separated by the modulation frequency. For the example in the figure, these three peaks would be at 9 Hz, 10 Hz and 11 Hz.

1.14 Basic Frequency Analysis

A propagating sound wave has been described either as an undefined disturbance as in Equation (1.18), or as a single frequency disturbance as given, for example, by Equation (1.22). Here it is shown how an undefined disturbance may be described conveniently as composed of narrow frequency bands, each characterised by a range of frequencies. There are various such possible representations and all are referred to broadly as spectra. It is customary to refer to spectral density level when the measurement band is one Hertz (Hz) wide, to 1/3-octave or octave band level when the measurement band is 1/3-octave or one octave wide, respectively, and to spectrum level for measurement bands of other widths. In air, sound is transmitted in the form of a longitudinal wave. To illustrate longitudinal wave generation, as well as to provide a model for the discussion of sound spectra, the example of a vibrating piston at the end of a very long tube filled with air is used, as illustrated in Figure 1.9.

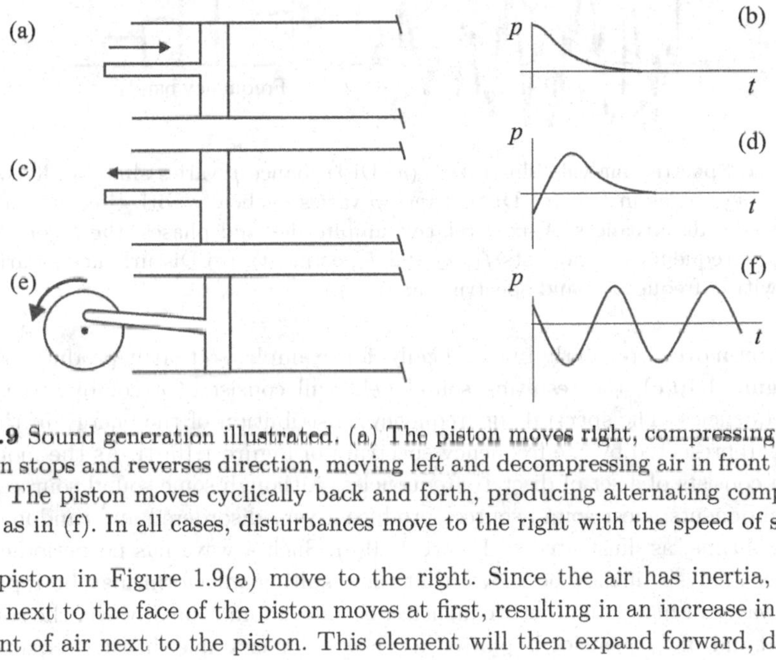

FIGURE 1.9 Sound generation illustrated. (a) The piston moves right, compressing air as in (b). (c) The piston stops and reverses direction, moving left and decompressing air in front of the piston, as in (d). (e) The piston moves cyclically back and forth, producing alternating compressions and rarefactions, as in (f). In all cases, disturbances move to the right with the speed of sound.

Let the piston in Figure 1.9(a) move to the right. Since the air has inertia, only the air immediately next to the face of the piston moves at first, resulting in an increase in the pressure in the element of air next to the piston. This element will then expand forward, displacing the next layer of air and compressing the next elemental volume. A pressure pulse is formed which travels along the tube with the speed of sound, c. Let the piston stop and subsequently move to the left. This results in the formation of a rarefaction next to the surface of the piston, which follows the previously formed compression down the tube. If the piston again moves to the right, the process is repeated with the net result being a "wave" of positive and negative pressure transmitted along the tube. If the piston moves with simple harmonic motion, a sine wave is produced; that is, at any instant, the pressure distribution along the tube will have the form of a sine wave, or at any fixed point in the tube, the pressure disturbance, displayed as a function of time, will have a sine wave appearance. Such a disturbance is characterised by a single frequency. The sound pressure variations at a single location as a function of time, together with the corresponding spectra, are illustrated in Figures 1.10(a) and (b), respectively. Although the sound pressure at any location varies sinusoidally with time, the particle motion is parallel to the direction of propagation of the wave, resulting in a longitudinal wave, as illustrated in Figure 1.2(a). Such a wave consists of compressions and rarefactions where the distance between particles is smaller in the compression part and larger in the rarefaction part.

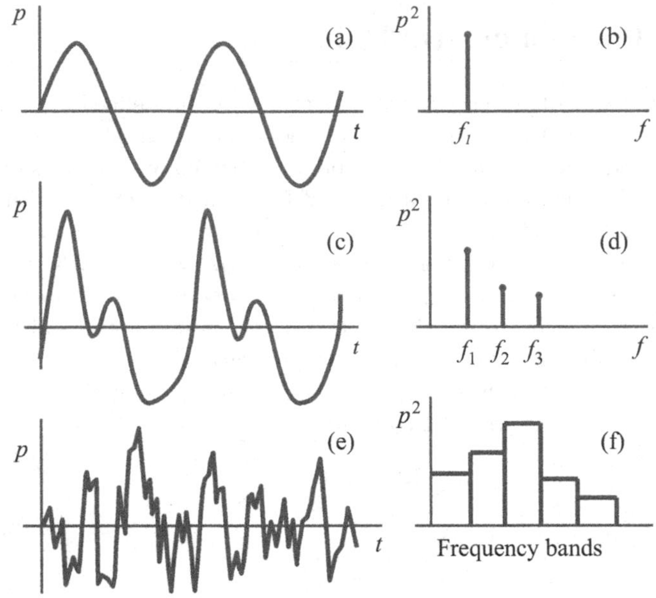

FIGURE 1.10 Spectral analysis illustrated. (a) Disturbance p varies sinusoidally with time t at a single frequency f_1, as in (b). (c) Disturbance p varies cyclically with time t as a combination of three sinusoidal disturbances of fixed relative amplitudes and phases; the associated spectrum has three single-frequency components f_1, f_2 and f_3, as in (d). (e) Disturbance p varies erratically with time t, with a frequency band spectrum as in (f).

If the piston moves irregularly but cyclically, for example, so that it produces the waveform shown in Figure 1.10(c), the resulting sound field will consist of a combination of sinusoids of several frequencies. The spectral (or frequency) distribution of the energy in this particular sound wave is represented by the frequency spectrum of Figure 1.10(d). As the motion is cyclic, the spectrum consists of a set of discrete frequencies. Although some sound sources have single-frequency components, most sound sources produce a very disordered and random waveform of pressure versus time, as illustrated in Figure 1.10(e). Such a wave has no periodic component, but by Fourier analysis, it may be shown that the resulting waveform may be represented as a collection of waves of many different frequencies. For a random type of wave, the sound pressure squared in a band of frequencies is plotted as shown, for example, in the frequency spectrum of Figure 1.10(f). Two special kinds of spectra are commonly referred to as white random noise and pink random noise. White random noise contains equal energy per Hertz and thus has a constant spectral density level. Pink random noise contains equal energy per measurement band and thus has an octave or 1/3-octave band level that is constant with frequency.

Frequency analysis is a process by which a time-varying signal is transformed into its frequency components. It can be used for quantification of a noise problem, as both criteria and proposed controls are frequency-dependent. When tonal components are identified by frequency analysis, it may be advantageous to treat these somewhat differently than broadband noise. Frequency analysis serves the important function of determining the effects of control, and it may aid, in some cases, in the identification of sources. Frequency analysis equipment and its use is discussed in Chapter 3 and more advanced frequency analysis is discussed in Chapter 12.

1.14.1 Frequency Bands

To facilitate comparison of measurements between instruments, frequency analysis bands have been standardised. International Standards Organisations have agreed on "preferred" frequency bands for sound measurement and by agreement, the *octave band* is the widest band usually considered for frequency analysis (ANSI/ASA S1.6, 2020; ANSI/ASA S1.11, 2019; IEC 61260-1, 2014; ISO 266, 1997). The upper-frequency limit of each octave band is approximately twice its lower-frequency limit and each band is identified by the geometric mean of the upper and lower frequency limits, called the band centre frequency. When more detailed information about a noise is required, standardised 1/3-octave band analysis may be used. The preferred frequency bands for octave and 1/3-octave band analysis are summarised in Table 1.2.

Reference to Table 1.2 shows that all information is associated with a band number, *BN*, listed in column one on the left. In turn the band number is related to the exact centre band frequencies, $f_{C(\text{exact})}$, of either the octaves or the 1/3-octaves listed in column two. The nominal band centre frequencies, f_C, represent a rounding up or down of the exact frequencies. The respective band limits are listed in columns five and six as the lower- and upper-frequency limits, f_ℓ and f_u, which apply to 1/3-octave bands. The lower octave band limit for an octave band is the lower limit in the line immediately above the particular octave band line and the upper limit is the upper limit in the line immediately below the particular octave band line. These observations may be summarised as:

$$f_{C(\text{exact})} = 10^{BN/10} = \sqrt{f_\ell f_u}; \quad f_\ell = f_{C(\text{exact})} \times 10^{-0.05}; \quad f_u = f_{C(\text{exact})} \times 10^{0.05} \tag{1.113}$$

In Table 1.2, it can be seen that ten times the logarithm of the exact band centre frequency is the band number in column one of the table. Consequently, as may be observed, the 1/3-octave centre frequencies repeat (with a different power of 10 multiplier) every decade in the table. The frequency band limits have been defined so that they are functions of the analysis band number, *BN*, and the ratios of the upper to lower frequencies, which, for base-10 filters (ANSI/ASA S1.11, 2019), are given by:

$$f_u/f_\ell = (10^{0.3})^{1/N}; \qquad N = 1, 3 \tag{1.114}$$

where $N = 1$ for octave bands and $N = 3$ for 1/3-octave bands.

The information provided thus far allows calculation of the bandwidth, Δf, of every band, using the following equation:

$$\Delta f = f_{C(\text{exact})} \frac{2^{1/N} - 1}{2^{1/(2N)}} = \begin{cases} 0.231563 f_{C(\text{exact})}; & \text{for 1/3-octave bands} \\ 0.707107 f_{C(\text{exact})}; & \text{for octave bands} \end{cases} \tag{1.115}$$

When logarithmic scales are used in plots, as will frequently be done in this book, the centre frequencies of the 1/3-octave bands between 12.5 Hz and 80 Hz inclusive (as an example) will lie, respectively, at 0.1, 0.2, 0.3, 0.4, 0.5, 0.6, 0.7, 0.8 and 0.9 of the distance on the scale between 10 and 100. The latter two numbers, in turn, will lie at 0.0 and 1.0, respectively, on the same scale.

Instruments are also available that provide analysis using bandwidths that are different to octave and fractional octave bands and usually much narrower (see Section 3.9). However, they do not enjoy the advantage of standardisation, so that the comparison of readings taken on such instruments may be difficult. One way to ameliorate the problem is to present such readings as mean levels per unit frequency. Data presented in this way are referred to as spectral density levels as opposed to band levels. In this case, the measured level is reduced by ten times the logarithm to the base ten of the bandwidth. For example, referring to Table 1.2, if the 500 Hz octave band, which has a bandwidth of $(707 - 353) = 354$ Hz, were presented in this way, the measured octave band level would be reduced by $10 \log_{10}(354) = 25.5$ dB to give an estimate of the spectral density level in the 500 Hz octave band. The problem is not entirely alleviated,

TABLE 1.2 Preferred octave and 1/3-octave frequency bands (IEC 61260-1, 2014; ISO 266, 1997; ANSI/ASA S1.11, 2019), where band limits are for base-10 filters

Band number	Exact centre frequency $f_{C(\text{exact})}$ (Hz)	Nominal octave band centre frequency (Hz)	Nominal 1/3-octave band centre frequency (Hz)	Band limits (Hz) Lower	Upper
−1	$10^{-1/10}$		0.8	0.71	0.89
0	$10^{0/10}$	1	1.0	0.89	1.12
1	$10^{1/10}$		1.3	1.1	1.4
2	$10^{2/10}$		1.6	1.4	1.8
3	$10^{3/10}$	2	2.0	1.8	2.2
4	$10^{4/10}$		2.5	2.2	2.8
5	$10^{5/10}$		3.15	2.8	3.5
6	$10^{6/10}$	4	4.0	3.5	4.5
7	$10^{7/10}$		5.0	4.5	5.6
8	$10^{8/10}$		6.3	5.6	7.1
9	$10^{9/10}$	8	8.0	7.1	8.9
10	$10^{10/10}$		10.0	8.9	11.2
11	$10^{11/10}$		12.5	11.2	14.1
12	$10^{12/10}$	16	16.0	14.1	17.8
13	$10^{13/10}$		20.0	17.8	22.4
14	$10^{14/10}$		25.0	22.4	28.2
15	$10^{15/10}$	31.5	31.5	28.2	35.5
16	$10^{16/10}$		40	35.5	44.7
17	$10^{17/10}$		50	44.7	56.2
18	$10^{18/10}$	63	63	56.2	70.8
19	$10^{19/10}$		80	70.8	89.1
20	$10^{20/10}$		100	89.1	112
21	$10^{21/10}$	125	125	112	141
22	$10^{22/10}$		160	141	178
23	$10^{23/10}$		200	178	224
24	$10^{24/10}$	250	250	224	282
25	$10^{25/10}$		315	282	355
26	$10^{26/10}$		400	355	447
27	$10^{27/10}$	500	500	447	562
28	$10^{28/10}$		630	562	708
29	$10^{29/10}$		800	708	891
30	$10^{30/10}$	1000	1000	891	1122
31	$10^{31/10}$		1250	1122	1413
32	$10^{32/10}$		1600	1413	1778
33	$10^{33/10}$	2000	2000	1778	2239
34	$10^{34/10}$		2500	2239	2818
35	$10^{35/10}$		3150	2818	3548
36	$10^{36/10}$	4000	4000	3548	4467
37	$10^{37/10}$		5000	4467	5623
38	$10^{38/10}$		6300	5623	7079
39	$10^{39/10}$	8000	8000	7079	8913
40	$10^{40/10}$		10000	8913	11220
41	$10^{41/10}$		12500	11220	14125
42	$10^{42/10}$	16000	16000	14125	17783
43	$10^{43/10}$		20000	17783	22387

as the effective bandwidth will depend on the sharpness of the filter cut-off, which is provided as a range of slopes in the standards (see ANSI/ASA S1.11 (2019)). Generally, the bandwidth is taken as lying between the frequencies, on either side of the passband, at which the signal is down 3 dB from the signal at the centre of the band. The spectral density level represents the energy level in a band one cycle wide, whereas by definition, a tone has a bandwidth of zero. Thus, for a pure tone, the same level will be measured, regardless of the bandwidth used for the measurement. When frequency analysis is done using a bandwidth of 1 Hz (for example, using FFT analysis as discussed in Chapter 12), the resulting band values obtained are referred to spectral density levels.

1.15 Doppler Shift

Doppler shift is the phenomenon whereby the frequency of sound experienced by an observer changes if one or both of the source and observer are moving relative to one another. To simplify the analysis, single frequency sound will be considered. However, the analysis can be applied to broadband sound by dividing the sound into narrow frequency bins using a Fourier transform as explained in Section 12.3. Each frequency bin can be treated separately and the results added together logarithmically to give a total result at each time step. However, the frequency resolution cannot be too fine or the time steps will be separated by too great an amount if the speed of the source in the direction of the observer is continually changing.

If the source is emitting sound of frequency, f_s, and moving with velocity, V_s, towards a stationary observer, the distance between successive maxima in the emitted sound wavefronts will be closer together than if the source were stationary. This can be explained by noting that the time between successive maxima emitted by a stationary source will be $1/f_s$. If the source is moving with velocity, V_s, towards the observer, then the time period between successive maxima will be reduced by V_s/λ, giving the resultant time period between successive maxima of $1/f_s - V_s/\lambda$. Thus, the new frequency experienced by the observer is given by:

$$f_s' = \frac{1}{\dfrac{1}{f_s} - \dfrac{V_s}{\lambda}} = \frac{f_s\lambda}{\lambda - f_s V_s} = \frac{f_s}{1 - V_s/c} \quad \text{(Hz)} \tag{1.116}$$

where λ is the wavelength of the emitted sound if the source is stationary and c is the speed of sound in the medium. If the source is moving away from the observer, the quantity, V_s, in Equation (1.116) is negative and the new frequency experienced by the observer is lower than the frequency corresponding to a stationary source.

For the case of a stationary source and an observer moving towards the source with velocity, V_o, the analysis is somewhat different, as the actual distance between pressure maxima (and hence the wavelength of the emitted sound) is unchanged by the motion of the observer. Instead, pressure maxima pass the observer at a faster speed, $c + V_o$, compared to the speed, c, when the observer is stationary. The new frequency is thus:

$$f_s' = \frac{c + V_o}{\lambda} = \frac{c}{\lambda} + \frac{V_o}{\lambda} = f_s + \frac{f_s V_o}{c} = f_s\left(1 + \frac{V_o}{c}\right) \quad \text{(Hz)} \tag{1.117}$$

If the source and receiver are stationary and the medium is moving, then there will be no Doppler shift in frequency. This is explained by noting that no matter how fast the medium is moving, the time period between adjacent maxima at the receiver will be the same as at the source, as the pressure maxima will pass the observer at the same rate as they are emitted from the source, resulting in the observed frequency being the same. However, the speed of sound in the medium will change to $c + V_o$, where V_o is the speed of the medium from the source to the receiver, so the time between the emission of a maximum from the source to the immission at the receiver of the same maximum will be changed, with the change the same for all maxima.

1.16 Impedance

In Section 1.4.7, the specific acoustic impedance for plane and spherical waves was introduced, and it was argued that similar expressions relating acoustic pressure and associated particle velocity must exist for waves of any shape. In free space and at large distances from a source, any wave approaches the propagation behaviour of a plane wave, and the characteristic impedance of the wave always tends to ρc.

Besides the specific acoustic impedance, two other kinds of impedance are commonly used in acoustics. These three kinds of impedance are summarised in Table 1.3 and their uses will be discussed in Sections 1.16.1, 1.16.2 and 1.16.3. All of the definitions suppose that with the application of a periodic force or pressure at some point in a dynamical system, a periodic velocity of fixed phase relative to the force or pressure will ensue. Note the role of cross-sectional area, S, in the definitions of the impedances shown in the table. In the case of mechanical impedance (radiation impedance) or ratio of force to velocity, the area, S, is the area of the radiating surface. For the case of acoustic impedance, the area, S, is the cross-sectional area of the duct containing sound.

TABLE 1.3 Three impedances used in acoustics

Impedance type	Definition	Dimensions
1. Mechanical impedance	$Z_M = F/u = pS/u$	$(\mathrm{MT^{-1}})$
2. Specific acoustic impedance	$Z_s = p/u$	$(\mathrm{MT^{-1}L^{-2}})$
3. Acoustic impedance	$Z_A = p/v = p/uS$	$(\mathrm{MT^{-1}L^{-4}})$

F = sinusoidally time-varying force $(\mathrm{MLT^{-2}})$
u = sinusoidally time-varying acoustic particle velocity $(\mathrm{LT^{-1}})$
p = sinusoidally time-varying acoustic pressure $(\mathrm{MT^{-2}L^{-1}})$
v = sinusoidally time-varying acoustic volume velocity $(\mathrm{L^3\ T^{-1}})$
S = area $(\mathrm{L^2})$

1.16.1 Mechanical Impedance, Z_M

The mechanical impedance is the ratio of a force to an associated velocity. This type of impedance is commonly used in acoustics to describe the radiation load (or radiation impedance) presented by a medium to a vibrating surface. In this case the impedance is usually referred to as radiation impedance, which is a special form of mechanical impedance. The associated force is that produced on the vibrating surface by the medium into which the surface is radiating as a result of the generation of the associated acoustic pressure field in the surrounding medium. The force is equal to the acoustic pressure adjacent to the vibrating surface multiplied by the area of the surface. The associated velocity is the acoustic particle velocity in the medium adjacent to the vibrating surface, which is the same as the vibration velocity of the surface. Radiation impedance will be encountered in Chapter 4 and is denoted Z_R.

1.16.2 Specific Acoustic Impedance, Z_s

The specific acoustic impedance is the ratio of acoustic pressure to associated particle velocity. It is important in describing the propagation of sound in free space and is continuous at junctions between media. In a medium of infinite extent, it is equal to the characteristic impedance, ρc, of the medium. The specific acoustic impedance is also important in describing the reflection and transmission of sound at an absorptive lining in a duct or on the wall or ceiling of a room and in describing reflection of sound by the ground. It will find use in Chapters 5 and 8. In Chapter 5, the characteristic impedance of the ground is denoted Z_m and is defined as the ratio of acoustic pressure, p, to particle velocity, u, in the ground.

1.16.3 Acoustic Impedance, Z_A

The acoustic impedance will find use in Chapter 8 in the discussion of sound propagation in reactive muffling devices, where the assumption is implicit that the propagating wavelength is long compared to the cross-sectional dimensions of the sound conducting duct. In the latter case, only plane waves propagate, and it is then possible to define a volume-velocity as the product of the duct cross-sectional area, S, and particle velocity. The volume velocity is continuous at junctions in a ducted system, as is the related acoustic pressure. Consequently, the acoustic impedance must also have the useful property that it is continuous at junctions in a ducted system (Kinsler et al., 1999). The characteristic acoustic impedance in a duct of cross-sectional area, S, is $\rho c/S$.

1.17 Flow Resistance

Porous materials are often used for the purpose of absorbing sound. Alternatively, it is the porous nature of many surfaces, such as grass-covered ground, that determines their sound reflecting properties. As discussion will be concerned with ground reflection in Chapter 5, with sound-absorption of porous materials as well as sound propagation through them in Chapter 6 and Appendix D, and with absorption of sound propagating in ducts lined with porous materials in Chapter 8, it is important to consider the property of porous materials that relates to their use for acoustical purposes.

A solid material that contains many voids is said to be porous. The voids may or may not be interconnected; however, for acoustical purposes, it is the interconnected voids that are important; the voids that are not connected are generally of little importance. The property of a porous material that determines its usefulness for acoustical purposes is the resistance of the material to induced flow through it, as a result of a pressure gradient. Flow resistance, an important parameter that is a measure of this property, is defined according to the following simple experiment (see Section D.1). A uniform layer of porous material of thickness, ℓ (m), and area, S (m^2), is subjected to an induced mean volume flow, V_0 (m^3/s), through the material, and the pressure drop, ΔP_s (Pa), across the layer is measured. Very low pressures and mean volume velocities are assumed (of the order of the particle velocity amplitude of a sound wave having a sound pressure level between 80 and 100 dB). The flow resistance of the material, R_f, is defined as the induced pressure drop across the layer of material divided by the resulting mean volume velocity, V_0, per unit area of the material:

$$R_f = \Delta P_s S / V_0 \tag{1.118}$$

The units of flow resistance are the same as for specific acoustic impedance, ρc; thus it is sometimes convenient to specify flow resistance in dimensionless form in terms of numbers of ρc units. The flow resistance of unit thickness of material is defined as the flow resistivity, R_1, which has the units Pa s m^{-2}, often referred to as MKS rayls per metre. Experimental investigation shows that porous materials of generally uniform composition may be characterised by a unique flow resistivity. Thus, for such materials, the flow resistance is proportional to the material thickness, ℓ, as:

$$R_f = R_1 \ell \tag{1.119}$$

Flow resistance characterises a layer of specified thickness, whereas flow resistivity characterises a bulk material in terms of resistance per unit thickness. For fibreglass and rockwool fibrous porous materials, which may be characterised by a mean fibre diameter, d, the following relation holds (Bies, 1988):

$$\frac{R_1 \ell}{\rho c} = 27.3 \left(\frac{\rho_B}{\rho_f}\right)^{1.53} \left(\frac{\mu}{d \rho c}\right) \left(\frac{\ell}{d}\right) \tag{1.120}$$

In the above equation, in addition to the quantities already defined, the gas density, ρ (= 1.206 kg/m^2 for air at 20°C), the porous material bulk density, ρ_B, and the fibre material density, ρ_f, have been introduced. The remaining variables are the speed of sound, c, of the gas and the dynamic gas viscosity, μ (1.84 × 10^{-5} Pa s for air at 20°C). The dependence of flow resistance on bulk density, ρ_B, and fibre diameter, d, of the porous material is to be noted. A decrease in fibre diameter results in an increase of flow resistivity and an increase in sound-absorption, so a useful fibrous material will have very fine fibres. Values of flow resistivity for some fibre glass and rockwool products have been measured and published (Bies and Hansen, 1979, 1980; Wang and Torng, 2001; Tarnow, 2002). For further discussion of flow resistance, its method of measurement and other properties of porous materials which follow from it, see Appendix D.

2

Human Hearing and Noise Criteria

LEARNING OBJECTIVES

In this chapter, the reader is introduced to:

- the anatomy of the ear;
- the response of the ear to sound;
- relationships between noise exposure and hearing loss;
- loudness measures;
- masking of some sound by other sound;
- various measures used to quantify occupational and environmental noise;
- various weighting networks for single number descriptions of the amplitude of complex sounds;
- hearing loss associated with age and exposure to noise;
- hearing damage risk criteria, requirements for speech recognition and alternative interpretations of existing data;
- hearing damage risk criteria and sound level versus exposure time trading rules;
- speech interference criteria for broadband noise and intense tones;
- psychological effects of noise as a cause of stress and effects on work efficiency;
- Noise Rating (NR), Noise Criteria (NC), Room Criteria (RC), Balanced Noise Criteria (NCB) and Room Noise Criteria (RNC) for ambient noise level specification; and
- environmental noise criteria.

2.1 Brief Description of the Ear

The comfort, safety and effective use of the human ear are the primary considerations motivating interest in the subject matter of this book; consequently, it is appropriate to describe that marvellous mechanism. The discussion will make brief reference to the external and the middle ears and extensive reference to the inner ear where the transduction process from incident sound to neural encoding takes place. This brief description of the ear will make extensive reference to Figure 2.1.

DOI: 10.1201/9780367814908-2

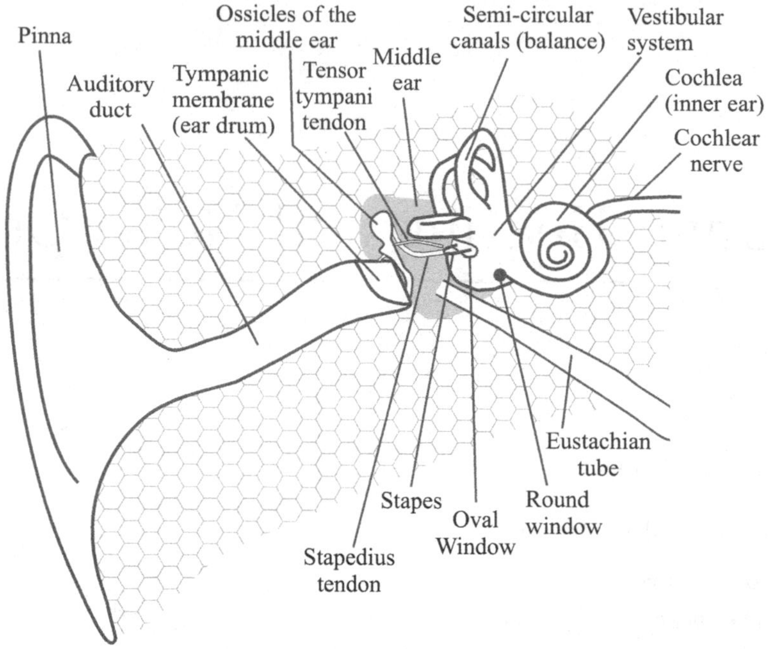

FIGURE 2.1 A representation of the pinna, middle and inner ear (right ear, face forward).

2.1.1 External Ear

The pinna, or external ear, will be familiar to the reader and no further reference will be made to it other than the following few words. As shown by Blauert (1983), the convolutions of the pinna give rise to multiple reflections and resonances within it, which are frequency and direction dependent. These effects and the location of the pinna on the side of the head make the response of the pinna directionally variable to incident sound in the frequency range of 3 kHz and above. For example, a stimulus in the latter frequency range is heard best when it is incident from the side of the head. If there were a mechanism tending to maintain levels in the ear within some narrow dynamic range, the variability in response resulting from the directional effects imposed by the pinna as the head moved around would be suppressed and would not be apparent to the listener. However, amplification through undamping provided by the active response of the outer hair cells, as will be discussed in Section 2.1.8.3, seems to be able to provide a mechanism for location of a noise source, whether or not there is a mechanism in the ear maintaining levels within a narrow dynamic range. Indeed, the jangling of keys is interpreted by a single ear in such a way as to infer the direction and general location of the source in space without moving the head.

2.1.2 Middle Ear

Sound enters the ear through the auditory duct (or ear canal), a more or less straight tube between 23 and 30 mm in length, at the end of which is the eardrum, a diaphragm-like structure known as the tympanic membrane. Sound entering the ear causes the eardrum to move in response to acoustic pressure fluctuations within the auditory canal and to transmit motion through a mechanical linkage provided by three tiny bones, called ossicles, to a second membrane at the oval window of the middle ear. Sound is transmitted through the oval window to the inner ear (see Figure 2.1).

The middle ear cavity is kept at atmospheric pressure by occasional opening, during swallowing, of the eustachian tube, also shown in Figure 2.1. If an infection causes swelling or mucus to block the eustachian tube, preventing pressure equalisation, the air in the middle ear will be gradually absorbed, causing the air pressure to decrease below atmospheric pressure and the tympanic membrane to implode. The listener then will experience temporary deafness.

Three tiny bones located in the air-filled cavity of the middle ear are identified in Figure 2.1 as the malleus (hammer), incus (anvil) and stapes (stirrup). They provide a mechanical advantage of about 3:1, while the relative sizes of the larger eardrum and smaller oval window provide a mechanical advantage of about 5:1, resulting in an overall mechanical advantage of 15:1. As the average length of the auditory canal is about 25 mm, the canal is resonant at about 4 kHz, giving rise to a further mechanical advantage about this frequency of the order of three.

The bones of the middle ear are equipped with a muscular structure (see Figure 2.1), which allows some control of the motion of the linkage, and thus transmission of sound to the inner ear. For example, a moderately loud buzz introduced into the earphones of a gunner may be used to induce tensing of the muscles of the middle ear, which effectively stiffens the ossicle linkage, and thus protects the inner ear from the loud percussive noise of firing. On the other hand, some individuals suffer from what is most likely a hereditary disease, which takes the form of calcification of the joints of the middle ear, rendering them stiff and the victim deaf. In this case, the cure may, in the extreme case, take the form of removal of part of the ossicles and replacement with a functional prosthesis.

A physician, who counted many miners among his patients, once told one of the authors that for those who had calcification of the middle ear, a prosthesis gave them good hearing. The calcification had protected them from noise-induced hearing loss for which they also received compensation.

No discussion of the function of the middle ear would be complete without mention of work by Bell (2014), who postulated that its action may be partly responsible for the response of some people to inaudible infrasound. Bell (2011) suggests that the physical gain control mechanism mentioned above that can protect the ear of a gunner from the loud noise of the gun firing is an acoustic reflex action that operates much more often. It is also initiated when we speak, when we shut our eyes tightly or when we touch our ears or face. Tensing of the middle ear muscles causes a pressure rise in the fluid of the cochlea, operating and implemented by flexure of the muscles supporting the ossicle linkage. Thus, tensing of middle ear muscles can result in feelings of pressure (or fullness) in the ears. If the pressure is excessive, it can cause vertigo and nausea. Tensing of the middle ear muscles may be caused by the presence of periodic infrasound, which can be well below the level of audibility, even though the sound pressure level is very high, as its frequency is too low to stimulate a nerve ending in the cochlea. However, the middle ear muscles could act to keep the ear drum in the middle of its operating range, causing them to pressurise the cochlea periodically in time with the stimulus. This may result in feelings of fullness in the ear and could also lead to fatigue as well as vertigo and nausea in some people. The conclusion that we can take from this is that infrasound may not need to be at audible levels to have adverse physiological effects on people, although this supposition is yet to be proven with a comprehensive study of many test subjects.

2.1.3 Inner Ear

The oval window at the entrance to the liquid-filled inner ear is connected to a small vestibule terminating in the semicircular canals and cochlea. The semicircular canals are concerned with balance and will be of no further concern here, except to remark that if the semicircular canals become involved, an ear infection can sometimes induce dizziness or vertigo.

In mammals, the cochlea is a small tightly rolled spiral cavity, as illustrated for humans in Figure 2.1. Internally, the cochlea is divided into an upper gallery (scala vestibuli) and a lower gallery (scala tympani) by an organ called the cochlear duct (scala media), or cochlear partition, which runs the length of the cochlea (see Figure 2.2). The two galleries are connected at the apex or apical end of the cochlea (the end furthest from the middle ear) by a small opening called the helicotrema. At the basal end of the cochlea, the upper and lower galleries terminate at the oval and round windows, respectively. The round window is a membrane-covered opening situated in the same general location of the lower gallery as the oval window in the upper gallery (see Figure 2.1). A schematic representation of the cochlea unrolled is shown in Figure 2.2 and a cross-sectional view is shown in Figure 2.3.

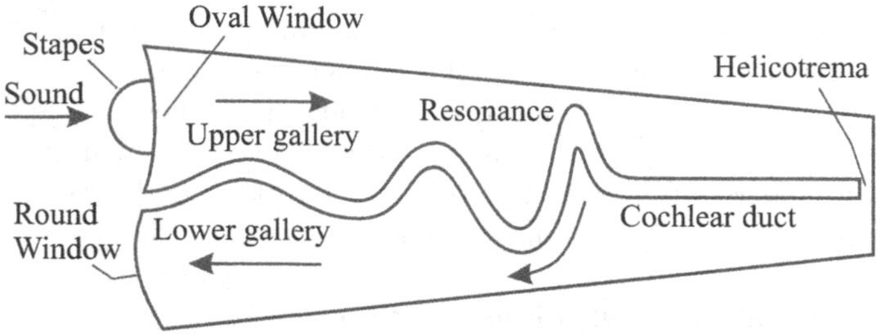

FIGURE 2.2 Schematic model of the cochlea (unrolled).

In humans, the average length of the cochlea from the basal end to the apical end is about 35 mm. The fluids within the cochlea, perilymph (sodium rich), which fills the upper and lower galleries, and endolymph (potassium rich), which fills the cochlear partition, are essentially salt water.

As will be shown, the central partition acts as a mechanical shunt between the upper and lower galleries. Any displacement of the central partition, which tends to increase the volume of one gallery, will decrease the volume of the other gallery by exactly the same amount. Consequently, it may be concluded that the fluid velocity fields in the upper and lower galleries are essentially the same but of opposite phase. For later reference, these ideas may be summarised as follows. Let subscripts 1 and 2 refer, respectively, to the upper and lower galleries; then the acoustic pressure, p, and volume velocity, v (particle velocity multiplied by the gallery cross-sectional area), may be written as:

$$p = p_1 = -p_2 \tag{2.1}$$

and

$$v = v_1 = -v_2 \tag{2.2}$$

2.1.4 Cochlear Duct or Partition

The cochlear duct (see Figure 2.3), which divides the cochlea into upper and lower galleries, is triangular in cross-section, being bounded on its upper side next to the scala vestibuli by Reissner's membrane, and on its lower side next to the scala tympani by the basilar membrane. On its third side, it is bounded by the stria vascularis, which is attached to the outer wall of the cochlea. The cochlear duct is anchored at its apical end to a bony ridge on the inner wall of the cochlear duct formed by the core of the cochlea. The auditory nerve is connected to the central partition through the core of the cochlea.

The closed three sides of the cochlear duct form a partition between the upper and lower galleries, and hence the alternative name of cochlear partition. It has been suggested that the potassium rich endolymph of the cochlear duct supplies the nutrients for the cells within the duct, as there are no blood vessels in this organ. Apparently, the absence of blood vessels avoids

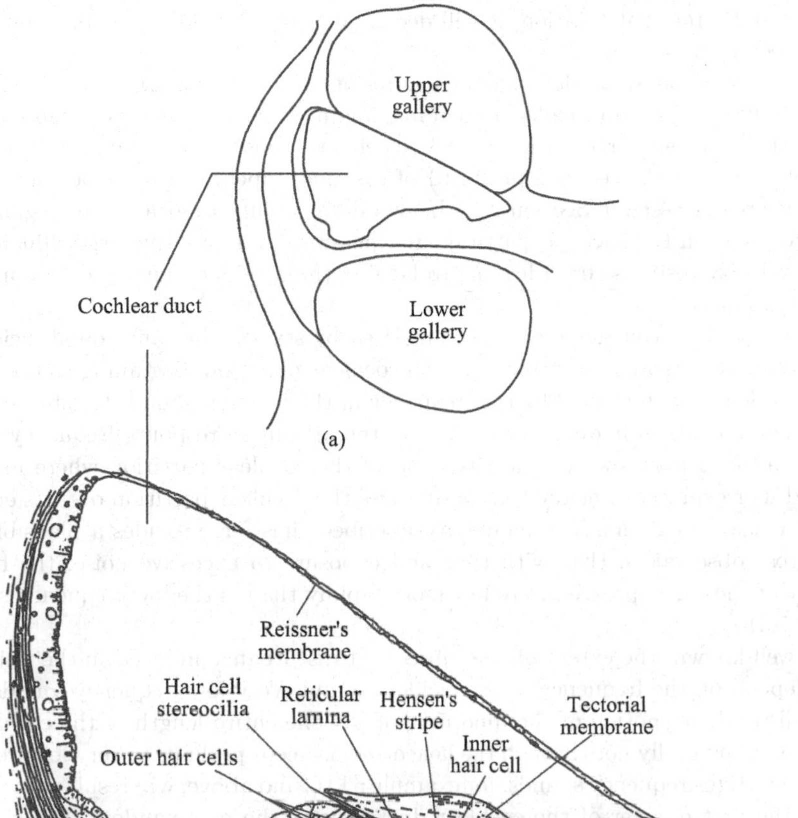

FIGURE 2.3 (a) Cross-section of the cochlea. (b) Cross-sectional detail of the organ of Corti.

the background noise which would be associated with flowing blood, because it is within the cochlear duct, in the organ of Corti, that sound is sensed by the ear.

The organ of Corti, shown in Figure 2.3(b), rests upon the basilar membrane next to the bony ridge on the inner wall of the cochlea and contains the sound-sensing hair cells. The sound-sensing hair cells, in turn, are connected to the auditory nerve cells (labelled as "cochlear nerve" in Figure 2.3) that pass through the bony ridge to the organ of Corti. The supporting basilar membrane attached, under tension, at the outer wall of the cochlea to the spiral ligament (see Figure 2.3) provides a resilient support for the organ of Corti.

The cochlear partition, the basilar membrane and upper and lower galleries form a coupled system, much like the flexible walled duct discussed in Section 1.4.4 on page 15. In this system, sound transmitted into the cochlea through the oval window proceeds to travel along the cochlear duct as a travelling wave with an amplitude that depends on the flexibility of the cochlear partition, which varies along its length. Depending on its frequency, this travelling wave will build to a maximum amplitude at a particular location along the cochlear duct as shown in

Figure 2.2 and after that location, it will decay quite rapidly. This phenomenon is analysed in detail in Section 2.1.8.1.

Thus, a tonal sound incident on the ear results in excitation along the cochlear partition that gradually increases up to a location of maximum response. The tone is sensed in the narrow region of the cochlear partition where the velocity response is a maximum. The ability of the ear to detect the pitch (see Section 2.2.6) of a sound appears to be dependent (among other things that are discussed in Section 2.1.6 below) on its ability to localise the region of maximum velocity response in the cochlear partition and possibly to detect the large shift in phase of the partition velocity response from leading to lagging the incident sound pressure in the region of maximum response.

The observations thus far may be summarised by stating that any sound incident upon the ear ultimately results in a disturbance of the cochlear partition, beginning at the stapes (basal) end, up to a length determined by the frequency of the incident sound. It is to be noted that all stimulus components of frequencies lower than the maximum response frequency result in some motion at all locations towards the basal end of the cochlear partition where high frequencies are sensed. For example, a heavy bass note drives the cochlear partition over its entire length to be heard at the apical end. The model, as described thus far, provides a plausible explanation for the gross observation that with time and exposure to excessive noise, the high-frequency sensitivity of the ear is progressively lost more rapidly than is the low-frequency sensitivity (see Section 2.5.3).

As is well-known, the extent of the subsequent disturbance induced in the fluid of the inner ear will depend on the frequency of the incident sound. Very low-frequency sounds, for example 50 Hz, will result in motion of the fluid over nearly the entire length of the cochlea. Note that such motion is generally not through the helicotrema except perhaps at very low frequencies, well below 50 Hz. High-frequency sounds, for example 6 kHz and above, will result in motion restricted to about the first quarter of the cochlear duct nearest the oval window. The situation for an intermediate audio frequency is illustrated in Figure 2.2. An explanation for these observations is proposed in Section 2.1.8.1

2.1.5 Hair Cells

The sound-sensing hair cells of the organ of Corti are arranged in rows on either side of a rigid triangular construction formed by the rods of Corti, sometimes called the tunnel of Corti. As shown in Figure 2.3(b), the hair cells are arranged in a single row of inner hair cells and three rows of outer hair cells. The hair cells are each capped with a tiny hair bundle, hence the name, called hair cell stereocilia, which are of the order of 6 or 7 μm in length.

The stereocilia of the inner hair cells are free standing in the cleft between the tectorial membrane above and the reticular lamina below, on which they are mounted. They are velocity sensors responding to any slight motion of the fluid that surrounds them (Bies, 1999). Referring to Figure 2.3(b), it may be observed that motion of the basilar membrane upward in the figure results in rotation of the triangular tunnel of Corti and a decrease of the volume of the inner spiral sulcus and an outward motion of the fluid between the tectorial membrane and the reticular lamina. This fluid motion is amplified by the narrow passage produced by Hensen's stripe and suggests its function (see Figure 2.3(b)).

The inner hair cells respond maximally at maximum velocity as the tectorial membrane passes through its position of rest. By contrast with the inner hair cells, the outer hair cells are firmly attached to and sandwiched between the radially stiff tectorial membrane and the radially stiff reticular lamina. The reticular lamina is supported on the rods of Corti, as shown in the Figure 2.3. The outer hair cells are capable of both passive and active response.

The capability of active response is referred to as motility. When active, the cells may extend and contract in length in phase with a stimulus up to a frequency of about 5 kHz. Since the effective hinge joint of the tectorial membrane is at the level of its attachment to the limbus, while the hinge joint for the basilar membrane is well below at the foot of the inner rods of Corti, any slight contraction or extension of the outer hair cells will result in associated motion of the basilar membrane and rocking of the tunnel of Corti.

Motility of the outer hair cells provides the basis for undamping and active response. In turn, undamping provides the basis for stimulus amplification. Undamping of a vibrating system requires that work must be done on the system so that energy is replaced as it is dissipated during each cycle. The motility of the outer hair cells provides the basis for undamping. A mechanism by which undamping is accomplished in the ear has been described by Mammano and Nobili (1993) and is discussed in Section 2.1.8.3.

2.1.6 Neural Encoding

As illustrated in Figure 2.3(b), the cochlear nerve is connected to the hair cells through the inner bony ridge on the core of the cochlea. The cochlear nerve is divided about equally between the afferent system which carries information to the brain and the efferent system which carries instructions from the brain to the ear. The cells of the afferent nervous system, which are connected to the inner hair cells, convert the analogue signal provided by the inner hair cells into a digital code, by firing once each cycle in phase, for excitation up to frequencies of about 5 kHz. At frequencies above about 5 kHz, the firing is random. As the sound pressure level increases, the number of neurons firing increases, so that the sound pressure level is encoded in the firing rate. Frequency information also is encoded in the firing rate up to about 5 kHz.

Pitch and frequency, though related, are not directly related (see Section 2.2.6). At frequencies higher than 5 kHz, pitch is associated with the location of the excitation on the basilar membrane. In Section 2.1.7 it will be shown that one way of describing the response of the cochlear partition is to model it as a series of independent short segments, each of which has a different resonance frequency. However, as also stated in Section 2.2.6, this is a very approximate model, due to the basilar membrane response being coupled with the response of the fluid in the cochlear duct and upper and lower galleries. When sound pressure in the upper gallery at a segment of the cochlear partition is negative, the segment is forced upward and a positive voltage is generated by the excited hair cells. The probability of firing of the attached afferent nerves increases. When the sound pressure at the segment is positive, the segment is pushed downward and a negative voltage is generated by the hair cells. The firing of the cells of the afferent nervous system is inhibited. Thus, in-phase firing occurs during the negative part of each cycle of an incident sound.

Neurons attached to the hair cells also exhibit resonant behaviour, firing more often for some frequencies than others. The hair cells are arranged such that the neuron maximum response frequencies correspond to basilar membrane resonance frequencies at the same locations. This has the effect of increasing the sensitivity of the ear.

The digital encoding favoured by the nervous system implies a finite dynamic range of discrete steps, which seems to be less than 170 counts per second in humans. The dynamic range of audible sound intensity, bounded by “just perceptible” at the lower end and “painful” at the higher end, is 10^{13}; thus, a logarithmic-type encoding of the received signal is required, which is related to the neuron firing rate. In an effort to provide an adequate metric for describing the intensity of sound, the decibel system has been adopted through necessity by the physiology of the ear. The digital encoding favoured by the nervous system implies a finite dynamic range of discrete steps, which seems to be less than 170 counts per second in humans. Thus, frequency analysis at the ear, rather than the brain, is essential, and this is done by the digital encoding including information about the response and relative phases of the response basilar membrane. Thus, the ear decomposes the sound incident upon it into frequency components and encodes the amplitudes of the components in rates of impulses forming information packets, which are transmitted to the brain for reassembly and interpretation.

In addition to the afferent nervous system, which conducts information from the ear to the brain, there also exists an extensive efferent nervous system of about the same size, which conducts instructions from the brain to the ear. A further distinction between the inner hair cells lying next to the inner rods of Corti and the outer hair cells lying on the opposite side of the rods of Corti (see Figure 2.3) can be observed. Whereas about 90% of the fibres of the afferent nerve connect directly with the inner hair cells, and only the remaining 10% connect with the more numerous outer hair cells, the efferent system connections seem to be about equally distributed between inner and outer hair cells. The efferent system is connected to the outer hair cells and to the afferent nerves of the inner hair cells (Spoendlin, 1975). Apparently, the brain is able to control the state (on or off) of the outer hair cells and similarly the state of the afferent nerves connected to the inner hair cells. As will be shown in Section 2.1.8.3, control of the state of the outer hair cells enables selective amplification.

It is suggested here that control of amplification may also allow interpretation of directional information imposed on the received acoustic signal by the directional properties of the external ear, particularly at high frequencies (see Section 2.1.8.3). In support of the latter view is the anatomical evidence that the afferent nerve system and density of hair cells, about 16 000 in number, is greatest in the basilar area of the cochlea nearest to the oval window, where high-frequency sounds are sensed. The connection of the efferent system to the afferent nerves of the inner hair cells suggests volume control to maintain the count rate within the dynamic range of the brain. In turn, this supports the suggestion that the inner hair cells are the sound detectors. The function of the outer hair cells will be discussed in Section 2.1.8.3.

2.1.7 Linear Array of Uncoupled Oscillators

Voldřich (1978) has investigated the basilar membrane in immediate post-mortem preparations of the guinea pig, and he has shown that rather than a membrane, it is accurately characterised as a series of discrete segments, each of which is associated with a radial fibre under tension. The fibres are sealed in between with a gelatinous substance of negligible shear viscosity to form what is referred to as the basilar membrane. The radial tension of the fibres varies systematically along the entire length of the cochlea from large at the basal end to small at the apical end. As the basilar membrane response is coupled with the fluid response in the cochlear duct, this is consistent with the observation that the location of maximum response of the cochlear partition varies in the audible frequency range from the highest frequency at the basal end to the lowest frequency at the apical end. This has led researchers in the past to state that the basilar membrane may be modelled as an array of linear, uncoupled oscillators. As the entire system is clearly coupled, this model is an approximate one only, but it serves well as an illustration.

Mammano and Nobili (1993) have considered the response of a segment of the central partition to an acoustic stimulus and have proposed the following differential equation describing the displacement, y, of the segment in response to two forces acting on the segment. One force, F_S, is due to motion of the stapes and the other force, F_B, is due to motion of all other parts of the membrane. In the following equation, m is the segment mass, k_s is the segment stiffness, z is a normalised longitudinal coordinate along the duct centre line from $z = 0$ at the stapes (basal end) to $z = 1$ at the helicotrema (apical end) and t is the time coordinate. The damping term, K, has the form given below in Equation (2.5). The equation of motion of a typical segment of the basilar membrane as proposed by Mammano and Nobili (1993) is:

$$m\frac{\partial^2 y}{\partial t^2} + K\frac{\partial y}{\partial t} + k_s y = F_S + F_B \tag{2.3}$$

which is similar in form to the equation of motion of a mass on a spring (see Equation (9.1)).

The total force $F_S + F_B$ is the acoustic pressure, p, multiplied by the area, $\ell_w \Delta z$, of the segment of the basilar membrane upon which it acts, where ℓ_w is the width of the membrane and Δz is the thickness of a segment in the direction along the duct centreline expressed as:

$$F_S + F_B = p\ell_w \Delta z \tag{2.4}$$

The damping term, K, expressed as an operator, has the following form:

$$K = C + \frac{\partial s(z)}{\partial z}\frac{\partial}{\partial z} \tag{2.5}$$

In Equation (2.5), the first term on the right-hand side, C, is the damping coefficient due to fluid viscosity and the second term provides an estimate of the contribution of longitudinal viscous shear. The quantity, $s(z)$, is the product of two quantities:

1. the effective area of the basilar membrane cross-section at location, z; and
2. the average shearing viscosity coefficient (≈ 0.015 kg m^{-1} s^{-1}) of a section of the organ of Corti at location, z.

In the formulation of Mammano and Nobili (1993), the first term on the RHS of Equation (2.5) is a constant when the cochlea responds passively, but is variable when the cochlea responds actively. In an active state, the variable, C, implies that Equation (2.3) is nonlinear.

The second term on the RHS of Equation (2.5) implies longitudinal coupling between adjacent segments and also implies that Equation (2.3) is nonlinear. However, it may readily be shown that the second term is negligible in all cases; thus, the term, K, will be replaced with the variable damping term, C, in subsequent discussion. Variable damping will be expressed in terms of damping and undamping as explained below (see Section 2.1.8.3).

When K is replaced with C, Equation (2.3) becomes the expression that formally describes the response of a simple one-degree-of-freedom linear oscillator for the case that C is constant or varies slowly (Equation (9.1)). It will be shown that in a passive state, C is constant and in an active state it may be expected that C varies slowly. In the latter case, the cochlear response is quasi-stationary.

It is proposed that the cochlear segments of Mammano and Nobili (1993) may be identified with a series of tuned mechanical oscillators. It is proposed to identify a segment of the basilar membrane, including each fibre that has been identified by Voldřich (1978) and the associated structure of the central partition, as parts of an oscillator.

Mammano and Nobili (1993) avoid discussion of nonlinearity when the cochlear response is active. Instead, they tacitly assume quasi-stationary response and provide a numerical demonstration that varying the damping in their equation of motion gives good results. Here, it will be explicitly assumed that slowly varying damping characterises cochlear response, in which case the response is quasi-stationary. The justification for the assumption of quasi-stationary response follows.

Quasi-stationary means that the active response time is long compared with the period of the lowest frequency that is heard as a tone. As the lowest audible frequency is, by convention, assumed to be 20 Hz, it follows that the active response time is longer than 0.05 seconds. As psychoacoustic response times seem to be of the order of 0.3 to 0.5 seconds, a quasi-stationary solution seems justified. This assumption is consistent with the observation that the efferent and afferent fibres of the auditory nerve are about equal in number and also with the observation that no other means of possible control of the outer hair cells has been identified.

The observation that longitudinal viscous shear forces may be neglected leads to the conclusion that each cochlear partition segment responds independently of any modal coupling between segments. Consequently, the cochlear partition may be modelled approximately as a series of modally independent linear mechanical oscillators that respond to the fluid pressure fields of the upper and lower galleries. Strong fluid coupling between any cochlear segment and all other segments of the cochlea accounts for the famous basilar membrane travelling wave discovered by Békésy and Wever (1960).

The modally independent segments of the cochlea will each exhibit their maximum velocity response at the frequency of undamped resonance for the corresponding mechanical oscillator. Thus, a frequency of maximum response (loosely termed resonance), which remains fixed at all sound pressure levels, characterises every segment of the cochlear duct.

The frequency of undamped resonant response will be referred to here as the characteristic frequency or the resonance frequency. The characteristic frequency has the important property that it is independent of the system damping (Tse et al., 1979). The amplitude of response, on the other hand, is inversely proportional to the system damping. Thus, variable damping provides variable amplification, but does not affect the characteristic frequency.

2.1.8 Mechanical Properties of the Central Partition

2.1.8.1 Basilar Membrane Travelling Wave

The pressure fields observed at any segment of the basilar membrane consist not only of contributions due to motion of the stapes and, as shown here, to motion of the round window but, very importantly, to contributions due to the motion of all other parts of the basilar membrane as well. Here, it is proposed that the upper and lower galleries may each be modelled as identical transmission lines coupled along their entire length by the central partition, which acts as a mechanical shunt between the galleries (Bies, 2000).

Introducing the acoustic pressure, p, volume velocity, v (particle velocity multiplied by the gallery cross-sectional area), defined by Equations (2.1) and (2.2), respectively, and the acoustical inductance, L_A, per unit length of either gallery, the equation of motion of the fluid in either gallery takes the following form:

$$\frac{\partial p}{\partial z} = L_A \frac{\partial v}{\partial t} \tag{2.6}$$

The acoustical inductance is an inertial term and is defined below in Equation (2.10). Noting that motion of the central partition, which causes a loss of volume of one gallery, causes an equal gain in volume in the other gallery, the following equation of conservation of fluid mass may be written for either gallery:

$$\frac{\partial v}{\partial z} = 2C_A \frac{\partial p}{\partial t} \tag{2.7}$$

where C_A is the acoustic compliance per unit length of the central partition and is defined below in Equation (2.16).

Equations (2.6) and (2.7) are the well-known transmission line equations due to Heaviside (Nahin, 1990), which may be combined to obtain the well-known wave equation:

$$\frac{\partial^2 p}{\partial z^2} = \frac{1}{c^2}\frac{\partial^2 p}{\partial t^2} \tag{2.8}$$

The phase speed, c, of the travelling wave takes the form:

$$c^2 = \frac{1}{2C_A L_A} \tag{2.9}$$

The acoustical inductance, L_A, per unit length of the central partition, is:

$$L_A = \frac{\rho}{S_g} \tag{2.10}$$

where ρ is the fluid density and S_g is the gallery cross-sectional area. Calculation of the acoustical compliance, C_A, per unit length of the central partition is discussed below. It will be useful for this purpose to introduce the velocity, u, of a segment of the basilar membrane, defined as:

$$u = \frac{\partial y}{\partial t} \tag{2.11}$$

Sinusoidal time dependence of amplitude, y_0, will also be assumed. Thus:

$$y = y_0 e^{j\omega t} \tag{2.12}$$

Introducing the mechanical compliance, C_M, of a segment of the basilar membrane, Equation (2.3) may be rewritten in the following form:

$$-\frac{ju}{C_M \omega} = F_S + F_B \tag{2.13}$$

where F_S and F_B are defined in Section 2.1.7.

The mechanical compliance, C_M, of the segment is defined as:

$$C_M = (k_s - m\omega^2 + jC\omega)^{-1} \tag{2.14}$$

where k_s and m are the stiffness and mass, respectively, of the segment of basilar membrane and C is the damping coefficient of the segment.

The acoustical compliance, C_A, of the basilar membrane per unit length is obtained by multiplying the mechanical compliance by the square of the area of the segment, upon which the total force acts and dividing by the length of the segment in the direction of the gallery centre-line. The expression for the acoustical compliance per unit length is related to the mechanical compliance as:

$$C_A = \ell_w^2 \Delta z C_M \tag{2.15}$$

Substitution of Equation (2.14) into Equation (2.15) gives the acoustical compliance as:

$$C_A = \frac{\ell_w^2 \Delta z}{(k_s - m\omega^2 + jC\omega)} \tag{2.16}$$

Substitution of Equations (2.10) and (2.16) into Equation (2.9) gives the following equation for the phase speed, c, of the travelling wave on the basilar membrane:

$$c = \sqrt{\frac{S_g(k_s - m\omega^2 + jC\omega)}{2\rho\ell_w^2 \Delta z}} \tag{2.17}$$

To continue the discussion, it will be advantageous to rewrite Equation (2.17) in terms of the following dimensionless variables, which characterise a mechanical oscillator. The undamped resonance frequency or characteristic frequency of a mechanical oscillator, ω_N, is related to the oscillator variables, stiffness, k_s, and mass, m, as (Tse et al., 1979):

$$\omega_N = \sqrt{k_s/m} \tag{2.18}$$

The frequency ratio, X, is defined as:

$$X = \omega/\omega_N \tag{2.19}$$

where ω is the stimulus frequency.

The critical damping ratio, ζ, defined as follows, will play a very important role in the following discussion (see Section 9.2.1, Equation (9.15)):

$$\zeta = \frac{C}{2m\omega_N} = \frac{C}{2\sqrt{k_s m}} \tag{2.20}$$

It will be convenient to describe the mass, m, of an oscillator as proportional to the mass of fluid in either gallery in the region of an excited segment, whereas k_s is the stiffness of the basilar membrane segment under consideration. The proportionality constant, α, is expected to be of the order of one.

$$m = \alpha \rho S_g \Delta z \tag{2.21}$$

Substitution of Equations (2.18) to (2.21) into Equation (2.17) gives the following equation for the speed of sound, which will provide a convenient basis for understanding the properties of the travelling wave on the basilar membrane:

$$c = \frac{\alpha S_g \omega_N}{\ell_w \sqrt{2}} \sqrt{1 - X^2 + \mathrm{j}2\zeta X} \tag{2.22}$$

At locations on the basal side of a location of maximum response along the cochlear partition, where frequencies higher than the stimulus frequency are sensed, the partition will be driven below its frequency of maximum response. In this region, the partition will be stiffness-controlled and wave propagation will take place. In this case, $X < 1$ and Equation (2.22) takes the following approximate form, which is real, confirming that wave propagation takes place.

$$c = \frac{\alpha S_g \omega_N}{\ell_w \sqrt{2}} \tag{2.23}$$

At distances on the apical side of a location of maximum response, the partition will be driven above the corresponding frequency of maximum response, the shunt impedance of the basilar membrane will be mass controlled and wave propagation in this region is not possible. In this case, when $X \gg 1$, Equation (2.22) takes the following imaginary form:

$$c = \frac{\alpha S_g \omega_N}{\ell_w \sqrt{2}} \mathrm{j}X \tag{2.24}$$

confirming that no real wave propagates. Any motion will be small and finally negligible, as it will be controlled by fluid inertia.

In the region of the cochlear partition that is in maximum response with the stimulus frequency, the motion will be large, and only controlled by the system damping. In this case, $X = 1$ and Equation (2.22) takes the following complex form:

$$c = \frac{\alpha S_g \omega_N}{\ell_w \sqrt{2}} (1 + \mathrm{j}) \sqrt{\zeta} \tag{2.25}$$

As shown in Equation (2.25), at a location of maximum response on the basilar membrane, the mechanical impedance becomes complex, having real and imaginary parts, which are equal. In this case, the upper and lower galleries are shorted together. At the location of maximum response at low sound pressure levels when the damping ratio, ζ, is small, the basilar membrane wave travels very slowly. Acoustic energy accumulates at the location of maximum response and is rapidly dissipated doing work transforming the acoustic stimulus into neural impulses for transmission to the brain. At the same time, the wave is rapidly attenuated and conditions for wave travel cease, so that the wave travels no further, as first observed by Békésy and Wever (1960). The model is illustrated in Figure 2.2, where motion is shown as abruptly stopping at about the centre of the central partition.

2.1.8.2 Energy Transport and Group Speed

In a travelling wave, energy is transported at the group speed. Lighthill (1991) has shown by analysis of Rhode's data that the group speed of a travelling wave on the basilar membrane tends to zero at a location of maximum response. Consequently, each frequency component of any stimulus travels to the location where it is resonant and there it slows down, accumulates and is rapidly dissipated doing work to provide the signal that is transmitted to the brain. The travelling wave is a marvellous mechanism for frequency analysis.

The group speed, c_g, is defined in Equation (1.34) in terms of the component frequency, ω, and the wavenumber, k. Rewriting Equation (1.34) in terms of the frequency ratio, X, given by Equation (2.19), gives the following expression for the group speed:

$$c_g = \omega_N \frac{\mathrm{d}X}{\mathrm{d}k} \tag{2.26}$$

The wavenumber is defined in Equation (1.25). Substitution of Equation (2.22) into Equation (1.25) gives an expression relating the wavenumber, k, to the frequency ratio, X, as:

$$k = \frac{\ell_w \sqrt{2}}{\alpha S_g} X(1 - X^2 + \mathrm{j}2\zeta X)^{-1/2} \tag{2.27}$$

Substitution of Equation (2.27) into Equation (2.26) gives, with the aid of a bit of tedious algebra, the following expression for the group speed:

$$c_g = \frac{\alpha S_g \omega_N}{\ell_w \sqrt{2}} \frac{(1 - X^2 + \mathrm{j}2\zeta X)^{3/2}(1 - \mathrm{j}\zeta X)}{(1 + \zeta^2 X^2)} \tag{2.28}$$

In Equation (2.28), the damping ratio, ζ, appears always multiplied by the frequency ratio, X. This has the physical meaning that the damping ratio is only important near a location of resonant response, where the frequency ratio tends to unity.

Furthermore, where the basilar membrane responds passively, the frequency ratio is small and the damping ratio then is constant, having its maximum value of 0.5 (see Section 2.1.8.4). It may be concluded that in regions removed from places of resonant response, the group speed varies slowly and is approximately constant.

As a stimulus component approaches a location of maximum response and at the same time the frequency ratio tends to 1, the basilar membrane may respond actively, depending on the level of the stimulus, causing the damping ratio to become small. At the threshold of hearing, the damping ratio will be minimal, of the order of 0.011. However, at sound pressure levels of the order of 100 dB, the basilar membrane response will be passive, in which case the damping ratio will be 0.5, its passive maximum value (see Section 2.1.8.4) (Bies, 1996).

When a stimulus reaches a location of maximum response, the frequency ratio, $X = 1$, and the group speed is controlled by the damping ratio. The damping ratio, in turn, is determined by the active response of the location stimulated, which, in turn, is determined by the level of the stimulus. For $X = 1$, the numerator of Equation (2.28) becomes small and dependent on the value of the damping ratio, ζ, indicating, as observed by Lighthill (1996) in Rhode's data, that the group speed tends to zero as the wave approaches a location of maximum response.

2.1.8.3 Undamping

It was shown in Section 2.1.6 that a voltage is generated at the outer hair cells in response to motion of the basilar membrane. In the same section, it was shown that the outer hair cells may respond in either a passive or an active state, presumably determined by instructions conveyed to the hair cells from the brain by the attached efferent nerves. In an active state the outer hair cells change length in response to an imposed voltage (Brownell et al., 1985) and an elegant mathematical model describing the biophysics of the cochlea, which incorporates this idea, has been proposed (Mammano and Nobili, 1993). In the latter model, the stereocilia of the outer hair cells are firmly embedded in the tectorial membrane and the extension and contraction of the outer hair cells result in a greater motion of the basilar membrane in the direction of the imposed motion. The associated rocking of the tunnel of Corti into and out of the sulcus increases the rate of flow through the cleft in which the inner hair cell cilia are mounted and thus amplifies their motion (velocity) relative to the surrounding fluid. In this model, the outer hair cells act upon the cochlea in a region of resonant response, resulting in amplification of as much as 25 dB at very low sound pressure levels.

The effect of the intervention of the outer hair cells is to undamp the cochlea in a region of resonant response when the stimulation is of low level. For example, at sound pressure levels corresponding to the threshold of hearing, undamping may be very large, so that the stimulated segment of the cochlea responds like a very lightly damped oscillator. Tuning is very sharp and the stimulus is greatly amplified. At increasing levels of stimulation undamping decreases, apparently to maintain the basilar membrane velocity response, at the location of maximum response, within a relatively narrow dynamic range. It is suggested here that this property may be the basis for interpretation of the distortions on an incident sound field imposed by the pinna. The latter distortions are associated with direction in the frequency range above 3 kHz. Thus, the direction of jangling keys may be determined with just one ear, without moving the head (see Section 2.1.1).

As shown in Equation (2.18) the frequency of maximum velocity response does not depend on the system damping. By contrast, as shown in Equation (9.18), the frequency of maximum displacement response of a linear mechanical oscillator is dependent on the system damping. As shown in the latter equation, when the damping of an oscillator is small, the frequency of maximum displacement response approaches that of the undamped resonance frequency (or the frequency of maximum response), but with increasing damping, the frequency of maximum displacement response shifts to lower frequencies, dependent on the magnitude of the damping ratio (see Equation (9.18)).

The inner hair cells, which are velocity sensors (Bies, 1999), are the cells that convert incident sound into signals that are transmitted by the afferent system to the brain, where they are interpreted as sound. Thus, inner hair cells are responsible for conveying most of the amplitude and frequency information to the brain. At low sound pressure levels, in a region of resonant response, the outer hair cells amplify the motion of the inner hair cells that sense the sound, by undamping the corresponding segments of the cochlea. Thus, the outer hair cells play the role of compressing the response of the cochlea (Bacon, 2006) so that our hearing mechanism is characterised by a huge dynamic range of up to 130 dB, which would not have been possible without some form of active compression.

At high sound pressure levels, undamping ceases, apparently to protect the ear. In summary, undamping occurs at relatively low sound pressure levels and within a narrow frequency range about the frequency of maximum response at a location of stimulation. At all other places on the cochlea, which do not respond to such an extent to the particular stimulus, and at all levels of stimulation, the cochlear oscillators are heavily damped and quite linear. Only in a narrow range of the location on the cochlea, where a stimulus is sensed in resonant response and at low sound pressure levels, is the cochlea nonlinear.

From the point of view of the engineer, it is quite clear that the kind of nonlinearity, which is proposed here to explain the observed nonlinear response of the cochlea, is opposite to that which is generally observed in mechanical systems. Generally, nonlinearity is observed at high levels of stimulation in systems that are quite linear at low levels of stimulation.

2.1.8.4 The Half-Octave Shift

Hirsh and Bilger (1955) first reported an observation that is widely referred to as the "half-octave shift". They investigated the effect on hearing levels at 1 kHz and 1.4 kHz of six subjects exposed to a 1 kHz tone for intervals of time ranging from 10 seconds to 4 minutes and for a range of sensation levels from 10 dB to 100 dB. Sensation levels are understood to mean sound levels relative to the hearing threshold of the individual test subject.

Reporting mean results of their investigation, Hirsh and Bilger (1955) found that for one-minute duration in each case, and for all sensation levels from 10 to 100 dB, the threshold shift at 1 kHz was essentially a constant 6 dB, but the threshold shift at 1.4 kHz was an increasing monotonic function of sensation level. At a sensation level of 60 dB, the shifts at the two frequencies were essentially the same, but at higher sensation levels, the shift at 1.4 kHz was greater than at 1 kHz. Many subsequent investigations have confirmed these results for other frequencies and for other species as well.

A result typical of such investigations has been used to construct Figure 2.4. In the figure, temporary threshold shift (TTS) in decibels is shown as a function of frequency in kHz, distributed on the abscissa generally as along the cochlear duct with high frequencies at the basal end on the left and low frequencies at the apical end on the right.

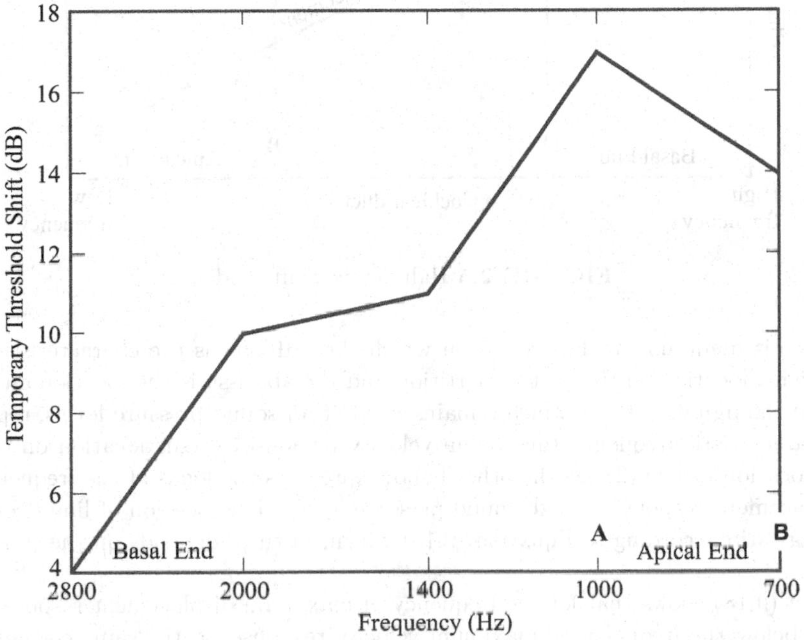

FIGURE 2.4 Typical half-octave shift due to exposure to a loud 700 Hz tone (from data published by Ward (1962, 1974)).

In the figure, TTS is shown after exposure to an intense 700 Hz tone. It is observed that an 11 dB shift at 700 Hz is associated with a larger shift of about 17 dB at 1 kHz, one half-octave higher in frequency. Significant hearing loss is also observed at even higher frequencies (reducing as the frequency increases), but essentially no loss is observed at frequencies below 700 Hz (Ward, 1962, 1974).

Crucial to the understanding of the explanation that will be proposed for the half-octave shift is the observation that the outer hair cells are displacement sensors and the inner hair cells, which provide frequency and amplitude information to the brain, are velocity sensors. For an explanation of Ward's data (Ward, 1962, 1974) shown in Figure 2.4, reference will be made to Figure 2.5. To facilitate the explanation proposed here for the half-octave shift, points **A** and **B** have been inserted in Figures 2.4 and 2.5. In Figure 2.4 they indicate the one half-octave above and the stimulus frequencies, respectively, while in Figure 2.5 they indicate locations on the cochlear duct corresponding to the places, respectively, where the one half-octave above and the stimulus frequencies are sensed. In the two figures, the "Apical end" and the "Basal end" have been inserted as a reminder that low frequencies are sensed at the apical end and high frequencies are sensed at the basal end of the cochlear duct.

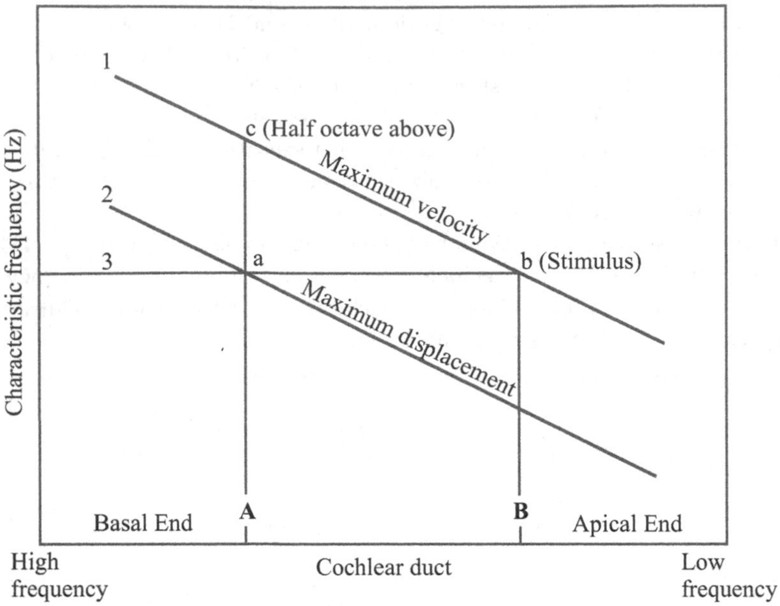

FIGURE 2.5 Half-octave shift model.

Reference is made now to Figure 2.5 on which the ordinate is the characteristic frequency associated with location on the central partition and the abscissa is the location on the central partition. In the figure, line (1), which remains fixed at all sound pressure levels, represents the locus of characteristic frequency (maximum velocity response) versus location on the cochlear duct. The location of line (2), on the other hand, represents the locus of the frequency of maximum displacement response at high sound pressure levels. The location of line (2) depends on the damping ratio according to Equation (9.18) which, in turn, depends on the sound pressure level.

Equation (9.18) shows that for the frequency of maximum displacement response to be one half-octave below the frequency of maximum velocity response for the same cochlear segment, the damping ratio must equal 0.5. In the figure, line (2) is shown at high sound pressure levels (> 100 dB) at maximum damping ratio and maximum displacement response. As the sound pressure level decreases below 100 dB, the damping decreases and line (2) shifts towards line (1) until the lines are essentially coincident at very low sound pressure levels.

Consider now Ward's investigation (Figure 2.4) with reference to Figure 2.5. Ward's 700 Hz loud exposure tone is represented by horizontal line (3), corresponding to exposure of the ear to a high sound pressure level for some period of time at the location of maximum displacement response at (a) and at the same time at the location of maximum velocity response at (b). The latter point (b) is independent of damping and independent of the amplitude of the 700 Hz tone, and remains fixed at location B on the cochlear partition. By contrast, the maximum displacement response for the loud 700 Hz tone is at a location on the basilar membrane where 700 Hz is half an octave lower than the characteristic frequency at that location for low-level sound. Thus, the maximum displacement response occurs at intersection (a) at location **A** on the cochlear partition, which corresponds to a normal low-level characteristic frequency of about 1000 Hz, which is one half of an octave above the stimulus frequency of 700 Hz.

The highest threshold shift, when tested with low-level sound, is always observed to be one half-octave higher than the shift at the frequency of the exposure tone at (b). Considering the active role of the outer hair cells, which are displacement sensors, it is evident that point (a) is now coincident with point (c) and that the greater hearing level shift is due to damage of the outer hair cells when they were excited by the loud tone represented by point (a). This damage may prevent the outer hair cells from performing their undamping action, resulting in an apparent threshold shift at the characteristic frequency for low-level sound (half an octave higher than the high-level sound used for the original exposure).

The lesser damage to the outer hair cells at frequencies higher than the frequency corresponding to one half-octave above the exposure tone may be attributed to the effect of being driven by the exposure tone at a frequency less than the frequency corresponding to the maximum velocity response. Estimation of the expected displacement response in this region on the basal side of point A on the cochlear duct, at the high damping ratio expected of passive response, is in reasonable agreement with this observation.

Here, a simple explanation has been proposed for the well-known phenomenon referred to as "the half-octave shift" (Bies, 1996).

2.1.8.5 Frequency Response

Although humans can hear down to frequencies of approximately 2 Hz, it is generally accepted that the frequency response of the central partition ranges from the 20 kHz at the basal end at the stapes to 20 Hz at the apical end at the helicotrema. Hearing outside of this frequency range is not dependent on achieving resonant response of the central partition.

To describe the frequency response along the central partition, it will be convenient to introduce the normalised distance, z, which ranges from 0 at the basal end of the basilar membrane to 1 at the apical end. The length of the basilar membrane in humans varies between about 33.5 mm and 36 mm (Keen, 1940). Based on work of Greenwood (1990), the following equation is proposed to describe the frequency response of the central partition:

$$f(z) = 165.4\left[10^{2.1(1-z)} - 0.88)\right] \quad \text{(Hz)} \tag{2.29}$$

The frequency response versus location from the stapes end is shown in Figure 2.6, where it is assumed that the length of the basilar membrane in a human ear is 35 mm (Greenwood, 1990). The dashed line in Figure 2.6 shows the frequency response where the constant term has been removed, and is written as:

$$f(z) = 165.4\left[10^{2.1(1-z)}\right] \quad \text{(Hz)} \tag{2.30}$$

Comparison of the frequency responses predicted by Equations (2.29) and (2.30) indicates that the difference between the two expressions is less than 10% for $z \leq 0.53$, or a distance from the stapes of $0.53 \times 35 = 18.6$ mm. Substitution of $z = 0.53$ in Equation (2.29) gives the predicted frequency response as 1460 Hz. Hence, for frequencies higher than about 1500 Hz, the relationship between frequency response and basilar membrane position will be log-linear.

2.1.8.6 Critical Frequency Band

A variety of psychoacoustic experiments has required for their explanation the introduction of the familiar concept of the bandpass filter. In the literature concerned with the ear, the band pass filter is given the name, "critical frequency band" (Moore, 1982). It will be useful to use the latter term in the following discussion in recognition of its special adaptations for use in the ear.

Of the 16 000 hair cells in the human ear, about 4000 are the sound sensing inner hair cells, suggesting the possibility of exquisite frequency discrimination at very low sound pressure levels when the basilar membrane is very lightly damped. On the other hand, as will be shown, frequency analysis may be restricted to just 35 distinct and non-overlapping critical bands and as has been shown, variable damping plays a critical role in the functioning of the basilar membrane. Further consideration is complicated by the fact that damping may range from very small to

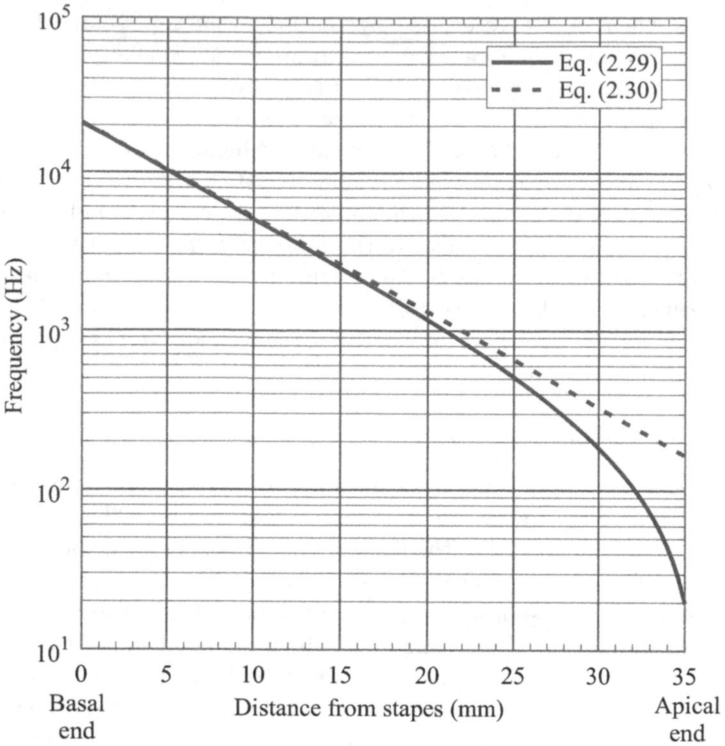

FIGURE 2.6 Frequency response of basilar membrane versus location (based on Figure 1 of Greenwood (1990)).

large with concomitant variation in frequency response of the segments of the basilar membrane. Clearly, active response plays a critical role in determining the critical bandwidth. For the case of 1 kHz and higher frequencies ($z \leq 0.6$), the derivative of Equation (2.30) may be written in the following differential form:

$$\frac{\Delta f}{f(z)} = 4.835\Delta z \tag{2.31}$$

In Equation (2.31), the critical bandwidth may be associated with Δf and the centre band frequency with $f(z)$. Moore (1982) has summarised the work of several authors (Scharf, 1970), who provided experimental determinations of critical bandwidth as a function of frequency. This summary was adapted to construct Figure 2.7.

Referring to Figure 2.7, it may be observed that the ratio of critical bandwidth to centre band frequency is constant in the frequency range above 1 kHz. Equation (2.31) shows that in this frequency range, each filter extends over a "constant length" of the central partition. A simple calculation using Equation (2.31), with information taken from Figure 2.7 and taking the average length of the basilar membrane as 34 mm (see Section 2.1.3), gives a value of about 1 mm for the "constant length" of the central partition (Moore, 1982). A maximum of 34 filters is suggested by this calculation, which is in very good agreement with experiments suggesting that the cochlear response may be described with about 35 critical bands (Moore, 1982). Each critical band is associated with a segment of the basilar membrane.

Critical bands are referred to in DIN 45631/A1:2010-03 (2010) for the assessment of time-varying loudness. The critical bandwidth of frequencies below about 500 Hz is 100 Hz so the lowest possible band spans the range from 0 to 100 Hz (Bark 0.5). Above 500 Hz, the bandwidth is about 20% of the centre frequency, f_c. The critical bands are defined in a similar way to

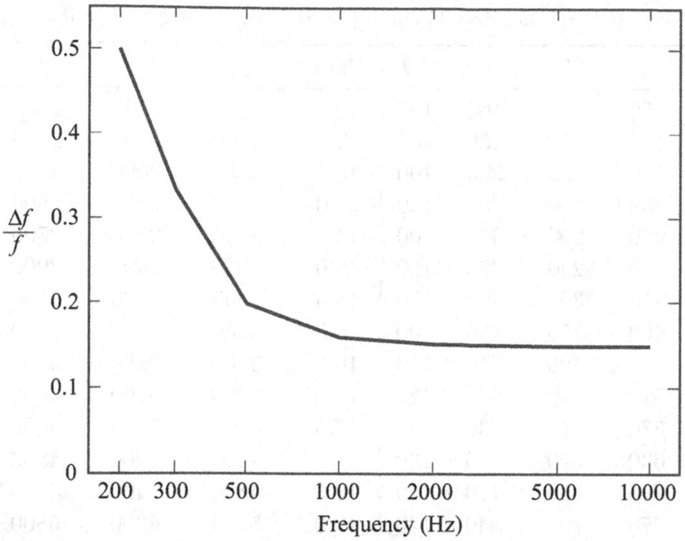

FIGURE 2.7 Normalised critical bandwidth as a function of centre band frequency based on Figure 3.6 of Moore (1982).

1/3-octave bands (see Section 1.14), but they are interpreted slightly differently. Rather than being considered as fixed, the critical bands can be thought of as varying along the Bark 0.5 to Bark 24 limits (Fastl and Zwicker, 2007), but defined entirely by an arbitrarily specified centre frequency and a corresponding bandwidth which is a function of frequency and given by Equation (2.32) and Figure 2.8 (IEC 61400-11 Ed.3.0, 2018). Thus, any frequency band can be defined as a fractional Bark, with its Bark value dependent on the chosen centre frequency, f_c, and the bandwidth determined by Equation (2.32). The bands corresponding to integer and half-integer Bark values between Bark 0.5 and Bark 24 (the upper limit) are listed in Table 2.1.

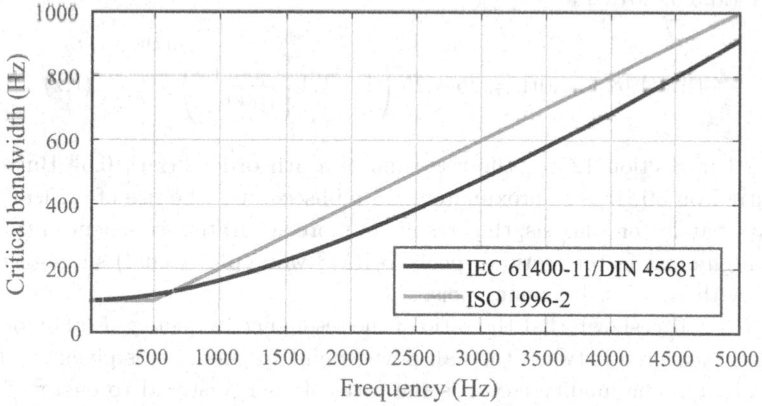

FIGURE 2.8 Critical bandwidths proposed in the IEC 61400-11 Ed.3.0 (2018), DIN 45681 (2005-08) (2005) and ISO 1996-2 (2017) standards.

Unlike 1/3-octave bands, which have specified centre frequencies and bandwidths, critical bands are defined by choosing the centre frequency, f_c, to be the frequency of interest, and once this has been specified, the critical bandwidth is defined using Equation (2.32) and Figure 2.8

TABLE 2.1 Critical frequency bands corresponding to integer and half-integer Bark values

Bark	f_c	f_ℓ	f_u	Δf	Bark	f_c	f_ℓ	f_u	Δf
0.5	50	0	100	100	12.5	1850	1720	2000	280
1.0	100	50	150	100	13.0	2000	1850	2150	300
1.5	150	100	200	100	13.5	2150	2000	2320	320
2.0	200	150	250	100	14.0	2320	2150	2500	350
2.5	250	200	300	100	14.5	2500	2320	2700	380
3.0	300	250	350	100	15.0	2700	2500	2900	400
3.5	350	300	400	100	15.5	2900	2700	3150	450
4.0	400	350	450	100	16.0	3150	2900	3400	500
4.5	450	400	510	110	16.5	3400	3150	3700	550
5.0	510	450	570	120	17.0	3700	3400	4000	600
5.5	570	510	630	120	17.5	4000	3700	4400	700
6.0	630	570	700	130	18.0	4400	4000	4800	800
6.5	700	630	770	140	18.5	4800	4400	5300	900
7.0	770	700	840	140	19.0	5300	4800	5800	1000
7.5	840	770	920	150	19.5	5800	5300	6400	1100
8.0	920	840	1000	160	20.0	6400	5800	7000	1200
8.5	1000	920	1080	160	20.5	7000	6400	7700	1300
9.0	1080	1000	1170	170	21.0	7700	7000	8500	1500
9.5	1170	1080	1270	190	21.5	8500	7700	9500	1800
10.0	1270	1170	1370	200	22.0	9500	8500	10500	2000
10.5	1370	1270	1480	210	22.5	10500	9500	12000	2500
11.0	1480	1370	1600	230	23.0	12000	10500	13500	3000
11.5	1600	1480	1720	240	23.5	13500	12000	15500	3500
12.0	1720	1600	1850	250	24.0	15500	13500	17500	4000

(IEC 61400-11 Ed.3.0, 2018)) as:

$$\text{Critical bandwidth} = 25 + 75\left(1 + 1.4\left[\frac{f_c}{1000}\right]^2\right)^{0.69} \quad \text{(Hz)} \tag{2.32}$$

As discussed in Section 12.2.1, the rise time of a 4th order "Bark 0.5" Butterworth band pass filter centred on 50 Hz is approximately 8.8 milliseconds. The use of critical bands, rather than 1/3-octave bands, for analysis, thus results in more accurate assessment of the potential for annoyance of high crest factor (ratio of peak to RMS value of a signal) short-term events that are associated with very low-frequency noise.

In Section 6.3.2, it is shown that the ratio of the resonance frequency, f, of an oscillator to the bandwidth, Δf, measured between the half power point frequencies, is given the name Quality Factor or simply, Q. The quality factor is the ratio of energy stored to energy dissipated and as damping is proportional to the energy dissipated, lower damping will result in higher Q and sharper tuning. Consequently, it is reasonable to conclude from Figure 2.7 that the sharpness of tuning of the ear is greatest above about 2000 Hz.

An example of a psychoacoustic experiment in which the critical frequency band plays an important role is the case of masking of a test tone with a second tone or band of noise (see Section 2.2.1). An important application of such investigations is concerned with speech intelligibility in a noisy environment.

The well-known phenomenon of beating between two pure tones of slightly different frequencies and the masking of one sound by another are explained in terms of the critical band. The critical bandwidth about a pure tone or a narrow band of noise is defined as the frequency band about the stimulus tone or narrow band of noise within which all sounds are summed together and without which all sounds are summed separately. This consideration suggests that the critical band is associated with a segment of the central partition. For example, energy summing occurs in the net response of a single stimulated segment.

Two pure tones will be heard as separate tones, unless their critical bands overlap, in which case they will be heard as one tone of modulated amplitude. This phenomenon is referred to as beating (see Section 1.12). In the case of masking of a test tone with a second tone or a narrow band of noise, only those frequency components of the masker that are in the critical band associated with the test tone will be summed with the test tone. Energy summing of test stimulus and masker components takes place at a location of resonance on the central partition.

2.1.8.7 Frequency Resolution

As shown in the discussion of spectra (see Section 1.14), noise of broad frequency content can best be analysed in terms of frequency bands, and within any frequency band, however narrow, there are always an infinite number of frequencies, each with an indefinitely small energy content. A tone, on the other hand, is characterised by a single frequency of finite energy content. The question is raised, "What bandwidth is equivalent to a single frequency?" The answer lies with the frequency analysing system of the ear, which is active and about which very little is known. As shown in Section 2.1.8.6, the frequency analysing system of the ear is based on a very clever strategy of transporting all components of a sound along the basilar membrane, without dispersion, to the places of resonance where the components are systematically removed from the travelling wave and reported to the brain.

As has been shown, the basilar membrane is composed of about 35 separate segments, which are capable of resonant response and which apparently form the mechanical basis for frequency analysis. The significance of a limited number of basic mechanical units or equivalently critical bands from which the ear constructs a frequency analysis, is that all frequencies within the response range (critical band) of a segment will be summed as a single component frequency. A well-known example of such summing, referred to as beating, was discussed in Sections 1.12 and 2.1.8.6.

2.2 Subjective Response to Sound Pressure Level

Often it is the subjective response of people to sound, rather than direct physical damage to their hearing, which determines the standard to which proposed noise control must be compared, and which will determine the relative success of the effort. For this reason, the subjective response of people to sound will now be considered, determined as means of large samples of the human population (Moore, 1982). The quantities of concern are loudness and pitch, as well as the masking of one type of sound by another. Sound quality, which is concerned with spectral energy distribution, will not be considered. Masking will be discussed first, followed by loudness and then pitch.

2.2.1 Masking

Masking is the phenomenon of one sound interfering with the perception of another sound. For example, the interference of traffic noise with the use of a mobile telephone on a busy street corner is probably well-known to everyone. Examples of masking are shown in Figure 2.9, in which is shown the effect of either a tone or a narrow band of noise on the threshold of hearing across the entire audible spectrum. The tone or narrow band of noise will be referred to as the masker.

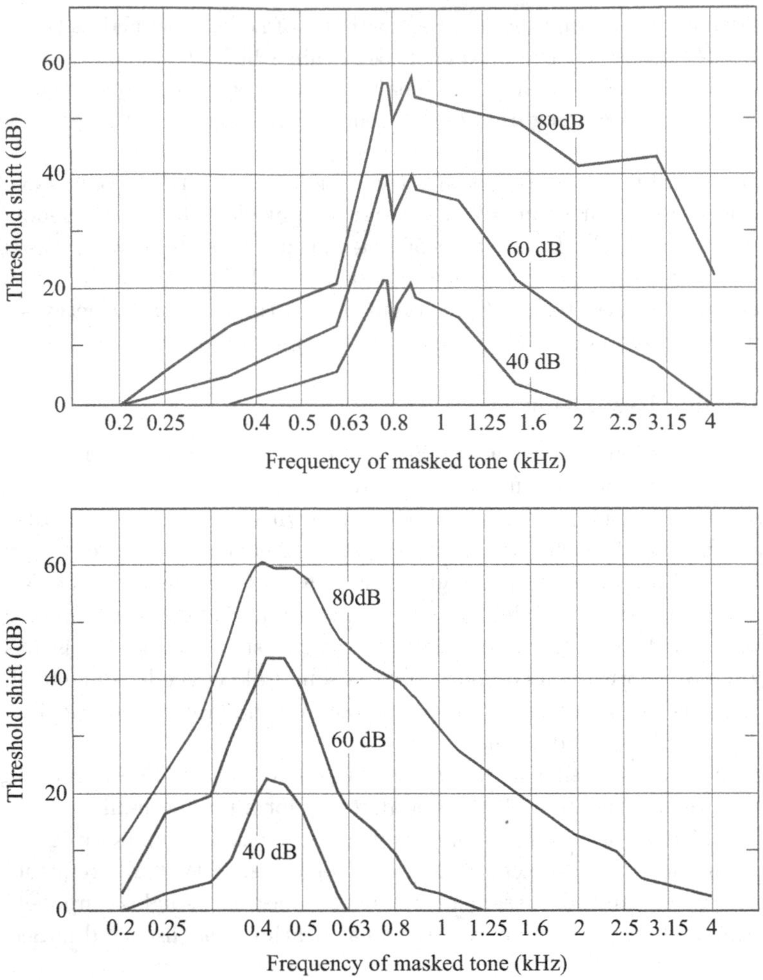

FIGURE 2.9 Example of masked audible spectra where the masker is either a tone or a narrow band of noise and the threshold shift is the amount by which the sound pressure level of a tone at any other frequency needs to be increased to be audible in the presence of the masker. The masker at three levels is: (a) 800 Hz tone; (b) a narrow band of noise 90 Hz wide centred on 410 Hz.

Referring to Figure 2.9, the following may be observed.

(a) The masker is an 800 Hz tone at three sound pressure levels. The masker at 80 dB has increased the level of detection of a 600 Hz tone by 25 dB and the level for detection of a 1100 Hz tone by 52 dB. The masker is much more effective in masking frequencies higher than itself than in masking frequencies lower than itself.

(b) The masker is a narrow band of noise 90 Hz wide centred at 410 Hz. The narrow band of noise masker is seen to be very much more effective in masking at high frequencies than at low frequencies, consistent with the observation in (a).

As shown in Figure 2.9, high frequencies are more effectively masked than are low frequencies. This effect is well-known and is always present, whatever the masker. The analysis presented here suggests the following explanation. The frequency component energies of any stimulus will each be transported essentially without loss at a relatively constant group speed, to a location of resonance on the basilar membrane. As a component approaches a location of resonance, the group speed of the component slows down and reaches a minimum at the location of resonance, where the component's energy is dissipated doing work to provide a stimulus, which is transmitted to the brain. Only in the region on the basilar membrane where the masker frequency is resonant will the masker and test tone components be summed, giving rise to high detection thresholds, except for the case of a tonal masker when the two frequencies coincide.

Evidently, the higher levels of threshold shift at high frequencies are due to the passage of the masker components through the places of resonance for high frequencies. It is suggested here that the most likely explanation is that the outer hair cells, which act to amplify a test stimulus at low levels, are inhibited by the high levels of excitation resulting from transmission of the masker. Consequently, the threshold level is elevated. By contrast, any residual components of the masker must decay very rapidly so that little or no masker is present on the apical side of the location of masker stimulation. The masking that is observed at low frequencies is due to the low-frequency response of the ear acting as a filter, which reduces the ability of the ear to perceive the higher frequency tone. In a linear filter, unique relations exist that can provide guidance, but for the ear, the system is active, which complicates the situation.

In all of the curves of Figure 2.9(a), where the masker is a tone, a small dip is to be noted when the variable tone approaches the masking tone. This phenomenon may be interpreted as meaning that the tones are close enough for their critical bands to overlap. In the frequency range of critical bandwidth overlap, one tone of modulated amplitude will be heard. For example, consider two closely spaced but fixed frequencies. As the phases of the two sound disturbances draw together, their amplitudes reinforce each other, and as they subsequently draw apart, until they are of opposite phase, their amplitudes cancel. The combined amplitude thus rises and falls, producing beats (see Section 1.12). Beats are readily identified, provided that the two tones are not greatly different in level; thus the dip is explained in terms of the enhanced detectability due to the phenomenon of beating. In fact, the beat phenomenon provides a very effective way of matching two frequencies. As the frequencies draw together the beating becomes slower and slower until it stops with perfect matching of tones. Pilots of propeller driven twin-engine aircraft use this phenomenon to adjust the two propellers to the same speed.

Reference is made now to Figure 2.10, where the effectiveness, as masker, of a tone and a narrow band of noise is compared. The tone is at 400 Hz and the band of noise is 90 Hz wide centred at 410 Hz. Both maskers are at 80 dB sound pressure level. It is evident that the narrow band of noise is more effective as a masker over most of the audio frequency range, except at frequencies above 1000 Hz where the tone is slightly more effective than a narrow band of noise.

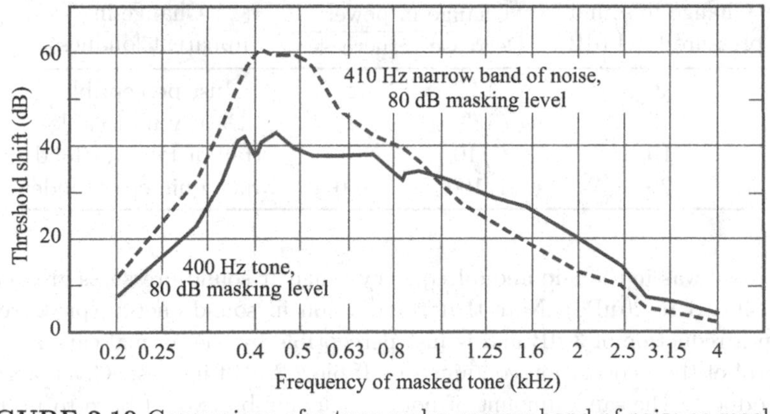

FIGURE 2.10 Comparison of a tone and a narrow band of noise as maskers.

It is of interest to note that the crossover, where the narrow band of noise becomes less effective as a masker, occurs where the ratio of critical bandwidth to centre band frequency becomes constant and relatively small (see Section 2.1.8.6 and Figure 2.7). In this range, the band filters are very sharply tuned. That is, the tone is more effective as a masker in the frequency range where the cochlear response is most sharply tuned, suggesting that the band pass filter is narrow enough to reject part of the narrow band of noise masker.

In the foregoing discussion of Figures 2.9 and 2.10, a brief summary has been presented of the effect of the masking of one sound by another. This information is augmented by reference to the work of Kryter (1970, pp. 30,33). Kryter has reviewed the comprehensive literature that was available to him and based on his review he has prepared the following summary of his conclusions.

- Narrowband noise causes greater masking around its frequency than does a pure tone of that frequency. This should be evident, since a larger portion of the basilar membrane is excited by the noise.
- Narrowband noise is more effective than pure tones in masking frequencies below 1000 Hz.
- A noise bandwidth is ultimately reached above which any further increase of bandwidth has no further influence on the masking of a pure tone at its frequency. This implies that the ear recognises certain critical bandwidths associated with the regions of activity on the basilar membrane.
- The threshold of the masked tone is normally raised to the level of the masking noise only in the critical bandwidth centred on that frequency.
- A tone, which is a few decibels above the masking noise, seems about as loud as it would sound if the masking noise were not present.

2.2.2 Loudness

The subjective response of a group of normal participantss to variation in sound pressure has been investigated (Stevens, 1957, 1972; Zwicker, 1958; Zwicker and Scharf, 1965). Table 2.2 summarises the results, which have been obtained for a single fixed frequency or a narrow band of noise containing all frequencies within some specified and fixed narrow range of frequencies.

TABLE 2.2 Subjective effect of changes in sound pressure level

Change in sound pressure level (dB)	Change in power		Change in apparent loudness
	Decrease	Increase	
3	1/2	2	Just perceptible
5	1/3	3	Clearly noticeable
10	1/10	10	Half or twice as loud
20	1/100	100	Much quieter or louder

The test sound was in the mid audio-frequency range at sound pressures greater than about 2×10^{-3} Pa (40 dB re 20 µPa). Note that a reduction in sound energy (pressure squared) of 50% results in a reduction of 3 dB and is just perceptible by the normal ear. The consequence for noise control of the information contained in Table 2.2 is of interest. Given a group of noise sources all producing the same amount of noise, their number would have to be reduced by a factor of 10 to achieve a reduction in apparent loudness of one-half. To decrease the apparent loudness by half again, that is to one-quarter of its original subjectively judged loudness, would require a further reduction of sources by another factor of 10. Alternatively, if we started with one trombone player behind a screen and subsequently added 99 more players, all doing their best, an audience out in front of the screen would conclude that the loudness had increased by a factor of four.

In contrast to what is shown in Table 2.2, an impaired ear with recruitment (see Section 2.5.3), in which the apparent dynamic range of the ear is greatly compressed, can readily detect small changes in sound pressure, so Table 2.2 does not apply to a person with recruitment. For example, an increase or decrease in sound power of about 10%, rather than 50% as in the table, could be just perceptible to a person with recruitment. It is expected that a person with recruitment may be more susceptible than a person with normal hearing to annoyance by an amplitude modulated sound.

It has been observed that outer hair cells are more sensitive than inner hair cells to excessive noise. It has also been observed that exposure to loud noise for an extended period of time will produce effects such as recruitment. These observations suggest that impairment of the outer hair cells is associated with recruitment. With time and rest, the ear will recover from the effects of exposure to loud noise if the exposure has not been too extreme. However, with relentless exposure, the damage to the hair cells will be permanent and recruitment may be the lot of their owner.

2.2.3 Comparative Loudness and the Phon

Variation in the level of a single fixed tone or narrow band of frequencies, and an average person's response to that variation, has been considered. Consideration will now be given to the comparative loudness of two sounds of different frequency content (ISO 226, 2023), as illustrated in Figure 2.11, where experimental results are summarised for the average perception of loudness by a large number of young test subjects. The results are derived from a considerable number of experiments spanning many years, with test subjects having undamaged normal hearing.

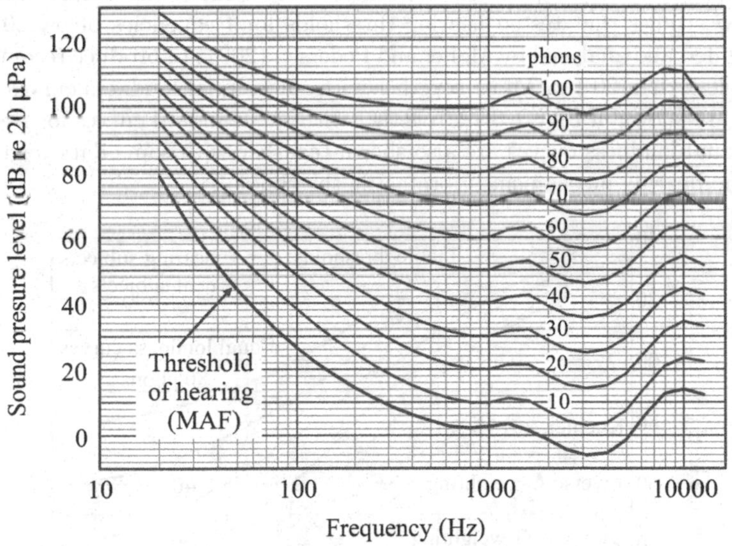

FIGURE 2.11 Equal loudness free-field frontal incidence contours in phons for tonal noise. MAF is the mean of the minimum audible field.

In the experiments represented by Figure 2.11, a subject was placed in a free field with sound frontally incident. The subject was presented with a 1 kHz tone used as a reference and alternately with a second sound used as a stimulus. The subject was asked to adjust the level of the stimulus until it sounded equally as loud as the reference tone. After the subject had adjusted a stimulus sound so that subjectively it seemed to be equally as loud as the 1 kHz tone, the sound pressure of the stimulus was recorded. Maps based on mean lines through the resulting data are shown in Figure 2.11. It is evident from the figure that the response of the ear, as subjectively reported, is both frequency and pressure-amplitude dependent.

The units used to label the equal-loudness contours in the figure are called phons. The lines in the figure are constructed so that all variable sounds of the same number of phons sound equally loud. In each case, the phon scale is chosen so that the number of phons equals the sound pressure level of the reference tone at 1 kHz. For example, according to Figure 2.11, a tonal stimulus of 40 phons at 31.5 Hz sounds equally loud as a 1000 Hz tone of 40 phons, even though the sound pressure level of the lower-frequency sound is about 49 dB higher. Humans are quite insensitive to sound at low frequencies. The bottom line in the figures represents the average threshold of hearing, or minimum audible field (MAF).

The phon has been defined so that for a tonal stimulus, every equal-loudness contour must pass through the point at 1000 Hz where the sound pressure level is equal to the corresponding phon number.

2.2.4 Low-Frequency Loudness

Many articles on hearing sensitivity in humans state that the range of human hearing is between 20 Hz and 20 kHz. However, this is not strictly correct as humans can detect much lower frequency sound via their vestibular system. Human response to sound has been measured at frequencies down to 2 Hz and the response is via the sensory apparatus of the inner ear that is associated with balance, rather than via tactile sensation.

As the annoyance threshold for low-frequency noise is very close to the hearing threshold for most people, it is of interest to examine the hearing threshold as a function of frequency below 20 Hz. The international standard, ISO 226 (2023) (see Figure 2.11), provides generally agreed on hearing threshold curves for frequencies between 20 Hz and 12500 Hz. Unfortunately, there have been only a few studies that tested hearing thresholds for frequencies below 20 Hz, and these have been compiled and analysed by Møller and Pedersen (2004) to produce recommended curves for frequencies less than 20 Hz. These together with some equal-loudness curves for frequencies below 1000 Hz are shown in Figure 2.12, together with the standard curves for frequencies above 20 Hz. Some equal-loudness curves are also shown together with some data for threshold levels that exclude the hearing mechanism (vibrotactile thresholds).

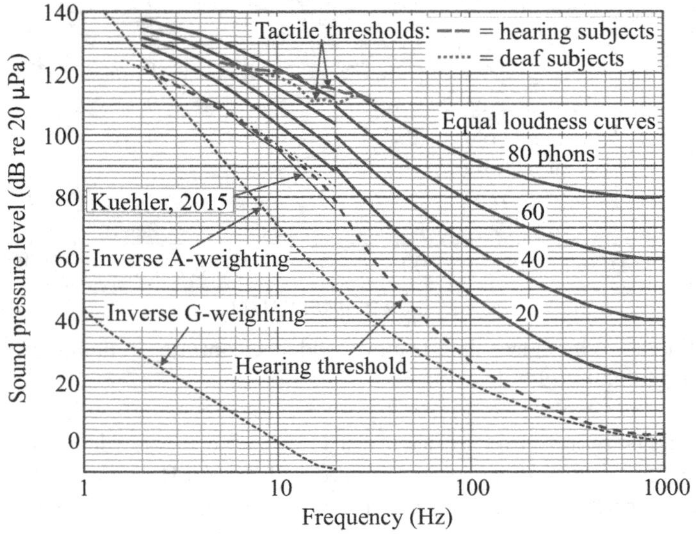

FIGURE 2.12 Low-frequency hearing thresholds and equal-loudness contours.

There are two characteristics of Figure 2.12 that are notable. The first is that the difference between the hearing threshold and an 80 phon sound becomes very small at low and infrasonic frequencies. The second characteristic is the discontinuity in the equal-loudness curves at 20 Hz. This is because the infrasound curves are based on fewer and different studies. The original curve suggested by Møller and Pedersen (2004) for the hearing threshold below 20 Hz is shown as the faint dotted line and is the average of ten past studies. Unfortunately, this curve does not meet the ISO 226 curve at 20 Hz, missing it by approximately 4 dB. On investigation of the data used by Møller and Pedersen (2004) to obtain the average, it is apparent that there are two studies for which the data match the ISO standard at 20 Hz and one of these is a study done by Watanabe and Møller (1990). Thus, when preparing Figure 2.12, it was decided to use the data from these latter two studies in the dashed curve shown in the figure for frequencies between 8 and 20 Hz so that the curve merges with the ISO standard at 20 Hz. The only significant difference between this curve and the one suggested by Møller and Pedersen (2004) is 4 dB at 20 Hz and 2 dB at 16 Hz. The hearing threshold curve below 20 Hz shown in Figure 2.12 is further justified by more recent threshold tests (Kuehler et al., 2015), shown as the faint solid line in the figure.

An inverse A-weighting curve (see Section 2.3) can be obtained by changing the sign of all of the A-weighting values, so the curve can then be directly related to equal-loudness curves. The inverse A-weighting curve is shown in Figure 2.12 to demonstrate that it is not a very good approximation of the perceived loudness of noise at frequencies below 20 Hz at any loudness level, nor is it a good approximation of the loudness of noise at a loudness level near our hearing threshold at any frequency. However, it is a reasonable approximation of the loudness of noise at a level of 60 phons above 20 Hz (but not necessarily representative of the annoyance of the noise, which is known to increase as the frequency decreases). The inverse G-weighting curve (see Section 2.3) is also shown in Figure 2.12 to demonstrate that it is a good approximation of the slope of the hearing threshold curve for noise at frequencies below 20 Hz, although it is not at all good for noise at frequencies above 20 Hz (see Figure 2.16). The slope of the hearing threshold curve below 20 Hz is 6 dB per octave, which implies that the hearing mechanism is responsive to the rate of change of the rate of change of pressure (that is, the acceleration of the pressure). Thus, a sharp impulse is readily detected by the hearing mechanism as a result of the rate-of-rate-of-change (i.e. double differential with respect to time) of its leading edge, and this is often closely correlated with the peak amplitude of the impulse (Swinbanks, 2015), indicating that the infrasonic sensitivity of the ear is related to the signal peak rather than its RMS value.

The thresholds and equal-loudness curves in Figures 2.11 and 2.12 represent the median of the population. One standard deviation is between 5 dB and 6 dB, so that if the hearing threshold in dB is normally distributed, then 2.5% of the population (representing two standard deviations on one side) would have a hearing threshold that was 10 to 12 dB less than that shown in Figures 2.11 and 2.12. The threshold curve in the infrasonic region of Figure 2.12 represents a sound pressure level of 97 dBG. If we subtract two standard deviations from this number, we arrive at 85 dBG which is the recognised hearing threshold used to evaluate noise dominated by infrasound. In fact, 2.5% of the population would have a lower hearing threshold than this and 0.15% would have a hearing threshold lower than 79 dBG.

It is important to note that the hearing thresholds in Figure 2.12 are for single frequency tonal noise. Møller and Pedersen (2004) state that 1/3-octave broadband, steady noise of a similar level is characterised by a similar hearing threshold. However, James (2012) states that the threshold of perception for a complex set of tones that are modulated in frequency and amplitude is likely to be much lower.

The mechanism by which our hearing system detects infrasound and very low-frequency sound is complex. von Gierke, H. E. and Nixon (1976) suggest that detection results from non-linearities of conduction in the middle ear, which generate higher frequency harmonics in the more audible higher frequency range. More recently, Salt and Lichtenhan (2014) point out that the outer hair cells (OHCs) in the cochlea are more sensitive than the inner hair cells (IHCs) to low-frequency sound and infrasound and respond at levels well below the threshold of hearing. The threshold of hearing is determined by the response of the inner hair cells (Salt and Hullar, 2010). The reason that OHCs are more sensitive to low-frequency sound and infrasound is that they are displacement sensors as a result of the ends of the stereocilia being attached to the tectorial membrane. On the other hand, the IHCs stereocilia are not attached at one end and so respond to the vibrations of the fluid near the tectorial membrane, resulting in them responding to velocity rather than displacement. They are also innervated by type I afferent nerve fibres and so are responsible for our hearing response.

Salt and Hullar (2010) estimate that at 5 Hz, the OHCs are 40 dB more sensitive than the IHCs. As Salt and Lichtenhan (2014) point out, the human hearing response to sound, as the frequency decreases from 500 Hz to 20 Hz, reduces at the rate of 18 dB per octave but below 10 Hz, the rate of sensitivity loss is much smaller and more like 12 dB/octave (see Figure 2.12). Salt and Hullar (2010) suggest that this may be due to stimulation of the IHCs that are responsible for hearing, by extracellular voltages generated by the OHCs. Stimulation of the IHCs at levels below the hearing threshold results in information transfer via pathways that do not involve conscious hearing, which may lead to various sensations, such as awakening from a deep sleep without having heard the noise that caused the awakening. According to Salt, this sort of stimulation could result in sleep disturbance and feelings of panic on awakening, with chronic sleep deprivation leading to blood pressure elevation and memory dysfunction.

The greatest response of the OHCs to low-frequency sound and infrasound occurs when stimulus in the mid-frequency range between 200 Hz and 2000 Hz is absent. This helps to explain why people in quiet rural environments may be more susceptible to wind farm noise than people in urban and suburban environments where the background sound pressure levels in the mid-frequency range are considerably higher (as found by Pedersen et al. (2009)). This idea is further supported by the work of Krahé (2010), who showed that low-frequency noise (between 20 Hz and 100 Hz) without the presence of significant noise above 100 Hz produced more stress than noise at the same level below 100 Hz, but which also included frequencies up to 1000 Hz. Thus, the noise with the higher A-weighted level was found to be less annoying, supporting the suggestion that the A-weighted level (see Section 2.3) is an inappropriate measure of the disturbance caused by low-frequency noise.

A particularly insidious effect of prolonged exposure to inaudible, periodic low-frequency noise and infrasound at levels well below the normal hearing threshold is that the sensitivity to audible noise of a person subjected to prolonged exposure over months and years can be increased substantially (Oud, 2013). This means that such a sensitised person may not find relief on leaving the vicinity of the low-frequency noise source such as a wind farm, as there are many other environmental noises to which they may have become sensitised (Oud, 2013).

As mentioned in Section 2.1.2, Bell (2014) has suggested that infrasound at levels below the threshold of hearing could cause the muscles of the middle ear to stiffen and relax in time with a periodic variation in infrasonic pressure, causing a feeling of fullness in the ear as well as vertigo and nausea in some people after prolonged exposure.

Some researchers have suggested that infrasound could be perceived through physiological mechanisms other than hearing. According to Dooley (2013), infrasound can be described as fluctuations or cyclic changes in the local barometric pressure, which are comparable to fluctuations in the surrounding barometric pressure experienced by an individual on a ship in high seas. The pressure fluctuations experienced by the individual on the ship occur due to changes in elevation as the ship moves between the crest and trough of ocean waves. Dooley (2013) proposed that this cyclic pressure variation may be the cause of motion sickness on ships as well as nausea in the vicinity of wind farms, as the amplitude of vertical motion in the 2 Hz to 4 Hz range required to produce seasickness in sensitive individuals corresponds to an atmospheric pressure variation that is similar to the levels of acoustic pressure variation experienced by people living in the vicinity of wind farms. The regular, periodic nature of these variations may explain why similar levels of randomly varying environmental infrasound do not result in motion sickness symptoms in sensitive individuals. Dooley also suggests that the problems of nausea are exacerbated when the sensation that is detected by one of our senses is not reflected by another. For the case of seasickness, the feeling of nausea improves when the sufferer comes onto the deck and can see the horizon, so that the sight sense input matches that of the vestibular system. With infrasound, there is no corresponding visual reference to satisfy the sight sense. In this case, Bray (2012) suggests that a person's mental construct of comfort and safety can be weakened or even destroyed, leading to conscious distress. Thus, it seems that regular, pulsing infrasound, even at inaudible levels, may be capable of affecting the vestibular system in some people, even though it may not be perceptible by their hearing mechanism.

2.2.5 Relative Loudness and the Sone

In the discussion in Section 2.2.3, the comparative loudness of a tone of variable frequency, compared to a reference tone, was considered and a system of equal-loudness contours was established. However, the labelling of the equal-loudness contours was arbitrarily chosen so that at 1 kHz the loudness in phons was the same as the sound pressure level of the reference tone at 1 kHz. This labelling provides no information about relative loudness; that is, how much louder is one sound than another of the same frequency but different level. In this section, the relative loudness of two sounds, as judged subjectively, will be considered. Reference to Table 2.2 suggests that an increase in sound pressure level of 10 decibels will result in a subjectively judged increase in loudness of a factor of two.

To take account of the information in the latter table, yet another unit, called the sone, has been introduced. The 40-phon contour of Figure 2.11 has been arbitrarily labelled "1 sone". Then the 50-phon contour of the figure, which, according to Table 2.2, would be judged twice as loud, has been labelled two sones, etc. The relation between the sone, S, and the phon, P, is summarised as:

$$\begin{cases} S = 2^{(P-40)/10} \\ P = 40 + 10\log_2 S = 40 + \dfrac{10\log_{10} S}{\log_{10} 2} \end{cases} \tag{2.33}$$

At levels of 40 phons and above, up to 100 phons, the preceding equation fairly well approximates subjective judgement of loudness. However, at levels of 100 phons and higher, the physiological mechanism of the ear begins to saturate, and subjective loudness will increase less rapidly than predicted by Equation (2.33). On the other hand, at levels below 40 phons, the subjective perception of increasing loudness will increase more rapidly than predicted by the equation. This is illustrated by the relationship between sones and phons in Figure 2.13. The definition of the sone is thus a compromise that works best in the mid-level range of ordinary experience, between extremely quiet (40 phons) and extremely loud (100 phons).

The loudness contours in Figure 2.11 are for pure tone sounds and are not necessarily applicable to complex sounds containing multiple tones or broadband noise. The procedure for calculating the loudness (in sones) of a noise spectrum was described in previous editions of this book, following the procedure outlined in ISO 532 (1975) and ANSI/ASA S3.4 (1980). The first of these standards, although superseded, represents a relatively simple way of calculating loudness. However, in recognition of the significant advances that have been made in recent years in evaluating the loudness of complex sounds, two more recent ISO standards have been developed (ISO 532-1, 2017; ISO 532-2, 2017) and ANSI/ASA S3.4 (1980) has been replaced with ANSI/ASA S3.4 (2020). These more recent standards are based on different and supposedly more accurate methods (Moore et al., 1997; Glasberg and Moore, 2006; Moore and Glasberg, 2007) than the earlier standards (ISO 532, 1975; ANSI/ASA S3.4, 1980).

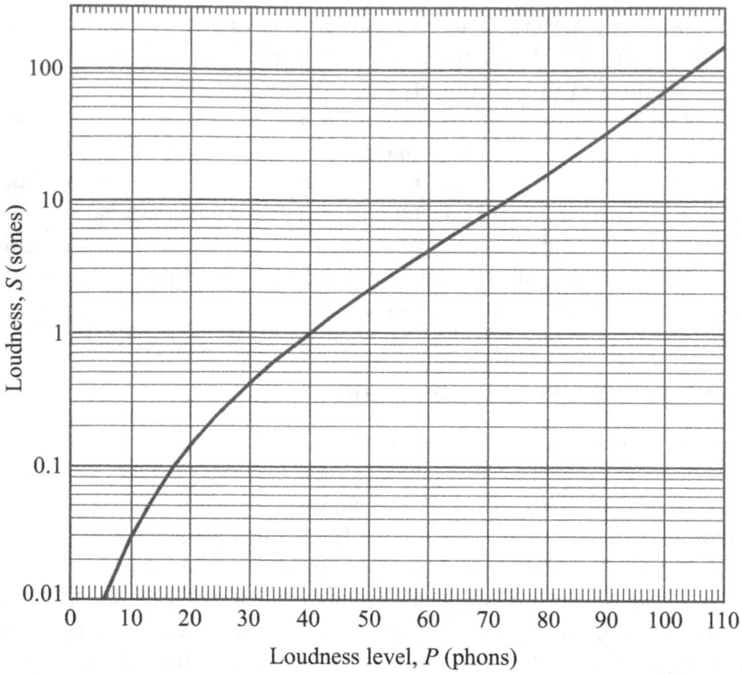

FIGURE 2.13 Relationship between phons and sones.

Each of the two ISO standards published in 2017 uses a different method to calculate loudness and the results from the application of each method can be significantly different. However, the committee responsible for the two standards could not agree on which method is more appropriate so they developed a standard for each method. According to Scheuren (2017), the Zwicker method, described in ISO 532-1 (2017), more closely approximates the earlier loudness contours described in ISO 226 (1987), whereas the Moore-Glasberg method, described in ISO 532-2 (2017) and Moore et al. (1997); Glasberg and Moore (2006); Moore and Glasberg (2007), more closely approximates the more recent loudness contours described in ISO 226 (2023). There is also some controversy (Scheuren, 2017) as to whether the equal loudness curves in Figure 2.11 (ISO 226, 2023) are accurate, particularly when loudness models using ISO 226 (1987) often agree better with subjective impressions than loudness models that use the more recent curves in ISO 226 (2023). Calculations made by Scheuren (2017) indicate that for broadband noise, the Zwicker loudness model (ISO 532-1, 2017) produces lower loudness levels for the same sound than the Moore-Glasberg method (ISO 532-2, 2017). The maximum difference corresponds to level differences of up to 5 dB in noise to produce the same loudness using both methods. Similarly, the earlier method outlined in ISO 532 (1975) produces lower sone values than those calculated using ISO 532-1 (2017) and ISO 532-2 (2017) for the same noise spectrum.

The more recent procedure outlined in ISO 532-1 (2017) is complex and the standard recommends use of the software that accompanies the standard and is available for download at no cost from http://standards.iso.org/iso/532/-1/ed-1/en and http://standards.iso.org/iso/532/-2/ed-1/en/, respectively. Of the two programs supplied, the file named "ISO_532-1_GUI.exe" is the easiest to use. It allows users to choose between entering sound pressure levels in 1/3-octave bands or providing a calibration wave file together with a wave file, which may be a recording of either a stationary or time-varying sound. Regardless of how the data are entered, the user has a choice of a diffuse field or a free field presented to the test subject.

The procedures outlined in ANSI/ASA S3.4 (2020) and ISO 532-2 (2017) can be used to calculate the loudness in phons and sones for both simple and complex sounds involving tones or broadband noise separately or involving the two noise types mixed. However, they are very complex to implement and the use of software that accompanies each standard is recommended.

The calculations described in ANSI/ASA S3.4 (2020) and ISO 532-2 (2017) are for sound presented to a single ear (monaural). However, the software that accompanies each standard allows binaural loudness to be calculated as well. The ISO 532-2 (2017) software allows different sounds to be presented to each ear, whereas the software that accompanies ANSI/ASA S3.4 (2020) assumes that the same sound is presented to each ear. When the same sound is presented to both ears, the single-ear loudness is multiplied by a factor of two in ANSI/ASA S3.4 (2020) and a factor of 1.5 in ISO 532-2 (2017). The difference arises as a result of work reported by Moore and Glasberg (2007), which suggested that the factor should be 1.5 rather than 2. As the ANSI/ASA S3.4 (2020) standard was published before Moore and Glasberg (2007), the associated software uses a factor of two to convert from monaural to binaural loudness.

The earlier ISO 532 (1975) standard has a much simpler procedure that results in a good estimate of the loudness of a sound represented by levels in octave or 1/3-octave bands and presented simultaneously to both ears (binaural). Of course, if tones are present along with the broadband noise, estimates of loudness will not be reliable and the calculated sone level will be too low. To calculate loudness using ISO 532 (1975), first the band loudness in sones is determined from the band sound pressure level presented to the subject for each octave or 1/3-octave band using Figure 2.14. For example, according to the figure, a 250 Hz octave band level of 50 dB has a loudness, S, of 1.8. Note that the calculation assumes that both ears are exposed to the noise.

Next, the composite loudness for all the bands is calculated using Equation (2.34). Note that the band with the highest index $S_{\max}$ must first be identified, excluded from the sum and added to the result of the sum after the result of the sum has been multiplied by the weighting, B. The following equation is used for the calculation, where the weighting B is equal to 0.3 for octave band and 0.15 for 1/3-octave band analysis, and the prime on the sum is a reminder that the highest-level band is omitted from the sum (Stevens, 1961):

$$L = S_{\max} + B\sum_i{}' S_i \quad \text{(sones)} \tag{2.34}$$

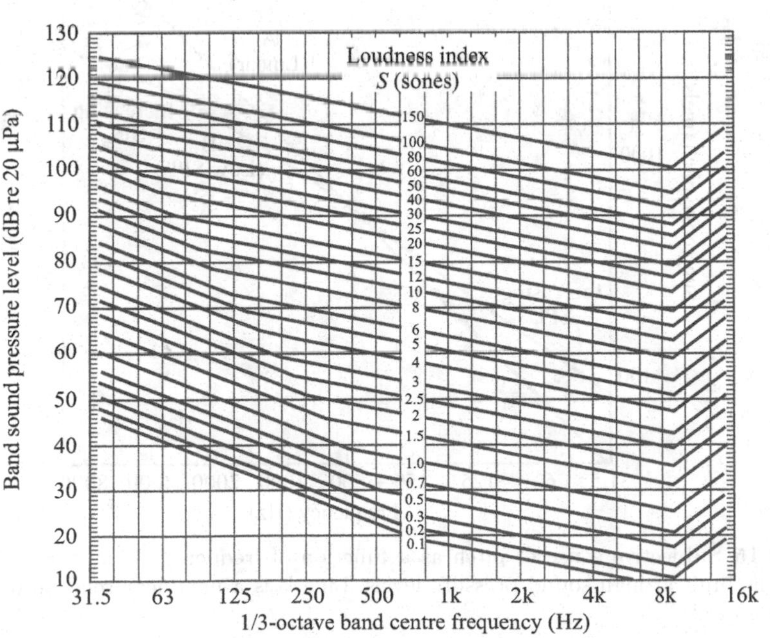

FIGURE 2.14 Relationship between loudness in sones, and band sound pressure level (octave or 1/3-octave).

When the composite loudness level, L (sones), has been determined using Equation (2.34), it may be converted back to phons using Equation 2.33. The latter number is also the sound pressure level for a 1 kHz tone.

2.2.6 Pitch

The lowest frequency that can be identified as a tone by a person with normal hearing is about 20 Hz. At lower frequencies, the individual pressure pulses are heard; the sound is that of a discrete set of events rather than a continuous tone. The highest frequency that a person can hear is very susceptible to factors such as age, health and previous exposure to high sound pressure levels. With acute hearing, the limiting frequency may be as high as 20 kHz, but normally the limit seems to be about 18 kHz. Pitch is the subjective response to frequency. Low frequencies are identified as "flat" or "low-pitched", while high frequencies are identified as "sharp" or "high-pitched". As few sounds of ordinary experience are of a single frequency (for example, the quality of the sound of a musical instrument is determined by the presence of many frequencies other than the fundamental frequency), it is of interest to consider what determines the pitch of a complex note.

If a sound is characterised by a series of integrally related frequencies (for example, the second lowest is twice the frequency of the lowest, the third lowest is three times the lowest, etc.), then the lowest frequency determines the pitch. Furthermore, even if the lowest frequency is removed, say by filtering, the pitch remains the same; the ear supplies the missing fundamental frequency. However, if not only the fundamental is removed but also the odd multiples of the fundamental as well, say by filtering, then the sense of pitch will jump an octave. The pitch will now be determined by the lowest frequency, which was formerly the second lowest. Clearly, the presence or absence of the higher frequencies is important in determining the subjective sense of pitch.

Pitch and frequency are not linearly related, and pitch is dependent on the sound level. The situation with regard to pitch is illustrated in Figure 2.15.

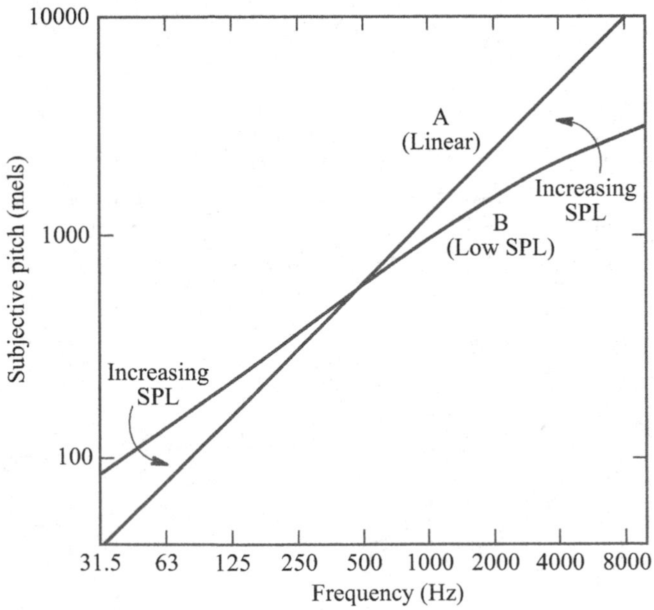

FIGURE 2.15 Subjective sense of pitch as a function of frequency. Line A is a linear relation observed in the limit at high sound pressure levels. Line B is a nonlinear relation observed at low sound pressure levels.

In the figure are shown two lines, A and B. Line A, with a slope of unity, illustrates a linear relationship between sense of pitch and frequency. However, an experimental study produced the empirical curve B, which describes the sense of pitch relative to frequency for tones of 60 dB re 20 µPa. The latter curve was obtained by presenting a reference 1 kHz tone and a variable tone, sequentially, to listeners who were asked to adjust the second tone until its pitch was twice that of the reference tone, half that of the reference tone, etc., until the curve shown could be constructed. The experimenters assigned the name "mel" to the units on the ordinate, such that the pitch of the reference tone was 1000 mels.

As mentioned previously, sense of pitch is also related to level. Our subjective response (curve B) tends to approach the linear response (curve A) at high sound pressure levels, with the crossover at about 500 Hz as indicated in the figure. It is worthy of note that the system tends to linearity at high sound pressure levels.

2.3 Weighting Networks

Attempts to present a single decibel number to describe the annoyance of environmental noise has led to the use of weighting networks, whereby the level of noise is adjusted as a function of frequency in an attempt to replicate how an average normal ear would hear. These weighting networks are more suitable for some types of noise than others. Most environmental and occupational noise measurements are taken using the A-weighting network. This is because the A-weighting curve is a good approximation of the ear response to sound that is typical of environmental noise produced by traffic and industrial facilities. The A-weighting curve also seems to be related to hearing damage risk in high noise level environments, even though the apparent loudness of high-level noise is closer to the C-weighting curve.

When the use of a weighting network proves desirable, Figure 2.16 shows the correction that is added to a linear reading to obtain the weighted reading for a particular frequency. For example, if the linear reading at 125 Hz were 90 dB re 20 µPa, then the A-weighted reading

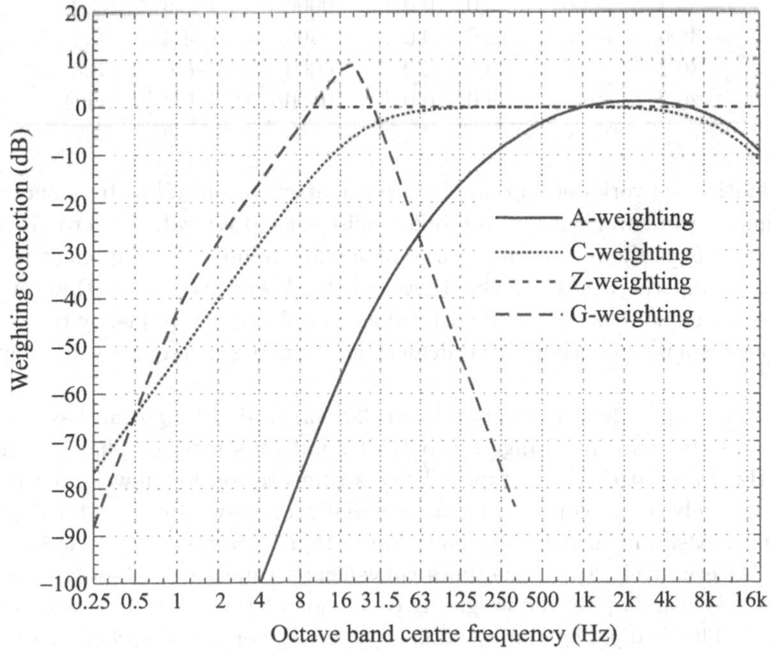

FIGURE 2.16 Various weighting curves used in the assessment of noise.

would be 74 dB(A). For convenience, corrections for the A-, C-, G- and Z-weighted networks at 1/3-octave band centre frequencies are listed in Table 2.3 (see also ANSI/ASA S1.42 (2020)).

TABLE 2.3 Weighting corrections (dB) at 1/3-octave band centre frequencies to be added to unweighted signal

Frequency (Hz)	A	C	G	Z	Frequency (Hz)	A	C	G	Z
0.25	−197.6	−77.0	−88.0	—	80	−22.5	−0.5	−36.0	0.0
0.315	−189.5	−73.0	−80.0	—	100	−19.1	−0.3	−44.0	0.0
0.4	−181.5	−69.0	−72.1	—	125	−16.1	−0.2	−52.0	0.0
0.5	−173.5	−65.0	−64.3	—	160	−13.4	−0.1	−60.0	0.0
0.63	−165.5	−60.9	−56.6	—	200	−10.9	0.0	−68.0	0.0
0.8	−157.4	−56.9	−49.5	—	250	−8.6	0.0	−76.0	0.0
1.0	−148.6	−52.5	−43.0	—	315	−6.6	0.0	−84.0	0.0
1.25	−140.6	−48.5	−37.5	—	400	−4.8	0.0	—	0.0
1.6	−132.6	−44.5	−32.6	—	500	−3.2	0.0	—	0.0
2.0	−124.6	−40.6	−28.3	—	630	−1.9	0.0	—	0.0
2.5	−116.7	−36.6	−24.1	—	800	−0.8	0.0	—	0.0
3.15	−108.8	−32.7	−20.0	—	1000	0.0	0.0	—	0.0
4.0	−100.9	−28.8	−16.0	—	1250	0.6	0.0	—	0.0
5.0	−93.1	−25.0	−12.0	—	1600	1.0	−0.1	—	0.0
6.3	−85.4	−21.3	−8.0	—	2000	1.2	−0.2	—	0.0
8.0	−77.8	−17.7	−4.0	—	2500	1.3	−0.3	—	0.0
10.0	−70.4	−14.3	0.0	0.0	3150	1.2	−0.5	—	0.0
12.5	−63.4	−11.2	4.0	0.0	4000	1.0	−0.8	—	0.0
16.0	−56.7	−8.5	7.7	0.0	5000	0.5	−1.3	—	0.0
20.0	−50.5	−6.2	9.0	0.0	6300	−0.1	−2.0	—	0.0
25.0	−44.7	−4.4	3.7	0.0	8000	−1.1	−3.0	—	0.0
31.5	−39.4	−3.0	−4.0	0.0	10000	−2.5	−4.4	—	0.0
40	−34.6	−2.0	−12.0	0.0	12500	−4.3	−6.2	—	0.0
50	−30.2	−1.3	−20.0	0.0	16000	−6.6	−8.5	—	0.0
63	−26.2	−0.8	−28.0	0.0	20000	−9.3	−11.2	—	0.0

The C-weighting network does not apply as much attenuation to low-frequency noise as does the A-weighting network and is meant for louder noise such as aircraft flyover noise near airports but it is also used to evaluate the importance of the low-frequency component in a broadband noise signal, by comparing the C-weighted level to the A-weighted level (Downey and Parnell, 2017). Low-frequency noise is commonly defined as noise having a frequency between 20 Hz and 200 Hz, although some researchers have defined it as having a frequency between 20 Hz and 160 Hz.

To assess infrasound (defined as noise below 20 Hz), the G-weighting has been developed, which only covers the frequency range up to 315 Hz but puts most emphasis on noise between 5 Hz and 40 Hz. In an attempt to standardise instrumentation for unweighted measurements, the Z-weighting has been developed, which is essentially a zero-weighting defined down to 10 Hz.

Most sound-measuring instruments have the option to apply various weighting networks electronically. Alternatively, it is sometimes convenient to measure 1/3-octave or octave band sound pressure levels and apply the weighting correction at the centre frequency of the band to the entire band. This is done by simply adding the correction in decibels (usually negative) to the measured unweighted level in the frequency band of interest. However, this latter method is not as accurate as applying the weighting network electronically to the signal being measured. This is because the same weighting is effectively applied to all frequencies in a band, which is only an approximation, as the correct weighting varies with frequency within each band.

2.4 Noise Measures

2.4.1 Equivalent Continuous Sound Pressure Level, L_{eq}

This level, denoted L_{eq}, is an energy-averaged level, which means that the sound pressure is squared prior to it being averaged. This is the level that is displayed in most digital sound level meters. The averaging time can be set by the user and the sound pressure level can be continuously updated by replacing the first sample in the average with the most recent sample. The Equivalent Continuous Sound Pressure Level, L_{eq}, averaged over time, T_e, may be written in terms of the instantaneous sound pressure level, $L_p(t)$, as:

$$L_{eq,T_e} = 10\log_{10}\left[\frac{1}{T_e}\int_0^{T_e} 10^{L_p(t)/10}\mathrm{d}t\right] \quad \text{(dB re 20 µPa)} \tag{2.35}$$

2.4.2 A-Weighted Equivalent Continuous Sound Pressure Level, L_{Aeq}

The A-weighted Equivalent Continuous Sound Pressure Level, L_{Aeq}, has a similar definition to the un-weighted continuous sound pressure level, defined in Section 2.4.1, except that the noise signal is A-weighted before it is squared and averaged. After A-weighting, the pressure squared is averaged (energy averaging). The A-weighted Equivalent Continuous Sound Pressure Level is used as a descriptor of both occupational and environmental noise and, for an average over time, T_e, it may be written in terms of the instantaneous A-weighted sound pressure level, $L_{pA}(t)$, as:

$$L_{Aeq,T_e} = 10\log_{10}\left[\frac{1}{T_e}\int_0^{T_e} 10^{L_{pA}(t)/10}\mathrm{d}t\right] \quad \text{(dB re 20 µPa)} \tag{2.36}$$

When the measured data are in the form of several measurements, each of which is averaged over a short time period, t_i, the integral of Equation (2.36) may be replaced with a sum. Thus:

$$L_{Aeq,T_e} = 10\log_{10}\frac{1}{T_e}\left[t_1 10^{L_{pA1}/10} + t_2 10^{L_{pA2}/10} + ... + t_m 10^{L_{pAm}/10}\right] \quad \text{(dB re 20 µPa)} \tag{2.37}$$

where L_{pAi} is the measured equivalent A-weighted sound pressure level for the ith measurement, t_i is the corresponding averaging time for measurement, i and $T_e = t_1 + t_2 + ... + t_m$. Equation (2.37) is referred to as decibel or logarithmic averaging.

For occupational noise, the most common descriptor is $L_{Aeq,8h}$, which implies a normalisation to 8 hours, even though the contributing noises may be experienced for more or less than 8 hours. Thus, for sound experienced over T_e hours:

$$L_{Aeq,8\mathrm{h}} = 10\log_{10}\left[\frac{1}{8}\int_0^{T_e} 10^{L_{pA}(t)/10}\mathrm{d}t\right] = L_{Aeq,T_e} + 10\log_{10}(T_e/8) \quad \text{(dB re 20 µPa)} \tag{2.38}$$

When the measured data are in the form of several measurements, each of which is averaged over a short time period, t_i, the integral of Equation (2.38) may be replaced with a sum. Thus:

$$L_{Aeq,8\mathrm{h}} = 10\log_{10}\frac{1}{8}\left[t_1 10^{L_{pA1}/10} + t_2 10^{L_{pA2}/10} + ... + t_m 10^{L_{pAm}/10}\right] \quad \text{(dB re 20 µPa)} \tag{2.39}$$

where L_{pAi} are the measured equivalent A-weighted sound pressure levels and t_i is the time, in hours, that a person spends at the ith location. The sum of t_1.......t_m does not necessarily have to equal 8 hours.

2.4.3 Noise Exposure Level, $L_{\mathrm{EX,8h}}$ or L_{EX} or $L_{ep'd}$

The quantity, daily noise exposure level (see ISO 9612 (2009)), is commonly characterised by one of three different symbols, as shown in the above header. However, the symbol, $L_{ep'd}$, is the most recent and is more commonly used. It represents noise exposure normalised to a nominal 8-hour working day. It is identical to $L_{Aeq,8h}$ of Equation (2.38) or (2.39) and can also be expressed in the form:

$$L_{\mathrm{EX,8h}} = L_{ep'd} = L_{Aeq,8\mathrm{h}} = L_{Aeq,T_e} + 10\log_{10}(T_e/8) \qquad (\mathrm{dB\ re\ 20\,\mu Pa}) \qquad (2.40)$$

where T_e is the exposure duration in hours per day.

If a worker is exposed to a different number of working days per week than the usual 5, the daily exposure level can be normalised to an equivalent 5 days per week of exposure using:

$$L_{\mathrm{EX,8h}} = 10\log_{10}\left[\frac{1}{5}\sum_{i=1}^{N} 10^{(L_{\mathrm{EX,8h}})_i/10}\right] \qquad (\mathrm{dB\ re\ 20\,\mu Pa}) \qquad (2.41)$$

where N is the number of days worked in a typical week. The number, "5", in the equation is the number of days to which the exposure is to be normalised and could be different to 5 if desired.

The measurement of occupational noise exposure is discussed in ANSI/ASA S12.19 (2020) and ISO 9612 (2009).

2.4.4 A-Weighted Sound Exposure, $E_{A,T}$

Environmental sound exposure may be quantified using the A-weighted Sound Exposure, $E_{A,T}$, defined as the time integral of the squared, instantaneous A-weighted sound pressure, $p_A^2(t)$ (Pa^2), over a particular time period, $T_e = t_2 - t_1$ (hours). The units are pascal-squared-hours (Pa^2.h) and the defining equation is:

$$E_{A,T} = \int_{t_1}^{t_2} p_A^2(t)\mathrm{d}t \qquad (2.42)$$

Using Equations (2.36) and (2.42), the relationship between the A-weighted Sound Exposure and the A-weighted Equivalent Continuous Sound Pressure Level, $L_{Aeq,T}$, can be shown to be:

$$E_{A,T} = 4T_e \times 10^{(L_{Aeq,T}-100)/10} \qquad (2.43)$$

2.4.5 A-Weighted Sound Exposure Level, L_{AE} or SEL

The A-weighted Sound Exposure Level is defined as ten times the logarithm of the time integral of the squared, instantaneous A-weighted sound pressure, $p_A^2(t)$, divided by the reference sound pressure squared, p_{ref}^2, over a particular time period, $T_e = t_2 - t_1$ (seconds). Thus:

$$L_{AE} = 10\log_{10}\left[\int_{t_1}^{t_2} \frac{p_A^2(t)}{p_{\mathrm{ref}}^2}\mathrm{d}t\right] = 10\log_{10}\left[\frac{E_{A,T}\times 3600}{p_{\mathrm{ref}}^2}\right] \qquad (\mathrm{dB\ re\ 20\,\mu Pa}) \qquad (2.44)$$

where the times t_1, t_2 and dt are in seconds (not hours as for A-weighted sound exposure in Equation (2.42)) and $p_{\mathrm{ref}} = 20\,\mu$Pa.

A "C-weighted" Sound Exposure Level, L_{CE}, is defined by substituting the C-weighted sound pressure level for the A-weighted level in Equation (2.44). These two exposure level quantities are mainly for the assessment of transient environmental noise, such as traffic noise, aircraft noise and train noise. When the event is a transient, the time interval, $t_1 - t_2$, must include the 10 dB down points as shown in Figure 2.17. The L_{AE} for any number (including zero) of transient events within the period of interest is simply obtained by making t_1 and t_2 define the beginning and end of the period of interest.

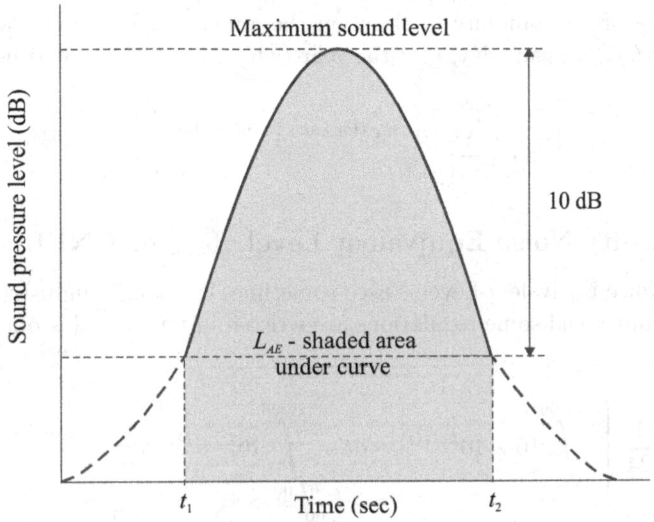

FIGURE 2.17 Sound exposure of a single event.

An Equivalent Continuous Sound Pressure Level for a nominal T_e = 8-hour working day may be calculated from $E_{A,8h}$ or L_{AE} using:

$$L_{Aeq,8h} = 10 \log_{10} \left(\frac{E_{A,8h}}{8 \times 4 \times 10^{-100/10}} \right) \quad \text{(dB re 20 µPa)} \tag{2.45}$$

$$L_{Aeq,8h} = 10 \log_{10} \left(\frac{1}{8 \times 3600} 10^{L_{AE,8h}/10} \right) = L_{AE,8h} - 44.6 \quad \text{(dB re 20 µPa)} \tag{2.46}$$

where $L_{AE,8h}$ is the value of L_{AE} for an eight hour exposure period.

2.4.6 Day-Night Average Sound Pressure Level, L_{dn} or DNL

Some standards regarding the intrusion of traffic noise into the community are written in terms of L_{dn}, which is defined as:

$$L_{dn} = 10 \log_{10} \frac{1}{24} \left[\int_{22:00}^{07:00} 10 \times 10^{L_{pA}(t)/10} \mathrm{d}t + \int_{07:00}^{22:00} 10^{L_{pA}(t)/10} \, \mathrm{d}t \right] \quad \text{(dB re 20 µPa)} \tag{2.47}$$

where t and dt are in hours. In calculations, hourly energy averages of $L_A(t)$ are usually used to give $L_{Aeq,1hr}$ and the integrals then become sums over the number of hours between the integration limits. That is, the integrals in Equation (2.47) are replaced with sums and dt is replaced with one.

For traffic noise, the Day-Night Average Sound Pressure Level for a particular vehicle class is related to the Sound Exposure Level by:

$$L_{dn} = L_{AE} + 10 \log_{10}(N_{\text{day}} + N_{\text{eve}} + 10 \times N_{\text{night}}) - 49.4 \quad \text{(dB re 20 µPa)} \tag{2.48}$$

where L_{AE} is the A-weighted Sound Exposure Level for a single vehicle pass-by, N_{day}, N_{eve} and N_{night} are the numbers of vehicles in the particular class that pass by in the daytime (0700 to 1900 hours), evening (1900 to 2200 hours) and nighttime (2200 to 0700 hours), respectively, and the normalisation constant, 49.4, is $10\log_{10}$ of the number of seconds in a day. To calculate the L_{dn} for all vehicles, the above equation is used for each class and the results added together logarithmically (see Section 1.11.3).

In many cases, regulations specify an allowed yearly or annual DNL which is the daily L_{dn} energy-averaged (see Equation (1.99)) over the 365 days of the year. Additionally, the long term annoyance response of a community is based on the annual DNL average, which is calculated from the values of L_{dni} for each day, i, of the year (where $i = 1,\ 2, ...365$) using:

$$\mathrm{DNL}_{\mathrm{annual}} = 10\log_{10}\left[\frac{1}{365}\left(\sum_{i=1}^{365} 10^{L_{dni}\ \text{(for each 24-hour day)}}\right)\right] \quad \text{(dB re 20 µPa)} \qquad (2.49)$$

2.4.7 Community Noise Equivalent Level, L_{den} or CNEL

The Community Noise Equivalent Level is used sometimes to quantify industrial noise and traffic noise in the community and some regulations are written in terms of this quantity, L_{den}, which is defined as:

$$L_{den} = 10\log_{10}\frac{1}{24}\left[\int_{22:00}^{07:00} 10\times 10^{L_{pA}(t)/10}\mathrm{d}t + \int_{07:00}^{19:00} 10^{L_{pA}(t)/10}\mathrm{d}t + \int_{19:00}^{22:00} 3\times 10^{L_{pA}(t)/10}\mathrm{d}t\right] \quad \text{(dB re 20 µPa)} \qquad (2.50)$$

where t and dt are in hours. In calculations, hourly energy energy averages of $L_{pA}(t)$ are usually used to provide $L_{Aeq,1\mathrm{hr}}$ and the integrals then become sums over the number of hours between the integration limits, with $L_{pA}(t)$ replaced by $L_{Aeq,1\mathrm{hr}}$.

For traffic noise, the Community Noise Equivalent Level for a particular vehicle class is related to the Sound Exposure Level by:

$$L_{den} = L_{AE} + 10\log_{10}(N_{\mathrm{day}} + 3N_{\mathrm{eve}} + 10N_{\mathrm{night}}) - 49.4 \quad \text{(dB re 20 µPa)} \qquad (2.51)$$

where L_{AE} is the A-weighted Sound Exposure Level for a single vehicle pass-by, the constant, $49.4 = 10\log_{10}$ (number of seconds in a day) and N_{day}, N_{eve} and N_{night} are the numbers of vehicles in the particular class that pass by in the daytime (0700 to 1900 hours), evening (1900 to 2200 hours) and nighttime (2200 to 0700 hours), respectively. To calculate the L_{den} for all vehicles, the above equation is used for each class and the results added together logarithmically (see Equation (1.99)).

2.4.8 Effective Perceived Noise Level, L_{EPN} or EPNL

This descriptor is used solely for evaluating aircraft noise. It is derived from the Perceived Noise Level, L_{PN}, which was introduced some time ago by Kryter (1959). It is a very complex quantity to calculate and is a measure of the annoyance of aircraft noise. It takes into account the effect of pure tones (such as engine whines) and the duration of each event.

The calculation procedure begins with a recording of the sound pressure level versus time curve, which is divided into 0.5-second intervals over the period that the aircraft noise exceeds background noise. Each 0.5-second interval (referred to as the kth interval) is then analysed to give the sound pressure level in that interval in 24 1/3-octave bands from 50 Hz to 10 kHz. A noisiness value (defined as a noy) for each 1/3-octave band is calculated using published tables (Edge and Cawthorn, 1976) or curves (Raney and Cawthorn, 1998). The total noisiness (in noys) corresponding to each time interval, t, is then calculated from the 24 individual 1/3-octave band noy levels using:

$$n_t = n_{\max} + 0.15\left(\sum_{i=1}^{24} n_i - n_{\max}\right) \quad \text{(noy)} \tag{2.52}$$

where n_i is the noy value for the ith 1/3-octave band and $n_{\max}$ is the maximum 1/3-octave band noy value for the time interval under consideration. The perceived noise level for each time interval is then calculated using:

$$L_{PN} = 40 + 33.22\log_{10} n_t \quad \text{(dB re 20 µPa)} \tag{2.53}$$

The next step is to calculate the tone-corrected perceived noise level (L_{PNT}) for each time interval. This correction varies between 0 dB and 6.7 dB, and it is added to the L_{PN} value. It applies whenever the level in any one band exceeds the levels in the two adjacent 1/3-octave bands. If two or more frequency bands produce a tone correction, only the largest correction is used. The calculation of the actual tone-correction is complex and is described in detail in the literature (Edge and Cawthorn, 1976). The maximum tone corrected perceived noise level over all time intervals is denoted $L_{PNT\max}$. Then next step in calculating L_{EPN} is to calculate the duration correction, D, which is usually negative and is given by Raney and Cawthorn (1998), corrected here, as:

$$D = 10\log_{10}\left(\sum_{k=i}^{i+2d} 10^{L_{PNT(k)}/10}\right) - 13 - L_{PNT\max} \quad \text{(dB)} \tag{2.54}$$

where $k = i$ is the time interval for which L_{PNT} first exceeds $L_{PNT\max}$ and d is the length of time in seconds that L_{PNT} exceeds $L_{PNT\max}$.

Finally, the effective perceived noise level is calculated using:

$$L_{EPN} = L_{PNT\max} + D \quad \text{(dB re 20 µPa)} \tag{2.55}$$

In a recent report (Yoshioka, 2000), it was stated that a good approximate and simple method to estimate L_{EPN} was to measure the maximum A-weighted sound pressure level, $L_{A\max}$, over the duration of the aircraft noise event (which lasts for approximately 20 seconds in most cases) and add 13 dB to obtain L_{EPN}.

2.4.9 Statistical Descriptors

Statistical descriptors are often used to characterise time-varying sound such as traffic noise and background noise in urban and rural environments. The most commonly used descriptors are L_{A10} for traffic noise and L_{A90} for background noise, where the subscript, A, refers to an A-weighted level. The quantity, L_{10}, is the sound pressure level that is exceeded 10% of the time. More generally, L_x is the sound pressure level that is exceeded x% of the time.

2.4.10 Other Descriptors, $L_{\max}$, L_{peak}, L_{Imp}

In a number of international acoustics standards, the terms, "maximum sound pressure level, $L_{\max}$" and "peak sound pressure level, L_{peak}" are used. These two quantities are very different. The maximum sound pressure level is the maximum level (in dB) recorded with a sound level meter response set to either "fast" or "slow" (see Section 3.2) and the weighting filter set to either

A-weighting or C-weighting (see Section 2.3) to produce L_{AFmax}, L_{ASmax}, L_{CFmax} or L_{CSmax}. Similar definitions hold for Z-weighting and G-weighting of the data and also for minimum sound pressure levels. On the other hand, L_{peak} is the maximum instantaneous sound level with no time constant and no RMS averaging as is done for the "fast" or "slow" responses. With old analog sound level meters, the effective time constant was approximately 50 µs (corresponding to a maximum frequency response of 20 kHz). However, with digital sound level meters that are now in common use, the sample rate for a meter response of 20 kHz is 48 kHz, which means that the samples are separated by approximately 20 µs and the peak sound pressure level will correspond to the sample with the highest level. Often, the peak sound pressure level is defined as a C-weighted peak, $L_{C\mathrm{peak}}$, which means that the sound pressure signal must pass through a C-weighting filter, which uses more than one sample to arrive at a result and effectively results in some averaging.

Another quantity sometimes referred to in international standards is the "impulse sound pressure level", L_{Imp}, which has a time constant of 35 ms. This quantity is used to assess occupational noise characterised by sudden bursts of sound energy such as punch press noise, hammering, pile driving, firearm or explosion noise. The impulse sound level is usually reported as a Z-weighted level (that is, no weighting). Sometimes impulse sound levels are labelled L_I, but this terminology is not used here due to likely confusion with the nomenclature for sound intensity level. It should be noted that the impulse time weighting is no longer required in sound level meters, as various investigations have concluded that it is not suitable for rating impulsive sounds with respect to their loudness – nor for determining the impulsiveness of a sound (IEC 61672-1, 2013).

There are a number of other descriptors used in the various standards, such as "long-time average A-weighted sound pressure level" or "long-term, time-averaged rating level", but these are all derived from the quantities mentioned in the preceding paragraphs and defined in the standards that specify them, so they will not be discussed further here. When applied to sound pressure levels, "average" usually implies logarithmic (or energy) average.

2.5 Hearing Loss

Hearing loss is generally determined using pure tone audiometry in the frequency range from about 100 Hz to 8 kHz, and is defined as the differences in sound pressure levels of a series of tones that are judged to be just audible compared with reference sound pressure levels that correspond to normal hearing for the same series of tones. It is customary to refer to hearing level which is the level at which the sound is just audible relative to the reference level when referring to hearing loss. However, the practice will be adopted here of always using the term hearing loss rather than the alternative term hearing level.

2.5.1 Threshold Shift

In Section 2.2.3 the sensitivity of the ear to tones of various frequencies was shown to be quite non-uniform. Equal loudness contours, measured in phons, were described, as well as the minimum audible field or threshold of hearing. The latter contours, and in particular the minimum audible field levels, were determined by the responses of a great many healthy young people, males and females in their 20s, who sat facing the source in a free field. When the subject had made the required judgement, that is, that two sounds were equally loud or the sound was just audible, the subject vacated the testing area, and the measured level of the sound in the absence of the subject was determined and assigned to the sound under test. In other words, the assigned sound pressure levels were the free-field levels, unaffected by diffraction effects due to the presence of the auditor. In Chapter 3, the problem of characterising the sensitivity of a microphone is discussed. It will be shown that diffraction effects, as well as the angle of incidence, very strongly affect the apparent sensitivity. Clearly, as the human head is much larger than any commercial microphone, the ear as a microphone is very sensitive to the effects mentioned. In fact, as mentioned in Sections 2.1.1 and 2.1.6, the ear and brain, in close collaboration, make use of such effects to gain source location information from a received signal. Thus, it is apparent that the sound pressure level at the entrance to the ear may be very different from the level of the freely propagating sound field in the absence of the auditor.

The threshold of audibility has been chosen as a convenient measure of the state of health of the auditory system. However, the provision of a free field for testing purposes is not always practical. Additionally, such a testing arrangement does not offer a convenient means for testing one ear at a time. A practical and much more convenient method of test is offered by the use of earphones. Such use forms the basis of pure tone audiometry. The assumption is then implicit that the threshold level determined as the mean of the responses of a great many healthy young people, males and females in their 20s, corresponds to the minimum audible field mentioned earlier. The latter interpretation will be put on published data for hearing loss based on pure tone audiometric testing. Thus, where the hearing sensitivity of a subject may be 20 dB less than the established threshold reference level, the practice is adopted in this chapter of representing such hearing loss as a 20 dB rise in the free-field sound pressure level which would be just audible to the latter subject. This method of presentation is contrary to conventional practice, but it better serves the purpose of illustrating the effect of hearing loss on speech perception.

2.5.2 Presbyacusis

It is possible to investigate the hearing sensitivity of populations of people who have been screened to eliminate the effects of disease and excessive noise. Hearing deterioration with age is observed in screened populations and is called presbyacusis. It is characterised by increasing loss with increasing frequency of the sound and increasing rate of loss with age. Men tend to lose hearing sensitivity more rapidly than women. There is evidence to show that hearing deterioration with age may also be race specific (Driscoll and Royster, 1984). Following the convention proposed in the preceding section, the effect of presbyacusis is illustrated in Figure 2.18 as a rise in the mean threshold of hearing level. For comparison, the range of quiet speech sounds is

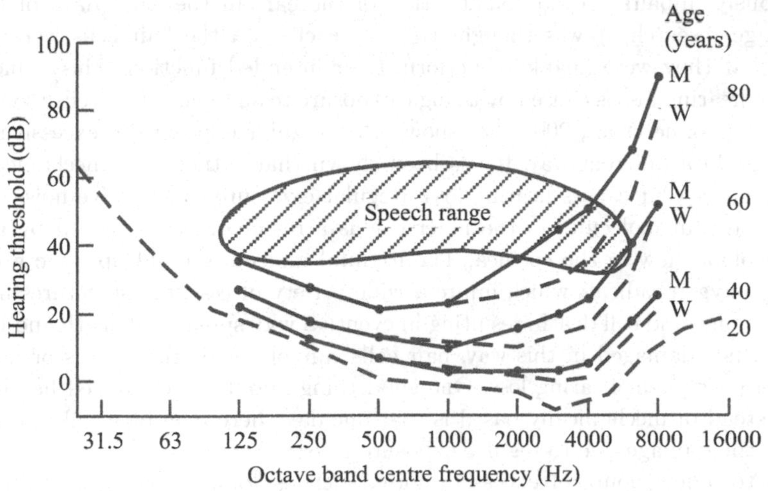

FIGURE 2.18 Threshold shift due to presbyacusis: M = men; W = women. Speech sounds: male, normal voice, at 1 m. Data from Smith et al. (2006).

also indicated in the figure. As the fricative parts of speech lie generally at the right and lower portion of the speech range, it is evident that old folks may not laugh as readily at the jokes, not because of a jaded sense of humour, but rather because they missed the punch lines.

2.5.3 Noise-Induced Hearing Loss

Hearing loss may be not only the result of advancing age but also the result of exposure to excessive noise. Loss caused by exposure to excessive noise usually occurs first in the frequency range from about 4000 to 6000 Hz, which is the range of greatest sensitivity of the human ear. Following the proposed method of presentation, the plight of women habitually exposed to excessive noise in a jute mill is illustrated in Figure 2.19. The dismal effect on their ability to understand speech is clearly illustrated.

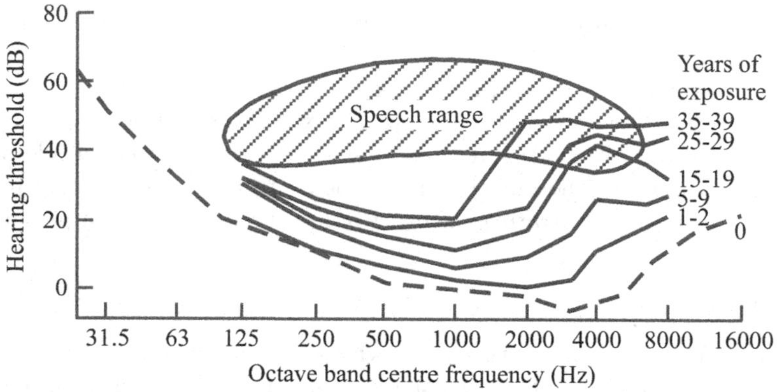

FIGURE 2.19 Threshold shift due to excessive noise exposure. Speech sounds: male, normal voice, at 1 m.

From the point of view of the noise control engineer interested in protecting the ear from damage, it is useful to note what is lost as a result of noise-induced damage to the ear. The outer hair cells are most sensitive to loud noise and may be damaged or destroyed before the inner hair cells suffer comparable damage. Damage to the outer hair cells, which are essential to good hearing, seriously impairs the normal function of the ear. In the early days of noise-induced hearing damage research, it was thought that stereocilia on the hair cells were mechanically damaged so that they were unable to perform their intended function. This is most likely the case where the hearing loss is caused by a single exposure to sudden, very intense sound. However, recent research (Bohne et al., 2007) has shown that regular exposure to excessive noise causes hearing damage in a different way. It has been shown that rather than mechanical damage, it is chemical damage that causes hearing loss. Regular exposure to excessive noise results in the formation of harmful molecules in the inner ear as a result of stress caused by noise-induced reductions in blood flow in the cochlea. The harmful molecules build up toxic waste products known as free oxygen radicals which injure a wide variety of essential structures in the cochlea causing cell damage and cell death, resulting in eventual widespread cell death and noise-induced hearing loss. Once damaged in this way, hair cells cannot repair themselves or grow back and the result is a permanent hearing loss. One good thing about noise-induced hearing loss being chemically instead of mechanically based is that one day there may be a pill that can be taken to ameliorate the damage due to regular exposure to excessive noise.

Exposure to a few hours of excessive noise, such as found in a typical night club, may produce a temporary loss of hearing sensitivity, which is often accompanied by a ringing in the ears, known as tinnitus, which persists after exposure to the noise ends. The temporary loss of hearing sensitivity is known as temporary threshold shift (TTS) or auditory fatigue. The onset of TTS and/or tinnitus after exposure to excessive sound levels is a good indicator that some permanent loss has occurred even though the tinnitus eventually goes away (sometime after the exposure ceases) and the loss is not noticeable to the individual. Noticeable permanent hearing loss and/or permanent tinnitus can develop as a result of many instances of temporary loss (with the anount of permanent loss increasing with each instance of TTS) or with even one instance of extremely excessive noise. Permanent hearing loss initially occurs in the frequency range from about 4000 to 6000 Hz, but as the hearing loss increases it spreads to both lower and higher frequencies. With increasing deterioration of hearing sensitivity, the maximum loss generally remains near 4000 Hz.

The first handicap due to noise-induced hearing loss to be noticed by the subject is usually some loss of hearing for high-pitched sounds, such as squeaks in machinery, bells, musical notes, etc. This is followed by a diminution in the ability to understand speech; voices sound muffled, and this is worse in difficult listening conditions. The person with noise-induced hearing loss complains that everyone mumbles. High-frequency consonant sounds of low intensity are missed, whereas vowels of low frequency and higher intensity are still heard. As consonants carry much of the information in speech, there is little reduction in volume but the context is lost. However, by the time the loss is noticed subjectively as a difficulty in understanding speech, the condition is far advanced. Fortunately, present-day hearing aids, which contain spectral shaping circuitry, can do much to alleviate this problem, although the problem of understanding speech in a noisy environment, such as a party, will still exist. A hearing aid, which makes use of a directional antenna worn as a band around one's neck, is also available to assist with directional selectivity of the hearing aid (Widrow, 2001).

It was suggested in Section 2.1.6 that the outer hair cell control of amplification would allow interpretation of directional information imposed on an auditory stimulus by the directional properties of the external ear. Apparently, outer hair cell loss may be expected to result in an ear unable to interpret directional information encoded by the pinna on the received acoustic stimulus. A person with outer hair cell loss may have the experience of enjoying seemingly good hearing and yet be unable to understand conversation in a noisy environment. For example, a hearing aid may adequately raise the received level to compensate for the lost sensitivity of the damaged ear, but it cannot restore the function of the outer hair cells and it bypasses the pinna altogether. It is to be noted that outer hair cell destruction may be well under way before a significant shift in auditory threshold and other effects, such as permanent tinnitus, are noticed. However, the ability of the ear to hear very low-level sound may be somewhat compromised by outer hair cell damage because, as explained in Section 2.1.8.4, the outer hair cells are responsible for the "undamping" action that decreases the hearing threshold of the inner hair cells.

With a single microphone hearing aid, all that a person may hear with severe outer hair cell loss in a generally noisy environment will be noise. In such a case, a microphone array system may be required, which will allow discrimination against a noisy background and detection of a source in a particular direction. Just such an array was reported by Widrow (2001).

Some people with hearing loss suffer an additional problem, known as recruitment, which is characterised by a very restricted dynamic range of tolerable sound pressure levels between loud enough and too loud. Here it is suggested that severe outer hair cell loss would seem to provide the basis for an explanation for recruitment. For example, it was suggested in Section 2.1.6 that the function of the outer hair cells is to maintain the response of the inner hair cells within a fairly narrow dynamic range. Clearly, if the outer hair cells cannot perform this function, the overall response of the ear will be restricted to the narrow dynamic range of the inner hair cells. For further discussion of recruitment, see Section 2.2.2.

It has been shown that the basilar membrane may be modelled approximately as a series of independent linear oscillators, which are modally independent but are strongly coupled through the fluid in the cochlear. It has been shown also that nonlinearity of response occurs at low to intermediate levels of stimulation in a region about resonant response, through variable damping. It is postulated that the efferent system controls the level of damping, based on cerebral interpretation of signals from the afferent system and that a time lag of the order of that typical of observed psychoacoustic integration times, which seem to range between 0.25 and 0.5 seconds, is required for this process (Moore, 1982).

It is postulated here that the ear's response is quasi-stationary, and thus the ear can only respond adequately to quasi-stationary sounds; that is, sounds that do not vary too rapidly in level. It is postulated that the ear will respond inadequately to non-stationary sounds. When the ear responds inadequately to sound of rapidly varying level, it may suffer damage by being tricked into amplifying stimuli that it should be attenuating, and thereby forced to contribute to its own destruction.

In Section 2.7, criteria are presented and their use is discussed for the purpose of the prevention of hearing loss among individuals who may be exposed to excessive noise. The latter criteria, which are widely accepted, make specific recommendations for exposure defined in terms of level and length of time of exposure, which should not be exceeded. The latter criteria are based on observed hearing loss among workers in noisy industrial environments.

It has been shown that exposure to loud sound of symphonic musicians often exceeds recommended maximum levels (Jansson and Karlsson, 1983) suggesting, according to the accepted criteria, that symphonic musicians should show evidence of hearing loss due to noise exposure. Karlsson et al. (1983) investigated the hearing of symphonic musicians and found no evidence of noise-induced hearing loss. Some more recent studies agreed with this conclusion, although others did not (see Zhao et al. (2010) for a review). One significant difference between symphonic music and industrial noise is that symphonic music is generally quasi-stationary whereas industrial noise is not quasi-stationary. The suggestion that the ear copes better with quasi-stationary sound than non-quasi-stationary sound, provides a possible explanation for the observation that symphonic music generally does not produce the hearing loss predicted using accepted criteria. The observations made here would seem to answer the question raised by Brüel (1977) when he asked "Does A-weighting alone provided an adequate measure of noise exposure for hearing conservation purposes?" The evidence presented here seems to suggest not (see Section 2.3 for an explanation of A-weighting).

In another interesting research finding, data have been presented in the USA (Royster et al., 1980) which show that hearing loss due to excessive noise exposure may be both race and sex specific. The study showed that, for the same exposure, white males suffered the greatest loss, with black males, white females and black females following with progressively less loss, in that order. The males tend to have the greatest loss at high frequencies, whereas the females tend to have a more uniform loss at all frequencies.

2.6 Hearing Damage Risk

The meaning of "damage risk" needs clarification in order to set acceptable noise levels to which an employee may be exposed. The task of protecting everyone from any change in hearing threshold over the entire audio-frequency range is virtually impossible and some compromise is necessary. The accepted compromise is that the aim of damage risk criteria must be to protect most individuals in exposed groups of employees against loss of hearing for everyday speech. Consequently, the discussion begins with the minimum requirements for speech recognition and proceeds with a review of what has been and may be observed.

The discussion will continue with a review of the collective experience on which a database of hearing level versus noise exposure has been constructed. It will conclude with a brief review of efforts to determine a definition of exposure, which accounts for both the effects of level and duration of excessive noise and to mathematically model the database ISO 1999 (2013) in terms of exposure so defined. The purpose of such mathematical modelling is to allow formulation of criteria for acceptability of variable level noise, which is not covered in the database. Criteria are designed to ensure exposed people retain the minimum requirements for speech recognition.

2.6.1 Requirements for Speech Recognition

For good speech recognition, the frequency range from 500 to 2000 Hz is crucial; thus, criteria designed to protect hearing against the loss of ability to recognise speech are concerned with protection for this frequency range. In the United States, loss for speech recognition purposes is assumed to be directly related to the arithmetic average of hearing loss in decibels at 500, 1000 and 2000 Hz. For compensation purposes, the 3000 Hz loss is included in the average. In Australia, a weighted arithmetic mean of loss in the frequency range from 500 to 6000 Hz is used as the criterion.

An arithmetic average of 25 dB loss defines the boundary between just adequate and inadequate hearing sensitivity for the purpose of speech recognition. For practical purposes, a hearing loss of 25 dB will allow speech to be just understood satisfactorily, while a loss of 92 dB is regarded as total hearing loss. If a person suffers a hearing loss between 25 dB and 92 dB, that person's hearing is said to be impaired, where the degree of impairment is determined as a percentage at the rate of 1.5 percentage points for each decibel loss above 25 dB.

An important part of any noise control program is the establishment of appropriate criteria for the determination of an acceptable solution to the noise problem. Thus, where the total elimination of noise is impossible, appropriate criteria provide a guide for determining how much noise is acceptable. At the same time, criteria provide the means for estimating how much reduction is required. The required reduction, in turn, provides the means for determining the feasibility of alternative proposals for control, and finally the means for estimating the cost of meeting the relevant criteria.

For industry, noise criteria ensure the following:

- that hearing damage risk to personnel is acceptably small;
- that reduction in work efficiency due to high sound pressure levels is acceptably small;
- that, where necessary, speech is possible; and
- that the sound pressure at plant boundaries is sufficiently small for noise levels in the surrounding community to be acceptable.

Noise criteria are also important for the design of assembly halls, classrooms, auditoria and all types of indoor facilities in which people congregate and seek to communicate, or simply seek rest and escape from excessive noise. Criteria are also essential for specifying acceptable environmental noise limits resulting from industrial, entertainment or transportation noise sources. Prior to discussing criteria, it is useful to first define the various noise measures that are used in standards and regulations to define acceptable noise limits.

2.6.2 Quantifying Hearing Damage Risk

In a population of people who have been exposed to excessive noise and who have consequently suffered an observable loss of hearing, it is possible to carry out retrospective studies to determine quantitative relationships between noise exposure and hearing threshold shift. Two such studies have been conducted (Burns and Robinson, 1970; Passchier-Vermeer, 1968, 1977), and these are referenced in ISO 1999 (2013). The standard states that neither of the latter studies forms part of its database. The International Standards Organisation document makes no reference to any other studies, including those used to generate its own database.

The standard provides equations for reconstructing its database and these form the basis for noise regulations around the world. It is important to note that studies to determine the quantitative relationship between noise exposure and hearing threshold shift are only feasible in situations where sound pressure levels are effectively "steady state", and it has only been possible to estimate these sound pressure levels and duration of exposure retrospectively. The International Standards Organisation database shows that permanent threshold shift is dependent on both the level of the sound and the duration of the exposure, but it cannot provide information concerning the effects of variable level sound during the course of exposure.

It is reasonable to assume that ageing makes some contribution to loss of hearing in people exposed to excessive noise (see Section 2.6.2). Consequently, in assessing the effects of excessive noise, it is common practice to compare a noise-exposed person with an unexposed population of the same sex and age when making a determination of loss due to noise exposure. As there is no way of directly determining the effect of noise alone, it is necessary for making any such comparison that some assumption be made as to how noise exposure and ageing collectively contribute to the observed hearing loss in people exposed to excessive noise over an extended time. An obvious solution, from the point of view of compensation for loss of hearing, is to suppose that the effects of age and noise exposure are additive on a decibel basis. In this case, the contribution due to noise alone is computed as the decibel difference between the measured threshold shift and the shift expected due to ageing, and this is the formulation adopted by ISO 1999 (2013) with a small correction (see Equation (2.56)). However, implicit in any assumption that might be made is some implied mechanism. For example, the proposed simple addition of decibel levels implies a multiplication of analogue effects. That is, it implies that damage due to noise and age is characterised by different mechanisms or damage to different parts of the hearing system and in this case it is reasonable to add the decibel hearing losses to obtain a total loss. However, if damage due to age and noise is to different parts of the hearing system, an addition of analogue effects may be more appropriate and an alternative interpretation of measured data may be possible (Kraak, 1981; Bies and Hansen, 1990).

2.6.3 International Standards Organisation Formulation

The International Standard, ISO 1999 (2013), and the American Standard, ANSI/ASA S3.44-1 (2020), provide the following empirical equation for the purpose of calculating the hearing threshold level, H', associated with age and noise of a noise-exposed population:

$$H' = H + N - HN/120 \qquad \text{(dB)} \tag{2.56}$$

H is the hearing threshold level associated with age and N is the actual or potential noise-induced permanent threshold shift, where the values of H, H' and N vary and are specific to the same fractiles of the population.

Only the quantities H and H' can be measured in noise-exposed and non-noise-exposed populations, respectively. The quantity N cannot be measured independently and thus is defined by Equation (2.56). It may be calculated using the empirical procedures provided by the Standard.

The values to be used in Equation (2.56) are functions of frequency, the duration of exposure, Θ (number of years), and the Equivalent Continuous A-weighted sound pressure level for a nominal eight-hour day, $L_{\text{EX,8h}}$, energy averaged over the duration of exposure, Θ (years). The energy average is obtained using Equation (1.99) with L_i replaced with $L_{\text{EX,8h},i}$ and N set equal to the number of work days in the period of Θ years. For exposure times between 10 and 40 years, the median (or 50% fractile) potential noise-induced permanent threshold shift values, N_{50} (meaning that 50% of the population will suffer a hearing loss equal to or in excess of this), are given by the following equation:

$$N_{50} = (u + v \log_{10} \Theta)(L_{\text{EX,8h}} - L_0)^2 \qquad \text{(dB)} \tag{2.57}$$

This equation defines the long-term relationship between noise exposure and hearing loss, where the empirical coefficients u, v and L_0 are listed in Table 2.4. If $L_{\text{EX,8h}} < L_0$, then $L_{\text{EX,8h}}$ is set equal to L_0 to evaluate Equation (2.56).

TABLE 2.4 Values of the coefficients u, v and L_0 used to determine the Noise Induced Permanent Threshold Shift (NIPTS) for the median value of the population, N_{50}

Frequency (Hz)	u	v	L_0 (dB)
500	−0.033	0.110	93
1000	−0.02	0.07	89
2000	−0.045	0.066	80
3000	0.012	0.037	77
4000	0.025	0.025	75
6000	0.019	0.024	77

For exposure durations, $\Theta < 10$ years, N_{50} can be calculated from the value for 10 years as:

$$N_{50} = \frac{\log_{10}(\Theta + 1)}{\log_{10}(11)} N_{50:\Theta=10} \tag{2.58}$$

where $N_{50:\Theta=10}$ is the value of N_{50} for an exposure time of 10 years. No procedures are offered by ISO 1999 (2013) for exposure times greater than 40 years due to lack of available data.

For other fractiles, Q, the threshold shift is given by:

$$N_Q = \begin{cases} N_{50} + kd_u; & 5 < Q < 50 \\ N_{50} - kd_L; & 50 < Q < 95 \end{cases} \tag{2.59}$$

The coefficient, k, is a function of the fractile, Q, and is given in Table 2.5. The parameters, d_u and d_L, can be calculated as:

$$d_u = (X_u + Y_u \log_{10} \Theta)(L_{\mathrm{EX,8h}} - L_0)^2 \tag{2.60}$$

$$d_L = (X_L + Y_L \log_{10} \Theta)(L_{\mathrm{EX,8h}} - L_0)^2 \tag{2.61}$$

If $L_{\mathrm{EX,8h}} < L_0$, then $L_{\mathrm{EX,8h}}$ is set equal to L_0 for the purposes of evaluating Equations (2.60) and (2.61). The coefficients, X_u, Y_u, X_L and Y_L, are listed in Table 2.6.

The threshold shift, H_{50}, for the 50% fractile due to age (Y years) alone is given in the standard as:

$$H_{50;\mathrm{Y}} = a(Y - 18)^2 + H_{50;18} \tag{2.62}$$

TABLE 2.5 Values of the multiplier k for each fractile, Q

$Q/100$		k
0.05	0.95	1.645
0.10	0.90	1.282
0.15	0.85	1.036
0.20	0.80	0.842
0.25	0.75	0.675
0.30	0.70	0.524
0.35	0.65	0.385
0.40	0.60	0.253
0.45	0.55	0.126
0.50		0.0

TABLE 2.6 Coefficients for use in calculating N_Q fractiles

Frequency (Hz)	X_u	Y_u	X_L	Y_L
500	0.044	0.016	0.033	0.002
1000	0.022	0.016	0.020	0.000
2000	0.031	−0.002	0.016	0.000
3000	0.007	0.016	0.029	−0.010
4000	0.005	0.009	0.016	−0.002
6000	0.013	0.008	0.028	−0.007

where $H_{50;18}$ is the 50% fractile hearing threshold shift for persons of the same sex with normal hearing, and this is taken as zero. For other fractiles, Q, the threshold shift is given by the following equations.

$$H_{Q;\mathrm{Y}} = \begin{cases} H_{50;\mathrm{Y}} + kS_u; & 5 < Q < 50 \\ H_{50;\mathrm{Y}} - kS_L; & 50 < Q < 95 \end{cases} \tag{2.63}$$

where k is given in Table 2.5, Q is the percentage of population that will suffer the loss, H_Q, and where:

$$S_u = b_u + 0.445H_{50;\mathrm{Y}} \tag{2.64}$$

$$S_L = b_L + 0.356H_{50;\mathrm{Y}} \tag{2.65}$$

Values of a, b_u and b_L differ for males and females and are listed in Table 2.7 as a function of the octave band centre frequency.

TABLE 2.7 Values of the parameters b_u, b_L and a used to determine, respectively, the upper and lower parts of the statistical distribution H_Q

Frequency (Hz)	Value of b_u		Value of b_L		Value of a	
	Males	Females	Males	Females	Males	Females
125	7.23	6.67	5.78	5.34	0.0030	0.0030
250	6.67	6.12	5.34	4.89	0.0030	0.0030
500	6.12	6.12	4.89	4.89	0.0035	0.0035
1000	6.12	6.12	4.89	4.89	0.0040	0.0040
1500	6.67	6.67	5.34	5.34	0.0055	0.0050
2000	7.23	6.67	5.78	5.34	0.0070	0.0060
3000	7.78	7.23	6.23	5.78	0.0115	0.0075
4000	8.34	7.78	6.67	6.23	0.0160	0.0090
6000	9.45	8.90	7.56	7.12	0.0180	0.0120
8000	10.56	10.56	8.45	8.45	0.0220	0.0150

2.6.4 United States Standard Formulation

The United States has also produced a standard for the assessment of hearing damage risk (ANSI/ASA S3.44-1, 2020). This standard is based on ISO 1999 (2013), but has some error corrections and two important additions. First, it suggests that exposure risk to tonal noise can be assessed by assuming that the tonal noise is equivalent to broadband noise at a level that is 5 dB(A) higher. Second, ANSI/ASA S3.44-1 (2020) allows for an exchange rate that is different to the equal energy exchange rate of 3 dB (see Section 2.6.7 for exchange rate discussions). The alternative exchange rate may be implemented by replacing $L_{\mathrm{EX,8h}}$ in Section 2.6.3 with an equivalent level, defined as EEH in the standard and defined as $L'_{Aeq,8h}$ in Section 2.6.7.

2.6.5 Alternative Formulations

The authors have demonstrated that an alternative interpretation of the International Standard ISO 1999 (2013) database is possible (Bies and Hansen, 1990). Alternatively, very extensive work carried out in Dresden, Germany, over a period of about two decades between the mid-1960s and mid-1980s, has provided yet a third interpretation of the existing database. These latter two formulations lead to the conclusion that for the purpose of determining hearing loss, noise exposure should be determined as an integral of the root mean square (RMS) pressure with time rather than the accepted integral of mean square pressure. This, in turn, leads to a 6 dB trading rule rather than the 3 dB trading rule that is widely accepted. Trading rules are discussed below in Section 2.6.7.

It has been shown (Macrae, 1991) that neither the formulation of Bies and Hansen nor the standard, ISO 1999 (2013), accounts for post-exposure loss observed in war veterans. Similarly, it may be shown that the formulation of the Dresden group (Kraak et al., 1977; Kraak, 1981) does not account for the observed loss. However, the formulation of Bies and Hansen (1990) as well as that of the Dresden group may be amended to successfully account for post-exposure loss for anyone (Bies, 1994).

2.6.5.1 Bies and Hansen Formulation

Bies and Hansen (1990) introduced sensitivity associated with age, *STA*, and with noise, *STN* (as amended by Bies (1994)), and they proposed that the effects of age and noise may be additive on a hearing sensitivity basis. They postulated the following relationship describing hearing loss, H', with increasing age and exposure to noise, which may be contrasted with the ISO 1999 (2013) formulation embodied in Equation (2.56):

$$H' = 10 \log_{10}(STA + STN) \quad \text{(dB)} \tag{2.66}$$

Additivity of effects on a sensitivity basis rather than on a logarithmic basis (which implies multiplication of effects) was proposed.

Hearing sensitivity associated with age is defined as:

$$STA = 10^{H/10} \quad \text{(dB)} \tag{2.67}$$

In the above equations, H is the observed hearing loss in a population unexposed to excessive noise, called presbyacusis, and is due to ageing alone. It may be calculated by using Equations (2.62) and (2.63).

Bies and Hansen (1990) proposed an empirically determined expression for the sensitivity to noise, *STN*. Their expression, modified according to Bies (1994), accounts for both loss at the time of cessation of exposure to excessive noise, $STN(Y_{ns})$ (where Y_{ns} (years) is the age when exposure to excessive noise stopped), and to post-exposure loss, M_c, (which is the additional loss to what would be expected due to aging alone) after exposure to excessive noise has stopped (Macrae, 1991).

Loss at the cessation of exposure is a function of the length of exposure, $\Theta = Y - 18$ (years), and the A-weighted sound pressure of the excessive noise, p_A (Pa). Here, Y is the age of the population and following the international standard, ISO 1999 (2013), it is assumed that exposure to excessive noise begins at age 18 years. The quantity *STN* is defined as zero when Θ is zero. Use of Equations (2.56), (2.66) and (2.67) gives the following expression for $STN(Y_{ns})$ in terms of N given by Equation (2.57) or (2.58) and H given by Equation (2.62):

$$STN(Y_{ns}) = 10^{H/10}\left(10^{(N-0.0083HN)/10} - 1\right) \quad \text{(dB)} \tag{2.68}$$

Hearing sensitivity, *STN*, associated with noise exposure is then:

$$STN = STN(Y_{ns}) + M_c(Y_{ns}, Y); \qquad Y > Y_{ns} \quad \text{(dB)} \tag{2.69}$$

The post-exposure term, M_c, has been determined empirically for one frequency (Bies, 1994) and may be expressed in terms of the age of the population, Y, and the age when exposure to excessive noise, Y_{ns}, stopped. The proposed post-exposure correction is based on data provided by Macrae (1991) and is limited to loss at 4 kHz as no information is available for other frequencies:

$$M_c = 0.0208\, Y_{ns}(Y - Y_{ns}) \quad \text{(dB)} \tag{2.70}$$

For the case of the reconstructed database of the International Standard, the quantity, M_c, is assumed to be zero, because the standard provides no post-exposure information.

Implicit in this formulation is the assumption that the A-weighted sound pressure, p_A, is determined in terms of the equivalent A-weighted sound pressure level averaged over a nominal 8-hour workday, where the prime is used to indicate an average of the RMS sound pressure, rather than the traditional energy average of sound pressure squared. Thus:

$$p_A = 10^{L'_{Aeq,8h}/20} \quad \text{(Pa)} \tag{2.71}$$

where

$$L'_{Aeq,8h} = 20\log_{10}\left[\frac{1}{8}\int_0^{T_e}\left[p_A^2(t)\right]^{1/2}\mathrm{d}t\right] = 20\log_{10}\left[\frac{1}{8}\int_0^{T_e}10^{L_A(t)/20}\mathrm{d}t\right] \quad \text{(dB re 20 µPa)} \tag{2.72}$$

which may be contrasted with the traditional Equation (2.38). Equation (2.72) implies that an equivalent sound pressure level may be calculated by integrating acoustic pressures rather than pressures squared as implied by Equation (2.38). This leads to a 6 dB trading rule for exposure time versus exposure level (see Section 2.6.7). In other words, for a fixed sound exposure, the sound pressure level may be increased by 6 dB(A) for each halving of the exposure time, rather than the 3 dB(A) that is implied by the averaging of squared acoustic pressures.

2.6.5.2 Dresden Group Formulation

The Dresden group investigated the relationship between noise exposure and hearing loss using retrospective studies of noise-exposed persons, temporary threshold shift investigations and animal experiments. Their major result supported by all three types of investigation describing the long-term effect of noise on the average hearing loss of an exposed population is summarised for the 4 kHz frequency below. The relationship describes exposure to all kinds of industrial and other noise including interrupted, fluctuating and impulsive noise with peak sound pressure levels up to 135 dB re 20 µPa. At higher levels, the observed loss seems to be dependent on pressure squared or energy input.

An A-weighted linear noise dose, B_e, is defined in terms of the total time, T_e, of exposure to noise in seconds as:

$$B_e = \int_0^{T_e}\left[p_A^2(t)\right]^{1/2}\mathrm{d}t \tag{2.73}$$

An age-related noise dose, B_a, in terms of the age of the person, T_a, in seconds is defined as:

$$B_a = 0.025(T_a - T_e) \tag{2.74}$$

The permanent threshold shift, H', is given by the following equation:

$$H' = k_f \log_{10}\left[\frac{B_e + B_a}{B_0}\right] \quad \text{(dB)} \tag{2.75}$$

The quantities of Equation (2.75), not already defined, are k_f, a coefficient specific for each audiometric frequency, with the value of 50 for 4 kHz and B_0, a critical noise dose used as a reference with the value of 2×10^7 Pa s.

Consideration of Equation (2.75) shows that if the term, B_e, associated with noise exposure is very much larger than the term, B_a, associated with age, then with cessation of exposure to noise, no further threshold shift should be observed until the term associated with age also becomes large. However, as pointed out above, Macrae (1991) has provided data showing the threshold shift continues post exposure and as suggested above, the expression given by Equation (2.75) may be corrected by the simple device of adding M_c, given by Equation (2.70).

2.6.6 Observed Hearing Loss

In Figure 2.20, the observed median loss in hearing at 4000 Hz is plotted as a function of the percentage risk of incurring that loss for a specified length of exposure at a specified sound pressure level. The presentation is based on published data (Smith et al., 2006; Burns and Robinson, 1970). Length of exposure is expressed in years, where it is assumed that a person would be exposed to the stated level for about 1900 hours during each year. In the figure, the curve labelled 80 dB(A) represents a lower bound for hearing loss that can be attributed to noise exposure; presumably all lower exposure levels would lie on the same curve because this loss is attributed to age and other causes.

Referring to Figure 2.20, it is evident that hearing deterioration is very rapid during the first 10 years and progressively more so as the exposure level rises above 80 dB(A). The data for percentage risk of developing a hearing loss refer to loss averaged arithmetically over 500 Hz, 1000 Hz and 2000 Hz, whereas the data for median loss (meaning that 50% of the population has this or greater loss) refer to loss measured at 4000 Hz. This choice of representation is consistent with the observation that noise-induced hearing loss always occurs first and proceeds most rapidly in the frequency range between 4000 Hz and 6000 Hz, and is progressively less at both lower and higher frequencies.

Inspection of Figure 2.20(a) shows that the 10-year exposure point for any given level always lies close to the 30-year exposure point for the level 10 dB lower. For example, the point corresponding to 30 years of exposure at a level of 80 dB(A) lies close to the point corresponding to 10 years at a level of 90 dB(A). This observation may be summarised by the statement that 30 years are traded for ten years each time the sound pressure level is increased by 10 dB. This observation, in turn, suggests that the metric proposed in Figure 2.20(b), which fairly well summarises the data shown in Figure 2.20(a), indicates that hearing loss is a function of the product of acoustic pressure and time, not pressure squared and time. This is equivalent to a trading rule that reducing the exposure level by 6 dB is equivalent to halving the time of exposure. Thus, a hearing deterioration index, *HDI*, is proposed, based on sound pressure, not sound pressure squared (or energy), which is the cumulative integral of the RMS sound pressure with time. Figure 2.20(b) shows that to avoid hearing impairment in 80% of the population, a strategy should be adopted that avoids acquiring a hearing deterioration index greater than 59 during a lifetime.

2.6.7 Occupational Noise Exposure Assessment

The measurement and assessment of occupational noise exposure is discussed in ANSI/ASA S12.19 (2020) and ISO 9612 (2009). In almost all current noise regulations worldwide, the assumption is implicit that hearing loss is a function of the integral of pressure squared with time as given by Equation (2.34). This implies that the relationship between exposure time and sound pressure level is such that for the same total noise exposure, a doubling of exposure time is equivalent to reducing the sound pressure level by 3 dB. This is referred to as the 3 dB trading

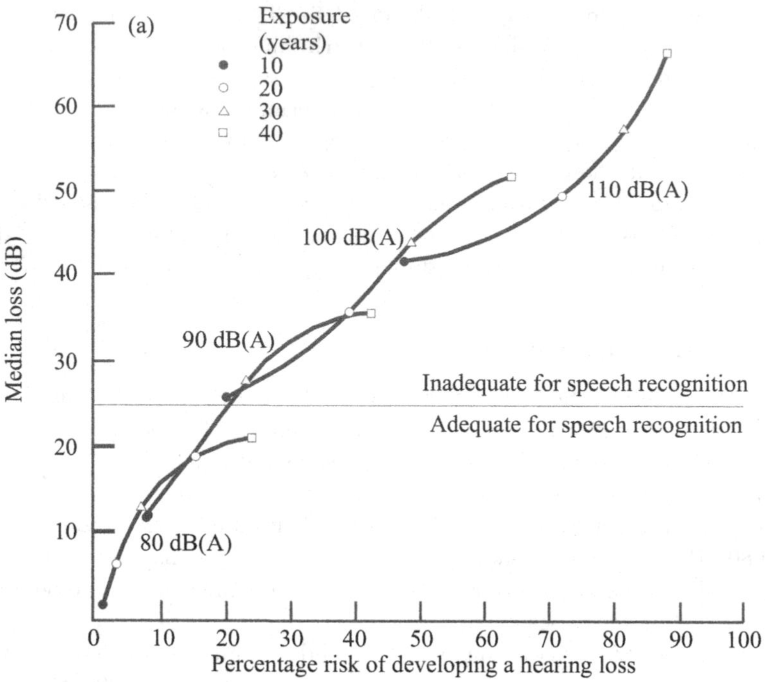

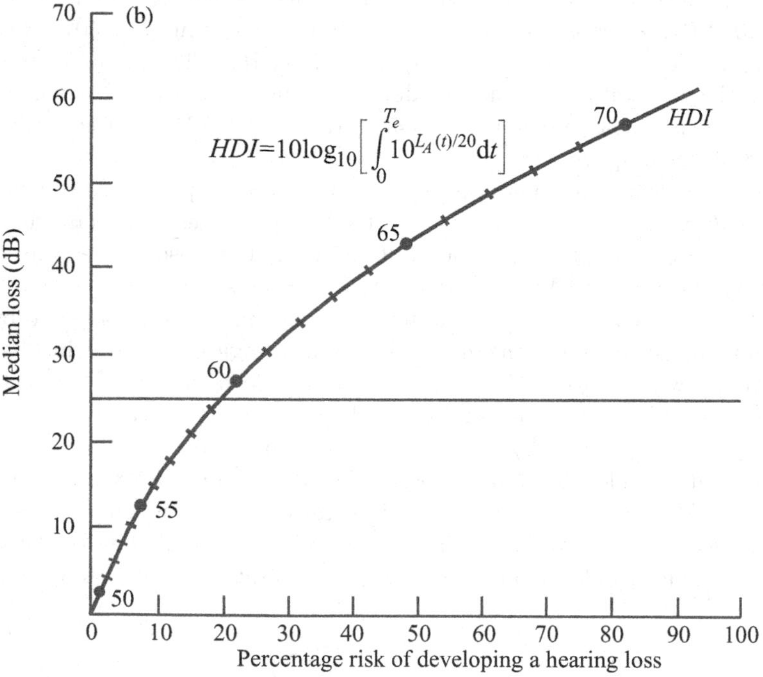

FIGURE 2.20 Hearing damage as a function of exposure. The % risk of developing a hearing loss (arithmetically averaged over 500 Hz, 1 kHz and 2 kHz) and the median loss (at 4 kHz) incurred with exposure are shown (a) as a function of the mean sound pressure level in the workplace (dB(A)) and exposure time (years); and (b) as a function of the hearing deterioration index, *HDI*. The quantity, $L_A(t)$, is the mean exposure level (dB(A)) over time, dt, and T_e is the exposure time in years.

rule. However, in the past some jurisdictions have used 4 dB or 5 dB instead of 3 dB and the discussion in Sections 2.6.5.1 and 2.6.5.2 implies that 6 dB could also be appropriate. Any trading rule may be represented by rewriting Equation (2.34) more generally as:

$$L'_{Aeq,8h} = \frac{10}{n} \log_{10} \left\{ \frac{1}{8} \int_0^{T_e} \left[10^{L_A(t)/10} \right]^n \mathrm{d}t \right\} \quad \text{(dB re 20 µPa)} \tag{2.76}$$

The prime has been used to distinguish the quantity from the traditional, energy-averaged $L_{Aeq,8h}$, defined in Equation (2.34). The value of n in Equation (2.76), corresponding to a trading rule of 3 dB is 1. If the observation that hearing loss due to noise exposure is a function of the integral of the RMS pressure with time, then $n = \frac{1}{2}$ and the trading rule is 6 dB(A). The relationship between n and the trading rule is:

$$n = 3/L \tag{2.77}$$

where L is the decibel trading level (or trading rule) that corresponds to a change in exposure by a factor of two for a constant exposure time. For a trading rule of 3, $L'_{Aeq,8h} = L_{Aeq,8\text{h}} = L_{\text{EX},8\text{h}}$ (see Equations (2.40) and (2.41)).

Introduction of a constant base level criterion, L_B, which $L'_{Aeq,8h}$ should not exceed, and use of Equation (2.77) allows Equation (2.76) to be rewritten in the following form:

$$L'_{Aeq,8h} = \frac{L}{0.3} \log_{10} \left\{ \frac{1}{8} \int_0^{T_e} \left[10^{0.3(L_A(t)-L_B)/L} \right] 10^{0.3L_B/L} \mathrm{d}t \right\} \quad \text{(dB re 20 µPa)} \tag{2.78}$$

Equation (2.78) may, in turn, be written as:

$$L'_{Aeq,8h} = \frac{L}{0.3} \log_{10} \left\{ \frac{1}{8} \int_0^{T_e} \left[10^{0.3(L_A(t)-L_B)/L} \right] \mathrm{d}t \right\} + L_B \quad \text{(dB re 20 µPa)} \tag{2.79}$$

Note that if discrete exposure levels were being determined with a sound level meter as described above, then the integral would be replaced with a sum over the number of discrete events measured for a particular person during a working day. For example, for a number of events, m, for which the ith event is characterised by an A-weighted sound pressure level of L_{Ai}, Equation (2.79) could be written as:

$$L'_{Aeq,8h} = \frac{L}{0.3} \log_{10} \left\{ \frac{1}{8} \sum_{i=1}^{m} \left[10^{0.3(L_{Ai}-L_B)/L} \right] \times t_i \right\} + L_B \quad \text{(dB re 20 µPa)} \tag{2.80}$$

When $L'_{Aeq,8h} = L_B$, reference to Equation (2.79) shows that the argument of the logarithm on the right-hand side of the equation must be one. Consequently, if an employee is subjected to higher levels than L_B, then to satisfy the criterion, the length of time, T_e, must be reduced to less than eight hours. Setting the argument equal to one, $L_A(t) = L_B = L'_{Aeq,8h}$ and evaluating the integral using the mean value theorem, the maximum allowed exposure time to an equivalent sound pressure level, $L'_{Aeq,8h}$, is:

$$T_a = 8 \times 10^{-0.3(L'_{Aeq,8h}-L_B)/L} \quad \text{(hours)} \tag{2.81}$$

In most developed countries, the equal energy trading rule is used with an allowable 8-hour exposure of 85 dB(A), which implies that in Equation (2.79), $L = 3$ and $L_B = 85$. In 2003, the European Parliament (European Parliament, 2003) decided that no employee is allowed to be exposed to $L_{Aeq,8\text{h}}$ levels greater than 87 dB(A) (measured with hearing protection such as ear muffs and/or ear plugs) and that in any environments where the L_{Aeq} levels exceeded 85 dB(A), hearing protection is mandatory.

If the number of hours of exposure is different to 8, then to find the actual allowed exposure time to the given noise environment, denoted $L'_{Aeq,T}$, the "8" in Equation (2.81) is replaced by the actual number of hours of exposure, T_e.

The daily noise dose (DND) is the percentage of the allowable daily exposure (ANSI/ASA S12.19, 2020), and is defined as equal to 100% × 8 hours divided by the allowed exposure time, T_a of Equation (2.81) with L_B set equal to 90. That is:

$$\text{DND} = 100 \times 10^{0.3(L'_{Aeq,8h}-90)/L} \quad (\%) \tag{2.82}$$

In most developed countries, the equal energy trading rule is used with an allowable 8-hour exposure of 85 dBA, which implies that in Equation (2.79), $L = 3$ and $L_B = 85$. In 2003, the European Parliament (European Parliament, 2003) decided that no employee is allowed to be exposed to $L_{Aeq,8\text{h}}$ levels greater than 87 dBA (measured with hearing protection included) and that in any environments where the L_{Aeq} levels exceeded 85 dBA, hearing protection is mandatory.

The US Federal Government has published regulations in two jurisdictions (OSHA, 1995; MSHA, 2021), in which they state that acceptable exposure levels correspond to using $L = 5$ and $L_B = 90$ and for levels above 85 dB(A) a hearing conservation program must be implemented and those exposed must be given hearing protection. More recently, the National Institute for Occupational Safety and Health (NIOSH) has recommended that $L = 3$ and $L_B = 85$ is more appropriate NIOSH (1998) but this has not been reflected in OSHA regulations.

2.7 Hearing Damage Risk Criteria

The sound pressure level below which damage to hearing from habitual exposure to noise should not occur in a specified proportion of normal ears is known as the hearing damage risk criterion (DRC). It should be noted that hearing damage is a cumulative result of level as well as duration, and any criterion must take both level and duration of exposure into account. Note that it is not just the workplace that is responsible for excessive noise. Many people engage in leisure activities that are damaging to hearing, such as going to nightclubs with loud music, shooting or jet ski riding. Also listening to loud music through headphones can be very damaging, especially for children.

2.7.1 Continuous Noise

A continuous eight-hour exposure each day to ordinary broadband noise of a level of 90 dB(A) results in a hearing loss of greater than 25 dB (arithmetically averaged over 0, 5, 1 and 2 kHz) for approximately 25% of people exposed for 30 years or more. This percentage is approximate only, as it is rare to get agreement between various surveys that are supposedly measuring the same quantity. This is still a substantial level of hearing damage risk. On the other hand, a criterion of 80 dB(A) for an eight-hour daily exposure would constitute a negligible hearing damage risk for speech. Therefore, to minimise hearing loss, it is desirable to aim for a level of 80 dB(A) or less in any industrial facility design. Limits higher than 80 dB(A) must be compromises between the cost of noise control, and the risk of hearing damage and consequent compensation claims.

Although an exposure to 80 dB(A) for eight hours per day would ensure a negligible hearing loss for speech due to noise exposure, a lower level would be required to ensure negligible hearing loss at all audible frequencies. One viewpoint is that 97% of the population should be protected from any measurable noise-induced permanent shift in hearing threshold at all frequencies, even after 40 years of exposure for eight hours per day for 250 days of the year. If we assume that, for about 10% of each eight-hour working day, a worker is out of the area of maximum noise (owing to visits to other areas) and, further, that he or she is exposed to sound pressure levels that are over 5 dB lower during the remaining sixteen hours of the day, then studies worldwide show that for 97% protection at all frequencies, the sound pressure level must not exceed 75 dB(A). If a worker is exposed to continuous noise for 24 hours per day, the level must not exceed 70 dB(A).

Another viewpoint is that it is only necessary to protect people from hearing damage for speech, and that to aim for the above levels is unnecessarily conservative and economically unrealistic. In 1974, having reviewed the published data, the Committee of American Conferences of Governmental Industrial Hygienists determined that an exposure level of 85 dB(A) during a working life would result in 90% of people suffering a hearing loss of less than 25 dB when arithmetically averaged over the frequencies, 0.5, 1 and 2 kHz.

Current standards in most countries now recommend 85 dB(A) as an acceptable level for eight hours of exposure, although most people agree that a level of 80 dB(A) is more desirable from the point of view of minimising hearing damage.

2.7.2 Impulse Noise

Industrial impulse noise is defined as a short-duration sound characterised by a shock front pressure waveform (i.e. virtually instantaneous rise), usually created by a sudden release of energy; for example, as encountered with explosives or gun blasts (Rice and Martin, 1973; Rice, 1974). Such a characteristic impulse pressure waveform is often referred to as a Friedlander wave and is illustrated in Figure 2.21.

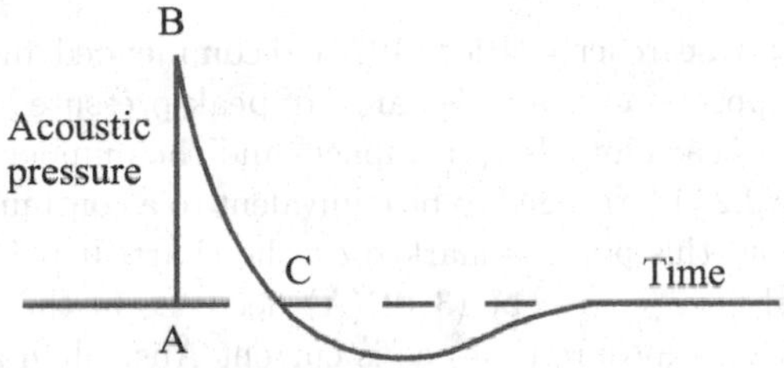

FIGURE 2.21 Idealised waveform of impulse noise.
Peak level = pressure difference AB; rise time = time difference AB; A duration = time difference AC.

This single-impulse waveform is typically generated in free-field environments, where sound-reflecting surfaces that create reverberation are absent. With gunfire, mechanically generated noise is also present in addition to the shock pulse, and in this case, the waveform envelope can take the form of an impact noise, which is discussed in Section 2.7.3.

The durations of impulsive noises may vary from microseconds up to 50 ms, although in confined spaces reverberation characteristics may cause the duration to extend considerably longer, in which case, the noise may be treated as an impact noise, as discussed in Section 2.7.3. For impulsive noise with no reverberation, an A-duration is defined as the time for the impulse acoustic pressure waveform to go from its peak to a zero value (see figure 2.21) and this may be used for the purpose of assessing hearing damage risk using Figure 2.23, or alternatively, following the procedure provided in MIL-STD-1474E (2015, pp. 45–46) to determine the percentage of allowed noise dose.

In general, people are not habitually exposed to impulsive noises. In fact, only people exposed to explosions such as quarry blasting or gunfire are exposed to impulse noises (as opposed to impact noises). Estimates of the number of pulses likely to be received on any one occasion vary between 10 and 100, although up to 1000 impulses may sometimes be encountered.

2.7.3 Impact Noise

Industrial impact noises are normally produced by non-explosive means, such as metal-to-metal impacts in industrial plant processes. In such cases, the characteristic shock front is not always present, and due to the reverberant industrial environments in which they are heard, the durations are often longer than those usually associated with impulse noise, producing a waveform like the one illustrated in Figure 2.22. The background noise present in such situations, coupled with the regularity with which impacts may occur, often causes the impacts to give the impression of running into one another. People in industry are often habitually exposed to such noises, and the number of impacts heard during an eight-hour shift usually runs into thousands.

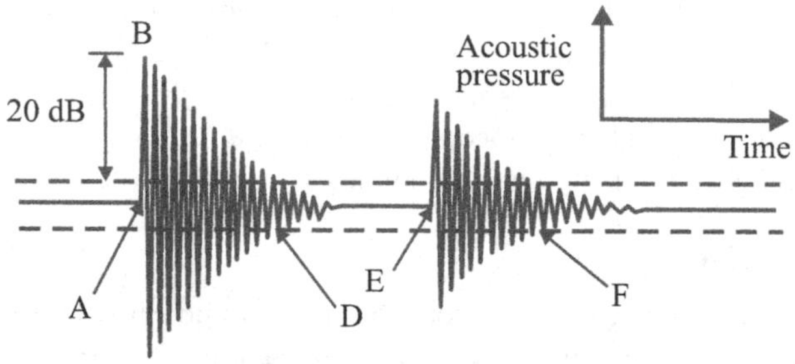

FIGURE 2.22 Idealised waveform of impact noise.
Peak level = pressure difference AB; rise time = time difference AB; B duration = time difference AD (+ EF when a reflection is present).

Figure 2.23 shows one researcher's (Rice, 1974) recommended impulse and impact upper bound criteria for daily exposure, over a wide range of peak pressure levels, as a function of the product of the B duration of each impulse (or impact) and the number of impulses (or impacts).

The criterion of Figure 2.23 is arranged to be equivalent to a continuous exposure to 90 dB(A) for an eight-hour period and this point is marked on the chart. It is interesting to note that if, instead of using the equal energy concept (3 dB(A) decrease in the allowable sound pressure level for each doubling of the exposure time) as is current Australian and European practice, a 5 dB(A) per halving of exposure is used (as is current US practice), the criteria for impulse and impact noise would essentially become one criterion.

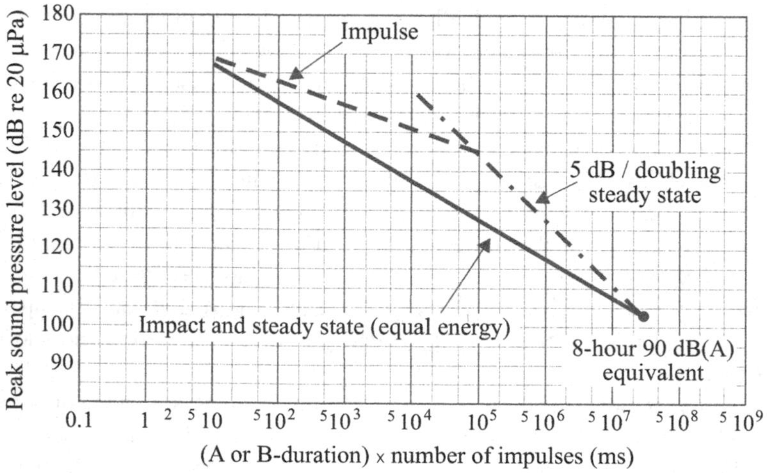

FIGURE 2.23 A comparison of impulse and impact damage risk criteria with steady-state criteria (after Rice (1974)).

The equivalent noise exposure corresponding to a particular A-duration or B-duration multiplied by the number of impacts or impulses, and a corresponding peak pressure level may be calculated using Figure 2.23. The first step is to calculate the product of the impulse A-duration or impact B-duration and number of impulses (impacts). This value is entered on the abscissa of Figure 2.23 and a vertical line is drawn until it intersects the appropriate curve. For impulse noise (usually less than 1000 impulses per exposure), the upper curve is used, while for impact noise the lower curve is used. From the point of intersection of the appropriate curve and the vertical line, a horizontal line is drawn to intersect the ordinate at the value of peak sound pressure level corresponding to an exposure level of $L_{Aeq,8\text{h}} = 90$ dB(A).

Alternatively, the peak level of an individual impact is entered on the ordinate and a horizontal line drawn until it intersects the lower curve. A vertical line is drawn downwards from the point of intersection. Where the vertical line intersects the abscissa indicates the product of B duration and number of impacts that will correspond to an exposure level of $L_{Aeq,8\text{h}} = 90$ dB(A). The allowed number of impacts is doubled for each 3 dB that the measured peak level is exceeded by the peak level corresponding to an exposure level of $L_{Aeq,8\text{h}} = 90$ dB(A). The same curve can be used for an allowed exposure level of $L_{Aeq,8\text{h}} = 85$ dB(A). In this case, the solid curve is shifted down by 5 dB(A) and the same procedure as just described is followed.

Due partly to the difficulty in measuring the duration of the impact noise and partly because the predictions are not considered very reliable, the preceding procedure has not seen much use. In fact, the ISO standard, ISO 1999 (2013), states, "The prediction method presented is based primarily on data collected with essentially broadband steady non-tonal noise. The application of the database to tonal or impulsive/impact noise represents the best available extrapolation." This implies that impact and impulse noise should be treated in a similar way to continuous noise in terms of exposure and that the best way of assessing its effect is to use an integrating sound level meter that has a sufficiently short time constant (20 µs) to record the energy contained in short impacts and impulses and include this energy with the continuous noise energy in determining the overall, energy-averaged, L_{Aeq}. However, some researchers in the audiometry community (von Gierke, H. E. et al., 1982) have suggested that impulse and impact noise should be assessed using a C-weighted energy-average or L_{Ceq} and this should be combined with the A-weighted level, L_{Aeq}, for continuous noise on a logarithmic basis to determine the overall level, which is then used in place of the continuous L_{Aeq} in the assessment of damage risk. Others feel that the impact noise should not be weighted at all, but this results in undue emphasis being placed on very low-frequency sound such as that produced by slamming a car door and is not recommended as a valid approach. With digital instrumentation, the peak sound pressure level is the highest absolute instantaneous value that is sampled. For an instrument with a sample rate of 48 kHz, this represents samples separated by approximately 20 µs. Instruments with a faster sampling rate may measure a higher peak for the same impact noise.

It is generally accepted in most standards that no one should be exposed to a peak sound pressure level that exceeds 140 dB, and for children the limit should be 120 dB. Some specify that this should be measured with a C-weighting network implemented on the measuring instrument and some specify that no weighting should be used. The latter specification is rather arbitrary as the measured level then depends on the upper and lower frequency range of the measuring instrument. For consistency, the sampling rate of the instrument measuring the peak sound pressure level should be 48 kHz and the peak level is then the highest absolute sample value that is recorded. A detailed discussion of the assessment of the hearing loss resulting from high intensity impulse noise can be found in Teague et al. (2016).

2.8 Implementing a Hearing Conservation Program

To protect workers in noisy industries from the harmful effects of excessive noise, it is necessary to implement a well-organised hearing conservation program. The key components of such a program include:

1. regular noise surveys of the work environment which includes:
    - making a preliminary general survey to determine the extent of any problems;
    - determining the sound power and directivity (or sound pressure at the operator locations) of noisy equipment;
    - identification, characterisation and ranking of noise sources;
    - identification of high sound pressure level areas and their contribution to worker exposures;
    - determination of individual worker exposures to noise using noise measurements and dosimeters (ANSI/ASA S12.19, 2020; ISO 9612, 2009)
    - prediction of the risk of hearing loss for individual or collective groups of workers using ISO 1999 (2013); and
    - identifying hearing conservation requirements.
2. regular audiometric testing of exposed workers to evaluate the program effectiveness and to monitor their temporary threshold shift (TTS) at the end of the work shift as well as permanent threshold shift (measured by testing after a quiet period) (see ANSI/ASA S3.6 (2018), ANSI/ASA S3.1 (2018), ISO 8253-1 (2010), ISO 8253-2 (2009), ISO 8253-3 (2022), IEC 60645-1 (2017), IEC 60645-3 (2020)) with the following notes:
    - elimination of temporary threshold shift will eliminate permanent threshold shift that will eventually occur as a result of sufficient incidences of TTS;
    - anyone with a permanent threshold shift in addition to the shift they had at the beginning of their employment should be moved to a quieter area and, if necessary, given different work assignments;
3. installation and regular monitoring of the effectiveness of noise control equipment;
4. consideration of noise in the specification of new equipment;
5. consideration of administrative controls involving the reorganisation of the workplace to minimise the number of exposed individuals and the duration of their exposure;
6. education of workers;
7. regular evaluation of the overall program effectiveness, including noting the reduction in temporary threshold shift in workers during audiometric testing;
8. careful record keeping, including noise data and audiometric test results, noise control systems purchased, instrumentation details and calibration histories, program costs and periodic critical analysis reports of the overall program; and
9. appropriate use of the information to:
    - inform workers of their exposure pattern and level;
    - act as a record for the employer;
    - identify operators whose exposure is above the legal limits;
    - identify areas of high sound pressure level;
    - identify machines or processes contributing to excessive sound pressure levels;

- indicate areas in which control is necessary;
- indicate areas where hearing protection must be worn prior to engineering noise controls being implemented;
- indicate areas where hearing protection must be worn even after engineering noise controls have been implemented; and
- identify the most appropriate locations for new machines and processes.

To be successful, a hearing conservation program requires:

- well defined goals and objectives;
- competent program management;
- commitment from management at the top of the organisation;
- commitment from the workers involved;
- adequate financial resources;
- access to appropriate technical expertise;
- good communication and monitoring systems; and
- a philosophy of continuous improvement.

2.9 Hearing Protection Devices

Many occupational noise guidelines recommend a maximum noise exposure limit of $L_{Aeq,8\text{h}} =$ 85 dB(A), which is typically measured using an instrument such as a sound level meter or a noise dosimeter. However, if an instrument is not available, a commonly used subjective technique to judge if the ambient sound pressure level is greater than 80 dB(A), and therefore consideration should be given to wearing hearing protection to lessen the likelihood of hearing damage, is if when two people standing about 1 m apart have to speak very loudly to communicate (see Webster (1970) and Section 2.10.1). In addition to the potential for hearing damage, there can be a reduction in the efficiency of workers at elevated sound pressure levels. The German standard, VDI 2058 Blatt 3 (2014), describes the mechanisms that lead to a decrease in work efficiency and an allocation of tasks of different complexity.

Personal hearing protection devices (HPDs), such as earmuffs and earplugs, reduce the noise exposure that a person receives, providing the device is fitted and used properly according to the manufacturer's instructions. The amount of noise reduction provided by the device is measured according to standards that typically involve conducting laboratory tests on untrained human subjects. The subjects are only provided with the instructions on the packaging, and the noise reduction provided by the hearing protection device is determined by measuring the difference in the subject's hearing threshold with and without wearing the device, and is termed the Real-Ear Attenuation at Threshold (REAT).

There are several accepted measurement standards and reporting schemes. Sections 2.9.1 to 2.9.5 describe the application of reported noise reduction metrics of hearing protection devices to the calculation of the sound pressure level that a wearer is likely to experience. If the reader intends to evaluate the noise reduction of hearing protection devices, they should refer to the appropriate testing standard, as each method is quite involved and beyond the scope of this book.

European and USA standards differ in terms of the quantity used to specify the acoustical performance of a hearing protector. Although both standards are similar in that they estimate a level experienced at the ear when wearing a particular hearing protection device, they differ in terms of the type of level that is estimated. European standards estimate an A-weighted overall "effective sound pressure level", $L'_{p,Ax}$, whereas USA standards estimate an A-weighted exposure level, referred to as an A-weighted "protected exposure level", $L_{\mathrm{prot},A}$, which is a sound pressure level normalised to an 8-hour exposure time (in other words, the estimated $L_{Aeq,8\mathrm{h}}$ at the ear).

2.9.1 Noise Reduction Rating, NRR

The Noise Reduction Rating (NRR) scheme is used in the USA and accepted in several other countries. This scheme provides a single number value in decibels of the noise reduction provided by a hearing protection device. The US Environment Protection Authority (EPA) has a legislative requirement that a single number NRR value is written on the packaging of hearing protection devices (EPA, 2020). However, it was found that the noise reduction that users were receiving in the workplace did not correlate with the attenuation indicated by the NRR value. Thus, it is recommended that the NRR value on the packaging should be derated when estimating the sound pressure level at the ears of the user. Calculation of appropriate derating factors is discussed in the following paragraphs.

The EPA proposed in 2009 (EPA, 2009) to update the labelling required on hearing protection devices with the Noise Level Reduction Statistic for use with A-weighting (NRS_{Ax}), which is described in Section 2.9.3, however these changes were not implemented, and the NRR value must be written on devices.

The US Occupational Safety and Health Administration (OSHA) and the National Institute for Occupational Safety and Health (NIOSH) have different methods for estimating the A-weighted exposure level at the ears when wearing hearing protection, called the A-weighted protected exposure level, $L_{\mathrm{prot},A}$, with different recommendations for derating the value of NRR. Note that OSHA is responsible for making regulations, and NIOSH is responsible for conducting research and making recommendations for the prevention of work-related illnesses but has no authority to make regulations.

The protected exposure level, $L_{\mathrm{prot},A}$, at the ears of the user is calculated by determing the A-weighted or C-weighted exposure level, $L_{Aeq,8\mathrm{h}}$ or $L_{Ceq,8\mathrm{h}}$, with no hearing protector in place (see Section 2.4.3), and subtracting a derated value of the hearing protection device's NRR to account for real-world performance. The $L_{Aeq,8\mathrm{h}}$ level (L_{Aeq} normalised to an 8 hour exposure) is referred to as exposure level, $L_{exp,A}$ in OSHA (2016), but will be written as $L_{Aeq,8\mathrm{h}}$ here to avoid confusion with the nomenclature in other parts of this book. Table 2.8 lists how to calculate the A-weighted protected exposure level, according to recommendations by OSHA (2016), and NIOSH (2018). The exposure level with no hearing protection is measured in decibels using either a C-weighting, given by $L_{Ceq,8\mathrm{h}}$ in units of dB re 20 µPa, or an A-weighting given by $L_{Aeq,8\mathrm{h}}$ in units of dB re 20 µPa. The derating factor, x, is listed in Table 2.9 and depends on the regulating authority and the type of hearing protection worn.

TABLE 2.8 Calculation of the A-weighted protected exposure level, $L_{\mathrm{prot},A}$, according to OSHA (2016) and NIOSH (1998) guidelines, based on either a C- or an A- weighted exposure level

Authority	C-weighting	A-weighting
OSHA	$L_{\mathrm{prot},A} = L_{Ceq,8\mathrm{h}} - [x \times \mathrm{NRR}]$	$L_{\mathrm{prot},A} = L_{Aeq,8\mathrm{h}} - [x \times (\mathrm{NRR} - 7)]$
NIOSH	$L_{\mathrm{prot},A} = L_{Ceq,8\mathrm{h}} - [x \times \mathrm{NRR}]$	$L_{\mathrm{prot},A} = L_{Aeq,8\mathrm{h}} - [\,(x \times \mathrm{NRR}) - 7]$

When earplugs and earmuffs are worn simultaneously, the OSHA (2016) guidelines suggest that there is only a 5 dB benefit in noise reduction, in addition to the reduction achieved using the device with the highest NRR value (NRR_h).

TABLE 2.9 Derating factor x to be applied to the manufacturer's NRR value according to OSHA and NIOSH recommendations

Authority	Protection type	Derating factor x
OSHA	All	0.5
NIOSH	Foam earplugs	0.5
NIOSH	All other earplugs	0.3
NIOSH	Earmuffs	0.75

2.9.2 Noise Reduction Rating Subjective Fit, NRR(SF)

Another noise metric that has been devised is called the Noise Reduction Rating Subjective Fit (NRR(SF)), which is intended to better represent the expected noise reduction achieved from practical use, and does not require derating. Some manufacturers of hearing protection devices will provide this information, and it is intended to be subtracted from the A-weighted exposure level with no hearing protection to provide the protected exposure level. The A-weighted protected exposure level is calculated as (NIOSH, 2005):

$$L_{\text{prot},A} = L_{Aeq,8\text{h}} - [\,\text{NRR(SF)}\,] \quad (\text{dB re } 20\,\mu\text{Pa}) \tag{2.83}$$

If the exposure level were measured using a C-weighting, then a 3 dB (previously 5 dB before re-evaluation of the database in 2017) adjustment is made to account for predicted differences in the A- and C- weighted levels, and so the protected exposure level is calculated as:

$$L_{\text{prot},A} = L_{Ceq,8\text{h}} - 3 - [\,\text{NRR(SF)}\,] \quad (\text{dBA re } 20\,\mu\text{Pa}) \tag{2.84}$$

If octave-band noise levels are measured, the recommended approach is to apply the A-weighting corrections, calculate the overall A-weighted noise level, L_{Aeq}, normalise this level to an 8-hour exposure using Equation (2.40) to obtain $L_{Aeq,8\text{h}}$, and use Equation (2.83) to calculate the protected A-weighted exposure level.

2.9.3 Noise Level Reduction Statistic, NRS_{Ax}

The US EPA proposed altering the labelling requirements on the packaging of hearing protectors, so that they display a calculated expected lower and upper range of the NRR value, if the device were worn by untrained and trained subjects, respectively (EPA, 2009). The noise reductions on the labels could be used directly with measured A-weighted sound exposure levels. The labels on hearing protection devices would have included two numbers called the Noise Level Reduction Statistic, NRS_{Ax}. The lower value indicates the expected noise reduction that would be achieved by $x = 80\%$ of the general population, and the higher value is the expected noise reduction by well-trained and motivated wearers for $x = 20\%$ of the population. The estimated A-weighted sound exposure level, $L_{\text{prot},Ax}$, at the ears of the person wearing hearing protection, for a protection performance of $x\%$, is calculated as (ANSI/ASA S12.68, 2020):

$$L_{\text{prot},Ax} = L_{Aeq,8\text{h}} - \text{NRS}_{Ax} \quad (\text{dBA re } 20\,\mu\text{Pa}) \tag{2.85}$$

where $L_{Aeq,8\text{h}}$ is the A-weighted sound exposure level (sound pressure level normalised to an 8-hour exposure) to which the person is exposed in their environment with no hearing protection, and which is used to assess their hearing damage risk.

More accurate calculation methods for predicting the A-weighted protected exposure level at the ear are described in ANSI/ASA S12.68 (2020) and ISO 4869-2 (2018).

2.9.4 Calculation of Effective A-Weighted Sound Pressure Level Using Assumed Protection Value, APV

The A-weighted "effective sound pressure level" is the level expected at the ear when a hearing protection device is fitted. It depends on the sound pressure level in the absence of the protection device and the extent of protection provided by the device. There is a statistical distribution of the amount of hearing protection provided to an individual that depends on factors such as the amount of attenuation provided by the device, how well the device is fitted on the person and the spectrum of the noise to which the person is exposed. We will begin the discussion by considering the extent of protection provided by a hearing protection device. The Assumed Protection Value (APV) is the attenuation of a hearing protection device, determined using laboratory tests described in ISO 4869-1 (2018), using a group of people to determine the mean sound attention. This mean is then adjusted by subtracting a coefficient, α, multiplied by the standard deviation from the mean attenuation, determined using ISO 4869-1 (2018), and is given by:

$$\text{APV}_{fx} = m_f - \alpha s_f \quad \text{(dB re 20 µPa)} \tag{2.86}$$

where f is the centre frequency of the octave band, x is the selected protection performance that is often set at $x = 84\%$, m_f is the mean sound attenuation determined using ISO 4869-1 (2018), s_f is the standard deviation of the measurements following ISO 4869-1 (2018) and α is a value that depends on the protection performance $x\%$, as listed in Table 2.10

TABLE 2.10 Values of α for a desired protection performance $x\%$

Protection performance ($x\%$)	Value of α	Protection performance ($x\%$)	Value of α
50	0.00	90	1.28
75	0.67	95	1.64
80	0.84	98	2.00
84	1.00		

On the ISO web site, an electronic spreadsheet is available with an example calculation of APV (ISO 4869-2 Spreadsheet, 2018). The APV values are used in the calculation of the overall A-weighted "effective sound pressure level", $L'_{p,Ax}$, at the wearer's ear, using the octave-band, High Medium Low (HML) and Single Number Rating (SNR) methods, described in the following subsections.

2.9.4.1 Octave Band Method

The standard ISO 4869-2 (2018) describes an octave-band calculation method for determining the A-weighted overall "effective sound pressure level", $L'_{p,Ax}$ at the centre of an acoustically transparent head, fitted with hearing protection, for a protection performance of $x\%$. The method gives results for any noise environment specified in terms of octave band sound pressure levels and uses these levels, together with the octave-band sound attenuation values for the hearing protector, to determine the A-weighted overall "effective sound pressure level". It is considered a reference method; however, it has inaccuracies, as it uses statistically determined mean sound attenuation values and standard deviations from tests on a group of test subjects, and does not use the measured sound attenuation values for the person in question.

The octave band method involves calculating the A-weighted overall "effective sound pressure level" by using the measured octave band sound pressure level of the noise, subtracting the Assumed Protection Value (APV) and the A-weighting correction value for the octave-band, and then logarithmically adding the result for each octave band to determine the overall A-weighted "effective sound pressure level" for a protection performance of $x\%$. Mathematically the procedure is expressed as"

$$L'_{p,Ax} = 10 \log_{10} \sum_{k=1}^{8} 10^{(L_{p,f(k)} - A_{f(k)} - \mathrm{APV}_{f(k),x})} \quad \text{(dBA re 20 µPa)} \tag{2.87}$$

where $L'_{p,Ax}$ is the A-weighted "effective sound pressure level" (expected at the ear) when wearing the hearing protection, rounded to the nearest integer, subscript $f(k)$ is the octave-band centre frequency where $f(1) = 63$ Hz, $\cdots f(8) = 8000$ Hz, $L_{p,f(k)}$ is the octave-band sound pressure level of the noise, $A_{f(k)}$ is the A-weighting value for the octave-band, as listed in Table 2.3, and $\mathrm{APV}_{f(k),x}$ is the Assumed Protection Value for the octave-band, $f(k)$, for a protection performance of $x\%$, which is typically set at $x = 84\%$.

An electronic spreadsheet is available on the ISO web site with an example calculation of protected level using the octave-band method (ISO 4869-2 Spreadsheet, 2018).

If the 63 Hz octave-band data for the noise level or the device attenuation are not available, then the summation in Equation (2.87) begins at 125 Hz. For the HML and SNR methods described below, the data at 63 Hz are not used, even if the data are available, and the calculations always begin at 125 Hz.

2.9.4.2 High, Medium, Low (HML) Method

The standard ISO 4869-2 (2018) describes a method to characterise the noise reduction of hearing protectors in high, H, medium, M, and low, L, frequency ranges. The method requires the attenuation values for the device measured in a laboratory according to ISO 4869-1 (2018), and the use of eight reference noise spectra that have spectral bias in the high, medium, and low frequency octave bands, to calculate the H, M, and L values in decibels. For example, some hearing protectors might be rated as having H=33 dB, M=31 dB and L=30 dB. The procedure to calculate the H, M, L values is involved and readers should refer to the ISO 4869-1 and -2 standards.

To estimate the A-weighted "effective sound pressure level" (expected at the ear), $L'_{p,Ax}$, (for a protection performance of $x\%$) using the HML ratings, requires measurement of both the A-weighted, L_{Aeq}, and C-weighted, L_{Ceq}, sound pressure levels to which the person is exposed. Note that in the ISO 4869-2 (2018) standard, the attenuation value for the 63 Hz octave band is excluded from the HML (and SNR) calculation methods. The steps to calculate the protected level are described below.

Step 1: Calculate the difference between the C- and A-weighted sound pressure levels as:

$$\Delta = L_{Ceq} - L_{Aeq} \quad \text{(dB)} \tag{2.88}$$

Step 2: If the value of $\Delta \leq 2$ dB, then calculate the Predicted Noise level Reduction (PNR) as:

$$\mathrm{PNR}_x = \mathrm{M}_x - \left[\frac{\mathrm{H}_x - \mathrm{M}_x}{4} \times (L_{Ceq} - L_{Aeq} - 2) \right] \quad \text{(dB)} \tag{2.89}$$

If the value of $\Delta > 2$ dB, then calculate the PNR as:

$$\mathrm{PNR}_x = \mathrm{M}_x - \left[\frac{\mathrm{M}_x - \mathrm{L}_x}{8} \times (L_{Ceq} - L_{Aeq} - 2) \right] \quad \text{(dB)} \tag{2.90}$$

Step 3: Calculate the A-weighted "effective sound pressure level" (expected at the ear) as:

$$L'_{p,Ax} = L_{Aeq} - \mathrm{PNR}_x \quad \text{(dBA re 20 µPa)}, \tag{2.91}$$

where the resulting $L'_{p,Ax}$ is rounded to the nearest integer.

An electronic spreadsheet is available on the ISO web site with an example deriving the HML values for a device based on measured results, and the application of the HML values to calculate the effective A-weighted protected level (ISO 4869-2 Spreadsheet, 2018).

2.9.4.3 Single Number Rating, SNR

The Single Number Rating (SNR) is used in countries affiliated with the European Union and is calculated using ISO 4869-2 (2018), and typically accompanies the HML rating. For example, a label on the package of the hearing protection device has written SNR = 37 dB, H = 36 dB, M = 35 dB, L = 34 dB. Tests are conducted at a number of laboratories that are independent of the manufacturers of the hearing protection device. The A-weighted "effective sound pressure level" (expected at the ear) is calculated by subtracting the SNR value from the measured C-weighted sound pressure level as:

$$L'_{p,Ax} = L_{Ceq} - \text{SNR} \quad (\text{dBA re } 20\,\mu\text{Pa}) \tag{2.92}$$

An electronic spreadsheet is available on the ISO web site with an example calculation using the SNR method (ISO 4869-2 Spreadsheet, 2018).

2.9.5 Sound Level Conversion, SLC_{80}

The Sound Level Conversion SLC_{80} rating system that is used in Australia and New Zealand, and defined in AS/NZS 1270 (2014), is based upon laboratory testing of untrained users to determine the threshold of hearing with and without the hearing protection device being assessed. The standard describes how to calculate the SLC_{80} value, which is an estimate of the amount of protection that could be achieved by 80% of users. A label on the packaging of the hearing protection device lists the SLC_{80} value followed by a class number; for example, SLC_{80} 27, Class 5. The SLC_{80} value indicates the amount of noise reduction and the Class is an integer from 1 to 5 that indicates the corresponding category for the measured SLC_{80} value, as shown in Table 2.11 (as defined in AS/NZS 1270 (2014), Appendix A). The table also lists the recommended maximum sound pressure level where the hearing protector may be used (as defined in AS/NZS 1269-3 (2016), Appendix A). The standard was developed based on work in the 1970's (Waugh, 1976).

TABLE 2.11 Classes of the SLC_{80} rating system and the recommended maximum sound pressure level of the environment where the hearing protection device will be used

Class	SLC_{80} range, dB	Max. SPL, dBA
1	10 to 13	Less than 90
2	14 to 17	90 to 95
3	18 to 21	95 to 100
4	22 to 25	100 to 105
5	26 or greater	105 to 110

The A-weighted protected level is calculated by subtracting the SLC_{80} value from the measured C-weighted exposure level with no hearing protection as (Williams and Dillon, 2005):

$$L_{\text{prot},A} = L_{Ceq,\text{8h}} - \text{SLC}_{80} \quad (\text{dBA re } 20\,\mu\text{Pa}) \tag{2.93}$$

The SLC_{80} method requires measurement of the $L_{Ceq,\text{8h}}$, which will typically be a few dB more than the corresponding $L_{Aeq,\text{8h}}$.

If the $L_{Aeq,8h}$ is greater than or equal to 110 dBA, or if the ambient noise is characterised by high levels of narrow band noise, or high levels of tonal noise, or high levels of low or high frequency components, then the octave-band method described in (AS/NZS 1270, 2014, Appendix A) should be used.

2.9.6 Standard Deviation

The various rating methods for hearing protection devices described in Sections 2.9.1, 2.9.2, 2.9.3, 2.9.4.3 and 2.9.5 have slightly different interpretations, which is puzzling as human ears are fairly consistent across jurisdictions. The rating of a hearing protection device is determined using a number of human test subjects in a laboratory environment that fit the device to themselves. Hence, the test is arguably attempting to simulate real-use of the device, and capture the inconsistencies of the measured noise reduction caused by sub-optimal fitting by untrained or trained users. One could argue that the test standards are more focused on examination of the variability caused by self-installation of the device, rather than measuring the noise reduction of the hearing protection device itself, which could be achieved using a standardised instrumented headform. However, as the intended use of the device is for humans that have a range of skill-levels, the standards arguably attempt to measure a statistical distribution of the expected noise reduction from operational use.

Each rating method described in Sections 2.9.1, 2.9.2, 2.9.3, 2.9.4.3 and 2.9.5 has an associated percentile value for which that percentage of users are likely to achieve the adjusted noise reduction, which is derived from measurements. Table 2.12 summarises the rating systems, the percentile of users likely to achieve the stated noise reduction and the number of standard deviations from the mean value. For example, the NRR rating is determined by measuring the mean noise reduction minus two standard deviations, which means that at least 97.7% of users should be able to achieve the measured noise reduction (Williams and Dillon, 2005).

TABLE 2.12 Comparison of noise rating schemes in terms of assumed percentile of population protected, and the statistic that is evaluated based on the mean and standard deviation (SD) values of attenuation

Rating scheme	Percentile of users	Statistic
NRR	98	mean $-2.00 \times$ SD
NRR(SF)	84	mean $-1.00 \times$ SD
$NRS_{A,80}$	80	mean $-0.84 \times$ SD
$NRS_{A,20}$	20	mean $+0.84 \times$ SD
SNR / HML	80	mean $-0.84 \times$ SD
SLC_{80}	80[a]	mean $-1.00 \times$ SD

[a] The SLC_{80} rating involves calculating the mean minus one standard deviation values of octave band attenuation, rather than evaluating an overall attenuation for subjects and then calculating the mean and standard deviations. It provides a protection rate of approximately 80% (Waugh, 1984).

2.9.7 Personal Attenuation Rating (PAR)

Fit testing systems are now available for measuring the actual performance of hearing protectors for individuals, producing a PAR expressed as an A-weighted noise reduction or insertion loss. In terms of predicting the noise level experienced at the ear, it is clear that application of an individual PAR will be much more accurate in assessing an individual's noise exposure that using a hearing protection device (HPD) rating, even if it has been somewhat derated, as described in previous paragraphs. This is because the performance of an HPD varies widely between individuals as well as the level of training in their use. A measured PAR can also indicate which individuals may benefit from more training in the use of their HPD. Comparison of PARs with various derating schemes used in various jurisdictions has shown that relying on HPD ratings for estimating the noise exposure of individuals can result in large errors in estimated exposure levels compared to what can be achieved using PARs (Murphy et al., 2022, 2023).

2.9.8 Degradation of Effectiveness from Short Lapses

It is important that hearing protection devices are always worn during exposure to high sound pressure levels, as even short periods of not wearing protection markedly reduces their effectiveness. Figure 2.24 shows the effective noise reduction of hearing protection devices that could provide nominal noise reductions of 30, 25, 20, 15, 10 dB, as a percentage of the time worn during exposure to loud noise. As an example, say a person is issued with a pair of high-grade earmuffs that are rated to provide 30 dB of noise reduction. However, the user only wears the device for 95% of the time during exposure to loud noise, which would be equivalent to having the earmuffs removed for only 24 minutes of an 8 hour shift. As shown in Figure 2.24 by the dotted curve, the effective noise reduction that the earmuffs provides is now only 13 dB, which is 17 dB less than expected! The MATLAB® script, `plot_effective_NR_vs_time.m`, used to plot Figure 2.24 is available for download from MATLAB scripts for ENC (2024).

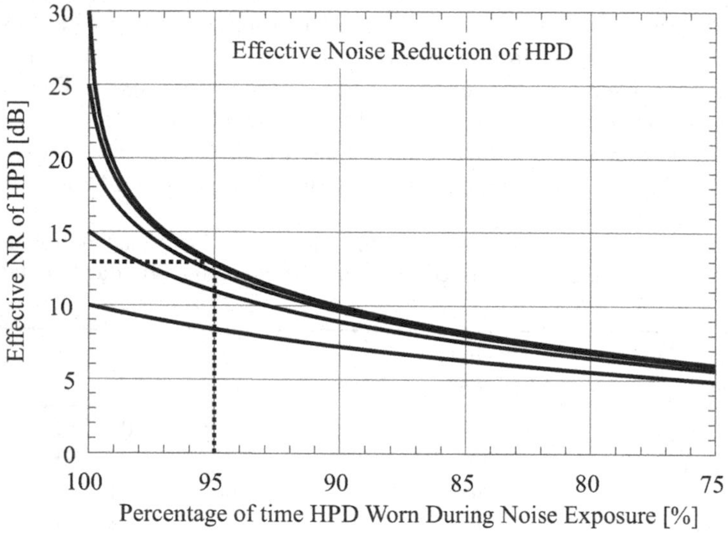

FIGURE 2.24 Effective noise reduction of hearing protection device as a percentage of the time worn during exposure to noise.

Groenewold et al. (2014) conducted an analysis of self-reporting use of hearing protectors and around 19000 corresponding audiograms over a 5-year period. They commented that it is preferable to reduce the noise exposure in the environment, rather than rely on the proper use of hearing protection.

2.9.9 Overprotection

Although hearing protection devices are intended to provide noise reduction for the wearer, one of the common complaints from people wearing hearing protection devices is that they have difficulty hearing another person speaking, they are unable to hear warning signals and they cannot get auditory feedback from task operations, leading to them feeling isolated from their environment. Sometimes workers will remove their hearing protection to communicate, and expose themselves to high sound pressure levels and significantly reduce their effective protection, as shown in Section 2.9.8. Hence, there can be instances where a hearing protecting device provides too much noise reduction. The standard BS EN 458 (2016) recommends that the "A-weighted equivalent sound pressure level effective to the ear" (or protected noise level) when wearing hearing protection should be between 75 and 80 dBA. A protected noise level less than 70 dBA is considered as overprotection.

Electronic hearing protection devices can provide high levels of passive noise reduction, and when the wearer wants to hear the ambient sounds such as a person speaking, the wearer presses a button that electronically transmits the outside noise into the earcup at a safe sound level. The noise reduction provided by these electronic hearing protection devices can be measured using the existing standards when operating in a passive mode.

Active noise reduction earmuffs present additional challenges for testing and can be evaluated using standards ISO 4869-6 (2019); BS EN 352-5 (2020). These devices can provide high levels of noise reduction at low frequencies, and moderate reduction of noise at mid- and high-frequencies that aid in improving speech intelligibility.

2.10 Speech Interference Criteria

In this section, the interfering effect of noise on oral communication is considered. Table 2.13 lists some of the significant frequency ranges that are of importance for these considerations.

TABLE 2.13 Significant frequency ranges for speech communication

	Approximate frequency range (Hz)
Range of hearing	16 to 20000
Speech range	200 to 6000
Speech intelligibility (containing the frequencies most necessary for understanding speech)	500 to 4000
Speech privacy range (containing speech sounds which intrude most objectionably into adjacent areas)	250 to 2500
Male voice (peak frequency of energy output)	350
Female voice (peak frequency of energy output)	700

2.10.1 Broadband Background Noise

Maintenance of adequate speech communication is often an important aspect of the problem of dealing with occupational noise. The degree of intelligibility of speech is dependent on the level of background noise in relation to the level of spoken words. Additionally, the speech level of a talker will depend on the talker's subjective response to the level of the background noise. Both effects can be quantified, as illustrated in Figure 2.25 (see also, ANSI/ASA S12.65 (2020)).

To enter Figure 2.25, the Speech Interference Level (SIL) is computed as the arithmetic average of the background sound pressure levels in the four octave bands, 500 Hz, 1000 Hz, 2000 Hz and 4000 Hz. Alternatively, the figure may be entered using the A-weighted scale, but with less precision. Having entered the figure, the voice level required for just adequate speech communication may then be determined for the various distances indicated on the abscissa. Each curve represents the combinations of talker/listener separation and background sound pressure level for which just-reliable face-to-face communication is possible for the voice level indicated on the curve. For example, for a speech interference level of 60 dB and distances between the

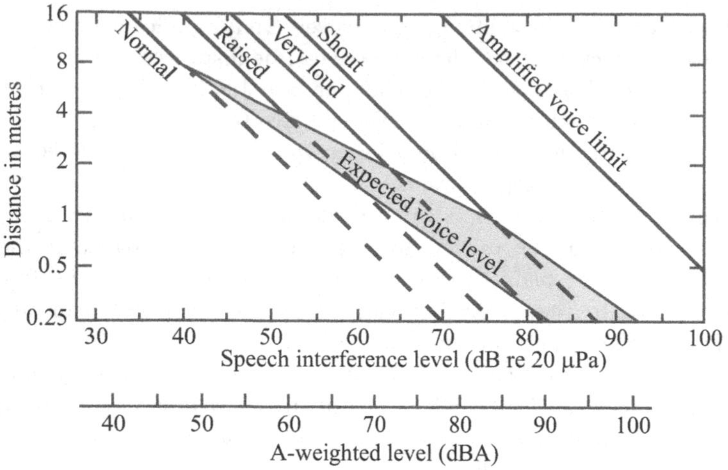

FIGURE 2.25 Rating noise with respect to speech interference.

talker and the listener less than 0.8 m, a normal speech level would seem to be adequate, while at 1.5 m, a raised voice level would be necessary for just adequate speech recognition.

The figure also shows as a shaded area, the voice level (in dBA - see lower abscissa scale) that the talker would automatically use (expected voice level) as a result of the background sound pressure level. The range of expected voice level represents the expected range in a talker's subjective response to the background noise and is independent of distance from the listener. If the talker is wearing an ear protection device such as ear plugs or earmuffs, the expected voice level will decrease by 4 dB.

For face-to-face communication with "average" male voices, the background sound pressure levels shown by the curves in Figure 2.25 represent upper limits for just acceptable speech communication, i.e. 95% sentence intelligibility, or alternatively 60% word-out-of-context recognition. For female voices, the Speech Interference Level, or alternatively the A-weighted level shown on the abscissa, should be decreased by 5 dB; that is, the scales should be shifted to the right by 5 dB as the generally higher frequency content of female voices makes them more difficult to hear in the presence of broadband background noise. The figure assumes no reflecting surfaces to reflect speech sounds. Where reflecting surfaces exist, the scale on the abscissa should be shifted to the right by 5 dB. Where the noise fluctuates greatly in level, the scale on the abscissa may be shifted to the left by 5 dB.

For industrial situations, where speech and telephone communication are important, such as in foreperson's offices or control rooms, an accepted criterion for background sound pressure level is 70 dBA.

2.10.2 Intense Tones

Intense tones may mask sounds associated with speech. The masking effect of a tone is greatest on sounds of frequency higher than the tone; thus low-frequency tones are more effective than high-frequency tones in masking speech sounds. However, tones in the speech range, which generally lies between 200 and 6000 Hz, are the most effective of all in interfering with speech recognition. Furthermore, as the frequencies, 500–5000 Hz are the most important for speech intelligibility, tones in this range are most damaging to good communication. However, if masking is required, then a tone of about 500 Hz, rich in harmonics, is most effective. For more on the subject of masking refer to Section 2.2.1.

2.10.3 Speech Intelligibility Index and Speech Transmission Index

There are two speech intelligibility measures in current use: the speech intelligibility index (SII) and the speech transmissibility index (STI), which both vary between 0 and 1, based on weighted contributions from a range of frequency bands that are present in speech. A value of zero implies that no part of the speech is intelligible (that is, no speech cues are received by the listener) while a value of 1 implies that the speech is completely intelligible (that is, all speech cues are received by the listener). Although intermediate values are not directly convertible to a speech intelligibility score, the STI or SII value is a direct measure of the proportion of speech cues that are received by the listener. When octave band analysis is used to calculate SII, the two measures give similar results in most environments and either is suitable for architectural acoustics purposes (Zhu et al., 2014). Two older measures of speech intelligibility (articulation index (AI) and rapid speech transmission index (RASTI)) are no longer in general use as they have been superseded by STI and SII. Thus, AI and RASTI will not be discussed here.

The SII value is calculated according to the American standard, ANSI/ASA S3.5 (2020) and the STI value is calculated according to the European standard, IEC 60268-16, 4th edn. (2020). Both measures are a function of the speech sound pressure level, the background noise sound pressure level and the extent of reverberation in the environment. Where desirable, SII calculations can also be done for a specific individual with a specified hearing threshold level.

Calculations of SII and STI are done at individual frequencies and the overall STI or SII value is determined by weighting and then summing the contributions from the individual frequencies. For calculating the SII value, the user is given 4 choices, which are in decreasing order of accuracy: 21 critical band (see Section 2.1.8.6) centre frequencies; 18 1/3-octave band centre frequencies; 17 equally contributing critical band centre frequencies; or 6 octave band centre frequencies. The octave band method is not recommended in situations where either the speech level or background noise level varies greatly as a function of frequency in any one octave. The STI calculation has no options and only uses spectrum levels at 7 octave band centre frequencies. Both STI and SII have their limitations, and these are discussed in detail in the above-mentioned standards.

It is often of interest to evaluate the intelligibility of speech over electronic communication systems such as telephones, public address systems and radio communication systems. The procedures used in this case rely on talkers and listeners communicating various words (ANSI/ASA S3.2, 2020) and are not related at all to the STI and SII procedures outlined in the following sections. Interested readers should consult the American standard, ANSI/ASA S3.2 (2020).

2.10.3.1 STI Calculation Procedure

The purpose of calculating an STI value for an indoor space is to provide a single number that is a measure of how intelligible speech will be in the space. However, the full STI method calculated as described in this section is rarely used as it is very time consuming. In most instances, it is sufficient to use the STIPA or STITEL methods outlined in Sections 2.10.3.2 and 2.10.3.3, respectively. The STI method, and its derivatives, STIPA or STITEL, are based on the finding determined from measurements on many people, that the most relevant information related to speech intelligibility are carried in the amplitude variations (modulation) of speech signals, which can be quantified as a function of the modulation frequency.

The calculation procedure for finding the STI for a particular space is outlined in detail in IEC 60268-16, 4th edn. (2020). A sound source is located at the intended talker location and an omni-directional microphone is located at the proposed listener location. A pink noise signal, approximately 10 s in length, of bandwidth one half of an octave and centred at an octave band centre frequency is fed into the sound source and is modulated sinusoidally at the modulation frequency, f_m (Hz), with a modulation depth of m_i (with m_i set equal to 1, if possible). The sound pressure level amplitude of the pink noise signal in terms of the overall A-weighted level must be decided upon at the beginning. Once this is done, the required level in each of the 7 octave bands used in the analysis is determined by adding the correction level in Table 2.14, to the required overall A-weighted level (different correction levels are used for males and females).

TABLE 2.14 Data for STI calculations

	Octave band centre frequency (Hz)							
	125	250	500	1000	2000	4000	8000	A-wt
Male correction level (dB)	2.9	2.9	−0.8	−6.8	−12.8	−18.8	−24.8	0.0
Female correction level (dB)	—	5.3	−1.9	−9.1	−15.8	−16.7	−18.0	0.0
Male weighting factor, α	0.085	0.127	0.230	0.233	0.309	0.224	0.173	—
Male weighting factor, β	0.085	0.078	0.065	0.011	0.047	0.095	—	—
Female weighting factor, α	—	0.117	0.223	0.216	0.328	0.250	0.194	—
Female weighting factor, β	—	0.099	0.066	0.062	0.025	0.076	—	—
Reception threshold intensity, $I_{rt,k}$	$10^{4.6}$	$10^{2.7}$	$10^{1.2}$	$10^{0.65}$	$10^{0.75}$	$10^{0.8}$	$10^{1.2}$	—

For the purposes of the following STI calculation procedure, the intensity at the listener in octave band, k is defined as:

$$I_k = 10^{L_k/10} \tag{2.94}$$

where L_k is the sound pressure level at the listener in the kth octave band, and this is the quantity that is measured at the listener location during an STI test.

The modulation depth of the intensity of the received signal, $m_o(k, f_m)$, for modulation frequency, f_m, and octave band, k, is determined at the listener microphone for each case by first deriving the intensity corresponding to each time sample, t_n, using Equation (2.94). Calculations are done for 7 octave bands (with centre frequencies from 125 Hz to 8 kHz) and for 14 modulation frequencies (equal to octave band centre frequencies from 0.63 Hz to 12.5 Hz) in each octave band.

The modulation depth for the received signal is then obtained for octave band, k, and modulation frequency, f_m, using:

$$m_o(k, f_m) = 2 \times \frac{\sqrt{\left[\sum\limits_{n=1}^{N} I_k(t_n)\sin(2\pi f_m t_n)\right]^2 + \left[\sum\limits_{n=1}^{N} I_k(t_n)\cos(2\pi f_m t_n)\right]^2}}{\sum\limits_{n=1}^{N} I_k(t_n)} \tag{2.95}$$

where t_n corresponds to time sample, n and N is the number of time samples, which need to include a whole number of periods at the modulation frequency, f_m. The calculation is repeated for each of the 14 modulation frequencies in each of the 7 octave bands.

The calculation of STI, requires calculation of the modulation transfer function for each octave band, k and modulation frequency, f_m, using:

$$m(k, f_m) = \frac{m_o(k, f_m)}{m_i(k, f_m)} \tag{2.96}$$

where m_i is the modulation depth of the signal fed into the loudspeaker source.

Auditory masking, which is the reduction in aural sensitivity at a particular frequency caused by a stronger, lower frequency sound, can be taken into account by modifying the value of $m(k, f_m)$ to obtain a corrected modulation transfer function, $m'(k, f_m)$, (or modulation transfer ratio) for octave band, k, and modulation frequency, f_m, as:

$$m'(k, f_m) = m(k, f_m) \times \frac{I_k}{I_k + I_{am,k} + I_{rt,k}} \tag{2.97}$$

The reception threshold intensity, $I_{rt,k}$, for each octave band, k, is provided in Table 2.14 and the masking intensity $I_{am,k}$, for octave band, k, is:

$$I_{am,k} = I_{k-1} \times a_{mf} \tag{2.98}$$

where I_k is the intensity of the signal in octave band, k (see Equation (2.94)), I_{k-1} is the intensity in the next lower octave band and a_{mf} is a level dependent auditory masking factor, dependent only on the level in the next lower adjacent octave band and is given in Table 2.15. The 125 Hz band is not masked at all.

TABLE 2.15 Auditory masking factor

Sound pressure level, L_p in octave band, $(k-1)$, (dB)	< 63	≥ 63 and < 67	≥ 67 and < 100	≥ 100
Auditory masking factor, a_{mf}	$10^{(L_p-130)/2}$	$10^{(1.8L_p-146.9)}$	$10^{(L_p-119.6)/2}$	0.1

The effective signal-to-noise ratio (SNR) for each modulation frequency, f_m, and each octave band, k, is calculated from the corrected modulation transfer function using:

$$\text{SNR}_{\text{eff}}(k, f_m) = 10 \log_{10} \left(\frac{m'(k, f_m)}{1 - m'(k, f_m)} \right) \tag{2.99}$$

where values of $\text{SNR}_{\text{eff}}(k, f_m)$ less than -15 dB are set equal to -15 dB and values of SNR greater than 15 dB are set equal to 15 dB. Each of the 98 values (14 modulation frequencies in 7 octave bands) is determined using a separate 10 second test.

The transmission index, $\text{TI}(k, f_m)$, is defined as:

$$\text{TI}(k, f_m) = \frac{\text{SNR}_{\text{eff}}(k, f_m) + 15}{30} \tag{2.100}$$

The transmission indices are then arithmetically averaged over the modulation frequencies for each octave band to obtain the octave band modulation transfer index, MTI_k for octave band, k using:

$$\text{MTI}_k = \frac{1}{n} \sum_{m=1}^{n} \text{TI}(k, f_m) \tag{2.101}$$

where m is the modulation frequency index and n is the number of modulation frequencies for each octave band.

The STI is calculated from the modulation transfer index using:

$$\text{STI} = \sum_{k=1}^{7} (\alpha_k \times \text{MTI}_k) - \sum_{k=1}^{6} \left(\beta_k \times \sqrt{\text{MTI}_k \times \text{MTI}_{k+1}} \right) \tag{2.102}$$

where the factors, α_k and β_k, are given in Table 2.14 for each octave band, k.

As the STI method involves a relatively long series of measurements, it is rarely used in practice. In fact, it is only used in situations where the difference between male and female voices is important. In other situations, the shorter, STIPA and STITEL methods, which are based on the STI method, are used. STEL is used for evaluating telecommunication channels and STIPA is used for assessing rooms for speech communication and evaluating PA systems.

2.10.3.2 STIPA Calculation Procedure

The STIPA procedure (IEC 60268-16, 4th edn., 2020) is based on the male speech spectrum (see Table 2.14) and allows simultaneous modulation and parallel processing of all frequency bands, where the STI method requires separate measurements for each modulation frequency and each frequency band. However, the STIPA method has a reduced ability to account for some forms of non-linear distortion. An additional simplification of the STIPA method is that only two modulation frequencies are used for each of the seven octave bands. Different modulation frequencies are used for each octave band (see Table 2.16), with the second modulation frequency 5 times the first. The modulation depth, m_i for each modulation frequency is 0.55 with a phase difference of 180° between the two components. The same general procedure is then used to calculate the STIPA as was done to calculate the STI.

TABLE 2.16 Modulation frequencies (Hz) for STIPA and STITEL methods

	Octave band centre frequency (Hz)						
	125	250	500	1000	2000	4000	8000
STIPA frequency 1 (Hz)	1.60	1.00	0.63	2.00	1.25	0.80	2.50
STIPA frequency 2 (Hz)	8.00	5.00	3.15	10.0	6.25	4.00	12.5
STITEL frequency (Hz)	1.12	11.33	0.71	2.83	6.97	1.78	4.53

2.10.3.3 STITEL Calculation Procedure

The STITEL method (IEC 60268-16, 4th edn., 2020) only uses one modulation frequency (see Table 2.16) applied to each octave band, which allows 100% modulation, thus increasing the SNR by 3 dB. The same general procedure is used to calculate the STITEL as was done to calculate the STI.

2.10.3.4 SII Calculation Procedure

Although the standard provides four different methods for calculating SII, only the octave band method is discussed here as that is the one most commonly used and which is also applicable in most situations. Similar procedures may be used for the other three methods (21 critical band centre frequencies, 18 1/3-octave band centre frequencies and 17 equally contributing critical band centre frequencies), and these are outlined in detail in ANSI/ASA S3.5 (2020). The basic formula for calculating SII for any of the 4 methods is:

$$\mathrm{SII} = \sum_{i=1}^{n} I_i A_i \tag{2.103}$$

where I_i is the band importance function and A_i is the band audibility function for frequency band, i, and n is the number of frequency bands (6 for octave band analysis).

The first step in determining the octave band SII is to determine the equivalent hearing threshold level, T_i', the equivalent noise spectrum level, N_i', and the equivalent speech spectrum level, E_i', for each octave band centre frequency, i. Spectrum level is the level in a frequency band that is 1 Hz wide. For example if the background noise level in an octave band is measured as N_B, then the spectrum level, N, of the noise at the band centre frequency is calculated as:

$$N = N_B - 10\log_{10}\Delta f \tag{2.104}$$

where Δf is the bandwidth (in Hz) of the octave band filter (see Table 2.17).

The equivalent hearing threshold level, T_i', is the arithmetically averaged hearing threshold levels, T_i across the group of listeners for which the SII calculations are to be done. The hearing threshold level is defined in ANSI/ASA S3.5 (2020) as the pure tone threshold level for the ear under test, minus the arithmetically averaged pure tone threshold level for a large number of ears from the general population within the age range of 18 to 30 years. In many SII calculations, T_i' is assumed to be 0 dB.

For calculations in a free field or in a space where reverberation is very small, the equivalent noise spectrum level, N_i', is calculated from the measured background noise levels in the speech environment in octave bands using Equation (2.104).

For calculations in a free field or in a space where reverberation is very small, the equivalent speech spectrum level, E_i', for octave band, i, can be calculated from the speech spectrum level, E_i that is measured at 1 m directly in front of the speaker's lips. Typical values for E_i (in dB/Hz), arithmetically averaged across a large group of adult male and female talkers are given in ANSI/ASA S3.5 (2020) and are listed in Table 2.17 for the 6 octave bands.

$$E_i' = E_i - 20 \log_{10} d \tag{2.105}$$

where d is the distance in metres between the speaker and listener. If the SII is to be determined with hearing protection on the listener, then the insertion gain, G_i for octave band, i, must be added to the RHS of Equation (2.105). Note that G_i is usually negative.

TABLE 2.17 Data for SII calculations

	Octave band centre frequency (Hz)					
	250	500	1000	2000	4000	8000
Bandwidth (Δf Hz)	177	354	707	1411	2825	5650
Band importance, I_i	0.0617	0.1671	0.2373	0.2648	0.2142	0.0549
$E_i = U_i$ for normal speech (dB/Hz)	34.75	34.27	25.01	17.32	9.33	1.13
E_i for raised speech (dB/Hz)	38.98	40.15	33.86	25.32	16.78	5.07
E_i for loud speech (dB/Hz)	41.55	44.85	42.16	34.39	25.41	11.39
E_i for shouting (dB/Hz)	42.50	49.24	51.31	44.32	34.41	20.72
Reference internal noise spectrum level, X_i (dB)	−3.90	−9.70	−12.50	−17.70	−25.90	−7.10
Free field to eardrum transfer function (dB)	1.00	1.80	2.60	12.00	14.30	1.80

For calculations in spaces where reverberation has an effect on the SII, ANSI/ASA S3.5 (2020) uses a similar measurement procedure as outlined in Section 2.10.3.1 for the calculation of STI. However, different modulation frequencies are used (0.5, 1.0, 1.5, 2.0, 3.0, 4.0, 6.0, 8.0 and 16.0 Hz) and a different symbol ($R_{f,i}$) is used for $\text{SNR}_{\text{eff}}(k, f_m)$ to denote the speech to noise ratio (or signal-to-noise ratio) for modulation frequency, f (see Equation (2.99)). As for the STI calculations, values of $R_{f,i}$ greater than 15 dB are set equal to 15 dB and values of $R_{f,i}$ less than -15 dB are set equal to -15 dB. The SII procedure requires calculation of the average apparent speech to noise ratio for each octave band, i by arithmetically averaging the values of $R_{f,i}$ over all 9 modulation frequencies to obtain R_i. In addition, the SII procedure requires an additional measurement of the spectrum level of the received signal, P_i with no modulation of the input signal. Using the values of R_i and P_i determined as described above, the values of E_i' and N_i' are given by:

$$N_i' = E_i' - R_i \tag{2.106}$$

$$E_i' = R_i + 10 \log_{10} \left[\frac{10^{0.1 P_i}}{1 + 10^{0.1 R_i}} \right] \tag{2.107}$$

Prior to calculating SII, there are some other variables that must be calculated. These are the equivalent internal noise spectrum, X_i', the equivalent disturbance spectrum level, D_i, the level distortion factor, L_i, and the audibility function, A_i. These variables are calculated for each octave band, i, using the following equations.

$$X_i' = X_i + T_i' \tag{2.108}$$

$$D_i = \max[N_i', X_i'] \tag{2.109}$$

$$L_i = 1 - (E_i' - U_i - 10)/160 \tag{2.110}$$

where X_i and U_i are provided in Table 2.17 for each octave band, i.

$$A_i = L_i(E_i' - D_i + 15)/30 \tag{2.111}$$

SII is then calculated using Equation (2.103), where I_i is provided in Table 2.17 and A_i is calculated using equation (2.111).

Computer programs to facilitate the calculations outlined in ANSI/ASA S3.5 (2020) are available at https://www.sii.to/index.html. However, these assume free field conditions and do not account for reverberation effects.

For 1/3 octave band calculations, the band centre frequencies in the bands between and including 160 Hz and 8000 Hz are used. However, the standard also allows users to specify the frequency range themselves or to use either octave band or critical band centre frequencies.

2.11 Psychological Effects of Noise

2.11.1 Noise as a Cause of Stress

Noise causes stress: the onset of loud noise can produce effects such as fear, and changes in pulse rate, respiration rate, blood pressure, metabolism, acuity of vision, skin electrical resistance, etc. However, most of these effects seem to disappear rapidly and the subject returns to normal, even if the noise continues, although there is evidence to show that prolonged exposure to excessive loud noise will result in permanently elevated blood pressure. Furthermore, Glass and Singer (1972) demonstrated that specific noise exposure scenarios increased negative after effects. These scenarios included exposure to predictable versus unpredictable noise and controllable versus uncontrollable noise. Exposure to unpredictable or uncontrollable noise resulted in reduced post-noise task performance and lowered tolerance for post-noise frustration.

2.11.2 Effect on Behaviour and Work Efficiency

Behavioural responses to workplace noise are usually explained in terms of arousal theory: there is an optimum level of arousal for efficient performance; below this level behaviour is sluggish and above it, tense and jittery. It seems reasonable to suppose, therefore, that noise improves performance when arousal is too low for the task, and impairs it when arousal is optimal or already too high. The complex task, multiple-action task or high repetition rate task is performed optimally under relatively quiet conditions, but performance is likely to be impaired under noisy conditions. Quiet conditions, on the other hand, are sub-optimal for the simple task, and performance is improved by the addition of noise. The important variable is the kind of task being performed and not the kind of noise present. To generalise, performance in doing complex tasks is likely to be impaired in the presence of noise and for simple tasks it is likely to be improved. However, various studies have shown that if the sound pressure level is far in excess of that required for the optimum arousal level for a particular task, workers become irritable as well as less efficient. This irritability usually continues for some time after the noise has stopped.

2.11.3 Effect on Sleep

Sleep is an active and restorative process necessary to facilitate daytime alertness and performance, quality of life and health (Basner and McGuire, 2018). During sleep, human sensory functions are still responsive to external noise, even if it is perceived in a non-conscious state (Muzet, 2007). However, noise exposure during sleep can have adverse impacts such as increasing the number of awakenings, modifying the distribution and duration of various sleep stages and modifying autonomic functions (heart rate, blood pressure, vasoconstriction and respiratory rate). Several experimental studies have demonstrated that reduced sleep can negatively impact immune function, glucose metabolism and memory consolidation. In the long-term, acute sleep loss increases the likelihood of cardio-metabolic conditions such as cardiovascular diseases, hypertension and diabetes.

To ensure a good night's sleep, continuous sound pressure levels exceeding 30 dBA or single event levels exceeding 45 dBA measured in a subject's place of sleep can disturb sleep (Berglund et al., 1999). The most recent guidelines for acceptable sound pressure levels external to residences are provided in World Health Organisation (2018). These guidelines recommend that traffic and railway noise, energy averaged over the nighttime hours of 10 pm to 7 am should be limited to 45 and dBA respectively, and aircraft noise should be limited to 40 dBA at receiver locations. This implies a sound reduction from outside to inside of approximately 15 dBA for road traffic and rail noise and 10 dBA for aircraft noise. Even lower levels than those recommended in World Health Organisation (2018) can cause sleep disturbance if the noise is tonal, amplitude modulated or in some cases, if it is randomly varying. Examples of such problem noise sources include industrial compressors and wind turbines.

2.12 Ambient Sound Pressure Level Specification

The use of a room or space for a particular purpose may, in general, impose a requirement for specification of the maximum tolerable background noise; for example, one would expect quiet in a church but not in an airport departure lounge. All have in common that a single number specification is possible.

The simplest way of specifying the maximum tolerable background noise is to specify the maximum acceptable A-weighted level. As the A-weighted level simulates the response of the ear at low levels and has been found to correlate well with subjective response to noise, such specification is often sufficient. Table 2.18 gives some examples of maximum acceptable A-weighted sound pressure levels and reverberation times in the 500 to 1000 Hz octave bands for unoccupied spaces. A full detailed list is published in AS/NZS 2107 (2016) and slightly different criteria are provided in ANSI/ASA S12.2 (2019). The values shown in Table 2.18 are for continuous background sound pressure levels within spaces, as opposed to single-event or intermittent sound pressure levels. In the table, the upper limit of the range of values shown is the maximum acceptable level and the lower limit is the desirable level. Recommended sound pressure levels for occupied spaces in vessels and offshore mobile platforms are listed for various spaces in AS 2254 (1988).

Broner and Leventhall (1983) conclude that the A-weighted measure is also acceptable for very low-frequency noise (20–90 Hz), although this conclusion is not universally accepted, especially in cases where people are trying to sleep. In these cases, the A-weighted level can often underestimate the disturbance effect. In such cases, very low A-weighted levels of noise can be very disruptive to sleep and this, in turn, can lead to serious health problems. Of course, some people are much more sensitive than others and the disturbance becomes more of a problem for these people as the time of exposure increases. This is why it is necessary to conduct human response experiments over long periods of time if useful data are to be obtained. It also helps to explain why there is such controversy over wind farm noise. Some people, who may not be very

TABLE 2.18 Recommended A-weighted ambient sound levels and reverberation times (arithmetic average of the 500 Hz and 1000 Hz octave bands) for different indoor spaces (space furnished but unoccupied, data mainly from AS/NZS 2107 (2016), where V = room volume in m^3)

Type of space	Recommended L_{Aeq} (dBA)	Recommended reverberation time at 500 to 1000 Hz (sec)
EDUCATIONAL BUILDINGS		
Art studio	40 – 45	<0.8
Assembly Hall (<250 m^3)	30 – 40	0.6 – 0.8
Assembly Hall (>250 m^3)	30 – 40	$0.367 \log_{10} V - 0.132$
Audio-visual areas	35 – 45	0.6 – 0.8
Churches	30 – 35	—
Computer rooms (teaching)	40 – 45	0.4 – 0.6
Computer rooms (working)	45 – 50	0.4 – 0.6
Conference rooms, seminar rooms	35 – 40	0.6 – 0.7
Corridors and lobbies	<50	<0.8
Drama studios	25 – 30	$0.367 \log_{10} V - 0.132$
Lecture theatres	30 – 35	$0.450 \log_{10} V + 0.065$
Libraries	40 – 50	<0.6
Music studios	30 – 35	$0.367 \log_{10} V - 0.132$
HEALTH BUILDINGS		
Emergency areas	40 – 45	<0.6
Consulting rooms	40 – 45	0.4 – 0.6
Intensive care wards	40 – 45	<0.6
Operating theatres	40 – 50	Short as possible
Ward bedrooms	35 – 40	0.4 – 0.7
INDUSTRIAL BUILDINGS		
Packaging and delivery	<60	Short as possible
Labs and test areas	40 – 50	<0.8
Foreperson's offices	45 – 50	0.4 – 0.7
SPORTS		
Indoor sports	<45	$0.745 \log_{10} V - 0.980$
OFFICE BUILDINGS		
Board and conference rooms	30 – 40	0.6 – 0.8
Professional and admin. offices	35 – 40	0.6 – 0.8
Executive offices	35 – 40	0.4 – 0.6
Open plan office areas	40 – 45	Short as possible
Reception areas	40 – 45	0.6 – 0.8
PUBLIC BUILDINGS		
Airport terminals	45 – 50	Short as possible
Cinemas	30 – 35	$0.367 \log_{10} V - 0.132$
Restaurants	40 – 50	Short as possible
Hotel bar	45 – 50	0.6 – 1.0
Shopping Malls and supermarkets	<55	Short as possible
RESIDENTIAL		
Houses and apartments – inner city (sleeping)	35 – 40	—
Houses and apartments – inner city (living)	35 – 45	—
Houses and apartments – suburban (sleeping)	30 – 35	—
Houses and apartments – suburban (living)	30 – 40	—
Houses in rural areas	25 – 30	—

sensitive, subject themselves to a short exposure (perhaps one night) and then declare that there is no problem, without realising that the problem may be much worse for sensitive people who are exposed for months. Thus, for some noise sources, regulations should consider low-frequency content and modulation. These can be quantified using various techniques and allowable limits can be set (Hansen et al., 2017).

A-weighted levels are also used for specifying acceptable occupational noise levels and also for environmental noise regulations. For impulsive environmental noise, characterised by very short duration impulses, the "C-weighted" sound exposure level, L_{CE}, defined in Section 2.4.5, is used. For transient environmental noise, the "A-weighted" sound exposure level, L_{AE} (see Section 2.4.5), is used. For traffic noise, L_{10} (see Section 2.4.9) or L_{dn} (see Section 2.4.6) are used. L_{dn} is also used for specifying acceptable sound pressure levels in houses. However, when it is desired to account for noise spectra in enclosed spaces that may be dominated by high or low-frequency components, noise weighting curves are often used to specify acceptable noise levels (see Section 2.12.2).

2.12.1 Ambient Sound Recommendations for Classrooms

Classrooms deserve special mention as they are where our children develop their language and learning skills. For children to understand the teacher, the teachers voice level should be 15 to 20 dBA above the background noise (Brill et al., 2018). As the A-weighted speech of a teacher could be as low as 55 dBA in a typical classroom, ANSI/ASA S12.60-1 (2015) recommends a maximum 1-hour energy averaged background noise level in an unoccupied classroom to be 35 dBA (or 55 dBC), which unfortunately seems to be lower than found in most classrooms in the USA Brill et al. (2018). Unfortunately, higher background noise levels in unoccupied classrooms is correlated with higher noise levels in occupied classrooms, as talkers need to raise their voice to be heard above the background noise and this exacerbates the intelligibility problem, especially when students need to communicate amongst themselves. To achieve the required background noise levels in unoccupied classrooms, attention must be paid to minimising:

- HVAC system noise;
- electronic teaching equipment noise (such as audio visual teaching equipment and charging stations with cooling fans); and
- sound leaking in from adjacent classrooms and corridors.

Suggested designs for walls and floor/ceiling assemblies that separate classrooms from one another are also provided in ANSI/ASA S12.60-1 (2015). Minimum STC ratings as specified by ANSI/ASA S12.60-1 (2015) for various adjacent spaces are:

1. 50 for other learning spaces, therapy and health-care rooms, and spaces requiring a high degree of privacy;
2. 53 for public toilets or showers;
3. 45 for corridors, staircases, offices or conference rooms; and
4. 60 for music performance spaces, auditoria, mechanical equipment rooms, gymnasia, swimming pools and cafes.

Impact sound isolation is also important and ANSI/ASA S12.60-1 (2015) recommends that floor/ceiling assemblies should have a minimum IIC rating of 45 (see Section 7.2.3).

Reverberation also affects speech intelligibility, so ANSI/ASA S12.60-1 (2015) recommends maximum unoccupied reverberation times (arithmetically averaged over the 500 Hz, 1000 Hz and 2000 Hz octave bands) of 0.6 seconds (but readily adaptable to 0.3 seconds) for classrooms with a volume less than 283 m^3 and 0.7 seconds for classrooms with a volume larger than 283 m^3.

2.12.2 Noise Weighting Curves

Although the specification of an A-weighted level is easy and convenient, it gives no indication of which frequency components may be the source of non-compliance. For most acoustic design purposes, it is more useful to make use of a weighting curve, which defines a spectrum of band levels in terms of a single number. Five currently used sets of single-number weighting curves are shown in Figures 2.26–2.30. These figures provide, respectively, noise rating (NR), noise criteria (NC), room criteria (RC), balanced noise criteria (NCB) and room noise criteria (RNC) weighting curves. A comparison of noise weighting curve values with A-weighted noise levels is provided in Section 2.13.2.

2.12.2.1 NR Curves

Noise Rating (NR) curves have been adopted by the International Standards Organisation and are intended for general use, particularly for rating environmental and industrial noise levels. They are also used in many cases by machinery manufacturers to specify machinery sound pressure levels.

The Noise Rating, NR, of any noise characterised in octave band levels may be calculated algebraically. The NR of a noise is equal to the highest octave band noise rating (NR_B) which

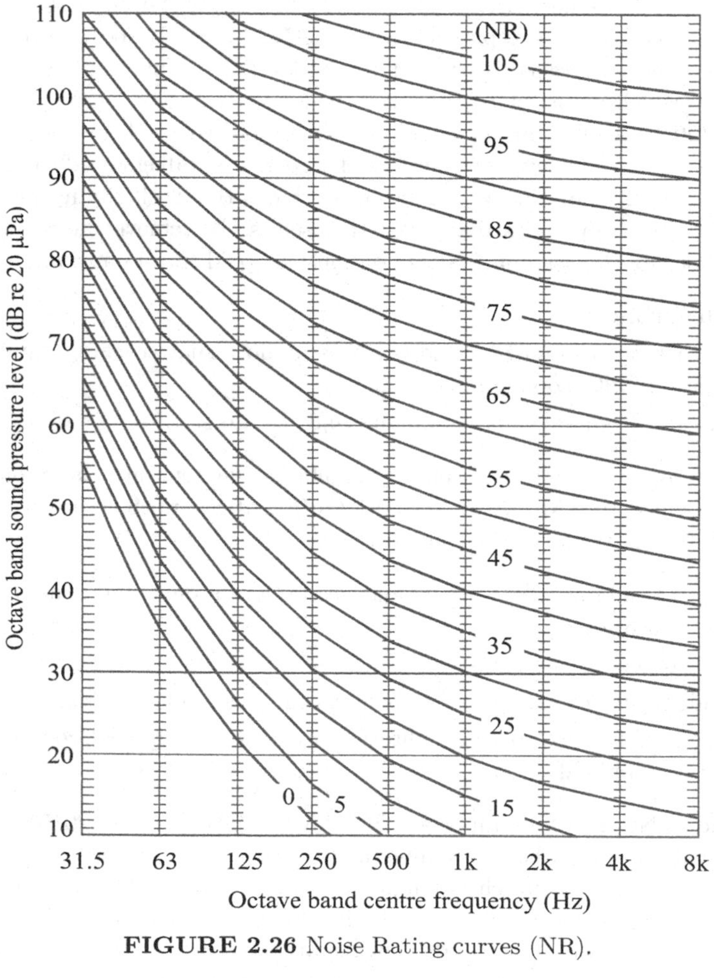

FIGURE 2.26 Noise Rating curves (NR).

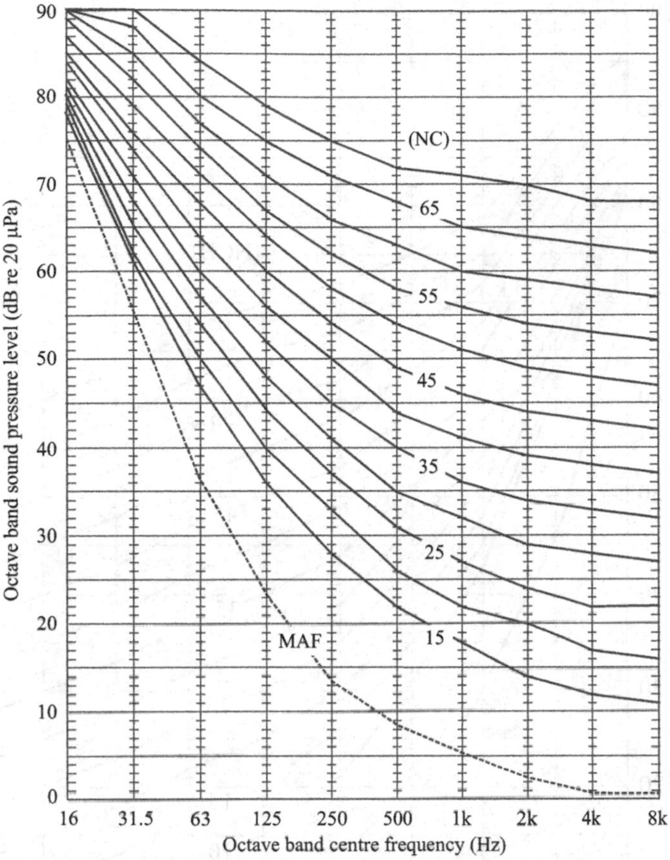

FIGURE 2.27 Noise Criteria curves (NC). "MAF" = "minimum audible field". At 16 Hz, NC60 = NC65 = NC70 = 90 dB.

is defined as:

$$\mathrm{NR}_B = \frac{L_{pB} - A_B}{B_B} \tag{2.112}$$

where A_B and B_B are as listed in Table 2.19 and L_{pB} is the octave band sound pressure level. However, the family of curves generated using Equation (2.114) and shown in Figure 2.26 is in more common use than the equation. By convention, the NR value of a noise is expressed as an integer.

TABLE 2.19 Coefficients used to calculate NR curve numbers for octave bands between 31.5 and 8000 Hz

Octave band centre frequency (Hz)	A_B	B_B	Octave band centre frequency (Hz)	A_B	B_B
31.5	55.4	0.681	1000	0.0	1.000
63	35.5	0.790	2000	−3.5	1.015
125	22.0	0.870	4000	−6.1	1.025
250	12.0	0.930	8000	−8.0	1.030
500	4.8	0.974			

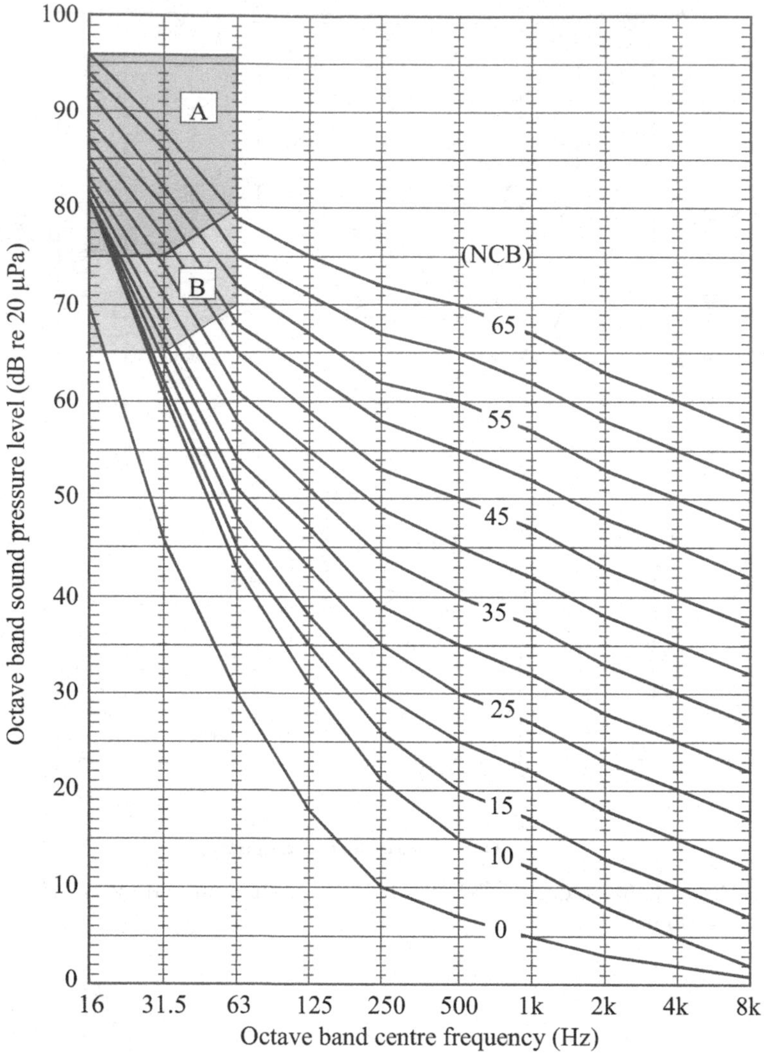

FIGURE 2.28 Balanced Noise Criterion curves (NCB). Regions A and B: noise-induced feelable vibration in lightweight structures, induced audible rattle in light fixtures, doors, windows, etc.; region A, high probability; region B, moderate probability.

To determine the NR rating of a noise, measured octave band sound pressure levels are plotted in Figure 2.26 and the rating is determined by the highest weighting curve which just envelopes the data. If the highest level falls between two curves, linear interpolation to the nearest integer value is used. Note that it is also possible to use 1/3-octave band data on 1/3-octave band NR curves, which are obtained by moving the octave band curves down by $10\log_{10}(3) = 4.77$ dB.

Specification of an NR number means that in no frequency band shall the octave band sound pressure in the specified space exceed the specified curve (tangent method). In practice, a tolerance of ±2 dB is implied. In specifications, an allowance of 2 dB above the curve is usually acceptable in any one octave band, provided that the levels in the two adjacent bands are less than 1 dB below the criterion curve.

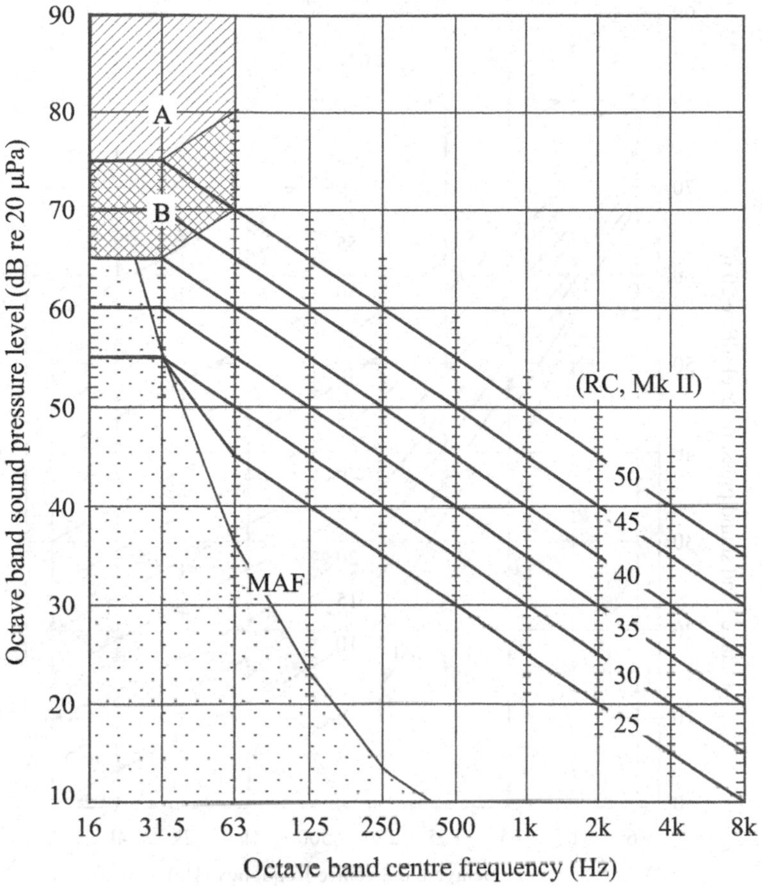

FIGURE 2.29 Room Criteria (RC Mark II) curves. Regions A and B: noise-induced feelable vibration in lightweight structures and induced rattles in light fixtures, doors, windows, etc. Region A, high probability; region B, moderate probability.

2.12.2.2 NC Curves

Noise Criteria curves (Figure 2.27) were developed in 1956 in response to the need for specification of just acceptable noise in occupied spaces with all systems running.

The original curves did not include the 16 Hz and 31.5 Hz octave bands and so did not account for low-frequency rumble noises. In the 1990s ANSI/ASA S12.2 (1995) replaced NC curves with NCB (see Section 2.12.2.3) and RC (see Section 2.12.2.4) curves which included the 16 Hz and 31.5 Hz octave bands. However, practitioners were generally unhappy with using NCB curves, so the more recent ANSI standard (ANSI/ASA S12.2 (2019)) reintroduced NC curves in place of NCB curves and extended the NC curves lower in frequency to include the 16 Hz and 31.5 Hz octave bands. The NC rating of a noise is determined in a similar way to the NR rating, except that Figure 2.27 is used instead of Figure 2.26. The NC rating, which is an integer value, corresponds to the curve that just envelopes the spectrum. No part of the spectrum may exceed the NC curve that describes it. Note that linear interpolation is used to generate curves corresponding to integer NC numbers between the 5 dB intervals shown in Figure 2.27.

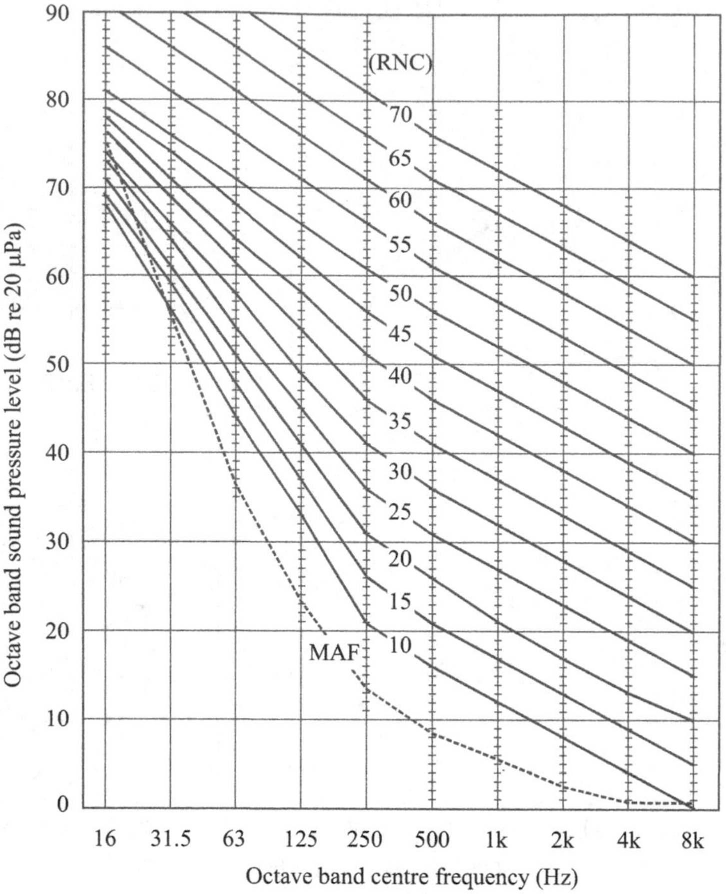

FIGURE 2.30 Room Noise Criterion curves (RNC). The lowest curve is the approximate threshold of hearing for octave band noise.

2.12.2.3 NCB Curves

Balanced Noise Criteria (NCB) curves (ANSI/ASA S12.2, 1995) are shown in Figure 2.28. NCB curves apply to occupied spaces with all systems running and are intended to replace the original NC curves. The designation number of an NCB curve is equal to the Speech Interference Level (SIL) of a noise with the same octave band levels as the NCB curve. The NCB or SIL of a noise is the arithmetic average of the 500 Hz, 1 kHz, 2 kHz and 4 kHz octave band decibel levels, calculated to the nearest integer. To determine whether the background noise is "rumbly", the octave band sound pressure levels of the measured noise are plotted on a chart containing a set of NCB curves. If any values in the 500 Hz octave band or lower exceed by more than 3 dB the curve corresponding to the NCB rating of the noise, then the noise is labelled "rumbly". To determine if the noise is "hissy", the NCB curve that is the best fit of the octave band sound pressure levels between 125 Hz and 500 Hz is determined. If any of the octave band sound pressure levels between 1000 Hz and 8000 Hz inclusive exceed this NCB curve, then the noise is rated as "hissy".

2.12.2.4 RC, Mark II Curves

Room Criteria, Mark II (RC) curves (see Figure 2.29) replace the original RC curves (see ANSI/ASA S12.2 (2019)) that, in turn, were developed to replace the original NC curves for rating only air conditioning noise in unoccupied spaces.

The difference between RC and RC Mark II curves is in the treatment of the 16 Hz octave band and the inclusion of the 8000 Hz octave band in the Mark II curves. Interest in the 31.5 Hz and 16 Hz bands stems from the fact that a level of the order of 65 dB or greater in these bands may result in noise-induced vibrations that can just be felt, especially in lightweight structures. Such vibration can also give rise to objectionable rattle and buzz in windows, doors and cabinets, etc.

For spectrum shapes that are ordinarily encountered, and if instrumentation does not provide levels in the 16 Hz octave band, the level in the 16 Hz band can be estimated from the difference in levels between the overall unweighted level and the 31.5 Hz octave band level. A difference of +4 dB or more is evidence of a level in the 16 Hz band equal to or greater than the level in the 31.5 Hz band.

The RC number is the arithmetic average of the 500 Hz, 1000 Hz and 2000 Hz octave band sound pressure levels, expressed to the nearest integer. If any octave band level below 500 Hz exceeds the RC curve by more than 5 dB, the noise is denoted "rumbly" (e.g. RC 29(R)). If any octave band level above 500 Hz exceeds the RC curve by more than 3 dB, the noise is denoted, "hissy". (e.g. RC 29(H)). If neither of the above occurs, the noise is denoted "neutral" (e.g. RC 29(N)). If the sound pressure levels in any band between and including 16 Hz to 63 Hz lie in the cross-hatched regions in Figure 2.29, perceptible vibration can occur in the walls and ceiling and rattles can occur in furniture. In this case, the noise is identified with "RV" (e.g. RC 29(RV)). The level, L_B, in the octave band of centre frequency, f, corresponding to a particular RC curve is given by:

$$L_B = \mathrm{RC} + \frac{5}{0.3}\log_{10}\left(\frac{1000}{f}\right) \quad \text{(dB re 20 µPa)} \tag{2.113}$$

RC curves represent the shape of the least objectionable noise spectrum. For this reason they are only defined up to an RC number of 50. By contrast, NR and NC curves are intended for specifying an upper bound of acceptability for background noise, and do not necessarily represent the least objectionable noise spectrum. Thus, in certain cases, such as in open plan offices, where it may be advantageous to introduce background noise to ensure speech privacy, noise with a spectrum shape of an NR or NC curve is unsatisfactory, being both too rumbly and too hissy. However, RC curves are suitable for specifying the introduction of acoustic "perfume", and noise with a spectrum of that shape has been found to be the least objectionable.

It is interesting that the "rumbly" or "hissy" labels discussed above are not used in ANSI/ASA S12.2 (2019) as they are inconsistent with the same definitions in the NCB rating scheme.

Acceptable RC ratings for background sound in rooms as a result of air conditioning noise are listed in Table 2.20.

2.12.2.5 RNC Curves

The RC and NCB curves have a number of limitations that can lead to undesirable results. The RC curves set criteria that are below the threshold of hearing to protect against large turbulence fluctuations that generate high levels of low-frequency noise for which the level can vary by up to 10 dB in synchronisation with fan surging. However, the RC curves could unnecessarily penalise a well designed HVAC system, such as may be used in a concert hall, requiring 10 dB or more of unnecessary noise attenuation at low frequencies. On the other hand, NCB curves are intended for well designed HVAC systems. However, they do not sufficiently penalise poorly designed systems that may be energy efficient, but are characterised by high levels of turbulence induced low-frequency noise.

Schomer (2000) proposed a new set of curves (room noise criteria or RNC) (see Figure 2.30) that are intended to address the limitations associated with the RC and NCB curves. The intention is for noise criteria to be above the threshold of hearing for well-behaved systems,

TABLE 2.20 Acceptable air conditioning sound pressure levels in various types of occupied space. Note that the spectrum shape of the noise should not deviate from an RC curve by more than 3 dB and should contain no easily distinguishable tonal components. Adapted from ASHRAE (2015)

Room type	Acceptable RC(N)	Room type	Acceptable RC(N)
Residence	25 – 35	Performing arts spaces	25
Hotel meeting rooms	25 – 35	Music practice rooms	30 – 35
Hotel suites	25 – 35	Laboratories	40 – 50
Other hotel areas	35 – 45	Churches	25 – 35
Offices and conf. rooms	25 – 35	Schools, lecture rooms	25 – 30
Building corridors	40 – 45	Libraries	30 – 40
Hospitals			
Private rooms	25 – 35	Indoor stadiums, gyms	40 – 50
Wards	30 – 40	Stadium with speech ampl.	45 – 55
Operating rooms	25 – 35	Courtrooms (no mics.)	25 – 35
Public areas	30 – 40	Courtrooms (speech ampl.)	30 – 40

while at the same time preventing a turbulence-producing, fan-surging HVAC system from being labelled acceptable.

The RNC rating system is now included in ANSI/ASA S12.2 (2019), and it is expected to receive more widespread use for rating air conditioning systems that are characterised by surging and/or turbulence (aerodynamic instability). Excessive surging is identified by observing the level variations in each of the 16 Hz, 31.5 Hz and 63 Hz octave bands with the meter response set to "fast" with Z-weighting. Excessive turbulence is identified by investigating the value of ($L_{\max} - L_{eq}$) or ($L_{10} - L_{eq}$) for the 125 Hz band by itself and the 16, 31.5 and 63 Hz bands combined to produce a single reading. If any of the differences exceed the values in Table 2.21, then the RNC rating scheme should be used to assess the noise impact.

TABLE 2.21 Screening criteria to determine the presence of large random fluctuations (ANSI/ASA S12.2, 2019)

	Combined 16, 31.5 and 63 Hz	125 Hz
$L_{\max} - L_{eq}$ (dB)	7	6
$L_{10} - L_{eq}$ (dB)	3.5	3

The RNC rating of a sound is obtained using the tangency method in much the same way as obtaining an NR number. That is, the RNC rating is the integer value of the highest RNC curve that intersects the plotted spectra. The complexity arises in the determination of what values to plot. These are the measured octave band values with correction terms added in the 31.5 Hz, 63 Hz and 125 Hz octave bands. It is the determination of the correction terms that is complex. For broadband noise radiated from an air conditioning duct without the presence of excessive turbulence or surging, the correction terms are zero. In cases where there is excessive turbulence, the correction at 31.5 Hz can be as high as a 4 dB increase and if surging is present it can be as high as 12 dB. The correction at 125 Hz is usually zero, except in the case of surging, it may be up to 2 dB. A straight line is drawn between the plotted corrected value at 31.5 Hz and the corrected value at 125 Hz to obtain the corrected value at 63 Hz. The 16 Hz value is not plotted on Figure 2.30 when determining the RNC rating for the noise for cases involving excessive turbulence or surging.

Briefly, the correction is calculated by taking a large number, $i = 1, N$ of octave band sound pressure level measurements, L_i, from 16 Hz to 8 kHz, over some reasonable time interval (for example, 20 seconds), using "fast" time weighting (125 ms time constant) and sampling intervals between 50 and 100 ms. The mean sound pressure level, L_m, is calculated by taking the arithmetic mean of all the measured dB levels in each octave band. The energy-averaged level, L_{eq}, is also calculated for each octave band using:

$$L_{eq} = 10\log_{10}\left[\frac{1}{N}\sum_{i=1}^{N}10^{L_i/10}\right] \quad \text{(dB re 20 µPa)} \tag{2.114}$$

The corrections Δ_{31} and Δ_{125} to be added to the 31.5 Hz and 125 Hz octave band measurements of L_{eq} are calculated for each of the two bands using:

$$\Delta_{31,125} = 10\log_{10}\left[\frac{1}{N}\sum_{i=1}^{N}10^{(L_i-L_m)/\delta}\right] + L_m - L_{eq} \quad \text{(dB)} \tag{2.115}$$

The quantity, δ, is set equal to 5 for calculations in the 31.5 Hz band and equal to 8 for calculations in the 125 Hz octave band and N is the number of measurements taken in each band. Note that for the 31.5 Hz correction term calculation, data for the 16 Hz, 31.5 Hz and 63 Hz bands are all included in calculating L_m and are also included in calculating each of the three terms in Equation (2.115). Before inclusion in Equation (2.115), the 16 Hz data have 14 dB subtracted from each measurement and the 63 Hz data have 14 dB added to each measurement. The noise is considered well behaved if the correction, Δ, for the 31.5 Hz octave band is less than 0.1 dB. In this case, the actual measured octave band data would be plotted on the set of RNC curves from 16 Hz to 8 kHz, with no correction applied to any band.

If no strong surging is present, then

$$\Delta_{125} = 0.118 \times \sigma_{125}^2 + L_m - L_{eq} \quad \text{(dB)} \tag{2.116}$$

where σ_{125}^2 is the standard deviation (in dB) of the 125 Hz octave band data.

Note that RNC values are reported, for example, as: RNC41-(63 Hz). In this example, the highest RNC curve intersected was the RNC 41 curve and this occurred at 63 Hz.

Judgement is often necessary for specifying a noise rating for a particular application. Consideration must be given to any unusual aspects, such as people's attitudes to noise, local customs and the need for economy. It has been found that there are differences in tolerance of noise from one climate to another. Suggested RNC and NC criteria for various unoccupied spaces are listed in Table 2.22. A more complete list is provided in ANSI/ASA S12.2 (2019).

2.12.3 Speech Privacy

When designing an office building, it is important to ensure that offices have speech privacy so that conversations taking place in an office cannot be heard in adjacent offices or corridors. Generally, the higher the background sound pressure levels from air conditioning and other mechanical equipment, the less one has to worry about speech privacy and the more flimsy can be the office partitions, as will become clear in the following discussion. One can deduce that speech privacy is likely to be a problem in a building with no air conditioning and no forced ventilation systems.

To avoid speech privacy problems between adjacent offices, it is important that the following guidelines are followed.

1. Use partitions or separation walls with adequate sound insulation (see Table 2.23). Sound insulation in this case is calculated as the arithmetically averaged, 1/3-octave band sound transmission loss values (see Chapter 7) of the separating partition in the frequency range, 160 Hz to 5000 Hz (ASTM E2638-10, 2017).

TABLE 2.22 Recommended NC and RNC weighting curves for different furnished (but unoccupied), data from ANSI/ASA S12.2 (2019)

Type of space	Recommended NC & RNC curve
EDUCATIONAL BUILDINGS	
Lecture theatres (<566 m^3)	25 – 30
Lecture theatres (≥566 m^3)	30 – 35
Open plan classrooms	25 – 30
HEALTH BUILDINGS	
Private rooms	25 – 30
Operating theatres	40 – 50
Laboratories	35 – 40
Corridors	35 – 45
Public areas	40 – 45
ENTERTAINMENT Live theatre	20 – 25
TV and broadcast studios	15 – 25
Premier movie theatre	20 – 25
Churches	30 – 35
Courtrooms	30 – 35
OFFICE BUILDINGS	
Board and conference rooms (small)	30 – 35
Board and conference rooms (large)	25 – 30
Professional and admin. offices	35 – 40
Executive offices	25 – 35
Open plan office areas	35 – 40
Reception areas	40 – 50
HOTELS & MOTELS	
Guest rooms	30 – 35
Meeting & banquet rooms	25 – 35
Service support areas	40 – 50
RESIDENTIAL	
Apartments	30 – 40
Bedrooms	25 – 30
Family rooms	30 – 40

TABLE 2.23 Speech Privacy Class (SPC) – average sound insulation plus average ambient noise (arithmetic average over the 1/3-octave bands 160 Hz to 5000 Hz)

Sound as heard by occupant	Speech Privacy Class (SPC)
Intelligible	70
Ranging between intelligible and unintelligible	75 – 80
Audible but not intrusive (unintelligible)	80 – 90
Inaudible	90

The American standard, ASTM E1110-06 (2019), provides a more reliable method than Table 2.23 for assessing speech privacy, and this is done in terms of a quantity defined as Articulation Class (AC), which is very different to the Articulation Index, which was superseded by the Speech Transmission Index discussed in Section 2.10.3.1.

The Articulation Class is defined as:

$$\mathrm{AC} = \sum_{i=23}^{37} A(f_i) \times w(f_i) \tag{2.117}$$

where i is the 1/3-octave band number, $A(f_i)$ and $w(f_i)$ are, respectively, the sound insulation in dB and the weighting factor in the 1/3-octave band with centre frequency, f_i. Weighting factors are provided in Table 2.24. The AC is reported to the nearest multiple of 10, usually exceeds 100 and values above 100 indicate increasing privacy and non-intelligibility. However, documented comparisons between AC ratings and subjective judgement of speech privacy are currently unavailable.

TABLE 2.24 Weighting factor, $w(f_i)$

Band number	Centre frequency (Hz)	Weighting factor	Band number	Centre frequency (Hz)	Weighting factor
23	200	0.12	31	1250	0.90
24	250	0.30	32	1600	1.11
25	315	0.30	33	2000	1.14
26	400	0.42	34	2500	1.02
27	500	0.42	35	3150	1.02
28	630	0.60	36	4000	0.72
29	800	0.60	37	5000	0.60
30	1000	0.72			

2. Ensure that there are no air gaps between the partitions and the permanent walls.

3. Ensure that the partitions extend all the way to the ceiling, roof or the underside of the next floor; it is common practice for the partitions to stop at the level of the suspended ceiling to make installation of ducting less expensive; however, this has a negative effect on speech privacy, as sound travels through the suspended ceiling along the ceiling backing space and back through the suspended ceiling into other offices or nearby spaces such as corridors. Alternatively, ceiling tiles with a high transmission loss (TL, see Chapter 7) as well as a high absorption coefficient could be used.

4. Ensure that acoustic ceiling tiles with absorption coefficients of at least 0.1 to 0.4 are used for the suspended ceiling to reduce the reverberant sound pressure level in the office spaces, and thus increase the overall reduction of sound transmitted from one office to another.

When speech privacy is essential, there are two alternative approaches that may be used at the building design stage. The first is to increase the sound insulation of the walls (for example, by using double stud instead of single stud walls, so that the same stud does not contact both leaves). The second approach is to add acoustic "perfume" to the corridors and offices adjacent to those where privacy is important. This "perfume" could be introduced using a random-noise generator, appropriate filter, amplifier and speakers, with the speakers mounted above the suspended ceiling. The filter would be adjusted to produce an overall noise spectrum (existing plus introduced noise), which followed the shape of one of the RC curves shown in Figure 2.29, although it is not usually desirable to exceed an RC of 30 for an executive office (see Table 2.20).

2.12.3.1 Measurement of Speech Privacy

The extent of speech privacy between two offices or between a space external to an office and the office can be measured as described in ASTM E2638-10 (2017) to produce a Speech Privacy Class (SPC) for use in Table 2.23. Essentially, the measurement process consists of using an omni-directional sound source located away from the room walls and floor in the room in which the talkers will be located and measuring the energy (or logarithmic) averaged sound pressure level in the room using either 5 microphone positions or a moving microphone. Several speaker locations should be used (usually at typical talker locations). A conventional speaker can also be used, but must be placed in the room corners and results obtained will be less reliable. The loudspeaker output should be the same for each measurement location. For each loudspeaker location inside the room, the sound pressure level should be measured at potential receiver locations external to the room, with the speaker being operated with the same power output as was done for the internal measurements. For each 1/3-octave band with centre frequency, f, the sound pressure levels at each measurement location inside the room are logarithmically averaged (see Equation (1.98)) over the N source and microphone positions to produce $L_s(f)$, and for each external receiver location the logarithmic average is over the n source positions to produce $L_r(f)$. Background levels are measured at each external receiver location in each of the 1/3-octave bands from 160 Hz to 5000 Hz to produce $L_b(f)$.

The Speech Privacy Class (SPC) at each receiver point is calculated using:

$$\mathrm{SPC} = LD_{\mathrm{average}} - L_{b,\mathrm{average}} \tag{2.118}$$

where

$$LD_{\mathrm{average}} = \sum_{f=160}^{5000} LD(f)/16 \; ; \quad L_{b,\mathrm{average}} = \sum_{f=160}^{5000} L_b(f)/16 \tag{2.119}$$

where, $LD(f) = L_s(f) - L_r(f)$.

Although the Articulation Index (AI) has been replaced by the Speech Transmissibility Index (STI) for assessing speech intelligibility, AI is still used in some standards for assessing speech privacy (see ASTM E1130-16 (2021)). AI can be derived from measurements that are similar to those used to derive SPC. However, instead of using the logarithmically averaged sound pressure levels in the source room, $L_s(f)$ is replaced with $L_{p1}(f)$, where $L_{p1}(f)$ is the sound pressure level at 1 m from the sound source in a free field (such as in an anechoic room or outdoors away from any reflecting surfaces) in a 1/3-octave band with a centre frequency of f. The Articulation Index is then calculated using:

$$\mathrm{AI} = \sum_{f=200}^{5000} \mathrm{SNR} \times w(f)/300 \tag{2.120}$$

where $w(f)$ can be obtained from Table 2.24 and $L_{\mathrm{voice}}(f)$ can be obtained from Table 2.25 for different voice levels. Also, $\mathrm{SNR} = L_{\mathrm{voice}}(f) - L_{p1}(f) + L_r(f) - L_b(f)$. If for any 1/3 octave band, f, SNR is less than 0, it is set equal to 0. If it exceeds 30, it is set equal to 30.

Values of AI range from 0.00 to 1.00. For speech to be audible but not understood (confidential), AI should be at or below 0.05. For open plan offices, for speech to be non-intrusive (but intelligible for eavesdroppers), the value of AI should be between 0.05 and 0.20. For values of AI greater than 0.20, speech may be readily understood and values greater than 0.30 would be generally unacceptable in an office environment.

An alternative rating, PI, often used in architectural situations is calculated directly from AI using

$$\mathrm{PI} = (1 - \mathrm{AI}) \times 100 \tag{2.121}$$

TABLE 2.25 Male speech peak levels, $L_{\text{voice}}(f)$ (dB)

1/3-octave centre frequency (Hz)	Casual voice	Normal voice	Raised voice
200	58	60	63
250	58	64	68
315	56	63	67
400	59	65	70
500	60	66	72
630	57	64	70
800	50	58	66
1000	49	57	65
1250	51	59	66
1600	51	56	63
2000	47	52	59
2500	45	53	59
3150	45	53	58
4000	47	50	56
5000	46	46	52
A-wt	63	69	76
Linear	67	73	78

Confidential speech privacy is achieved for PI values greater than 95, while non-intrusive speech, which is normal in open plan offices is characterised by PI values between 80 an 95 (ASTM E1130-16, 2021). For healthcare facilities, ANSI/ASA S12.70 (2016) specifies a PI value of 95 or greater for enclosed to enclosed spaces and also for enclosed to open spaces. For open spaces, the requirement is for PI to be 85 or greater.

2.13 Environmental Noise Criteria

Comprehensive documents (Berglund et al., 1999; Hurtley, 2009; World Health Organisation, 2018), which address many environmental noise issues, have been published by Stockholm University and the World Health Organisation. These are recommended as excellent sources for detailed information. Here, the discussion is limited to an overview of the assessment of environmental noise impacts.

Acceptable environmental (or community) sound pressure levels are almost universally specified in terms of A-weighted sound pressure levels. Standards exist (ANSI/ASA S12.9-1 (2018), ANSI/ASA S12.9-2 (2018), ANSI/ASA S12.9-3 (2018), ANSI/ASA S12.9-4 (2021), ANSI/ASA S12.9-5 (2020), ANSI/ASA S12.9-7 (2020), ASTM E1686-16 (2016), ISO 1996-1 (2016), ISO 1996-2 (2017), AS 1055.1 (1997), AS 1055 (1978) and AS 1055.3 (1997)) that specify how to measure and assess environmental noise. The ideas in these standards are summarised below.

A comprehensive review of the effect of vehicle noise regulations on road traffic noise, changes in vehicle emissions over the 30 years prior to 2001 and recommendations for consideration in the drafting of traffic noise regulations has been provided by Sandberg (2001).

2.13.1 A-Weighting Criteria

To minimise annoyance to neighbouring residents or to occupiers of adjacent industrial or commercial premises, it is necessary to limit intrusive noise. The choice of limits is generally determined by noise level criteria at the property line or plant boundary. Criteria may be defined in noise-control legislation or regulations. Typical specifications (details may vary) might be written in terms of specified A-weighted equivalent sound pressure levels, L_{Aeq} (dBA), and might state: Noise emitted from non-domestic premises is excessive if the sound pressure level at the measurement place (usually, the noise sensitive location nearest to the noise emitter) for a period during which noise is emitted from the premises:

1. exceeds by more than 5 dBA the background sound pressure level at that place (usually defined as the L_{90} level – see Section 2.4.9); and
2. exceeds the maximum permissible sound pressure level prescribed for that time of the day.

Permissible plant-boundary sound pressure levels generally are dependent on the type of area in which the industrial premises are located, and the time of day. Premises located in predominantly residential areas and operating continuously (24 hours per day) face restrictive boundary sound pressure level limits, as it is generally accepted that people are 10–15 dBA more sensitive to intrusive noise between 10 pm and 7 am.

Applicable standards, such as many of those listed at the beginning of this section, are guides to the assessment of environmental noise measurement and annoyance and do not have any legal force. They are prepared from information gained primarily from studies of noise generated in industrial, commercial and residential locations. In general, they (1) are intended as a guide for establishing sound pressure levels that are acceptable in the majority of residential areas; and (2) are a means for establishing the likelihood of complaints of noise nuisance at specific locations. The general method of assessment involves comparison of noise levels measured in dBA, with acceptable levels. For steady noise, the measured level is the average of the meter reading on a standard sound level meter with the A-weighting network switched in. For fluctuating or cyclic noise, the Equivalent Continuous A-weighted Sound Pressure Level, L_{Aeq}, must be determined. This can be done using a statistical noise analyser (ASTM E1503-22, 2022) or an integrating/averaging sound level meter.

Recommended allowable noise levels are calculated using Table 2.26 with a base level of 40 dB(A). Table 2.26 is used to find an adjustment, ΔL, which may be positive or negative, to be added to the measured L_{Aeq} noise level to obtain a value for comparison with the 40 dB(A) base level. The total adjustment, ΔL, is obtained by arithmetically adding the appropriate adjustment from each of the three categories identified in bold font. For the character category, for a noise that contains both impulsiveness and tones, the maximum allowable adjustment is 10 dB for AS 1055 (1978) and ASTM E1686-16 (2016) and 12 dB for ISO 1996-1 (2016). When the adjusted measured L_{Aeq} level exceeds the base sound pressure level of 40 dBA, a guide to probable complaints is provided in Table 2.27. These guidance levels are no longer included in the most recent version of AS 1055 (2018), but they remain useful for illustrating concepts.

2.13.1.1 Low-Frequency Noise

Although ISO 1996-1 (2016) states that the A-weighted sound level is not an adequate guide to the assessment of community annoyance when the noise is dominated by low frequencies, no quantitative guidance is provided regarding an appropriate adjustment to the measured A-weighted sound pressure level.

2.13.1.2 Transportation Noise

Although dBA levels are generally used to characterise industrial noise, L_{dn} or L_{den} levels (see Sections 2.4.6 and 2.4.7) are generally used to characterise transportation noise (World Health Organisation, 2018). Procedures for calculating the percentage of people who will be highly

TABLE 2.26 Adjustments to be added to the measured sound pressure level prior to comparison with the 40 dBA base level (adapted from AS 1055 (1978), ASTM E1686-16 (2016) and ISO 1996-1 (2016))

	Adjustment (dBA)		
	AS 1055 (1978)	ASTM E1686-16 (2016)	ISO 1996-1 (2016)
Character of the sound			
Tones readily detectable	+5	+5	+3 to +6
Impulsive noise readily detectable	+5	+5	+5
Highly impulsive	—	—	+12
Intermittency readily detectable	+5	—	—
Aircraft noise	—	—	+5 to +8
Railroad noise	—	—	−3 to −6
Time of day			
Evening (6 pm to 10 pm)	+5	+5	+5
Nighttime (10 pm to 7 am)	+10	+10	+10
Weekend daytime	—	—	+5
Neighbourhood			
Rural and outer suburban areas with negligible traffic	0	+10	—
General suburban areas with infrequent traffic	−5	+5	—
General suburban areas with medium density traffic or suburban areas with some commerce or industry	−10	0	—
Areas with dense traffic or some commerce or industry	−15	−5	—
City or commercial areas or residences bordering industrial areas or very dense traffic	−20	−10	—
Predominantly industrial areas or extremely dense traffic	−25	—	—

The neighbourhood adjustments listed under ASTM E1686-16 (2016) correspond to a base level of 50 dBA, so they are in effect the same as the adjustments listed under AS 1055 (1978).

The base level of 40 dBA (or 50 dBA in ASTM E1686-16 (2016)) is at the lower end of the range suggested in current standards.

TABLE 2.27 Estimated public reaction to noise when the adjusted measured sound pressure level exceeds the base sound pressure level (40 dB(A)), by an amount, ΔL (see Table 2.26 for adjustments to the measured level

ΔL (dBA)	Public reaction	Expression of public reaction in a residential situation
0–5	Marginal	From no observed reaction to sporadic complaints
5–10	Little	From sporadic complaints to widespread complaints
10–15	Medium	From sporadic complaints to threats of community action
15–20	Strong	From widespread complaints to threats of community action
20–25	Very strong	From threats of community action to vigorous community action
25 and over	Extreme	Immediate direct community and personal action

annoyed by a particular L_{dn} or L_{den} level corresponding to aircraft, road vehicle and railway vehicle noise are provided in ISO 1996-1 (2016) and ANSI/ASA S12.9-4 (2021).

2.13.1.3 Tonality

There are three different ways generally accepted for identifying tonality (AS 1055, 2018; ANSI/ASA S12.9-5, 2020; IEC 61400-11 Ed.3.0, 2018; ISO/PAS 20065, 2016). The simplest method (ANSI/ASA S12.9-3, 2018; AS 1055, 2018) will be outlined here. This involves obtaining 1/3-octave band sound pressure levels, L_{eq}, energy averaged over the measurement interval (usually 10 or 15 minutes) and comparing the level in each 1/3-octave band with the levels in the two adjacent bands. Tones are labelled as readily detectable if the L_{eq} in the 1/3-octave band of interest exceeds the arithmetic average of the L_{eq} in the two adjacent 1/3-octave bands by the frequency-dependent values listed below (ANSI/ASA S12.9-3, 2018; AS 1055, 2018).

1. 15 dB for frequencies between and including 31.5 Hz and 125 Hz.
2. 8 dB for frequencies between and including 160 Hz and 400 Hz.
3. 5 dB for frequencies between and including 500 Hz and 16000 Hz.

When tonal noise is amplitude modulated, the associated sidebands will increase the adjacent 1/3-octave band noise levels, so in this case it is strongly recommended that the simplest method not be used. A detailed treatment of more complex procedures for tonal noise assessment may be found in (Hansen et al., 2017, pp. 351–362).

2.13.1.4 Impulsiveness

There are two generally accepted ways for identifying the presence of impulsiveness (AS 1055, 2018; ANSI/ASA S12.9-5, 2020), which is defined as the sudden onset of a noise. The onset rate (OR) in dB/s is defined as the rate of increase of the sound pressure level following the onset of impulsiveness. The onset of an impulse is defined as the time at which the rate of increase of the sound pressure level reaches or exceeds 10 dB/s. Cessation of the impulse occurs when the sound pressure level no longer increases at a rate equal to or greater than 10 dB/s. The onset rate is the average rate of increase in sound pressure level (determined by a least squares fit of the data) between the onset and cessation of the impulse. The prominence, P of an impulse is defined as $P = 3\log_{10}(\text{OR}) + 2\log_{10}(\text{LD})$, where LD is the difference in sound pressure level between the level at the onset of the impulse and the cessation of the impulse.

2.13.1.5 Intermittency

A procedure for identifying intermittency is not addressed in AS 1055 (2018), but subjectively it would be applied to any noise that comes and goes many times during a 15-minute measurement period.

2.13.2 Comparison of Noise Weighting Curves with dBA Specifications

For the majority of occupied spaces, advisory limits can be placed on maximum permissible background sound pressure levels, but recommended levels will vary slightly depending on the source of information.

As mentioned earlier, when an attempt was made to use NC rating curves as a guide for deliberate spectral shaping of background noise, the result was unsatisfactory. The internationally accepted NR curves do not provide an improvement in this respect. For example, if the NR weighting curves 15 to 50 are taken as background noise spectra, then A-weighted according to Table 2.3, and their overall equivalent A-weighted levels determined as described in Section 2.3, it is found that in all cases, the low-frequency bands dominate the overall level. As A-weighting accords with subjective response, one would intuitively expect trouble with a spectrum which emphasises a frequency range that ordinarily contributes little to an A-weighted level. In this respect, the RC weighting curves are much more satisfactory. When A-weighted, it is found that neither the high- nor the low-frequency extremes dominate; rather, the mid-frequency range contributes most to the computed equivalent A-weighted level.

Considering noise only in the 500 Hz, 1000 Hz and 2000 Hz bands, the dBA equivalent to any of the noise rating (NR, NC, RC, NCB or RNC) curves is obtained by adding between 5 and 10 dB to the rating number. For example, RC 20 and NC 20 are both equivalent to between 25 and 30 dBA. In terms of perception, RC 20 is "very quiet", RC 30 is "quiet", RC 40 is "moderately noisy" and RC 50 is "noisy". RC curves do not go beyond 50, but if they did, RC 60 would be regarded as "very noisy".

2.14 Environmental Noise Surveys

To document existing environmental noise, one or more ambient sound surveys must be undertaken and if a new facility is being planned, the calculated emissions of the facility must be compared with existing sound pressure levels to assess the potential noise impact. Existing noise regulations must be met, but experience has shown that noise problems may arise, despite compliance with all applicable regulations.

When undertaking a noise survey to establish ambient sound pressure levels, it is important to exclude transient events and noise sources, such as insect noise, which may not provide any masking of noise from an industrial facility. Similarly, high noise levels resulting from weather conditions favourable to sound propagation must be recognised and corresponding maximum sound pressure levels determined. To this end, ISO 1996-2 (2017) gives guidance regarding the minimum required wind speed component in the direction from the source to the receiver for various times of day and night, together with the associated fraction of cloud cover. During the day, the required downwind component ranges from 0.9 to 2.7 m/s whilst at night it ranges between 0.1 and 0.5 m/s. If the facility is to operate 24 hours a day, the daily noise variation with time must be understood using data collected by continuous monitoring in all four seasons. For installations that shut down at night, measurements can be limited to daytime and one nighttime period. In situations where there are significant levels of low-frequency noise, special care must be taken in shielding microphones (see ANSI/ASA S12.9-7 (2020)) from any wind, as wind becomes more and more of a problem as the frequency becomes lower and lower (below about 200 Hz), resulting in measured levels being potentially much higher than the actual noise levels. The quantity commonly used to characterise existing background sound pressure levels is L_{A90}, which is the level exceeded 90% of the time.

For low density vehicular traffic, railway noise and aircraft noise, the sound exposure level, L_{AE}, is recorded for each event (one vehicle pass-by = one event) for the time, T_e (seconds), that the noise is within 10 dB of its maximum value. The L_{AE} values for each event, i, in the time period of interest are converted to L_{Aeq} values of interest using Equation (2.46) as a basis:

$$L_{Aeq,i} = 10\log_{10}\left(\frac{1}{T_e}10^{L_{AE,i}/10}\right) \quad \text{(dB re 20 µPa)} \tag{2.122}$$

The overall L_{Aeq} for the total time period of interest is obtained by logarithmically averaging the individual $L_{Aeq,i}$ values using Equation (1.99).

Existing standards (ASTM E1686-16 (2016), ISO 1996-1 (2016) and ISO 1996-2 (2017)) should always be followed for general guidance.

2.14.1 Measurement Locations

Residential areas closest to the noise source are usually chosen as measurement locations, but occasionally it is necessary to take measurements at other premises such as offices and churches. If the receiver locations are above the planned site or at the other side of a large body of water, measurements may be needed as far as 2 km away from the noise source. Sometimes it is possible to take measurements on residential properties; at other times, it may be necessary to use a laneway. Whatever locations are used, it is important that they are as far from the main residence access road as the residences themselves and definitely not near the side of the road, as the resulting percentile levels and equivalent sound pressure levels will not represent actual levels at the front or rear of a residence. Noise emission codes and zoning regulations specify allowable levels at the property line of the noise generator, so it is often necessary to take measurements at a number of locations on the property boundary of the noise generator.

2.14.2 Duration of the Measurement Survey

Current practice commonly involves both continuous unattended monitoring over a two-week period and periodic 10- to 15-minute attended sampling. Most jurisdictions require unattended monitoring to occur for a minimum of two weeks and where there is a large seasonal variation in ambient sound pressure levels, additional two-week sampling periods are required to ensure that the season with minimum sound pressure levels is sampled. If both attended and unattended monitoring are used concurrently and the origins of noise events are recorded during the attended sampling periods, the two techniques provide complementary information that can be combined to obtain a good understanding of the ambient noise environment. It is recommended that at least one continuous monitor (preferably 3 or 4) be used at the most critical locations and that regular attended sampling be undertaken at a number of locations, including the critical locations.

Environments that are dominated by high-density daily traffic noise are generally very similar from day to day (with typical standard deviations of less than 1 dB for the same time of day). However, in the early morning hours, weather conditions become more important than traffic volume and minimum levels vary in accordance with sound propagation conditions. Conversely, measurements in urban and rural environments, not dominated by traffic, are not very repeatable and can be characterised by standard deviations higher than 5 dB. These data are often affected by wind and temperature gradients, as well as by the wind generating noise indirectly, such as by causing leaves in trees to rustle.

2.14.3 Measurement Parameters and Procedures

The statistical measure, L_{A10} (see Section 2.4.9), is primarily used for assessing traffic noise; L_{A50} is a useful measure of the audibility of noise from a planned facility; and L_{A90} is used to classify and characterise residential area environments. In many cases, L_1 is used as a measure of the peak sound pressure level, but it often underestimates the true peak by as much as 20 dB. In most cases, the energy-averaged Equivalent A-weighted Sound Pressure Level, L_{Aeq}, is the preferred measure. To gain a good understanding of the environmental noise character and the important contributors to it, it is often useful to express the statistical measures mentioned above in unweighted octave or 1/3-octave bands. More detail on measurement parameters and procedures is provided in ANSI/ASA S12.9-1 (2018). Requirements for long-term average measurements using unattended instrumentation for environmental noise assessment, noise prediction validation and regulation compliance, are detailed in ANSI/ASA S12.9-2 (2018). Measurement procedures for attended measurements for the purpose of measuring the contribution of a specific sound source to a specific measurement location (in terms of L_{Aeq}) are detailed in ANSI/ASA S12.9-3 (2018) and ANSI/ASA S12.18 (2019). When undertaking measurements, it is important that environmental conditions are taken into account so that the worst-case situation can be recorded. This means that the receiver should be downwind of the source (within $\pm 45°$) or if there is little or no wind, there should be a temperature inversion (ANSI/ASA S12.18, 2019). In order to obtain the noise contribution from an existing noise source, the total L_{Aeq} with the noise source off must be measured and then logarithmically subtracted from the L_{Aeq} measured with the noise source operating using Equation (1.102). It is not scientifically justifiable to subtract statistical levels (such as L_{A90}) from one another in this way. For cases involving transportation noise, it is often preferable to use quantities such as L_{dn} or L_{den} (ISO 1996-1, 2016).

When environmental noise due to a particular noise source is measured, the following items should be included in the measurement report:

- reference to the appropriate noise regulation document;
- date and time of measurement;
- details of measurement locations;
- weather conditions (wind speed and direction, relative humidity, temperature and recent precipitation);
- operating conditions of the sound source (e.g. % load, etc.), description of the noise source and its condition, and any noticeable characteristics such as tones, modulation or vibration;
- instrumentation used and types of measurements recorded (e.g. spectra, L_{A10}, L_{Aeq});
- total levels of noise due to all other sources;
- measured data or results of any calculations pertaining to the noise source being measured;
- any calculation procedures used for processing the measurements;
- results and interpretation; and
- any other information required by the noise regulation document.

2.14.4 Measurement Uncertainty

When undertaking noise measurements, it is important to include an uncertainty estimate in the results. In the following descriptions, the measured sound pressure level, L_{eq} may apply to an un-weighted octave band measurement or an overall measurement that may or may not be A-weighted. The averaging time required to obtain an L_{eq} measurement may range from 10 minutes to a year, depending on the situation being investigated. Short term measurements as defined in (ISO 1996-2, 2017) may range from 10 minutes to several hours, but the measurement time should be selected to cover all variations in noise emission. Long term measurements may be required to take into account different seasons and may involve short periods of one or two weeks during each season.

Uncertainty arises from a number of causes, and it is usually expressed in terms of a quantity called expanded uncertainty, u_e. The expanded uncertainty means that there is a 95% probability that the measured value will be within $\pm\, u_e$ of the true value. The expanded uncertainty is related to the overall standard uncertainty, $u_{s,\text{tot}}$, by $u_e = 2u_{s,\text{tot}}$. Environmental noise measurements suffer from the following uncertainties (ISO 1996-2, 2017):

1. measurement system accuracy, $u_{s,1} = u_{s,\text{slm}}$;
2. source sound power output, $u_{s,2} = u_{s,\text{sou}}$;
3. meteorological conditions deviating from the assumed conditions, $u_{s,3} = u_{s,\text{met}}$;
4. selection of receiver location, $u_{s,4} = u_{s,\text{loc}}$; and
5. residual noise, $u_{s,5} = u_{s,\text{res}}$ (or noise not associated with the noise source being investigated).

The overall standard uncertainty, $u_{s,\text{tot}}$, in the measured sound pressure level is obtained by summing the squares of the product of each individual standard uncertainty, u_i, listed above and the corresponding sensitivity coefficient, c_i as follows:

$$u_{s,\mathrm{tot}} = \sqrt{\sum_{i=1}^{5}(c_i u_{s,i})^2} \qquad \text{(dB)} \tag{2.123}$$

where c_i represents the importance of component, $u_{s,i}$ to the overall uncertainty (see above list). Values of c_i are usually set equal to 1 to reflect that all variables have an equal effect on the uncertainty. However, if the residual noise level (or noise not associated with the noise source under consideration), L_{res}, can be measured, the sensitivity coefficients, $c_1 = c_{slm}$ and $c_5 = c_{\mathrm{res}}$, for the measured overall sound pressure level, L'_{eq}, and the residual sound pressure level, L_{res}, respectively, can be calculated using (ISO 1996-2, 2017):

$$c_1 = c_{slm} = \frac{1}{1 - 10^{-(L'_{eq} - L_{\mathrm{res}})/10}} \tag{2.124}$$

and

$$c_5 = c_{\mathrm{res}} = \frac{-10^{-(L'_{eq} - L_{\mathrm{res}})/10}}{1 - 10^{-(L'_{eq} - L_{\mathrm{res}})/10}} \tag{2.125}$$

The standard uncertainty, $u_{s,1} = u_{s,\mathrm{slm}}$, for the measurement system is 0.5 dB for a Class 1 system (see Section 3.2) and 1.5 dB for a Class 2 system (ISO 1996-2, 2017). The standard uncertainty for the source sound power output is usually combined with the standard uncertainty for meteorological conditions and is determined by taking N_m individual measurements, $L_{eq,i}$, $i = 1, \ldots N_m$, when meteorological conditions are within the required window. The average L_{eq} is obtained using:

$$L_{eq} = 10\log_{10}\left(\frac{1}{N_m}\sum_{i=1}^{N_m} 10^{L_{eq,i}/10}\right) \qquad \text{(dB re 20 µPa)} \tag{2.126}$$

and the estimate of the standard deviation due to variations in source power output and meteorological conditions (items 2 and 3 in the above list) is:

$$s_{2,3} = \sqrt{\frac{1}{N_m - 1}\sum_{i=1}^{N_m}\left(10^{L_{eq,i}/10} - 10^{L_{eq}/10}\right)^2} \qquad (\mathrm{Pa}^2) \tag{2.127}$$

Thus, the combined uncertainty due to the source power output variability and the meteorological conditions is (ISO 1996-2, 2017):

$$u_{s,2,3} = 10\log_{10}\left(10^{L_{eq}/10} + s_{2,3}\right) - L_{eq} \qquad \text{(dB)} \tag{2.128}$$

Alternatively, the uncertainties due to source variability and meteorological conditions could be determined separately. The uncertainty due to source power output variability, $u_{s,2} = u_{s,\mathrm{sou}}$, could be determined by taking measurements sufficiently close to the source that meteorological effects are negligible. In this case, the standard uncertainty according to ISO 1996-2 (2017), due to source power output variability is:

$$u_{s,2} = u_{s,\mathrm{sou}} = \sqrt{\frac{\sum_{i=1}^{N_m}\left(L_{eq,i} - L_m\right)^2}{N_m - 1}} \qquad \text{(dB)} \tag{2.129}$$

where L_m is the arithmetic mean of the N_m L_{eq} measurements and $L_{eq,i}$ represents the energy-averaged sound pressure level for one full cycle of operation of the source. It is unclear why ISO 1996-2 (2017) suggests using the arithmetic mean and the form of Equation (2.129), when it seems more appropriate to use the energy average of Equation (2.126) and the form of Equation (2.128), which is suggested by the same standard for the calculation of the combined source and meteorological uncertainty.

The value to be used for the standard uncertainty due to meteorological effects alone, for meteorological conditions favourable for propagation and for source–receiver distances less than 400 m, is $u_{s,3} = u_{s,\text{met}} = 2$ dB (ISO 1996-2, 2017) unless better information is available. For distances, $d > 400$ m:

$$u_{s,3} = u_{s,\text{met}} = (1 + d/400) \qquad \text{(dB)} \tag{2.130}$$

where d is in metres. When calculating the uncertainty for transportation noise, other variables need to be included in the calculation of the effect of the source uncertainty, and these are discussed in detail in ISO 1996-2 (2017).

The uncertainty due to receiver location, $u_{s,4} = u_{s,\text{loc}}$, takes into account the effect of any reflecting surfaces on the measured sound pressure level. When no reflecting surfaces (other than the ground) are closer to the receiver than twice the distance from the sound source to the microphone location or if the reflecting surface is sufficiently small to have a negligible effect on the measured sound pressure level, $u_{s,4} = u_{s,\text{loc}} = 0$ dB. However, if measuring near a building, or other reflecting surface, it is recommended to mount the microphone flush with the reflecting surface and subtract 5.7 dB from the measured result (ISO 1996-2, 2017). In this case, $u_{s,4} = u_{s,\text{loc}} = 0.4$ dB. For a microphone NOT mounted flush with the reflecting surface, but at a perpendicular distance of d from a point, O, on the surface, the correction is 3 dB, provided that the following criteria are met (see ISO 1996-2 (2017, App. B)).

1. 5.4 m $\leq d \leq 0.05L$ for a point source, and 1.6 m $\leq d \leq 0.1L$ for an extended source, where L is the distance from the source to point O on the reflecting surface.
2. The distance from the edge of the reflecting surface to the location where the perpendicular from the microphone to the reflecting surface strikes the surface, must be greater than $4d$ for a surface vertical edge and $2d$ for a horizontal edge.

For frequencies below 4 kHz, a 13 mm (half-inch) diameter microphone should be used, but for higher frequencies a 6 mm (quarter-inch) diameter microphone should be used (ISO 1996-2, 2017). It is also possible to position the microphone at distances from the reflecting surface that are greater than 1.6 m for an extended source and 5.4 m for a point source, in which case 3 dB is subtracted from the measured result and $u_{s,4} = u_{s,\text{loc}} = 0.4$ dB (ISO 1996-2, 2017).

If the residual sound pressure level can be measured by removing or turning off the sound source under investigation, then the residual noise uncertainty, $u_{s,5} = u_{s,\text{res}}$, is the same as for the overall sound pressure level measurement (0.5 dB for a Class 1 measurement system and 1.5 dB for a Class 2 system – see Section 3.2).

Following calculation of the standard uncertainties and the overall standard uncertainty using Equation (2.123), the expanded uncertainty, u_e, can be calculated. The expanded uncertainty, u_e, corresponding to 95% confidence limits, represents a 95% chance that the actual value will be within $\pm\, u_e$ of the measured value. For a normal distribution and 95% confidence limits (usually used for measurements), $u_e = 1.96u_{s,\text{tot}} \approx 2u_{s,\text{tot}}$.

2.14.5 Noise Impact

To calculate the overall noise impact of an industry on the surrounding community, the number of people exposed to various sound pressure levels is used to arrive at a single noise exposure index called the Total Weighted Population (or TWP). This quantity is calculated using day-night average sound pressure levels (L_{dn}), weighting factors to weight higher levels as more important, and the number of people exposed to each level as:

$$\mathrm{TWP} = \sum_i W_i P_i \tag{2.131}$$

where P_i is the number of people associated with the ith weighting factor which, in turn, is related to a particular L_{dn} level as defined in Table 2.28.

TABLE 2.28 Annoyance weighting factors corresponding to values of L_{dn}

Range of L_{dn} (dB)	W_i	Range of L_{dn} (dB)	W_i	Range of L_{dn} (dB)	W_i
35–40	0.01	55–60	0.18	75–80	1.20
40–45	0.02	60–65	0.32	80–85	1.70
45–50	0.05	65–70	0.54	85–90	2.31
50–55	0.09	70–75	0.83		

The relative impact of one particular noise environment may be compared with another by comparing the Noise Impact Index for each environment, defined as:

$$\mathrm{NII} = \frac{\mathrm{TWP}}{\sum_i P_i} \tag{2.132}$$

3

Instrumentation for Noise Measurement and Analysis

LEARNING OBJECTIVES

In this chapter, the reader is introduced to:

- acoustic instrumentation;
- condenser, electret and piezoelectric microphones;
- microphone sensitivity, definition and use;
- acoustic instrumentation calibration;
- sound level meters;
- noise dosimeters;
- recording of acoustical data;
- sound intensity analysers and particle velocity sensors;
- statistical noise analysers;
- frequency spectrum analysers; and
- instrumentation for sound source localisation.

3.1 Microphones

A wide variety of transduction devices have been demonstrated over the years for converting sound pressure fluctuations to measurable electrical signals, but only two such devices are commonly used for precision measurement (Starr, 1988; Bruce, 1988). As this chapter is concerned with the precision measurement of sound pressure level, the discussion will be restricted to these two types of transducers. The most commonly used sound pressure transducer for precision measurement is the condenser microphone. To a lesser extent piezoelectric microphones are also used. Both microphones are used because of their very uniform frequency response and their long-term sensitivity stability. The condenser microphone generally gives the most accurate and consistent measure but it is much more expensive to construct than is the piezoelectric microphone. The condenser microphone is available in two forms, which are either externally polarised by application of a bias voltage in the power supply or pre-polarised internally by use of an electret. The externally polarised microphone is sensitive to dust and moisture on its diaphragm, but it is capable of reliable operation at elevated temperatures and generally has a lower electrical noise floor than its electret counterpart. The pre-polarised type is not nearly as sensitive to dust and

DOI: 10.1201/9780367814908-3

moisture and is the microphone of choice in current instrumentation for accurate measurement of sound. Low cost electret microphones are generally not temperature compensated and do not have a very flat frequency response. However, they are the most commonly used microphone in active noise control systems and headsets. Both forms of condenser microphone are relatively insensitive to vibration. Specifications for laboratory standard microphones are provided in ANSI/ASA S1.15-1 (2020) and IEC 61043 (2017) and their pressure sensitivity calibration, which requires 3 laboratory standard microphones or two microphones and a sound source, is discussed in detail in ANSI/ASA S1.15-2 (2020).

The piezoelectric microphone is less sensitive to dust and moisture but it can be damaged by exposure to elevated temperatures and, in general, it tends to be quite microphonic; that is, a piezoelectric microphone may respond about equally well to vibration and sound whereas a condenser microphone will respond well to sound and effectively not at all to vibration.

MEMS microphones are also available (Piper et al., 2015), and these are excellent for constructing low-cost arrays for beam forming and acoustic holography. Their main limitation is their relatively high electronic noise floor of about 25–30 dBA which makes them unsuitable for low level sound measurements.

3.1.1 Condenser Microphones

A condenser microphone consists of a diaphragm that serves as one electrode of a condenser, and a polarised backing plate, parallel to the diaphragm and separated from it by a very narrow air gap, which serves as the other electrode. The condenser is polarised by means of a bound charge, so that small variations in the air gap due to pressure-induced displacement of the diaphragm result in corresponding variations in the voltage on the condenser.

The bound charge on the backing plate may be provided either by means of an externally supplied bias voltage of the order of 200 V, or by use of an electret, which forms either part of the diaphragm or the backing plate. Details of the electret construction and its use are discussed in the literature (Frederiksen et al., 1979). For the purpose of the present discussion, however, the details of the latter construction are unimportant. In either case, the essential features of a condenser microphone and a sufficient (but simplified) representation of its electrical circuit for the present purpose are provided in Figure 3.1. A more detailed treatment is provided by Brüel and Kjær (1996).

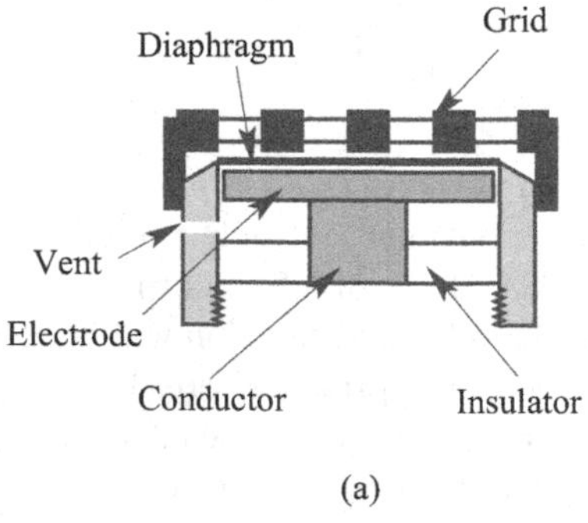

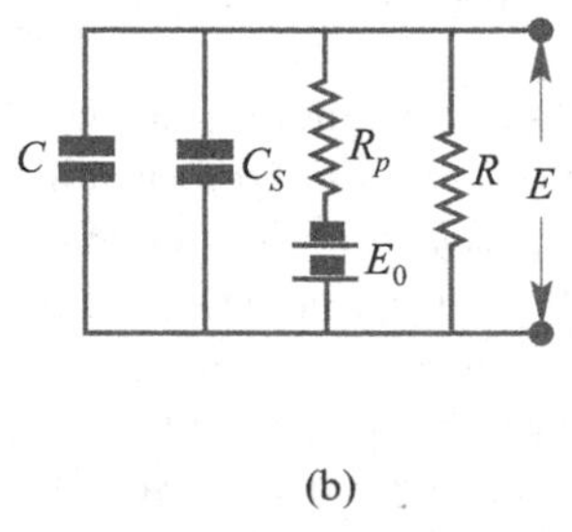

FIGURE 3.1 A schematic representation of a condenser microphone and equivalent electrical circuit.

Referring to Figure 3.1, the bound charge, Q, may be supplied by a DC power supply of voltage, E_0, through a very large resistor, R_p. Alternatively, the branch containing the DC supply and resistor, R_p, may be thought of as a schematic representation of the electret. The microphone response voltage is detected across the load resistor, R. A good signal can be obtained at the input to a high internal impedance detector, even though the motion of the diaphragm is only a small fraction of the wavelength of light.

An equation relating the output voltage of a condenser microphone to the diaphragm displacement will be derived. It is first observed that the capacitance of a condenser is defined as the stored charge on it, Q, divided by the resulting voltage across the capacitance. Using this observation, it can be seen by reference to Figure 3.1, where C is the capacitance of the microphone and C_S is the stray capacitance of the associated circuitry, that for the diaphragm at rest with a DC bias voltage of E_0:

$$\frac{Q}{C+C_s} = E_0 \tag{3.1}$$

The microphone capacitance is inversely proportional to the spacing at rest, h, between the diaphragm and the backing electrode. If the microphone diaphragm moves a distance, x, inward (negative displacement, positive pressure) so that the spacing becomes $h - x$, the microphone capacitance will increase from C to $C + \delta C$ and the voltage across the capacitors will decrease in response to the change in capacitance by an amount E to $E_0 + E$. Thus:

$$E = -\frac{Q}{C + \delta C + C_s} + E_0 \tag{3.2}$$

The microphone capacitance is inversely proportional to the spacing between the diaphragm and the backing electrode, thus:

$$\frac{C + \delta C}{C} = \frac{h}{h - x} \tag{3.3}$$

Equation (3.3) may be rewritten as:

$$\delta C = C\left(\frac{1}{1 - x/h} - 1\right) \tag{3.4}$$

Substitution of Equation (3.4) into Equation (3.2) and use of Equation (3.1) gives:

$$E = -\frac{Q}{C}\left[\frac{1 - x/h}{1 + (C_s/C)(1 - x/h)} - \frac{1}{1 + C_s/C}\right] \tag{3.5}$$

Equation (3.5) may be rewritten as:

$$E = -\frac{Q}{C}\left[\frac{1 - x/h}{1 + (C_s/C) - (C_s/C)(x/h)} - \frac{1}{1 + C_s/C}\right] \tag{3.6}$$

The empirical coefficient, K_1, is now introduced and defined as:

$$K_1 = \frac{1}{Ch} \tag{3.7}$$

By design, $C_s/C << 1$ and $x/h << 1$; thus in a well-designed microphone the product, $(C_s/C)(x/h)$, in Equation (3.6) is negligible compared to unity and thus may be omitted. Introducing Equation (3.7), Equation (3.6) takes the following approximate form:

$$E \approx -K_1 Qh\left[\frac{1}{1 + C_s/C} - \frac{1}{1 + C_s/C} - \frac{x/h}{1 + C_s/C}\right] \approx \frac{K_1 Qx}{1 + C_s/C} \tag{3.8}$$

Equation (3.8) reflects the fact that a positive pressure, which causes a negative displacement x (inward motion), results in a negative value of the induced voltage E.

Reference to Equation (3.7) shows that the coefficient, K_1, depends on two parameters:

1. the capacitance of the device; and
2. the spacing at rest between the microphone diaphragm and the backing electrode.

Both of these parameters are generally very difficult to determine by design, and consequently, the coefficient must be determined by calibration. For good linear response, the capacitance ratio C_s/C must be kept as small as possible, and similarly, the microphone displacement relative to the condenser diaphragm backing electrode spacing, x/h, must be very small. Thus, in a well-designed microphone, the term, $(C_s/C)(x/h)$, in Equation (3.6) is negligible compared to unity.

The vent shown in Figure 3.1 is an essential element of a microphone in that it allows the mean pressure to be equalised on both sides of the microphone diaphragm. Without the presence of this vent, any gradual changes in atmospheric pressure would act to push the diaphragm one way or the other and drastically change the frequency response of the microphone. Unfortunately, the presence of the vent limits the low-frequency response of the microphone so that below the cut-on frequency caused by the vent, the microphone will be insensitive to acoustic pressure. The cut-on frequency is proportional to the size of the vent; the larger the vent the higher will be the cut-on frequency, which is defined as (Olsen, 2005):

$$f_{co} = \frac{1}{2\pi R_v C_{dc}} \quad \text{(Hz)} \tag{3.9}$$

where R_v is the resistance of the vent and for a vent consisting of a tube of radius, a, and length, ℓ, it is given by (Beranek, 1954):

$$R_v = \frac{8\mu\ell}{\pi a^4} \tag{3.10}$$

where μ is the dynamic viscosity for air and is equal to 1.84×10^{-5} N-s/m^2 at 20° C. The dynamic viscosity varies with absolute temperature, T, in Kelvin such that:

$$\mu \propto T^{0.7} \tag{3.11}$$

A typical value for R_v is 36×10^9 N-s/m^5. The quantity, C_{dc}, in Equation (3.9) is the compliance of the diaphragm–cavity system and is the reciprocal of the sum of the reciprocals of the compliance, C_c, of the cavity behind the microphone diaphragm and the compliance, C_d, of the diaphragm, as these latter two compliances act together in parallel. Thus:

$$\frac{1}{C_{dc}} = \frac{1}{C_c} + \frac{1}{C_d} = \frac{C_d + C_c}{C_d C_c} \tag{3.12}$$

A typical value for the diaphragm compliance, C_d, is 0.3×10^{-12} m^5/N (Brüel and Kjær, 1996). The compliance of the cavity of volume V_c (m^3) is given by:

$$C_c = \frac{V_c}{\gamma P} \tag{3.13}$$

where $\gamma = 1.4$ is the ratio of specific heats for air and P is the mean atmospheric pressure (Pa). A typical value for C_c is 2.8×10^{-12} (Brüel and Kjær, 1996), which thus defines a typical value for V_c of approximately 4×10^{-10} m^3 or 0.4 mm^3.

3.1.2 Piezoelectric Microphones

A sketch of the essential features of a typical piezoelectric microphone and a schematic representation of its electrical circuit are shown in Figure 3.2. In this case, sound incident upon the diaphragm tends to stress or unstress the piezoelectric element which, in response, induces a

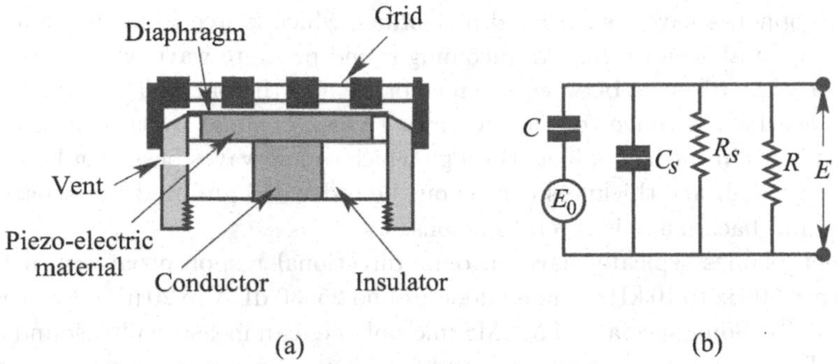

FIGURE 3.2 A schematic representation of a piezoelectric microphone and equivalent electrical circuit.

bound charge across its capacitance. The effect of the variable charge is like that of a voltage generator, as shown in the circuit.

For a piezoelectric microphone, the equivalent circuit of which is shown in Figure 3.2, the microphone output voltage is given by:

$$E = \frac{Q}{C + C_s} = \frac{Q/C}{1 + C_s/C} \tag{3.14}$$

As $E_0 = Q/C$:

$$E = \frac{E_0}{1 + C_s/C} \tag{3.15}$$

The voltage E_0 generated by the piezoelectric crystal is proportional to its displacement, x. Thus:

$$E_0 = K_2 x \tag{3.16}$$

Substituting Equation (3.16) into (3.15) gives:

$$E = \frac{K_2 x}{1 + C_s/C} \tag{3.17}$$

3.1.3 MEMS Microphones

Very small Micro-Electro Mechanical Systems (MEMS) microphones are packaged as an electronic chip, intended to be soldered on to a printed circuit board (PCB). MEMS microphones have dimensions around 3 × 2 × 1 mm, and operate the same way as the condenser microphone described in Section 3.1.1. The microphone element is constructed on a silicon wafer, using manufacturing processes similar to integrated circuits. Figure 3.3 shows two configurations of MEMS microphone with an input port for sound to enter the device (a) through the bottom, and (b) through the top.

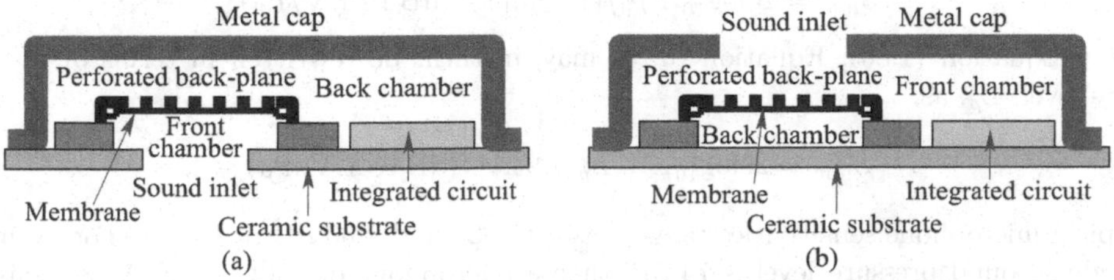

FIGURE 3.3 A schematic representation of a MEMS condenser microphone with (a) bottom port and (b) top port.

MEMS microphones have a suspended membrane, which is free to move about a fixed back-plane, forming a variable capacitor. An incoming sound pressure wave will cause the membrane to move and vary the distance between the membrane and the back-plane, which in turn varies the capacitance, which is converted into a varying voltage signal by the on-board electronics. The perforated holes in the back-plane, through which sound waves pass, can be less than 10 µm in diameter, the membrane thickness can be on the order of 1 µm, and the separation between the membrane and back-plane is several microns.

MEMS microphones typically have an omni-directional response pattern, a flat frequency response between 50 Hz to 10 kHz, a noise floor around 25–30 dBA re 20 µPa, and can measure up to 130 dB re 20 µPa. Some specialised MEMS microphones can measure ultrasound at frequencies up to 80 kHz. Their low-cost, compact size and low power consumption, requiring around 20 µA to 120 µA and only 1.5 V to 3.6 V, makes them ideal for consumer electronics. They are also used in acoustic instrumentation, such as microphone arrays and acoustic cameras.

Some MEMS microphones output analogue signals, and others output a digital stream. There are two common digital communication protocols: pulse density modulation (PDM) or inter-IC sound (I^2S), which are used to transport a digital stream to a micro-processor that can store the audio data.

3.1.4 Pressure Response

Equations (3.8) and (3.17) show that the response voltage, E, for both condenser and piezo-electric microphones, is essentially linearly related to the displacement, x. In either case, an acoustic wave of acoustic pressure, p, incident upon the surface of the microphone diaphragm of area, S, will induce a mean displacement, x, determined by the compliance, K_3, of the diaphragm. Thus, the relation between acoustic pressure and mean displacement is:

$$x = -K_3 S p \tag{3.18}$$

Substitution of Equation (3.18) into either Equation (3.8) or Equation (3.17) results in a relation, formally the same in either case, between the induced microphone voltage and the incident acoustic pressure. Thus, for either microphone:

$$E = -\frac{KSp}{(1 + C_s/C)} \tag{3.19}$$

In Equation (3.19) the coefficient, K, must be determined by calibration.

3.1.5 Microphone Sensitivity and Dynamic Range

It is customary to express the sensitivity of microphones in decibels relative to a reference level. Following accepted practice, the reference voltage, E_{ref}, is set equal to 1 V and the reference pressure, p_0, is set equal to 1 Pa. The following equation may then be written for the sensitivity of a microphone:

$$S_{\text{mic}} = 20\log_{10}[Ep_0/(E_{\text{ref}}p)] \quad \text{(dB re 1 V/Pa)} \tag{3.20}$$

Using Equation (1.86), Equation (3.20) may, in turn, be rewritten in terms of the sound pressure level, L_p, as:

$$S_{\text{mic}} = 20\log_{10}E - L_p + 94 \quad \text{(dB re 1 V/Pa)} \tag{3.21}$$

Typical microphone sensitivities range between −25 and −60 dB re 1V/Pa. For example, if the incident sound pressure level is 74 dB, then a microphone of −30 dB re 1V/Pa sensitivity will produce a voltage of −50 dB re 1V. The voltage thus produced is $E = 10^{-50/20} = 3.16$ mV.

The dynamic range of typical high-quality condenser microphones is between 100 and 120 dB. The dynamic range in dB is the maximum sound pressure level that can be measured without distortion minus the minimum that can be measured without undue influence from electronic noise.

3.1.6 Field Effects and Calibration

Reference to Equation (3.19) shows that the output voltage of a microphone is directly proportional to the area of the diaphragm. Thus, the smaller a microphone of a given type, the smaller will be its sensitivity. If low sound pressure levels are to be measured, the output voltage may be rather small, and the magnitude of the gain that is possible in practice is limited by the internal noise of the amplification devices. These considerations call for a microphone with a large diaphragm, which will produce a corresponding relatively large output voltage.

In the design of a good microphone one is concerned with uniform frequency response, besides general sensitivity as discussed above, and the demand for high-frequency response recommends against a microphone with a large diaphragm. The problem is that, at high frequencies, the wavelength of sound becomes very small. Thus, for any diaphragm, there will be a frequency at which the diameter of the diaphragm and the wavelength of sound are comparable. When this happens, large diffraction effects begin to take place, resulting in the microphone response to the incident sound becoming quite irregular and also sensitive to the direction of the incident sound. This may be undesirable when the direction of incidence is unknown. Unfortunately, the problem of diffraction cannot be avoided; in fact diffraction effects begin to be apparent when the wavelength of the measured sound is still as much as 10 times the diameter of the microphone, and this is always well within the expected measurement range of the instrument. At still higher frequencies and shorter wavelengths the response of the microphone becomes quite sensitive to the angle of incidence of the measured sound, as illustrated in Figure 3.4. In the figure, the increase or decrease in sound pressure level due to the presence of the microphone, relative to the sound pressure level that would exist in the absence of the microphone, is shown. Thus, for a microphone characterised by the diffraction effects shown in the figure, and for a microphone diameter to wavelength ratio of 0.63, a normally incident sound will produce a sound pressure at the microphone which is 8 dB higher than the same sound will produce at grazing

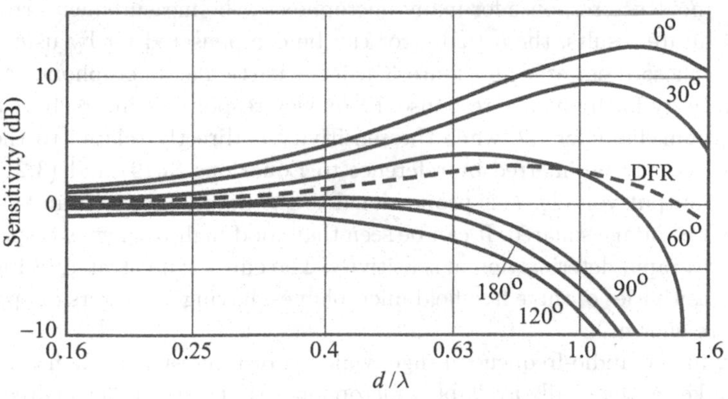

FIGURE 3.4 Microphone free-field correction: the sound pressure level on the diaphragm relative to the free-field level in the absence of the microphone, as a function of angle of incidence. Angles are measured relative to the normal to the diaphragm. DFR is the diffuse-field response, i.e. sound pressure level on the diaphragm in a diffuse field (random-incidence calibration); adapted from Brüel and Kjær (1973).

incidence. Essentially, what is affected is the phase as well as the amplitude of the sound pressure distribution over the diaphragm of the microphone.

Since the problem of diffraction cannot be avoided for practical reasons of sensitivity, as has been explained, it is necessary to take account of the expected angle of incidence of sound upon a microphone during use. This has led to the so-called free-field calibration, which is a function of angle of incidence. Alternatively, if sound is assumed to be incident from all directions at once, then the properly weighted average of the free-field calibrations produces the random-incidence calibration. Such a weighted average is shown in Figure 3.4 as the diffuse-field response (DFR).

Yet another possibility exists. If a uniform pressure is imposed on the diaphragm (done in practice electrostatically), then a pressure response can be determined. The latter response is affected by the design of the backing electrode, the size and shape of the backing cavity, and the mass and tension of the diaphragm. It is thus possible to shape the high-frequency pressure response to compensate for the effective increase in pressure at the diaphragm due to diffraction. For example, if the pressure-response roll-off is shaped by design to just compensate for the increase of the free-field pressure at normal incidence, giving a combined response which is fairly flat for normally incident sound, then a free-field microphone is produced. Alternatively, the pressure response roll-off can be adjusted to compensate for the free-field increase at any angle of incidence, or it can be adjusted to compensate for the random incidence response, resulting in a random-incidence or diffuse-field microphone.

In summary, three types of microphone are recognised. These are free field, random incidence and pressure sensitive. Microphones are commonly sold as (1) free field, generally meaning flat response for normal incidence, or (2) as random incidence, meaning generally flat response in a diffuse sound field characterised by sound of equal intensity incident from all directions. Alternatively, some sound level meters provide a filter network to allow simulation of diffuse field response for any particular microphone. In some applications where diffraction is not present (such as when the microphone is mounted flush with the wall of a duct to measure turbulent pressure fluctuations), the pressure-response calibration only is used.

Both free-field and diffuse-field microphones are used for industrial noise measurement. When it is obvious from which direction the noise is coming, then a free-field microphone is used and pointed directly at the source. When it is not obvious where the noise is coming from or when there are noise sources all around the microphone, then best results are obtained by using a random-incidence microphone and pointing it straight upwards. In this case, most sound will then be incident at 90° to the microphone axis, and as the diffuse-field response is reasonably close to the 90° incidence response for many microphones, the measurement error is minimised.

For more accurate results, the actual error can be compensated for by using the characteristics, similar to those illustrated in Figure 3.4, for a particular microphone. All microphones must trade sensitivity for frequency response. Frequency response is inversely related to the microphone diaphragm diameter, D, while the sensitivity is directly related to the fourth power of the diameter, as may be inferred by reference to Equations (3.19) and (3.21), where it can be seen that the output voltage, E is proportional to the diaphragm area and the sensitivity is proportional to the voltage squared. It can be seen that good high-frequency response is obtained at the expense of a rapid deterioration in sensitivity. The effect is illustrated in Figure 3.5, where the respective sensitivities of three free-field microphones, having diameters of approximately 25, 12 and 6 mm, are shown.

Fortunately, in the audio-frequency range, which is of interest here, it has been possible to produce, and make commercially available, microphones that have sufficient frequency response and sensitivity for most purposes and amplifiers are available that can adequately cope with their small output signals. Recent advances in microphone technology have resulted in 12 mm microphones that are 10 dB more sensitive than shown in Figure 3.5. Thus, 25 mm diameter microphones are not used very often any more as 12 mm microphones with similar sensitivity and better frequency response are available.

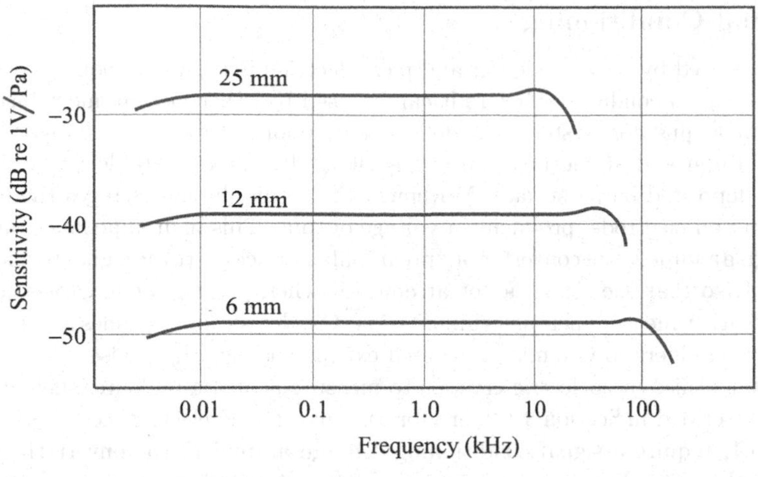

FIGURE 3.5 Free-field condenser microphone sensitivities for three different diaphragm diameters.

3.1.7 Microphone Accuracy

The accuracy of a microphone is dependent on the frequency of the sound being measured, the quality of the microphone and the accuracy of the calibrator. The accuracy and frequency response of modern digital instrumentation to which microphones are attached, is usually well in excess of the microphone accuracy and thus need not be considered here.

In a calibration laboratory, it is possible to calibrate microphones to an accuracy of 0.05 dB. However, outside of a calibration laboratory, a microphone (and the attached amplifier and meter) are usually calibrated at a single frequency by placing on the microphone, a sound source (pistonphone) that generates a known sound pressure level at the microphone diaphragm, as described in Section 3.4. This is always done before and after each measurement. To ensure that the microphones retain their performance over the entire frequency range for which they are rated, it is important that they are calibrated at other frequencies (usually in a calibration laboratory) at least once per year.

3.1.8 Infrasound Sensors

The measurement of infrasound has become increasingly of interest, as more and more wind farms come online and more and more people are suffering adverse health effects, even though it is has not been proven that such adverse effects are related to infrasound. Infrasound is also used in the study of volcanoes, tornadoes and earthquakes, as well as long-distance detection of incidents involving atomic bomb testing. Condenser microphones are available that measure down to 0.5 Hz, but in many cases their sensitivity is insufficient for many purposes and they are adversely affected by even small air movements. At the very low-frequency end, even primary and secondary windscreens are incapable of reducing wind noise to acceptable levels, even when the wind is blowing at a speed of only a few km/hr. Higher sensitivity microphones are limited to a 5 Hz lower frequency limit. However, sensors designed just to measure frequencies below 20 Hz in the presence of significant wind are available (Williams, 2015). These sensors can measure accurately over a frequency range from 0.02 Hz to 100 Hz, which may be extended via calibration. Wind noise is made insignificant by venting the sensor with 4 ports separated by 90° around the cylindrical transducer. These 4 ports are attached to 15 m long garden soaker hoses and act to cancel noise that is uncorrelated between the four inputs, including atmospheric turbulence.

3.1.9 Signal Conditioning

The signal generated by both condenser and piezoelectric microphones requires passing through an appropriate signal conditioner before it can be used by other measurement instrumentation, such as a data acquisition system. Condenser microphones produce a change in capacitance across a diaphragm and an electrode, which is charged by a very stable DC voltage or by an electret layer deposited on its surface. Movement of the diaphragm changes the capacitance of the diaphragm and electrode, producing a voltage output. This high impedance output must be passed through an impedance converter or "preamplifier" to convert the signal to a low impedance voltage output, so that the signal is not attenuated when passing along cables connecting the microphone to other measurement instrumentation. Condenser microphones are usually mounted directly on preamplifiers to minimise unwanted extraneous electrical noise.

Sensors that utilise piezoelectric crystals to measure acoustic pressure, such as piezoelectric microphones described in Section 3.1.2, or vibration using piezoelectric accelerometers described in Section 9.5.1, require a signal conditioning "charge amplifier" to convert the charge signal developed by the piezoelectric crystal into a voltage signal that can be recorded by a data acquisition system. During a measurement period, care has to be taken not to move or vibrate the cable that connects the piezoelectric sensor to the charge amplifier, as this movement can generate spurious signals called tribo-electric noise. Further, it is difficult to make good quality low-level measurements using long cables as they can be affected by electrical noise picked up by the cable. A work-around is to place the charge-amplifier close to the piezoelectric sensor, and then use long co-axial cables to transmit voltage signals to the data acquisition system.

3.1.9.1 IEPE Sensors

An enhancement to sensors that use piezoelectric crystals, is to place a signal conditioning amplifier inside the sensor package, which converts the charge signal to a low impedance voltage signal that can be transmitted easily along lengthy co-axial cables. The instrumentation, such as a data acquisition system, connected to the other end of the cable connected to the sensor, must provide a small, constant, electrical current, usually between 2 mA to 20 mA, to power the amplifier built into the sensor. This results in a positive DC bias voltage, typically between 8 V and 12 V, at the output from the sensor, which is removed by the data acquisition system (or external amplifier).

Piezoelectric sensors that include an inbuilt amplifier, which is powered by a constant current, are generally called Integrated Electronic Piezoelectric (IEPE), or Constant Current Line Drive (CCLD) sensors. Several manufacturers have trademarked names for this type of technology such as ICP® Integrated Circuit-Piezoelectric, a trademark of PCB Piezotronics, Inc., IsoTron® by Endevco, Inc., DeltaTron® by Bruel and Kjaer, and Piezotron® by Kistler Inc. One of the differences between manufacturers of IEPE sensors is the amount of electrical current necessary to power the sensor. Typically between 2 mA and 20 mA is required, but it is important to check the current required by the sensor, and that the data acquisition system (or external IEPE amplifier) is able to supply the required current.

3.1.9.2 TEDS Sensors

Many modern acoustic and vibration sensors, such as microphones, pressure transducers, accelerometers, and force transducers, have an in-built electronic chip (or EPROM), called transducer electronic data sheet (TEDS), which stores information about the sensor, such as the manufacturer, sensor type, serial number, calibration factors, calibration dates, and other relevant data. When TEDS equipped sensors are connected to data acquisition devices that can read the TEDS data, it helps expedite the setup of the instrumentation, and ensures that the calibration factors are correctly entered into the data acquisition system so that data are recorded with the correct units, and the serial number of the sensor is also recorded, which is beneficial for traceability. TEDS sensors can also be connected to instrumentation that does not read the TEDS data, and the sensor will function normally.

The TEDS standard was originally defined in IEEE Std 1451.4-2004 (2004), and has since been adopted broadly by International standards organisations in ISO/IEC/IEEE 21451-4 (2010); ISO/IEC/IEEE 21450 (2010). These standards have been adopted by many instrumentation manufacturers, thereby enabling the transfer of data between instrumentation manufactured by any of these equipment manufacturers.

There are two classes of TEDS devices:

- Class 1: uses the same wires for communicating the analogue signal from the sensor, and digital signals when communicating the TEDS information. The TEDS information is accessed by applying a negative bias voltage to the sensor, which activates the TEDS chip and enables digital communication.
- Class 2: uses separate wires for the analogue signals from the sensor, and digital signals when communicating the TEDS information.

In addition to the wired sensors described previously, wireless acoustic and vibration sensors are available, and the standard IEEE Std 1451.5 (2007) established the communication protocol for TEDS capable wireless sensors.

The TEDS chip has no adverse effect on the operation or performance of the sensor. Although the chip does add a small amount of mass to the sensor, this added mass can be compensated for in the calibration factor, or for very light-weight sensors, the TEDS chip can be removed from the sensor and embedded in the connector.

3.2 Sound Level Meters (SLMs)

The most common and convenient instrument for measuring the sound pressure level of a sound field is a sound level meter. Currently available sound level meters are all digital with dynamic ranges up to 100 dB. They consist of an analogue to digital converter connected to a digital signal processor board that converts the time varying signal to whatever is specified by the user through the GUI (graphical user interface). Sound level meters range from the very simple to the very complex. At the very simple end there may be no user adjustments other than an on-off switch and they usually have a digital readout showing a single number A-weighted sound pressure level that updates every second or so. Slightly better sound level meters have a sensitivity adjustment that allows calibration and they display the A-weighted sound pressure level, L_{Aeq}, energy-averaged over a specified time. Top of the range sound level meters can display narrow-band spectra, octave and 1/3-octave band spectra, as well as L_{eq}, L_{Aeq} and statistical quantities such as L_{10} (sound pressure level exceeded 10% of the time) and L_{90}. They also have A-, B-, C-, G-, and Z-weighted internal filter networks and an all-pass (linear) arrangement without weighting, as well as some adjustment that allows the user to select the measurement range so that the instrument can be optimised to measure low level or high level noises. These sound level meters are particularly useful for assessing noise for control purposes, especially if the sound pressure level is non-steady or fluctuates with time. In this latter case, it is essential that the sound level meter is able to measure L_{eq} and L_{Aeq} (see Sections 2.4.1 and 2.4.2).

The dynamic range (difference between the maximum and minimum measurable sound pressure levels) of a sound level meter as well as its electronic noise floor are two quantities used to distinguish better equipment. These quantities are also characteristic of the microphone chosen for the task. When high sound pressure levels are to be measured (such as above 140 dB), a low sensitivity microphone should be chosen and when low sound pressure levels are to be measured (such as below 30 dB), a high sensitivity microphone should be used so that the noise floor of the sound level meter or the microphone is not a problem.

Most sound level meters have a choice of two meter response times. One, labelled "fast", has a time constant of 125 ms and is designed to approximate the response of the ear. The other, labelled "slow", has a time constant of 1 second, and while not simulating the response of the ear, it was useful in older sound level meters for determining mean levels when the measured sound fluctuated continuously and violently during the course of measurement and the sound level meter did not have the ability to calculate L_{eq}. Some sound level meters have a third response referred to as "impulse" response (with a standard time constant of 35 ms) for the measurement of impulsive-type noises such as drop forges. Many sound level meters with an impulse response characteristic also have a capability to measure peak sound pressure levels. The peak sound pressure level is different to the fast, slow and impulse measurements in that it is the maximum sound pressure level sampled by the D/A converter in the sound level meter, prior to any filtering. On the other hand, the fast, slow and impulse measurements are averages of sound pressure squared over the specified averaging time of 0.125 s, 1.0 s and 35 ms respectively, with the result converted to decibels using Equation (1.87). For a sampling rate of 50 kHz, which is typical of a sound level meter, the samples would be separated by 20 µs. A-weighted peak levels should not be used due to the A-weighted filtering affecting the measurement of the true peak (ISO 1996-1, 2016). The maximum sound pressure level, L_{MAX}, is the maximum sound pressure level obtained when using either the fast or slow response time of the sound level meter. Thus, it is very different to the peak level. Similarly, the minimum sound pressure level, L_{MIN}, is the minimum sound pressure level obtained when using either the fast or slow response time of the sound level meter. Measurement standards and statutory regulations generally specify which meter response should be used for particular measurements. Note that if recorded time series data are analysed in software, the weighting filter (A, C, G or Z) must be applied prior to implementation of the fast, slow or impulse time constant.

3.3 Classes of Sound Level Meter

The various classes of sound level meter are discussed in IEC 61672-1 (2013). Earlier standards published by other organisations are out of date and do not reflect current sound level meter technology. The standard divides sound level meters into two classes: Class 1 or Class 2, in order of decreasing accuracy. These classes are described as:

Class 1: precision sound level meter, intended for laboratory use or for field use where accurate measurements are required;

Class 2: general-purpose sound level meter, intended for general field use and for recording sound pressure level data for later frequency analysis.

3.4 Sound Level Meter Calibration

The battery voltage should be checked regularly and the sound level meter and microphone calibration should be checked and adjusted according to the operating instructions. It is recommended that the instrument be overhauled and sent to a calibration facility for full calibration every two years.

Strictly, calibration is an end-to-end check of an instrument over its entire useful frequency range, and is generally a time-consuming procedure involving specialised equipment dedicated to that purpose. Reference checks at one or more selected frequencies, quite often loosely called calibration, may be used in the field and are strongly recommended to be undertaken both immediately before and immediately after every measurement set.

3.4.1 Electrical Calibration

An externally or internally generated electrical signal of known amplitude and frequency is injected into the microphone amplifier circuit. If the measured signal level is incorrect, slight compensation can be made to the amplification by adjustment of a preset control. Although this calibration can check the amplifier, and the weighting networks and filters, the microphone sensitivity is not checked. Thus, it is important to supplement this form of calibration with regular acoustic calibration.

3.4.2 Acoustic Calibration

Acoustic calibration (or reference checking) is recommended to be carried out in the field before and after each measurement set. Acoustic calibrators used during field measurements are intended to calibrate not just the microphone but the entire measurement system. During calibration, a tonal acoustic signal of known sound pressure level is applied to the microphone and the meter reading is compared with the reference level. Any error outside the quoted calibrator tolerance may be adjusted by the preset gain control (if available).

Most devices use 1000 Hz as the calibration frequency, as the same result is obtained with the A-weighting network switched in or out. Calibrators operating at other frequencies (for example, 250 Hz) require all weighting networks to be switched out during the calibration process. The noise-generating device is usually called a "pistonphone" and care should be taken to ensure that there is a good seal between the microphone housing and calibrator cavity (usually achieved with an "O-ring" on the calibrator part that slips over the microphone). Pistonphones produce sound pressure levels at nominally 94 dB or 124 dB (Brüel and Kjær) or 114 dB (GRAS, PCB, Rion).

During calibration, the sensitivity of the sound level meter is adjusted to make it read a value equal to the sound pressure level generated by the pistonphone calibrator. Calibration is accurate to ±0.5 dB for most instruments intended for use in the field (general purpose sound level meters) and ±0.2 dB for precision instruments for laboratory use. Such calibration is limited to one or a few discrete frequencies, so that the procedure can only serve as a spot check at these frequencies. Large errors may indicate poor cable connections or damage to either the sound level meter or the calibrator, and in the latter two cases, both should be returned to the manufacturer for checking.

3.4.3 Measurement Accuracy

The accuracy of measurements taken with a sound level meter depends on the calibration accuracy as well as the flatness of the frequency response of the sound level meter electronics and microphone. So it is important to examine the calibration chart supplied with the instrument to determine the accuracy over various frequency ranges. In practice, an accuracy of ± 0.5 dB is usually achievable with a Class 1 sound level meter and repeatability should be ± 0.1 dB provided that the noise source does not vary and measurements are taken indoors in a stable environment with no air movement or temperature variation. Thus, when checking to determine whether an item of equipment meets its specified noise level, which could be sound power level or sound pressure level at a specified location relative to the noise source, it is unrealistic to expect to be able to measure with an accuracy better than ± 0.5 dB with a Class 1 sound level meter and ± 1dB with a Class 2 sound level meter. In many cases of outdoor or on-site sound pressure level measurements, it is difficult to justify reporting results with any more precision than 1 dB, as sound fields are rarely constant in amplitude and sound pressure levels will generally be continually varying with time.

3.5 Noise Measurements Using Sound Level Meters

Accurate measurement of sound pressure levels requires careful attention to detail and the elimination of possible sources of measurement error. It should be noted that the following considerations are quite general and may apply whether or not measurements are made with a sound level meter. The main sources of error are listed below.

3.5.1 Microphone Mishandling

If the microphone or its stand is placed on a surface which is vibrating, then the diaphragm may vibrate and thus the microphone may produce spurious signals that are misinterpreted as sound pressure levels. Care must be taken to keep the microphone away from sources of vibration. Exposure to extreme sound pressure levels (see manufacturer's data), or shocks due to dropping the microphone, may cause a significant change in calibration.

3.5.2 Sound Level Meter Amplifier Mishandling

The amplifier of the sound level meter must be protected from magnetic fields, intense sound and vibration; otherwise, it may itself act as a microphone and give an erroneous output.

Some sound level meters have a choice of dynamic range for the instrument. In this case, it is important to make sure that the instrument range is selected to best cover the range in sound pressure levels to be measured.

3.5.3 Microphone and Sound Level Meter Response Characteristics

The manufacturer's data should be consulted when a noise measurement is to be made, to ensure that both the pressure amplitude (minimum as well as maximum) and frequency of the signal are within the specified ranges of the microphone and SLM.

3.5.4 Background Noise

Background noise is the factor most likely to affect sound pressure level readings. To determine whether or not background noise may interfere with measurements, the source to be measured should be turned off (if possible). If the background noise is less than 10 dB below the total level when the source is turned on, a correction must be made to each reading as described and summarised in Section 1.11.5. If the difference between the total sound pressure level with the sound source in operation and the background sound pressure level in any octave or 1/3-octave measurement band is less than 3 dB, then meaningful acoustic measurements in that band probably cannot be made.

3.5.5 Wind Noise

Even light winds blowing past a microphone produce spurious readings, particularly at frequencies below 200 Hz. These effects are best reduced by fitting a porous acoustic foam windscreen over the microphone. Air movement within some industrial plants can be enough to warrant taking such a precaution. However, most commercially available foam microphone wind screens are inadequate when wind speeds exceed 20 km/hr (5.6 m/s). If it is necessary to take noise measurements when the wind is stronger than this and one can be sure that noise due to wind rustling in nearby vegetation or other objects is not going to be a problem, then an additional screen is needed. This may take the form of a hollow secondary wind screen made of a porous material skin mounted on a wire frame, covered with fur-like material and about 400 to 500 mm in diameter (see IEC 61400-11 Ed.3.0 (2018) for material specifications). If desired, the secondary wind screen may be replaced or enhanced with a cubic enclosure of dimensions approximately 1 metre × 1 metre × 1 metre, made of shade cloth and open at the top with the microphone in the centre of the cube. An alternative to the shade cloth cube is a 2 m diameter dome tent with the impervious nylon replaced by porous acoustic material. Infrasound measurements, using low-frequency microphones (and not specialised infrasound sensors), are difficult without all of the above-mentioned wind screens, even in light winds. In many cases, infrasound measurements must be made with micro-porous hoses attached to the four orthogonal ports of an infrasound sensor. This arrangement takes advantage of the differing coherence lengths of turbulent pressure fluctuations and sound, so that sound pressure fluctuations are reinforced as they travel along inside the tube while turbulent pressure fluctuations are not. A detailed analysis of the mechanisms and control of wind noise on microphones is provided in Lyons et al. (2021).

Whenever foam windscreens are used, note should be taken of the microphone manufacturer's advice regarding the slight effect on the microphone sensitivity at high frequencies. A useful study of the effect of wind and windscreens on environmental noise measurements was reported by Hessler et al. (2008). Some instrument manufacturers provide instruments that compensate their output to account for the attenuation of their windscreens (which is frequency dependent and greater at high frequencies than at low frequencies).

3.5.6 Temperature

The temperature-sensitive elements in the measurement system are the batteries and the microphone. As the temperature increases, the life of the batteries decreases. Some microphones are particularly insensitive to temperature. The best grade of condenser microphone typically retains its calibration within 0.5 dB over a range from −40°C to 150°C. Piezoelectric microphones are also quite stable over a range of −40°C to 60°C. They may be damaged, however, at higher temperatures.

3.5.7 Humidity and Dust

Condenser microphones in particular, are sensitive to both humidity and dust on the diaphragm, both of which cause a high level of self-noise in the instrument. When not in use, the equipment should be stored in a dry place, and the microphone protected with a desiccator cap. A better plan might be to remove and store the microphone in a tight enclosure containing a desiccant. Alternatively, condenser microphones that use an electret, referred to commercially as pre-polarised, do not suffer these disadvantages and no special precautions are necessary for their care and use, except that they should not be subjected to rainfall or immersed in water, or have their diaphragms scratched or contaminated with dirt.

As mentioned in Section 3.1.1, condenser microphones have a small hole in the side so that the pressure acting on the back of the microphone diaphragm can be equalised with the pressure (usually atmospheric) acting on the front of the microphone diaphragm. The diameter of this hole controls the lower frequency limit of the microphone. The smaller the hole diameter, the lower will be the lower frequency limit, which is why the hole diameter is usually very small. If care is not taken to ensure the microphone stays clean when used in the field, it is possible that this hole can become blocked with dirt. Any change in atmospheric pressure after blockage can have a severe impact on the microphone calibration and frequency response, as the microphone diaphragm can be pushed against one of its limits, causing the microphone to appear to be faulty. If the microphone is then air freighted back to the manufacturer, it is quite likely that the reduced atmospheric pressure in the aircraft will result in the blocked hole being cleared as a result of a large pressure difference between the back of the diaphragm and the front allowing air in the cavity behind the diaphragm to escape through the hole and thus unblock it. So the microphone is likely to be returned to the user with the embarrassing message that there is nothing wrong with it. This scenario has happened!

3.5.8 Reflections from Nearby Surfaces

Nearby objects, which may affect the sound radiation of a source, must often remain undisturbed for the purposes of the measurement; for example, while assessing operator sound pressure levels. At other times their presence may be highly undesirable, such as when assessing the true acoustic power output of a particular machine. Such possible effects may be expected when an object's dimensions are comparable with or larger than the wavelength of the sound being measured.

The possible effect on the sound field may be judged by reference to Table 4.1 and the accompanying discussion of Section 4.9.1 of Chapter 4. When it is desirable to avoid the effect on sound measurements of nearby objects which cannot be removed, the objects should be covered with sound-absorptive material, the thickness of which should increase as the frequency of interest becomes lower. For example, a minimum thickness of 100 mm would be necessary to measure down to 200 Hz.

3.5.9 Time-Varying Sound

Most industrial noise is not steady (although it may be continuous), but rather fluctuates significantly in level over a short period of time. Traffic noise, which is the predominant source of environmental noise in urban areas, is similarly characterised.

Time-varying sound is usually described statistically, either in terms of L_{eq}, the equivalent continuous sound pressure level for a given time period, or in terms of L_x, the sound pressure level that is exceeded for x% of the time. Both L_{eq} and L_x are generally expressed in dBA units. The two most commonly used exceedance levels are L_{10} and L_{90}. The quantity L_{10}, the level exceeded for 10% of the time, is a measure of the higher level (and hence more intrusive) components of noise, whereas L_{90} is a measure of the background or residual level (also referred to as ambient level).

For an older-style sound level meter with an analogue meter display and without the capability to display statistical quantities, both L_{10} and L_{90} may be estimated, with the meter set to "fast" response. The quantity L_{10} corresponds approximately to the average of the maximum pointer deflections, and L_{90} to the average of the minima. Both quantities may be weighted (see Section 2.3) or unweighted, with the use of A-weighting the most common. In cases where the fluctuating component of a sound pressure level is relatively small, that is $L_{10} - L_{90}$ does not exceed 10 dB, L_{eq} corresponds approximately to the average of the maximum pointer deflections with the sound level meter set to a "slow" response.

3.5.10 Sound Pressure Level Measurement

The type of noise measurement made will depend on the purpose of the measurement. For example, is it desired to establish dBA levels to determine whether allowable criteria have been exceeded? Or are octave band levels needed for comparison with model predictions or for determining machine noise reduction requirements? Or are narrow band measurements needed for the purpose of identifying noise sources?

The type of noise measurement made will also depend on the type of noise source; steady state and impulsive noises require different types of measurement. These measurements and the appropriate instrumentation are summarised in Table 3.1 for various noise source types.

TABLE 3.1 Noise types and their measurement Noise characteristics are classified according to the way they vary with time. Constant noise remains within ±5 dB of the mean level for a long time. Constant noise which starts and stops is called intermittent. Fluctuating noise varies significantly but has a constant long term average ($L_{Aeq,T}$). Impulse noise lasts for less than one second

	Characteristics	Type of source	Type of measurement	Type of instrument	Remarks
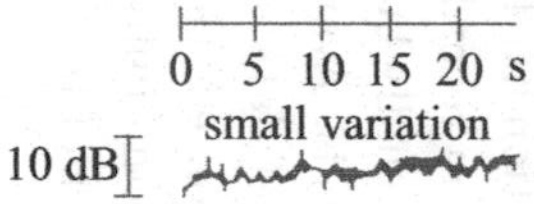	Constant, continuous noise.	Pumps, electric motors, gearboxes, conveyors.	L_{Aeq}.	Sound level meter.	Octave or 1/3-octave analysis if noise is excessive.
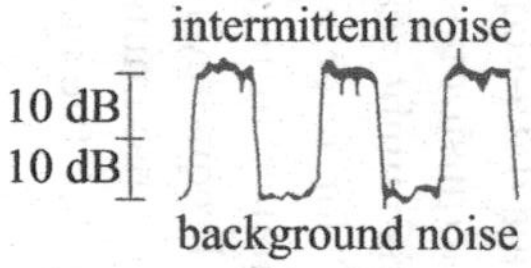	Constant but intermittent noise.	Air compressor, automatic machinery during a work cycle.	L_{Aeq} and exposure time.	Sound level meter. Integrating sound level meter.	Octave or 1/3-octave analysis if noise is excessive.
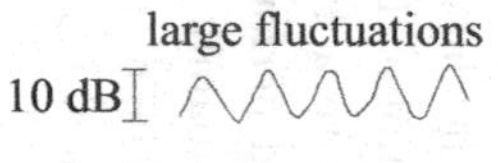	Periodically fluctuating noise.	Mass production, surface grinding.	L_{Aeq} and noise dose.	Sound level meter. Integrating sound level meter.	Octave or 1/3-octave analysis if noise is excessive.
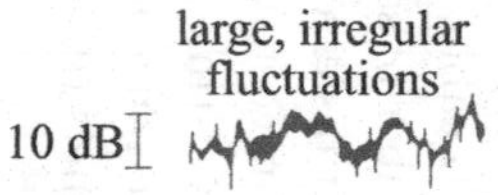	Fluctuating non-periodic noise.	Manual work, grinding, welding, component assembly.	L_{Aeq} or noise dose, statistical analysis.	Noise dosimeter. Integrating sound level meter.	Long term measurement usually required.
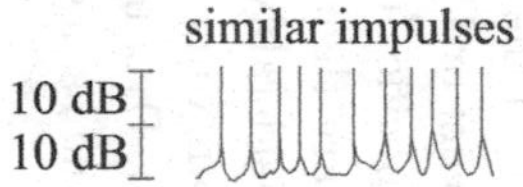	Repeated impulses.	Automatic press, pneumatic drill, riveting.	L_{Aeq} or noise dose and "Impulse" level L_{imp}. Check "peak" value.	Impulse sound level meter or SLM with "peak" hold.	Difficult to assess. More harmful to hearing than it sounds.
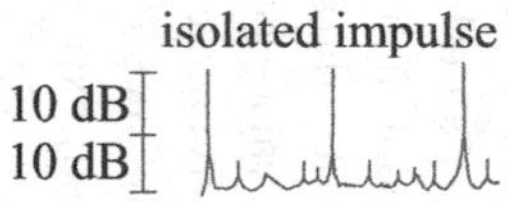	Single impulse.	Hammer blow, material handling, punch press.	L_{Aeq} and L_{peak}	Impulse sound level meter or SLM with "peak" hold.	Difficult to assess. Very harmful to hearing.

3.6 Data Loggers

These instruments (sometimes referred to as statistical analysers) measure the distribution of fluctuating noise with time for the purpose of community noise assessment and hearing damage risk assessment. In addition to providing energy-averaged sound pressure levels (L_{eq}), they readily provide statistical quantities such as L_{10}, L_{50} and L_{90}, which are, respectively, the levels exceeded 10%, 50% and 90% of the time and which are usually A-weighted. Provided that the data loggers are well protected or hidden to avoid theft or vandalism, they can often be left in the field unattended, gathering data for weeks or months at a time.

3.7 Personal Sound Exposure Meters

Personal sound exposure meters have replaced the majority of equipment used for the assessment of noise hazard to hearing in industry, as they provide a simple, unified and accurate measure of noise exposure. They are designed to be carried by individuals who are exposed to varying levels of high-intensity noise; thus they are small in size and simple to operate, with a microphone that can clip on to the lapel of a coat or shirt. Required specifications are outlined in IEC 61252 (2017) and ANSI/ASA S1.25 (2020).

Personal sound exposure meters (sometimes referred to as dosimeters) measure the cumulative noise dose in terms of A-weighted sound energy over a period of a normal working day for the purpose of assessing hearing damage risk. The instruments perform the necessary integration to determine L_{Aeq,T_e} over T_e hours according to Equation (2.36) (see Sections 2.4.1 and 2.4.2). The L_{Aeq,T_e} measurement can be standardised to the eight-hour equivalent, $L_{Aeq,8h}$, using Equation (2.38) and in many cases, this is also done by the instrument. In addition, most instruments provide a measure of noise exposure level, $L_{EX,\mathrm{8h}}$, (see Section 2.4.3) and sound exposure level, L_{AE} or SEL (see Section 2.4.5), as well as a % measure of the allowable noise dose. To obtain the latter quantity, the user can enter their choice of trading rule (or exchange rate), base level (or criterion level) and threshold level below which noise exposure is not recorded (see Section 2.6.7 for an explanation of trading rule and base level). The technical specifications of dosimeters are the same as for Class 2 sound level meters and are provided in the standard, IEC 61672-1 (2013). Personal sound exposure meters that satisfy either IEC 61252 (2017) or ANSI/ASA S1.25 (2020) are suitable for use in any country and all should have settings that allow them to comply with European (ISO 9612, 2009) and American (OSHA, 1995) regulations. The upper limit for some personal sound exposure meters is 130 dBA but 140 dBA is more desirable, especially when the user is exposed to impact noise.

Many (but not all) personal sound exposure meters are capable of providing the user with information in the form of L_{Aeq} or any statistical parameter (such as L_{10} or L_{90}) as well as the peak level over any predetermined increment of time (such as 1 second) and thus provide a time history of the noise exposure. The accuracy of this instrumentation is usually $\pm$ 0.5 dBA.

3.8 Data Acquisition and Recording

Most recordings made at the time of writing are via a data-acquisition hardware module attached to a PC via an ethernet or USB cable. The data-acquisition system records data in blocks, each of which contain a specific number of samples. This number can usually be specified by the user. The recorded data are then transferred to a PC for later analysis. Available data-acquisition hardware varies considerably in cost and performance in terms of data storage size, frequency response, dynamic range and real-time capability. Frequency response is the maximum frequency that can be recorded. Dynamic range is the ratio of maximum to minimum signal level that can be recorded, expressed as decibels (dB) and related to the number of bits in the A/D converter.

Good systems usually have at least 24 bits and systems with up to 32 bits are available. Real-time capability is related to the speed of the data-acquisition system and is the highest frequency that can be continuously recorded with no data loss between record blocks. It is also important to make sure that data-acquisition systems include low pass anti-aliasing filtering to prevent aliasing of high-frequency data into the relatively low audio range (see Chapter 12).

Some data-acquisition systems have an external clock input which allows the analyser to sample at an externally determined rate. This facility is useful for monitoring rotating machinery noise and vibration where speed variations exist over short periods of time. The external clock signal may be derived from a marker on the rotating shaft (using an optical tachometer). The result of using this external clock is that the rotating machinery harmonics always appear at the same frequency and are not "washed out" in the averaging process. Aspects of frequency analysis are discussed in more detail in Chapter 12.

To make full use of the dynamic range of any recording device, it is necessary to record with as high an input signal level as possible without overloading the data-acquisition system. Most instruments have a "gain" control that can be adjusted to obtain he required signal level. For impact or impulse type noise (see Section 2.7.2) an impulse sound level meter should be used to check for the maximum signal level and then the input level to the data-acquisition system should be adjusted accordingly.

As a means of minimising storage requirements, most data-acquisition systems record data in binary form and proprietary software (usually supplied with the data-acquisition system) is required to convert the data to wav files, which can then be analysed using filtering or spectrum analysis software.

When recording a signal for subsequent analysis, it is recommended that calibration tones be introduced at the microphone, while keeping careful track of all gains and subsequent gain changes in the recording system. This will allow determination of recorded levels with minimum likelihood of significant error. Alternatively, the average signal level could be manually logged at the time of recording, allowing the playback levels to be adjusted until their average is equal to the average level that was manually logged. However, this method is less precise than recording a calibration tone on the recording device. It is good practice to record at the beginning of a recording the following information:

- time and date;
- data-acquisition system sample rate;
- number of samples in each data record;
- time to acquire each data record;
- microphone position;
- unweighted sound pressure level measured with a sound level meter;
- calibration level;
- attenuator or gain settings;
- a description of the test configuration; and
- weather conditions.

A redundancy of recorded information can be very valuable when something goes wrong unexpectedly during a recording session that cannot conveniently be repeated. Care should also be taken when adjusting amplifier gains between calibration and recording of the signal of interest. Some instruments have dynamic range changes in 10 dB increments. However, it is sometimes found that the gain is not exactly 10 dB, which results in a calibration error. This error can be measured and thus compensated for by using a constant pink or white noise input and averaging for 30 seconds with different attenuator settings.

3.9 Spectrum Analysers

Spectrum analysers, which consist of a dedicated unit that does data acquisition as well as signal processing, are no longer used due to the availability of powerful PCs that can process data acquired with a data-acquisition unit as described in Section 3.8.

When it is desirable to have a detailed description of sound pressure levels as a function of frequency, spectral analysis software is used. This sort of analysis is especially useful for the analysis of signals containing pure tone components. Often, the frequency of a pure tone gives an indication of its origin, and such information indicates areas for the initiation of noise control. Spectral analysis software first divides the time series data into a number of segments called records (typically 1024 or 2048 samples in length, depending on the frequency resolution required and the sample rate). Each record is then processed using the Fast Fourier Transform (FFT), implemented as a Digital Fourier Transform (DFT) algorithm, to convert the time domain data into the frequency domain. The data can be down-sampled by a factor of m if the maximum frequency of interest is a factor of m less than provided by the data-acquisition system. Software is also available for obtaining octave and 1/3-octave band levels from a time-series data file. In many cases, sound level meters can also provide octave and 1/3-octave band levels, as well as overall A-weighted levels.

The frequency resolution provided by DFT analysis is a function of the number of samples in the record being processed and the sampling rate. Table 12.2 shows the relationships between the various quantities used in spectral analysis.

Most spectrum analysis software packages have very powerful options for presentation of the frequency (or time) domain data, including a “zoom” capability where a small part of the spectrum may be analysed in detail. A “waterfall” display capability may also be included, which allows the changes in the frequency spectrum to be observed as a function of time. Various averaging options are also offered for power spectra (pressure squared spectra), such that a number of frequency spectra may be averaged to give a more representative result. Multi-channel analysers (with two-channel being the most common) offer the further capability of investigation of the interrelationships between different input signals.

Spectrum analysis apps are also available for smart phones. However, the accuracy of the result is dependent on the quality of the microphone that is embedded in the smart phone. The anti-aliasing filter requirement may also not be satisfied by a smart phone app.

3.9.1 Spectrograms

The use of spectrograms as a means of analysis is becoming more common with the advent of more powerful personal computers allowing analysis, recognition and classification of patterns appearing in spectrograms. Spectrograms are two dimensional plots of sound pressure level (or acceleration), with frequency on the x-axis and time on the y-axis. Sound pressure level can be represented on the graph by a grey scale or a colour scale. For a grey scale, white represents the lowest sound pressure level, while black represents the highest sound pressure level. For the colour scale, the colours of the rainbow are used to represent sound pressure level, with violet representing the lowest level and red representing the highest level. Spectrograms allow the tracking of the variation of tonal noise over time as well as allowing the time variation of the spectral energy distribution to be observed on just one graph.

3.10 Sound Intensity Meters

The direct measurement of the sound intensity of propagating sound waves is possible using two carefully matched microphones and band limited filters. The filter bandwidth is generally not in excess of one octave.

The measurement of sound intensity requires the simultaneous determination of sound pressure and particle velocity. The determination of sound pressure is straightforward, but the determination of particle velocity presents some difficulties; thus, there are two principal techniques for its determination and consequently the measurement of sound intensity. Either the acoustic pressure and particle velocity are measured directly ($p-u$ method) or the acoustic pressure is measured simultaneously at two closely spaced points. In this latter case, the mean sound pressure is calculated by arithmetically averaging the two sound pressure measurements and the particle velocity is calculated by taking the difference between the two measurements and dividing the result by the microphone spacing and some constants ($p-p$ method). In either case, the acoustic pressure and particle velocity are multiplied together to produce the instantaneous intensity and the time-averaged intensity (see Section 1.8).

Errors inherent in intensity measurements and limitations of instrumentation are discussed in Sections 3.10.1.1, 3.10.2.1 and by Fahy (1995).

3.10.1 Sound Intensity by the $p-u$ Method

For the $p-u$ method, the acoustic particle velocity is measured directly. The most common method for measuring particle velocity directly in air is by employing a "Microflown" particle velocity sensor (de Bree et al., 1996; Druyvesteyn and de Bree, 2000) which uses a measure of the temperature difference between two resistive sensors spaced 40 µm apart to estimate the acoustic particle velocity. The temperature difference between the two sensors is caused by the transfer of heat from one sensor to the other by convection as a result of the acoustic particle motion. This in turn leads to a variation in resistance of the sensor, which can be detected electronically. To get a temperature difference which is sufficiently high to be detected, the sensors are heated with a DC current to a temperature of about 500 K. The sensor consists of two cantilevers of silicon nitride (dimensions 800 × 40 × 1 µm) with an electrically conducting platinum pattern, used as both the sensor and heater, placed on them. The base of the sensor is silicon, which allows it to be manufactured using the same wafer technology as used to make integrated circuit chips. Up to 1000 sensors can be manufactured in a single wafer. The sensitivity of the sensors (and signal-to-noise ratio) can be increased by packaging of the sensor in such a way that the packaging increases the particle velocity near the sensor.

The spacing of 40 µm between the two resistive sensors making up the particle velocity sensor is an optimal compromise. Smaller spacing reduces the heat loss to the surroundings so that more of the heat from one sensor is convected to the other, making the device more sensitive. On the other hand, as the sensors come closer together, conductive heat flow between the two of them becomes an important source of heat loss and thus measurement error.

It has been shown (Jacobsen and Liu, 2005) that measurement of sound intensity using the "Microflown" sensor to determine the particle velocity directly is much more accurate than the indirect method involving the measurement of the pressure difference between two closely spaced microphones, which is described in Section 3.10.2. Jacobsen and Liu (2005) also showed that the "Microflown" sensor was more accurate than microphones for acoustic holography which involves using the measurement of acoustic pressure OR particle velocity in a plane located in front of a noise source to predict the acoustic pressure AND particle velocity in another plane.

Particle velocity sensors such as the "Microflown" are more useful than microphones for measuring particle velocity as they are directional, which makes them less susceptible to background noise, and they are more sensitive. When used close to a reasonably stiff structure that is radiating noise, their superiority over microphones is even more apparent. This is because the pressure associated with any background noise is approximately doubled at the surface, whereas the particle velocity associated with the background noise will be close to zero where it is reflected. It is possible to purchase a small sound intensity probe (12 mm diameter) that includes a microphone and a "Microflown" and this can be used to directly measure sound intensity in one direction. Microflown probes containing one pressure sensor and three particle velocity sensors are also available for measuring 3-D sound intensity, rather than just the component in a single direction.

The active and reactive intensities for a harmonic sound field are given by Equations (1.76) and (1.77) in Chapter 1, respectively.

3.10.1.1 Accuracy of the $p-u$ Method

With sound intensity measurements, there are systematic errors, random errors and calibration uncertainty. The calibration uncertainty for a typical $p-u$ probe is approximately ± 0.5 dB. However, the systematic error arises from the inaccuracy in the model that compensates for the phase mismatch between the velocity sensor and the pressure sensor in the probe. This model is derived by exposing the $p-u$ probe to a sound field where the relative amplitudes and phases between the acoustic particle velocity and pressure are well-known (Jacobsen and de Bree, 2005).

Although the phase error is not as important for a $p-u$ probe as it is for a $p-p$ probe (as the $p-p$ probe also relies on the phase between the two microphones to estimate the particle velocity), it is still the major cause of uncertainty in the intensity measurement, especially for reactive fields where the reactive intensity amplitude is larger or comparable with the active intensity. Reactive intensity fields occur close to noise-radiating structures and in reverberant acoustic fields (such as found in rooms with reflective surfaces). For such fields, the errors in sound intensity measurement using a $p-u$ probe increase as the phase between the acoustic pressure and particle velocity increases. The Microflown handbook gives the systematic error for intensity measurements using a $p-u$ probe at a particular frequency as:

$$\text{Error (dB)} = 10\log_{10}\left[1+\left|\frac{\hat{I}-I}{I}\right|\right] = 10\log_{10}(1+\beta_e\tan\beta_f) \tag{3.22}$$

where I is the actual active intensity at the $p-u$ probe location, $\hat{I}$ is the intensity measured by the probe, β_e (in radians) is the phase calibration error, which for a typical $p-u$ probe is of the order of 2.5° or 0.044 radians and β_f is the phase between the acoustic pressure and particle velocity in the sound field. It can be seen that this systematic error will exceed 1 dB if the phase between the acoustic pressure and particle velocity exceeds 80°. In practice, if the sound field reactivity is too high, the probe can be moved further from the noise source and the reactivity due to the near field will decrease. Of course, if the source is located in a small room with reflective surfaces, the reactivity of the reverberant field will increase as the distance from the source increases, so the optimum probe location would be where the combined reactivity of the source near field and the room reverberant field is a minimum.

As sound reflected from a single surface and sound from sources other than the one being measured do not usually increase the reactivity of the sound field, they will not influence the accuracy of intensity measurements made with the $p-u$ probe, in contrast to their significant influence on intensity measurements made with the $p-p$ probe (see Section 3.10.2.1). Note that very reactive fields are unlikely to occur except at low frequencies, whereas the phase error in the $p-p$ probe affects the accuracy of the intensity measurement over the entire audio frequency range. Thus, the $p-u$ probe usually gives a more accurate value of sound intensity at high frequencies and in fact is useful up to 20 kHz whereas the limit of commercially available $p-p$ probes is 10 kHz.

Estimates of the magnitude of random errors are very difficult to make and no estimates have been reported in the literature.

3.10.2 Sound Intensity by the $p-p$ Method

For the $p-p$ method, the determination of acoustic pressure and acoustic particle velocity are both made using a pair of high quality condenser microphones. Requirements that should be met by instruments using this method of sound intensity measurement are provided in IEC 61043 (2017). The microphones are generally mounted side by side or facing one another and separated by a fixed distance (6 mm to 50 mm), depending on the frequency range to be investigated. A signal proportional to the particle velocity at a point midway between the two microphones and along the line joining their acoustic centres is obtained using the finite difference in measured pressures to approximate the pressure gradient while the mean is taken as the pressure at the midpoint.

The useful frequency range of $p - p$ intensity meters is largely determined by the selected spacing between the microphones used for the measurement of pressure gradient, from which the particle velocity may be determined by integration. The spacing must be sufficiently small to be much less than a wavelength at the upper frequency bound so that the measured pressure difference approximates the pressure gradient. On the other hand, the spacing must be sufficiently large for the phase difference in the measured pressures to be determined at the lower frequency bound with sufficient precision to determine the pressure gradient with sufficient accuracy.

Clearly, the microphone spacing must be a compromise and a range of spacings is usually offered to the user as illustrated in Figure 3.6. The assumed positive sense of the determined intensity is in the direction of the centre line from microphone 1 to microphone 2. For convenience, where appropriate in the following discussion, the positive direction of intensity will be indicated by unit vector $\boldsymbol{n}$, and this is in the direction from microphone 1 to microphone 2.

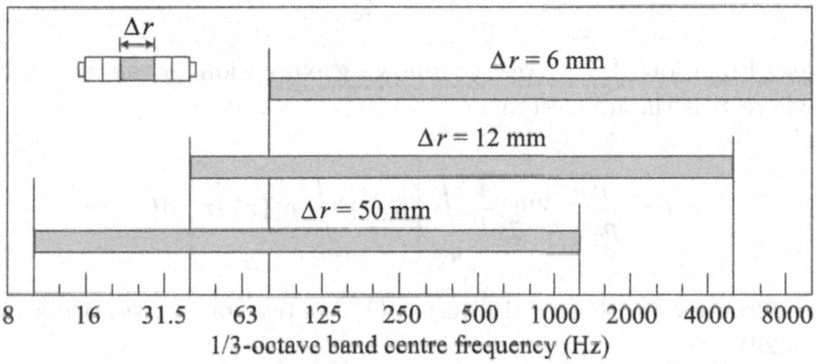

FIGURE 3.6 Microphone spacing requirements for 1 dB accuracy for sound intensity measurements in a free field for a total phase mismatch in probe and analyser of 0.1 degrees. If the phase mismatch is 0.05 degrees, the lower frequency limit can be shifted one octave lower in all three cases.

Taking the gradient of Equation (1.11) in the direction of unit vector $\boldsymbol{n}$, and using Equation (1.12), gives the equation of motion relating the pressure gradient to the particle acceleration. That is:

$$\boldsymbol{n}\frac{\partial p}{\partial n} = -\rho\frac{\partial \boldsymbol{u_n}}{\partial t} \tag{3.23}$$

where $\boldsymbol{u_n}$ is the component in direction, $\boldsymbol{n}$, of particle velocity, $\boldsymbol{u}$; p and $\boldsymbol{u}$ are both functions of the vector location, $\boldsymbol{r}$, and time, t; and ρ is the density of the acoustic medium. The normal component of particle velocity, u_n, is obtained by integration of Equation (3.23), where the assumption is implicit that the particle velocity is zero at time $t = -\infty$:

$$\boldsymbol{u_n}(t) = -\frac{\boldsymbol{n}}{\rho}\int\limits_{-\infty}^{t}\frac{\partial p(\tau)}{\partial n}\mathrm{d}\tau \tag{3.24}$$

The integrand of Equation (3.24) is approximated using the finite difference between the pressure signals p_1 and p_2 from microphones 1 and 2, respectively, and Δ is the separation distance between them:

$$\boldsymbol{u_n}(t) = -\frac{\boldsymbol{n}}{\rho\Delta}\int\limits_{-\infty}^{t}[p_2(\tau) - p_1(\tau)]\,\mathrm{d}\tau \tag{3.25}$$

The pressure midway between the two microphones is approximated as the mean:

$$p(t) = \frac{1}{2}[p_1(t) + p_2(t)] \tag{3.26}$$

Thus, the instantaneous intensity in direction, $\boldsymbol{n}$, at time, t, is approximated as:

$$\boldsymbol{I_i}(t) = \frac{\boldsymbol{n}}{2\rho\Delta}[p_1(t) + p_2(t)]\int\limits_{-\infty}^{t}[p_1(\tau) - p_2(\tau)]\,\mathrm{d}\tau \tag{3.27}$$

For stationary sound fields, the instantaneous intensity can be obtained from the product of the signal from one microphone and the integrated signal from a second microphone in close proximity to the first (Fahy, 1995):

$$\boldsymbol{I_i}(t) = \frac{\boldsymbol{n}}{\rho\Delta}p_2(t)\int\limits_{-\infty}^{t}p_1(\tau)\mathrm{d}\tau \tag{3.28}$$

The time-average of Equation (3.28) gives the following expression for the time-averaged intensity in direction $\boldsymbol{n}$ (where $\boldsymbol{n}$ is the unit vector):

$$\boldsymbol{I} = \frac{\boldsymbol{n}}{\rho\Delta}\lim_{T\to\infty}\frac{1}{T}\int\limits_{0}^{T}\left[p_2(t)\int\limits_{-\infty}^{t}p_1(\tau)\mathrm{d}\tau\right]\mathrm{d}t \tag{3.29}$$

Commercial instruments with digital filtering (1/3-octave or octave) are available to implement Equation (3.29).

As an example, consider two harmonic pressure signals from two closely spaced microphones:

$$p_i(t) = |\hat{p}_i|\mathrm{e}^{j(\omega t+\beta_i)} = |\hat{p}_i|\cos(\omega t + \beta_i) + \mathrm{j}|\hat{p}_i|\sin(\omega t + \beta_i); \qquad i = 1,2 \tag{3.30}$$

Substitution of the real components of these quantities in Equation (3.27) gives the instantaneous intensity as:

$$\begin{aligned}\boldsymbol{I_i}(t) &= \frac{\boldsymbol{n}}{2\rho\Delta}\Big[|\hat{p}_1|\cos(\omega t + \beta_1) + |\hat{p}_2|\cos(\omega t + \beta_2)\Big] \times \left[\frac{|\hat{p}_1|}{\omega}\sin(\omega t + \beta_1) - \frac{|\hat{p}_2|}{\omega}\sin(\omega t + \beta_2)\right]\\ &= \frac{\boldsymbol{n}}{4\rho\omega\Delta}\Big[|\hat{p}_1|^2\sin(2\omega t + 2\beta_1) - |\hat{p}_2|^2\sin(2\omega t + 2\beta_2) + 2|\hat{p}_1||\hat{p}_2|\sin(\beta_1 - \beta_2)\Big]\end{aligned} \tag{3.31}$$

where the first two terms in brackets represent the time varying reactive intensity and the last term represents the active intensity. Taking the time average of Equation (3.31) results in sine terms containing ωt becoming zero and what is left is the active intensity as follows:

$$\boldsymbol{I} = \frac{\boldsymbol{n}|\hat{p}_1||\hat{p}_2|}{2\rho\omega\Delta}\sin(\beta_1 - \beta_2) \tag{3.32}$$

If the argument of the sine is a small quantity, then Equation (3.32) becomes approximately:

$$\boldsymbol{I} = \frac{\boldsymbol{n}|\hat{p}_1||\hat{p}_2|}{2\rho\omega\Delta}(\beta_1 - \beta_2); \qquad (\beta_1 - \beta_2) \ll 1 \tag{3.33}$$

This equation also follows directly from Equation (1.73), where the finite difference approximation is used to replace $\partial\beta_p/\partial r$ with $(\beta_1 - \beta_2)/\Delta$ and $|\hat{p}|^2$ is approximated by $|\hat{p}_1||\hat{p}_2|$.

The first two terms of the right-hand side of Equation (3.31) describe the reactive part of the intensity. If the phase angles β_1 and β_2 are not greatly different, for example, the sound pressures $|\hat{p}_1|$ and $|\hat{p}_2|$ are measured at points that are closely spaced compared to a wavelength, the magnitude of the reactive component of the intensity is approximately:

$$\boldsymbol{I_r} = \frac{1}{4\omega\rho\Delta}\left[|\hat{p}_1|^2 - |\hat{p}_2|^2\right] \tag{3.34}$$

Equation (3.34) also follows directly from the last (reactive) term in curly brackets of Equation (1.72), where $|\hat{p}|$ is replaced by $(|\hat{p}_1| + |\hat{p}_2|)/2$ and $\partial p/\partial r$ is replaced with the finite difference approximation, $(|\hat{p}_1| - |\hat{p}_2|)/\Delta$.

Measurement of the sound intensity in a harmonic stationary sound field can be made with only one microphone, a phase measurement system and a stable reference signal if the microphone can be located sequentially at two suitably spaced points. The single reference signal can be the driving signal for the acoustic or vibration source (such as the input to the amplifier driving a loudspeaker) that is generating the sound field for which the intensity is to be determined. Indeed, the 3-D sound intensity vector field can be measured by automatically traversing, stepwise, a single microphone over an area of interest, measuring the acoustic pressure at each step and also the phase between the microphone signal and the reference signal.

Use of a single microphone for sound intensity measurements eliminates problems associated with microphone, amplifier and integrator phase mismatch as well as enormously reducing diffraction problems encountered during the measurements. Although useful in the laboratory, this technique is difficult to implement in the field as it is often difficult to obtain the stable reference signal of exactly the same frequency as the sound or vibration field. In general, the determination of the total instantaneous intensity vector, $\boldsymbol{I_i}(t)$, requires the simultaneous determination of three orthogonal components of particle velocity. Current instrumentation is available to do this using a single $p-u$ probe (one microphone and one particle velocity sensor), a 3-D $p-u$ probe (one microphone and three particle velocity sensors), a single $p-p$ probe (two microphones) or a 3-D $p-p$ probe (six microphones).

Note that with a $p-p$ probe, the finite difference approximation used to obtain the acoustic particle velocity has problems at both low and high frequencies, which means that at least three different microphone spacings are needed to cover the audio frequency range. At low frequencies, the instantaneous pressure signals at the two microphones are very close in amplitude and a point is reached where the precision in the microphone phase matching is insufficient to accurately resolve the difference. At high frequencies the assumption that the pressure varies linearly between the two microphones is no longer valid. The $p-u$ probe suffers from neither of these problems, as it does not use an approximation to the sound pressure gradient to determine the acoustic particle velocity.

3.10.2.1 Accuracy of the $p-p$ Method

The accuracy of the $p-p$ method is affected by both systematic and random errors. The systematic error stems from the amplitude sensitivity difference and phase mismatch between the microphones and is a result of the approximations inherent in the finite difference estimation of particle velocity from pressure measurements at two closely spaced microphones.

The error due to microphone phase mismatch and the associated finite difference approximation can be expressed in terms of the difference, δ_{pI}, between sound pressure and intensity levels measured in the sound field being evaluated and the difference, δ_{pIO}, between sound pressure and intensity levels measured by the instrumentation in a specially controlled uniform pressure field in which the phase at each of the microphone locations is the same and for which the intensity is zero.

The quantity, δ_{pIO}, known as the "Residual Pressure–Intensity Index", is a measure of the accuracy of the phase matching between the two microphones making up the sound intensity probe and the higher its value, the higher is the quality of the instrumentation (that is, the better is the microphone phase matching).

The quantity, δ_{pI}, known as the "Pressure–Intensity Index' for a particular sound field, is defined as the difference in dB between the measured intensity and the intensity that would characterise a plane wave having the measured sound pressure level, corrected by the term, $10\log_{10}(\rho c/400)$. Thus:

$$\delta_{pI} = L_p - L_I = 10\log_{10}\left(\frac{\rho c}{400}\right) - 10\log_{10}\left(\frac{\beta_f \lambda}{2\pi\Delta}\right) \tag{3.35}$$

where β_f is the actual phase difference between the sound field at the two microphone locations, Δ is the microphone spacing and λ is the wavelength of the sound. The first term on the right of Equation (3.35) accounts for the difference in reference levels for sound intensity and sound pressure. If noise is coming from sources other than the one being measured or there are reflecting surfaces in the vicinity of the noise source, the Pressure–Intensity Index will increase and so will the error in the intensity measurement. Other sources or reflected sound do not affect the accuracy of intensity measurements taken with a $p-u$ probe, but the $p-u$ probe is more sensitive than the $p-p$ probe in reactive sound fields (such as the near field of a source, see Jacobsen and de Bree (2005)).

The normalised systematic error in sound intensity due to microphone phase mismatch is a function of the actual phase difference, β_f (radians), between the two microphone locations in the sound field and the phase mismatch error, β_s (radians). The normalised error is given by Fahy (1995) as:

$$e_\beta(I) = \beta_s/\beta_f \tag{3.36}$$

The Pressure–Intensity Index, δ_{pI}, may be written in terms of this error as:

$$\delta_{pIO} - \delta_{pI} = 10\log_{10}|1 + (1/e_\beta(I))| = 10\log_{10}|1 + (\beta_f/\beta_s)| \quad \text{(dB)} \tag{3.37}$$

A normalised error of Equation (3.36) equal to 0.25 corresponds to a sound intensity error of approximately 1 dB ($= 10\log_{10}(1 + 0.25)$), and a difference, $\delta_{pIO} - \delta_{pI} = 7$ dB.

A normalised error equal to 0.12 corresponds to a sound intensity error of approximately 0.5 dB and a difference, $\delta_{pIO} - \delta_{pI} = 10$ dB. The pressure intensity index will be large (leading to relatively large errors in intensity estimates) in near fields and reverberant fields and this can extend over the entire audio frequency range. The phase mismatch (in radians) between the microphones in the $p-p$ probe is related to the Residual Pressure–Intensity Index (which is often supplied by the $p-p$ probe suppliers) by:

$$\beta_s = k\Delta 10^{-\delta_{pIO}/10} \tag{3.38}$$

Phase mismatch also distorts the directional sensitivity of the $p-p$ probe so that the null in response of the probe (often used to locate noise sources) is changed from the 90° direction to β_m, as given by (Fahy, 1995):

$$\beta_m = \cos^{-1}(\beta_s/k\Delta) \tag{3.39}$$

where k is the wavenumber at the frequency of interest and Δ is the spacing between the two microphones in the $p-p$ probe.

In state of the art instrumentation, microphones are available that have a phase mismatch of less than 0.05° (0.0009 radians). In cases where the phase mismatch is larger than this, the instrumentation sometimes employs phase mismatch compensation in the signal processing path. The error due to amplitude mismatch is zero for perfectly phase matched microphones.

However, for imperfectly phase matched microphones, the error is quite complicated to quantify and depends on the characteristics of the sound field being measured.

Fahy (1995) shows that random errors in intensity measurements add to the uncertainty due to systematic errors and in most sound fields where the signals received by the two microphones are random and have a coherence close to unity, the normalised random error is given by $(BT_A)^{-1/2}$, corresponding to an intensity error of:

$$e_r(I) = 10\log_{10}\left(1 + (BT_A)^{-1/2}\right) \quad \text{(dB)} \tag{3.40}$$

where B is the bandwidth of the measurement in Hz and T_A is the effective averaging time, which may be less than the measurement time unless real-time processing is performed. The coherence of the two microphone signals will be less than unity in high-frequency diffuse fields or where the microphone signals are contaminated by electrical noise, unsteady flow or cable vibration. In this case, the random error will be greater than indicated by Equation (3.40).

Finally, there is an error due to instrument calibration, which adds to the random errors. This is approximately ± 0.2 dB for one particular manufacturer, but the reader is advised to consult calibration charts that are supplied with the instrumentation.

3.10.3 Frequency Decomposition of the Intensity

It is often necessary to decompose the intensity signal into its frequency components. This may be done either directly or indirectly, as described in the following sections.

3.10.3.1 Direct Frequency Decomposition

$p - u$ Probe

For this probe type, the frequency distribution of the mean intensity may be obtained by passing the two output signals (p and u) through identical bandpass filters prior to performing the time averaging. Separate filters are needed for each frequency band of interest.

$p - p$ Probe

For this probe type, the frequency distribution may be determined by passing the two signals through appropriate identical bandpass filters, either before or after performing the sum, difference and integration operations of Equation (3.27) and then time averaging the resulting outputs.

3.10.3.2 Indirect Frequency Decomposition

Determination of the intensity using the indirect frequency decomposition method is based on Fourier analysis of the two probe signals (either the $p - u$ signals or the $p - p$ signals).

$p - u$ Probe

Fahy (1995) (pp. 95–97) shows that for a $p - u$ probe, the intensity as a function of frequency is given by the single-sided cross-spectral density, G_{Dpu} (see Section 12.3.12), between the two signals as:

$$\boldsymbol{I_D}(f_n) = \text{Re}\{G_{Dpu}(f_n)\} \tag{3.41}$$

$$\boldsymbol{I_{Dr}}(f_n) = -\text{Im}\{G_{Dpu}(f_n)\} \tag{3.42}$$

where $\boldsymbol{I_D}(f_n)$ represents the real (or active) time-averaged intensity power spectral density (see Section 12.3.2) at frequency f_n and $\boldsymbol{I_{Dr}}(f_n)$ represents the amplitude of the reactive component.

The cross-spectral density of two signals is defined as the product of the complex instantaneous spectral density of one signal with the complex conjugate of the complex instantaneous spectral density of the other signal (see Section 12.3.12). Thus, if $G_{Dp}(f_n)$ and $G_{Du}(f_n)$ represent the complex single-sided spectral densities of the pressure and velocity signals, respectively, then the associated cross-spectral density is given by:

$$G_{Dpu}(f_n) = G^*_{Dp}(f_n)G_{Du}(f_n) \tag{3.43}$$

where the * represents the complex conjugate.

For single-frequency signals and harmonics, the cross-spectral density function, $G_{Dpu}(f_n)$, may be replaced with the cross-spectrum, $G_{pu}(f_n)$, obtained by multiplying the cross-spectral density by the bandwidth of each FFT filter (or the frequency resolution) used to obtain the cross-spectrum.

$p - p$ Probe

For the $p - p$ probe, (Fahy, 1995, p. 97) shows that the mean active intensity spectral density, $\boldsymbol{I_D}(f_n)$, and amplitude, $\boldsymbol{I_{Dr}}(f_n)$, of the reactive intensity spectral density in direction, $\boldsymbol{n}$, at frequency, f_n, are:

$$\boldsymbol{I_D}(f_n) = -\frac{\boldsymbol{n}}{\rho f_n \Delta}\mathrm{Im}\{G_{Dp1p2}(f_n)\} \tag{3.44}$$

$$\boldsymbol{I_{Dr}}(f_n) = \frac{\boldsymbol{n}}{2\rho f_n \Delta}\left[G_{Dp1p1}(f_n) - G_{Dp2p2}(f_n)\right] \tag{3.45}$$

where G_{Dp1p2} is the cross-spectral density of the two pressure signals, and G_{Dp1p1} and G_{Dp2p2} represent the auto-spectral densities (see Section 12.3.12). The intensity spectrum is obtained using the same equations with the subscript, D, removed, so that cross-spectra and auto-spectra are used in place of cross- and auto-spectral densities, respectively.

For the case of a stationary harmonic sound field, it is possible to determine the sound intensity by using a single microphone and the indirect frequency decomposition method just described by taking the cross-spectrum (as the sound is tonal) between the microphone signal and a stable reference signal (referred to as A) of the same frequency, for two locations p_1 and p_2 of the microphone. The reference signal, A for example, could be the electrical signal driving a loudspeaker producing the sound being measured or an accelerometer on a vibrating surface radiating the sound being measured. Thus, the effective cross-spectrum G_{p1p2} for use in the preceding equations can be calculated as:

$$G_{p1p2}(f_n) = \frac{G^*_{p1}(f_n)G_A(f_n)G^*_{p2}(f_n)G_{p2}(f_n)}{G_{p2}(f_n)^*G_A(f_n)} = \frac{G_{p1A}(f_n)G_{p2p2}(f_n)}{G_{p2A}(f_n)} \tag{3.46}$$

Equations (3.41) to (3.46) are easily implemented using commercially available FFT software with the signals from two carefully phase-matched microphones processed using using a standard two-channel data acquisition system. Most FFT software allows selection of the bandwidth over which the sound intensity is averaged, including standard 1/3-octave and octave bands. If the microphones are not well matched, measurements are still possible (Fahy, 1995).

Most spectrum analysis software has the ability to determine cross-spectra as well as a cross-spectral densities, with the cross-spectral density being useful for calculating the intensity for broadband noise sources and the cross-spectrum being useful for calculating the intensity of tonal or harmonic noise sources.

3.11 Sound Source Localisation

The identification of the actual location of sources of sound on vibrating structures or surfaces in air flows (aerodynamic sound) is of great interest to any noise control engineer. There is a range of instrumentation available for this purpose and practically all of it is quite expensive.

There are four main techniques for locating sound sources: near-field acoustic holography (NAH), statistically optimised near-field acoustic holography (SONAH), beamforming (BF) and direct sound intensity measurement. Instrumentation using each of these techniques is discussed below. The physical principles underpinning each technique are discussed and a limited amount of mathematical analysis is presented to illustrate the fundamental principles of each technique. More detail is provided in the references.

It is worth noting that none of the techniques to be discussed can distinguish between a sound wave radiated directly from a source and a wave reflected from an object or surface. Thus, reflecting surfaces can often be identified as noise sources. In addition, to determine whether the sound is coming from in front or behind the array, the array must be either non-planar or made up of microphones in two parallel planes with associated more complex signal processing.

3.11.1 Near-Field Acoustic Holography (NAH)

Acoustic holography is a method for obtaining the distribution of sound intensity close to the surface of a noise-radiating structure. The method is based on the measurement of the amplitude and phase of a sound field at many locations on a plane at some distance from the sound sources, but in their near field so that evanescent waves contribute significantly to the microphone signals. The measurements are then used to predict the complex acoustic pressure and particle velocity on a plane that approximates the surface of the source (the prediction plane).

Multiplying these two quantities together gives the sound intensity as a function of location in the prediction plane, which allows direct identification of the relative strength of the acoustic radiation from various areas of a vibrating structure. The theoretical analysis that underpins this technique is described in detail in a book devoted just to this topic (Williams, 1999) and so it will only be summarised here.

Acoustic holography may involve only sound pressure measurements in the measurement plane (using phase matched microphones), and these are used to predict both the acoustic pressure and particle velocity in the plane of interest. The product of these two predicted quantities is then used to determine the sound intensity on the plane of interest which is usually adjacent to the sound radiating structure. Alternatively, only acoustic particle velocity may be measured in the measurement plane (using the "Microflown") and then these measurements may be used to predict the acoustic pressure as well as the acoustic particle velocity in the plane of interest. Finally, both acoustic pressure and acoustic particle velocity may be measured in the measurement plane. Then the acoustic particle velocity measurement is used to predict the acoustic particle velocity in the plane of interest and the acoustic pressure measurement is used to predict the acoustic pressure in the plane of interest.

Jacobsen and Liu (2005) showed that measurement of both acoustic pressure and particle velocity gave the best results, followed by measurement of only particle velocity, with the measurement of only acoustic pressure being a poor third. The reason that using like quantities to predict like quantities gives the best results is intuitively obvious, but the reasons that the particle velocity measurement gives much better results than the pressure measurement need some explanation. The reasons are summarised below:

- Particle velocity decays more rapidly than sound pressure towards the edges of the measurement region and has a larger dynamic range. This means that spatial windowing (similar to the time domain windowing discussed in Section 12.3.5, but in this case, in the spatial domain), which is a necessary part of the application of the technique to finite size structures, does not have such an influence when particle velocity measurements are used.
- Predicting particle velocity from acoustic pressure measurements results in amplification of high spatial frequencies, which increases the inherent numerical instability of

the prediction. On the other hand, predicting acoustic pressure from acoustic particle velocity results in a reduction of the amplitudes of high spatial frequencies.

- Phase mismatch between transducers in the measurement array has a much greater effect on the prediction of acoustic particle velocity from acoustic pressure measurements than vice versa.

The practical implementation of acoustic holography is complex and expensive, involving many sensors whose relative phase calibration must be accurately known, and whose relative positions must be known to a high level of accuracy. However, commercially available instrumentation exists, which does most of the analysis transparently to the user, so it is relatively straightforward to use.

3.11.1.1 Summary of the Underlying Theory

Planar near field acoustic holography involves the measurement of the amplitudes and relative phases of either or both of the acoustic pressure and acoustic particle velocity at a large number of locations on a plane located in the near field of the radiating structure on which it is desired to locate and quantify noise sources. The measurement plane must be in the near field of the structure and as close as practical to it, but no closer than the microphone spacing. As the measurement surface is a plane, the noise source map will be projected on to a plane that best approximates the structural surface. If the structural surface is curved, it is possible to use a curved sensor array but the analysis becomes much more complicated. The resolution (or accuracy of noise source location) of this technique is equal to the spacing of the measurement array from the noise-radiating surface at low frequencies and at frequencies above the frequency where this distance is equal to half a wavelength, the resolution is half a wavelength. The following analysis is for a measurement array of infinite extent, with microphones sufficiently close together that the measured sound pressure on the measurement plane is effectively continuous. The effect of deviating from this ideal situation by using a finite size array and a finite number of sensors will be discussed following the analysis.

For the analysis, it is assumed that the plane encompassing the radiating structure is in the $x - y$ plane at a z coordinate, $z_p \leq 0$, and the measurement plane is located at $z_m > 0$ and is parallel to the plane at z_p. The wavenumber transform corresponding to complex pressure measurements made on the measurement plane is a transform from the spatial domain to the wavenumber domain and may be written as:

$$K(k_x, k_y) = \int_{-\infty}^{\infty}\int_{-\infty}^{\infty} p(x, y, z_m)\mathrm{e}^{\mathrm{j}(k_x x + k_y y)}\mathrm{d}x\,\mathrm{d}y \tag{3.47}$$

where k_x and k_y are the wavenumber components in the x and y directions and their values cover the wavenumber range from $-\infty$ to $+\infty$. Of course, in practice the finite size of the array and the physical spacing of the microphones will limit the wavenumber spectrum and hence the frequency range that can be covered. The acoustic pressure is given by the inverse transform as:

$$p(x, y, z_m) = \frac{1}{(2\pi)^2}\int_{-\infty}^{\infty}\int_{-\infty}^{\infty} K(k_x, k_y)\mathrm{e}^{-\mathrm{j}(k_x x + k_y y)}\mathrm{d}k_x\,\mathrm{d}k_y \tag{3.48}$$

From the preceding two equations, it can be seen that the wavenumber spectrum, $K(k_x, k_y)$, for any given value of k_x and k_y may be interpreted as the amplitude of a plane wave provided that:

$$k_x^2 + k_y^2 \leq k^2 \tag{3.49}$$

The plane wave propagates in the $(k_x,\ k_y,\ k_z)$ direction and k_z must satisfy:

$$k_x^2 + k_y^2 + k_z^2 = k^2 = (\omega/c)^2 \tag{3.50}$$

If Equation (3.49) is not satisfied, then the particular component of the wavenumber spectrum represents an evanescent wave whose amplitude decays exponentially with distance from the sound source.

The wavenumber transform in the prediction plane, $z = z_p$, can be calculated by multiplying Equation (3.48) with an exponential propagator defined as:

$$G(z_p, z_m, k_x, k_y) = \mathrm{e}^{-\mathrm{j}k_z(z_p - z_m)} \tag{3.51}$$

After the multiplication of the wavenumber transform by G in Equation (3.51), the inverse transform (see Equation (3.48)) is taken to obtain the sound pressure in the prediction plane.

The acoustic particle velocity in the prediction plane is calculated from the wavenumber transform of Equation (3.47) by multiplying it by the *pu* propagator defined as:

$$G_{pu}(z_p, z_m, k_x, k_y) = \frac{k_z}{\rho c k}\mathrm{e}^{-\mathrm{j}k_z(z_p - z_m)} \tag{3.52}$$

where

$$k_z = \begin{cases} \sqrt{k^2 - k_x^2 - k_y^2}; & \text{for } k_x^2 + k_y^2 \le k^2 \\ -\mathrm{j}\sqrt{k_x^2 + k_y^2 - k^2}; & \text{for } k_x^2 + k_y^2 > k^2 \end{cases} \tag{3.53}$$

Taking the inverse transform, as for the acoustic pressure, it is also possible to complete the entire process using particle velocity measurements instead of pressure measurements. In this case, the particle velocity is substituted for the acoustic pressure in Equation (3.47) to give the particle velocity wavenumber transform and then Equation (3.48) gives the particle velocity on the LHS instead of the acoustic pressure. The propagator in Equation (3.51) is then multiplied with the particle velocity wavenumber transform and then the inverse transform is taken as in Equation (3.48) to give the particle velocity on the prediction plane. The acoustic pressure on the prediction plane can be calculated by multiplying the particle velocity wavenumber transform by the following predictor, prior to taking the inverse transform.

$$G_{up}(z_p, z_m, k_x, k_y) = \frac{\rho c k}{k_z}\mathrm{e}^{-\mathrm{j}k_z(z_p - z_m)} \tag{3.54}$$

Implementation of the above procedure in practice, requires the simplifications of a finite size measurement plane (which must be slightly larger than the radiating structure being analysed) and a finite spacing between measurement transducers. The adverse effect of the finite size measurement plane is minimised by multiplying the wavenumber transform by a spatial window that tapers towards the edge of the array so less and less weighting is placed on the measurements as one moves from the centre to the edge of the array. To prevent wrap around errors, the array used in the wavenumber transform is larger than the measurement array and all points outside the actual measurement array are set equal to zero.

The effect of the finite spacing between sensors in the measurement array is to limit the ability of the array to sample high spatial frequency components, which are a way of representing the sound field variation, with the components being more prominent if the sound field varies strongly as a function of location. All components that exist must be properly sampled to enable accurate location of noise sources. Thus, if the array is incapable of sampling high-frequency spatial components, the need to sample them can be avoided by moving the array to some distance from the noise-radiating structure to a location where the high spatial frequency components are negligible, but not so far that the evanescent modes are so low in intensity that they cannot be measured.

The other effect of finite microphone spacing is to limit the upper frequency for the measurement as a result of the spatial sampling resolution. In theory, this spacing should be less than half a wavelength but, in practice, good results are obtained for spacings less than about one quarter of a wavelength. The lower frequency ability of the measurement is limited by the array size, which should usually be at least a wavelength, but in some special cases, if the sound pressure at the edges of the array has dropped off sufficiently, an array size of 1/3 of a wavelength can be used.

To be able to calculate the finite Fast Fourier Transforms, the measurement grid for NAH must be uniform; that is, all sensors must be uniformly spaced.

3.11.2 Statistically Optimised Near-Field Acoustic Holography (SONAH)

The SONAH method (Hald et al., 2007) for locating sound sources on a noise-radiating structure is a form of near-field acoustic holography in which the FFT calculation is replaced with a least squares matrix inversion. The advantage of this method is that the measurement array does not need to be as large as the measurement source and the measurement sensors need not be regularly spaced. In addition, the sound field can be calculated on a surface that matches the contour of the noise-radiating surface. However, the computer power needed for the inversion of large matrices can be quite large. The noise source location resolution is similar to that for NAH. When inverting large matrices using singular value decomposition, large errors can result if the matrix condition number is large. The condition number is defined as the ratio of the relative error in the input data to the relative error in the solutions. A high condition number results in small errors in the input data of the matrix producing large errors in the solution. For SONAH, any condition number larger than about 10 is probably too large. Condition numbers are usually provided by matrix inversion software as one of the outputs. In SONAH, the matrix condition number can be reduced by using a smaller microphone spacing and in many cases it can be reduced by rearranging the order of the rows in the matrix (pivoting), which is the same as changing the order of the equations that are being solved.

The requirements for maximum microphone spacing are similar to NAH and typical maximum operating frequencies of commercially available SONAH systems range from 1 kHz to 6 kHz. The typical dynamic range (difference in level between the strongest and weakest sound sources that can be detected) is 15 to 20 dB. The array size sets the lower frequency limit of the measurement but the requirements for SONAH are much less stringent than those for NAH. It is possible to undertake measurements down to frequencies for which the array size is 1/8 of a wavelength.

The analysis begins by representing the sound field as a set of plane evanescent wave functions defined as:

$$\Phi_k(x, y, z) = \mathrm{e}^{-\mathrm{j}(k_x x + k_y y + k_z(z_p - z_m))} \tag{3.55}$$

where k_z is defined by Equation (3.53). In practice, the sound field at any location, (x, y, z), is represented by N plane wave functions, chosen to cover the wavenumber spectrum of interest, so that:

$$p(x, y, z) \approx \sum_{n=1}^{N} a_n \Phi_{kn}(x, y, z) \tag{3.56}$$

The coefficients, a_n, are determined by using the pressure measurements over L locations, $(x_\ell, y_\ell, z_\ell, \ell = 1, ..., L)$, in the measurement array so that:

$$p(x_\ell, y_\ell, z_\ell) \approx \sum_{n=1}^{N} a_n \Phi_{kn}(x_\ell, y_\ell, z_\ell); \qquad \ell = 1, ..., L \tag{3.57}$$

The plane wave functions corresponding to all the measurement points, L, can be expressed in matrix form as:

$$\boldsymbol{A}_m = \begin{bmatrix} \Phi_{k1}(x_1, y_1, z_1) & \Phi_{k2}(x_1, y_1, z_1) & \dots & \Phi_{kN}(x_1, y_1, z_1) \\ \Phi_{k1}(x_2, y_2, z_2) & \Phi_{k2}(x_2, y_2, z_2) & \dots & \Phi_{kN}(x_2, y_2, z_2) \\ \vdots & \vdots & \vdots & \vdots \\ \Phi_{k1}(x_L, y_L, z_L) & \Phi_{k2}(x_L, y_L, z_L) & \dots & \Phi_{kN}(x_L, y_L, z_L) \end{bmatrix} \tag{3.58}$$

Then the pressure at the measurement locations may be written as:

$$\boldsymbol{p}_m = \boldsymbol{A}_m \boldsymbol{a} \tag{3.59}$$

where

$$\boldsymbol{p}_m = \begin{bmatrix} p(x_1, y_1, z_1) \\ p(x_2, y_2, z_2) \\ \vdots \\ p(x_L, y_L, z_L) \end{bmatrix}; \quad \boldsymbol{a} = \begin{bmatrix} a_1 \\ a_2 \\ \vdots \\ a_N \end{bmatrix} \text{ and } \boldsymbol{\phi}(x, y, z) = \begin{bmatrix} \Phi_{k_1}(x, y, z) \\ \Phi_{k_2}(x, y, z) \\ \vdots \\ \Phi_{k_N}(x, y, z) \end{bmatrix} \tag{3.60}$$

Equation (3.56) may then be used with Equation (3.60) to write an expression for the pressure at any other point not in the measurement array (and usually on the surface of the noise source being examined) as:

$$p(x, y, z) = \boldsymbol{a}^{\mathrm{T}} \boldsymbol{\phi}(x, y, z) \tag{3.61}$$

The vector $\boldsymbol{a}$ is found by inverting Equation (3.59), which includes the measured data. As the $\boldsymbol{A}_m$ matrix is non-square, a pseudo-inverse is obtained. Thus:

$$\boldsymbol{a} = \left(\boldsymbol{A}_m^{\mathrm{H}} \boldsymbol{A}_m + \alpha \boldsymbol{I}\right)^{-1} \boldsymbol{A}_m^{\mathrm{H}} \boldsymbol{p}_m \tag{3.62}$$

where $\boldsymbol{I}$ is the identity matrix, $\boldsymbol{A}_m^{\mathrm{H}}$ is the Hermitian transpose of $\boldsymbol{A}_m$ (see Equations (A.21) and (A.22)) and α is the regularisation parameter, which is usually selected with the following equation to give a value close to the optimum:

$$\alpha = \left(1 + \frac{1}{2(kd)^2}\right) \times 10^{\mathrm{SNR}/10} \tag{3.63}$$

where SNR is the signal-to-noise ratio for the measured data and $d = z_m - z_p$ is the distance between the sound source and the measurement plane. The particle velocity is obtained using the same procedure and the same equations as above, except that Equation (3.55) is replaced with:

$$\Phi_k(x, y, z) = \frac{k_z}{\rho} cke^{-\mathrm{j}(k_x x + k_y y + k_z(z_p - z_m))} \tag{3.64}$$

This measurement technique is accurate in terms of quantifying the sound intensity as a function of location on a noise emitting structure, and it has the same resolution and frequency range as NAH. No particle velocity measurement is needed.

3.11.3 Helmholtz Equation Least Squares Method (HELS)

The HELS method for locating sound sources (Wu, 2000a,b; Isakov and Wu, 2002) is very similar to the SONAH method discussed above, except that, instead of using plane wave functions to describe the sound field as in Equation (3.55), the HELS method uses spherical wave functions. The analysis is quite a bit more complicated than it is for the SONAH method and as this method does not produce any better results than the SONAH method, it will not be discussed further here.

3.11.4 Beamforming

Beamforming is the process of determining the direction from which sound is coming by measuring the amplitude and phase of the sound pressure over a planar or spherical or linear array of many microphones. These measurements are used to maximise the total summed output of the array for sound coming from a specified direction, while minimising the response due to sound coming from different directions. By adding an adjustable delay when processing the electronic signal from each microphone, it is possible to "steer" the array so that the direction of maximum response can be varied. In this way, the relative intensity of sound coming from different directions can be determined and the data analysed to produce a map of the relative importance of different parts of a sound source to the total far field sound pressure level. This principle underpins the operation of a device known as an acoustic camera. Unlike NAH and SONAH, beamforming operates in the far field of the sound source, and it is more accurate at higher frequencies.

Typical frequency ranges and recommended distances between the array and the radiating noise source for various commercially available array types are listed in Table 3.2. The dynamic range of a beamforming measurement varies from about 6 dB for ring arrays to up to 15 dB for spiral arrays. However, spiral arrays have a disadvantage of poor depth of field, so that it is more difficult to focus the array on the sound source, especially if the sound source is non-planar. Poorly focussed arrays used on more than one sound source existing at different distances from the array but in a similar direction can result in sources cancelling one another so that they disappear from the beamforming image altogether.

TABLE 3.2 Beamforming array properties. The backward attenuation represents how much a wave is attenuated if it arrives from behind the array

Array name	Array size	Number of microphones	Distance from source (m)	Frequency range (Hz)	Backward attenuation
Star	3×2 m arms	36	3–300	100–7000	21 dB
Ring	0.35 m dia	32	0.35–1.5	400–20000	0 dB
Ring	0.75 m dia	48	0.7–3	400–20000	0 dB
Ring	1.4 m dia	72	2.5–20	250–20000	0 dB
Cube	0.35 m across	32	0.3–1.5	1000–10000	20 dB
Sphere	0.35 m dia	48	0.3–1.5	1000–10000	20 dB

When a spherical array is used inside an irregular enclosure such as a car passenger compartment, it is necessary to use a CAD model of the interior of the enclosure and accurately position the beamforming array within it. Then the focus plane of the array can be adjusted in software for each direction to which the array is steered.

Beamformers have the disadvantage of poor spatial resolution of noise source locations, especially at low frequencies. For a beamforming array of largest dimension, D, and distance from the source, L, the resolution (or accuracy with which a source can be located) is given by:

$$\text{Res} = 1.22\frac{L}{D}\lambda \tag{3.65}$$

For acceptable results, the array should be sufficiently far from the source that it does not subtend an angle greater than 30° in order to cover the entire source. In general, the distance of the array from the sound source should be at least the same as the array diameter. A big advantage of the beamforming technique is that it can image distant sources as well as moving sources. It is also possible to get quantitative measures of the sound power radiated by the source (Hald et al., 2007).

One disadvantage of beamforming compared to NAH is that it is not possible to distinguish between sound radiated directly by the source and sound reflected from the source (as the measurements are made in the far field of the source). Also, for planar arrays, one cannot distinguish between sound coming from in front of or behind the array.

Beamforming can give erroneous results in some situations. For example, if the array is not focussed at the source distance, the source location will not be clear and sharp – it can look quite fuzzy. If two sources are at different distances from the array, it is possible that neither will be identified.

Beamforming array design is also important as there is a trade-off between depth of focus of the array and its dynamic range. The spiral array has the greatest dynamic range (up to 15 dB) but a very small depth of focus whereas the ring array only has a dynamic range of 6 dB but a large depth of focus, allowing the array to focus on noise sources at differing distances and not requiring such precision in the estimate of the distance of the noise source from the array. The dynamic range is greatest for broadband noise sources and least for low-frequency and tonal sources. More detailed analysis of various aspects affecting the beamformer performance is provided by Christensen and Hald (2004) and Johnson and Dudgeon (1993).

3.11.4.1 Summary of the Underlying Theory

Beamforming theory is complicated so only a brief summary will be presented here. For more details, the reader is referred to Bai et al. (2013); Christensen and Hald (2004); Johnson and Dudgeon (1993).

There are two types of beamforming: infinite-focus distance and finite-focus distance. For the former, plane waves are assumed and for the latter, spherical waves are assumed to originate from the focal point of the array.

In essence, infinite-focus beamforming in the context of interest here is the process of summing the signals from an array of microphones and applying different delays to the signals from each microphone so that sound coming from a particular direction causes a maximum summed microphone response and sound coming from other directions causes no response. Of course in practice, sound from any direction will still cause some response but the principle of operation is that these responses will be well below the main response due to sound coming from the direction of interest.

It is also possible to scale the beamformer output so that a quantitative measure of the active sound intensity at the surface of the noise radiator can be made (Hald, 2004). Consider a planar array made up of L microphones at locations (x_ℓ, y_ℓ; $\ell = 1, ..., L$) in the $x - y$ plane. If the measured pressure signals, p_ℓ, are individually delayed and then summed, the output of the array is:

$$p(\boldsymbol{n}, t) = \sum_{\ell=1}^{L} w_\ell p_\ell(t - \Delta_\ell(\boldsymbol{n})) \tag{3.66}$$

where w_ℓ is the weighting coefficient applied to the pressure signal, p_ℓ, and its function is to reduce the importance of the signals coming from the array edges, which in turn reduces the amplitudes of side lobes in the array response. Side lobes are peaks in the array response in directions other than the design direction and serve to reduce the dynamic range of the beamformer.

The quantity, $\boldsymbol{n}$, in Equation (3.66) is the unit vector in the direction of maximum sensitivity of the array and the time delays, Δ_ℓ, are chosen to maximise the array sensitivity in direction, $\boldsymbol{n}$. This is done by delaying the signals associated with a plane wave arriving from direction, $\boldsymbol{n}$, so that they are aligned in time before being summed. The time delay, Δ_ℓ, is the dot product of the unit vector, $\boldsymbol{n}$, and the vector, $\boldsymbol{r}_\ell = (x_\ell, y_\ell)$, divided by the speed of sound, c. That is:

$$\Delta_\ell = \frac{\boldsymbol{n}\cdot\boldsymbol{r}_\ell}{c} \tag{3.67}$$

If the analysis is done in the frequency domain, the beamformer output at angular frequency, ω, is:

$$P(\boldsymbol{n},\omega) = \sum_{\ell=1}^{L} w_\ell P_\ell(\omega)\mathrm{e}^{-\mathrm{j}\omega\Delta_\ell(\boldsymbol{n})} = \sum_{\ell=1}^{L} w_\ell P_\ell(\omega)\mathrm{e}^{-\mathrm{j}\boldsymbol{k}\cdot\boldsymbol{r}_\ell} \tag{3.68}$$

where $\boldsymbol{k} = -k\boldsymbol{n}$ is the wavenumber vector of a plane wave incident from the direction, $\boldsymbol{n}$, which is the direction in which the array is focussed.

Finite-focus beamforming using a spherical wave assumption to locate the direction of a source, and its strength at a particular distance from the array (array focal point) and follows a similar but slightly more complex analysis than outlined above for infinite-focus beamforming. For the array to focus on a point source at a finite distance, the various microphone delays should align in time, so that the microphone signals corresponding to a spherical wave emitted from the focus point are in-phase. Equation (3.68) still applies, but the delay, Δ_ℓ, is defined as:

$$\Delta_\ell = \frac{|\boldsymbol{r}| - |\boldsymbol{r} - \boldsymbol{r}_i|}{c} \tag{3.69}$$

where $\boldsymbol{r}$ is the vector location of the source from an origin point in the same plane as the array, $\boldsymbol{r}_\ell$ is the vector location of microphone, ℓ, in the array with respect to the same origin and $|\boldsymbol{r}-\boldsymbol{r}_\ell|$ is the scalar distance of microphone, ℓ, from the source location.

More complex beamforming analyses applicable to aero-acoustic problems, where the array is close to the source and there is a mean flow involved, are discussed by Brooks and Humphreys (2006).

3.11.5 Direct Sound Intensity Measurement

The use of a stethoscope, which is essentially a microphone, to manually scan close to the surface of an item of equipment to locate noise sources is well-known. However, in the presence of significant levels of background noise, this method is no longer effective. Manually scanning a particle velocity sensor such as the "Microflown" to measure the normal acoustic particle velocity over a surface has a number of advantages;

1. the particle velocity signal is larger than the pressure signal close to a source;
2. background noise reflected from the surface being scanned produces close to zero particle velocity at the surface, whereas the acoustic pressure is approximately doubled; and
3. the particle velocity sensor is directional (in contrast to the omni-directional nature of a microphone), thus further reducing the influence of background noise.

There is also equipment available that transforms the particle velocity signal to an audible signal by feeding it into a head set. This crude method of source location in the presence of high levels of background noise seems to be very effective (de Bree and Druyvesteyn, 2005; de Vries and de Bree, 2008), even though in theory, the relative sound power radiated by the various locations on a surface or structure can only be determined if the near field sound intensity is measured.

If more accurate measurements are needed, then a small intensity probe made up of a miniature microphone and a "Microflown" transducer can be used to scan the surface over which the noise source identification and quantification is required. The scan should be as close as possible to the surface and as far as is practicable, it should follow the surface contour. It is quite feasible for this scan to be done manually by taking intensity measurements adjacent to a large number of points on the noise-radiating surface and having the intensity probe stationary for each measurement. A less accurate measurement is to manually scan the intensity probe over an imaginary planar surface adjacent to the noise-radiating surface.

One significant advantage of the direct measurement of sound intensity adjacent to a noise-radiating surface is the relatively large dynamic range (difference in dB between maximum and minimum measurable intensities) that can be achieved. A dynamic range between 30 and 60 dB is common compared to 20 dB for NAH and SONAH using microphones and 40 dB for NAH and SONAH using particle velocity sensors. The dynamic range of beamforming measurements varies from 6 to 15 dB (Heilmann et al., 2008).

Another advantage of the direct intensity measurement method is the wide bandwidth possible (20 Hz to 20 kHz) compared to 200 Hz to 2 kHz for planar holography and 2 kHz to 10 kHz for beamforming arrays. However, direct measurement of sound intensity close to a surface does have some problems due the dominance of the reactive sound field in that region. This means that any errors in the phase matching between pressure and particle velocity sensors can have a relatively large impact on the accuracy of the intensity measurement. For this reason, intensity measurements can only be made accurately if the phase difference between the acoustic pressure and particle velocity is less than 85°, which corresponds to a reactivity of 10 dB. The reactivity for a harmonic sound field is defined as:

$$\text{Reactivity} = 10\log_{10}(\boldsymbol{I}/\boldsymbol{I}_r) \tag{3.70}$$

where $\boldsymbol{I}$ and $\boldsymbol{I}_r$ are defined in Equations (1.76) and (1.77), respectively. If the reactivity is too high, the intensity probe must be moved further from the noise-radiating structure.

4

Sound Sources and Sound Power

LEARNING OBJECTIVES

In this chapter, the reader is introduced to:

- the simple source as a source of spherical waves and a fundamental building block for acoustical analysis;
- the dipole source, directional properties and modes of generation;
- the quadrupole source, its various forms, directional properties and some modes of generation;
- line sources and their uses for modelling;
- the piston in an infinite wall, far-field and near-field properties, radiation load, and uses for modelling;
- the incoherent plane radiator;
- directional properties of sources and the concept of directivity;
- effects of reflection;
- radiation impedance of a source and what influences it;
- the relation between sound power and sound pressure;
- radiation field, near field and far field of a sound source;
- sound power determination using sound intensity measurements, surface vibration measurements and sound pressure measurements, both in the laboratory and in the field; and
- some uses of sound power information.

4.1 Introduction

Sources of sound generation are generally quite complicated and often beyond the capability of ordinary analysis if a detailed description is required. Fortunately, a detailed description of the noise-generation mechanism is often not necessary for noise control purposes and in such cases, very gross simplifications will suffice. For example, a source of sound of characteristic dimensions much less than the wavelengths it generates can often be approximated as a simple point source of zero dimension. In this case, the properties of the idealised point source of sound will provide a sufficient description of the original sound source.

Alternatively, familiarity with the properties of common idealised sources may provide the means for identification of the noise source. For example, in a duct, noise that increases in level

DOI: 10.1201/9780367814908-4

with the sixth power of the flow speed can readily be identified as due to a dipole source, and most probably originates at some obstruction to the flow in the duct (Dowling and Ffowcs-Williams, 1983). For these reasons, as well as others to be mentioned, it is worthwhile considering the properties of some idealised sound sources.

In the second part of this chapter, it will be shown how the quantities, sound pressure, sound intensity and sound power are related. Sound pressure is the quantity most directly related to the response of people and things to airborne sound, and this is the quantity that is most often to be controlled. However, sound pressure is not always the most convenient descriptor for a noise source, since it will depend on distance from the source and the environment in which the sound and measurement position are located. A better descriptor is usually the sound power radiated by the source (see Section 1.9).

In many cases, sound power level information is sufficient and no additional information is required for the specification of sound pressure levels. In other cases, information about the directional properties of the source may also be required. For example, the source may radiate much more effectively in some directions than in others. However, given both sound power level and directivity index (see Section 4.8), the sound pressure level at any position relative to the source may then be determined. In fact, a moment's reflection will show that specification of sound pressure level could be quite insufficient for a source with complicated directivity; thus the specification of sound power level for sources is preferred for many noise control problems.

The sound field generated by a source can be defined in terms of the source properties and the environment surrounding the source. The sound field, in turn, produces a radiation load on the source that affects both: (1) the sound power generated by the source; and (2) the relationship between sound power and sound pressure. In this chapter, the discussion will include the experimental determination of source sound power using sound pressure or sound intensity measurements and the use of sound power information for the determination of sound pressure levels in the free field.

When a noise source radiates into free space, one is interested in directivity as well as sound power level. However, if a source radiates into a reverberant field, directivity information is lost, and sound power level is all that one need know about the source. This matter will be considered in detail later in this chapter. Thus, for the present, it will be sufficient to point out that sound power level information is useful for the following purposes:

- it allows comparison of the noise-producing properties of different machines;
- it allows verification that the noise produced by a particular machine meets specifications for noise control purposes; and
- it provides a means for predicting expected sound pressure levels in reverberant spaces, and in the free field when directivity information is also known.

4.2 Simple Source

A simple source is a source of sound that radiates uniformly in all directions and which is very much smaller than the wavelength of the radiated sound. The source could be a small speaker mounted in one wall of a very small rigid box, but provided that the wavelength it generates is large compared to any of the dimensions of the box, it will qualify as a simple source. That is, the small speaker-box source will generate an acoustic field similar to that generated by a small pulsating sphere. Thus, the class of simple sources may be thought of as those sources that have the properties of a small pulsating sphere, the diameter of which is very much less than any wavelength it generates. Radiation of sound waves from such a source is illustrated in Figure 4.1.

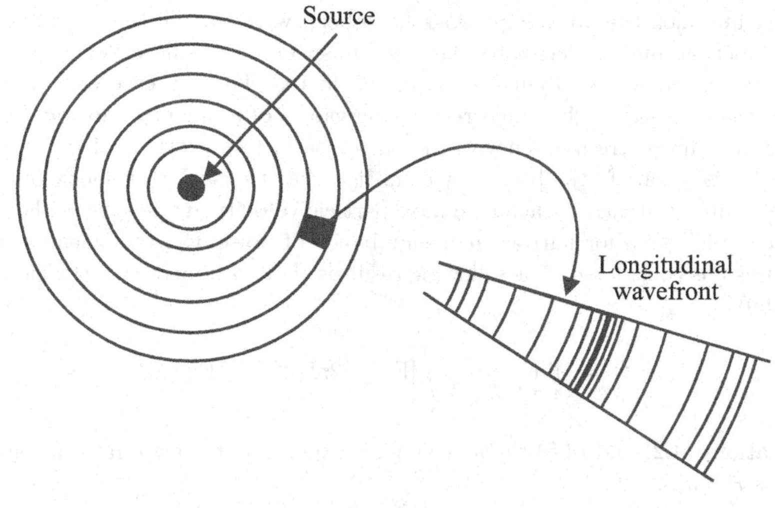

FIGURE 4.1 Sound radiation from a simple source.

4.2.1 Pulsating Sphere

In Section 1.5.3, an outward-travelling spherical wave solution to the wave equation was considered. Clearly, as the particle velocity of an outward-travelling spherical wave can be made to match the surface velocity of a pulsating sphere, the radiated field of the pulsating sphere can be determined from the spherical wave acoustic potential function given by Equation (1.42).

According to Section 1.8.2, the sound intensity is obtained as the long-time average of the product of the acoustic pressure and particle velocity. Use of Equations (1.43), (1.44) and (1.67) gives the following equation for the intensity:

$$I = \lim_{T_A \to \infty} \frac{\rho c}{T_A} \int_0^{T_A} \frac{\mathrm{F\,F'} + r\mathrm{F'F'}}{r^3} \mathrm{d}t \tag{4.1}$$

However, integration over an encompassing sphere of surface area $4\pi r^2$ (effectively multiplying the intensity by the surface area), shows that if the radiated power, W, is to be constant, independent of radius, r, any term in the integral of Equation (4.1) must be a constant when multiplied by r^2. If a term still contains a dependence on r after this multiplication, it must be set equal to 0 or else the power, W, will not be independent of radius. Thus:

$$\lim_{T_A \to \infty} \frac{1}{T_A} \int_0^{T_A} \mathrm{F\,F'd}t = 0 \tag{4.2}$$

The intensity of the diverging spherical wave is therefore:

$$I = \rho c \frac{\langle \mathrm{F'F'} \rangle}{r^2} \tag{4.3}$$

where

$$\langle \mathrm{F'F'} \rangle = \lim_{T_A \to \infty} \frac{1}{T_A} \int_0^{T_A} \mathrm{F'F'd}t \tag{4.4}$$

The symbol $\langle\ \rangle$ indicates a time-averaged quantity (see Section 1.6).

In the form in which the above equations have been written, the functional dependence of the potential function and its derivative has been unspecified to allow for the possibility that a narrow band of noise, rather than a tone, is considered. For the case of a narrow band of noise, for example a 1/3-octave band, corresponding values of frequency and wavelength may be estimated by using the centre frequency of the narrowband (see Section 1.14).

The next step is to satisfy the boundary condition that the surface velocity of the pulsating small spherical source matches the acoustic wave particle velocity at the surface. For the purpose of providing a result useful for narrow frequency bands of noise, the root mean square particle velocity at radius r is considered. The following result is then obtained with the aid of Equations (1.44) and (1.59):

$$\langle u^2 \rangle = \lim_{T_A \to \infty} \frac{1}{T_A r^4} \int_0^{T_A} [\mathrm{F\,F} + 2r\mathrm{F\,F'} + r^2\mathrm{F'F'}]\mathrm{d}t \tag{4.5}$$

Using Equations (4.2) and (4.5) the following equation may be written for a spherical source having a radius $r = a$:

$$\langle u^2 \rangle = \frac{\langle \mathrm{F\,F} \rangle}{a^4} + \frac{\langle \mathrm{F'F'} \rangle}{a^2} \tag{4.6}$$

The RMS volume flux Q_{RMS} of fluid produced at the surface of the pulsating spherical source may be defined as:

$$Q_{\mathrm{RMS}} = 4\pi a^2 \sqrt{\langle u^2 \rangle} \tag{4.7}$$

Substitution of Equation (4.6) in Equation (4.7) gives:

$$Q_{\mathrm{RMS}} = 4\pi \sqrt{\langle \mathrm{F\,F} \rangle + \langle \mathrm{F'F'} \rangle a^2} \tag{4.8}$$

The particle velocity is now matched to the surface velocity of the pulsating spherical source. The assumption is made that the following expression, which holds exactly for a pure tone or single frequency, also holds for a narrow band of noise:

$$\langle \mathrm{F\,F} \rangle k^2 = \langle \mathrm{F'F'} \rangle \tag{4.9}$$

The coefficient, k, is the wavenumber, equal to ω/c, where for a narrow band of noise the centre frequency in radians per second is used. Substitution of Equation (4.9) into Equation (4.8) gives:

$$Q_{\mathrm{RMS}} = 4\pi \frac{\sqrt{\langle \mathrm{F'F'} \rangle (1 + k^2a^2)}}{k} \tag{4.10}$$

Finally, substituting Equation (4.10) into Equation (4.3) gives the intensity of a simple source in terms of the mean square flux of the source:

$$I_M = \frac{Q_{\mathrm{RMS}}^2 k^2 \rho c}{(4\pi r)^2 (1 + k^2a^2)} \tag{4.11}$$

Note that for a very small source ($a \ll \lambda$), the term, k^2a^2 may be ignored. As the simple source is also known as a monopole, the subscript "M" has been added to the intensity, I. The power radiated by a simple or monopole source is given by the integral of the intensity over the surface of any convenient encompassing sphere of radius r; that is, multiplication of Equation (4.11) by the expression for the surface area of a sphere gives, for the radiated sound power of a monopole:

$$W_M = \frac{Q_{\mathrm{RMS}}^2 k^2 \rho c}{4\pi (1 + k^2a^2)} \tag{4.12}$$

An expression for the mean square pressure radiated by a simple source may be obtained using Equations (1.78) and (4.11).

$$\langle p^2 \rangle = \frac{(Q_{\rm RMS} k \rho c)^2}{(4\pi r)^2 (1 + k^2 a^2)} \tag{4.13}$$

Using Equations (4.12) and (4.13), the following may be written for the sound pressure at distance r from the source in terms of the source sound power:

$$\langle p^2 \rangle = \frac{W_M \rho c}{4\pi r^2} \tag{4.14}$$

The RMS flux $Q_{\rm RMS}$ must be limited by the source volume to less than $2\sqrt{2}\omega\pi a^3/3$, where a is the radius of the source, as the maximum pulsation amplitude of a small source is physically limited to about half of the source radius. The mean square acoustic pressure at distance r from the source is obtained by substituting $Q_{\rm RMS} < 2\sqrt{2}\omega\pi a^3/3$ and Equation (1.30) into Equation (4.13) and must also be limited. Thus:

$$\langle p^2 \rangle < \left[\frac{\rho c^2 \sqrt{2}}{6} \frac{a}{r} \frac{k^2 a^2}{\sqrt{1 + k^2 a^2}} \right]^2 \tag{4.15}$$

The *directivity* of a monopole source is the same in all directions, so it is referred to as an omnidirectional source and the sound pressure level measured at a fixed distance from the source will not vary with angular location around the source.

4.2.2 Fluid Mechanical Monopole Source

An example of a fluid mechanical monopole source is the inlet or the exhaust of a reciprocating pump or the exhaust of a reciprocating engine. In either case, the source dimension is small compared to the wavelength of radiated sound. Referring to Equation (4.12), the following proportional relationships will be assumed:

$$Q_{\rm RMS} \propto SU; \quad k \propto U/(\sqrt{S}c) \tag{4.16a,b}$$

where S is the inlet or exhaust cross-sectional area, U is the local mean flow speed, $k = \omega/c$ is the wavenumber, c is the speed of sound and $\sqrt{S}$ is a characteristic length. Substitution of Equations (4.16a and b) into Equation (4.12) and introduction of the Mach number, $M = U/c$, gives the following result:

$$W_M \propto (S\rho U^4/c) \propto (S\rho U^3)M \tag{4.17a,b}$$

Equation (4.17a) shows that the sound power of a fluid mechanical monopole source may be expected to increase with the fourth power of the stream speed while Equation (4.17b) shows that the sound power is proportional to the quantity $S\rho U^3$, which is a measure of the convected kinetic energy or the stream power, and the Mach number, M, which for values less than one, may be interpreted as a measure of the source efficiency.

4.3 Dipole Source

If two simple sources of equal strength are placed close together and arranged to be always of exactly opposite phase, then as one produces a net outflow, the other produces an exactly opposite inflow. Thus, at some distance from the sources, there is no net fluid flux. However, being of opposite sign there is a net thrust exerted on the fluid from the positive to the negative source, which fluctuates in direction. It is the time rate of change of this force on the fluid that is important for noise production. Such an arrangement becomes a dipole source in the limit as the separation distance between the sources is made indefinitely small; that is, very much less than the radiated wavelength. A familiar example of a dipole source is one of the prongs of a tuning fork. A less obvious example is the fluctuating flow over a spoiler introduced into an air stream.

4.3.1 Pulsating Doublet or Dipole (Far-Field Approximation)

Consider two simple pulsating sources of opposite phase placed at $y = h$ and $y = -h$, and determine the resulting acoustic potential function at the general point R(x, y), as shown in Figure 4.2.

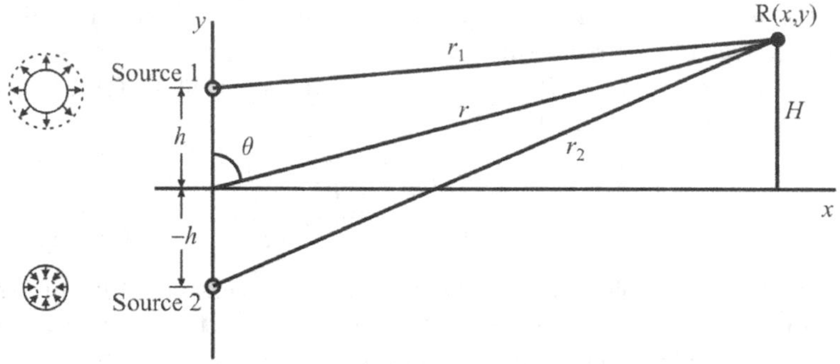

FIGURE 4.2 Geometry of a dipole source.

The potential ϕ at (x, y) in the far field of the dipole is the sum of the potentials due to sources 1 and 2:

$$\phi = \phi_1 + \phi_2 \tag{4.18}$$

The form of ϕ_1 and ϕ_2 is that given by Equation (1.47a), with r replaced, respectively, with r_1 and r_2.

The distances r_1 and r_2 may be rewritten in terms of the distance r measured from the origin to the observation point O. That is:

$$r_1 = r\sqrt{1 + (h/r)^2 - 2(h/r)\cos\theta} \tag{4.19}$$

Thus,

$$r_1 \approx r - h\cos\theta \tag{4.20}$$

and, similarly

$$r_2 \approx r + h\cos\theta \tag{4.21}$$

In writing Equations (4.20) and (4.21) it has been assumed that:

$$(h/r) \ll 1 \tag{4.22}$$

Substitution of Equations (4.20) and (4.21) into Equation (4.18) and use of Equation (1.47a) results in the following equation, to first order in small quantities:

$$\phi = \frac{\mathrm{F}(ct - r_1)}{r_1} - \frac{\mathrm{F}(ct - r_2)}{r_2} = \frac{\mathrm{F}(ct - r + h\cos\theta)}{r} - \frac{\mathrm{F}(ct - r - h\cos\theta)}{r} \tag{4.23}$$

Using a Taylor series,

$$\mathrm{F}(x + \Delta) = \mathrm{F}(x) + \frac{\mathrm{F}'(x)}{1!}\Delta + \frac{\mathrm{F}''(x)}{2!}\Delta^2 + ... \tag{4.24}$$

To expand Equation (4.23), again to first order, the following equation is obtained:

$$\phi = \frac{1}{r}\left[\mathrm{F}(ct-r) + \mathrm{F}'(ct-r)h\cos\theta + ... - \mathrm{F}(ct-r) + \mathrm{F}'(ct-r)h\cos\theta - ....\right] \tag{4.25}$$

Equation (4.25) reduces to the following expression for the dipole source potential function, valid for distances greater than about one wavelength:

$$\phi = 2\mathrm{F}'(ct-r)(h/r)\cos\theta \tag{4.26}$$

Equations (1.11) and (1.12) of Section 1.4.6 and Equation (4.26) are used to obtain expressions for the acoustic pressure p and particle velocity u, as:

$$p = \rho\frac{\partial\phi}{\partial t} = 2\rho c\frac{h}{r}\mathrm{F}''(ct-r)\cos\theta \tag{4.27}$$

and

$$u = -\frac{\partial\phi}{\partial r} = \frac{2h}{r^2}\left[\mathrm{F}'(ct-r) + r\mathrm{F}''(ct-r)\right]\cos\theta \tag{4.28}$$

Division of Equation (4.27) by Equation (4.28) and use of the following relationships, produces Equation (1.45), which is formally the same as obtained for spherical waves.

$$\langle \mathrm{F}'\mathrm{F}'\rangle k^2 = \langle \mathrm{F}''\mathrm{F}''\rangle$$

$$\mathrm{F}'(ct-r) = \mathrm{j}kA\mathrm{e}^{\mathrm{j}k(ct-r)} \tag{4.29}$$

$$\mathrm{F}''(ct-r) = -k^2A\mathrm{e}^{\mathrm{j}k(ct-r)}$$

Thus, for narrow bands of noise or single frequencies, Equation (1.50) also holds for dipoles. At large distances from the source, the characteristic impedance for dipole waves becomes simply ρc.

Using exactly the same procedures and argument used in Section 4.2.1, the dipole field intensity I_D may be shown to be:

$$I_D = \rho c\frac{h^2 Q_{\mathrm{RMS}}^2 k^4}{(2\pi r)^2(1+k^2a^2)}\cos^2\theta \tag{4.30}$$

where Q_{RMS} is the RMS strength of each monopole source making up the dipole and a is the radius of each of these sources. Finally, integration of the intensity over an encompassing sphere of radius, r, gives for the radiated power of a dipole:

$$W_D = \rho c\frac{k^4h^2Q_{\mathrm{RMS}}^2}{3\pi(1+k^2a^2)} \tag{4.31}$$

The mean square sound pressure at distance, r, and angular direction, θ, measured from the dipole axis, is obtained by substituting into Equation (4.27) the relations, $\mathrm{F}''(ct-r) = -k^2\mathrm{F}(ct-r) = \mathrm{j}k\mathrm{F}'(ct-r)$, and Equation (4.10) to give:

$$\langle p^2\rangle = \left[\frac{hQ_{\mathrm{RMS}}k^2\rho c\cos\theta}{2\pi r}\right]^2\frac{1}{(1+k^2a^2)} = 3W_D\frac{\rho c}{4\pi r^2}\cos^2\theta \tag{4.32}$$

The characteristic figure-of-eight radiation pattern of the pressure field of the dipole source is illustrated in Figure 4.3. In the plane normal to the axis of the dipole, the sound pressure is nil,

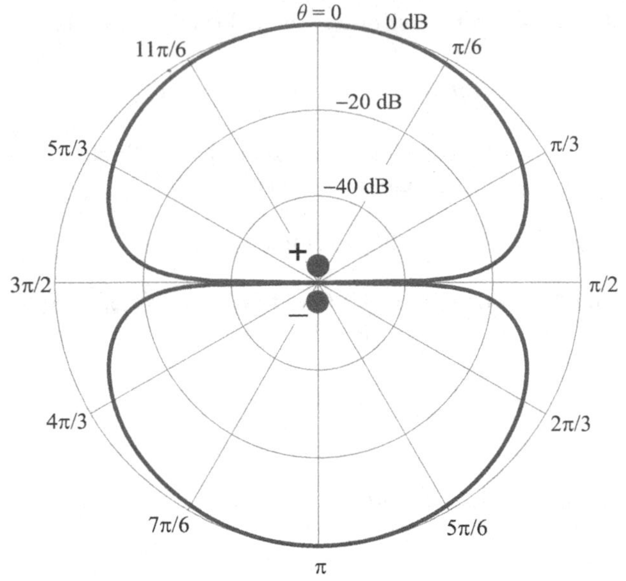

FIGURE 4.3 Radiation pattern of a dipole source. The axis of the source is vertical. The figure indicates the relative intensity of the sound pressure in the direction, θ. In particular, a null in intensity is observed at $\theta = \pi/2$.

while along the axis of the dipole, the sound pressure is maximum and of opposite phase in opposite directions from the centre. The MATLAB® script, `dipole_spl_vs_angle.m`, can be used to plot the directivity of a dipole source, and is available for download from www.causalsystems.com.

It is instructive to compare the radiation efficiencies of the dipole and monopole sources. Using Equations (4.12) and (4.31), their ratio can be determined as:

$$\frac{W_D}{W_M} = \frac{4}{3}(kh)^2 \tag{4.33}$$

As the quantity kh, which is a measure of the sources' separation distance-to-wavelength ratio, is small, it can be seen that the dipole is much less efficient as a radiator of sound than either of its constituent simple sources taken separately. This is because the presence of the second source, of opposite phase and in close proximity, tends to unload the first source so that it produces less sound. This effect is more pronounced at low frequencies, because the wavelength becomes larger at low frequencies, which results in the source separation distance, measured in terms of wavelengths, becoming smaller.

In Equations (4.30) and (4.31), the RMS source strength Q_{RMS} is multiplied by the sources' separation distance h (see Figure 4.2). In a mathematical sense, the product hQ_{RMS} may be set equal to the RMS dipole strength D_{RMS}, which remains constant while h and Q_{RMS} are allowed to become indefinitely small and large, respectively.

If the separation distance, $2h$, between the two out-of-phase monopole sources making up the dipole is not small compared to the distance of the receiver from the dipole centre, or is not small compared to a wavelength, then the equations derived in this section are no longer valid. In this case, the problem would need to be treated as two monopole sources and the sound pressure and relative phase at a receiver location may be determined by adding the complex sound pressures (taking relative phase into account) arriving from each of the two monopole sources. That is, Equation (4.18) may be used, with ϕ replaced by p.

4.3.2 Pulsating Doublet or Dipole (Near Field)

It is of interest to consider the near-field sound pressure in the vicinity of a dipole, as it will be useful in the discussion of the oscillating sphere in Section 4.3.3. Beginning with Equation (1.49), and the definition for flux, Q, from a single pulsating sphere of radius, a, as:

$$Q = Q_0 \mathrm{e}^{\mathrm{j}\omega t} = 4\pi a^2 u = 4\pi a^2 u_0 \mathrm{e}^{\mathrm{j}\omega t} \tag{4.34a,b,c}$$

the following can be written for the source flux amplitude in terms of the constant, A, of Equation (1.49):

$$Q_0 = 4\pi a^2 \left[\frac{A}{a^2} + \frac{jkA}{a}\right] \mathrm{e}^{-\mathrm{j}ka} = 4\pi A(1 + \mathrm{j}ka)\mathrm{e}^{-\mathrm{j}ka} \tag{4.35}$$

Thus:

$$A = \frac{Q_0}{4\pi}\left[\frac{1}{1+\mathrm{j}ka}\right] \mathrm{e}^{\mathrm{j}ka} \tag{4.36}$$

Defining a Green's function between a source point, $\boldsymbol{r}_0$, and a measurement point, $\boldsymbol{r}$, as:

$$G(\boldsymbol{r}, \boldsymbol{r}_0, \omega) = \frac{1}{4\pi|\boldsymbol{r} - \boldsymbol{r}_0|}\mathrm{e}^{-\mathrm{j}k|\boldsymbol{r}-\boldsymbol{r}_0|} \tag{4.37}$$

and using Equations (1.48b) and (4.36), the following expression (which is accurate in both the near and far fields) may be written for the acoustic pressure at any time, t, radiated by the dipole to some location $\boldsymbol{r}$:

$$\begin{aligned} p(\boldsymbol{r}, \theta, t) &= \frac{\mathrm{j}k\rho c Q_0}{(1+\mathrm{j}ka)}\left[-G(r_1, h, \omega) + G(r_2, -h, \omega)\right] \mathrm{e}^{\mathrm{j}(\omega t + ka)} \\ &= \frac{\mathrm{j}k\rho c Q_0}{(1+\mathrm{j}ka)}\left[-G(r - h\cos\theta, h, \omega) + G(r + h\cos\theta, -h, \omega)\right] \mathrm{e}^{\mathrm{j}(\omega t + ka)} \end{aligned} \tag{4.38}$$

where h is half the distance between the two sources making up the dipole, which is centred at the origin. If the distance between the two sources making up the dipole is small compared to a wavelength, Equation (4.38) may be written using Taylor's series (see Equation (4.24)) as:

$$p(\boldsymbol{r}, \theta, t) = \frac{\mathrm{j}k\rho c Q_0 2h\cos\theta}{(1+\mathrm{j}ka)}\left[\frac{\partial}{\partial r}G(\boldsymbol{r}, \boldsymbol{r}_0, \omega)\right]_{\{\boldsymbol{r}_0 = 0\}} \mathrm{e}^{\mathrm{j}(\omega t + ka)} \tag{4.39}$$

Using Equation (4.37), Equation (4.39) may be written as:

$$p(r, \theta, t) = \frac{k^2 \rho c Q_0 h\cos\theta}{2\pi r(1+\mathrm{j}ka)}\left[1 - \frac{\mathrm{j}}{kr}\right] \mathrm{e}^{\mathrm{j}(\omega t - k(r-a))} = \frac{B\cos\theta}{kr}\left[1 - \frac{\mathrm{j}}{kr}\right] \mathrm{e}^{\mathrm{j}(\omega t - k(r-a))} \tag{4.40}$$

where $r = |\boldsymbol{r} - \boldsymbol{r}_0|$, with $\boldsymbol{r}_0 \approx 0$ and:

$$B = \frac{\rho c h Q_0 k^3}{2\pi(1+\mathrm{j}ka)} \tag{4.41}$$

Using Equations (1.11) and (1.12) allows the following to be written for the particle velocity:

$$u(\boldsymbol{r}, \theta, t) = \frac{B\cos\theta}{kr\rho c}\left[1 - \frac{2}{(kr)^2} - \frac{2\mathrm{j}}{kr}\right] \mathrm{e}^{\mathrm{j}(\omega t - k(r-a))} \tag{4.42}$$

Inspection of Equations (4.40) and (4.42) shows that, for small values of kr (that is, close to the source), the acoustic pressure and particle velocity are virtually in quadrature (90° out of phase), suggesting a source that radiates very poorly.

In the far field, where kr is large, the mean square pressure can be derived from Equation (4.40) as:

$$\langle p^2 \rangle = \left[\frac{hQ_{\mathrm{RMS}} k^2 \rho c\cos\theta}{2\pi r}\right]^2 \frac{1}{(1+k^2a^2)} \tag{4.43}$$

which is identical to Equation (4.32).

4.3.3 Oscillating Sphere

An oscillating sphere (see Figure 4.4) acts in a similar way to a dipole in terms of its sound radiation. It is interesting to consider Equation (4.28) and note that it is of the form, $u = u_0 \cos\theta e^{j\omega t}$. The latter equation, however, is just the expression for the normal component of velocity over the surface of a sphere, which vibrates with velocity amplitude, u_0, along the dipole axis. For example, in Figure 4.4 the pair of sources might be replaced with a small sphere of radius a at the origin, which is made to vibrate up and down along the y-axis. The net force F exerted on the surrounding medium by the sphere vibrating along the y-axis (see Figure 4.4) is determined by integrating the y-component of the force due to the acoustic pressure over the surface of the sphere. However, to proceed, expressions for the acoustic pressure and particle velocity in the near field of the source are required.

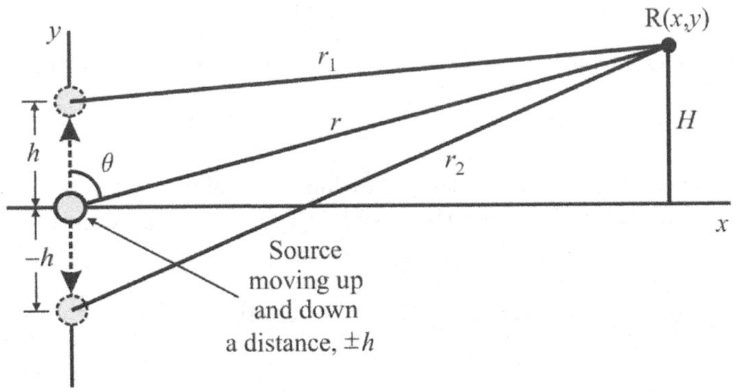

FIGURE 4.4 Geometry of an oscillating sphere source.

Slight rearrangement of the expression given by Dowling and Ffowcs-Williams (1983, Eq. (2.39)), changing the sign to account for a slightly different general solution to the wave equation given by Equation (1.43), and use of Equation (4.37), results in the following for the sound pressure radiated by an oscillating sphere:

$$p(\boldsymbol{r}, \theta, t) = \frac{j4\pi k\rho c u_0 a^3 \cos\theta}{2(1 + jka) - k^2a^2} \left[\frac{\partial}{\partial r} G(\boldsymbol{r}, \boldsymbol{r}_0, \omega)\right]_{\boldsymbol{r}_0=0} e^{j(\omega t + ka)} \tag{4.44}$$

Differentiating the expression in brackets, allows Equation (4.44) to be rewritten as:

$$p(\boldsymbol{r}, \theta, t) = \frac{k^2\rho c u_0 a^3 \cos\theta}{r[2(1 + jka) - k^2a^2]} \left[1 - \frac{j}{kr}\right] e^{j(\omega t - k(r-a))} \tag{4.45}$$

For a compact sphere ($ka \ll 1$), Equation (4.45) may be written as:

$$p(\boldsymbol{r}, \theta, t) = \frac{k^2\rho c u_0 a^3 \cos\theta}{2r(1 + jka)} \left[1 - \frac{j}{kr}\right] e^{j(\omega t - k(r-a))} \tag{4.46}$$

Comparing Equations (4.40) and (4.46) for a compact oscillating sphere, which is much smaller than a wavelength, shows that a compact oscillating sphere, of radius, a, and oscillating with a velocity amplitude of u_0, radiates the same sound field as a dipole made up of two simple sources separated by $2h$, and each having a flux amplitude given by:

$$Q_0 = \pi a^3 u_0 / h \tag{4.47}$$

If the oscillating sphere is not compact, the analysis becomes much more complicated, and this is beyond the scope of this text.

Using Equations (1.11), (1.12), (4.46) and (4.47) allows the following to be written for the particle velocity radiated by an oscillating sphere:

$$u(\boldsymbol{r},\theta,t) = \frac{B\cos\theta}{kr\rho c}\left[1 - \frac{2}{(kr)^2} - \frac{2\mathrm{j}}{kr}\right]\mathrm{e}^{\mathrm{j}(\omega t - k(r-a))} \tag{4.48}$$

It can be seen that this equation is identical to Equation (4.42) for a dipole. However, in this case,

$$B = \frac{\rho c a^3 u_0 k^3}{2(1+\mathrm{j}ka)}; \qquad |B| = \frac{\rho c a^3 u_0 k^3}{2\sqrt{(1+k^2a^2)}} \tag{4.49}$$

According to Equation (1.67) of Section 1.8.1, the sound intensity is determined as the long-time average of the product of the real parts of the acoustic pressure given by Equation (4.40), and the particle velocity given by Equation (4.42), where the second term in brackets is small compared to the other two term for large r. In the far field (that is, more than a wavelength from the source), the expression for the sound intensity is:

$$I_D = \frac{|B|^2\cos^2\theta}{2\rho c(kr)^2} \tag{4.50}$$

where the factor of 2 arises from the time average being $\sqrt{2}$ smaller than the amplitude for a single frequency waveform.

The sound power radiated by the small vibrating sphere may be determined as the long-time average of the real part of the force acting to move the sphere multiplied by the real part of the velocity of the sphere. However, the power radiated must equal the power flow to the far field; thus, an alternative means for determining the power flow due to vibration of the small sphere is to integrate the expression for the far-field intensity over the surface of a sphere, with spherical coordinates, $(r,\ \theta,\ \psi)$. Using either method, the expression obtained for the power radiated is:

$$W_D = \iint\limits_S \frac{|B|^2\cos^2\theta}{2\rho c(kr)^2}\mathrm{d}S = \frac{|B|^2}{2\rho c k^2}\int_0^{2\pi}\int_0^{\pi}\frac{\cos^2\theta\, r^2\sin\theta}{r^2}\mathrm{d}\theta\mathrm{d}\psi = \frac{2\pi|B|^2}{3\rho c k^2} \tag{4.51}$$

The amplitude of the oscillating force acting on the small vibrating sphere of radius, a may be determined by integrating the component of the acoustic pressure given by Equation (4.46) over the surface of the sphere. Note that the expression for the acoustic pressure has the form $p = p_0\cos\theta$. The force along the principal axis of the dipole (y-axis in Figure 4.4) is:

$$F = \iint\limits_S p\cos\theta\,\mathrm{d}S = \iint\limits_S p_0\cos\theta\cos\theta\,\mathrm{d}S = a^2p_0\int_0^{2\pi}\int_0^{\pi}\cos^2\theta\sin\theta\,\mathrm{d}\theta\mathrm{d}\psi = \frac{4}{3}\pi a^2 p_0 \tag{4.52}$$

Making use of Equation (4.46) evaluated at $r = a$, the dummy variable p_0 may be replaced and the following expression obtained:

$$F = \frac{4\pi a B}{3k}\left[1 - \frac{\mathrm{j}}{ka}\right]\mathrm{e}^{\mathrm{j}\omega t} \tag{4.53}$$

The above equation may be solved for B using $F = |F|\mathrm{e}^{\mathrm{j}\omega t}$ giving the following expression:

$$|B| = \frac{3k^2}{4\pi\sqrt{(1+k^2a^2)}}|F| \tag{4.54}$$

Substitution of Equation (4.54) into (4.51) gives the following expression for the sound power radiated as a result of the force, F, acting on a small vibrating rigid body:

$$W_D = \frac{3\pi f^2 |F|^2}{2\rho c^3 (1 + k^2 a^2)} \tag{4.55}$$

As the analysis assumed a compact source (wavelength of radiated sound is large compared to the radiating object), the k^2a^2 term in the denominator may be neglected.

It is not necessary that the compact source be spherical. It may be of any shape and an example is the tooth of a circular saw turning at high speed, which may be successfully modelled as such a source (Bies, 1992). In this case, the quantity, a, would be a characteristic source dimension. For a rectangular-shaped block, it is not clear whether the "characteristic dimension" is half the length in the direction of flow or a different dimension. Equation (4.55) has significance for noise generation and control as it shows that the radiated power is only dependent on the force exerted on the object.

For an oscillating sphere of radius, a, moving with a velocity amplitude of u_0, Equations (4.49) and (4.54) may be combined to give the following expression for the force term in Equation (4.55):

$$|F| = 2\rho\omega\pi a^3 u_0 / 3 \tag{4.56}$$

4.3.4 Fluid Mechanical Dipole Source

As mentioned earlier, turbulent flow over a small obstruction in an air stream provides a good example of a fluid mechanical dipole source of sound. Consideration of dipole noise generation in a fluid stream suggests the following proportional relationships. Referring to Equation (4.31):

$$Q \propto h^2 U \propto SU; \quad k \propto U/(hc) \tag{4.57a,b}$$

where c is the speed of sound, $k = \omega/c$ is the wavenumber, U is the mean stream speed and $h \propto \sqrt{S}$ is a characteristic dipole source dimension. Substitution of Equations (4.57a and b) into Equation (4.31) and introduction of the Mach number $M = U/c$ gives:

$$W_D \propto S\rho U^6/c^3 \propto S\rho U^3 M^3 \tag{4.58}$$

Equation (4.58) shows that the sound power of a dipole source in a fluid stream tends to increase with the sixth power of the fluid stream speed. However, the idling noise of a saw tooth on a moving saw is dipole in nature but, due to the effect of the blade on the radiation impedance, the sound power of the saw is proportional to the fifth power of the saw speed, not the sixth power as predicted by Equation (4.58). See Section 4.10 for more information on the effect of the source radiation impedance on its radiated sound power.

Equation (4.58) shows that the radiated sound power is proportional to the stream power, $S\rho U^3/2$, and the cube of the Mach number, M. As in the case of the monopole source, the Mach number term may be interpreted as an efficiency factor for values of Mach number less than one. This interpretation shows that the dipole source is much less efficient in converting stream power to noise than is the monopole source. However, the efficiency of the dipole source increases much more rapidly with flow speed than does the monopole source.

4.4 Quadrupole Source (Far-Field Approximation)

It was shown earlier that a dipole could be thought of as a sphere vibrating along an axis coincident with the dipole. Similarly, any multipole source may be thought of as a sphere vibrating in appropriate modal response. For these cases, exact solutions in spheroidal coordinates are possible (Morse, 1948), which describe the entire acoustic field. For the quadrupole source considered here, the analysis is restricted to the far field (see Section 4.12), where simpler approximate expressions apply.

If two dipole sources are combined, then there will be no net flux of fluid, and no net unbalanced force, but in general, there will remain a net shear force on the fluid. In the extreme, the net shear force may reduce to a local stress on the fluid. Again, it is the time rate of change of shear force or stress that is important in producing sound. As fluids can be expected to support such forces poorly, it will come as no surprise that quadrupole sources are relatively poor radiators of sound. However, they do play a dominant part in the mixing region when a fluid jet is introduced into a quiescent atmosphere.

Referring to Figure 4.5, where the coordinates are indicated, it may be shown in the same way as was done for the dipole that the potential function of the generalised quadrupole has the following form:

$$\phi = \frac{4hL}{r}\mathrm{F}''(ct-r)H(\theta,\psi,\alpha) \tag{4.59}$$

where

$$H(\theta,\psi,\alpha) = \cos\theta(\sin\theta\sin\psi\sin\alpha + \cos\theta\cos\alpha) \tag{4.60}$$

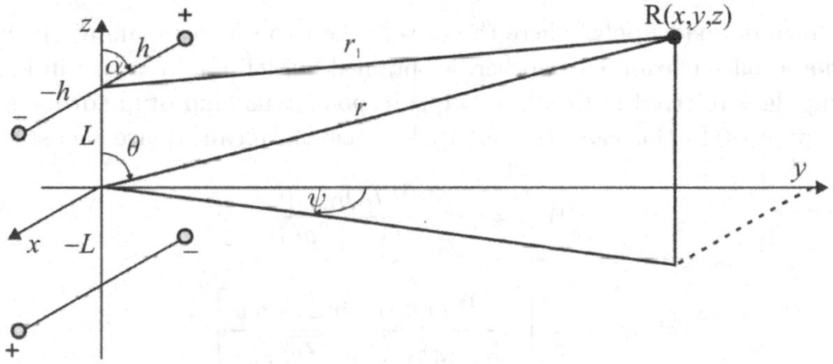

FIGURE 4.5 Geometry of a quadrupole source. α is the angle between the vertical z-axis and the axis of the top dipole or bottom dipole, which together comprise the quadrupole source. θ is the angle that the line joining the centre of the quadrupole source to the receiver makes with the $x-z$ plane. ψ is the angle that the line joining the centre of the quadrupole source to the receiver makes with the $y-z$ plane.

Expressions for the acoustic pressure, p, and particle velocity, u, in the far field immediately follow from Equation (4.59), using Equations (1.11) and (1.12):

$$p = \frac{4hL\rho c}{r}\mathrm{F}'''(ct-r)H(\theta,\psi,\alpha) \tag{4.61}$$

$$u = \frac{4hL}{r^2}\left[\mathrm{F}''(ct-r) + r\mathrm{F}'''(ct-r)\right]H(\theta,\psi,\alpha) \tag{4.62}$$

An expression for the near-field acoustic pressure at receiver location, $\mathrm{R}(x,y,z)$, may be derived following the procedure outlined by Morse and Ingard (1968), and remembering that positive time dependence is used in this text, whereas negative time dependence is used by Morse and Ingard. The result for the near-field pressure is:

$$p = -4\mathrm{j}\frac{k\rho c Q_0 Lh\cos\theta}{\sqrt{1+k^2a^2}}\left[\frac{\partial}{\partial r}G(\boldsymbol{r},\boldsymbol{r}_0,\omega)\right]_{r_0=0}\left[\sin\psi\cos\theta + \sin\psi\sin\theta + \cos\psi\right]\mathrm{e}^{\mathrm{j}\omega t} \tag{4.63}$$

where $r = |\boldsymbol{r} - \boldsymbol{r}_0|$.

Expressions for the intensity, I_q, radiated power, W_q, and alternative expressions for the mean square acoustic pressure $\langle p^2 \rangle$ are, respectively:

$$I_q = \rho c \left[\frac{k^3 h L Q_{\mathrm{RMS}}}{\pi r} \right]^2 \frac{[H(\theta, \psi, \alpha)]^2}{(1 + k^2 a^2)} \tag{4.64}$$

$$W_q = \rho c \frac{(2k^3 h L Q_{\mathrm{RMS}})^2}{5\pi(1 + k^2 a^2)} \left(\cos^2 \alpha + \frac{1}{3} \sin^2 \alpha \right) \tag{4.65}$$

$$\langle p^2 \rangle = \left[\rho c \frac{k^3 h L Q_{\mathrm{RMS}}}{\pi r} \right]^2 \frac{[H(\theta, \psi, \alpha)]^2}{(1 + k^2 a^2)} \tag{4.66}$$

or

$$\langle p^2 \rangle = \rho c \frac{5 W_q [H(\theta, \psi, \alpha)]^2}{4\pi r^2 \left(\cos^2 \alpha + \frac{1}{3} \sin^2 \alpha \right)} \tag{4.67}$$

4.4.1 Lateral Quadrupole

The common form of quadrupole, where the axes of the two dipoles making up the quadrupole are parallel to and off-set from one another, is obtained by defining $\alpha = \pi/2$ in Figure 4.5. This type of quadrupole is referred to as a lateral quadrupole. This kind of quadrupole characterises the jet-mixing process. In this case, the radiated power and mean square acoustic pressure are:

$$W_{lat} = \rho c \frac{[2k^3 h L Q_{\mathrm{RMS}}]^2}{15\pi(1 + k^2 a^2)} \tag{4.68}$$

$$\langle p_{lat}^2 \rangle = \left[\rho c \frac{k^3 h L Q_{\mathrm{RMS}} \sin 2\theta \sin \psi}{2\pi r \sqrt{(1 + k^2 a^2)}} \right]^2 \tag{4.69}$$

or

$$\langle p_{lat}^2 \rangle = \rho c \frac{15 W_{lat} \sin^2 2\theta \sin^2 \psi}{16\pi r^2} \tag{4.70}$$

The sound pressure level radiation pattern for the lateral quadrupole source is illustrated in Figure 4.6 for $\psi = \pi/2$. The MATLAB® script, `quadrupole_lateral_spl_vs_angle.m`, can be used to plot the directivity of a lateral quadrupole source, and is available for download from MATLAB scripts for ENC (2024).

4.4.2 Longitudinal Quadrupole

If $\alpha = 0$ in Equations (4.60) to (4.67) a longitudinal (or axial) quadrupole is obtained. In this case, the radiated power is:

$$W_{long} = \rho c \frac{[2k^3 h L Q_{\mathrm{RMS}}]^2}{5\pi(1 + k^2 a^2)} \tag{4.71}$$

and the radiated mean square acoustic pressure is:

$$\langle p_{long}^2 \rangle = 5 \rho c \frac{W_{long} \cos^4 \theta}{4\pi r^2} \tag{4.72}$$

The sound pressure level radiation pattern for the longitudinal quadrupole source is illustrated in Figure 4.7. The MATLAB® script, `quadrupole_longitudinal_spl_vs_angle.m`, can be used to plot the directivity of a longitudinal quadrupole source, and is available for download from MATLAB scripts for ENC (2024).

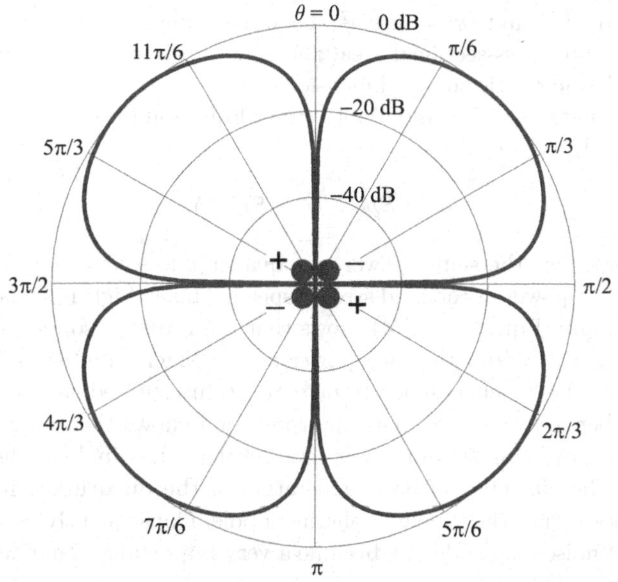

FIGURE 4.6 Radiation pattern of a lateral quadrupole source, where the relative intensity of the sound pressure in the direction, θ, is shown for $\psi = \pi/2$.

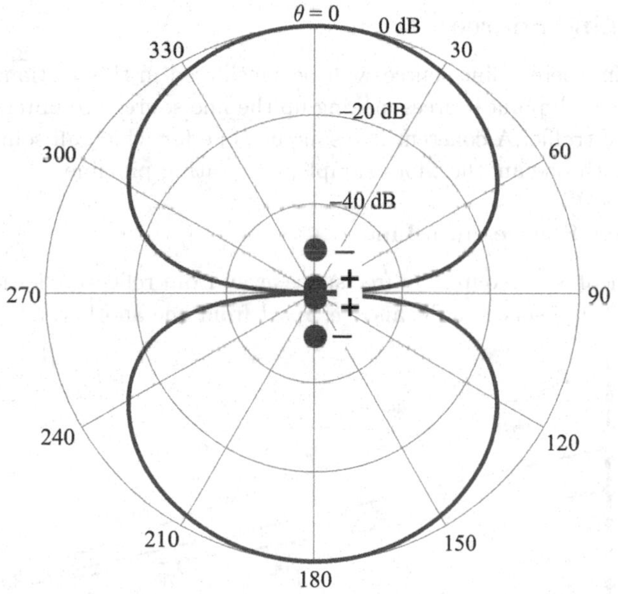

FIGURE 4.7 Radiation pattern of a longitudinal quadrupole source, where the relative intensity of the sound pressure in the direction, θ, is shown for $\psi = \pi/2$.

4.4.3 Fluid Mechanical Quadrupole Source

To investigate the parametric dependence of the radiated sound power on the parameters characteristic of a jet, the following proportional relationships will be introduced into Equation (4.67).

$$Q \propto Uh^2; \quad k \propto U/(hc); \quad h \approx L \tag{4.73a,b}$$

where c is the speed of sound, $k = \omega/c$ is the wavenumber, U is the mean stream speed, $S \propto L$ or h, is the stream cross-sectional area, and L and h, which will be assumed to be equal, both represent the characteristic source dimension.

Substitution of Equations (4.73a and b) into either Equation (4.68) or (4.71) and introduction of the Mach number, $M = U/c$, gives:

$$W \propto S\rho U^8/c^5 \propto S\rho U^3 M^5 \tag{4.74}$$

Equation (4.74) shows that the sound power of a quadrupole source in a fluid stream tends to increase with the eighth power of the fluid stream speed. This is Lighthill's famous prediction for jet noise (Lighthill, 1952). Equation (4.74) shows that the radiated sound power is proportional to the stream power, $S\rho U^3$, and the fifth power of the Mach number, M. As in the case of the monopole source, the Mach number term may be interpreted as an efficiency factor for values of Mach number less than one. This interpretation shows that the quadrupole source is much less efficient in converting stream power to noise than the dipole or the monopole sources. On the other hand, the efficiency of noise generation of the quadrupole increases much more rapidly with flow speed than the dipole or the monopole. Consequently, what might have been a negligible source of noise may suddenly become a very important source with only a relatively small increase in flow speed.

4.5 Line Source

4.5.1 Infinite Line Source

Both coherent and incoherent line sources will be considered in this section. An incoherent line source is one for which all point sources making up the line source are unrelated to one another; for example, a line of traffic. A coherent line source is one for which all sources have a constant phase relationship with one another; for example, a vibrating pipeline.

4.5.1.1 Incoherent Sources in a Line

Consider a row of incoherent sources in free space away from reflecting surfaces, each radiating a sound of power, W, and spaced a distance, b, apart from one another, as shown in Figure 4.8.

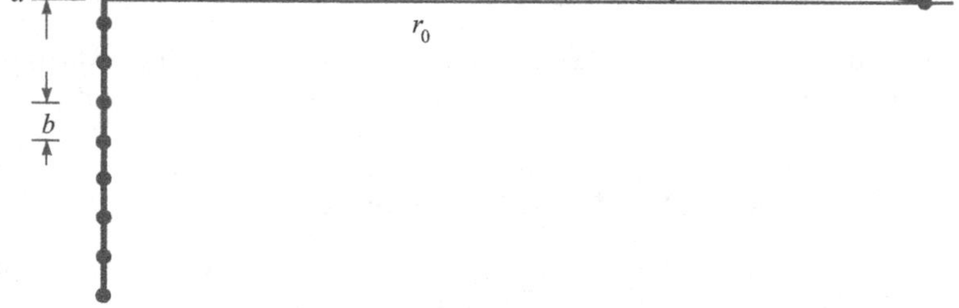

FIGURE 4.8 Geometry for a line source.

Since the sources are assumed to be incoherent, that is of random phases, the mean square sound pressure of each will add to give the total mean square sound pressure that will be measured at an observation point, R, at normal distance, r_0, from the row of sources. Let the nearest source to location R be numbered zero, with subsequent sources above and below, as shown in the figure, numbered with positive or negative integer values, n. Use of Equation (4.14) gives for the mean square sound pressure at the observation point, R:

$$\langle p^2 \rangle = \rho c \frac{W}{4\pi} \sum_{n=-\infty}^{\infty} 1/r_n^2 \tag{4.75}$$

where

$$r_n^2 = r_0^2 + (nb + d)^2 \tag{4.76}$$

Referring to Figure 4.8, the distance, d, introduced in Equation (4.76) is less than, or at most equal to, $0.5b$ in magnitude, and takes on a positive value if the zeroth source lies above the normal, r_0, to the line of sources and a negative value if it lies below (see Figure 4.8). For a distance $r_0 > 3b$, Equation (4.75) reduces to Equation (4.77) below, with an error of less than 1 dB.

$$\langle p^2 \rangle \approx \rho c \frac{W}{4\pi} \sum_{n=-\infty}^{\infty} [r_0^2 + n^2 b^2]^{-1} \tag{4.77}$$

which can be expressed as the following without any further increase in error:

$$\langle p^2 \rangle \approx \rho c \frac{W}{4 b r_0} \coth \left[\frac{\pi r_0}{b} \right] \tag{4.78}$$

Equation (4.78) can be simplified to the following equation, without any increase in error for $r_0 > 3b$.

$$\langle p^2 \rangle \approx \rho c \frac{W}{4 b r_0} \tag{4.79}$$

According to Equation (4.79), the corresponding sound pressure level at the observation point is:

$$L_p = L_W - 6 - 10 \log_{10} r_0 - 10 \log_{10} b + 10 \log_{10}(\rho c/400) \qquad \text{(dB re 20 µPa)} \tag{4.80}$$

where L_W is the sound power level (dB re 10^{-12} W) of each source. At 20°C, where $\rho c = 414$ (SI units), there would be an error of approximately 0.1 dB if the last term in Equation (4.80) is omitted. If r_0 is less than b/π, then the receiver distance is close enough to the nearest point source for the sound pressure to be calculated with an error of less than 1 dB by considering this source exclusively. Thus, in this case:

$$\langle p^2 \rangle = \rho c \frac{W}{4\pi r_0^2} \tag{4.81}$$

4.5.1.2 Coherent Sources in a Line

In the preceding discussion, the assumption was implicit that, because the sources were incoherent, they tended not to interfere with each other, so that the calculation could proceed on the assumption that each source radiated independently of the other and the effect at any point of observation of all the sources was obtained as a simple sum of their energy contributions. However, in general, the problem can be more complicated, as sources that are coherent do tend to affect one another when placed in close proximity; that is, the energy that a source radiates will depend on the radiation impedance presented to it and the radiation impedance will be affected by the presence of nearby coherent sources. When a line of coherent point sources is considered, the sound pressure contributions from all sources add at any point of observation. In this case, the phase distribution of the sources along the line, as well as the phase shifts introduced in propagation from the sources to the point of observation, must be taken into account in determining the sum of the contributions. In general, the problem can become quite complicated. However, for the present purpose, it will be assumed that all sources radiate in phase and a great simplification is then possible.

When the point of observation r_0 satisfies the condition that $r_0/\lambda \gg 1$, then a line of coherent sources radiates like a continuous source of cylindrical waves. Consequently, if the sources are coherent and all radiate in phase with one another, then the mean square sound pressure at an observation point at normal distance, r_0, may be calculated by assuming that radiation is essentially in the form of cylindrical waves, corresponding to the surface of a cylinder of radius r_0 and area per unit length of $2\pi r_0$. The mean square sound pressure at r_0 is then:

$$\langle p^2 \rangle = \rho c \frac{W}{2\pi b r_0} \tag{4.82}$$

where W/b is effectively the sound power per unit length of the line source. The corresponding sound pressure level is given by:

$$L_p = L_W - 8 - 10\log_{10} r_0 - 10\log_{10} b + 10\log_{10}(\rho c/400) \qquad \text{(dB re } 20\,\mu\text{Pa)} \tag{4.83}$$

According to Equations (4.80) and (4.83), the sound pressure level decreases at the rate of 3 dB for each doubling of distance from a line of coherent or incoherent sources, and the coherent sources produce 2 dB less sound pressure level for the same radiated sound power.

4.5.2 Finite Line Source

4.5.2.1 Incoherent Sources in a Line

If the number of sources, each radiating a sound power of W and each spaced b apart from one another, is finite and they are arranged in a line of length, D, as illustrated in Figure 4.8, then the summation of Equation (4.75) may be expressed approximately as an integral. In this case, the sound power per unit length becomes W/b and Equation (4.81) (for incoherent sources) takes the following form (Rathe, 1969):

$$\langle p^2 \rangle = \int_{x_1}^{x_2} \frac{W}{b} \frac{\rho c}{4\pi r^2} \mathrm{d}x \tag{4.84}$$

Reference to Figure 4.8 allows the following relationships to be determined:

$$x_1 = r_0 \tan \alpha_\ell; \qquad x_2 = r_0 \tan \alpha_u; \qquad r = r_0/\cos\alpha;$$
$$x = r_0 \tan\alpha; \qquad \mathrm{d}x/\mathrm{d}\alpha = r_0 \sec^2\alpha; \qquad \text{and} \qquad \mathrm{d}x = r_0\,\mathrm{d}\alpha \sec^2\alpha$$

Substitution of these definitions into Equation (4.84) gives the following expression for the mean square sound pressure at the receiver (for an incoherent line source):

$$\langle p^2 \rangle = \rho c \int_{\alpha_\ell}^{\alpha_u} \frac{W}{b} \frac{\cos^2\alpha}{4\pi r_0^2} r_0 \sec^2\alpha \mathrm{d}\alpha \tag{4.85}$$

or

$$\langle p^2 \rangle = \rho c \frac{W}{4\pi r_0 b}[\alpha_u - \alpha_\ell] \tag{4.86}$$

Equation (4.86) can also be written as:

$$\langle p^2 \rangle = \rho c \frac{W_t}{4\pi r_0 D}[\alpha_u - \alpha_\ell] \tag{4.87}$$

where W_t is the total sound power radiated by the line source.

4.5.2.2 Coherent Sources in a Line

For a coherent source of finite length, the sound pressure level will be the same as for a finite length incoherent source for large distances from the source compared with the source length ($r_0 > D/2$) (Equation (4.87)). Close to a coherent source of finite length D ($r_0 < D/10$ from the source), the sound pressure level (normal to the centre of the source) will be the same as for an infinite coherent source (Equation (4.82)). In between, the sound pressure level will be between the two levels calculated using Equations (4.82) and (4.87) and interpolation can be used.

The preceding analysis is based on the assumption that the line source is radiating into free space away from any reflecting surfaces. If reflecting surfaces are present, then they must be accounted for as described in Section 4.9.

4.6 Piston in an Infinite Baffle

The classical circular piston source in an infinite rigid baffle has been investigated by many authors; for example, Kinsler et al. (1999), Pierce (1981) and Meyer and Neumann (1972). Other authors have contributed as well and will be mentioned during the course of the following discussion, which takes a different approach to previously published analyses. Readers interested in the analysis of a rectangular-shaped piston in an infinite baffle can find it in Beranek and Mellow (2012), Chapter 13.17 and in Mellow and Kärkkäinen (2016).

A piston is defined as a plane surface vibrating so that all parts of the surface vibrate in phase with the same amplitude. A baffle is defined as a plane, rigid, thin wall of infinite extent, with a hole containing the piston with an airtight seal around its edges. A baffle may be considered infinite in extent for our purposes here if its smallest dimension is less than three wavelengths of the sound being considered.

4.6.1 Far Field

The circular piston in an infinite rigid baffle, illustrated in Figure 4.9, is of interest because it has relatively simple geometry; it serves conveniently as an introduction to the behaviour of all

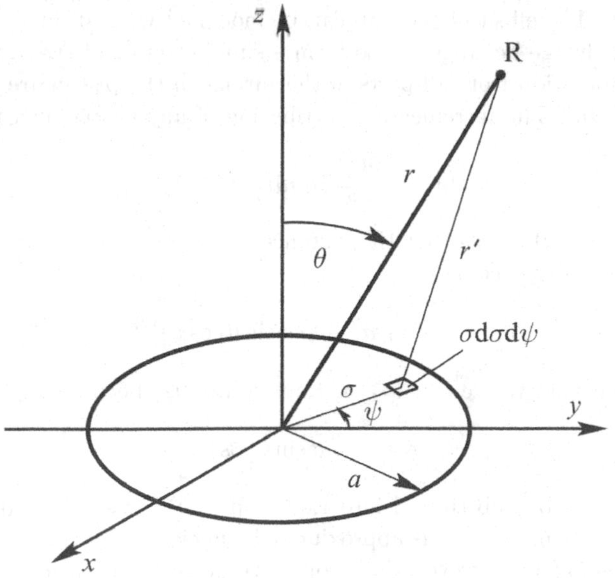

FIGURE 4.9 Piston source in an infinite baffle. The piston of radius a lies in the $x-y$ plane and vibrates vertically parallel to the z-axis with velocity amplitude, U.

radiating surfaces, and it can be approximated by a speaker in a wall. The requirement that the baffle be infinite means that edges are far enough removed for diffraction effects originating there to be ignored. Alternatively, the edges of the baffle might be covered with a sound-absorbent material with the same effect. The presence of the baffle implies that all of the sound power radiated by the piston is radiated into the hemispherical half-space bounded by the plane of the baffle. The piston, generally restricted to a circle of radius, a, is assumed to vibrate with uniform normal velocity of amplitude, U, and frequency, ω (rad/s).

A solution for the sound field generated by a piston in an infinite baffle will be constructed as an integral of tiny, simple point sources, each of which has a radius of a_i. The analysis begins with recognising that ka is much less than unity, which allows the use of Equation (4.36) to evaluate the constant, A of Equation (1.49) as:

$$A = \frac{Q_0}{4\pi} \tag{4.88}$$

Substituting Equation (4.88) into Equation (1.49), letting $r = a_i$ and assuming that $ka_i \ll 1$ gives:

$$u = \frac{(1 + \mathrm{j}ka_i)}{a_i^2}\frac{Q_0}{4\pi}\mathrm{e}^{\mathrm{j}(\omega t - ka_i)} \approx \frac{Q_0}{4\pi a_i^2}\mathrm{e}^{\mathrm{j}(\omega t - ka_i)} \tag{4.89}$$

As shown in Equation (4.34a), the quantity Q_0 may be interpreted as the strength amplitude or volume-velocity amplitude of a source. With this interpretation, substitution of Equation (4.88) into Equation (1.48) gives an expression for the sound pressure at location, R, in the acoustic field in terms of the source strength, Q_0:

$$p = \frac{\mathrm{j}\rho ckQ_0}{4\pi r}\mathrm{e}^{\mathrm{j}(\omega t - kr)} \tag{4.90}$$

Suppose that the piston in an infinite baffle is composed of a uniform distribution of an infinite number of simple sources, such as given by Equation (4.90), and each source has an elemental surface area of $\sigma \mathrm{d}\sigma \mathrm{d}\psi$ (see Figure 4.9). The amplitude of the strength of the source corresponding to the elemental area is $Q_0 = 2U\sigma \mathrm{d}\sigma \mathrm{d}\psi$, where U is the piston velocity amplitude. The factor of 2 is introduced because radiation is only occurring into half-space, due to the presence of the rigid wall and piston. The effect of the wall can be modelled with an image source, so that the actual source and its image act together to form a simple source of strength, $Q_0 = 2U\sigma \mathrm{d}\sigma \mathrm{d}\psi$. It is supposed that radiation from all parts of the surface of the piston are in phase; thus all of the sources are coherent. The incremental contribution of any source at a point of observation at distance r' is then:

$$\mathrm{d}p = \frac{\mathrm{j}\rho ckU}{2\pi r'}\sigma \mathrm{d}\sigma \mathrm{d}\psi \mathrm{e}^{\mathrm{j}(\omega t - kr')} \tag{4.91}$$

Reference to Figure 4.9 shows that the distance, r', is given by the following expression in terms of the centre-line distance, r.

$$r' = [r^2 + \sigma^2 - 2r\sigma \sin\theta \cos\psi]^{1/2} \tag{4.92}$$

For the case of the far field where $r \gg a \geq \sigma$, Equation (4.92) becomes approximately:

$$r' \approx r - \sigma \sin\theta \cos\psi \tag{4.93}$$

Equation (4.93) may be substituted into Equation (4.91), where the further simplification may be made that the denominator is approximated sufficiently by the first term on the right-hand side of Equation (4.93). However, the second term must be retained in the exponent, which reflects the fact that small variations in relative phase of the pressure contributions arriving at the observation point have a very significant effect on the sum.

Summing contributions given by Equation (4.91) from all points over the surface of the piston to obtain the total pressure, p, at the far-field observation point, O, and using Equation (4.93) results in the following integral expression:

$$p = \frac{\mathrm{j}\rho ck}{2\pi r} U \mathrm{e}^{\mathrm{j}(\omega t - kr)} \int_0^a \sigma \mathrm{d}\sigma \int_0^{2\pi} \mathrm{e}^{\mathrm{j}k\sigma \sin\theta \cos\psi} \mathrm{d}\psi \tag{4.94}$$

Integration of Equation (4.94) begins by noting that the second integral (divided by 2π) is a Bessel function of the first kind of order zero and argument, $k\sigma \sin\theta$ (McLachlan, 1941). Thus, the equation becomes:

$$p = \frac{\mathrm{j}\rho ck}{r} U \mathrm{e}^{\mathrm{j}(\omega t - kr)} \int_0^a \sigma \mathrm{J}_0(k\sigma \sin\theta) \mathrm{d}\sigma \tag{4.95}$$

Integration of Equation (4.95) gives the following useful result:

$$p = \frac{\mathrm{j}\rho ck}{2\pi r} \mathrm{F}(q) \mathrm{e}^{\mathrm{j}(\omega t - kr)} \tag{4.96}$$

where

$$\mathrm{F}(q) = U\pi a^2 [\mathrm{J}_0(q) + \mathrm{J}_2(q)] = U\pi a^2 2 \frac{\mathrm{J}_1(q)}{q} \tag{4.97}$$

and where

$$q = ka \sin\theta; \qquad \text{and} \qquad k = \omega/c = 2\pi f/c \tag{4.98a,b}$$

The quantities $\mathrm{J}_0(q), \mathrm{J}_1(q)$ and $\mathrm{J}_2(q)$ are Bessel functions of the first kind of order 0, 1 and 2, respectively, and argument, q. The real part of the pressure given by Equation (4.96) may be written as:

$$p = -\frac{\rho\omega}{2\pi r} \mathrm{F}(q) \sin(\omega t - kr) \tag{4.99}$$

The acoustic intensity in the far field is related to the acoustic pressure by Equation (1.78) and can be evaluated as:

$$I = \frac{\rho ck^2}{8\pi^2 r^2} \mathrm{F}^2(q) \tag{4.100}$$

The corresponding sound pressure level, L_p, is given by:

$$L_p = 10\log_{10}(\rho cI) + 94 = 20\log_{10}\left[\frac{\rho ck}{2\sqrt{2}\,\pi r}\mathrm{F}(q)\right] + 94 \quad \text{(dB re 20 µPa)} \tag{4.101}$$

Consideration of Equations (4.96), (4.99), (4.100) and (4.101) shows that as they are all functions of $\mathrm{F}(q)$, and in some directions, θ, there may be nulls in the sound field, according to Equation (4.98a). For example, reference to Figure 4.10, where $\mathrm{F}(q)/U\pi a^2$ is plotted as a function of q, shows that for q equal to 3.83, 7.0, etc., the function F is zero. On the other hand, for values of q less than 3.83, the function, F, becomes large. Generally, the sound radiated to the far field tends to beam on-axis and, in appropriate cases, to exhibit side lobes, depending on the frequency of sound radiation. This is illustrated in Figure 4.11, which shows the far-field sound pressure level distribution around a piston source, calculated using Equation (4.101) for three different ka values. The MATLAB® script, `radiation_pattern_baffled_piston.m`, can be used to plot the directivity of an oscillating piston in an infinite plane baffle, and is available for download from MATLAB scripts for ENC (2024).

Thus, in the far field, it can be seen from Figure 4.11 that, for ka values greater than about 3.8, the sound radiation from a piston in an infinite baffle is quite directional, and a directivity index, as discussed in Section 4.8, can be defined. In the near field, by contrast, a directivity index may make little sense, because of the presence of local maxima and minima in the pressure field.

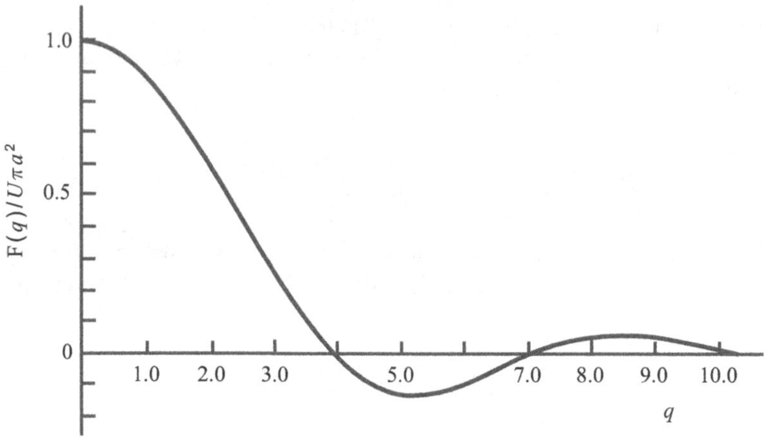

FIGURE 4.10 Far-field directivity function for a piston source in an infinite baffle. The piston has a radius of a and vibrates with a velocity amplitude, U. The abscissa is $q = ka\sin\theta$, where k is the wavenumber and θ is the angle of observation relative to the piston axis (see Figure 4.9).

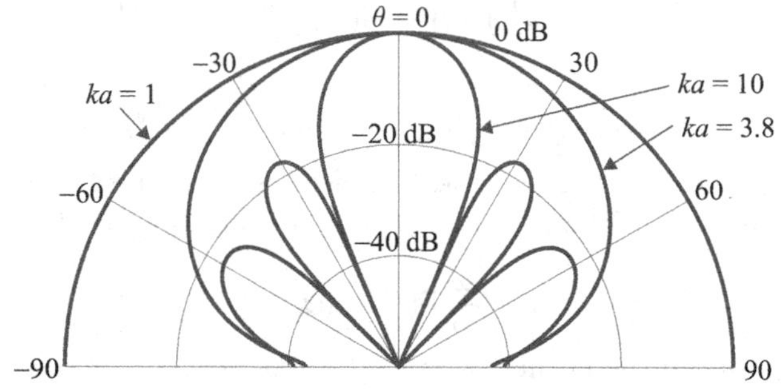

FIGURE 4.11 Radiation pattern of a piston source in an infinite baffle for three different frequencies (ka values).

4.6.2 Near Field On-Axis

In general, the near field of a piston source is quite irregular, being characterised by a complicated pattern of local pressure maxima and minima (Meyer and Neumann, 1972). Consequently, no attempt will be made here to calculate the near field except along the axis of symmetry, where the integration is simple and the result is instructive. Referring to Figure 4.9, let the point of observation lie at distance, r, on the axis of symmetry, z. In this case, set $\theta = 0$ in Equation (4.92) and differentiate, giving the following result:

$$r'\mathrm{d}r' = \sigma\mathrm{d}\sigma \tag{4.102}$$

Substitution of Equation (4.102) in Equation (4.91) and expressing the pressure on-axis at a distance, r, from the piston of radius, a, as an integral of the pressure contributed by all points on the surface of the piston, gives the following result:

$$p = \mathrm{j}\frac{\rho ckU}{2\pi}\int_0^{2\pi}\mathrm{d}\psi\int_r^{\sqrt{r^2+a^2}}\mathrm{e}^{\mathrm{j}(\omega t-kr')}\mathrm{d}r' \tag{4.103}$$

Carrying out the integration of Equation (4.103) gives the following expression for the pressure on-axis in the near field of the piston:

$$p = -\rho c U \left[\mathrm{e}^{\mathrm{j}(\omega t - k\sqrt{r^2+a^2})} - \mathrm{e}^{\mathrm{j}(\omega t - kr)} \right] \tag{4.104}$$

Introducing the following identity:

$$\mathrm{e}^{-\mathrm{j}\alpha} - \mathrm{e}^{-\mathrm{j}\beta} = -\mathrm{j}2\sin\frac{\alpha-\beta}{2}\mathrm{e}^{-\mathrm{j}(\alpha+\beta)/2} \tag{4.105}$$

allows Equation (4.104) to be rewritten as:

$$p = \mathrm{j}2\rho c U \mathrm{e}^{\mathrm{j}\omega t}\sin\left[\frac{k}{2}\left(\sqrt{r^2+a^2}-r\right)\right]\mathrm{e}^{-\mathrm{j}\dfrac{k}{2}(\sqrt{r^2+a^2}+r)} \tag{4.106}$$

Equation (4.106) shows that the pressure amplitude has zeroes (nulls) on-axis for values of axial distance, r, which satisfy the following condition, where n is the number of nulls, counting from the furthest towards the surface of the piston:

$$r = \frac{(a/\lambda)^2 - n^2}{2n/\lambda} \tag{4.107}$$

Equation (4.107) shows that the number of nulls, n, on-axis is bounded by the size of the ratio of piston radius to wavelength, a/λ, because only positive values of distance, r, are admissible. The equation also shows that there can only be nulls on-axis when $a/\lambda > 1$.

Letting r be large allows the argument of the sine function of Equation (4.106) to be expressed in series form as in the following equation:

$$\frac{k}{2}\left(\sqrt{r^2+a^2}-r\right) = \frac{kr}{2}\left[1+\frac{1}{2}\left(\frac{a}{r}\right)^2 + ... - 1\right] \approx \frac{ka^2}{4r} \tag{4.108}$$

Replacing the sine function in Equation (4.106) with its argument given by Equation (4.108) gives the following equation, valid for large distances, r:

$$p = \mathrm{j}\rho c\frac{Uka^2}{2r}\mathrm{e}^{\mathrm{j}[\omega t-(k/2)(\sqrt{r^2+a^2}+r)]} \tag{4.109}$$

In the far field of a source, the sound pressure decreases inversely with distance r; thus, Equation (4.109) describes the sound field in the far field of the piston source.

The sound field in front of the piston, which is characterised by the presence of maxima and minima, is called the geometric near field. The bound between the far field and the geometric near field cannot be defined precisely but may be taken as lying beyond the null on-axis furthest from the surface of the piston (see Equation (4.107)). Thus setting $n = 1$ in Equation (4.107) corresponding to the null furthest from the piston, readily leads to the following condition for a point of observation to lie in the far field (see Section 4.12):

$$r \gg a^2/2\lambda \tag{4.110}$$

4.6.3 Radiation Load of the Near Field

As mentioned earlier, the near field of a piston source can be quite complicated, but it is possible to compute rather simply the load that the acoustic field imposes on the vibrating piston, and this information is often quite useful. For example, the net time-varying force divided by the corresponding time-varying piston velocity can be computed and this quantity, known as the radiation impedance, is useful, as will be shown in Section 4.15.

Integration of Equation (4.91) over the surface, S of the piston gives the pressure, p_s, acting on surface element, $\mathrm{d}S'$ as a result of contributions from all other elements on the surface. Further integration of this pressure over the surface gives the force on the piston.

Referring to Figure 4.12 for definition of the coordinates, the following integral expression for the force on the piston is obtained:

$$F_R = \iint_S p_s \mathrm{d}S' = \frac{\mathrm{j}\rho c k U \mathrm{e}^{\mathrm{j}\omega t}}{2\pi} \int_0^a r\mathrm{d}r \int_0^{2\pi} \mathrm{d}\phi \int_0^a \sigma \mathrm{d}\sigma \int_0^{2\pi} \frac{\mathrm{e}^{-\mathrm{j}kh}}{h} \mathrm{d}\psi \tag{4.111}$$

where $h = \sqrt{\sigma^2 + r^2 - 2r\sigma\cos(\phi - \psi)}$.

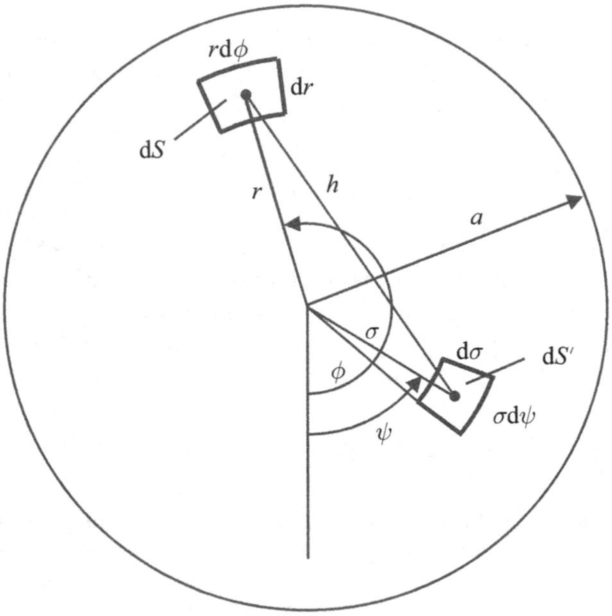

FIGURE 4.12 Piston coordinates and arrangement for calculating the radiation load.

Carrying out the integration and dividing by the piston velocity $U\mathrm{e}^{\mathrm{j}\omega t}$ gives the following expression for the radiation impedance:

$$Z_R = \frac{F_R}{U\mathrm{e}^{\mathrm{j}\omega t}} = \rho c \pi a^2 [R_R(2ka) + \mathrm{j}X_R(2ka)] \tag{4.112}$$

The radiation impedance is a complex quantity with a real part, $\rho c \pi a^2 R_R$, and a mass reactive part, $\rho c \pi a^2 X_R$, where R_R and X_R are defined by the following equations:

$$R_R(z) = \frac{z^2}{2 \times 4} - \frac{z^4}{2 \times 4^2 \times 6} + \frac{z^6}{2 \times 4^2 \times 6^2 \times 8} - \ldots \tag{4.113}$$

$$X_R(z) = \frac{4}{\pi}\left[\frac{z}{3} - \frac{z^3}{3^2 \times 5} + \frac{z^5}{3^2 \times 5^2 \times 7} - \ldots\right] \tag{4.114}$$

where $z = 2ka$.

The real part, R_R, may be used to calculate the radiated power, given the piston velocity amplitude, U; that is:

$$W = \frac{1}{2}\text{Re}\{Z_R\}U^2 = \frac{1}{2}\rho c R_R \pi a^2 U^2 \tag{4.115}$$

Values for the real term, R_R, and mass reactance term, X_R, are plotted in Figure 4.13 as solid lines, for the piston mounted in an infinite baffle (Equations (4.113) and (4.114)). For comparison, the analytically more complex case of a piston mounted in the end of a tube (Levine and Schwinger, 1948) is represented by dashed lines. Reference to the figure shows that the mass reactance term in both cases passes through a maximum, then tends to zero with increasing frequency, while the real term approaches and then remains approximately unity with increasing frequency. The figure shows that the effect of the baffle at low frequencies is to double the real term (increase it by 3 dB) and increase the mass reactance by $\sqrt{2}$, while at frequencies above $ka = 5$, the effect of the baffle is negligible.

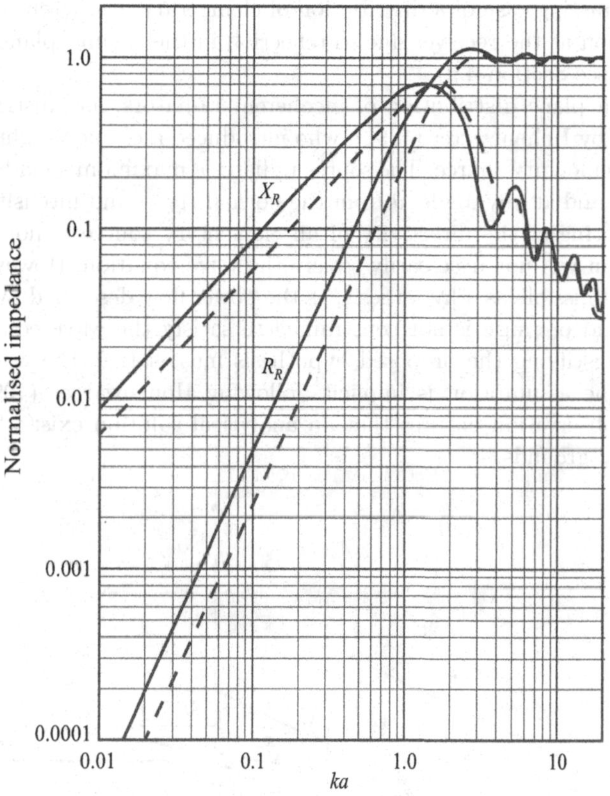

FIGURE 4.13 Radiation impedance functions for a piston source mounted either in an infinite baffle (solid line) or in the end of a long tube (dashed line). The piston has a radius of a and the impedance has the form: $Z_R = \rho c \pi a^2[R_R(2ka) + jX_R(2ka)]$.

Figure 4.13 also shows that, in the limit of high frequencies, the radiation impedance of the piston becomes the same as that of a plane wave, and although the pressure distribution over the surface of the piston is by no means uniform, the power radiated is the same as would be radiated by a section of a uniform plane wave of area πa^2 (Meyer and Neumann, 1972).

The behaviour of the radiation impedance shown in Figure 4.13 is typical of all sound radiators. For example, the open end of a duct will have a similar radiation impedance, with the result that at low frequencies, a sound wave propagating within the duct will be reflected back into the duct as though it were reflected from the mass reactance at the free end. Resonance in a musical instrument is achieved in this way. However, at high frequencies, where the impedance becomes essentially that of a freely travelling plane wave, there will be little or no reflection at the open end.

4.7 Incoherent Plane Radiator

4.7.1 Single Wall

The piston in an infinite baffle, which was considered in Section 4.6, may be considered as an example of a plane distribution of coherent radiators; that is, all parts of the piston vibrate with the same amplitude and phase. By contrast, the wall of a building housing noisy machinery may be modelled as a distribution of elementary incoherent radiators. Openings such as windows or open doors in walls of buildings may be similarly modelled. Consequently, it will be of interest to consider the contrasting case of a distribution of elementary incoherent radiators. In all such cases, the contribution at the receiver due to reflection in the ground plane must be taken into account separately (see Section 4.9).

Radiation from a plane distribution of incoherent radiators, as illustrated in Figure 4.14, has been considered by Hohenwarter (1991), who introduced a cosine weighting on the radiation directivity of each elementary source. The source radiates a maximum sound intensity, I_0, normal to the plane surface and at any angle, θ from the normal the sound intensity, $I = I_0 \cos\theta$. Note that arbitrary weighting of the directivity of an elementary source is not admissible, because the resulting equation will not necessarily satisfy the wave equation. However, weighting of the energy radiated is admissible as a hypothesis, as the power flow described by the sound intensity or mean square sound pressure is not constrained to satisfy the wave equation. Of course, an analytical solution justifying the proposed hypothesis must satisfy the wave equation, but for the purpose here, the assumption is implicit, following Hohenwarter (1991), that a solution in agreement with observation ensures that an analytical solution exists. The analysis is thus justified on empirical grounds.

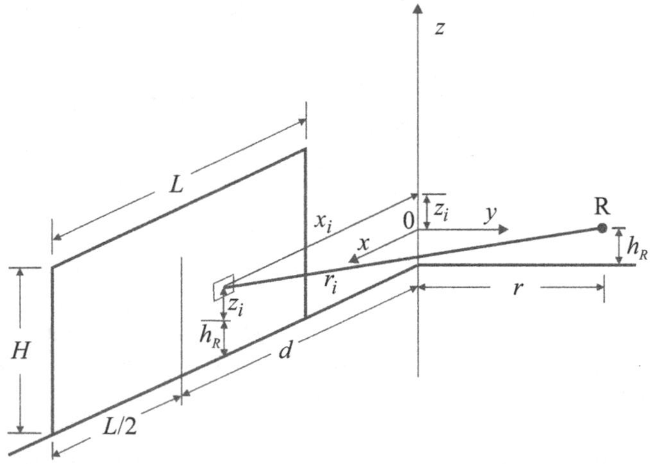

FIGURE 4.14 Rectangular noise source of length, L, and height, H, showing general field observation point, R, and associated coordinates.

Use of the cosine-weighted directivity for each elementary source both greatly simplifies the integration, which is otherwise difficult, and provides an answer in better agreement with observation than an analysis based on unweighted sources. For example, the weighted-source model predicts a finite sound pressure level as one approaches such a source (wall of a building), whereas the unweighted model predicts a sound pressure level increasing without bound.

The plane incoherent source is assumed to be located in a large plane baffle (or wall) in the $x - z$ plane at $y = 0$. The wall is assumed to be much greater than a wavelength in dimension so that sound is effectively excluded from the rear side. The axis origin is shown in the figure and the receiver, R, is located at a normal distance of r from the plane surface at coordinates, $(x,\ y,\ z) = (0, r, 0)$. The source of height, H and width, L, has its centre located at coordinates, $(d,\ 0,\ (H/2 - h_R))$ where h_R is the height of the receiver above the bottom of the plane source.

Applying Hohenwarter's cosine weighting function to the elemental source, i, shown in Figure 4.14 implies that the sound pressure squared at receiver location, R, due to the elemental source, i, is obtained by multiplying the squared pressure, calculated by assuming that the source radiates uniformly in all directions, with the directivity factor, D_i, given by the following equation:

$$D_i = r(x_i^2 + z_i^2 + r^2)^{-1/2} = r/r_i \tag{4.116}$$

The quantities, x_i and z_i are dependent on the location of element i and r_i and r are dependent on both the location of the elemental source, i, and the receiver location and are defined in Figure 4.14.

Using Equation (4.14) including the cosine weighted directivity factor D_i, and replacing the "4" with a "2" to account for radiation only into half-space, the mean square sound pressure at the receiver location, R, due to elementary source, i, located at a distance, r_i, is:

$$\langle p_i^2 \rangle = \frac{W_i \rho c D_i}{2\pi r_i^2} \tag{4.117}$$

The sound power, W_i, radiated by an elementary source, i, of dimensions $\mathrm{d}x \times \mathrm{d}z$ is defined as:

$$W_i = \frac{W \mathrm{d}x \mathrm{d}z}{HL} \tag{4.118}$$

where W is the total power radiated by the plane source. Substituting Equations (4.118) and (4.116) for W_i and D_i, respectively, into Equation (4.117) gives:

$$\langle p_i^2 \rangle = \frac{\rho c W}{2\pi HL} \frac{r}{r_i^3} \mathrm{d}x \mathrm{d}z = \frac{\rho c W}{2\pi HL} [x_i^2 + z_i^2 + r^2]^{-3/2} r \mathrm{d}x \mathrm{d}z \tag{4.119}$$

Summing the contributions to the receiver mean square pressure of each elemental source over the area of the wall gives the following integral expression for the total mean square sound pressure at the observation point:

$$\langle p^2 \rangle = \frac{\rho c W}{2\pi HL} \int_{-h_R}^{H-h_R} \mathrm{d}z \int_{d-L/2}^{d+L/2} [x^2 + z^2 + r^2]^{-3/2} r \mathrm{d}x \tag{4.120}$$

Integration of the above expression gives the following result, where the dimensionless variables that have been introduced are $\alpha = H/L$, $\beta = h_R/L$, $\gamma = r/L$ and $\delta = d/L$.

$$\begin{aligned}\langle p^2 \rangle = \frac{\rho c W}{2\pi HL} \Bigg[& \tan^{-1} \frac{(\alpha - \beta)(\delta + 1/2)}{\gamma\sqrt{(\alpha - \beta)^2 + (\delta + 1/2)^2 + \gamma^2}} \\ & + \tan^{-1} \frac{\beta(\delta + 1/2)}{\gamma\sqrt{\beta^2 + (\delta + 1/2)^2 + \gamma^2}} - \tan^{-1} \frac{(\alpha - \beta)(\delta - 1/2)}{\gamma\sqrt{(\alpha - \beta)^2 + (\delta - 1/2)^2 + \gamma^2}} \\ & - \tan^{-1} \frac{\beta(\delta - 1/2)}{\gamma\sqrt{\beta^2 + (\delta - 1/2)^2 + \gamma^2}} \Bigg]\end{aligned} \tag{4.121}$$

As there are no special assumptions that restrict the use of the above equation, it may be used to predict the mean square sound pressure at any point in the field away from the surface of the plane source. For further discussion, see Hohenwarter (1991).

The mean square sound pressure on the axis of symmetry, $\delta = 0$, $\beta = 0.5\alpha$, is also the maximum mean square sound pressure in a plane parallel to the plane of the source, determined by making γ a constant value. Equation (4.121) takes the following form on the axis of symmetry:

$$\langle p^2 \rangle = \frac{2\rho c W}{\pi H L} \tan^{-1}\left[\frac{HL}{2r\sqrt{H^2 + L^2 + 4r^2}}\right] \tag{4.122}$$

At large distances from the source, so that r is very much larger than the dimensions H and L, Equation (4.122) reduces to the following approximate form:

$$\langle p^2 \rangle = \frac{\rho c W}{2\pi r^2} \tag{4.123}$$

The reason for the "2" in the denominator, rather than the "4" shown in Equation (4.14) is that the plane source is mounted in a baffle (or wall) that is much larger than a wavelength and is thus radiating into a hemispherical space instead of a spherical space. Equation (4.123) shows that at large distances from the source, the mean square sound pressure is solely a function of the total radiated sound power and the distance from the plane of the source to the receiver, r, and not of any of the other variables. The form of Equation (4.123) is the same as that of a simple monopole source radiating into half-space (see Equation (4.14)) so that r may be taken as the distance from the centre of the source to the observation point and the source may be treated like a point source.

Close to the wall, as r tends to zero, Equation (4.122) leads to the following asymptotic form, which is consistent with expectation based on the assumed model:

$$\langle p^2 \rangle = \frac{\rho c W}{HL} \tag{4.124}$$

Equations (4.122) and (4.123) have been used to construct Figure 4.15, which shows the difference one might expect in sound pressure level at a particular distance, r, from the source when comparing the field levels due to a point and plane source of the same power output radiating into a hemispherical half-space.

Close to the source the sound pressure level due to the point source will be greater, whereas at large distances, where the ratio $r/\sqrt{HL}$ is large, the sound pressure levels due to the point and plane sources will be similar and will decrease at the rate of 6 dB per doubling of distance from the source. It can be seen that as the aspect ratio of the lane source becomes larger, it requires a greater distance from the source before the plane source appears as a point source to an observer. All curves represent panels with identical total areas. The MATLAB® script, `single_wall_spl_vs_angle.m`, can be used to plot the directivity and sound pressure level contour map of an incoherent plane radiator wall, and is available for download from MATLAB scripts for ENC (2024).

The side of a building which houses noisy machinery, or an open window, is often modelled as an incoherent plane radiator. Although this is not strictly correct, experimental data show that acceptable results are obtained when 1/3-octave or wider frequency bands are used for the analysis. The sound power radiated by a wall can be calculated in one of two ways. The first method uses calculated values for the interior sound pressure level and measured wall noise reduction properties as described in Section 7.2.2. The second method uses measured wall vibration levels, as described in Section 4.15.

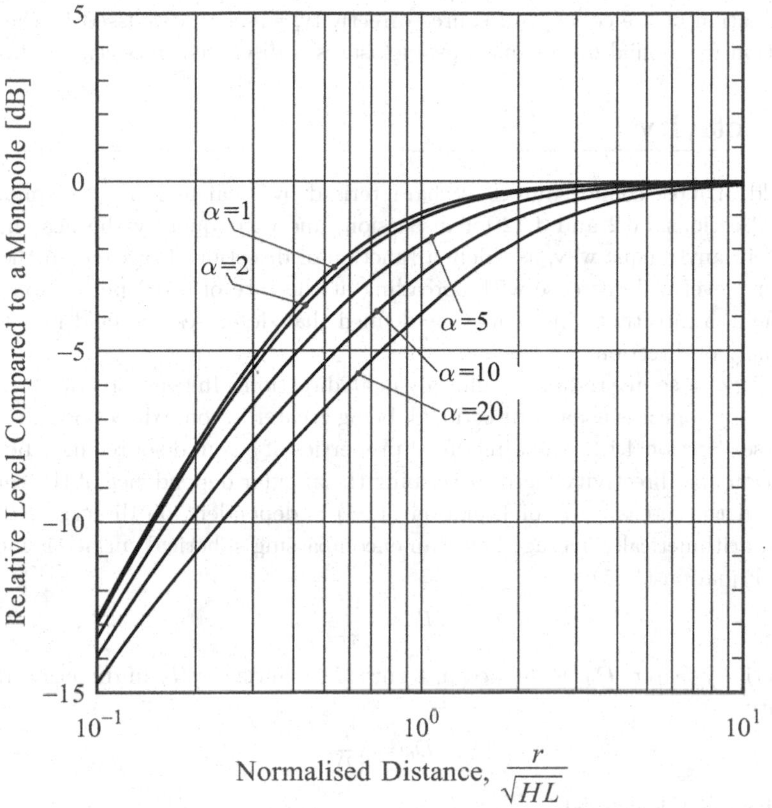

FIGURE 4.15 Relative levels of point monopole (Equation (4.123)) and plane (Equation (4.122)) sources on the axis of symmetry. The plane source is of length, L, and height, H, and the observation point is at normal distance, r, from the centre of the plane source. If $L < H$, then $\alpha = H/L$, otherwise if $L > H$, $\alpha = L/H$. The normalised distance from the monopole is calculated using the dimensions, H and L, of the plane source with which it is being compared, so the actual distance from the source is the normalised distance multiplied by $\sqrt{HL}$ for both the plane source and the point source. The curves are independent of the panel area.

4.7.2 Several Walls of a Building or Enclosure

The sound pressure level at a specified distance, r, from a building due to radiation by the walls and roof is calculated, in practice, by assuming that the sound energy is uniformly distributed over the area of an encompassing hemispherical surface of radius, r. Thus:

$$L_p = L_{Wt} - 10\log_{10} S + 10\log_{10}\frac{\rho c}{400} \quad \text{(dB re } 20\,\mu\text{Pa)} \tag{4.125}$$

where S is the surface area of the hemisphere of radius r, and L_{Wt} is the total radiated sound power level, calculated on the basis of the total sound power radiated by all of the exterior walls and the roof of the building. The origin of the last term in the equation can be explained by reference to Equation (1.92). It is usually less than 0.2 dB and is often ignored. The sound propagation effects discussed in Chapter 5 would also have to be included in practice. The effect of differences in power radiated by the walls and roof is taken into account by adding a correction to the sound pressure level, L_p, radiated in each direction. For a rectangular building where five directions are of concern, the correction is the difference between L_{Wi} and $L_{Wt} - 7$ ($= L_{Wt} - 10\log_{10} 5$) where L_{Wi} is the sound power level radiated by the ith surface. This leads to the introduction of the concept of source directivity, which is discussed in the next section. Sound radiation from buildings or machine enclosures is discussed in Section 7.4.2.

4.8 Directivity

The near field of most sound sources is characterised by local maxima and minima in sound pressure (see Sections 4.6.2 and 4.12 for discussion) and consequently the near field cannot be characterised in any unique way, as solely a function of direction. However, in the far field, the sound pressure level will decrease with spreading at the rate of 6 dB per doubling of distance and in this field, a directivity index may be defined that describes the field in a unique way as a function solely of direction.

The simple point source radiates uniformly in all directions. In general, however, the radiation of sound from any source is usually directional, being greater in some directions than in others. In the far field (see Section 4.12), the directional properties of a sound source may be quantified by the introduction of a directivity factor describing the angular dependence of the sound intensity. For example, if the intensity, I, of Equation (1.84) is dependent on direction, then the mean intensity, $\langle I \rangle$, arithmetically averaged over an encompassing spherical surface is introduced and, according to Equation (1.85):

$$\langle I \rangle = \frac{W}{4\pi r^2} \tag{4.126}$$

The directivity factor, D_θ, is defined in terms of the intensity I_θ in direction (θ, ψ) and the mean intensity:

$$D_\theta = \frac{I_\theta}{\langle I \rangle} \tag{4.127}$$

The directivity index is defined as:

$$\mathrm{DI} = 10 \log_{10} D_\theta \tag{4.128}$$

Alternatively, making use of Equations (4.126) and (4.127), it can be written as:

$$\mathrm{DI} = 10 \log_{10} I_\theta - 10 \log_{10} W + 10 \log_{10} 4\pi r^2 \qquad \text{(dB)} \tag{4.129}$$

In general, the directivity index is determined by measurement of the intensity I_θ at distance r and angular orientation (θ, ψ) from the source centre. Alternatively, the sound pressure level, L_p, may be measured instead of intensity where Equations (1.78), (1.87) and (1.89) have been used to rewrite Equation (4.129) in the following useful form:

$$\mathrm{DI} = L_p - L_W + 20 \log_{10} r - 10 \log_{10}(\rho c/400) + 11 \qquad \text{(dB)} \tag{4.130}$$

4.9 Reflection Effects

The presence of a reflecting surface near to a source will affect the sound radiated and the apparent directional properties of the source. Similarly, the presence of a reflecting surface near to a receiver will affect the sound received by the receiver. In general, a reflecting surface will affect not only the directional properties of a source but also the total power radiated by the source (Bies, 1961). As the problem can be complicated, the simplifying assumption is often made, and will be made here, that the source is of constant power output; thus only the case of constant power sources will be considered in the following sections. Other source types are discussed in Section 4.10.

4.9.1 Simple Source Near a Reflecting Surface

The concept of directivity may be used to describe the radiation from a simple source in the proximity of one or more bounding planes when it may be assumed that:

- the distance between the source and the reflecting plane is small compared with the distance from the source to the observation point;
- the sound power of the source may be assumed to be constant and unaffected by the presence of the reflecting plane.

For example, a simple source on the ground plane or next to a wall will radiate into the resulting half-space. If the simple source is assumed to be a constant power source, then in this case, it may be represented by modifying Equation (4.14) and using Equation (1.78) to give:

$$W = I\frac{4\pi r^2}{D} = \langle p^2 \rangle \frac{4\pi r^2}{\rho c D} \tag{4.131}$$

The intensity, I, is independent of angle in the restricted region of propagation, and the directivity factor, D, takes the value listed in Table 4.1. For example, the value of D for the case of a simple source next to a reflecting wall is 2, showing that all of the sound power is radiated into the half-space defined by the wall.

TABLE 4.1 Directivity factors for a simple source near reflecting surfaces (a constant power source is assumed – see text)

Situation	Directivity factor, D	Directivity index, DI (dB)
Free space	1	0
Centred in a large flat surface	2	3
Centred at the edge formed by the junction of two large flat surfaces	4	6
At the corner formed by the junction of three large flat surfaces	8	9

If the distance, a, between the source and a reflecting surface is large in comparison, or comparable with the distance, b, between the source and receiver point and the source is radiating broadband noise, then the effective sound pressure squared is calculated by adding the separate sound pressure squared contributions due to the source and its image in the reflecting plane, with the image position taken as far behind the reflecting surface as the source is in front, and with the image power equal to the source power (see Figure 4.16). The image source is the apparent location of the source behind the reflection plane when viewed from the receiver point.

If the noise is tonal in nature, then sound pressures of the direct and reflected waves must be added taking into account the phase shift on reflection and relative phase shift due to the differences in lengths of the two propagation paths.

4.9.2 Observer Near a Reflecting Surface

When the receiver is located close, of the order of one-tenth of a wavelength or less, to a reflecting surface, the sound pressure of the source image (or reflected wave, which can also be modelled using a receiver image instead of the source image) is added to that at the receiver due to the direct sound wave, i.e. pressure doubling occurs with an apparent increase in sound pressure level

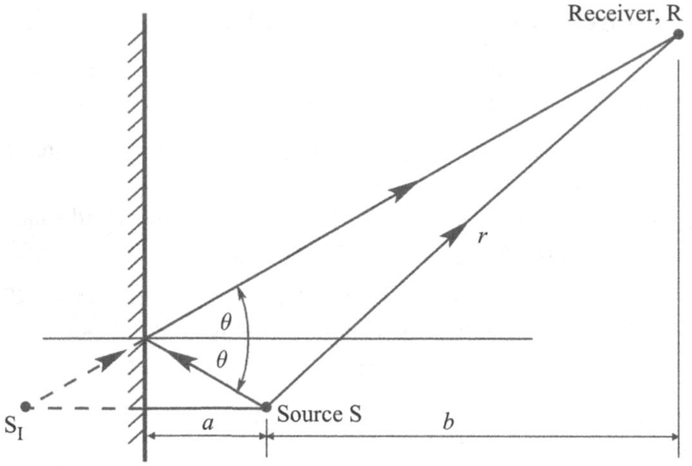

FIGURE 4.16 Geometry illustrating reflection from a plane rigid surface.

of 6 dB. On the other hand, when the receiver is located further than one-tenth of a wavelength from a reflecting surface, then the path difference between the direct and reflected waves is usually sufficiently large for most practical sound sources radiating non-tonal noise for the two waves to combine with random phase; the squared pressures, $\langle p^2 \rangle$, will add with an apparent increase in sound pressure level of approximately 3 dB, provided that the source–receiver distance is much larger that the distance between the source and the reflecting surface.

If the noise is tonal in nature, then sound pressures of the direct and reflected waves must be added taking into account the phase shift on reflection and relative phase shift due to the differences in lengths of the two propagation paths.

4.9.3 Observer and Source Both Close to a Reflecting Surface

For both source and receiver located close to the same reflecting plane, where their distances h_S and h_R from the plane are small compared to the distance d between them (see Figure 4.17), the path difference between the direct and reflected waves is relatively small. In this case, if there is no turbulence, no temperature gradients in the medium in which the wave is propagating, and the real part of the characteristic impedance of the reflecting plane, Z_s, satisfies the relation

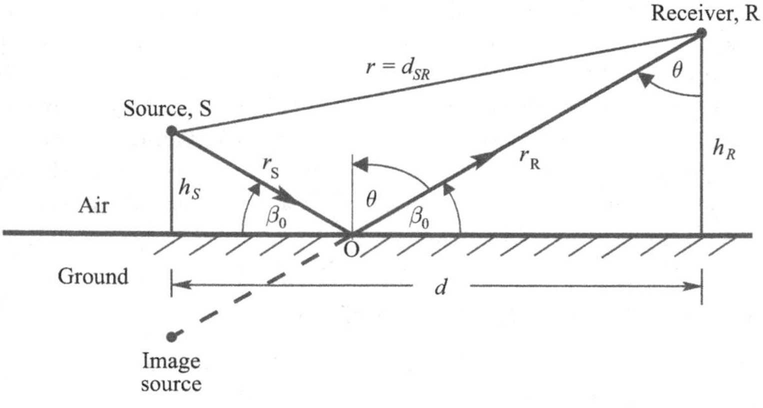

FIGURE 4.17 Geometry illustrating reflection and transmission above a reflecting plane.

$\text{Re}\{Z_s\}/\rho c > d/(h_S + h_R)$, the combination of the direct and reflected waves is coherent, and they will be essentially in-phase with the same amplitude at the receiver. This means that the resulting sound pressure amplitude will be the sum of the individual sound pressure amplitudes corresponding to the direct and reflected waves. This is in contrast to the usual case where incoherent combination of the direct and reflected waves means that the phase relationship is random, so the total sound pressure is the square root of the sum of the squared pressures of the direct and reflected waves.

Coherent addition of source and image thus results in a 6 dB rather than a 3 dB increase in sound pressure level over that for a free field. In this case only, all directivity factors shown in Table 4.1 should be multiplied by two. For the case where the impedance of the reflecting plane does not satisfy the preceding relationship, reference should be made to Chapter 5.

4.10 Radiation Impedance

When the sound power level of a source is specified, unless otherwise stated, the assumption is implicit that the radiation impedance presented to the source is the same as it would be in a free unbounded space, commonly referred to as "free field". The radiation impedance is analogous to the load impedance presented to a generator in the more familiar case of electrical circuit theory. In the latter case, as is well-known, the power delivered by a generator to a load depends on both the load impedance and the generator internal impedance. For the case of acoustic sources, the internal impedance is seldom known, although for vibrating and radiating structures the assumption is commonly made that they are constant volume-velocity (or infinite internal impedance) sources. The meaning of this is that the motion of the vibrating surface is assumed to be unaffected by the acoustic radiation load, and this is probably a good approximation in many cases.

Most aerodynamic sources are, however, not well represented as constant volume-velocity sources. In these cases, the problem becomes quite complicated, although such sources are often approximated as constant acoustic-pressure sources, which means that the acoustic pressure at the source, rather than the source volume-velocity, is unaffected by the acoustic radiation load.

The radiation impedance presented to a source in a confined space or in the presence of nearby reflectors will seldom be the same as that presented to the same source by a free field. For example, if a sound source is placed in a highly reverberant room, as will be discussed in Section 6.4, the radiation impedance will strongly depend on both position and frequency, but in an indeterminate way. However, if the source is moved about to various positions, the arithmetic average of the radiation impedance will tend to that of a free field. Thus, in using sound power level as a descriptor, it is tacitly assumed that, on average, the source will be presented with a free field, and on average this assumption seems justified.

Some special situations in which the free-field assumption is not valid, however, are worth mentioning. If an omnidirectional constant volume-velocity source is placed next to a large reflecting surface such as a wall, then the source will radiate twice the power that it would radiate in free space. This may be understood by considering that the pressure everywhere would be doubled, and thus the intensity would be four times what it would be in free space. As the surface area of integration is reduced by a factor of two from a sphere to a hemisphere encompassing the source, the integral of the intensity, (which is equal to the power radiated), is double what it would be in free space. Thus, the sound power radiated by an omnidirectional constant volume-velocity source will be least in the free field, 3 dB greater when the source is placed next to a reflecting surface, 6 dB greater when the source is placed at the junction of two surfaces and 9 dB greater when the source is placed in the corner at the junction of three surfaces. The corresponding intensity of sound at some distance r from the surfaces will be 6, 12 and 18 dB, respectively, greater than in a free field. A typical constant volume-velocity source would be a loudspeaker mounted on one wall of a small, airtight enclosure.

A constant acoustic-pressure source would radiate half the sound power that it radiates in a free field, as it is radiating into half-space and the sound pressure is the same as if it were radiating into a full hemispherical space. A typical constant acoustic-pressure source would be an aerodynamic source.

A constant power source would radiate the same power into free space, as it would when placed near to a reflecting surface. Thus, the sound intensity would be doubled as the same amount of acoustic power as was radiated into a spherical space is radiated into a hemispherical space.

A restriction in application of these ideas must be mentioned. For a source to be close to a reflecting surface it must be closer than one-tenth of a wavelength of the emitted sound. At distances greater than one-tenth of a wavelength the effect on radiated sound power rapidly diminishes until, for a band of noise at a distance of one-half wavelength, the effect is negligible. For tones, the effect persists to somewhat greater distances than a half wavelength.

If the source is sufficiently far away (half wavelength for octave or 1/3-octave band sources and two wavelengths for pure tone sources) that its sound power is not at all affected by the reflecting surface, or if the source is a constant power source, then the sound pressure level (and directivity D) at any location can be estimated by adding the sound field of an image source to that of the source with no reflector present, as discussed in Section 4.9.

Making reference to the discussion of reflection effects in Section 4.9, and noting that for the cases considered in Table 4.1 and for a constant volume-velocity source located closer than one-tenth of a wavelength to the reflecting surface(s), the following equation may be written for the intensity, I:

$$I = W_0 D^2 / 4\pi r^2 = \langle p^2 \rangle / \rho c \tag{4.132}$$

where W of Equation (4.126) has been replaced with $W_0 D$, W_0 is the power radiated in the free field and D takes the values shown in Table 4.1.

The internal impedance of a practical source may not be large, and it will never be infinite. The case of infinite impedance (i.e. a constant volume-velocity source) may be taken as defining the upper bound on radiated power. For example, a loudspeaker, which, unless backed by a small, airtight enclosure, is not in general a constant volume-velocity source. When placed in a corner in a room, it will produce more sound power and a higher sound pressure and sound intensity than when placed in the free field, but the increase in intensity at some reference point on axis will be less than 18 dB.

4.11 Relation between Sound Power and Sound Pressure

The sound pressure level produced by a source at a specified location may be calculated in terms of the specified sound power level and directivity. Use of Equation (4.130), and the observation that in the SI system of units the characteristic impedance, ρc, is approximately 414, allows the following equation to be written relating sound pressure level, sound power level and directivity factor:

$$L_p = L_W + 10\log_{10} D - 10\log_{10} S + 0.15 \quad \text{(dB re } 20\,\mu\text{Pa)} \tag{4.133}$$

where

$$S = 4\pi r^2 \quad (\text{m}^2) \qquad \text{and} \qquad 10\log_{10}\left(\frac{\rho c}{400}\right) = 0.15 \text{ (dB)} \tag{4.134}$$

Equation (4.133) relates the sound pressure level, L_p, at a point to the sound power level, L_W, of a source, its directivity factor, D, dependent on direction from the source, and the distance, r, from the source to the measurement point. The equation holds as long as the measurement point is in the far field of the source. The far field will be discussed in Section 4.12.

If sound power is not constant and the alternative case is considered of a constant volume-velocity source, radiating equally well in all directions and located within one-tenth of a wavelength of a reflecting surface, then, as discussed earlier, W in Equation (4.133) is replaced with $W_0 D$, where W_0 is the sound power that the source would radiate in the free field. In this case, the relation between sound pressure level L_p and sound power level L_{W0} becomes:

$$L_p = L_{W0} + 20\log_{10} D - 10\log_{10} S + 0.15 \qquad \text{(dB re 20 µPa)} \tag{4.135}$$

4.12 Radiation Field of a Sound Source

The sound field radiated by a source in a free field may be divided into three regions: the hydrodynamic near field, the geometric (or Fresnel) near field and the far field.

In general, the hydrodynamic near field is considered to be that region immediately adjacent to the vibrating surface of the source, extending outward a distance much less than one wavelength. This region is characterised by fluid motion that is not directly associated with sound propagation. For example, local differences in phase of the displacement of adjacent parts of a vibrating surface will result in fluid motion tangential to the surface, if the acoustic wavelength is long compared with the separation distance between the adjacent parts. The acoustic pressure will be out of phase with local particle velocity. As sound propagation to the far field is associated with the in-phase components of pressure and particle velocity, it follows that measurements of the acoustic pressure amplitude in the near field give an inaccurate indication of the sound power radiated by the source.

The sound field adjacent to the hydrodynamic near field is known as the geometric near field. In this region, interference between contributing waves from various parts of the source leads to sound pressure levels that do not necessarily decrease monotonically at the rate of 6 dB for each doubling of the distance from the source; rather, relative maxima and minima are to be expected. This effect is greater for pure tones than it is for bands of noise. However, in the geometric near field, the particle velocity and acoustic pressure of the contributing waves from the various parts of the source are in phase, as for waves in the far field.

The possibility of determining the sound power radiated by an extended source from pressure measurements made in the geometric near field has been investigated for the case of the baffle mounted piston considered in Section 4.6, using a computer simulation (Bies and Bridges, 1993). It was shown that a simple sound level meter can provide an accurate determination of the radiated sound power even in the geometric near field of the source, where close investigation of the field with an intensity meter would suggest that such determination might not be possible. In fact, in the case investigated, the intensity meter would appear to provide no special advantage. While the reported study was not exhaustive, it is probably indicative and thus it may be concluded that radiated sound power can be calculated from a sufficient number of sound pressure measurements made in the geometric near field. The problem becomes the determination of a sufficient number of measurements, for which there is no general rule.

The region of the sound field extending beyond the geometric near field to infinity is known as the far field, where sound pressure levels decrease monotonically at the rate of 6 dB for each doubling of the distance from the source (for exceptions see Chapter 5). In the far field, the source directivity is well defined. The far field is characterised by the satisfaction of three criteria, written as (Bies, 1976):

$$r \gg \lambda/(2\pi), \qquad r \gg \ell, \qquad r \gg \pi\ell^2/(2\lambda) \tag{4.136}$$

where r is the distance from the source to the measurement position, λ is the wavelength of radiated sound and ℓ is the characteristic source dimension. The characteristic source dimension is the square root of the area of a radiating surface. If the source is a 3-D object, the characteristic dimension is the square root of the area facing the receiver of interest. The "much greater than" criterion in the above three expressions refers to a factor of three or more. More generally, defining $\gamma = 2r/\ell$, and $\kappa = \pi\ell/\lambda$, the above criteria reduce to:

$$\gamma \gg 1/\kappa, \quad \gamma \gg 2, \quad \gamma \gg \kappa \tag{4.137}$$

These criteria are used to construct Figure 4.18.

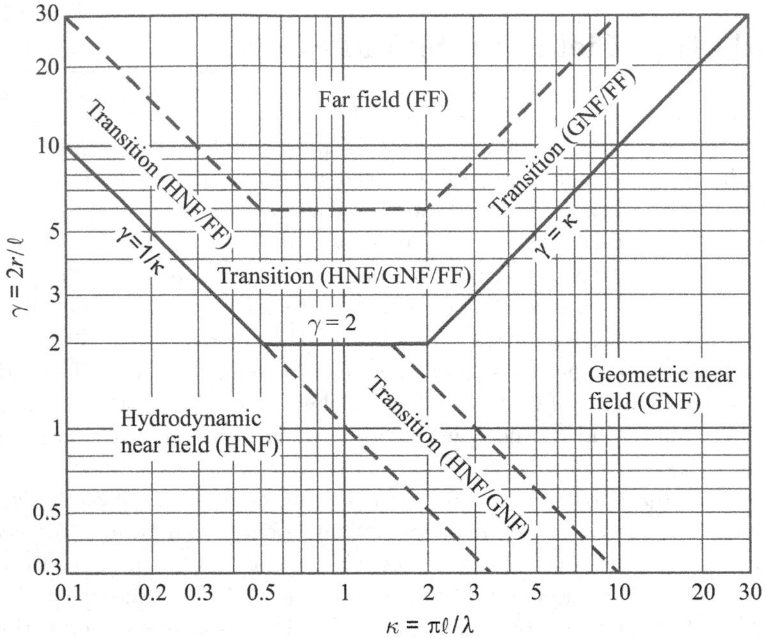

FIGURE 4.18 Radiation field of a source.

The criteria given by Equations (4.136) or (4.137) define the bound between the geometric near field and the transition to the far field, as shown in Figure 4.18. It should be pointed out, however, that while satisfaction of the inequality given by the equations is sufficient to ensure that one is in the far field, it may not always be a necessary condition. For example, a very large pulsating sphere has only a far field.

4.12.1 Free-Field Simulation in an Anechoic Room

One way to produce a free field for the study of sound radiation is to construct a room that absorbs all sound waves which impinge upon the walls. Such a room is known as an anechoic room, and a sound source placed in the room will produce a sound field similar in all respects to the sound field that would be produced in a boundary-free space, except that its extent is limited by the room boundaries. The walls of an anechoic room are constructed from wedges that point into the interior of the room. The wedges should be at least 1 m long and approximately 150 mm square at the end that is attached to a solid wall (usually 300 mm-thick concrete to minimise the intrusion of exterior noise into the room). If measurements are required below 100 Hz, then the wedges need to be longer than 1 m. Thus, for measurements down to 50 Hz, the wedges would need to be 2 m long. The wedge material should be an acoustically absorbing material with high absorption coefficients over the frequency range of interest. Many anechoic rooms use rockwool, fibreglass or acoustic foam. If either of the former two are used, they should be encapsulated in an acoustically transparent thin cloth to prevent fibres escaping. A disadvantage of acoustic foam is that it may not satisfy fire regulations as it emits toxic gas when ignited. The required acoustic specifications of anechoic rooms used for sound power measurements are listed in ISO 3745 (2017).

Figure 4.18 has special significance for the use of anechoic rooms for the purpose of simulating a free field. For example, if the characteristic length of a test source is 0.2 m and the maximum distance from the source that can be accommodated within an anechoic room is 2 m, then the far field of the source is limited to values of $\gamma(= 2r/\ell) \leq 20$. When this bound is entered in Figure 4.18, a region above the upper dashed line in the figure is defined in which values of κ and γ may be chosen to ensure that far field is achieved. Similarly, an upper bound may be placed on κ, which is dependent on the absorptive properties of the walls at high frequencies, which, in turn, sets the lower bound for λ.

Inspection of Figure 4.18 shows that the region of the far field and even of the geometric near field may be quite limited in a given anechoic room. Indeed, for some sources, it may not even be possible to achieve far-field conditions in a given room. In these instances, estimates of sound power levels will have to be made from sound pressure or intensity measurements in the geometric near field. However, any directivity information inferred from measurements made in the geometric near field must be treated with caution, especially if the source is characterised by one or more tones.

4.12.2 Sound Field Produced in a Non-Anechoic Room

If the sound source is located in a room or enclosure that does not have highly absorptive walls, the sound field radiated by the source will have superimposed upon it, additional sound waves caused by reflection of the original sound waves from the room surfaces. The sound field generated by reflection from the room surfaces is called the reverberant field, and at a sufficient distance from the sound source, the reverberant field may dominate the direct field generated by the source. In a reverberant field, many reflected wavetrains are usually present and the average sound pressure squared reaches a level (in the region where the reverberant field dominates) that is essentially independent of distance from the source. The reverberant field is called a diffuse field if a great many reflected wavetrains cross from all possible directions and the sound energy density is very nearly uniform throughout the field. A room in which this is the case is known as a reverberant room. Such a room has boundaries that are acoustically very hard, resulting in a reverberant field that dominates the whole room except for a small region near the source. More will be said about reverberant spaces in the following sections, and in Chapter 6.

4.13 Determination of Sound Power Using Sound Intensity Measurements

The recommended practice for the direct measurement of sound intensity, for the purpose of determining the radiated sound power level generated by a noise source, is described in various standards (ANSI/ASA S12.12, 2020; ISO 9614-1, 1993; ISO 9614-2, 1996; ISO 9614-3, 2002). The measurement of sound intensity provides a means for directly determining the magnitude and direction of the acoustic energy flow at any location in space. Measuring and arithmetically averaging the sound intensity over an imaginary surface enclosing a machine allows determination of the total acoustic power radiated by the machine.

To measure sound power, a test surface of area, S, which entirely encloses the sound source, is set up. If the source is mounted on the floor, the measurement surface only needs to include the region above the floor. The sound field is then sampled using an intensity probe scanned over a single test surface, or where a pure tone sound source is measured, a single microphone may be scanned over two test surfaces as described following Equation (3.34). By either method, the sound intensity, I_n is determined, which is the time-averaged sound intensity component normal to the test surface. Once the sound intensity has been determined over the enclosing test surface, either by using a number of point measurements or by slowly scanning the intensity probe and arithmetically averaging the sound intensity values, the sound power may be determined using:

$$W = I_n S \tag{4.138}$$

Details concerning the use of intensity meters and various procedures for measuring sound intensity are discussed in Section 3.10.

Theoretically, sound power measurements made using sound intensity can be conducted in the near field of a machine, in the presence of reflecting surfaces and near other noisy machinery. Any energy from these sources flowing into the test surface at one location will flow out at another location provided that there is a negligible amount of acoustic absorption enclosed by the test surface. Because sound intensity is a vector quantity, when it is averaged over a test surface only the net outflow of energy through that surface is measured. However, if the reactive field (see Section 1.8.1) associated with reflecting surfaces, or the near field of a sound source, is 10 dB or more greater than the active field, or if the contribution of other nearby sound sources is 10 dB or more greater than the sound pressure level of the source under investigation, then in practice, reliable sound intensity measurements cannot be made with currently available precision instrumentation. Such instrumentation consists of two phase matched microphones with an accurately measured separation distance. This is because in this situation the acoustic pressures measured by the two microphones are relatively large, but the difference between the levels at the two microphones is relatively small. Sound intensity measurements and the corresponding precision that may be expected are discussed in Section 3.10.

The availability of the intensity meter and some experience with its use has shown that, in general, the radiation of sound from a source is much more complicated than might be supposed. It is not uncommon for a vibrating surface to exhibit areas of sound-absorption as well as radiation. Thus, a map of the acoustic power radiation can be quite complicated, even for a relatively simple source. Consequently, the number of measurements required to determine the field may be quite large, and herein lies the difficulty in the use of the sound intensity meter. When using the intensity meter to determine the net power transmission away from a source, one must be sure to make a sufficient number of measurements on the test surface enclosing the source to adequately describe the resulting sound field. Stated differently, the sound intensity meter may provide too much detail when a few naively conducted pressure measurements may provide an adequate result.

It is suggested that measurements using an intensity meter be made in a free or semi-free field where possible. Alternatively, intensity measurements can be made close to a radiating surface in a reverberant field (ANSI/ASA S12.12, 2020). Errors inherent in sound intensity measurements and limitations of instrumentation are discussed in Sections 3.10.1.1 and 3.10.2.1 and by Fahy (1995).

4.13.1 Uncertainty in Sound Power Determined Using Sound Intensity Measurements

The expanded measurement uncertainty for the sound power level determined using sound intensity measurements is given by:

$$u_e = 2\sqrt{s_R^2 + s_{\text{omc}}^2} \quad \text{(dB)} \tag{4.139}$$

There is a 95% probability that the actual sound power level will be within $\pm u_e$ dB of the average of the measured results.

The estimate of the operating and mounting standard deviation, s_{omc}, may be obtained by taking at least 6 sound power measurements using the same test equipment and the same measurement site. The mean and estimated standard deviation may be calculated using Equations (2.126) and (2.127), where the measured sound power level, L_W, is substituted for L_{eq}.

Estimates of the reproducibility standard deviation, s_R, of the sound power level for three intensity measurement methods for various precision levels and frequency ranges are listed in Table 4.2 (ISO 9614-1, 1993; ISO 9614-2, 1996; ISO 9614-3, 2002).

TABLE 4.2 Estimate of the reproducibility standard deviation, s_R (dB) for sound intensity measurement of sound power

Octave band centre frequency (Hz)	1/3-octave band centre frequency (Hz)	s_R (dB) Precision grade 1	Engineering grade 2	Survey grade 3
63–125	50–160	2.0	3.0	
250–500	200–630	1.5	2.0	
1000–4000	800–5000	1.0	1.5	
—	6300	2.0	2.5	
Overall A-weighted		1.0	1.5	4

Using point measurements, it is possible to obtain a precision grade of 1, provided the instructions in Appendices A and B of ISO 9614-1 (1993) are followed. Using scanning procedures described in ISO 9614-2 (1996), the best that can be obtained is a precision grade of 2. However, the more recent ISO 9614-3 (2002) provides procedures to obtain a precision grade of 1. The survey precision level (grade 3) only applies to overall A-weighted levels and can be obtained with much simpler procedures than required for grades 1 and 2.

4.14 Determination of Sound Power Using Sound Pressure Measurements

There are a number of accepted methods for the determination of sound power based on sound pressure measurements made in the vicinity of a source. The choice of method is dependent to a large degree on the precision required, the mobility of the source, the presence of other noise sources if the source to be tested cannot be moved, and the expected field location of the source with respect to reflecting surfaces such as floors and walls.

Each method discussed in the paragraphs to follow is based on sound pressure measurements. The discussion begins with the most accurate method and ends with the least accurate; the latter method is used when the source to be tested cannot be moved and other immobile noise sources are in the near vicinity.

Most of the methods discussed in the following sections are discussed in great detail in American and International standards. These standards and their ranges of applicability are listed in Table 4.3.

4.14.1 Measurement in Free or Semi-Free Field

The determination of the sound power radiated by a machine in the presence of one or more plane-reflecting surfaces requires that any reverberant sound (i.e. sound returned to the machine from other than the plane reflecting surfaces considered in the measurement) is negligible. ISO 3745 (2017) describes appropriate measurement arrangements in anechoic rooms for precision measurements. ISO 3744 (2010) describes similar but less precise measurements in the open

TABLE 4.3 Standards and their ranges of applicability for sound power determination from sound pressure measurements

Standard designation (book section)	Precision	Test environment	Source volume	Noise character	Sound power data
ISO 3741 ANSI S12.51 ANSI S12.58 (Sect. 4.14.2.2)	Grade 1	Qualified reverberation room	Less than 2% room volume	Steady, impulsive, fluctuating, tonal, broadband	A-weighted, 1/3-octave, octave
ISO 3743-1 ANSI S12.53-1 (Sect. 4.14.3.2)	Grade 2	Hard-walled test room	Less than 2.5% room volume	Steady, impulsive, fluctuating, broadband	A-weighted, octave
ISO 3743-2 ANSI S12.53-2 (Sect. 4.14.2.1)	Grade 2	Special reverberation test room	Less than 1% room volume	Steady, fluctuating, tonal, broadband, narrowband	A-weighted, octave
ISO 3744 ANSI S12.54 (Sect. 4.14.1)	Grade 2	Outdoors or large room	None	Any	A-weighted, 1/3-octave, octave
ISO 3745 ANSI S12.55 (Sect. 4.14.1)	Grade 1	Anechoic or semi-anechoic	Less than 0.5% room volume	Any	A-weighted, 1/3-octave
ISO 3746 ANSI S12.56 (Sect. 4.14.3.4)	Grade 3	None	Unrestricted	Any	A-weighted
ISO 3747 ANSI S12.7 (Sect. 4.14.3.2)	Grade 2 Grade 3	Sufficiently reverberant	Unrestricted	Steady, impulsive, fluctuating, broadband	A-weighted, octave
ISO/TS 7489-1 (Sect. 4.15)	Grade 3	None	Unrestricted	Any	A-weighted
ISO/TS 7489-2 (Sect. 4.15)	Grade 2	None	Unrestricted	Any	A-weighted, 1/3-octave, octave

(such as in a car park with no obstructions or in a large field) or in large rooms, which are not necessarily anechoic.

The sound power of a machine is determined by the integration of the sound pressure squared over a hypothetical spherical surface surrounding it. The centre of the sphere should be the acoustic centre of the machine and a good approximation to this is generally taken as the geometrical centre of the machine. The sphere should be chosen such that its surface is in the radiation far field of the source (where the sound intensity level is related directly to the sound pressure level and where sound power levels and directivity information can be obtained), or at least in the geometric near field where sound power levels but not reliable directivity information can be obtained, but not in the hydrodynamic near field (Bies and Bridges, 1993). A better approximation of the acoustic centre of the sound source is the location from which the inverse distance law applies (Jacobsen et al., 2004). This location can be found by measuring the sound pressure level, L_{p1} and L_{p2} at two locations at different distances, r_1 and r_2 from the source, in the far field of the source and in the same angular location with respect to the geometric centre of the source. Setting $r_2 > r_1$ and $r_2 - r_1 = d$, we can write:

$$\frac{r_2}{r_1} = 10^{(L_{p1}-L_{p2})/20} \quad \text{and} \quad r_2 - r_1 = d \tag{4.140}$$

from which we obtain:

$$r_1 = \frac{d}{10^{(L_{p1}-L_{p2})/20} - 1} \quad \text{(m)} \tag{4.141}$$

If measurements are made in an anechoic room, then, according to the standards, the surface of the sphere should be at least one-quarter of a wavelength of sound away from the anechoic room walls and the sphere should have a radius of at least twice the major machine dimensions, but not less than 0.6 m. A more accurate means for determining the required radius is to use Figure 4.18. For example, if the major machine dimension is 2 m and the frequency of interest is 125 Hz ($\lambda = 2.75$ m) then $\kappa = 2.28$ and from Figure 4.18, γ ranges from 1.3 to 2.6 for the geometric near field and is equal to 7.5 at the beginning of the far field. This corresponds to a distance, r, from the source ranging between 1.3 and 2.6 m for the geometric near field and a distance, r, greater than 7.5 m for the far field. It can be seen that for most machines, anechoic rooms of very large size are implied if measurements are to be made in the far field of the source.

For large items of equipment (such as a gas turbine, see ISO 10494 (2018)), the excessively large size of an appropriate anechoic room strongly recommends consideration of the anechoic space above a reflecting plane, as would be provided outdoors with the machine mounted on the ground away from any other significant reflectors. Alternatively, such a space can be provided indoors, as described in ISO 3744 (2010). Carrying the consideration further, the mounting arrangement of the machine in use might suggest that its radiated sound power be determined in the presence of a reflecting floor and possibly one or two walls. The measurement surface can be chosen as a parallelepiped that approximates the shape of the machine under test rather than a hemisphere (ISO 3744, 2010). In some cases, it may be necessary to measure in the near field of the noise source as described in Section 4.14.3.4. In other cases, where reflecting surfaces in addition to the floor are present, it may be preferable to use sound intensity methods to determine the sound power of the equipment (see Section 4.13).

The integration of the sound intensity over the encompassing spherical surface is achieved by determining time-average squared sound pressures at a discrete set of measurement points arranged to uniformly sample the integration surface, arithmetically averaging the results over all measurement locations and then dividing the average by ρc. The number of measurement points, N, and their recommended coordinates, relative to the acoustic centre of the noise source, were first suggested in the 1977 version of ISO 3745. These coordinate locations were updated in the 2003 version and again in the 2012 version, when the number of measurement points was doubled from 20 to 40. For precision measurements in an anechoic room, ISO 3745 (2017) recommends that only measurements in a fully anechoic room or in an anechoic room with a hard floor (semi-anechoic room) be undertaken. For both configurations, the number of measurement points surrounding the source is recommended to be 40. For less precise ("engineering") measurements, ISO 3744 (2010) suggests that 20 measurement locations be used for both the fully anechoic space and the semi-anechoic space and also includes locations for cases where measurements are made on a quarter sphere and a one-eighth sphere, corresponding to the source being mounted at the junction of two or three reflecting planes, respectively.

The sound pressure level, L_p, averaged over the N locations on the measurement surface is:

$$L_p = 10 \log_{10} \left[\frac{1}{N} \sum_{1}^{N} 10^{(L_{pi}/10)} \right] \quad \text{(dB re 20 µPa)} \tag{4.142}$$

The sound power is calculated using:

$$L_W = L_p + 20 \log_{10} r + C + C_1 + C_2 + C_3 \qquad \text{(dB re } 10^{-12}\text{ W)} \tag{4.143}$$

where the coefficient, C, is listed in Table 4.4 for various source mounting configurations, along with the recommended number of measurement locations according to ISO 3744 (2010) and ISO 3745 (2017). The coefficient, $C_1 = 10\log_{10}[400/(\rho c)]$ is included in ISO 3745 (2017) to account for the different reference levels used for sound power and sound pressure as well as the presence of ρc in the equation for converting sound pressure squared to sound intensity (see equation (1.78)).

TABLE 4.4 Values of the coefficient, C, used in Equation (4.143), together with the required number of measurement locations on the measurement surface surrounding the source

Source location	Coefficient, C (dB)	Number of measurement locations	
		ISO 3744 (2010)	ISO 3745 (2017)
1. Near junction of three planes	2	6	—
2. Near junction of two planes	5	9	—
3. Near a single plane	8	20	40
4. Free field	11	20	40

The coefficient, C_2, is included in ISO 3745 (2017) to account for meteorological conditions at the place of measurement being different to reference conditions and although small, it can be calculated using:

$$C_2 = -10\log_{10}\left(\frac{P_0}{101325}\right) + 15\log_{10}\left(\frac{273+T}{296}\right) \qquad \text{(dB)} \tag{4.144}$$

where P_0 and T are, respectively, the atmospheric pressure and temperature existing at the measurement time and place.

The coefficient, $C_3 = A_a$, is included in ISO 3745 (2017) to account for air absorption and this also is usually very small. However, it can be calculated using Equation (5.68), where d_{SR} is the measurement radius (distance of the measurement surface from the acoustic centre of the source) in metres:

If the measurement surface is in the far field, the directivity index corresponding to measurement location, i, can be determined as:

$$\text{DI} = L_{pi} - L_p \qquad \text{(dB)} \tag{4.145}$$

4.14.2 Measurement in a Diffuse Field

To provide a diffuse sound field, a test room is required that has adequate volume (see below) and suitable shape (to ensure uniform spacing of resonant modes), and whose boundaries over the frequency range of interest can be considered acoustically hard (to ensure adequate diffusion). The volume of the room should be large enough so that the number of normal acoustic modes (see Section 6.3.1) in the octave or 1/3-octave frequency band of interest is enough to provide a satisfactory state of sound diffusion. Various standards state that at least 20 acoustic modes are required in the lowest frequency band used. This implies a minimum room volume of $1.3\lambda^3$, if measuring in octave bands, and $4.6\lambda^3$, if measuring in 1/3-octave bands, where λ is the wavelength of sound corresponding to the lowest band centre frequency.

If the machine emits sound containing discrete frequencies or narrowband components, it is necessary to use a rotating sound diffuser (Ebbing, 1971; Lubman, 1974; Bies and Hansen, 1979). When a rotating diffuser is used, the lowest discrete frequency that can be reliably measured is given by the relation:

$$f_q = 2000(T_{60}/V)^{1/2} \qquad \text{(Hz)} \tag{4.146}$$

In Equation (4.146), T_{60} is the time (in seconds) required for a 1/3-octave band of noise (with centre frequency f_q) to decay by 60 dB after the source is shut off (see Sections 6.5.1 and

6.5.2), and V is the room volume (m^3). The rotating sound diffuser should have a swept volume equal to or greater than the cube of the wavelength of sound at the lowest frequency of interest. Procedures to be followed for the measurement of tonal noise in reverberant test chambers are described in ISO 3741 (2010).

The shape of the test room for both broadband and tonal measurements should be such that the ratio of any two dimensions is not equal to, or close to, an integer. Ratios of height, width and length of 2:3:5 and 1:1.260:1.588 have been found to give good results (ISO 3741, 2010, Annex A). The absorption coefficient of the test room should be less than 0.06 in all test frequency bands, although when noise containing pure tone components is measured, it is advantageous to increase the absorption coefficient in the low frequencies to 0.15 (to increase the modal overlap – see Sections 6.3.2 and 6.3.3, and thus reduce the sound pressure level variation as a function of frequency). For test purposes, the machine should be placed at a distance of at least one-quarter wavelength away from all surfaces not associated with the machine.

There are two methods for determining the sound power of a machine in a reverberant room: the substitution method (ISO 3743-2, 2018) and the absolute method (ISO 3741, 2010). Both methods require the determination of the space-average sound pressure level produced in the room by the machine. This may be measured using a microphone travelling on a linear traverse across the room, or a circular traverse on a rotating boom. The traverse length should be at least 1.5λ or 3 m, whichever is greatest. Alternatively, the microphone may be sequentially moved from point to point over a set of discrete measurement positions. In each case, an energy-averaged sound pressure level is determined (see Equation (4.142)). The microphone should at all times be kept at least one-half of a wavelength away from reflecting surfaces, and out of any region in which the direct field from the sound source dominates the reverberant sound field.

The number of discrete microphone positions required is at least six for broadband noise when a rotating sound diffuser is used. The microphone positions should be at least one-half of a wavelength of sound apart. When a continuous microphone traverse is used, the equivalent number of discrete microphone positions is equal to twice the traverse length divided by the wavelength of sound corresponding to the centre frequency of the measurement band.

If no rotating diffuser is used, or if the machine radiates discrete frequency components, more microphone positions are required to obtain the same measurement accuracy. The measurement errors associated with various frequency bandwidths and numbers of microphone positions have been well documented (Beranek, 1988; ISO 3741, 2010; ISO 3743-2, 2018).

4.14.2.1 Substitution Method

With this method (ISO 3743-2, 2018), the sound source to be tested is placed in the room and the space-average sound pressure level, L_p, is determined for each frequency band using energy averaging (see Equation (4.142)). The sound source is then replaced with another sound source of known sound power output, L_{WR} (the reference sound source). The space-averaged sound pressure level, L_{pR}, produced by the reference source in the room is determined using energy averaging. The sound power level, L_W, for the test source is then calculated using the following relation:

$$L_W = L_{WR} + (L_p - L_{pR}) \qquad (\text{dB re } 10^{-12}\,\text{W}) \tag{4.147}$$

Reference sound sources are available from manufacturers of acoustic measurement instrumentation.

4.14.2.2 Absolute Method

For this method (ISO 3741, 2010; ISO 3743-2, 2018), the sound-absorbing properties of the room are determined in each measurement band from measurements of the frequency band reverberation times, T_{60}, of the room (see Sections 6.5.1 and 6.5.2). The steady-state space-average sound pressure level, L_p, produced by the noise source is also determined for each frequency band (see Equation (4.142)). The sound power level, L_W, produced by the source is then calculated in each frequency band using the following equation (Beranek, 1988):

$$L_W = L_p + 10\log_{10} V - 10\log_{10} T_{60} + 10\log_{10}[1 + S\lambda/(8V)] - 13.9 \qquad (\text{dB re } 10^{-12}\,\text{W}) \quad (4.148)$$

where the constant, "13.9 dB", has been calculated for an atmospheric pressure of 101.325 kPa and a temperature of 20°C, using Equations (6.43) and (6.55) as a basis (with $(1-\bar{\alpha}) \approx 1$, where $\bar{\alpha}$ is the arithmetically averaged Sabine absorption coefficient of the room surfaces – see Section 6.4.2). In Equation (4.148), V is the volume of the reverberant room, S is the total area of all reflecting surfaces in the room and λ is the wavelength of sound at the band centre frequency. The fourth term on the right of the equation is not derived from Equation (6.55) and represents a correction ("Waterhouse correction") to account for the measurement bias resulting from the space-averaged sound pressure level, L_p (see Equation (4.142)), excluding measurements of the sound field closer than $\lambda/4$ to any room surface (Waterhouse, 1955). All the other terms in the equation can be derived directly from Equations (6.43) and (6.55), where the contribution of the direct field is considered negligible and $S\bar{\alpha} \approx S\bar{\alpha}/(1-\bar{\alpha})$.

4.14.3 Field Measurement (ISO 3744, 2010)

If the machine is mounted in an environment such that the conditions of free or semi-free field can be met, then sound power measurements can be made using the appropriate method of Section 4.14.1. However, when these conditions are not satisfied, and it is inconvenient or impossible to move the machine to be tested, less precise sound power measurements can be made with the machine on site, using one of the methods described in this section.

Most rooms in which machines are installed are neither well damped nor highly reverberant; the sound field in such rooms is said to be semi-reverberant. For the determination of sound power in a semi-reverberant room, no specific assumptions are made concerning the room, except that the room should be large enough so that measurements can be made in the far field of the source and not too close to the room boundaries. The microphone should at all times be at least one-half of a wavelength away from any reflecting surfaces or room boundaries not associated with the machine. The machine should be mounted in its normal position, which will typically include the hard floor but may also include mounting in the corner of a room or at the junction between the floor and a wall. Unusually long or narrow rooms will generally degrade the results obtained and are best avoided if possible. Where the conditions mentioned can be satisfied there are three alternative measurement procedures, as described in Sections 4.14.3.1, 4.14.3.2 and 4.14.3.3. Each of these methods is discussed in Appendix A of ISO 3744 (2010).

4.14.3.1 Semi-Reverberant Field Measurements Using a Reference Source to Determine Room Absorption

To compute the sound power level, the total room absorption (see Section 6.6.1 and ISO 3747 (2010)) of the test room must be determined. This is usually done in octave or 1/3-octave bands. For this purpose, a reference sound source of known sound power output in octave or 1/3-octave bands is used. The reference sound source is placed on the floor away from the walls or any other reflecting surfaces, and a hypothetical hemispherical test surface surrounding the reference source is chosen. The radius, r, of the test surface should be large enough for the test surface to be in the far field of the reference source. Reference to Figure 4.18 shows that this condition is easily satisfied for a typical fan-based reference source, which has a characteristic dimension, ℓ, of the order of 0.1 m. For example, at 500 Hz the distance, r, should be greater than 0.3 m. For best results, the test surface should lie in the region about the test source, where its direct field and the reverberant field are about equal (see Section 6.4.4). Measurements on the surface of the test hemisphere in each octave or 1/3-octave band (see Table 4.4) allow determination of the reference source average sound pressure level, L_{pR} (see Equation (4.142)), due to the combination of direct and reverberant sound fields. At the radius, r, of the test hemisphere, the sound pressure level, L_{p2}, due to the direct sound field of the reference source only, is calculated using the reference source sound power levels, L_{WR}, and the following equation obtained by setting DI = 3 dB in Equation (4.130):

$$L_{p2} = L_{WR} - 20\log_{10} r - 8 + 10\log_{10}\frac{\rho c}{400} \quad \text{(dB re 20 µPa)} \tag{4.149}$$

The total room absorption may be expressed in terms of the room constant, R_c. The expression for determining the reciprocal room constant factor, $4/R_c$ (see Section 6.4.4), of the test room is obtained by using the expression relating the sound pressure in a room to the sound power of a sound source derived in Chapter 6, Equation (6.44), as:

$$L_{WR} = L_{pR} - 10\log_{10}\left[\frac{D}{4\pi r^2} + \frac{4}{R_c}\right] - 10\log_{10}\frac{\rho c}{400} \quad \text{(dB re } 10^{-12}\text{ W)} \tag{4.150}$$

For a measurement over a hemispherical surface of area S_H above a hard floor, $D/(4\pi r^2) = 1/S_H$, or $S_H = 2\pi r^2$, as $D = 2$ (see Table 4.1). Assuming that the reference source radiates the same sound power into the room as it does into a free field, the values of L_{WR} in Equations (4.149) and (4.150) are equal. Noting that $20\log_{10} r + 8$ in Equation (4.149) may be replaced with $10\log_{10} S_H$, the following equation is obtained:

$$L_{p2} + 10\log_{10} S_H = L_{pR} - 10\log_{10}\left[\frac{1}{S_H} + \frac{4}{R_c}\right] \quad \text{(dB)} \tag{4.151}$$

or

$$L_{pR} - L_{p2} = 10\log_{10}\left[1 + \frac{4S_H}{R_c}\right] \quad \text{(dB)} \tag{4.152}$$

Rearranging gives an expression for the room constant factor, $4/R_c$, as:

$$4/R_c = [10^{(L_{pR}-L_{p2})/10} - 1]/S_H \quad (\text{m}^{-2}) \tag{4.153}$$

The room constant, R_c, is discussed in Chapter 6, but for the present purpose it may be taken as the total Sabine absorption, $S\bar{\alpha}$, in the room, measured in units of area. Alternative methods for determining the room constant, R_c, are given in Section 6.6.

The noise source under test is now operated in the room and the mean square sound pressure level, L_p, over a test surface of radius, r, and centre at the acoustic centre of the noise source (see Section 4.14.1) is determined. The radius of the test hemisphere should be chosen large enough for the sound pressure level measurements to be made at least in the geometric near field of the source but preferably in the far field. The sound power level may be computed using Equation (4.150) by replacing L_{WR} with L_W and L_{pR} with L_p.

Table 4.1 provides a guide for the choice of directivity factor D in Equation (4.150). For example, if the noise source is mounted on a hard reflecting plane surface then the test surface should be a hemisphere, and D in Equation (4.150) should take the value 2.

4.14.3.2 Semi-Reverberant Field Measurements Using a Reference Source Substitution

If the machine to be tested is located in a hard-walled room and on a hard floor, and in addition, it can be moved, the measurement of L_W is simplified. In this case, the substitution method described in Section 4.14.2.1 can be employed, as described in ISO 3743-1 (2010). The average sound pressure level L_p (see Equation (4.142)) is determined over a test hemisphere surrounding the machine. The machine is replaced by the reference sound source and the average sound pressure level L_{pR} is determined over the same test hemisphere. The sound power output of the machine is then calculated using the following expression:

$$L_W = L_{WR} + (L_p - L_{pR}) \qquad (\text{dB re } 10^{-12}\,\text{W}) \tag{4.154}$$

where L_W is the sound power level of the machine and L_{WR} is the sound power level of the reference source. Note that the test hemisphere should be located such that its centre is the acoustic centre of the machine (see Section 4.14.1), and its surface is in the far field, or at least the geometric near field, of both the machine and the reference sound source. Any reflecting surfaces present should not be included within the test hemisphere.

If the machine is too large to be moved, this method can also be used as described in ISO 3747 (2010). In this case, the reference source is placed on top of the machine under test or as close as possible to it. Additionally, the machine under test may be located in the position in which it is used, which may be near one or more reflecting surfaces in addition to the floor.

4.14.3.3 Semi-Reverberant Field Measurements Using Two Test Surfaces

Alternatively, instead of using a reference sound source, the sound power level of any source may be determined by taking sound pressure measurements on two separate test surfaces having different radii. The discussion of Section 4.14.1 may be used as a guide in choosing the test surfaces and measurement locations. In all cases, the centres of the test surfaces are at the acoustic centre of the noise source. The test surface areas are given by $4\pi r^2/D$, where D is given in Table 4.1, and r is the test surface radius. The radii of the test surfaces are such that they are in the far field, or at least in the geometric near field of the source.

The measurement procedure assumes that the background sound pressure levels produced by other machines at the microphone measurement positions make a negligible contribution to the measurements associated with the test machine. This implies that the sound pressure levels at the measurement positions produced by the machine under test are at least 10 dB above the background sound pressure level. If this is not the case, then the sound pressure level data must be corrected for the presence of background noise. All measurements must be repeated with the test machine turned off and the background levels measured. The background levels must be logarithmically subtracted from the test measurement levels, using the method and corresponding limitations outlined in Section 1.11.5.

Let L_{p1} be the average sound pressure level (see Equation (4.142)) measured over the smaller test surface of area S_1 and L_{p2} be the average sound pressure level measured over the larger test surface of area S_2. The sound pressure levels, L_{p1} and L_{p2}, are calculated using Equation (4.142). Since both sets of measurements should give the same result for the sound power, Equation (4.150) gives the following expression relating the measured quantities:

$$L_{p1} - 10\log_{10}\left[\frac{D}{4\pi r_1^2} + \frac{4}{R_c}\right] = L_{p2} - 10\log_{10}\left[\frac{D}{4\pi r_2^2} + \frac{4}{R_c}\right] \qquad (\text{dB}) \tag{4.155}$$

The unknown quantity, R_c, is called the room constant, and it is evaluated by manipulating the preceding equation and substituting $1/S_1$ for $D/4\pi r_1^2$ and $1/S_2$ for $D/4\pi r_2^2$. Taking antilogs of Equation (4.155) gives:

$$10^{(L_{p1}-L_{p2})/10} = \frac{[1/S_1 + 4/R_c]}{[1/S_2 + 4/R_c]} \tag{4.156}$$

Rearranging gives:

$$\frac{4}{R_c} = \frac{(1/S_1) - (1/S_2)[10^{(L_{p1}-L_{p2})/10}]}{10^{(L_{p1}-L_{p2})/10} - 1} \qquad (\text{m}^{-2}) \tag{4.157}$$

or

$$\frac{1}{S_2} + \frac{4}{R_c} = \frac{(1/S_1 - 1/S_2)}{(10^{(L_{p1}-L_{p2})/10} - 1)} \quad (\text{m}^{-2}) \tag{4.158}$$

Substituting Equation (4.158) into (4.150) gives the following equation (Diehl, 1977):

$$L_W = L_{p2} - 10\log_{10}[S_1^{-1} - S_2^{-1}] + 10\log_{10}[10^{(L_{p1}-L_{p2})/10} - 1] - 10\log_{10}\frac{\rho c}{400} \; (\text{dB re } 10^{-12}\,\text{W}) \tag{4.159}$$

4.14.3.4 Near-Field Measurements

The previous methods outlined for the on-site measurement of sound power radiated by a machine have all required that the room in which the machine is situated is large enough for the measurements to be made in the far field of the source. The previous methods have also assumed that the background sound pressure levels produced by other machines have either made a negligible contribution to the measurements or that a correction can be made for their contribution; that is, it has been assumed that the machine under test has produced a sound pressure level at least 3 dB higher than the background sound pressure level at each microphone measurement position.

In some cases, the above assumptions may not be valid. For example, the room in which the machine is situated may be too small for a far field to exist. Alternatively, if the room is of adequate size, the background sound pressure levels produced by other machines in the room may be too high to allow valid measurements to be made at a sufficient distance from the test machine to be in its far field. For cases such as these, an alternative, but less accurate, procedure may be used to estimate the sound power radiated by the machine. This procedure relies on sound pressure level measurements made close to the machine surface (Diehl, 1977; ISO 3746, 2010) at a number of points on a hypothetical test surface surrounding the machine (which is usually mounted on a reflective surface such as a hard floor).

The test surface usually conforms approximately to the outer casing of the machine so that its area is easy to calculate (in many cases, a parallelepiped may be used), and it is sufficiently close to the machine for the measurements not to be affected too much by nearby reflecting surfaces or background noise. If background noise is a problem it must be accounted for by measuring the sound pressure levels with the machine under test turned on and then with it turned off. The level with it turned off is then subtracted from the level with the machine on, as illustrated by Equation (1.102) in Chapter 1.

The effect of nearby reflecting surfaces can be minimised by placing sound-absorbing material on them (e.g. 50 mm thick glass-fibre blanket). The standard and generally accepted distance between the test surface and machine surface is 1 m, but may need to be less in some cases, at the expense of reduced accuracy in the estimation of the radiated sound power.

The average sound pressure level, L_p, over the test surface is found by measuring the sound pressure level, L_{pi}, at a number of equally spaced, discrete points over the surface, and then using Equation (4.142). The number of measurement positions, N, is determined by the irregularity of the acoustic field and size of the machine, and should be sufficient to take any irregularities into account. Suitable measurement locations are discussed in ISO 3746 (2010) and involve between 5 and 16 locations for a typical machine. Once determined, the value of L_p is used to determine the sound power level of the machine (estimated to the nearest 1 dB) using the following equation:

$$L_W = L_p + 10\log_{10} S - \Delta_1 - \Delta_2 \qquad (\text{dB re } 10^{-12}\,\text{W}) \tag{4.160}$$

where S is the area of the test surface, Δ_1 and Δ_2 are correction terms to account for the effects of the reverberant and near fields, respectively, and it is assumed that $\rho c \approx 400$.

In the near field of a machine, sound propagation will not necessarily be normal to the arbitrarily chosen measurement surface. As the integration process implied by Equation (4.160) implicitly assumes propagation normal to the measurement surface, the correction factor Δ_2 is introduced to account for possible tangential sound propagation. Values of Δ_2 are given in Table 4.5 as a function of the ratio of the area of the measurement surface S divided by the area of the smallest parallelepiped, S_m, which just encloses the source.

TABLE 4.5 Correction factor, Δ_2, due to near-field effects (Jonasson and Eslon, 1981)

Ratio of test surface area to machine surface area S/S_m	Near-field correction factor Δ_2 (dB)
1–1.1	3
1.1–1.4	2
1.4–2.5	1
2.5–∞	0

The correction factor, Δ_1, has been suggested to account for the absorption characteristics of the test room. Values of Δ_1 typical of production rooms are given in Table 4.6 for various ratios of the test room volume, V, to area, S, of the test surface.

TABLE 4.6 Value of room effect correction factor, Δ_1 (Pobol, 1976)

Characteristics of production or test room	Ratio of room volume to test surface area, V/S (m)			
Usual production room without highly reflective surfaces	20–50	50–90	90–3000	Over 3000
Room with highly reflective surfaces, with no sound-absorbing treatment	50–100	100–200	200–600	Over 600
Δ_1 (dB)	3	2	1	0

In Table 4.6, the distance from the test surface to the machine surface is approximately 1 m. Note that in deriving Equation (4.160), ρc has been assumed to be equal to 400.

A more accurate method of determining Δ_1 is to measure the average sound pressure level (see Equation (4.142)) on two imaginary test surfaces surrounding the machine, which correspond roughly to the shape of the machine. The smaller test surface is usually displaced about 1 m from the machine surface. The two test surfaces should be sufficiently far apart that the average sound pressure level over one surface differs from that measured over the other surface by at least 0.5 dB. Values of Δ_1 as a function of the area ratio of the two test surfaces and the difference in average sound pressure level measured over each surface may be estimated using Figure 4.19, or calculated as described in the text to follow.

Assuming that the smallest measurement surface of area, S_1, is sufficiently large that the correction term, $\Delta_2 = 0$ in Equation (4.160), substituting S_1 for S and L_{p1} for L_p in Equation (4.160) and setting the RHS of Equation (4.160) equal to the RHS of Equation (4.159), the following result is obtained:

$$L_W = L_{p1} + 10\log_{10} S_1 - \Delta_1 = L_{p2} - 10\log_{10}[S_1^{-1} - S_2^{-1}] + 10\log_{10}[10^{(L_{p1}-L_{p2})/10} - 1] \quad \text{(dB re } 10^{-12}\,\text{W)} \quad (4.161)$$

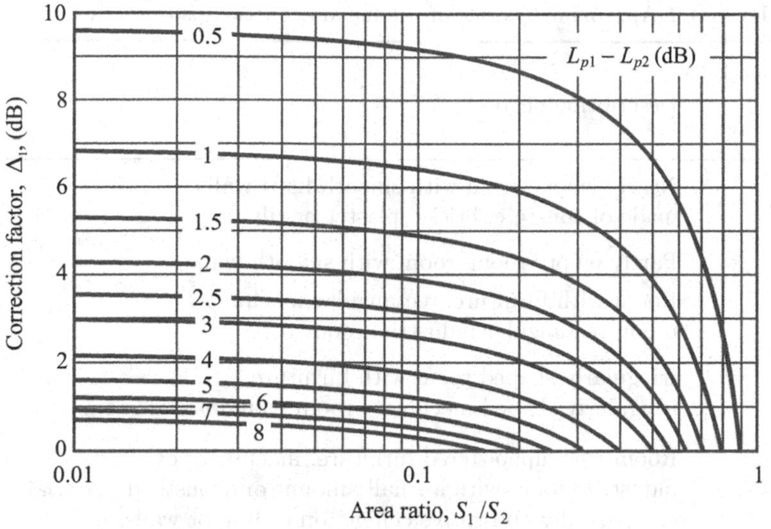

FIGURE 4.19 Graph for estimating the correction factor, Δ_1.

Rearranging Equation (4.161) gives:

$$\Delta_1 = L_{p1} - L_{p2} - 10\log_{10}[10^{(L_{p1}-L_{p2})/10} - 1] + 10\log_{10}[1 - S_1/S_2] \quad \text{(dB)} \qquad (4.162)$$

If the machine being tested is sufficiently small that Δ_2 cannot be neglected, then Equation (4.162) becomes:

$$\begin{aligned}\Delta_1 = L_{p1} - L_{p2} + \Delta_{2,\text{large}} - \Delta_{2,\text{small}} - 10\log_{10}[10^{(L_{p1}-L_{p2}+\Delta_{2,\text{large}}-\Delta_{2,\text{small}})/10} - 1]\\ + 10\log_{10}[1 - S_1/S_2] \quad \text{(dB)}\end{aligned} \qquad (4.163)$$

In Equations (4.162) and (4.163), the subscripts 1 and 2 on quantities, L_p and S, refer, respectively, to the measurement surfaces near and remote from the machine. The quantity, $\Delta_{2,\text{small}}$ refers to the near-field correction for the measurement surface closest to the machine $\Delta_{2,\text{large}}$ refers to the near-field correction for the measurement surface furthest from the machine.

Alternatively, Δ_1 may be estimated using only one measurement surface, the mean acoustic absorption of the room surfaces and the following procedure. In this case, Equation (4.150) (with $D/4\pi r^2 = 1/S_1$) may be set equal to Equation (4.160), noting that Δ_2 can be eliminated as it would be the same near-field correction in both equations. Thus:

$$L_p + 10\log_{10} S_1 - \Delta_1 \approx L_p - 10\log_{10}\left[\frac{1}{S_1} + \frac{4}{R_c}\right] \quad \text{(dB)} \qquad (4.164)$$

Rearranging Equation (4.164) and use of Equation (6.45) (see Section 6.4.4) gives the following expression for the correction term Δ_1:

$$\Delta_1 = 10\log_{10}\left[1 + \frac{4S_1(1-\bar{\alpha})}{(S_R\bar{\alpha})}\right] \quad \text{(dB)} \qquad (4.165)$$

where S_1 is the area of the measurement surface, S_R is the total area of all the room surfaces and $\bar{\alpha}$ is the mean acoustic Sabine absorption coefficient for the room surfaces (see Chapter 6 for further discussion and measurement methods). Representative values of $\bar{\alpha}$ are included in Table 4.7.

TABLE 4.7 Approximate value of the mean acoustic absorption coefficient, $\bar{\alpha}$

Mean acoustic absorption coefficient, $\bar{\alpha}$	Description of room
0.05	Nearly empty room with smooth hard walls made of concrete, bricks, plaster or tile
0.1	Partly empty room, room with smooth walls
0.15	Room with furniture, rectangular machinery room, rectangular industrial room
0.2	Irregularly shaped room with furniture, irregularly shaped machinery room or industrial room
0.25	Room with upholstered furniture, machinery or industrial room with a small amount of acoustical material, e.g. partially absorptive ceiling, on ceiling or walls
0.35	Room with acoustical materials on both ceiling and walls
0.5	Room with large amounts of acoustical materials on ceiling and walls

In making near-field measurements, a possible source of error is associated with the microphone response. To avoid such an error, the microphone directional response variation should be less than 1 dB over the maximum angle between the line joining the microphone to the furthest point on the machine surface and the normal to the measurement surface. This angle is approximately 60° for a source dimension of 3 m and a measurement surface 1 m from the source. Thus, for a 12 mm microphone, the error due to microphone directional response should be negligible up to frequencies of about 3.15 kHz, as shown by reference to Figure 3.4.

4.14.4 Measurement of the Total Sound Power of Multiple Sources Covering a Large Area

In large industrial facilities, it is often not practical to measure the sound power of individual sound sources due to interference from other sources. However, a sound power estimation of all sources is often needed in order to determine the contribution of a particular industrial facility to community noise levels (see Chapter 5). According to ISO 8297 (2021), to undertake a measurement of the overall sound power level of a facility, the first step is to determine the test facility area, S_p. Second, draw a measurement contour around the facility (see Figure 4.20) and determine the corresponding enclosed area, $S_m = a \times b$ and perimeter, $L = 2a + 2b$.

The third step is to specify microphone locations on the contour for provision of the sound pressure level data. The distance, d_i of each microphone position, i, on the measurement contour from the nearest point on the industrial facility perimeter is used to calculate the average microphone distance, $\bar{d} = (1/N)\sum_{i=1}^{N} d_i$, where N is the number of microphone locations. The average distance, $\bar{d}$, must lie between the maximum of $0.05\sqrt{S_p}$ or 5 m and the minimum of $0.5\sqrt{S_p}$ or 35 m. Within this range it should be as large as possible, whilst maintaining background noise levels more than 6 dB below facility levels at all microphone positions. The microphones should be omni-directional and equally spaced, at a distance apart of $D_m \leq 2\bar{d}$.

The fourth step in the process is to determine an average height, H, of all the significant noise sources and then specify a measurement microphone height, h, of the maximum of $h = 5$ m or $h = H + 0.025\sqrt{S_m}$ (m).

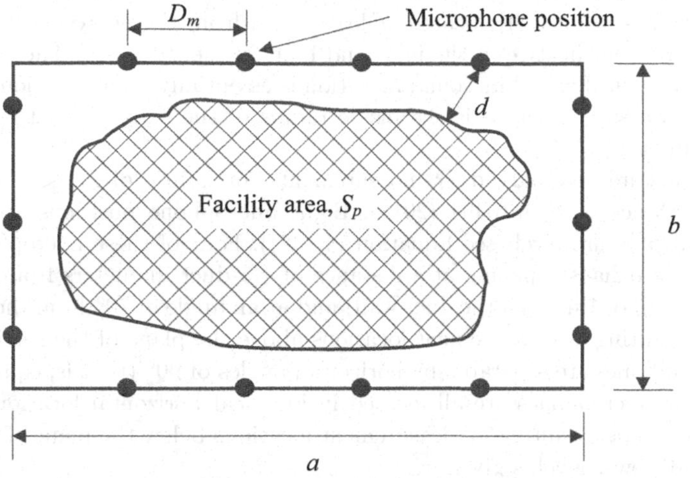

FIGURE 4.20 Measurement contour and microphone measurement locations.

The fifth step is to measure the time-averaged sound pressure levels, $L_{p,i}$ at each microphone location and logarithmically average them using Equation (4.142) to obtain $\bar{L}_p$.

The sixth step is to calculate the correction terms to be used in Equation (4.166) to calculate the sound power level from the average sound pressure level. These terms, which are all in dB, are $\Delta L_S = 10\log_{10}(S_m + hL)$ for area, $\Delta L_F = \log_{10}\left[\bar{d}/(4\sqrt{S_p})\right]$ to account for the proximity of the microphones to the test facility and $\Delta L_a = 0.5\alpha_a\sqrt{S_m}$ to account for the effect of atmospheric sound absorption (see Section 5.3.2 for the calculation of α_a). Note that the term, ΔL_S, in ISO 8297 (2021), uses $2S_m$ in place of S_m, but this is likely an error as the effective measurement surface consists of a top surface of area S_m and an edge surface of hL but no bottom surface. The term, ΔL_F, is very similar to the term, Δ_2 discussed in Section 4.14.3.4.

Finally, the sound power level in 1/3-octave or octave bands may be calculated from mean sound pressure levels in 1/3-octave or octave bands using:

$$L_W = \bar{L}_p + \Delta L_S + \Delta L_F + \Delta L_a \qquad \text{(dB)} \tag{4.166}$$

The expanded uncertainty inherent in this method is a function of the mean distance, $\bar{d}$, of the microphones from the facility perimeter and the area, S_p, occupied by the facility. Suggested values are listed in Table 4.8 (ISO 8297, 2021). More details about this measurement procedure are provided in ISO 8297 (2021).

TABLE 4.8 Uncertainty in measured L_W

$\bar{d}/\sqrt{S_p}$	u_e (dB)
0.05	+3.0, –3.5
0.1	+2,5, –2.5
0.2	+2.0, –2.5
0.5	+1.5, –2.0

4.14.5 Gas Turbine Exhaust Sound Power Measurement

Gas turbines are becoming an increasing community noise problem as more and more are being constructed to compensate for variability in sun and wind power sources and to replace coal fired power stations. Thus, it is of interest to explore how to measure the sound power of such a source. The community noise problem produced by gas turbines is generally low-frequency in nature, as turbines are usually located at least 1 km from the nearest resident and exhaust mufflers are much more effective in the mid- and high-frequency range. The turbine exhaust is always orientated vertically, and its sound radiation is essentially omnidirectional for frequencies corresponding to $ka < 1$, where a is the inside radius of the exhaust duct cross-section for a circular section duct.

Surprisingly accurate sound power measurements for values of $ka < 1$ can be made by following the guidance of ISO 10494 (2018). Eight microphone positions are used to obtain an average sound pressure level (see Equation (4.142)). Four of these microphones are located around the turbine exhaust opening at a distance of 1.5 duct diameters from the central duct axis (but a minimum of 1 m from the duct wall) and at an angle of 70° from the duct axis at the exhaust outlet, resulting in measurement locations above the plane of the vertical duct exhaust outlet. The microphones are separated by horizontal angles of 90° (that is, equally spaced). The second set of four microphones are all located in identical horizontal locations at an angle of 110° from the duct axis, resulting in measurement locations below the plane of the duct exhaust outlet. The sound power level is given by:

$$L_W = L_{pm} + 10\log_{10}(S_1 + S_2) \qquad \text{(dB re } 10^{-12}\text{W)} \tag{4.167}$$

where L_{pm} is the average sound pressure level measured by the eight microphones, given by:

$$L_{pm} = 10\log_{10}\left(\frac{1}{8}\sum_{i=1}^{8} 10^{L_{pi}/10}\right) \qquad \text{(dB re } 20\,\mu\text{Pa)} \tag{4.168}$$

The measurement surface area, S_1, is the area of the hemispherical measurement surface above the outlet of the vertically oriented duct and S_2 is the area of the hemispherical measurement surface below the outlet, for which the area occupied by the duct is subtracted. For circular section ducts, these areas are given by:

$$S_1 = 2\pi(r_i + D_m)^2 \qquad (\text{m}^2) \tag{4.169}$$

and

$$S_2 = 2\pi(r_i + D_m)\sqrt{(r_i + D_m)^2 - r_0^2} \qquad (\text{m}^2) \tag{4.170}$$

where

$$D_m = \begin{cases} 1; & \text{if } r_i \leq 0.5\text{m} \\ 2r_i; & \text{if } r_i > 0.5\text{m} \end{cases} \tag{4.171}$$

where r_i is the inside radius and r_o is the outside radius of the circular section exhaust duct.

For values of $ka > 1$, the duct outlet directivity must be taken into account, as discussed in Section 8.19. Rectangular section ducts are treated slightly differently (see ISO 10494 (2018)). In some cases, turbine inlet and casing noise are also important and ISO 10494 (2018) provides procedures for determining these quantities. The turbine casing sound power is determined by taking sound pressure measurements on an enveloping surface above a reflecting plane (see Sections 4.14.1 and 4.14.3.4), while the inlet sound power is measured in a similar way to the exhaust sound power.

4.14.6 Wind Turbine Sound Power Measurements

With the increasing use of renewable energy systems, the requirement to measure wind turbine sound power is becoming more and more common. This measurement is discussed separately here because it cannot be done using the procedures discussed in previous sections due to the large size of the source and the potential effect of wind and turbine induced turbulence on the measurements. A standard has been developed to guide practitioners in this measurement, but it is important to realise that the standard (IEC 61400-11 Ed.3.0, 2018) is based on the assumption that the source directivity is constant in the vertical direction. That is, it is assumed that in the far field, the sound pressure level generated by the turbine will remain constant as the receiver ascends vertically in an arc such that the distance between the centre of the rotor and the receiver remains constant. Making this assumption allows the sound power level of a turbine to be determined by making sound pressure level measurements near or on the ground surrounding the turbine at a specified constant distance from the turbine, such that the measurements are made in the far field. In an acknowledgement that the assumption of uniform directivity in the vertical direction may not be accurate, the standard refers to the resulting sound power calculation as "apparent sound power".

Wind turbine sound power estimations are based on sound pressure level measurements in octave or 1/3-octave bands at locations on a ground mounted board, 1 m in diameter and at least 12 mm thick for a plywood board or 2.5 mm thick for a metal board. The microphone is placed horizontally on the board and covered with a hemispherical primary wind screen of 90 mm diameter open-cell foam (usually provided by the microphone manufacturer). It is also recommended that the microphone and primary windscreen be covered with a secondary windscreen having a diameter of at least 450 mm. The secondary windscreen should be a wire frame covered with 25 mm thick open-cell foam (IEC 61400-11 Ed.3.0, 2018, p. 50) with a thin layer of synthetic fur, especially designed to minimise wind noise. It is important that the insertion loss of the windscreen be measured before use, as described in IEC 61400-11 Ed.3.0 (2018, p. 50)

A total of 4 ground-based measurement positions around the turbine are used, with one (referred to as the reference position) immediately downwind (within ±15°) from the turbine at a horizontal distance of $H + D/2$ from the tower centreline, where H is the height of the turbine rotor centre above the ground and D is the diameter of the rotor. Reported sound power data are based solely on measurements at the single reference position. Data from the other three positions may be used in propagation analyses in directions other than downwind from the turbine.

Overall A-weighted sound pressure levels, as well as 1/3-octave band levels in bands with centre frequencies from 20 Hz to 10 kHz, are separated into wind speed "bins". Wind speeds corresponding to the centre of each wind speed bin are separated by a maximum of 1 m/s increments of wind speed from the turbine cut-in wind speed to the maximum operational wind speed, measured at the turbine nacelle height. The A-weighted levels in each 1/3 octave band are determined either by filtering with the measuring instrument or by applying the centre frequency corrections of Table 2.3 to the unfiltered levels. The measured 1/3-octave band sound pressure levels are corrected by an amount, Δ, which is determined by first adding together logarithmically, the measured 1/3-octave A-weighted levels (see Equation (1.98)). The resulting total A-weighted level, L_{Aeq}, is compared to the overall A-weighted level, $L_{Aeq,o}$, measured by the instrument. A correction, Δ, determined using Equation (4.172), is then added to each 1/3-octave band sound pressure level measurement to obtain the "normalised" level to be used in calculating the average sound pressure level for a particular 1/3-octave band and wind speed bin.

$$\Delta = L_{Aeq,o} - L_{Aeq} \quad \text{(dB)} \tag{4.172}$$

Thus, the resulting "normalised" 1/3-octave band sound pressure level for band, i, is given by:

$$L_{Aeq,n,i} = L_{Aeq,i} + \Delta \quad \text{(dB re 20 µPa)} \tag{4.173}$$

Normalised sound pressure level data in a given wind speed bin, k, corresponding to a bin centre wind speed, U and 1/3-octave band, i, are averaged to obtain the total sound pressure level, $L_{U,T,i,k}$, which includes the background noise as well as the turbine noise. The same measurements and analyses are repeated with the turbine switched off to obtain background noise levels, $L_{U,B,i,k}$. The sound pressure level, $L_{U,c,i,k}$, resulting from the wind turbine alone is then calculated using Equation (1.102). IEC 61400-11 Ed.3.0 (2018) also provides a procedure to estimate the uncertainty of the result. The turbine "apparent" A-weighted sound power level, $L_{WA,i,k}$, for 1/3-octave band, i and wind speed bin, k, is calculated from $L_{U,c,i,k}$ using:

$$L_{WA,i,k} = L_{U,c,i,k} - 6 + 10\log_{10}[4\pi r_1^2] \qquad (\text{dB re } 10^{-12}\,\text{W}) \tag{4.174}$$

where r_1 is the distance from the centre of the turbine rotor to the measurement location. More details on the procedure and uncertainty analyses are provided in IEC 61400-11 Ed.3.0 (2018, Sect. 9.2).

4.14.7 Road Traffic Noise Measurement

The measurement of road traffic noise and other transportation noise is a special case, as the noise sources are moving and many different types of source are contributing to the resulting noise levels at receiver locations. Methods for measuring environmental noise resulting from road traffic are discussed at length in ISO 362-1 (2015), ISO 362-2 (2009), ISO 362-3 (2016) and ISO 16254 (2016). Procedures are provided for measuring stationary or slowly-moving vehicles (ISO 16254, 2016) for the purpose of detectability by pedestrians, as well as for accelerating and constant speed vehicles (ISO 362-1, 2015; ISO 362-2, 2009) for the purpose of assessing the radiated noise under typical urban traffic conditions. Measurements are undertaken on a flat test track under specified acceleration and vehicle mass conditions. The test program outlined in ISO 362-1 (2015); ISO 362-2 (2009) is complex and difficult to adequately summarise here. Procedures are also available for testing vehicles indoors (ISO 362-3, 2016).

4.14.8 Specialist Procedures for Other Noise Sources

International standards exist for the sound power measurement of a number of specific sources. These methods are generally minor variations of the methods already discussed in Section 4.14 and are summarised in Table 4.9.

4.14.9 High-Frequency Correction for Sound Power Level Measurements

Sound power levels determined by outdoor measurements outlined in this chapter are based on the assumption that air absorption is negligible. However, for sound containing significant energy at frequencies above about 10 kHz, this is no longer true and the measured sound power level must be corrected to allow for the effects of air absorption of sound, which is discussed in detail in Section 5.3.2. For sound power measurements made outdoors or in an anechoic room, according to ISO 3744 (2010), using a measurement radius greater than 2 m, the measured result for 1/3-octave bands of 10 kHz and above and for the octave band of 16 kHz must be increased by $\alpha_a r$, where r is the measurement radius and α_a is calculated as described in Section 5.3.2. For procedures involving source substitution, reverberation times or measurements on two test surfaces, the air absorption is already taken into account.

4.14.10 Uncertainty in Sound Power Levels Determined Using Sound Pressure Level Measurements

According to ISO 3741 (2010); ISO 3743-2 (2018); ISO 3744 (2010); ISO 3745 (2017); ISO 3746 (2010) and ISO 3746 (2010), arithmetic means and standard deviations of sound pressure levels are used as a basis of uncertainty estimation for sound power measurements. In many cases, the symbol, s, is used in place of σ to indicate that the quantity is an estimate of the standard deviation.

TABLE 4.9 Measurement standards for determining sound power and sound pressure level at the operator location for specific noise sources

Standard number	Sound source description
IEC 60704-2-1 (2020)	Vacuum cleaners[1]
IEC 60704-2-2 (2009)	Fan heaters[1]
IEC 60704-2-3 (2017)	Dishwashers[1]
IEC 60704-2-4 (2011)	Washing machines and spin extractors[1]
IEC 60704-2-5 (2014)	Electric thermal storage room heaters[1]
IEC 60704-2-6 (2012)	Tumble dryers[1]
IEC 60704-2-7 (2020)	Fans[1]
IEC 60704-2-8 (2020)	Electric shavers[1]
IEC 60704-2-9 (2003)	Electric hair care appliances[1]
IEC 60704-2-10 (2011)	Electric cooking ranges, ovens, grills, microwave ovens, any combination of these[1,2]
IEC 60704-2-11 (1998)	Electrically-operated food preparation[1]
IEC 60704-2-13 (2016)	Range hoods, other cooking fume extractors[1]
IEC 60704-2-14 (2019)	Refrigerators, freezers, frozen-food storage cabinets[1]
IEC 60704-2-15 (2008)	Household food waste disposers[1]
IEC 60704-2-16 (2019)	Washer-dryers[1]
IEC 60076-10 (2016)	Electric power transformers[1]
IEC 60076-10-1:2016+AMD1:2020CSV (2016)	Electric power transformers[1]
ANSI/ASA S12.10-1 (2020)	IT & telecommunications equipment[1,2]
ANSI/ASA S12.14 (2020)	Audible public and outdoor warning devices[1]
ANSI/ASA S12.15 (2016)	Portable and fixed electric power tools, Gardening appliances[1]
ASTM E1265-04 (2021)	Pneumatic exhaust silencers[5]
ASTM F1334-18 (2021)	Vacuum cleaners[1]
ASTM F2544-11 (2021)	Central vaccuum power units[1]
ISO 1680 (2013)	Rotating electrical machines[1,3]
ISO 2922 (2020)	External noise emitted by vessels on inland waterways and harbours[4]
ISO 2923 (1997)	Shipboard noise[3]
ISO 3381 (2021)	Railway vehicle interior[2,3]
ISO 5129 (2013)	Aircraft interior[2,3]
ISO 5130 (2019)	Stationary road vehicles[3]
ISO 5135 (2020)	Air exit devices in air flow systems[1]
ISO 5136 (2003)	Fans in ducts[1]
ISO 7779 (2018)	IT & telecommunications equipment[1,2]
ISO 9645 (1990)	2-wheeled mopeds[3]
ISO 10302-1 (2011)	Small air moving devices[1]
ISO 13475-1 (1999)	Audible public and outdoor warning devices[1]
ISO 13475-2 (2000)	Audible public and outdoor warning devices[1]

Measured quantities

1. Sound power level
2. Sound pressure level at the operator location
3. Sound pressure level at locations other than the operator location
4. A-weighted sound exposure level
5. Insertion Loss

The total standard deviation, σ_{tot}, of the measured data is made up of two components, σ_{omc}, corresponding to variability in the operating and mounting conditions, and σ_R, corresponding to the reproducibility of the measurement caused by uncertainties resulting from the use of different instrumentation, different measurement facilities and different implementations of the measurement procedure (ISO 3741, 2010). The expanded uncertainty, u_e, according to ISO 3741 (2010) is equal to $2\sigma_{\text{tot}}$, so that there is a 95% chance that the true value lies within $\pm u_e$ of the measured value. The total standard deviation, σ_{tot}, may be calculated using (ISO 3741, 2010):

$$\sigma_{\text{tot}} = \sqrt{\sigma_{\text{omc}}^2 + \sigma_R^2} \quad \text{(dB)} \tag{4.175}$$

where

$$\sigma_{\text{omc}} = \sqrt{\frac{1}{N_m - 1}\sum_{i=1}^{N_m}(L_{p,i} - L_{p,\text{av}})^2} \quad \text{(dB)} \tag{4.176}$$

where $L_{p,i}$ is the sound pressure level (usually an L_{eq} measurement, energy-averaged over at least 1 minute), measured at a prescribed location corresponding to the highest sound pressure level measurement, corrected for background noise, for the ith repetition of the prescribed operating and mounting conditions. Alternatively, all measurement positions could be used and in this case, $L_{p,i}$ (corrected for background noise according to Equation (1.102)) would be the arithmetic average of the sound pressure level over all measurement locations for condition, i. The average sound pressure level is the arithmetic (not energy) average of all sound pressure levels (N_m) used in Equation (4.176) and:

$$L_{p,\text{av}} = \frac{1}{N_m}\sum_{i=1}^{N_m} L_{p,i} \quad \text{(dB)} \tag{4.177}$$

Typical values of σ_{omc} vary from 0.5 dB for cases characterised by a well-defined measurement procedure, which includes specification of mounting conditions, and a small variability of sound power level over time, to 2 dB for cases characterised by a large variability in sound power level caused, for example, by material flowing into and out of the machine.

Typical upper bound values of σ_R are given in the ISO standards listed above for various methods of sound power measurement and are listed in Tables 4.10, 4.11 and 4.12. If it is suspected that these values are too high, then it may be necessary to test the machine in several different test facilities to obtain a more accurate estimate of σ_R.

TABLE 4.10 Upper bound for σ_R (dB) for various sound power measurement procedures

	Octave band centre frequency (Hz)				Overall
	125	250	500–4000	8000	A-wt
ISO 3743-1 (2010) Section 4.14.3.2	3.0	2.0	1.5	2.5	1.5
ISO 3743-2 (2018) Section 4.14.2.1	5.0	3.0	2.0	3.0	2.0
ISO 3746 (2010) Section 4.14.3.4	—	—	—	—	3.0 (tones not significant) 4.0 (discrete tones present)
ISO 3747 (2010) Section 4.14.3.2	—	—	—	—	1.5 (Engineering – grade 2) 4.0 (Survey – grade 3)

It is also possible to estimate the uncertainty in sound power level associated with a batch of the same equipment items (ISO 7574-1, 1985). As before, the expanded uncertainty is given by $2\sigma_{\text{tot}}$, where $\sigma_{\text{tot}} = \sqrt{\sigma_P^2 + \sigma_R^2}$ and $\sigma_P = \sqrt{\sigma'^2 + \sigma_{\text{omc}}^2/N_m}$ and σ' is the standard deviation of the mean sound power level measured for a number of different machines from the same batch.

TABLE 4.11 Upper bound for σ_R (dB) for various sound power measurement procedures

	1/3-octave band centre frequency (Hz)					Overall
	50–80	100–160	200–315	400–5000	6300–10000	A-wt
ISO 3741 (2010) Section 4.14.2.2	3.9	3.0	2.0	1.5	3.0	0.5
ISO 3744 (2010) Section 4.14.1 Section 4.14.3.1 Section 4.14.3.2 Section 4.14.3.3	—	3.0	2.0	1.5	2.5	1.5

TABLE 4.12 Upper bound for σ_R (dB) for measurements in anechoic and hemi-anechoic rooms

	1/3-octave band centre frequency (Hz)					Overall
	50–80	100–630	800–5000	6300–10000	12500–20000	A-wt
ISO 3745 (2017) Section 4.14.1 anechoic room	2.0	1.0	0.5	1.0	2.0	0.5
ISO 3745 (2017) Section 4.14.1 hemi-anechoic room	2.0	1.5	1.0	1.5	2.0	0.5

4.15 Determination of Sound Power Using Surface Vibration Measurements

Vibration measurements to determine the sound power radiated by a vibrating surface are sometimes used in the following situations:

1. background noise is too high to use sound pressure or sound intensity measurements;
2. structurally radiated noise is to be separated from aerodynamic noise
3. relative contributions to the overall radiated noise are to be determined for various structural components in close proximity.

The sound power radiated by a machine surface can be estimated from a determination of the mean square vibration velocity averaged over the surface (Takatsubo et al., 1983). The measurements are usually made using an accelerometer, an integrating circuit, a bandpass filter and a sound level meter or vibration meter. In many cases, the integrating circuit and bandpass filter are incorporated in the sound level or vibration meter. Generally, five to ten measurements distributed at random, but not too close to the edges of the radiating surface, are sufficient. However, referring to Equation (4.115) for a plane piston noise source, a similar expression can be derived by replacing $(U^2/2)$ with the surface mean square velocity, $\langle v^2 \rangle_{S,t}$, replacing the piston surface area, πa^2, with the noise-radiating area of the surface, S_p, and renaming the real part of the function, R_R, as the surface radiation efficiency, σ. With these changes, Equation (4.115) becomes:

$$W = \langle v^2 \rangle_{S,t} S_p \rho c \sigma \tag{4.178}$$

Taking logarithms to the base 10 of both sides of Equation (4.178) and using octave or 1/3-octave band measurement of the space-and-time-averaged mean square surface velocity, $\langle v^2 \rangle_{S,t}$, the level of sound power radiated in each of the 1/3-octave or octave bands can be computed using the following equation:

$$L_W = 10\log_{10}\langle v^2\rangle_{S,t} + 10\log_{10} S_p + 10\log_{10}\sigma + 146 \qquad (\text{dB re } 10^{-12}\text{ W}) \tag{4.179}$$

where S_p is the area of the noise-radiating plane surface. The quantity, σ, is a measure of the efficiency of radiation, called the radiation efficiency or radiation factor. Since it can be larger than unity, it is sometimes called "radiation ratio". The radiation efficiency may be determined in a laboratory, away from other noise sources, by measuring the radiated sound power and then using Equation (4.178) (ISO 7849-2, 2009).

For simply supported, rectangular, flat panels of dimensions $\ell_1 \times \ell_2$, with $\ell_1 \leq \ell_2$ and perimeter, P, the following procedure is suggested for calculating the radiation efficiency, which is a function of the ratio of the excitation frequency, f to the panel critical frequency, f_c (see Equations (7.3) and (7.4)) and fundamental resonance frequency, $f_{1,1}$ (see Equation (7.46)).

$$\sigma = 4S_pf^2/c^2; \qquad f < f_{1,1} \qquad (\text{Vér, 2006}) \tag{4.180}$$

$$\sigma = \left(\frac{2c^2}{f_c^2 S_p}\delta_1 + \frac{Pc}{f_c S_p}\delta_2\right)\gamma; \qquad 2f_{1,1} < f < xf_c \qquad \begin{matrix}\text{(Maidanik (1962)}\\ \text{corrected according to}\\ \text{Crocker and Price (1969))}\end{matrix} \tag{4.181}$$

$$\sigma = 0.5\sqrt{2\pi f\ell_1/c}; \qquad f = f_c \qquad (\text{Leppington et al., 1982}) \tag{4.182}$$

$$\sigma = (1 - f_c/f)^{-1/2}; \qquad f \gg f_c \qquad \begin{matrix}\text{(Maidanik (1962) and}\\ \text{Leppington et al. (1982))}\end{matrix} \tag{4.183}$$

where x is a value between 0.7 and 0.99 and:

$$\delta_1 = \begin{cases}\dfrac{4}{\pi^4}\dfrac{(1-2\xi_c^2)}{\xi_c(1-\xi_c^2)^{1/2}}; & f < f_c/2 \qquad (\text{Maidanik, 1962})\\ 0; & f > f_c/2\end{cases} \tag{4.184}$$

$$\delta_2 = \frac{1}{4\pi^2}\left[\frac{(1-\xi_c^2)\log_e\left(\dfrac{1+\xi_c}{1-\xi_c}\right) + 2\xi_c}{(1-\xi_c^2)^{3/2}}\right] \qquad \begin{matrix}\text{(Maidanik (1962) and}\\ \text{Leppington et al. (1982))}\end{matrix} \tag{4.185}$$

and where

$$\xi_c = (f/f_c)^{1/2} \tag{4.186}$$

For simply supported panels, γ takes the value of 1 while for clamped edge panels γ takes the value 2. All other conditions lie between these extremes.

Leppington et al. (1982) showed that the term containing δ_1 in Equation (4.181) is invalid and should be omitted for all frequencies. This correction is included in Figure 4.21.

It can be seen from the preceding equations that there are gaps in the predictions between the lowest panel resonance and twice this frequency, as well as on either side of the critical frequency, although there is an estimate at exactly the critical frequency. The first mentioned gap is filled by drawing a straight line on a $\log_{10}\sigma$ versus $\log_{10}(f/f_c)$ plot between σ at $f_{1,1}$ and σ at $2f_{1,1}$. For the gap above the critical frequency, Equation (4.183) defines valid values of frequency to be such that $f \gg f_c$, although "$\gg$" is not defined. Similarly for frequencies below f_c, the upper valid value of f is that frequency, f, corresponding to the quantity, $k\ell_1[(f_c/f)^{1/2} - 1]$, no longer being "large" (Leppington et al., 1982), where "large" is undefined. Thus, in the construction of

Figure 4.21 to follow, the calculated radiation efficiency at the critical frequency has been set as the maximum value over the entire frequency range and then the values of radiation efficiency in the vicinity of the critical frequency are adjusted so that the curves above and below the critical frequency meet smoothly at the critical frequency. The approximations so obtained are sufficiently accurate for sound power calculations. Note the upper practical limit of $\log_{10}\sigma = 5$ dB, as experimentally measured flat panel radiation efficiencies above this value have not been reported in the literature.

For panels supported at the edge and on intermediate battens (or studs), the perimeter, P, in Equation (4.181) is the overall length of the panel perimeter plus twice the length of all the studs. The area, S_p, is the area of the entire panel.

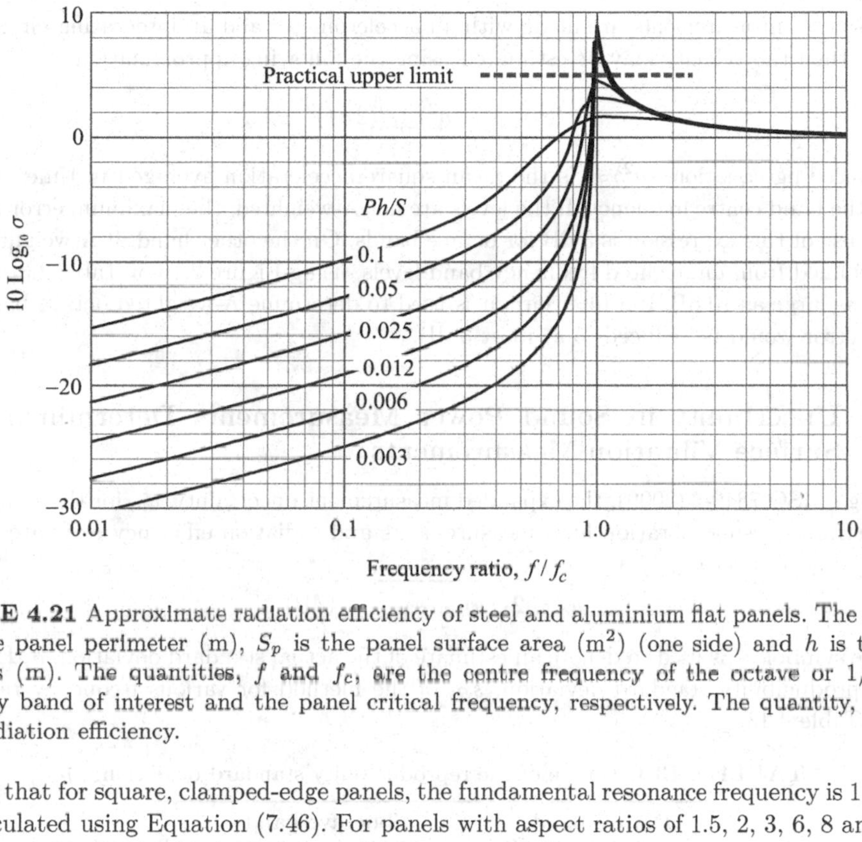

FIGURE 4.21 Approximate radiation efficiency of steel and aluminium flat panels. The quantity, P, is the panel perimeter (m), S_p is the panel surface area (m^2) (one side) and h is the panel thickness (m). The quantities, f and f_c, are the centre frequency of the octave or 1/3-octave frequency band of interest and the panel critical frequency, respectively. The quantity, σ, is the panel radiation efficiency.

Note that for square, clamped-edge panels, the fundamental resonance frequency is 1.83 times that calculated using Equation (7.46). For panels with aspect ratios of 1.5, 2, 3, 6, 8 and 10 the factors are 1.89, 1.99, 2.11, 2.23, 2.25 and 2.26, respectively.

Between the lowest order modal resonance and twice that frequency, the radiation efficiency is found by interpolating linearly (on a log σ versus log f plot).

For a thick plate ($h > 0.3c_B/f$), where c_B is the plate bending wave speed (see Equation (7.1)), the radiation efficiency for excitation by a vibration field is (Vér, 2006, p. 511):

$$\sigma = \begin{cases} 0.45\sqrt{Pf/c}; & f \leq (f_c + 5c/P) \\ 1; & f \gg (f_c + 5c/P) \end{cases} \tag{4.187}$$

In the preceding equations, the quantities P and S_p are the panel perimeter and area, respectively. The panel is assumed to be isotropic of uniform thickness, h, and characterised by longitudinal wave speed, c_{LI}. Values of c_{LI} for various materials are given in Appendix C.

Radiation efficiencies for other structures (e.g. I-beams, pipes, etc.) are available in the published literature (Vér, 2006; Wallace, 1972; Lyon, 1975; Jeyapalan and Richards, 1979; Anderton and Halliwell, 1980; Richards, 1980; Jeyapalan and Halliwell, 1981; Sablik, 1985).

In cases for which it is not possible to determine a radiation efficiency, ISO 7849-1 (2009) suggests that it be set equal to one and that the A-weighted vibration velocity be used in Equation (4.179) to calculate an upper bound to the radiated sound power level. The A-weighted vibration velocity, $\langle v^2 \rangle_{S,t}$, can be obtained using an accelerometer, integrating the signal and then passing it through an A-weighting network. The success of this scheme depends on the observation that for high frequencies, the radiation efficiency is approximately unity, and only for low frequencies is it less than unity and uncertain. The A-weighting process minimises the importance of inefficient, low-frequency vibration in terms of its contribution to the overall radiated sound power. Thus, the use of A-weighted vibration measurements and setting $\sigma = 1$ in Equation (4.179) generally allows identification of dominant noise sources on many types of machines and an estimate of the radiated sound power.

If vibration measurements are made with an accelerometer and an integrating circuit is not available, then the velocity may be estimated using the following approximation:

$$\langle v^2 \rangle_{S,t} = \langle a^2 \rangle_{S,t} / (2\pi f)^2 \tag{4.188}$$

In the preceding equation, $\langle a^2 \rangle_{S,t}$ is the mean square acceleration averaged in time and space and f is the band centre frequency. If the levels are not A-weighted, the maximum error resulting from the use of this expression is 3 dB for octave bands. On the other hand, if A-weighted levels are determined from unweighted frequency band levels, using Figure 2.16 or Table 2.3, the error could be as large as 10 dB. If a filter circuit is used to determine A-weighted octave band levels then the error would be reduced to at most 3 dB.

4.15.1 Uncertainty in Sound Power Measurements Determined Using Surface Vibration Measurements

According to ISO 7849-2 (2009), the expanded measurement uncertainty, u_e, for the sound power level determined using vibration level measurements and radiation efficiency estimates is given by:

$$u_e = 2\sqrt{s_R^2 + s_{\text{omc}}^2} \quad \text{(dB)} \tag{4.189}$$

where the symbol, s, is used to denote an estimate of the actual standard deviation, σ. Estimates of the reproducibility standard deviation, s_R, of the method for various frequency ranges are listed in Table 4.13.

TABLE 4.13 Estimate of the reproducibility standard deviation, s_R

Octave band centre frequency (Hz)	1/3-octave band centre frequency(Hz)	s_R
125	100–160	3
250	200–315	2
500–4000	400–5000	1.5
8000	6300–10000	2.5
Overall A-weighted		1.5

The estimate, s_{omc}, of the operating and mounting standard deviation may be obtained by taking at least 6 sound power measurements using the same test equipment and the same measurement site (ISO 7849-2, 2009). The mean and estimated standard deviation may then be calculated using Equations (2.126) and (2.127), where the measured sound power level, L_W, is substituted for L_{eq}. Alternatively, they may be calculated as described in Appendix G of ISO 3741 (2010), where arithmetic, rather than logarithmic averaging of sound pressure levels is used, even though using arithmetic instead of logarithmic averaging is difficult to justify scientifically (see Section 4.14.10).

There is a 95% probability that the actual sound power level will be within $\pm u_e$ dB of the average of the measured results.

4.16 Uses of and Alternatives to Sound Power Information

The location of a machine with respect to a large reflecting surface such as a floor or wall of a room may affect its sound power output. Thus, the interpretation of sound power information must include consideration of the mounting position of the machine (e.g. on a hard floor or wall, at the junction of a wall and floor or in the corner of a room). Similarly, such information should be contained in any sound power specifications for new machines. The sound power output of a machine may be used to estimate sound pressure levels generated at various locations; for example, in a particular room. In this way, the contribution made by a particular machine to the overall sound pressure level at a particular position in a room may be determined. Means for estimating sound pressure levels from sound power level information are required, dependent on the situation, and these will now be considered.

4.16.1 Far Free Field

When the sound pressure level, L_p, is to be estimated in the far-free-field of the source (see Figure 4.18), the contribution of the reverberant sound field to the overall sound field will be assumed to be negligible for two particular cases. The first is if the machine is mounted in the open, away from any buildings. The second is if the machine is in a large room, which has had its boundaries (not associated with the mounting of the machine) treated with acoustically absorptive material, and where the position at which L_p is to be estimated is not closer than one-half of a wavelength to any room boundaries. In this situation, directivity information about the source is useful.

An expression for estimating L_p at a distance, r, from a source for the preceding two cases is obtained by rearrangement of Equation (4.130). The directivity index, DI, that appears in Equation (4.130) is given in Table 4.1. If no directivity information exists, then the average sound pressure level to be expected at a distance, r, from the source can be estimated by assuming a uniform sound radiation field.

4.16.2 Near Free Field

The near-free-field of a sound source (geometric or hydrodynamic) is generally quite complicated, and cannot be described by a simple directivity index. Thus, estimates of the sound pressure level at fixed points near a sound source are based on the simplifying assumption that the sound source has a uniform directivity pattern. This is often necessary as, for example, the operator's position for most machines is usually in the near field. Rough estimates of the sound pressure level at points on a hypothetical surface of area, S, conforming to the shape of the surface of the machine, and at a specified short distance from the machine surface, can be made using Equation (4.160). Referring to the latter equation, if the contribution due to the reverberant field can be considered negligible, then $\Delta_1 = 0$.

4.16.3 Sound Pressure Levels at Operator Locations

In many cases, where the machine operator must work in the near vicinity of a machinery noise source, it is more useful to provide sound pressure levels at the operator location, rather than machine sound power levels. These can be determined directly by measurement (ISO 11202, 2010; ISO 11204, 2010) or from the sound power level (ISO 11203, 1995). The standard, ISO 11203

(1995), requires the use of a rearranged Equation (4.160) (and the associated parallelepiped measurement surface located at the same distance from the machine surface as the operator location) but without the two correction terms, Δ_1 and Δ_2. The same approach applies to the calculation of sound pressure levels at other locations close to the machinery noise source.

5

Outdoor Sound Propagation and Outdoor Barriers

LEARNING OBJECTIVES

In this chapter, the reader is introduced to:

- reflection and transmission at a plane two media interface;
- sound propagation outdoors, ground reflection, atmospheric effects, barrier effects, vegetation effects, methods of prediction; and
- application of sound propagation models.

5.1 Introduction

When calculating the sound pressure level at a receiver in the community as a result of a noise source, it is necessary to consider the effects of the ground and the atmosphere on the propagation of the noise. Treatment of this complex problem is considered here and is based on many references given in the following text.

To begin, we will analyse the relatively simple case of sound reflection and transmission at a plane interface between two different media, such as the air and ground. Later, this analysis will be applied to sound travelling in air for cases where the source/receiver distance is much greater than the heights of the source and receiver above the ground. This latter analysis will be used as a basis for understanding the various noise propagation models to be discussed in the remainder of this chapter. However, it is worth remembering that community sound pressure levels predicted using a propagation model may not necessarily be representative of the annoyance of the noise, nor may it be a very accurate representation of the range of sound pressure levels that will be experienced, due to all the uncertainties and parameter variations involved. As will be seen in following sections, at best we may expect an uncertainty in the predicted sound pressure level at residences of ±3 to ±4 dBA for any specified atmospheric condition. In addition to the sound pressure level, the annoyance of the noise will depend upon its frequency distribution, its variability with time and the presence or otherwise of tones.

5.2 Reflection and Transmission at a Plane Interface between Two Different Media

Sound rays may impinge on the ground as a result of the sound source directing some energy towards the ground or as the result of atmospheric refraction, which causes the sound ray path

DOI: 10.1201/9780367814908-5

to curve downwards, so that it eventually strikes the ground, even if not initially directed at the ground. When sound impinges on a porous surface such as the ground, some sound energy is reflected and some is transmitted into the ground as indicated in Figure 5.1. The proportion of reflected energy to transmitted energy is a function of the difference in characteristic impedance between the two media. When this difference is small, the transmitted wave amplitude and energy is greater than the reflected wave amplitude and energy and when this difference is large, the reflected wave energy is greater than the transmitted wave energy. The larger the difference in characteristic impedance, the larger will be the difference between transmitted and reflected wave energies and amplitudes. The ratio of transmitted to reflected wave amplitudes is also a function of the angle of incidence of the incident wave. The closer this angle is the normal to the surface, the greater will be the transmitted energy and wave amplitude.

Sound transmission over a porous ground surface is most accurately modelled assuming the reflection of a spherical wave at a plane (that is, flat) reflecting surface. Unfortunately, the associated calculations are very complicated. However, the resulting equation for the spherical wave amplitude reflection coefficient, Q, may be expressed in terms of the much simpler plane-wave amplitude reflection coefficient, R_p (see Equation (5.23)). Consequently, it will be convenient to consider first, the reflection of a plane wave at a plane interface between two media.

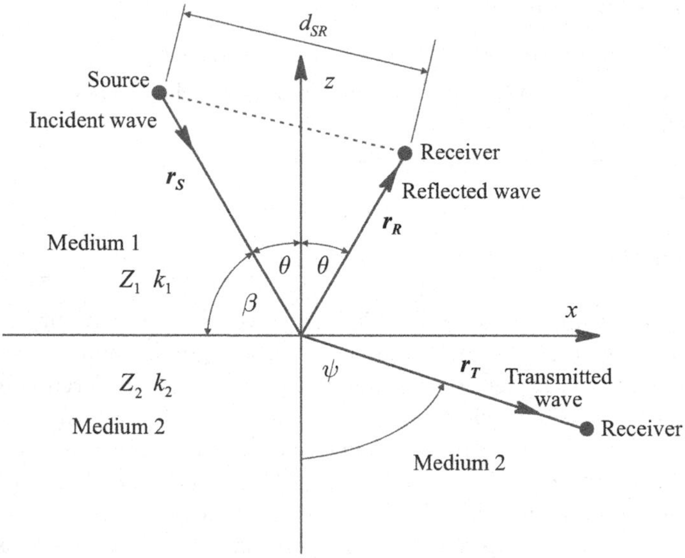

FIGURE 5.1 Geometry illustrating reflection and transmission at the plane interface between two acoustic media.

In the literature, one of three assumptions is commonly made, often without comment, when considering the reflection of sound at an interface between two media; for example, at the ground surface. Either it is assumed that the second medium is locally reactive, so that the response of any point on the surface is independent of the response at any other point in the second medium; or it is assumed that the surface of the second medium is modally reactive, where the response of any point on the surface is dependent on the response of all other points on the surface of the second medium. Alternatively, in media in which the sound wave attenuates as it propagates, the response at any point on the surface will depend only on the response of nearby points, within an area whose size depends on the extent of attenuation. In this latter case, the surface will be referred to as extensively reactive or as a case of extended reaction. A criterion given by Equation (5.19) for determining how a porous surface, for example the ground, should be treated is discussed in Section 5.2.2.

5.2.1 Porous Ground

When one of the media, such as the ground, is described as porous and the other medium is a gas, such as air, which penetrates the pores of the porous medium, then the term "porous" has the special meaning that sound is transmitted through the pores and not the structure, which is generally far less resilient than the gas in the pores. In such a case, the acoustic properties associated with a porous medium are determined by the combined properties of a rigid, gas-filled structure, which may be replaced with a fictitious gas of prescribed properties for the purposes of analysis, as described in Appendix D.

For the case of the ground, which is well modelled as a porous medium (here indicated by subscript, m), the characteristic impedance, Z_m, and the propagation coefficient, k_m (both complex), may be calculated from a knowledge of the ground surface flow resistivity, R_1, in MKS rayls/m as described in Appendix D. Values of flow resistivity, R_1, for various ground surfaces are given in Table 5.1. The values were obtained from ANSI/ASA S1.18 (2018), Nordtest (1999), Embleton et al. (1983) and Attenborough et al. (2007).

In Appendix D, it is shown that both the characteristic impedance and the propagation coefficient may be expressed as functions of the dimensionless scaling parameter, $\rho f/R_1$, where ρ is the density of the gas in the pores and f is the frequency of the sound being considered.

As can be seen from Table 5.1, there are great variations in measured flow resistivity data from different sources. For this reason, some propagation models classify ground surfaces into 8 types labelled A–H, as listed in Table 5.2 (see Plovsing (2006a)).

In general, a wavenumber (or propagation coefficient) may be complex, where the real part is associated with the wave speed and the imaginary part is associated with the rate of sound propagation loss. When propagation loss is negligible, the wavenumber takes the form given by Equation (1.25). Alternatively, when sound propagation loss is not negligible, as in the case of propagation in a porous medium, the wavenumber, k_m, takes the form, $k_m = \omega/c_m - \mathrm{j}\alpha_m$, where c_m is the wave speed in the porous medium and α_m is the propagation loss factor (see Appendix D).

5.2.2 Plane Wave Reflection and Transmission

The reflection and transmission of a plane sound wave at a plane interface between two media will be considered. As illustrated in Figure 5.1, the interface is assumed to be flat and the incident, reflected and transmitted waves are assumed to have plane wavefronts. The plane interface is assumed to lie along the abscissa at $z = 0$ and the angles of incidence and reflection, θ, and transmission, ψ, are measured from the normal to the plane of the interface.

The vectors, $\boldsymbol{r_I}, \boldsymbol{r_R}$, and $\boldsymbol{r_T}$, indicate directions of wave travel and progress of the incident, reflected and transmitted waves, respectively. Medium 1 lies above and medium 2 lies below the x-axis. Both media extend away from the interface by an infinite distance and have characteristic impedances, Z_1 and Z_2, and propagation coefficients (complex wavenumbers) k_1 and k_2, any or all of which may be complex. For the infinitely extending media, the characteristic impedances are equal to the normal impedances, Z_{N1} and Z_{N2}, respectively, at the interface.

Referring to Figure 5.1 the component propagation coefficients are defined as:

$$k_{1x} = k_1 \sin\theta, \qquad k_{1z} = k_1 \cos\theta \tag{5.1}$$

$$k_{2x} = k_2 \sin\psi, \qquad k_{2z} = k_2 \cos\psi \tag{5.2}$$

The time dependent term, $\mathrm{e}^{\mathrm{j}\omega t}$, may be suppressed, allowing the sound pressure of the propagating incident wave of amplitude A_I to be written as:

$$p_I = A_I \mathrm{e}^{-\mathrm{j}k_1 r_S} \tag{5.3}$$

TABLE 5.1 Flow resistivities measured for some common ground surfaces

Ground surface type	Flow resistivity, R_1 (kPa s/m^2)
Dry snow, newly fallen 0.1 m over about 0.4 m older snow	10–30
Sugar snow	25–50
Soft forest floor with blueberry greens and moss	40
Forest floor covered by weeds	63–100
Pine or hemlock forest floor	20–80
Soft forest floor covered with pine needles	160
Sandy forest floor	630–2000
Dense shrubbery, 20 cm high	100
Soil and bark, sparse vegetation	100
Peat or turf area, homogeneous organic material	100
Soil covered with leaves and twigs	160–250
Soil mixed with sawdust	250
Relatively dense soil sparsely covered by grass and other low greens	630
Short grass, green moss and blueberry greens	40
Rough grassland and pasture	100–300
Grass, soccer field	630–2000
Lawn, moderately stepped on	160–250
Lawn, seldom stepped on	250
Agricultural field	160–250
Hard soil	400–2000
Soil, exposed and rain packed	4000–8000
Wet, sandy loam	1500
Moistened sand	500
Bare sandy plain	250–500
Dry sand	60–140
Sandy silt, hard packed by vehicles	800–2500
Quarry dust, hard packed by vehicles	5000–20000
Mixed paving stones and grass	630–2000
Old gravel field with sparse vegetation	2000
Gravel road, stones and dust	2000
Gravel parking lot	630–2000
Asphalt sealed by dust and light use	30000
Concrete	20000

where $r_S = |\boldsymbol{r_S}|$ is the distance travelled by the sound ray from the source to the point of reflection on the ground.

Reference to Figure 5.1 shows that the z-component of the incident and transmitted waves travels in the negative direction and on reflection in the positive direction, whereas the x-component travels in the positive direction in all cases. Taking note of these observations, multiplying $k_1 r_S$ by 1, where $1 = \cos^2\theta + \sin^2\theta$, and using Equation (5.1) allows Equation (5.3) to be rewritten for a single sound source, as:

$$p_I = A_I \mathrm{e}^{-\mathrm{j}(k_{1x}x - k_{1z}z)} \tag{5.4}$$

According to the laws of reflection, the angle of reflection must be equal to the angle of incidence. Thus, expressions for the sound pressure of the propagating, reflected and transmitted

TABLE 5.2 Flow resistivities for some common ground surface types (to be used with the ISO 9613-2 and NMPB-2008 propagation models)

Ground surface class	Value of G recom-mended by ISO[1]	Value of G recom-mended by EU[2] for ISO 9613-2	Value of G recom-mended by EU[3] for NMPB-2008	Repres-entative flow resist-ivity, R_1 (kPa s/m^2)	Ground surface description
A	1	1	1	12.5	Very soft (snow or moss)
B	1	1	1	31.5	Soft forest floor
C	1	0.8	1	80	Uncompacted, loose ground
D	1	0.6	1	200	Normal uncompacted ground (pastures, forest floors)
E	0	0.4	0.7	500	Compacted fields, lawns and gravel
F	0	0.2	0.3	2000	Compacted dense ground (gravel road, parking lot)
G	0	0	0	20000	Asphalt, concrete
H	0	0	0	200000	Water

[1]ISO 9613-2 (1996)
[2]European Commission (2010a, p. 30)
[3]Kephalopoulos et al. (2012, p. 86)

waves may be written as:

$$p_R = A_R \mathrm{e}^{-\mathrm{j}(k_{1x}x + k_{1z}z)} \tag{5.5}$$

and

$$p_T = A_T \mathrm{e}^{-\mathrm{j}(k_{2x}x - k_{2z}z)} \tag{5.6}$$

Continuity of pressure at the interface requires that at $z = 0$:

$$p_I + p_R = p_T \tag{5.7}$$

Substitution of Equations (5.4), (5.5) and (5.6) into Equation (5.7) gives:

$$A_I \mathrm{e}^{-\mathrm{j}k_{1x}x} + A_R \mathrm{e}^{-\mathrm{j}k_{1x}x} = A_T \mathrm{e}^{-\mathrm{j}k_{2x}x} \tag{5.8}$$

Equation (5.8) must be true for all x; thus it must be true for $x = 0$, which leads to the following conclusion:

$$A_I + A_R = A_T \tag{5.9}$$

Substitution of Equation (5.9) in Equation (5.8) gives:

$$(A_I + A_R)(\mathrm{e}^{-\mathrm{j}k_{1x}x} - \mathrm{e}^{-\mathrm{j}k_{2x}x}) = 0 \tag{5.10}$$

The amplitudes, A_I and A_R, may be positive and non-zero; thus, the second term of Equation (5.8) must be zero for all values of x. It may be concluded that:

$$k_{1x} = k_{2x} \tag{5.11}$$

Using Equations (5.1) and (5.2) and introducing the index of refraction, $n = c_2/c_1$, Equation (5.11) becomes Snell's law of refraction:

$$\frac{k_1}{k_2} = n = \frac{\sin\psi}{\sin\theta} \tag{5.12}$$

Continuity of the particle velocity (= sound pressure divided by impedance, see Equation (1.21)) at the interface requires that at $z = 0$, the normal component of the particle velocity in medium 1 is equal to that in medium 2. Thus:

$$\frac{p_I - p_R}{Z_{N1}} \cos\theta = \frac{p_T}{Z_{N2}} \cos\psi \tag{5.13}$$

where Z_{N1} and Z_{N2} are the specific normal impedances at the surfaces of media 1 and 2, respectively.

Substitution of Equations (5.4), (5.5), (5.6) and (5.9) into Equation (5.13) leads to:

$$A_I \left[\frac{\cos\theta}{Z_{N1}} - \frac{\cos\psi}{Z_{N2}}\right] = A_R \left[\frac{\cos\theta}{Z_{N1}} + \frac{\cos\psi}{Z_{N2}}\right] \tag{5.14}$$

For later reference, when considering reflection from the surface of the ground and without limiting the generality of the equations, it will be convenient to consider medium 1 as air of infinite extent having a normal impedance of Z_{N1} (at the ground interface) equal to its characteristic impedance, ρc, and propagation coefficient (wavenumber), $k_1 = k = \omega/c$. Similarly, the porous ground will be considered as extending infinitely in depth and having a normal impedance of Z_{N2} (at the ground interface) equal to its characteristic impedance, Z_m, and propagation coefficient, $k_2 = k_m$ (see Equations (D.9) and (D.10)). On making the indicated substitutions, the complex amplitude reflection coefficient for plane waves, $A_R/A_I = R_p$, may be written as:

$$R_p = \frac{A_R}{A_I} = \frac{Z_{N2}\cos\theta - Z_{N1}\cos\psi}{Z_{N2}\cos\theta + Z_{N1}\cos\psi} = \frac{Z_m\cos\theta - \rho c\cos\psi}{Z_m\cos\theta + \rho c\cos\psi} = |R_p|\mathrm{e}^{\mathrm{j}\alpha_p} \tag{5.15a,b,c,d}$$

where the phase of the reflected wave relative to the incident wave is:

$$\alpha_p = \tan^{-1}[\mathrm{Im}\{R_p\}/\mathrm{Re}\{R_p\}] \tag{5.16}$$

Equation (5.15) only applies to infinitely extending media or media that extend for a sufficient distance that waves reflected from any termination back towards the interface have negligible amplitude on arrival at the interface. The equation applies for the interface between a fluid and a porous surface such as the ground (characterised by very different wavenumbers and characteristic impedances) as well as between two fluids, such as two different density air layers where the wavenumbers and characteristic impedances are not very different. When the two characteristic impedances are very different, the transmitted wave amplitude is much smaller than the reflected wave amplitude, whereas when the two characteristic impedances are of a similar magnitude, the transmitted wave amplitude is much greater than the reflected wave amplitude, as can be seen by inspection of Equations (5.12), (5.15b) and (5.21).

Equation (5.12) may be used to show that:

$$\cos\psi = \sqrt{1 - \left(\frac{k}{k_m}\right)^2 \sin^2\theta} \tag{5.17}$$

Examination of Equation (5.17) for the case, $k_m = k_2 \gg k_1 = k$, suggests that for this case, the angle, ψ, tends to zero and Equation (5.15) reduces to the following form:

$$R_p = \frac{Z_{N2}\cos\theta - Z_{N1}}{Z_{N2}\cos\theta + Z_{N1}} = \frac{Z_m\cos\theta - \rho c}{Z_m\cos\theta + \rho c} \tag{5.18a,b}$$

which is the equation for a sound wave in a fluid impinging on a locally reactive porous surface, such as the ground. Use of Equation (5.17) and reference to Appendix D, Equation (D.23), gives the following criterion for the porous surface to be essentially locally reactive ($k_m > 100k$):

$$\rho f < 10^{-3} R_1 \tag{5.19}$$

Equations (5.15) and (5.18) can be evaluated, as discussed in Section 5.2.1, for any value of flow resistivity of the ground. Typical flow resistivities for various ground surfaces are listed in Tables 5.1 and 5.2.

The attenuation (or reflection loss, A_{rf} in dB) of a sound wave in a fluid incident on a porous surface, such as the ground, is related to the reflection coefficient as:

$$A_{rf} = -20 \log_{10} |R_p| \tag{5.20}$$

A_{rf} is plotted in Figure 5.2 for reflection of a sound wave in a fluid from a locally reactive porous surface, for various values of the dimensionless parameter $\rho f / R_1$, where ρ is the fluid density, R_1 is the flow resistivity of the porous surface and f is the tonal frequency, or the centre frequency of the measurement band. Alternatively, if Equation (5.19) is not satisfied (that is, local reaction cannot be assumed for the porous surface), then R_p should be calculated using Equation (5.15). Figure 5.2 is used to determine the decrease in energy, A_{rf} (dB), of the reflected sound wave on reflection from a porous surface such as the ground.

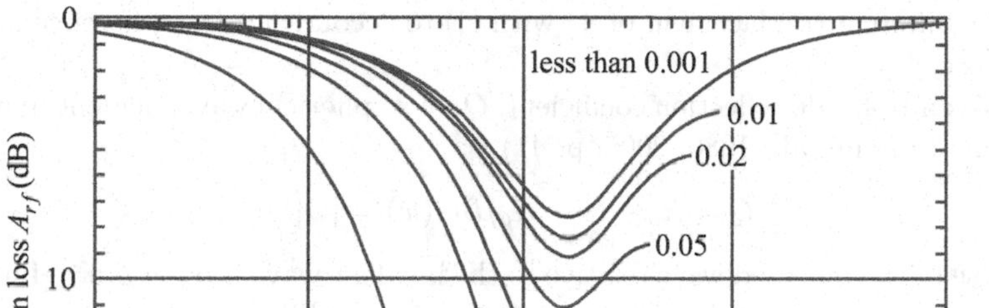

FIGURE 5.2 Reflection loss, A_{rf} (dB), as a function of reflection angle β (where β is measured from the horizontal as shown in Figures 4.17 and 5.1, and $\beta = 90-\theta$ degrees), surface flow resistivity R_1, fluid density ρ and frequency f. Curves are truncated when β reaches 90° or the reflection loss exceeds 20 dB. The surface is assumed to be locally reactive.

The ratio of the amplitude of the transmitted wave to the amplitude of the incident wave may readily be determined by use of Equation (5.15). Also of interest is the sound power transmission coefficient, τ_p, which is a measure of the energy incident at the interface that is transmitted into the second medium. Thus:

$$\tau_p = \frac{W_T}{W_I} = \frac{I_T S}{I_I S} = \frac{I_T}{I_I} = \frac{|Z_{N1}|\,|p_T|^2}{|Z_{N2}|\,|p_I|^2} \tag{5.21}$$

where S represents a surface area, W_T and W_I are, respectively, the transmitted and incident sound powers and I_T and I_I are, respectively, the transmitted and incident sound intensities.

Multiplying the left-hand side of Equation (5.13) with the left-hand side of Equation (5.7) and repeating the operation with the right-hand sides to obtain a new equation and then using Equations (5.7), (5.12) and (5.15b) gives the following expression for the sound power transmission coefficient:

$$\tau_p = \frac{(1 - |R_p|^2)\cos\theta}{\sqrt{1 - n^2 \sin^2\theta}} \tag{5.22}$$

5.2.3 Spherical Wave Reflection at a Plane Interface Between Two Media

The problem of determining the complex amplitude reflection coefficient for a spherical wave incident on a plane interface between two media, which is produced by a point source above the interface, has been considered by Rudnik (1957) and by Attenborough et al. (1980). In the following discussion the air above is considered as medium 1 and the porous ground as medium 2. The air above is characterised by air density, ρ, propagation coefficient, $k = \omega/c$, and characteristic impedance, ρc, while the porous ground is characterised by density, ρ_m, propagation coefficient, k_m, and characteristic impedance, Z_m. In general, the listed variables of the two media may be either real or complex, but in the case of the ground and the air above, only the variables associated with the ground will be considered complex. Where the ground may be characterised by an effective flow resistivity, R_1 (see Table 5.1), the complex quantities, ρ_m, k_m and Z_m may be calculated by reference to Appendix D. In the following analysis positive time dependence, $\mathrm{e}^{\mathrm{j}\omega t}$, has been assumed to be consistent with all other parts of the book. However, in the original references, negative time dependence, $\mathrm{e}^{-\mathrm{i}\omega t}$, has been assumed, which results in a sign change for the imaginary parts of a complex quantity. Thus, Equation (5.33) used in this book is slightly different (replacement of +j with –j) to account for the difference in the assumed time dependence.

The complex amplitude reflection coefficient. Q, of a spherical wave incident upon a reflecting surface (Attenborough et al., 1980, 2007, p. 41) is:

$$Q = R_p + (1 - R_p)BG(w) = |Q|\mathrm{e}^{\mathrm{j}\alpha_s} \tag{5.23a,b}$$

where the phase of the reflected wave relative to the incident wave is $\alpha_s = \tan^{-1}[\mathrm{Im}\{Q\}/\mathrm{Re}\{Q\}]$, and R_p is the plane wave complex amplitude reflection coefficient, which, in turn, is given by either Equation (5.15) or (5.18), as appropriate.

The total time varying sound pressure for a frequency with a wavenumber, k, generated at a receiver by a source over a reflecting plane is:

$$p_t = p_D + p_R = \frac{A}{d_{SR}}\mathrm{e}^{\mathrm{j}(\omega t - k d_{SR})} + \frac{A}{r_S + r_R} Q \mathrm{e}^{\mathrm{j}(\omega t - k(r_S + r_R))} \tag{5.24}$$

where the distances, d_{SR} and r_R, are defined in Figures 5.1 and 4.17, and A is an arbitrary constant dependent on the sound power of the source and various propagation attenuation factors, as explained in Section 5.3. If we assume that propagation attenuation factors other than the ground can be accounted for separately, we may write Equation (5.24) as a ratio of the mean square sound pressure for a spherically symmetric sound at a source–receiver distance, d_{SR}, to the mean square sound pressure at 1 m from the source as:

$$\begin{aligned}\frac{\langle p^2\rangle_{d_{SR}}}{\langle p^2\rangle_{1\mathrm{m}}} &= \frac{1}{d_{SR}^2}\left|1 + \left(\frac{d_{SR}}{r_S + r_R}\right) Q\mathrm{e}^{-\mathrm{j}k(r_S + r_R - d_{SR})}\right|^2 \\ &= \frac{1}{d_{SR}^2} + \frac{|Q|^2}{(r_S + r_R)^2} + \frac{2|Q|}{d_{SR}(r_S + r_R)}\cos[\alpha_s - k(r_S + r_R - d_{SR})] \quad (\mathrm{Pa}^2)\end{aligned} \tag{5.25}$$

where the pressure at 1 m, $\langle p^2\rangle_{1\mathrm{m}}$, is that due to the direct (non-reflected) field only, and it is equal to d_{SR}^2 times the direct field sound pressure at a distance of d_{SR} m. For the general case that the reflecting interface is extensively reactive, B in Equation (5.23) is defined as:

$$B = \frac{B_1 B_2}{B_3 B_4 B_5} \tag{5.26}$$

For most ground surfaces $B \approx 1$. However, for the interface with other porous media, the following expressions may be used to calculate the components of B (Attenborough et al., 1980).

$$B_1 = \left[\cos\theta + \frac{\rho c}{Z_m}\left(1 - \frac{k^2}{(k_m)^2}\sin^2\theta\right)^{1/2}\right]\left[1 - \frac{k^2}{(k_m)^2}\right]^{1/2} \tag{5.27}$$

$$B_2 = \left[\left(1 - \frac{\rho^2}{(\rho_m)^2}\right)^{1/2} + \frac{\rho c}{Z_m}\left(1 - \frac{k^2}{(k_m)^2}\right)^{1/2} \cos\theta + \left(1 - \left(\frac{\rho c}{Z_m}\right)^2\right) \sin\theta \right]^{1/2} \tag{5.28}$$

$$B_3 = \cos\theta + \frac{\rho c}{Z_m}\left(1 - \frac{k^2}{(k_m)^2}\right)^{1/2} \left[1 - \frac{\rho^2}{(\rho_m)^2}\right]^{-1/2} \tag{5.29}$$

$$B_4 = \left[1 - \frac{k^2}{(k_m)^2} \sin\theta\right]^{1/2} \tag{5.30}$$

$$B_5 = \left[1 - \frac{\rho^2}{(\rho_m)^2}\right]^{3/2} [2\sin\theta]^{1/2} \left[1 - \left(\frac{\rho c}{Z_m}\right)^2\right]^{1/2} \tag{5.31}$$

$$B_6 = 1 + \left[\frac{\rho c}{Z_m}\left(1 - \frac{k^2}{(k_m)^2}\right)^{1/2} \cos\theta + \left(1 - \frac{\rho c}{Z_m}\right)^{1/2} \sin\theta\right] \left[1 - \frac{\rho^2}{(\rho_m)^2}\right]^{-1/2} \tag{5.32}$$

where B_3 and B_6 are used in Equation (5.33) to follow, θ is defined in Figure 5.1, and Z_m, ρ_m and k_m are defined in Appendix D.

The complex argument, w, of $G(w)$ in Equation (5.23) is referred to as the square root of the numerical distance and is calculated using: (Attenborough et al., 1980)

$$w = \frac{1}{\sqrt{2}}(1 - \mathrm{j})[k(r_S + r_R)]^{1/2} \frac{B_3}{B_6^{1/2}} \tag{5.33}$$

where r_S and r_R are defined in Figures 5.1 and 4.17. Equation (5.33) differs slightly from that given in Attenborough et al. (1980) (the use of $1-\mathrm{j}$ instead of $1+\mathrm{j}$ on the RHS), which accounts for the choice of positive time dependence in defining the contributing parameters that are used (see the text preceding Equation (5.23)).

When Equation (5.19) is satisfied and the porous surface is essentially locally reactive, such as typical ground surfaces, then $k_m \gg k$ and $Z_m \gg \rho c$ and the following simplifications are possible:

$$B_1 = B_3 = \cos\theta + \frac{\rho c}{Z_m} \tag{5.34}$$

$$B_2 = (1 + \sin\theta)^{1/2} \tag{5.35}$$

$$B_4 = 1 \tag{5.36}$$

$$B_5 = (2\sin\theta)^{1/2} \tag{5.37}$$

$$B_6 = B_2^2 \tag{5.38}$$

To find the spherical wave reflection coefficient for a locally reactive surface, Equation (5.18b) is used to evaluate R_p. For a locally reacting ground surface, Equations (5.34) to (5.38) may be used to calculate the B coefficients used in Equations (5.26) and (5.33). For many ground surfaces, it is preferable to assume extended reaction and use the extended reaction equations for B and w, as these will produce more accurate results for approximately locally reacting ground as well as extended reacting ground.

The term, $G(w)$, in Equation (5.23) is defined as (Attenborough et al., 1980):

$$G(w) = 1 + \mathrm{j}\sqrt{\pi} w g(w) \tag{5.39}$$

where

$$g(w) = g(w_r, \mathrm{j}w_i) = \mathrm{e}^{-w^2} \mathrm{erfc}(-\mathrm{j}w) \tag{5.40}$$

where $g(w) = g(w_r, \mathrm{j}w_i)$ is the scaled complementary error function, erfc( ) is the complementary error function (Abramowitz and Stegun, 1965), and w, which is given by Equation (5.33), is a complex number.

The most recent published procedure for calculating $g(w)$ is provided in Annex D of ANSI/ASA S1.18 (2018).

$$g(w) = \frac{h}{\pi(-\mathrm{j}w)} + \frac{2h(-\mathrm{j}w)}{\pi}\sum_{n=1}^{N}\frac{\mathrm{e}^{-n^2h^2}}{n^2h^2 - w^2} - \epsilon P - \mathrm{E}(h) \tag{5.41}$$

where

$$\epsilon = \begin{cases} 1; & \text{if } \mathrm{Im}\{w\} < \pi/h \\ \dfrac{1}{2}; & \text{if } \mathrm{Im}\{w\} = \pi/h \\ 0; & \text{if } \mathrm{Im}\{w\} > \pi/h \end{cases} \tag{5.42}$$

and

$$P = \frac{2\mathrm{e}^{-w^2}}{\mathrm{e}^{2\pi(-\mathrm{j}w)/h} - 1} \tag{5.43}$$

The error bound can be estimated from Annex D in ANSI/ASA S1.18 (2018):

$$E(h) \leq \frac{2|-\mathrm{j}w|\mathrm{e}^{-(\pi^2/h^2)}}{\sqrt{\pi}\left(1 - \mathrm{e}^{-(\pi^2/h^2)}\right)|\mathrm{Im}\{w\}^2 - \pi^2/h^2|} \tag{5.44}$$

If $h = 0.5$, then summing the series to $N = 14$ will produce an error of $|\mathrm{E}(h)| < 10^{-15}$ (ANSI/ASA S1.18, 2018).

When $\mathrm{Re}\{w\} \approx nh$, where n is a positive integer, Equation (5.41) cannot produce a valid result (ANSI/ASA S1.18, 2018) and the following equation should be used.

$$g(w) = \frac{2h(-\mathrm{j}w)}{\pi}\sum_{n=0}^{N}\frac{\mathrm{e}^{-(n+\frac{1}{2})^2h^2}}{(n+\frac{1}{2})^2h^2 - w^2} + \epsilon P' + \mathrm{E}(h) \tag{5.45}$$

where

$$P' = \frac{2\mathrm{e}^{-w^2}}{\mathrm{e}^{2\pi(-\mathrm{i}w)/h} + 1} \tag{5.46}$$

Note the $n = 0$ lower limit on the sum, in contrast to Equation (5.41).

Equation (5.45) cannot produce a valid result if $\mathrm{Re}\{w\} = (n + 1/2)h$, so the following procedure can be used to evaluate $g(w)$ (ANSI/ASA S1.18, 2018):

1. if $0.1 < \varphi\,[\mathrm{Re}\{w\}/h] < 0.9$, use Equation (5.41);
2. otherwise, use Equation (5.45)

The function $\varphi\,[\mathrm{Re}\{w\}/h]$ means that the quantity to be used in the evaluation is the non-integer part of $[\mathrm{Re}\{w\}/h]$. For example, if $[\mathrm{Re}\{w\}/h] = 4.3$, then $\varphi\,[\mathrm{Re}\{w\}/h] = 0.3$.

The above procedure is only valid if $\mathrm{Im}\{w\} > 0$. If the imaginary part of w is negative (which seems to be always when positive time dependence is used), replace w in Equations (5.41) to (5.46) with $w_a = -w$. Then calculate $g(w_a)$. After calculating $g(w_a)$, calculate $g(w)$ using:

$$g(w) = 2\exp(-w_a^2) - g(w_a) \tag{5.47}$$

The attenuation (or reflection loss, A_{rf} in dB) of a sound wave incident on the ground is related to the reflection coefficient as:

$$A_{rf} = -20\log_{10}|Q| \tag{5.48}$$

5.2.4 Effects of Turbulence

Turbulence in the acoustic medium containing the direct and ground-reflected waves has a significant effect on the effective surface spherical-wave-amplitude reflection coefficient. This effect will now be discussed with reference to sound propagation outdoors over the ground. Local turbulence near the ground is especially important because it introduces variability of phase between the reflected sound and the direct sound from the source to the receiver. Variability in phase between the direct and reflected sound determines whether the two sounds, the direct and the reflected sound, should be considered as adding coherently or incoherently. Coherent reflection requires minimal variability of phase and can result in constructive or destructive interference, in which case the variation in level can be very large, while incoherent reflection, associated with a large variability of phase, can result in at most a 3 dB variation in observed level.

Solar-driven local air currents near the ground, which result as the ground heats up during the day relative to the cooler air above, will cause local convection and turbulence near the ground of the kind being considered here. Sound of wavelength of the order of or less than the turbulence scale will be observed to warble strongly only a short distance away from the source when observed across a paved car-parking lot. The model proposed here suggests that coherent reflection should be observed more often at night than during the day.

The effect of atmospheric turbulence on sound propagation over an acoustically smooth surface has been investigated by Clifford and Lataitis (1983) and by Raspet et al. (1995). The presence or absence of turbulence may be included by a generalisation of their results to give the following general expression for the mean square sound pressure at a receiver located at a distance, d_{SR}, from the source for a spherically symmetric sound source having unit far field mean square sound pressure at 1 m:

$$\begin{aligned}\langle p^2\rangle_t &= \frac{1}{d_{SR}^2}\left|1+\left(\frac{d_{SR}}{r_S+r_R}\right)T_t Q \mathrm{e}^{-\mathrm{j}k(r_S+r_R-d_{SR})}\right|^2 \\ &= \frac{1}{d_{SR}^2}+\frac{|Q|^2}{(r_S+r_R)^2}+\frac{2|Q|}{d_{SR}(r_S+r_R)}T_t\cos[\alpha_s-k(r_S+r_R-d_{SR})] \quad (\mathrm{Pa}^2)\end{aligned} \tag{5.49}$$

which is different to Equation (5.25) only by inclusion of the term, T_t and where the sign of k is opposite to that in the work by Clifford and Lataitis (1983) and Raspet et al. (1995) to reflect that positive time dependence is used in this book, whereas negative time dependence is used in the above-mentioned references. The modulus of the spherical wave reflection coefficient, |Q| and the phase angle, α_s, represent the complex conjugate of Q', which, in turn, is defined by Equation (5.23) and the following text. The distances, d_{SR}, r_S and r_R are shown in Figures 4.17 and 5.1.

The exact solution (Clifford and Lataitis, 1983; Raspet et al., 1995) for the term T_t that appears in the above equation is very complicated. However, if it is assumed that the lateral correlation of the wave parameters between the direct and reflected paths is approximately zero, then we obtain the following approximate expression:

$$T_t = \exp\left\{-4\alpha\pi^{5/2}\langle n_1^2\rangle 10^3\Phi\right\} \tag{5.50}$$

where $\exp\{x\} = \mathrm{e}^x$ and

$$\Phi = 0.001\frac{dL_0}{\lambda^2} \tag{5.51}$$

In Equation (5.51), d is the horizontal distance between the source and receiver (see Figure 4.17), L_0 is the scale of the local turbulence and λ is the wavelength of the sound under consideration. A value of L_0 of about 1 to 1.2 m is suggested if this quantity is unknown or cannot be measured conveniently. When Φ is greater than 1, incoherent reflection can be expected and when Φ is less than 0.1, coherent reflection can be expected.

In Equation (5.50), α has the value 0.5 when $d/k \gg L_0^2$ and the value 1 when $d/k \ll L_0^2$. For values of d/k and L_0 which do not satisfy either of the preceding conditions, use of the exact formulation for T_t may be used (Raspet et al., 1995). However, the exact solution is extremely complicated on the one hand and on the other it is reasonable to expect that α will be bounded by its extreme values, 0.5 and 1. Consequently, as the far field is generally satisfied by the value $\alpha = 0.5$, the latter value is used here to evaluate the exponent of the right-hand side of Equation (5.50).

The term $\langle n_1^2 \rangle$ is the mean square of the fluctuations in the speed of sound relative to its mean value in the absence of turbulence. A value of 5×10^{-6} is suggested (Raspet and Wu, 1995). Evaluation of the exponent allows Equation (5.50) to be rewritten in terms of the figure of merit, Φ, introduced in Equation (5.51) as:

$$T_t = \mathrm{e}^{-0.17\Phi} \tag{5.52}$$

As the figure of merit, Φ, approaches 0, indicating little or no turbulence, the term T_t approaches 1, resulting in coherent ground reflection and a large variation in attenuation as a function of distance and frequency. On the other hand, as the figure of merit becomes large, indicating a large turbulence effect, the term T_t approaches 0, and the direct and ground-reflected waves are combined incoherently, resulting in an expected sound pressure level increase due to ground reflection of 3 dB or less.

5.3 Sound Propagation Outdoors – General Concepts

The problems arising from sound propagation outdoors may range from relatively simple to very complex, depending on the nature of the source and distribution of the affected surrounding areas. If the source is composed of many individual component sources, as would often be the case with an industrial plant, and the surrounding area is extensively affected, then the use of a computer to carry out the analysis associated with level prediction is essential and a number of schemes have been designed for the purpose. These schemes generally rely heavily on empirical information determined from field surveys but gradually empiricism is being replaced with well-established analysis based on extensive research. At present, the most successful schemes rely on a mixture of both theory and empiricism. In all cases, the method of predicting community sound pressure levels as a result of one or more noise sources proceeds as follows.

1. Determine sound power levels, L_W, of all sources.
2. For a given environment calculate the individual components of attenuation, A_{Ei}, (see below) for all sources $i = 1$ to N.
3. Compute the resulting sound pressure levels at selected points in the environment for each of the individual sources for each octave (or 1/3-octave band).
4. Calculate the sound pressure level at the receiver due to all sound sources for each octave (or 1/3-octave) frequency band. This is done by calculating the octave (or 1/3-octave) band sound pressure levels produced by each of the individual sources at the receiver location. The sound pressure levels due to each source are then added logarithmically, as described in Section 1.11.3 for each octave or 1/3-octave band.
5. The A-weighted sound pressure level in each octave (or 1/3-octave) band is calculated by arithmetically adding the A-weighting correction (dB) to each octave (or 1/3-octave) band, using Table 2.3. The overall A-weighted sound pressure level at the receiver location is then calculated by logarithmically adding all of the octave (or 1/3-octave) band A-weighted levels (see Section 1.11.3).

The expression relating sound pressure level, L_{pi}, at a receiver location produced by a single source, i, of sound power level, L_{Wi}, is:

$$L_{pi} = L_{Wi} - A_{\text{div},i} + \text{DI}_{Mi} - A_{Ei} \qquad \text{(dB)} \tag{5.53}$$

where A_{Ei} is an attenuation factor (see Equation (5.55)) for propagation from source, i, to the receiver, and $A_{\text{div},i}$ is a geometric divergence factor, which is dependent on the type of source and accounts for spherical spreading as the sound propagates away from the source.

The quantity, DI_{Mi}, is a directivity index, which accounts for directional properties of source, i. In many cases, the directivity of a source is not known, so the DI_M term is often ignored. If the directivity is known, then the combination of the directivity and the ground effect can be difficult to establish accurately, except in cases where the directivity varies in a horizontal plane parallel to the ground rather than in a vertical plane perpendicular to the ground. When the directivity varies in a plane perpendicular to the ground, the effect of the ground plane becomes more complicated. For cases of highly directional sources that radiate most of their sound power in a direction away from the ground, the effect of the ground plane may be a negligible addition to the directivity effect, due to there being negligible sound reflected in the ground plane.

In deriving Equation (5.53) from Equations (1.78) and (1.85), it has been assumed that $\rho c \approx 400$ (SI units), so that $10 \log_{10}(\rho c/400) \approx 0$.

In the following analysis, it will be assumed that $\text{DI}_M = 0$ or alternatively, the sound power was derived only from sound pressure measurements near the sound source (but in the far field of the source), in the direction of the receiver so that:

$$L_{pi} = L_{Wi} - A_{\text{div},i} - A_{Ei} \qquad \text{(dB)} \tag{5.54}$$

where

$$A_{Ei} = [A_a + A_g + A_b + A_{\text{met}} + A_{\text{misc}}]_i \tag{5.55}$$

The attenuation term, A_{Ei}, is made up of components, A_a, A_g, A_b, A_{met} and A_{misc}, which refer, respectively, to attenuation due to absorption by the atmosphere, effects due to the ground, effects of barriers and obstacles, meteorological effects such as wind and temperature gradients and other miscellaneous effects such as reflections from vertical surfaces, source height, in-plant screening and vegetation screening. In the more advanced propagation models, the ground effect and barrier effect are combined together and then further combined with the effects of atmospheric refraction caused by vertical wind and temperature gradients as well as the effects of atmospheric turbulence. Note that the ground effect is usually negative, resulting in an increased sound pressure level at the receiver. For the simpler models, the attenuation components of A_{Ei} are simply added together arithmetically when carrying out the computations of step 3 above. That is, if two individual attenuations are 3 dB and 2 dB, respectively, the overall attenuation due to the two effects is 5 dB.

Enclosed sound sources may be included by using Equations (7.126) and (7.127) of Chapter 7 to calculate the sound pressure level, L_{p1}, energy-averaged over the external enclosure surface of area S_E, where L_W of Equation (7.126) is the sound power level of the enclosed sources. The quantity L_W needed for Equation (5.53) is the sound power level radiated by the enclosure walls, which is equal to $(L_{p1} + 10 \log_{10} S_E)$, and the quantity DI_M is the directivity index of the enclosure in the direction of the receiver. Again, this can be included in the sound power level data if the latter data are determined solely from sound pressure level measurements made near the sound source and in the direction of the receiver. The relevant equation is Equation (4.143), where only sound pressure level measurements on the side of the sound source closest to the receiver are used to determine the average L_p of Equation (4.142).

Each propagation model to be discussed in the following sections uses the same Equation (5.54) for calculating the sound at a single receiver location due to a single source. However, the models differ in the way they calculate some of the attenuation terms in Equation (5.55) and how they combine the attenuation terms together. However, the terms, $A_{\text{div},i}$ and A_a, are calculated in the same way in all models and so they will be treated separately first of all. Also, some general observations will be made about the other attenuation effects just mentioned.

5.3.1 Geometric Divergence, A_{div}

The spherical divergence factor, A_{div}, is the same for all sound propagation models and characterises the reduction in sound pressure level with distance from the source, as a result of the sound wave spreading out spherically into an ever increasing volume. Referring to Equation (4.14) for a monopole source, putting the equation in the form of Equation (5.53) and setting the terms DI_M and A_{Ei} equal to zero, it is readily determined that, for a simple monopole source radiating in free space away from any reflective boundaries:

$$A_{\text{div}} = 10\log_{10} 4\pi + 20\log_{10} d_{SR} \tag{5.56}$$

where d_{SR} is the distance from the source to the point of observation. Alternatively, if the source is on a wall and thus radiating hemispherically into half-space then:

$$A_{\text{div}} = 10\log_{10} 2\pi + 20\log_{10} d_{SR} \tag{5.57}$$

Note that the effect of the ground (including hard ground) is taken into account by the A_g term, Thus, Equation (5.57) should not be used to account for a hard ground. However, Equation (5.57) can be used if the source is mounted on hard ground, but in this case, the A_g term is set equal to zero. If both the A_g term and Equation (5.57) were used together, the ground effect would be erroneously taken into account twice.

Referring to Equation (4.86), it is readily determined that for a line source:

$$A_{\text{div}} = 10\log_{10}(4\pi d_{SR} D/\alpha) \tag{5.58}$$

where D is the length of the line source, d_{SR} is the distance from the source centre to the point of observation and α is the angle subtended by the source at the point of observation (equal to $\alpha_u - \alpha_\ell$), as shown in Figure 4.8.

For the case of a wall, which may be modelled as a plane incoherent radiator and which is large compared to the wavelength of the radiated sound, reference to Equation (4.121) shows that A_{div} has the following form:

$$A_{\text{div}} = 10\log_{10} 2\pi + 20\log_{10} d_{SR} - 10\log_{10} F(\alpha,\beta,\gamma,\delta) \tag{5.59}$$

The dimensionless variables of Equation (5.59) are $\alpha = H/L$, $\beta = h_R/L$, $\gamma = r/L$ and $\delta = d/L$, where h_R, L, r and d are defined in Figure 4.14 and d_{SR} is the distance from the centre of the wall to the receiver. For the case of the near field, where $(\gamma/10) < (\alpha,\ \beta,\ \text{and}\ \delta)$, the function, F, takes the following form:

$$\begin{aligned} F = \frac{\gamma^2}{\alpha}\Bigg[&\tan^{-1}\frac{(\alpha-\beta)(\delta+1/2)}{\gamma\sqrt{(\alpha-\beta)^2+(\delta+1/2)^2+\gamma^2}} \\ &+\tan^{-1}\frac{\beta(\delta+1/2)}{\gamma\sqrt{\beta^2+(\delta+1/2)^2+\gamma^2}} - \tan^{-1}\frac{(\alpha-\beta)(\delta-1/2)}{\gamma\sqrt{(\alpha-\beta)^2+(\delta-1/2)^2+\gamma^2}} \\ &-\tan^{-1}\frac{\beta(\delta-1/2)}{\gamma\sqrt{\beta^2+(\delta-1/2)^2+\gamma^2}}\Bigg] \end{aligned} \tag{5.60}$$

Alternatively, in the case of the far field, where $\gamma/10 > \alpha, \beta, \delta$, the function, $F = 1$, and the corresponding term in Equation (5.59) is zero.

5.3.2 Atmospheric Absorption, A_a

The attenuation, A_a, due to atmospheric absorption is given by:

$$A_a = \alpha_a d_{SR} \qquad \text{(dB)} \tag{5.61}$$

where d_{SR} is the distance (m) travelled by the sound wave from the source to the receiver.

The attenuation, α_a in dB per metre of travel through the atmosphere for a pure tone frequency, f, is given by the standards, ANSI/ASA S1.26 (2019) and ISO 9613-1 (1993) as:

$$\begin{aligned}\alpha_a = 8.686 f^2 \Bigg\{ & \left[1.84\times 10^{-11}\left(\frac{P_a}{P_0}\right)^{-1}\left(\frac{T}{T_r}\right)^{1/2}\right] \\ & + \left(\frac{T}{T_r}\right)^{-5/2} \times \left[0.01275\,\mathrm{e}^{-2239.1/T}\left(\frac{f_{ro}}{f_{ro}^2+f^2}\right)\right. \\ & \left.+0.1068\,\mathrm{e}^{-3352/T}\left(\frac{f_{rN}}{f_{rN}^2+f^2}\right)\right]\Bigg\} \qquad \text{(dB/m)}\end{aligned} \tag{5.62}$$

where T is the ambient atmospheric temperature in Kelvin (273.15 + °C), T_r is the reference temperature in Kelvin (293.15), P_a is the ambient atmospheric pressure and P_0 is the reference atmospheric pressure (101325 Pa). The relaxation frequencies for oxygen and nitrogen, f_{ro} and f_{rN}, respectively are:

$$f_{ro} = \left(\frac{P_a}{P_0}\right)\left[24 + \left(\frac{(4.04\times 10^4 h_M)(0.02+h_M)}{0.391+h_M}\right)\right] \qquad \text{(Hz)} \tag{5.63}$$

and

$$f_{rN} = \left(\frac{T}{T_r}\right)^{-1/2}\left(\frac{P_a}{P_0}\right)\left\{9 + 280 h_M\,\mathrm{e}^{-4.170\left[(T/T_r)^{-1/3}-1\right]}\right\} \qquad \text{(Hz)} \tag{5.64}$$

where h_M is the % molar concentration of water vapour which may be calculated from the % relative humidity, h_{rel}, by:

$$h_M = h_{\mathrm{rel}} \times 10^V \left(\frac{P_a}{P_0}\right)^{-1} \tag{5.65}$$

where

$$\begin{aligned}V =& 10.79586\left[1-\left(\frac{273.16}{T}\right)\right] - 5.02808\log_{10}\left(\frac{T}{273.16}\right) \\ & +1.50474\times 10^{-4}\left\{1-10^{-8.29692[(T/273.16)-1]}\right\} \\ & +0.42873\times 10^{-3}\left\{-1+10^{4.76955[1-(273.16/T)]}\right\} - 2.2195983\end{aligned} \tag{5.66}$$

or V can be calculated to a sufficiently close approximation using:

$$V = -6.8346\left(\frac{273.16}{T}\right)^{1.261} + 4.6151 \tag{5.67}$$

where 273.16 Kelvin = 0.01°C is the triple point isotherm temperature (ISO 9613-1, 1993).

Calculated values of the absorption rate, $m = 1000\alpha_a$ for a range of atmospheric temperatures and relative humidities are listed in Table 5.3 for nominal octave band centre frequencies. Values for other temperatures, atmospheric pressures and relative humidities may be calculated using the standards, ISO 9613-1 (1993) and ANSI/ASA S1.26 (2019), which both provide values of m corresponding to pure tones at the exact 1/3-octave band centre frequencies, which are slightly different to the nominal frequencies in Table 1.2. The absorption rate for the 1/3-octave

TABLE 5.3 Attenuation due to atmospheric absorption

Relative humidity %	Temperature °C	Octave band centre frequency (Hz) 63	125	250	500	1000	2000	4000
		m (dB per 1000 m)						
25	15	0.2	0.6	1.2	2.4	6.4	21.9	74.7
	20	0.2	0.7	1.4	2.5	5.5	16.9	59.4
	25	0.2	0.7	1.6	2.9	5.4	14.0	46.9
	30	0.2	0.6	1.8	3.5	6.0	12.7	38.4
	35	0.2	0.6	1.8	4.2	7.1	12.8	33.3
	40	0.1	0.5	1.8	4.8	8.7	14.3	31.1
50	15	0.1	0.5	1.2	2.2	4.2	10.8	36.5
	20	0.1	0.4	1.3	2.7	4.7	9.9	29.7
	25	0.1	0.4	1.3	3.2	5.7	10.2	25.8
	30	0.1	0.3	1.3	3.6	7.0	11.6	24.6
	35	0.1	0.3	1.2	3.7	8.4	14.1	25.6
	40	0.1	0.3	1.0	3.6	9.5	17.5	29.0
75	15	0.1	0.4	1.1	2.4	4.1	8.5	25.1
	20	0.1	0.3	1.1	2.8	5.1	9.0	22.2
	25	0.1	0.3	1.0	3.0	6.3	10.5	21.5
	30	0.1	0.2	0.9	3.0	7.4	13.0	23.0
	35	0.1	0.2	0.8	2.9	8.2	16.1	26.8
	40	0.0	0.2	0.7	2.7	8.5	19.3	32.6

band centre frequency may be used as an approximate representation of the absorption rate arithmetically averaged over a 1/3-octave band (ISO 9613-1, 1993; ANSI/ASA S1.26, 2019). A more accurate estimate of the total air absorption over a frequency band for a path of length, d_{SR}, is (Joppa et al., 1996; ANSI/ASA S1.26, 2019):

$$A_a = \alpha_a d_{SR} \left[1.0 + (B_r^2/10)(1.0 - 0.2303\alpha_a d_{SR})\right]^{1.6} \quad \text{(dB)} \tag{5.68}$$

where α_a is the atmospheric absorption coefficient (dB/m) at the band centre frequency, $B_r = (f_u - f_\ell)/f_C = 2^{1/(2b)} - 2^{-1/(2b)}$, $1/b$ is the band fraction $= 1/3$ for 1/3-octave bands and 1 for octave bands, f_u is the band upper limit, f_ℓ is the band lower limit and f_C is the band centre frequency. For 1/3-octave bands, $B_r \approx 0.23156$ and for octave bands, $B_r \approx 0.7071$.

5.3.3 Ground Effect, A_g

The effect of the ground is generally to increase the sound pressure level at the receiver location. However, in some cases, destructive interference between the direct and ground-reflected waves results in a decreased sound pressure level at the receiver compared to the sound pressure level with no ground effect included.

There are four waves, listed below, which are considered to contribute to the ground effect. The simplest sound propagation models only include the first reflected wave type, as they only use the plane wave reflection coefficient (not the spherical wave reflection coefficient). Models that use the spherical wave reflection coefficient automatically include all wave types.

- **Reflected wave** – this is the wave that is reflected from the ground and is included in all propagation models. It combines at the receiver with the direct wave from the source and can either reinforce or interfere with the direct wave, respectively, increasing or decreasing the sound pressure level as a function of frequency, depending on the relative phase between the direct and reflected wave and whether or not atmospheric effects influence the coherence between the two waves. The location of specular reflection for a flat ground surface can be found by assuming an image source exists directly beneath the actual source and as far below the ground surface as the actual source is above it. A straight line is then drawn between the image source and the receiver. The point of specular reflection is where this line intersects the ground surface, as shown by point O in Figure 4.17. This wave is accounted for entirely by the plane wave reflection coefficient.

- **Ground wave** – this is an additional contribution to the reflected wave field arising from diffusion of the image source as a result of spherical wave incidence on the ground. This wave is taken into account by use of the spherical wave reflection coefficient in the more advanced propagation models (Attenborough et al., 2007, p. 42).

- **Surface wave** – this is effectively a 2-D ground propagating wave, as its amplitude decays exponentially with height above the ground and with the inverse of the square root of distance from the source, rather than directly with distance as other components do. This wave is not affected by atmospheric refraction due to wind and temperature gradients and is more likely to exist at low frequencies. Its properties are discussed in detail in (Attenborough et al., 2007, p. 50–54). The surface wave travels slower than the speed of sound in air and is only important for propagation over thin, soft surfaces over a hard backing such as snow on frozen ground, such that the magnitude of the imaginary surface impedance, $|\mathrm{Im}\{Z_m\}|$ is greater than the magnitude of the resistive impedance, $|\mathrm{Re}\{Z_m\}|$. Even with this requirement, the sound wave must be travelling at close to grazing incidence so $\theta > 70°$, where θ is defined in Figure 5.1. In practice, at long distances the surface wave is difficult to detect, as most outdoor ground surfaces produce substantial attenuation of waves propagating adjacent to the surface. The contribution of the surface wave is approximately accounted for in Equation (5.23) for the spherical wave reflection coefficient (Attenborough et al., 2007, p. 42).

- **Creeping wave** – this wave only occurs in situations where there is no direct transmission of sound, such as along a convex surface or upward refracting atmosphere (downwind and/or normal temperature profile) (Embleton, 1996). As sound transmission in an upward refracting atmosphere is not usually important from the environmental noise perspective, it will not be considered further here.

For cases where there is incoherent addition between the direct and ground-reflected waves, the ground effect can be calculated directly from the reflection loss, A_{rf}, defined in Equations (5.20) or (5.48), corresponding to the assumption of plane wave reflection or spherical wave reflection, respectively, as:

$$
\begin{aligned}
A_g &= -10\log_{10}\left(\frac{\langle p_t^2\rangle}{\langle p_{\mathrm{dir}}^2\rangle}\right) = -10\log_{10}\left[1+\left(\frac{d_{SR}}{r_S+r_R}\right)^2 10^{-A_{rf}/10}\right] \quad \text{(dB)}\\
&= -10\log_{10}\left[1+\left(\frac{d_{SR}}{r_S+r_R}\right)^2 |R_p|^2\right] \quad \text{for plane wave reflection}\\
&= -10\log_{10}\left[1+\left(\frac{d_{SR}}{r_S+r_R}\right)^2 |Q|^2\right] \quad \text{for spherical wave reflection}
\end{aligned}
\tag{5.69a,b,c,d}
$$

where p_t is the total sound pressure squared at the receiver due to both direct and ground-reflected waves and p_{dir} is the sound pressure squared due to the direct wave only.

Note that for hard surfaces such as asphalt and concrete, the reflection coefficient is close to 1 so the reflection loss, A_{rf}, is close to zero, resulting in a ground effect, A_g, of -3 dB for the case of incoherent addition between the direct and ground-reflected waves.

Equations (5.69b,c,d) are simplifications of the actual situation, as there is generally some degree of coherence between the direct and ground-reflected waves. Thus, it is generally more representative to calculate the ground effect for single frequencies using coherent addition of the direct and ground-reflected waves, and then logarithmically average the results (see Equation (4.142)) for a number of single frequencies to obtain an octave band value. When coherent addition is used to combine the direct and ground-reflected waves to find the ground effect at frequency, f, the difference in distance travelled by the two waves has to be taken into account in addition to the reflection coefficient. This is because the difference in distance usually has a significant effect on the phase between the direct and ground-reflected waves, even though the difference in distance travelled usually only has a very small effect on the relative amplitudes of the two waves. To recap, the ground effect is the difference between the direct wave sound pressure level and the sound pressure level of the combined direct and ground-reflected waves. This effect can be calculated using the spherical wave reflection coefficient, Q (defined in Equation (5.23)), Equation (5.25), together with coherent addition of the sound pressure levels of the direct and ground-reflected waves to give:

$$\begin{aligned}A_g &= -10\log_{10}\left|1+\left(\frac{d_{SR}}{r_S+r_R}\right)Q\mathrm{e}^{-\mathrm{j}k(r_S+r_R-d_{SR})}\right|^2\\ &= -10\log_{10}\left[1+\left(\frac{d_{SR}}{r_S+r_R}\right)^2|Q|^2+\left(\frac{2d_{SR}}{r_S+r_R}\right)|Q|\cos\left[\alpha_s-k(r_S+r_R-d_{SR})\right]\right] \quad \text{(dB)}\end{aligned} \tag{5.70}$$

where r_S, r_R and d_{SR} are defined in Figure 4.17 and where the complex spherical wave reflection coefficient, Q, may be written as:

$$Q = |Q|\mathrm{e}^{\mathrm{j}\alpha_s} \quad \text{and} \quad \alpha_s = \tan^{-1}\left[\frac{\mathrm{Im}\{\mathrm{Q}\}}{\mathrm{Re}\{\mathrm{Q}\}}\right] \tag{5.71}$$

The complex plane wave reflection coefficient, R_p (defined in Equation (5.15) or Equation (5.18)), can also be used in place of Q in Equation (5.70).

For the case involving turbulence, the situation is a little more complex. Equation (5.49), for the total sound pressure at the receiver in the presence of turbulence for a ray with unit mean square sound pressure at 1 m, can be used together with Equation (5.70) to obtain:

$$A_g = -10\log_{10}\left[1+\left(\frac{d_{SR}}{r_S+r_R}\right)^2|Q|^2+\left(\frac{2d_{SR}}{r_S+r_R}\right)|Q|T_t\cos\left[\alpha_s-k(r_S+r_R-d_{SR})\right]\right] \quad \text{(dB)} \tag{5.72}$$

where T_t is defined in Equation (5.50) and Q is the complex spherical wave reflection coefficient. The turbulence parameter, T_t only appears in the third term in brackets in Equation (5.72), as turbulence only affects the phase, but not the amplitude, of the rays arriving at the receiver. For strong turbulence, the term, T_t, approaches 0 (see Equations (5.51) and (5.52)), and Equation (5.72) becomes the same as Equation (5.69), which is for incoherent combination of direct and ground-reflected waves. For weak or no turbulence, T_t is equal to 1 and Equation (5.72) becomes equal to Equation (5.70), which is for coherent combination of direct and ground-reflected waves. For values of T_t between 0 and 1, the combination of direct and ground-reflected waves lies between incoherent ($T_t = 0$) and fully coherent ($T_t = 1$).

The attenuation, A_g, averaged over a 1/3-octave or octave band, is calculated by averaging logarithmically the attenuations calculated at a number of different frequencies (at least 10) equally spaced throughout the band (see Equation (8.225)). Band-averaged attenuation is probably more useful for noise predictions in practice, as atmospheric turbulence and undulating ground will result in considerable fluctuations in the single frequency values. However, almost all propagation models are based on calculations only at centre frequencies of octave or 1/3-octave bands. Alternatively, A_g can be calculated as described in the various propagation models, each of which treat the calculation of the ground effect slightly differently, as discussed in Sections 5.5, 5.6 and 5.7.

5.3.4 Meteorological Effects, A_{met}

Meteorological effects are primarily a result of a vertical sonic gradient caused by the variation of wind speed and temperature with altitude above the ground. Turbulence is not included, as it does not affect the amplitude of waves arriving at the receiver; it only affects their relative phases, which indirectly affects the total sound pressure level. However, this turbulence effect is included in the ground effect.

The vertical sonic gradient causes the sound rays to bend towards or away from the ground, depending on whether the resulting vertical sound speed gradient is positive or negative, and this causes significant variations in the sound pressure level arriving at the receiver.

Wind shear is the common name for a vertical wind speed profile in which the wind speed and direction changes with height above the ground. Due to surface friction effects, the wind is almost always characterised by an increasing speed as the distance from the ground increases. Generally, wind shear is greater at lower ground level wind speeds and the wind shear variation with altitude is usually greater at night.

In the direction upwind from the source (that is, the direction into the wind from the source), sound rays are diffracted upwards away from the ground and may result in a *shadow zone*. In this zone, the only sound in the mid- and high-frequency ranges that reaches a receiver is what is scattered due to atmospheric turbulence or due to other obstacles impacted by the sound ray, although this may not be the case for low-frequency sound and infrasound (see Section 5.3.3). Thus, in the upwind direction, the sound experienced at a receiver is usually less than would be expected in the absence of wind.

In the direction downwind from the source, sound rays are refracted towards the ground so that in this direction the sound pressure level at large distances from the source is likely to be greater than would be expected in the absence of wind. At a sufficient distance from the source, there will be multiple rays striking the ground, which have all been reflected from the ground one or more times. At distances greater than the shortest distance from the source that multiple reflections occur, there will be less attenuation of sound with distance than might otherwise be expected, and this is the main reason that most noise propagation models are not considered reliable at distances greater than 1000 m from the source.

In the early hours of the morning when there is a clear, cloud-free sky, the atmospheric temperature usually increases with altitude close to the ground, instead of the more normal condition where it decreases with altitude. The layer of air in which the temperature is increasing is known as the *inversion layer*, and this is usually just a few hundred metres thick. The effect of an atmospheric temperature inversion is to cause sound rays to bend towards the ground in much the same way as they do when propagating downwind. Conversely, the more normal (especially during the daytime) non-inverted atmospheric temperature profile results in sound rays that are refracted upwards, away from the ground, thus resulting in reduced sound pressure levels at the receiver.

In the following subsections, procedures for calculating the attenuation due to meteorological effects are outlined. We begin with estimations of bounds on the possible variations in sound pressure level that may result from meteorological effects. Subsequent subsections outline procedures for calculating meteorological effects more accurately as a function of the vertical atmospheric sonic gradient resulting from vertical wind and temperature gradients.

5.3.4.1 Uncertainty Bounds

If the weather conditions are not accurately known, then it is generally assumed that meteorological effects will result in a sound pressure level variation about the predicted level at the receiver, as shown in Table 5.4.

TABLE 5.4 Variability in sound pressure level predictions due to meteorological influences, A_{met} (dB)

Octave band centre frequency (Hz)	Distance from the source (m)			
	100	200	500	1000
		A_{met} (dB)		
63	+1	+4, −2	+7, −2	+8, −2
125	+1	+4, −2	+6, −4	+7, −4
250	+3, −1	+5, −3	+6, −5	+7, −6
500	+3, −1	+6, −3	+7, −5	+9, −7
1000	+7, −1	+11, −3	+12, −5	+12, −5
2000	+2, −3	+5, −4	+7, −5	+7, −5
4000	+2, −1	+6, −4	+8, −6	+9, −7
8000	+2, −1	+6, −4	+8, −6	+9, −7

The range of values in Table 5.4 are primarily due to variations in vertical wind shear and temperature gradients which cause variations in the vertical sound speed gradient.

5.3.4.2 Overview of Methods Used in Standard Propagation Models

When predicting outdoor sound using a standard propagation model, it is usual to include downwind or temperature inversion meteorological conditions, corresponding to CONCAWE category 5 or 6 (see Section 5.5.4). The downwind condition implies that the wind direction makes an angle of less than 45° to the line joining the source to the receiver.

The noise propagation model, CONCAWE, takes the vertical sound speed gradient into account indirectly by using atmospheric stability classes, as discussed in Section 5.5. The ISO model discussed in Section 5.6 also takes the vertical sound speed gradient into account indirectly, but instead of using atmospheric stability classes, it calculates the sound pressure level at the receiver for "worst-case" atmospheric conditions (downward refracting atmosphere - see ISO 9613-2 (1996)). The NMPB-2008 model discussed in Section 5.7 provides calculations for two different stability conditions: a "neutral" atmosphere (Pasquill stability category, D) and a downward refracting atmosphere. More complex propagation models such as Nord2000 and Harmonoise exist, which can calculate receiver sound pressure levels for any specified sound speed gradient. However, they are not discussed here and the reader is referred to Hansen et al. (2017) for a detailed description of these models.

5.3.4.3 Methods Using Linear Sonic Gradient Estimates

The attenuation due to meteorological effects can be estimated more accurately than is possible using procedures in standard propagation models, by estimating the actual sonic gradient and then using the estimate to calculate the propagation path and propagation time of the direct and ground-reflected sound waves that arrive at the receiver from the source. The use of the radius of curvature of the sound ray to calculate propagation path lengths and relative phases (or propagation times) is not used in any standard noise propagation models. However, it is outlined here, as it results in a more accurate combination at the receiver location, of the direct and ground-reflected waves from the source, than is possible using procedures described in Section 5.3.4.2, especially when the two waves are coherent or partially coherent. When this approach is used, the ground and meteorological effects are combined, as the ground reflection induces a phase shift in addition to the phase shift due to the propagation distance, so the two effects cannot be treated independently.

As sound propagation in a downward refracting atmosphere (downwind or atmospheric temperature inversion or both) is of most importance in community noise estimates, this meteorological condition will be discussed first. In a downward refracting atmosphere (convex shape of the sound wave propagation path), one or more ground-reflected rays will arrive at the receiver in addition to the direct ray with no ground reflection.

It has been established (Rudnick, 1947) that the radius of curvature of a sound ray propagating in the atmosphere is dependent on the vertical gradient of the speed of sound (sonic gradient), which can be caused either by an atmospheric vertical wind or temperature gradient or by both. It has also been shown (Piercy et al., 1977) that refraction of a sound ray due to either a vertical wind gradient or a vertical temperature gradient produces equivalent acoustic effects, which are essentially additive. A sound ray travelling at an angle, ψ_S, above (or below) the line parallel to the ground plane will have a curved path with a radius of curvature, R_c at height, z, given by the following equation (de Jong, R. and Stusnik, 1976). When R_c is positive the sound rays are curved downward and when R_c is negative the sound rays are curved upward:

$$R_c = \frac{c}{\left[\dfrac{\mathrm{d}c}{\mathrm{d}z}\right]\cos\psi_S} \tag{5.73}$$

The sonic gradient, $\mathrm{d}c/\mathrm{d}z$, results from vertical atmospheric wind and temperature gradients, and it can also vary with altitude, as wind and temperature gradients may vary with altitude.

In the following analysis in this section, it will be assumed that the wind contribution to the sound speed gradient adds linearly to the temperature gradient contribution. The total vertical gradient can thus be expressed as:

$$\frac{\mathrm{d}c}{\mathrm{d}z} = \frac{\mathrm{d}\boldsymbol{U}}{\mathrm{d}z} + \left[\frac{\partial c}{\partial z}\right]_T \tag{5.74}$$

where the term, $\left[\frac{\partial c}{\partial z}\right]_T$, on the right-hand side is evaluated by assuming a linear vertical temperature profile. The sonic gradient due to the vertical atmospheric temperature profile can be calculated from a knowledge of the vertical atmospheric temperature profile and the use of Equation (1.8). The latter equation may be rewritten more conveniently for the present purpose as:

$$c = c_{00}\sqrt{T/273} \tag{5.75}$$

where T is the temperature in Kelvin and $c_{00} = 331.3$ m/s is the speed of sound at sea level, one atmosphere pressure and 0°C.

The vertical sound speed gradient resulting from the atmospheric temperature gradient is found by differentiating Equation (5.75) with respect to z to give:

$$\left[\frac{\partial c}{\partial z}\right]_T = \frac{\mathrm{d}T}{\mathrm{d}z}\frac{\partial c}{\partial T} = \frac{\mathrm{d}T}{\mathrm{d}z}T^{-1/2}\frac{c_{00}}{2\sqrt{273}} = 10.025\frac{\mathrm{d}T}{\mathrm{d}z}\left[T_0 + 273\right]^{-1/2} = A_m \tag{5.76a,b,c,d}$$

In Equation (5.76), $\mathrm{d}T/\mathrm{d}z$ is the vertical temperature gradient (in °C/m) and T_0 is the ambient temperature in °C at 1 m height. Typical values of A_m range from -0.1 to $+0.1$ s^{-1}.

Only a light downwind condition is needed to counteract the effect of the usual atmospheric temperature gradient in non-inverted conditions. This can be verified using Equation (5.76c), which gives for a ground temperature of 20°C and an atmospheric temperature gradient of −3°C per 300 m, a sonic gradient of −5.86 m/sec per 1000 m, which represents the gradient associated with a light surface wind (less than 1 m/s at 10 m – see Equation (5.79) and Figure 5.3).

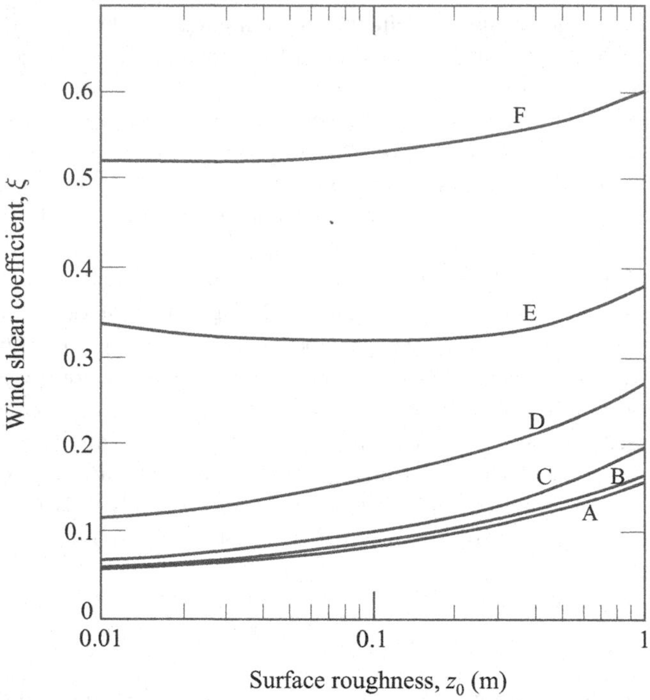

FIGURE 5.3 Estimates of the wind shear coefficient as a function of surface roughness, based on averaged measurements between 10 m and 100 m above the ground (Irwin, 1979). The exponent is a function of Pasquill stability category, A–F, (see Table 5.11) and surface roughness.

Two procedures, outlined below, may be used to calculate the component of the sonic gradient due to the wind, $\mathrm{d}\boldsymbol{U}/\mathrm{d}z$. This gradient is then added to the sonic gradient due to the atmospheric temperature gradient of Equation (5.76) and the total sonic gradient, $\mathrm{d}c/\mathrm{d}z$, is then used with Equation (5.73) to obtain the radius of curvature of the sound ray travelling in a circular arc from the source to the receiver, which is an approximation of the true ray path, as it is based on the assumption that the sonic gradient does not vary in the height band through which the sound ray passes. The results obtained using this approximation are generally only useful for overall A-weighted calculations at source/receiver distances less than 1000 m. For octave band data and for distances greater than 1000 m, it is necessary to calculate the curved ray path incrementally as discussed in Section 5.3.4.4.

The wind gradient is a vector. Hence $\boldsymbol{U}$ is the velocity component of the wind in the direction from the source to the receiver and measured at height z above the ground. The velocity component, $\boldsymbol{U}$, is positive when the wind is blowing in the direction from the source to the receiver and negative for the opposite direction.

Two procedures are outlined in this section for estimating the propagation path lengths and propagation times of the direct and ground reflected waves, using a linearisation of the vertical wind profile. The linearised wind speed profile is used to calculate a linearised sonic gradient wind contribution, $\mathrm{d}\boldsymbol{U}/\mathrm{d}z$, which is added to the sonic gradient due to the vertical atmospheric temperature gradient of Equation (5.76) to obtain a total linear sonic gradient. The linearisation of the vertical wind speed profile, and hence making constant the vertical sonic gradient due to the wind, is necessary in order to use a circular arc with a fixed radius to describe the entire propagation path. Although the circular arc path is only an approximation to the true ray path, its use greatly simplifies the problem. In both procedures used to linearise the vertical wind speed profile, the linearisation is achieved by using the nonlinear equations to calculate the wind speed gradient at a height midway between the source and receiver heights (Plovsing, 2006b) and then using this gradient for the entire propagation path. Although this linearisation approach provides better estimates of meteorological effects than used in standard propagation models, large errors are still possible, so this linearisation approach is only considered useful for overall A-weighted calculations at source/receiver distances less than 1000 m. A more accurate but more complex approach may be used for octave band data and for distances greater than 1000 m. This approach, which is outlined in Section 5.3.4.4, only requires piece-wise linearisation of the wind speed profile over very small layers of altitude, with each layer being characterised by a unique wind speed gradient, resulting in a propagation path represented by a series of circular arcs with different radii for propagation through each altitude layer. Two procedures for linearising the wind speed gradient, based on a power law and a logarithmic representation of the vertical wind speed profile, are discussed in the following few pages.

Use of a Linearised Power Law Wind Speed Profile

This approach uses a power law representation of the vertical wind speed profile and then linearises it to obtain the wind speed gradient at a specified height, z. The gradient at one specific height is used to determine the radius of curvature of a circular sound ray that extends from the source to the receiver. The height often chosen is the height of the line joining the source and receiver at a location halfway between the source and receiver (Plovsing, 2006b).

The power law representation for the wind speed profile is:

$$U(z) = U_0 \left(\frac{z}{z_r}\right)^{\xi} \tag{5.77}$$

In Equation (5.77), ξ is the wind shear coefficient, which is a function of surface roughness and atmospheric stability (see Figure 5.3), although the value for neutral atmospheric conditions is often used for all atmospheric conditions without justification. $U(z)$ is the wind speed at height z and U_0 is the wind speed at some reference height, z_r, normally taken as 10 m, although a height of 5 m is sometimes used.

Values of ξ for various ground surfaces and neutral (adiabatic) atmospheric conditions (Pasquill stability category D, see Table 5.10), obtained from various sources including Davenport (1960), are listed in Table 5.5. The last 4 values in the table are inconsistent with curve "D" in Figure 5.3 so they may not be valid.

The surface roughness, z_0, can be obtained from Tables 5.6 and 5.7 or it can be estimated using measurements of the wind speed at two different heights in a stable atmosphere and using Equation (5.82) as a basis to give (IEC 61400-11 Ed.3.0, 2018):

$$z_0 = \exp\left(\frac{U(z)\log_e z_r - U_0 \log_e z}{U(z) - U_0}\right) \tag{5.78}$$

To obtain the wind gradient (and hence the sonic gradient due to the wind), Equation (5.77) is differentiated with respect to height, giving the following expression for the expected wind gradient at height, z.

$$\frac{\mathrm{d}\boldsymbol{U}}{\mathrm{d}z} = \xi \frac{U(z)}{z} \tag{5.79}$$

Equations (5.76c) and (5.79) are substituted into Equation (5.74) to obtain the total sonic gradient.

TABLE 5.5 Values of the wind shear coefficient, ξ, for a neutral (adiabatic) atmosphere (Pasquill stability category D)

Type of ground surface	ξ
Very smooth (mud flats, ice)	0.08
Snow over short grass	0.11
Swampy plain	0.12
Sea	0.12
Lawn grass, 1 cm high	0.13
Desert	0.14
Snow cover	0.16
Thin grass, 10 cm high	0.19
Grass airfield	0.21
Thick grass, 10 cm high	0.24
Countryside with hedges	0.29
Thin grass, 50 cm high	0.36
Beet field	0.42
Thick grass, 50 cm high	0.43
Grain field	0.52

TABLE 5.6 Estimates of surface roughness for various ground surface types (Davenport, 1960; Wieringa, 1980, 1992), where the parameter, x is the distance of travel of the sound wave across the surface and the parameter, h is the average height of the crop, bushes or obstacles

Surface type	Roughness, z_0 (m)
Still water or calm open sea unobstructed downwind for 5 km	0.0002
Open terrain with a smooth surface such as concrete runways in airports, mowed grass	0.0024
Mud flats, snow; no vegetation, no obstacles	0.005
Open flat terrain; grass, few isolated obstacles	0.03
Low crops, occasional large obstacles; x/h >20	0.10
Agricultural land with some houses and 8-meter tall sheltering hedgerows within a distance of about 500 meters	0.10
High crops, scattered obstacles, $15 < x/h < 20$	0.25
Parkland, bushes, numerous obstacles, $x/h \approx 10$	0.50
Regular large obstacle coverage (suburb, forest)	0.50–1.0
City centre with high- and low-rise buildings	≥ 2

TABLE 5.7 Alternative estimates of surface roughness for various ground surface types (IEC 61400-11 Ed.3.0, 2018)

Surface type	Roughness, z_0 (m)
Water, snow or sand	0.0001
Open flat land, mown grass, bare soil	0.01
Farmland with some vegetation	0.05
Suburbs, towns, forests, many trees & bushes	0.3

The height, z, is fixed to a height midway between the source and receiver heights (Plovsing, 2006b), resulting in a linear sonic gradient. Equation (5.79) is then used with Equations (5.73), (5.74) and (5.76c) to calculate the radius of curvature due to the combined effects of atmospheric wind and temperature gradients.

The angle, ψ_S, at which the ray actually leaves the source, is determined iteratively, starting with $\psi_S = 0$, calculating the corresponding value of R_c using Equations (5.73)–(5.76) and (5.79) and then calculating the distance, d_c, from the source at which the ray is at a height equal to the receiver height. Referring to Figure 5.4, we may write the following for d_c:

$$d_c = 2R_c \sin(\psi_S + \varphi) \cos \varphi \tag{5.80}$$

where

$$\varphi = \arctan \left[(h_S - h_R)/d_c \right] \tag{5.81}$$

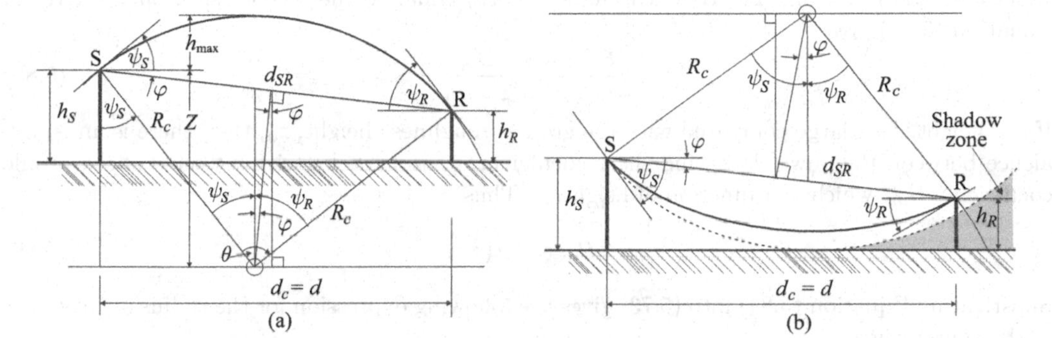

FIGURE 5.4 Geometry for a sound ray originating at source, S, and arriving at receiver, R: (a) positive atmospheric vertical sonic gradient; (b) negative atmospheric vertical sonic gradient showing shadow zone.

As d_c appears on the left-hand side of Equation (5.80) and also in the equation for φ, and we eventually require that $d_c \approx d$, it is more efficient to replace the estimate, d_c, of the horizontal distance from the source to the receiver with the actual distance, d in Equation (5.81).

The above equations are also valid for the receiver being higher than the source, in which case, φ is negative. The value of ψ_S is incremented by small amounts until $d_c \approx d$ to the required accuracy (usually a few percent).

Use of a Linearised Logarithmic Wind Speed Profile

A second approach to calculating the vertical wind speed profile assumes that it is logarithmic and represented by Equation (5.82).

$$U(z) = U_0 \left(\frac{\log_e[(z/z_0) + 1]}{\log_e[(z_r/z_0) + 1]} \right) \tag{5.82}$$

where

z_0 = roughness length (see Table 5.6 and Table 5.7);
z = height at which wind speed is to be determined; and
z_r = anemometer height corresponding to the measured wind speed, U_0.

In the approach described here, Equation (5.82) is also used without justification for other atmospheric conditions, which can sometimes lead to substantial errors.

The overall sound speed profile is obtained by combining Equation (5.82) for the wind speed profile and Equation (5.76) for the atmospheric temperature profile to obtain the following equation for the sound speed, $c(z)$, at height, z.

$$c(z) = B_m \log_e \left(\frac{z}{z_0} + 1 \right) + A_m z + c_0 \tag{5.83}$$

where c_0 is the speed of sound at ground level ($z = 0$). The speed of sound, $c_0 = c$ is calculated using Equation (5.75), with the value of T corresponding to the temperature at ground level. Equation (5.83) is then differentiated to obtain the sonic gradient as:

$$\frac{\mathrm{d}c}{\mathrm{d}z} = A_m + \frac{B_m}{z + z_0} \tag{5.84}$$

where the first term on the right of the equation is the sonic gradient contribution due the atmospheric temperature gradient and the second term is the contribution due to the vertical wind speed gradient. The wind speed gradient is also given by the power law representation in Equation (5.79). Setting the RHS of Equation (5.79) equal to the second term on the RHS of Equation (5.84) gives:

$$\frac{\xi U(z)}{z} = \frac{B_m}{z + z_0} \tag{5.85}$$

If the height, z, is large compared with the ground roughness height, z_0, then there is an equivalence between the power law wind shear coefficient, ξ, and the logarithmic wind speed profile coefficient, B_m, which is a function of height, z. Thus:

$$B_m \approx \xi U(z) \tag{5.86}$$

Substituting Equation (5.84) into (5.73) gives the following expression for the radius of curvature of the sound ray.

$$\frac{1}{R_c} = \frac{1}{c} \left[A_m + \frac{B_m}{z + z_0} \right] \cos \psi_S \tag{5.87}$$

If A_m and B_m can be evaluated, then it is possible to use Equation (5.87) to find the radius of curvature of a ray at any height, z, and Equation (1.8) to find the corresponding speed of sound, c.

The coefficient, A_m, is the gradient of the speed of sound with height due to the atmospheric temperature profile and is given by:

$$A_m = \left[\frac{\partial c}{\partial z} \right]_T \tag{5.88}$$

where $\left[\frac{\partial c}{\partial z} \right]_T$ is given by Equation (5.76).

The coefficient, A_m, can also be determined by fitting a straight line to the measured atmospheric temperature profile. Typical values of A_m range from -0.1 to $+0.1$ s^{-1}.

The coefficient, B_m, in Equation (5.83) is a function of the wind speed gradient and arises directly from Equation (5.82):

$$B_m = \frac{U_0}{\log_e \left(\frac{z_r}{z_0} + 1 \right)} \tag{5.89}$$

Typical values of B_m range from -1 to $+1$ ms^{-1}, although higher values are possible at night in the presence of a "low-level jet". The above equations for $c(h)$ and B_m are strictly only valid for neutral atmospheric conditions. If we substitute Equation (5.86) for B_m in Equation (5.89), we can obtain an approximate equivalence of wind shear coefficient, ξ and surface roughness, z_0. Thus:

$$\xi U(z) \approx \frac{U_0}{\log_e \left(\frac{z_r}{z_0} + 1 \right)} \tag{5.90}$$

Setting $z = z_r = 10$ m results in $U(z) = U_0$ and this leads to:

$$\xi \approx \frac{1}{\log_e\left(\dfrac{10}{z_0}+1\right)} \tag{5.91}$$

In cases where the atmosphere is stable as often occurs at night, it is better to use the power law representation for the wind speed gradient, which means replacing the first term in Equation (5.83) with the right-hand side of Equation (5.77), with ξ determined from Figure 5.3 or by curve fitting SODAR (SOnic Detection And Ranging) data.

A more accurate means for calculating the radius of curvature of the sound ray involves using Figure 5.5 to obtain the following equation for ψ_S in terms of R_c.

$$\psi_S = -\varphi + \arcsin\left(\frac{d}{2R_c\cos\varphi}\right) \tag{5.92}$$

where

$$\varphi = \arctan\left(\frac{h_S - h_R}{d}\right) \tag{5.93}$$

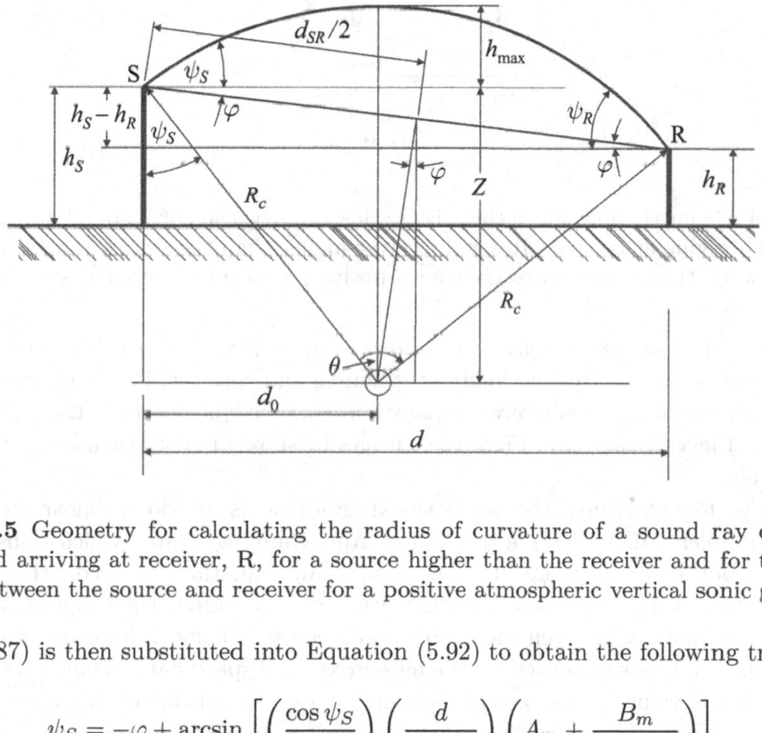

FIGURE 5.5 Geometry for calculating the radius of curvature of a sound ray originating at source, S, and arriving at receiver, R, for a source higher than the receiver and for the maximum ray height between the source and receiver for a positive atmospheric vertical sonic gradient.

Equation (5.87) is then substituted into Equation (5.92) to obtain the following transcendental equation:

$$\psi_S = -\varphi + \arcsin\left[\left(\frac{\cos\psi_S}{c}\right)\left(\frac{d}{2\cos\varphi}\right)\left(A_m + \frac{B_m}{(z+z_0)}\right)\right] \tag{5.94}$$

where z_0 is defined in Tables 5.6 and 5.7, and z is the mean height above the ground of a straight line drawn between the source and receiver (usually equal to $(h_S + h_R)/2$). Once the value of ψ_S is obtained by solving Equation (5.94) iteratively, Equation (5.87) or (5.92) may be used to obtain R_c.

Equation (5.94) has been derived on the assumption that the sonic gradient is constant and equal to the actual gradient (calculated by assuming a logarithmic wind speed profile) for the mean height above the ground of the straight line between source and receiver, resulting in a circular ray path. In some situations, better results may be obtained by using the actual gradient calculated from the logarithmic wind speed profile for a slightly different height. However, the height above the ground of the straight line between source and receiver at the mid point between them has been found to give good results in most cases (Plovsing, 2006b, 2014). If the assumption of constant sonic gradient as a function of altitude is not met, the ray path will not be circular and the procedures outlined in Section 5.3.4.4 must be used.

Equations (5.92), (5.93) and (5.94) also apply for cases of the receiver higher or lower than the source and also for the cases where the maximum height of the sound ray above the reference plane is not between the source and receiver. These cases are illustrated in Figures 5.6, 5.7 and 5.8.

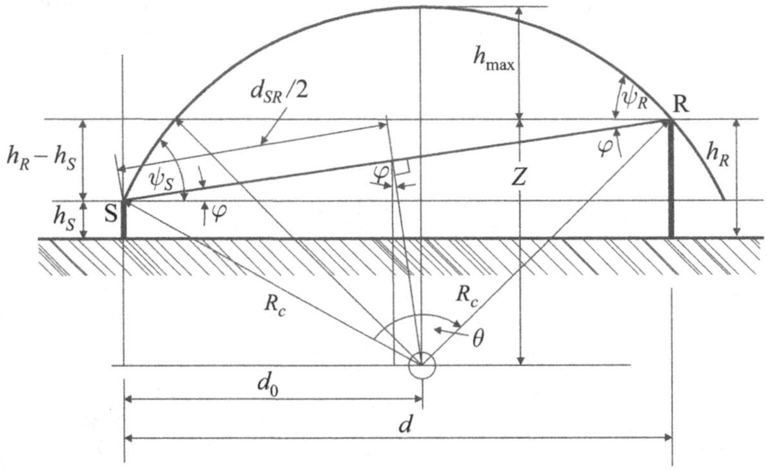

FIGURE 5.6 Geometry for calculating the radius of curvature of a sound ray originating at source, S, and arriving at receiver, R, for a receiver higher than the source and for the maximum ray height between the source and receiver for a positive atmospheric vertical sonic gradient.

The angle, φ, is positive when $h_S > h_R$ and negative when $h_S < h_R$. It represents the angle between the horizontal and the line joining the source and receiver. For cases where the source is high compared to the source/receiver separation distance, the centre of the circular arc may be above ground level as shown in Figure 5.9. It can be shown that the same equations apply in this case as well.

A third means for obtaining the wind speed gradient is to do a linear fit to Doppler SODAR or Doppler LIDAR (LIght Detection And Ranging) data, which consists of wind speed as a function of altitude. If the power law equation is used, the sonic gradient is calculated using Equation (5.79), with the measured wind speed at a height, $z = (h_S + h_R)/2$, midway between the source and receiver heights. If a logarithmic wind speed profile is assumed, a linear fit is used for the measured wind speed data to obtain the wind speed gradient and thus a value for the second term on the right-hand side of Equation (5.84), which provides a value for B_m as a linear function of z. This result is then used to estimate the sonic gradient at a height midway between the source and receiver heights (Plovsing, 2006b).

A SODAR system uses a vertically oriented sound transmitter and receiver to determine wind speed profiles and turbulence measurements over a 3000 m range, beginning at the lowest possible altitude of 45 m above ground level. The LIDAR system uses a laser transmitter and receiver to measure wind speed as a function of altitude, usually between 40 m minimum and 750 m maximum above the ground.

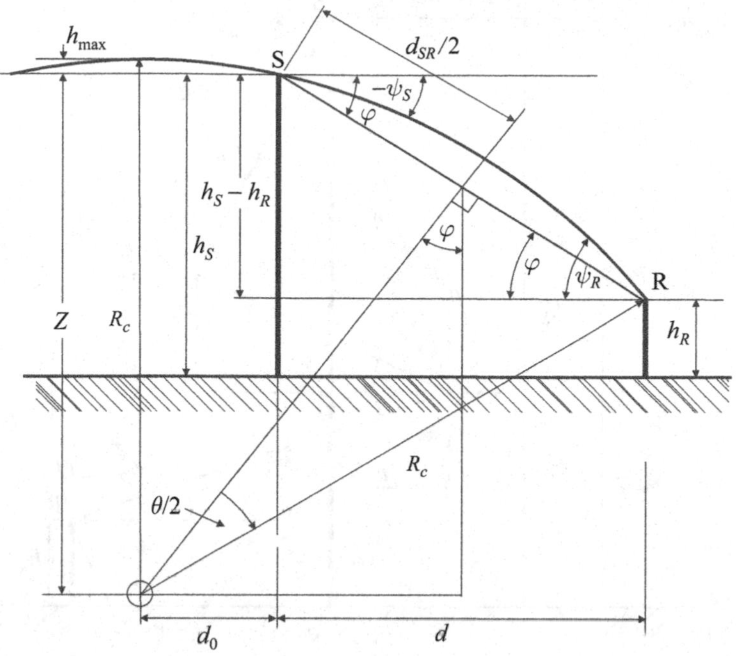

FIGURE 5.7 Geometry for calculating the radius of curvature of a sound ray originating at source, S, and arriving at receiver, R, for a source higher than the receiver and for the maximum ray height *not* between the source and receiver for a positive atmospheric vertical sonic gradient.

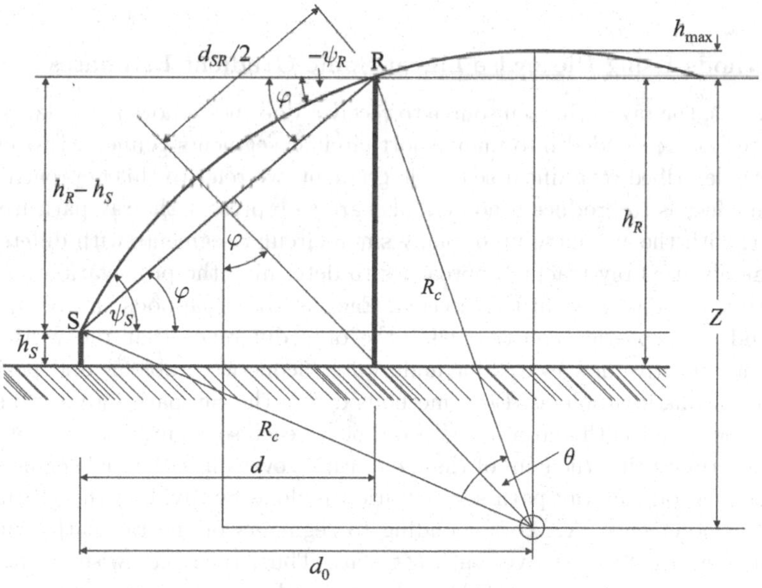

FIGURE 5.8 Geometry for calculating the radius of curvature of a sound ray originating at source, S, and arriving at receiver, R, for a receiver higher than the source and for the maximum ray height *not* between the source and receiver for a positive atmospheric vertical sonic gradient.

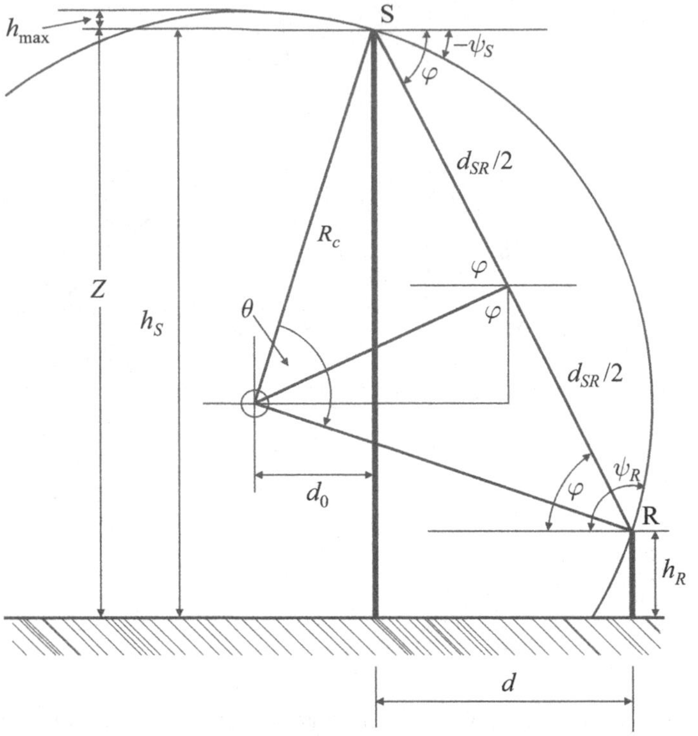

FIGURE 5.9 Geometry illustrating the arrangement for determining h_{max} and Z when the source/receiver distance is comparable to the source height for a positive atmospheric vertical sonic gradient.

5.3.4.4 Methods Using Piecewise Linear Sonic Gradient Estimates

In most situations, the ray path from source to receiver does not follow a circular path, although the actual path can be divided into many short circular segments connected together. As each segment can be described by a linearised sonic gradient, we refer to this approach as "piecewise linear", as the effect is to produce a non-circular arc to represent the ray path from the source to the receiver, with the arc made up of many small circular segments with different radii.

Two commonly used ray tracing approaches to determine the propagation path length and propagation time are discussed in this section. Each of these methods can be applied to both underwater and atmospheric acoustics, with the only difference that the vertical coordinate system is defined positive downwards for underwater acoustics problems and positive upwards for atmospheric acoustic problems. Both methods divide the ray path into small segments and then calculate the path length and propagation time for each segment, with the total path length and propagation time being the sum of these quantities over all individual segments.

In the first method, the ray path segmentation is done by dividing the altitude of the ray path into small increments, Δz, corresponding to segments of the ray path, with a constant sonic gradient assumed to exist over each segment. Thus, each piecewise segment of the ray path is circular and the coordinates of the end of a circular segment can be calculated from the coordinates of the beginning of the segment using simple linear relationships, with the radius of curvature of the segment being constant over the segment length. The second method uses Euler's method (which can also be replaced by the second-order Runge-Kutta method) to calculate the location of the end point of each segment based on the coordinates of the beginning point and the sonic gradient in the vertical and horizontal directions. Although the first method seems simpler, the ray segment length, Δr, becomes larger as the ray approaches its peak height and smaller as it approaches the ground, requiring the altitude step size, Δz to be adaptive so it continually adjusted to maintain Δr within a specified range. Inplementing the adaptation in computer code means that calculations made using the first method take longer to achieve the same accuracy.

Both proposed ray tracing methods require as inputs, the sonic gradient, $\mathrm{d}c_i/\mathrm{d}z$, as a function of altitude and range. For upward refraction, the sonic gradient is negative, so the radius of curvature is such that its centre is above the ray path. This situation is represented with a negative value for $R_{c,i}$. For downward refraction, the centre of curvature of the ray is below the ray path and the radius of curvature is represented as a positive value. The vertical sonic gradient to be used for each ray path segment can be calculated using either Equation (5.74) or (5.84), or it may be estimated for each height from vertical wind speed data, usually obtained using either a SODAR or LIDAR system. For both ray tracing methods, the beginning of the second segment is the end point of the first segment and the coordinates of the end of the second segment are calculated as for the first segment. This process repeats until the entire ray path is determined. Variations in topography between the source and receiver affect the value of the sonic gradient and this can be taken into account by specifying a range-dependent vertical sonic gradient, rather than specifying a horizontal sonic gradient. The topography details can be conveniently sourced from a geophysical archive data delivery system (Intergovernmental Committee on Surveying and Mapping, 2016). The vertical sound speed gradient for ridges and hills can then be estimated by applying a topographical effect multiplier to the velocity profile (Ngo and Letchford, 2009; Stull, 1988). This factor is a function of ridge type and location along the ridge and different formulations are available in various standards, including ASCE/SEI 7-05 (2013); AS/NZS 1170-2 (2021); EN1991-1-4 (2010); AIJ-RLB (2004).

The ray calculation for both ray tracing methods begins by specifying the initial conditions, which include the ray launch angle, ψ_0, which is the angle between the horizontal and the ray path tangent at the source location, the source height, $z_0 = h_S$, the sound speed, c_0, at the source location and the vertical sound speed gradient, $\mathrm{d}c/\mathrm{d}z$, at the source location. Method 1 steps in the vertical z-direction using a step size of Δz, while method 2 steps along the ray path using a step size of Δr along the path of total length, r. A numerical code, dependent on the method used, can then be implemented to calculate the value of the end point of the first segment based on the coordinates of the beginning location (the source) and the local sonic gradient, $\mathrm{d}c/\mathrm{d}z$.

For both ray tracing methods, the launch angle, ψ_0, must be found that corresponds to the sound ray landing closest to the receiver. To do this the ray tracing procedure is undertaken for ray launch angles at the source, ψ_0, beginning with $-89.9°$ and incrementing with a reasonable increment such as one degree until the optimum angle, ψ_0, is found that corresponds to the ray path that lands closest to the receiver. If the distance between the landing point of this path and the receiver is not within the required tolerance, the launch angle increment is reduced (usually by half) and the ray tracing procedure is repeated for a small range of angles surrounding the optimum angle that was found with the larger increment.

Method 1: This method involves stepping along the ray path at fixed vertical step sizes, Δz, so that the coordinates of the end of one segment are used as the coordinates of the beginning of the next segment. The method is based on TU Delft (2020) and begins with defining Δz and Δx in the coordinate system shown in Figure 5.10. The launch angles, ψ_i and ψ_{i+1} (radians), at the two ends of the segment represent the angles between the horizontal and the tangent to the ray path at each end of the ray path segment. If the angle is above the horizontal, it is positive, otherwise it is negative. The change in elevation, Δz_i, range Δx_i and length, Δr_i, for the piecewise segment, i, can then be calculated using trigonometry and Figure 5.10,

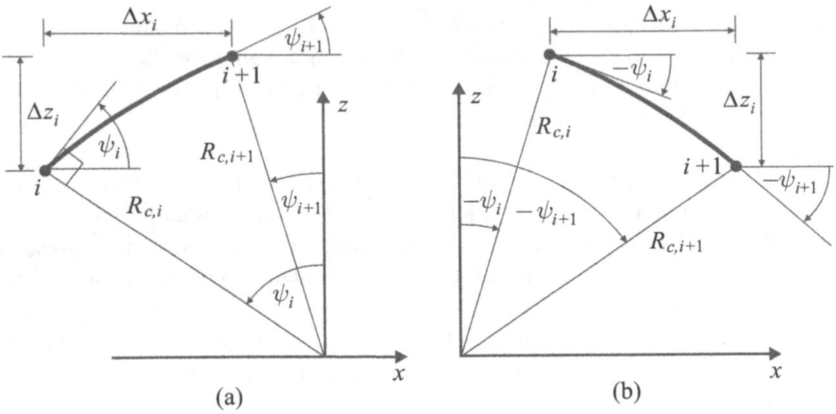

FIGURE 5.10 Diagram used to calculate the change in elevation of a piecewise ray segment with known launch and landing angles, ψ_i and ψ_{i+1}, respectively: (a) positive Δz and ψ; (b) negative Δz and ψ.

according to:

$$\Delta z_i = R_{c,i} \left(\cos \psi_{i+1} - \cos \psi_i\right) \tag{5.95}$$

$$\Delta x_i = -R_{c,i} \left(\sin \psi_{i+1} - \sin \psi_i\right) \tag{5.96}$$

$$\Delta r_i = R_{c,i}(\psi_i - \psi_{i+1}) \tag{5.97}$$

where it is assumed that the step size is sufficiently small so that $R_{c,i} \approx R_{c,i+1}$.

Note that the sign on the right side of Equation (5.95) is different to the sign used in underwater acoustics literature, as the sign of Δz is dependent on the coordinate system definition. For example, moving from point i to $i+1$ in Figure 5.10b would correspond to a positive or negative Δz_i in the context of underwater or atmospheric acoustics, respectively. The sign of Δx_i is always positive, which is the same for underwater and atmospheric acoustics. Equation (5.95) can rearranged to provide an expression for ψ_{i+1} as a function of the launch angle, ψ_i, the radius of curvature, $R_{c,i}$, of the sound ray at location i, and the chosen (thus known) step size, Δz_i, in the vertical direction.

$$\psi_{i+1} = \pm \arccos \left(\cos \psi_i + \Delta z_i / R_{c,i}\right) \tag{5.98}$$

where the sign of ψ_i and the sign of the solution for ψ_{i+1} must both be chosen to be the same as the sign of Δz_i; that is, ψ_i and ψ_{i+1} is chosen positive when Δz_i is positive. When the ray strikes the ground ($z_{i+1} \leq 0$), the sign of Δz_{i+1} is changed from negative to positive and when the ray experiences a tuning point at its maximum height, the sign of of Δz_{i+1} is changed from positive to negative.

As discussed on page 276, the increment, Δz_i is generally not the same for each segment. For example, the step size could be halved each time the segment length, Δr, exceeded a specified amount or doubled if the segment length became less than a specified amount. The radius of curvature, $R_{c,i}$, of the sound ray at location, i, is based on the speed of sound, c_i for segment, i and Equation (5.73). Thus:

$$R_{c,i} = \frac{c_i}{\left[\dfrac{\mathrm{d}c_i}{\mathrm{d}z}\right] \cos \psi_i} \tag{5.99}$$

The subscripts i and $i+1$ in Equation (5.98) refer to the corresponding points on the curve in Figure 5.10 and $i = 0$ represents the source location. Stepping from point, i, to point, $i+1$ with a vertical step size of Δz_i can be described as an iteration step, where Δz_i is positive in

Figure 5.10(a) and negative in Figure 5.10(b). The x and z-axes shown in Figure 5.10 are provided to show the x and z directions and the angles subtended from the vertical by the ends of a segment. These axes have their origin at the centre of the circle of which the ray path arc segment is a part and the origin location is different for each ray path segment.

To implement this method, the z-coordinate is set equal to zero on the ground immediately beneath the source and the sign of the increment, Δz_0 is set positive for positive ψ_0 and negative for negative ψ_0. The source, of height, h_S, is defined as located at $x = 0$. Thus, $(x_0,\ z_0) = (0,\ h_S)$. The z-coordinate at the beginning of the ith segment is:

$$z_i = z_0 + \sum_{k=1}^{i} \Delta z_k \tag{5.100}$$

where the first segment is identified with a subscript, $i = 0$, so for this segment no terms in the sum are used.

If the ith ray segment strikes the ground ($z_{i+1} \leq 0$), the sign of Δz_{i+1} is changed from negative to positive, resulting in the sound ray being automatically reflected. The coordinates of the reflection point, $(\tilde{x},\ 0)$, may be set equal to $(x_{i+1},\ 0)$ if the step size is sufficiently small or $\tilde{x}$ may be determined more accurately using linear interpolation, so that:

$$\tilde{x} = \frac{x_{i+1}z_i - x_i z_{i+1}}{z_i - z_{i+1}} \tag{5.101}$$

The new x-coordinate for the end of ray path segment, i, where the segment strikes the ground is then $x_{i+1} = \tilde{x}$, the new z-coordinate is $z_{i+1} = 0$ and the new ray path length, $\Delta\tilde{r}_i$, is:

$$\Delta\tilde{r}_i = \sqrt{(\tilde{x} - x_i)^2 + z_i^2} \tag{5.102}$$

The maximum ray height can be determined when the value of ψ_{i+1} of Equation (5.98) becomes undefined (as the arccos argument is greater than 1 or less than -1), at which point the sign of Δz_i is changed in Equation (5.98) to calculate a new ψ_{i+1}, resulting in the calculation proceeding in the opposite z-direction. If the ray strikes the ground before the value of ψ_{i+1} becomes undefined, the maximum ray height reached by the ray is the source height. The x-direction is always positive (to the right); thus, the sign of Δx is never flipped. The MATLAB® script, `propagation_ray_tracing.m`, for implementing this method can be found on the website listed in Appendix F.

The procedure described in the preceding paragraphs is also applicable to sound propagation over ground that is not flat. In this case the ground is divided into segments so that the ground under each segment has a uniform slope defined by equation (5.104). However, in this case, the z-coordinate of the ground, $z_{g,i}$, corresponding to x_i, must be defined for the ends of each segment, i. In many cases, such as uniformly sloping ground, the z-coordinate of the ground may be expressed as a function of x. The test of whether the ray strikes the ground between x_i and x_{i+1} is when $z_{i+1} \leq z_{g,i+1}$. The coordinates of the reflection point, $(\tilde{x},\ \tilde{z})$, may be set equal to $(x_{i+1},\ z_{i+1})$ if the step size is sufficiently small or they may be determined more accurately by linear interpolation using Figure 5.11, which assumes that the ground slope between locations i and $i+1$ is uniform and the curved ray path may be approximated as a straight line, as shown in the figure. This assumption is usually sufficiently accurate in locating the ground reflection point for reasonably small step sizes.

Referring to Figure 5.11, the intersection point, $(\tilde{x},\ \tilde{z})$, between the sound ray and the ground is:

$$\tilde{x} = \frac{(z_i x_{i+1} - z_{i+1}x_i + x_i z_{g,i+1} - x_{i+1}z_{g,i})}{z_i - z_{i+1} - z_{g,i} + z_{g,i+1}}; \quad \tilde{z} = \frac{(z_i z_{g,i+1} - z_{i+1}z_{g,i})}{z_i - z_{i+1} - z_{g,i} + z_{g,i+1}} \tag{5.103a,b}$$

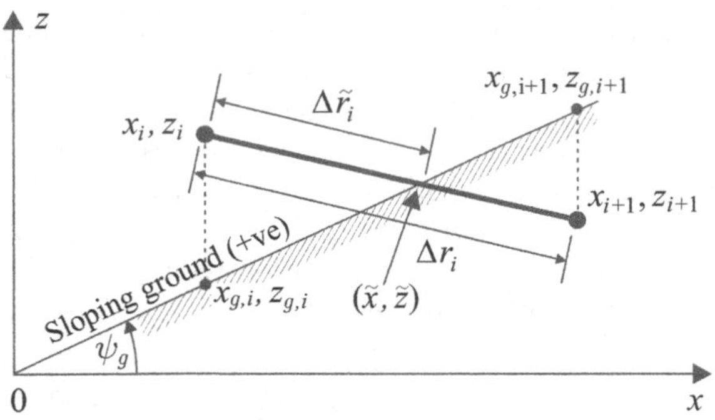

FIGURE 5.11 Geometry used to find the ground intersection point, $(\tilde{x},\ \tilde{z})$, for Method 1 and to calculate the reduced step size for Method 2, so that the end of the ith segment lands on the ground.

and the new ray path length, $\Delta\tilde{r}_i$ for the ith segment is given by Equation (5.102). The calculation of total ray path lengths and propagation times is discussed in Section 5.3.4.5. When the ground is sloping at a ground reflection location between x_i and x_{i+1} the launch angle, ψ_{i+1}, calculated using equation (5.98) must have twice the slope angle, $2 \times \psi_g$, added to it, where:

$$\tan\psi_g = (z_{g,i+1} - z_{g,i})/(x_{g,i+1} - x_{g,i}) \tag{5.104}$$

Method 2: This method involves stepping along the ray path using fixed ray-path step sizes, Δr, instead of the vertical step sizes, Δz_i, used in Method 1 above. Similar to Method 1, the coordinates of the end of one segment are used as the coordinates of the beginning of the next segment. The method has been implemented in the well-known Bellhop code (Porter, 2011) and the derivation of the governing equations is provided by Jensen et al. (2011), with the final results repeated below as Equations (5.105) and (5.106) (Jensen et al., 2011, p. 161).

$$\frac{\mathrm{d}x}{\mathrm{d}r} = c\chi, \quad \frac{\mathrm{d}\chi}{\mathrm{d}r} = -\frac{1}{c^2}\frac{\partial c}{\partial x} \tag{5.105}$$

$$\frac{\mathrm{d}z}{\mathrm{d}r} = c\zeta, \quad \frac{\mathrm{d}\zeta}{\mathrm{d}r} = -\frac{1}{c^2}\frac{\partial c}{\partial z} \tag{5.106}$$

The coordinate system labels chosen here are different to those used by Jensen et al. (2011), but they are consistent with earlier sections in this book, where x and z are the horizontal and vertical coordinates, respectively, and r is the distance along a ray path, so it it is a function of x and z. The variable, χ, represents the fraction of the speed of sound that is equivalent to the ratio of the change in the x-coordinate to the change in the r-coordinate as we move along the ray path. Similarly for ζ, except that the z-coordinate replaces the x-coordinate.

The initial conditions to begin ray tracing are:

$$x = x_0, \quad \chi = \frac{\cos\psi_0}{c_0} \tag{5.107}$$

$$z = h_S, \quad \zeta = \frac{\sin\psi_0}{c_0} \tag{5.108}$$

where c_0 is the speed of sound at the source location and ψ_0 is the ray launch angle at the source. This is the angle between the horizontal and the tangent to the ray path at the source. Angles above the horizontal are positive and angles below the horizontal are negative.

As Equations (5.105) and (5.106) take the standard form given in Equation (5.109), where $y = f(x, z, \chi, \zeta)$, they can be solved by numerical discretisation using Euler's method.

$$y' = f(r, y) \tag{5.109}$$

Thus, Equations (5.105) to (5.106) can be discretised as:

$$\chi_{i+1} = \chi_i - \Delta r \left(\frac{1}{c_i^2} \frac{\partial c_i}{\partial x} \right) \tag{5.110}$$

$$\zeta_{i+1} = \zeta_i - \Delta r \left(\frac{1}{c_i^2} \frac{\partial c_i}{\partial z} \right) \tag{5.111}$$

$$x_{i+1} = x_i + \Delta r \, c_i \chi_i \tag{5.112}$$

$$z_{i+1} = z_i + \Delta r \, c_i \zeta_i \tag{5.113}$$

where Δr is the arc length of one segment of the path of total length, r, and $\partial c_i / \partial x$ and $\partial c_i / \partial z$ are the sound speed gradients for segment, i in the x and z directions, respectively. In practice, it is more efficient to set $\partial c_i / \partial x = 0$ and take into account the range dependence of the sonic gradient by specifying different vertical sonic gradients as a function of range, x. As higher-order stepping schemes yield better accuracy (Chapra and Canale, 2011), the second order Runge-Kutta method is more commonly implemented (rather than the Euler method) in computer codes, such as the Bellhop codes (Porter, 2011).

Equations (5.110) to (5.113) allow sound ray paths to be traced accurately, including turning points. The point at which the ray impacts the ground may be calculated using Figure 5.11 in a similar way as was done for Method 1. For flat, horizontal ground, for which the ground coordinate, $z_{g,i} = 0$ for all segments, i, the condition $z_{i+1} < 0$ can be checked at each iteration step. At the point where this condition is met, z_{i+1} is set equal to 0 and the reduced step size, $\Delta \tilde{r}_i$, is found using Equation (5.113) as:

$$\Delta \tilde{r}_i = \frac{-z_i}{c_i \zeta_i} \tag{5.114}$$

For a ground surface that is not flat, the z-coordinate of the ground, $z_{g,i}$, corresponding to x_i, must be defined for each segment, i. In many cases, such as uniformly sloping ground, the z-coordinate of the ground may be expressed as a function of x. The test of whether the ray strikes the ground between x_i and x_{i+1} is when $z_{i+1} \le z_{g,i+1}$. For this case, a reduced step size, $\Delta \tilde{r}_i$, must be found by considering the slope of both the ground surface and the ray trajectory at the point where the sound ray just penetrates the ground surface. The intersection point, $(\tilde{x}, \tilde{z})$, between the sound ray and the ground can then be found by linear interpolation using Figure 5.11, which assumes that the ground slope between locations i and $i+1$ is uniform, as shown in the figure. This assumption is usually accurate for reasonably small step sizes. The intersection point coordinates are given by Equations (5.103a) and (5.103b), and the reduced step size is then:

$$\Delta \tilde{r}_i = \sqrt{(\tilde{x} - x_i)^2 + (\tilde{z} - z_i)^2} \tag{5.115}$$

Equations (5.103a), (5.103b) and (5.115) can also be used for flat ground instead of Equation (5.114), in which case, $z_{g,i} = z_{g,i+1}$ and therefore the denominators of Equations (5.103a) and (5.103b) become simpler.

When the sound ray is reflected at the boundary, the launch angle from the boundary is 180 degrees minus the incident angle and the conditions associated with the reflection point are defined as:

$$\chi_{i+1} = \chi_i - \Delta \tilde{r}_i \left(\frac{1}{c_i^2} \frac{\partial c_i}{\partial x} \right) \tag{5.116}$$

$$\zeta_{i+1} = -\zeta_i + \Delta \tilde{r}_i \left(\frac{1}{c_i^2} \frac{\partial c_i}{\partial z} \right) \tag{5.117}$$

The equations presented in both methods outlined above can be used to generate a sound ray map, which can be used to provide a two-dimensional contour plot of the sound field. To trace a single sound ray path from the source to a receiver, an iterative process can be used to obtain the angle, ψ_S, at which the ray arriving at the receiver left the source. The procedure involves setting ψ_0 equal to $-89.9°$ at the source location and then incrementing ψ_0 by a small amount, $\Delta\psi_0$, until the ray arrives at the receiver within a specified tolerance. Then ψ_S is set equal to the angle, ψ_0 that corresponds to this condition. Alternatively, ψ_S could be set equal to the value of ψ_0 that results in the ray striking the ground closer to the receiver than the rays corresponding to the next higher and next lower ψ_0 increment.

The calculation of total ray path lengths and propagation times, which are needed to determine the sound pressure level at the receiver are discussed in Section 5.3.4.5. The ray path length, r, is used to calculate the propagation loss due to geometric divergence, using equations (5.56) to (5.59), where d_{SR} is replaced with r. For cases where the ground reflected ray is coherent with the direct ray, the ray path propagation times are also needed so that the phase differences between the propagation paths contributing to the sound pressure level at the receiver can be calculated.

5.3.4.5 Calculation of Ray Path Lengths and Propagation Times

The ray path length is needed to calculate the attenuation, due to geometrical spreading, of the propagating wave for each path that leads to the receiver. These paths include the direct path with no ground reflection, the path with one ground reflection and the paths with more than one ground reflection. Paths with more than one ground reflection will only exist beyond a certain distance from the source. This distance is dependent on the source and receiver heights, the ground absorption properties and meteorological effects. Standard noise propagation models such as CONCAWE and ISO 9613-2 only consider the direct wave and the wave that has been reflected only once from the ground. However, in the discussion in the following sections, we will consider waves that have undergone more than one reflection as well.

The propagation time, τ, for each wave arriving at the receiver must be known if the different waves arriving at the receiver are to be combined coherently (as for a tonal sound) or partially coherently (see Appendix E), as the difference in propagation times defines the relative phase between the arriving waves. The combination of sound pressures due to all waves arriving at the receiver to obtain a total sound pressure level is discussed at the end of Sections 5.3.4.6 and 5.3.4.7.

Linearised wind speed profile

The expression for the radius of curvature of the sound ray can be used to calculate the angles at which the sound ray leaves the source, follows a circular ray path and arrives at the receiver, using Equation (5.92) (see Figure 5.12).

The included angle, θ, for the curved path between the source, S, and receiver, R, is:

$$\theta = \psi_S + \psi_R \tag{5.118}$$

The horizontal distance, d_0, of the centre of the circular sound ray path from the source, S, is:

$$d_0 = \sqrt{R_c^2 - (R_c - h_{\max})^2} = R_c \sin\psi_S \tag{5.119a,b}$$

and ψ_S is defined by Equation (5.92). Note that d_0 will be negative if $\psi_S < 0$ (corresponding to the ray leaving the source at an angle below the horizontal). Negative d_0 implies that the centre of the arc is on the opposite side of the source to the receiver.

For distances, d_1, less than d_0 from the source, the height, h_1, of the ray at any specified distance, d_1, is:

$$h_1 = \sqrt{R_c^2 - (d_0 - d_1)^2} - R_c \cos\psi_S \tag{5.120}$$

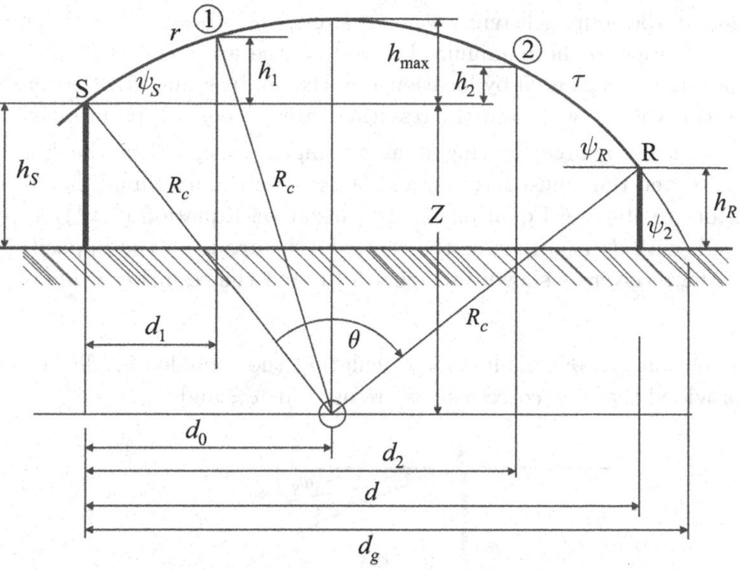

FIGURE 5.12 Geometrical parameters for calculating the height and centre of curvature of a curved sound ray beginning at point, S, and ending at point, R.

For distances, d_2, greater than d_0 from the source, the height, h_2, of the ray at any specified distance, d_2, is:

$$h_2 = \sqrt{R_c^2 - (d_2 - d_0)^2} - R_c \cos \psi_S \tag{5.121}$$

The ray variables that need to be calculated in order to determine the contribution of the direct and ground-reflected rays from the same source, arriving at the receiver, are listed below. In the following items, the source may be the actual noise source or the top of a barrier and the receiver may be the actual receiver or the top of a barrier.

- The propagation time, τ, along the sound ray from the source to the receiver.
- The propagation distance, r, along the sound ray from the source to the receiver.

The distance, r, travelled by a direct ray to the receiver over an arc of a circle of radius, R_c, with an included angle of θ radians (see Equation (5.118)), is:

$$r = R_c \theta \tag{5.122}$$

To be able to calculate the angle, θ, it is necessary to determine the location, (d_0, Z), of the centre of curvature of the ray path relative to the source, as shown in Figures 5.5 to 5.9. From the figures:

$$d_0 = R_c \sin \psi_S \quad \text{and} \quad Z = R_c \cos \psi_S \tag{5.123a,b}$$

There are three different situations for calculating an arc length and all arc lengths are calculated from the originating point of the arc at the source or the top of a barrier. The different situations for calculating arc lengths are listed below.

1. Arc begins at the source, leaving it at an angle, ψ_S, above the horizontal, and has an end point after the maximum height of the arc. Referring to Figure 5.12, and in particular to location 2 on the arc, the angle, ψ_R (radians), for use in Equation (5.118) for calculating θ, and hence the arc length, is:

$$\psi_R = \arcsin \frac{d - d_0}{R_c} \tag{5.124}$$

2. Arc begins at the source, leaving it at an angle, ψ_S, above the horizontal, and has an end point prior to the maximum height that the arc could reach if it continued. This situation is represented by location 1 on the arc in Figure 5.12. Equation (5.124) applies in this case as well, and the resulting value of ψ_R will be negative.
3. Arc begins at the source, leaving it at an angle, $-\psi_S$, below the horizontal (see Figure 5.13), and thus must have an end point after the maximum height of the arc. This case occurs if ψ_S of Equation (5.92) is negative. Equation (5.124) applies in this case as well, with d_0 negative, as the centre of the arc is on the opposite side of the source to the receiver as shown in Figure 5.13, and the resulting value of ψ_R will be positive.

In all cases, Equation (5.118) applies for calculating the included angle, θ, for use in Equation (5.122), provided that the correct signs are used for ψ_S and ψ_R.

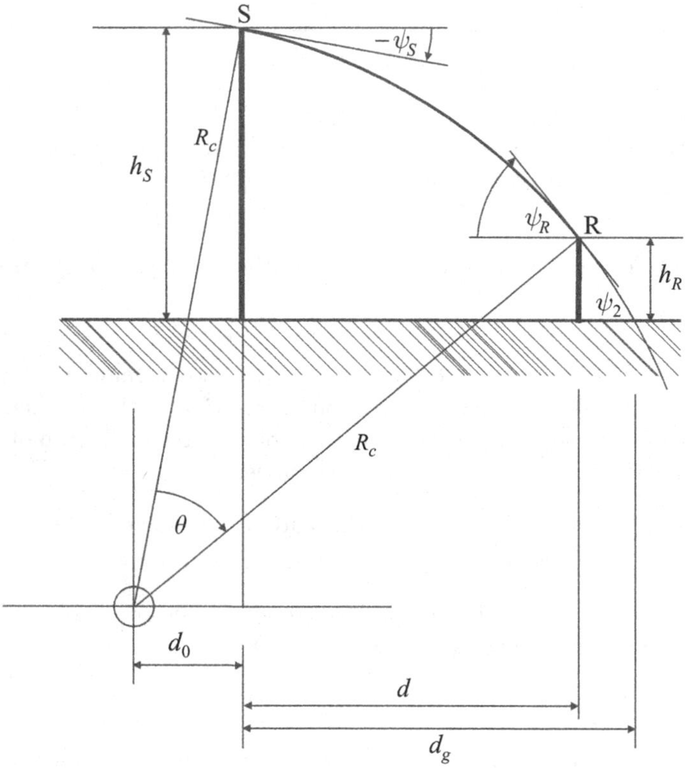

FIGURE 5.13 Geometrical parameters for a direct ray from source, S, to receiver, R, leaving the source at an angle, $-\psi_S$, below the horizontal.

The calculation of the arc length from a source point to the receiver point can be done by substituting Equation (5.118) into Equation (5.122), with ψ_S defined by Equation (5.92) and ψ_R defined by Equation (5.124). The radius of curvature for use in Equation (5.124) is given by Equation (5.87). However, it is preferable to use iteration to find R_c. In this case, ψ_S is incremented from $\psi_S = -89.9$ degrees and the radius of curvature, R_c, for each value of ψ_S is calculated using Equations (5.87), (5.88) and (5.89). Iteration continues until the value of d, calculated using Equation (5.92), is equal to the actual horizontal distance between the source and receiver (within an acceptable error).

To calculate the propagation time, τ, from the source to the receiver, it is necessary to take into account the variation in sound speed over different parts of the arc due to them being at differing altitudes. The simplest way to do this is to divide the arc up into a number of segments of length, Δr_i, each with a particular mid-point height, z_i above the ground and associated constant sound speed, c_i, which is calculated using Equation (5.83) or obtained from measured data. The total propagation time, τ, is calculated by summing the propagation times, τ_i for each of the N segments of the propagation path as follows, where the segment nearest the source corresponds to $i = 0$:

$$\tau = \sum_{i=0}^{N-1} \tau_i = \sum_{i=0}^{N-1} \frac{\Delta r_i}{c_i} \tag{5.125a,b}$$

Typically, the above process may be repeated using progressively smaller values of Δr_i until the difference between successive repeats is less than the required accuracy.

At the point of reflection, the reflection coefficient must be calculated using Equations (5.15c), (5.18b) or (5.89b). The sound pressure amplitude must be multiplied by the modulus of the complex reflection coefficient and the difference (phase of reflected ray minus phase of incident ray) must be added to the phase difference between the direct and reflected rays (reflected minus direct) at the receiver.

Piecewise linearised wind speed profile

When the approach to calculating the propagation path outlined in Section 5.3.4.4 is used, the total distance travelled along the ray path, or the total ray path length, r, is obtained by summing the lengths of all of the path segments over the number of segments, N. Thus:

$$r = \sum_{i=0}^{N-1} \Delta r_i \tag{5.126}$$

where for Method 1, Δr_i is calculated using Equation (5.97) (or Equation (5.102) for a segment that is incident on the ground), and for Method 2, $\Delta r_i = \Delta r$ (or it is calculated using Equation (5.115) for a segment that is incident on the ground). Smaller step sizes in both methods leads to greater accuracy.

The propagation time for propagation over segment, r_i for Method 1 is given by the ith term in Equation (5.125b). When using Method 1, outlined in Section 5.3.4.4, the propagation time τ_i associated with ray path segment, i, whose tangent at the segment beginning and end points subtends angles to the horizontal, $\psi_{1,i}$ and $\psi_{2,i}$, respectively, is:

$$\tau_i = \int_{\psi_{1,i}}^{\psi_{2,i}} \frac{R_{c,i}}{c_i} \mathrm{d}\psi = \int_{\psi_{1,i}}^{\psi_{2,i}} \left(\frac{\partial c}{\partial z} \cos\psi \right)^{-1} \mathrm{d}\psi \tag{5.127a,b}$$

$$= \left| \frac{2}{\partial c / \partial z} \left[\operatorname{arctanh}(\tan(\psi_{2,i}/2)) - \operatorname{arctanh}(\tan(\psi_{1,i}/2)) \right] \right|$$

where

$$\operatorname{arctanh}(x) = \frac{1}{2} \log_e \left(\frac{1+x}{1-x} \right) \tag{5.128}$$

As described in the previous section, the total propagation time, τ, from source to receiver is calculated by summing the propagation times, τ_i, for each segment, i using Equation (5.125b).

The ray propagation time can also be expressed in terms of an integral over the ray of total length, r. This approach can be used with both Methods 1 and 2 (see Section 5.3.4.4). In this case, the total propagation time τ associated with the ray path under consideration is:

$$\tau = \int_0^r \frac{1}{c_r} \, \mathrm{d}r \tag{5.129}$$

where c_r is the speed of sound for segment dr. This integral can be solved using the trapezoidal rule, which, for segment, i, gives the propagation time as:

$$\tau_i = (r_{i+1} - r_i) \frac{1/c_{i+1} + 1/c_i}{2} \tag{5.130}$$

where $r_0 = 0$, $r_i = \sqrt{(x_{i+1} - x_i)^2 + (z_{i+1} - z_i)^2}$ and $r_{i+1} = \sqrt{(x_{i+2} - x_{i+1})^2 + (z_{i+2} - z_{i+1})^2}$, as r_i and r_{i+1} define the start and end points, respectively, of segment i and the first segment corresponds to $i = 0$. As described in the previous section, the total propagation time, τ, is calculated by summing the propagation times, τ_i, for each segment, i using Equation (5.125b).

The use of propagation times and propagation path lengths to calculate the sound pressure level at the receiver due to the combination of all contributing ray paths is discussed at the end of Sections 5.3.4.6 and 5.3.4.7.

5.3.4.6 Ground-Reflected Rays – Single Ground Reflection

If the source/receiver separation distance is sufficiently small compared to the source and receiver heights, there will only be a single ground-reflected ray arriving at the receiver. In this case, for a ray reflected from the ground, the following additional parameters are needed.

- The location of the point of reflection relative to the source position.
- The propagation distance, r_S, along the sound ray from the source to the reflection point.
- The propagation distance, r_R, along the sound ray from reflection point to receiver.
- Ground reflection angle, ψ_2 (grazing angle), for a ray.

The CONCAWE, ISO 9613-2 and the NMPB-8 models only consider a single ground reflection and do not consider the effect of multiple ground reflections. For these cases, the ground-reflection is located on the line joining the ground at the base of the source with the ground at the base of the receiver. For flat ground, the point on this line is located a horizontal distance, d_S, from the source defined by:

$$d_S = h_S d/(h_S + h_R) \tag{5.131}$$

The angle of incidence and reflection (for calculating the ground attenuation) is defined as (see Figure 4.17):

$$\beta_0 = \tan^{-1}(h_S/d_S) \tag{5.132}$$

For calculations outlined in Section 5.3.4.3, a circular ray path of constant radius is assumed for propagation between the source and receiver. For this case, the propagation path of the reflected ray from the source to the ground reflection point is calculated as in the previous procedures by setting the receiver location at the ground reflection point. Similarly, the propagation path from the ground reflection point to the receiver is calculated as in the previous procedures, by setting the source location at the ground reflection point. Although the ground reflection point is unknown, we do know that it is a point where the angle of incidence, ψ_R, of the ray from the source is equal to the angle of reflection, ψ_S, of the ray that begins at the ground reflection point and ends at the receiver.

The angle of incidence, ψ_{R2}, is defined by Equation (5.124) and the angle of reflection, ψ_{S2} for the ground-reflected ray is obtained from Equation (5.119b). If ψ_{S2} is greater than ψ_{S2}, the reflection point is moved incrementally closer to the receiver. If ψ_{S2} is less than ψ_{S2}, the reflection point is moved incrementally closer to the source. In practice, the movement of the reflection point is achieved by increasing or decreasing the ray launch angle, ψ_S. For each new reflection point location, the procedures described in Sections 5.3.4.3 or 5.3.4.4 are repeated. This continues, using decreasing increments for ψ_S, until the calculated angles of incidence, ψ_{S2}, and reflection, ψ_{R2}, at the ground only differ by a small, acceptable amount.

The preceding analysis in this section assumes that there is flat ground between the source and receiver. If the ground is sloping (where the slope, ψ_g, is positive if the ground under the source is lower than the ground under the receiver and negative if the ground under the receiver is lower), the calculation of the angles of incidence and reflection become slightly more complex. In this case, the angle of incidence, ψ_{R2}, is replaced with $\psi_{R2} + \psi_g$ and the angle of reflection, ψ_{S2}, is replaced with $\psi_{S2} - \psi_g$, where ψ_g is the angle defining the slope of the ground and ψ_{S2} and ψ_{R2} are calculated as for flat ground.

For the piecewise linearised wind speed profile calculations outlined in Section 5.3.4.4, the ray path consists of small circular segments, each of which may have a unique radius of curvature. As discussed at the end of Section 5.3.4.4, the direct ray path is identified as the one corresponding to the launch angle, ψ_S, that results in the ray landing closest to the receiver with no ground reflection. The single ground-reflected ray path to be used in the calculations of sound pressure level at the receiver is the one that arrives closer to the receiver than any other ground reflected ray.

Once the path lengths, propagation times and sound pressures for the direct and ground-reflected ray paths, which both contribute to the sound pressure level at the receiver, have been determined, they must be combined together to give the total sound pressure level. The two rays may be considered incoherent or partially coherent if bands of noise are considered and coherent or partially coherent if tones are considered.

For the incoherent case, Equation (1.98) may be used with two terms, L_1 and L_2 on the RHS of the equation. At the ground reflection point, the reflection loss, A_{rf}, calculated using Equations (5.20) or (5.48), must be determined and subtracted from the sound pressure level contribution of the reflected ray at the receiver, prior to it being logarithmically added to the sound pressure level contribution of the direct ray, to obtain the total sound pressure level.

For the coherent case, Equation (1.95) may be used, with the relative phase, $\beta_2 - \beta_1$, (phase of reflected ray minus that of direct ray) between the two sound rays calculated from the frequency and propagation time difference, $\Delta\tau$, as:

$$\beta_2 - \beta_1 = 2\pi f \Delta\tau \quad \text{(radians)} \tag{5.133}$$

At the point of reflection, the complex reflection coefficient must be calculated using Equations (5.15c), (5.18b) or (5.89b). The sound pressure amplitude for the reflected ray arriving at the receiver must be multiplied by the modulus of the complex reflection coefficient, $|Q|$, corresponding to the ground reflected ray 2 and defined by Equation (5.23). Furthermore, the phase of the ground-reflected ray minus the phase of incident ray must be calculated to produce a phase shift of ϑ, due to reflection from the ground, which must be added to the phase difference due to propagation time calculated using Equation (5.133).

The partially coherent case, which may be applied to broadband or tonal noise, is discussed at length in Appendix E.

5.3.4.7 Ground-Reflected Rays – Multiple Ground Reflections

If ray paths having more than one ground reflection are to be considered, the piecewise ray tracing method of Section 5.3.4.4 is a better choice than the method described in Section 5.3.4.3, as the latter method becomes much more difficult to program and is also less accurate. In addition, the former method is easily adapted to a non-flat ground profile. As the number of ground reflections being considered increases, it likely that the increment size for Δz or Δr of the piecewise method will need to be reduced.

The calculation procedure begins by defining the maximum allowable distance between the receiver and the nearest landing point of a ray. Next, the beginning ray launch angle relative to the horizontal is set, and this is usually −89° (negative values are for angles below the horizontal). The ray tracing calculations are repeated with increasing increments of launch angle until a ray lands either side of the receiver for two consecutive launch angle values. If the distance between the receiver and one of the two landing locations is within the maximum allowable distance defined in the first step, then the path length and propagation time for the ray corresponding to the nearest landing location is recorded. If neither landing location is within the required tolerance, then the angular increment for ψ_S is halved and the process is repeated, beginning with the angular increment corresponding to the ray with the landing location closest to and on the source side of the receiver. This procedure is repeated until a ray is found with its landing location closer to the receiver than the maximum allowable distance.

The next step is to repeat the above procedure with the additional requirement that the ray must undergo a ground reflection between the source and receiver. Then the procedure is repeated for two ground reflections, then three ground reflections, etc, until a point is reached at which the required number of ground reflections between the source and receiver cannot be found. This could occur as early as the two ground reflection case. Of course, there are computational techniques that result in faster solution times, but description of these is outside the scope of this textbook.

To increase the accuracy of the solution, linear interpolation using the rays landing on either side of the receiver for each case can be used to find more accurate propagation times and phase differences between the various ray paths. Once the path lengths, propagation times and sound pressures for each path contributing to the sound pressure level at the receiver have been determined, they must be combined together to give the total sound pressure level due to all ray paths. The various contributions may be considered incoherent or partially coherent if bands of noise are considered and coherent or partially coherent if tones are considered.

For the incoherent case, Equation (1.98) may be used with the number of terms on the RHS of the equation equal to the number of ray paths contributing at the receiver. For ray paths involving one or more ground reflections, the reflection loss, A_{rf} (dB), at each ground reflection point, calculated using Equations (5.20) or (5.48), must be determined and the arithmetic sum of the reflection losses (in dB) for any one path must be subtracted from the sound pressure level contribution (in dB) at the receiver for that path, prior to being logarithmically added to the sound pressure level contributions of the other paths to obtain the total sound pressure level at the receiver.

For the coherent case, Equations (1.96) and (1.87) may be used, with the relative phases between each possible pairing combination, $(i,\ k)$, of the various sound rays calculated from the frequency and propagation time differences, $(\tau_i - \tau_k)$ seconds, as:

$$\beta_i - \beta_k = 2\pi f(\tau_i - \tau_k) \ \text{(radians)} \ = 360 f(\tau_i - \tau_k) \ \text{(degrees)} \tag{5.134}$$

At the point of reflection, the complex reflection coefficient must be calculated using Equations (5.15c), (5.18b) or (5.89b). The sound pressure amplitude for a particular ray arriving at the receiver must be multiplied by the modulus of each complex reflection coefficient corresponding to the ground reflections for the ray being considered. Furthermore, the phase of the reflected ray minus the phase of incident ray for each ground reflection must be calculated and added together to produce a total phase shift due to ground reflections of ϑ_i for path, i. Thus, equation (5.134) must be modified as follows.

$$\beta_i - \beta_k = 2\pi f(\tau_i - \tau_k) + \vartheta_i - \vartheta_k \ \text{(radians)} \tag{5.135}$$

The partially coherent case, which may be applied to broadband or tonal noise, is discussed at length in Appendix E.

5.3.4.8 Low-Level Jets (LLJs)

Low-level jets are typically a nighttime phenomenon (Baas et al., 2009), characterised by a maximum in the wind speed profile, which is typically located 100 to 500 m above the ground. They are also characterised by a much higher than expected wind gradient (wind shear) near the ground, which results in much more severe than expected curvature of the sound rays due to atmospheric refraction. This results in the distance from the source, at which multiple ground reflections begin, to be significantly reduced (see Section 5.3.4.7), and thus reduces the applicability of noise propagation models to shorter propagation distances than one would normally expect. The presence of an LLJ will result in considerably higher than expected sound pressure levels downwind of a noise source.

LLJs are more prevalent near flat terrain and near coastal areas and can have a significant effect on noise propagation from off-shore turbines. An LLJ is defined by Bass (2011), and it occurs when the following criteria are satisfied (see Figure 5.14).

$$\frac{U_{\mathrm{max}} - U_{\mathrm{min}}}{U_{\mathrm{min}}} > 1.25 \tag{5.136}$$

$$U_{\mathrm{max}} - U_{\mathrm{min}} > 2 \text{ (m/s)}$$

$$4 < U_{\mathrm{max}} < 14 \text{ (m/s)}$$

$$80 < z_{\mathrm{jmax}} < 400 \text{ m}$$

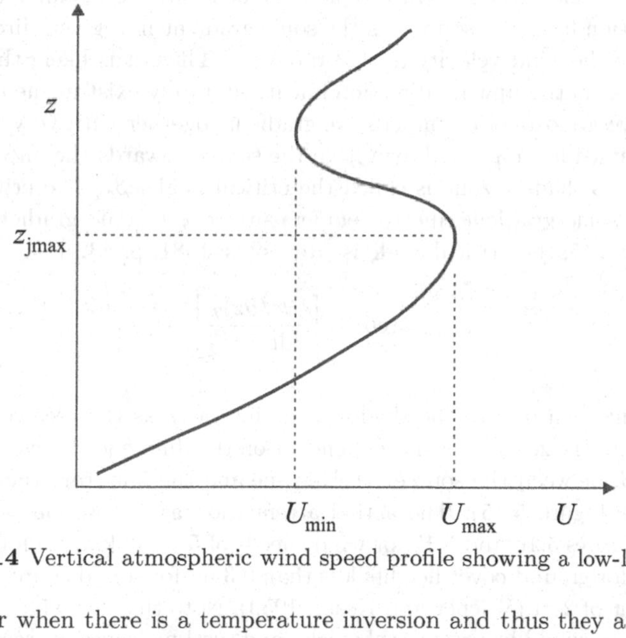

FIGURE 5.14 Vertical atmospheric wind speed profile showing a low-level jet (LLJ).

LLJs only occur when there is a temperature inversion and thus they are a nighttime phenomenon, reaching their maximum strength in the early hours of the morning, around 5 to 6 am.

5.3.4.9 Attenuation in the Shadow Zone (Negative Sonic Gradient)

Most of the time, for the purpose of assessing the impact of environmental noise, it is necessary to calculate sound pressure levels at the receiver, which correspond to propagation in a neutral or downward refracting atmosphere, as these conditions contribute most to observed sound pressure levels. However, at times it is of interest to determine sound pressure levels corresponding to an upward refracting atmosphere, which occurs when the wind is blowing from the receiver towards the source and/or there is no atmospheric temperature inversion close to the ground. The zone where the receiver is relatively isolated from the noise source due to the upward refracting atmosphere is referred to as the "shadow zone", which is defined as that region where direct sound cannot penetrate as a result of the upward refracting atmosphere. Of course, a small amount of sound will always transmit to the shadow zone as a result of scattering from air turbulence or objects, so that one might expect an increase in attenuation in the shadow zone to be limited to about 30 dB. To create a shadow zone, a negative sonic gradient must exist, as this causes the sound rays emitted by the source to have a radius of curvature, R_c, that is finite. By convention, in a negative sonic gradient, the radius of curvature is negative (sound rays curved upwards), whereas in a positive sonic gradient, the radius of curvature is positive (sound rays curved downwards). For the case of a negative sonic gradient and corresponding negative radius of curvature, R_c, the distance, x, between the source (height, h_S) and receiver (height, h_R) beyond which the receiver will be in the shadow zone is:

$$x = \sqrt{-2R_c}\left[\sqrt{h_S} + \sqrt{h_R}\right] \tag{5.137}$$

Note that a shadow zone can also exist in the presence of no wind when there is a negative temperature gradient, and it will be symmetrical around the source. This is because the sonic gradient due to temperature (in an atmosphere with a normal negative temperature gradient) produces around the source, a shadow zone that does not vary in extent with angle, whereas the sonic gradient due to the wind produces a shadow zone that is directional, with a maximum extent towards the upwind direction. In the presence of wind, the distance to the shadow zone will vary with direction from the source, as the sonic gradient in a given direction is dependent on the component of the wind velocity in that direction. Thus, it is likely that in most cases, a shadow zone will exist in the upwind direction but it will rarely exist in the downwind direction (only with a negative atmospheric temperature gradient together with very light or no wind).

The angle subtended from a line drawn from the source towards the oncoming wind beyond which there will be no shadow zone is called the critical angle, β_c. The critical angle depends on the ratio of the sonic gradient due to temperature to the sonic gradient due to wind. As illustrated in Figure 5.15, the critical angle is (Reynolds, 1981, p. 432):

$$\beta_c = \cos^{-1}\left[\frac{(\partial c/\partial z)_T}{\mathrm{d}U/\mathrm{d}z}\right] \tag{5.138}$$

The actual attenuation due to the shadow zone increases as the receiver moves away from the source further into the zone. It is also dependent on the difference between the critical angle, β_c, and the angle, β, between the source/receiver line and the line from the source towards the oncoming wind (see Figure 5.15). The actual attenuation as a function of β and β_c may be determined using Figures 5.16 and 5.17 for wind speeds of 5 to 30 km/hr, for octave bands from 250 Hz to 4000 Hz, for ground cover heights less than 0.3 m, for sound source heights of 3 to 5 m and a receiver height of 2 m (Wiener and Keast, 1959). Note that for wind speeds greater than 30 km/hr, noise due to wind blowing over obstacles and rustling leaves in trees usually dominates other environmental noise, unless the latter is particularly severe. At frequencies below 250 Hz, other ground effects limit the attenuation in the shadow zone, as discussed in Section 5.3.3.

The attenuation for a particular value of β is calculated using Figure 5.16. This is only the attenuation due to a negative vertical sonic gradient (upwind propagation and/or negative temperature gradient close to the ground). The attenuation cannot exceed the value given by Figure 5.17, which is a function of $\beta_c - \beta$.

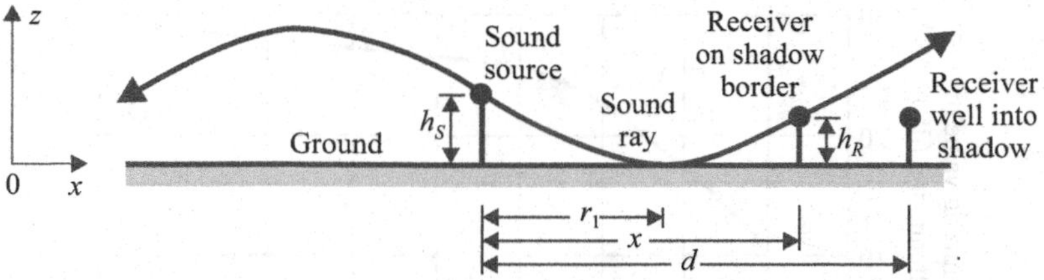

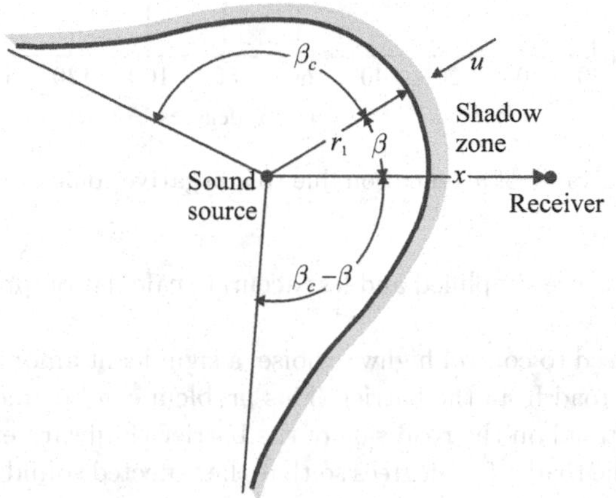

FIGURE 5.15 Illustration of the shadow zone and the limiting angle, β_c, beyond which the shadow zone does not exist. β is the angle subtended between the wind direction and the source/receiver line.

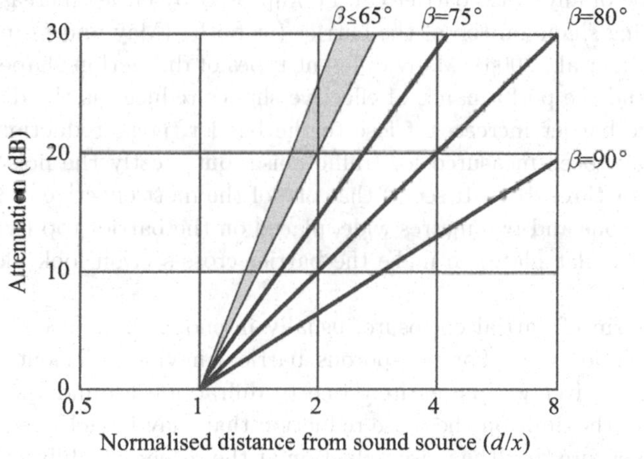

FIGURE 5.16 Attenuation in the shadow zone due to the negative vertical sonic gradient for various angles, β, between the wind direction and the line joining the source and receiver.

5.3.5 Barrier Effects, A_b

Barriers are often interposed between an outdoor sound source and receiver to reduce sound level experienced at the receiver. Thus, calculating the effect of a barrier is an important part of any propagation modelling process. In this section we discuss some general aspects of barrier attenuation and how the attenuation effect of barriers may be calculated accurately. Nevertheless,

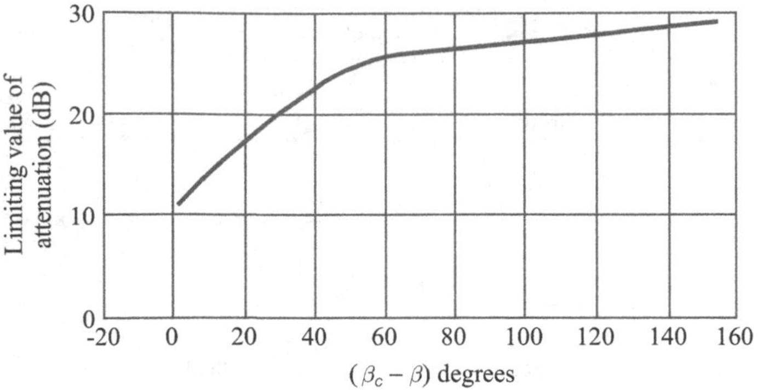

FIGURE 5.17 Limiting value of attenuation due to a negative sonic gradient in the shadow zone.

some propagation models use simplified and less accurate calculation procedures as will be found in Section 5.4.

When barriers are used to control highway noise, a significant amount of sound is reflected to the opposite side of the road from the barrier. This problem can be ameliorated by using a layer of sound absorptive material on the road side of the barrier (Fujiwara et al., 1977b) or by tilting the barrier away from the road a few degrees so that the reflected sound is directed upwards. The tilt angle should be sufficient so that the reflected rays arrive well above any sensitive receiver location on the opposite side of the road. This height can be calculated by remembering that the angle between the normal to the barrier surface and the incident sound is the same as that between the normal and the reflected sound (but on the other side of the normal – see Figure 5.1).

The performance of any noise barrier can be improved by either increasing its height or by installing a diffracting shape on top of the barrier (or both) (May and Osman, 1980; Ekici and Bougdah, 2003; Egan et al., 2006). Many different types of diffraction shape and size have been tested in the past and the performance of effective shapes reduces as the distance of the source or receiver from the barrier increases. Close to the barrier, noise reduction increases of up to four or five dBA have been measured for traffic noise, but mostly the noise reduction increase is restricted to two or three dBA. It seems that one of the most effective of the simple shapes is a flat plate, between one and two metres wide, placed on the barrier top (with sound absorbing material on top of the flat plate) to make the barrier cross-section look like a T (Cohn et al., 1994).

Barriers are a form of partial enclosure, usually intended to reduce the direct sound field radiated in one direction only. For non-porous barriers having sufficient surface density, the sound reaching the receiver will be entirely due to diffraction around the barrier boundaries. Since diffraction sets the limit on the noise reduction that may be achieved, the barrier surface density is chosen to ensure that the noise reduction at the receiver is diffraction limited. For this purpose, the barrier surface density will usually exceed 20 kg/m^2. There are many proprietary designs for barriers; typical barriers are built of lightweight concrete blocks, but asbestos board, cement board, sheet metal, glass, fibreglass panels and high-density plastic sheeting have also been used.

The attenuation due to barriers is frequency dependent. This is why attenuation values are usually reported in octave bands or 1/3-octave bands so the results can be used in noise propagation models. However, when comparing the improvement in performance that may be achieved by different treatments to the top of the barrier, a single number descriptor of the measured difference is useful (BS EN 1793-4, 2015). This number is an average over the eighteen

1/3-octave bands from 100 Hz to 5000 Hz, and it is called the diffraction index difference, $\text{DI}_{\Delta\text{DI}}$:

$$\text{DI}_{\Delta\text{DI}} = -10\log_{10}\left(\frac{\sum\limits_{i=1}^{18} 10^{L_i/10}10^{-\Delta_{\text{DI}i}/10}}{\sum\limits_{i=1}^{18} 10^{L_i/10}}\right) \tag{5.139}$$

where L_i is the A-weighted sound pressure level of the noise in the ith 1/3-octave band and $\Delta_{\text{DI}i}$ is the increase in barrier attenuation compared to the case with no treatment of the barrier top.

In the following sections, we will restrict our discussion to straight barriers with no top or face treatment. Barriers that have diffraction shapes mounted on their top are best analysed using numerical methods.

5.3.5.1 Diffraction at the Edge of a Thin Sheet

In general, sound diffracted around a barrier may take a number of paths. To make the problem tractable, the idealised case of sound diffraction at the straight-edge of a thin semi-infinite, acoustically opaque, plane barrier will first be considered, and the basis for all subsequent calculations will be established. In some practical applications of barriers, such as those along highways, the barrier face is tilted slightly so that reflected waves are directed upwards to minimise reflections between barriers on opposite sides of the highway, which can increase noise levels on the receiver side of each barrier. The analysis presented in this section also applies to this case.

The discussion will make reference to Figure 5.18, in which a point source model is illustrated. The model may also be used for a line source that lies parallel to the top edge of the barrier. The following analysis makes use of a 3-D coordinate system, with the x-axis normal to the barrier face in the horizontal plane, the y-axis parallel to the barrier edge in the horizontal plane and the z-axis parallel to the barrier face in the vertical plane, with coordinates, ($x = 0$, $z = z_{bg}$) on the ground at the base of the barrier. The $y = 0$ location can be chosen to be any point along the x-axis, the $x = x_b = 0$ location is the face of the barrier nearest the source and the $z = 0$ location is arbitrary. The coordinates of the two ends of the barrier and the source and receiver are specified relative to the ($x = 0$, $y = 0$, $z = 0$) location. The source is located at (x_S, y_S, z_S) (where x_S must be negative). Similarly, the receiver location is at (x_R, y_R, z_R) (where x_R must be positive). The location where the direct ray passes over the barrier is at x_b, y_b, z_b, where $x_b = 0$ and y_b is found by using similar triangles associated with the view looking down on top of the barrier and including the height difference between the source and barrier in the distance

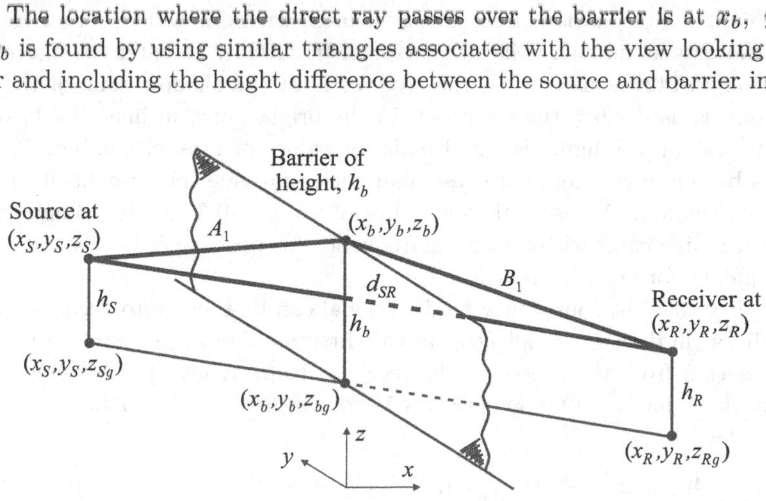

FIGURE 5.18 Geometry for determining the Fresnel number for the sound propagation path over a barrier, showing ground heights, z_{Sg}, z_{bg} and z_{Rg} beneath the source, barrier and receiver respectively.

calculations. The result is:

$$y_b = \frac{y_R\sqrt{(x_S - x_b)^2 + (z_S - z_b)^2} + y_S\sqrt{(x_R - x_b)^2 + (z_R - z_b)^2}}{\sqrt{(x_S - x_b)^2 + (z_S - z_b)^2} + \sqrt{(x_R - x_b)^2 + (z_R - z_b)^2}} \tag{5.140}$$

If the barrier top is a constant height above flat ground, then $z_b = h_b + z_{bg}$, where h_b is the height of the barrier above the ground under the barrier. However, if the z- coordinates of the two ends (1 and 2) of the barrier are different ($z_{b1} \neq z_{b2}$), iteration is required to find y_b, as in this case, z_b depends on y_b and vice versa. However, the iteration converges quickly if the initial value of y_b is chosen to be halfway between y_{b1} and y_{b2}. As z_b is a function of the y_b location along the barrier top, proportionality can be used to derive the relationship to find z_b:

$$z_b = \begin{cases} z_{b1} + \dfrac{(z_{b2} - z_{b1})(y_b - y_{b1})}{(y_{b2} - y_{b1})}; & z_{b2} > z_{b1} \\[2ex] z_{b2} + \dfrac{(z_{b1} - z_{b2})(y_{b2} - y_b)}{(y_{b2} - y_{b1})}; & z_{b1} > z_{b2} \\[2ex] h_b + z_{bg} = z_{b1} = z_{b2}; & z_{b1} = z_{b2} \end{cases} \tag{5.141a,b,c}$$

The lengths, A_1 and B_1, are introduced as two segments of the shortest path over the barrier from the source to the receiver. The length, d_{SR}, is the straight-line distance from the source to the receiver. In terms of these path lengths, the Fresnel number, N_1, is defined by:

$$N_1 = \pm(2/\lambda)(A_1 + B_1 - d_{SR}) \tag{5.142}$$

where λ is the wavelength of the centre frequency of the narrow band of noise considered; for example, a 1/3-octave or octave band of noise. The lengths, d_{SR}, A_1 and B_1 shown in the figure, are calculated using:

$$\begin{aligned} d_{SR} &= [(x_S - x_R)^2 + (y_S - y_R)^2 + (z_S - z_R)^2]^{1/2} \\ A_1 &= [(x_S - x_b)^2 + (y_S - y_b)^2 + (z_S - z_b)^2]^{1/2} \\ B_1 &= [(x_R - x_b)^2 + (y_R - y_b)^2 + (z_R - z_b)^2]^{1/2} \end{aligned} \tag{5.143a,b,c}$$

In Figure 5.19, an attenuation factor, Δ_b, associated with diffraction at an edge is plotted as a function of the Fresnel number (Maekawa, 1968, 1977, 1985). To enter the figure, the positive sign of Equation (5.142) is used when the receiver is in the shadow zone of the barrier, and the negative sign is used when the receiver is in the bright zone, in line-of-sight of the source. The horizontal scale in the figure is logarithmic for values of Fresnel number, N, greater than one, but it has been adjusted for values less than one to provide the straight-line representation shown. The attenuation, Δ_b, is 0 dB for values of $N \leq -0.3$. As frequency, $f \propto |N|$, the barrier attenuation increases with increasing frequency for positive N values and decreases with increasing frequency for negative N values.

The possibility that the source may be directional can be taken into account by introducing directivity indices (in dB), $D_{\theta B}$ and $D_{\theta R}$, in the direction from the source to the barrier edge and in the direction from the source to the receiver, respectively. The attenuation, $A_{b,i}$, of a single sound path, i, due to diffraction over the barrier is given by the following equation (Kurze and Beranek, 1988):

$$A_{b,i} = \Delta_{b,i} + 20\log_{10}[(A_i + B_i)/d_{SRi}] + D_{\theta R} - D_{\theta B} \quad \text{(dB)} \tag{5.144}$$

As an alternative to using Figure 5.19, Kurze and Anderson (1971) proposed the expression:

$$\Delta_{b,i} = 5 + 20\log_{10}\frac{\sqrt{2\pi N_i}}{\tanh\sqrt{2\pi N_i}} \quad \text{(dB)} \tag{5.145}$$

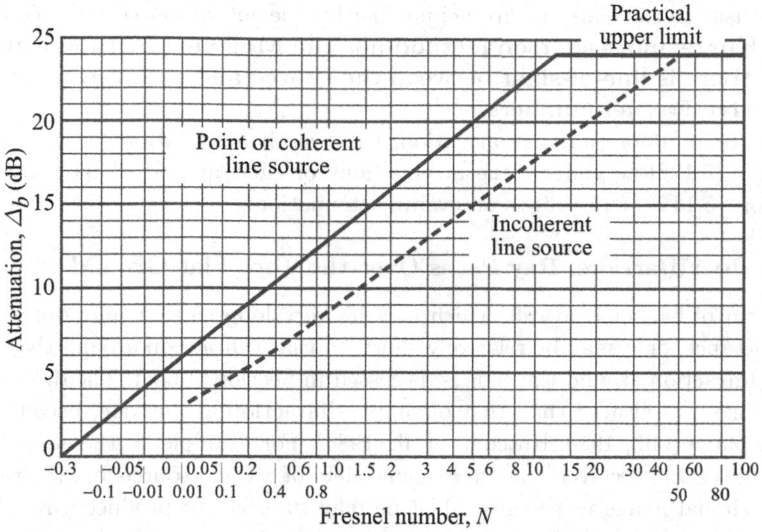

FIGURE 5.19 Sound attenuation by a semi-infinite screen in free space. If the receiver has direct line-of-sight to the source, then N is set negative.

where N_i is the Fresnel number for path, i over the barrier. Equation (5.145) is a very good approximation to the point source curve in Figure 5.19, for $N_i > 0.5$. Below $N_i = 0.5$, the amount by which Maekawa's curve exceeds the Kurze and Anderson formula gradually increases to a maximum of 1.5 dB at $N_i = 0.1$, and then gradually decreases again for smaller N_i.

A correction to Equation (5.145) was proposed by Menounou (2001) to make it more accurate for locations of the source or receiver close to the barrier or for the receiver close to the boundary of the bright and shadow zones. The more accurate equation for any particular path, i, over the barrier is:

$$A_{b,i} = \mathrm{IL}_s + \mathrm{IL}_b + \mathrm{IL}_{sb} + \mathrm{IL}_{sp} + D_{\theta R} - D_{\theta B} \tag{5.146}$$

where

$$\mathrm{IL}_s = 20 \log_{10} \frac{\sqrt{2\pi N_i}}{\tanh\sqrt{2\pi N_i}} - 1 \tag{5.147}$$

$$\mathrm{IL}_{sp} = \begin{cases} 3 \text{ dB} & \text{for plane waves} \\ 10 \log_{10} \left(1 + (A_1 + B_1)/d_{SR}\right); & \text{for coherent line source} \\ 10 \log_{10} \left((A_1 + B_1)^2/d_{SR}^2 + (A_1 + B_1)/d_{SR}\right); & \text{for point source} \end{cases} \tag{5.148}$$

$$\mathrm{IL}_b = 20 \log_{10} \left[1 + \tanh\left(0.6 \log_{10} \frac{N_2}{N_i}\right)\right] \tag{5.149}$$

$$\mathrm{IL}_{sb} = \left(6 \tanh\sqrt{N_2} - 2 - \mathrm{IL}_b\right)\left(1 - \tanh\sqrt{10 N_i}\right) \tag{5.150}$$

The term represented by Equation (5.150) should only be calculated when N_i is very small and positive. The quantity, N_2, is the Fresnel number calculated for a ray travelling from the image source to the receiver where the image source is generated by reflection from the barrier (not the ground). Thus, the image source will be on the same side of the barrier as the receiver, it will be at the same height as the source and it will be at the same distance from the barrier as the source. The distance, d_{SR}, used in Equation (5.142) to calculate the Fresnel number, N_2, is the straight-line distance between the image source on the receiver side of the barrier (resulting from reflection from the barrier face) and receiver. The distance $(A_1 + B_1)$ in Equation (5.148) is the same as used to calculate the Fresnel number for the actual source and receiver. Note that **neither the Kurze and Anderson method nor the Menonou method can be used for cases where there is line-of-sight between the source and receiver, corresponding to zero or negative Fresnel numbers.**

Note that for outdoor sound propagation, the overall attenuation must be calculated by using Equation (5.178) to convert the attenuations of each individual path calculated using either Equation (5.144) or (5.146) to an overall attenuation.

5.3.5.2 Outdoor Barriers, Ray Paths Over the Top, Flat Ground

A common form of barrier is a wall, which may be very long so that only diffraction over the top is of importance, or it may be relatively short, so that diffraction around the ends is also of importance. Diffraction around the ends is discussed in Section 5.3.5.4. The diffraction problem is generally more complicated than that of simple diffraction at an edge, because of reflection in the ground plane (Gill, 1980; Hutchins et al., 1984). For example, in the absence of a barrier between the source and receiver, the direct sound field of a tonal source and the reflected sound field from its virtual image in the ground plane may interfere to produce a relative minimum in the sound field at the position of the receiver. The introduction of a barrier may effectively prevent such interference, with the result that the placement of a barrier may result in a net gain in level at the point of observation. However, introduction of a barrier introduces four new sound ray paths over the top of the barrier (see Figure 5.20), which can also interfere on arrival at the receiver. To make the problem tractable, it will be assumed here that rays combining at the receiver consist of 1/3-octave or octave bands of noise and the paths that they travel through the atmosphere are sufficiently different so that their contributions may be combined incoherently at the receiver. For the purpose of evaluating the expected outdoor barrier attenuation for a particular frequency band, the centre frequency of the band will be used in Figure 5.19 and Equations (5.144) to (5.150).

Diffraction over the top of a very long wall (from now on referred to as a barrier), as illustrated in Figure 5.20(a) for flat ground, will be considered first. Referring to the figure, the four possible diffraction paths over the top of the barrier are SBR, SP_SBR, SBP_RR and SP_SBP_RR. As indicated in the figure, placement of an image source and receiver in the diagram is helpful in determining the path lengths that involve one or two ground reflections. These images are placed as far beneath the ground plane as the source and receiver are above it.

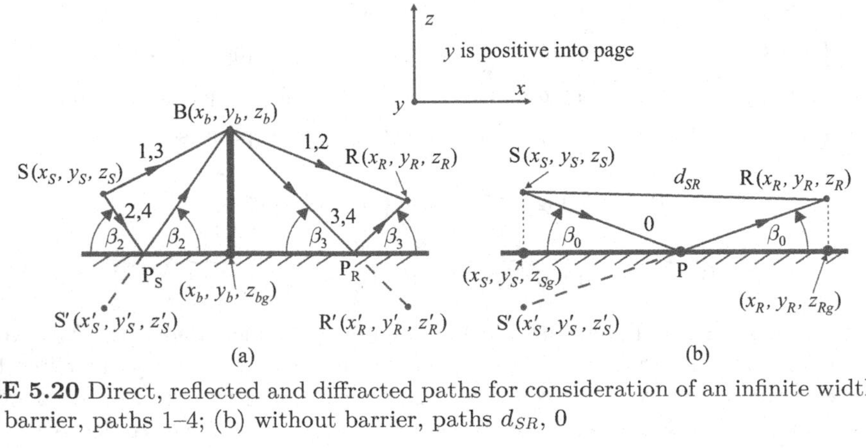

FIGURE 5.20 Direct, reflected and diffracted paths for consideration of an infinite width barrier: (a) with barrier, paths 1–4; (b) without barrier, paths d_{SR}, 0

To minimise confusion in the analyses to follow, the following numbers are used to identify the various paths from source to receiver around and over an intervening barrier, as well as the angles made with the ground for the ground-reflected paths.

0. Ground-reflected path in the absence of the barrier.
1. Direct path from the source to the receiver (over the top of the barrier in the presence of a barrier).
2. Ground-reflected path from the source to the barrier top on the source side of the barrier and direct path from the barrier top to the receiver on the receiver side.
3. Direct path from the source to the barrier top on the source side of the barrier and ground-reflected path from the barrier top to the receiver on the receiver side.
4. Source - receiver path with a ground reflection on both sides of the barrier.
5. Direct path from the source to the receiver around the right hand end of the barrier, when looking at the barrier from the source (see Section 5.3.5.4).
6. Direct path from the source to the receiver around the left hand end of the barrier.
7. Ground-reflected path from the source to the receiver around the right hand end of the barrier.
8. Ground-reflected path from the source to the receiver around the left hand end of the barrier.

When paths involving one or more ground reflections are considered, the straight-line distance, d_{SR}, used in Equation (5.142) is the distance between the effective source and effective receiver. The coordinate system used in the discussion in the following sections was explained in Section 5.3.5.1 and Figure 5.20. For a wave reflected from the ground on the source side of the barrier, the effective source location is the image source location, which is as far below the ground as the source is above it, that is $(x'_S,\ y'_S,\ z'_S)$, where $x'_S = x_S$, $y'_S = y_S$, $z'_S = 2z_{Sg} - z_S$ and z_{Sg} is the coordinate of the ground beneath the source. For a wave reflected on the receiver side of the barrier, the effective receiver location is the image receiver location at $(x'_R,\ y'_R,\ z'_R)$, where $x'_R = x_R$, $y'_R = y_R$ and $z'_R = 2z_{Rg} - z_R$. For flat ground, the reflected ray paths pass over the top of the barrier at the same location as the direct ray path at $(x_b,\ y_b,\ z_b)$ and the ground vertical z-coordinates beneath the source, barrier and receiver are set equal, so that $z_{Sg} = z_{Rg} = z_{bg}$.

For the analysis outlined in this and future subsections, the Cartesian coordinate system is defined such that the y-axis is parallel to the barrier face and the x-axis is perpendicular to the barrier face. When undertaking an analysis for flat ground, the x, y and z-coordinates of the source, receiver and the tops of the two ends of the barrier, as well as the z-coordinate for the ground are necessary inputs. For sloping ground, additional required inputs are the z-coordinates of the ground beneath the source, barrier and receiver. For double barriers, additional inputs are the x and z-coordinates of the two ends of the second diffraction edge, with the additional input for double barriers, the z-coordinate of the ground under the second diffraction edge.

The zero location for the x-coordinate is the barrier face nearest the source, the zero locations for the y and z-coordinates can be chosen to be any convenient location. The x-coordinate for the source must be negative and the x-coordinate for the receiver must be positive.

For the ray path over the top of the barrier, the quantities, A_1 and B_1, for use in Equation (5.142) are calculated using Equation (5.143).

For the ground-reflected ray on the side of the source, the quantity, A_1, for use in Equation (5.142) is:

$$A_1 = \left[(x_S - x_b)^2 + (y_S - y_b)^2 + (z'_S - z_b)^2\right]^{1/2} \tag{5.151}$$

and for the ground-reflected ray on the side of the receiver the quantity, B_1, for use in Equation (5.142) is:

$$B_1 = [(x_R - x_b)^2 + (y_R - y_b)^2 + (z'_R - z_b)^2]^{1/2} \tag{5.152}$$

where x_b is the barrier x-coordinate, y_b is defined by Equation (5.140) and z_b is defined by Equation (5.141). The coordinate locations are illustrated in Figure 5.18.

In the absence of a barrier, there are only two paths between the source and receiver, as shown in Figure 5.20(b). Thus, if the ground plane is hard and essentially totally reflective, and octave bands of noise are considered so that contributions from the various paths may be added incoherently, the noise reduction due to the barrier is calculated as follows.

1. Making use of the image source and image receiver, respectively, as indicated in Figure 5.20(a), the expected reduction in level is determined using Equation (5.144) or (5.145) for each of the four paths.
2. The process is repeated for the two paths shown in Figure 5.20(b), again assuming total incoherent reflection. It is also assumed that the total power radiated by the source is not affected by the barrier.
3. The noise reduction due to the barrier (overall attenuation, A_b) is determined by combining the results of steps 1 and 2 above using Equation (1.105).
4. If the ground is not acoustically hard but is somewhat absorptive, as is generally the case, then the dB attenuation due to reflection must be added arithmetically to the dB attenuation due to the barrier for each path that includes a reflection. One of the paths over the top of the barrier includes two reflections, so for this path, two reflection losses (in dB) must be added arithmetically to the barrier attenuation.

The angle, β ($= 90-\theta$, see Figure 5.1), that a ground-reflected wave makes with the horizontal is important for calculating the loss due to ground absorption. In the absence of the barrier, this angle will be defined as β_0, as shown in Figure 5.20(b). In the presence of a barrier, the angle made on the source side for a ground-reflected wave (travelling over the top of the barrier) will be denoted β_2 and the angle made on the receiver side will be denoted β_3, as shown in Figure 5.20(a). The angle corresponding to the ground reflection for a wave travelling around the right-hand side of the barrier will be denoted β_7 (see Figure 5.23) and the angle corresponding to the ground reflection for a wave travelling around the left-hand side of the barrier will be denoted β_8 (similar figure to Figure 5.23, but on the other side of the barrier).

Referring to Figures 5.18 and 5.20 for flat ground, $z_{Sg} = z_{Rg} = z_{bg}$. Thus:

$$\beta_0 = \arctan[(z_S + z_R - 2z_{Sg})/d_{SR}] \tag{5.153}$$

$$\beta_2 = \arctan\left[\frac{(z_b + z_S - 2z_{bg})}{\sqrt{(x_S - x_b)^2 + (y_S - y_b)^2}}\right] \quad \text{and} \quad \beta_3 = \arctan\left[\frac{(z_b + z_R - 2z_{bg})}{\sqrt{(x_R - x_b)^2 + (y_R - y_b)^2}}\right] \tag{5.154}$$

In the absence of the barrier, the direct path length, d_{SR}, from source to receiver is given by Equation (5.143a) and the path length, d_{SPR}, of the ground reflected ray (see Figure 5.20(b)) is:

$$d_{\text{SPR}} = [(x_S - x'_R)^2 + (y_S - y'_R)^2 + (z_S - z'_R)^2]^{1/2} \tag{5.155}$$

5.3.5.3 Outdoor Barriers, Ray Paths Over the Top, Sloping Ground

Here, sloping ground in the direction between the source and receiver only is considered. Ground slopes in other directions are not considered. When the ground is at different heights beneath the source, barrier and receiver, a first approximation is to assume a uniform ground slope between the source and barrier, and another uniform ground slope between the barrier and receiver.

Direct ray path. Equation (5.143) for flat ground may also be used to calculate the path lengths, A_1, B_1 and d_{SR}, for the direct ray over the top of the barrier. For a barrier with a sloping top edge, where the heights of the top of each end are different, the coordinates of the point (x_b, y_b, z_b), where the source/receiver ray path passes over the top of the barrier, can be calculated for the direct ray path using the same equations as used for flat ground (Equations (5.140) and (5.141)).

Reflected ray paths – source side of barrier. For sloping ground, the x-coordinate of the image source is not the same as that of the source, and the same applies to the image receiver, as illustrated in Figure 5.21. Thus, the y_b value where the diffracted ray is incident on the barrier top will be slightly different for reflected ray paths. However, Equation (5.140) can still be used, but for a ray reflected on the source side, x_S must be replaced with x'_S in the equation and for a ray reflected on the receiver side, x_R must be replaced with x'_R. For a ray reflected on both sides of the barrier, both replacements must be made.

Equation (5.143), with y_b calculated as described above, can be used for the calculation of reflected ray path lengths, but for a ground reflection on the source side, x_S, y_S and z_S are replaced with x'_S, y'_S and z'_S in the equation for A_1. The calculation of the coordinates of the image source for sloping ground is described below.

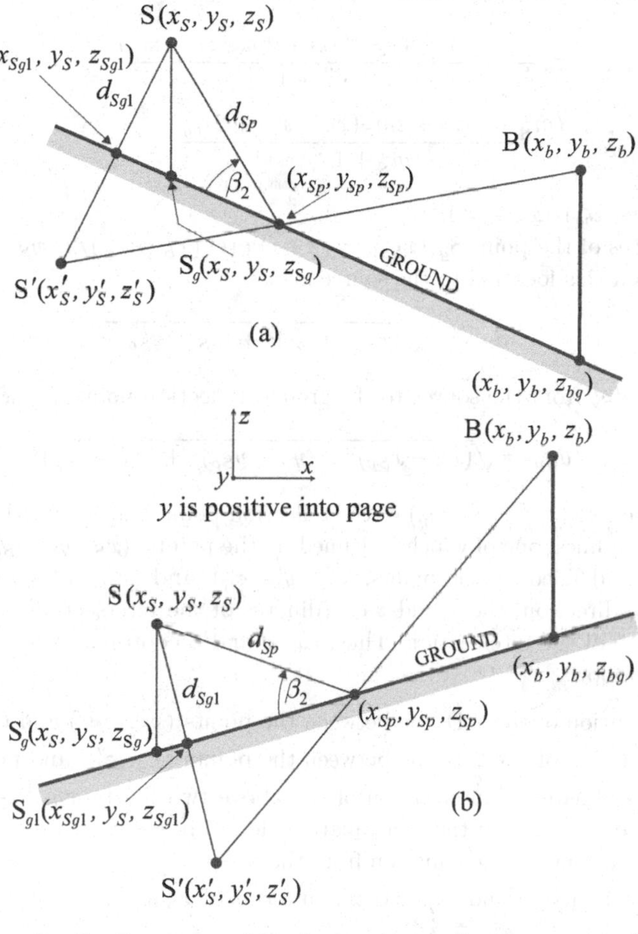

FIGURE 5.21 Ray paths for the source side of a barrier on sloping ground: (a) negative slope; (b) positive slope. The x-coordinate is horizontal, the y-coordinate is perpendicular to the page and the z-coordinate is vertical. S_g is the ground location directly below the source and S_{g1} is the location where the normal line from the source strikes the ground. For flat ground, $S_g = S_{g1}$.

In the figure, it can be seen that for uniformly sloping ground, the true image source location is along a line passing through $S_{g1}(x_{Sg1},\ y_S,\ z_{Sg1})$. The image source location is found by implementing the following steps.

1. Find the equation of the line ($z = m_S x + b$) representing the ground surface from the source, (x_S, y_S, z_{Sg}) to the barrier, (x_b, y_b, z_{bg}). This will be a line in the $x - z$ plane.
2. Find the equation describing the normal to the ground line that passes through (x_S, y_S, z_S). This will also be a line in the $x - z$ plane.
3. Find the coordinates of intersection, (x_{Sg1}, y_S, z_{Sg1}) by inserting x_{Sg1} for x and z_{Sg1} for z in the two equations found in steps one and two and solving for x_{Sg1} by subtracting one equation from the other.
4. Find the coordinates of the image source using $x'_S = 2x_{Sg1} - x_S$, $y'_S = y_S$ and $z'_S = 2z_{Sg1} - z_S$

The coordinates of the image source are therefore:

$$x'_S = x_b + \frac{(1 - m_S^2)(x_S - x_b) + 2m_S(z_S - z_{bg})}{m_S^2 + 1}; \qquad y'_S = y_S$$
$$z'_S = \frac{(m_S^2 - 1)z_S + 2m_S(x_S - x_b) + 2z_{bg}}{m_S^2 + 1} \tag{5.156}$$

where $m_S = (z_{Sg} - z_{bg})/(x_S - x_b)$.

The coordinates of the point $S_{g1}(x_{Sg1},\ y_S\ z_{Sg1})$ are: $(x_S + x'_S)/2$, y_S, $(z_S + z'_S)/2$ and the distance, d_{Sg1} from this location to the source is:

$$d_{Sg1} = \sqrt{(x_S - x_{Sg1})^2 + (z_S - z_{Sg1})^2} \tag{5.157}$$

and the distance, d_{Sp} from the source to the ground reflection point, (x_{Sp}, y_{Sp}, z_{Sp}), is:

$$d_{Sp} = \sqrt{(x_S - x_{Sp})^2 + (y_S - y_{Sp})^2 + (z_S - z_{Sp})^2} \tag{5.158}$$

where the coordinates, (x_{Sp}, y_S, z_{Sp}) of the reflection point are evaluated by finding the intersection point of two lines, one of which is defined by the points, (x_S, y_S, z_{Sg}), and (x_b, y_b, z_{bg}), with the other line defined by the points, (x'_S, y'_S, z'_S), and (x_b, y_b, z_b). As the ground does not slope in the y-direction, the x and z coordinates of the intersection point are independent of the y-coordinate of the intersection. Thus, the x and z coordinates of the intersection point may be found as follows.

1. Find the equation of the 2-D line between the points (x_S, z_{Sg}) and (x_b, z_{bg})
2. Find the equation of the 2-D line between the points (x'_S, z'_S) and (x_b, z_b).
3. Find the coordinates of intersection of the above two lines, (x_{Sp}, z_{Sp}) by inserting x_{Sp} for x and z_{Sp} for z in the two equations found in steps one and two and solving for x_{Sp} by subtracting one equation from the other.
4. Use proportionality to find y_{Sp} and z_{Sp} in terms of x_{Sp}.

The coordinates of the reflection point on the source side of the barrier are therefore:

$$x_{Sp} = x_b + \frac{(z_b - z_{bg})(x'_S - x_b)(x_{Sg1} - x_b)}{(x'_S - x_b)(z_{Sg1} - z_{bg}) + (z_b - z'_S)(x_{Sg1} - x_b)}$$
$$y_{Sp} = y_b + \frac{(x_{Sp} - x_b)(y'_S - y_b)}{(x'_S - x_b)}; \qquad z_{Sp} = z_b + \frac{(x_{Sp} - x_b)(z'_S - z_b)}{(x'_S - x_b)} \tag{5.159}$$

where the y_{Sp} and z_{Sp} coordinates are obtained from x_{Sp} using proportionality. The reflection angle, β_2 is:

$$\beta_2 = \sin^{-1}(d_{Sg1}/d_{Sp}) \tag{5.160}$$

Reflected ray paths – receiver side of barrier. A similar approach to that just discussed may be used to determine the ray path lengths on the receiver side of the barrier (see Figure 5.22). Equation (5.143) still applies, but for a ground reflection on the receiver side, x_R, y_R and z_R are replaced with x'_R, y'_R and z'_R in the equation for B_1 and in the equation for d_{SR}. For the ray reflected on both sides of the barrier, the distance, A_1, corresponding to reflection on the source side and the distance, B_1, corresponding to reflection on the receiver side are both used. In this case, the distance, d_{SR} in Equation (5.143) is found by replacing x_S, y_S, z_S, x_R, y_R and z_R, with x'_S, y'_S, z'_S, x'_R, y'_R and z'_R, respectively. In addition, the value of y_b to be used in Equation (5.140) can still be calculated using Equation (5.143), but x_R must be replaced with x'_R.

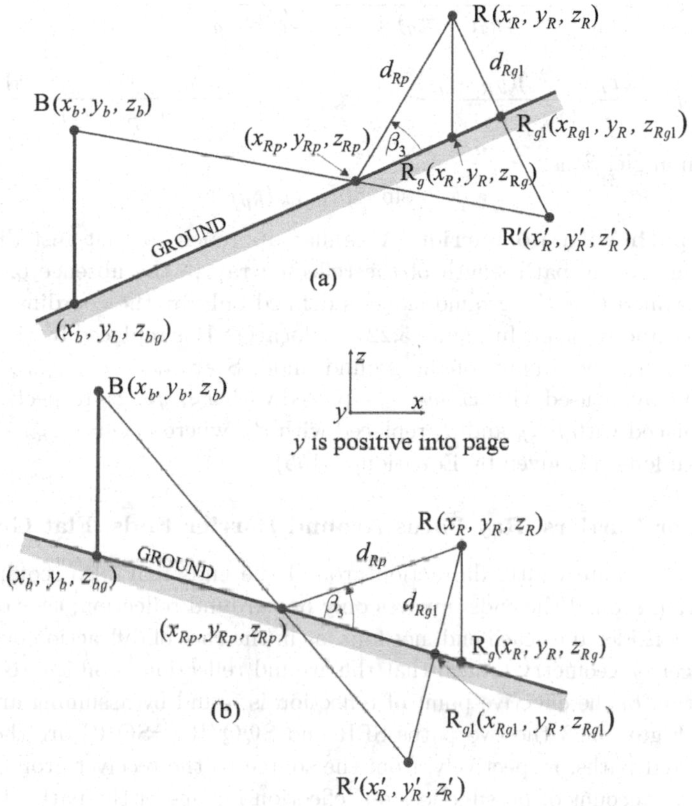

FIGURE 5.22 Ray paths for the receiver side of a barrier on sloping ground: (a) positive slope; (b) negative slope. The x-coordinate is horizontal, the y-coordinate is perpendicular to the page and the z-coordinate is vertical. R_g is the ground location directly below the receiver and R_{g1} is the location where the normal line from the receiver strikes the ground. For flat ground, $\mathrm{R}_g = \mathrm{R}_{g1}$.

The calculation of the image source location has been discussed in earlier paragraphs and the image receiver location is found in a similar way, resulting in the following equations.

$$x'_R = x_b + \frac{(1-m_R^2)(x_R - x_b) + 2m_R(z_R - z_{bg})}{m_R^2+1}; \qquad y'_R = y_R$$
$$z'_R = \frac{(m_R^2-1)z_R + 2m_R(x_R - x_b) + 2z_{bg}}{m_R^2+1} \tag{5.161}$$

where $m_R = (z_{Rg} - z_{bg})/(x_R - x_b)$.

The coordinates of the point $R_{g1}(x_{Rg1},\ y_R,\ z_{Rg1})$ are: $(x_R + x'_R)/2,\ y_R,\ (z_R + z'_R)/2$ and the distance, d_{Rg} from this location to the receiver is:

$$d_{Rg1} = \sqrt{(x_R - x_{Rg1})^2 + (z_R - z_{Rg1})^2} \tag{5.162}$$

and the distance, d_{Rp} from the receiver to the ground reflection point, $(x_{Rp},\ y_{Rp}\ z_{Rp})$, is:

$$d_{Rp} = \sqrt{(x_R - x_{Rp})^2 + (y_R - y_{Rp})^2 + (z_R - z_{Rp})^2} \tag{5.163}$$

where the reflection point is calculated in a similar way to the reflection point on the source side of the barrier. Thus:

$$\begin{aligned} x_{Rp} &= x_b + \frac{(z_b - z_{bg})(x'_R - x_b)(x_{Rg1} - x_b)}{(x'_R - x_b)(z_{Rg1} - z_{bg}) + (z_b - z'_R)(x_{Rg1} - x_b)} \\ y_{Rp} &= y_b + \frac{(x_{Rp} - x_b)(y'_R - y_b)}{(x'_R - x_b)}; \qquad z_{Rp} = z_b + \frac{(x_{Rp} - x_b)(z'_R - z_b)}{(x'_R - x_b)} \end{aligned} \tag{5.164}$$

and the reflection angle, β_3 is:

$$\beta_3 = \sin^{-1}(d_{Rg1}/d_{Rp}) \tag{5.165}$$

Reflected ray path with no barrier. A similar approach to that just discussed may be used to determine the ray path length of the reflected ray in the absence of the barrier. For this case, it is assumed that the ground slope is defined only by the coordinates of the ground beneath the source and receiver. In Figure 5.22, the location B is replaced by the source location, $S(x_S,\ y_S,\ z_S)$ and the coordinates of the ground under S are $(x_S,\ y_S,\ z_{Sg})$. Then Equations (5.161) to (5.165) may be used with $x_b,\ y_b,\ z_b$ replaced with $x_S,\ y_S,\ z_S$ respectively, z_{bg} replaced with z_{Sg}, m_R replaced with m_{SR} and β_3 replaced with β_0, where $m_{SR} = (z_{Rg} - z_{Sg})/(x_R - x_S)$. The reflected path length is given by Equation (5.155).

5.3.5.4 Outdoor Barriers, Ray Paths Around Barrier Ends, Flat Ground

When a barrier is of finite width, diffraction around the ends may also require consideration. However, diffraction around the ends involves only one ground reflection; thus only two possible paths need to be considered at each end, not four, as in the case of diffraction over the top. If the source-barrier-receiver geometry is such that the ground reflection is on the receiver side of the barrier, the location of the effective point of reflection is found by assuming an image receiver, R′. Referring to Figure 5.23 the two paths SPR and SQQ″R (=SQR′) are the shortest direct and ground-reflected paths, respectively, from the source to the receiver around one end of the wall. Again, taking account of possible loss on reflection for one of the paths, the contributions over the two paths are determined using Figures 5.18 and 5.19 and Equation (5.144), where for the non-reflected ray, paths A_1 and B_1 are replaced by paths, A_5 = SP and B_5 = PR, with the z-coordinate at the base of the barrier equal to z_{bg}. For the ray reflected from the ground on the receiver side of the barrier, paths A_1 and B_1 are replaced by paths, A_7 = SQ and B_7 = QQ″R and for a ground reflection on the source side, paths A_1 and B_1 are replaced by paths, A_7 = SQ″Q and B_7 = QR, respectively.

Referring to Figure 5.23, for definition of the symbols, and defining h'_b as the height of point P above the ground and h''_b as the height of point Q above the ground, Equations (5.166) and (5.168) may be derived for ground reflection on the receiver side of the barrier, using 3-D similar triangles. Equation (5.167) can be derived using a similar figure to Figure 5.23, but with the ground reflection on the source side of the barrier.

$$h'_b = z_P - z_{bg} = \frac{(z_S - z_{bg})\sqrt{(x_R - x_b)^2 + (y_{b1} - y_R)^2} + (z_R - z_{bg})\sqrt{(x_S - x_b)^2 + (y_{b1} - y_S)^2}}{\sqrt{(x_S - x_b)^2 + (y_{b1} - y_S)^2} + \sqrt{(x_R - x_b)^2 + (y_{b1} - y_R)^2}} \tag{5.166}$$

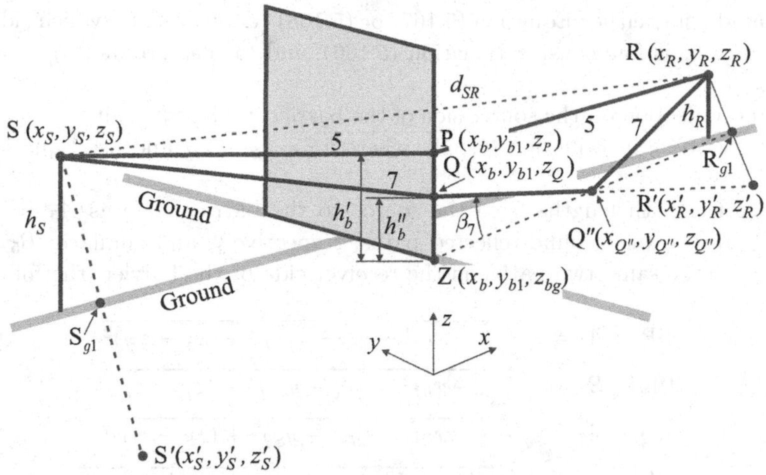

FIGURE 5.23 Geometry for diffraction around one edge of a finite width barrier on flat or sloping ground, with the ground reflection on the receiver side of the barrier. S and S′ are source and image source, respectively, R is the receiver location, P is defined by the shortest path from S to R around the barrier end, Q and Q″ are defined by the shortest reflected path from S to R around the barrier end. Point Z is where the end of the barrier meets the ground and point R_{g1} is where the normal from the receiver strikes the ground, which is directly under the receiver for flat ground.

$$h''_b = z_Q - z_{bg} = \frac{(z_R - z_{bg})\sqrt{(x'_S - x_b)^2 + (y_{b1} - y'_S)^2} + (z'_S - z_{bg})\sqrt{(x_R - x_b)^2 + (y_{b1} - y_R)^2}}{\sqrt{(x'_S - x_b)^2 + (y_{b1} - y'_S)^2} + \sqrt{(x_R - x_b)^2 + (y_{b1} - y_R)^2}} \tag{5.167}$$

$$h''_b = z_Q - z_{bg} = \frac{(z_S - z_{bg})\sqrt{(x'_R - x_b)^2 + (y_{b1} - y'_R)^2} + (z'_R - z_{bg})\sqrt{(x_S - x_b)^2 + (y_{b1} - y_S)^2}}{\sqrt{(x_S - x_b)^2 + (y_{b1} - y_S)^2} + \sqrt{(x'_R - x_b)^2 + (y_{b1} - y'_R)^2}} \tag{5.168}$$

where for flat ground, $x'_S = x_S$, $y'_S = y_S$, $z'_S = 2z_{bg} - z_S$, $x'_R = x_R$, $y'_R = y_R$, $z'_R = 2z_{bg} - z_R$. Also, z_{bg} is the z-coordinate of the ground under the barrier (often set equal to zero for convenience) and y_{b1} is the y-coordinate of the right-hand end of the barrier when observing the barrier from the source. Note that all variables in the preceding equations are coordinates, not distances, so they can have negative values.

Equation (5.167) applies if the ground reflection is on the source side of the barrier and Equation (5.168) applies if the ground reflection is on the receiver side. If the calculation of h''_b returns a negative value, then the ground reflection point is on the other side of the barrier to what was assumed in the calculation and the alternative equation must be used.

To find the path lengths, we need to find the coordinates, ($x_{Q''}$, $y_{Q''}$, $z_{Q''}$), of the ground reflection point, Q″, which is the intersection of the lines Z – R_{g1} and Q – R′. For flat ground, $z_{Q''} = z_{bg}$ and the intersection coordinates, ($x_{Q''}$, $y_{Q''}$) can be found using the procedure outlined in Section 5.3.5.5 with the ground coordinates, z_{Rg1} set equal to z_{bg} in Equations (5.175) and (5.176) to give:

$$\begin{aligned} x_{Q''} &= x_b + \frac{(z_{bg} - z_Q)(x'_R - x_b)}{(z'_R - z_Q)} = x_b + (x'_R - x_b)\left(1 - \frac{h_R}{h_R + h''_b}\right) \\ y_{Q''} &= y_{b1} + \frac{(z_{bg} - z_Q)(y'_R - y_{b1})}{(z'_R - z_Q)} = y_{b1} + (y'_R - y_{b1})\left(1 - \frac{h_R}{h_R + h''_b}\right) \end{aligned} \tag{5.169}$$

where h''_b is found using either Equation (5.167) or (5.168) (depending on which side the ground reflection occurs), h'_b is found using Equation (5.166), and for flat ground, $y'_R = y_R$ and $h_R = z_R - z_{bg} = z_{bg} - z'_R$.

For a ground reflection on the source side of the barrier, replace h_R with h_S, x_R with x_S, y_R with y_S and z_R with z_S in Equation (5.169), where $h_S = z_S - z_{bg}$ and use Equation (5.167) to calculate h''_b.

Referring to the path lengths from the source to the barrier edge as A_5 (= SP) and A_7 (= SQ) for the direct and ground-reflected paths, respectively, and similarly B_5 (= PR) and B_7 (= QQ″R) for the same two paths on the receiver side of the barrier, the following can be written:

$$\mathrm{SP} = A_5 = \sqrt{(x_S - x_b)^2 + (y_S - y_{b1})^2 + (z_S - z_P)^2} \tag{5.170}$$

$$\mathrm{PR} = B_5 = \sqrt{(x_R - x_b)^2 + (y_R - y_{b1})^2 + (z_R - z_P)^2} \tag{5.171}$$

$$\mathrm{SQ} = A_7 = \sqrt{(x_S - x_b)^2 + (y_{b1} - y_S)^2 + (z_S - z_Q)^2} \tag{5.172}$$

$$\mathrm{QQ''R} = B_7 = \sqrt{(x'_R - x_b)^2 + (y'_R - y_{b1})^2 + (z'_R - z_Q)^2} \tag{5.173}$$

The quantity, d_{SR}, is the same as it is for diffraction over the top of the barrier (see Figure 5.23). If the source/receiver geometry is such that the point of reflection for the reflected ray is on the source side of the barrier, then expressions for A_5 and B_5 remain the same, but A_7 and B_7 are calculated by interchanging the subscripts, R and S, in Equations (5.169), (5.172) and (5.173) and interchanging the R and S on the LHS of Equations (5.172) and (5.173).

Equation (5.142) may then be used to calculate the Fresnel number and thus the noise reduction corresponding to each of the two paths around the end of the barrier.

To allow calculation of the reflection loss of the ground-reflected ray using Figure 5.2, the angle, β_7 shown in Figure 5.23, which the reflected wave makes with the ground, must be determined. This angle may be calculated using:

$$\beta_7 = \arcsin\left[\frac{(z_R - z_{bg})}{\sqrt{(x_R - x_{Q''})^2 + (y_R - y_{Q''})^2 + (z_R - z_{bg})^2}}\right] \tag{5.174}$$

For a wave reflected on the source side of the barrier, z_R is replaced with z_S, x_R with x_S and y_R with y_S, with the same substitutions made in Equation (5.169) for calculating $x_{Q''}$ and $y_{Q''}$. For diffraction around end 2 (left-hand end when looking from the source to the barrier), we replace y_{b1} with y_{b2} in Equations (5.167) to (5.174).

5.3.5.5 Outdoor Barriers, Ray Paths Around Barrier Ends, Sloping Ground

For sloping ground, the equations describing the location of point P on the edge of the barrier and the height, h'_b are the same as for the flat ground case (Equation (5.166)). Although the locations of points Q and Q'' are different (see Figure 5.23), the equations describing the height, h''_b are identical to Equations (5.167) and (5.168), with different equations for the image source coordinates, (x'_S, y'_S and z'_S) and image receiver coordinates, (x'_R, y'_R and z'_R). For sloping ground, these coordinates are defined by Equations (5.156) and (5.161), respectively. Equation (5.167) applies if the ground reflection is on the source side of the barrier and Equation (5.168) applies if the ground reflection is on the receiver side. If the calculation of h''_b returns a negative value, then the ground reflection point is on the other side of the barrier to what was assumed in the calculation and the alternative equation must be used. A third possibility, which may occur if the ground slope is too great or if there is a particular combination of different slopes on each side of the barrier, is that the two calculations of h''_b both return either a positive or a negative number. In this case there is no ground reflected wave.

The path lengths for rays travelling around the edge of the barrier may be calculated using Equations (5.170) to (5.173), where Equation (5.166) is used to calculate h'_b and Equation (5.167) or (5.168) is used to calculate h_b''. The reflection angle for rays reflected on the receiver side of the barrier may be calculated using Equation (5.174).

The coordinates, $(x_{Q''}, y_{Q''}, z_{Q''})$, of the ground reflection point for rays travelling around the end of a barrier may be found using the following steps.

1. Write a 3-D equation for each of the two lines, Q – R′ and Z – R_{g1} in the form $(x-x_1)/(x_2-x_1) = (y-y_1)/(y_2-y_1) = (z-z_1)/(z_2-z_1) = \Lambda$ and $(x-x_3)/(x_4-x_3) = (y-y_3)/(y_4-y_3) = (z-z_3)/(z_4-z_3) = \Gamma$, respectively, where the subscripts 1, 2, 3 and 4 represent the coordinates at Q, R′, Z and R_{g1}, respectively.
2. Both lines pass through the point, $(x_{Q''}, y_{Q''}, z_{Q''})$ so set the two line equations (in x and z only or y and z only as x and y together yield a trivial solution) equal at that point and solve for Λ.
3. Use the above equations and set $x = x_{Q''}$, $y = y_{Q''}$, $z = z_{Q''}$ to obtain $x_{Q''} = x_1 + \Lambda(x_2 - x_1)$, $y_{Q''} = y_1 + \Lambda(y_2 - y_1)$ and $z_{Q''} = z_1 + \Lambda(z_2 - z_1)$.
4. Substitute the actual coordinates at the points, Q, R′, Z and R_{g1} to obtain the following equations.

$$x_{Q''} = x_b + \Lambda(x'_R - x_b); \quad y_{Q''} = y_{b1} + \Lambda(y'_R - y_{b1}); \quad z_{Q''} = z_Q + \Lambda(z'_R - z_Q) \tag{5.175}$$

where the coefficient, Λ is:

$$\Lambda = \frac{(x_{Rg1} - x_b)(z_{bg} - z_Q)}{(z'_R - z_Q)(x_{Rg1} - x_b) - (x'_R - x_b)(z_{Rg1} - z_{bg})} \tag{5.176}$$

where z_Q is calculated using Equation (5.167) or (5.168) and z_{bg} is the coordinate of the ground under the barrier. The coordinates of the point $(x_{Rg1}, y_{Rg1}, z_{Rg1})$ are $(x_R + x'_R)/2$, y_R, $(z_R + z'_R)/2$, respectively and the coordinates, x'_R, y'_R and z'_R are calculated using Equation (5.161).

For ground reflection on the receiver side of the barrier, the ground reflection angle, β_7, may be calculated using:

$$\beta_7 = \arcsin\left[\frac{\sqrt{(z_R - z_{Rg1})^2 + (x_R - x_{Rg1})^2}}{\sqrt{(x_R - x_{Q''})^2 + (y_R - y_{Q''})^2 + (z_R - z_{Q''})^2}}\right] \tag{5.177}$$

For ground reflection on the source side of the barrier, Equations (5.175) to (5.177) may be used, provided that z_R is replaced with z_S, x_R with x_S, y_R with y_S, x_{Rg1} with x_{Sg1}, y_{Rg1} with y_{Sg1} and z_{Rg1} with z_{Sg1}. For calculating h_b'' for this case, Equation (5.167) is used and for calculating the coordinates, x'_S, y'_S and z'_S, Equation (5.156) is used.

The same procedures and equations apply for rays travelling around side 2 (left-hand side) of the barrier, except that in all equations, y_{b1} is replaced by y_{b2}.

5.3.5.6 Combining Contributions From All Paths Around a Barrier

For a finite-length wall, eight separate paths should be considered and the results combined to determine the expected noise reduction provided by the placement of the barrier. In practice, however, not all paths will be of importance. In summary, if there are n_A paths around the barrier, then the overall noise reduction due to the barrier is calculated using Equation (1.105). That is:

$$A_b = 10\log_{10}\left[1 + 10^{-(A_{rf,w}/10)}\right] - 10\log_{10}\sum_{i=1}^{n_A} 10^{-(A_{b,i}+A_{rf,i})/10} \tag{5.178}$$

where the reflection loss, $A_{rf,i}$, due to the ground is added arithmetically to each path, i that involves a ground reflection. The subscript i refers to the ith path around the barrier and the subscript w refers to the ground-reflected path in the absence of the barrier.

For ground that is not uniform between the source and receiver, the reflection loss at the point of the ground corresponding to specular reflection is used. For plane wave reflection, A_{rf} (equal to $A_{rf,i}$ or $A_{rf,w}$ in Equation (5.178)) is given by:

$$A_{rf} = -20\log_{10}|R_p| \tag{5.179}$$

where R_p is defined in Equation (5.15) for extended reactive ground and in Equation (5.18) for locally reactive ground. If the more complex and more accurate spherical wave reflection model is used (see Sections 5.2.3 and 5.2.4):

$$A_{rf} = -20\log_{10}|Q| \tag{5.180}$$

It is interesting to note that the first term in Equation (5.178) is equivalent to the attenuation due to ground effects in the absence of the barrier, multiplied by −1. For the case of source distributions other than those considered, the simple strategy of dividing the source into a number of equivalent line or point sources, which are then each treated separately, may be used. Implicit in this approach is the assumption that the parts are incoherent, consistent with the analysis described here.

5.3.5.7 Thick Barriers

Existing buildings may sometimes serve as barriers (Tocci and Marcus, 1978). In this case, it is possible that a higher attenuation than that calculated using Equation (5.144) may be obtained due to double diffraction at the two edges of the building. This has the same effect as using two thin barriers placed a distance apart equal to the building thickness (ISO 9613-2, 1996). Double barriers are also discussed by Foss (1979). The effect of the double diffraction is to add an additional attenuation, ΔC, to the noise reduction achieved using a thin barrier (Fujiwara et al., 1977a) located at the centre of the thick barrier:

$$\Delta C = K\log_{10}(2\pi b/\lambda) \tag{5.181}$$

where b is the barrier thickness, λ is the wavelength at the band centre frequency, and K is a coefficient that may be estimated using Figure 5.24.

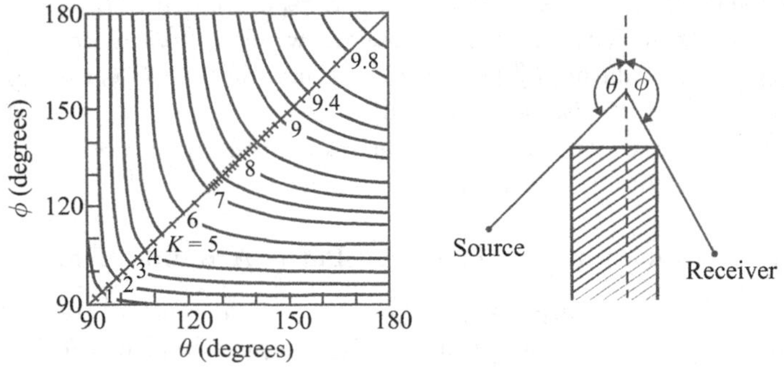

FIGURE 5.24 Finite width barrier correction factor, K (Equation (5.181)).

Note that to use Equation (5.181), the condition $b > \lambda/2$ must be satisfied. Otherwise, the barrier may be assumed to be thin.

Similar results are obtained for soil mounds, with the effective barrier width being the width of the top of the soil mound. Any trees planted on top of the soil mound are not considered to contribute significantly to the barrier attenuation and can be ignored.

Double diffraction edge barriers - paths over the top. Thick barriers can also be modelled as a double diffraction edge, as shown in Figure 5.25, where the two diffraction edges are assumed to be parallel.

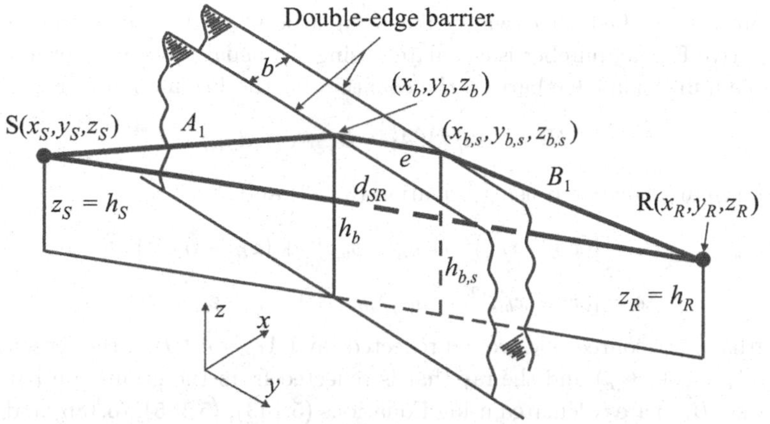

FIGURE 5.25 Geometry for double edge diffraction (single, thick barrier or two thin, parallel barriers).

The y and z-locations where the path passes over the edge nearest the source are given by Equations (5.140) and (5.141), respectively. The y-location where the ray passes over the second diffraction edge (the one nearest the receiver) can be derived from Equation (5.140), where the coordinates, x_b, y_b and z_b are replaced by $x_{b,s} = x_b + b$, $y_{b,s}$ and $z_{b,s}$, respectively, where:

$$y_{b,s} = \frac{y_R\sqrt{(x_S - x_{b,s})^2 + (z_S - z_{b,s})^2} + y_S\sqrt{(x_R - x_{b,s})^2 + (z_R - z_{b,s})^2}}{\sqrt{(x_S - x_{b,s})^2 + (z_S - z_{b,s})^2} + \sqrt{(x_R - x_{b,s})^2 + (z_R - z_{b,s})^2}} \tag{5.182}$$

If the barrier diffraction edge nearest the receiver is a constant height above the ground and the z-coordinate of the ground under the barrier is $z_{bg,s}$, then $z_{b,s} = z_{bg,s} + h_{b,s}$, where $h_{b,s}$ is the height above the ground of the diffraction edge nearest the receiver. However, if the z-coordinates of the two ends (3 and 4) of the diffraction edge nearest the receiver are different ($z_{b3} \neq z_{b4}$), iteration is required to find $y_{b,s}$ and $z_{b,s}$, as in this case, $z_{b,s}$ depends on $y_{b,s}$ and vice versa. However, the iteration converges quickly if the initial value of $y_{b,s}$ is chosen to be halfway between the y-coordinates of the two ends, $y_{b1,s}$ and $y_{b2,s}$. The general expression for $z_{b,s}$ is:

$$z_{b,s} = \begin{cases} z_{b1,s} + \dfrac{(z_{b2,s} - z_{b1,s})(y_{b,s} - y_{b1,s})}{(y_{b2,s} - y_{b1,s})}; & z_{b2,s} > z_{b1,s} \\[2ex] z_{b2,s} + \dfrac{(z_{b1,s} - z_{b2,s})(y_{b2,s} - y_{b,s})}{(y_{b2,s} - y_{b1,s})}; & z_{b1,s} > z_{b2,s} \\[2ex] h_{b,s} + z_{bg,s} = z_{b1,s} = z_{b2,s}; & z_{b1,s} = z_{b2,s} \end{cases} \tag{5.183a,b,c}$$

The $y_{b,s}$ value will be slightly different for reflected ray paths when the ground is sloping (ground under the source, barrier and receiver is at different heights), as explained in Section 5.3.5.3. However, Equations (5.140) and (5.182) for flat ground can still be used for sloping ground, except for a ray reflected on the source side, x_S must be replaced with x'_S in the two equations and, for a ray reflected on the receiver side, x_R must be replaced with x'_R. For a ray reflected on both sides of the barrier, both replacements must be made.

Important note. If there is line-of-sight from the top of the barrier nearest the source to the receiver, the double diffraction edge should be replaced with a single diffraction edge at the same location as the edge nearest the source. Similarly, if there is line-of-sight from the source to the top of the barrier nearest the receiver, the double diffraction edge should be replaced with a single diffraction edge at the same location as the edge nearest the receiver. In either of these cases, the barrier attenuation is calculated as for a single-diffraction-edge barrier. If both of the above conditions are satisfied, there will be line of sight between the source and receiver and the associated negative Fresnel number is calculated using the usual procedure for a double barrier.

For a double-diffraction-edge barrier, the equation for the Fresnel number is:

$$N_1 = \pm(2/\lambda)(A_1 + B_1 + e - d_{SR}) \tag{5.184}$$

where A_1 and d_{SR} are calculated using Equation (5.143) and

$$B_1 = [(x_R - x_{b,s})^2 + (y_R - y_{b,s})^2 + (z_R - z_{b,s})^2]^{1/2} \tag{5.185}$$

$$e = [(x_b - x_{b,s})^2 + (y_b - y_{b,s})^2 + (z_b - z_{b,s})^2]^{1/2} \tag{5.186}$$

The path lengths of the source side ground reflected ray $(A_{1,r} + e + B_1)$, the receiver side ground reflected ray $(A_1 + e + B_{1,r})$ and the ray that is reflected from the ground on both sides of the barrier $(A_{1,r} + e + B_{1,r})$ are calculated using Equations (5.143), (5.185), (5.186) and the following two equations.

$$A_{1,r} = [(x'_S - x_b)^2 + (y'_R - y_b)^2 + (z'_s - z_b)^2]^{1/2} \tag{5.187}$$

$$B_{1,r} = [(x'_R - x_{b,s})^2 + (y'_R - y_{b,s})^2 + (z'_R - z_{b,s})^2]^{1/2} \tag{5.188}$$

In the equations, the image source and receiver locations are the same as for a single diffraction edge (Equations (5.156) and (5.161)). The ground-reflection location for the ray reflected on the source side of the barrier is given by Equation (5.159), and the ground-reflection location for reflection on the receiver side can be derived from Equation (5.164) by replacing the subscript, b, with the subscript, b, s. Note that the equations for sloping ground also apply to flat ground, although for flat ground, the much simpler equations in Section 5.3.5.2 may also be used, except in Equation (5.152), x_b, y_b and z_b must be replaced with $x_{b,s}$, $y_{b,s}$ and $z_{b,s}$, respectively. The path lengths of Equations (5.157) to (5.159) on the source side of the barrier and the corresponding reflection angle, β_2, of Equation (5.160) are calculated using the same equations as for the single diffraction case, using the coordinates of the diffraction edge nearest the source. For the receiver side, the Equations (5.162) to (5.165) for the single edge barrier are used with the subscript, b, replaced with the subscript, b, s, where the s part of the subscript refers to the coordinates of the diffraction edge nearest the receiver.

Double diffraction edge barriers - non-reflected paths around the ends. The non-ground-reflected path lengths around the ends of a double-diffraction-edge barrier are calculated in a similar way to the paths for a single-diffraction-edge barrier. The z-coordinate location, z_P, where the direct (non-ground reflected) ray passes around the right-hand diffraction edge nearest the source is given by Equation (5.166). The equation for the z-coordinate location, z_{Ps}, where the direct ray passes around the diffraction edge nearest the receiver (edge s) may be derived from Equation (5.166) by replacing x_b with $x_{b,s}$, y_{b1} with $y_{b1,s}$, z_P with $z_{P,s}$ and h'_b with $h'_{b,s}$. Thus:

$$\begin{aligned} h'_{b,s} &= z_{P,s} - z_{bg,s} \\ &= \frac{(z_S - z_{bg,s})\sqrt{(x_R - x_{b,s})^2 + (y_{b1,s} - y_R)^2} + (z_R - z_{bg,s})\sqrt{(x_S - x_{b,s})^2 + (y_{b1,s} - y_S)^2}}{\sqrt{(x_S - x_{b,s})^2 + (y_{b1,s} - y_S)^2} + \sqrt{(x_R - x_{b,s})^2 + (y_{b1,s} - y_R)^2}} \end{aligned} \tag{5.189}$$

where $x_{b,s} = x_b + b$, $y_{b1,s}$ and $z_{P,s} = h'_{b,s} + z_{bg,s}$ are the x, y and z-coordinates of where the ray passes end 1 (right-hand end) of the diffraction edge nearest the receiver and $z_{bg,s}$ is the z-coordinate of the ground under the barrier diffraction edge nearest the receiver.

The location on end 1 of the barrier where the ray passes the diffraction edge nearest the source is obtained using Equation (5.166). The ray path length, $\mathrm{SPP_sR}$, for the non-reflected ray passing the right-hand end is:

$$\begin{aligned}\mathrm{SPP_sR} =& \sqrt{(x_S - x_b)^2 + (y_S - y_{b1})^2 + (z_S - z_P)^2} \\ &+ \sqrt{(x_R - x_{b,s})^2 + (y_R - y_{b1,s})^2 + (z_R - z_{P,s})^2} \\ &+ \sqrt{(x_{b,s} - x_b)^2 + (y_{b1,s} - y_{b1})^2 + (z_{P,s} - z_P)^2}\end{aligned} \tag{5.190}$$

The x-coordinate where the ray passes around end 2 (left-hand end when looking from the source to the barrier) nearest the source is x_b and the x-coordinate where the ray passes the left-hand diffraction edge nearest the receiver is $x_{b,s}$. The height, $h'_{b2,s}$, on the barrier edge where the non-ground-reflected ray from source and receiver passes the diffraction edge nearest the receiver at end 2 of the barrier is calculated by substituting y_{b2} for y_{b1}, $y_{b2,s}$ for $y_{b1,s}$, $h'_{b2,s}$ for $h'_{b,s}$, z_{P2} for z_P and $z_{P2,s}$ for $z_{P,s}$ in Equations (5.189) and (5.190), where the subscript 2 refers to the left hand end of the barrier.

Double diffraction edge barriers - ground-reflected paths around the ends. The path lengths for ground-reflected rays travelling around the ends of a double-diffraction-edge barrier are calculated in a similar way to the path lengths for a single-diffraction-edge barrier, but as can be seen from Figure 5.26, the situation is rather more complicated.

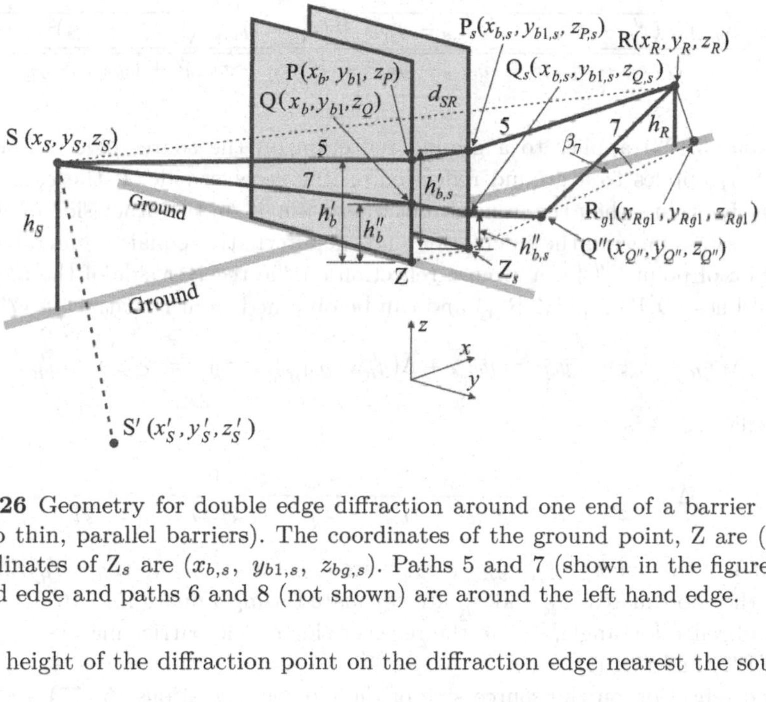

FIGURE 5.26 Geometry for double edge diffraction around one end of a barrier (single, thick barrier or two thin, parallel barriers). The coordinates of the ground point, Z are (x_b, y_{b1}, z_{bg}) and the coordinates of Z_s are ($x_{b,s}$, $y_{b1,s}$, $z_{bg,s}$). Paths 5 and 7 (shown in the figure) are around the right hand edge and paths 6 and 8 (not shown) are around the left hand edge.

First, the height of the diffraction point on the diffraction edge nearest the source is calculated.

$$\begin{aligned}h''_b &= z_Q - z_{bg} \\ &= \frac{(z_R - z_{bg})\sqrt{(x'_S - x_b)^2 + (y_{b1} - y'_S)^2} + (z'_S - z_{bg})\sqrt{(x_R - x_b)^2 + (y_{b1} - y_R)^2}}{\sqrt{(x'_S - x_b)^2 + (y_{b1} - y'_S)^2} + \sqrt{(x_R - x_b)^2 + (y_{b1} - y_R)^2}}\end{aligned} \tag{5.191}$$

$$h_b'' = z_Q - z_{bg}$$
$$= \frac{(z_S - z_{bg})\sqrt{(x_R' - x_b)^2 + (y_{b1} - y_R')^2} + (z_R' - z_{bg})\sqrt{(x_S - x_b)^2 + (y_{b1} - y_S)^2}}{\sqrt{(x_S - x_b)^2 + (y_{b1} - y_S)^2} + \sqrt{(x_R' - x_b)^2 + (y_{b1} - y_R')^2}} \tag{5.192}$$

The coordinates, x_S', y_S' and z_S' are calculated using Equation (5.156) and the coordinates, x_R', y_R' and z_R' are calculated using Equation (5.161). Equation (5.191) applies if the ground reflection is on the source side of the diffraction edge nearest the source and Equation (5.192) applies if the ground reflection is on the receiver side. If the calculation of h_b'' returns a negative value, then the ground reflection point is on the other side of the diffraction edge to what was assumed in the calculation and the alternative equation must be used.

Next, the height of the diffraction point around the diffraction edge nearest the receiver is calculated. This is done by replacing x_b, y_{b1} and z_{bg} in Equations (5.191) and (5.192) with $x_{b,s}$, $y_{b1,s}$ and $z_{bg,s}$, respectively, to give the results, $h_{b,s}''$ and $z_{Q,s}$. Thus:

$$h_{b,s}'' = z_{Q,s} - z_{bg,s}$$
$$= \frac{(z_R - z_{bg,s})\sqrt{(x_S' - x_{b,s})^2 + (y_{b1,s} - y_S')^2} + (z_S' - z_{bg,s})\sqrt{(x_R - x_{b,s})^2 + (y_{b1,s} - y_R)^2}}{\sqrt{(x_S' - x_{b,s})^2 + (y_{b1,s} - y_S')^2} + \sqrt{(x_R - x_{b,s})^2 + (y_{b1,s} - y_R)^2}} \tag{5.193}$$

$$h_{b,s}'' = z_{Q,s} - z_{bg,s}$$
$$= \frac{(z_S - z_{bg,s})\sqrt{(x_R' - x_{b,s})^2 + (y_{b1,s} - y_R')^2} + (z_R' - z_{bg,s})\sqrt{(x_S - x_{b,s})^2 + (y_{b1,s} - y_S)^2}}{\sqrt{(x_S - x_{b,s})^2 + (y_{b1,s} - y_S)^2} + \sqrt{(x_R' - x_{b,s})^2 + (y_{b1,s} - y_R')^2}} \tag{5.194}$$

where Equation (5.193) applies to a ground reflection on the source side of the barrier and Equation (5.194) applies to a ground reflection on the receiver side. If the calculation of $h_{b,s}''$ returns a negative value, then the ground reflection point is on the other side of the diffraction edge to what was assumed in the calculation and the alternative equation must be used.

The location of point Q″ for a ground reflection on the receiver side of the barrier is at the intersection of lines (Q_s'R) and (Z_sR_{g1}) and can be obtained from Equation (5.175) as:

$$x_{Q''} = x_{b,s} + \Lambda(x_R' - x_{b,s}); \quad y_{Q''} = y_{b1,s} + \Lambda(y_R' - y_{b1,s}); \quad z_{Q''} = z_{Q,s} + \Lambda(z_R' - z_{Q,s}) \tag{5.195}$$

where the coefficient, Λ is:

$$\Lambda = \frac{(x_{Rg1} - x_{b,s})(z_{bg,s} - z_{Q,s})}{(z_R' - z_{Q,s})(x_{Rg1} - x_{b,s}) - (x_R' - x_{b,s})(z_{Rg1} - z_{bg,s})} \tag{5.196}$$

The coordinates of the point $(x_{Rg1}, y_{Rg1}, z_{Rg1})$ are $(x_R + x_R')/2$, y_R, $(z_R + z_R')/2$, respectively, $y_R' = y_R$ and the coordinates, x_R' and z_R' are calculated using Equation (5.161).

The ground reflection angle, β_7, on the receiver side of the barrier may be calculated using Equations (5.177) and (5.195).

For ground reflection on the source side of the barrier, Equations (5.177) and (5.195) may also be used, provided that z_R is replaced with z_S, x_R with x_S, y_R with y_S, x_{Rg1} with x_{Sg1}, y_{Rg1} with y_{Sg1}, z_{Rg1} with z_{Sg1}, $x_{b,s}$ with x_b, $z_{bg,s}$ with z_{bg} and $z_{Q,s}$ with z_Q, where z_Q is calculated using Equation (5.191).

If the h_b'' calculation for the diffraction edge on the source side indicates that the ground reflection is on the receiver side of that edge, and the $h_{b,s}''$ calculation for the diffraction edge on the receiver side indicates that the ground reflection is on the source side of that edge, then the ground-reflection location is between the two diffraction edges. In this case, the coordinates of the ground reflection location are ($x_{Q''}$, $y_{Q''}$, $z_{Q''}$), where:

$$x''_Q = x_b + \frac{(x_{b,s} - x_b)(z_{bg} - z_{Q''})}{(z_{bg} - z_{bg,s})}; \quad y_{Q''} = y_{b1} + \frac{(y_{b1,s} - y_{b1})(x_b - x_{Q''})}{(x_b - x_{b,s})} \tag{5.197}$$

For ground reflection between the two diffraction edges, the reflection angle, β_{7i} for the incident ray from the source side is:

$$\beta_{7i} = \arcsin\left[\frac{(z_Q - z_{Q''})}{\sqrt{(x_b - x_{Q''})^2 + (y_{b1} - y_{Q''})^2 + (z_Q - z_{Q''})^2}}\right] - \beta_g \tag{5.198}$$

The reflection angle, β_{7r} for the reflected ray travelling to the receiver side is:

$$\beta_{7r} = \arcsin\left[\frac{(z_{Q,s} - z_{Q''})}{\sqrt{(x_{b,s} - x_{Q''})^2 + (y_{b1,s} - y_{Q''})^2 + (z_{Q,s} - z_{Q''})^2}}\right] + \beta_g \tag{5.199}$$

where the ground slope, β_g (positive if the ground under the barrier nearest the receiver is higher than the ground under the diffraction edge nearest the source and negative otherwise) is:

$$\beta_g = \arcsin\left[\frac{z_{bg,s} - z_{bg}}{\sqrt{(x_{b,s} - x_b)^2 + (z_{bg,s} - z_{bg})^2}}\right] \tag{5.200}$$

The value of $z_{Q''}$ is adjusted iteratively and the calculations of Equations (5.197) to (5.199) repeated until $\beta_{7i} \approx \beta_{7r}$ to an acceptable accuracy.

The source / receiver path length for the ray reflected from the ground on the receiver side of the barrier is:

$$\begin{aligned} \text{SQQ}_s\text{Q}''\text{R} = &\sqrt{(x_S - x_b)^2 + (y_S - y_{b1})^2 + (z_S - z_Q)^2} \\ &+ \sqrt{(x_{b,s} - x_b)^2 + (y_{b1,s} - y_{b1})^2 + (z_{Q,s} - z_Q)^2} \\ &+ \sqrt{(x_{b,s} - x_{Q''})^2 + (y_{b1,s} - y_{Q''})^2 + (z_{Q,s} - z_{Q''})^2} \\ &+ \sqrt{(x_R - x_{Q''})^2 + (y_R - y_{Q''})^2 + (z_R - z_{Q''})^2} \end{aligned} \tag{5.201}$$

For a ray reflected from the ground on the source side of the barrier, the source / receiver path length is:

$$\begin{aligned} \text{SQ}''\text{QQ}_s\text{R} = &\sqrt{(x_S - x_{Q''})^2 + (y_S - y_{Q''})^2 + (z_S - z_{Q''})^2} \\ &+ \sqrt{(x_b - x_{Q''})^2 + (y_{b1} - y_{Q''})^2 + (z_Q - z_{Q''})^2} \\ &+ \sqrt{(x_{b,s} - x_b)^2 + (y_{b1,s} - y_{b1})^2 + (z_{Q,s} - z_Q)^2} \\ &+ \sqrt{(x_R - x_{b,s})^2 + (y_R - y_{b1,s})^2 + (z_R - z_{Q,s})^2} \end{aligned} \tag{5.202}$$

For a ray reflected from the ground between the two barrier diffraction edges, the source / receiver path length is:

$$\begin{aligned} \text{SQQ}''\text{Q}_s\text{R} = &\sqrt{(x_S - x_b)^2 + (y_S - y_{b1})^2 + (z_S - z_Q)^2} \\ &+ \sqrt{(x_b - x_{Q''})^2 + (y_{b1} - y_{Q''})^2 + (z_Q - z_{Q''})^2} \\ &+ \sqrt{(x_{b,s} - x_{Q''})^2 + (y_{b1,s} - y_{Q''})^2 + (z_{Q,s} - z_{Q''})^2} \\ &+ \sqrt{(x_R - x_{b,s})^2 + (y_R - y_{b1,s})^2 + (z_R - z_{Q,s})^2} \end{aligned} \tag{5.203}$$

To obtain results for barrier end 2, y_{b2}, $y_{b2,s}$ and β_8 are substituted in place of y_{b1}, $y_{b1,s}$ and β_7, respectively, in Equations (5.191) to (5.203). All equations in this section apply to both flat and sloping ground between the source and receiver, with different slopes permitted on either side of the barrier as well as between the two diffraction edges.

The calculated total barrier attenuation in any octave band for the double diffraction case should not exceed 25 dB. For a single barrier, it should not exceed 20 dB.

5.3.5.8 Shielding by Terrain

For outdoor sound propagation over undulating or mountainous terrain, the equivalent barrier effect due to the terrain is calculated using the geometry shown in Figure 5.27. The equivalent Fresnel number is:

$$N = \pm\frac{2}{\lambda}[A + B + C - d_{SR}] \tag{5.204}$$

where the negative sign is used if there is direct line-of-sight between the source and receiver and the positive sign is used if there is no direct line-of-sight.

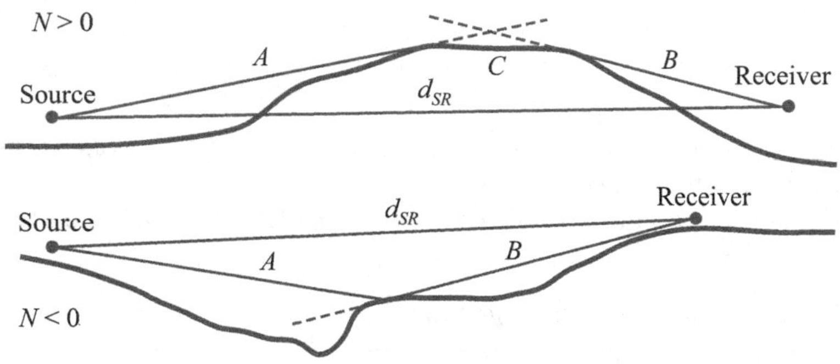

FIGURE 5.27 Geometry for calculation of barrier effect due to terrain.

5.3.5.9 Effects of Wind and Temperature Gradients on Barrier Attenuation

The attenuation of a barrier outdoors is affected by vertical atmospheric wind and temperature gradients. This effect can be taken into account approximately as follows for diffraction over the top of a barrier (Tonin, 1985).

The effect of a positive vertical wind or temperature gradient (that is wind speed or temperature increasing with altitude) is to effectively move both the source and receiver higher up and further from the barrier, as shown in Figure 5.28.

The radius of curvature, R_c, of the curved sound ray shown in Figure 5.28 is given by Equation (5.73). If R_c in Equation (5.73) is positive, then rays are curved downwards as shown in Figure 5.28, resulting in less attenuation. If R_c is negative then the converse is true, resulting in more attenuation.

Referring to Figure 5.28(a), the effective source position for a positive sonic gradient is given by ℓ'_S, which is the horizontal distance of the effective source from the barrier, and h'_S, which is the increase in height of the effective source above the actual source. Both ℓ'_S and h'_S may be

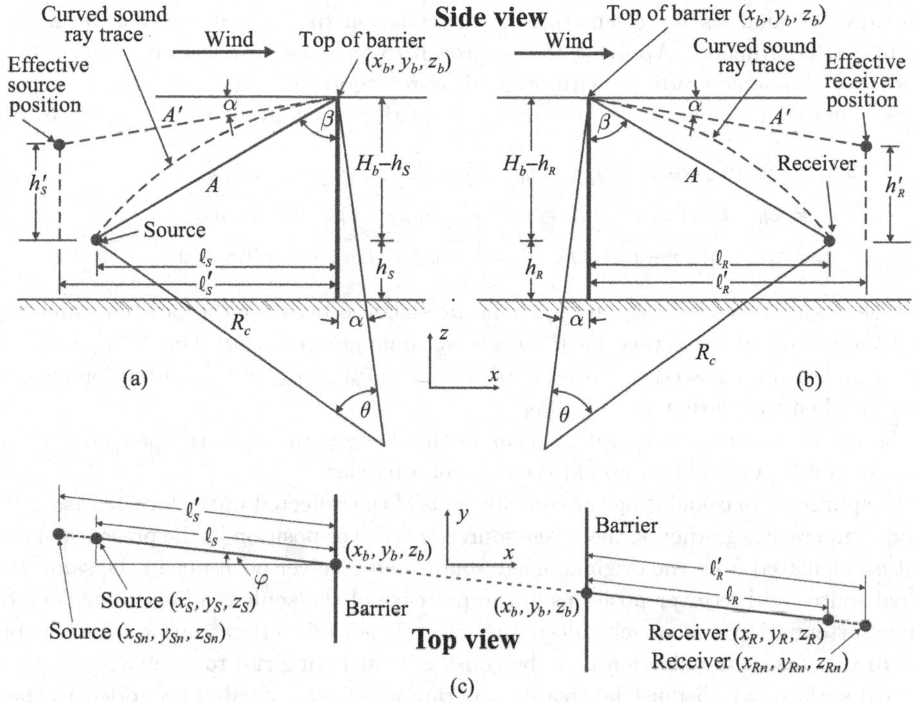

FIGURE 5.28 Geometry for locating the effective source and receiver positions, assuming a positive vertical sonic gradient: (a) side view showing effective source position; (b) side view showing effective receiver position; (c) top view showing a positive angle, φ. If $y_S < y_b$, φ is negative.

calculated given the radius of curvature, R_c, using the following equations (Tonin, 1985):

$$
\begin{aligned}
\ell'_S &= R_c\theta\cos\alpha \\
h'_S &= H_b - h_S - R_c\theta\sin\alpha \\
\alpha &= \tfrac{1}{2}(\pi - \theta) - \beta \\
\beta &= \cos^{-1}[(H_b - h_S)/A] \\
\theta &= \pm\cos^{-1}[1 - (A^2/2R_c^2)]; \qquad |R_c| > A/2 \\
A &= \sqrt{(x_S - x_b)^2 + (y_S - y_b)^2 + (z_S - z_b)^2}
\end{aligned}
\tag{5.205}
$$

where H_b is the barrier height, and A is the distance from the actual source to the barrier top. The angle θ must have the same sign as R_c, where R_c is positive for a positive sonic gradient and negative for a negative sonic gradient.

The quantities, ℓ'_R and h'_R in Figure 5.28(b) may be found by replacing the subscript, S with R in Equation (5.205). For a double barrier, the value of H_b used in the calculation of ℓ'_R and h'_R is the height above the ground of the diffraction edge nearest the receiver.

The effect of a negative sonic gradient (that is wind speed or temperature decreasing with altitude) is to effectively move the source and receiver lower and closer to the barrier. Note that wind speed is measured in the direction from the source to the receiver, so that wind blowing from the receiver to the source will have a negative speed (and usually a corresponding negative sonic gradient). The analysis is very similar to that for the positive sonic gradient, resulting in identical equations to Equation (5.205), except that the sign of R_c and θ are both negative. This results in negative values for h'_S and h'_R in Equation (5.205) resulting in the effective source and receiver positions being *below* the actual source and receiver positions respectively. In addition, the negative value for θ results in a large α, which results in $\ell'_S < \ell_S$ and $\ell'_R < \ell_R$.

The previous analysis for calculating the attenuation due to barriers has been based on the use of 3-D coordinates. Applying this approach to the case of refraction caused by an atmospheric vertical sonic gradient, with original source coordinates (x_S, y_S, z_S), results in new source coordinates, (x_{Sn}, y_{Sn}, z_{Sn}), where, for a positive sonic gradient (see Figure 5.28(c)):

$$\begin{aligned} x_{Sn} &= x_b - \ell'_S \cos\varphi; \quad y_{Sn} = y_b + \ell'_S \sin\varphi; \quad z_{Sn} = z_S \pm h'_S \\ x_{Rn} &= x_b + \ell'_R \cos\varphi; \quad y_{Rn} = y_b - \ell'_R \sin\varphi; \quad z_{Rn} = z_R \pm h'_R \\ \varphi &= \tan^{-1}|(y_S - y_R)/(x_R - x_S)| = \tan^{-1}|(y_S - y_b)/(x_b - x_S)| \end{aligned} \quad (5.206)$$

where in the expressions for z_{Sn} and z_{Rn}, the plus sign is used for the positive sonic gradient case and the negative sign is used for the negative sonic gradient case. For the case of a double barrier, x_b and y_b are replaced with the coordinates, $x_{b,s}$ and $y_{b,s}$, of the diffraction edge nearest the receiver when calculating x_{Rn} and y_{Rn}.

As the ray curvature is assumed to occur in the $x - y$ plane only, the preceding analysis is not applied to waves travelling around the sides of a barrier.

For the purposes of calculating the contributions of the reflected waves for the case of positive wind and temperature gradients, the image source or receiver position in the presence of the sonic gradient is calculated from the original image source and receiver positions in the same way that the actual source and receiver positions in the presence of the sonic gradient are calculated.

Barrier attenuation is adversely affected by reflections from vertical surfaces such as building façades. In this case, the reflection must be considered as giving rise to an image source behind the vertical surface, which must be treated separately and its contribution added to the sound level at the receiver (see Section 1.11.3) due to the non-reflected wave.

5.3.5.10 Barrier Insertion Loss (IL) Measurement

The measurement of barrier IL is discussed in ANSI/ASA S12.8 (2020) and ISO 10847 (1997). The ANSI standard is much more prescriptive and detailed than the ISO standard. Essentially, there are two methods for measuring the barrier IL: the direct method for which the receiver microphone remains in the same location with and without the barrier present; and the indirect method for which no data are available prior to installation of the barrier and such data must be obtained at a different but acoustically "equivalent" site. In both cases, it is necessary to use a reference microphone in addition to the receiver microphone, where the reference microphone is in a location close to the barrier that is unaffected by the barrier. The reference location should be at least 15 m from the sound source and at least 1.5 m directly above the barrier if the barrier is more than 15 m from the sound source. If the barrier is closer than 15 m to the sound source, then the reference microphone should be sufficiently high that the angle between the line joining the top of the barrier to the sound source and the line joining the reference microphone to the sound source is at least 10°.

For overall A-weighted values of IL, it is important that the sound source used for the measurements is the same as the sound source for which the barrier is designed. For 1/3-octave or octave band measurements it is possible to use an artificial sound source. The barrier insertion loss is determined using:

$$\mathrm{IL} = L_{p,\mathrm{ref,a}} - L_{p,\mathrm{ref,b}} + L_{p,\mathrm{b}} - L_{p,\mathrm{a}} \quad (5.207)$$

where $L_{p,\mathrm{ref,a}}$ is the sound pressure level at the reference microphone after insertion of the barrier, $L_{p,\mathrm{ref,b}}$ is the sound pressure level at the reference microphone before insertion of the barrier, $L_{p,\mathrm{a}}$ is the sound pressure level at the receiver microphone after insertion of the barrier and $L_{p,\mathrm{b}}$ is the sound pressure level at the receiver microphone before insertion of the barrier. The receiver microphones for the before and after cases must be in the same locations relative to reflecting surfaces. If this is not possible due to the presence of building façades, then for the situation where a building façade would influence the results, the microphone should be mounted on a rigid flat plate of minimum dimensions 1 m × 1 m and a correction of 6 dB should be subtracted from this microphone level. It is also important that meteorological and ground conditions be the same for the before and after measurements.

5.3.6 Miscellaneous Effects, A_{misc}

Miscellaneous effects include effects such as those due to vegetation, reflections from vertical surfaces, source height and in-plant screening. These effects are treated differently by the various noise propagation models and details of the treatment by each model can be found in the section, later in this chapter, which discusses the particular model.

5.3.7 Low-Frequency Noise and Infrasound

As low-frequency noise and infrasound are not attenuated significantly by atmospheric absorption nor by ground reflections, and mid- and high-frequency noise is significantly attenuated, then at distances of one or more kilometres from a typical broadband noise source, the noise spectrum tends to be dominated by low-frequency energy and infrasound, in contrast to the spectrum in the vicinity of the source. When there is a strong downwind component from the source to the receiver and/or a temperature inversion, there will be a minimum distance beyond which more than one sound ray will reach the receiver (see Section 5.3.4.7). In addition, as a result of more and more rays affecting the sound pressure level as the distance from the source increases, the sound pressure levels no longer decrease at a rate of 6 dB for each doubling of distance from the source. It has been suggested that the rate becomes closer to 3 dB per doubling of distance once the minimum distance of the receiver from the source for more than one ray arriving is exceeded. Thus, the reasons why it seems that infrasound and low-frequency noise decays at a slower rate than mid- and high-frequency sound are listed below.

1. Low-frequency noise and infrasound are insignificantly attenuated by ground reflections.
2. Low-frequency noise and infrasound are insignificantly attenuated by air absorption.
3. Multiple reflections (see Section 5.3.4.7) may begin to occur in a downwind direction or in conditions of atmospheric temperature inversion close to the ground, after about one kilometre from a typical sound source. This effect is more apparent for low-frequency noise and infrasound, which is attenuated by atmospheric absorption or by ground reflection to a much lesser extent than mid- to high-frequency noise.

5.3.8 Impulse Sound Propagation

The calculation of the community sound exposure level resulting from bird scarers, gun shots, mine explosions and military weapons is discussed in detail in ANSI/ASA S12.17 (2011), ANSI/ASA S2.20 (2020) and ISO 17201-3 (2019). Exposure to impulse noise is quantified in terms of sound exposure level and in ANSI/ASA S12.17 (2011), the C-weighted sound exposure level is used (see Section 2.4.5). The expression for an engineering estimate of the C-weighted sound exposure level, L_{CE} is (ANSI/ASA S12.17, 2011):

$$L_{CE} = A - Gd_{SR} + C - C_b \tag{5.208}$$

where the sound source is considered to be omni-directional and where:

$$C_b = \text{smaller of} \begin{cases} 18 + 10b/(M)^{1/3} \\ 30b/(M)^{1/3} \end{cases} \tag{5.209}$$

where b is the burial depth of the explosive (m), M is the equivalent mass of TNT (kg), d_{SR} is the source/receiver distance (km) and the constants, A and G are given in Table 5.8 for various ground surfaces. For explosions other than military-style guns, $C = 8.2 \log_{10} M$ and the directivity correction is zero.

TABLE 5.8 Values of the coefficients, A and G for various ground surfaces.

Type of ground surface	A	G
General, undefined	102.3	31.7
Water	111.1	22.4
Arid land	106.1	32.8
Dense forest	95.9	32.3

For military-style guns, $C = Y + B \log_{10}(16M) + D$, where values of Y, B and D (directivity correction) are given for various guns in ANSI/ASA S12.17 (2011).

The standard deviation in L_{CE} is:

$$s = 5 + 2.9 \log_{10} d_{SR} + 0.28 d_{SR} \quad \text{for distances, } 1\text{ km} \leq d_{SR} \leq 30\text{ km} \tag{5.210}$$

For explosions at mines and quarries:

$$L_{CE} = 99.1 - 29 \log_{10} \left(\frac{d_{SR}}{M^{1/3}} \right) - 0.025 \left(\frac{d_{SR}}{M^{1/3}} \right) + C_b \tag{5.211}$$

The effect of explosion height on the C-weighted sound exposure level for above-ground explosions is discussed in detail in ANSI/ASA S2.20 (2020).

5.4 Propagation Models in General Use

Since the early 1970s, a number of propagation models of varying complexity have been developed and validated to varying degrees. The implementation of the various models in commercial software has not been consistent, so that the results obtained for the same model from various commercial software are not identical for the more complex models, which leads to confusion and doubt about the results obtained using these models. Perhaps for this reason, most practitioners in the past have used the simpler, but less accurate and less reliable models.

In an attempt to improve model accuracy and consistency, the European Commission of the European Union, in the context of the European Environmental Noise Directive 2002/49/EC (END), decided to prepare Common NOise aSSessment methOdS (CNOSSOS-EU) for road, railway, aircraft and industrial noise in order to improve the reliability and the comparability of results obtained by different organisations and different commercial software packages. CNOSSOS investigated three commonly used and complex noise propagation models in detail (Harmonoise, NMPB-2008 and ISO 9613-2), as described in European Commission (2010b), Kephalopoulos et al. (2012) and European Commission (2010a), respectively. In 2015, CNOSSOS published software modules, algorithms and documentation for implementation of all three models, noting that the model based on the NMPB algorithm had been selected as the preferred one.

In the sections to follow, two of the three CNOSSOS models as well as one other well-known and extensively used propagation model will be discussed in order of increasing complexity. The models to be discussed include the CONCAWE model, the ISO 9613-2 model and the NMPB-2008 model. All models have limitations, as it is impossible to accurately model the ground or atmospheric conditions, which can vary significantly, in both space and time, between any given source and receiver locations. In addition, for the purposes of the analysis it is assumed that the ground between the source and receiver is flat. These limitations result in uncertainty limits being placed on the predictions. Although predictions for a large number of receiver locations are generally better on average than indicated by the uncertainty limits, the predictions at any single location are well described by the uncertainty limits. This means that all noise predictions at receiver locations should be accompanied by uncertainty estimates and the upper level of the uncertainty estimate should not be permitted to exceed the allowable level. In fact, Verheijen et al. (2011) showed that the standard models, which are used in many current studies, can substantially underestimate sound pressure levels for locations at large distances during temperature inversions. Uncertainty is discussed in Section 5.9.

5.5 CONCAWE Noise Propagation Model

CONCAWE is a noise propagation model developed for estimating environmental noise levels radiated by petroleum and petrochemical complexes to surrounding communities (Manning, 1981). Since that time, the original version or its modified form (Marsh, 1982) has been used in almost all commercially available software for calculating the level of noise radiated into surrounding communities by any sound source.

The equation used in the CONCAWE model is a derivative of Equation (5.54) and may be written for the ith source producing a sound pressure level, L_{pik}, at the kth community location as:

$$L_{pik} = L_{Wi} + \mathrm{DI}_{ik} - A_{Eik} \qquad \text{(dB)} \tag{5.212}$$

where L_{pik} is the octave band sound pressure level at community location, k, due to the ith source, and L_{Wi} is the sound power radiated by the ith source. DI_{ik} is the directivity index of source, i, in the direction of community location, k. DI_{ik} is usually assumed to be 0 dB, unless specific source directivity information is available.

Calculations are done in octave bands from 63 Hz to 8 kHz and the overall A-weighted sound pressure level is calculated by applying the A-weighting correction (see Table 2.3) to each octave band level and then summing the levels logarithmically (see Section 1.11.3). C-weighted sound pressure levels can be calculated in a similar way.

The coefficient, A_{Eik}, is the attenuation experienced by a sound pressure disturbance travelling from source, i, to community location, k, and is given by:

$$A_{Eik} = (K_1 + K_2 + K_3 + K_4 + K_5 + K_6 + K_7)_{ik} \tag{5.213}$$

Each of these attenuation factors are discussed in the following paragraphs.

5.5.1 Spherical Divergence, K_1

Sound sources are treated as point sources and the attenuation as a result of the sound waves spreading out as they travel away from the sound source is given by Equation (5.56).

5.5.2 Atmospheric Absorption, K_2

This is discussed in detail in Section 5.3.2 and is the same for all propagation models. Note that $K_2 = A_a$, and is calculated using Equation (5.61).

5.5.3 Ground Effects, K_3

For a hard surface such as asphalt, concrete or water, $K_3 = -3$ dB. For all other surfaces a set of empirical curves is used (see Figure 5.29). However, these curves were developed for noise sources close to the ground and positive values imply that there is coherent or partially coherent combination, causing destructive interference of the direct and ground-reflected rays arriving at the receiver. This situation is uncommon, especially for long-distance propagation and high sound sources and thus will often result in over-predicting the attenuation effect of the ground, resulting in higher predicted sound pressure levels at the receiver. The error will become larger as distance from the source increases. A conservative approach is to use the hard ground value for K_3 (-3 dB) for relatively hard ground surfaces and a value of 0 dB for soft ground surfaces.

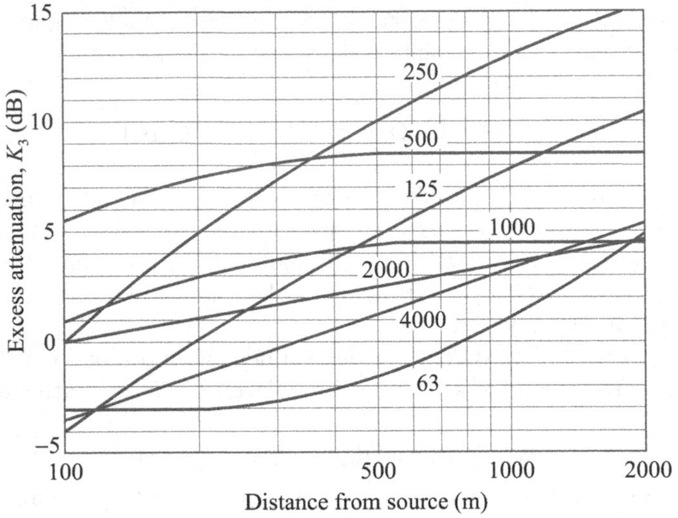

FIGURE 5.29 Attenuation due to the ground. The octave band centre frequency (Hz) corresponding to each curve is indicated on the figure.

5.5.4 Meteorological Effects, K_4

Accounting for meteorological effects is perhaps the most difficult of all the propagation attenuations. In this procedure, meteorological effects have been graded into six categories based on a combined vertical wind and temperature gradient. In Table 5.9, incoming solar radiation is defined for use in Table 5.10. In Table 5.10, the temperature gradient and wind speed combination is coded in terms of Pasquill stability category A–G. Category A represents a strong lapse condition (large temperature decrease with height). Categories E, F and G, on the other hand, represent a weak, moderate and strong temperature inversion, respectively, with the strong inversion being that which may be observed early on a clear morning. Thus, category G represents very stable atmospheric conditions while category A represents very unstable conditions and category D represents neutral atmospheric conditions. The wind speed in this table is not a vector as the direction is not relevant for the purpose of assessing atmospheric stability.

TABLE 5.9 Daytime incoming solar radiation (full cloud cover is 8 octas, half cloud cover is 4 octas, etc.)

Latitude of sun	Cloud cover (octas)	Incoming solar radiation
$< 25°$	0–7	slight
25°–45°	<4	moderate
$> 45°$	5–7	moderate
$> 45°$	<4	strong

TABLE 5.10 CONCAWE determination of Pasquill stability category from meteorological information (full cloud cover is 8 octas, half cloud cover is 4 octas, etc.). Wind speed is measured to the nearest 0.5 m/s and is measured at a height of 10 m above the ground. Category G is restricted to nighttime with less than 1 octa of cloud cover and a wind speed less than 0.5 m/s (after Marsh (1982)). In this table, the direction of the wind is not relevant.

Wind speed m/s	Daytime incoming solar radiation strong	moderate	slight	Over-cast	1 hour before sunset or after sunrise	Nighttime cloud cover (octas) 0–3	4–7	8
<2	A	A–B	B	C	D	F or G	F	D
2.0–2.5	A–B	B	C	C	D	F	E	D
3.0–4.5	B	B–C	C	C	D	E	D	D
5.0–6.0	C	C–D	D	D	D	D	D	D
>6.0	C	D	D	D	D	D	D	D

The Pasquill stability category is combined with the magnitude of the wind vector component in the direction from the source to the receiver using Table 5.11 to determine one of the six meteorological categories (CAT) for which attenuations, $A_{\text{met}} = K_4$, were experimentally determined from surveys of three European process plants (Manning, 1981). Attenuations are shown in Figures 5.30(a)–(g) for the octave bands from 63 Hz to 4 kHz. The wind speed used in Tables 5.10 and 5.11 is measured at a height of 10 m above ground level and in Table 5.11 (but NOT in Table 5.10) it is the component of the wind vector in the direction from the source to the receiver.

TABLE 5.11 CONCAWE determination of meteorological category (negative values represent the receiver upwind of the source and positive values are for the receiver downwind of the source). Category 4 is the one corresponding to minimal meteorological influence (after Marsh (1982)). The wind speed, v is the component of the wind velocity vector in the direction from the source to the receiver. Attenuation curves for each meteorological category (CAT) are shown in Figure 5.30.

Meteorological category	Pasquill stability category A, B	C, D, E wind speed, v (m/s)	F, G
1	$v < -3.0$	—	—
2	$-3.0 < v < -0.5$	$v < -3.0$	—
3	$-0.5 < v < +0.5$	$-3.0 < v < -0.5$	$v < -3.0$
4	$+0.5 < v < +3.0$	$-0.5 < v < +0.5$	$-3.0 < v < -0.5$
5	$v > +3.0$	$+0.5 < v < +3.0$	$-0.5 < v < +0.5$
6	—	$v > 3.0$	$+0.5 < v < +3.0$

5.5.5 Source Height Effects, K_5

If Figure 5.29 is used to calculate the ground effect, then for source heights greater than 2 m, an additional correction, K_5, must be added to K_3. The correction, K_5, is given by:

$$K_5 = \begin{cases} (K_3 + K_4 + 3) \times (\gamma - 1) \quad \text{(dB)}; & \text{if}(K_3 + K_4) > -3 \text{ dB} \\ 0 \quad \text{(dB)}; & \text{if}(K_3 + K_4) \leq -3 \text{ dB} \end{cases} \tag{5.214}$$

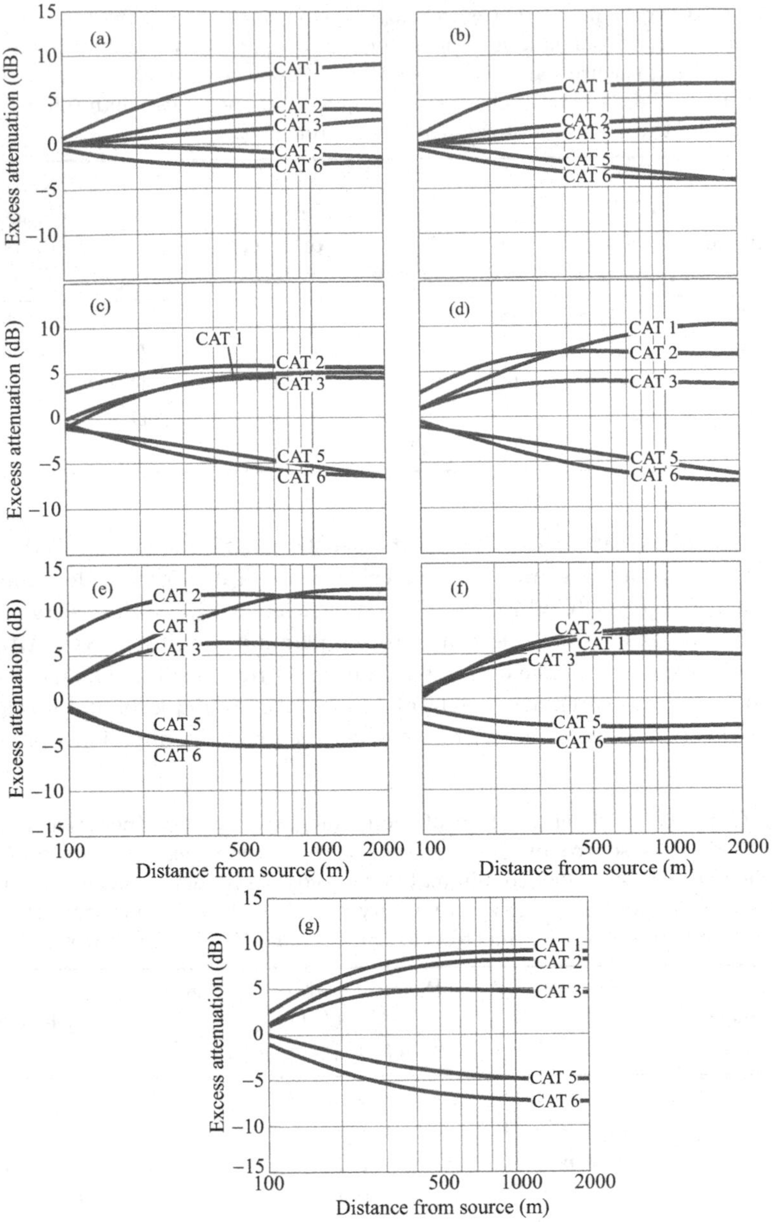

FIGURE 5.30 CONCAWE meteorological curves for various octave bands. (a) 63 Hz, (b) 125 Hz, (c) 250 Hz, (d) 500 Hz, (e) 1000 Hz, (f) 2000 Hz, (g) 4000 Hz.

where γ has a maximum value of 1 and is given by:

$$\gamma = 1.08 - 0.478(90 - \theta) + 0.068(90 - \theta)^2 - 0.0029(90 - \theta)^3 \tag{5.215}$$

Note that K_5 is always negative, which means that it acts to reduce the attenuation. The angle, θ, is in degrees and is defined in Figures 5.1 and 4.17. If propagation is to a receiver located on a hillside, or across a valley floor, the value of K_5 should be reduced (made more negative) by up to 3 dB to account for multiple reflections from the hillside.

5.5.6 Barrier Attenuation, K_6

Barriers are any obstacles that represent interruptions of the line-of-site from the sound source to the community location. In the CONCAWE model, barriers are modelled as thin screens and the corresponding attenuation is calculated using a procedure according to Maekawa (1968, 1977), as described in Section 5.3.5.1, with allowance made for bending of sound over the barrier as a result of atmospheric wind and temperature gradients, as described in Section 5.3.5.9.

5.5.7 In-Plant Screening, K_7

Manning (1981) found that in-plant screening was only significant for large petrochemical plants and for these, a value for K_7 was difficult to estimate, due to the limited amount of data available. However, a conservative estimate would be to set $K_7 = 0$.

5.5.8 Vegetation Screening, K_v

Marsh (1982) suggested that the following equation could be used to estimate the attenuation of sound (in an octave band of centre frequency, f) as a result of travelling a distance, r, through vegetation such as a forest. Marsh did point out that this equation was likely to underestimate the attenuation in European forests.

$$K_v = 0.01rf^{1/3} \tag{5.216}$$

For sound propagating a distance, r, through long grass or shrubs, the attenuation may be calculated using:

$$K_v = (0.18 \log_{10} f - 0.31)r \tag{5.217}$$

5.5.9 Limitations of the CONCAWE Model

The limitations of the CONCAWE model are:

- The model has been validated only for distances from the source to the receiver between 100 m and 2000 m.
- The model was developed using empirical data from petrochemical plants with noise sources less than 20 m in height. When used with the hard ground option, it has been shown to overpredict wind farm noise (Evans and Cooper, 2011).
- The model is only valid for wind speeds less than 7 m/s.
- The model is only valid for octave band analysis for octave band centre frequencies ranging from 63 Hz to 4000 Hz. Of course, it is also valid for calculating overall A-weighted sound pressure levels over the frequency range covered by the 63 Hz to 4000 Hz octave bands.
- The attenuation due to the ground is based on a very simplified model derived from experimental data taken at two airfields. Its accuracy is questionable for propagation over other types of ground surface.
- The model used to calculate the attenuation due to vegetation is not reliable.
- Most software packages that implement the CONCAWE model only implement its standard form, in which case, the following aspects discussed in this section are excluded.

1. Vegetation effects, and if they are included, they are limited to forests.
2. Ground-reflected rays are not included when considering barrier attenuation; only sound rays travelling over the top of the barrier with no ground reflection are included.
3. Reduced attenuation of barriers in the presence of a downward refracting atmosphere (downwind and/or temperature inversion) is not taken into account.
4. Terrain shielding is not included.

A detailed statistical analysis has been carried out on comparisons between predictions and measurements with the 95% confidence limits shown in Table 5.12.

TABLE 5.12 95% confidence limits of the CONCAWE model, which is representative of the expected reliability of a single calculation (after Marsh (1982)). The average of many measured values minus the predicted value for a range of meteorological conditions is shown in brackets, indicating the extent of bias (or error in the model in terms of predicting average values)

Meteorological category	dB(A)	Octave band centre frequency (Hz) 63	125	250
2	6.8 (0.5)	5.4 (0.1)	5.4 (0.1)	9.1 (2.0)
3	6.9 (0.6)	5.0 (0.0)	6.2 (0.5)	9.4 (1.6)
4	5.7 (0.5)	4.8 (0.3)	6.5 (0.8)	8.7 (−1.2)
5	4.7 (0.0)	3.9 (−0.1)	5.4 (0.0)	8.4 (−2.3)
6	4.5 (0.5)	5.2 (−0.8)	6.1 (−0.3)	6.7 (−1.7)

Meteorological category	Octave band centre frequency (Hz) 500	1000	2000	4000
2	9.4 (2.2)	7.8 (2.2)	9.8 (−0.2)	12.4 (0.4)
3	10.1 (0.4)	8.5 (0.8)	8.5 (0.8)	9.4 (0.4)
4	9.8 (−0.2)	6.6 (0.1)	5.6 (1.4)	6.7 (0.2)
5	8.1 (0.4)	5.2 (−0.6)	5.6 (0.9)	6.7 (−0.9)
6	9.3 (1.2)	4.9 (−0.2)	5.5 (0.1)	8.2 (−0.9)

Table 5.12 also shows in brackets the average of many observed minus predicted values, indicating the extent of bias in the model. These limits are for use of the model over the range from 100 m to 2000 m. For smaller distances, the confidence limits would be smaller than shown in the table, but for larger distances over 1000 m, they are larger (Marsh, 1982). The confidence limits are smaller for meteorological categories 5 and 6, which represent downwind propagation. Marsh points out that these numbers may be even smaller if the analysis used to estimate the uncertainty were limited to a range of ±45° from the directly downwind direction.

Marsh (1982) also showed that the bias in the CONCAWE model was small, as evidenced by comparing the mean of a large number of observed minus predicted measurements and calculations, as shown in Table 5.12.

5.6 ISO 9613-2 (1996) Noise Propagation Model

The ISO 9613-2 (1996) standard is a noise propagation model that recommends calculations be done in octave bands with centre frequencies between 63 Hz and 8 kHz, and that for each octave band, the same basic equation as used by the CONCAWE model (Equation (5.212)) be used. This equation is used to calculate the sound pressure level, L_{pik}, at location, k, due to the sound power, L_{Wi}, radiated by sound source, i, with a directivity index, DI_{ik}, in the direction of the receiver location:

$$L_{pik} = L_{Wi} + \mathrm{DI}_{ik} - A_{Eik} \qquad \text{(dB)} \tag{5.218}$$

The overall A-weighted sound pressure level is calculated by applying the A-weighting correction (see Table 2.3) to each octave band level and then summing the levels logarithmically (see Section 1.11.3). The overall C-weighted sound pressure level can be calculated by following a similar procedure.

In the standard, the attenuation A_{Eik} is given by:

$$A_{Eik} = [A_{\mathrm{div}} + A_a + A_g + A_b + A_{\mathrm{misc}}]_{ik} \tag{5.219}$$

The attenuation component, A_{div}, is calculated assuming a point source using Equation (5.56) and the component, A_a, is calculated using Equation (5.61), as described in Section 5.3.2. Both of these attenuation effects are applied to the final total calculated sound pressure level at the receiver for each sound source.

The attenuation components, A_g and A_b, are equivalent to the CONCAWE attenuation components, K_3 and K_6, respectively, but are calculated differently. The ISO model does not explicitly include an attenuation due to meteorological effects as its calculation of A_g is for downwind sound propagation only and this would be the worst-case situation for most cases. However, it is really an oversimplification of a complex problem and may not be as accurate as the much earlier CONCAWE model, described above. The attenuation components, A_g and A_b, are discussed separately in Sections 5.6.1 and 5.6.3, respectively.

The attenuation component, A_{misc}, is made up of the following subcomponents.

- Attenuation, A_{site}, due to propagation of sound through an industrial site on its way from the source to the receiver (identical to the quantity, K_7, in the CONCAWE model).
- Attenuation, A_h, due to propagation of the sound through a housing estate on its way from the source to the receiver.
- Attenuation, (A_f), due to propagation of sound through foliage. This similar to the quantity, K_v, which is estimated in Section 5.5.8. However, the calculation of A_f relies on a different procedure, which is outlined in Section 5.6.4 below.

5.6.1 Ground Effects, A_g

For the ISO 9613-2 (1996) method, the space between the source and receiver is divided into three zones, source, middle and receiver zones. The source zone extends a distance of $30h_S$ from the source towards the receiver and the receiver zone extends $30h_R$ from the receiver towards the source, where h_S and h_R are defined in Figure 4.17, page 216. The middle zone includes the remainder of the path between the source and receiver and will not exist if the source/receiver separation (projected onto the ground plane) is less than $d = 30h_S + 30h_R$, where d is defined in Figure 4.17. The acoustic properties of each zone are quantified using the parameter, G (see Table 5.2). This parameter has a value of 0.0 for hard ground, a value of 1.0 for soft (or porous) ground and for a mixture of hard and soft ground it is equal to the fraction of ground that is soft. It is assumed that for downwind propagation, most of the ground effect is produced by the ground in the vicinity of the source and receiver so the middle part does not contribute much to the overall value of A_g.

The total attenuation due to the ground is the sum of the attenuations for each of the three zones. That is:

$$A_g = A_S + A_{mid} + A_{Rec} \tag{5.220}$$

Values for each of the three quantities on the right-hand side of Equation (5.220) may be calculated using Table 5.13.

TABLE 5.13 Octave band ground attenuation contributions, A_S, A_{Rec} and A_{mid} (ISO 9613-2 (1996))

	Octave band centre frequency (Hz)			
	63	125	250	500
A_S (dB)	-1.5	$-1.5 + G_S a_S$	$-1.5 + G_S b_S$	$-1.5 + G_S c_S$
A_{Rec} (dB)	-1.5	$-1.5 + G_R a_R$	$-1.5 + G_R b_R$	$-1.5 + G_R c_R$
A_{mid} (dB)	$-3q$	$-3q(1 - G_m)$	$-3q(1 - G_m)$	$-3q(1 - G_m)$
	Octave band centre frequency (Hz)			
	1000	2000	4000	8000
A_S (dB)	$-1.5 + G_S d_S$	$-1.5(1 - G_S)$	$-1.5(1 - G_S)$	$-1.5(1 - G_S)$
A_{Rec} (dB)	$-1.5 + G_R d_R$	$-1.5(1 - G_R)$	$-1.5(1 - G_R)$	$-1.5(1 - G_R)$
A_{mid} (dB)	$-3q(1 - G_m)$	$-3q(1 - G_m)$	$-3q(1 - G_m)$	$-3q(1 - G_m)$

In Table 5.13, G_S, G_R and G_m are the values of G (Table 5.2) corresponding to the source zone, the receiver zone and the middle zone, respectively. The quantity, A_{mid}, is zero for source/receiver separations of less than $d = 30h_S + 30h_R$, and for greater separation distances it is calculated using the lines labelled A_{mid} in Table 5.13 with:

$$q = 1 - \frac{30(h_S + h_R)}{d} \tag{5.221}$$

The coefficients a_S, b_S, c_S and d_S and the coefficients a_R, b_R, c_R and d_R of Table 5.13 may be calculated using the following equations. In each equation, $h_{S,R}$ is replaced with h_S for calculations of A_S and by h_R for calculations of A_{Rg}.

$$a_S,\ a_R = 1.5 + 3.0\mathrm{e}^{-0.12(h_{S,R}-5)^2}(1 - \mathrm{e}^{-d/50}) + 5.7\mathrm{e}^{-0.09h_{S,R}^2}\left(1 - \mathrm{e}^{-2.8\times10^{-6}\times d^2}\right) \tag{5.222}$$

$$b_S,\ b_R = 1.5 + 8.6\mathrm{e}^{-0.09h_{S,R}^2}(1 - \mathrm{e}^{-d/50}) \tag{5.223}$$

$$c_S,\ c_R = 1.5 + 14.0\mathrm{e}^{-0.46h_{S,R}^2}(1 - \mathrm{e}^{-d/50}) \tag{5.224}$$

$$d_S,\ d_R = 1.5 + 5.0\mathrm{e}^{-0.9h_{S,R}^2}(1 - \mathrm{e}^{-d/50}) \tag{5.225}$$

5.6.2 Meteorological Effects, A_{met}

The ISO model is intended to provide octave band and A-weighted overall community sound pressure level predictions for downwind conditions, which are considered to be "worst-case". So there is no specific allowance for including meteorological conditions in the octave band calculations. However, a correction term for meteorological effects on the overall A-weighted sound pressure level is provided. This term reduces the calculated A-weighted sound pressure level for long-time energy averages of several months to a year. Essentially, the correction is to allow for the fact that downwind propagation does not occur 100% of the time. The corrections are only to be subtracted from the overall A-weighted level calculations and they are not to be included for locations closer to the source than ten times the sum of the source and receiver heights. The correction to account for downwind propagation not occurring 100% of the time is:

$$A_{\text{met}} = A_0[1 - 10(h_S + h_R)/d] \qquad \text{(dB)} \tag{5.226}$$

where h_S and h_R are the source and receiver heights, respectively, and d is the horizontal distance between the source and receiver. The quantity, A_0, depends on local meteorological statistics and varies between 0 and 5 dB with values over 2 dB very rare. The standard offers no other procedure for calculating A_0. The value of A_{met} calculated using this procedure is intended only as a correction to the A-weighted sound pressure level, which is why it contains no frequency-dependent terms.

5.6.3 Barrier Attenuation, A_b

For an obstacle to be classified as a barrier for which the calculations in this section are valid, it must satisfy the following conditions.

- Obstacle surface density must be greater than 10 kg/m^2.
- No cracks or gaps that would allow sound to travel through.
- The obstacle length normal to the line between the sound source and receiver should be greater than a wavelength at the frequency corresponding to the lower frequency limit of the octave band centre frequency of interest.

Obstacles that fulfil the above conditions are replaced for the purposes of the calculation with an equivalent flat rectangular panel with height above the ground equal to the average height of the obstacle.

As the barrier interrupts the ground-reflected wave, then for downwind propagation (the only condition addressed by ISO 9613-2 (1996)), the ground effect term is replaced in the barrier calculation by an expression that includes a new ground reflection term. For paths, i, $i = 1 - 4$, over the top of the barrier (see Figures 5.20(a) and 5.25), the octave band attenuation, $A_{b,t}$, is given by:

$$A_{b,t} = -10\log_{10}\left[10^{-D_{z1}/10} + 10^{-(D_{z2}+A_{rf2})/10} + 10^{-(D_{z3}+A_{rf3})/10} + 10^{-(D_{z4}+A_{rf4})/10}\right] - A_g \tag{5.227}$$

For paths, i, $i = 5 - 8$, around the two ends of the barrier (see Figures 5.23 and 5.26 which show paths five and seven around the right hand end):

$$A_{b,s} = -10\log_{10}\left[10^{-(D_{z5})/10} + 10^{-(D_{z6})/10} + 10^{-(D_{z7}+A_{rf7})/10} + 10^{-(D_{z8}+A_{rf8})/10}\right] \tag{5.228}$$

where paths six and eight are not shown in the figures, but they are identical to paths five and seven, respectively, except they are paths around the left hand edge (when looking from the source to the barrier).

The total attenuation, A_b, due to the barrier is:

$$A_b = -10\log_{10}\left[10^{-A_{b,t}} + 10^{-A_{b,s}}\right] \tag{5.229}$$

where A_b is set $= 0$ if $A_b < 0$.

In Equations (5.227) and (5.228), the attenuation, D_{z1}, refers to the non-reflected path over the top, D_{z2}, D_{z3} and D_{z4} refer to the three ground-reflected paths over the top that have a ground-reflection loss, respectively, of A_{rf2}, A_{rf3} and A_{rf4} dB (which consists of two reflection losses added arithmetically). The attenuations, D_{z5} and D_{z6} refer to the two non-reflected paths around the ends of the barrier, and D_{z7} and D_{z8} refer to the two ground-reflected paths around the ends of the barrier (one at each end) that have a ground reflection loss of A_{rf7} and A_{rf8} dB, respectively. Any of the terms may be omitted from Equations (5.227) and (5.228) if it is not desired to consider a particular path. If diffraction around the ends of the barrier is to be ignored, then $A_b = A_{b,t}$.

The attenuation, D_{zi}, $i = 1 - -8$, for any propagation path, z_i over and around the ends of the barrier is given by:

$$D_{zi} = 10\log_{10}\left[3 + (C_2/\lambda)C_3K_{\text{met}}\Delta z_i\right] \quad \text{(dB)} \tag{5.230}$$

where λ is the wavelength of sound at the octave band centre frequency, $C_2 = 20$ if ground-reflected paths over and around the barrier are included in the calculation, and $C_2 = 40$ if ground-reflected paths are calculated separately. Thus, if $C_2 = 20$, then the terms corresponding to i values of 2, 3, 4, 7 and 8 must be excluded from Equations (5.227) and (5.228). The paths around the barrier and the calculation of their lengths are described in Section 5.3.5. Usually, $C_2 = 20$ is chosen in Equation (5.230), so that the only terms to be used in Equations (5.227) and (5.228) are those corresponding to $i = 1,\ 5$ and 6.

In Equation (5.227), the A_g term is not included when the barrier is shielding multiple sources and is a building more than 10 m high, or if the sound sources are high (ISO 9613-2, 1996). It is important to note that the A_g term is still included in Equation (5.219), which is used to calculate the total attenuation due to all effects.

In Equation (5.227), the coefficient, $C_3 = 1$ for single edge diffraction (Figure 5.18). For double edge diffraction (Figure 5.25):

$$C_3 = \frac{1 + (5\lambda/e)^2}{(1/3) + (5\lambda/e)^2} \tag{5.231}$$

The quantity, Δz_i, is the difference between the line-of-sight distance between source and receiver and the path length over the top or around the side of the barrier, for path, i.

The path length difference, Δz_1 for use in Equation (5.230) for diffraction over the top of the barrier may be calculated using Figure 5.18 for single diffraction (a thin obstacle) or Figure 5.25 for double diffraction (a wide obstacle). If the barrier does not interrupt the line-of-sight between the source and receiver, Δz_i is negative. Otherwise, Δz_i is positive.

From Figure 5.18 for single diffraction, the path difference, Δz_1, is given by:

$$\Delta z_1 = \pm(A_1 + B_1 - d_{SR}) \tag{5.232}$$

where

$$\begin{aligned} d_{SR} &= \left((x_R - x_S)^2 + (y_R - y_S)^2 + (z_R - z_S)^2\right)^{1/2} \\ A_1 &= \left((x_S - x_b)^2 + (y_S - y_b)^2 + (z_S - z_b)^2\right)^{1/2} \\ B_1 &= \left((x_R - x_b)^2 + (y_R - y_b)^2 + (z_R - z_b)^2\right)^{1/2} \end{aligned} \tag{5.233}$$

From Figure 5.25 for double edge diffraction, the path difference, Δz_1, for the two edges of heights, z_b and $z_{b,s}$, respectively, is given by:

$$\Delta z_1 = \pm(A_1 + B_1 + e - d_{SR}) \tag{5.234}$$

or, alternatively:

$$\begin{aligned} \Delta z_1 = \pm\Big[(x_S - x_b)^2 + (z_S - z_b)^2 + (y_S - y_b)^2\Big]^{1/2} + \Big[(x_R - x_{b,s})^2 + (z_R - z_{b,s})^2 \\ + (y_R - y_{b,s})^2\Big]^{1/2} + \Big[(x_b - x_{b,s})^2 + (z_b - z_{b,s})^2 + (y_b - y_{b,s})^2\Big]^{1/2} - d_{SR} \end{aligned} \tag{5.235}$$

The term, $K_{\rm met}$, in Equation (5.230) is a meteorological correction factor for downwind propagation and is given by:

$$K_{\rm met} = \exp\left[-\frac{1}{2000}\sqrt{\frac{A_1 B_1 d_{SR}}{2(A_1 + B_1 + e - d_{SR})}}\right] \tag{5.236}$$

where the dimension, e, is zero for single edge diffraction, all dimensions are in metres and $\exp(x) = {\rm e}^x$. For diffraction around the vertical edge of a screen, $K_{\rm met}$ is set equal to 1.

The upper allowed limit of the attenuation, D_{z1}, for the diffracted wave over the barrier top is 20 dB for single-edge diffraction and 25 dB for double-edge diffraction. If ground-reflected rays are treated separately, the upper diffraction limit is not applied to those rays; it is only applied to the non-reflected ray over the barrier top, prior to combining the attenuations of the various paths using Equation (5.227). No upper limit is applied to diffracted rays around the barrier ends. If there is line-of-sight between the source and receiver, the barrier attenuation according to ISO 9613-2 (1996) is set equal to 0 dB.

5.6.4 Vegetation Screening, A_f

ISO 9613-2 (1996) gives the attenuation values in Table 5.14 for sound propagation through dense foliage. For distances less than 20 m, the values given are absolute dB. For distances between 20 and 200 m, the values given are dB/m and for distances greater than 200 m, the value for 200 m is used.

TABLE 5.14 Octave band attenuation, A_f, due to dense foliage (after ISO 9613-2 (1996))

	Octave band centre frequency (Hz)							
	63	125	250	500	1000	2000	4000	8000
A_f (dB) for 10m $\leq r_f \leq$20 m	0	0	1	1	1	1	2	3
A_f (dB/m) for 20m $\leq r_f \leq$200 m	0.02	0.03	0.04	0.05	0.06	0.08	0.09	0.12

The distance of travel through the foliage is not equal to the extent of the foliage between the source and receiver. It depends on the height of the source and receiver and the radius of curvature of the propagating ray as a result of wind and temperature gradients. ISO 9613-2 (1996) recommends that a radius of 5 km be used for downwind propagation. The centre (always below the ground plane) of the circular arc, representing the sound ray path from the source to the receiver, is easily found, using a scaled drawing, as the intersection of two lines of length equal to 5 km, with one line intersecting the source location and the other intersecting the receiver location. The distance $r_f = r_1 + r_2$, where r_1 and r_2 are defined in Figure 5.31.

FIGURE 5.31 Path lengths for sound propagation through foliage.

5.6.5 Industrial Equipment Screening, A_{site}

In the absence of measured data, Table 5.15 may be used as a guide for estimating attenuation through process equipment in an industrial facility. To calculate the number of metres, $r_1 + r_2$, of travel of the sound wave through the industrial facility, the same approach is used as that described in Section 5.6.4 for attenuation through vegetation (see Figure 5.31).

TABLE 5.15 Octave band attenuation, A_{site}, due to process equipment (after ISO 9613-2 (1996))

	Octave band centre frequency (Hz)							
	63	125	250	500	1000	2000	4000	8000
A_{site} (dB/m)	0	0.015	0.025	0.025	0.02	0.02	0.015	0.015

5.6.6 Housing Screening, A_h

The overall A-weighted attenuation due to propagation through rows of houses may be calculated approximately using:

$$A_h = 0.1Br_b - 10\log_{10}[1 - (P/100)] \quad \text{(dB)} \tag{5.237}$$

where A_h should not exceed 10 dBA, B is the density of buildings along the path (total plan area of buildings divided by the total ground area (including that occupied by the buildings), r_b is the distance that the curved sound ray travels through the houses and P is the percentage ($\leq 90\%$) of the length of housing facades relative to the total length of a road or railway or industrial facility contributing significantly to the sound pressure level at the receiver. The second term in Equation (5.237) is only used if there are well defined rows of houses perpendicular to the direction of sound propagation. The second term may also not exceed the calculated Insertion Loss for a continuous barrier the same height as the building facades (see Chapter 7). The quantity, r_b is calculated in the same way as r_f ($= r_1 + r_2$) in Figure 5.31 for travel through foliage, except that the foliage is replaced by houses.

Note that if A_h of Equation (5.237) is non zero, the ground effect, A_g, through the built up area is set equal to zero, unless the ground effect with the buildings removed (and calculated using Equation (5.220)) is greater than the first term in Equation (5.237) for A_h. In that case, the ground effect for the ground without buildings is substituted for the first term in Equation (5.237). As the ground effect is frequency dependent, the overall A-weighted ground effect must be calculated before it can be compared with the first term in Equation (5.237). This is done by A-weighting and then logarithmically adding the A-weighted octave band sound pressure levels with no ground effect using Equation (1.98) to obtain the overall A-weighted sound pressure level at the receiver. The same process is done again with the ground effect included in each octave band to obtain the overall A-weighted sound pressure level at the receiver with the ground effect included. Subtracting the A-weighted level with the ground effect included from the level with the ground effect excluded gives the A-weighted ground effect (which may be negative; that is, it may result in an increased sound pressure level at the receiver).

5.6.7 Effect of Reflections Other than Ground Reflections

If the receiver is close to the wall of a building, the expected sound pressure level will be increased as a result of reflection from the wall. ISO 9613-2 (1996) provides a means for calculating this effect and the procedure is summarised below.

As can be seen from inspection of Figure 4.17, if a reflecting plane is present, sound arrives at the receiver by a direct path and also by a reflected path. If the reflecting plane is not the ground but is a reflecting surface such as a tank or wall of a house, then provided the reflecting surface satisfies the conditions below, it can be taken into account using the procedure outlined here.

- The magnitude of the surface absorption coefficient, α_r, is less than or equal to 0.8.
- The surface is sufficiently large in extent such that the following equation is satisfied:

$$\frac{1}{\lambda} > \left[\frac{2}{(L_{\min}\cos\theta)^2}\right]\left[\frac{r_S r_R}{r_S + r_R}\right] \tag{5.238}$$

  In Equation (5.238), λ is the wavelength of the incident sound (at the 1/3-octave or octave band centre frequency of interest), $L_{\min}$ is the minimum dimension (width or height) of the reflecting surface and r_S, r_R and θ are defined in Figure 4.17, page 216, where the ground is replaced by the reflecting surface.

The sound pressure level at the receiver location is then determined by calculating the sound pressure levels due to the direct and reflected waves separately (using all the attenuation parameters for each of the direct and reflected waves) and then adding the two results logarithmically (see Section 1.11.3). The sound power used for the reflected wave, L_{Wr}, is derived from the sound power level of the source, L_W, which is the sound power level used for the direct wave. Thus:

$$L_{Wr} = L_W + 10\log_{10}(1-\alpha_r) + \text{DI}_r \quad \text{(dB)} \tag{5.239}$$

where DI_r is the source directivity index in the direction of the reflecting surface and α_r is the absorption coefficient of the surface. If measured absorption coefficient data are unavailable, then ISO recommends using the values in Table 5.16.

TABLE 5.16 Estimates of sound-absorption coefficient (derived from ISO 9613-2 (1996))

Object	α_r
Flat, hard walls with no openings	0.0
Walls of building with windows and small additions or bay	0.2
Walls of building with 50% openings, installation or pipes	0.6
Cylinders with hard surfaces (tanks, silos)[a]	$\left[1 - \dfrac{D\sin(90-\theta)}{2r_S}\right]$ where D=cylinder diameter (m) r_S is the distance from the source to the cylinder centre and the cylinder centre is at location O in Figure 4.17

[a]Applies only if the cylinder is much closer to the source than it is to the receiver.

A number of adjustments were suggested during investigations of the most appropriate EU noise model (European Commission, 2010a). The most appropriate ones are listed below.

- Overall G values are to be weighted by the relative extents of the different ground types between source and receiver. The values proposed for G in the ISO 9613-2 (1996) standard should preferably be replaced by those values in Table 5.17, which are more recent recommendations for the NMPB-2008 propagation model. However, these new G values have not been included in the ISO 9613-2 (1996) standard and are only provided here for information and to allow for the possibility of obtaining more accurate predictions than expected using the G values recommended in ISO 9613-2 (1996).

TABLE 5.17 Values for the parameter, G

Surface description	G
Very soft	1
Uncompacted, loose ground	0.8
Normal uncompacted ground	0.6
Compacted field, lawns, gravel	0.4
Compacted dense ground (unpaved road)	0.2
Hard and very hard (asphalt, concrete, water)	0

- Section 7.3.2, which is an alternative method for calculating A-weighted sound pressure levels at the receiver, is to be entirely disregarded.
- Under Section 7.4:
    - Equation (12) in the standard (see Equation (5.227) above) shall be replaced with:

$$A_b + A_g = \max(A_g, A_b) \tag{5.240}$$

    - The value of D_z calculated using Equation (14) in ISO 9613-2 (1996) (D_{zi} in Equation (5.230) above) cannot be less than 0.
    - The quantity, C_2, in Equation (5.230) above is always equal to 20.

5.6.8 Limitations of the ISO 9613-2 Model

The ISO 9613-2 model, when applied to the prediction of noise in surrounding communities, has the following uncertainties and issues associated with it.

- It has only been validated for distances between source and receiver of less than 1 km.
- The model is only valid for octave band analysis for octave band centre frequencies ranging from 63 Hz to 4000 Hz. Of course, it is also valid for calculating overall A-weighted sound pressure levels over the frequency range covered by the 63 Hz to 4000 Hz octave bands.
- Downwind propagation is assumed by the ISO model, but only wind speeds between 1 m/s and 5 m/s (measured between 3 m and 11 m above the ground) are valid.
- Significant deviations from the ISO model may be expected for wind speeds above the 5 m/s limit (Kalapinski and Pellerin, 2009).
- The ISO model has only been validated for source heights <30 m above the ground.
- No allowance is made for the attenuation effects of scattering due to atmospheric turbulence.
- The ISO standard states that the expected accuracy of sound pressure level predictions is ±3 dB for source heights less than 30 m and for distances between the source and receiver of less than 1000 m. It is expected that the errors will increase as the source/receiver distance increases. Also, as the source height increases above 30 m, the calculation of the attenuation due to the ground becomes less accurate. Accuracy and prediction uncertainty are discussed in more detail in Section 5.9.
- The ISO model assumes that sound propagation is from point sources, which makes the calculated sound pressure levels close to large sound sources subject to greater errors than those estimated above.

5.7 NMPB-2008 Noise Propagation Model

The NMPB-2008 model (sometimes called the NMPB-Routes-2008 model) is a revised French model that was originally intended for the prediction of the propagation of road traffic noise. However, it has been selected by the European Union as the preferred model for propagation from all noise sources, including industrial sources such as wind turbines (European Commission, 2015; Kephalopoulos et al., 2012; Dutilleux et al., 2010). The calculation procedure is implemented for the centre frequencies of the octave bands from 63 Hz to 8 kHz. The propagation effect is considered to be the same for all frequencies in any one octave band.

The recommended calculation procedure to determine the sound pressure level at a receiver due to all relevant sound sources is as follows.

- Determine equivalent point sources for all noise sources.
- Determine the sound power level, L_{Wi}, for the ith source in the direction of the kth receiver. This will then include the directional properties of the source.
- Determine all propagation paths between the ith source and the receiver, including reflected and diffracted paths.
- For each propagation path calculate:
    - the attenuation for neutral (or homogeneous) (no wind or temperature gradients) atmospheric conditions;
    - the attenuation for downward refracting (favourable) atmospheric conditions;

  - the percentage of time that favourable conditions occur; and
  - the long-term sound pressure level, arriving at the receiver via each path.
- Determine the total sound pressure level arising from combining the sound pressure levels from each contributing path. This is done by assuming incoherent addition of contributions as discussed in Section 1.11.3.

The basic equation for calculating the octave band sound pressure level, L_{pik}, at location, k, due to the octave band sound power, L_{Wi}, radiated by sound source, i, with a directivity index in the direction of the receiver location of DI_{ik}, is similar to Equation (5.212) for the CONCAWE model and is given by:

$$L_{pik} = L_{Wi} - A_{\mathrm{div},ik} + \mathrm{DI}_{ik} - A_{Eik} \quad \text{(dB)} \tag{5.241}$$

where the directivity index, DI_{ik}, is often assumed to be 0 dB or is taken into account in the sound power measurement.

The total octave band sound pressure level due to all sound sources is calculated by incoherent addition (see Section 1.11.3) of the sound pressure level generated by each source. This total is denoted L_H for neutral or homogeneous atmospheric conditions for which no vertical sonic gradient exists, and L_F for downward refracting conditions for which the speed of sound increases with increasing height above the ground.

The long-term average octave band sound pressure level, L_{LT}, is obtained by logarithmically summing the sound pressure levels due to the two atmospheric conditions, with the sound pressure levels weighted by the fraction of the time that each condition is deemed to exist.

$$L_{LT} = 10\log_{10}\left[w10^{(L_F/10)} + (1-w)10^{(L_H/10)}\right] \quad \text{(dB)} \tag{5.242}$$

where $w \leq 1.0$ is the fraction of time that downward refraction conditions exist and the remaining time is allocated to the homogeneous atmospheric condition.

The overall A-weighted sound pressure level is calculated by applying the A-weighting correction (see Table 2.3) to each octave band level and then summing the levels logarithmically (see Section 1.11.3). The overall C-weighted sound pressure level can be calculated by following a similar procedure.

The attenuation, A_{Eik}, is defined by:

$$A_{Eik} = [A_a + A_{g+b} + A_r]_{ik} \quad \text{(dB)} \tag{5.243}$$

where A_{g+b} is the only term affected by wind and temperature gradients and atmospheric turbulence, and the term, A_{misc}, used in the ISO 9613-2 model is excluded from this model.

The attenuation component, $A_{\mathrm{div},ik}$, is the attenuation calculated assuming a point source using Equation (5.56) and the component, A_a, is calculated using Equation (5.61) as described in Section 5.3.2. Both of these attenuation effects are applied to the final total calculated sound pressure level at the receiver for each sound source.

The attenuation component, A_r, only applies to rays that are reflected from vertical (or near vertical, less than 15° from vertical) surfaces such as buildings. The attenuation is calculated as described in Section 5.7.1.7 and is applied to the ray prior to its sound pressure level at the receiver being incoherently summed with the sound pressure levels from non-reflected rays. The term, A_r, does not include ground reflections as these are included in the A_{g+b} term, which is discussed in Section 5.7.1. This model does not take into account the effect of multiple ground reflections when the receiver is at a large distance from the source in comparison with the source or receiver height.

5.7.1 Ground, Barrier and Terrain Attenuation, A_{g+b}

In comparing Equations (5.243) with (5.219), it is noted that in the NMPB-2008 model, the attenuation due to the ground is combined with that due to barriers, whereas these two attenuations are treated separately in the ISO 9613-2 model. Two atmospheric conditions are considered in the calculation of the A_{g+b} term – a neutral atmosphere characterised by no refraction and a downward refracting atmosphere (referred to in the NMPB documentation as "favourable"). For each atmospheric condition, three different situations are considered in the NMPB model.

1. Ground reflection with no diffraction, so that $A_{g+b} = A_{g,H}$ for a homogeneous atmosphere (no sonic gradient) and $A_{g+b} = A_{g,F}$ for a downward refracting atmosphere.
2. Diffraction over a barrier with no ground effect so that $A_{g+b} = A_{b,H}$ for a homogeneous atmosphere and $A_{g+b} = A_{b,F}$ for a downward refracting atmosphere.
3. Diffraction over or around a barrier with the ground effect included, so that $A_{g+b} = A_{g+b,H}$ for a homogeneous atmosphere and $A_{g+b} = A_{g+b,F}$ for a downward refracting atmosphere.

The calculation of the quantities, $A_{g,H}$, $A_{g,F}$, $A_{b,H}$, $A_{b,F}$, $A_{g+b,H}$ and $A_{g+b,F}$, is discussed in the following pages.

5.7.1.1 Mean Ground Plane

Before continuing, it is necessary to define a mean ground plane and the heights of the source and receiver above this mean plane, as illustrated in Figure 5.32. The equivalent height of a ground point is the height above the mean plane. If a height is negative, it is set equal to zero. Real heights above the real ground are denoted with the symbol, h, and heights above the mean ground plane are denoted with the symbol, z.

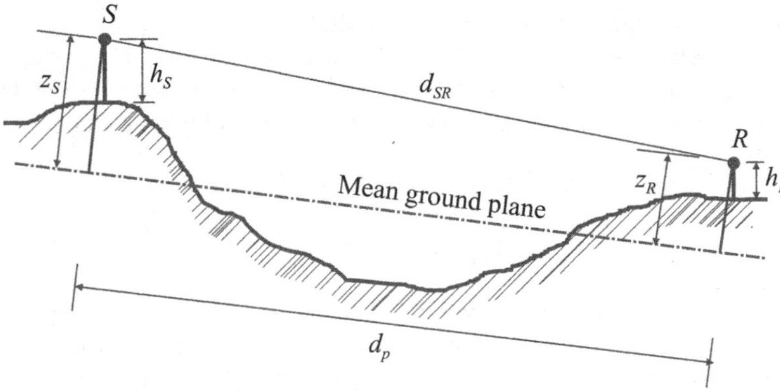

FIGURE 5.32 Source and receiver heights above the ground and the mean ground plane.

The location of the mean plane shown in Figure 5.32 is calculated as follows. The cross-section of the topography between the source, S, and receiver, R, is defined by a set of discrete points (x_k, z_k); $k = 1, 2, ..., n$, where x is the coordinate in the horizontal direction and z is the coordinate in the vertical direction, *referenced to an arbitrary origin.* For convenience, the origin is usually chosen as the point on the ground directly beneath the source so that this point corresponds to $k = 1$. The distance in the x-direction between data points should be 1 m or less. Each discrete point is joined to the next by a straight line to form a polyline, the segments of which are defined by (European Commission, 2015, p. 27):

$$z_k = a_k x_k + b_k \tag{5.244}$$

where

$$a_k = \frac{z_{k+1} - z_k}{x_{k+1} - x_k} \quad \text{and} \quad b_k = \frac{z_k x_{k+1} - z_{k+1} x_k}{x_{k+1} - x_k} \tag{5.245}$$

The equation of the mean line is:

$$z = ax + b; \quad x \epsilon [x_1, ... x_n] \tag{5.246}$$

where

$$a = \frac{3[2A - B(x_n + x_1)]}{(x_n - x_1)^3} \tag{5.247}$$

and

$$b = \frac{2B(x_n^3 - x_1^3)}{(x_n - x_1)^4} - \frac{3A(x_n + x_1)}{(x_n - x_1)^3} \tag{5.248}$$

where

$$A = \frac{2}{3} \sum_{k=1}^{n-1} a_k \left(x_{k+1}^3 - x_k^3\right) + \sum_{k=1}^{n-1} b_k \left(x_{k+1}^2 - x_k^2\right) \tag{5.249}$$

and

$$B = \sum_{k=1}^{n-1} a_k \left(x_{k+1}^2 - x_k^2\right) + 2 \sum_{k=1}^{n-1} b_k \left(x_{k+1} - x_k\right) \tag{5.250}$$

5.7.1.2 Ground Effect Factor, G

The ground effect is a function of the flow resistivity of the ground surface, where the extent of different surfaces between the source and receiver must be taken into account by simple averaging. The purpose of this section is to introduce the factor, G, which is used to characterise the flow resistivity of the ground surface in a similar way as done in the ISO 9613-2 model. Values of G to be used in the NMPB-2008 model for various ground surfaces are listed in Table 5.2.

To account for varying ground surfaces between the source and receiver, a quantity, G_{path}, is defined and used in place of G in the following analysis.

$$G_{\text{path}} = \frac{\sum_{i=1}^{\ell} G_i d_i}{\sum_{i=1}^{\ell} d_i} \tag{5.251}$$

where ℓ is the number of different ground surfaces between the source and receiver, G_i is the value of G for the ith ground surface and d_i is the length of the ith ground surface projected on the horizontal plane. The distance, d_p, between the source and receiver along the mean ground plane is:

$$d_p = \sqrt{(z_{mn} - z_{m1})^2 + (x_1 - x_n)^2} \tag{5.252}$$

where z_{m1} is the mean ground plane vertical coordinate at the source location, x_1 is the ground plane horizontal coordinate at the source location, z_{mn} is the mean ground plane vertical coordinate at the receiver location and x_n is the ground plane horizontal coordinate at the receiver location.

Some adjustment is necessary for relatively small source/receiver distances, so a new quantity, G'_{path}, to replace G_{path} is defined as:

$$G'_{\text{path}} = \begin{cases} G_{\text{path}} \dfrac{d_p}{30(z_S + z_R)} + G_S \left(1 - \dfrac{d_p}{30(z_S + z_R)}\right); & d_p \le 30(z_S + z_R) \\ G_{\text{path}}; & d_p > 30(z_S + z_R) \end{cases} \tag{5.253}$$

where G_S is the value of G in the vicinity of the source.

5.7.1.3 Ground Effect With No Diffraction: Homogeneous Atmosphere

The ground effect with no diffraction condition can be divided into two categories. The first is for a homogeneous atmosphere (discussed in this section) and the second is for a downward diffracting atmosphere (discussed in Section 5.7.1.4).

The attenuation, $A_{g,H}$, due to the ground effect for homogeneous atmospheric conditions is calculated as (European Commission, 2015, p. 34):

$$A_{g,H} = \max \begin{cases} -10\log_{10}\left[\dfrac{4k^2}{d_p^2}\left(z_S^2 - z_S\sqrt{\dfrac{2C_f}{k}} + \dfrac{C_f}{k}\right)\left(z_R^2 - z_R\sqrt{\dfrac{2C_f}{k}} + \dfrac{C_f}{k}\right)\right] \\ A_{g,H,\min} \end{cases} \tag{5.254}$$

where the wavenumber, $k = 2\pi f/c$, f is the octave band centre frequency and c is the speed of sound in air (= 343 m/s at 20°C), and:

$$A_{g,H,\min} = -3(1 - G_m) \tag{5.255}$$

The quantity, C_f, is defined as:

$$C_f = d_p \frac{1 + 3wd_p\exp(-\sqrt{wd_p})}{1 + wd_p} \tag{5.256}$$

where

$$w = 0.0185 \frac{f^{2.5}G_w^{2.6}}{f^{1.5}G_w^{2.6} + 1.3 \times 10^3 f^{0.75}G_w^{1.3} + 1.16 \times 10^6} \tag{5.257}$$

and where for a homogeneous atmosphere and no diffraction:

$$G_w = G_m = G'_{\text{path}} \tag{5.258}$$

If $G'_{\text{path}} = 0$, then $A_{g,H} = -3$ dB.

5.7.1.4 Ground Effect for a Downward Refracting Atmosphere and No Diffraction

For a downward refracting atmosphere and for $G'_{\text{path}} \neq 0$, Equation (5.254) can be used but with $A_{g,H}$ replaced with $A_{g,F}$ on the LHS of the equation and with $A_{g,H,\min}$ replaced with $A_{g,F,\min}$ on the RHS of the equation. Also, the quantities, z_S and z_R in Equation (5.253) are replaced by $z_S + \delta z_S + \delta z_T$ and $z_R + \delta z_R + \delta z_T$, respectively, where δz_S and δz_R account for the bending of the sound rays due to refraction and δz_T accounts for the effect of atmospheric turbulence. These quantities are defined as (European Commission, 2015, p. 35):

$$\delta z_S = a_0 \left(\frac{z_S}{z_S + z_R}\right)^2 \frac{d_p^2}{2}; \qquad \delta z_R = a_0 \left(\frac{z_R}{z_S + z_R}\right)^2 \frac{d_p^2}{2} \tag{5.259a,b}$$

$$\delta z_T = 6 \times 10^{-3} \left(\frac{d_p}{z_S + z_R}\right) \tag{5.260}$$

where $a_0 = 2 \times 10^{-4}$ m^{-1} is the inverse of the radius of curvature of the sound ray (assumed here to be 5 km). This corresponds to an assumed mean value of the sound speed gradient of $B_m = 2c \times 10^{-4}$ m^{-1} $= 0.07s^{-1}$. This value of a_0 is an assumed quantity for the purposes of this analysis and is considered to represent a typical downward refracting atmosphere but is not representative of what occurs in the presence of a low jet.

For the downward refracting case, the lower bound, $A_{g,F,\min}$, for Equation (5.254) is defined as:

$$A_{g,F,\min} = \begin{cases} -3(1 - G_m); & \text{if } d_p \le 30(z_S + z_R) \\ -3(1 - G_m)\left\{1 + 2\left[1 - \dfrac{30(z_S + z_R)}{d_p}\right]\right\}; & \text{if } d_p > 30(z_S + z_R) \end{cases} \tag{5.261}$$

If $G'_{\text{path}} = 0$, then $A_{g,F} = A_{g,F,\min}$.

In Equation (5.257) for the downward refracting case in the absence of diffraction, the quantities, G_w and G_m, are re-defined as:

$$\begin{cases} G_w = G_{\text{path}} \\ G_m = G'_{\text{path}} \end{cases} \tag{5.262}$$

5.7.1.5 Diffraction with No Ground Effect

For pure diffraction with no ground effects, the attenuation is (European Commission, 2015, p. 37):

$$A_b = \begin{cases} 10C_h \log_{10}\left(3 + \dfrac{40}{\lambda}C''\delta\right); & \text{if } \dfrac{40}{\lambda}C''\delta \geq -2 \\ 0; & \text{if } \dfrac{40}{\lambda}C''\delta < -2 \end{cases} \tag{5.263}$$

where

$$C_h = \min \begin{cases} fh_0/250 \\ 1 \end{cases} \tag{5.264}$$

and $h_0 = \max(z_{\text{BR}}, z_{\text{BS}})$ (see Figure 5.33), λ is the wavelength at the octave band centre frequency, δ is the path difference between direct and diffracted sound rays (see Figures 5.34 and 5.35), C'' is a coefficient to take into account multiple diffractions and $C'' = 1$ for a single diffraction, as shown in Figures 5.34(a) and (b) and 5.35(a), (b) and (c). For multiple diffractions, as shown in Figures 5.34(c), (d) and (e) and 5.35(d), (e) and (f):

$$C'' = \frac{1 + (5\lambda/e)^2}{1/3 + (5\lambda/e)^2} \tag{5.265}$$

If $A_b < 0$, then A_b is set equal to 0. The path difference, δ, for various relationships between source, receiver and diffraction edges is illustrated in Figure 5.34 for a homogeneous atmosphere

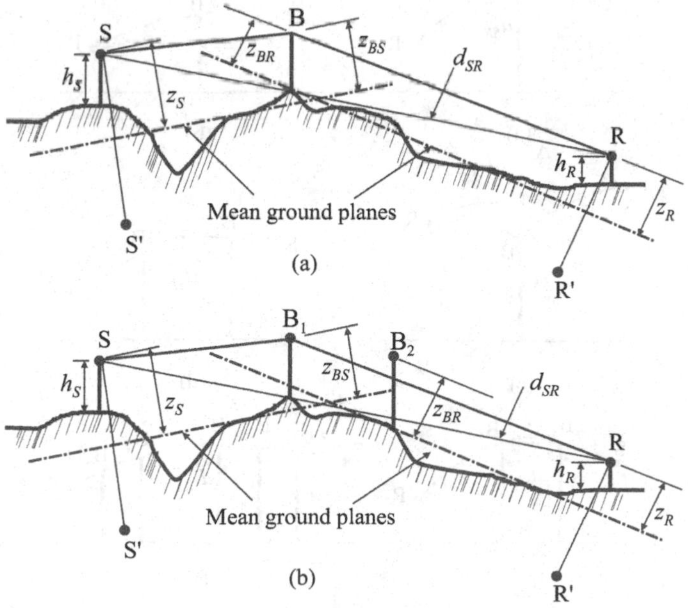

FIGURE 5.33 Geometry for diffraction over single and double diffracting edges. For more than one diffracting edge, the ground effect between two adjacent diffracting edges is ignored as shown in part (b).

(no vertical sonic gradient and straight sound rays) and in Figure 5.35 for a downward refracting atmosphere (positive vertical sonic gradient and curved sound rays).

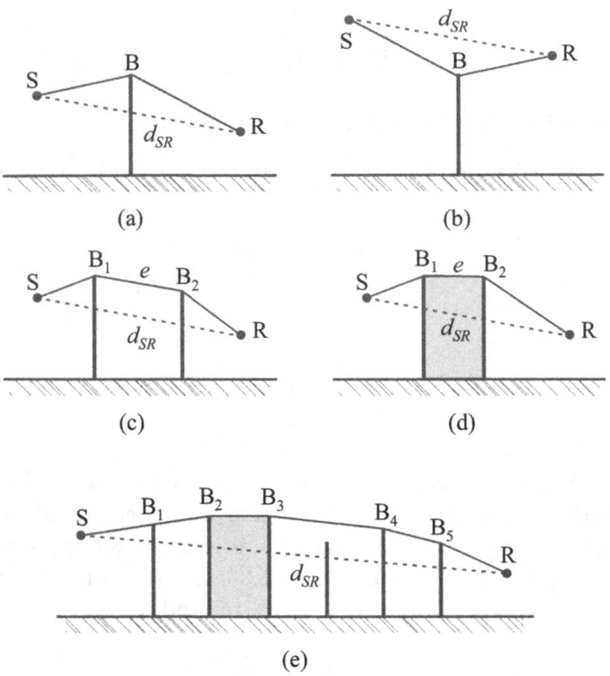

FIGURE 5.34 Various geometries for determining the path length differences between direct and diffracted rays for a homogeneous atmosphere. In part (e), $e = B_1B_2 + B_2B_3 + B_3B_4 + B_4B_5$.

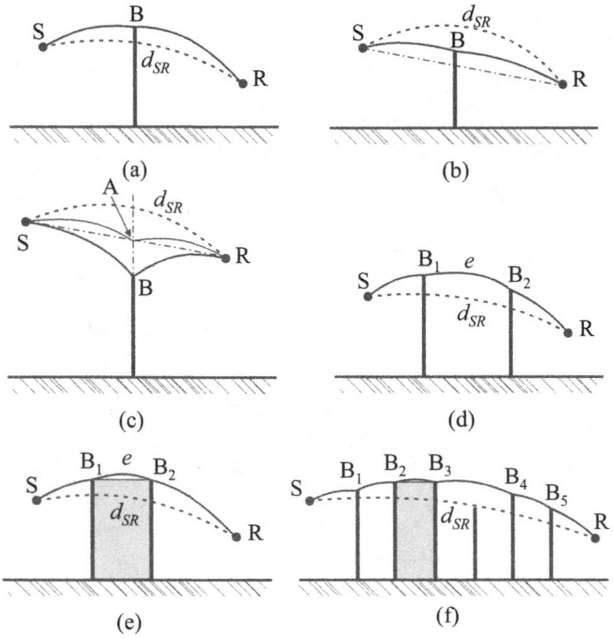

FIGURE 5.35 Various geometries for determining the path length differences between direct and diffracted rays for a downward refracting atmosphere. In part (f), e is the sum of the curved path lengths, $B_1B_2 + B_2B_3 + B_3B_4 + B_4B_5$.

The path differences, δ, for use in Equation (5.263) are defined for a homogeneous atmosphere, represented by the various parts of Figure 5.34, as:

$$\delta = \begin{cases} \mathrm{SB} + \mathrm{BR} - d_{SR} & \text{Figure 5.34(a)} \\ -(\mathrm{SB} + \mathrm{BR} - d_{SR}) & \text{Figure 5.34(b)} \\ \mathrm{SB}_1 + e + \mathrm{B}_2\mathrm{R} - d_{SR} & \text{Figure 5.34(c)} \\ \mathrm{SB}_1 + e + \mathrm{B}_2\mathrm{R} - d_{SR} & \text{Figure 5.34(d)} \\ \mathrm{SB}_1 + \mathrm{B}_1\mathrm{B}_2 + \mathrm{B}_2\mathrm{B}_3 + \mathrm{B}_3\mathrm{B}_4 + \mathrm{B}_4\mathrm{B}_5 + \mathrm{B}_5\mathrm{R} - d_{SR} & \text{Figure 5.34(e)} \end{cases} \tag{5.266}$$

The path differences, δ, for use in Equation (5.263) are defined for a downward refracting atmosphere, represented by the various parts of Figure 5.35 as:

$$\delta = \begin{cases} \mathrm{SB} + \mathrm{BR} - d_{SR} & \text{Figure 5.35(a)} \\ \mathrm{SB} + \mathrm{BR} - d_{SR} & \text{Figure 5.35(b)} \\ 2\mathrm{SA} + 2\mathrm{AR} - \mathrm{SB} - \mathrm{BR} - d_{SR} & \text{Figure 5.35(c)} \\ \mathrm{SB}_1 + e + \mathrm{B}_2\mathrm{R} - d_{SR} & \text{Figure 5.35(d)} \\ \mathrm{SB}_1 + e + \mathrm{B}_2\mathrm{R} - d_{SR} & \text{Figure 5.35(e)} \\ \mathrm{SB}_1 + \mathrm{B}_1\mathrm{B}_2 + \mathrm{B}_2\mathrm{B}_3 + \mathrm{B}_3\mathrm{B}_4 + \mathrm{B}_4\mathrm{B}_5 + \mathrm{B}_5\mathrm{R} - d_{SR} & \text{Figure 5.35(f)} \end{cases} \tag{5.267}$$

For the curved sound ray case, all path lengths are along the curved paths and are thus greater than the straight-line distance between the two points joined by the curved ray. For the multiple diffraction cases, involving more than two diffraction edges, a convex hull is drawn and any diffraction edges below the hull outline are ignored as shown in Figures 5.34(e) and 5.35(f).

The path lengths in Equation (5.267) are all curved, as shown in Figure 5.35. In part (c), point A is the point of intersection of the straight line between the source and receiver and the vertical extension of the diffracting obstacle.

If the path length difference, δ, is less than $-\lambda/20$ for any of the three cases shown in Figures 5.34(b) and 5.35(b) and (c), then the diffraction effect is ignored and the attenuation due to diffraction is 0 dB.

The definition of the path length difference, δ, for the general case of n diffraction edges on the convex hull between the source and receiver is:

$$\delta = \mathrm{SB}_1 + \sum_{i=1}^{i=n-1} \mathrm{B}_i\mathrm{B}_{i+1} + \mathrm{B}_n\mathrm{R} - d_{SR} \tag{5.268}$$

where the paths are straight lines for the homogeneous atmosphere case and curved lines for the downward refracting case. The radius of curvature, R_c, of the curved rays is given by:

$$R_c = \max(1000, 8d_{SR}) \tag{5.269}$$

where in this particular equation, d_{SR} is the straight line distance between the source and receiver. This value of R_c is different to that used for the case of no diffraction (see line following Equation (5.260)). However, no explanation for this discrepancy is provided in European Commission (2015) and is difficult to justify.

The length, ℓ_c, of a curved ray path between any point B_i and B_{i+1} is:

$$\ell_c = 2R_c \arcsin\left(\frac{\mathrm{B}_i\mathrm{B}_{i+1}}{2R_c}\right) \tag{5.270}$$

where $\mathrm{B}_i\mathrm{B}_{i+1}$ is the straight-line distance between points B_i and B_{i+1}. The length, ℓ_c, of the curved ray path is all that is needed to calculate δ with Equation (5.268), which is then used with Equation (5.263) to calculate A_b.

5.7.1.6 Diffraction with Ground Effect

When the ground effect is included with the diffraction effect the following quantities are calculated.

- Attenuation, A_{g,SB_1}, due to the ground between the source and the closest diffraction edge. A new mean ground plane must be calculated for the ground between the source and the nearest diffraction edge, in order to calculate the quantity, z_S and locate the image source, S′, as shown in Figure 5.33(a). The image source coordinates may be obtained using Equation (5.156), where m_S is the slope of the mean ground plane.
- Attenuation, A_{g,SB_n}, due to the ground between the receiver and the closest diffraction edge. A new mean ground plane must be calculated for the ground between the receiver and the nearest diffraction edge, in order to calculate the quantity, z_R and locate the image receiver, R′, as shown in Figure 5.33(a). The image receiver coordinates may be obtained using Equation (5.161), where m_R is the slope of the mean ground plane.
- Diffraction between the source, S, and the receiver, R, by considering the path difference, δ, between the direct path from the source to the receiver and the path over all obstacles between the source and receiver.
- Diffraction between the image source, S′, and the receiver, R, by considering the path difference, δ, between the direct path from the image source to the receiver and the path over all obstacles between the image source and receiver.
- Diffraction between the source, S, and the image receiver, R′, by considering the path difference, δ, between the direct path from the source to the image receiver and the path over all obstacles between the source and image receiver.

The calculation of the attenuation, A_{g+b}, due to diffraction with ground reflections only takes into account the ground between the source and its nearest diffraction edge, B_1, and the ground between the receiver and its nearest diffraction edge, B_n (see Figure 5.33). A_{g+b} is calculated as:

$$A_{g+b} = A_{b(\mathrm{SR})} + \Delta_{g(\mathrm{SB}_1)} + \Delta_{g(\mathrm{B}_n\mathrm{R})} \tag{5.271}$$

where B_1 is the top of the diffraction edge nearest the source and B_n is the top of the diffraction edge nearest the receiver and:

$$\Delta_{g(\mathrm{SB}_1)} = -20\log_{10}\left[1 + \left(10^{-A_{g(\mathrm{SB}_1)}/20} - 1\right) \times 10^{-(A_{b(\mathrm{S'R})} - A_{b(\mathrm{SR})})/20}\right] \tag{5.272}$$

and

$$\Delta_{g(\mathrm{B}_n\mathrm{R})} = -20\log_{10}\left[1 + \left(10^{-A_{g(\mathrm{B}_n\mathrm{R})}/20} - 1\right) \times 10^{-(A_{b(\mathrm{SR'})} - A_{b(\mathrm{SR})})/20}\right] \tag{5.273}$$

- $A_{g(\mathrm{SB}_1)}$ is the attenuation due to the ground between the source and diffraction edge, B_1, calculated as described in Sections 5.7.1.2, 5.7.1.3 and 5.7.1.4, but with the following conditions:
  - z_R is replaced with z_{BS} – see Figure 5.33;
  - The height, z_S, is calculated relative to the mean ground plane representing the ground between the source and nearest diffraction edge as shown in Figure 5.33(a);
  - the value of G_{path} of Equation (5.251) is calculated only for the ground between the source and the nearest diffraction edge;
  - in homogeneous atmospheric conditions, G_w of Equation (5.257) is replaced by G'_{path} of Equation (5.253);

- in downward refracting atmospheric conditions, G_w of Equation (5.257) is replaced by G_{path}; and
- in both homogeneous and downward refracting atmospheric conditions, G_m of Equations (5.255) and (5.261) is replaced by G'_{path}.

- $A_{g(\text{B}_n\text{R})}$ is the attenuation due to the ground between the receiver and diffraction edge, B_n, calculated as described in Section 5.7.1.2, but with the following conditions:
  - z_S is replaced with z_{BR} – see Figure 5.33;
  - The height, z_R, is calculated relative to the mean ground plane representing the ground between the receiver and nearest diffraction edge as shown in Figure 5.33;
  - the value of G_{path} is calculated only for the ground between the receiver and the nearest diffraction edge;
  - as the source has been replaced with the diffraction edge nearest to the receiver, the G'_{path} correction is not used;
  - in both homogeneous and downward refracting atmospheric conditions, G_w of Equation (5.257) is replaced by G_{path}; and
  - in both homogeneous and downward refracting atmospheric conditions, G_m of Equations (5.255) and (5.261) is replaced by G_{path}.
- $A_{b(\text{SR})}$ is the attenuation due to pure diffraction between S and R, calculated as described in Section 5.7.1.5, Equation (5.263).
- $A_{b(\text{S}'\text{R})}$ is the attenuation due to pure diffraction between the image source, S′ and R, calculated as described in Section 5.7.1.5, where the image source location is shown in Figure 5.33 and calculated using Equation (5.156).
- $A_{b(\text{SR}')}$ is the attenuation due to pure diffraction between S and the image receiver, R′, calculated as described in Section 5.7.1.5, where the image receiver location is shown in Figure 5.33 and calculated using Equation (5.161).

The ray from the image source to the image receiver is ignored in this model, probably as its contribution is considered negligible in almost all situations.

5.7.1.7 Vertical Edge Diffraction with Ground Effect

In some cases, it is of interest to be able to calculate the sound pressure level arriving at a receiver after diffraction around a vertical edge, such as the side of a building, which lies between the source and receiver. The sound pressure level due to a path from the source to the receiver around a vertical edge is then logarithmically added to the sound pressure level at the receiver due to other paths (including other diffracted paths) to obtain the total sound pressure level at the receiver, using Equation (1.11.3). In this case, the attenuation is still calculated using Equation (5.243). However, the term, A_{div}, is calculated from the direct distance, d_{SR}, while the terms, A_a and A_g, are calculated using the total length of the propagation path around the side of the obstacle. The term, A_{g+b}, is calculated using:

$$A_{g+b} = A_g + A_{b(\text{SR})} \tag{5.274}$$

where $A_{b(\text{SR})}$ is calculated using Equation (5.263) with the path difference, δ, being the difference between the direct path from the source to the receiver and the path around the diffraction edge.

5.7.2 Reflections from Vertical Surfaces

A vertical surface that does not interrupt the line of sight between the source and receiver, may result in the generation of a ray reflected from its surface that will contribute to the total sound pressure level at the receiver. A potentially reflective surface is considered to be vertical if its slope is less than 15°. The obstacle is ignored if any dimension is less than 0.5 m. The sound pressure arriving at the receiver due to the reflected ray is added incoherently to the ray arriving via the non-reflected path. The reflected ray will exist if it is possible to find a location on the surface where the angle between the incident ray and the surface normal is the same as the angle between the reflected ray that arrives at the receiver and the surface normal. The additional attenuation due to reflection must be added to the attenuations due to spherical spreading, scattering and atmospheric absorption. The resulting sound pressure level at the receiver is then added incoherently to the sound pressure level due to other ray paths. If both a ground-reflected and direct ray are reflected from a vertical surface, then each of these is treated separately and the contribution of each at the receiver is added incoherently at the receiver. The additional attenuation due to absorption by the vertical surface is:

$$A_r = -10\log_{10}(1 - \alpha_r) \tag{5.275}$$

where $0 < \alpha_r < 1$ is the absorption coefficient of the surface.

5.7.3 Limitations of the NMPB-2008 Model

1. The model is valid for octave band analysis for octave band centre frequencies ranging from 63 Hz to 8000 Hz. It is also valid for calculating overall A-weighted sound pressure levels over the frequency range covered by the 63 Hz to 8000 Hz octave bands.
2. NMPB-2008 is valid only for point sources. Where necessary, other sources are modelled as collections of point sources with the same total sound power output. However, even large sources appear as point sources at distances greater than about 2 source dimensions, so this should not result in significant errors in almost all situations.
3. The terrain between source and receiver is approximated by one or two straight lines, which may not model actual situations accurately. However, this approach is superior to that used by the CONCAWE and ISO models described in Sections 5.5 and 5.6.
4. The geometry used for the terrain effects is based on 2-D rather than 3-D modelling which may be inaccurate in some cases.
5. Different estimates are used for the radius of curvature of the sound ray paths for diffracted and non-diffracted cases (Equation (5.269) versus 5 km).
6. Any barriers are generally assumed to be infinite in length, with their face normal to the line between the source and receiver. However, if the effect of finite width on the barrier attenuation needs to be taken into account, then refraction around the vertical edges of the barrier may be included.
7. Only two atmospheric sonic profiles are considered — one with zero velocity gradient and one with a downward refracting gradient that produces a radius of curvature of the sound rays emitted from the source, given by Equation (5.269).
8. The model does not account for multiple ground reflections for the downward refracting atmosphere case, so its accuracy may not be sufficient for large distances between the source and receiver.
9. Only reflecting surfaces angled less than 15° from the vertical are considered.
10. No uncertainty data are provided. Accuracy and prediction uncertainty are discussed in Section 5.9.

5.8 Required Input Data for the Various Propagation Models

One of the most difficult aspects of applying the various noise propagation models in practice is obtaining the required input data. The required data becomes more detailed as the propagation model becomes more complex. In addition, if sound pressure level predictions are required for other than worst-case downwind propagation, then the specification of atmospheric wind and temperature gradients as well as turbulence is needed for the more complex models. No models allow for different atmospheric conditions to exist at different locations along the path from the source to the receiver. In hilly terrain, it is quite possible that non-uniform atmospheric conditions will exist and this will result in a degree of uncertainty in the final predicted sound pressure levels. Each of the models discussed in this chapter will now be considered in terms of the input data needed to enable the model to be applied.

5.8.1 CONCAWE

The input data needed for implementation of a CONCAWE propagation model are listed below.

- Height of the source, h_S, and receiver, h_R, above a horizontal reference line passing through the ground at the base of the source or receiver.
- Horizontal distance, d_{SR}, between the source and receiver.
- Cloud cover (octas).
- Wind speed (in the direction from the source to the receiver) at 10 m above the ground.
- Average atmospheric temperature and relative humidity between the source and receiver.
- Whether the ground is hard (concrete, asphalt, water, packed soil) or other.
- Location, length and height of any barriers between the source and receiver. This includes terrain and man-made barriers, and includes the difference in path lengths between the straight-line source/receiver distance and the source/receiver path via the top of the barrier.

5.8.2 ISO 9613-2

The input data needed for implementation of an ISO 9613-2 propagation model are listed below.

- Height of source, h_S, and receiver, h_R, above a horizontal reference line passing through the ground at the base of the source or receiver.
- Horizontal distance, d, between source and receiver.
- Average atmospheric temperature and relative humidity between the source and receiver.
- Whether the ground is hard (concrete, asphalt, water, packed soil) or soft (anything else). For this purpose, three assessments are needed.
  - One for ground extending from the source in the direction of the receiver for a distance of 30 times the source height.
  - One for ground extending from the receiver in the direction of the source for a distance of 30 times the receiver height.
  - One for the ground in between, which is not included in the two ground sections near the source and receiver described above.
- Location, length and height of any barriers between the source and receiver. This includes terrain and man-made barriers, and includes the difference in path lengths

between the straight-line source/receiver distance and the source/receiver path via the top of the barrier, for both single and double diffraction situations. If multiple barriers exist, only the two most significant are considered.

- Length of the source/receiver ray path that travels through foliage or housing estates, based on a radius of curvature of the sound ray of 5 km (see Equation (5.122)).
- Location and absorption coefficient of any vertically reflecting objects that could affect the sound pressure level at the receiver.

5.8.3 NMPB-2008

The input data needed for implementation of an NMPB-2008 propagation model are listed below.

- Average atmospheric temperature and relative humidity between the source and receiver.
- Category (A–G, see Table 5.2) of ground sections between the source and receiver and extent of each ground section type so a weighted average value of G can be calculated for use in the model.
- Terrain profile from the source to the receiver represented as a series of joined straight lines with defined end points. Equivalent barriers to represent any hills between the source and receiver.
- Heights, h_S, of the source and h_R of the receiver above the local ground immediately below.
- Distance, d_{SR}, between source and receiver.
- Fraction of the time that downward refracting atmospheric conditions exist.
- Location, length and height of any man-made barriers between the source and receiver.
- Location and absorption coefficient of any vertically reflecting objects that could affect the sound pressure level at the receiver.

5.9 Propagation Model Prediction Uncertainty

The accuracy of noise propagation modelling is often the subject of discussions in court, especially when predicted sound pressure levels are close to allowable sound pressure levels. When testing a prediction model with measured data, uncertainty in the comparison is a function of the uncertainty in the prediction model results and the uncertainty in the measured results. The uncertainty in the predicted results is a function of the accuracy of the propagation model in representing all of the physical parameters that influence the sound pressure level at the receiver. This type of uncertainty is often expressed as an accuracy and is an estimate based on experience. It is referred to in the statistics literature as a type B standard uncertainty. The uncertainty in the measured results is discussed in detail in Section 2.14.4, and as it is based on the statistics of measured data, it is referred to as a Type A standard uncertainty. Uncertainty is most conveniently expressed as an expanded uncertainty, u_e. These uncertainty types are discussed in the following sections.

5.9.1 Type A Standard Uncertainty

This type of uncertainty estimate is usually associated with repeated measurements to obtain a number of measurement sets, each of which is characterised by a mean and standard deviation. The means of all of the data sets are assumed to follow a normal distribution, characterised by its own mean and standard deviation. However, the standard deviation, u_s, of the means of all data sets can be calculated from the standard deviation, σ, of a single data set using:

$$u_s = \frac{\sigma}{\sqrt{N}} \tag{5.276}$$

where N is the number of measurements making up the data set. To avoid confusion, u_s is referred to as the standard uncertainty. For N_m individual measurements, $L_{pi,\ i=1,...N_m}$, for which the mean is L_{pm}, the standard deviation is defined as:

$$\sigma = \sqrt{\frac{\sum_{i=1}^{N_m}(10^{L_{pi}/10} - 10^{L_{pm}/10})^2}{(N_m - 1)}} \tag{5.277}$$

where $L_{pm} = 10\log_{10}\Big(\frac{1}{N_m}\sum_{i=1}^{N_m} 10^{L_{pi}/10}\Big)$ dB. For outdoor sound pressure level measurements, $L_{p,i}$ is usually an L_{eq} measurement, energy-averaged over 10 minutes.

The terms, "accuracy" and "range" do not apply to this type of uncertainty. This type of uncertainty could be applied to a series of sound pressure level measurements at a single location to determine the uncertainty in the estimate of the mean sound pressure level (see Section 2.14.4).

5.9.2 Type B Standard Uncertainty

Type B uncertainties are not based on statistics but usually on judgement and experience and, in the context of this chapter, they apply to the sound pressure levels predicted using a propagation model. This type of uncertainty usually assumes a rectangular distribution of values between an upper and a lower limit (range) or in terms of an accuracy. The terms, "accuracy" and "range", are used interchangeably and are expressed as $\pm x$ dB. In many cases, the expected occurrence of all values between the maximum and minimum limits is considered equally likely, so the distribution is rectangular rather than normal. In this case, the standard uncertainty, u_s, is related to the accuracy, x, by $u_s = x/\sqrt{3}$, and this is what is used in assessing the 95% confidence limits for a sound propagation model.

In cases where the occurrence of values increases in likelihood near the centre of the range, a normal distribution may be more appropriate and in this case, the standard uncertainty, u_s, is related to the accuracy or range, $\pm x$, by $u_s = x/2$, where x (dB) is half the difference between the maximum and minimum values of the sound pressure level.

5.9.3 Combining Standard Uncertainties

Type B standard uncertainties can be combined by summing the squares of each standard uncertainty element and then taking the square root of the result, provided that all quantities contribute equally to the quantity for which the overall standard uncertainty is to be determined. If the quantities that are represented by the various uncertainties contribute in different amounts to the quantity for which the overall uncertainty is to be determined, the individual uncertainties must be weighted by their relative contribution (see Equation (2.123)). For standard uncertainty, $u_{s,i}$, this weighting is referred to in ISO 1996-2 (2017) as a fractional quantity called the sensitivity index, c_i.

For N sound sources contributing different amounts to the total sound pressure level at a receiver, where the contribution from source, i, is represented by a standard uncertainty, $u_{s,i}$ (dB), the overall uncertainty, u_{tot}, is given by:

$$u_{s,\text{tot}} = \frac{\sqrt{\sum_{i=1}^{N}\left(u_{s,i} \times 10^{L_i/10}\right)^2}}{\sum_{i=1}^{N} 10^{L_i/10}} \quad \text{(dB)} \tag{5.278}$$

where L_i is the sound pressure level contribution at the receiver from source, i. Equation (5.278) shows that as the number of sources contributing significantly to the sound pressure levels at a receiver increases, the standard uncertainty of the sound pressure level prediction at a specific receiver will decrease.

The individual standard uncertainties in Equation (5.278), for the case of sound source, i, are themselves made up of two standard uncertainties: that due to the uncertainty in the sound power level data, $u_{s,\text{source}}$, (see Section 4.14.10), and that due to the uncertainty associated with the propagation model, $u_{s,\text{prop}}$. Thus:

$$u_{s,i} = \sqrt{u^2_{s,\text{source}} + u^2_{s,\text{prop}}} \quad \text{(dB)} \tag{5.279}$$

The uncertainty associated with the propagation model can itself be made up of a number of terms, with each term contributing to one of the attenuation effects that make up the propagation model. There would be a term for each of the following:

- meteorological effects;
- ground effects;
- atmospheric absorption;
- vegetation;
- source height;
- in-plant screening;
- barriers;
- terrain.

These uncertainties would be combined as indicated by Equation (2.123), with sensitivity coefficients, c_i, set equal to 1 in the absence of any other information. Usually propagation model uncertainty is expressed as an overall uncertainty (or accuracy) based on experience, as it is simpler and avoids the necessity of estimating the individual uncertainties associated with the items in the above list.

5.9.4 Expanded Uncertainty

The expanded uncertainty, u_e, is calculated by multiplying the standard uncertainty, u_s, by a factor to account for the uncertainty confidence limits. For a confidence limit of 95% (that is, a 95% chance that the true sound pressure level, corresponding to the meteorological and ground conditions assumed in the model, will be in this range), the factor is 1.96 if a two-sided normal distribution of the standard error is assumed. However, if the standard uncertainty is related to a one-sided distribution, so that it reflects the chance that a predicted value will be greater than the actual value (or alternatively, less than the actual value), then the expanded uncertainty for a 95% confidence limit is calculated by multiplying the standard uncertainty by 1.65 (instead of 1.96), which makes the expanded uncertainty very close numerically to the accuracy or range of possible values. The expanded uncertainty for a two-sided distribution is the quantity that should be used in assessing the uncertainty associated with acoustic predictions and measurements. The expanded uncertainty for a 95% confidence limit may be interpreted as there being a 95% chance that the difference between the predicted (or measured) sound pressure level and the actual sound pressure level at any time for which the model meteorological conditions (usually downwind or temperature inversion) are satisfied, will be between $\pm u_e$.

Generally, most noise models allow noise predictions to be made for the worst-case meteorological conditions that are expected to occur on a regular basis. However, the best accuracy (or expanded uncertainty) that can realistically be expected for sound pressure level predictions at distances of 100 m or more from a noise source is ±3 dB. However, an expanded uncertainty of

±4 dB is probably more realistic when uncertainties in the source sound power levels are taken into account. Some practitioners claim an accuracy of better than ±2 dB, but there are insufficient data available to confirm that. The difficulty in obtaining accurate predictions is mainly associated with the variability of the atmospheric wind and temperature profiles over time and geographic location.

There is another type of uncertainty called model bias, which is the difference between predictions and measured data averaged over many different locations. This result is also expressed as a 95 percentile, but it is inappropriate for use in a noise level prediction report as it does not indicate the error that could exist at a single location. Rather it tells us that, although in some locations the predictions may be high, in others they will be low, so that the average is usually less than the uncertainty for a single location.

The most extensive study of propagation model prediction uncertainty was done by Marsh (1982) for the CONCAWE model for source/receiver separations spanning distances of 200 to 2000 m. His results are summarised in Table 5.12 for the various meteorological categories discussed in Section 5.5. For downwind propagation only, the expanded uncertainty (for 95% confidence limits) for the CONCAWE model is approximately ±4.5 dBA.

The ISO 9613-2 model (see Section 5.6) includes an estimate of the accuracy of the predicted overall A-weighted sound pressure level. For a mean source and receiver height less than 5 m ($[h_S + h_R]/2 < 5$ m), the estimated accuracy is ±3 dB. For a mean source and receiver height between 5 m and 30 m, the estimated accuracy is ±1 dB for source/receiver distances of less than 100 m, and ±3 dB for source/receiver distances between 100 m and 1000 m. It is expected that for higher sources such as wind turbines, the upper distance limits mentioned above may be extended further, which implies that the propagation models are more accurate for higher sound sources, as the distance from the source at which multiple ground reflections begin to occur increases with increasing difference between the source and receiver height. Perhaps the greatest uncertainty lies in the input data used in the prediction models, especially meteorological data, including atmospheric wind and temperature variation as a function of height above the ground.

In summary, for most A-weighted environmental noise predictions, it would be wise to suggest that the variation between prediction and measurement for any particular location in a downward refracting atmosphere is of the order of ± 4 to 5 dBA.

6

Sound in Enclosed Spaces

LEARNING OBJECTIVES

In this chapter, the reader is introduced to:

- wall-cavity modal coupling, when it is important and when it can be ignored;
- the simplifying assumption of locally reactive walls;
- three kinds of rooms: Sabine rooms, flat rooms and long rooms;
- Sabine and statistical absorption coefficients;
- low-frequency modal description of room response;
- high-frequency statistical description of room response;
- transient response of Sabine rooms and reverberation decay;
- reverberation time calculations;
- the room constant and its determination;
- porous, metamaterial, micro-perforate and panel sound absorbers; and
- applications of sound-absorption.

6.1 Introduction

Sound in an enclosed space is strongly affected by the reflective properties of the enclosing surfaces and to the extent that the enclosing surfaces are reflective, the shape of the enclosure also affects the sound field. When an enclosed space is bounded by generally reflective surfaces, multiple reflections will occur, and a reverberant field will be established in addition to the direct field from the sound source. Thus, at any point in such an enclosure, the overall sound pressure level is a function of the energy contained in the direct and reverberant fields.

In general, the energy distribution and variation with frequency of a sound field in an enclosure with reflective walls is difficult to determine with precision. Fortunately, average quantities are often sufficient and procedures have been developed for determining these quantities. Accepted procedures divide the problem of describing a sound field in a reverberant space into low- and high-frequency ranges, loosely determined by the ratio of a characteristic dimension of the enclosure to the wavelength of the sound considered. For example, the low-frequency range might be characterised by a ratio of less than 10 while the high-frequency range might be characterised by a ratio of greater than 10; however, precision will be given to the meaning of these concepts in the discussion in Section 6.3.

DOI: 10.1201/9780367814908-6

6.1.1 Wall-Interior Modal Coupling

A complication that arises from consideration of sound in an enclosure is that coupling between acoustic modes in the enclosed space (cavity modes) and vibration modes in the enclosure boundaries (wall modes) generally cannot be ignored. For example, in the low-frequency range of the relatively lightweight structures that characterise aircraft and automobiles, coupled wall-cavity modes may be dominant. In such cases, the sound field in the enclosed space cannot be considered in isolation; coupling between the modes in the wall and the cavity must be considered and the wall modes thus become equally as important as the cavity modes in determining the acoustic field in the enclosure (Pan and Bies, 1990a,b,c). Generally, for the lightweight enclosure cases cited above, the low-frequency range extends over most of the audio-frequency range.

It has become traditional to begin a discussion of room acoustics with the assumption that the walls of the enclosure are locally reactive (Kuttruff, 2009) or effectively infinitely stiff; the alternative that the walls may be bulk reactive, meaning that wall-cavity mode coupling is important, seems rarely to have been considered for the case of sound in rooms. However, it has been shown that when a sound field is diffuse, meaning that the sound energy travels in all directions within the enclosed space with equal probability, the modal response of a bounding surface will be similar to that of a surface which is locally reactive (Pan and Bies, 1988). Thus, in the frequency range in which the sound field may be assumed to be diffuse, the assumption of locally reactive walls gives acceptable results. However, in the low-frequency range, where the sound field will not be diffuse, cavity-wall modal coupling can be expected to play a part in the response of the room. Modal coupling will affect the resulting steady-state sound field levels as well as the room reverberation time (Pan and Bies, 1990b,c,d). See also Sections 6.5 and 6.6.2.

In the high-frequency range, the concept of a locally reactive boundary is of great importance, as it serves to uncouple the cavity and wall modes and greatly simplify the analysis (Morse, 1939). Locally reactive means that the response to an imposed force at a point is determined by local properties of the surface at the point of application of the force and is independent of forces applied at other points on the surface. That is, the modal response of the boundary plays no part in the modal response of the enclosed cavity.

6.1.2 Sabine Rooms

When the reflective surfaces of an enclosure are not too distant from one another, and none of the dimensions are so large that air absorption becomes important, the sound energy density of a reverberant field will tend to uniformity throughout the enclosure. However, this does not mean that the sound pressure level will be uniform throughout the enclosure.

Generally, reflective surfaces will not be too distant, as intended here, if no enclosure dimension exceeds any other dimension by more than a factor of about three. As the distance from the sound source increases in this type of enclosure, the relative contribution of the reverberant field to the overall sound field will increase until it dominates the direct field (Embleton, 1988; Smith, 1971). This kind of enclosed space has a generally uniform reverberant energy density field, characterised by a mean sound pressure and standard deviation (see Section 6.4). Such a space has been studied extensively and will be the principal topic of this chapter because it characterises rooms used for assembly and general living. For convenience, this type of enclosed space will be referred to as a Sabine enclosure, named after the man who initiated an investigation of the acoustical properties of such rooms (Sabine, 1993).

All enclosures exhibit low- and high-frequency response and generally all such response is of interest. However, only the high-frequency sound field in an enclosure exhibits those properties that are amenable to Sabine-type analysis; the concepts of the Sabine room are thus strictly associated only with the high-frequency response.

The number of acoustic resonances in an enclosure increases very rapidly as the frequency of excitation increases. Consequently, in the high-frequency range, the possible resonances become so numerous that they cannot be distinguished from one another. Thus, one observes a generally uniform sound field in the regions of the reverberant field not in the vicinity of the source. In this frequency range, the resulting sound field is essentially diffuse and may be described in statistical terms or in terms of average properties.

In the discussion of high-frequency response in Sabine type rooms, the acoustic power transmission into the reverberant sound field has traditionally been treated as a continuum, injected from some source and continually removed by absorption at the boundaries. The sound field is then described in terms of a simple differential equation. The concept of Sabine absorption is introduced and a relatively simple method for its measurement is obtained. This development, which will be referred to as the classical description, is introduced in Section 6.5.1.

In Section 6.5.2 an alternative analysis, based on a modal description of the sound field, is introduced. It is shown that with appropriate assumptions, the formulations of Norris-Eyring and Millington-Sette are obtained.

6.1.3 Flat and Long Rooms

Enclosed spaces are occasionally encountered in which some of the bounding surfaces may be relatively remote or highly absorptive, and such spaces are also of importance. For example, lateral surfaces may be considered remote when the ratio of enclosure width-to-height or width-to-length exceeds a value of about three. Among such possibilities are flat rooms, characteristic of many industrial sites in which the side walls are remote or simply open, and long rooms such as corridors or tunnels. These two types of enclosure, which have been recognised and have received attention in the technical literature, are discussed in Section 6.9.

6.2 Low Frequencies

In the low-frequency range, an enclosure sound field is dominated by standing waves at certain characteristic frequencies. Large spatial variations in the reverberant field are observed if the enclosure is excited with pure tone sound, and the sound field in the enclosure is said to be dominated by resonant or modal response.

When a source of sound in an enclosure is turned on, the resulting sound waves spread out in all directions from the source. When the advancing sound waves reach the walls of the enclosure they are reflected, generally with a small loss of energy, eventually resulting in waves travelling around the enclosure in all directions. If each path that a wave takes is traced around the enclosure, there will be certain paths of travel that repeat upon themselves to form normal modes of vibration, and at certain frequencies, waves travelling around such paths will arrive back at any point along the path in phase. Amplification of the wave disturbance will result and the normal mode will be resonant. When the frequency of the source equals one of the resonance frequencies of a normal mode, resonance occurs and the interior space of the enclosure responds strongly, being only limited by the absorption present in the enclosure.

A normal mode has been associated with paths of travel that repeat upon themselves. Evidently, waves may travel in either direction along such paths so that, in general, normal modes are characterised by waves travelling in opposite directions along any repeating path. As waves travelling along the same path, but in opposite directions, produce standing waves, a normal mode may be characterised as a system of standing waves, which, in turn, is characterised by nodes (locations of minimum response) and antinodes (locations of maximum response). At locations where the oppositely travelling waves arrive with their acoustic pressures 180 degrees out-of-phase, pressure cancellation will occur, resulting in a sound pressure minimum called a node. Similarly, at locations where the oppositely travelling waves arrive with their acoustic pressures in-phase, pressure amplification will occur, resulting in a sound pressure maximum called an antinode.

In an enclosure at low frequencies, the number of resonance frequencies within a band, such as an octave or 1/3-octave, will be small. Thus, at low frequencies, the response of a room as a function of frequency and location will be quite irregular; that is, the spatial distribution in the reverberant field will be characterised by sound pressure nodes and antinodes.

6.2.1 Rectangular Rooms

If a source, radiating slowly-increasing, single-frequency sound, is placed in a room of rectangular shape, the sound pressure level at any location (other than at a node in the room for that frequency) will at first rapidly increase, momentarily reach a maximum at resonance, then rapidly decrease. The process repeats with each room resonance. The measured frequency response of a 180 m^3 rectangular reverberation room is shown in Figure 6.1 for illustration. The sound pressure was measured in a corner of the room (where there are no pressure nodes) while the frequency of the source (placed at an opposite corner) was very slowly increased.

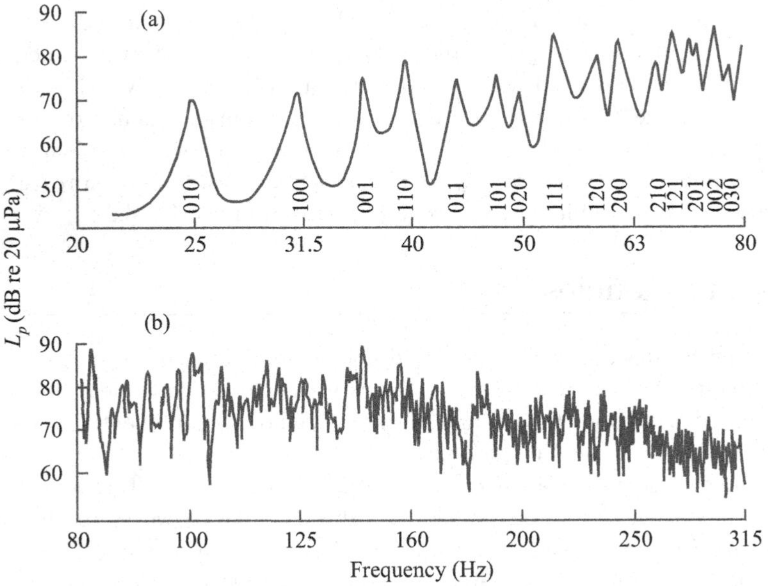

FIGURE 6.1 Measured frequency response of an 180 m^3 rectangular room. (a) In this frequency range, room resonances are identified by mode numbers. (b) In this frequency range, peaks in the room response cannot be associated with room resonances identified by mode numbers.

Consideration of a rectangular room provides a convenient model for understanding modal response and the placement of sound-absorbents for sound control. The mathematical description of the modal response of the rectangular room, illustrated in Figure 6.3, takes on a particularly simple form; thus it will be advantageous to use the rectangular room as a model for the following discussion of modal response. It is emphasised that modal response is by no means peculiar to rectangular or even regular-shaped rooms. Modal response characterises enclosures of all shapes. Splayed, irregular or odd numbers of walls will not prevent resonances and accompanying pressure nodes and antinodes in an enclosure constructed of reasonably reflective walls; nor will such peculiar construction necessarily result in a more uniform distribution in frequency of the resonances of an enclosure than would occur in a rectangular room of appropriate dimensions (see Section 6.3.1). However, it is simpler to calculate the resonance frequencies and mode shapes for rectangular rooms.

For sound of frequency, ω radians/s, in a rectangular enclosure, a standing wave solution for the acoustic potential function takes the following simple form (see Section 1.4.6):

$$\phi = X(x)Y(y)Z(z)\mathrm{e}^{\mathrm{j}\omega t} \tag{6.1}$$

Substitution of Equation (6.1) into the wave equation (Equation (1.16)), use of $k^2 = \omega^2/c^2$ and rearrangement gives:

$$\frac{X''}{X} + \frac{Y''}{Y} + \frac{Z''}{Z} = -k^2 \tag{6.2}$$

Each term of Equation (6.2) on the left-hand side is a function of a different independent variable, whereas the right-hand side of the equation is a constant. It may be concluded that each term on the left must also equal a constant; that is, Equation (6.2) takes the form:

$$k_x^2 + k_y^2 + k_z^2 = k^2 \tag{6.3}$$

This implies the following:

$$X'' + k_x^2 X = 0 \tag{6.4}$$

$$Y'' + k_y^2 Y = 0 \tag{6.5}$$

$$Z'' + k_z^2 Z = 0 \tag{6.6}$$

Solutions of Equations (6.4), (6.5) and (6.6) are as:

$$X = A_x\mathrm{e}^{\mathrm{j}k_x x} + B_x\mathrm{e}^{-\mathrm{j}k_x x} \tag{6.7}$$

$$Y = A_y\mathrm{e}^{\mathrm{j}k_y y} + B_y\mathrm{e}^{-\mathrm{j}k_y y} \tag{6.8}$$

$$Z = A_z\mathrm{e}^{\mathrm{j}k_z z} + B_z\mathrm{e}^{-\mathrm{j}k_z z} \tag{6.9}$$

Boundary conditions will determine the values of the constants. For example, if it is assumed that the walls are essentially rigid, so that the normal particle velocity, u_x, at the walls is zero, then, using Equations (1.11), (6.1) and (6.7), the following is obtained:

$$u_x = -\partial\phi/\partial x \tag{6.10}$$

and

$$-\,\mathrm{j}k_x YZ\mathrm{e}^{\mathrm{j}\omega t}[A_x\mathrm{e}^{\mathrm{j}k_x x} - B_x\mathrm{e}^{-\mathrm{j}k_x x}]_{x=0,L_x} = 0 \tag{6.11}$$

Since:

$$\mathrm{j}k_x YZ\mathrm{e}^{\mathrm{j}\omega t} \neq 0 \tag{6.12}$$

then:

$$[A_x\mathrm{e}^{\mathrm{j}k_x x} - B_x\mathrm{e}^{-\mathrm{j}k_x x}]_{x=0,L_x} = 0 \tag{6.13}$$

First, consider the boundary condition at $x = 0$. This condition leads to the conclusion that $A_x = B_x$. Similarly, it may be shown that:

$$A_i = B_i; \qquad i = x,\ y,\ z \tag{6.14}$$

Next consider the boundary condition at $x = L_x$. This second condition leads to the following equation:

$$\mathrm{e}^{\mathrm{j}k_x L_x} - \mathrm{e}^{-\mathrm{j}k_x L_x} = 2\mathrm{j}\sin(k_x L_x) = 0 \tag{6.15}$$

Similar expressions follow for the boundary conditions at $y = L_y$ and $z = L_z$. From these considerations it may be concluded that the k_i are defined as:

$$k_i = n_i \frac{\pi}{L_i}; \qquad n_i = 0, \ \pm 1, \ \pm 2, \ ...; \qquad i = x, \ y, \ z \tag{6.16}$$

Substitution of Equation (6.16) into Equation (6.3) and use of $k^2 = \omega^2/c^2$ leads to the following useful result:

$$f_n = \frac{\omega_n}{2\pi} = \frac{c}{2}\sqrt{\left[\frac{n_x}{L_x}\right]^2 + \left[\frac{n_y}{L_y}\right]^2 + \left[\frac{n_z}{L_z}\right]^2} \quad \text{(Hz)} \tag{6.17}$$

In this equation, the subscript, n, on the frequency variable, f, indicates that the particular solutions or "eigen" frequencies of the equation are functions of the particular mode numbers, n_x, n_y and n_z.

Following Section 1.4.6 and using Equation (1.12), the following expression for the acoustic pressure is obtained:

$$p = \rho \frac{\partial \phi}{\partial t} = \mathrm{j}\omega\rho X(x)Y(y)Z(z)\mathrm{e}^{\mathrm{j}\omega t} \tag{6.18}$$

Substitution of Equations (6.14) and (6.16) into Equations (6.7), (6.8) and (6.9) and, in turn, substituting these altered equations into Equation (6.18) gives the following expression for the acoustic pressure for mode (n_x, n_y, n_z) in a rectangular room with rigid walls:

$$p = \hat{p}\cos\left[\frac{\pi n_x x}{L_x}\right]\cos\left[\frac{\pi n_y y}{L_y}\right]\cos\left[\frac{\pi n_z z}{L_z}\right]\mathrm{e}^{\mathrm{j}\omega t} \tag{6.19}$$

where $\hat{p}$ is the acoustic pressure amplitude of the standing wave.

In Equations (6.17) and (6.19), the mode indices, n_x, n_y and n_z, have been introduced. These indices take on all positive integer values, plus zero. There are three types of normal modes of vibration in a rectangular room, which have their analogues in enclosures of other shapes. They may readily be understood as:

1. axial modes for which only one modal index is not zero;
2. tangential modes for which one modal index is zero; and
3. oblique modes for which no modal index is zero.

Axial modes correspond to wave travel back and forth parallel to an axis of the room. For example, the $(n_x, 0, 0)$ mode in the rectangular room of Figure 6.3 corresponds to a wave travelling back and forth parallel to the x-axis. Such a system of waves forms a standing wave having n_x nodal planes normal to the x-axis and parallel to the end walls. This may be verified by using Equation (6.19). The significance for noise control is that only sound-absorption on the walls normal to the axis of sound propagation, where the sound is multiply reflected, will significantly affect the energy stored in the mode. Sound-absorptive treatment on any of the other walls would have only a small effect on an axial mode. The significance for sound coupling is that a speaker placed in the nodal plane of any mode will couple at best very poorly to that mode. Thus, the best place to drive an axial mode is to place the sound source on the end wall where the axial wave is multiply reflected; that is, at a pressure antinode.

Tangential modes correspond to waves travelling parallel to two opposite walls of an enclosure while successively reflecting from the other four walls. For example, the (n_x, n_y, 0) mode of a rectangular enclosure corresponds to a wave travelling around the room parallel to the floor and ceiling. In this case, the wave impinges on all four vertical walls and absorptive material on any of these walls would be most effective in attenuating this mode. Absorptive material on the floor or ceiling would be less effective.

Oblique modes correspond to wave travel oblique to all room surfaces. For example, the (n_x, n_y, and n_z) mode in the rectangular room of Figure 6.3 would successively impinge on all six walls of the enclosure. Consequently, absorptive treatment on the floor, ceiling or any wall would be equally effective in attenuating an oblique mode.

The various mode types described above are illustrated in Figure 6.2.

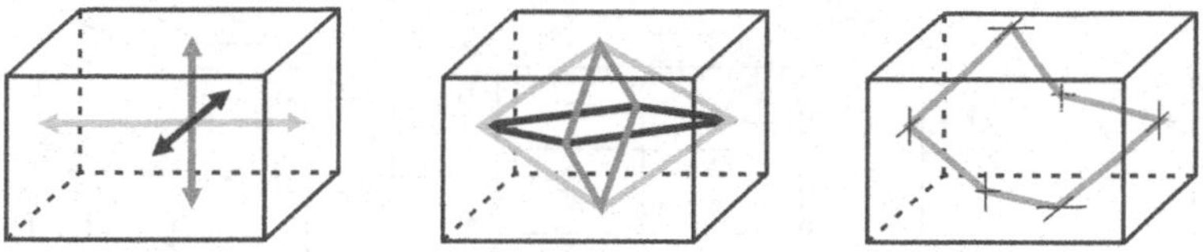

FIGURE 6.2 Illustration of axial, tangential and oblique modes in a rectangular room.

The first few lowest order modes are illustrated in Figure 6.3 for 1D, 2D and 3D rooms, where the plus and minus signs in the figures represent the relative phase of the acoustic sound pressure in each location. A plus sign represents a 180° phase difference to a minus sign. In practice, even a small amount of absorption due to the wall material and the presence of air will make this phase difference slightly different to 180°.

For the placement of a speaker to drive a room, it is of interest to note that every mode of vibration has a pressure antinode at the corners of a room. This may be verified by using Equation (6.19). A corner is a good place to drive a rectangular room when it is desirable to introduce sound. It is also a good location to place absorbents to attenuate sound and to sample the sound field for the purpose of determining room frequency response.

In Figure 6.1, the first 15 room resonant modes have been identified using Equation (6.17). Reference to the figure shows that of the first 15 lowest order modes, seven are axial modes, six are tangential modes and two are oblique modes. Reference to the figure also shows that as the frequency increases, the resonances become too numerous to identify individually and in this range, the number of axial and tangential modes will become negligible compared to the number of oblique modes. It may be useful to note that the frequency at which this occurs is about 80 Hz in the reverberation room described in Figure 6.1 and this corresponds to a room volume of about 2.25 cubic wavelengths. As the latter description is non-dimensional, it is probably general; however, a more precise boundary between low- and high-frequency behaviour will be given in Section 6.3.

In a rectangular room, for every mode of vibration for which one of the modal indices is odd, the sound pressure is zero at the centre of the room, as shown by consideration of Equation (6.19); that is, when one of the modal indices is odd, the corresponding term in Equation (6.19) is zero at the centre of the corresponding coordinate (room dimension). Consequently, the centre of the room is a very poor place to couple, either with a speaker or an absorber, into the modes of the room. Consideration of all the possible combinations of odd and even in a group of three modal indices shows that only one-eighth of the modes of a rectangular room will not have nodes at the centre of the room. At the centre of the junction of two walls, only one-quarter of the modes of a rectangular room will not have nodes, and at the centre of any wall only half of the modes will not have nodes.

6.2.2 Cylindrical Rooms

The analysis of cylindrical rooms follows the same procedure as for rectangular rooms except that the cylindrical coordinate system is used instead of the Cartesian system. The result of this analysis is the following expression for the resonance frequencies of the modes in a cylindrical

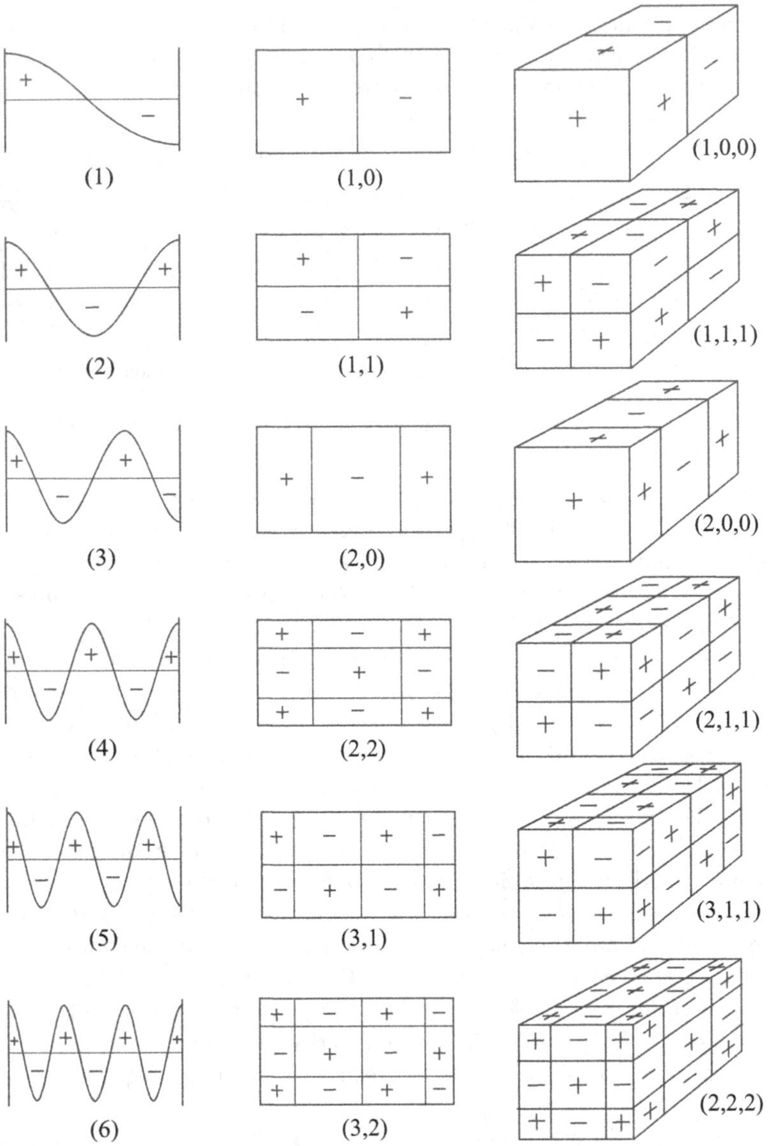

FIGURE 6.3 Sound pressure mode shapes for 1-D, 2-D and 3-D rooms, where the mode numbers correspond to the number of nodal points, nodal lines or nodal planes respectively (theoretical zero sound pressure). (a) 1-D room, modes (1), (2), (3), (4), (5) and (6); (b) 2-D room, modes (1,0), (1,1), (2,0), (2,2), (3,1) and (3,2); 3-D room modes (1,0,0), (1,1,1), (2,0,0), (2,1,1), (3,1,1), (2,2,2).

room. Values of ψ_{mn} for the first few modes are given in Table 6.1.

$$f(n_z, m, n) = \frac{c}{2}\sqrt{\left(\frac{n_z}{\ell}\right)^2 + \left(\frac{\psi_{m,n}}{a}\right)^2} \tag{6.20}$$

In Equation (6.20), n_z, is the number (varying from 0 to ∞) of nodal planes normal to the axis of the cylinder, ℓ is the length of the cylinder and a is its radius. The characteristic values, ψ_{mn}, are functions of the modal indices m, n, where m is the number of diametral pressure nodes and n is the number of circumferential pressure nodes (excluding the boundary surface).

TABLE 6.1 Values of $\psi_{m,n}$

$m \backslash n$	0	1	2	3	4
0	0.0000	1.2197	2.2331	3.2383	4.2411
1	0.5861	1.6971	2.7172	3.7261	4.7312
2	0.9722	2.1346	3.1734	4.1923	5.2036
3	1.3373	2.5513	3.6115	4.6428	5.6623
4	1.6926	2.9547	4.0368	5.0815	6.1103

6.3 Boundary between Low-Frequency and High-Frequency Behaviour

Referring to Figure 6.1, where the frequency response of a rectangular enclosure is shown, it can be observed that the number of peaks in response increases rapidly with increasing frequency. At low frequencies, the peaks in response are well separated and can be readily identified with resonant modes of the room. However, at high frequencies, so many modes may be driven in strong response at once that they tend to interfere, so that at high frequencies, individual peaks in response cannot be associated uniquely with individual resonances. In this range, statistical analysis is appropriate.

Clearly, a need exists for a frequency bound that defines the crossover from the low-frequency range, where modal analysis is appropriate, to the high-frequency range where statistical analysis is appropriate. Reference to Figure 6.1 provides no clear indication of a possible bound, as a continuum of gradual change is observed. However, analysis does provide a bound, but to understand the determination of the bound, called here the crossover frequency, three separate concepts are required: modal density, modal damping and modal overlap. These concepts will be introduced in Sections 6.3.1, 6.3.2 and 6.3.3 and then used to define the crossover frequency.

6.3.1 Modal Density

The approximate number of modes, N, which may be excited in the frequency range from zero up to f Hz, is given by the following expression for a rectangular room (Roe, 1941):

$$N = \frac{4\pi f^3 V}{3c^3} + \frac{\pi f^2 S}{4c^2} + \frac{fL}{8c} \tag{6.21}$$

In Equation (6.21), c is the speed of sound, V is the room volume, S is the room total surface area and L is the total perimeter of the room, which is the sum of lengths of all edges. For a rectangular room, L is four times the sum of the length, width and breadth of the room. It has been shown (Morse and Ingard, 1968) that Equation (6.21) has wider application than for rectangular rooms; to a good approximation it describes the number of modes in rooms of any shape, with the approximation improving as the irregularity of the room shape increases. It should be remembered that Equation (6.21) is an approximation only and the actual number of modes fluctuates above and below the prediction of this equation as the frequency gradually increases or decreases.

For the purpose of estimating the number of modes that, on average, may be excited in a narrow frequency band, the derivative of Equation (6.21), called the modal density, is useful. The expression for the modal density is:

$$\frac{\mathrm{d}N}{\mathrm{d}f} = \frac{4\pi f^2 V}{c^3} + \frac{\pi f S}{2c^2} + \frac{L}{8c} \tag{6.22}$$

which also applies approximately to rooms of any shape, including cylindrical rooms of radius a and length, ℓ, where $V = \pi a^2 \ell$, $S = 2\pi a(a + \ell)$ and $L = 4\pi a + 4\ell$ (Garret, 2020, p. 643).

Consideration of Equation (6.22) shows that, at low frequencies, the number of modes per unit frequency that may be excited will be very small but, as the modal density increases quadratically with increasing frequency, at high frequencies the number of modes excited will become very large. Thus, when a room is excited with a narrow band of noise at low frequencies, one can expect large spatial fluctuations in sound pressure level, as observed in Figure 6.1, but at high frequencies, the fluctuations become small and the reverberant field approximates uniformity throughout the room. The number of oblique modes in a room of any shape is described approximately by the cubic term of Equation (6.21), although the linear and quadratic terms also contribute a little to the number of oblique modes (Morse and Ingard (1968), pp. 586), with these latter contributions becoming steadily less important as the frequency increases. Similarly, the number of tangential modes is dominated by the quadratic term with the linear term also contributing. The number of axial modes is actually 4 times the linear term in Equation (6.21) (Morse and Ingard, 1968), but this latter term has also been modified by negative contributions from oblique and tangential modes. Thus, it is evident that at high frequencies, the number of oblique modes will far exceed the number of tangential and axial modes and to a good approximation at high frequencies, the latter two mode types may be ignored.

6.3.2 Modal Damping and Bandwidth

Referring to Figure 6.1, it may be observed that the recorded frequency response peaks in the low-frequency range have finite widths, which may be associated with the response of the room that was investigated. A bandwidth, Δf, may be defined and associated with each mode, being the frequency range about resonance over which the sound pressure squared is greater than or equal to half the same quantity at resonance. The lower and upper frequencies bounding a resonance and defined in this way are commonly known as the half-power points. The corresponding response at the half-power points is down 3 dB from the peak response. Referring to Figure 6.4, the corresponding bandwidths are easily determined where individual resonances may be identified.

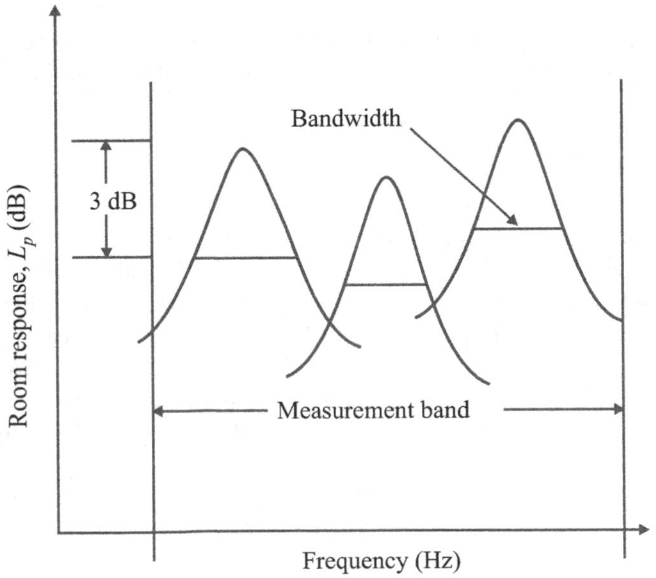

FIGURE 6.4 Three modes in a specified frequency range with a modal overlap of 0.6.

The bandwidth, Δf, is dependent on the damping of the mode; the greater the modal damping, the larger will be the bandwidth. For acoustical spaces such as rooms the modal damping is commonly expressed in terms of the damping factor (similar to the critical damping ratio), which is a viscous based quantity and proportional to particle velocity, whereas for structures, modal damping is commonly expressed in terms of a modal loss factor, η, which is a hysteretic based quantity and proportional to displacement. Alternatively, damping in structures may be viscously based as well and may be expressed in terms of the damping ratio, ζ, commonly used to describe damping in mechanical systems. These quantities are related to each other and to the energy-based quality factor, Q, of the resonant mode, and the logarithmic decrement, δ, by the following relations (see Table 9.3 for a more complete list of damping measures):

$$\Delta f/f = 1/Q = \eta = \frac{2\zeta}{\sqrt{1-\zeta^2}} = \delta/\pi \tag{6.23}$$

The quality factor, Q, is discussed in Section 8.8.2.2, the critical damping ratio, ζ, is discussed in Section 9.2.1 and the logarithmic decrement, δ, is discussed in Section 9.8. Here the modal loss factor, η, is presented as an energy-based quantity by its relation to the quality factor, Q. The loss factor, η, is sometimes used in acoustics as a viscous based damping quantity. More usually, it has meaning as a structural loss factor based on a hysteretic damping effect in a structural member. For a solid material, it is defined in terms of a complex modulus of elasticity $E' = E(1 + \mathrm{j}\eta)$ where E is Young's modulus. This use of the loss factor is discussed in Section 9.2.1. As may be observed by reference to Equation (6.23), when the modal loss factor, η, is small, which is true for most practical cases, the implication is that the critical damping ratio is also small and $\eta \approx 2\zeta$.

At low frequencies, individual modal bandwidths can be identified and measured directly. At high frequencies, where individual modes cannot be identified, the average bandwidth may be calculated from a measurement of the decay time (see Section 6.5.1) using the following equation (Embleton, 1988):

$$\Delta f = 2.20/T_{60} \tag{6.24}$$

6.3.3 Modal Overlap

Modal overlap, M, is calculated as the product of the average bandwidth, given by either Equation (6.23) or (6.24) and the modal density, given by Equation (6.22). The expression for modal overlap is:

$$M = \Delta f \frac{\mathrm{d}N}{\mathrm{d}f} \tag{6.25}$$

The modal overlap is a measure of the extent to which the resonances of a reverberant field cover the range of all possible frequencies within a specified frequency range. The concept is illustrated for a hypothetical case of a low modal overlap of 0.6 in Figure 6.4. In the figure, three resonant modes, their respective bandwidths and the frequency range of the specified measurement band are indicated.

6.3.4 Crossover Frequency

There are two criteria commonly used for determining the crossover frequency. The chosen criterion depends on whether the room is excited with bands of noise or with pure tones.

If room excitation with 1/3-octave, or wider bands of noise, is to be considered, then the criterion for statistical (high-frequency) analysis is that there should be a minimum of between 3 and 6 modes resonant in the frequency band. The exact number required is dependent on the modal damping and the desired accuracy of the results. More modes are necessary for low values of modal damping and/or if high accuracy is required. If room excitation with a pure tone or a very narrow band of noise is of concern, then the criterion for reliable statistical analysis is that the modal overlap should be greater than or equal to 3.

6.4 High Frequencies, Statistical Analysis

At high frequencies, the sound field in a reverberant space may be thought of as composed of plane waves travelling in all directions with equal probability. If the mean square pressure amplitude of any travelling plane wave is on average the same, independent of direction, then the sound field is said to be diffuse. In a reverberant space, the sound field may be considered to be diffuse when the modal overlap is three or greater, in which case, the sound field steady-state frequency response is essentially a random phenomenon. For excitation with bands of noise, the parameters describing the field are essentially predictable from the room reverberation time (see Section 6.5, Section 6.6.2, and Schroeder (1969)).

The concept of diffuse field implies that the net power transmission along any coordinate axis is negligibly small; that is, any power transmission is essentially the same in any direction. The reverberant sound field may be considered to consist of constant mean energy density throughout the room. However, the concept does not imply a sound field where the sound pressure level is the same throughout. Even in a perfectly diffuse sound field, the sound pressure level will fluctuate over time at any given location in the room and the long-time-averaged sound pressure level will also fluctuate from point to point within the room. The amount of fluctuation is dependent on the product of the measurement bandwidth, B (in Hz), and the reverberation time, T_{60}, averaged over the band (see Sections 6.5 and 6.6.2). The expected standard deviation, which describes the spatial fluctuations in a diffuse sound field, is given by the following approximate expression (Lubman, 1969):

$$\sigma \simeq 5.57(1 + 0.238BT_{60})^{-1/2} \quad \text{(dB)} \tag{6.26}$$

6.4.1 Effective Intensity in a Diffuse Field

In a diffuse (reverberant) field, sound propagation in all directions is equally likely and consequently the intensity at any point in the field is zero, as it is a vector quantity. However, an effective intensity associated with power transmission in a specified direction can be defined. An expression for the effective intensity will now be derived in terms of the reverberant field sound pressure.

Consider sound energy in a reverberant field propagating along a narrow column of circular cross-section, as illustrated in Figure 6.5.

Let the column just encompass a small spherical region in the field. The ratio of the volume of the spherical region to the cylindrical section of the column that just encompasses the spherical region is:

$$\frac{4\pi r^3}{3} \times \frac{1}{2\pi r^3} = \frac{2}{3} \tag{6.27}$$

Equation (6.27) shows that the spherical region occupies two-thirds of the volume of the encompassing cylinder.

Referring to Figure 6.5, consider the convergence on the spherical region of sound from all directions. Let the intensity of any incident sound beam of cross-sectional area, dS, be I. The time for the beam to travel through the spherical region (length of the encompassing cylinder) is $2r/c$; thus the incremental contribution per unit area to the energy E in the spherical region due to any beam is:

$$\Delta E = \frac{2}{3} I \frac{2r}{c} \tag{6.28}$$

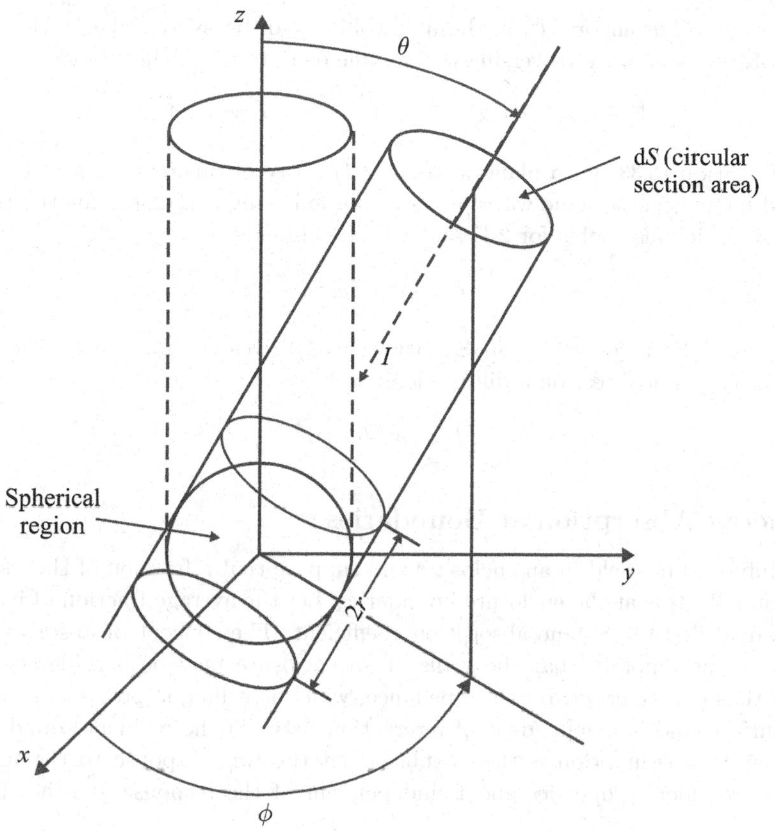

FIGURE 6.5 Geometrical arrangement for determining effective intensity in a diffuse field.

The total energy is obtained by integrating the incremental energy contribution per unit area of the sphere over the area of the sphere. The incremental area of sphere for use in integration is:

$$\mathrm{d}S = r^2 \sin\theta \mathrm{d}\theta \mathrm{d}\phi \tag{6.29}$$

Thus:

$$E = \frac{4I}{3c} \int_0^{2\pi} \mathrm{d}\phi \int_0^{\pi} r^3 \sin\theta \mathrm{d}\theta = \frac{16 I \pi r^3}{3c} \tag{6.30}$$

Let the time-averaged acoustic energy density be ψ at the centre of the region under consideration; then the total energy (energy density $\times$ volume) in the central spherical region is:

$$E = \frac{4\pi r^3 \psi}{3} \tag{6.31}$$

Combining Equations (6.30) and (6.31) gives, for the effective intensity, I, in any direction in terms of the time-averaged energy density, ψ:

$$I = \psi c/4 \tag{6.32}$$

To obtain an expression for the energy density in terms of mean square sound pressure, we consider a plane wave of unit cross-sectional area travelling unit distance to form a volume, $V = 1$ m^3. The length of time, t, for the plane wave to travel unit distance is $1/c$, where c is the speed of sound. The energy, E, in the unit volume can be expressed in terms of the sound intensity, I, of the plane wave traversing the volume or in terms of the energy density as:

$$E = I \times 1 \times 1 \times t = I \times 1 \times 1/c = \psi \times 1 \times 1 \times 1 \tag{6.33}$$

Thus, from Equation (6.33) for a plane wave, $\psi = I/c$. Use of this relation and Equation (1.78) for the sound intensity of a plane wave provides the following expression for the time-averaged energy density, which also holds for 2-D and 3-D sound fields:

$$\psi = \langle p^2 \rangle / (\rho c^2) \tag{6.34}$$

Substitution of Equation (6.32) in Equation (6.34) gives the following expression for the effective intensity in one direction a diffuse field:

$$I = \langle p^2 \rangle / (4\rho c) \tag{6.35}$$

6.4.2 Energy Absorption at Boundaries

Consider a diffuse sound field in an enclosure and suppose that a fraction of the incident energy is absorbed on reflection at the enclosure boundaries. Let the average fraction of incident energy absorbed be $\bar{\alpha}$, called the Sabine absorption coefficient. The concept of absorption coefficient follows from the assumption that the walls of an enclosure may be considered to be locally reactive and thus characterised by an impedance, which is a unique property exhibited by the wall at its surface and is independent of interaction between the incident sound and the wall anywhere else. The assumption is then explicit that the wall response to the incident sound depends solely on local properties and is independent of the response at other points on the surface.

The locally reactive assumption has proven very useful for architectural purposes but is apparently of very little use in predicting interior noise in aircraft and vehicles of various types. In the latter cases, the modes of the enclosed space couple with modes in the walls, and energy stored in vibrating walls contributes very significantly to the resulting sound field. In such cases, the locally reactive concept is not even approximately true, and neither is the concept of Sabine absorption that follows from it.

6.4.3 Air Absorption

In addition to energy absorption on reflection, some energy is absorbed by the air during propagation between reflections. Generally, propagation loss due to air absorption is negligible, but at frequencies above 500 Hz, especially in large enclosures, it makes a significant contribution to the overall loss.

Air absorption may be taken into account as follows. As shown in Section 6.5.3, the mean distance, Λ, travelled by a plane wave in an arbitrarily shaped enclosure between reflections is called the mean free path and is given by:

$$\Lambda = \frac{4V}{S} \tag{6.36}$$

where V is the room volume and S is the room surface area (Kuttruff, 2009). It will now be assumed that the fraction of propagating sound energy lost due to air absorption between reflections is linearly related to the mean free path. If the fraction lost is not greater than 0.4, then the error introduced by this approximation is less than 10% (0.5 dB).

At this point, many authors write $4m'V/S = \alpha_{\text{air}}$ for the contribution due to air absorption, and they provide tables of values of the coefficient, m', as a function of temperature and relative humidity. Here, use will be made of values for air absorption m already given in Table 5.3 for sound propagating outdoors. In a distance of one mean free path the attenuation of sound is:

$$\frac{4mV}{S}10^{-3} = -10\log_{10} \mathrm{e}^{-4m'V/S} \tag{6.37}$$

Let:

$$4m'V/S = \alpha_{\mathrm{air}} \tag{6.38}$$

Then:

$$\alpha_{\mathrm{air}} = 4mV/(S \times 10^4 \log_{10} \mathrm{e}) = 9.21 \times 10^{-4} mV/S \tag{6.39}$$

Using the above relation, the total mean absorption coefficient, including air absorption, may be written as:

$$\bar{\alpha} = \bar{\alpha}_{wcf} + 9.21 \times 10^{-4} mV/S \tag{6.40}$$

where α_{wcf} is the absorption coefficient due to absorption by the room surfaces, m is the atmospheric absorption coefficient, $m = 1000\alpha_a$, and α_a is defined by Equation (5.62).

Equations (6.32) and (6.35) may be used to write for the power, W_a, or rate of energy absorbed:

$$W_a = \psi S c \bar{\alpha}/4 = \langle p^2 \rangle S \bar{\alpha}/(4\rho c) \tag{6.41}$$

where $\bar{\alpha}$ is defined by Equation (6.40).

6.4.4 Steady-State Response

At any point in a room, the sound field is a combination of the direct field radiated by the source and the reverberant field. Thus, the total sound energy measured at a point in a room is the sum of the sound energy due to the direct field and that due to the reverberant field.

Using Equation (4.14) and introducing the directivity factor, $D_{\theta,\phi}$ (see Section 4.8), the sound pressure squared due to the direct field at a point in the room at a distance, r, and in a direction, (θ, ϕ), from the source may be written as:

$$\langle p^2 \rangle_D = W \rho c D_{\theta,\phi}/4\pi r^2 \tag{6.42}$$

The quantity $D_{\theta,\phi}$ is the directivity factor of the source in direction (θ, ϕ), ρ is the density of air (kg/m^3), c is the speed of sound in air (m/s) and W is the sound power, in watts, radiated by the source. In writing Equation (6.42), it is assumed that the source is sufficiently small or r is sufficiently large for the measurement point to be in the far field of the source.

Consider that the direct field must be once reflected to enter the reverberant field. The fraction of energy incident at the walls, which is reflected into the reverberant field, is $(1 - \bar{\alpha})$. Using Equation (6.41) and setting the power absorbed equal to the power introduced, W, the sound pressure squared due to the reverberant field may be written as:

$$\langle p^2 \rangle_R = 4W\rho c(1 - \bar{\alpha})/(S\bar{\alpha}) \tag{6.43}$$

The sound pressure level at any point due to the combined effect of the direct and reverberant sound fields is obtained by adding together Equations (6.42) and (6.43). Thus, using Equations (1.86), (1.89) and (1.92):

$$L_p = L_W + 10\log_{10}\left[\frac{D_\theta}{4\pi r^2} + \frac{4}{R_c}\right] + 10\log_{10}\left[\frac{\rho c}{400}\right] \tag{6.44}$$

At 20°C, where $\rho c = 414$ (SI units), there would be an error of approximately 0.1 dB if the last term in Equation (6.44) is omitted. In common industrial spaces, which have lateral dimensions much greater than their height, Equation (6.44) underpredicts reverberant field sound pressure levels (see Section 6.9) close to the noise source and overpredicts levels far away from the source (Hodgson, 1994b). The predictions of sound levels in these types of space are discussed in Section 6.9. Equation (6.44) has been written in terms of the room constant, R_c, where the room constant is:

$$R_c = \frac{S\bar{\alpha}}{1-\bar{\alpha}} \tag{6.45}$$

The sound pressure levels due to the direct and reverberant field are equal when the terms in Equation (6.44), $D_\theta 4/\pi r^2$ and $4/R_c$. The corresponding distance, r is called the hall radius, r_h.

6.5 Transient Response

If sound is introduced into a room, the reverberant field level will increase until the rate of sound energy introduction is just equal to the rate of sound energy absorption. If the sound source is abruptly shut off, the reverberant field will decay at a rate determined by the rate of sound energy absorption. The time required for the reverberant field to decay by 60 dB, called the reverberation time, is the single most important parameter characterising a room for its acoustical properties. For example, a long reverberation time may make the understanding of speech difficult but may be desirable for organ recitals. The reverberation time, T_{60}, is typically found by measuring the time taken for the sound field to decay by an amount of 20 dB or 30 dB, beginning at a level of 5 dB below the steady-state level before the source is turned off and ending at a level of 25 or 35 dB below the steady-state level, respectively (see Section 6.6.2 and ISO 3382-2 (2008)). The result is then multiplied by a factor of three or two respectively, to obtain the 60 dB decay time, T_{60}. The reason for measuring the 20 or 30 dB decay times instead of the 60 dB decay time is that it is difficult for the level to remain 10 dB above the noise floor after it has decayed by 65 dB from the steady-state level.

As the reverberation time is directly related to the energy dissipation in a room, its measurement provides a means for the determination of the energy absorption properties of a room. Knowledge of the energy absorption properties of a room, in turn, allows estimation of the resulting sound pressure level in the reverberant field when sound of a given power level is introduced. The energy absorption properties of materials placed in a reverberation chamber may be determined by measurement of the associated reverberation times of the chamber, with and without the material under test in the room. The Sabine absorption coefficient, which is assumed to be a property of the material under test, is determined in this way and standards (ASTM C423-17, 2017; ISO 354, 2003) are available that provide guidance for conducting these tests.

In Sections 6.5.1 and 6.5.2, two methods will be used to characterise the transient response of a room. The classical description, in which the sound field is described statistically, will be presented first and a second method, in which the sound field is described in terms of modal decay, will then be presented. The second method provides a description in better agreement with experiment than does the classical approach (Bies, 1995).

6.5.1 Classical Description

At high frequencies, the reverberant field may be described in terms of a simple differential equation, which represents a gross simplification of the physical process, but nonetheless gives generally useful results.

Using Equation (6.41) and the observation that the rate of change of the energy stored in a reverberant field equals the rate of supply, W_0, less the rate of energy absorbed, W_a (Equation (6.41)), gives the following result:

$$W = V\partial\psi/\partial t = W_0 - \psi Sc\bar{\alpha}/4 \tag{6.46}$$

Introducing the dummy variable:

$$X = [4W_0/S c\bar{\alpha}] - \psi \tag{6.47}$$

and using both parts of Equation (6.46), we can write:

$$\frac{\mathrm{d}X}{\mathrm{d}t} = -\frac{\partial \psi}{\partial t} = \frac{-1}{V}[W_0 - \psi S c\bar{\alpha}/4] \tag{6.48}$$

$$\frac{1}{X} = \frac{\psi S c\bar{\alpha}}{4}[W_0 - \psi S c\bar{\alpha}/4]^{-1} \tag{6.49}$$

Thus:

$$\frac{1}{X}\frac{\mathrm{d}X}{\mathrm{d}t} = -\frac{S c\bar{\alpha}}{4V} \tag{6.50}$$

Integrating Equation (6.50) gives:

$$X = X_0 \mathrm{e}^{-S c\bar{\alpha} t/(4V)} \tag{6.51}$$

where X_0 is the initial value.

Two cases will be considered. Suppose that initially, at time zero, the sound field is nil and a source of power, W_0, is suddenly turned on. The initial conditions are time $t = 0$ and sound pressure $\langle p_0^2 \rangle = 0$. Use of Equation (6.34) and substitution of Equation (6.47) into Equation (6.51) gives the following expression for the resulting reverberant field at any later time t:

$$\langle p^2 \rangle = \frac{4W_0 \rho c}{S\bar{\alpha}}\left(1 - \mathrm{e}^{-S c\bar{\alpha} t/(4V)}\right) \tag{6.52}$$

Alternatively, consider that a steady-state sound field has been established when the source of sound is suddenly shut off. In this case, the initial conditions are time $t = 0$, sound power $W_0 = 0$, and sound pressure $\langle p^2 \rangle = \langle p_0^2 \rangle$. Again, use of Equation (6.34) and substitution of Equation (6.47) into Equation (6.51) gives, for the decaying reverberant field at some later time t:

$$\langle p^2 \rangle = \langle p_0^2 \rangle \mathrm{e}^{-S c\bar{\alpha} t/(4V)} \tag{6.53}$$

Taking logarithms to the base ten of both sides of Equation (6.53) gives:

$$L_{p0} - L_p = 1.0857 S c\bar{\alpha} t/V \tag{6.54}$$

Equation (6.54) shows that the sound pressure level decays linearly with time and at a rate proportional to the Sabine absorption $S\bar{\alpha}$. It provides the basis for the measurement of the total Sabine absorption coefficient, $\bar{\alpha}$, which is defined in Equation (6.40).

Sabine introduced the reverberation time, T_{60} (seconds), as the time required for the sound energy density level to decay by 60 dB from its initial value (see Section 6.5). He showed that the reverberation time, T_{60}, was related to the room volume, V, the total wall area including floor and ceiling, S, the speed of sound, c, and an absorption coefficient, $\bar{\alpha}$, which was characteristic of the room and generally a property of the bounding surfaces and the air in the room. Sabine's reverberation time equation, which follows from Equations (6.53) and (6.54) with $L_{p0} - L_p = 60$, may be written as:

$$T_{60} = \frac{55.26V}{S c\bar{\alpha}} \tag{6.55}$$

It is interesting to note that the effective Sabine absorption coefficient used to calculate reverberation times in spaces such as typical concert halls or factories is not the same as that measured in a reverberation room (Hodgson, 1994a; Kuttruff, 1994), which can lead to inaccuracies in predicted reverberation times. For this reason, it is prudent to follow the advice given in the Section 6.5.2.

It is often more useful to calculate decay rate (dB/s) rather than reverberation time. Using Equations (6.39) and (6.55), the decay rate, D (dB/s), is then given by:

$$D = \frac{60}{T_{60}} = \frac{60Sc\bar{\alpha}}{55.25V} = \frac{60Sc\bar{\alpha}_{wcf}}{55.25V} + \frac{0.921 \times 60Sc\bar{\alpha}_{\text{air}}}{1000 \times 55.25V} = \frac{60Sc\bar{\alpha}_{wcf}}{55.25V} + \frac{mc}{1000} \tag{6.56}$$

where m is the atmospheric absorption coefficient, $m = 1000\alpha_a$ (dB/1000 m), α_a is defined by Equation (5.62) and α_{wcf} is the absorption coefficient due to absorption by the room surfaces.

6.5.2 Modal Description

The discussion thus far suggests that the reverberant field within a room may be thought of as composed of the excited resonant modes of the room. This is still true even in the high-frequency range where the modes may be so numerous and close together that they tend to interfere and cannot be identified separately. In fact, if any enclosure is driven at a frequency slightly off-resonance and the source is abruptly shut off, the frequency of the decaying field will be observed to shift to that of the driven resonant mode as it decays (Morse, 1948). In general, the reflection coefficient, β_e (the fraction of incident energy that is reflected), characterising any surface is a function of the angle of incidence. It is related to the corresponding absorption coefficient, α (the fraction of incident energy that is absorbed), as:

$$\alpha + \beta_e = 1 \tag{6.57}$$

Note that the energy reflection coefficient, β_e, referred to here is the modulus squared of the pressure amplitude reflection coefficient, $|R_p|^2$, discussed in Section 5.2.

When a sound field decays, all of the excited modes decay at their natural frequencies (Morse, 1948). This implies that the frequency content of the decaying field may be slightly different to that of the steady-state field, as the decay of the sound field is modal decay (Larson, 1978). In a reverberant field in which the decaying sound field is diffuse, it is also necessary to assume that scattering of sound energy continually takes place between modes, so that even though the various modes decay at different rates, scattering ensures that they all contain about the same amount of energy, on average, at any time during the decay process. Effectively, in a Sabine room, all modes within a measurement band will decay, on average, at the same rate, because energy is continually scattered from the more slowly decaying modes into the more rapidly decaying modes.

Let $\langle p^2(t)\rangle$ be the mean square band sound pressure level at time, t, in a decaying field and $\langle p_k^2(0)\rangle$ be the mean square sound pressure level of mode, k, at time, $t = 0$. The decaying field may be expressed in terms of modal mean square pressures, $\langle p_k^2(0)\rangle$, mean energy reflection coefficients, $\beta_{e,k}$, and modal mean free paths, Λ_k, as:

$$\langle p^2(t)\rangle = \sum_{k=1}^{N} \langle p_k^2(0)\rangle \beta_{e,k}^{(ct/\Lambda_k)} \tag{6.58}$$

where N is the number of modes within a measurement band and $\beta_{e,k}$, are the energy reflection coefficients for mode k, given by:

$$\beta_{e,k} = \prod_{i=1}^{n} [\beta_{e,ki}]^{S_i/S_k} \tag{6.59}$$

where $\beta_{e,ki}$, is the energy reflection coefficient of surface, i of area, S_i, encountered by a wave travelling around a modal circuit associated with mode k. (Morse and Bolt, 1944). The S_k are the sums of the areas of the reflecting surfaces encountered in one modal circuit of mode, k. The symbol, $\prod_{i=1}^{n}$, represents the product of the n reflection coefficients where n is either a multiple of the number of reflections in one modal circuit or a large number.

The modal mean free path, Λ_k, is the mean distance between reflections of a sound wave travelling around a closed modal circuit and, for a rectangular room, it is given by the following equation (Larson, 1978):

$$\Lambda_k = \frac{2f_k}{c}\left[\frac{n_x}{L_x^2} + \frac{n_y}{L_y^2} + \frac{n_z}{L_z^2}\right]^{-1} \tag{6.60}$$

where f_k is the resonance frequency given by Equation (6.17) for mode, k, of a rectangular enclosure, which has the modal indices, n_x, n_y and n_z.

The assumption will be made that the energy in each mode is on average the same, so that in Equation (6.58), $p_k(0)$ may be replaced with $p_0/\sqrt{N}$, where p_0 is the measured initial sound pressure in the room when the source is shut off. Equation (6.58) may be rewritten as:

$$\langle p^2(t)\rangle = \langle p_0^2\rangle \frac{1}{N}\sum_{k=1}^{N} \mathrm{e}^{(ct/\Lambda_k)\log_\mathrm{e}(1-\alpha_k)} \tag{6.61}$$

A mathematical simplification is now made. In the above expression, the modal mean free path length, Λ_k, is replaced with the mean of all of the modal mean free paths, $4V/S$, and the modal mean absorption coefficient, α_k, is replaced with the area weighted mean statistical absorption coefficient, $\bar{\alpha}_{st}$, for the room (see Section 6.7 and Appendix D). The quantity, V, is the total volume and S is the total wall, ceiling and floor area of the room. In exactly the same way as Equation (6.55) was derived from Equation (6.53), the well-known reverberation time equation of Norris-Eyring may be derived from Equation (6.61) as:

$$T_{60} = -\frac{55.25V}{Sc\log_\mathrm{e}(1-\bar{\alpha}_{st})} \tag{6.62}$$

This equation is often preferred to the Sabine equation by many who work in the field of architectural acoustics, as some authors suggest that it is more appropriate for rooms that are reasonably absorptive and that it gives results that are closer to measured data (Neubauer, 2001). However, Beranek and Hidaka (1998) obtained good agreement between measured and predicted reverberation times in concert halls using the Sabine relation. Of course, if sound-absorption coefficients measured in a reverberation chamber are to be used to predict reverberation times, then the Sabine equation must be used as the Norris-Eyring equation is only valid if statistical absorption coefficients are used (see Appendix D).

Note that air absorption must be included in $\bar{\alpha}_{st}$ in a similar way as it is included in $\bar{\alpha}$ (Equation (6.40)). It is worth careful note that Equation (6.62) is a predictive scheme based on a number of assumptions that cannot be proven, and consequently inversion of the equation to determine the statistical absorption coefficient $\bar{\alpha}_{st}$ is not recommended.

With a further simplification, the famous equation of Sabine is obtained. When $\bar{\alpha}_{st} < 0.4$ an error of less than 0.5 dB is made by setting $\bar{\alpha}_{st} \approx -\log_\mathrm{e}(1-\bar{\alpha}_{st})$ in Equation (6.62), and then by replacing $\bar{\alpha}_{st}$ with $\bar{\alpha}$, Equation (6.55) is obtained. Bies (1995) showed that the above equality of the Sabine and statistical absorption coefficients is accurate provided that edge diffraction of the absorbing material being tested in the reverberation room is taken into account by appropriately increasing the effective area of absorbing material.

Alternatively, if in Equation (6.61), the quantity, $(1-\alpha_k)$, is replaced with the modal energy reflection coefficients $\beta_{e,k}$ and these, in turn, are replaced with a mean value, called the mean statistical reflection coefficient, $\bar{\beta}_{e,st}$, the following equation of Millington and Sette is obtained:

$$T_{60} = -\frac{55.25V}{Sc\log_\mathrm{e}\bar{\beta}_{e,st}} \tag{6.63}$$

The quantity, $\bar{\beta}_{e,st}$, may be calculated using Equation (6.59) but with changes in the meaning of the symbols. The term, $\beta_{e,k}$, is replaced with $\bar{\beta}_{e,st}$, which is now to be interpreted as the area-weighted geometric-mean of the random incidence energy reflection coefficients, $\beta_{e,i}$, for all of the room surfaces; that is:

$$\bar{\beta}_{e,st} = \left[\prod_{i=1}^{n} \beta_{e,i}^{S_i/S} \right] \tag{6.64}$$

The quantity $\beta_{e,i}$ is related to the statistical absorption coefficient, α_i, for surface, i, of area, S_i, by $\beta_{e,i} = 1 - \alpha_i$. It is of interest to note that although taken literally, Equation (6.64) would suggest that an open window having no reflection would absorb all of the incident energy and there would be no reverberant field. The interpretation presented here suggests that an open window must be considered as only a part of the wall in which it is placed and the case of total absorption will never occur. Alternatively, reference to Equation (6.58) shows that if any term $\beta_{e,i}$ is zero, it simply does not appear in the sum and thus will not appear in Equation (6.63) which follows from it.

6.5.3 Empirical Description

For calculating reverberation times in rooms for which the distribution of absorption is non-uniform (such as rooms with large amounts of absorption on the ceiling and floor and little on the walls), Fitzroy (1959) proposed the following empirical equation:

$$T_{60} = \frac{0.16V}{S^2} \left[\frac{-S_x}{\log_e(1 - \bar{\alpha}_{xst})} + \frac{-S_y}{\log_e(1 - \bar{\alpha}_{yst})} + \frac{-S_z}{\log_e(1 - \bar{\alpha}_{zst})} \right] \tag{6.65}$$

where V is the room volume (m^3), S_x, S_y and S_z are the total areas (m^2) of two opposite parallel room surfaces, $\bar{\alpha}_{xst}$, $\bar{\alpha}_{yst}$ and $\bar{\alpha}_{zst}$ are the arithmetically averaged statistical absorption coefficients of a pair of opposite room surfaces (see Equation (6.85)) and S is the total room surface area.

Neubauer (2001) presented a modified Fitzroy equation, which he called the Fitzroy-Kuttruff equation, and which gave more reliable results than the original Fitzroy equation. In fact, this equation has been shown to be even more accurate than the Norris-Eyring equation for architectural spaces with non-uniform sound-absorption. The Fitzroy-Kuttruff equation is:

$$T_{60} = \left(\frac{0.32V}{S^2} \right) \left(\frac{L_z(L_x + L_y)}{\bar{\alpha}_w} + \frac{L_x L_y}{\bar{\alpha}_{cf}} \right) \tag{6.66}$$

where L_x, L_y and L_z are the room dimensions (m) and:

$$\bar{\alpha}_w = -\log_e(1 - \bar{\alpha}_{st}) + \left[\frac{\beta_{e,w}(\beta_{e,w} - \bar{\beta}_{e,st}) S_w^2}{(\bar{\beta}_{e,st} S)^2} \right] \tag{6.67}$$

$$\bar{\alpha}_{cf} = -\log_e(1 - \bar{\alpha}_{st}) + \left[\frac{\beta_{e,cf}(\beta_{e,cf} - \bar{\beta}_{e,st}) S_{cf}^2}{(\bar{\beta}_{e,st} S)^2} \right] \tag{6.68}$$

where $\beta_e = (1 - \alpha)$ is the energy reflection coefficient, for which the subscript, w, refers to the walls and the subscript, cf, refers to the ceiling and floor. $\bar{\alpha}_{st}$ and $\bar{\beta}_{e,st}$ represent the arithmetic mean over the six room surfaces of the surface averaged statistical absorption and reflection coefficients respectively.

Equations (6.55), (6.62), (6.63), (6.65) and (6.66) for reverberation time are all based on the assumption that the room dimensions satisfy the conditions for Sabine rooms (see Section 6.1.2) and that the absorption is reasonably well distributed over the room surfaces. However, in practice this is not often the case and for rooms that do not meet this criterion, Kuttruff (1994) has proposed that Equation (6.55) be used except that $\bar{\alpha}$ should be replaced with α defined as:

$$\alpha = -\log_e(1-\bar{\alpha}_{st})\left[1+0.5\gamma^2\log_e(1-\bar{\alpha}_{st})\right] + \frac{\sum_{i=1}^{n}\beta_{e,i}(\beta_{e,i}-1+\bar{\alpha}_{st})S_i^2}{S^2(1-\bar{\alpha}_{st})^2} \tag{6.69}$$

In Equation (6.69), n is the number of room surfaces (or part room surfaces if whole surfaces are subdivided), $\bar{\alpha}_{st}$ is the statistical absorption coefficient, area averaged over all room surfaces (see Section 6.7.7, Equation (6.86)) and $\beta_{e,i}$ is the statistical energy reflection coefficient of surface, i, of area S_i. The first term in Equation (6.69) accounts for room dimensions that exceed the Sabine room criterion. The quantity γ^2 is the variance of the distribution of path lengths between reflections, divided by the square of the mean free path length. It has a value of about 0.4, provided that the room shape is not extreme. The second term in Equation (6.69) accounts for non-uniform placement of sound-absorption.

Neubauer (2000) provided an alternative modified Fitzroy equation for flat and long rooms as:

$$T_{60} = \left[\frac{-0.126S_x}{\log_e(1-\bar{\alpha}_{stx})P_x} - \frac{0.126S_y}{\log_e(1-\bar{\alpha}_{sty})P_y} - \frac{0.126S_z}{\log_e(1-\bar{\alpha}_{stz})P_z}\right]^{1/2} \tag{6.70}$$

where P_x and P_y are the total perimeters for each of the two pairs of opposite walls and P_z is the total perimeter of the floor and ceiling. Similar definitions apply for S_x, S_y and S_z and also for $\bar{\alpha}_{stx}$, $\bar{\alpha}_{sty}$ and $\bar{\alpha}_{stz}$. Note that for a cubic room, Equation (6.70) may be used with the exponent, (1/2), replaced by (1/3).

6.5.4 Mean Free Path

When air absorption was considered in Section 6.4.3, the mean free path was introduced as the mean distance travelled by a sound wave between reflections, and frequent reference has been made to this quantity in subsequent sections. Many ways have been demonstrated in the literature for determining the mean free path and two will be presented in this section.

The classical description of a reverberant space, based on the solution of a simple differential equation presented in Section 6.5.1, leads directly to the concept of mean free path. Let the mean free path be Λ, then in a length of time equal to Λ/c all of the sound energy in the reverberant space will be once reflected and reduced by an amount (one reflection), $e^{-\bar{\alpha}}$. If the energy stored in volume V was initially $V\langle p_0^2\rangle/\rho c^2$ and at the end of time $t = \Lambda/c$ it is $V\langle p^2\rangle/\rho c^2$, then according to Equation (6.53):

$$V\frac{\langle p^2\rangle}{\rho c^2} = V\frac{\langle p_o^2\rangle}{\rho c^2}e^{-\bar{\alpha}} = V\frac{\langle p_o^2\rangle}{\rho c^2}e^{-S\bar{\alpha}\Lambda/(4V)} \tag{6.71}$$

Consideration of Equation (6.71) shows that the mean free path, Λ, is given by Equation (6.36).

Alternatively, a modal approach to the determination of the mean free path may be employed, using modal indices, n_x, n_y and n_z, respectively. Consideration in this case will be restricted to rectangular enclosures for convenience. For this purpose the following quantities are defined:

$$X = \frac{n_x c}{2f_n L_x},\ Y = \frac{n_y c}{2f_n L_y} \text{ and } Z = \frac{n_z c}{2f_n L_z} \tag{6.72}$$

Substitution of Equations (6.72) into Equation (6.17) gives the following result:

$$1 = X^2 + Y^2 + Z^2 \tag{6.73}$$

Letting $a_1 = L_y \times L_z$, $a_2 = L_z \times L_x$, $a_3 = L_x \times L_y$ and $V = L_x \times L_y \times L_z$, multiplying the numerator of the reciprocal of Equation (6.60) by V and the denominator by $L_x \times L_y \times L_z$ and use of Equation (6.72) gives the following result:

$$V/\Lambda_i = a_1 X + a_2 Y + a_3 Z \tag{6.74}$$

An average value for the quantity V/Λ_i may be determined by summing over all possible values of Λ_i. When the modal density is large, it may be assumed that sound is incident from all directions, and it is then possible to replace the sum with an integral. Introducing the following spherical coordinates:

$$X = \sin\phi\cos\theta; \quad Y = \sin\phi\sin\theta; \quad Z = \cos\phi \tag{6.75}$$

Substituting Equation (6.75) into Equation (6.74) and forming the integral, the following result is obtained:

$$V/\Lambda = \frac{2}{\pi}\int_0^{\pi/2} \mathrm{d}\phi \int_0^{\pi/2} (a_1\sin^2\phi\cos\theta + a_2\sin^2\phi\sin\theta + a_3\cos\phi\sin\phi)\mathrm{d}\theta \tag{6.76}$$

Carrying out the indicated integration gives for the mean free path, Λ, the result given previously by Equation (6.36).

6.6 Measurement of the Room Constant

Measurements of the room constant, R_c, given by Equation (6.45), or the related Sabine absorption, $S\bar{\alpha}$, may be made using either a reference sound source or by measuring the reverberation time of the room in the frequency bands of interest. These methods are described in Sections 6.6.1 and 6.6.2. Alternatively, yet another method is offered in Section 4.14.3.1.

6.6.1 Reference Sound Source Method

The reference sound source is placed at a number of positions chosen at random in the room to be investigated, and sound pressure levels are measured at a number of positions in the room for each source position. In each case, the measurement positions are chosen to be remote from the source, where the reverberant field of the room dominates the direct field of the source. The number of measurement positions for each source position and the total number of source positions used are usually dependent on the irregularity of the measurements obtained. Generally, four or five source positions with four or five measurement positions for each source position are sufficient, giving a total number of measurements between 16 and 25. The room constant, R_c, for the room is then calculated using Equation (6.44) rearranged as:

$$R_c = 4 \times 10^{(L_W - L_p)/10} \tag{6.77}$$

In writing Equation (6.77), the direct field of the source has been neglected, following the measurement procedure proposed above, and it has been assumed that $\rho c = 400$. In Equation (6.77), L_p is the logarithmic average of all the N sound pressure level measurements and is calculated using Equation (4.142).

6.6.2 Reverberation Time Method

The second method is based on a measurement of the room reverberation time, which is the time (seconds) for the sound in the room to decay by 60 dB after the sound source is turned off.

When measuring reverberation time in a room, the source of sound is usually a speaker driven by a random noise generator in series with a bandpass filter. When the sound is turned off, the room rate of decay can be measured simply by using a sound level meter attached to a level recorder (or data acquisition system and computer) as illustrated in Figure 6.6.

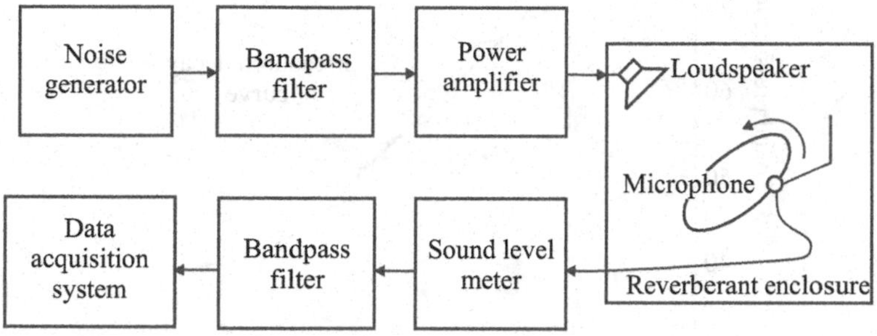

FIGURE 6.6 Equipment for reverberation time (T_{60}) measurement.

Alternatively, there are many acoustic instruments such as spectrum analysers that can calculate the reverberation time internally for all 1/3-octave bands simultaneously. In this case, the "data-acquisition system" box, "bandpass filter" box and the "sound level meter" box are replaced with a "sound analyser" box. However, it is important to ensure that the signal level in each band is at least 45 dB above any background noise. If it is less, the reverberation time results will be less accurate.

The reverberation time, T_{60}, in each frequency band is determined as $60/D$, where D is the sound pressure level decay rate (dB/s) obtained using the level recorder or the spectrum analyser. According to Equation (6.54), the recorded level in decibels should decay linearly with time. The slope, generally measured as the best straight line fit to the recorded decay between 5 dB and 35 dB down from the initial steady-state level, is used to determine the decay rate. The time for the sound to decay 30 dB (from −5 dB to −35 dB below the level at the time the sound source was switched off) is estimated from the slope of the straight line fitted to the trace of sound pressure level versus time. The reverberation time, T_{60}, for a 60 dB decay, is found by multiplying the 30 dB decay result by 2. For the example shown in Figure 6.7, the reverberation time, $T_{60} = (1.77 - 0.14) \times 2 = 3.3$ seconds. The 35 dB down level should be at least 10 dB above the background noise level for any 1/3-octave or octave band measurement.

When measuring reverberation time in a room or laboratory, it is important to use a number of different microphone and loudspeaker locations and obtain an average value. At this point, a word of caution is in order. When processing the data, determine average decay rates not decay times, even though ISO 354 (2003) suggests that similar results are obtained by arithmetically averaging reverberation times! To average decay rates we take the reciprocal of the average of reciprocal T_{60} values, as illustrated in Equation (6.78):

$$\frac{1}{T_{60}} = \frac{1}{N} \sum_{i=1}^{N} \frac{1}{T_{60_i}} \tag{6.78}$$

Once found, the average reverberation time, T_{60}, is used in Equation (6.55), rearranged as follows, to calculate the room absorption. Thus:

$$S\bar{\alpha} = (55.25V)/(cT_{60}) \quad (\text{m}^2) \tag{6.79}$$

The standard deviation of the reverberation time measurements, to be used in uncertainty estimates for the case of the 30 dB decay time measurement, T_{30} is (ISO 3382-2, 2008):

$$\sigma_{(T_{30})} = 0.55T_{30}\sqrt{\frac{1 + (1.52/n)}{NBT_{30}}} \tag{6.80}$$

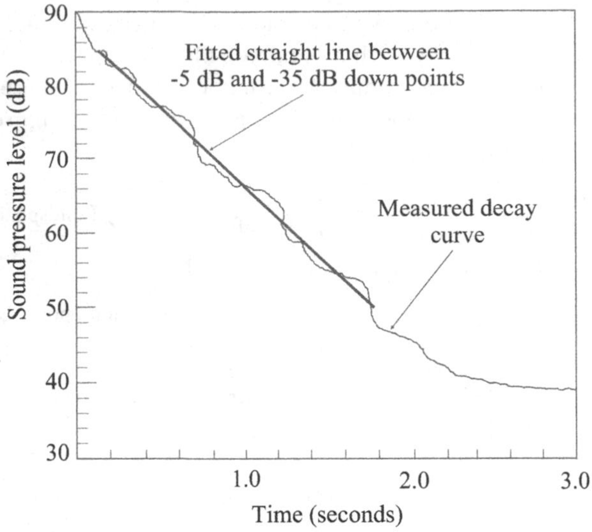

FIGURE 6.7 Measurement of reverberation time.

where n is the number of measurements at each speaker/microphone location pair, N is the number of different microphone/speaker location pairs and B is the bandwidth in Hz (usually 1/3-octave or octave, see Equation (1.115)).

ASTM E2235-04 (2020) provides procedures for measuring the average room absorption that are slightly different to what has been discussed. Rather than average reciprocals of reverberation times, ASTM E2235-04 (2020) recommends averaging the sound pressure levels to obtain $L_{p,i}$, using Equation (4.142), at each time increment, i on the decay curve, where time is calculated from when the sound source is first switched off. This produces a similar result to averaging reciprocal reverberation times but relies on there being no variation in source sound power output between tests, so this approach is not recommended here. ASTM E2235-04 (2020) also suggests that the user can choose between logarithmic and arithmetic averaging the sound pressure levels at each time increment, i. However, logarithmic averaging is the choice that is technically correct.

To determine reverberation decay rates (or reverberation times), ASTM E2235-04 (2020) uses the portion of the decay curve between 5 dB down and 35 dB down from the steady-state level occurring prior to the sound source being turned off (for laboratory test facilities) and between 5 dB down and 25 dB down when measurements are undertaken in the field, where background noise levels are usually higher. The 30 dB decay time, T_{30} is multiplied by two to obtain T_{60} and the 20 dB decay time, T_{20}, is multiplied by three.

When observing reverberation decay curves (average sound pressure level versus time) it will be noted that for almost any room, two different slopes will be apparent. The steeper slope occurs for the initial 7 to 10 dB of decay, the exact number of dB being dependent on the physical characteristics of the room and contents. When this initial slope is extrapolated to a decay level of 60 dB, the corresponding time is referred to as the early decay time (EDT). The slope of the remainder of the decay curve, when extrapolated to 60 dB, corresponds to what is commonly referred to as the reverberation time (RT). The ratio of EDT to RT as well as the absolute values of the two quantities are widely used in the design of architectural spaces. See also Beranek (1962), Mackenzie (1979), Cremer and Müller (1982) and Egan (1987).

6.7 Porous Sound Absorbers

6.7.1 Measurement of Absorption Coefficients

There are two measures of absorption coefficient that are used to assess the ability of a material to absorb randomly incident sound: the statistical absorption coefficient, α_{st}, and the Sabine absorption coefficient, $\bar{\alpha}$. The statistical absorption coefficient is calculated from the normal specific acoustic impedance, which is measured using an impedance tube as described in Appendix D. The statistical absorption coefficient has a maximum theoretical value of just over 0.95, and it is always less than the Sabine absorption coefficient for the same material, which can have maximum values up to about 1.3 (meaning that in principal, there is more energy absorbed than is incident on the test material). This peculiar phenomenon arises from the sample of absorbing material distorting the sound field in the reverberation room, causing the sound field to diffract towards the sample, so that more energy falls on its surface than fell on the floor in the absence of the material. Sabine absorption coefficients for materials are generally measured in a laboratory using a reverberant test chamber. Procedures and test chamber specifications are described in various standards (ASTM C423-17, 2017; ISO 354, 2003). The material to be tested is placed in a reverberant room and the reverberation time, T_{60}, is measured. The test material is removed and the reverberation time, T'_{60}, of the room containing no test material is measured next. Provided that the absorption of the reverberation room in the absence of the test material is dominated by the absorption of the walls, floor and ceiling, the reverberation times are related to the test material absorption coefficient, $\bar{\alpha}$, by the following equation (derived directly from Equation (6.55)):

$$\bar{\alpha} = \frac{55.3V}{Sc}\left[\frac{1}{T_{60}} - \frac{(S'-S)}{S'T'_{60}}\right] \quad (\text{m}^2) \tag{6.81}$$

The quantity, S', is the total area of all surfaces in the room including the area covered by the material under test. Equation (6.81) is written with the implicit assumption that the surface area, S, of the test material is large enough to measurably affect the reverberation time, but not so large as to seriously affect the diffusivity of the sound field, which is basic to the measurement procedure. The standards recommend that S should be between 10 and 12 m^2 with a length-to-breadth ratio between 0.7 and 1.0. The tested material should also be more than 1 m from the nearest room edge and the edge of the test specimen should not be parallel to the nearest room edge. Absorption coefficients for materials other than flat sheets can be determined in a similar way to flat sheets, but there should a sufficient number of test specimens to achieve a total value of $S\alpha$ between 1 and 12 m^2.

In many cases, the absorption of a reverberation room is dominated by things other than the room walls, such as loudspeakers at low frequencies, stationary and rotating diffuser surfaces at low- and mid-frequencies and air absorption at high frequencies. For this reason, the contribution of the room to the total absorption is often considered to be the same with and without the presence of the sample. In this case, the Sabine absorption coefficient of the sample may be written as:

$$\bar{\alpha} = \frac{55.3V}{Sc}\left[\frac{1}{T_{60}} - \frac{1}{T'_{60}}\right] \quad (\text{m}^2) \tag{6.82}$$

Equation (6.82) is what appears in most current standards (see ISO 354 (2003)), even though its accuracy is questionable (Equation (6.81) is more accurate). Note that the reverberation times used in Equations (6.81) and (6.82) are averages of many measured reverberation times. Refer to Section 6.6.2 for procedures to be used to obtain average values.

The absorption coefficient, $\bar{\alpha}$, of a material sample can also be measured using a reference sound source of known sound power level, L_W, by measuring the space-averaged sound pressure level in the room with and without the sample present and using Equations (6.77) and (6.45) to give:

$$\bar{\alpha} = \frac{S'\bar{\alpha}_2 - S'\bar{\alpha}_1 + S\bar{\alpha}_1}{S} \tag{6.83}$$

where S is the area of test material, $\bar{\alpha}_1$ is the space-averaged absorption coefficient in the room without the test material present and $\bar{\alpha}_2$ is the space-averaged absorption coefficient in the room with the test material present.

The measured value of the Sabine absorption coefficient is dependent on the sample size, sample distribution and the properties of the room in which it is measured. Because standards specify the room characteristics and sample size and distribution for measurement, similar results can be expected for the same material measured in different laboratories (although even under these conditions significant variations have been reported). However, these laboratory-measured values are used to calculate reverberation times and reverberant sound pressure levels in auditoria and factories that have quite different characteristics, which implies that in these cases, values of reverberation time, T_{60}, and reverberant field sound pressure level, L_p, calculated from measured Sabine absorption coefficients are approximate only. This is why many practitioners prefer to do their calculations of reverberation time in building spaces that are not very reverberant using statistical absorption coefficients and Equation (6.62), rather than using Sabine absorption coefficients and Equation (6.55).

Statistical absorption coefficients may be estimated from impedance tube measurements, as discussed in Appendix D. A list of Sabine absorption coefficients selected from the literature is included in Table 6.2 for various materials. The approximate nature of the available data makes it desirable to either use manufacturer's data or take measurements (if possible).

6.7.2 Single Number Descriptors for Absorption Coefficient

Relevant standards (ASTM C423-17, 2017) require that absorption coefficients be measured in 1/3-octave bands and that octave band data are found by arithmetically averaging the absorption coefficients for the three 1/3-octave bands included in the particular octave band. The octave band values so obtained and rounded to the nearest 0.05 are also referred to as Practical Sound Absorption Coefficients, α_p (ISO 11654, 1997). Mid-point values are rounded up; for example 0.855 is rounded to 0.86. If the sound absorption of a particular sample is required to be reported in m^2, the absorption coefficient is multiplied by the area of the sample.

For the purposes of quickly comparing the relative absorption effectiveness of different products, it is useful for the absorption characteristics of a material to be described by a single number. There are three different descriptors commonly used: the Weighted sound absorption, α_w, described in the International Standard, ISO 11654 (1997); the Noise Reduction Coefficient or NRC (now outdated) and Sound Absorption Average (SAA), which are both described in the American standard, ASTM C423-17 (2017).

The α_w value is obtained by plotting the measured Practical Sound Absorption Coefficient data in Figure 6.8 and shifting the reference curve shown in the figure up or down until the sum of the "unfavourable deviations"$\leq$ 1. An unfavourable deviation occurs at a particular frequency when the measured value is less than the value of the reference curve.

The standard, ISO 11654 (1997), also requires the reporting of a shape indicator whenever a Practical Sound Absorption Coefficient value exceeds the final reference curve by more than. If this occurs at 250 Hz, the shape indicator is L. If it occurs at 500 Hz or 1 kHz, the shape indicator is M. If it occurs at 2 kHz or 4 kHz, the shape indicator is H.

NRC is defined as:

$$\text{NRC} = \frac{(\bar{\alpha}_{250} + \bar{\alpha}_{500} + \bar{\alpha}_{1000} + \bar{\alpha}_{2000})}{4} \tag{6.84}$$

where each octave band absorption coefficient is rounded to the nearest 0.05 and represents the average of the three included 1/3-octave bands. Again, mid-point values are rounded up.

TABLE 6.2 Sabine absorption coefficients for some commonly used materials

Material	Octave band centre frequency (Hz)						
	63	125	250	500	1000	2000	4000
Concert hall seats							
Unoccupied – heavily upholstered seats (Beranek and Hidaka, 1998)		0.65	0.76	0.81	0.84	0.84	0.81
Unoccupied – medium upholstered seats		0.54	0.62	0.68	0.70	0.68	0.66
Unoccupied – light upholstered seats		0.36	0.47	0.57	0.62	0.62	0.60
Unoccupied – very light upholstered seats		0.35	0.40	0.41	0.38	0.33	0.27
Unoccupied – average well-upholstered seating areas	0.28	0.44	0.60	0.77	0.84	0.82	0.70
Unoccupied – leather-covered upholstered seating areas		0.40	0.50	0.58	0.61	0.58	0.50
Unoccupied – metal or wood seats		0.15	0.19	0.22	0.39	0.38	0.30
Unoccupied – concert hall, no seats halls lined with thin wood or other materials <3 cm thick		0.16	0.13	0.10	0.09	0.08	0.08
Halls lined with heavy materials		0.12	0.10	0.08	0.08	0.08	0.08
100% occupied audience (orchestra and chorus areas) – upholstered seats	0.34	0.52	0.68	0.85	0.97	0.93	0.85
Audience, per person seated $S\bar{\alpha}(\mathrm{m}^2)$		0.23	0.37	0.44	0.45	0.45	0.45
Audience, per person standing $S\bar{\alpha}(\mathrm{m}^2)$		0.15	0.37	0.43	0.44	0.44	0.43
Wooden pews – 100% occupied		0.57	0.61	0.75	0.86	0.91	0.86
Wooden chairs – 100% occupied		0.60	0.74	0.88	0.96	0.93	0.85
Wooden chairs – 75% occupied		0.46	0.56	0.65	0.75	0.72	0.65
Walls							
Acoustic plaster, 10 mm thick sprayed on solid wall		0.08	0.15	0.30	0.50	0.60	0.70
Hard surfaces (brick walls, plaster, hard floors, etc.)		0.02	0.02	0.03	0.03	0.04	0.05
Gypsum board on 50 × 100 mm studs		0.29	0.10	0.05	0.04	0.07	0.09
Plaster, gypsum or lime, smooth finish;							
on brick,		0.013	0.015	0.02	0.03	0.04	0.05
on concrete block,		0.012	0.09	0.07	0.05	0.05	0.04
on lath		0.014	0.10	0.06	0.04	0.04	0.03
Solid timber door		0.14	0.10	0.06	0.08	0.10	0.10

Cont. on next page

Material	Octave band centre frequency (Hz)						
	63	125	250	500	1000	2000	4000
Acoustic material							
Fibreglass or rockwool blanket							
16 kg/m^3, 25 mm thick		0.12	0.28	0.55	0.71	0.74	0.83
16 kg/m^3, 50 mm thick		0.17	0.45	0.80	0.89	0.97	0.94
16 kg/m^3, 75 mm thick		0.30	0.69	0.94	1.0	1.0	1.0
16 kg/m^3, 100 mm thick		0.43	0.86	1.0	1.0	1.0	1.0
24 kg/m^3, 25 mm thick		0.11	0.32	0.56	0.77	0.89	0.91
24 kg/m^3, 50 mm thick		0.27	0.54	0.94	1.0	1.0	1.0
24 kg/m^3, 75 mm thick		0.28	0.79	1.0	1.0	1.0	1.0
24 kg/m^3, 100 mm thick		0.46	1.0	1.0	1.0	1.0	1.0
48 kg/m^3, 50 mm thick		0.3	0.8	1.0	1.0	1.0	1.0
48 kg/m^3, 75 mm thick		0.43	0.97	1.0	1.0	1.0	1.0
48 kg/m^3, 100 mm thick		0.65	1.0	1.0	1.0	1.0	1.0
60 kg/m^3, 25 mm thick		0.18	0.24	0.68	0.85	1.0	1.0
60 kg/m^3, 50 mm thick		0.25	0.83	1.0	1.0	1.0	1.0
Polyurethane foam, 27 kg/m^3 15 mm thick		0.08	0.22	0.55	0.70	0.85	0.75
Floors							
Wood platform with large space beneath		0.40	0.30	0.20	0.17	0.15	0.10
Wood floor on joists		0.15	0.11	0.10	0.07	0.06	0.07
Concrete or terrazzo		0.01	0.01	0.01	0.02	0.02	0.02
Concrete block painted		0.01	0.05	0.06	0.07	0.09	0.08
Linoleum, asphalt, rubber or cork tile on concrete		0.02	0.03	0.03	0.03	0.03	0.02
Varnished wood joist floor		0.15	0.12	0.10	0.07	0.06	0.07
Carpet, heavy, on concrete		0.02	0.06	0.14	0.37	0.60	0.65
Carpet, heavy, on 1.35 kg/ m^2 hair felt or foam rubber		0.08	0.24	0.57	0.69	0.71	0.73
Carpet, 5 mm thick, on hard floor		0.02	0.03	0.05	0.10	0.30	0.50
Carpet, 6 mm thick, on underlay		0.03	0.09	0.20	0.54	0.70	0.72
Cork floor tiles (3-4 inch thick) – glued down		0.08	0.02	0.08	0.19	0.21	0.22
Glazed tile/marble		0.01	0.01	0.01	0.01	0.02	0.02
Ceilings							
13 mm mineral tile direct fixed to floor slab		0.10	0.25	0.70	0.85	0.70	0.60
13 mm mineral tile suspended 500 mm below ceiling		0.75	0.70	0.65	0.85	0.85	0.90
Curtains							
Light velour, 338 g/m^2							
hung flat on wall		0.03	0.04	0.11	0.17	0.24	0.35
hung in folds on wall		0.05	0.15	0.35	0.40	0.50	0.50
Medium velour, 475 g/m^2 draped to half area		0.07	0.31	0.49	0.75	0.70	0.60
Heavy velour, 610 g/m^2 draped to half area		0.14	0.35	0.55	0.72	0.70	0.65

Cont. on next page

Material	Octave band centre frequency (Hz)						
	63	125	250	500	1000	2000	4000
Glass							
Glass, heavy plate		0.18	0.06	0.04	0.03	0.02	0.02
Ordinary window		0.35	0.25	0.18	0.12	0.07	0.04
Other							
Stage openings		0.30	0.40	0.50	0.60	0.60	0.50
Water (surface of pool)		0.01	0.01	0.01	0.015	0.02	0.03
Orchestra with instruments on podium, 1.5 m^2 per person		0.27	0.53	0.67	0.93	0.87	0.80

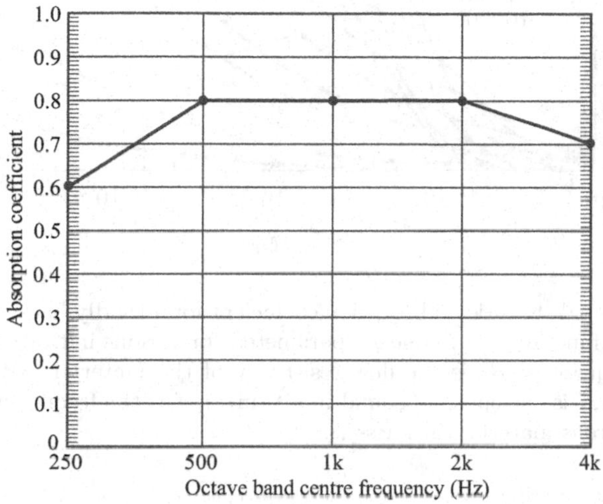

FIGURE 6.8 Absorption coefficient reference curve.

The Sound Absorption Average (SAA) is the arithmetic average of the twelve 1/3-octave band absorption coefficients (rounded to the nearest 0.01) in the frequency range between and including 200 Hz to 2500 Hz. Again, mid-point values are rounded up.

6.7.3 Porous Liners

Where manufacturer's data are not available, the statistical absorption coefficient for a porous blanket of thickness, ℓ, backed by a cavity of depth, L, may be calculated as outlined in Appendix D. Some results of such calculations are shown in Figure 6.9. Implicit in the calculations is the assumption that sound propagation within the porous material is predominantly normal to the surface. This condition is sufficiently well satisfied if the porous material is fairly dense. Alternatively, the porous material could be contained between solid partitions arranged to prevent lateral propagation.

The calculated statistical absorption coefficient is optimum when the total flow resistance, R_f, through the material is between $2\rho c$ and $5\rho c$ (see Appendix D for a discussion of flow resistance). This is shown in Figure 6.9, where it can be seen that a porous liner as little as one-tenth of a wavelength thick will give a statistical absorption coefficient of about 0.92. This is close to the maximum that is theoretically possible.

The performance of a porous blanket material can be improved by mounting it so that an air gap exists between the material and the hard backing wall. Calculations (see Appendix D)

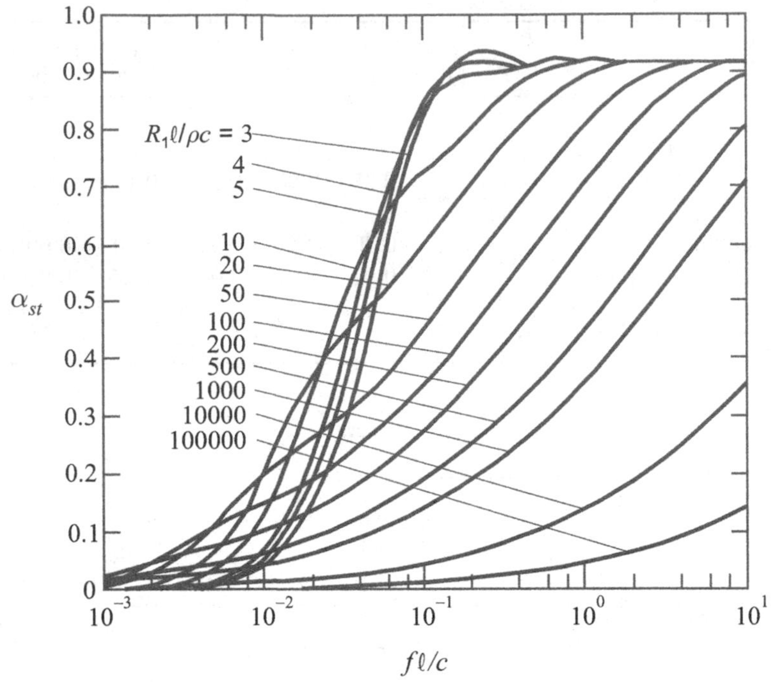

FIGURE 6.9 Calculated statistical absorption coefficient for a rigidly backed porous material in a reverberant field, as a function of a frequency parameter, for various indicated values of normalised flow resistance. The quantity R_1 is the flow resistivity of the material (MKS rayls/m), ρ is the density of air (kg/m^3), c is the speed of sound in air (m/s), f is the frequency (Hz) of the incident sound and ℓ is the porous material thickness (m).

show that the optimum air gap depth is equal to the thickness of the porous liner and that the absorption will be the same as that which would characterise a porous layer equal in thickness to the air gap plus liner. To achieve this optimum, it would be necessary to place partitions in the air gap to prevent lateral sound propagation in the gap parallel to the wall.

In practice, it is often necessary to protect a porous liner from contamination by moisture, dust, oil, chemicals, etc., which would render it useless. For this reason, the material is often wrapped in a thin, limp, impervious blanket made of polyurethane, polyester, aluminium or PVC. The effect of the introduction of a wrapping material is generally to improve the low-frequency absorption at the expense of the high-frequency absorption. However, if the wrapping material is sufficiently thin (for example, 20 µm thick polyurethane) then the effect is not measurable. The effect can be calculated as shown in Appendix D.

6.7.4 Porous Liners with Perforated Panel Facings

When mechanical protection is needed for a porous liner, it may be covered using perforated wood, plastic or metal panels. If the open area provided by the perforations is greater than about 20%, the expected absorption is entirely controlled by the properties of the porous liner, and the panel has no significant effect.

Alternatively, if the facing panel has an open area of less than 20%, then the frequency at which absorption is maximum may be calculated by treating the facing as an array of Helmholtz resonators in a manner similar to that used in Section 8.8.2.

This procedure leads to the following approximate equation:

$$f_{\max} = \frac{c}{2\pi}\left[\frac{P/100}{L[w + 0.85d(1 - 0.22d/q)]}\right]^{1/2} \tag{6.85}$$

In Equation (6.85), P is the percentage open area of the panel, L is the depth of the backing cavity including the porous material layer, w is the panel thickness, d is the diameter of the perforations and q is the spacing between hole centres. If the porous material fills the entire cavity so that the thickness of the porous material is also L, then the speed of sound c should be replaced with $0.85c$ to account for isothermal rather than adiabatic propagation of sound in the porous material at low frequencies. The condition $f_{\max}L/c < 0.1$ must be satisfied for Equation (6.85) to give results with less than 15% error. Measured data for a panel of 10% open area are presented in Figure 6.11 (see Davern (1977) for additional data).

If a perforated facing is used over a porous liner wrapped in an impervious membrane, care must be taken to ensure that the facing and liner are not in contact; otherwise, the sound-absorbing properties of the combination will be severely degraded. Non-contact is usually achieved in practice by placing a 12 mm square wire-mesh spacer between the liner and the perforated sheet.

6.7.5 Micro-Perforated Panels and Sheets

Micro-perforated panels (MPPs) can be used as sound absorbers when it is necessary to achieve a sound-absorbing surface in the absence of porous acoustic materials such as fibreglass, rockwool or polyester. The advantage of the micro-perforated panel absorber is that it can be made of any material and can thus function in environments unsuitable for traditional porous acoustic materials. Although micro-perforated absorbers were developed in the late 1960s (Maa, 1998), they were not widely used until the mid to late 2000s, due to the difficulty and expense in producing absorbers that have an acceptable performance over a wide frequency range. More recently, absorbers have been developed that are not prohibitively costly and which, through the use of multiple layers and novel designs, can achieve the required performance.

The theory underpinning the acoustic performance of micro-perforated panels in absorbing sound was published by Maa (1998), who showed that the main disadvantage of these absorbers was the limited frequency bandwidth (approximately two octaves) over which good performance could be achieved. Since then, the theoretical work has been extended to include configurations with multiple layers of micro-perforated material (Cobo and Simon, 2019). Multiple layers of micro-perforated material extend the frequency range over which the construction is efficient, with three layers achieving good performance over four octaves, with statistical absorption coefficients between 0.8 and 0.95 over the frequency range 1000 Hz to 6000 Hz, decreasing to 0.25 at 300 Hz (Cobo and Simon, 2019). The cost of manufacturing MPPs has been reduced significantly through the use of 3D printing and the drilling of epoxy substrates, as opposed to the traditional methods of laser drilling, micro-punching or etching of steel or aluminium substrates. However, the use of SLS (selective laser sintering) using a CO_2 laser to sinter powdered polymer materials, which is an advanced form of 3D printing, has been shown to be cost-effective in producing multi-layer MPPs (Yang et al., 2020).

6.7.6 Acoustic Metamaterials

Acoustic metamaterials (AMMs or AMs) are usually artificial composites that possess properties that do not exist naturally. They rely on an array of sub-wavelength unit cells to achieve high levels of sound-absorption or low levels of sound transmission using material thicknesses that are substantially less than conventional materials which achieve similar acoustic absorption performance. A review of a range of various AMM structures is provided by Liao et al. (2021) and a general overview is provided by Cummer et al. (2016).

Acoustic metamaterials often consist of an array of small (compared to a wavelength of sound) resonators, with each resonator tuned to a different frequency for control of a band of noise or to a single frequency for tonal control. For the control of sound in mufflers the array could consist of one wall of the duct supporting an array of 1/4-wavelength tubes or Helmholtz resonators. For increasing the transmission loss at the coincidence frequency (see Section 7.2.1) of a wall made from a sandwich panel, resonators could be placed between the two skins as shown in Figure 6.10(b) (Liu et al., 2018). For increasing the low-frequency absorption capability of acoustic foam, steel masses could be embedded in the foam to form resonators, with the foam acting as the stiffness as shown in Figure 6.10(c) (Fuller and Mathur, 2017). Alternatively, micro-perforated sheets could be placed between layers of acoustic foam as shown in Figure 6.10(d) (Fuller and Mathur, 2017). Configurations (c) and (d) are used to increase the low-frequency sound-absorption capability of acoustic foam.

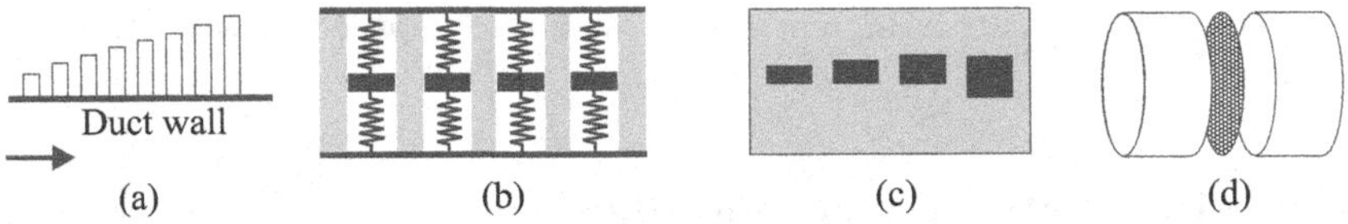

FIGURE 6.10 Various metamaterial configurations. (a) 1/4-wave tubes mounted on a flow duct; (b) resonators embedded between the two skins of a composite panel to increase the sound transmission loss at coincidence (with the sandwich material shown as grey shading); (c) Steel masses embedded in acoustic foam to increase the sound-absorption at low frequencies; (d) micro-perforated sheets between acoustic foam layers.

Care must be taken in the use of metamaterials in flow ducts such as mufflers, so that unwanted sound generation caused by interaction of flow with the metamaterial, is avoided (D'Elia et al., 2020).

Metamaterials have also been used to focus sound as well as act as a collimator so that sound from a source such as a loudspeaker is restricted to a specific direction. This has been done using small brick elements, which consist of folded passages that delay the passage of sound and which can be combined into a metamaterial sheet that can focus and bend an incoming sound field (Memoli et al., 2019). Metamaterials made up of similar brick elements, are also being used to reduce sound transmission between spaces while maintaining air flow for ventilation.

6.7.6.1 Layered Fibrous Material

A separate class of metamaterials is characterised by two or more layers of similar material but with different thicknesses, densities and flow resistivities (Ingard, 1994; Lambert, 2006). The less porous face, corresponding to the higher flow resistivity and smaller thickness, is placed closest to the acoustic field. The theoretical prediction of the performance of these materials can be obtained using the approach outlined in Section D.5.2 to obtain the normal impedance and Equations (D.44) and (D.50) to convert normal impedance to statistical absorption coefficient. Such layered materials can significantly increase the low-frequency sound absorption performance for a fixed total thickness of material, although in most cases, the high-frequency performance is reduced. In one optimised example provided by Lambert (2006), the sound absorption coefficient for a 100 mm thick layer of fibrous material was increased from 0.44 to 0.66 at 250 Hz and from 0.78 to 0.95 at 500 Hz, with little change at higher frequencies, which greatly increased the NRC value (see Equation (6.84)). For the previous example, the fibrous material had a flow resistivity of 7030 MKS rayls/m and the 1 mm-thick covering layer had a flow resistivity of 74600 MKS rayls/m.

Yet another class of metamaterials consists of layers of sound absorbing material separated by very thin steel or aluminium plates which may or may not be perforated. These constructions can also enhance the low-frequency sound-absorption performance of fibrous or foam acoustic material without significantly increasing the total thickness.

6.7.6.2 Porous Concrete

Porous concrete is a special type of meta-material that uses pores in the concrete to absorb sound. The porosity of concrete can be increased by a number of methods listed below (Tie et al., 2020).

- Using lightweight aggregate such as expanded shale.
- Using a mixture of small and large aggregate to increase the pore number and size.
- Adding crumbed rubber.
- Using building construction waste such as recycled concrete with a particle size greater than 5 mm.
- Using foam with or without fine aggregate.
- Using Hemp concrete made from a mixture of hemp stems with a mineral binder such as lime or cement.

The density of normal concrete is approximately 2300 kg/m^3, whereas the density of porous concrete can be as low as 300-500 kg/m^3. Generally, reducing the density of concrete as a result of more pores and/or pores of larger size, results in higher sound-absorption coefficient values. It is possible to achieve NRC values greater than 0.5 with a concrete density of 500 kg/m^3.

6.7.6.3 Functionally Graded Materials

Functionally graded materials are a special type of metamaterial in which the properties of the material gradually change throughout a thickness or length. Such materials can also be used for acoustic cloaking (Sanders et al., 2018) with one particular application directed towards hiding underwater objects from sonar. This is done by arranging the material properties so that incident sound is refracted around the object in such a way that the magnitude and phase of the downstream sound field is the same as it would have been without the object in place. This also implies that the amplitude of reflected sound will be negligible so the object essentially disappears.

6.7.7 Sound-Absorption Coefficients of Materials in Combination

When a surface is characterised by several different materials having different values of the absorption coefficient, it is necessary to determine an average Sabine absorption coefficient. For example, for a room characterised by q different materials, the average absorption coefficient will be the area-weighted average absorption of the q different materials calculated as:

$$\bar{\alpha} = \frac{\sum\limits_{i=1}^{q} S_i \bar{\alpha}_i}{\sum\limits_{i=1}^{q} S_i} \tag{6.86}$$

where S_i are the areas and $\bar{\alpha}_i$ are the Sabine absorption coefficients of the q materials.

6.8 Panel Sound Absorbers

Figure 6.9 indicates that porous liners are not very effective in achieving low-frequency absorption. When low-frequency absorption is required, panel absorbers are generally used. These consist of flexible panels mounted over an air space; for example, on battens (or studs) attached to a solid wall. Alternatively, such panels may be mounted on a suspended ceiling. In any case, to be effective, the panels must couple with and be driven by the sound field. Acoustic energy is then dissipated by flexure of the panel. Additionally, if the backing air space is filled with a porous material, energy is also dissipated in the porous material. Maximum absorption occurs at the first resonance of the coupled panel-cavity system. A typical example of absorption coefficients for a panel absorber is shown in Figure 6.11.

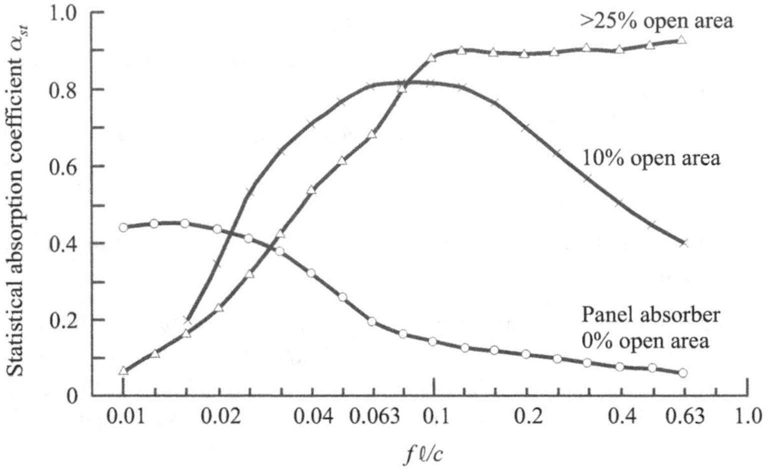

FIGURE 6.11 Effect of perforations on the sound-absorption of a panel backed by a porous liner. The panel surface weight is 2.5 kg/m^2 and its thickness is 3 mm. The porous liner is 50 mm thick and about $5\rho c$ flow resistance (see Appendix D).

It is customary to assign a Sabine absorption coefficient to a resonant panel absorber, although the basis for such assignment is clearly violated by the mode of response of the panel absorber; that is, the absorption is not dependent on local properties of the panel but is dependent on the response of the panel as a whole. Furthermore, as the panel absorber depends on strong coupling with the sound field to be effective, the energy dissipated is very much dependent on the sound field and thus on the rest of the room in which the panel absorber is used. This latter fact makes the prediction of the absorptive properties of panel absorbers difficult.

Two methods will be described for estimating the Sabine absorption of panel absorbers. One is empirical and is based on data measured in auditoria and concert halls and must be used with caution while the other is based on analysis, but requires considerable experimental investigation to determine all of the required parameters.

6.8.1 Empirical Method

The essence of an empirical prediction scheme (Hardwood Plywood Manufacturers' Association, 1962) for flexible panel absorbers, which has been found useful in auditoria and concert halls is contained in Figures 6.12 and 6.13. First, the type of Sabine absorption curve desired is selected from curves A to J in Figure 6.12. The solid curves are for configurations involving a blanket (25 mm thick and flow resistance between $2\rho c$ and $5\rho c$) in the air gap behind the panel, while

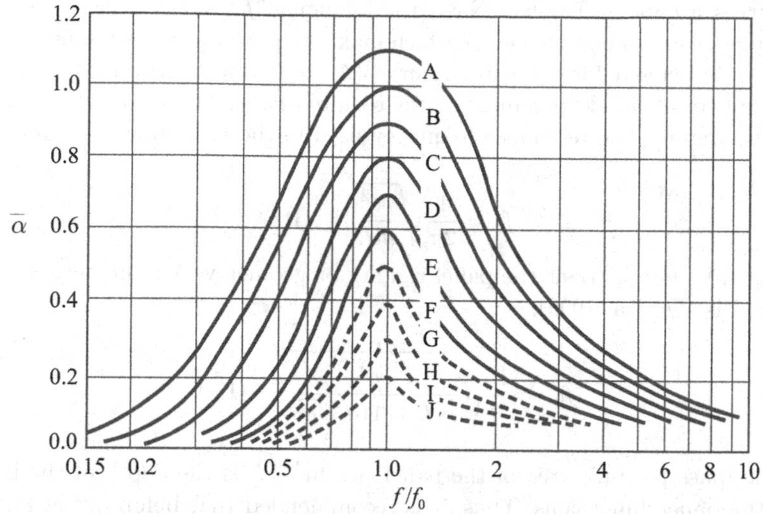

FIGURE 6.12 Sabine absorption coefficients for resonant plywood panels. The panel configurations corresponding to the curves labelled A–J may be identified using Figure 6.13. Dashed curves (G–J) represent configurations with no absorptive material in the cavity behind the panel. Configurations A–F require a sound-absorbing blanket between the panel and backing wall. The blanket must not contact the panel and should be between 10 and 50 mm thick and consist of glass or mineral fibre with a flow resistance between 1000 and 2000 MKS rayls. Panel supports should be at least 0.4 m apart.

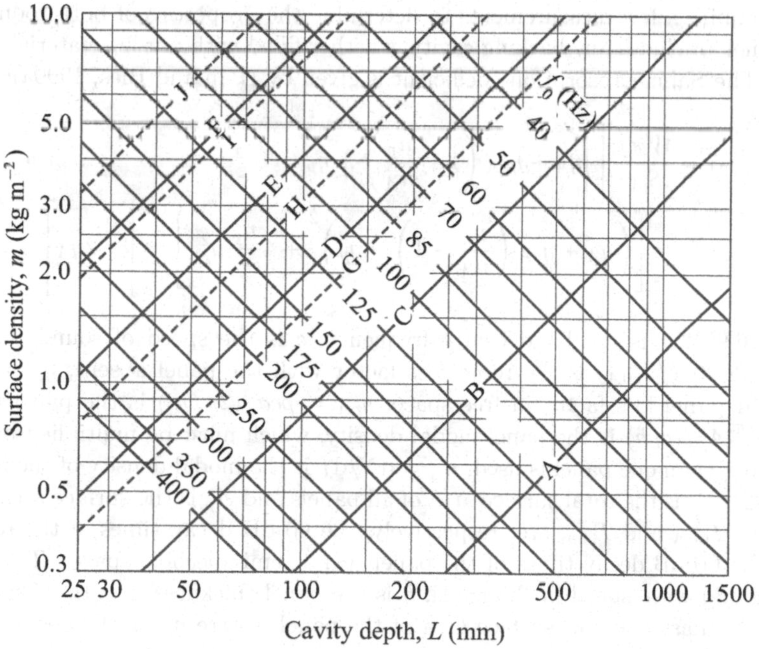

FIGURE 6.13 Design curves for resonant plywood panels, to be used in conjunction with Figure 6.12. The quantity, f_0, is the frequency at which maximum sound absorption is required.

the dashed curves are for no blanket. Next, the frequency f_0, which is the fundamental panel resonance frequency and the frequency at which maximum absorption is required, is determined and Figure 6.13 is entered for the chosen curve (A to J) and the desired frequency f_0. The required air gap (mm) behind the panel and the required panel surface density (kg/m^2) are read directly from the figure. The resonance frequency used in the preceding procedure is calculated using:

$$f_0 = \frac{1}{2\pi}\sqrt{\frac{\rho c^2}{mL}} \quad \text{(Hz)} \tag{6.87}$$

which does not take into account the panel rigidity or geometry. A more accurate equation for a plywood panel is (Sendra, 1999):

$$f_0 = \frac{1}{2\pi}\sqrt{\frac{\rho c^2}{mL + 0.6L\sqrt{ab}}} \quad \text{(Hz)} \tag{6.88}$$

where m is the mass per unit area of the panel (kg/m^2), L is the depth of the backing cavity and a, b are the panel dimensions. Thus, it is recommended that before using Figure 6.13, the frequency of maximum desired absorption be multiplied by the ratio:

$$\sqrt{\frac{m}{m + 0.6\sqrt{ab}}} \tag{6.89}$$

6.8.2 Analytical Method

The Sabine absorption coefficient of n resonant panels of total surface area, S, and individual surface area, A_p, may be calculated using Equation (6.90). In this case, the absorption coefficient is explicitly a function of the properties of the room as well as the properties of the panel, and consequently fairly good results can be expected. On the other hand, the price paid for good results will be quite a few measurements to determine the properties of both room and panels. The term "panel" includes any backing cavity, whether filled with porous material or not, as the case may be. The Sabine absorption coefficient is given by (Pan and Bies, 1990a):

$$\begin{aligned}\bar{\alpha} = \frac{4V\pi f}{S_p c}\Bigg[\Bigg|\eta_A + \eta_{pA}\left(1 + \frac{n_p}{n_A}\right) + \eta_p \\ - \left[\left(\eta_A + \eta_{pA}\left(\frac{n_p}{n_A} - 1\right) - \eta_p\right)^2 + 4\frac{n_p}{n_A}\eta_{pA}^2\right]^{1/2}\Bigg| - 2\eta_A\Bigg]\end{aligned} \tag{6.90}$$

In Equation (6.90), f is the band centre frequency, c is the speed of sound, V is the room volume, $\eta_A = 2.20/(fT_{60A})$ is the room loss factor with the panel absent, $\eta_p = 2.20/(fT_{60p})$ is the mounted panel loss factor in free space, $\eta_{pA} = \rho c\sigma/(2\pi f m)$ is the panel radiation loss factor, $n_p = \sqrt{3}A_p/(c_L h)$ is the panel modal density, which must be multiplied by the number of panels if more than one panel is used, $n_A = \mathrm{d}N/\mathrm{d}f$ is the modal density of the room given by Equation (6.22), S_p is the total surface area of all panels and A_p is the surface area of one panel. The quantities, T_{60A} and T_{60p}, are, respectively, the 60 dB decay times of the room (without panels) and the 60 dB decay times of the panels (in a semi-anechoic space). The quantity, c_L, is the longitudinal wave speed in the panel, h is the panel thickness, ρ is the density of air and $m = \rho_m h$ is the mass per unit surface area of the panel, where ρ_m is the density of the panel material.

The equations used for calculating the panel radiation efficiency, σ, in previous editions of this text were for excitation of the panel by an external force. However, it is more accurate to use the radiation efficiency of a panel excited non-resonantly by an external diffuse sound field. In this case, the radiation efficiency is (Davy, 2009a):

$$\sigma = \log_e\left(\frac{1+\sqrt{1+q^2}}{F+\sqrt{F^2+q^2}}\right) + \frac{1}{B}\log_e\left(\frac{H+\sqrt{H^2+q^2}}{F+\sqrt{F^2+q^2}}\right) \tag{6.91}$$

where $k = 2\pi/\lambda$, $H = \left[0.67\sqrt{2ka/\pi} - 0.124\right]^{-1}$, $q = \pi/(2k^2a^2)$, $F = 1.3\sqrt{\pi/(2ka)}$, $a = 2A_p/P$, A_p is the plate area, P is the plate perimeter and $B = (H/F) - 1$. If $B = 0$ or if B is close to 0, the second term on the RHS of Equation (6.91) is replaced with $F/\sqrt{H^2+q^2}$.

6.9 Flat and Long Rooms

Many enclosures have dimensions that are not conducive to the establishment of a reverberant sound field of the kind that has been the topic of discussion thus far and was first investigated by Sabine. Other types of enclosure (flat and long rooms) are considered briefly in this section and their investigation is based on work of Kuttruff (1985, 1989).

Reflections at the boundaries of either flat rooms or long rooms produce a reverberant field in addition to the direct field of the source. In contrast to Sabine-type rooms discussed earlier, where the reverberant field could be considered to be characterised by a constant mean energy density (level) throughout the room, the non-Sabine-type rooms considered here are characterised by a reverberant field that decays as the distance from the source increases. However, as in the case of Sabine-type rooms, it is useful to separately identify the direct and reverberant fields, because the methods of their control differ. For example, where the direct field is dominant, the addition of sound-absorption will be of little value.

Examples of enclosures of the type to be considered here, called flat rooms, are often encountered in factories in which the height, though it may be large, is much smaller than any of the lateral dimensions of the room. Open plan offices provide other familiar examples. For analytical purposes, such enclosures may be considered as contained between the floor and a parallel ceiling but of infinite extent and essentially unconstrained in the horizontal directions except close to the lateral walls of the enclosure. In the latter case, use of the method of images is recommended but is not discussed here (Elfert, 1988).

Examples of long rooms are provided again by factories in which only one horizontal dimension, the length, may be very much greater than either the height or width of the room. Other examples are provided by corridors and tunnels. Enclosed roadways, which are open above, may be thought of as corridors with completely absorptive ceilings and thus also may be treated as long rooms. As with flat rooms, the vertical dimension may be very large; that is, many wavelengths long. The horizontal dimension normal to the long dimension of the room may also be very large. The room cross-section is assumed to be constant and sufficiently large in terms of wavelengths that, as in the case of the Sabine rooms considered earlier, the sound field may be analysed using geometrical analysis.

Reflection at a surface may be quite complicated; thus to proceed, the problem of describing reflection will be simplified to one or the other of two extremes; that is, specular or diffuse reflection. Specular reflection is also referred to in the literature as geometrical reflection. Consideration of specular reflection may proceed by the method of images. In this case, the effect of reflection at a flat surface may be simulated by replacement of the bounding surface with the mirror image of the source. Multiple reflections result in multiple mirror image sources symmetrically placed. Diffuse reflection occurs at rough surfaces where an incident wave is scattered in all directions. For the cases considered here, it will be assumed that the intensity, $I(\theta)$, of scattered sound follows Lambert's rule taken from optics. In this case:

$$I(\theta) \propto \cos\theta \tag{6.92}$$

where θ is the angle subtended by the scattered ray relative to the normal to the surface.

Diffuse reflection at a surface is wavelength dependent; an observation that follows from the consideration that surface roughness is characterised by some size distribution. If the wavelength is large compared to the characteristic dimensions of the roughness, the reflection will be essentially specular, as the roughness will impose only negligible phase variation on the reflected wave at the surface. Alternatively, if the wavelength is small compared to the smallest size of the roughness dimensions, then the reflection, though it may be complicated, must again be specular. In the range where the wavelength is comparable to the surface roughness, the reflection will be diffuse.

In the discussion to follow, specular reflection will be mentioned as a reference case and also as an introduction to the more complicated diffuse reflection cases to follow. However, the discussion will be concerned principally with diffuse reflection based on the following observation. The floor of a furnished open plan office or the ceiling of a factory with extensive fittings such as piping and conduits may be thought of as a rough surface. Here the simple assumption will be made that sound scattering objects may be considered as part of the surface on which they rest so that the surface with its scatterers may be replaced with an effective diffusely reflecting surface. The use of this concept considerably reduces the complexity of the problem and makes tractable what may be an otherwise intractable problem. However, simplification is bought at the price of some empiricism in determining effective energy reflection coefficients for such surfaces and predictions can only provide estimates of average room sound pressure levels. Limited published experimental data suggest that measurement may exceed prediction by at most 4 dB with proper choice of reflection coefficients (Kuttruff, 1985).

The discussions of the various room configurations in the sections to follow are based on theoretical work undertaken by Kuttruff (1985). As an alternative for estimating sound pressure levels and reverberation times in non-Sabine rooms, there are various ray tracing software packages available that work by following the path of packets of sound rays that emanate from the source in all possible directions and eventually arrive at the receiver (specified as a finite volume) after various numbers of reflections from various surfaces. The principles underlying this technique are discussed by many authors including Krokstad et al. (1968), Naylor (1993), Lam (1996), Bork (2000), Keränen et al. (2003) and Xiangyang et al. (2003). There are also empirical models based on experimental data that allow the prediction of sound pressure levels in typical workshops as a function of distance from a source of known sound power output (Heerema and Hodgson, 1999; Hodgson, 2003).

In the following subsections, the term, energy reflection coefficient, β_e, is used in place of the pressure reflection coefficient amplitude squared, $|R_p|^2$, to simplify the notation in the equations to follow. However, the two quantities are identical.

6.9.1 Flat Room with Specularly Reflecting Floor and Ceiling

The flat room with specularly reflecting floor and ceiling will be encountered rarely in practice, because the concept really only applies to empty space between two relatively smooth reflecting surfaces. For example, a completely unfurnished open plan office might be described as a room of this type. The primary reason for its consideration is that it serves as a convenient reference for comparison with rooms that are furnished and with rooms which have diffuse reflecting surfaces. It also serves as a convenient starting point for the introduction of the concepts used later.

A source of sound placed between two infinite plane parallel reflecting surfaces will give rise to an infinite series of image sources located along a line through the source, which is normal to the two surfaces. If the source is located at the origin and the receiver is located at $\boldsymbol{r} = \boldsymbol{r}_0$ and each is located midway between the two reflecting surfaces, then the line of image sources will take the form illustrated in Figure 4.8 where, referring to the figure, $d = 0$ and the separation distance b between adjacent image sources in the figure is now the distance, a, between the reflecting planes; that is, the height of the room. The effective distance from the nth image to the receiving point will be represented by r_n, where the index, n, represents the number of reflections required to produce the image.

The source is assumed to emit a band of noise so that the source and all of its images may be considered as incoherent. In this case, summation at the point of observation may be carried out on an energy basis; sound pressures squared may be added without consideration of phase. It will be assumed that the surfaces below and above have uniform energy reflection coefficients, $\beta_{e,1}$ and $\beta_{e,2}$, respectively, which are independent of angle of incidence, and that the sound power of the source is W. The mean square sound pressure observed at the receiving point, $\boldsymbol{r}$, consists of the direct field, given by Equation (4.14) and shown as the first term in brackets on the right-hand side, and the reverberant field, given by the summation, where i is the image order, in the following expression (Kuttruff, 1985):

$$\langle p^2(\boldsymbol{r})\rangle = \frac{W\rho c}{4\pi}\left[\frac{1}{r^2} + \sum_{i=1}^{\infty}\left(\frac{1/\beta_{e,1} + 1/\beta_{e,2}}{r_{2i-1}^2} + \frac{2}{r_{2i}^2}\right)(\beta_{e,1}\beta_{e,2})^i\right] \tag{6.93}$$

For $r_n = r_{2i}$ or r_{2i-1}, that is, $n = 2i$ or $(2i-1)$:

$$r_n^2 = r^2 + (na)^2 \tag{6.94}$$

Equation (6.93) has been used to construct Figure 6.14 where the direct field and the reverberant field terms are plotted separately as a function of normalised distance, r/a, from the source for several values of the energy reflection coefficients $\beta_{e,1} = \beta_{e,2} = \beta_e$.

The figure shows that at large distances, the reverberant field may exceed the direct field when the reflection coefficient is greater than one-third. This may readily be verified by setting the direct field term of Equation (6.93) equal to the far-field reverberant field term of the same equation with $\beta_{e,1} = \beta_{e,2} = \beta_e$.

Two limiting cases are of interest. If the distance between the receiver and the source is large so that $r \gg a$, then Equation (6.93) becomes, using Equation (6.94) and the well-known expression for the sum of an infinite geometric series, in the limit:

$$\langle p^2(\boldsymbol{r})\rangle = \frac{W\rho c}{4\pi r^2}\left[1 + \frac{\beta_{e,1} + \beta_{e,2} + 2\beta_{e,1}\beta_{e,2}}{1 - \beta_{e,1}\beta_{e,2}}\right] \tag{6.95}$$

Equation (6.95) shows that the sound field, which includes both the direct field and the reverberant field, decays with the inverse square of the distance from the source. Equation (6.95) also shows that the reverberant field sound pressure may be greater than or less than the direct field at large distances from the source, depending on the values of the energy reflection coefficients $\beta_{e,1}$ and $\beta_{e,2}$.

If the distance between the source and receiver is small and the energy reflection coefficients approach unity, then (Kuttruff, 1985):

$$\langle p^2(\boldsymbol{r})\rangle = \frac{W\rho c}{4\pi}\left[\frac{1}{r^2} + \frac{\pi^2}{3a^2}\right] \tag{6.96}$$

In Equation (6.96), the first term in brackets on the right-hand side is the direct field term and the second term is the reverberant field term. Equation (6.96) shows that in the vicinity of the floor and ceiling and in the limiting case of a very reflective floor and ceiling, the direct field is dominant to a distance of $r = (\sqrt{3}/\pi)a \simeq 0.55a$ or slightly more than half the distance from the floor to the ceiling.

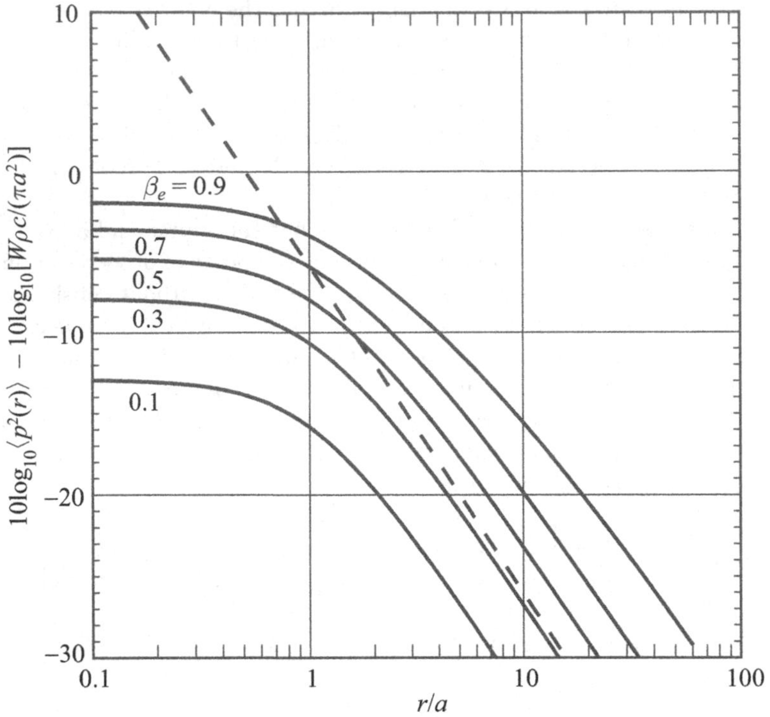

FIGURE 6.14 Direct and reverberant sound fields in a flat room of height, a, with specularly reflecting floor and ceiling, as a function of the normalised distance from the source to the receiver. The reverberant field contribution is shown as a function of the energy reflection coefficient, β_e, assumed the same for floor and ceiling. The direct field is shown by the dashed diagonal straight line.

6.9.2 Flat Room with Diffusely Reflecting Floor and Ceiling

The case of a flat room, which is furnished or occupied with machinery, or objects that tend to scatter incident sound, is considered in this section. A surface with scattering objects located upon it is replaced by an effective diffusely scattering surface for the purpose of the analysis. This approach, while obviously well suited for consideration of the floor of a furnished open plan office, may also be applied to the ceiling of a factory, which may be characterised by large open beams, conduits, pipes, corrugations, etc. Determination of effective surface reflection coefficients then becomes a problem, and generally an empirical approach will be necessary. For noise control purposes, the proposed model is useful in spite of the mentioned limitation, as it provides the basis for consideration of the effectiveness of the introduction of measures designed to reduce the floor and/or ceiling reflection coefficients.

Based on the proposed model, Kuttruff (1985) derived expressions for the sound intensities $I_1(\boldsymbol{r}')$ and $I_2(\boldsymbol{r}'')$, which characterise the contributions from the floor (surface 1) and ceiling (surface 2). The calculation of the mean square sound pressure at an observation point located by vector, $\boldsymbol{r}$, requires that the quantity, $I_1(\boldsymbol{r}')$, be integrated over surface 1 (floor) and $I_2(\boldsymbol{r}'')$ be integrated over surface 2 (ceiling) to determine their respective contributions to the reverberant field. The total acoustic field is obtained by summing the reverberant field and the direct field contributions on an energy basis as the simple sum of mean square sound pressures. The direct field is given by Equation (4.14). For source and receiver at height h from surface 1 (floor), the calculation of the reverberant part of the acoustic field proceeds as:

$$\langle p^2(\boldsymbol{r})\rangle_R = \frac{\rho c}{\pi}\left[\beta_{e,1} h \int\limits_{S_1} \frac{I_1(\boldsymbol{r}')}{R_1^3}\mathrm{d}S + \beta_{e,2}(a-h)\int\limits_{S_2}\frac{I_2(\boldsymbol{r}'')}{R_2^3}\mathrm{d}S\right] \tag{6.97}$$

where

$$R_1 = \left[|\boldsymbol{r}-\boldsymbol{r}'|^2 + h^2\right]^{1/2} \tag{6.98}$$

and

$$R_2 = \left[|\boldsymbol{r}-\boldsymbol{r}''|^2 + (a-h)^2\right]^{1/2} \tag{6.99}$$

Here $\boldsymbol{r}'$ and $\boldsymbol{r}''$ are, respectively, vector locations in surfaces S_1 and S_2, $\boldsymbol{r}$ is the vector from the source to the receiver and a is the distance from the floor to the ceiling.

Kuttruff (1985) shows how expressions for I_1 and I_2 may be obtained, which when substituted in Equation (6.97) allow solution for several special cases of interest. Equation (6.97), and all of the special cases that follow, may be compared with Equation (6.44) for the Sabine room. In the following analysis, the source will be located at the origin of the vector coordinate, $\boldsymbol{r}$, and $r = |\boldsymbol{r}|$.

For the case that the energy reflection coefficients of the bounding surfaces (floor and ceiling) are the same ($\beta_{e,1} = \beta_{e,2} = \beta_e$), Equation (6.97) may be simplified and the solution for the reverberant field contribution takes the following form (Kuttruff, 1985):

$$\langle p^2(\boldsymbol{r})\rangle_R = \frac{W\rho c\beta_e}{\pi a^2}\int\limits_0^\infty \frac{\mathrm{e}^{-z}}{1-\beta_e z\mathrm{K}_1(z)}\mathrm{J}_0(rz/a)\, z\,\mathrm{d}z \tag{6.100}$$

where $\mathrm{J}_0(rz/a)$ is a zero order Bessel Function of the first kind with argument rz/a and $\mathrm{K}_1(z)$ is a Modified Bessel Function of the second kind (Gradshteyn and Ryshik, 1965). In general, sufficient accuracy is achieved in evaluation of Equation (6.100) by use of the following approximation; otherwise, the equation must be evaluated numerically. The following approximation holds, where the empirical coefficient, Γ, is evaluated according to Equation (6.101), using $\mathrm{K}_1(1) = 0.6019$ so that the two sides of the expression are exactly equal for $z = 1$:

$$[1-\beta_e z\mathrm{K}_1(z)]^{-1} \approx 1 + \frac{\beta_e}{1-\beta_e}\mathrm{e}^{-\Gamma z} \tag{6.101}$$

where

$$\Gamma = \log_e\left[\frac{1-0.6019\beta_e}{(1-\beta_e)0.6019}\right] \tag{6.102}$$

Substitution of Equation (6.101) into Equation (6.100) and integrating gives the following closed-form approximate solution:

$$\langle p^2(\boldsymbol{r})\rangle_R \approx \frac{W\rho c\beta_e}{\pi a^2}\left[\left[1+\frac{r^2}{a^2}\right]^{-3/2} + \frac{\beta_e(\Gamma+1)}{1-\beta_e}\left[(\Gamma+1)^2 + \frac{r^2}{a^2}\right]^{-3/2}\right] \tag{6.103}$$

Equation (6.103) allows consideration of two limiting cases. For $r = 0$, the equation reduces to the following expression:

$$\langle p^2(\boldsymbol{r})\rangle_R = \frac{W\rho c\beta_e}{\pi a^2}\left[1+\frac{\beta_e}{(\Gamma+1)^2(1-\beta_e)}\right] \tag{6.104}$$

Equation (6.104) shows that the reverberant field is bounded in the vicinity of the source and thus will be dominated by the source near (direct) field. For $r \gg a$, Equation (6.103) takes the following approximate form:

$$\langle p^2(\boldsymbol{r})\rangle_R \approx \frac{W\rho c\beta_e}{\pi a^2}\left[1+\frac{(\Gamma+1)\beta_e}{(1-\beta_e)}\right]\left[\frac{a}{r}\right]^3 \tag{6.105}$$

Equation (6.105) shows that the reverberant field decreases proportionally to the cube of the distance, r, from the source, whereas the direct field decreases as the square of the distance from the source. Thus, the direct field will be dominant at a large distance from the source.

In Figure 6.15 the direct field term, calculated using Equation (4.14), and the reverberant field term, calculated using Equation (6.105), are plotted separately as a function of normalised distance, r/a, from the source, to illustrate the points made here.

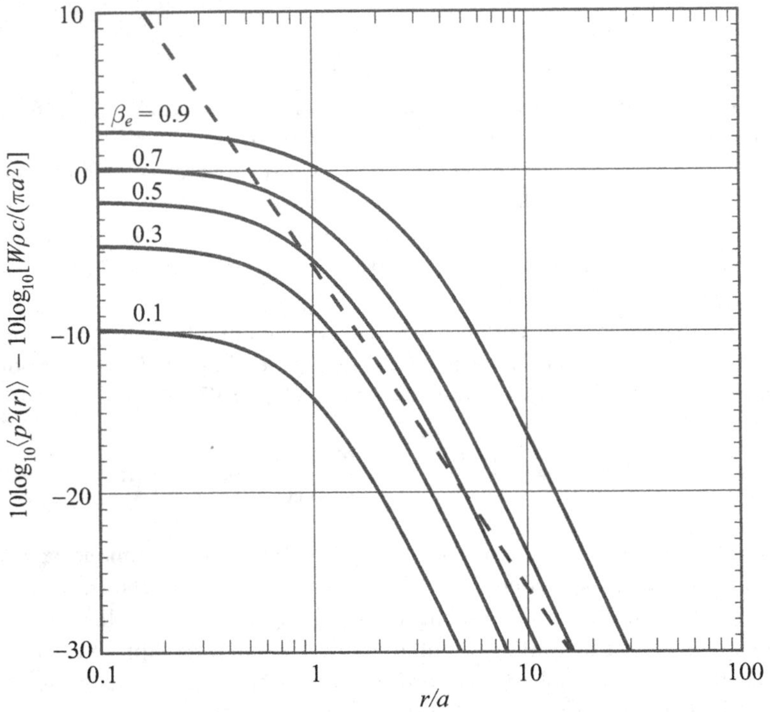

FIGURE 6.15 Direct and reverberant sound fields in a flat room of height, a, with diffusely reflecting floor and ceiling, as a function of the normalised distance from the source to the receiver. The reverberant field contribution is shown as a function of the energy reflection coefficient, β_e, assumed the same for floor and ceiling. The direct field is indicated by the dashed diagonal straight line.

When the direct and reverberant fields are equal at large distance from the source, a second hall radius (see Section 6.4.4) is defined. Setting Equation (6.105) equal to Equation (4.14) for the direct field, the second hall radius, r_{h2}, may be calculated as:

$$r_{h2} \approx 4a\beta_e \left[1 + \frac{(\Gamma + 1)\beta_e}{1 - \beta_e}\right] \tag{6.106}$$

When the energy reflection coefficients of the bounding surfaces (floor and ceiling) of a flat room are unequal, Equation (6.100) must be replaced with a more complicated integral equation, which shows the dependency on energy reflection coefficients $\beta_{e,1}$ and $\beta_{e,2}$ associated with bounding surfaces 1 (floor) and 2 (ceiling). In this case, it will be convenient to introduce the geometric mean value, $\beta_{e,g}$, of the reflection coefficients of the floor and ceiling:

$$\beta_{e,g} = \sqrt{\beta_{e,1}\beta_{e,2}} \tag{6.107}$$

As shown by Kuttruff (1985), the mean square sound pressure associated with the reverberant field in a flat room with unequal diffuse reflection coefficients given by:

$$\langle p^2(\boldsymbol{r})\rangle_R = \langle p^2(\boldsymbol{r})\rangle_{R1} + \frac{W\rho c}{2\pi a^2}\int_0^\infty \frac{\beta_{e,1}\mathrm{e}^{-2zh/a} + \beta_{e,2}\mathrm{e}^{-2z(1-h/a)} - 2\beta_{e,g}\mathrm{e}^{-z}}{1-(\beta_{e,g}z\mathrm{K}_1(z))^2}\mathrm{J}_0(rz/a)\,z\,\mathrm{d}z \tag{6.108}$$

where $\mathrm{J}_0(rz/a)$ is a zero order Bessel Function of the first kind with argument rz/a, $\mathrm{K}_1(z)$ is a Modified Bessel Function of the second kind and $\langle p^2(\boldsymbol{r})\rangle_{R1}$ is calculated using Equation (6.100) with $\beta_e = \beta_{e,g}$. The source is assumed to be located at the origin of coordinates and the receiver is located at the position given by the vector, $\boldsymbol{r}$. In general, Equation (6.108) will require numerical integration to obtain a solution. However, some special cases are of interest, which each allow a relatively simple closed-form solution, and these will be considered. If the ceiling is removed so that $\beta_{e,2} = 0$, and if the source and receiver are both at the same height, h, and separated by distance, r, then Equation (6.108) reduces to the following form (Kuttruff, 1985; Chien and Carroll, 1980):

$$\langle p^2(\boldsymbol{r})\rangle_R = \frac{W\rho c\beta_{e,1}h}{\pi}\left[4h^2 + r^2\right]^{-3/2} \tag{6.109}$$

which could be used to describe propagation of sound along a road between rows of high-rise buildings.

Some results of an investigation into the variation in height of the source and the receiver obtained using Equation (6.108) are shown in Figure 6.16. The energy reflection coefficient of the floor is $\beta_{e,1} = 0.9$ and the energy reflection coefficient of the ceiling is $\beta_{e,2} = 0.1$. Both surfaces are assumed to be diffuse reflectors. The results of the calculations shown in the figure suggest that when the receiver and source are very close together (curve $r = 0$), the reverberant field will become very large, near either the floor or the ceiling. As the receiving point is moved away from the source, the reverberant field becomes fairly uniform from the floor to the ceiling when the distance from the source to the receiver is equal to the room height a (curve $r = a$). Finally, the reverberant field tends to a distribution at large distances in which the level is about 6 dB higher at the ceiling than at the floor (curve $r \gg a$).

If the source and receiver are located at distance, r, apart and both are midway between the two bounding planes so that $h = a/2$, then introducing the arithmetic mean value of the energy reflection coefficient, $\beta_{e,a} = (\beta_{e,1} + \beta_{e,2})/2$, Equation (6.108) takes the following form:

$$\langle p^2(\boldsymbol{r})\rangle_R = \langle p^2(\boldsymbol{r})\rangle_{R1} + \frac{W\rho c}{\pi a^2}(\beta_{e,a} - \beta_{e,g})\int_0^\infty \frac{\mathrm{e}^{-z}}{1-(\beta_{e,g}z\mathrm{K}_1(z))^2}\mathrm{J}_0(rz/a)z\,\mathrm{d}z \tag{6.110}$$

where $\mathrm{J}_0(rz/a)$ is a zero order Bessel Function of the first kind with argument rz/a, $\mathrm{K}_1(z)$ is a Modified Bessel Function of the second kind and $\langle p^2(\boldsymbol{r})\rangle_{R1}$ is calculated using Equation (6.100) with $\beta_e = \beta_{e,g}$. A comparison between measured and predicted values using Equation (6.110) shows generally good agreement, with the theoretical prediction describing the mean of the experimental data (Kuttruff, 1985).

The second integral of Equation (6.110) may be evaluated using approximations similar to those used in deriving Equation (6.108) with the following result:

$$\langle p^2(\boldsymbol{r})\rangle_R = \frac{W\rho c}{\pi a^2}\left[\beta_{e,a}\left[1+\frac{r^2}{a^2}\right]^{-3/2} + \beta_{e,g}^2\frac{(\Gamma+1)}{1-\beta_{e,g}}\left[\Gamma+1+\frac{r^2}{a^2}\right]^{-3/2}\right] \tag{6.111}$$

Consideration of Equation (6.111) shows that as Γ is of the order of unity (see Equation (6.102)), then for a large separation distance between the source and the receiver, so that $r \gg a$, the quantity $\langle p_2(\boldsymbol{r})\rangle_R$ decreases as the inverse cube of the separation distance or as $(a/r)^3$. Comparison

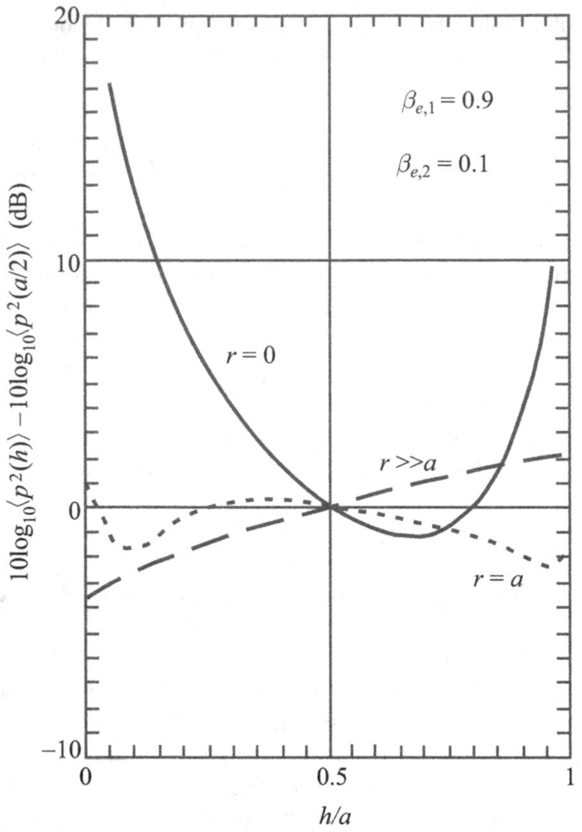

FIGURE 6.16 Reverberant sound field in a flat room of height, a, with diffusely reflecting floor and ceiling at three distances of the receiver from the source. The reflection coefficients of the floor and ceiling are, respectively, 0.9 and 0.1. The sound pressure is shown relative to that halfway between floor and ceiling, as a function of the normalised distance, h/a, from the surface of the greater reflection coefficient.

of Equations (6.111) and (6.103) shows that they are similar. Indeed, it is found that for most values of the two energy reflection coefficients, the use of the mean energy reflection coefficient, $\beta_{e,g}$, in Equation (6.103) will give a sufficiently close approximation to the result obtained using $\beta_{e,1}$ and $\beta_{e,2}$ in Equations (6.108) and (6.111).

6.9.3 Flat Room with Specularly and Diffusely Reflecting Boundaries

An open plan office might best be characterised as a flat room with a diffusely reflecting floor and a specularly reflecting ceiling. Indeed, a flat room characterised by one specularly reflecting boundary and one diffuse reflecting boundary may be the most common case. However, the cases already considered allow a simple extension to this case, when it is observed that the room and sound source form an image room with image source when reflected through the specularly reflecting bounding surface. Thus, a room of twice the height of the original room and with equally reflecting diffuse bounding surfaces is formed. The strength of the image source is exactly equal to the source strength and the double room is like a larger room with two identical sources symmetrically placed. The reflection loss at the specularly reflecting surface may be taken into account by imagining that the double room is divided by a curtain having a transmission loss equal to the energy reflection coefficient of the geometrically reflecting bounding surface. Thus, any ray that crosses the curtain is reduced by the magnitude of the energy reflection coefficient.

Using the analysis that has been outlined (Kuttruff, 1985), the following solution is obtained, which again requires numerical integration for the general case, but which also has a useful closed-form approximate solution for a source and receiver height, h, above the floor and floor to ceiling spacing of a:

$$\langle p^2(\boldsymbol{r})\rangle_R = \frac{W\rho c}{4\pi a^2}\left[2\beta_{e,1}\int_0^{\infty}\frac{\left(e^{-zh/a}+\beta_{e,2}e^{-z(2-h/a)}\right)^2}{1-\beta_{e,1}\beta_{e,2}2zK_1(2z)}J_0\left(\frac{r}{a}z\right)z\mathrm{d}z+\beta_{e,2}\left[4[1-h/a]^2+[r/a]^2\right]^{-1}\right] \tag{6.112}$$

Equation (6.113) that follows and Table 6.3 that provides calculated values of $\pi a^2\langle p^2(0)\rangle_R/(W\rho c)$ for various values of $\beta_{e,1}$ and $\beta_{e,2}$ allow construction of approximate solutions.

Table 6.3 provides estimates of the local reverberant field, $r \approx 0$, while for large values, $r \gg a$ and for $h \neq 0$ and $h \neq a$, the approximate solution for Equation (6.112) is:

$$\langle p^2(\boldsymbol{r})\rangle_R = \frac{W\rho c}{\pi a^2}\left[\frac{\beta_{e,1}(1+\beta_{e,2})}{1-\beta_{e,1}\beta_{e,2}}\left[(1-\beta_{e,2})\frac{h}{a}+2\beta_{e,2}+\beta_{e,1}\beta_{e,2}(1+\beta_{e,2})\Gamma\right]\left(\frac{a}{r}\right)^3+\frac{\beta_{e,2}}{4}\left(\frac{a}{r}\right)^2\right] \tag{6.113}$$

where Γ is defined by Equation (6.102).

TABLE 6.3 Calculated values of $\pi a^2\langle p^2(0)\rangle_R/(W\rho c)$

$\beta_{e,2}$	$\beta_{e,1}$				
	0.1	0.3	0.5	0.7	0.9
0.1	0.078	0.184	0.291	0.398	0.507
0.3	0.133	0.253	0.376	0.502	0.634
0.5	0.190	0.324	0.467	0.619	0.784
0.7	0.247	0.399	0.566	0.754	0.973
0.9	0.304	0.476	0.675	0.916	1.244

In Figure 6.17, the normalised mean square sound pressure associated with the reverberant field for a sound source placed at height, $h = a/2$, has been plotted as a function of the normalised distance, r/a, according to Equation (6.112) for two cases. In both cases, $\beta_{e,1}$ refers to the diffuse reflecting surface while $\beta_{e,2}$ refers to the specularly reflecting surface; for example, the furnished floor, surface 1, and smooth ceiling, surface 2, of an open plan office. In case (a), $\beta_{e,1} = 0.9$ and $\beta_{e,2} = 0.1$, while in case (b), $\beta_{e,1} = 0.1$ and $\beta_{e,2} = 0.9$. The figure shows that in case (a), the local reverberant field is about 2 dB higher, but decreases more rapidly with distance from the source than in case (b).

6.9.4 Long Room with Specularly Reflecting Walls

A long room with a constant rectangular cross-section of height, a, and width, b, and with specularly reflecting walls is considered. It is convenient to introduce the coordinate, r, along the central axis of symmetry (long axis of the room). The reflection coefficients of all four walls are assumed to be the same and a point source placed at the origin is separated from the receiving point, also assumed to be located on the central axis of symmetry of the room, by distance, r.

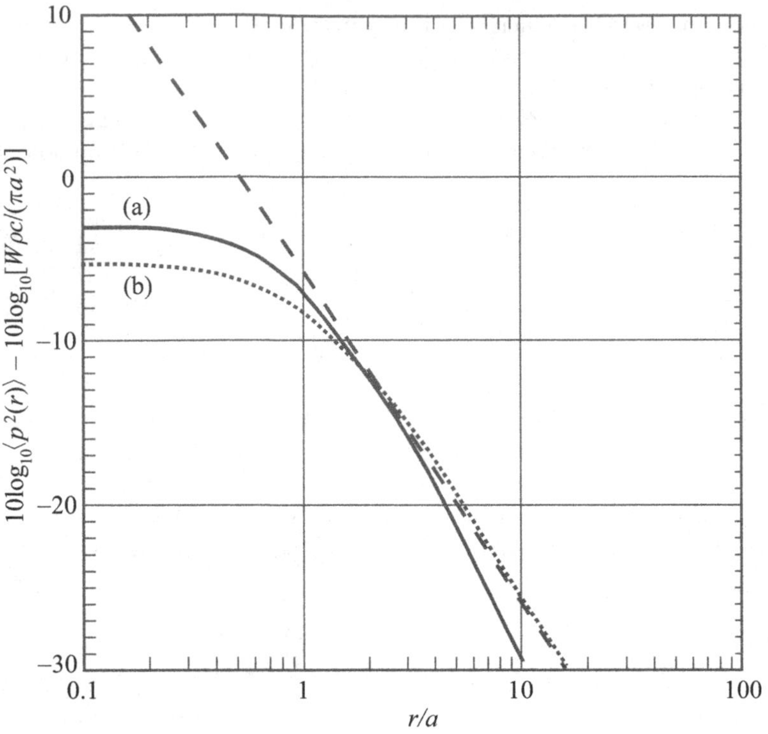

FIGURE 6.17 Reverberant sound field in a flat room of height, a, at $a/2$ with diffusely reflecting floor of energy reflection coefficient $\beta_{e,1}$ and specularly reflecting ceiling of energy reflection coefficient $\beta_{e,2}$. (a) $\beta_{e,1} = 0.9$, $\beta_{e,2} = 0.1$; (b) $\beta_{e,1} = 0.1$, $\beta_{e,2} = 0.9$. The direct field is indicated by the dashed diagonal straight line.

More general cases of long rooms with geometrically reflecting walls have been discussed in the literature (Cremer et al., 1982) and will not be discussed here. Rather, only this special case will be considered as a reference for the discussion of long rooms with diffusely reflecting walls.

Multiple reflections will produce two infinite series of image sources, which lie on vertical and horizontal axes through the source and which are normal to the long room central axis of symmetry. The source is assumed to emit a band of noise of power, W, and the source and its images are assumed to be incoherent. Summing the contributions of the source and its images on a pressure squared basis leads to the following expression for the mean square sound pressure at the receiving point:

$$\langle p^2(\boldsymbol{r})\rangle = \frac{W\rho c}{4\pi}\left[\frac{1}{r^2} + \sum_{m=1}^{\infty}\sum_{n=1}^{\infty}\frac{4\beta_e^{m+n}}{(ma)^2+(nb)^2+r^2} + \sum_{n=1}^{\infty}\frac{2\beta_e^n}{(nb)^2+r^2} + \sum_{m=1}^{\infty}\frac{2\beta_e^m}{(ma)^2+r^2}\right] \tag{6.114}$$

Equation (6.114) has been used to construct Figure 6.18 for the case of a square cross-section, long room of width, b, equal to height, a, for some representative values of energy reflection coefficient, β_e, which is assumed to be the same for all four walls.

A circular cross-section room of the same cross-sectional area as a square-section room is also approximately described by Figure 6.18.

In Equation (6.114), the first term in brackets represents the direct field while the next three terms represent the reverberant field due to the contributions of the four lines of image sources extending away from each wall.

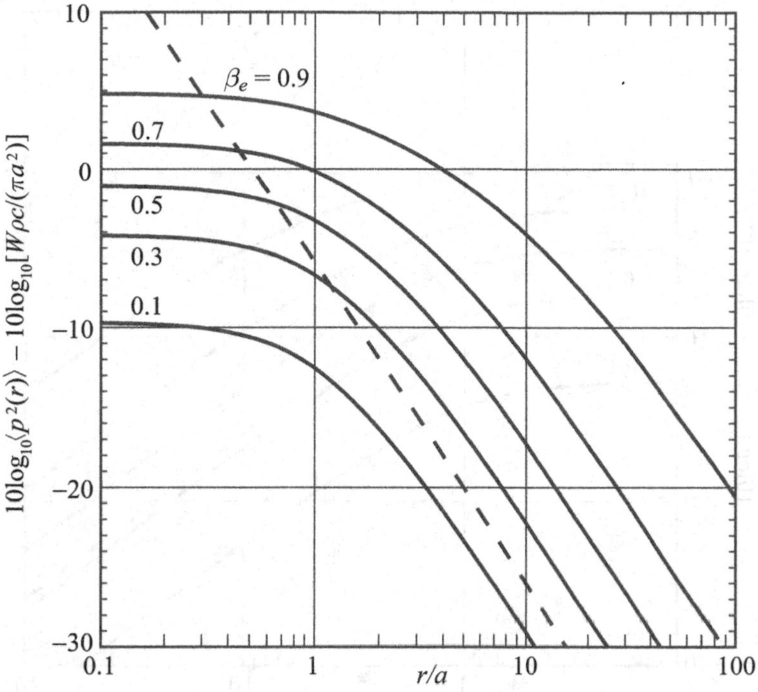

FIGURE 6.18 Direct and reverberant sound fields in a square cross-section long room of height and width, a, with specularly reflecting floor, walls and ceiling, as a function of the normalised distance from a point source to the receiver. The reverberant field is represented as solid lines for various values of the energy reflection coefficient, β_e, and the direct field is indicated by the dashed diagonal straight line.

In the limit of a very large distance, r, so that $a/r \approx b/r \approx 0$, the double sum of Equation (6.114) can be written in closed form as:

$$\langle p^2(\boldsymbol{r})\rangle \approx \frac{W\rho c}{4\pi r^2}\left[1 + \frac{4\beta_e}{(1-\beta_e)^2}\right] \tag{6.115}$$

If in the case considered of a long room of square cross-section of height and width, a, the point source is replaced with an incoherent line source perpendicular to the axis of symmetry and parallel to two of the long walls of the room, then Equation (6.114) takes the following simpler form, where the power per unit length, W', is defined so that $W'a = W$, the power of the original point source:

$$\langle p^2(\boldsymbol{r})\rangle = \frac{W'\rho c}{4}\left[\frac{1}{r} + \sum_{n=1}^{\infty} \frac{2\beta_e^n}{[(na)^2 + r^2]^{1/2}}\right] \tag{6.116}$$

Equation (6.116) has been used to construct Figure 6.19.

In Equation (6.116), the first term on the right-hand side is the direct field contribution and the second term is the reverberant field contribution. In the equation, it has been assumed that the line source emits a band of noise and thus the contributions of the source and its images add incoherently as the sum of squared pressures. Again, in the limit of very large distance, r, so that $a/r \approx 0$, the sum can be written in closed form as:

$$\langle p^2(\boldsymbol{r})\rangle = \frac{W'\rho c}{4r}\left[1 + \frac{2\beta_e}{1-\beta_e}\right] \tag{6.117}$$

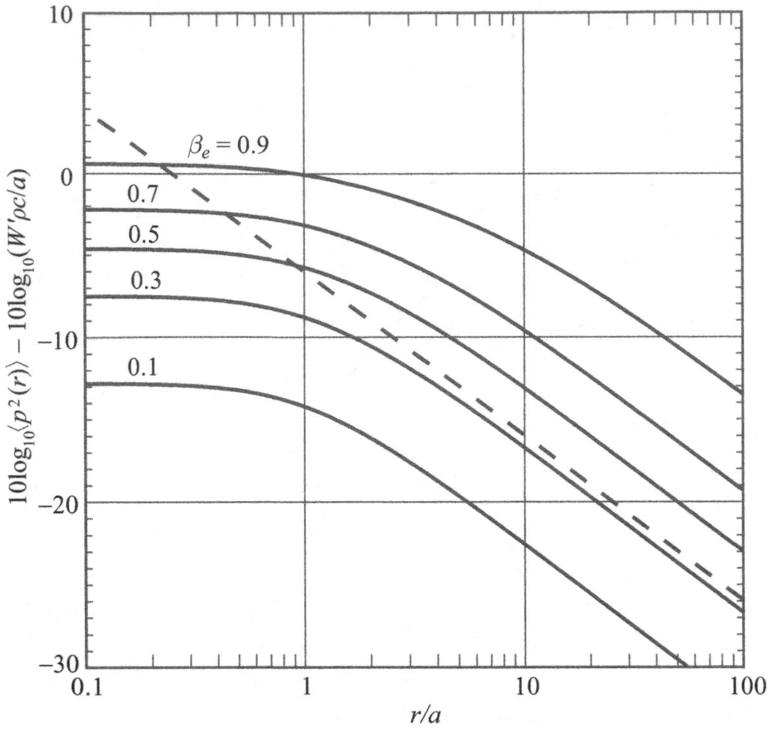

FIGURE 6.19 Direct and reverberant sound fields in a square cross-section, long room of height and width, a, with specularly reflecting floor, walls and ceiling, as a function of the normalised distance from a line source to the receiver. The reverberant field is represented as solid lines for various values of the energy reflection coefficient, β_e, and the direct field is indicated by the dashed diagonal straight line.

Equation (6.117) shows that both the direct field (first term in brackets) and the reverberant field (second term in brackets) decay at the same rate with increasing distance, r. The equation also shows, by comparison of the two terms, that the direct field is equal to or greater than the reverberant field when the energy reflection coefficient $\beta_e < 1/3$.

6.9.5 Long Room: Circular Cross-Section and Diffusely Reflecting Wall

After some extensive mathematics, Kuttruff (1989) derives the following equation for the reverberant field in a long room of circular cross-section, radius a, and diffusely reflecting wall:

$$\langle p^2(\boldsymbol{r})\rangle_R = \frac{2W\rho c\beta_e}{\pi^2 a^2}\int_0^\infty \frac{[\xi \mathrm{K}_1(\xi)]^2}{1-\beta_e\lambda(\xi)}\cos\left(\frac{\xi r}{a}\right)\mathrm{d}\xi \tag{6.118}$$

Numerical integration of Equation (6.118) gives the result shown in Figure 6.20. For comparison, the contribution of the direct field is also shown in the figure.

In Equation (6.118), the function, $\lambda(\xi)$, is defined in terms of the modified Bessel function, $\mathrm{I}_2(2\xi)$, and the modified Struve function, $\mathrm{L}_{-2}(2\xi)$, as (Abramowitz and Stegun, 1965):

$$\lambda(\xi) = \pi\xi\left[\mathrm{I}_2(2\xi) - \mathrm{L}_{-2}(2\xi)\right] \tag{6.119}$$

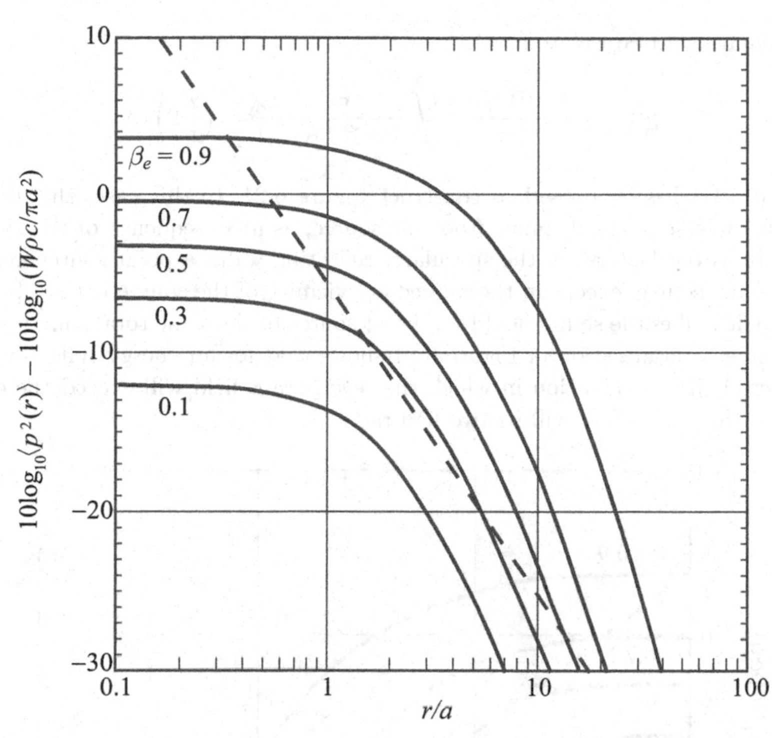

FIGURE 6.20 Direct and reverberant sound fields in a circular cross-section, long room of diameter, $2a$, with diffusely reflecting walls, as a function of normalised distance from a point source to the receiver. The reverberant field is represented as solid lines for various values of the energy reflection coefficient, β_e, and the direct field is indicated by the dashed diagonal straight line.

Equation (6.119) may be approximated as:

$$\lambda(\xi) \approx \left[1 + \frac{4}{3}\xi^2\right]^{-1} \tag{6.120}$$

Note that a square cross-section room of the same cross-sectional area as a circular-section room is also approximately described by Figure 6.20. Comparison of Figure 6.20 with Figure 6.18 shows that the diffuse reflected reverberant field decreases more rapidly than does the specularly reflected reverberant field. It may be shown that at large distances, the reverberant sound field decreases as $(r/a)^3$ or at the rate of 9 dB per doubling of distance. Reference to the figure shows that here again two hall radii (distance at which the direct and reverberant fields are equal) may be defined.

6.9.6 Long Room with Rectangular Cross-Section

A rectangular cross-section, long room of height, a, with a diffusely reflecting floor and ceiling and width, b, with specularly reflecting side walls is considered.

The energy reflection coefficients of the floor and ceiling are β_e, while the energy reflection coefficients of the side walls are unity; that is, they are assumed to reflect incident sound without loss. A line source is assumed, which lies perpendicular to the room axis midway between the floor and ceiling and parallel to them. The source has a sound power of W' per unit length and is assumed to radiate incoherently. Kuttruff (1989) gives the following expression for the reverberant sound field (see Figure 6.21):

$$\langle p^2(\boldsymbol{r})\rangle_R = \frac{2W'\rho c\beta_e}{\pi a}\int_0^{\infty}\frac{\mathrm{e}^{-\xi}}{1-\beta_e\xi K_1(\xi)}\cos\left(\frac{r}{a}\xi\right)\mathrm{d}\xi \tag{6.121}$$

Equation (6.121) has been used to construct Figure 6.21. In this case, the direct field diminishes as the inverse of the distance from the source, as a consequence of the assumption of a line source. However, because of the specularly reflecting walls, a point source and its images will look like a line source, except in the immediate vicinity of the source ($r/a \leq 1$), so that the expression also describes the sound field for a point source of the same total sound power output as the line source. Consideration of Figure 6.21 shows that for an energy reflection coefficient, $\beta_e \geq 0.47$, there will exist a region in which the reverberant field will exceed the direct sound field, so that in this case, there will be two hall radii.

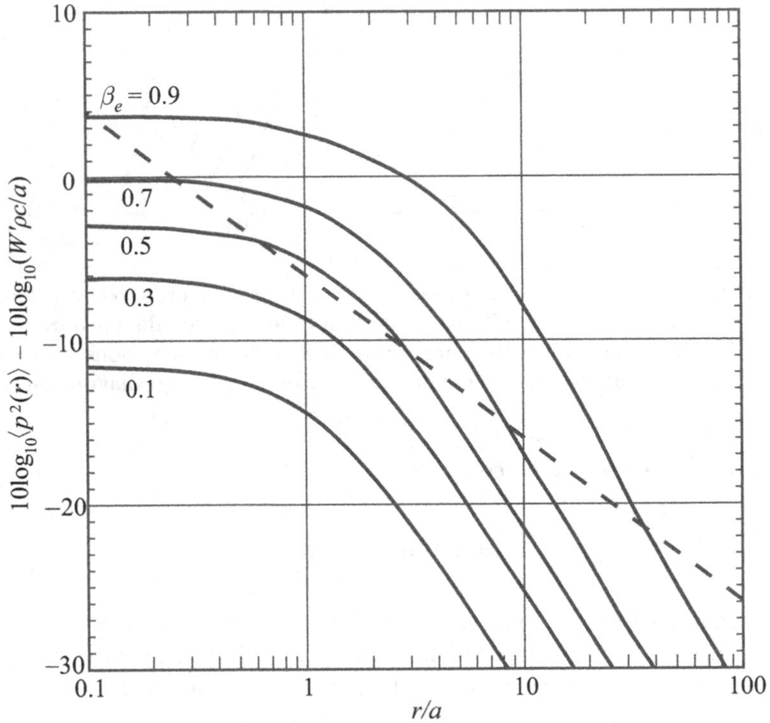

FIGURE 6.21 Direct and reverberant sound fields in a rectangular-section, long room of height, a, and width, b, with specularly reflecting walls and diffusely reflecting floor and ceiling as a function of the normalised distance from a line source (perpendicular to the room axis and parallel to the floor and ceiling) to the receiver. The reverberant field is represented as solid lines for various values of the energy reflection coefficient, β_e, of the floor and ceiling and the direct field is indicated by the dashed diagonal straight line. The energy reflection coefficient of the specularly reflecting walls is assumed to be unity.

6.10 Applications of Sound Absorption

6.10.1 Relative Importance of the Reverberant Field

Consideration will now be given to determining when it is appropriate to treat surfaces in a room with acoustically-absorbing material. The first part of the procedure is to determine whether the reverberant sound field dominates the direct sound field at the point where it is desired to reduce the overall sound pressure level, because treating reflecting surfaces with acoustically-absorbing material can only affect the reverberant sound field. At locations close to the sound source (for example, a machine operator's position) it is likely that the direct field of the source will dominate, so there may be little point in treating a factory with sound-absorbing material to protect operators from noise produced by their own machines. However, if an employee is affected by noise produced by other machines some distance away, then treatment may be appropriate.

For the case of a Sabine room, the relative strength of the reverberant sound field may be compared with the direct field produced by a machine at a particular location by comparing the direct and reverberant field terms of the argument of Equation (6.44); that is, $4/R$ and $D_\theta/4\pi r^2$. For the cases of flat rooms or long rooms, see the discussion of Section 6.9. When the reverberant sound field dominates, for example in the Sabine room when $4/R$ is much larger than $D_\theta/4\pi r^2$, then the introduction of additional absorption may be useful.

6.10.2 Reverberation Control

If the reverberant sound field dominates the direct field, then the sound pressure level will decrease if absorption is added to the room or factory. The decrease in reverberant sound pressure level, ΔL_p, to be expected in a Sabine room for a particular increase in sound-absorption, expressed in terms of the room constant, R (see Equation (6.45)), may be calculated by using Equation (6.44) with the direct field term set equal to zero. The following equation is thus obtained, where R_{orig} is the original room constant and R_{final} is the room constant after the addition of sound-absorbing material.

$$\Delta L_p = 10 \log_{10} \left[\frac{R_{\text{final}}}{R_{\text{orig}}} \right] \tag{6.122}$$

Referring to Equation (6.45) for the definition of the room constant, R, it can be seen from Equation (6.122) that if the original room constant, R_{orig}, is large, then the amount of additional absorption to be added must be very large so that $R_{\text{final}} \gg R_{\text{orig}}$ and ΔL_p is significant and worth the expense of the additional absorbent. Clearly, it is more beneficial to treat hard surfaces such as concrete floors, which have small Sabine absorption coefficients, because this will have the greatest effect on the room constant.

To affect as many room modes as possible, it is better to distribute any sound-absorbing material throughout the room, rather than having it only on one surface. However, if the room is very large compared to the wavelength of sound considered, distribution of the sound-absorbing material is not so critical, because there will be many more oblique modes than axial or tangential modes in a particular frequency band. As each mode may be assumed to contain approximately the same amount of sound energy, then the larger percentage of sound energy will be contained in oblique modes, as there are more of them. Oblique modes consist of waves reflected from all bounding surfaces in the room and thus treatment anywhere will have a significant effect, although distribution of the sound-absorbing material equally between the room walls, floor and ceiling will be more effective than having it on only one of those surfaces.

The energy reflection coefficients, β_e, used in the discussion of flat rooms and long rooms (see Section 6.9), are related to absorption coefficients, α, by the relation, $\alpha + \beta_e = 1$. Thus, an equivalent room constant, R, can be calculated and subsequently used with Equation (6.122) to determine the effect on sound pressure levels of adding sound-absorbing material to these rooms. For optimum results, the sound-absorbing material should be added to the floor or ceiling of flat rooms and to the floor, ceiling and side walls of long rooms.

Reverberation control in classrooms is also very important so that students can properly understand their teacher. Guidelines for modifying classrooms to achieve the required reverberation times by adding sound-absorbing material are provided in Appendix C of the freely available standard, ANSI/ASA S12.60-1 (2015), which also discusses absorptive material types, mounting arrangements and optimal locations of the added material. It has also been shown that classrooms have acceptable acoustics for listening if the clarity, C_{50}, value measured at the centre of the classroom is greater than 3 dB (Astolfi et al., 2022). Clarity for speech purposes is defined as:

$$C_{50} = 10\log_{10}\left[\frac{\text{energy arriving within 50 ms of direct sound}}{\text{energy arriving later than 50 ms of direct sound}}\right] \tag{6.123}$$

The quantity in the numerator includes the direct sound energy, which is measured by integrating the acoustic pressure squared over the 50 ms measurement period that follows the arrival of the direct sound.

Clarity may be measured using a filtered impulse signal (such as a gunshot) or a filtered MLS signal (see Section 12.3.18), where the filter range corresponds to the speech range needed for intelligibility (octave bands between and including 500 Hz to 4000 Hz). The single clarity measure is an arithmetic average of the clarity for each of the four octave bands between 500 and 4000 Hz.

7

Partitions, Enclosures and Indoor Barriers

LEARNING OBJECTIVES

In this chapter, the reader is introduced to:

- sound transmission through partitions and the importance of bending waves;
- transmission loss, its calculation and its measurement for single (isotropic and orthotropic) and double panels;
- enclosures for keeping sound in and out;
- barriers for the control of sound out of doors and indoors; and
- pipe lagging.

7.1 Introduction

In many situations, for example, where plant or equipment already exists, it may not be feasible to modify the characteristics of the noise source to reduce its radiated sound power. In these cases, a possible solution to a noise problem is to modify the acoustic transmission path or paths between the source of the noise and the observer. In such a situation, the first task for noise control purposes is to determine the transmission paths and order them in relative importance. For example, on close inspection it may transpire that, although the source of noise is readily identified, the important acoustic radiation originates elsewhere, such as from structures mechanically connected to the source. In this case, structure-borne sound is more important than the airborne component. In considering enclosures for noise control one must always guard against such a possibility; if structure-borne sound is the problem, an enclosure to contain airborne sound can be completely useless.

In this chapter, the control of airborne sound is considered (the control of structure-borne sound will be considered in Chapter 9). Control of airborne sound takes the form of interposing a barrier to interrupt free transmission from the source to the observer; thus the properties of materials and structures, which make them useful for this purpose, will first be considered and the concept of transmission loss will be introduced. Various means for calculating the transmission loss, over the audio frequency range, of single, double and triple walls and roofing constructed from a wide range of materials will be described, and this will include estimates of flanking effects in practical structures. Both flat panels and corrugated building panels will be considered.

DOI: 10.1201/9780367814908-7

Complete and partial enclosures will then be analysed and means for estimating their effectiveness as both machine enclosures and personnel enclosures will be outlined. Finally, lagging for the containment of noise in conduits such as air ducts and pipes will be considered and means will be provided for estimating their effectiveness.

7.2 Sound Transmission through Partitions

7.2.1 Bending Waves

Solid materials are capable of supporting shear as well as compressional stresses, so that in solids, shear and torsional waves as well as compressional (longitudinal) waves may propagate. In the audio-frequency range in thick structures, for example in the steel beams of large buildings, all three types of propagation may be important, but in the thin structures of which wall panels are generally constructed, purely compressional wave propagation is of negligible importance. Rather, audio-frequency sound propagation through panels and thus walls is primarily through the excitation of bending waves, which are a combination of shear and compressional waves.

In the discussion to follow, both isotropic and orthotropic panels will be considered. Isotropic panels are characterised by uniform stiffness and material properties, whereas orthotropic panels are usually characterised by a stiffness that varies with the direction of bending wave travel (for example, a corrugated or ribbed steel panel).

Bending waves in thin panels, as the name implies, take the form of waves of flexure propagating parallel to the surface, resulting in normal displacement of the surface. The speed, c_B, of bending wave propagation for an isotropic panel is given by:

$$c_B = (B\omega^2/m)^{1/4} \qquad \text{(m/s)} \tag{7.1}$$

The bending stiffness per unit width, B, is defined as:

$$B = EI'/(1-\nu^2) = Eh^3/[12(1-\nu^2)] \qquad (\text{kg m}^2\text{s}^{-2}) \tag{7.2}$$

where ω is the angular frequency (rad/s), h is the panel thickness (m), ρ_m is the material density, $m = \rho_m h$ is the surface density (kg/m^2), E is Young's modulus (Pa), ν is Poisson's ratio and $I' = h^3/12$ is the cross-sectional second moment of area per unit width (m^3), computed for the panel cross-section about the panel's neutral axis.

When the lateral dimension of a panel equals a multiple of half of a bending wavelength, a resonant vibration mode is excited and the frequency at which this occurs is called a resonance frequency. For any panel, there is an infinite number of resonance frequencies; however, in practice we are only usually interested in resonance frequencies that lie below about 10 kHz.

As shown in Equation (7.1), the speed of propagation of bending waves increases with the square root of the excitation frequency; thus there exists, for any panel capable of sustaining shear stress, a critical frequency (sometimes called the coincidence frequency) at which the speed of bending wave propagation is equal to the speed of acoustic wave propagation in the surrounding medium. The frequency for which airborne and solid-borne wave speeds are equal, the critical frequency, is given by the following equation which is also illustrated in Figure 7.1:

$$f_c = \frac{c^2}{2\pi}\sqrt{\frac{m}{B}} \qquad \text{(Hz)} \tag{7.3}$$

where c is the speed of sound in air.

Substituting Equation (7.2) into (7.3), and using Equations (1.1) and (1.4) for the longitudinal wave speed in a two-dimensional solid (or panel), the following equation is obtained for the critical frequency:

$$f_c = 0.55c^2/(c_{LI}h) \tag{7.4}$$

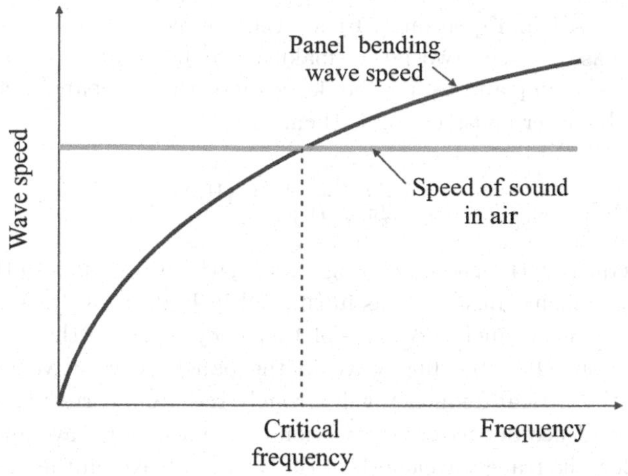

FIGURE 7.1 Illustration of critical frequency.

Here, the longitudinal wave speed, c_{LI}, for thin plates (see Section 1.4.4) is given by:

$$c_{LI} = \sqrt{E/[\rho_m(1-\nu^2)]} \tag{7.5}$$

Representative values of $\sqrt{E/\rho_m}$, the longitudinal wave speed in thin rods, are given in Appendix C. Using Equations (7.2) and (7.5), the longitudinal wave speed may be written as:

$$c_{LI} = \frac{\sqrt{12}}{h}\sqrt{\frac{B}{m}} \tag{7.6}$$

For a panel made of two layers of different materials bonded firmly together (such as gypsum board and particle board bonded together to make a wall panel, as illustrated in Figure 7.2), the bending stiffness and surface mass in the preceding equation must be replaced with an effective bending stiffness, B_{eff}, and surface mass, m_{eff}.

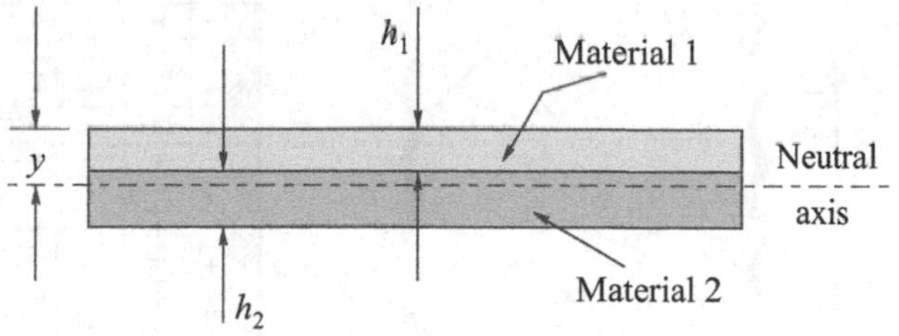

FIGURE 7.2 Composite material notation.

The effective bending stiffness may be calculated as:

$$\begin{aligned} B_{eff} &= \frac{E_1 h_1}{12(1-\nu_1^2)}\left[h_1^2 + 12(y - h_1/2)^2\right] \\ &+ \frac{E_2 h_2}{12(1-\nu_2^2)}\left[h_2^2 + 12\left(y - (2h_1 + h_2)/2\right)^2\right] \end{aligned} \tag{7.7}$$

where the neutral axis location (see Figure 7.2) is given by:

$$y = \frac{E_1 h_1 + E_2(2h_1 + h_2)}{2(E_1 + E_2)} \tag{7.8}$$

The surface mass used in Equation (7.6) for the double layer construction is simply the sum of the surface masses of the two layers making up the composite construction. That is, $m_{eff} = \rho_1 h_1 + \rho_2 h_2$, where ρ_1 and ρ_2 are the densities of the two panel materials. The critical frequency of this double layer construction is then:

$$f_c = \frac{c^2}{2\pi}\sqrt{\frac{m_{eff}}{B_{eff}}} \quad \text{(Hz)} \tag{7.9}$$

At the critical frequency, the panel bending wavelength corresponds to the trace wavelength of an acoustic wave at grazing incidence, as illustrated in Figure 7.3(a). A sound wave incident from any direction at grazing incidence, and of frequency equal to the critical frequency, will strongly drive a corresponding bending wave in the panel. Alternatively, a panel excited in bending (flexure) at the critical frequency will strongly radiate a corresponding acoustic wave.

As the angle of incidence between the direction of the acoustic wave and the normal to the panel becomes smaller, the trace wavelength of the acoustic wave on the panel surface becomes longer, as shown in Figure 7.3(b). Thus, for any given angle of incidence (measured from the normal to the panel surface) smaller than grazing incidence, there will exist a frequency (which will be higher than the critical frequency) at which the bending wavelength in the panel will match the acoustic trace wavelength on the panel surface. This frequency is referred to as a coincidence frequency and must be associated with a particular angle of incidence or radiation of the acoustic wave. The coincidence frequency for grazing incidence (incident angle of 90° from the panel normal) is equal to the critical frequency. Thus, in a diffuse field, in the frequency range about and above the critical frequency, a panel will be strongly driven and will radiate sound well. However, the response is a resonance phenomenon, being strongest in the frequency range about the critical frequency and strongly dependent on the damping in the system. This phenomenon is called coincidence, and it is of great importance in the consideration of transmission loss.

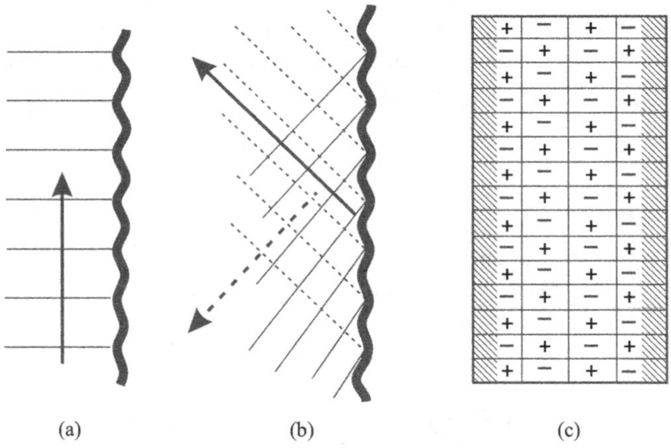

FIGURE 7.3 Coupling of the acoustic field and the panel flexural wave. The horizontal lines represent the maxima of longitudinal acoustic wavefronts. The edge view of the panel shows panel bending waves. (a) At the critical frequency, the radiated grazing acoustic wave has the same wavelength as the panel bending wave and the panel radiates. (b) Above the critical frequency, for every frequency, there is a radiation angle at which the radiated acoustic wave has the same trace wavelength on the panel as the panel bending wave and the panel radiates. (c) At frequencies less than the critical frequency, the disturbance is local and the panel does not radiate except at the edges (shaded area on the figure). The + and − signs indicate areas vibrating in opposite phase and the lines represent vibration nodes on the panel.

At frequencies less than the critical frequency (lowest critical frequency for orthotropic panels), the panel bending wavelength is shorter than the airborne acoustic wavelength and wave coupling is not possible (Cremer et al., 1988, 2005). In this case, an infinite panel is decoupled from an incident sound field. However, in finite panels, radiation coupling occurs at the edges (as illustrated in Figure 7.3(c)) and at stiffeners, where the disturbance is not matched by a compensating disturbance of opposite sign. At these places of coupling, the panel radiates sound or, alternatively, it is driven by an incident sound field. Between the edges, radiation from one cell (represented by a + or − sign) is cancelled by out-of-phase radiation from an adjacent cell.

An important concept, which follows from the preceding discussion, is concerned with the difference in sound fields radiated by a panel excited by an incident acoustic wave and one excited by a mechanical, localised force. In the former case, the structure will be forced to respond in modes that are characterised by bending waves having wavelengths equal to the trace wavelengths of the incident acoustic field. Thus, at excitation frequencies below the structure critical frequency, the modes that are excited by an incident acoustic field will not be resonant, because the structural wavelength of the resonant modes will be smaller than the wavelength in the adjacent medium. Lower order modes will be excited by an acoustic field at frequencies above their resonance frequencies. As these lower order modes are more efficient than the unexcited higher order modes that are resonant at the excitation frequencies, the level of radiated sound will be higher than it would be for a resonantly excited structure (resulting from mechanical excitation) having the same mean square velocity levels at the same excitation frequencies. As excitation of a structure by a mechanical force results in resonant structural response, it can be concluded that sound radiation from an acoustically excited structure will be greater than that radiated by a structure excited mechanically to the same vibration level (McGary, 1988). A useful item of information that follows from this conclusion is that structural damping will only be effective for controlling mechanically excited structures because it is only the resonant structural response that is significantly influenced by damping.

As the bending wave speed (which is a function of panel bending stiffness) plays a very important role in the transmission of sound, the difference in bending stiffness between isotropic and orthotropic panels is of importance. Unless the panel is essentially isotropic in construction, the bending stiffness will be variable and dependent on the direction of wave propagation. For example, ribbed or corrugated panels commonly found in industrial constructions are orthotropic, being stiffer along the direction of the ribs than across the ribs. The consequence is that orthotropic panels are characterised by a range of bending wave speeds (dependent on the direction of wave propagation across the plate) due to the two different values of the cross-sectional second moment of area per unit width, I'. By contrast, isotropic panels are characterised by a single bending wave speed, given by Equation (7.1).

For wave propagation along the direction of ribs or corrugations the bending stiffness per unit width may be calculated by referring to Figure 7.4:

$$B = \frac{Eh}{(1-\nu^2)\ell}\sum_{i=1}^{N} b_i\left(z_i^2 + \frac{h^2+b_i^2}{24} + \frac{h^2-b_i^2}{24}\cos 2\theta_i\right) \quad (\mathrm{kg\,m^2s^{-2}}) \qquad (7.10)$$

The summation is taken over all segments (total, N) in width ℓ and the distances z_i are distances of the centre of the segment from the neutral axis of the entire section. The location of the neutral axis is found by selecting any convenient reference axis, such as one that passes through the centre of the upper segment (see Figure 7.4). The distance of segment i from the reference axis is z_{ri}. Then the neutral axis location from the reference axis location is given by:

$$z_n = \frac{\sum_{i=1}^{N} z_{ri} b_i h_i}{\sum_{i=1}^{N} b_i h_i} \qquad (7.11)$$

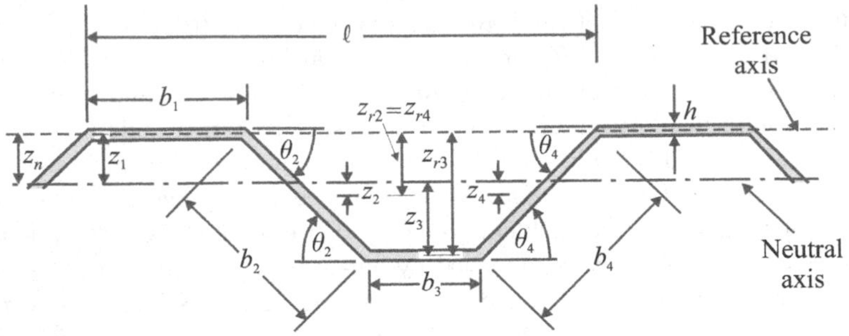

FIGURE 7.4 A typical cross-section of a ribbed panel.

For the special case of a sinusoidally-shaped corrugated panel with a peak to trough height of d and thickness of h, the bending stiffness is, $B = 0.13Ehd^2/(1-\nu^2)$ (Garifullin et al., 2021).

For wave propagation across the corrugations in a stiffened panel, the bending stiffness per unit width will be similar to that for an isotropic panel; that is:

$$B = \frac{Eh^3}{12(1-\nu^2)} \quad (\text{kg}\,\text{m}^2\text{s}^{-2}) \tag{7.12}$$

Note that Equation (7.12) follows from Equation (7.10) if in the latter equation z_n and θ_n are set equal to zero (see Figure 7.4). The derivation of Equation (7.10) is predicated on the assumption that the wavelength of any flexural wave will be long compared to any panel dimension. Thus at high frequencies, where a flexural wavelength may be of the order of a characteristic dimension of the panel structure (for example b_n in Figure 7.4), the bending stiffness will approach that for an isotropic panel, as given by Equation (7.2).

Although for an isotropic panel there exists just one critical frequency, for orthotropic panels the critical frequency is dependent on the direction of the incident acoustic wave. However, as shown in Equation (7.3), the range of critical frequencies is bounded at the lower end by the critical frequency corresponding to a wave travelling in the panel stiffest direction (e.g. along the ribs for a corrugated panel), and at the upper end by the critical frequency corresponding to a wave propagating in the least stiff direction (e.g. across the ribs of a corrugated panel). For the case of an orthotropic panel, characterised by an upper and lower bound of the bending stiffness, B per unit width, a range of critical frequencies will exist. The response will now be strong over this frequency range, which effectively results in a strong critical frequency response occurring over a much more extended frequency range than for the case of the isotropic panel.

As an interesting example, consider sound incident on one side of a floor or roof containing parallel rib stiffeners. At frequencies above the critical frequency, there will always be angles of incidence of the acoustic wave for which the projection of the acoustic wave on the structure will correspond to multiples of the rib spacing. If any one of these frequencies corresponds to a frequency at which the structural wavelength is equal to a multiple of the rib spacing, then a high level of sound transmission may be expected.

Another mechanism that reduces the transmission loss of ribbed or corrugated panels at some specific high frequencies is the resonant behaviour of the panel sections between the ribs. At the resonance frequencies of these panels, the transmission loss is markedly reduced.

7.2.2 Transmission Loss, TL (or Sound Reduction Index, R)

When sound is incident upon a wall or partition, some of it will be reflected and some will be transmitted through the wall. The fraction of incident energy that is transmitted is called the transmission coefficient, τ. The transmission loss, TL (referred to as the sound reduction index,

R, in ISO standards), is in turn defined in terms of the transmission coefficient as:

$$\text{TL} = -10\log_{10}\tau \qquad \text{(dB)} \tag{7.13}$$

In general, the transmission coefficient, and thus the transmission loss, will depend on the angle of incidence of the incident sound. Normal incidence, diffuse field (random) incidence and field incidence transmission loss (denoted TL_N, TL_d and TL, respectively) and corresponding transmission coefficients (denoted τ_N, τ_d and τ_F, respectively) are terms commonly used. These terms and their meanings will be described in Section 7.2.6. Field incidence transmission loss, TL, is the transmission loss commonly observed in testing laboratories and in the field, and reported in tables.

The transmission loss of a partition is usually measured in a laboratory by placing the partition in an opening between two adjacent reverberant rooms designed for such tests. Noise is introduced into one of the rooms, referred to as the source room, and some of the sound energy passes through the test partition into the second room, referred to as the receiver room. The resulting mean space-average sound pressure levels (well away from the sound source) in the source and receiver rooms are measured (see Equation (4.142)) and the difference in levels, called the noise reduction, NR, is determined.

The receiver room constant is determined either by use of a standard sound power source or by measurements of the reverberation decay, as discussed in Section 6.6.2. The Sabine absorption in the room, including loss back through the test partition, is thus determined. However, sound that escapes through the partition into the source room causes reverberation in the source room, resulting in sound transmitted back into the receiver room, thus resulting in errors in the reverberation time measurement (see Sections 6.5 and 6.6.2). These errors can be minimised by adding sound-absorbing material to the source room until there is no further effect on the receiver room reverberation times (ASTM E90-09, 2016). This effect is particularly noticeable when the measured TL is less than 10 dB (Bies and Davies, 1977). The sound-absorbing material is only present for the reverberation time measurements and must be removed when the steady-state difference (NR) in sound pressure levels between the two rooms is determined.

An expression for the field incidence transmission loss in terms of these measured quantities can then be derived using the analysis of Section 6.4, as will now be shown.

The power transmitted through the wall is given by the effective intensity in a diffuse field (see Section 6.4.1) multiplied by the area, A, of the panel and the fraction of energy transmitted, τ; thus, using Equation (6.35) we may write for the power transmitted:

$$W_t = \frac{\langle p_i^2 \rangle A\tau}{4\rho c} \tag{7.14}$$

The sound pressure level in the receiver room (from Equation (6.43)) is:

$$\langle p_r^2 \rangle = \frac{4W_t\rho c(1-\bar{\alpha})}{S\bar{\alpha}} = \frac{\langle p_i^2 \rangle A\tau(1-\bar{\alpha})}{S\bar{\alpha}} \tag{7.15}$$

and the noise reduction is thus given by:

$$\text{NR} = 10\log_{10}\frac{\langle p_i^2 \rangle}{\langle p_r^2 \rangle} = \text{TL} - 10\log_{10}\frac{A(1-\bar{\alpha})}{S\bar{\alpha}} \qquad \text{(dB)} \tag{7.16}$$

In reverberant test chambers used for transmission loss measurement, $\bar{\alpha}$ is always less than 0.1; thus $S\bar{\alpha}/(1-\bar{\alpha})$ may be approximated as $S\bar{\alpha}$. Equation (7.16) may then be rearranged to give the following expression, which is commonly used for the laboratory measurement of sound transmission loss, TL, (ASTM E90-09, 2016) or sound reduction index, R, (ISO 10140-2, 2021):

$$\text{TL} = R = \text{NR} + 10\log_{10}(A/S\bar{\alpha}) \qquad \text{(dB)} \tag{7.17}$$

In Equation (7.17), $S\bar{\alpha}$ is the Sabine absorption of the receiving room, including losses through the test partition (measured as described in Section 6.5), and A is the area of the test partition. S and $\bar{\alpha}$ are, respectively, the receiving room total surface area, including that of the test partition, and the mean Sabine absorption coefficient (including the test partition).

When conducting a transmission loss test in a laboratory, great care must be taken to ensure that all other acoustic transmission paths are negligible; that is, "flanking paths" must contribute an insignificant amount to the total energy transmitted. The test procedures for standardised tests in certified acoustic test facilities as well as for testing in the field are described in relevant standards publications (ISO 10140-2, 2021; ASTM E90-09, 2016; ISO 16283-1, 2014; ASTM E336-19, 2019). Care should also be taken to subtract out any background noise from sound pressure level measurements using Equation (1.102).

The sound transmission loss of a partition may also be determined using a single reverberant room as the source room and a not too reverberant space (preferably free field) as the receiving room. In this case, the sound power incident on the partition may be determined using Equation (7.14) with the quantity, τ, excluded and the transmitted power may be determined by measuring the average of the active sound intensity very close (500 to 100 mm) to the panel on the receiving room side. The transmitted power is then determined by multiplying the average sound intensity by the panel surface area, the transmission coefficient is determined as the ratio of the transmitted to incident power and the transmission loss is then determined using Equation (7.13). This latter method of transmission loss measurement is more accurate than the sound pressure measurement method and is gradually becoming more accepted. It is described in detail in three ISO standards and one ASTM standard (ISO 15186-1, 2000; ISO 15186-2, 2003; ISO 15186-3, 2002; ASTM E2249-19, 2019).

7.2.2.1 Measurement of Transmission Loss Outside of a Laboratory

In many cases, it is desirable to measure the sound transmission between different spaces in a building such an apartment or office building. In this case, we use a different descriptor to quantify the transmission loss, and this is the quantity, D, defined in ISO 16283-1 (2014) as the difference $(L_{p1} - L_{p2})$ in space and time averaged sound pressure levels (see Equation (4.142)) between two spaces, 1 and 2. The standardised level difference, D_{nT}, is:

$$D_{nT} = D + 10\log_{10}(2T_{60}) \qquad \text{(dB)} \tag{7.18}$$

where T_{60} is the reverberation time (see Sections 6.5 and 6.6.2) of the receiver room (see ISO 3382-2 (2008) for T_{60} measurements outside of a laboratory). The apparent sound reduction index, R', is:

$$R' = D + 10\log_{10}[A/(S\bar{\alpha}/10)] \qquad \text{(dB)} \tag{7.19}$$

where A is the area of the partition and $S\bar{\alpha}$ is the room absorption (room surface area × arithmetically averaged Sabine sound absorption coefficients of the room surfaces).

As an alternative, the apparent transmission loss (ATL) is defined in ASTM E336-19 (2019) by replacing R' in Equation (7.19) with ATL. However, ASTM E336-19 (2019) uses the symbol NR, which is defined in the same way as D; that is, NR $= L_{p1} - L_{p2}$. ASTM E336-19 (2019) also defines the term, NNR as the normalised noise reduction, which is equivalent to D_{nT} in Equation (7.18).

7.2.2.2 Single Number Ratings for Transmission Loss of Partitions

STC Rating:

In practice, it is desirable to characterise the transmission loss of a partition with a single number descriptor to facilitate comparison of the performance of different partitions. For this reason, a single number ASTM rating scheme (ASTM E413-22, 2022) called STC (or Sound Transmission

Class) has been introduced. To determine the STC rating for a particular partition, a curve fitting technique is used to fit the measured or calculated 1/3-octave transmission loss (TL) data for the partition. Figure 7.5 shows a typical STC contour.

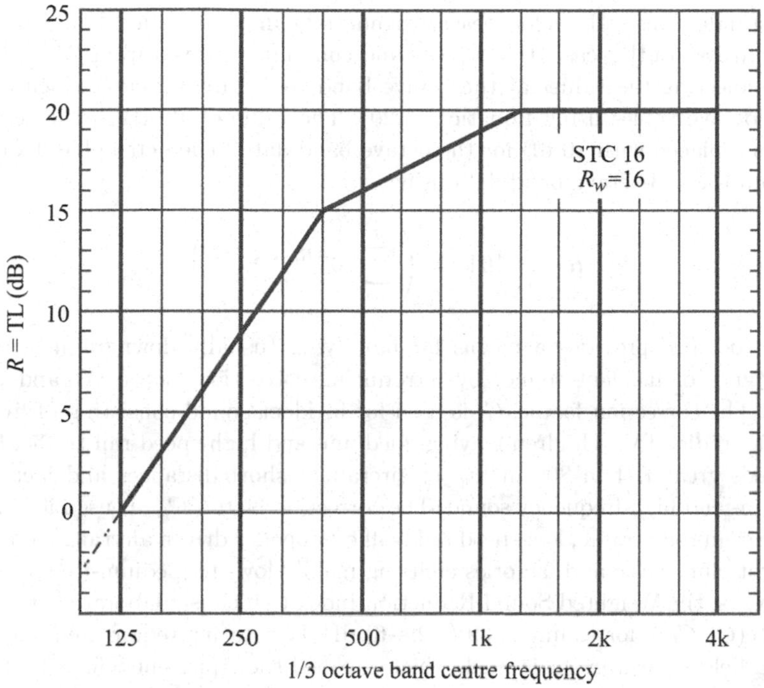

FIGURE 7.5 Example STC contour.

STC contours consist of a horizontal segment from 1250 to 4000 Hz, a middle segment increasing by 5 dB from 400 to 1250 Hz and a low-frequency segment increasing by 15 dB from 125 to 400 Hz. The STC rating of a partition is determined by plotting the 1/3-octave band TL (rounded up or down to the nearest integer dB) of the partition and comparing it with the STC contours. The STC contour is shifted vertically downwards in 1 dB increments from a large value until the following criteria are met.

1. The TL data are never more than 8 dB below the STC contour in any 1/3-octave band.
2. The sum of the deficiencies of the TL data below the STC contour over the 16 1/3-octave bands does not exceed 32 dB. Only data that are below the STC contour are included in this calculation.

When the STC contour is shifted to meet these criteria, the STC rating is given by the integer TL value of the contour at 500 Hz.

R_w Rating:
The ISO method for determining a single number to describe the sound transmission loss characteristics of a construction is outlined in ISO 717-1 (2020). Different terminology is used to that in the method described in ASTM E413-22 (2022); otherwise, the methods are very similar. The ISO standard uses Sound Reduction Index (R) instead of sound transmission loss and Weighted Sound Reduction Index (R_w) instead of Sound Transmission Class (STC). The shape of the contour for 1/3-octave band data is identical to that shown in Figure 7.5, except that the straight line at the low-frequency end continues down to 100 Hz and at the upper-frequency end, the line terminates at 3150 Hz. In addition, there is no requirement to satisfy criterion number

1 listed above (the 8 dB criterion). However, measured TL values are rounded to the nearest 0.1 dB (rather than the 1 dB for STC) when calculating the deficiencies below the R_w contour, but as with the STC rating, the contour values are integer only. Figure 7.5 shows a typical R_w contour.

The ISO standard also allows for measurements to be made in octave bands between 125 Hz and 2 kHz inclusive. In this case, the octave band contour is derived from the 1/3-octave band contour by connecting the values at the octave band centre frequencies, which are calculated from the 1/3-octave values using Equation (7.20). The value of 32 dB in the second criterion listed above is replaced with 10 dB for the octave band data. The octave band values of R are calculated from the 1/3-octave band data using:

$$R_{\rm oct} = 10\log_{10}\left(\sum_{i=1}^{3} 10^{-R_{i,(1/3-\rm oct)}}\right) \tag{7.20}$$

The ISO method also provides a means for modifying (usually downgrading) the R_w value for different types of incident sound, by introducing correction factors, C and C_{tr}, that are added to R_w. The correction factor, C, is used for incident sound consisting of living activities (talking, music, radio, TV), children playing, medium- and high-speed rail traffic, highway road traffic at speeds greater than 80 km/hr, jet aircraft at short distances and factories emitting mainly medium- and high-frequency sound. The correction factor, C_{tr}, is used for incident sound consisting of urban road traffic, low-speed rail traffic, propeller driven aircraft, jet aircraft at long distances, night club music and factories emitting mainly low- to medium-frequency sound. For building elements, the Weighted Sound Reduction Index (which is a laboratory measurement) is written as R_w (C; C_{tr}), for example, 39 (−2; −6) dB. For stating requirements or performance of buildings, a field measurement is used, which is called the Apparent Sound Reduction Index, R'_w, and it is written as a sum with a spectral correction term, such as $R'_w + C_{tr} > 47$ (for example), where the measurements are conducted in the field according to ISO 16283-1 (2014) or ISO 16283-3 (2016).

The correction terms C and C_{tr} are calculated from values in Table 7.1 and the following equations:

$$C = -10\log_{10}\sum_{i=1}^{N} 10^{(L_{i,1}-R_i)/10} - R_w \tag{7.21}$$

$$C_{tr} = -10\log_{10}\sum_{i=1}^{N} 10^{(L_{i,2}-R_i)/10} - R_w \tag{7.22}$$

where L_{i1} and L_{i2} are listed for 1/3-octave or octave bands in Table 7.1, R_i is the transmission loss or sound reduction index for frequency band i and N is the number of bands used to calculate R_w (octave or 1/3-octave). Although the table shows values in the frequency range from 50 Hz to 5000 Hz, the standard frequency range usually used is 100 Hz to 3150 Hz. In this case (and for the case of the frequency range from 50 Hz to 3150 Hz), the octave and 1/3-octave band values in the table for L_{i1} (only) must be increased (made less negative) by 1 dB. When the expanded frequency range is used, the calculation of R_w is unchanged but the values of C and C_{tr} are different and indicated by an appropriate subscript; for example, $C_{50-3150}$ or $C_{50-5000}$ or $C_{100-5000}$.

ASTC, R'_w and $D_{nT,w}$ Ratings:
The STC and R_w ratings are intended to be applied to laboratory measurements of wall, roof and floor constructions where the influence of flanking transmission is minimal. However, when transmission loss measurements are undertaken outside of the laboratory in existing buildings, the results are influenced by flanking transmission, which influences the difference in average

TABLE 7.1 Correction terms for Equations (7.21) and (7.22)

Band centre frequency	$L_{i,1}$		$L_{i,2}$	
	1/3-octave	octave	1/3-octave	octave
50	−41		−25	
63	−37	−32	−23	−18
80	−34		−21	
100	−30		−20	
125	−27	−22	−20	−14
160	−24		−18	
200	−22		−16	
250	−20	−15	−15	−10
315	−18		−14	
400	−16		−13	
500	−14	−9	−12	−7
630	−13		−11	
800	−12		−9	
1000	−11	−6	−8	−4
1250	−10		−9	
1600	−10		−10	
2000	−10	−5	−11	−6
2500	−10		−13	
3150	−10		−15	
4000	−10	−5	−16	−11
5000	−10		−18	

sound pressure levels (see Equation (4.142)) measured between two rooms separated by a partition, with one of the rooms excited by a loudspeaker. The difference, D, in measured average sound pressure levels, L_{p1} and L_{p2} in rooms 1 and 2 (with the sound source located in room 1), is then used to calculate an apparent transmission loss (ATL), defined in (ASTM E336-19, 2019), or an apparent sound reduction index, R', defined in ISO 16283-1 (2014), or a standardised level difference, $D_{nT,w}$, also defined in ISO 16283-1 (2014). Equations to calculate these quantities are provided in Section 7.2.2.1.

The 1/3-octave ATL values are then used to calculate the ASTC rating by following the same procedure used to find the STC rating from measured values of TL. Similarly, the R' values are used to calculate the weighted apparent sound reduction index, R'_w, and the D_{nT} values are used to calculate the weighted standardised level difference, $D_{nT,w}$, using the same procedure as used to find R_w from R values.

OITC Rating:
Another method commonly used for rating the sound transmission of walls is the Outdoor–Indoor Transmission Class (OITC). This method (ASTM E1332-22, 2022) was developed in 1990 to provide a single number for the assessment of the transmission through building façades, of external traffic noise and aircraft noise, which contains significant levels of low-frequency components. This is the reason that the OITC rating system emphasises low frequencies. The same 1/3-octave band sound transmission loss data are used as used for calculation of the STC rating, except the OITC method requires data in the 80 Hz and 100 Hz 1/3-octave bands as well. The OITC rating method is not intended for use in rating interior walls. The OITC rating number is calculated using:

$$\text{OITC} = 100.13 - 10\log_{10}\left[\sum_{f=80\text{Hz}}^{f=4\text{kHz}} 10^{(LA_f - \text{TL}_f)/10}\right] \tag{7.23}$$

where TL_f is the measured 1/3-octave band sound transmission loss and the frequency-dependent coefficient, LA_f, is listed in Table 7.2.

TABLE 7.2 Values of LA_f as a function of 1/3-octave band centre frequency

1/3-octave band centre frequency (Hz)	LA_f	1/3-octave band centre frequency (Hz)	LA_f	1/3-octave band centre frequency (Hz)	LA_f
80	80.5	315	87.4	1250	89.6
100	82.9	400	88.2	1600	89.0
125	84.9	500	89.8	2000	89.2
160	84.6	630	89.1	2500	88.3
200	86.1	800	89.2	3150	86.2
250	86.4	1000	89.0	4000	85.0

Data from ASTM E1332-22 (2022).

The ISO approach to a single number rating for façades is the same as the R'_w method described in the previous section.

7.2.2.3 Uncertainty in TL and R Measurements

There are two approaches (ASTM E90-09, 2016; ISO 12999-1, 2020) recommended in international standards for estimating uncertainty in 1/3-octave band TL and R measurements. Each will be discussed in the following paragraphs.

Uncertainty in TL according to ISO 12999-1 (2020)
The ISO approach to calculating uncertainty in R, R', R_w and R'_w values covers three situations and provides tabulated data that are typical of each situation and that can be used for reporting results. These situations (A, B and C) are summarised below.

A. Measurements are undertaken in a number of different test facilities that qualify to undertake the measurements according to ISO 10140-2 (2021).
B. Measurements are undertaken in the same test facility by different measurement personnel.
C. Measurements are repeated in the same test facility by the same personnel.

Standard uncertainties, u_s, which can be used for all three situations, are provided in Tables 2 and 3 in ISO 12999-1 (2020) for both R and R_w. In addition, 95% confidence limits are provided for situation A and, in the absence of inter-laboratory data for a particular test specimen, these values are those that ISO 12999-1 (2020) requires to be reported. For situations B and C, the standard uncertainties, u_s, provided in Tables 2 and 3 in ISO 12999-1 (2020) can be used to calculate the expanded uncertainty, u_e with 95% confidence limits using:

$$u_e = k_f u_s \tag{7.24}$$

where $k_f = 1.96$ for a two-sided test, implying that there is a 95% certainty that the actual value will lie in the range, $R \pm u_e$, where R is the measured value. For a one-sided test, $k_f = 1.65$ for a 95% certainty that the measured value conforms with a requirement. That is, when it is to be determined whether a measured value is smaller than a requirement, the value of u_e for a one-sided test is added to the measured data and when it is to be determined whether a measured value is larger than a requirement, then the value of u_e is subtracted from the measured data.

Different values of k_f correspond to different values of uncertainty confidence limits (ISO 12999-1, 2020). For example, for 90% confidence limits, $k_f = 1.65$ for a two-sided test and $k_f = 1.28$ for a one-sided test.

Uncertainty in TL according to ASTM E90-09 (2016)

The ASTM approach allows uncertainty estimates to be undertaken for a single laboratory, based on the variation in sound pressure level measurements made in the source and receiver room and decay rate measurements made in the receiver room. In both standards, uncertainties are expressed in terms of 95% confidence limits; that is, there is a 95% confidence that the actual value of TL or R is within the range of the measured value $\pm$ (the uncertainty level in dB), expressed as TL $\pm\,\Delta$TL or $R \pm \Delta R$ or $R_w \pm \Delta R_w$.

The overall expanded uncertainty, $u_e = \Delta\mathrm{TL}$, corresponding to 95% confidence limits, is a function of the uncertainties in the individual quantities used in the calculation of TL. Thus, for each 1/3 octave band with a centre frequency of f, the 95% confidence limit range for the TL is the expanded uncertainty, $\mathrm{TL}_f \pm \Delta\mathrm{TL}_f$ (dB), where:

$$\Delta\mathrm{TL}_f = u_e(f) = \sqrt{\Delta L_{pS,f}^2 + \Delta L_{pR,f}^2 + 18.86(\Delta D_f/D_f)^2} \quad \text{(dB)} \tag{7.25}$$

and where the decay rate, D_f, for the 1/3-octave band with a centre frequency of f is: $D_f = 60/T_{60,f}$ dB/s and $T_{60,f}$ is the measured reverberation time for the 1/3-octave band with a centre frequency of f (see Section 6.5).

The individual uncertainties are calculated using:

$$\Delta X_f = \sqrt{\frac{a^2}{N-1}\sum_{i=1}^{N}\left[X_{i,f} - \overline{X}_f\right]^2} \tag{7.26}$$

where $X = L_{pS}$, L_{pR} or D, N is the number of measurements and a is a constant, listed in Table 7.3, which is dependent on the number of microphone positions used for the measurements (minimum allowed is 4 in each room).

TABLE 7.3 Values of the coefficient, a

Number of microphone positions, N	Coefficient a	Number of microphone positions, N	Coefficient a	Number of microphone positions, N	Coefficient a
4	1.591	8	0.836	12	0.635
5	1.241	9	0.769	13	0.604
6	1.050	10	0.715	14	0.577
7	0.925	11	0.672	14	0.554

The maximum allowed values of $\Delta\mathrm{TL}_f$ are listed in Table 7.4.

The average, $\overline{X}_f$, is an arithmetic average of decay rates or sound pressure levels. The average is calculated using:

$$\overline{X}_f = \frac{1}{N}\sum_{i=1}^{N} X_{i,f} \tag{7.27}$$

If uncertainty values calculated using Equation (7.25), exceed the maximum allowed values, the number of microphone positions used for the measurements must be increased until the calculated values are less than or equal to the allowed maxima. It is also possible to use continuous microphone traverses rather than single microphone locations. This requires determining the minimum number of microphone locations, N_{min}, that will satisfy the maximum allowed values listed in Table 7.4. Once this has been determined, the required minimum radius, r_{min}, of a circularly rotating microphone is the larger of $r_{\mathrm{min}} = 1.2$ m and $r_{\mathrm{min}} = N_{\mathrm{min}}\lambda/(4\pi)$ and the required length, L_{min} of a linear traverse is $L_{\mathrm{min}} = (N_{\mathrm{min}} - 1)\lambda/2$ (ASTM E492-09, 2016).

TABLE 7.4 Maximum allowed uncertainty level, ΔTL_f

1/3-octave band centre frequency, f (Hz)	95% confidence limit, ΔTL_f (dB)	1/3-octave band centre frequency, f (Hz)	95% confidence limit, ΔTL_f (dB)
80	6	200, 250	2
100	4	315–4000	1
125, 160	3		

7.2.3 Impact Isolation Measurement According to ASTM Standards

7.2.3.1 Laboratory Measurements

The ability of a construction such as a floor or ceiling to prevent transmission of impact noise such as footsteps is quantified in terms of its impact isolation, which is measured in the laboratory according to ASTM E492-09 (2016) and ASTM E989-21 (2021) using a standard tapping machine. A standard tapping machine can be purchased from suppliers of acoustic instrumentation and should conform to the specifications in Appendix F of ISO 10140-5 (2021) and repeated in ASTM E492-09 (2016). Basically, such a machine consists of five standard hammers, equally spaced 100 mm apart in a line, which bang on the floor sequentially, by falling from a height of 40 mm in a direction perpendicular to the floor, with 0.1 seconds between successive impacts. The hammers weigh 0.5 kg each and consist of steel cylinders, 30 mm in diameter and with a radius of curvature of 500 mm on the hardened steel end that strikes the floor. To measure the impact isolation of a floor in a laboratory test facility, the standard tapping machine is placed on it and the average sound pressure levels, $L_{p,k}$, in the room on the opposite side of the floor or ceiling are measured in 1/3-octave or octave bands (more commonly 1/3-octave bands) for 4 different locations ($k = 1, 4$) of the tapping machine. The average sound pressure level, $L_{p,k}$, in the receiving room for each tapping machine location, k is:

$$L_{p,k} = 10 \log_{10} \left(\frac{1}{N} \sum_{i=1}^{N} 10^{L_{p,i}/10} \right) \tag{7.28}$$

The number of microphone positions required to obtain an acceptable accuracy for $L_{p,k}$ is discussed in Section 7.2.3.4.

The sound pressure level, $L_{p,O}$, averaged over all 4 tapping machine locations is:

$$L_{p,O} = 10 \log_{10} \left(\frac{1}{4} \sum_{i=1}^{4} 10^{L_{p,k}/10} \right) \tag{7.29}$$

The normalised impact sound pressure level, L_n, for each measurement band is then calculated using:

$$L_n = L_{p,O} + 10 \log_{10} (S\bar{\alpha}/10) \tag{7.30}$$

The quantity, $S\bar{\alpha}$, applies to the room in which the sound measurements are made and may be determined from the room reverberation time using Equation (6.55) and/or ASTM E2235-04 (2020).

From these measurements a single number Impact Insulation Class (IIC) may be determined (ASTM E989-21, 2021). This is done in a similar way to determining the STC rating of a wall. The normalised impact sound pressure levels, L_n, for each 1/3-octave band between 100 Hz and 3150 Hz, are rounded up or down to the nearest decibel and plotted on a set of axes similar to those used in Figure 7.6.

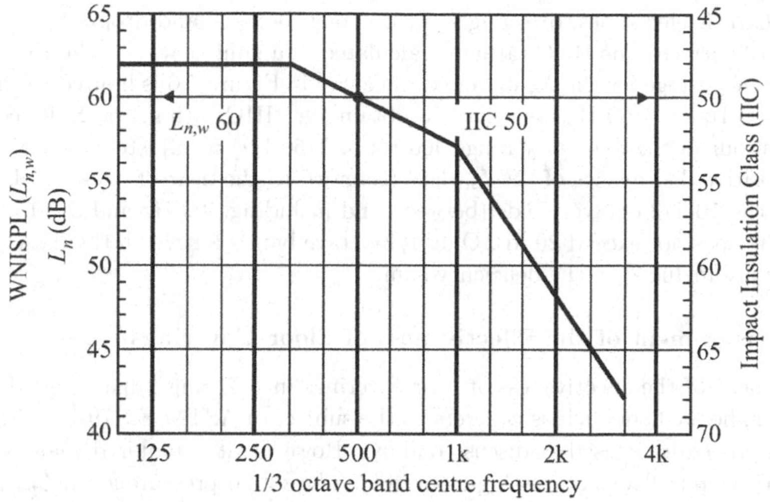

FIGURE 7.6 Example IIC and $L_{n,w}$ contour. The $L_{n,w}$ calculation is the ISO equivalent to the IIC calculation, and it is described in Section 7.2.4.1

The IIC curve shown in the figure is started at a low L_n level (or high value of IIC) and then shifted vertically upwards (decreasing the IIC value) in 1 dB increments until the following conditions are met:

1. the 1/3-octave L_n data of normalised measured sound pressure levels (rounded to the nearest decibel) measured in the room on the opposite side of the floor or ceiling to the tapping machine is never more than 8 dB above the IIC contour in any 1/3-octave band; and
2. the sum of the positive deficiencies of the L_n data (rounded to the nearest decibel) above the IIC contour over the 16 1/3-octave bands is as large as possible but does not exceed 32 dB. Only 1/3-octave bands for which the data are above the IIC curve are included in the deficiency sum. Note that the lower the IIC contour on the figure, the higher (and better) will be the IIC.

When the IIC contour has been adjusted to meet the above criteria, the IIC is the integer value of the contour at 500 Hz on the right of the figure or the value on the left ordinate (L_n) subtracted from 110. Note that there is no octave band calculation specified to calculate an octave band IIC rating.

ASTM has also introduced low- and high-frequency ratings referred to as LIIC (ASTM E3207-21, 2021) and HIIC (ASTM E3222-20a, 2020), respectively. The purpose of the low-frequency rating is to allow different constructions to be compared with respect to their performance in reducing low-frequency impact noise caused primarily by footfall on lightweight structures.

The LIIC rating is calculated using:

$$\mathrm{LIIC} = 190 - 20\log_{10}\left[\sum_{f=50}^{f=80} 10^{L_n(f)/10}\right] \tag{7.31}$$

where, for LIIC only, L_n values are rounded to the nearest 0.1 dB and the sum is over the three 1/3-octave bands with centre frequencies, 50, 63 and 80 Hz.

The purpose of the HIIC rating is to allow comparison of the relative performance of structures and floor coverings in terms of their ability to reduce the transmission of impact noise arising from hard-heeled shoes, dragging furniture, dog toenails, and dropping objects on flooring with a hard surface. The HIIC rating is calculated in a similar way to the IIC rating except that the frequency range for the L_n data and the curve in Figure 7.6 is limited to the 1/3-octave bands from 400 Hz to 3150 Hz inclusive. To obtain the HIIC rating, the HIIC contour (same as the IIC contour in the frequency range 400 Hz to 3150 Hz) is adjusted vertically so that the sum of the positive deficiencies of the L_n data (rounded to the nearest decibel) above the HIIC contour over the 10 1/3-octave bands (between and including 400 Hz and 3150 Hz) is as large as possible but does not exceed 20 dB. Only 1/3-octave bands for which the data are above the HIIC contour are included in the deficiency sum.

7.2.3.2 Measurement of the Effectiveness of Floor Coverings

The measurement of the effectiveness of floor coverings in reducing impact sound transmission through a bare, heavy floor such as concrete, is the subject of ASTM E2179-21 (2021). A similar measurement procedure is used as discussed above. However, it is required that the covering be installed on a concrete floor in a testing laboratory. The sound pressure level, $L_{p,O}$, in the room averaged over the room volume for four tapping machine locations is calculated using Equation (7.29) for the bare concrete floor. The same procedure is then done for the case with the floor covering installed over the concrete slab to produce $L_{p,c}$. The quantity, $L_{n,\text{ref},c}$, which is the normalised impact sound pressure level for the reference floor with the floor covering under test, is calculated using:

$$L_{n,\text{ref},c} = L_{p,c} - L_{p,O} + L_{n,\text{ref}} \tag{7.32}$$

where values of $L_{n,\text{ref}}$ are provided in Table 7.5.

TABLE 7.5 Values of the normalised impact sound pressure level, $L_{n,\text{ref}}$, assumed for a reference concrete floor

Frequency (Hz)	$L_{n,\text{ref}}$ (dB)	Frequency (Hz)	$L_{n,\text{ref}}$ (dB)	Frequency (Hz)	$L_{n,\text{ref}}$ (dB)	Frequency (Hz)	$L_{n,\text{ref}}$ (dB)
100	67.0	250	69.0	630	71.0	1600	72.0
125	67.5	315	69.5	800	71.5	2000	72.0
160	68.0	400	70.0	1000	72.0	2500	72.0
200	68.5	500	70.5	1250	72.0	3150	72.0

The impact insulation class, IIC_c for the 1/3-octave band values, $L_{n,\text{ref},c}$, is then calculated as described in Section 7.2.3.1 by using $L_{n,\text{ref},c}$ values in place of L_n values. The improvement in insulation class due to the floor covering is then:

$$\Delta\text{IIC} = \text{IIC}_c - 28 \tag{7.33}$$

where 28 = IIC rating for the reference concrete floor.

The high-frequency improvement in insulation class is calculated in the same way, except that the frequency range is limited to the 1/3-octave bands between and including 400 Hz and 3150 Hz. Thus:

$$\Delta\text{HIIC} = \text{HIIC}_c - 28 \tag{7.34}$$

where HIIC_c is calculated using values of $L_{n,\text{ref},c}$, calculated using Equation (7.32), in the frequency range 400 Hz to 3150 Hz.

7.2.3.3 Field Measurements

It is also possible to measure the impact isolation via the floor/ceiling between two rooms in the field outside of a test laboratory, over the same frequency range, using a similar procedure as used in the laboratory setting (ASTM E1007-19, 2019). It is important that the sound pressure levels in the receiving room due to acoustic transmission from the room containing the tapping machine are at least 10 dB less than the acoustic levels in the receiving room resulting from the structural vibration generated by the tapping machine. This can be checked by using a loudspeaker to generate the same acoustic levels in the tapping machine room and then measuring the resulting levels in the receiving room. Flanking transmission should NOT be suppressed, which is opposite to the laboratory test requirement.

The main differences between the laboratory and field measurement processes lie in the reporting of the measured data. That is, for field measurements outside of a laboratory, where flanking transmission may play a role, the reported quantities are ISPL, ISR, ANISPL and AIIC. ISPL is the space-averaged sound pressure level in the receiving room, calculated using:

$$\mathrm{ISPL} = 10 \log_{10} \left(\frac{1}{N} \sum_{i=1}^{N_m} 10^{L_{p,i}/10} \right) \tag{7.35}$$

where $N_m = 4N$ and N is the number of microphone locations used for one tapping machine location. ISR is the value of IIC calculated using ISPL in place of L_n. A low-frequency rating, LIR, analogous to ISR, may be calculated using ISPL values in place of L_n values in Equation (7.31). A high-frequency, HIR, analogous to ISR, may be calculated as for the calculation of HIIC, except that ISPL values are used instead of L_n values.

AIIC is the value of IIC calculated using ANISPL in place of L_n where ANISPL is the absorption normalised impact sound pressure level, calculated using Equation (7.30) with L_n replaced with ISPL. An analogous high-frequency rating using only the 1/3-octave bands between and including 400 Hz and 3150 Hz is AHIR (ASTM E3222-20a, 2020).

NISR (only valid for rooms of volume less than 150 m^2) is the value of IIC calculated using the reverberation time normalised impact sound pressure level (RTNISPL) in place of L_n. RTNISPL is defined for each 1/3-octave band as:

$$\mathrm{RTNISPL} = \mathrm{ISPL} - 10 \log_{10} (2T_{60}) \tag{7.36}$$

where T_{60} is the reverberation time of the receiving room in the 1/3-octave band under test (see Section 6.5 and 6.6.2). All measurements of sound pressure level in the receiving room must be adjusted to account for background noise levels using Equation (1.102). NISR provides a measure of the isolation of the impact sound produced by a standard tapping machine in a receiving room, as if the receiving room had a reverberation time of 0.5 s.

NHIR is the high-frequency rating analogous to NISR, calculated in the same way as NISR but with the frequency range restricted to 400 Hz to 3150 Hz. In other words, NHIR is the value of HIIC calculated using RTNISPL in place of L_n.

7.2.3.4 Uncertainty According to ASTM E492-09 (2016)

As discussed in Section 5.9.4, the expanded uncertainty, u_e for a 95% confidence interval implies that there is a 95% chance that the measured value of L_n will be within $\pm u_e$ of the true value. For the laboratory measurement case, the expanded uncertainty, u_e, for each 1/3-octave band is:

$$u_e = \Delta L_n = k_c \left[s_0^2 + s_f^2/N \right]^{1/2} \quad \text{(dB)} \tag{7.37}$$

The standard, (ASTM E492-09, 2016) uses a coverage factor value of $k_c = 1.6$, but for a 2-sided normal distribution, this corresponds to confidence limits of approximately 90%. If confidence limits of 95% are used, then $k_c = 2$ (or more accurately, 1.96). Alternatively, if all values within a range were considered to be equally likely (that is, a rectangular distribution), then a 95% confidence limit would correspond to $k_c = 1.65$. Similarly, if a one-sided uncertainty test were required, such as checking whether a measurement met a minimum requirement, then u_e, calculated with $k_c = 1.65$, would be added to the best estimate found by measurement, and the result compared with the requirement.

The estimate in Equation (7.37), s_0, of the actual standard deviation, σ_0, of the mean sound pressure level, based on only the four tapping machine locations is:

$$s_0 = \sqrt{\frac{1}{3}\sum_{k=1}^{4}(L_{p,k} - L_{p,O})^2} \quad \text{(dB)} \tag{7.38}$$

where $L_{p,k}$ and $L_{p,O}$ are defined in Equations (7.28) and (7.29), respectively. The estimated standard deviation, $s_{f,k}$, of the measured sound pressure level in the receiving room for tapping machine location, k is (ASTM E492-09, 2016):

$$s_{f,k} = \sqrt{\frac{1}{N-1}\sum_{i=1}^{N}(L_{p,i} - L_{p,Ak})^2} \quad \text{(dB)} \tag{7.39}$$

where N is the number of sound pressure level measurements corresponding to the kth tapping machine position and $L_{p,Ak}$ is the arithmetic average of the sound pressure level measurements in the receiver room for tapping machine location, k, given by:

$$L_{p,Ak} = \frac{1}{N}\sum_{i=1}^{N} L_{p,i} \quad \text{(dB)} \tag{7.40}$$

Although implementation of the ASTM E492-09 (2016) procedure uses arithmetic sums in Equations (7.39) and (7.40), it is more accurate to use logarithmic sums (as in Equations (2.127) and (2.126)), so that we are applying statistics to energy quantities (pressure squared) rather than sound pressure levels in decibels.

A requirement of ASTM E492-09 (2016) is that the 95% confidence limits, as_f, should be less than 3 dB for 1/3-octave bands from 100 to 400 Hz and less than 2.5 dB for bands ranging from 500 Hz to 3150 Hz for each location of the tapping machine. If not, the number of measurements, N must be increased until the criteria are satisfied. It is also possible to use continuous microphone traverses rather than single microphone locations. This requires determining the minimum number of microphone locations, N_{min}, that will satisfy the maximum allowed values of as_f. Once this has been determined, the required minimum radius, r_{min}, of a circularly rotating microphone is the larger of $r_{\text{min}} = 1.2$ m and $r_{\text{min}} = N_{\text{min}}\lambda/(4\pi)$ and the required length, L_{min} of a linear traverse is $L_{\text{min}} = (N_{\text{min}} - 1)\lambda/2$ (ASTM E492-09, 2016). Values of the coefficient, a, are listed in Table 7.3.

The reproducibility, R (dB), corresponding to the 95% confidence limit that the absolute difference between two field test results, by two different test agencies will be less than the value specified, was determined by doing a round robin test (ASTM E1007-19, 2019). The round robin test consisted of six different groups testing the same floor/ceiling assembly in six different multi-family residential buildings (one for each group). The results are listed in Table 7.6.

7.2.4 Impact Isolation Measurement According to ISO Standards

7.2.4.1 Laboratory Measurements

The laboratory measurement procedure specified in ISO 10140-3 (2021) is very similar to that specified in ASTM E492-09 (2016), and Equation (7.30) is used to find the normalised impact sound pressure level, L_n. In addition, ISO 10140-3 (2021) allows the reporting of octave band

TABLE 7.6 Reproducibility, R (95% confidence limits), estimates for field measurements of ISPL and ANISPL for a specific floor/ceiling assembly (ASTM E1007-19, 2019)

1/3-octave band centre frequency (Hz)	R (dB) for ANISPL	R (dB) for ISPL	1/3-octave band centre frequency (Hz)	R (dB) for ANISPL	R (dB) for ISPL
80	3.4	3.3	1000	1.8	2.4
100	8.7	8.7	1250	1.5	2.0
125	3.2	2.5	1600	1.5	2.2
160	2.0	1.7	2000	1.9	2.8
200	2.8	2.7	2500	2.9	3.9
250	3.0	2.5	3150	3.8	4.7
315	1.8	1.9	4000	2.6	3.7
400	1.5	1.7	5000	4.4	5.4
500	1.5	1.6	AIIC	2.1	—
630	2.6	2.2	ISR	—	3.1
800	1.6	2.0			

data, which, for each octave band, is the logarithmic sum (not average) of the values for the three 1/3 octave bands that make up the octave band. That is:

$$L_{n,\text{oct}} = 10 \log_{10} \left(\sum_{i=1}^{3} 10^{L_{n,i,(1/3-\text{oct})}} \right) \tag{7.41}$$

ISO 10140-3 (2021) also allows extension of the measured data in the high-frequency range to include the 4000 Hz and 5000 Hz 1/3-octave bands and extension of the low-frequency range to include the 50 Hz, 63 Hz and 80 Hz 1/3-octave bands, if it is considered necessary. However, these bands are not used in the calculation of the single integer number rating, $L_{n,w}$. The $L_{n,w}$ value is calculated using Figure 7.6, and the $L_{n,w}$ is defined as the L_n axis value at the 500 Hz point on the $L_{n,w}$ contour. Additionally, the L_n data used to determine $L_{n,w}$ using the 32 dB deficiency criterion are rounded to the nearest 0.1 dB rather than the 1 dB used for IIC calculations. To find the value of the $L_{n,w}$ contour corresponding to the 1/3-octave band L_n data, the contour shown in Figure 7.6 starts at a low value of L_n and then is shifted upwards in increments of 1 dB until the arithmetic sum of the differences between the 1/3-octave band data and the contour is as large as possible but less than 32 dB. Only differences for which the data are higher on the graph in Figure 7.6 than the $L_{n,w}$ contour, are included in the sum, so the number of values used in the sum could be less than the total number of 1/3-octave bands (16). The criterion for IIC calculations, which states that the L_n data are never more than 8 dB above the IIC contour in any 1/3-octave band, does not need to be satisfied for the $L_{n,w}$ calculation. The data in the extended low-frequency range, although not used for calculating $L_{n,w}$, are used for calculating a spectral adaptation term (see 7.2.4.4), which is reported along with $L_{n,w}$. The low and high-frequency data in the extended ranges mentioned above may be used to compare different products but are not used in the $L_{n,w}$ calculation. The $L_{n,w}$ value is always within a few dB of 110−IIC, with the difference caused by the difference in rounding the L_n values (0.1 dB for $L_{n,w}$ calculations and 1 dB for IIC calculations) used for the calculations.

The standard, ISO 717-2 (2020), also allows the use of octave band measurements of L_n (from 125 Hz to 2 kHz) to calculate a single number, $L_{n,w}$. In this case, the octave band $L_{n,w}$ contour (with octave L_n values calculated from 1/3-octave values using Equation (7.41)) is adjusted in 1 dB increments until the sum of the octave band deficiencies of the measured data above the $L_{n,w}$ contour in the 5 relevant octave bands is as large as possible but no more than 10 dB. As for the 1/3-octave calculation, data points that are below the $L_{n,w}$ contour are not included in the sum. The value of $L_{n,w}$ so obtained using the octave band data is then reduced by 5 dB (to account for the sum of Equation (7.41)).

7.2.4.2 Measurement of the Effectiveness of Floor Coverings

The measurement of the effectiveness of floor coverings in reducing impact sound transmission through a bare, heavy floor such as concrete, as well as three types of light-weight floor, is also covered in ISO 717-2 (2020). The floor separating the two test chambers must be one of the four types mentioned above, depending on the type of floor for which data on the improvement offered by the floor covering are required. The procedure for testing the effectiveness of the covering on a concrete floor is identical to that for IIC testing (including the reference values in Table 7.5), except that ISO 717-2 (2020) uses different symbol nomenclature to ASTM E2179-21 (2021) and calculates a weighted reduction, ΔL_w, in $L_{n,w}$, instead of an increase, ΔIIC, in IIC, as a result of the floor covering. Thus, for a floor covering over a heavy concrete floor:

$$\Delta L_w = 78 - L_{n,r,w} \qquad \text{(dB)} \tag{7.42}$$

where $L_{n,r,w}$ is calculated using $L_{n,\text{ref},c}$ values in place of L_n values in Figure 7.6 and $L_{n,\text{ref},c}$ is calculated using Equation (7.32).

To calculate the value of ΔL_w for a covering over one of the three standard lightweight floors, types 1, 2 and 3, as specified in ISO 10140-5 (2021), the reference floor values of $L_{n,\text{ref}}$ are replaced with different values listed in ISO 717-2 (2020) and the value of 78 in Equation (7.42) is replaced with 72 for floor types 1 and 2 and 75 for floor type 3.

7.2.4.3 Field Measurements

The procedure outlined in ISO 16283-2 (2020) for field measurements is similar to the ASTM approach described in Section 7.2.3.3, except that the ISO method can also be used for impact isolation of walls between rooms as well as floors between building levels. ISO 16283-2 (2020) uses Equation (7.36) to find the standardised impact sound pressure level, where RTNISPL in that equation is replaced with L'_{nT} and ISPL is replaced with L_i which is the same as ISPL and is calculated using Equation (7.35). The normalised impact sound pressure level, L'_n, measured in the field outside of a laboratory, is calculated using Equation (7.30) with L_n replaced with L'_n and $S\bar{\alpha}$ is replaced with $0.16V/T_{60}$ (see Equation (6.55), where $c = 343$ m/s). In addition, a standardised impact sound pressure level, L'_{nT}, is calculated using Equation (7.36) with RTNISPL replaced with L'_{nT} and ISPL replaced with L_i. The prime is used to identify quantities that apply to field measurements only.

A corresponding weighted normalised impact sound pressure level, $L'_{n,w}$ for field data, is calculated by plotting values of L'_n on Figure 7.6 and adjusting the $L'_{n,w}$ contour until the sum of the deficiencies of the L'_n values above the $L'_{n,w}$ contour over the 16 1/3-octave bands is as large as possible but does not exceed 32 dB (ISO 717-2, 2020). Similarly the Weighted Standardised Impact Sound Pressure Level, $L'_{nT,w}$, is calculated by plotting values of L'_{nT} on Figure 7.6 and adjusting the $L'_{nT,w}$ contour until the sum of the deficiencies of the L'_{nT} values above the $L'_{nT,w}$ contour over the 16 1/3-octave bands is as large as possible but does not exceed 32 dB. Note that the value assigned to the final $L'_{n,w}$ contour is the L'_n value of the $L'_{n,w}$ contour at 500 Hz and similarly for the final $L'_{nT,w}$ contour.

To make the treatment here consistent with ISO 717-2 (2020), the variables, L_n and $L_{n,w}$, apply to laboratory measurements where flanking transmission is negligible, whereas the same quantities with a prime, L'_n and L'_{nT}, apply to field measurements where flanking transmission is significant. For field measurements, two additional quantities, $L'_{n,w}$ and $L'_{nT,w}$ also apply and for these two quantities, there are no equivalent laboratory measurement quantities.

7.2.4.4 Additional Impact Spectrum Adaptation Term

The $L_{n,w}$, or $L'_{n,w}$, rating has been shown to be adequate for wooden floors or concrete floors with carpet covering or floating concrete floors. However, it has been found to be inadequate for bare concrete and timber joist floors. Therefore a spectral adaptation term, C_I, calculated to the nearest 1 dB, has been introduced (ISO 717-2, 2020), so that a requirement may be written as the sum of $L'_{n,w}$ and C_I, where C_I is slightly positive for timber joist floors, ranges from -15 to 0 dB for bare concrete and is approximately zero for massive floors. To obtain C_I, 1/3-octave band measured data for L_n, L'_n or L'_{nT} over the frequency range, 100 Hz to 2500 Hz are summed logarithmically (see Equation (1.98)) to produce $L_{n,sum}$, $L'_{n,sum}$ or $L'_{nT,sum}$. Depending on the quantity used as a basis, the spectral adaptation term (expressed as an integer) is then calculated using one of the following equations.

$$C_I = -15 + \begin{cases} L_{n,sum} - L_{n,w} \\ L'_{n,sum} - L'_{n,w} \\ L'_{nT,sum} - L'_{nT,w} \end{cases} \tag{7.43}$$

If desired, the frequency range for the spectral adaptation term can be extended at the low-frequency end to include the 80 Hz, the 63 Hz + 80 Hz or the 50 Hz + 63 Hz + 80 Hz bands, respectively (ISO 717-2, 2020). In these cases, the spectral adaptation term is written as $C_{I,80-2500}$, $C_{I,63-2500}$ or $C_{I,50-2500}$, respectively. Note that the frequency range is also extended when calculating $L_{n,sum}$, $L'_{n,sum}$ or $L'_{nT,sum}$.

7.2.4.5 Uncertainty According to ISO 12999-1 (2020)

The ISO approach for uncertainty in impact isolation measurements covers the latter two (items B and C) of the three situations described in Section 7.2.2.3 for sound transmission. For situations B and C, the standard uncertainties, u_s, provided in Tables 2 and 3 in ISO 12999-1 (2020) can be used to calculate the expanded uncertainty, u_e with 95% confidence limits using:

$$u_e = k_f u_s \tag{7.44}$$

where $k_f = 1.96$ for a two-sided test, implying that there is a 95% certainty that the actual value will lie in the range, $L_n \pm u_e$, where L_n is the measured value. For a one-sided test, $k_f = 1.65$ for a 95% certainty that the measured value conforms with a requirement. That is, when it is to be determined whether a measured value is smaller than a requirement, the value of u_e for a one-sided test is added to the measured data and when it is to be determined whether a measured value is larger than a requirement, then the value of u_e is subtracted from the measured data. Standard uncertainties that can be used for situations B and C are provided in Tables 4 and 5 in ISO 12999-1 (2020) for L_n L'_n, $L_{n,w}$ and $L'_{n,w}$. Standard uncertainties for the reduction in impact sound pressure level due to floor coverings are provided in Tables 6 and 7 in ISO 12999-1 (2020).

7.2.5 Recommended Sound and Impact Isolation Values for Apartment and Office Buildings

In previous subsections, we have described how to measure and quantify the sound transmission between spaces in an acoustic test laboratory or in a building. However, it is useful to know what values of the relevant quantities are acceptable. As a guide, the Association of Australasian Acoustical Consultants suggested values and associated building star ratings that are summarised in Table 7.7. Anything less than 4 stars is generally considered unacceptable.

TABLE 7.7 Recommended values for the weighted standardised level difference, $D_{nT,w}$, and the Weighted Standardised Impact Sound Pressure Level, $L'_{nT,w}$, between tenancies in an apartment or office building

Star rating	★★	★★★	★★★★	★★★★★	★★★★★★
$D_{nT,w} + C_{tr}$ greater than	35	40	45	50	55
$L'_{nT,w}$ less than	65	55	50	45	40

7.2.6 Panel Transmission Loss (or Sound Reduction Index) Estimates

It will be instructive to consider the general behaviour of the field incidence transmission loss of a single uniform partition (isotropic panel) over the broad audio-frequency range. An illustration of typical behaviour is shown in Figure 7.7(a), in which various characteristic frequency ranges are indicated.

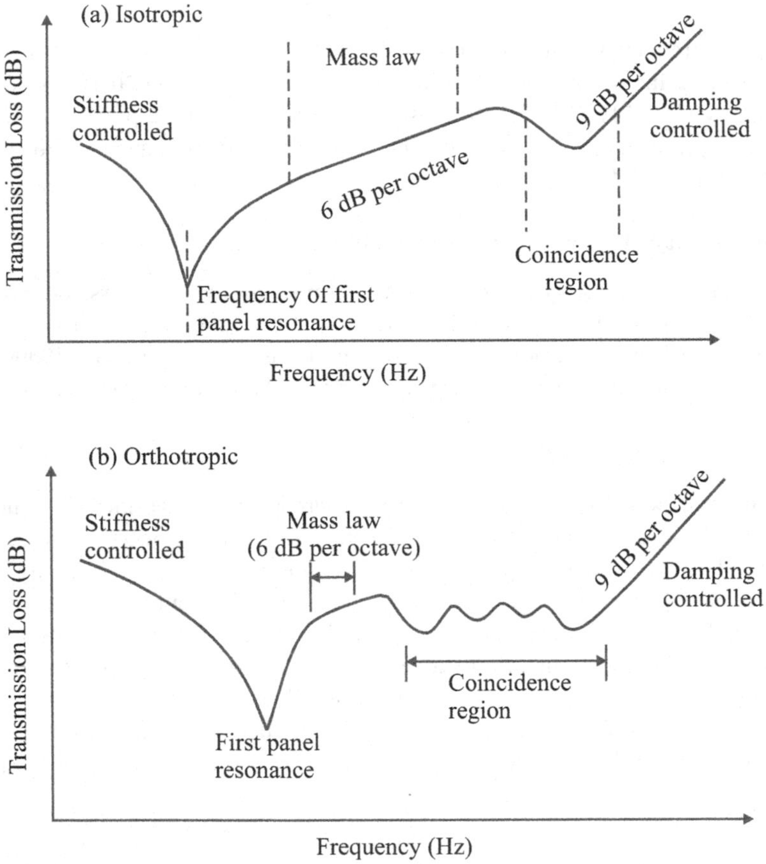

FIGURE 7.7 Typical single panel transmission loss as a function of frequency: (a) isotropic panel characterised by a single critical frequency; (b) orthotropic panel characterised by a critical frequency range.

At low frequencies, the transmission loss is controlled by the stiffness of the panel. At the frequency of the first panel resonance, the transmission of sound is high and, consequently, the transmission loss passes through a minimum determined in part by the damping in the system. Subsequently, at frequencies above the first panel resonance, a generally broad frequency range is encountered, in which transmission loss is controlled by the surface density of the panel. In this frequency range (referred to as the mass law range, due to the approximately linear dependence of the transmission loss on the mass of the panel), the transmission loss increases with frequency at the rate of 6 dB per octave. Ultimately, however, at still higher frequencies in the region of the critical frequency, coincidence is encountered. Finally, at very high frequencies, the transmission loss again rises, being damping controlled, and gradually approaches an extension of the original mass law portion of the curve. The rise in this region is of the order of 9 dB per octave.

The transmission loss of orthotropic panels is strongly affected by the existence of a critical frequency range, rather than a single critical frequency. In this case, the coincidence region may extend over two decades for common corrugated or ribbed panels. Figure 7.7(b) shows a typical transmission loss characteristic of orthotropic panels. This type of panel should be avoided where noise control is important, although it can be shown that damping can improve the performance of the panel slightly, especially at high frequencies.

The resonance frequencies of a simply supported rectangular isotropic panel of width, ℓ_1, length, ℓ_2, and bending stiffness B per unit width may be calculated using the following equation:

$$f_{i,n} = \frac{\pi}{2}\sqrt{\frac{B}{m}}\left[\frac{i^2}{\ell_1^2} + \frac{n^2}{\ell_2^2}\right] \quad \text{(Hz);} \quad i, n = 1, 2, 3, ... \tag{7.45}$$

The lowest order (or fundamental) frequency corresponds to $i = n = 1$. For an isotropic panel, Equation (7.5) can be substituted into Equation (7.45) to give:

$$f_{i,n} = 0.453 c_{LI} h \left[\frac{i^2}{\ell_1^2} + \frac{n^2}{\ell_2^2}\right] \tag{7.46}$$

The resonance frequencies of a simply supported rectangular orthotropic panel of width ℓ_1 and length ℓ_2 are (Hearmon, 1959):

$$f_{i,n} = \frac{\pi}{2m^{1/2}}\left(\frac{B_a i^4}{\ell_1^4} + \frac{B_b n^4}{\ell_2^4} + \frac{2B_{ab} i^2 n^2}{\ell_1^2 \ell_2^2}\right)^{1/2}; \quad i, n = 1, 2, 3, ... \tag{7.47}$$

where m is the surface density (mass per unit area) of the panel (kg/m^2), and

$$B_{ab} = 0.5(B_a \nu + B_b \nu + Gh^3/3) \tag{7.48}$$

In the preceding equations, $G = E/[2(1 + \nu)]$ is the material modulus of rigidity, E is Young's modulus, ν is Poisson's ratio and B_a and B_b are the bending stiffnesses per unit width in directions, ℓ_1 (x-direction in Figure 7.8) and ℓ_2 (y-direction in Figure 7.8), respectively, calculated according to Equations (7.12) and (7.10) respectively.

The following behaviour is especially to be noted. A very stiff construction tends to move the first resonance to higher frequencies but, at the same time, the frequency of coincidence tends to move to lower frequencies. Thus, the extent of the mass law region depends on the stiffness of the panel. For example, steel-reinforced concrete walls of the order of 0.3 m thick exhibit coincidence at about 60 Hz, and this severely limits the transmission loss of such massive walls. On the other hand, a lead curtain wall exhibits coincidence well into the ultrasonic frequency range, and its large internal damping greatly suppresses the first resonance, so that its behaviour is essentially mass-law controlled over the entire audio-frequency range.

The transmission coefficient for a wave incident on a panel surface is a function of the bending wave impedance, Z, which for an infinite isotropic panel is (Cremer, 1942):

$$Z = \mathrm{j}2\pi f m \left[1 - \left(\frac{f}{f_c}\right)^2 (1 + \mathrm{j}\eta)\sin^4\theta\right] \tag{7.49}$$

where η is the panel loss factor (see Equation (6.23) and Appendix C) and m is the panel surface density (kg/m^2). For an infinite orthotropic panel the bending wave impedance is (Hansen, 1993):

$$Z = \mathrm{j}2\pi fm \left[1 - \left[\frac{f}{f_{c1}}\cos^2\vartheta + \frac{f}{f_{c2}}\sin^2\vartheta\right]^2 (1+\mathrm{j}\eta)\sin^4\theta\right] \tag{7.50}$$

where f_{c1} and f_{c2} are, respectively, the lowest and highest critical frequencies of the panel (see Equation (7.3)) and ϑ is the angle of incidence with respect to the axis about which the panel is least stiff to a bending moment (see Figure 7.8). For example, for a corrugated panel, it is with respect to the axis parallel to the corrugations.

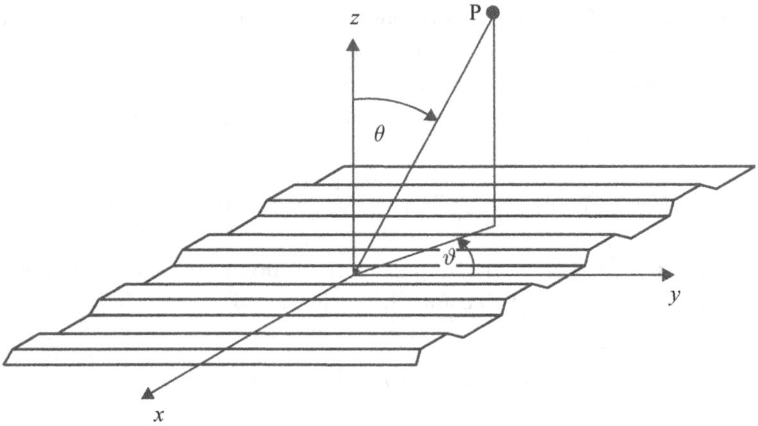

FIGURE 7.8 Geometry of a corrugated panel.

For a panel of infinite extent, the transmission coefficient at an angle $(\theta,\ \vartheta)$ to the normal to the panel surface is given by Cremer (1942) as:

$$\tau(\theta,\vartheta) = \left|1 + \frac{Z\cos\theta}{2\rho c}\right|^{-2} \tag{7.51}$$

The transmission coefficient for normal incidence, τ_N, is found by substituting $\theta = 0$ in Equation (7.51). The diffuse field transmission coefficient, τ_d, is found by determining a weighted average for $\tau(\theta,\ \vartheta)$ over all angles of incidence using the following relationship:

$$\tau_d = \frac{1}{\pi}\int_0^{2\pi} \mathrm{d}\vartheta \int_0^{\pi/2} \tau(\theta,\vartheta)\cos\theta\sin\theta\mathrm{d}\theta \tag{7.52}$$

The $\cos\theta$ term accounts for the projection of the cross-sectional area of a plane wave that is incident upon a unit area of wall at an angle, θ, to the wall normal. The $\sin\theta$ term arises because the width of the annulus between θ and $\theta + \Delta\theta$ is $2\pi\sin\theta\Delta\theta$.

For isotropic panels, Equation (7.52) can be simplified to:

$$\tau_d = \int_0^1 \tau(\theta)\mathrm{d}(\sin^2\theta) \tag{7.53}$$

and for corrugated panels, Equation (7.52) becomes:

$$\tau_d = \frac{2}{\pi}\int_0^{\pi/2} \mathrm{d}\vartheta \int_0^1 \tau(\theta,\vartheta)\mathrm{d}(\sin^2\theta) \tag{7.54}$$

as τ is a function of ϑ as well as θ.

In practice, panels are not of infinite extent and results obtained using the preceding equations do not agree well with results measured in the laboratory. However, it has been shown that good comparisons between prediction and measurement can be obtained if the upper limit of integration of Equation (7.53) is changed so that the integration does not include angles of θ between some limiting angle and 90°. Davy (1990) has shown that this limiting angle θ_L is dependent on the size of the panel as:

$$\theta_L = \cos^{-1}\sqrt{\frac{\lambda}{2\pi\sqrt{A}}} \tag{7.55}$$

where A is the area of the panel and λ is the wavelength of sound at the frequency of interest.

In a later paper, Davy (2009b), with a slight modification according to Davy (2010), suggested that the calculation of the limiting angle should be redefined as:

$$\cos^2\theta_L = \begin{cases} 0.9; & \text{if } \dfrac{\lambda}{8\pi A/P} > 0.9 \\ \dfrac{\lambda}{8\pi A/P}; & \text{if } \dfrac{\lambda}{8\pi A/P} \leq 0.9 \end{cases} \tag{7.56}$$

where A is the panel area and P is its perimeter.

Introducing the limiting angle, θ_L, allows the field incidence transmission coefficient, τ_F, of isotropic panels to be defined as:

$$\tau_F = \int_0^{\sin^2\theta_L} \tau(\theta)\mathrm{d}(\sin^2\theta) \tag{7.57}$$

Hansen (1993) has shown that the same reasoning is valid for orthotropic panels as well, giving:

$$\tau_F = \frac{2}{\pi}\int_0^{\pi/2} \mathrm{d}\vartheta \int_0^{\sin^2\theta_L} \tau(\theta,\vartheta)\mathrm{d}(\sin^2\theta) \tag{7.58}$$

Substituting Equation (7.49) or (7.50) into (7.51), then into (7.57) or (7.58), respectively, and performing the numerical integration allows the field incidence transmission coefficient to be calculated as a function of frequency for any isotropic or orthotropic panel, for frequencies above 1.5 times the first resonance frequency of the panel. At lower frequencies, the infinite panel model used to derive the equations is not valid and a different approach must be used as discussed in Section 7.2.6.1.

Third-octave band results are obtained by arithmetically averaging the τ_F results over a number of frequencies (at least 20) in each band. The field incidence transmission loss can then be calculated by substituting τ_F for τ in Equation (7.13). Results obtained by this procedure generally agree well with measurements made in practice.

To reduce the extent of the numerical calculations, considerable effort has been made by various researchers to simplify the above equations by making various approximations. At frequencies below $f_c/2$ in Equation (7.49), or below $f_{c1}/2$ in Equation (7.50), the quantities in brackets in Equations (7.49) and (7.50) are in each case approximately equal to 1, giving for both isotropic and orthotropic panels:

$$Z = j2\pi fm \tag{7.59}$$

Substituting Equation (7.59) into (7.51) and the result into Equation (7.13) gives the following expression for the mass law transmission loss of an infinite isotropic or orthotropic panel subject to an acoustic wave incident at angle θ to the normal to the panel surface:

$$\mathrm{TL}_\theta = 10\log_{10}\left[1+\left(\frac{\pi f m}{\rho c}\cos\theta\right)^2\right] \tag{7.60}$$

Normal Incidence TL (TL_N) is obtained by substituting $\theta = 0$ in Equation (7.60).

7.2.6.1 Sharp's Prediction Scheme for Isotropic Panels

Sharp (1973) showed that good agreement between prediction and measurement in the mass law range is obtained for single panels by using a constant value for θ_L equal to about 78°. In this case, the field incidence transmission loss, TL, is related to the normal incidence transmission loss, TL_N, for predictions in 1/3-octave bands, for which $\Delta f/f = 0.23156$, by Sharp (1978):

$$\mathrm{TL} = \mathrm{TL}_N - 10\log_{10}\left(1.5+\log_\mathrm{e}\frac{2f}{\Delta f}\right) = \mathrm{TL}_N - 5.6 \qquad \text{(dB)} \tag{7.61}$$

Note that the mass law predictions assume that the panel is limp. As panels become thicker and stiffer, their mass law performance drops below the ideal prediction, so that in practice, very few constructions will perform as well as the mass law prediction.

Substituting Equation (7.60) with $\theta = 0$ into (7.61) and rearranging gives the following for the field incidence transmission loss in the mass-law frequency range below $f_c/2$ for isotropic panels or $f_{c1}/2$ for orthotropic panels:

$$\mathrm{TL} = 10\log_{10}\left[1+\left(\frac{\pi f m}{\rho c}\right)^2\right] - 5.6 \qquad \text{(dB)} \tag{7.62}$$

Many authors (including Sharp (1978)) do not include the "1" in Equation (7.62) as it is much smaller than the second term in brackets for most practical constructions when $f > 200$ Hz. However, it is included in the equations in this book for completeness so that the equations are applicable to all single panel structures.

Equation (7.62) is not valid for frequencies below 1.5 times the first panel resonance frequency, but above this frequency, it agrees reasonably well with measurements taken in 1/3-octave bands. For octave band predictions, for which $\Delta f/f = 0.707$, the 5.6 in Equation (7.62) should be replaced with 4.0. Alternatively, better results are usually obtained for the octave band transmission loss, TL_o, by logarithmically averaging the predictions, TL_1, TL_2 and TL_3, for the three 1/3-octave bands included in each octave band as:

$$\mathrm{TL}_o = -10\log_{10}\frac{1}{3}\left[10^{-\mathrm{TL}_1/10}+10^{-\mathrm{TL}_2/10}+10^{-\mathrm{TL}_3/10}\right] \qquad \text{(dB)} \tag{7.63}$$

For frequencies equal to or higher than the critical frequency, Sharp gives the following equation for an isotropic panel:

$$\mathrm{TL} = 10\log_{10}\left[1+\left(\frac{\pi f m}{\rho c}\right)^2\right] + 10\log_{10}\left[2\eta f/(\pi f_c)\right] \qquad \text{(dB)} \tag{7.64}$$

Equation (7.64) is only used until a frequency is reached at which the calculated TL is equal to that calculated using the mass law expression given by Equation (7.62) (see Figure 7.9(a)). Values for the panel loss factor, η, which appears in the above equation, are listed in Appendix C. Note that the loss factors listed in Appendix C are not solely for the material but include the effects of typical support conditions found in wall structures. The upper limit is applicable to support

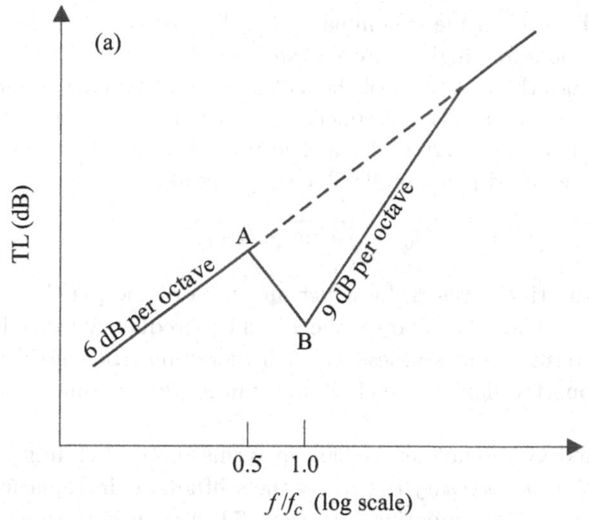

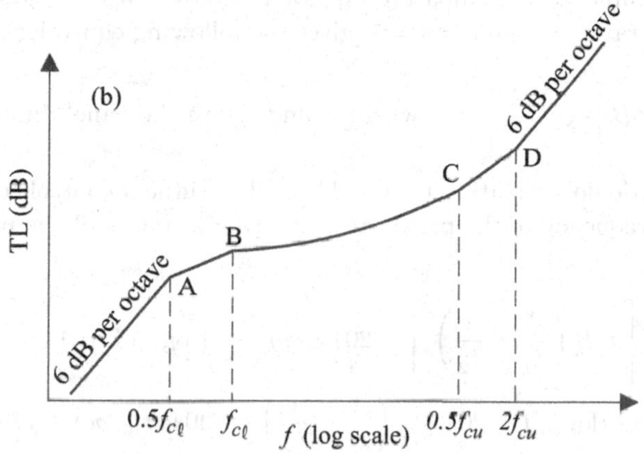

FIGURE 7.9 Design charts for estimating the transmission loss of a single panel. See Appendix C for values of $\rho(= m/h)$ and η for typical materials. See the text for parameter definitions.
(a) A design chart for an isotropic panel.
The points on the chart are calculated as follows:
point A: $\text{TL} = 20\log_{10} f_c m - 54$ (dB)
point B: $\text{TL} = 20\log_{10} f_c m + 10\log_{10}\eta - 44$ (dB)
(b) A design chart for an orthotropic (or ribbed) panel, with critical frequencies f_{c1} and f_{c2}, and small damping. For a well damped panel, see the discussion in the last paragraph of Section 7.2.6.5.
The points in the chart are calculated as follows:
point A: $\text{TL} = 20\log_{10} f_c m - 54$ (dB)
Between and including points B and C:
$\text{TL} = 20\log_{10} f + 10\log_{10} m - 10\log_{10} f_{c\ell} - 20\log_{10}[\log_e(4f/f_{c\ell})] - 13.2$ (dB)
Point D: $\text{TL} = 10\log_{10} m + 15\log_{10} f_{cu} - 5\log_{10} f_{c\ell} - 17$ (dB).
$f_{c\ell}$ is the critical frequency for wave propagation along the stiffest direction of the corrugated panel and f_{cu} is the critical frequency for wave propagation along the least stiff direction of the corrugated panel (see Equation (7.3)).

conditions found in normal building structures, while the lower limit is for the unsupported material or, in the case of metals, for a welded structure.

The transmission loss between $0.5f_c$ and f_c is approximated by connecting with a straight line the points corresponding to $0.5f_c$ and f_c on a graph of TL versus $\log_{10}$ (frequency).

The preceding prediction scheme is summarised in Figure 7.9(a), where a method for estimating the transmission loss for single isotropic panels is illustrated. The lowest valid frequency for this scheme is 1.5 times the frequency of the first panel resonance. Occasionally, it may be of interest to be able to predict the TL at frequencies below this and, for this purpose, we adapt the analysis of Fahy and Gardonio (2007) for a rigid panel on flexible supports. They define the resonance frequency of the rigid panel on flexible supports as:

$$f_0 = \sqrt{s/m} \quad \text{(Hz)} \tag{7.65}$$

where s and m are, respectively, the stiffness per unit area of the panel support, and the mass per unit area of the panel. They then express the TL in the frequency range below the first panel resonance frequency in terms of the stiffness, s. Their model may be considered to be equivalent to that for a simply supported flexible panel vibrating in its first resonant mode (not necessarily at the resonance frequency).

To re-write Fahy and Gardonio's expression in terms of the bending stiffness of a simply supported flexible panel, it is necessary to express the stiffness, s, in Equation (7.65) in terms of the panel bending stiffness, B. Comparing Equation (7.65) with Equation (7.45), evaluated for the first mode of vibration where $i = n = 1$, gives the following equivalence between B and s:

$$s = \pi^4 B \left[\frac{1}{\ell_1^2} + \frac{1}{\ell_2^2}\right]^2; \quad \text{where } \ell_1 \text{ and } \ell_2 \text{ are the panel dimensions.} \tag{7.66}$$

Thus, Fahy and Gardonio's equation for the TL in the stiffness controlled region below half of the first resonance frequency of the panel can be written in terms of the panel bending stiffness and dimensions as:

$$\begin{aligned} \text{TL} &= 20\log_{10}\left[\pi^4 B \left(\frac{1}{\ell_1^2} + \frac{1}{\ell_2^2}\right)^2\right] - 20\log_{10} f - 20\log_{10}(4\pi\rho c) \\ &= 20\log_{10} B - 20\log_{10} f + 20\log_{10}\left(\frac{1}{\ell_1^2} + \frac{1}{\ell_2^2}\right)^2 - 20\log_{10}(\rho c) + 17.8 \quad \text{(dB)} \end{aligned} \tag{7.67}$$

In the vicinity of the lowest panel resonance frequency, provided the loss factor, $\eta \gg \rho c/(2\pi f m)$, the following expression may be used to calculate the TL of the panel:

$$\text{TL} = 20\log_{10} f_{1,1} + 20\log_{10} m + 20\log_{10}\eta - 20\log_{10}(\rho c/\pi) \quad \text{(dB)} \tag{7.68}$$

where $f_{1,1}$ is defined by Equation (7.45) with $i = n = 1$. Equation (7.68) can be used to estimate the panel TL over the frequency range from $0.5 f_{1,1}$ to $1.5 f_{1,1}$. If the loss factor, $\eta \ll \rho c/(2\pi f m)$, then the TL in this frequency range is set equal to 0.

7.2.6.2 Davy's Prediction Scheme for Isotropic Panels

A prediction scheme for the frequency range above $1.5 f_{1,1}$, which is claimed to be more accurate and which allows variation of the limiting angle as a function of frequency to be taken into account according to Equation (7.55), has been proposed by Davy (1990).
In the frequency range below $0.95 f_c$:

$$\text{TL} = 20\log_{10}(a) - 10\log_{10}\left[\log_e\left(\frac{1 + a^2}{1 + a^2\cos^2\theta_L}\right)\right]; \tag{7.69}$$

where $\cos^2\theta_L$ is given by Equation (7.56) and

$$a = \frac{\pi f m}{\rho c} \tag{7.70}$$

In the frequency range above $1.05f_c$:

$$\mathrm{TL} = 20\log_{10}\left(\frac{\pi f m}{\rho c}\right) + 10\log_{10}\left[\left(\frac{2\eta}{\pi}\right)\left(\frac{f}{f_c} - 1\right)\right]; \tag{7.71}$$

In the frequency range around the critical frequency ($0.95f_c < f < 1.05f_c$):

$$\mathrm{TL} = 20\log_{10}\left(\frac{\pi f m}{\rho c}\right) + 10\log_{10}\left[\frac{2\eta\Delta_b}{\pi}\right]; \tag{7.72}$$

where η is the panel loss factor (upper limit of the range in Appendix C) and Δ_b is the ratio of the filter bandwidth to the filter centre frequency used in the measurements. For a 1/3-octave band, $\Delta_b = 0.23156$ and for an octave band, $\Delta_b = 0.7071$.

In the frequency range $1.05f_c < f < 2f_c$, the larger of the two values calculated using Equations (7.71) and (7.72) is used, while in the range $0.5f_c < f < 0.95f_c$, the larger of the two values calculated using Equations (7.69) and (7.72) is used. Note that Equation (7.71) is the same as Equation (7.64) except for the "−1" in the argument of Equation (7.71). Also, Equation (7.72) is the same as (7.64) (with $f = f_c$), except for the Δ_b term in Equation (7.72).

It seems that Equation (7.72) agrees better with experimental results when values for the panel loss factor, η, towards the high end of the expected range are used, whereas Equation (7.64) is in better agreement when small values of η are used. It is often difficult to decide which equation is more nearly correct because of the difficulty in determining a correct value for η. Ranges of η for some materials are given in Appendix C. Most loss factors listed include a contribution from the mounting of the panel. However, in practice, loss factors at the upper end of the range should be used, as the mounting contribution can be quite significant for building structures.

The Davy method for calculating TL values is generally more accurate at low frequencies while the Sharp method gives better results around the critical frequency of the panel.

7.2.6.3 ISO 12354-1 (2017) Prediction Scheme for Isotropic Panels

An alternative method for calculating the R value (equivalent to TL) for single panels is provided by the European standard, ISO 12354-1 (2017).

The field incidence transmission coefficient, of Equation (7.13) for 1/3-octave band centre frequency, f, is defined as:

$$\tau = \begin{cases} \left(\dfrac{\rho c}{\pi f m}\right)^2 \left(2\sigma_f + \dfrac{(\ell_1+\ell_2)^2}{\ell_1^2+\ell_2^2}\sqrt{\dfrac{f_c}{f}}\dfrac{\sigma^2}{\eta_{tot}}\right); & f < f_c \\ \left(\dfrac{\rho c}{\pi f m}\right)^2 \left(\dfrac{\pi\sigma^2}{2\eta_{tot}}\right); & f \approx f_c \\ \left(\dfrac{\rho c}{\pi f m}\right)^2 \left(\dfrac{\pi f_c \sigma^2}{2 f \eta_{tot}}\right); & f > f_c \end{cases} \tag{7.73}$$

where σ is the radiation efficiency for free bending waves, σ_f is the radiation efficiency for forced transmission, ℓ_1 and ℓ_2 are the lengths of the sides of the rectangular wall and η_{tot} is the combined loss factor of the panel material and the loss factor due to the panel mounting (measured with the panel supported on all four sides according to its intended use).

The radiation efficiency for forced (non-resonant) waves (with $\ell_1 \geq \ell_2$) is given by (ISO 12354-1, 2017) as:

$$\sigma_f = 0.5\left[\log_e\left(k\sqrt{\ell_1\ell_2}\right) - \Lambda\right] \tag{7.74}$$

where

$$\Lambda = -0.964 - \left(0.5 + \frac{\ell_2}{\pi\ell_1}\right)\log_e\left(\frac{\ell_2}{\ell_1}\right) + \frac{5\ell_2}{2\pi\ell_1} - \frac{1}{4\pi\ell_1\ell_2k^2} \tag{7.75}$$

If the calculated value of σ_f exceeds 2, it is set equal to 2.

The radiation efficiency, σ_f, can also be calculated using Equation (6.91) (Davy, 2009a). The difference in results between the two methods is usually less than 2 dB.

7.2.6.4 Thickness Correction for Isotropic Panels

When the thickness of the panel exceeds about 1/6 of the bending wavelength, a correction is needed for the high-frequency transmission loss (Ljunggren, 1991). This is in the form of the maximum allowed transmission loss which the prediction result cannot exceed. This is given by:

$$\begin{aligned}\mathrm{TL}_{\mathrm{max}} &= 20\log_{10}\left(\frac{mc_{LI}}{4\rho ch}\right) + 10\log_{10}\eta + 17 \\ &= 20\log_{10}\left(\frac{mc_{LI}}{h}\right) + 10\log_{10}\eta - 47.4 \qquad \text{(dB at 20°C)}\end{aligned} \tag{7.76}$$

and is implemented in the frequency range defined by:

$$f > \sqrt{\frac{B}{h^4 m}} \tag{7.77}$$

where B is the bending stiffness defined in Equation (7.2).

7.2.6.5 Orthotropic Panels

Below half the first critical frequency, the transmission loss may be calculated using Equation (7.61). In the frequency range between the lowest critical frequency and half the highest critical frequency, the following relationship gives reasonably good agreement with experimental results:

$$\tau_F = \frac{\rho c}{2\pi^2 f m}\frac{f_{c\ell}}{f}\left[\log_e\left(\frac{4f}{f_{c\ell}}\right)\right]^2 \tag{7.78}$$

This equation is an approximation to Equation (7.54) in which Equation (7.51) is substituted with $\eta = 0$ and has been derived by Heckl (1960). Equation (7.78) can be rewritten in terms of transmission loss using Equation (7.13) (with $\rho c = 414$) as:

$$\begin{aligned}\mathrm{TL} =& 20\log_{10} f + 10\log_{10} m - 10\log_{10} f_{c\ell} \\ &- 20\log_{10}\left[\log_e\left(\frac{4f}{f_{c\ell}}\right)\right] - 13.2 \text{ (dB)} \qquad f_{c\ell} \le f < 0.5 f_{cu}\end{aligned} \tag{7.79}$$

Above $2f_{cu}$, the TL is given by (Heckl, 1960):

$$\mathrm{TL} = 20\log_{10} f + 10\log_{10} m - 5\log_{10} f_{c\ell} - 5\log_{10} f_{cu} - 23 \qquad \text{(dB)} \tag{7.80}$$

Between $0.5f_{cu}$ and $2f_{cu}$, the TL is estimated by connecting the points $0.5f_{cu}$ and $2f_{cu}$, with a straight line on a graph of TL versus $\log_{10}$ (frequency). Between $f_{c\ell}/2$ and $f_{c\ell}$, the TL is also found in the same way. Note that although Equations (7.78) to (7.80) do not include the limiting angle as was done for isotropic panels, they provide reasonably accurate results and are satisfactory for most commonly used orthotropic building panels. Nevertheless, there are two important points worth noting when using the above prediction schemes for orthotropic panels.

1. Particularly for small panels, the transmission loss below about $0.7f_{c1}$ is underestimated, the error becoming larger as the frequency becomes lower or the panel becomes smaller.

2. For common corrugated panels, there is nearly always a frequency between 2000 and 4000 Hz where there is a dip of up to 5 dB in the measured transmission loss curve, which is not predicted by theory. This corresponds either to an air resonance between the corrugations or one or more mechanical resonances of the panel. Work reported by Windle and Lam (1993) indicates that the air resonance phenomenon does not affect the TL of the panel and that the dips in the measured TL curve correspond to a few resonances in the panel which seem to be more easily excited than others by the incident sound field.

The transmission loss for a single orthotropic panel may be calculated using Figure 7.9(b). If the panel is heavily damped, then the transmission loss will be slightly greater (by about 1 to 4 dB) at higher frequencies, beginning with 1 dB at 500 Hz and increasing to 4 dB at 4000 Hz for a typical corrugated building panel.

7.2.7 Sandwich Panels

In the aerospace industry, sandwich panels are used due to their high stiffness and light weight. Thus, it is of great interest to estimate the transmission loss of such structures. These structures consist of a core of paper honeycomb, aluminium honeycomb or foam. The core is sandwiched between two thin sheets of material commonly called the "laminate", which is usually aluminium on both sides or aluminium on one side and paper on the other. One interesting characteristic of these panels is that in the mid-frequency range it is common for the transmission loss of the aluminium laminate by itself to be greater than the honeycomb structure (Nilsson, 2001). Panels with thicker cores perform better than thinner panels at high frequencies but more poorly in the mid-frequency range. The bending stiffness of the panels is strongly frequency-dependent. However, once a model enabling calculation of the stiffness as a function of frequency has been developed, the methods outlined in Section 7.2.6 may be used to calculate the transmission loss (Nilsson, 2001). Loss factors, η, for these panels when freely suspended are frequency-dependent and are usually in the range 0.01 to 0.03. However, when included in a construction such as a ship's deck, the loss factors are much higher as a result of connection and support conditions and can range from 0.15 at low frequencies to 0.02 at high frequencies (Nilsson, 2001).

7.2.8 Double Wall Transmission Loss

When a high transmission loss structure is required, a double wall or triple wall is less heavy and more cost-effective than a single wall. Design procedures have been developed for both types of wall. However, the present discussion will be focussed mainly on double wall constructions. A single stud construction is illustrated in Figure 7.10. For a more thorough discussion of transmission loss, consideration of triple wall constructions and for some experimental data for wood stud walls, the reader is referred to the published literature (Sharp, 1973, 1978; Brekke, 1981; Bradley and Birta, 2001; Davy, 2010; Davy et al., 2012).

For best results, the two panels of the double wall construction must be both mechanically and acoustically isolated from one another as much as possible. Mechanical isolation may be accomplished by mounting the panels on separate staggered studs or by resiliently mounting the panels on common studs. Resilient mounting can be achieved using rubber strips or flexible steel sections for studs (see Figure 7.12). Acoustic isolation is generally accomplished by providing as wide a gap between the panels as possible and by placing sound-absorbing material in the gap. For best results, the panels should be isotropic.

The Institute for Research in Construction within the Canadian National Research Council has supported an extensive assessment program of measurements of various types of double wall construction, using both wooden and steel studs to separate and support the two wall leaves

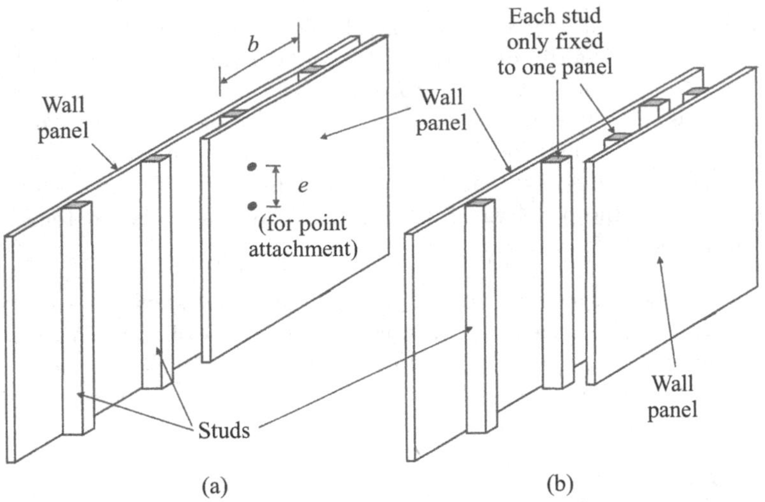

FIGURE 7.10 Double wall constructions. (a) Single row of studs separated by a distance, b. This figure also shows the distance, e, which is part of the discussion later in this section. (b) Staggered stud wall with each stud only attached to one of the two wall panels.

or panels. Some of these results have been reported by Bradley and Birta (2000); Quirt et al. (2008); Quirt and Nightingale (2008); Warnock (2008). The latter report contains equations for estimating the STC value for three different double wall constructions containing type TC steel studs (see Figure 7.12). These equations are listed below.

- Non-load-bearing steel studs or load-bearing steel studs with resilient metal channels and sound-absorbing material almost filling the entire cavity.

$$\text{STC} = 13.5 + 11.4(\log_{10} m_1 + \log_{10} m_2) + 82.6d + 8.5b \qquad \text{(dB)}$$

  where m_1 and m_2 are the masses per unit area of the two panels attached to the studs (kg/m^2), d is the cavity depth (separation distance between the two panels) in metres, and b is the stud spacing in metres.

- Non-load-bearing steel studs or load-bearing steel studs with resilient metal channels and no sound-absorbing material in the cavity.

$$\text{STC} = -18.8 + 17.55(\log_{10} m_1 + \log_{10} m_2) + 165.0d + 10.0b \qquad \text{(dB)}$$

- Staggered steel studs with sound-absorbing material in the cavity.

$$\text{STC} = 17.7 + 14.54(\log_{10} m_1 + \log_{10} m_2) + 23.0d + 27.0t \qquad \text{(dB)}$$

  where t is the insulation thickness in metres.

The flexibility of steel studs reduces the extent of structure-borne sound transmission through a wall construction resulting in higher TL values for walls with steel studs compared to walls with wooden studs (except in the low-frequency range below a few hundred Hz).

In the following subsections, we will outline three different models that are currently being used to estimate the transmission loss (TL) of double walls: the Sharp model, the Davy model and the European standard model. The Davy model is the most recent and perhaps the most accurate. It should be pointed out that these models estimate the transmission loss for the panel construction being considered and do not take into account the transmission of sound from one space to another via "flanking" paths; that is, paths that are not directly through the wall under consideration (see Quirt et al. (2008) and Warnock (2008) for examples of such paths).

7.2.8.1 Sharp Model for Double Wall TL

In Section 7.2.6, it was shown that the transmission loss of a single isotropic panel is determined by two frequencies, namely the lowest order panel resonance frequency, $f_{1,1}$, and the coincidence frequency, f_c. The double wall construction introduces three new important frequencies. The first is the lowest order acoustic resonance, the second is the lowest order structural resonance and the third is a limiting frequency related to the gap between the panels. The lowest order acoustic resonance, f_2, replaces the lowest order panel resonance of the single panel construction (below which the following procedure cannot be used) and may be calculated as:

$$f_2 = c/(2L) \tag{7.81}$$

where c is the speed of sound in air and L is the longest cavity dimension.

The lowest order structural resonance may be approximated by assuming that the two panels are limp masses connected by a massless compliance, which is provided by the air in the gap between the panels.

Introducing the empirical constant, 1.8, the following expression (Fahy, 1985) is obtained for the mass-air-mass resonance frequency, f_0, for panels that are large compared to the width of the gap between them:

$$f_0 = \frac{1}{2\pi}\left(\frac{1.8\rho c^2(m_1+m_2)}{dm_1m_2}\right)^{1/2} \quad \text{(Hz)} \tag{7.82}$$

where m_1 and m_2 are, respectively, the surface densities (kg/m^2) of the two panels and d is the gap width (m). The empirical constant, 1.8, has been introduced by Sharp (1973) to account for the "effective mass" of the panels being less than their actual mass.

Finally, a limiting frequency, f_ℓ, which is related to the gap width d (m) between the panels, is defined as:

$$f_\ell = c/(2\pi d) \simeq 55/d \quad \text{(Hz)} \tag{7.83}$$

The frequencies, f_2, f_0 and f_ℓ, given by Equations (7.81) to (7.83) for the two-panel assembly, are important for determining the transmission behaviour of the double wall. Note that f_ℓ is equal to the lowest cavity resonance frequency for wave propagation in the cavity normal to the plane of the panels, divided by π. The critical frequencies, f_{c1} for panel one and f_{c2} for panel two are also important and are calculated using Equation (7.3).

For double wall constructions, with the two panels completely isolated from one another both mechanically and acoustically, the expected transmission loss is given by the following equations (Sharp, 1978):

$$\text{TL} = \begin{cases} \text{TL}_M; & f \le f_0 \\ \text{TL}_1 + \text{TL}_2 + 20\log_{10}(2kd); & f_0 < f < f_\ell \\ \text{TL}_1 + \text{TL}_2 + 6; & f \ge f_\ell \end{cases} \tag{7.84}$$

where $k = 2\pi f/c$. The quantities, TL_1, TL_2 and TL_M, are calculated by replacing m in Equation (7.64) with the values for the respective panel surface densities, m_1 and m_2, and the total surface density, $M = m_1 + m_2$, respectively.

Equation (7.84) is formulated on the assumption that standing waves in the air gap between the panels are prevented, so that airborne coupling is negligible. To ensure such decoupling, sound-absorbing material is usually placed in the gap. The density of material ought to be chosen to be high enough that the total flow resistance through it is of the order of $3\rho c$ or greater (see Appendix D). When installing a porous material, care should be taken that it does not form a mechanical coupling between the panels of the double wall; thus an upper bound on total flow resistance of $5\rho c$ is suggested or, alternatively, the material can be attached to just one wall without any contact with the other wall. Generally, the sound-absorbing material should be as thick as possible, with a minimum thickness of $15/f$ (m), where f is the lowest frequency of interest.

The transmission loss predicted by Equation (7.84) is difficult to realise in practice. The effect of connecting the panels to supporting studs at points (using spacers), or along lines, is to provide a mechanical bridge for the transmission of structure-borne sound from one panel to the other. Above a certain frequency, called the bridging frequency, such structure-borne conduction limits the transmission loss that can be achieved, to much less than that given by Equation (7.84). Above the bridging frequency, which lies above the structural resonance frequency, f_0, given by Equation (7.82), and below the limiting frequency, f_ℓ, given by Equation (7.83), the transmission loss increases at the rate of 6 dB per octave increase in frequency.

As the nature of the attachment of a panel to its supporting studs determines the efficiency of conduction of structure-borne sound from the panel to the stud and vice versa, it is necessary to distinguish between two possible means of attachment (line and point) and, in the double panel wall under consideration, four possible combinations of the two attachment types. A panel attached directly to a supporting stud generally will make contact along the length of the stud. Such support is called line support and the spacing between studs, b, is assumed regular. Alternatively, the support of a panel on small spacers mounted on the studs is called point support; the spacing, e, between point supports is assumed to form a regular square grid. The dimensions b and e are important in determining transmission loss.

In the following discussion, it is assumed that the two panels are numbered, so that the critical frequency of panel 1 is always less than or at most equal to the critical frequency of panel 2. With this understanding, four combinations of panel attachment are possible as follows: line–line, line–point, point–line and point–point. Of these four possible combinations of panel support, point–line will be excluded from further consideration, as the transmission loss associated with it is always inferior to that obtained with line–point support. In other words, for best results the panel with the higher critical frequency should be point supported if point support of one of the panels is considered.

In the frequency range above the bridging frequency and below about one-half of the critical frequency of panel 2 (the one with the higher critical frequency), the expected transmission loss for the three cases (for adequate sound-absorbing material in the cavity), is as follows (see Figure 7.11).

For line–line support (Sharp, 1973):

$$\mathrm{TL} = 20\log_{10} m_1 + 10\log_{10}(f_{c2}b) + 20\log_{10} f + 20\log_{10}\left(1 + \frac{m_2 f_{c1}^{1/2}}{m_1 f_{c2}^{1/2}}\right) - 72 \quad \text{(dB)} \qquad (7.85)$$

For point–point support:

$$\mathrm{TL} = 20\log_{10} m_1 + 20\log_{10}(f_{c2}e) + 20\log_{10} f + 20\log_{10}\left(1 + \frac{m_2 f_{c1}}{m_1 f_{c2}}\right) - 93 \quad \text{(dB)} \qquad (7.86)$$

For line–point support (panel two point supported):

$$\mathrm{TL} = 20\log_{10} m_1 + 20\log_{10}(f_{c2}e) + 20\log_{10} f + 10\log_{10}[1 + 2X + X^2] - 93 \quad \text{(dB)}$$
$$\text{where } X = \frac{77.7 m_2}{m_1 e\sqrt{f_{c1} f_{c2}}} \qquad (7.87)$$

Based on limited experimental data, Equation (7.85) seems to give very good comparison between prediction and measurement, whereas Equation (7.86) seems to give fair comparison. For line–point support, the term X is generally quite small, so that the term in Equation (7.87) involving it may generally be neglected. Based on limited experimental data, Equation (7.87)

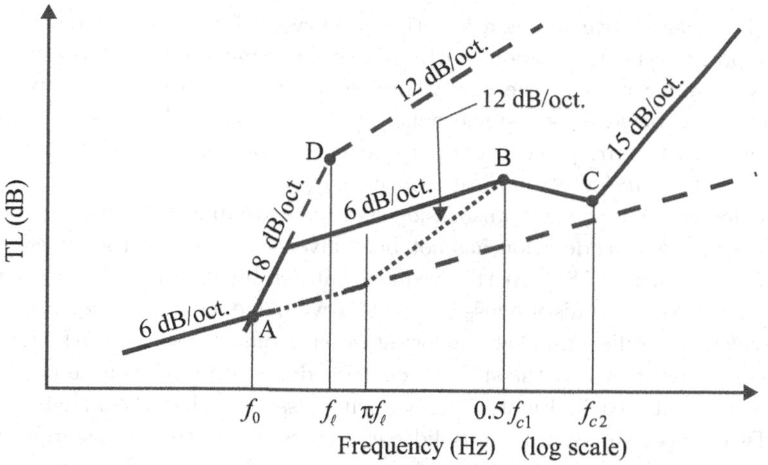

FIGURE 7.11 A design chart for estimating the transmission loss of a double panel wall, based on Sharp (1973). The upper dashed line is used when point B is higher than shown in the figure. The lower dashed line indicates the mass-law TL but is not used in these TL calculations. The panels are assumed to be numbered so that the critical frequency, f_{c1}, of panel 1 is always less than or equal to the critical frequency, f_{c2}, of panel 2, i.e. $f_{c1} \leq f_{c2}$; m_1 and m_2 (kg m^{-2}) are the respective panel surface densities; d (m) is the spacing between panels; b (m) is the spacing between line supports; and e (m) is the spacing of an assumed square grid between point supports. c and c_{LI} (m/s) are, respectively, the speed of sound in air and in the panel material, h is the panel thickness and η_1 and η_2 are the loss factors for panels 1 and 2 respectively.

Calculate the points, A, B, C and D in the chart and then construct a line with a slope of −6 dB/octave from the left side of B to where it intersects the line from A to D.

Point A:

$f_0 = 80.4\sqrt{(m_1 + m_2)/(dm_1m_2)}$ (Hz); $\quad \text{TL}_A = 20\log_{10}(m_1 + m_2) + 20\log_{10} f_0 - 48$ (dB)

Point B: $f_{c1} = 0.55c^2/(c_{LI1}h_1)$ (Hz)

The transmission loss, TL_B, at point B is equal to TL_{B1} if no sound-absorptive material is placed in the cavity between the two panels; otherwise, TL_B is equal to TL_{B2}, provided sufficient absorption is achieved:

$$\text{TL}_{B1} = \text{TL}_A + 20\log_{10}(f_{c1}/f_0) + 20\log_{10}(f_{c1}/f_\ell) - 22 \quad \text{(dB)}$$

(a) Line–line support

$$\text{TL}_{B2} = 20\log_{10} m_1 + 10\log_{10} b + 20\log_{10} f_{c1} + 10\log_{10} f_{c2} + 20\log_{10}\left[1 + \frac{m_2 f_{c1}^{1/2}}{m_1 f_{c2}^{1/2}}\right] - 78 \quad \text{(dB)}$$

(b) Line–point support (f_{c2} is the critical frequency of the point supported panel):

$$\text{TL}_{B2} = 20\log_{10} m_1 e + 20\log_{10} f_{c1} + 20\log_{10} f_{c2} - 99 \quad \text{(dB)}$$

(c) Point–point support:

$$\text{TL}_{B2} = 20\log_{10} m_1 e + 20\log_{10} f_{c1} + 20\log_{10} f_{c2} + 20\log_{10}\left[1 + \frac{m_2 f_{c1}^{1/2}}{m_1 f_{c2}^{1/2}}\right] - 105 \quad \text{(dB)}$$

Point C:

(a) $f_{c2} \neq f_{c1}$; $\quad \text{TL}_C = \text{TL}_B + 6 + 10\log_{10}\eta_2 + 20\log_{10}\left(\frac{f_{c2}}{f_{c1}}\right)$ (dB)

(b) $f_{c2} = f_{c1}$; $\quad \text{TL}_C = \text{TL}_B + 6 + 10\log_{10}\eta_2 + 5\log_{10}\eta_1$ (dB)

Point D: $f_\ell = 55/d$ (Hz)

The final TL curve for sound-absorbing material in the cavity is the solid line in the figure. The dotted line deviation between f_0 and $0.5f_{c1}$ replaces the solid line in this range when there is no sound-absorbing material in the cavity. The lower dashed line represents the mass-law TL that would be achieved at frequencies below $f_c/2$ with no cavity between the panels, where f_c is the critical frequency corresponding to the total panel thickness.

seems to predict greater transmission loss than observed. The observed transmission loss for point–point support seems to be about 2 dB greater than that predicted for line–point support.

If there is no absorption in the cavity, limited experimental data (Sharp, 1973) indicate that the double wall behaves as a single panel of mass equal to the sum of the masses of the individual panels up to a frequency of the first cavity resonance of πf_ℓ. Above this frequency, the TL increases at 12 dB/octave until it reaches $0.5 f_{c1}$.

A method for estimating the transmission loss for a double panel wall is outlined in Figure 7.11. In the figure, consideration has not been given explicitly to the lowest order acoustic resonance, f_2, of Equation (7.81). At this frequency, it can be expected that somewhat less than the predicted mass-law transmission loss will be observed, dependent on the cavity damping that has been provided. In addition, below the lowest order acoustic resonance, the transmission loss will again increase, as shown by the stiffness controlled portion of the curve in Figure 7.7.

The procedure outlined in Figure 7.11 explicitly assumes that the inequality, $Mf > 2\rho c$, is satisfied. Two curves are shown; the solid curve corresponds to the assumption of sufficient sound-absorbing material between the panels to suppress the acoustic resonances in the cavity and prevent acoustic coupling between the panels; the dotted (not dashed) line corresponds to no sound-absorbing material in the cavity and, in Figure 7.11, it is only different to the solid curve in the frequency range between f_0 and 0.5 f_{c1}. Of course, the TL at point B is different for the two cases, but the curves for the two cases are constructed in the same way except for the frequency range between f_0 and $0.5 f_{c1}$. In some cases, such as double glazed window constructions, it is only possible to put sound-absorbing material in the cavity around the perimeter of the construction, so that the transparency of the construction is unaffected. Provided that this material is at least 50 mm thick and consists of fibreglass or rockwool of sufficient density, it will have almost as good an effect as if the material were placed in the cavity between the two panels. However, in these cases, the TL in the frequency range between f_0 and πf_ℓ will be slightly less than predicted.

7.2.8.2 Davy Model for Double Wall TL

The equations outlined in the previous section for a double wall are based on the assumption that the studs connecting the two leaves of the construction are infinitely stiff. This is an acceptable assumption if wooden studs are used but not if metal studs (typically thin-walled channel sections with the partition leaves attached to the two opposite flanges) are used (Davy, 1990).

Davy (1990, 1991, 1993, 1998) presented a method for estimating the transmission loss of a double wall that takes into account the compliance, C_M (reciprocal of the stiffness), of the studs. This prediction procedure is more complicated than the one just discussed and has been considerably modified since its introduction in the early 1990s and earlier editions of this textbook. The latest version is described in Davy et al. (2012, 2019), and this is the version presented here.

In the following analysis, the panels are numbered so that $f_{c1} \leq f_{c2}$. The transmission loss is found by calculating the transmission coefficients, τ_{Fa} and τ_{Fc}, due to airborne and structure-borne sound respectively, and then substituting these values into Equation (7.88):

$$\mathrm{TL} = -10 \log_{10}(\tau_{Fa} + \tau_{Fc}) \qquad \text{(dB)} \tag{7.88}$$

In the most recent Davy model (Davy et al., 2019), the mass-air-mass resonance frequency is calculated in a different way to how it is done in the Sharp model. First, the fundamental resonance frequencies of the panels on either side of the cavity are calculated by treating the leaves of the wall as rectangular plates that are simply supported on two opposite edges (the edges attached to the studs) and free on the other two edges. The expression used for the resonance frequency of the first bending mode for leaf i is (Leissa, 1969, p. 55):

$$f_i = \frac{\pi}{2b^2}\sqrt{\frac{E_i h_i^2}{12\rho_{mi}(1-\nu_i^2)}}, \quad ....i = 1, 2 \tag{7.89}$$

where b is the spacing between the studs and ρ_{mi} is the density of leaf, i.

When the two leaves making up the double wall are identical, the Davy model calculates and uses a modified value, f_0' for the mass-air-mass resonance frequency:

$$f_0' = \sqrt{\frac{\rho c^2(m_1+m_2)}{4\pi^2 d m_1 m_2} + f_1^2} \tag{7.90}$$

When the panels are not identical, the calculation is slightly more complex. In this case:

$$f_0' = \sqrt{-0.5p + 0.5\sqrt{p^2-4q}} \tag{7.91}$$

where $p = -(f_1^2 + f_2^2 + f_{a1}^2 + f_{a2}^2)$ and $q = f_1^2 f_2^2 + f_1^2 f_{a2}^2 + f_2^2 f_{a1}^2$ and

$$f_{ai} = \frac{1}{2\pi}\sqrt{\frac{\rho c^2}{d m_i}} \tag{7.92}$$

Davy et al. (2019) also proposes various multipliers for f_1 and f_2 that allow his theory to fit experimental data. For all thicknesses of the steel studs greater than 0.5 mm, the best multiplier is 1.7 and for steel studs with a thickness less than 0.5 mm, the best multiplier is 0.8 (Davy, 2020). Davy et al. (2019) indirectly measured loss factors for gypsum plaster board that ranged from 0.03 to 0.04. These were the factors that resulted in best agreement with his TL model in the vicinity of the panel critical frequencies.

Below the adjusted mass-air-mass resonance frequency, f_0', the double wall behaves like a single wall of the same mass and the single wall procedures may be used to estimate the TL below and including the frequency, $f = 2f_0'/3$. In this case, the TL is given by Equations (7.69) and (7.70), with the mass, m replaced with $m_1 + m_2$. Between $f = 2f_0'/3$ and f_0', a straight line is drawn on a plot of TL versus $\log_{10} f$ between the TL calculated at $f = 2f_0'/3$ and the TL calculated at $f = f_0'$. The TL at $f = f_0'$ is calculated using Equations (7.88), (7.95) (or (7.93)) and (7.100).

Above f_0', the transmission from one leaf to the other consists of airborne energy through the cavity and structure-borne energy through the studs. For point connections, the structure-borne sound transmission coefficient for all frequencies equal to and above f_0' is (Davy et al., 2012):

$$\tau_{Fc} = \frac{16 n \rho^2 c^4 Q R}{\pi^5 f^2\left[(m_1 f_{c2} + m_2 f_{c1})^2 + 64 f^2 m_1^2 m_2^2 c^4 C_M^2\right]} \tag{7.93}$$

where n is the number of point connections per square metre, C_M is the stud compliance (m/N) and

$$Q = \begin{cases} 1+e & f < f_{c1} \\ e & f \geq f_{c1} \end{cases} \quad \text{and} \quad R = \begin{cases} 1+r & f < f_{c2} \\ r & f \geq f_{c2} \end{cases} \tag{7.94}$$

where $e = \pi f_{c1}\sigma_1/(4f\eta_1)$, $r = \pi f_{c2}\sigma_2/(4f\eta_2)$, f_{c1} and f_{c2} are the critical frequencies of panels 1 and 2 respectively, η_1 and η_2 are the loss factors, σ_1 and σ_2 are the radiation efficiencies, calculated as described in Section 6.8.2, and m_1 and m_2 are the masses per unit area. When calculating panel radiation efficiencies, the perimeter, P is the overall length of the panel perimeter plus twice the length of all the studs. A line of point connections can be treated as a stud. When wall leaves are nailed or screwed to the studs, point connections may be assumed. When the wall leaves are bonded to the studs or bonded plus nailed, then line connections may be assumed.

For line connections, the structure-borne sound transmission coefficient for all frequencies equal to and above f_0' is (Davy et al., 2012):

$$\tau_{Fc} = \frac{8\rho^2 c^3 Q R}{\left[g^2 + \left(8\pi f^{3/2} m_1 m_2\, c\, C_M - g\right)^2\right] b\pi^3 f^2} \tag{7.95}$$

where b is the spacing between the studs, $f_{c1} \leq f_{c2}$, $e = \dfrac{\sigma_1}{2\eta_1}\sqrt{\dfrac{f_{c1}}{f}}$, $r = \dfrac{\sigma_2}{2\eta_2}\sqrt{\dfrac{f_{c2}}{f}}$ and these values are used together with Equation (7.94) to calculate Q and R. Also:

$$g = m_1(f_{c2})^{\frac{1}{2}} + m_2(f_{c1})^{\frac{1}{2}} \tag{7.96}$$

Davy (2020) has found that for wooden studs, the point connection model works best in all cases and the line connection model should not be used. For steel studs where the leaves are screwed to the studs, the line connection model should be used at frequencies for which the screw spacing is less than one-quarter of the leaf bending wavelength and the point connection model should be used at frequencies for which the screw spacing is greater than one-quarter of the leaf bending wavelength. The leaf bending wavelength may be calculated using $\lambda_B = c_B/f$, where c_B is defined in equation (7.1).

Based on all of his and his co-workers work on developing empirical compliance models, Davy (2020) recommends that the most recent model, as described in (Davy et al., 2019), is the one that should be used for all steel studs, even though the model was derived only using steel C-section studs, 92 mm wide. There are two different models provided; one for point connections and one for line connections.

For the line connection model, the compliance, C_M is given by:

$$C_M = Af^{x_f} m_r^{x_m} b^{x_b} g^{x_g} S^{x_S} \tag{7.97}$$

where m_r is defined as:

$$m_r = \frac{m_1 m_2}{m_1 + m_2} \tag{7.98}$$

where m_1 and m_2 are the mass per unit areas of leaves 1 and 2, respectively, b is the stud spacing, g is the gauge of steel used in the stud and S is the area of one side of the wall.

For the point connection model, the compliance, C_M is given by:

$$C_M = Af^{x_f} m_r^{x_m} n^{x_n} g^{x_g} S^{x_S} \tag{7.99}$$

where n is the number of point connections (number of screw connections) per square metre, Values of the constant, A and exponents corresponding to Equations (7.97) and (7.99) are listed in Table 7.8.

TABLE 7.8 Values of the coefficients in Equations (7.97) and (7.99) for TC type studs for two different frequency ranges

Coefficient	Line connection		Point connection	
	63 Hz–250 Hz	250 Hz–5000 Hz	63 Hz–250 Hz	250 Hz–5000 Hz
A	6.07×10^{-4}	2.58×10^{-4}	4.06×10^{-5}	4.94×10^{-7}
x_f	−1.04	−1.52	−0.76	−1.16
x_m	−1.4	−1.12	−1.96	−1.18
x_b	0.0	−0.257	0.0	0.0
x_n	0.0	0.0	0.0	0.747
x_g	0.666	1.52	1.68	2.49
x_S	0.0	0.0	0.0	0.355

Davy et al. (2012) investigated the effect of using different steel stud sections (see Figure 7.12). They concluded that the compliance corresponding to these stud types can be obtained by multiplying the value of C_M for TC type studs by the factors listed in Table 7.9, which indicates that AWS section studs will result in the highest TL values.

FIGURE 7.12 Cross-sections of various steel stud types. The wall panels are attached to the top and bottom of the stud sections shown in the figure.

TABLE 7.9 Factor by which to multiply the TC compliance to obtain the compliance for the other stud sections illustrated in Figure 7.12

Cross-section	AWS	LR	S	O
Multiplying factor	3.39	1.96	0.98	0.73

The field incidence transmission coefficient for airborne sound transmission through a double panel (each leaf of area A), for frequencies between f_0' and $0.9f_{c1}$ (where f_{c1} is the lower of the two critical frequencies corresponding to the two panels), is:

$$\tau_{Fa} = \frac{1-\cos^2\theta_L}{\left[\dfrac{m_2^2+m_1^2}{2m_1m_2} + a_1a_2\bar{\alpha}\cos^2\theta_L\right]\left[\dfrac{m_2^2+m_1^2}{2m_1m_2} + a_1a_2\bar{\alpha}\right]} \tag{7.100}$$

where

$$a_i = \left[\frac{\pi f m_i}{\rho c}\right]; \qquad i = 1,2 \tag{7.101}$$

and the limiting angle, θ_L, is defined in Equation (7.55). Davy (1998) states that the limiting angle should not exceed 80°, but in a later paper (Davy, 2009b), he changed this to 61° and in an even later paper (Davy, 2010), he suggested replacing the equation from (Davy, 2009b) with the following:

$$\cos^2\theta_L = \begin{cases} 0.9; & \text{if } \dfrac{\lambda}{8\pi A/P} > 0.9 \\ \dfrac{\lambda}{8\pi A/P}; & \text{if } \cos^2(61°) \le \dfrac{\lambda}{8\pi A/P} \le 0.9 \\ \cos^2(61°); & \text{if } \dfrac{\lambda}{8\pi A/P} < \cos^2(61°) \end{cases} \tag{7.102}$$

where $\lambda = c/f$ is the wavelength of sound in air, A is the panel area and P is its perimeter.

In Equations (7.100) and (7.101), m_1, m_2 are the surface densities of panels 1 and 2 and $\bar{\alpha}$ is the cavity absorption coefficient, generally taken as the minimum of 1.0 or kd, for a cavity filled with sound-absorbing material, such as fibreglass or rockwool at least 50 mm thick. For cavities containing no sound-absorbing material, a value between 0.1 and 0.15 may be used for $\bar{\alpha}$ (Davy, 1998), but it should not exceed kd.

At frequencies above $0.9f_{c1}$, the following equations may be used to estimate the field incidence transmission coefficient for airborne sound transmission (Davy, 2010):

$$\tau_{Fa} = \frac{\pi(\eta_2\xi_1 + \eta_1\xi_2)}{2\hat{a}_1^2\hat{a}_2^2\eta_1\eta_2(q_1^2+q_2^2)\bar{\alpha}^2} \tag{7.103}$$

$$\hat{a}_i = \frac{\pi f m_i}{\rho c}; \qquad \xi_i = \left(\frac{f}{f_{ci}}\right)^{\frac{1}{2}}; \qquad i = 1,2 \tag{7.104}$$

$$q_1 = \eta_1\xi_2 + \eta_2\xi_1 \tag{7.105}$$

$$q_2 = 2(\xi_1 - \xi_2) \tag{7.106}$$

The quantities η_1 and η_2 are the loss factors of the two panels and f is the 1/3-octave band centre frequency. Equation (7.106) was incorrect in Davy's earlier paper and consequently also in previous editions of this book. However, the corrected version has since been published by Davy (2010) and is reflected here.

The overall transmission coefficient is:

$$\tau_F = \tau_{Fa} + \tau_{Fc}; \qquad f \geq f_0' \tag{7.107}$$

The value of τ_F from Equation (7.107) is then used in Equation (7.13) to calculate the transmission loss (TL). The quantity, f_0', is defined by Equation (7.91).

The calculation of TL using the Davy model is summarised in Table 7.10.

TABLE 7.10 Summary of TL calculations using the Davy model (TL $= -10\log_{10}\tau$)

Frequency range	$\tau = \tau_{Fa} + \tau_{Fc}$	Comments
$1.5f_1 < f \leq 2f_0'/3$	τ is calculated using Equations (7.69), (7.70) and (7.89)–(7.92)	m is the total mass/unit area of the two panels making up the wall
$2f_0'/3 < f < f_0'$	Straight line on a log-log plot between $2f_0'/3$ and f_0'	—
$f_0' < f < 0.9f_{c1}$	Equation (7.93)+Equation (7.100) Equation (7.95)+Equation (7.100)	Point connections Line connections
$0.9f_{c1} < f$	Equation (7.93)+Equation (7.103) Equation (7.95)+Equation (7.103)	Point connections Line connections

7.2.8.3 Model from ISO 12354-1 (2017)

This is a simple model that uses as the basis the model for a single panel wall discussed in Section 7.2.6.3. The increase in STC or R_w as a result of adding the second leaf is given in Table 7.11, where R_w (see Section 7.2.2.2) is the weighted sound reduction index for the leaf

TABLE 7.11 Increase in R_w or STC as a result of converting an existing single panel wall to a double panel wall (ISO 12354-1, 2017). For resonance frequencies below 200 Hz, the minimum value of ΔR_w is 0 dB. Values of ΔR_w for intermediate frequencies can be obtained by linear interpolation on a graph of R_w or TL versus log(frequency)

Mass-air-mass resonance frequency, f_0'', (Hz)	Increase, ΔR_w, in R_w or STC (dB)
≤80	$35 - R_w/2$
100	$32 - R_w/2$
125	$30 - R_w/2$
160	$28 - R_w/2$
200	−1
250	−3
315	−5
400	−7
500	−9
630–1600	−10
>1600	−5

with the greater mass per unit area. It is assumed that the cavity between the two leaves is filled with sound-absorbing material and the mass-air-mass resonance frequency, f_0'', is given by:

$$f_0'' = 53.3 \left(\frac{m_1 + m_2}{d m_1 m_2} \right)^{1/2} \tag{7.108}$$

where d is the panel separation gap in m and m_1 and m_2 are the masses per unit area of the two panels in kg/m^2.

7.2.8.4 Stud Spacing Effect in Walls with Wooden Studs

An important point regarding stud walls with gypsum board leaves is that a stud spacing of between 300 and 400 mm has been shown (Rindel and Hoffmeyer, 1991) to severely degrade the performance of the double wall in the 160 and 200 Hz 1/3-octave bands by up to 13 dB. Other stud spacings (even 100 and 200 mm) do not result in the same performance degradation, although smaller stud spacings improve the low-frequency performance (below 200 Hz) at the expense of a few dB loss at all frequencies between 250 and 2000 Hz.

It is important not to use walls of the same thickness (and material) as this greatly accentuates the dip in the TL curve at the critical frequency. This is also important for double glazing constructions. As an aside, one problem with double glazing is that it can suffer from condensation, so if used, drainage holes are essential.

7.2.8.5 Staggered Studs

A staggered stud arrangement is commonly used to obtain a higher transmission loss. In this arrangement, studs of a common wall are alternately displaced. Panels on opposite sides are then supported on alternate studs. The only common support between opposite panels is at the perimeter of the common wall, for example, at the base and top.

For the purpose of calculating expected transmission loss, the staggered stud construction could be modelled as a perimeter-supported double wall, using the double wall TL models discussed previously for line–line support, with a stud spacing equal to the panel width. However, the introduction of studs improves the structure-borne coupling and degrades the transmission loss which is obtained. However, if care is taken to ensure that at least one of the panels is very well damped, even higher transmission loss may be obtained with staggered studs than obtained without the addition of damping material. Thus, if neither of the two panels making up the double wall construction is well damped, the expected transmission loss for a double wall on staggered studs will lie between that of perimeter mounting and line–line support given by Equation (7.85). Alternatively, if at least one panel is very well damped then the double wall may be modelled as perimeter supported and a slightly higher transmission loss can be expected than predicted.

7.2.8.6 Panel Damping

A simple means for achieving the very high panel damping alluded to above is to construct a thick panel of two thin panels glued together at spots in a regular widely spaced grid. Slight motion of the panels results in shear between the panels in the spaces between attachments, resulting in very effective panel damping due to the shearing action, which dissipates energy in the form of heat. This mechanism can be considered to approximately double the loss factor of the base panels. Alternatively, the panels could be connected together with a layer of visco-elastic material to give a loss factor of about 0.2.

When glass is the material used for the wall or for a window, damping can be increased by using laminated glass, which is a sandwich of two layers of glass separated by a plastic sheet.

sound-absorbing material may also be placed around the perimeter of the cavity between two glass walls to increase acoustic absorption without affecting the transparency of the glass.

7.2.8.7 Effect of the Flow Resistance of the Cavity Sound-Absorbing Material

Although the Sharp theory discussed in the preceding sections does not specify any properties of the sound-absorbing material that is placed in the cavity of a double wall, experimental work has shown that the type of material used is important. Ideally, the material should have a value of $R_1\ell/(\rho c)$ between 2 and 10, where R_1 is the flow resistivity of the material, ℓ is the material thickness and ρ and c are, respectively, the density of air and the speed of sound in air. This should rule out the use of low-density fibreglass (such as insulation batts used in house ceilings), as well as typical polyester blankets. In fact, polyester blankets are likely to be somewhat ineffective.

7.2.8.8 Multi-Leaf and Composite Panels

A multi-leaf panel, for the purposes of the following text, is a panel made up of two or more leaves of the same material, which are connected together in one of three ways: rigid, which is essentially glued very firmly; flexible, which is glued or nailed together at widely separated spots (0.3 to 0.6 m apart); and visco-elastic, which is connected together with visco-elastic material such as silicone rubber (in the form of silastic, for example). For the latter two constructions, the flexibility in the connections between the panels means that they essentially act separately in terms of bending waves propagating through them. Thus, it makes sense to use the lowest critical frequency of the individual leaves for any TL calculations (as the thickness of each leaf does not have to be the same as any other). It is understood that this is an approximation only as one might expect differences in measured TL depending on where the thickest leaf is located amongst the various leaves. Thus, the TL for single and double-panel walls (where each wall may be made up of multiple leaves connected together as described above), is calculated following the procedures outlined previously, with the critical frequency calculated using the thickness and mass of the thickest leaf in each wall and the TL then calculated using this critical frequency together with the total mass per unit area of the entire panel including all the leaves that make it up. The loss factor used in the calculations is that described in Section 7.2.8.6.

When the leaves are connected rigidly together with glue covering the entirety of each leaf, the panel may be considered to act as a single leaf panel of thickness equal to the total thickness of all the leaves and mass equal to the total mass (including glue) of all the leaves.

A composite panel is defined here as a two-leaf panel whose leaves consist of different materials that are bonded rigidly together. The effective stiffness of the panel is calculated using Equation (7.7) and the critical frequency is calculated using Equation (7.9). Then the TL of the single or double panel may be calculated by following the procedures in the previous sections.

It is possible to have a multi-leaf panel made up of composite leaves, where each leaf consists of two layers, each of a different material, bonded rigidly together. In this case, the effective bending stiffness and mass of each leaf are calculated first and the construction is then treated as a multi-leaf construction described above, except that each leaf will be a composite of two rigidly bonded layers.

7.2.8.9 TL Properties of Some Common Stud Wall Constructions

Bradley and Birta (2001) reported on a series of TL measurements made on a range of double walls with wooden studs, staggered studs and resilient steel studs. Based on their measurements they found the following.

- An OITC rating of 25 was measured for a "base" double wall, consisting of a single 13 mm thick gypsum board leaf on the inside screwed to 140 mm deep wooden studs, with an outside leaf consisting of 11 mm thick oriented strand board (OSB) sheathing which was screwed to the other side of the studs.

- A wall consisting of panels rigidly connected to the stud connector is characterised by a low-frequency structural resonance that results in a poor TL value (approaching 10 dB for the "base" wall) at a frequency about double the mass-air-mass resonance frequency for this construction.
- The frequency of the structural resonance mentioned above decreases with increasing stud spacing and increasing gap between the two leaves making up the wall. Changing the stud spacing from 406 mm to 610 mm for the base wall increased the OITC rating from 25 to 31, mainly as a result of shifting the structural resonance to a lower frequency. However, there was no improvement in the 80 Hz and lower 1/3-octave bands.
- The effects of the structural resonance can be negated by using resilient steel studs or staggered studs. Replacing the wooden studs with resilient steel studs increased the OITC rating for the "base" wall to 32.
- The effect of adding mass to either the inside or outside leaf via a second layer is to increase the OITC by approximately $10\log_{10}(m_{tot}/m_{or})$, where m_{tot} is the total wall mass per unit area after the addition of the additional leaf and m_{or} is the original mass per unit area.

7.2.9 Triple Wall Sound Transmission Loss

Very little work has been done in this area, but work reported by Tadeu and Mateus (2001) indicates that for double and triple glazed windows with the same total weight of glazing and total air gap, nothing is gained by using triple glazing over double glazing. However, this is because the cut-on frequency above which 3-D reflections occur in the cavity is above the frequency range of interest for typical panel separations (d m) used in windows (30 to 50 mm). The cut-on frequency is given by the following equation:

$$f_{co} = c/(2d) \tag{7.109}$$

Note that the poorest performance is achieved with panes of glass separated by 10 to 30 mm (Tadeu and Mateus, 2001). Above the cut-on frequency, it is possible to achieve a marked improvement with a triple panel wall (Brekke, 1981).

Sharp (1973) reported that for constructions of the same total mass and total thickness, the double wall construction has better performance for frequencies below $4f_0$, whereas the triple wall construction performs better at frequencies above $4f_0$, where f_0 is the double panel resonance frequency defined by Equation (7.82), using the total distance between the two outer panels as the air gap and the two outer panels as the masses, m_1 and m_2. Below f_0, the two constructions will have the same transmission loss and this will be the same as for a single wall of the same total mass.

For the idealised case of no mechanical connections between panels, the sound transmission loss for a triple panel construction has a similar form to that for a double wall construction, given by Equation (7.84). That is (Sharp, 1973; Long, 2014):

$$\mathrm{TL} = \begin{cases} \mathrm{TL}_M; & f \le f_{01} \\ \mathrm{TL}_1 + \mathrm{TL}_2 + \mathrm{TL}_3 + 20\log_{10}(2kd_1) + 20\log_{10}(2kd_2) & f_{02} \ge f < f_\ell \\ \mathrm{TL}_1 + \mathrm{TL}_2 + \mathrm{TL}_3 + 12; & f \ge f_\ell \end{cases} \tag{7.110}$$

where f_ℓ is given by Equation (7.83). The two mass-air-mass resonance frequencies, f_{01} and f_{02}, are given by (Vinokur, 1996):

$$f_{01} = \frac{1}{2\pi}\sqrt{1.8\rho c^2}\sqrt{a - \sqrt{a^2 - 4q}} \quad \text{(Hz)} \tag{7.111}$$

$$f_{02} = \frac{1}{2\pi}\sqrt{1.8\rho c^2}\sqrt{a + \sqrt{a^2 - 4q}} \quad \text{(Hz)} \tag{7.112}$$

where

$$a = \frac{m_1 + m_2}{m_1 m_2 d_1} + \frac{m_2 + m_3}{m_2 m_3 d_2} \tag{7.113}$$

and

$$q = \frac{(m_1 + m_2 + m_3)}{m_1 m_2 m_3 d_1 d_2} \tag{7.114}$$

where m_1, m_2 and m_3 are the surface densities (kg/m^2) of panels 1, 2 and 3, respectively, d_1 is the gap (m) between panels 1 and 2 and d_2 is the gap (m) between panels 2 and 3.

The TL between f_{01} and f_{02} is found by drawing a straight line on a plot of TL (dB) versus log(frequency), between the TL at f_{01} and the TL at f_{02}. However, it is important to note that the TL calculations of Equation (7.110) are for an idealised situation where there are no mechanical connections between the panels and so the TL so calculated represents a theoretical maximum. This maximum will not be achieved in practice as panels have to be connected in some way for structural integrity, even if it is only at the edges.

7.2.10 Sound-Absorptive Linings

When an enclosure is to be constructed, some advantage will accrue by lining the walls with a porous material. The lining will prevent reverberant sound build-up, which would lessen the effectiveness of the enclosure for noise reduction, and at high frequencies it will increase the transmission loss of the walls. The transmission loss of a porous lining material is discussed in Appendix D. Calculated transmission loss values for a typical blanket of porous material are given in Table 7.12.

7.2.11 Common Building Materials

Some examples of results of transmission loss (field incidence) tests on conventional building materials and structures (including panels, doors and windows) are listed in Table 7.13. More

TABLE 7.12 Calculated transmission loss (TL) values (dB) for fibreglass or rockwool porous acoustic materials (see Appendix D). L, M, H labels in brackets next to the TL values refer to low, mid and high-frequency ranges according to Figure D.4. The air density, $\rho = 1.206$ kg/m^3 and the speed of sound, $c = 343$ m/s. Unfortunately, the TL does not scale with thickness. In most cases the reflection loss from Figure D.6 in Appendix D will have to be added to these values.

Flow resistivity R_1 (Rayls/m)	Bulk density ρ_B (kg/m^3)	Thickness (m)	Octave band centre frequency (Hz)						
			63	125	250	500	1000	2000	4000
4×10^3	20	0.05	1(L)	1(L)	1(L)	1(L)	2(M)	3.5(M)	5.5(M)
2×10^4	48	0.05	3(L)	6.5(L)	8(M)	8(M)	9(M)	11(M)	15(M)
2×10^4	48	0.1	7(L)	8(L)	10(M)	11(M)	14(M)	20(M)	30(H)
5.5×10^4	100	0.05	8(M)	9(M)	10(M)	11(M)	13(M)	15(M)	23(H)

TABLE 7.13 Representative values of airborne sound transmission loss for some common structures and materials

Panel construction	Thick-ness (mm)	Surface weight (kg/m^2)	Octave band centre frequency (Hz)							
			63	125	250	500	1000	2000	4000	8000
Panels of sheet materials										
1.5 mm lead sheet	1.5	17	22	28	32	33	32	32	33	36
3 mm lead sheet	3	34	24	30	31	27	38	44	33	38
20 g aluminium sheet, stiffened	0.9	2.5	8	11	10	10	18	23	25	30
6 mm steel plate	6	50	—	27	35	41	39	39	46	—
22 g galvanised steel sheet	0.55	6	3	8	14	20	23	26	27	35
20 g galvanised steel sheet	0.9	7	3	8	14	20	26	32	38	45
18 g galvanised steel sheet	1.2	10	8	13	20	24	29	33	39	44
16 g galvanised steel sheet	1.6	13	9	14	21	27	32	37	43	42
18 g fluted steel panels stiffened at edges, joints scaled	1.2	39	25	30	20	22	30	28	31	31
Corrugated asbestos sheet, stiffened and sealed	6	10	20	25	30	33	33	38	39	42
Chipboard sheets on wood framework	19	11	14	17	18	25	30	26	32	38
Fibreboard on wood framework	12	4	10	12	16	20	24	30	31	36
Plasterboard sheets on wood framework	9	7	9	15	20	24	29	32	35	38
2 layers 13 mm plaster board	26	22	—	24	29	31	32	30	35	—
Plywood sheets on wood framework	6	3.5	6	9	13	16	21	27	29	33
Plywood sheets on wood framework	12	7	—	10	15	17	19	20	26	—
Hardwood (mahogany) panels	50	25	15	19	23	25	30	37	42	46
Woodwork slabs, unplastered	25	19	0	0	2	6	6	8	8	10
Woodwork slabs, plastered (12 mm on each face)	50	75	18	23	27	30	32	36	39	43

Cont. on next page

Panel construction	Thick-ness (mm)	Surface weight (kg/m^2)	Octave band centre frequency (Hz)							
			63	125	250	500	1000	2000	4000	8000
Panels of sheet materials(Cont.)										
Plywood	6	3.5	—	17	15	20	24	28	27	—
Plywood	9	5	—	7	13	19	25	19	22	—
Plywood	18	10	—	24	22	27	28	25	27	—
Lead vinyl curtains	3	7.3	—	22	23	25	31	35	42	—
Lead vinyl curtains	2	4.9	—	15	19	21	28	33	37	—
Panels of sandwich construction										
Machine enclosure panels										
16 g steel + damping with 100 mm of glass-fibre	100	25	20	21	27	38	48	58	67	66
As above, but covered by 22 g perforated steel	100	31	25	27	31	41	51	60	65	66
As above, but 16 g steel replaced with 5 mm steel plate	100	50	31	34	35	44	54	63	62	68
1.5 mm lead between two sheets of 5 mm plywood	11.5	25	19	26	30	34	38	42	44	47
9 mm asbestos board between two sheets of 18 g steel	12	37	16	22	27	31	27	37	44	48
Compressed straw between two sheets of 3 mm hardboard	56	25	15	22	23	27	27	35	35	38
Single masonry walls										
Single leaf brick, plastered on both sides	125	240	30	36	37	40	46	54	57	59
Single leaf brick, plastered on both sides	255	480	34	41	45	48	56	65	69	72
Single leaf brick, plastered on both sides	360	720	36	44	43	49	57	66	70	72
Solid breeze or clinker, plastered (12 mm both sides)	125	145	20	27	33	40	50	58	56	59

Panel construction	Thickness (mm)	Surface weight (kg/m^2)	Octave band centre frequency (Hz)							
			63	125	250	500	1000	2000	4000	8000
Single masonry walls (Cont.)										
Solid breeze or clinker blocks, unplastered	75	85	12	17	18	20	24	30	38	41
Hollow cinder concrete blocks, painted (cement base paint)	100	75	22	30	34	40	50	50	52	53
Hollow cinder concrete blocks, unpainted	100	75	22	27	32	32	40	41	45	48
Thermalite blocks	100	125	20	27	31	39	45	53	38	62
Glass bricks	200	510	25	30	35	40	49	49	43	45
Plain brick	100	200	—	30	36	37	37	37	43	—
Aerated concrete blocks	100	50	—	34	35	30	37	45	50	—
Aerated concrete blocks	150	75	—	31	35	37	44	50	55	—
Double masonry walls										
280 mm brick, 56 mm cavity, strip ties, outer faces plastered to thickness of 12 mm	300	380	28	34	34	40	56	73	76	78
280 mm brick, 56 mm cavity, expanded metal ties, outer faces plastered to thickness of 12 mm	300	380	27	27	43	55	66	77	85	85
Stud partitions										
50 mm × 100 mm studs, 12 mm insulating board both sides	125	19	12	16	22	28	38	50	52	55
50 mm × 100 mm studs, 9 mm plasterboard and 12 mm plaster coat both sides	142	60	20	25	28	34	47	39	50	56

Cont. on next page

Panel construction	Thickness (mm)	Surface weight (kg/m^2)	Octave band centre frequency (Hz)							
			63	125	250	500	1000	2000	4000	8000
Stud partitions (Cont.)										
Gypsum wall with steel studs and 16 mm-thick panels each side										
Empty cavity, 45 mm wide	75	26	—	20	28	36	41	40	47	—
Cavity, 45 mm wide, filled with fibreglass	75	30	—	27	39	46	43	47	52	—
Empty cavity, 86 mm wide	117	26	—	19	30	39	44	40	43	—
Cavity, 86 mm wide, filled with fibreglass	117	30	—	28	41	48	49	47	52	—
Gypsum wall, 16 mm leaves, 200 mm cavity with no sound-absorbing material and no studs	240	23	—	33	39	50	64	51	59	—
As above with 88 mm sound-absorbing material	240	26	—	42	56	68	74	70	73	—
As above but staggered 4-inch studs	240	30	—	35	50	55	62	62	68	—
Gypsum wall, 16 mm leaves, 100 mm cavity, 56 mm thick sound-absorbing material, single 4-inch studs with resilient metal channels on one side to attach the panel to the studs	140	28	—	25	40	48	52	47	52	—
Single glazed windows										
Single glass in heavy frame	4	10	—	20	22	28	34	29	28	—
Single glass in heavy frame	6	15	17	11	24	28	32	27	35	39
Single glass in heavy frame	8	20	18	18	25	31	32	28	36	39
Single glass in heavy frame	9	22.5	18	22	26	31	30	32	39	43
Single glass in heavy frame	16	40	20	25	28	33	30	38	45	48
Single glass in heavy frame	25	62.5	25	27	31	30	33	43	48	53
Laminated glass	13	32	—	23	31	38	40	47	52	57

Panel construction	Thick-ness (mm)	Surface weight (kg/m^2)	Octave band centre frequency (Hz)							
			63	125	250	500	1000	2000	4000	8000
Doubled glazed windows										
2.44 mm panes, 7 mm cavity	12	15	15	22	16	20	29	31	27	30
9 mm panes in separate frames, 50 mm cavity	62	34	18	25	29	34	41	45	53	50
6 mm glass panes in separate frames, 100 mm cavity	112	34	20	28	30	38	45	45	53	50
6 mm glass panes in separate frames, 188 mm cavity	200	34	25	30	35	41	48	50	56	56
6 mm glass panes in separate frames, 188 mm cavity with absorbent blanket in reveals	200	34	26	33	39	42	48	50	57	60
6 mm and 9 mm panes in separate frames, 200 mm cavity, absorbent blanket in reveals	215	42	27	36	45	58	59	55	66	70
3 mm plate glass, 55 mm cavity	63	25	—	13	25	35	44	49	43	—
6 mm plate glass, 55 mm cavity	70	35	—	27	32	36	43	38	51	—
6 mm and 5 mm glass, 100 mm cavity	112	34	—	27	37	45	56	56	60	—
6 mm and 8 mm glass, 100 mm cavity	115	40	—	35	47	53	55	50	55	—
Doors										
Flush panel, hollow core, normal cracks as usually hung	43	9	1	12	13	14	16	18	24	26
Solid hardwood, normal cracks as usually hung	43	28	13	17	21	26	29	31	34	32
Typical proprietary "acoustic" door, double heavy sheet steel skin, absorbent in air space, and seals in heavy steel frame	100	—	37	36	39	44	49	54	57	60
2-skin metal door	35	16	—	26	26	28	32	32	40	—
Plastic laminated flush wood door	44	20	—	14	18	17	23	18	19	—
Veneered surface, flush wood door	44	25	—	22	26	29	26	26	32	—
Metal door; damped skins, absorbent core, gasketing	100	94	—	43	47	51	54	52	50	—

Cont. on next page

Panel construction	Thickness (mm)	Surface weight (kg/m²)	Octave band centre frequency (Hz)							
			63	125	250	500	1000	2000	4000	8000
Doors (Cont.)										
Metal door; damped skins, absorbent core, gasketing	180	140	—	46	51	59	62	65	62	—
Metal door; damped skins, absorbent core, gasketing	250	181	—	48	54	62	68	66	74	—
Two 16 g steel doors with 25 mm sound-absorbing material on each, and separated by 180 mm air gap	270	86	—	50	56	59	67	60	70	—
Hardwood door	54	20	—	20	25	22	27	31	35	—
Hardwood door	66	44	—	24	26	33	38	41	46	—
Floors										
T & G boards, joints scaled	21	13	17	21	18	22	24	30	33	63
T & G boards, 12 mm plasterboard ceiling under, with 3 mm plaster skin coat	235	31	15	18	25	37	39	45	45	48
As above, with boards "floating" on glass-wool mat	240	35	20	25	33	38	45	56	61	64
Concrete, reinforced	100	230	32	37	36	45	52	59	62	63
Concrete, reinforced	200	460	36	42	41	50	57	60	65	70
Concrete, reinforced	300	690	37	40	45	52	59	63	67	72
126 mm reinforced concrete with "floating" screed	190	420	35	38	43	48	54	61	63	67
200 mm concrete slabs	200	280	—	34	39	46	53	59	64	65
As above, but oak surface	212	282	—	34	41	46	55	64	70	—
As above, but carpet + hair felt underlay, no oak surface	200	281	—	34	36	46	55	66	72	—
Gypsum ceiling, mounted resiliently, and vinyl finished wood joist floor with glass-fibre insulation and 75 mm plywood	318	—	—	30	36	45	52	47	65	—

comprehensive lists have been published in various handbooks, and manufacturers generally supply data for their products. Where possible, manufacturer's data should be used; otherwise, the methods outlined in Section 7.2.6 or values in Table 7.13 may be used.

7.3 Noise Reduction versus Transmission Loss

When a partition is placed between two rooms and one room contains a noise source that affects the other room (receiver room), the difference in sound pressure level between the two rooms is related to the TL of the partition by Equation (7.16). When there are paths, other than through the partition, for the sound to travel from one room to the other, the effective transmission loss of the panel in terms of the sound reduction from one room to the other will be affected. These alternative transmission paths could be through doors, windows or suspended ceilings. If the door or wall forms part of the partition, then the procedures for calculating the effective transmission loss are discussed in Section 7.3.1. If the transmission path is around the wall, then the effective transmission loss of this path needs to be calculated or measured in the laboratory according to such standards as ISO 10140-4 (2021) or ISO 10848-1 (2017) and normalised to the area of the wall (which is done automatically if the procedures in the standards are followed). In this case, the effective transmission loss of the partition, including the flanking paths, is calculated as described in Section 7.3.2.

7.3.1 Combined Transmission Loss

The wall of an enclosure may consist of several elements, each of which may be characterised by a different transmission loss coefficient. For example, the wall may be constructed of panels of different materials, it may include permanent openings for passing materials or cooling air in and out of the enclosure, and it may include windows for inspection and doors for access. Each such element must be considered in turn, in the design of an enclosure wall, and the transmission loss of the wall determined as an overall area weighted average of all of the elements. The composite transmission coefficient is calculated as:

$$\tau = \frac{\sum_{i=1}^{q} S_i \tau_i}{\sum_{i=1}^{q} S_i} \tag{7.115}$$

where S_i is the surface area (one side only), and τ_i is the transmission coefficient of the ith element. The transmission coefficient of any element may be determined, given its transmission loss, TL, as:

$$\tau = 10^{(-\mathrm{TL}/10)} \tag{7.116}$$

The overall transmission coefficient is then calculated using Equation (7.115) and, finally, the overall transmission loss is calculated using Equation (7.13).

If a wall or partition consists of just two elements, then Figure 7.13 is useful. The figure shows the transmission loss increment to be added to the lesser transmission loss of the two elements to obtain an estimate of the overall transmission loss of the two-element structure. The transmission loss increment, δTL, is plotted as a function of the ratio of the area of the lower transmission loss element divided by the area of the higher transmission loss element with the difference, ΔTL, between the transmission losses of the two elements as a parameter.

It can be seen from Figure 7.13 that low transmission loss elements within an otherwise very high transmission loss wall can seriously degrade the performance of the wall. Thus, the transmission loss of any penetrations, such as doors, windows, ductwork or electrical and plumbing

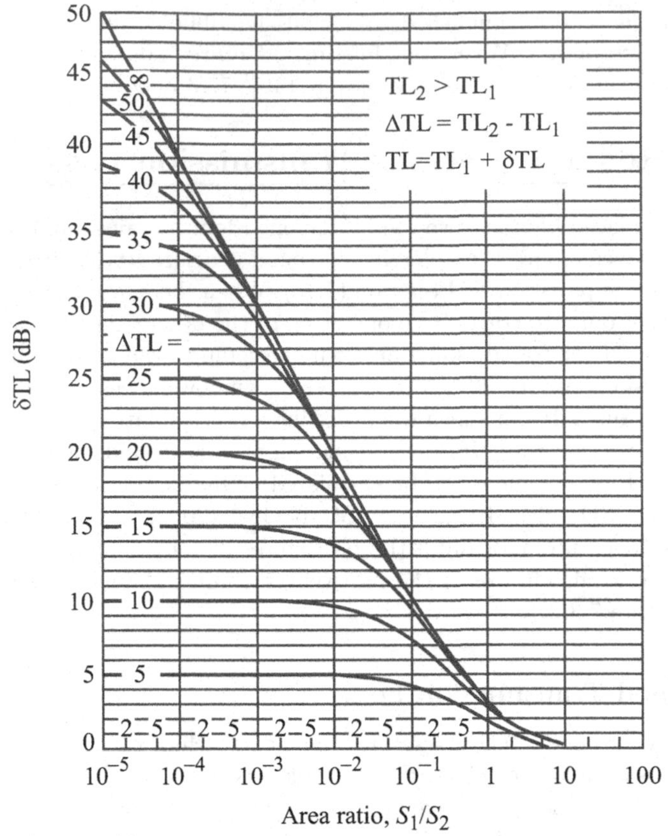

FIGURE 7.13 Scheme for estimation of the transmission loss of a two-element structure.

services, must be commensurate with the transmission loss of the whole structure. In practice, this generally means that the transmission loss of such things as doors, windows and access and ventilation openings should be kept as high as possible, and their surface areas as small as possible.

7.3.2 Flanking Transmission

The effective transmission loss of a partition, including the effects of flanking transmission, is given by (ISO 12354-1, 2017):

$$\text{TL}_{\text{overall}} = -10\log_{10}\left(10^{-\text{TL}_{\text{flank}}/10} + 10^{-\text{TL}/10}\right) \qquad \text{(dB)} \tag{7.117}$$

where TL_{flank} is the combined effective TL of all the flanking paths normalised to the area of the partition and TL is the transmission loss of the partition. When measuring the flanking path effects, the following quantity is reported:

$$D_{n,f} = L_{p1} - L_{p2} - 10\log_{10}\left(\frac{S\bar{\alpha}}{10}\right) \tag{7.118}$$

where $S\bar{\alpha}$ is the absorption area of the receiving test chamber and L_{p1} and L_{p2} are the sound pressure levels in the source and receiving rooms, respectively, with the receiver level due only to the flanking path or paths. An example of a flanking path into a room, which would reduce the effective TL of a partition, is a suspended ceiling that connects the two rooms separated by the partition. The Flanking path, $D_{n,f}$ would be measured separately in a test facility.

The flanking transmission loss, TL_{flank}, or Flanking Sound Reduction Index, R_{flank}, is calculated from the normalised sound pressure level difference quantity, $D_{n,f}$, measured in the test facility as:

$$\text{TL}_{\text{flank}} = D_{n,f} - 10\log_{10}\left(\frac{10}{A}\right) = L_{p1} - L_{p2} + 10\log_{10}\left(\frac{A}{S\bar{\alpha}}\right) \tag{7.119}$$

which is the same as Equation (7.17), where A is the area of the field mounted partition, which is not necessarily the same as the area of the partition used in the test facility. If the field situation flanking condition exactly matches the laboratory situation, then the result of Equation (7.119) may be used with Equation (7.117) to estimate an effective TL for use in a field installation. However, if the field situation is different to the laboratory configuration, then some adjustment needs to be made to Equation (7.119).

One example given in ISO 12354-1 (2017) is for a suspended ceiling where the ceiling dimensions for the installation are different to those used in the laboratory measurement. In this case, the quantity $D_{n,f}$ in Equation (7.119) is replaced by $D_{n,s}$, defined by:

$$D_{n,s} = D_{n,f} + 10\log_{10}\left(\frac{h_{pl}\ell_{pl}}{h_{\text{lab}}\ell_{\text{lab}}}\right) + 10\log_{10}\left(\frac{S_{cs,\text{lab}}S_{cr,\text{lab}}}{S_{cs}S_{cr}}\right) + C_\alpha \tag{7.120}$$

where h_{pl} and h_{lab} are the heights of the space above the suspended ceiling in the actual installation and in the laboratory, respectively, ℓ_{pl} and ℓ_{lab} are the thicknesses of the partition where it connects to the suspended ceiling in the actual installation and in the laboratory, respectively, S_{cs} and $S_{cs,\text{lab}}$ are the areas of the suspended ceiling in the source room in the actual installation and in the laboratory, respectively, S_{cr} and $S_{cr,\text{lab}}$ are the areas of the suspended ceiling in the receiver room in the actual installation and in the laboratory, respectively, and C_α is defined as follows. For no absorption in the space above the suspended ceiling, or if sound-absorbing material exists and the condition, $f \leq 0.015c/t_a$ (where t_a is the thickness of the absorbing material) is satisfied, then $C_\alpha = 0$. For absorption in this space, where the preceding condition is not satisfied:

$$C_\alpha = \begin{cases} 10\log_{10}\left[\dfrac{h_{\text{lab}}}{h_{pl}}\sqrt{\dfrac{S_{cs}S_{cr}}{S_{cs,\text{lab}}S_{cr,\text{lab}}}}\right]; & 0.015\dfrac{c}{t_a} < f < \dfrac{0.3c}{\min(h_{\text{lab}}, h_{pl})} \\[2ex] 10\log_{10}\left[\left(\dfrac{h_{\text{lab}}}{h_{pl}}\right)^2\sqrt{\dfrac{S_{cs}S_{cr}}{S_{cs,\text{lab}}S_{cr,\text{lab}}}}\right]; & f \geq \dfrac{0.3c}{\min(h_{\text{lab}}, h_{pl})} \end{cases} \tag{7.121}$$

Note that for an ISO test facility, $S_{cs,\text{lab}} = S_{cr,\text{lab}} = 20$ m^2 and $h_{\text{lab}} = 0.7$ m.

7.4 Enclosures

7.4.1 Noise Inside Enclosures

The use of an enclosure around a noise source will produce a reverberant sound field within it, in addition to the existing direct sound field of the source. Both the reverberant and direct fields will contribute to the sound radiated by the enclosure walls as well as to the sound field within the enclosure.

Equation (6.44) of Chapter 6 may be used to estimate the sound pressure level at any location within the enclosure, but with the restriction that the accuracy of the calculation will be impaired if the location considered is less than one-half of a wavelength from the enclosure or machine surfaces.

7.4.2 Noise Outside Enclosures

The sound field immediately outside of an enclosure consists of two components. One component is due to the internal reverberant field and the other is due to the direct field of the source. The transmission coefficient corresponding to the reverberant field is τ, while that corresponding to the direct field is τ_θ, where θ is the angle that the line, joining the source in the enclosure to the receiver location, makes with the normal to the enclosure wall through which it passes. The corresponding transmission losses TL and TL_θ were discussed on page 424. The expression for the total radiated sound power is:

$$W_t = S_E \langle p_1^2 \rangle / (\rho c) = W\tau_\theta + W(1-\bar{\alpha}_i)[S_E/(S_i\bar{\alpha}_i)]\tau \tag{7.122}$$

In Equation (7.122), the subscript i indicates quantities interior to the enclosure and the subscript, 1, indicates a quantity adjacent to the outside surface of the enclosure. In writing Equation (7.122), the external radiated sound power, given by the integral over the external surface, S_E, of the sound intensity, $\langle p^2 \rangle/\rho c$ (see Equation (1.78)), has been set equal to the fraction of the source sound power, W, transmitted by the direct field, plus the fraction transmitted by the reverberant field. In the latter case, the power contributed by the reverberant field is determined by use of Equations (6.35) and (6.43). The quantity, S_i, is the enclosure internal surface area, including any machine surfaces.

Consider the transmission of the direct sound field, and suppose that the line joining the sound source to the receiver location is normal to the wall through which it passes, with a corresponding transmission coefficient, τ_θ, which can be estimated by making use of Equation (7.60). From Equation (7.61), $\text{TL}_N - \text{TL} = \text{TL}_{(\theta=0)} - \text{TL} = 5.6$ dB. Use of Equation (7.13) leads to the conclusion, based on the above result, that $\tau_N = \tau 10^{-5.6/10} = 0.28\tau$. However, from Equation (7.60), $\tau_\theta \approx \tau_N / \cos^2\theta$. The mean square sound pressure adjacent to the outside of the enclosure can be written using Equation (4.14), with the surface area $4\pi r^2$ replaced by the enclosure outer surface area, S_E. Thus:

$$S_E \langle p_1^2 \rangle / (\rho c) = W\tau_E \tag{7.123}$$

where $W\tau_E$ is the sound power radiated by the enclosure walls and roof, and:

$$\tau_E = \tau[K + S_E(1-\bar{\alpha}_i)/(S_i\bar{\alpha}_i)] \tag{7.124}$$

where

$$K \approx \tau_N/(\tau\cos^2\theta) = (10^{-5.6/10})/\cos^2\theta \tag{7.125}$$

For angles of $\theta < 20°$, $K \approx 0.3$. Values of K greater than 2 may not be realistic and could overestimate the transmission loss of the direct field component, especially at frequencies in the vicinity of and above the panel critical frequency.

Taking logarithms to the base ten of both sides of Equation (7.123) and assuming $\rho c = 400$ gives:

$$L_{p1} = L_W - \text{TL} - 10\log_{10} S_E + C \tag{7.126}$$

where TL is the field incidence transmission loss and L_{p1} is the average sound pressure level (see Equation (4.142)) immediately outside of the enclosure. Values of the coefficient, C, may be calculated using:

$$C = 10\log_{10}[K + S_E(1-\bar{\alpha}_i)/(S_i\bar{\alpha}_i)] \quad \text{(dB)} \tag{7.127}$$

If the enclosure is located outdoors, the following expression gives a reasonable approximation to the sound pressure level, L_{p2}, to be expected at a point some distance, r, from the enclosure:

$$L_{p2} = L_{p1} + 10\log_{10} S_E + 10\log_{10}(D_\theta/4\pi r^2) \quad \text{(dB)} \tag{7.128}$$

In Equation (7.126), the distance from the enclosure to the measurement position, r (m), is assumed to be large compared with the relevant enclosure wall area, S_w ($r > 3S_w$), and D_θ is the directivity factor for the enclosure. Normally, for an enclosure on a hard floor, $D_\theta = 2$ (see Table 4.1). The quantity, $L_{p1} + 10\log_{10} S_E$, represents the sound power radiated by the external surface of the enclosure.

If the enclosure is located indoors in a room, then the reverberant sound field due to the enclosing room must be considered. In this case, the sound pressure level, L_{p2}, at a position in the room is derived using the results of Chapters 5 and 6, as follows. The sound power, W_t, radiated by the enclosure into the room can be written as:

$$W_t = \langle p_1^2 \rangle [S_E/(\rho c)] \quad \text{(W)} \tag{7.129}$$

or

$$L_{Wt} = L_{p1} + 10\log_{10} S_E \quad \text{(dB)} \tag{7.130}$$

where S_E is the total external surface area of the enclosure and the assumption has been made that $\rho c \approx 400$. In this case, use of Equation (6.44) gives:

$$L_{p2} = L_{Wt} + 10\log_{10}\left[\frac{D_\theta}{4\pi r^2} + \frac{4(1-\bar{\alpha})}{S\bar{\alpha}}\right] \quad \text{(dB)} \tag{7.131}$$

where $\bar{\alpha}$ is the average Sabine absorption coefficient (see Equation (6.86)) and S is the total area of all room surfaces.

Substituting Equation (7.130) into Equation (7.131) gives the following for L_{p2}:

$$L_{p2} = L_{p1} + 10\log_{10} S_E + 10\log_{10}\left[\frac{D_\theta}{4\pi r^2} + \frac{4(1-\bar{\alpha})}{S\bar{\alpha}}\right] \quad \text{(dB)} \tag{7.132}$$

The noise reduction due to the enclosure may now be calculated. The sound pressure level at a position in the room with no enclosure is obtained using Equation (6.44) as:

$$L'_{p2} = L_W + 10\log_{10}\left[\frac{D_\theta}{4\pi r^2} + \frac{4(1-\bar{\alpha})}{S\bar{\alpha}}\right] \quad \text{(dB)} \tag{7.133}$$

Assuming that the enclosure has not altered the directivity characteristics of the noise source in the enclosure, the noise reduction is given by $\text{NR} = L'_{p2} - L_{p2}$. Thus:

$$\text{NR} = L_W - L_{p1} - 10\log_{10} S_E \quad \text{(dB)} \tag{7.134}$$

Substituting Equation (7.126) for L_{p1} into Equation (7.134) provides the following expression for the noise reduction, which holds for an enclosure located outdoors or inside a building:

$$\text{NR} = \text{TL} - C \quad \text{(dB)} \tag{7.135}$$

The quantity C may be determined using Equation (7.127).

The previous analysis assumes that the TL of the building is the same for all four walls and roof and that the receiver is in a direction normal to the centre of one wall or roof. Most situations are considerably more complicated but, nevertheless, are still amenable to logical analysis. To account for different transmission coefficients for different wall and roof sections that contribute to noise at a particular receiver, an overall average field incidence transmission coefficient, τ_a, and corresponding transmission loss, TL_a, must be calculated that takes into account the different transmission coefficients of the different wall and roof components as well as the directivity of sound radiation from each wall or roof that contributes to the sound pressure level at the receiver. The average value of TL is calculated for each wall (usually four) and roof using the transmission coefficients and areas of the different sections (such as doors, windows and wall parts) making up the wall, using Equation (7.115). The average TL values for each wall that contributes to the sound pressure level at the receiver must then be modified using the directivity properties of an incoherent panel source to account for a receiver not being located in a direction normal to the centre of the wall. The new overall TL value (TL_a) for all walls contributing to the sound pressure level at the receiver and to be used in Equation (7.135) in place of TL, can be calculated by first adjusting each wall TL value to account for the direction of the receiver with respect to the normal to the centre of the wall and then combining the contributions from each wall or roof that contributes to the sound pressure level at the receiver, to obtain TL_a. Thus:

$$\text{TL}_a = -10\log_{10}\left[\frac{\sum_{i=1}^{N}\left[\dfrac{\langle p_i^2\rangle}{\langle p_{n,i}^2\rangle}\times S_i\times 10^{-\text{TL}_i/10}\right]}{\sum_{i=1}^{N}\left[\dfrac{\langle p_i^2\rangle}{\langle p_{n,i}^2\rangle}\times S_i\right]}\right] \quad \text{(dB)} \qquad (7.136)$$

where $\langle p_i^2\rangle, \ldots. i = 1, N$ is the mean square sound pressure in the direction of the receiver from the centre of the ith wall or roof, where N is usually 1 or 2 or at most 3 (see Equation (1.87)), S_i is the area of the ith wall or roof, TL_i is the overall TL value for the ith wall and $\langle p_{n,i}^2\rangle$ is the mean square sound pressure at a location normal to the centre of the ith wall (see Equation (4.122)).

The value of the coefficient, K in Equation (7.125) must also be adjusted to account for the difference between TL_a and the TL of the wall section through which the direct field passes. Assuming that the direct line of sight from the source in the enclosure to the receiver outside the enclosure is through wall section, n, the required value of K_a to be used in place of K in Equation (7.127) to calculate C_a is:

$$K_a = \left(\frac{\tau_{N,j}}{\tau_j}\right)\left(\frac{\tau_j}{\tau_a}\right)\left(\frac{1}{\cos^2\theta}\right) = 10^{-5.6/10}\left(\frac{10^{-\text{TL}_j/10}}{10^{-\text{TL}_a/10}}\right)\left(\frac{1}{\cos^2\theta}\right) \qquad (7.137)$$

where θ is the angle between the line joining the source and receiver, and the normal to the section of wall through which the line of sight from source to receiver passes. The subscript, j, refers to the section of wall through which the line of sight from source to receiver passes. The coefficients, C_a and TL_a, replace the coefficients C and TL in Equation (7.135) to calculate the enclosure noise reduction for the more complex case of the receiver not being located normal to the centre of one wall or roof.

7.4.3 Personnel Enclosures

For personnel enclosures, the noise source is external and the purpose of the enclosure is to reduce levels within. Suppose that the enclosure is located within a space in which the reverberant field is dominant. The sound pressure level in the reverberant field, removed at least one half of a wavelength from the walls of the enclosure, is designated L_{p1} (dB). Use will be made of Equation (6.41), in which the mean absorption coefficient $\bar{\alpha}$ is replaced with the field-incidence wall transmission coefficient, τ. The power transmission W_i into the enclosure through the external walls of surface area, S_E, is that associated with the sound intensity vector that is normal to the wall surface and is:

$$W_i = S_E\langle p_1^2\rangle\tau/(4\rho c) \qquad (7.138)$$

Taking logarithms to the base ten of both sides of the equation, and noting that the numerical value of ρc is approximately 400, Equation (7.138) may be rewritten as:

$$L_{Wi} = L_{p1} + 10\log_{10} S_E - \text{TL} - 6 \quad \text{(dB)} \qquad (7.139)$$

To estimate the space averaged sound pressure level, L_{pi}, within the enclosure, use is made of Equation (6.44). In the latter equation, the direct field term is replaced with the reciprocal of the external surface area, S_E, and Equation (6.44) is rewritten as:

$$L_{pi} = L_{Wi} + 10\log_{10}\left[\frac{1}{S_E} + \frac{4(1-\bar{\alpha}_i)}{S_i\bar{\alpha}_i}\right] \quad \text{(dB)} \tag{7.140}$$

Substitution of the above equation into Equation (7.139), and use of Equation (7.126), leads to the following result:

$$\text{NR} \approx L_{p1} - L_{pi} = \text{TL} - C \quad \text{(dB)} \tag{7.141}$$

The coefficient, C, is calculated using Equation (7.127).

The problem is not so simple when the direct field of the source is dominant at one or more walls of the enclosure. In this case, for the purposes of estimating the sound field incident on the exterior of each of the enclosure walls, it is necessary to treat the enclosure as a barrier and calculate the Fresnel number associated with the source and the centre of the enclosure wall (see Figure 7.14 and Section 5.3.5.1). The Fresnel number, N_1, for each surface is:

$$N_1 = \pm(2/\lambda)(A + B - d_{SR}) \tag{7.142}$$

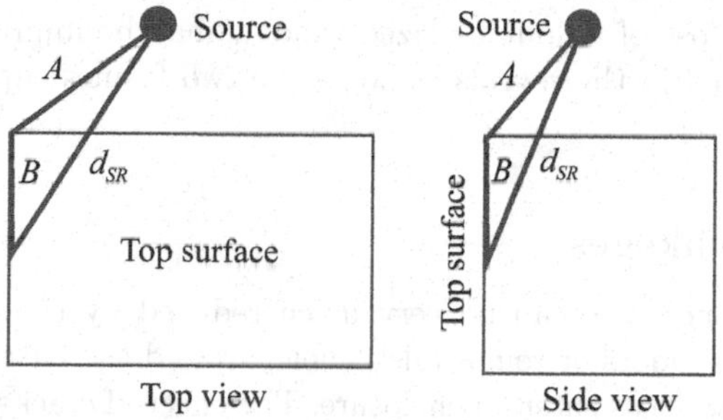

FIGURE 7.14 Fresnel number calculation to determine the barrier effect of the enclosure in terms of calculating the approximate average of the direct field noise level incident on the exterior surface of the wall or roof of a personnel enclosure.

The sound power level introduced into the interior of the enclosure by any side or top surface of the enclosure exposed to the direct field from the sound source can be estimated approximately using Equation (7.139). However, for this case S_E is the area and TL is the transmission loss of the side or top surface under consideration. The sound pressure level, $L_{p1,m}$, adjacent to the outside centre of any side or top surface, m, and associated only with the direct field, can be calculated from the sound power level, L_W, of the source with a correction for spherical spreading and the barrier effect, $A_{b,m}$. Thus:

$$L_{p1,m} \approx L_W - 20\log_{10} d_{SR} - 8 - A_{b,m} \quad \text{(dB)} \tag{7.143}$$

where it is assumed that the sound pressure at the centre of any side or top surface is a good approximation of the average sound pressure level over the entire surface. The variables, A, B and d_{SR} are defined in Figure 7.14 and the Fresnel number, calculated using Equation (7.142), is used together with Figure 7.14 and Equation (5.144) to determine $A_{b,m}$.

The total sound power level, L_{Wi}, for use in Equation (7.140) to calculate the space average sound pressure level, L_{pi} inside the enclosure is calculated from the sound power levels, $L_{Wi,m}$, introduced by each of the five enclosure surfaces as:

$$L_{Wi} = 10\log_{10}\sum_{m=1}^{5} 10^{L_{Wi,m}/10} \quad \text{(dB)} \tag{7.144}$$

The value of L_{Wi} calculated using Equation (7.144) is then used with Equation (7.140) to find L_{pi} which is then used with Equation (7.141) to find NR, except that for this case, $L_{p1} \approx L_W - 20\log_{10} d - 8$ (dB), where d is the distance from the source to the location of interest without the enclosure.

7.4.4 Enclosure Windows

Inspection windows are usually double glazed. However, the use of double glazing may show no improvement over single glazing at low frequencies where the interaction of the mass of the glass panes and stiffness of the air trapped between them can produce a series of resonances. The lowest frequency corresponding to a resonance of this type and a corresponding poor value of transmission loss is given approximately by Equation (7.82) (see Appendix C for glass properties). There are other acoustic resonances of the cavity that are also important, but these usually have higher resonance frequencies (see Equations (6.17) and (7.81)).

Reference to Figure 7.11 shows that the glass thickness and pane separation should be chosen so that f_0 (see Equation (7.82)) is well below the frequency range in which significant noise reduction is required. For example, a pane thickness of 6 mm and a separation of 150 mm gives $f_0 = 78$ Hz. Good transmission loss should not be expected at frequencies below about $1.15 f_0$ or 90 Hz. The transmission loss of a double-glazed window may be improved by placing a blanket of porous acoustic material in the reveals between the two frames supporting the glass (Quirt, 1982).

7.4.5 Enclosure Leakages

The effectiveness of an enclosure can be very much reduced by the presence of air gaps. Air gaps usually occur around doors or removable panels, around the base of the enclosure where it meets the floor or where services enter an enclosure. The effect of cracks (or slits) around doors or around the base of a machine enclosure can be calculated with the help of Figure 7.15, which gives the transmission coefficient of a crack (or slit) as a function of frequency and width. If the crack is between one plane surface and another plane surface normal to it, the effective crack width must be doubled (because of reflection – for example, a crack under a door) before using Figure 7.15. However, note that the effective area of the crack is not doubled when overall TL values are calculated using Equation (7.115). Once the transmission coefficient, τ, has been determined for a particular frequency, the procedure outlined in Section 7.3.1 is used for estimating the average value of τ for the enclosure wall or cover.

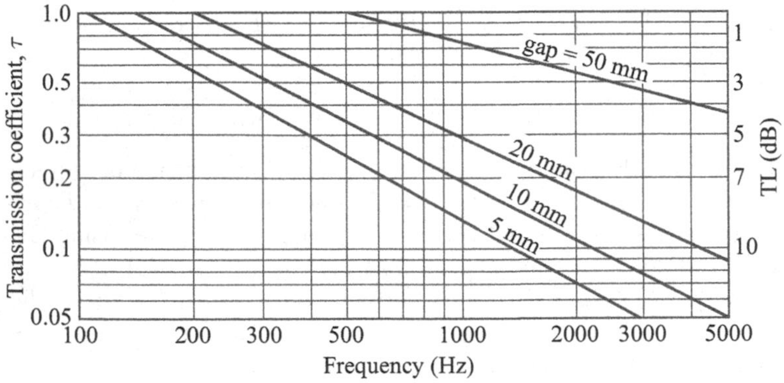

FIGURE 7.15 Transmission coefficients of long narrow cracks.

The importance of small air gaps is illustrated by reference to Figures 7.13 and 7.15. For example, consider a door with a 20 mm air gap beneath it, which for a typical door results in an area ratio S_1/S_2 of 0.01, where S_1 is the area (in the plane of the door) of the gap at the bottom of the door and S_2 is the area of the door. The effective width of a 20 mm gap under a door is 40 mm due to reflection on the floor under the gap. Thus, the transmission coefficient for a 20 mm gap under a door is 1.0 for frequencies below about 400 Hz. For $S_1/S_2 = 0.01$, Figure 7.13 shows that the best transmission loss that can be achieved at frequencies below 400 Hz is 20 dB.

Gomperts and Kihlman (1967) presented the following expression for calculating the transmission coefficient of a slit:

$$\tau = \frac{mk\ell_w \cos^2(k\alpha)}{2n^2\left\{\dfrac{\sin^2(k\ell_d + 2k\alpha)}{\cos^2(k\alpha)} + \dfrac{k^2\ell_w^2}{2n^2}\left[1 + \cos(k\ell_d + 2k\alpha)\cos(k\ell_d)\right]\right\}} \tag{7.145}$$

where

$$\alpha = \frac{\ell_w}{\pi}\left[\log_e\left(\frac{8}{k\ell_w}\right) - 0.57722\right] \tag{7.146}$$

where k is the wavenumber, ℓ_w is the height or width of the slit, ℓ_d is the depth of the slit (or thickness of the wall) and the length of the slit must exceed a wavelength at the frequency of interest. If $\alpha < 0.0$, set $\alpha = 0.0$. The coefficient, m, is equal to 4 for normally incident plane waves and is equal to 8 for a diffuse field incident on the slit. The coefficient, n, is equal to one if the slit is in the middle of a wall and is equal to 0.5 if the slit is between one plane surface and another plane surface normal to it, which takes into account the effect of the reflected path of sound travelling through the slit. Thus, for this model, the slit area is not doubled for a slit between one plane surface and and another plane surface normal to it. Using Equation (7.145), it will be found that the slit transmission coefficient exceeds 1.0 most of the time, which indicates that the incident sound energy is concentrated in the vicinity of the slit as a result of its presence. As the slit becomes wider, this effect reduces, as the sound energy concentration in the slit vicinity reduces.

An alternative, much simpler expression for the transmission coefficient of a long narrow slit ($k\ell_w \ll 2\pi$ or $k\ell_w < 0.5$) in the centre of a wall has been presented by Mechel (1986) as:

$$\tau = \frac{2\ell_w}{k\ell_d^2} \tag{7.147}$$

For a slit near the edge of a wall, τ in Equation (7.147) should be multiplied by 2.

For doors with no seals a leak height, ℓ_w, of 1.2 mm around the full perimeter of the door is usually used with Equations (7.115), (7.145), and (7.147) to calculate the effect on the overall TL of a wall with a door. For a door with weather stripping, a leak height of 0.5 mm is used and for a door with magnetic seals a leak height of 0.2 mm is used.

7.4.6 Enclosure Access and Ventilation

Most enclosures require some form of ventilation. They may also require access for passing materials in and out. Such necessary permanent openings must be treated with some form of silencing to avoid compromising the effectiveness of the enclosure. In a good design, the acoustic performance of access silencing will match the performance of the walls of the enclosure. Techniques developed for the control of sound propagation in ducts may be employed for the design of silencers (see Chapter 8).

If ventilation for heat removal is required but the heat load is not large, then natural ventilation with silenced air inlets at low levels close to the floor and silenced outlets at high levels, well above the floor, will be adequate. If forced ventilation is required to avoid excessive heating, then the approximate amount of air flow needed is:

$$V = \frac{H}{\rho C_p \Delta T} \tag{7.148}$$

where V is the volume (m^3/s) of airflow required, H is the heat input (W) to the enclosure, ΔT is the temperature differential (°C) between the external ambient and the maximum allowable internal temperature of the enclosure, ρ is the gas (air) density (kg/m^3), and C_p is the specific heat of the gas (air) in SI units (1010 m^2s^{-2} °C^{-1}).

If a fan is provided for forced ventilation, the silencer will usually be placed external to the fan so that noise generated by the fan will be attenuated as well. When high volumes of air flow are required, the noise of the fan should be considered very carefully, as this noise source is quite often a cause of complaint. As fan noise generally increases with the fifth power of the blade tip speed, large slowly rotating fans are always to be preferred over small high-speed fans.

7.4.7 Enclosure Vibration Isolation

Any rigid connection between the machine and enclosure must be avoided. If at all possible, all pipes and service ducts passing through the enclosure wall should have flexible sections to act as vibration breaks; otherwise, the pipe or duct must pass through a clearance hole surrounded by a packing of mineral wool closed by cover plates and mastic sealant.

It is usually advisable to mount the machine on vibration isolators (see Chapter 9), particularly if low-frequency noise is the main problem. This ensures that little energy is transmitted to the floor. If this is not done, there is a possibility that the floor surrounding the enclosure will re-radiate the energy into the surrounding space, or that the enclosure will be mechanically excited by the vibrating floor and act as a noise source.

Sometimes it is not possible to mount the machine on vibration isolators. In this case, excitation of the enclosure can be avoided by mounting the enclosure itself on vibration isolators. Great care is necessary when designing machinery vibration isolators, to ensure that the machine will be stable and that its operation will not be affected adversely. For example, if a machine must pass through a system resonance when running up to speed, then "snubbers" can be used to prevent excessive motion of the machine on its isolation mounts. Isolator design is discussed in Chapter 9.

7.4.8 Enclosure Resonances

Two types of enclosure resonance will be considered. The first is a mechanical resonance of the enclosure panels, while the second is an acoustic resonance of the air space between an enclosed machine and the enclosure walls. At the frequencies of these resonances, the noise reduction due to the enclosure is markedly reduced from that calculated without regard to resonance effects.

The lowest order enclosure panel resonance is associated with a large loss in enclosure effectiveness at the resonance frequency. Thus, the enclosure should be designed so that the resonance frequencies of its constituent panels are not in the frequency range in which appreciable sound attenuation is required. Only the lowest order, first few, panel resonances are of concern here. The panels may be designed in such a way that their resonance frequencies are higher than or lower than the frequency range in which appreciable sound attenuation is required. Additionally, the panels should be well-damped.

If the sound source radiates predominantly high-frequency noise, then an enclosure with low resonance frequency panels is recommended, implying a massive enclosure. On the other hand, if the sound radiation is predominantly low-frequency in nature, then an enclosure with a high resonance frequency is desirable, implying a stiff but not massive enclosure. The discussion of

Sections 7.2.6 and 7.2.8 may be used as a guide to resonance frequency and transmission loss calculations.

The resonance frequency of a panel may be increased by using stiffening ribs, but the increase that may be achieved is generally quite limited. For stiff enclosures with high resonance frequencies, materials with large values of Young's modulus to density ratio, E/ρ, are chosen for the wall construction to ensure large values of the longitudinal wave speed, c_{LI} (see Section 1.4.4). For massive enclosures with low resonance frequencies, small values of E/ρ are chosen. Values of E and ρ for common panel materials are listed in Appendix C. In practice, stiff enclosures will generally be restricted to small enclosures.

If a machine is enclosed, reverberant build-up of the sound energy within the enclosure will occur unless adequate sound-absorption is provided. The effect will be an increase of sound pressure at the inner walls of the enclosure over that which would result from the direct field of the source. A degradation of the noise reduction expected of the enclosure is implied.

In close-fitting enclosures, noise reduction may be degraded by yet another resonance effect. At frequencies where the average air spacing between a vibrating machine surface and the enclosure wall is an integral multiple of half wavelengths of sound, strong coupling will occur between the vibrating machine surface and the enclosure wall, resulting in a marked decrease in the enclosure wall transmission loss.

The effect of inadequate absorption in enclosures is very noticeable. Table 7.14 shows the reduction in performance of an ideal enclosure with varying degrees of internal sound-absorption. The sound power of the source is assumed constant and unaffected by the enclosure. "Percent" refers to the fraction of internal surface area that is treated. A negative noise reduction represents a sound pressure level increase.

TABLE 7.14 Enclosure noise reduction as a function of percentage of internal surface covered with sound-absorptive material

Percent sound-absorbing material	10	20	30	50	70
Noise reduction (dB)	−10	−7	−5	−3	−1.5

For best results, the internal surfaces of an enclosure are usually lined with glass or mineral fibre or open-cell polyurethane foam blanket. Typical values of absorption coefficients are given in Table 6.2.

Since the absorption coefficient of absorbent lining is generally highest at high frequencies, the high-frequency components of any noise will be subject to the highest attenuation. Some improvement in low-frequency absorption can be achieved by using a thick layer of lining. See Section 6.7.3 for a discussion of suitable linings and their containment for protection from contamination with oil or water, which impairs their acoustical absorption properties.

7.4.9 Close-Fitting Enclosures

The cost of acoustic enclosures of any type is proportional to size; therefore there is an economic incentive to keep enclosures as small as possible. Thus, because of cost or limitations of space, a close-fitting enclosure may be fitted directly to the machine which is to be quietened, or fixed independently of it but so that the enclosure internal surfaces are within, say, 0.5 m of major machine surfaces.

When an enclosure is close fitting, the panel resonance frequency predicted by Equation (7.46) may be too low; in fact, the resonance frequency will probably be somewhat increased due to the stiffening of the panel by the enclosed air volume. Thus, an enclosure designed to be massive with a low resonance frequency may not perform as well as expected when it is close fitting. Furthermore, system resonances will occur at higher frequencies; some of these modes of vibration will be good radiators of sound, producing low noise reductions, and some will be poor radiators, little affecting the noise reduction. The magnitude of the decrease in noise reduction caused by these resonances may be controlled to some extent by increasing the mechanical damping of the wall. Thus, if low-frequency sound (less than 200 Hz) is to be attenuated, the close-fitting enclosure should be stiff and well-damped, but if high-frequency sound is to be attenuated, the enclosure should be heavy and highly absorptive but not stiff. Doubling of the volume of a small enclosure will normally lead to an increase in noise reduction of 3 dB at low frequencies, so that it is not desirable to closely surround a source, such as a vibrating machine, if a greater volume is possible.

Generally, if sufficient space is left within the enclosure for normal maintenance on all sides of the machine, the enclosure need not be regarded as close fitting. If, however, such space cannot be made available, it is usually necessary to upgrade the transmission loss of an enclosure wall by up to 10 dB at low frequencies (less at high frequencies), to compensate for the expected degradation in performance of the enclosure due to resonances.

The analysis of close-fitting enclosures is treated in detail in the literature (Tweed and Tree, 1976; Byrne et al., 1988).

7.4.10 Partial Enclosures

In many situations, where easy and continuous access to parts of a machine is necessary, a complete enclosure may not be possible, and a partial enclosure must be considered (Alfredson and Scow, 1976). However, the noise reductions that can be expected at specific locations from partial enclosures are difficult to estimate and will depend on the particular geometry. Estimates of the sound power reduction to be expected from a partial enclosure, which is lined with a 50 mm thick layer of sound-absorbing material, are presented in Figure 7.16.

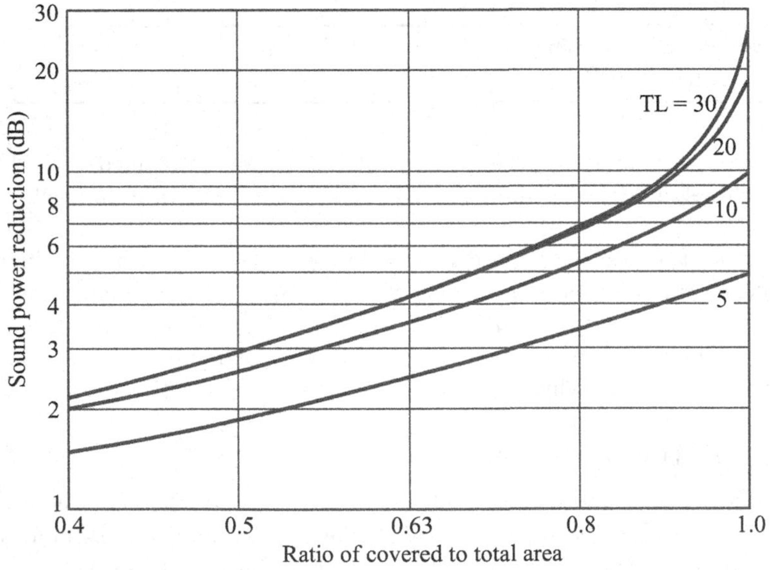

FIGURE 7.16 Sound attenuation due to a partial enclosure. The transmission loss (TL) of the enclosure wall is shown parametrically.

Figure 7.16 shows the sound power reduction to be expected for various degrees of enclosure, with transmission loss through the enclosure walls and roof as a parameter. The figure shows fairly clearly that the enclosure walls should have a transmission loss of about 20 dB, and the most sound power reduction that can be achieved for a realistic partial enclosure (90% closed) is about 10 dB. However, sound pressure levels in some directions may be reduced more, especially in areas immediately behind solid parts of the enclosure.

7.4.11 Enclosure Performance Measurement

The Insertion Loss (IL) of an enclosure may be determined by measuring the sound power level of the internal noise source with and without the enclosure present using the methods discussed in Section 4.14. The IL (dB) is then the arithmetic difference between the two levels and may be expressed in 1/3-octave bands, octave bands or overall A-weighted IL (dB). Guidelines for measuring the enclosure IL are provided in ISO 11546-1 (1995) for laboratory measurements and ISO 11546-2 (1995) for field measurements. The noise source for the performance measurement should be the same source as the one for which the operational enclosure is to be used. If such a source is not available, a constant-volume-velocity source should be used (see Section 4.10). The standards mentioned above provide a design of such a source which has a low output at low frequencies, making it unsatisfactory for testing in an environment outside of a laboratory. An alternative constant-volume-velocity source is a loudspeaker with a small backing cavity. To ensure that the source has the same output with and without the enclosure present, a microphone can be inserted into the backing cavity, at the same time making sure that the airtight property of the enclosure is maintained. The output of the amplifier driving the loudspeaker can then be adjusted so that the microphone output is the same for both test scenarios. A faster approach is to leave the amplifier settings unchanged between tests and then adjust the measured sound power level by the difference in sound pressure levels recorded by the microphone in the loudspeaker backing cavity for the two cases (with and without the test enclosure present).

7.5 Indoor Barriers

In Chapter 5, only the use of barriers outdoors was considered or, more explicitly, the situation in which the contribution of the source direct field to the overall sound pressure level was much larger than any reverberant field contribution. In this section, the effect of placing a barrier in a room where the reverberant sound field and reflections from other surfaces cannot be ignored will be considered.

The following assumptions are implicit in the calculation of the insertion loss for an indoor barrier:

- The transmission loss of the barrier material is sufficiently large that transmission through the barrier can be ignored. A transmission loss of at least 20 dB in the frequency range of interest is recommended.
- The sound power radiated by the source is not affected by the insertion of the barrier.
- The receiver is in the shadow zone of the barrier; that is, there is no direct line of sight between source and receiver.
- Interference effects between waves diffracted around the side of the barrier, waves diffracted over the top of the barrier and reflected waves are negligible. This implies octave band analysis, as frequency averaging minimises interference effects that occur at single frequencies .

The barrier insertion loss, IL, in dB is (Moreland and Minto, 1976):

$$\mathrm{IL} = 10\log_{10}\left(\frac{D}{4\pi r^2} + \frac{4}{S_0\bar{\alpha}_0}\right) - 10\log_{10}\left(\frac{DF}{4\pi r^2} + \frac{4K_1K_2}{S(1-K_1K_2)}\right) \tag{7.149}$$

where D is the source directivity factor in the direction of the receiver (for an omnidirectional source on a hard floor, $D = 2$); r is the distance between source and receiver in the absence of the barrier; $S_0\bar{\alpha}_0$ is the absorption of the original room before inserting the barrier, where S_0 is the total original room surface area and $\bar{\alpha}_0$ is the room mean Sabine absorption coefficient; S is the open area between the barrier perimeter and the room walls and ceiling in the plane of the barrier; and F is the diffraction coefficient given by the following equation:

$$F = \sum_i \frac{1}{3 + 10N_i} \tag{7.150}$$

where N_i is the Fresnel number for diffraction around the ith edge of the barrier and is given by Equation (5.142), and K_1 and K_2 are dimensionless numbers given by:

$$K_1 = \frac{S}{S + S_1\bar{\alpha}_1}; \qquad K_2 = \frac{S}{S + S_2\bar{\alpha}_2} \tag{7.151}$$

In Equation (7.151), S_1 and S_2 are the total room surface areas on sides 1 and 2, respectively, of the barrier; that is, $S_1 + S_2 = S_0$+(area of two sides of the barrier). The quantities, $\bar{\alpha}_1$ and $\bar{\alpha}_2$, are the mean Sabine absorption coefficients associated with areas S_1 and S_2, respectively.

When multiple barriers exist, as in an open plan office, experimental work (West and Parkin, 1978) has shown the following statements to be true in a general sense (test screens were 1.52 m high by 1.37 m wide).

1. No difference in attenuation is obtained when a 300 mm gap is permitted between the base of the screen and the floor;
2. When a number of screens interrupt the line-of-sight between source and receiver, an additional attenuation of up to 8 dBA over that for a single screen can be realised;
3. Large numbers of screens remove wall reflections and thus increase the attenuation of sound with distance from the source;
4. For a receiver immediately behind a screen, a local shadow effect results in large attenuation, even for a source a large distance away. This is in addition to the effect mentioned in (2) above;
5. For a screen less than 1 m from a source, floor treatment has no effect on the screen's attenuation;
6. A maximum improvement in attenuation of 4–7 dB, as frequency is increased from 250 Hz to 2 kHz, can be achieved by ceiling treatment. However, under most conditions, this greater attenuation can only be achieved at the higher frequencies; and
7. Furnishing conditions are additive; that is, the attenuations measured under two different furnishing conditions are additive when the two conditions coexist.

7.6 Pipe Lagging

Radiation from the walls of pipes or air conditioning ducts is a common source of noise. The excitation usually arises from disturbed flow through valves or dampers, in which case, it is preferable to reduce the excitation by treatment or modification of the source. However, as treatment of the source is not always possible, an alternative is to acoustically treat the walls of the pipe or duct to reduce the transmitted noise. For ventilation ducts, the most effective solution is to line the duct internally with acoustic absorbent, whereas with pipework, an external treatment is normally used. The former treatment is discussed in Section 8.10.3 and the latter treatment is discussed here.

7.6.1 Porous Material Lagging

The effect of wrapping a pipe with a layer of porous absorbent material may be calculated by taking into account sound energy loss due to reflection at the porous material surfaces and loss due to transmission through the material. Methods for calculating these losses are outlined in Appendix D. The procedures outlined in Appendix D only include the effects of the pipe fundamental "breathing" mode on the sound radiation. Kanapathipillai and Byrne (1991) showed that pipe "bending" and "ovaling" modes are also important and the sound radiation from these is influenced in a different way by the lagging. Indeed, for a pipe vibrating in these latter modes, the insertion loss of a porous lagging is negative at low frequencies, and a sound increase is observed, partly as a result of the effective increase in sound radiating area. In practice, it may be assumed that below 250 Hz, there is nothing to be gained in terms of noise reduction by lagging pipework with a porous acoustic blanket.

7.6.2 Impermeable Jacket and Porous Blanket Lagging

Several theories have been advanced for the prediction of the noise reduction (or insertion loss) resulting from wrapping a pipe first with glass-fibre and then with a limp, massive jacket. Unfortunately, none of the current theories reliably predicts measured results; in fact, some predicted results are so far removed from reality as to be useless. Predictions based on the theory presented in the following paragraphs have been shown to be in best agreement with results of experiment, but even these predictions can vary by up to 10 dB from the measured data. For this reason, the analysis is followed by Figure 7.17, which shows the measured insertion loss for a pipe lagging configuration that was commonly used in the past; that is 50 mm of 70–90 kg m^{-3} glass-fibre or rockwool covered with a lead–aluminium jacket of 6 kg/m^2 surface density. However, lead is an unpopular material due to its toxicity, so the lead–aluminium jacket has been replaced with a thicker, single-layer, aluminium or stainless steel jacket or a double layer consisting of one layer of aluminium and one layer of a high density composite film, which is bonded to the aluminium using a viscoelastic film adhesive. One example of the latter construction is called Muffl-Jac®, which consists of a 0.5 mm-thick sheet of aluminium bonded to a 1.5 to 2.3 mm-thick layer of high density composite film; the total surface density of the jacket being approximately 5 kg/m^2. Of course the jacket needs to be separated from the pipe with a layer of rockwool (usually 100 mm thick and density between 60 and 100 kg/m^3) or acoustic foam (usually 50 mm thick) to obtain the insertion loss values shown in the figure. Data for this latter jacket were measured according to ASTM E1222-90 (2016) and are also shown in Figure 7.17. As results for frequencies below 500 Hz are generally unreliable, ASTM E1222-90 (2016) recommends that data in this frequency range not be measured. Another example of a commercially available pipe lagging material is called Soundlag 4525C®, which consists of a thin aluminium foil bonded to a mass loaded vinyl blanket, which, in turn, is bonded to a corrugated acoustic foam sheet to vibration isolate the vinyl blanket from the pipe. Data for this construction, which also has a surface density of 5 kg/m^2, are also shown in Figure 7.17.

Before proceeding with the analysis it should be noted that, where possible, manufacturer's data rather than calculations should be used. Guidelines for obtaining measured IL data are provided in ASTM E1222-90 (2016). In practice, it has been found that at frequencies below about 250 Hz, the insertion loss resulting from this type of treatment is either negligible or negative. It has also been found that porous acoustic foam gives better results than fibreglass or rockwool blanket because of it greater compliance. However, note that the following analysis applies to fibreglass or rockwool blankets only.

For the purposes of the analysis, the frequency spectrum is divided into three ranges (Hale, 1978) by two characteristic frequencies, which are the ring frequency, f_r, and the critical frequency, f_c, of the jacket ($f_r < f_c$). The jacket is assumed to be stiff and at the ring frequency,

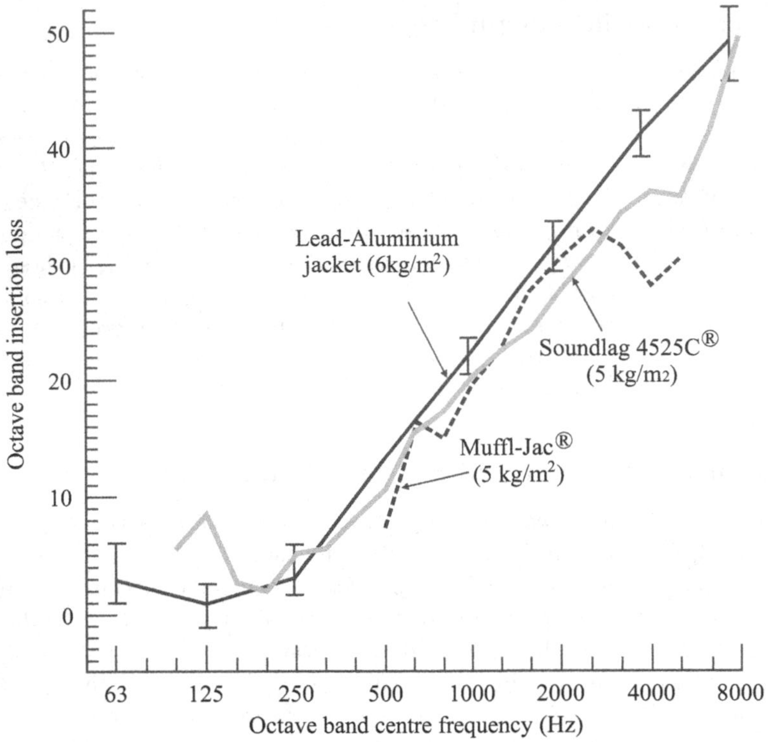

FIGURE 7.17 Typical pipe lagging insertion loss for some pipe lagging jackets. The solid black curve applies to a construction consisting of a 50 mm glass fibre, density 70–90 kg/m^3, covered with a lead/aluminium jacket of 6 kg/m^2 surface density. The I symbols represent variations in measured values for three pipe diameters (75 mm, 150 mm and 360 mm) for the lead aluminium jacket construction. The dashed curve represents the Muffl-Jac® construction and the solid grey curve represents the Soundlag 4525C® construction.

the circumference of the jacket (of diameter, d m) is one longitudinal wavelength long; thus $f_r = c_{LI}/\pi d$. The critical frequency given by Equation (7.3) is discussed in Section 7.2.1. For jackets made of two layers (such as lead and aluminium), Equations (7.6) to (7.9) must be used to calculate f_c and c_{LI}.

In the low-frequency range below the jacket ring frequency, f_r, the insertion loss is:

$$\text{IL} = 10\log_{10}[1 - 0.012X_r \sin 2C_r + (0.012X_r \sin C_r)^2] \quad (7.152)$$

where

$$X_r = [1000(m/d)^{1/2}(\xi_r - \xi_r^2)^{1/4}] - [2d/(\ell\xi_r)] \quad (7.153)$$

and

$$C_r = 30\xi_r\ell/d \quad (7.154)$$

where $\xi_r = f/f_r$, m is the jacket surface density (kg m^{-2}), ℓ is the absorptive material thickness (m), f_r is the jacket ring frequency (Hz), c_{LI} is the longitudinal wave speed (thin panel) in the jacket material, and f is the octave or 1/3-octave band centre frequency.

In the high-frequency range above the critical frequency, f_c, of the jacket (see Equation (7.3)) the insertion loss is:

$$\text{IL} = 10\log_{10}[1 - 0.012X_c \sin 2C_c + (0.012X_c \sin C_c)^2] \quad (7.155)$$

where

$$X_c = [41.6(m/h)^{1/2}\xi_c(1 - 1/\xi_c)^{-1/4}] - [258h/(\ell\xi_c)] \tag{7.156}$$

and

$$C_c = 0.232\xi_c\ell/h \tag{7.157}$$

The quantity, h, is the jacket thickness (m), $\xi_c = f/f_c$, and f_c is the jacket critical frequency (Hz).

In the mid-frequency range between f_r and f_c:

$$\text{IL} = 10\log_{10}[1 - 0.012X_m \sin 2C_c + (0.012X_m \sin C_c)^2] \tag{7.158}$$

where

$$X_m = [226(m/h)^{1/2}\xi_c(1 - \xi_c^2)] - [258h/(\ell\xi_c)] \tag{7.159}$$

As an alternative to the preceding prediction scheme, a simpler formula (Michelsen et al., 1980) is offered, which provides an upper bound to the expected insertion loss at frequencies above 300 Hz. That is:

$$\text{IL} = \frac{40}{1 + 0.12/D}\log_{10}\frac{f\sqrt{m\ell}}{132} \tag{7.160}$$

where D is the pipe diameter (m) and $f \geq 120/\sqrt{m\ell}$. The preceding equations are based on the assumption that there are no structural connections between the pipe and the jacket. If solid spacers are used to support the jacket, then the insertion loss will be substantially less.

In recent times, the use of acoustic foam in place of rockwool and fibreglass has become more popular. Manufacturers of the pre-formed foam shapes (available for a wide range of pipe diameters) claim superior performance over that achieved with the same thickness of rockwool. An advantage of the foam (over rockwool or fibreglass) is that it doesn't turn to powder when applied to a pipe experiencing relatively high vibration levels. However, acoustic foam is much more expensive than rockwool or fibreglass.

8

Muffling Devices

LEARNING OBJECTIVES

In this chapter, the reader is introduced to:

- noise reduction, NR, and transmission loss, TL, of muffling devices;
- diffusers as muffling devices;
- classification of muffling devices as reactive and dissipative;
- acoustic impedance for analysis of reactive mufflers;
- acoustic impedance of orifices and expansion chambers;
- analysis of several reactive muffler types;
- pressure loss calculations for reactive and dissipative mufflers;
- lined duct mufflers and splitter mufflers as dissipative mufflers;
- design and analysis of lined ducts;
- duct break-out noise transmission calculations;
- lined plenum attenuators;
- water injection for noise control of exhausts; and
- directivity of exhaust stacks.

8.1 Introduction

Muffling devices are commonly used to reduce noise associated with internal combustion engine exhausts, high-pressure gas or steam vents, compressors and fans. These examples lead to the conclusion that a muffling device allows the passage of fluid while at the same time restricting the free passage of sound. Muffling devices might also be used where direct access to the interior of a noise containing enclosure is required, but through which no steady flow of gas is necessarily to be maintained. For example, an acoustically treated entry way between a noisy and a quiet area in a building or factory might be considered as a muffling device.

Muffling devices may function in any one or any combination of three ways: they may suppress the generation of noise; they may attenuate noise already generated; and they may carry or redirect noise away from sensitive areas. Careful use of all three methods for achieving adequate noise reduction can be very important in the design of muffling devices.

8.2 Measures of Performance

There are a number of metrics that can be used to evaluate the acoustic performance of a muffler (also commonly known as a silencer). Two terms, insertion loss, IL, and transmission loss, TL,

DOI: 10.1201/9780367814908-8

are commonly used to describe the effectiveness of a muffling system. These terms are similar to the terms noise reduction, NR, and transmission loss, TL, respectively, introduced in Chapter 7 in connection with sound transmission through a partition. These three measures of performance are defined as follows:

Insertion Loss (IL) is defined as the reduction in radiated sound power level due to the replacement (insertion) of a section of duct with the proposed muffler and is calculated as:

$$\mathrm{IL} = L_{W\,:\,\mathrm{before}} - L_{W\,:\,\mathrm{after}} \quad \mathrm{(dB)} \tag{8.1}$$

where $L_{W\,:\,\mathrm{before}}$ is the sound power level exiting the duct without the muffler installed, and $L_{W\,:\,\mathrm{after}}$ is the sound power level after the muffler has been installed by replacing a section of the duct. IL can be expressed as (Munjal et al., 2006, Eq. (9.6), p. 284):

$$\mathrm{IL} = L_b - L_a = 10\log\left[\frac{p_b^2}{p_a^2}\right] \quad \mathrm{(dB)} \tag{8.2}$$

where L_a, L_b are the sound pressure levels measured at the same location relative to the exhaust outlet, with the muffler installed, and when an equivalent straight section of duct is used in place of the muffler. If the muffler is placed at the end of the duct, it may change the directivity of the emitted noise so the sound pressure squared may need to be averaged at a number of locations (all at the same distance from the exhaust outlet) before and after installation of the muffler and these average values used in Equation (8.2). If the locations over which the sound pressures squared are averaged are not all at the same distance from the exhaust outlet, then each pressure squared value must be multiplied by the square of the distance from the centre of the exhaust outlet before averaging. It is also important to ensure that sound pressure measurements are not taken in the near field of the exhaust outlet, which usually means taking measurements at a distance from the exhaust outlet of at least half a wavelength at the lowest frequency of interest. Furthermore, it is important that the microphones used for the measurements are not subject to any significant exhaust flow as this will affect measurements, especially in the low-frequency range.

Transmission Loss (TL) is the difference between the sound power level *incident* on the muffler ($L_{W:\mathrm{incident}}$) and the sound power level propagating downstream after the muffler ($L_{W:\mathrm{transmitted}}$), when the system has an infinite (anechoic) end condition (Munjal, 2014, p. 55). When a reactive muffler is installed in a duct, it tends to reflect sound power back upstream and the sound pressure level can increase upstream of the muffler compared to a system without the muffler installed. The calculation of TL uses the sound power *incident* on the muffler, and not the total sound power that exists upstream of the muffler, to quantify the baseline acoustic performance. Note that the total power consists of both upstream and downstream travelling power and the power of interest for the determination of TL is the downstream travelling component. A technique is described in ASTM E2611-19 (2019) that can be used to calculate the incident and reflected acoustic field, by processing the signals from two microphones placed along the duct. Although this standard is aimed towards determining the TL through a sound-absorbing material when plane waves propagate along the duct, the algorithm and method can be applied to measuring the incident acoustic power on a muffler, and the transmitted power down a duct. This method is discussed in more detail in Section D.4.3. The sound power that is *transmitted* into an anechoic termination after the muffler is used to evaluate TL. TL is expressed mathematically as:

$$\mathrm{TL} = L_{W:\mathrm{incident}} - L_{W:\mathrm{transmitted}} \quad \mathrm{(dB)} \tag{8.3}$$

The TL is not equal to the NR or the IL, as the muffler can affect the sound power generated by the sound source (and thus the sound power incident on the muffler), in addition to the transmitted sound power, as explained below.

Noise Reduction (NR) (also known as Level Difference) is the difference between the sound pressure level measured upstream and downstream of the muffler, at a distance of several duct diameters from the muffler (Munjal, 2014, p. 56; Munjal et al., 2006, Eq. (9.9), p. 285) and is calculated as:

$$\mathrm{NR} = \mathrm{LD} = L_{p,\mathrm{upstream}} - L_{p,\mathrm{downstream}} \quad \mathrm{(dB)} \tag{8.4}$$

The measurement of NR does not require anechoic duct terminations and does not require knowledge of the source impedance. It is the easiest of the three metrics to measure in practice, but it is also the least useful in terms of estimating the effect that the muffler will have on sound pressure levels emitted from the duct exit.

Some people consider that the most useful metric of a muffler's performance is IL, because it is the difference between the acoustic power from the exit of the exhaust before and after installation of the muffler. However, if there is significant sound radiation from the engine, then attempts to measure IL could be misleading, as exhaust sound power measurements could be contaminated with engine noise. Although it can be easy to measure IL, it is hard to accurately predict, as it depends on the source and termination impedances, which can be difficult to estimate, although not too difficult to measure if a similar system is available with or without a muffler installed (see ASTM E2611-19 (2019) and Appendix D).

Muffling devices make use of one or the other or a combination of two effects in their design. Either, sound propagation may be prevented (or strongly reduced) by reflection or suppression, or sound may be dissipated. Muffling devices based on reflection or source sound power output suppression are called reactive devices and those based on dissipation are called dissipative devices.

The performance of reactive devices is dependent on the impedances of the source and termination (outlet). In general, a reactive device will strongly affect the generation of sound at the source. This has the effect that the TL and IL of reactive devices may be very different. As IL is the quantity directly related to the reduction of noise exiting the end of the duct, it will be used here to describe the performance of reactive muffling devices in preference to TL; however, TL will be also considered for some simple reactive devices.

The performance of dissipative devices, on the other hand, by the very nature of the mode of operation, tends to be independent of the effects of source and termination impedance. Provided that the TL of a dissipative muffler is at least 5 dB, it may be assumed that the IL and the TL are the same. This assertion is justified by the observation that any sound reflected back to the source through the muffler will be reduced by at least 10 dB and is thus small and generally negligible compared to the sound introduced. Consequently, the effect of the termination impedance on the source must also be small and negligible.

8.3 Design for a Required Performance

Mufflers are invariably part of an industrial air handling system or exhaust system, such as that attached to a gas turbine. Almost all systems containing mufflers to reduce radiated noise also have maximum allowed back pressure requirements, and all mufflers contribute to this back pressure. Thus, any muffler design must not only address the allowed sound power to be radiated from the duct exit; it must also address the pressure loss problem, and care must be taken to ensure that the entire duct system does not exceed the allowed pressure loss. For this reason, the assessment of the performance of duct elements such as bends and plenum chambers is included in this chapter in addition to mufflers, and this performance assessment includes pressure loss calculations as well as self-noise generation due to air or exhaust gas flowing through the system.

All of the performance calculations outlined in this chapter are estimates only, and depending on the application, a factor of safety should always be applied so that the calculated muffler IL is always greater than needed, and the pressure loss, as well as the self-noise generation, is always less than the maximum allowed. Factors of safety usually range from 1.1 in non-critical applications to 1.5 in critical applications. For example, a factor of safety of 1.5 implies that the calculation of the system IL should be 50% greater than needed (in dB), so that if 20 dB is needed, the design goal should be 30 dB. Factors of safety may be increased beyond the values mentioned above in situations where the characteristics of the flow entering the muffler or duct bend and/or the flow temperature are not known accurately.

A muffler is often installed on an engine or item of equipment to reduce the exhaust noise at some distance from the exhaust outlet, so that it is less than a specified sound pressure level. The calculation methods in this chapter can aid in the acoustic design of a muffler. However, there is no unique optimal design for a muffler, as there are often a number of competing requirements. These can include the requirement that the muffler should provide high levels of noise attenuation, but not impose significant back-pressure or occupy a large space; the muffler should be lightweight, but not radiate sound from its structure.

There are also a number of practical requirements for a muffler that include:

- An adequate IL, such that the noise emanating from the outlet of the exhaust duct is at least 5 dB lower than other noise sources, to ensure that the usually lower frequency exhaust noise is not objectionable. The topic of perceivable noise is discussed in Section 2.2 and for further information, the reader should consult Lyon (2000); Fastl and Zwicker (2007).
- An acoustic performance that is suitable for an engine's range of operating conditions, including the expected variation in engine speed, exhaust gas temperature, and exhaust gas flow speed.
- An acoustic performance that does not degrade over time.
- A minimal increase in back-pressure, as any increase adversely affects the maximum power developed by a power source as well as the volumetric efficiency.
- Minimal radiated sound from the muffler exterior surface (called "break-out noise").
- Minimal noise generated by the flow of gas through the muffler.
- Materials used to construct the muffler that are robust to vibration, able to withstand temperature gradients, resistant to corrosion, not excessively expensive, not large in size, not made from flammable materials, and not able to ignite flammable materials on contact (such as dry grass if driven over by a car).

Table 10.20 lists estimated IL values for small, medium and large mufflers (from Joint Departments of the Army, Air Force, and Navy USA (1983), Table 3-2). It is suggested that for a piston-engine application, the volume of the muffler should ideally be 10-times the cylinder volume of the engine to minimise power loss and provide around 20 dB of IL, which corresponds to the "medium size" muffler in Table 10.20 (Munjal, 2014, p. 332). Often space is limited and a medium or large muffler cannot be installed. In this case, it is still necessary that the muffler volume be at least 3-times the cylinder volume of the engine (Kampichler et al., 2010, p. 563).

8.4 Diffusers as Muffling Devices

A commonly used device, often associated with the design of dissipative mufflers for the reduction of high-pressure gas exhaust noise, is a gas diffuser. When properly designed, this device can very effectively suppress the generation of noise. Alternatively, if attention is not given to the design, it can become a serious source of noise.

Aerodynamic sources were considered in Chapter 4, where it was shown that a fluctuating force or stress on a moving fluid can produce a dipole or quadrupole source of sound. As these sources were shown to generate sound power proportional to the sixth and eighth power, respectively, of the stream speed, it should be an aim of the diffuser design to minimise the generation of fluctuating forces and stresses. For this reason, a diffuser should have as its primary function the reduction of the pressure gradient associated with the exhaust. Thus, while the pressure loss may be fixed, the gradient may be reduced by extending the length over which the pressure loss is produced. This is quite often accomplished by providing a tortuous path for the exhausting gas, as illustrated in Figure 8.1.

In the mixing region between the exhausting gas and the ambient or slowly moving air, very large shear stresses can occur, giving rise to quadrupole noise generation. Such noise can become very serious at high discharge rates because of its eighth power dependence on stream speed. Thus, a second function of a diffuser should be to reduce the shear in the mixing region between the exhausting gas and ambient or slowly moving air in the neighbourhood of the exhaust stream. This is quite often accomplished by breaking the exhaust stream up into many small streams. A common design, illustrated in Figure 8.1, is to force the exhaust to pass through the myriad holes in a perforated plate, perhaps incorporated in the muffler as a closed perforated cylinder. Such a design can also accomplish the task of reducing the pressure gradient.

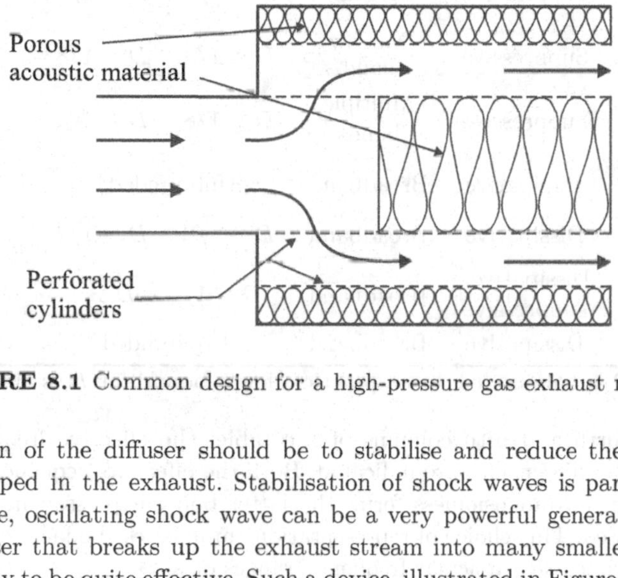

FIGURE 8.1 Common design for a high-pressure gas exhaust muffler.

A third function of the diffuser should be to stabilise and reduce the magnitude of any shock waves developed in the exhaust. Stabilisation of shock waves is particularly important because an unstable, oscillating shock wave can be a very powerful generator of noise. In this connection, a diffuser that breaks up the exhaust stream into many smaller streams has been found experimentally to be quite effective. Such a device, illustrated in Figure 8.2, has been shown to accomplish by itself, without any additional muffling, a 10 dB insertion loss in broadband noise in a steam-generating plant blow-out operation.

8.5 Classification of Muffling Devices

As an aid in the discussion that follows, it is convenient to classify commonly used muffling devices in some systematic way. A classification suitable for the present purpose is presented in Table 8.1, in which seven types of devices are identified in column one, and additionally classified according to the mechanism by which they function, in column two. Thus, under the heading "Mechanism", in column two, the devices are identified either as suppressive (reactive) or dissipative.

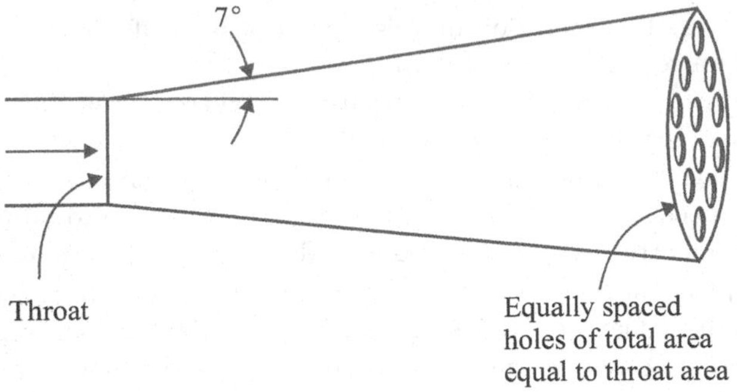

FIGURE 8.2 Noise suppression diffuser.

TABLE 8.1 Classification of muffling devices, where "length" is the axial dimension, "width" is the transverse dimension and f, c, λ and ℓ are, respectively, frequency, speed of sound, wavelength of sound and critical dimension of the device

Device	Mechanism	Effective frequency range	Critical dimensions $D = f\ell/c = \ell/\lambda$		Dependence of performance on end conditions
			length	width	
1. Lumped element	Suppressive	Band	$D < 1/8$	$D < 1/8$	Critical
2. Side branch resonator	Suppressive	Narrow band	$D \leq 1/4$	$D < 1/8$	Critical
3. Transmission line	Suppressive	Multiple bands	$D > 1/8$	$D < 1/4$	Critical
4. Lined duct	Dissipative	Broadband	Unbounded[a]		Slightly dependent
5. Lined bend	Dissipative	Broadband	$D > 1/2$	$D > 1/2$	Not critical
6. Plenum chamber	Dissipative/ suppressive	Broadband	$D > 1$	$D > 1$	Not critical
7. Water injection	Dissipative	Broadband	Unbounded		Not critical

[a]Theoretically, D is unbounded, but a practical lower bound for D is about 1/4.

In the third, fourth and fifth columns of the table, the effective frequency range of the device and the critical dimensions are indicated. Both the effective frequency range and critical dimensions are given in dimensionless form, the latter being expressed in wavelengths of the sound to be attenuated. This choice of representation emphasises the fact that, in principle, the devices are not restricted to particular frequency ranges or sizes.

In practical designs, it will generally be found that reactive devices are favoured for the control of low-frequency noise, since they tend to be more compact than dissipative devices for the same attenuation. Conversely, dissipative devices are generally favoured for the attenuation of high-frequency noise, due to their simpler and cheaper construction and greater effectiveness over a broad frequency range.

The critical dimensions of columns four and five in Table 8.1 are arbitrary; however, there are possible exceptions. Thus, the width of the side branch resonator shown as not exceeding one-eighth could be increased to perhaps one-half for a special design, and similar comments are true of the other dimensions shown. However, in these cases, the lumped element analysis used in this chapter to determine the resonator dimensions for a particular design frequency may not be very accurate and some *in situ* adjustment should be allowed for. Side branch resonators also exhibit noise attenuations at frequencies that are multiples of the design frequency.

The last column of the table shows the expected dependence of performance, of the devices listed, on the characteristics of the source and termination.

8.6 Acoustic Impedance

Various types of impedance were discussed in Section 1.16. In the discussion of muffling devices and sound propagation in ducts, the most commonly used and useful type of impedance is acoustic impedance. This is an important parameter in the discussion of reactive muffling devices, where the assumption is implicit that the muffler cross-dimensions are small compared to a wavelength (see Equation (8.228)), so that only plane waves propagate. In duct systems where there are changes in the cross-sectional area, the volume velocity, which is defined as the product of the particle velocity and the duct cross-sectional area, is unaltered by the changes in area.

Further, as was discussed in Section 1.16, the acoustic impedance relates the acoustic pressure to the associated volume velocity. At a junction between two ducts, the acoustic pressure will also be continuous (Kinsler et al., 1999); thus the acoustic impedance has the useful property that it is continuous at junctions in a ducted system.

As will be shown, a knowledge of the acoustic impedance of sources would be useful, but such information is generally not available. However, those devices characterised by fixed cyclic volume displacement, such as reciprocating pumps and compressors and internal combustion engines, can be approximated as constant-acoustic-volume-velocity sources of infinite internal acoustic impedance (as the velocity of the mechanism producing the noise is unaffected by the acoustic field loading). On the other hand, other devices characterised by an induced pressure rise, such as centrifugal and axial fans and impeller-type compressors and pumps, are probably best described as constant-acoustic-pressure sources of zero internal impedance (as the sound pressure they produce is independent of the acoustic field loading). The case for the engine exhaust is a bit better, since the impedance for a simple exhaust pipe has been modelled as a vibrating piston at the end of a long pipe, analysed theoretically and verified by measurement (see Chapter 4, Figure 4.13), and such a termination is the most common case. A loudspeaker backed by a small air-tight cavity may be approximated as a constant-volume-velocity source at low frequencies where the wavelength is much greater than any cavity dimension. See Section 8.9.11 for further discussion about techniques to measure the acoustic impedances of acoustic sources and duct terminations.

8.7 Lumped Element Devices

The discussion of lumped element devices and the side branch resonator will be carried out in terms of acoustical circuits, which are analogues of electrical circuits. For this purpose, it will be necessary to develop expressions for the acoustical analogues of electrical inductances and capacitances. Using electrical analogies, voltage drop is the analogue of acoustic pressure, and current is the analogue of acoustic volume velocity. Thus, the acoustic pressure in a volume or across an orifice is the analogue of the voltage across a capacitance or an inductance, respectively. See Section 8.8.1 for a further discussion on acoustic-to-electrical analogies.

In electrical networks, one can calculate electrical power as the product of voltage and current. Referring to the definition of acoustic volume velocity in Section 8.6 and sound intensity given by Equations (1.78) and (1.79), the acoustic power transmission in a duct is proportional to the mean square pressure and particle velocity.

8.7.1 Impedance of an Orifice or a Short Narrow Duct

In the following subsections, the impedance equations will be derived using lumped element analysis followed by transmission line analysis.

8.7.1.1 Lumped Element Analysis

In the analysis in this section, the slug of fluid vibrating in the orifice, whose cross-section is illustrated in Figure 8.3, is assumed to be much smaller in length than a wavelength of sound and as such it may be treated as a lumped element. Consideration of the acoustically induced cyclic flow through an orifice in a thin plate, shows that the acoustic energy will pass through the orifice largely in the form of kinetic energy. There will be a small volume of fluid, somewhat larger than the actual orifice, that will participate in the induced motion. The bounds of this volume are obviously quite arbitrary, but the volume can be imagined as a little cylinder of air of cross-section equal to the area of the orifice and some effective length ℓ_e. Its motion will be governed by Newton's second law, written as:

$$-\frac{\partial p}{\partial x} = \rho\left[\frac{\partial u}{\partial t} + U\frac{\partial u}{\partial x}\right] \tag{8.5}$$

In Equation (8.5), p (Pa) is the acoustic pressure, x (m) is the displacement coordinate along the orifice axis (vertical direction in Figure 8.3), u (m/s) is the particle velocity, ρ (kg/m^3) is the density of the gas, t (s) is time and U (m/s) is any convection velocity in the direction (vertical direction in Figure 8.3) of sound propagation through the orifice of cross-sectional area, S.

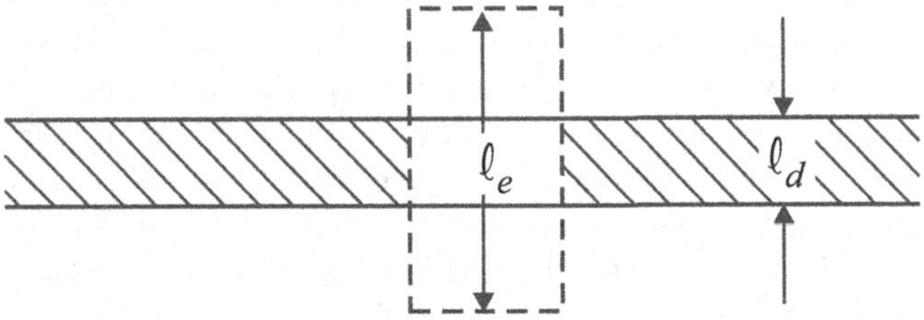

FIGURE 8.3 Schematic representation of acoustically induced flow through an orifice, of depth, ℓ_d, shown in cross-section. A cylinder of equivalent inertia of area S and effective length ℓ_e is indicated. The effective length is the orifice length plus an amount to account for the fluid on either side of the orifice also being affected by the motion of the fluid in the orifice.

To proceed, a solution is assumed for Equation (8.5) of the form:

$$u = u_0 \mathrm{e}^{\mathrm{j}(\omega t - kx)} \tag{8.6}$$

where ω (rad/s) is the angular frequency, $k = \omega/c = 2\pi/\lambda$ is the wavenumber (m^{-1}), c (m/s) is the speed of sound, λ (m) is the wavelength of sound and j is the imaginary number, $\sqrt{-1}$. It will be assumed that sound propagation and direction of mean flow are the same. The case for mean flow opposite to the direction of sound propagation is not analysed in detail here, but would predict an increasing end correction, $(1+M)$, in the last term of Equation (8.10) below (instead of the $(1-M)$ term) with a corresponding change in Equation (8.7) that follows. Substitution of Equation (8.6) into Equation (8.5) results in the following expression:

$$-\frac{\partial p}{\partial x} = \rho\left[\mathrm{j}\omega u + U\left[\frac{1}{u_0}\frac{\partial u_0}{\partial x} - \frac{\mathrm{j}\omega}{c}\right]u\right] \tag{8.7}$$

Equation (8.7) may be simplified by making use of the definition of the Mach number for the convection velocity, $M = U/c$, and the following approximations:

$$-\frac{\partial p}{\partial x} \approx \frac{p}{\ell_e} \tag{8.8}$$

$$\frac{1}{u_0}\frac{\partial u_0}{\partial x} \approx \frac{1}{u_0}\frac{u_0}{\ell_e} \approx \frac{1}{\ell_e} \tag{8.9}$$

Adding a resistive acoustic impedance, R_A, Equation (8.7) can be used to write:

$$Z_A = R_A + \mathrm{j}X_A = R_A + \frac{p}{Su} = R_A + \mathrm{j}\frac{\rho c}{S}k\ell_e(1-M) = R_A + \mathrm{j}\frac{\rho\omega\ell_e}{S}(1-M) \tag{8.10}$$

Reference to Figure 8.3 shows that the effective length, ℓ_e, is made up of three parts: the length, ℓ_d, of the orifice and an end correction at each end. The end corrections are also a function of the Mach number, M, of the flow through the orifice. Since the system under consideration is symmetric, the end correction for each end without flow is ℓ_0, so the effective length is:

$$\ell_e \approx \ell_d + 2\ell_0(1-M)^2 \tag{8.11}$$

The "no flow" end correction, ℓ_0, will be considered in detail later in the next section.

For the case where the length, ℓ_d, of the orifice is short compared to the effective length, ℓ_e, substitution of Equation (8.11) into Equation (8.10) gives the following result:

$$Z_A \approx R_A + \mathrm{j}\frac{\rho c}{S}k2\ell_0(1-3M+3M^2-M^3) \tag{8.12}$$

The first term (the real term) in Equations (8.10) and (8.12) can be evaluated using Equation (8.34).

For tubes of very small diameter (such that radius, $a < 0.002/\sqrt{f}$), Beranek (1954) indicates that the imaginary term in Equations (8.10) and (8.12) should be multiplied by 4/3 and the real part is given by Equation (8.37). For slits of cross-sectional area, S, the imaginary term in Equations (8.10) and (8.12) should be multiplied by 6/5, with ℓ_e set equal to the depth of the slit plus two end corrections (see Equation (8.39)). The real impedance is given by Equation (8.38).

Equation (8.12) describes experimental data quite well. The results of experiments indicate that the convection velocity, $\boldsymbol{U}$, could be a steady superimposed flow through the hole, a steady flow at grazing incidence to the hole or the particle velocity itself at high sound pressure levels. Equations (8.12) and (8.34) show that, for zero mean flow or low sound pressure levels, the reactive part of acoustic impedance is essentially inductive and the total acoustic impedance is:

$$Z_A \approx R_A + \mathrm{j}\frac{\rho\omega\ell_e}{S} = R_A + \mathrm{j}\frac{\rho ck\ell_e}{S} = R_A + \mathrm{j}X_A \tag{8.13}$$

However, for high sound pressure levels, or in the presence of a significant mean flow, the resistive part of the acoustic impedance becomes important, as shown in Equation (8.34). From Equation (8.13), it can be seen that the acoustical inductance, analogue of the electrical inductance, is $\rho\ell_e/S$, where S is the duct cross-sectional area.

8.7.1.2 Transmission Line Analysis

The following alternative approach is presented as it shows how lumped element analysis is really a first approximation to the more general transmission-line analysis. Consider a plane acoustic wave that propagates from the left in a duct of cross-sectional area, S, and open end, as shown in Figure 8.4.

At the open end on the right, a wave will be reflected to the left, back along the duct towards the source. The approximation is now made that the acoustic pressure is essentially zero at a point distant from the duct termination by an amount equal to the end correction ℓ_0, and for convenience, the origin of coordinates is taken at this point. Referring to Equation (1.35) the equation for waves travelling in both directions in the duct is:

$$\phi = B_1\mathrm{e}^{\mathrm{j}(\omega t - kx)} + B_2\mathrm{e}^{\mathrm{j}(\omega t + kx + \theta)} \tag{8.14}$$

Assuming that the wave is reflected at the open end on the right in Figure 8.4, with negligible loss of energy so that the amplitude of the reflected wave, B_2, is essentially equal to the amplitude

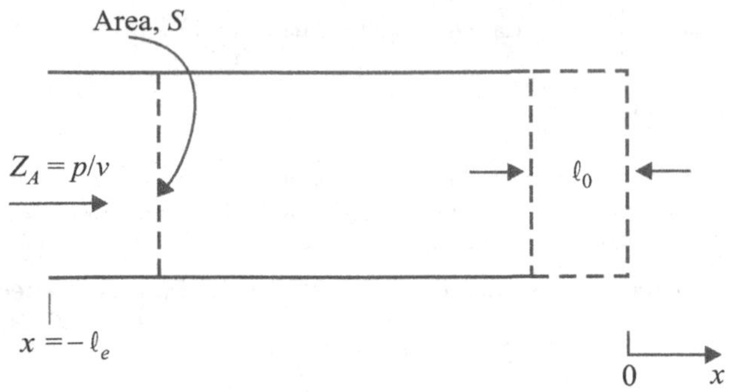

FIGURE 8.4 Schematic representation of an acoustic transmission line with an open end: Z_A is the acoustic impedance at the cross-section of area S, and ℓ_0 is the end correction.

of the incident wave, B_1, then $B_1 = B_2 = B$. At $x = 0$ the acoustic pressure and thus the velocity potential, ϕ, is required to be zero; thus, $\theta = \pi$. Use of Equation (1.12) gives the acoustic pressure as:

$$p = \rho \frac{\partial \phi}{\partial t} = \mathrm{j}\omega \rho B \mathrm{e}^{\mathrm{j}\omega t} \left[\mathrm{e}^{-\mathrm{j}kx} - \mathrm{e}^{\mathrm{j}kx}\right] \tag{8.15}$$

Similarly, use of Equation (1.11) gives for the particle velocity at any point, x, in the duct:

$$u = -\frac{\partial \phi}{\partial x} = \mathrm{j}kB\mathrm{e}^{\mathrm{j}\omega t} \left[\mathrm{e}^{-\mathrm{j}kx} + \mathrm{e}^{\mathrm{j}kx}\right] \tag{8.16}$$

The reactive part of the acoustic impedance at x, looking to the right, is:

$$\mathrm{j}X_A = \frac{p}{v} = \frac{p}{Su} = -\frac{\rho c}{S} \frac{\mathrm{e}^{\mathrm{j}kx} - \mathrm{e}^{-\mathrm{j}kx}}{\mathrm{e}^{\mathrm{j}kx} + \mathrm{e}^{-\mathrm{j}kx}} \tag{8.17}$$

or

$$\mathrm{j}X_A = -\mathrm{j}\frac{\rho c}{S} \tan kx \tag{8.18}$$

This expression can be used to estimate the acoustic impedance looking into a tube from the left at $x = -\ell_e$ as shown in Figure 8.4.

$$R_A + \mathrm{j}X_A = R_A + \mathrm{j}\frac{\rho c}{S} \tan k\ell_e \tag{8.19}$$

For small ℓ_e, Equation (8.13) follows as a first approximation. If the mean flow speed through the tube is non-zero and has a Mach number, M, and the resistive component, R_A of the impedance is included, Equation (8.19) becomes:

$$Z_A = R_A + \mathrm{j}\frac{\rho c}{S} \tan(k\ell_e(1 - M)) \tag{8.20}$$

where R_A is given by Equation (8.34). The effective length, ℓ_e, is used to describe the length of the tube because the actual length is increased by an end effect at each end of the tube. The amount by which the effective length exceeds the actual length is called the "end correction" (see Section 8.7.1.4). Generally, one end correction (ℓ_0) is added for each end of the tube so that the effective length is $\ell_e = \ell_d + 2\ell_0(1 - M)^2$, where ℓ_d is the physical length of the tube.

8.7.1.3 Impedance of a Perforated Plate

The acoustic impedance corresponding to a single hole in a perforated plate may be calculated using Equations (8.20) and (8.11), where ℓ_d is the thickness of the perforated plate. The acoustic impedance for $N = (P_{\mathrm{open}}/100) \times (S_p/S)$ holes is obtained by dividing the acoustic impedance for one hole by N, so that for a perforated plate of area S_p, the acoustic impedance due to the holes is:

$$Z_{Ah} = \frac{100}{P_{\text{open}} S_p} (R_A S + \text{j}\rho c \tan(k\ell_e(1 - M))) \tag{8.21}$$

where R_A is the acoustic resistance, defined in Equation (8.34), S is the area of a single hole and P_{open} is the % open area of the perforated plate, given by:

$$P_{\text{open}} = 100NS/S_p = 100N\pi a^2/S_p \tag{8.22}$$

where N is the total number of holes in area, S_p. The Mach number, M, refers to the speed of the flow either through the holes or across the face of the plate. The Mach number of the grazing flow across the plate is positive if it is on the same side of the holes as the side at which the acoustic impedance looking into the holes is required and negative if it is on the opposite side. The Mach number of the through flow is positive if the flow is in the same direction as the direction in which the acoustic impedance is required.

A more accurate expression for the acoustic impedance of a perforated plate involves including the mass of the plate between the holes and calculating the acoustic impedance assuming the plate mass impedance acts in parallel with the hole acoustic impedance as suggested by Bolt (1947). The acoustic impedance of the plate of mass per unit area, m, and area, S_p, is:

$$Z_{Ap} = \text{j}\omega m/S_p \tag{8.23}$$

Thus, the acoustic impedance of the plate mass in parallel with the holes in the plate is:

$$Z_A = \frac{Z_{Ah}}{1 + Z_{Ah}/Z_{Ap}} = \frac{\dfrac{100}{P_{\text{open}} S_p}(R_A S + \text{j}\rho c \tan(k\ell_e(1 - M))R_A S)}{1 + \dfrac{100}{\text{j}\omega m P_{\text{open}}}(R_A S + \text{j}\rho c \tan(k\ell_e(1 - M)))} \tag{8.24}$$

Note that for standard thickness (less than 3 mm) perforated plates, it is sufficiently accurate to replace the tan functions in the preceding equation by their arguments.

8.7.1.4 End Correction

Reference to Figure 8.3 shows that the end correction accounts for the mass reactance of the medium (air) just outside of the orifice or at the termination of an open-ended tube. The mass reactance, however, is just the reactive part of the radiation impedance presented to the orifice, treated here as a vibrating piston. As shown in Chapter 4, the radiation impedance of any source depends on the environment into which the source radiates. Consequently, the end correction will depend on the local geometry at the orifice termination.

Circular orifice in a baffle. In general, for a circular orifice of radius, a, centrally located in a baffle in a long tube with a circular cross-section such that the ratio, ξ, of orifice diameter to tube diameter is less than 0.6, the end correction *without* either through or grazing flow for each side of the hole is (Bolt et al., 1949):

$$\ell_0 = \frac{8a}{3\pi}(1 - 1.25\xi) \tag{8.25}$$

As ξ tends to zero, the value of the end correction tends to the value for a piston in an infinite baffle, as may be inferred from the discussion in Section 4.6. In particular, using Equations (4.112) and (4.114), substituting $k = \omega/c$ and converting the piston radiation impedance to acoustic impedance by dividing by the area squared, it can be shown that the imaginary part of Equation (4.112) is the same as the RHS of Equation (8.13).

When there is flow of Mach number, M, either through the hole (positive M in the direction of sound propagation) or past the hole (grazing flow with positive M in any direction), the end correction ℓ_0 must be multiplied by $(1 - M)^2$. End corrections for quarter-wavelength tubes (QWTs) and Helmholtz resonators are discussed in Section 8.8.2.1.

Non-circular Orifice. For orifices not circular in cross-section, an effective radius, a, can be defined, provided that the ratio of orthogonal dimensions of the orifice is not much different from unity. In such cases,

$$a \approx 2S/P_D \tag{8.26}$$

where S (m^2) is the cross-sectional area of the orifice and P_D (m) is its perimeter. Alternatively, if the orifice is of cross-sectional area, S, and aspect ratio (major dimension divided by minor orthogonal dimension) n, the effective radius, a, may be determined using the following equation:

$$a = K\sqrt{S/\pi} \tag{8.27}$$

where K is plotted in Figure 8.5 as a function of the aspect ratio, n.

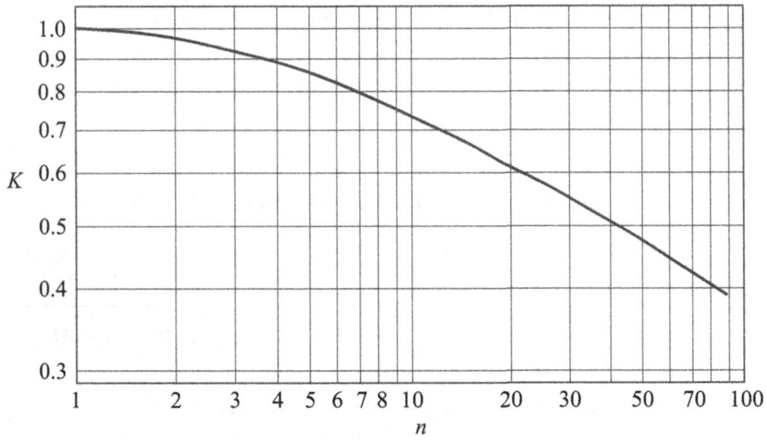

FIGURE 8.5 Ratio, K, of the effective radius for a non-circular orifice to the radius of a circular orifice of the same area. The non-circular orifice aspect ratio is n.

Unflanged tube. For tubes that are unflanged and look into free space, as will be of concern later in discussing engine exhaust tailpipes, the end correction *without* either through or grazing flow is (Beranek (1954), p. 133):

$$\ell_0 = 0.61a \tag{8.28}$$

Perforated plate. For holes separated by a distance q (centre to centre, where $q > 2a$) in a perforated plate, the end correction for each side of a single hole, *without* either through or grazing flow is:

$$\ell_0 = \frac{8a}{3\pi}(1 - 0.43a/q) \tag{8.29}$$

The effective length, ℓ_e, of the holes is derived from Equations (8.11) and (8.29) as:

$$\ell_e = \ell_d + \left[\frac{16a}{3\pi}\left(1 - 0.43\frac{a}{q}\right)\right](1 - M)^2 \tag{8.30}$$

An alternative expression for the effective length, which may give slightly better results than Equation (8.30), for grazing flow across the holes, and which only applies for flow speeds such that $u_\tau/(\omega d) > 0.03$, is (Dickey and Selamet, 2001):

$$\ell_e = \ell_d + \left[\frac{16a}{3\pi}\left(1 - 0.43\frac{a}{q}\right)\right]\left[0.58 + 0.42\mathrm{e}^{-23.6(u_\tau/(\omega d_h) - 0.03)}\right] \tag{8.31}$$

where d_h is the diameter of the holes in the perforated plate, and the friction velocity, u_τ, is:

$$u_\tau = 0.15U[Re]^{-0.1} \tag{8.32}$$

"*Re*" is the Reynolds number of the grazing flow given by:

$$Re = \frac{U d_h \rho}{\mu} \tag{8.33}$$

where μ is the dynamic gas viscosity (1.83×10^{-5} kg m^{-1} s^{-1} for air at 20°C), ρ is the gas density (1.206 kg/m^3 for air at 20°C) and U is the mean flow speed (m/s).

TABLE 8.2 Summary of end correction equations for no flow. To include flow, the RHS of the equations is multiplied by $(1 - M^2)$

Element type	Relevant equations
Circular orifice in a baffle (one side)	(8.25)
One end of a circular tube flush with a baffle	(8.25)
Unflanged tube	(8.28)
Non-circular orifice or tube additional equations	(8.26) and (8.27)
Perforated plate (one side)	(8.29), (8.30) and (8.31)

8.7.1.5 Acoustic Resistance

The acoustic resistance, R_A (kg m^{-4}s^{-1}), of an orifice or a tube of length, ℓ_d (m), cross-sectional area, S (m^2), and internal cross-sectional perimeter, D (m), may be calculated using the following equation:

$$R_A = \frac{\rho c}{S}\left[\frac{ktD\ell_d}{2S}\left[1 + (\gamma - 1)\sqrt{\frac{5}{3\gamma}}\right] + 0.288kt\log_{10}\left[\frac{4S}{\pi h^2}\right] + \epsilon\frac{Sk^2}{2\pi} + 0.7|M|\right] \tag{8.34}$$

The derivation of Equation (8.34) was done with reference principally to Morse and Ingard (1968). In Equation (8.34), ρc is the characteristic acoustic impedance for air (414 MKS rayls for air at room temperature), γ is the ratio of specific heats (1.40 for air), $k (= \omega/c)$ is the wavenumber (m^{-1}), ω is the angular frequency (rad/s), c is the speed of sound (m/s), M is the Mach number of any mean flow through the tube or orifice or across the face of the orifice and:

$$t = \sqrt{2\mu/\rho\omega} \tag{8.35}$$

where μ is the dynamic viscosity of the gas and ρ is the gas density.

The first term in Equation (8.34) accounts for attenuation along the length of the tube. This term is generally negligible except for small tubes or high frequencies. On the other hand, because the term depends on the length ℓ_d, it may become significant for very long tubes. It was derived using Equations (6.4.5), (6.4.31) and (9.1.12) from Morse and Ingard (1968).

The second term in Equation (8.34) accounts for viscous loss at the orifice or tube entry, and is a function of the quantity h. For orifices in thin plates of negligible thickness ($\ell_d \approx 0$), h is either the half plate thickness or the viscous boundary layer thickness t, given by Equation (8.35), whichever is bigger. Alternatively, if the orifice is the entry to a tube ($\ell_d \gg 0$), then h is the orifice edge radius or the viscous boundary layer thickness, whichever is bigger. This term was derived using Equations (6.4.31) and (9.1.23) from Morse and Ingard (1968).

The third term in Equation (8.34) accounts for radiation loss at the orifice or tube exit and was derived assuming that the tube exit radiates sound as a piston (see Equation (4.112)). For tubes that radiate into spaces of diameter much less than a wavelength of sound (for example, an expansion chamber in a duct) the parameter ϵ may be set equal to zero. For tubes that terminate in a well-flanged exit or radiate into free space from a very large plane wall or baffle, ϵ should be set equal to 1. Alternatively, for tubes that radiate into free space, but without a flange at their exit, ϵ should be set equal to 0.5. When a Helmholtz resonator or quarter-wave tube (QWT) is attached as a side-branch to a main duct, the radiation impedance is greater than radiation into a free space, and ϵ can be set equal to 2. This agrees with Equation (2.149) in Munjal (2014), when the unconventional definition of acoustic impedance used in that reference is taken into account.

The fourth and last term of Equation (8.34) accounts for mean flow through the orifice and is often an important term in the presence of a mean flow. It is valid only for Mach numbers less than about 0.2. This term is provided in Sullivan (1979b) and is valid for grazing flow past the orifice in any direction. If flow is through the orifice in any direction, the $0.7M$ term should be replaced with $2.7M$ (Sullivan, 1979b). For high sound pressure levels in the absence of a mean flow, the mean flow Mach number, M, may be replaced with the Mach number corresponding to the particle velocity amplitude and the $2.7M$ term used.

Alternatively, for grazing flow across a plate with circular holes, each having a cross-sectional area, S, with a speed such that $u_\tau/(\omega d_h) \geq 0.05$, Dickey and Selamet (2001) give the following expression for the acoustic resistance:

$$R_A = \frac{\rho c k d_h}{S}\left[-0.32 + \frac{9.57 u_\tau}{\omega d_h}\right] \tag{8.36}$$

where u_τ is defined by Equation (8.32) and d_h is the hole diameter.

For tubes that have a very small diameter ($d_h < 0.004/\sqrt{f}$), Beranek (1954) gives the following expression for the acoustic resistance:

$$R_A = \frac{8\pi\mu\ell_d}{S^2} \tag{8.37}$$

where μ is the dynamic viscosity for air (1.84×10^{-5} kg m^{-1} s^{-1} at 20°C), which varies with absolute temperature, T, in Kelvin as $\mu \propto T^{0.7}$.

For slits with an opening height, ℓ_w (much smaller than the slit width, a), the acoustic resistance is given by (Beranek, 1954):

$$R_A = \frac{12\mu\ell_w}{\ell_d^2 S} \tag{8.38}$$

where ℓ_d is the effective depth of the slit. The effective depth of the slit is the actual depth plus two end corrections, ℓ_0, (one for each end), given by (Morse and Ingard, 1968, p. 488):

$$\ell_0 = (2 + \pi)\ell_w/\pi \tag{8.39}$$

For situations where there is grazing flow past the orifice and a small bias flow either into or out of the orifice, expressions for the acoustic impedance can be found in Sun et al. (2002). Lee and Ih (2003) compare many expressions for the acoustic impedance of an orifice with grazing-flow and through-flow.

8.7.2 Impedance of a Volume

From the definition of electrical capacitance as charge per unit induced voltage, it may be concluded that a volume should be the acoustical analogue of the electrical capacitor. To begin, consider the adiabatic compression of a gas contained in a volume, V. It has been shown both experimentally and theoretically that the heat conduction of air is so low that adiabatic compression is a very good approximation for compression in volumes of dimensions very much less than a wavelength of sound (Daniels, 1947; Golay, 1947). Thus:

$$p/P_0 + \gamma(\Delta V/V) = 0 \tag{8.40}$$

In this equation, p is the acoustic pressure, P_0 is the absolute pressure (atmospheric pressure if the volume is open to the atmosphere), γ is the ratio of specific heats, V is the volume and ΔV is the incremental change in volume.

It will be assumed that the volume flow (or current) v into the volume is:

$$v = v_0 \mathrm{e}^{\mathrm{j}\omega t} \tag{8.41}$$

then, as a decrease in volume is achieved by a positive pressure and associated volume flow:

$$\Delta V = -\int v dt = -v/\mathrm{j}\omega \tag{8.42}$$

Substitution of Equation (8.42) into Equation (8.40) and noting that the impedance is in the positive x-direction (see Figure 8.6) leads to the following expressions for the acoustic impedance (assuming the acoustic resistance is negligible for this case):

$$Z_A \approx \mathrm{j}X_A = \frac{p}{v} = -\mathrm{j}\frac{\gamma P_s}{V\omega} = -\mathrm{j}\frac{\rho c^2}{V\omega} \tag{8.43}$$

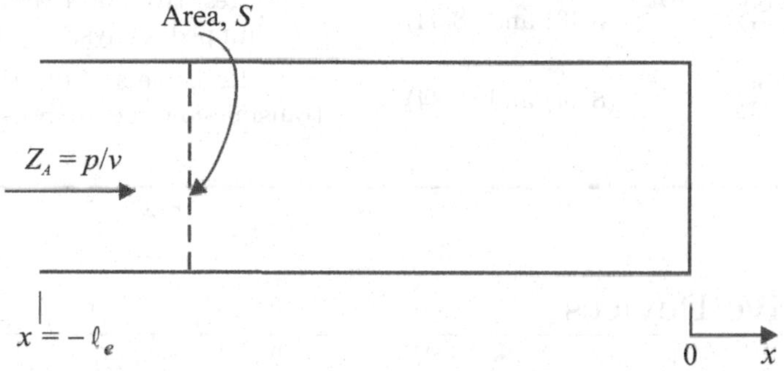

FIGURE 8.6 Schematic representation of an acoustic volume with a closed end: Z_A is the acoustic impedance at the cross-section of area S.

The last two alternate forms of Equation (8.43) follow directly from the relationship between the speed of sound, gas density and compressibility. Consideration of the alternate forms shows that the acoustical capacitance, the analogue of electrical capacitance, is either $V/\gamma P_s$ or $V/\rho c^2$.

Equation (8.43) may be shown to be a first approximation to the more general transmission-line analysis represented by Equation (8.44). Referring to the closed end duct shown in Figure 8.6, and taking the coordinate origin at the closed end, the analysis is the same as was followed in deriving Equation (8.18), except that as the end is now closed, the reflected wave is in phase with the incident wave resulting in pressure doubling and zero particle velocity at the reflecting end (x=0). The reactive acoustic impedance looking right, into the tube, can be derived using the same procedure as for an open tube and including a resistance term, the acoustic impedance is written as:

$$Z_A = R_A + \mathrm{j}X_A = R_A - \mathrm{j}\frac{\rho c}{S}\cot(kx) \tag{8.44}$$

This expression can be used to estimate the acoustic impedance looking into a tube closed at the opposite end, and of cross-sectional area, S, and effective length, ℓ_e, by replacing x with the effective length, $-\ell_e$, which includes an end correction so it is a bit larger than the physical length of the tube. For small ℓ_e, Equation (8.44) reduces to Equation (8.43) as a first approximation, where $V = S\ell_e$. For the situation of flow past the end of the tube, the acoustic impedance may be written as:

$$Z_A = R_A - \mathrm{j}\frac{\rho c}{S}\cot(k\ell_e(1-M)) \tag{8.45}$$

Note that in most cases, the resistive acoustic impedance associated with a volume is considered negligible, except in the case of a QWT.

TABLE 8.3 Summary of impedance equations without flow and with flow of Mach number of M

Element type	Relevant equations	Comments
Circular orifice or open tube	(8.10), (8.12) and (8.34)	Resistive and reactive, lumped analysis, flow of Mach number, M
Circular orifice or open tube	(8.20) and (8.34)	Resistive and reactive, transmission line analysis, flow of Mach number, M
Perforated plate	(8.20) and (8.34)	Resistive and reactive, transmission line analysis, flow of Mach number, M
Closed tube or volume, V	(8.43) and (8.34)	Resistive and reactive, lumped analysis, no flow
Closed tube or volume, V	(8.44) and (8.34)	Resistive and reactive, transmission line analysis, no flow

8.8 Reactive Devices

Commercial mufflers for internal combustion engines are generally of the reactive type. These devices are designed, most often by cut and try, to present an essentially imaginary input acoustic impedance to the noise source in the frequency range of interest. The input power and thus the radiated sound power is then reduced to some acceptably low level.

The subject of reactive muffler design, particularly as it relates to automotive mufflers, is difficult, although it has received much attention in the literature (Jones, 1984; Davies, 1992a,b, 1993). Consideration of such design will be limited to three special cases. Some attention will also be given in this section to the important matters of pressure loss. Flow-generated noise, which is often important in air conditioning mufflers but is often neglected in muffler design, will also be discussed.

8.8.1 Acoustical Analogues of Kirchhoff's Laws

The following analyses of reactive devices are based on acoustical analogies of the well-known Kirchhoff laws of electrical circuit analysis. Referring to Section 8.6, it may be observed that the acoustic volume velocity is continuous at junctions and is thus the analogue of electrical current. Similarly, acoustic pressure is the analogue of voltage. Table 8.4 lists analogies between acoustical and electrical concepts.

TABLE 8.4 Analogies between acoustical and electrical concepts

Acoustical	Electrical
Pressure	Voltage
Volume velocity	Current
Mass	Inductance
Compliance	Capacitance
Resistance	Resistance

Thus, in acoustical terms, the Kirchhoff current and voltage laws may be stated as:

1. the algebraic sum of acoustic volume velocities at any instant and at any location in the system must be equal to zero (that is, the volume velocity flowing into a junction must equal that flowing out) (see Figure 8.7(a)); and
2. the algebraic sum of the acoustic-pressure losses around any closed loop in the system at any instant must be equal to zero (which implies that the pressure drop across any element that connects two parts of a circuit must be the same) (see Figure 8.7(b)).

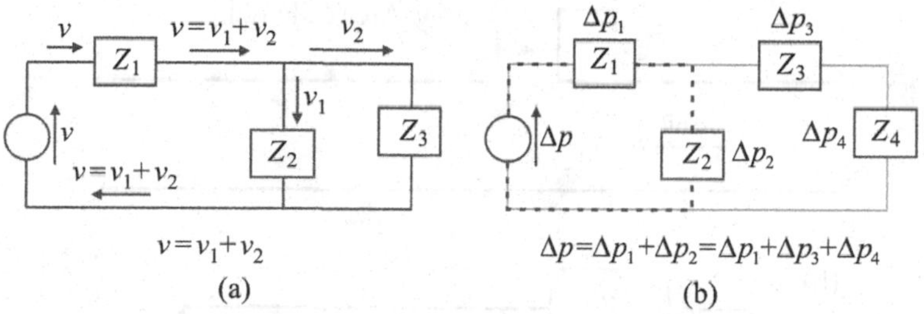

FIGURE 8.7 Illustration of Kirchhoff laws. (a) Current (or acoustic volume velocity) law and (b) voltage (or acoustic pressure) law.

In the remainder of the discussion on reactive mufflers, the subscript, "A", which has been used to denote acoustic impedance as opposed to specific acoustic impedance, will be dropped to simplify the notation and impedance will mean acoustic impedance unless otherwise indicated.

8.8.2 Side Branch Resonator

A particularly useful device for suppressing pure tones of constant frequency, such as might be associated with constant speed pumps or blowers, is the side branch resonator. The side branch resonator functions by placing a very low impedance in parallel with the impedance of the remainder of the line at its point of insertion. It is most effective when its internal resistance is low, and it is placed at a point in the line where the impedance of the tone to be suppressed is real. This point will be considered further later.

The side branch resonator may take the form of a short length of pipe, for example, a quarter-wavelength tube (QWT), whose length (approximately a quarter of the wavelength of sound at the frequency of interest plus an end correction of approximately 0.3 × the pipe internal diameter) may be adjusted to tune the device to maximum effectiveness. The QWT diameter should be relatively constant along its length, with no step changes, as these will compromise the performance. The impedance of the QWT with grazing flow of Mach number, M, is:

$$Z_{\text{side}} = -\frac{\mathrm{j}\rho c}{S}\cot[k\ell_e(1-M)] + R_{\text{side}} \quad (= 0 + R_{\text{side}} \quad \text{if } \ell_e = \lambda/4) \tag{8.46}$$

where $R_{\text{side}} = R_A$ can be calculated using Equation (8.34).

The resistive impedance of the entrance to the QWT, R_{side}, is a function of the grazing flow speed, and considerable research has been reported for the impedance of small-diameter holes with grazing flow, such as flow over perforated plates or flow through perforated tubes. However, there is no expression available for large-diameter holes, although Equation (8.34) may be used to obtain an approximate estimate. It has been found experimentally that the acoustic response varies considerably for variations in hole diameter (Peat et al., 2003; Karlsson and Åbom, 2010), side-branch geometry (Knotts and Selamet, 2003; Howard and Craig, 2014b) and ratio of cross-sectional areas between the main duct and side-branch resonator. For the simplest case, where there is no flow, it is assumed that there is no additional impedance at the QWT entrance ($Z_{\text{side}} = 0$). This enables comparison of the theoretical predictions with the finite element analysis results. For further discussion the reader is referred to Howard and Craig (2014b).

An alternative type of side branch resonator is the Helmholtz resonator, which consists of a connecting orifice or tube (neck) and a backing volume, as indicated schematically in Figure 8.8(a), with an equivalent acoustical circuit shown in Figure 8.8(b).

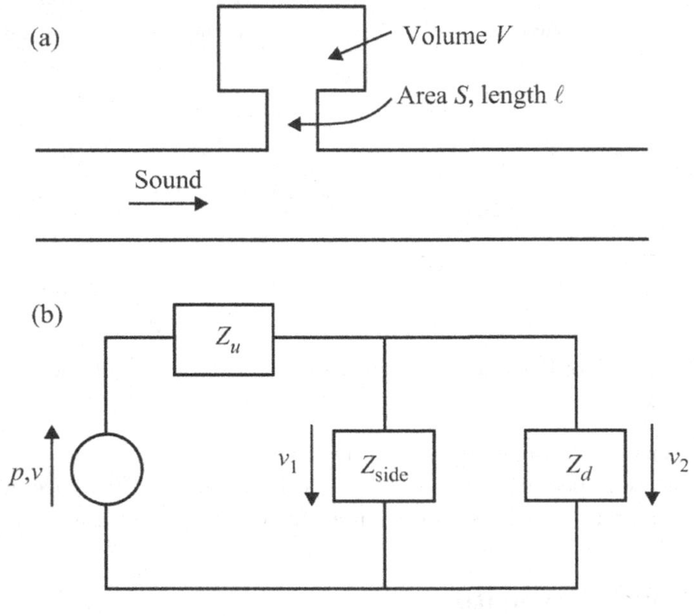

FIGURE 8.8 Side branch resonator: (a) acoustical system; (b) equivalent acoustical circuit.

The side branch Helmholtz resonator appears in the equivalent acoustical circuit of Figure 8.8(b) as a series acoustic impedance, Z_{side}, in parallel with the downstream duct impedance, Z_d. The quantity, Z_u, is the acoustic impedance of the duct upstream of the resonator. The series acoustic impedance, Z_{side}, consists of a capacitance, Z_C, associated with the resonator cavity, an inductance, Z_t, associated with the resonator connecting orifice or tube and a resistance, R_{side}, in series as shown in the following equation:

$$Z_{\text{side}} = \frac{\mathrm{j}\rho\omega\ell_e}{S} - \frac{\mathrm{j}\rho c^2}{V\omega} + R_{\text{side}} \tag{8.47}$$

The electrical capacitance analogy simulates the potential energy in the resonator volume, as a result of the fluctuating sound pressure in the volume. The electrical inductance analogy simulates the kinetic energy in the resonator neck, as a result of the fluctuating volume velocity in the neck. The electrical resistance analogy simulates energy loss, which occurs mainly in the neck due to frictional effects along the walls and at the ends where the cross-section changes rapidly.

The side branch has a resonance or minimum impedance when the sum of the capacitive and inductive impedances is equal to zero. Using lumped element analysis and setting the inductive term given by Equation (8.13) equal to the negative of the capacitive term given by Equation (8.43) gives the following expression for the frequency, ω_0 (rad/s), of the series resonance for the Helmholtz resonator:

$$\omega_0 = c\sqrt{S/\ell_e V} \tag{8.48}$$

where S and ℓ_e are the cross-sectional area and effective length (actual length plus two end corrections), respectively, of the connecting neck and V is the volume of the cavity.

Equation (8.48) has been shown to become inaccurate when the resonator cross-sectional or length dimensions exceed 1/16 of a wavelength or when the aspect ratio of the resonator cavity dimensions is much different to unity. Panton and Miller (1975) provided a more accurate expression for cylindrical Helmholtz resonators that takes into account axial wave propagation in the resonator chamber, but still requires that the neck dimensions are small compared to a wavelength of sound. Their expression is:

$$\omega_0 = c\sqrt{\frac{S}{\ell_e V + 0.33\ell_V^2 S}} \tag{8.49}$$

where ℓ_V is the cylindrical cavity length, which is concentric with the cylindrical neck.

Equation (8.49) is still not as accurate as really needed. However, Li (2003) gives a more accurate expression for the resonance frequency as:

$$\omega_0 = c\sqrt{-\frac{3\ell_e S_c + \ell_c S}{2\ell_e^3 S_c} + \sqrt{\left(\frac{3\ell_e S_c + \ell_c S}{2\ell_e^3 S_c}\right)^2 + \frac{3S}{\ell_e^3 S_c \ell_c}}} \tag{8.50}$$

In using Equation (8.50), there are no restrictions on the cavity or neck lengths, although the cavity diameter must be less than a wavelength at the resonance frequency.

If the orifice is offset from the centre of the cavity, the above equations will significantly overestimate the resonator resonance frequency. In the presence of grazing flow of Mach number, M, Equations (8.47) to (8.50) can be used by substituting $(1 - M)\ell_e$ for ℓ_e in all places.

8.8.2.1 End Corrections for a Helmholtz Resonator Neck and Quarter-Wavelength Tube

To calculate the effective length of a Helmholtz resonator neck or QWT, it is necessary to determine the end correction at each end of the neck. For a Helmholtz resonator, one end looks into the resonator volume and the other looks into the main duct on which the resonator is mounted. For the QWT, there is only the end correction for the end looking into the main duct.

For a cylindrical Helmholtz resonator where the cavity is concentric with the neck (of radius, a), the end correction for the end of the neck connected to the cavity is given by Selamet and Ji (2000), for configurations where $\xi < 0.4$, as:

$$\ell_0 = 0.82a(1 - 1.33\xi) \tag{8.51}$$

which is very similar to Equation (8.25) for an orifice in an anechoically terminated duct, and where ξ is the ratio of the neck diameter to cavity diameter.

The end correction for the neck end attached to the main duct is difficult to determine analytically. However, Ji (2005) gives the following equations, based on a Boundary Element Analysis.

$$\ell_0 = a \begin{cases} (0.8216 - 0.0644\xi - 0.694\xi^2); & \xi \leq 0.4 \\ (0.9326 - 0.6196\xi); & 0.4 < \xi \leq 1.0 \end{cases} \tag{8.52}$$

where in this case, ξ is the ratio of neck diameter to main duct diameter. In the presence of a mean grazing flow of Mach number, M, in the duct, the right-hand-side of Equation (8.52) as well as (8.53) below should be multiplied by $(1 - M)^2$.

Dang et al. (1998) undertook a series of measurements using a closed end tube attached to a duct and derived an empirical expression as follows, which applies to a QWT and also the neck duct interface for a Helmholtz resonator. This expression agrees remarkably well over the range $0 < \xi \leq 1.0$, with the expression derived by Ji (2005) and represented in Equation (8.52).

$$\ell_0 = a\left(\frac{1.27}{1 + 1.92\xi} - 0.086\right); \qquad 0.2 < \xi < 5.0 \tag{8.53}$$

where a is the radius of the QWT or radius of the neck of the Helmholtz resonator.

As an alternative, Kurze and Riedel (2013)(p. 287) suggest the following expression for the end-correction:

$$\ell_0 = \pi a \left(1 - 1.47\xi^{0.5} + 0.47\xi^{1.5}\right) \tag{8.54}$$

The end-correction length that should be added is the subject of ongoing research and da Silva and Nunes (2016) present a summary of various expressions.

For cases where the Helmholtz resonator is a narrow, rectangular slit of opening height, ℓ_w, opening width, a and depth, ℓ_d, the effective length, ℓ_0, to be added to each end to obtain the total depth is (Smits and Kosten, 1951):

$$\ell_0 = \left[\frac{1}{\pi} + \frac{2}{\pi}\log_e\left(\frac{2a}{\ell_w}\right)\right]\ell_w \tag{8.55}$$

Comparison of Equation (8.55) with experimental data for various slit heights is provided by Sakamoto et al. (2022), who show that Equation (8.55) is reasonably accurate for slit heights between 0.3 and 1 mm, with the accuracy decreasing with increasing slit heights above 1 mm.

8.8.2.2 Quality Factor of a Helmholtz Resonator and Quarter-Wavelength Tube (QWT)

A quality factor, Q, may be defined for any resonance, as the ratio of energy stored divided by the energy dissipated during one cycle:

$$Q = 2\pi E_s / E_D \tag{8.56}$$

Energy stored, E_s, is proportional to the square root of the product of the capacitative and inductive impedances, while energy dissipated, E_D, is proportional to the resistive impedance. This leads to the following expression for the series circuit quality factor, Q, for a Helmholtz resonator (where it is assumed that $k\ell_e$ is small so that $k\ell_e = \tan(k\ell_e)$):

$$Q = \frac{1}{R_{\text{side}}}\sqrt{Z_C Z_t} \approx \frac{\rho c}{R_{\text{side}}}\sqrt{\frac{\ell_e}{SV}} \tag{8.57}$$

where ℓ_e, S, V, Z_C and Z_t are the effective length of the neck, cross-sectional area of the neck, volume of the resonator chamber, impedance of the resonator cavity and inductive impedance of the resonator throat, respectively.

In most instances, the acoustic resistance term, R_{side}, is dominated by the resistance of the neck of the resonator and may be calculated using Equation (8.34). However, the placement of acoustically absorptive material in the resonator cavity or, especially in the neck, will greatly increase the acoustic resistance. This will broaden the bandwidth over which the resonator will be effective at the expense of reducing the effectiveness in the region of resonance.

For a QWT of cross-sectional area, S, and effective length, ℓ_e, the resonance frequency and quality factor are given by (Beranek (1954), pp. 129 and 138):

$$\omega_0 = \frac{\pi c}{2\ell_e}; \quad Q = \frac{\pi \rho c}{6 R_{\text{side}} S} \tag{8.58}$$

The preceding discussion for Helmholtz resonators and QWTs is applicable at low frequencies for which no dimension of the resonator exceeds a quarter of a wavelength. It is also only applicable for resonator shapes that are not too different to a sphere or a cube. Howard et al. (2000) showed, using finite element analysis, that the resonance frequencies of Helmholtz resonators are dependent on the resonator aspect ratio. Resonators (both QWT and Helmholtz) have many additional resonances at frequencies higher than the fundamental given by Equations (8.48) and (8.58). At each of these frequencies, the resonator IL is substantial. This multi-resonance nature is useful in the design of large industrial resonator mufflers, which are considered later in this chapter.

The quality factor, Q, which is proportional to the resonator bandwidth, is a function of the ratio of the resonator (or QWT) neck cross-sectional area to the cross-sectional area of the duct on which it is installed. The larger this ratio, the smaller will be the quality factor. According to Equation (6.23), the smaller the quality factor, the larger will be the frequency bandwidth over which the resonator will provide significant noise reduction. However, even for area ratios approaching unity, the quality factor is relatively high and the corresponding frequency bandwidth is small, resulting in noise attenuation that is dependent on the frequency stability of the tonal noise source. This, in turn, is dependent on the stability of the physical parameters on which the noise depends, such as duct temperature or rotational speed of the equipment generating the noise. Thus, large changes in the effectiveness of the resonator can become apparent as these physical quantities vary. This problem could be overcome by using a control system to drive a moveable piston to change the volume of the resonator so that the tone is maximally suppressed. Alternatively, the limited bandwidth problem could be partly overcome by using two or more resonators tuned to slightly different frequencies, and physically separated by at least one wavelength of sound at the frequency of interest (Ihde, 1975).

8.8.2.3 Insertion Loss Due to Side Branch

Referring to Figure 8.8(b), the following three equations may be written using the acoustical analogues of Kirchhoff's current and voltage laws (see Section 8.8.1). First, the pressure drop around one of the loops (the outside loop, for example) is equal to the pressure across the source, p (see Figure 8.8(b)). Thus:

$$p = v_2 Z_d + (v_1 + v_2) Z_u \tag{8.59}$$

Second, the volume-velocity flow, v, into the junction to the right of Z_u must equal the total flow, $v_1 + v_2$, out of the junction. Thus:

$$v = v_1 + v_2 \tag{8.60}$$

Third, the pressure drop across Z_{side} must equal that across Z_d, so that:

$$v_1 Z_{\text{side}} = v_2 Z_d \tag{8.61}$$

The details of the analysis to follow depend on the assumed acoustic characteristics of the sound source. To simplify the analysis, the source is modelled as either a constant-acoustic-volume-velocity (infinite internal impedance) source or a constant-acoustic-pressure (zero internal impedance) source (see Sections 8.6 and 8.7 for a discussion of impedances). In the much of the following reactive muffler analysis, two different source types will be considered. One is a constant-volume-velocity-amplitude source (henceforth referred to a a constant-volume-velocity source) for which the volume velocity amplitude of the source is independent of the acoustic load seen by the source. The second source type is a constant-acoustic-pressure source (henceforth referred to as a constant-acoustic-pressure source) for which the acoustic pressure generated by the source is independent of the acoustic load seen by the source.

For a constant-volume-velocity source, consider the effect on the power transmission to the right in the duct of Figure 8.8 when the side branch resonator is inserted into the system. Initially, before insertion, the power transmission is proportional to the supply volume-velocity squared, $|v|^2$, whereas after insertion the power transmission is proportional to the load volume-velocity squared, $|v_2|^2$. Insertion loss, IL, is defined as a measure of the decrease in transmitted power in decibels. A large, positive IL corresponds to a large decrease.

$$\mathrm{IL} = 20\log_{10}\left|\frac{v}{v_2}\right| \tag{8.62}$$

Using the definition of IL given by Equation (8.62), use of Equations (8.60) and (8.61) gives the following expression for the side-branch IL:

$$\mathrm{IL} = 20\log_{10}\left|1 + \frac{Z_d}{Z_{\mathrm{side}}}\right| \tag{8.63}$$

Alternatively, the source, such as a fan, can be modelled as one of constant-acoustic-pressure, which produces the same acoustic pressure in the medium surrounding it, regardless of the acoustic impedance provided to the source by its surroundings. In this case, by similar reasoning, before insertion of the side branch, the acoustic power transmission is proportional to the pressure loss squared across Z_u and Z_d, which is $|p|^2$. After insertion of the side branch, the acoustic power transmission is proportional to the pressure loss squared across Z_d, which is $|v_2 Z_d|^2$. The Insertion Loss is thus defined as:

$$\mathrm{IL} = 20\log_{10}\left|\frac{p}{v_2 Z_d}\right| \tag{8.64}$$

Combining Equations (8.59), (8.61) and (8.64) gives the following expression for the IL of a side branch resonator for a constant-acoustic-pressure source:

$$\mathrm{IL} = 20\log_{10}\left|1 + \frac{Z_u(Z_d + Z_{\mathrm{side}})}{Z_{\mathrm{side}} Z_d}\right| \tag{8.65}$$

To maximise the IL for both types of sound source, the magnitude of Z_{side} must be made small while at the same time, the magnitude of Z_d (as well as Z_u for a constant-acoustic-pressure source) must be made large. The side-branch impedance, Z_{side}, is made small by making the side branch resonant (zero reactive impedance, such as a uniform tube, one-quarter of a wavelength long), and the associated resistive impedance as small as possible (rounded edges, no sound-absorptive material). The quantities Z_d and Z_u are made large by placing the side branch at a location on the duct where the internal acoustic pressure is a maximum. This can be accomplished by *placing the side branch at an odd multiple of quarter wavelengths (minus the duct end correction) from the duct exhaust*, and as close as possible to the noise source (but no closer than about three duct diameters for fan noise sources to avoid undesirable turbulence effects). Because acoustic-pressure maxima in the duct are generally fairly broad in the duct axial direction, the resonator position need not be precise. On the other hand, the frequency of maximum attenuation is very sensitive to resonator physical dimensions, and it is wise to allow some means of fine-tuning the resonator on-site (such as a moveable piston to change the effective volume or length).

One should be prepared for loss in performance with an increase in mean flow through the duct, because the resistive part of the side branch resonator impedance will usually increase, as indicated by Equation (8.34). For example, a decrease in IL from 30 dB to 10 dB at resonance, with flow speeds above about 40 m/s, as a result of the increased damping caused by the flow, has been reported in the literature (Gosele, 1965). Meyer et al. (1958) measured an increase of 20% in the resonance frequency in the presence of a flow rate of 70 m/s.

The value of the resistance, R_{side}, which appears in Equations (8.57) and (8.58), is difficult to calculate accurately and control by design, although Equation (8.34) provides an adequate approximation to the actual value in most cases. As a rule of thumb, the related quality factor, Q, for a side branch resonator may be expected to range between 10 and 100 with a value of about 30 being quite common.

Alternatively, as shown in Equation (6.23), if the sound pressure within the resonator (preferably at the closed end of a QWT or on the far wall of a Helmholtz resonator) can be measured with a microphone while the resonator is driven by an external variable frequency source, the quality factor Q may be readily determined. This measurement will usually require mounting the resonator on the wall of a duct or large enclosure and introducing the sound into the duct or enclosure using a speaker backed by an additional small enclosure. The frequency input is varied until the lowest frequency peak is found and then the frequency is varied on either side of the peak frequency until the two frequency locations are found at which the microphone signal is 3 dB less than the peak signal. The bandwidth, Δf, is the difference between the two frequencies corresponding to the 3 dB down level and f is the frequency corresponding to the peak level. The two quantities, Δf and f are then used with Equation (6.23) to find the quality factor.

The *in situ* quality factor can also be determined by mounting the resonator on the duct to be treated and then varying its volume around the design volume (Singh et al., 2006).

It should be noted that the presence of a mean flow has the effect of increasing the resonator damping and increasing its resonance frequency.

8.8.2.4 Transmission Loss Due to Side Branch

Sometimes it is useful to be able to compare the TL of various mufflers, even though it is not directly related to the noise reduction at the duct exit as a result of installing the muffler, as explained in Section 8.2. The TL of a side branch may be calculated by referring to Figure 8.8a and the harmonic solution of Equation (1.54), which is the wave equation. It will also be assumed that all duct and side branch dimensions are sufficiently small that only plane waves propagate (see Equation (8.228)), that the downstream duct diameter is equal to upstream duct diameter and that each has a cross-sectional area denoted S_d. TL is defined in Equation (7.13), where the transmission coefficient, τ is the ratio of transmitted power to incident power, As we are only considering plane waves, this may be written as:

$$\text{TL} = -10\log_{10}\left|\frac{p_T}{p_I}\right|^2 = 10\log_{10}\left|\frac{A_I}{A_T}\right|^2 \tag{8.66}$$

When a plane wave propagates down the duct from the left and encounters a side branch, a transmitted wave and a reflected wave are generated. The total sound pressure in the duct to the left of the resonator is the sum of the incident pressure, p_I, and the pressure, p_R, reflected by the impedance mismatch in the duct caused by the side-branch resonator. From Equation (1.54), this pressure may be written using complex constants, A_I and A_R as:

$$p_{\text{inlet}} = A_I \mathrm{e}^{\mathrm{j}(\omega t - kx)} + A_R \mathrm{e}^{\mathrm{j}(\omega t + kx + \beta_1)} \tag{8.67}$$

In the absence of any reflected wave from the downstream end of the duct, the transmitted pressure may be written as:

$$p_T = A_T \mathrm{e}^{\mathrm{j}(\omega t - kx + \beta_2)} \tag{8.68}$$

where β_1 and β_2 represent arbitrary phase angles that have no effect at all on the TL. Equations (1.11) and (1.12) may be used with the preceding two equations to write the following for the acoustic volume velocities (particle velocity multiplied by the duct cross-sectional area) in the same locations:

$$v_{\text{inlet}} = \frac{S_d}{\rho c}\left[A_I \mathrm{e}^{\mathrm{j}(\omega t - kx)} - A_R \mathrm{e}^{\mathrm{j}(\omega t + kx + \beta_1)}\right] \tag{8.69}$$

$$v_T = \frac{S_d A_T}{\rho c} \mathrm{e}^{\mathrm{j}(\omega t - kx + \beta_2)} \tag{8.70}$$

where S_d is the diameter of the duct on which the side branch is mounted.

If the duct axial coordinate, x, is set equal to zero at the location in the duct corresponding to the centre of the side branch resonator, the total acoustic pressure at the entrance to the side branch, p_{side}, is equal to the acoustic pressure in the duct at $x = 0$, so that at this location for continuity reasons, the sum of acoustic pressures due to the incoming wave (incident plus reflected) is equal to the acoustic pressure of the outgoing (or transmitted) wave:

$$p_{\text{side}} = p_T = A_T \mathrm{e}^{\mathrm{j}(\omega t + \beta_2)} = A_I \mathrm{e}^{\mathrm{j}(\omega t)} + A_R \mathrm{e}^{\mathrm{j}(\omega t + \beta_1)} \tag{8.71}$$

At the junction of the side branch and duct, there must be continuity of volume velocity so that at $x = 0$, the incoming volume-velocity from the left is equal to the sum of that moving in the duct to the right and that entering the side branch. The incoming and outgoing volume velocities in the duct are given by Equations (8.69) and (8.70), respectively, while the side-branch volume-velocity at $x = 0$ is:

$$v_{\text{side}} = \frac{p_{\text{side}}}{Z_{\text{side}}} = \frac{p_T}{Z_{\text{side}}} \tag{8.72}$$

where Z_{side} is the acoustic impedance of the side branch.

Thus, using Equations (8.68) to (8.72), continuity of volume velocity at the side branch junction can be written as $v_{\text{inlet}} = v_T + v_{\text{side}}$, or:

$$\frac{S_d}{\rho c}\left[A_I \mathrm{e}^{\mathrm{j}(\omega t - kx)} - A_R \mathrm{e}^{\mathrm{j}(\omega t + kx + \beta_1)}\right] = \frac{S_d A_T}{\rho c} \mathrm{e}^{\mathrm{j}(\omega t - kx + \beta_2)} + \frac{A_T}{Z_{\text{side}}} \mathrm{e}^{\mathrm{j}(\omega t + \beta_2)} \tag{8.73}$$

As $x = 0$ and $\mathrm{e}^{\mathrm{j}\omega t}$ appears in all terms, we can simplify Equation (8.73) and use Equation (8.71) to rewrite the result as:

$$\frac{2 A_I S_d}{\rho c} = A_T \mathrm{e}^{\mathrm{j}\beta_2}\left(\frac{2 S_d}{\rho c} + \frac{1}{Z_{\text{side}}}\right) \tag{8.74}$$

Further rearranging gives:

$$\left|\frac{A_I}{A_T}\right| = \left|1 + \frac{\rho c}{2 S_d Z_{\text{side}}}\right| \tag{8.75}$$

Thus, the TL of the side-branch resonator of impedance, Z_{side}, is:

$$\mathrm{TL} = 20 \log_{10}\left|1 + \frac{\rho c}{2 S_d Z_{\text{side}}}\right| \tag{8.76}$$

The side-branch acoustic impedance, Z_{side}, may be calculated using lumped analysis, and is found by adding together Equations (8.13) and (8.43). However, more accurate results are obtained if the resonator is treated as a 1-D transmission line so that wave motion is allowed in the axial direction. In this case, the impedance is calculated using Equation (D.91) in Appendix D, where all the specific acoustic impedances are replaced with acoustic impedances. The impedance, Z_m, corresponds to the air in the neck and is given by $\rho c/S$ where S is the cross-sectional area of the neck, and the impedance, Z_L, is the load impedance, which is the impedance looking into the resonator chamber, given by the imaginary part of Equation (8.45) with $M = 0$, as only the zero flow condition is being considered, and the acoustic resistance of the chamber is considered negligible. In this case, Equation (D.91) can be written in terms of the overall acoustic impedance of the side branch, Z_{side}, the acoustic impedance of the resonator cavity, Z_c, the length of the cavity in the resonator axial direction, ℓ_c, the volume of the resonator cavity, V, the effective length, ℓ_e, cross-sectional area of the resonator neck, S, and the cross-sectional area, S_c, of the resonator cavity in the plane normal to the resonator axis as:

$$Z_{\text{side}} = \frac{\rho c}{S}\frac{Z_c S/\rho c + \mathrm{j}\tan(k\ell_e)}{1 + \mathrm{j}(SZ_c/\rho c)\tan(k\ell_e)} \tag{8.77}$$

Substituting Equation (8.45) for Z_c into Equation (8.77), with $R_A = M = 0$, gives:

$$Z_{\text{side}} = \frac{\rho c}{S}\frac{\dfrac{-\mathrm{j}S}{S_c\tan(k\ell_c)} + \mathrm{j}\tan(k\ell_e)}{1 + \mathrm{j}\dfrac{-\mathrm{j}S\tan(k\ell_e)}{S_c\tan(k\ell_c)}} = \mathrm{j}\frac{\rho c}{S}\frac{(S_c/S)\tan(k\ell_e)\tan(k\ell_c) - 1}{(S_c/S)\tan(k\ell_c) + \tan(k\ell_e)} = \mathrm{j}\frac{\rho c}{S}\frac{Z_{\text{num}}}{Z_{\text{den}}} \tag{8.78}$$

Substituting Equation (8.78) into Equation (8.75) gives:

$$\left|\frac{A_I}{A_T}\right| = \left|1 - \mathrm{j}\frac{S}{2S_d}\frac{Z_{\text{den}}}{Z_{\text{num}}}\right| \tag{8.79}$$

Thus:

$$\text{TL} = 10\log_{10}\left[1 + \left(\frac{S}{2S_d}\frac{Z_{\text{den}}}{Z_{\text{num}}}\right)^2\right] \tag{8.80}$$

which is valid for frequencies above any axial resonances in the cavity and up to the resonance frequency of the first cross mode in the resonator cavity.

We can take into account dissipation by including the resistance term, R_A, as an addition to Equation (8.77) to produce:

$$Z_{\text{side}} = \mathrm{j}\frac{\rho c}{S}\frac{Z_{\text{num}}}{Z_{\text{den}}} + \frac{\rho c}{S}R_{\text{rat}} = \frac{\rho c}{S}\left[\mathrm{j}\frac{Z_{\text{num}}}{Z_{\text{den}}} + R_{\text{rat}}\right] \tag{8.81}$$

where $R_{\text{rat}} = R_A S/(\rho c)$. Equation (8.75) then becomes:

$$\left|\frac{A_I}{A_T}\right| = \left|1 + \frac{\rho c}{2S_d Z_{\text{side}}}\right| = \left|1 + \frac{S}{2S_d}\left(\frac{Z_{\text{den}}}{\mathrm{j}Z_{\text{num}} + Z_{\text{den}}R_{\text{rat}}}\right)\right| \tag{8.82}$$

Thus, Equation (8.80) becomes:

$$\text{TL} = 10\log_{10}\left\{\left[1 + \frac{S}{2S_d}\left(\frac{Z_{\text{den}}^2 R_{\text{rat}}}{(R_{\text{rat}}Z_{\text{den}})^2 + Z_{\text{num}}^2}\right)\right]^2 + \left[\frac{S}{2S_d}\left(\frac{Z_{\text{den}}Z_{\text{num}}}{(R_{\text{rat}}Z_{\text{den}})^2 + Z_{\text{num}}^2}\right)\right]^2\right\} \tag{8.83}$$

Equations (8.80) and (8.83) also apply to a QWT resonator by setting $S = S_c$. Using trigonometric identities for the tan function, Equation (8.80) for a QWT then becomes:

$$\text{TL} = 10\log_{10}\left[1 + \left(\frac{S_c}{2S_d}\tan[k(\ell_e + \ell_c)]\right)^2\right] \tag{8.84}$$

which is the TL of a QWT resonator of length $\ell_e + \ell_c$ and the side branch cross-sectional area, $S = S_c$. A procedure similar to that used to obtain Equation (8.83) can be used to modify Equation (8.84) to obtain the equivalent expression including the acoustic resistance term. Thus:

$$\text{TL} = 10\log_{10}\left\{\left[1 + \frac{S}{2S_d}\left(\frac{R_{\text{rat}}}{R_{\text{rat}}^2 + \cot^2[k(\ell_e + \ell_c)]}\right)\right]^2 + \left[\frac{S}{2S_d}\left(\frac{\cot[k(\ell_e + \ell_c)]}{R_{\text{rat}}^2 + \cot^2[k(\ell_e + \ell_c)]}\right)\right]^2\right\} \tag{8.85}$$

If S_c approaches infinity, then there is effectively no cavity on the end and Equation (8.80) reduces to the TL for a side branch consisting of an open-ended tube as:

$$\mathrm{TL} = 10\log_{10}\left[1 + \left(\frac{S}{2S_d}\cot(k\ell_e)\right)^2\right] \tag{8.86}$$

where the top and bottom lines of Equation (8.78) were multiplied by S/S_c to obtain the result.

The axial resonance frequencies of the Helmholtz resonator can be determined by realising that they occur when the denominator in Equation (8.80) is zero and the TL is infinite. Thus, the resonance frequencies are those that satisfy:

$$(S_c/S)\tan(k\ell_e)\tan(k\ell_c) = 1 \tag{8.87}$$

where k is the wavenumber defined as $k = 2\pi f/c$. Note that in the presence of grazing flow of Mach number, M, Equations (8.77) to (8.87) can be used by substituting $(1 - M)\ell_e$ for ℓ_e in all places. Also, the resistance term in Equation (8.83) results in the TL being somewhat less than infinity at resonance.

8.8.3 Resonator Mufflers

Resonator mufflers are used in industry where large, low-frequency noise reductions are needed and also in applications where it is not possible to use porous sound-absorbing material in the muffler (due to possible contamination of the air flow or contamination of the sound-absorbing material by particles or chemicals in the air flow).

8.8.3.1 Resonator Mufflers For Tonal Control

Resonator mufflers can be used for controlling the radiation of low-frequency tonal sound from the discharge of a duct, such as a tone at the blade pass frequency of a fan in a duct. In this case, one or more QWTs of different lengths are attached to the side-wall of the duct. The QWTs are usually cylindrical in shape with a diameter at least half that of the duct on which they are mounted. The reason for using more than one QWT is that the temperature in the duct may change, or the fan speed may change in response to different fluid loadings on the fan or different throughput requirements. These changes will alter the wavelength of the tone and thus the optimal length of the QWT. One way of extending the effective frequency range of a fixed-length QWT is to add some sound absorbing material to the inside of the tube, at the expense of reducing the noise reduction at the design frequency. Alternatively, to cope with a range of different wavelengths corresponding to different fan operating conditions or gas temperatures, a single QWT could be used together with a microphone located in the wall of the duct downstream of the QWT. A control system connected to a motor and gearbox could adjust the location of a plunger that can move axially to change the length of the QWT to minimise the microphone output at the frequency of the tone to be controlled. Howard and Craig (2014a) describe implementation of such a system to reduce tonal exhaust noise from a variable-speed diesel engine. As the standing wave in a duct at any particular frequency has a sound pressure null at the open end of the duct, the ideal location of a QWT is at one quarter or three quarters of a wavelength from the open end of the duct. The latter length is usually preferred to minimise any local phenomena occurring at the duct discharge.

8.8.3.2 Resonator Mufflers for Broadband Noise Control

Resonator mufflers for broadband noise control consist of a mix of QWTs and Helmholtz resonators, tuned to cover the frequency range of interest and attached to the walls of the duct through which the sound is propagating. Helmholtz resonators have a lower Q (and hence act over a broader frequency range) than QWTs, but their performance in terms of IL is not as good. Thus, in practice, many mufflers are made up of a combination of these two types of resonators, with the QWTs tuned to tonal noise or frequency ranges where the greatest noise reduction is needed. To ensure good frequency coverage, and to allow for possible variations in tonal frequencies, changes in gas temperature, and resonator manufacturing errors, Helmholtz resonators are installed that cover a range of resonance frequencies about those of the QWTs.

It is important that adjacent resonators are not tuned to identical or near identical frequencies, so they do not interact and substantially reduce the muffler performance at that frequency. In many cases, the duct cross-sectional area, through which the sound is travelling, is so large that the resonator muffler is constructed using splitters to contain the resonators. This requires dividing the duct cross-section into a number of parallel sections using dividing walls and including resonators in each dividing wall as shown in Figure 8.9.

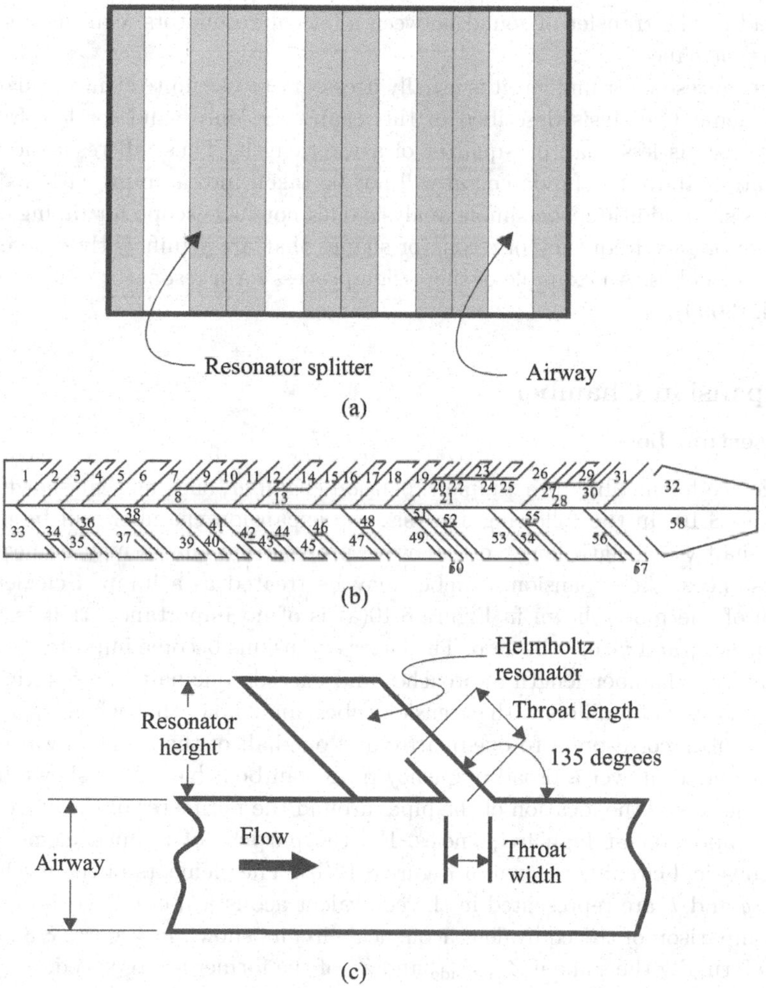

FIGURE 8.9 Resonator muffler details: (a) cross-section of splitter muffler (splitters are shaded); (b) detail of a splitter containing 58 resonators; (c) detail of a single resonator.

Note that the resonators are usually angled towards the flow direction, as shown in Figure 8.9(c), to avoid filling up with particles from the air flow, and to minimise the generation of tonal noise due to vortex shedding from the edge of the resonator inlet (Singhal, 1976; Ingard,

2010). To ensure that the pressure drop due to insertion of the muffler is not too great, it is usually sized so that the total open cross-sectional area between splitters is the same as the inlet duct cross-sectional area.

There are two types of resonator muffler for broadband noise control – those that contain no sound-absorbing material at all and those that include sound-absorbing material in the resonator chambers. Those that contain no sound-absorbing material at all have resonators with a relatively high Q, and many resonators are needed to cover a reasonable frequency range. A small amount of sound-absorbing material in the resonator chambers will produce a muffler with much more uniform attenuation characteristics as a function of frequency than a muffler without sound-absorbing material. However, the peak attenuation at some frequencies may not be as high as achieved by a resonator muffler with no sound-absorbing material, for the case where an excitation frequency corresponds very closely to one of the side-branch resonances.

In constructing resonator mufflers, it is important that the walls of the resonator are made using sufficiently thick material, or they are sufficiently stiff, so that the vibration of the internal walls does lead to the transfer of sound between adjacent resonators, causing a degradation of the muffler performance.

In designing a resonator muffler, it is usually necessary to use finite element analysis software. The lumped element analysis described in this chapter is only valid for low frequencies and resonator dimensions less than one-quarter of a wavelength. Thus, all resonances of the side-branch resonators above the fundamental will not be taken into account with a simple lumped element analysis. In addition, the simple analysis does not have scope for taking account of the effect on the resonance frequency of resonator shapes that are significantly different to straight tubes, spheres or cubes. An example of the design process for a resonator muffler is described in Howard et al. (2000).

8.8.4 Expansion Chamber

8.8.4.1 Insertion Loss

A common device for muffling is a simple expansion chamber, such as that shown diagrammatically in Figure 8.10. In the following analysis, the expansion chamber will be assumed to be less than one-half wavelength long, so that wave propagation effects may be neglected. Under these circumstances, the expansion chamber may be treated as a lumped element device and the extension of the pipes shown in Figure 8.10(a) is of no importance; that is, the extension lengths shown as x_1 and y_1 may be zero. The latter extensions become important at mid- to high frequencies, as the chamber length approaches and exceeds one-half wavelength. Section 8.9.8 discusses the analysis of mufflers with extension tubes, and shows that when the lengths, x_1 and y_1, of the two tubes correspond to one-quarter and one-half of the acoustic wavelength respectively, then attenuation over a broad frequency range can be achieved, as shown in Figure 8.22. Similarly, the details of the location of the pipes around the perimeter of the expansion chamber only become important at high frequencies. For the purpose of lumped element analysis, an equivalent acoustical circuit is shown in Figure 8.10(b). The elements of the mechanical system labelled a, b, g and L are represented in the equivalent acoustical circuit as impedances Z_a, Z_b and Z_L. A comparison of the equivalent acoustical circuits shown in Figures 8.8 and 8.10 shows that they are formally the same if Z_u, Z_{side} and Z_d of the former are replaced, respectively, with Z_a, Z_b and Z_L of the latter. For a constant-volume-velocity source, the expression for IL for the expansion chamber is formally identical to Equation (8.63).

If it is assumed that pipe (g) is not part of the muffler, but part of the original load so that it contributes to Z_L, then, according to Equation (8.63):

$$\text{IL} = 20 \log_{10} \left| 1 + \frac{Z_L}{Z_b} \right| \tag{8.88}$$

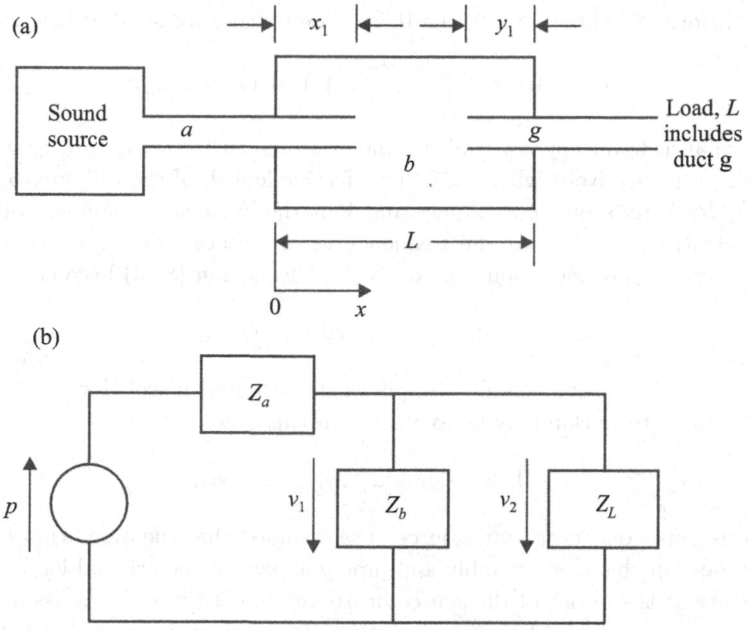

FIGURE 8.10 Expansion chamber muffler: (a) acoustical system; (b) equivalent acoustical circuit.

Reference to Figure 8.10 shows that the impedance, Z_b, is capacitative, and may be calculated using Equation (8.43). For a muffler terminated by a tube open at both ends, the impedance, Z_L, can be written as the sum of the tube resistance, R_L, given by Equation (8.34), and the inductance, given by Equation (8.18). Dividing the result by Z_b (see Equation (8.43)) gives:

$$\frac{Z_L}{Z_b} = \mathrm{j}\frac{\omega V}{\rho c^2}\left[R_L + \mathrm{j}\frac{\rho c}{S_{\text{duct}}}\tan k\ell_e\right] \tag{8.89}$$

Here V (m^3) is the volume of the expansion chamber, S_{duct} (m^2) is the cross-sectional area of the tailpipe (g) and ℓ_e (m) is its effective length, ρ (kg/m^3) is the gas density and c (m/s) is the speed of sound.

Expansion chamber mufflers may also be characterised by a resonance frequency, ω_0 radians/s, and a quality factor, Q. The resonance frequency occurs when the capacitative impedance, Z_b, of the expansion chamber volume plus the inductive impedance of the tailpipe is equal to zero. These impedances are, respectively, $Z_b = -\mathrm{j}\rho c^2/(V\omega_0)$ (see Equation (8.43)) and $\text{Imag}\{Z_L\} = \mathrm{j}(\rho c/S_{\text{duct}})k\ell_e$ (see Equation (8.13), which applies for small $k\ell_e$ so that $k\ell_e \approx \tan(k\ell_e)$). Setting $Z_b = -\text{Imag}\{Z_L\}$, gives:

$$\omega_0 = c\sqrt{\frac{S_{\text{duct}}}{V\ell_e}} \tag{8.90}$$

Equation (8.57) can be used to define a quality factor, where the capacitance is simulated by the expansion chamber and the inductance by the tailpipe, so that for small $k\ell_e$:

$$Q \approx \frac{\rho c}{R_L}\sqrt{\frac{\ell_e}{S_{\text{duct}}V}} \tag{8.91}$$

where R_L may be evaluated using Equation (8.34).

Equations (8.90) and (8.91) may be used to rewrite Equation (8.89), for small values of $k\ell_e$ as:

$$\frac{Z_L}{Z_b} \approx -\left[\frac{\omega}{\omega_0}\right]^2 + \mathrm{j}\frac{\omega}{\omega_0 Q} \tag{8.92}$$

Using Equations (8.88) and (8.92), the IL may be written for small values of $k\ell_e$ as:

$$\text{IL} \approx 10\log_{10}\left[\left(1-(\omega/\omega_0)^2\right)^2 + Q^{-2}(\omega/\omega_0)^2\right] \tag{8.93}$$

where the resonance frequency, ω_0 (rad/s), and the quality factor, Q, are given by Equations (8.90) and (8.91), respectively, where ℓ_e is the effective length of the tailpipe, S_{duct} is its cross-sectional area, R_L is its acoustic resistance and V is the expansion chamber volume. Reference to Equation (8.93) shows that, at the frequency of resonance when $\omega = \omega_0$, the IL becomes negative and a function of the quality factor, so that Equation (8.93) becomes:

$$\text{IL}|_{\omega=\omega_0} = -20\log_{10} Q \tag{8.94}$$

In this case, the expansion chamber amplifies the radiated noise! However, well above resonance, appreciable attenuation may be expected. In this case:

$$\text{IL} \approx 40\log_{10}(\omega/\omega_0), \qquad \omega \gg \omega_0 \tag{8.95}$$

For the constant-acoustic-pressure source, it is assumed that the pipe (a) of Figure 8.10(a) is part of the expansion chamber assembly and pipe g is part of the original load. Thus, constant-acoustic pressure at the outlet of the source or at the inlet to pipe (a) is assumed, so that the pressure drop, p, across the load prior to insertion of the expansion chamber is the same as the constant-acoustic pressure supplied by the source. In this case, the expression for the IL of the expansion chamber assembly takes the following form:

$$\text{IL} = 20\log_{10}\left|\frac{p}{v_2 Z_L}\right| \tag{8.96}$$

where $v_2 Z_L$ is the pressure drop across the load after insertion of the expansion chamber and p is the pressure drop across the load prior to installation of the expansion chamber.

Using circuit analysis (acoustical analogies of Kirchhoff's circuit laws discussed in Section 8.8.1) and Figure 8.10, we can write for a constant-acoustic-pressure source:

$$p = v_2(Z_L + Z_a) + v_1 Z_a \tag{8.97}$$

and:

$$v_1 Z_b = v_2 Z_L \tag{8.98}$$

Solving Equations (8.96) to (8.98) gives:

$$\text{IL} = 20\log_{10}\left|1 + \frac{Z_a(Z_L + Z_b)}{Z_b Z_L}\right| \tag{8.99}$$

The results of an experiment with an expansion chamber muffler driven by a constant-volume-velocity source (loudspeaker with backing enclosure) are shown in Figure 8.11, where the effect of flow is also shown. The experiment has been carried out to frequencies well above the upper bound, about 220 Hz, for which the analysis is expected to apply.

Thus, the decrease in IL at about 400 Hz must be accounted for by transmission-line theory, which is not considered here. However, 4-pole analysis does allow extension of the TL calculations to higher frequencies (see Sections 8.9.7 and 8.9.8), although the diameter of the muffler elements must be sufficiently small that only plane waves propagate (see Equation (8.228)).

In Figure 8.11, the data below about 200 Hz generally confirm the predictions of Equation (8.93) if the resonance frequency is assumed to be about 50 Hz and the quality factor without flow to be about 10. The latter magnitude is quite reasonable. The data show further that the quality factor Q decreases with flow, in part because the inductances of Figure 8.10

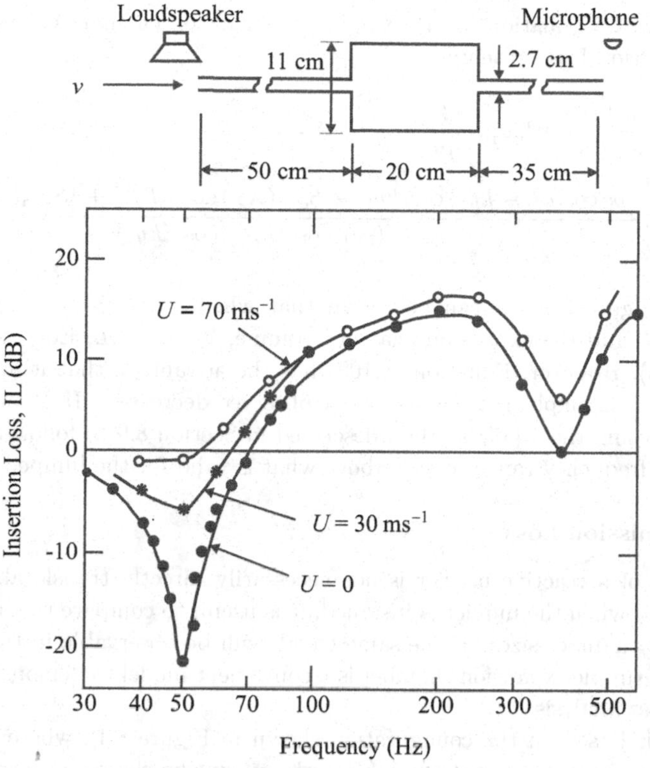

FIGURE 8.11 Experimental evaluation of a cylindrical, expansion-chamber muffler. Near the resonance the IL is increased by steady airflow; if the flow velocity is large enough, the resonance (and negative IL) may be eliminated.

decrease and in part because the corresponding resistances increase, as generally predicted by the considerations of Section 8.7.1.5 and shown in Equation (8.34). Note that the performance of the device improved with flow.

When wave propagation along the length of the expansion chamber is considered, the analysis becomes more complicated. It will be sufficient to state the result of the analysis for the case of zero mean flow. As before, the simplifying assumption is made that the chamber is driven by a constant-volume-velocity source. A further simplification is made that the outlet pipe is much less than one-half wavelength long, so that propagational effects may be neglected. The expansion chamber has a length, L, as shown in Figure 8.10(a) and a cross-sectional area of S_{exp}. Söderqvist (1982) provides the following expression (slightly rearranged) for the ratio of the volume velocity, v, at the expansion chamber inlet to the pressure, p_L, at the outlet as:

$$\frac{v}{p_L} = \frac{\left[\mathrm{j}S_{\text{exp}}Z_L \sin(kL) + \rho c \cos(kL - ky_1)\cos(ky_1)\right]}{Z_L \rho c \cos(kx_1)\,\cos(ky_1)} \tag{8.100}$$

Using Equation (8.20), assuming that there is no flow and the tailpipe is short so that $k\ell_e$ is small and $\tan(k\ell_e) \approx k\ell_e$, the impedance, Z_L, pf the tailpipe (load) may be written as:

$$Z_L = R_L + \mathrm{j}X_L \approx R_L + \mathrm{j}\frac{\rho c k \ell_e}{S_{\text{duct}}} \tag{8.101}$$

where R_L is defined by Equation (8.34), S_{exp} is the expansion chamber cross-sectional area and $k = \omega/c$. The insertion loss is then:

$$\text{IL} = 10\log_{10}\left|\frac{v}{v_L}\right|^2 = 10\log_{10}\left|\frac{vZ_L}{p_L}\right|^2$$
$$= 10\log_{10}\left[\frac{\left[\rho c\cos(kL - ky_1)\cos(ky_1) - S_{\text{exp}}(X_L)\sin(kL)\right]^2 + \left[S_{\text{exp}}(R_L)\sin(kL)\right]^2}{(\rho c)^2\cos^2(kx_1)\cos^2(ky_1)}\right] \tag{8.102}$$

Referring to Figure 8.10, it can be shown that when $x_1 = y_1 = 0$ and kL is small so that $\sin(kL) \approx kL$ and the expansion chamber volume, $V = S_{\text{exp}}L$, Equation (8.102) reduces to Equation (8.93). However, Equation (8.102) has the advantage that it applies over a wider frequency range. For example, it accounts for the observed decrease in IL shown in Figure 8.11 at 400 Hz. Analysis using the 4-pole method described in Section 8.9 to follow allows extension of the valid analysis frequency range to well above what is valid for the lumped element approach.

8.8.4.2 Transmission Loss

Although the TL of a reactive muffler is not necessarily directly translatable to the NR that will be experienced when the muffler is installed, it is useful to compare the TL performance for various expansion chamber sizes, as the same trends will be observable in their noise reduction performance. The simple expansion chamber is a convenient model to demonstrate the principles of transmission loss analysis.

The analysis is based on the configuration shown in Figure 8.12 where it is assumed that the inlet and discharge tubes do not extend into the expansion chamber. There will exist a right travelling wave in the inlet duct, which will be denoted p_I and a left travelling wave reflected from the expansion chamber inlet, denoted p_R. There will also be a right and a left travelling wave in the expansion chamber, denoted p_A and p_B, respectively. In the exit pipe, there will only be a right travelling wave, p_T, as an anechoic termination will be assumed. It will also be assumed that all muffler dimensions are sufficiently small that only plane waves will be propagating (see Equation (8.228)), and that the exit pipe diameter is equal to the inlet pipe diameter and each has a cross-sectional area denoted S_{duct}. The cross-sectional area of the expansion chamber is S_{exp}. TL is defined as the ratio of transmitted power to incident power and as we are only considering plane waves, this may be written as:

$$\text{TL} = 10\log_{10}\left|\frac{p_I}{p_T}\right|^2 = 10\log_{10}\left|\frac{A_I}{A_T}\right|^2 \tag{8.103}$$

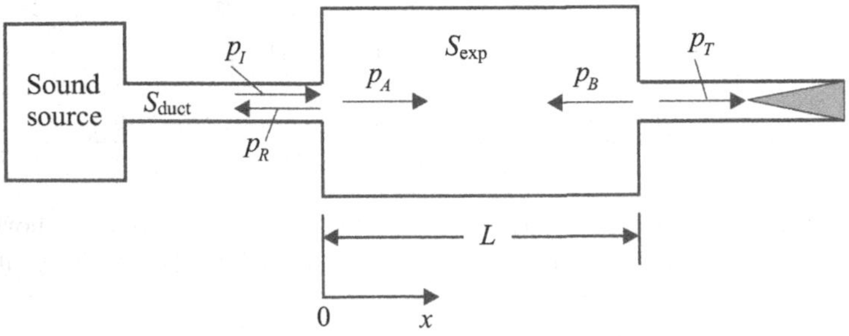

FIGURE 8.12 Expansion chamber muffler arrangement for TL analysis.

The total sound pressure in the inlet pipe, which is the sum of the incident pressure, p_I, and the pressure, p_R, reflected from the expansion chamber entrance, may be written using the harmonic pressure solution to the wave equation (see Equation (1.54)) as:

$$p_{\text{inlet}} = A_I \mathrm{e}^{\mathrm{j}(\omega t - kx)} + A_R \mathrm{e}^{\mathrm{j}(\omega t + kx + \beta_1)} \tag{8.104}$$

The total sound pressure in the expansion chamber may be written in terms of the right travelling wave and the reflected left travelling wave as:

$$p_{\text{exp}} = A_A \mathrm{e}^{\mathrm{j}(\omega t - kx + \beta_2)} + A_B \mathrm{e}^{\mathrm{j}(\omega t + kx + \beta_3)} \tag{8.105}$$

The total sound pressure in the exit pipe may be written as:

$$p_T = A_T \mathrm{e}^{\mathrm{j}(\omega t - kx + \beta_4)} \tag{8.106}$$

Equations (1.11) and (1.12) may be used with the preceding three equations to write the following for the acoustical particle velocities in the same locations:

$$u_{\text{inlet}} = \frac{1}{\rho c}\left[A_I \mathrm{e}^{\mathrm{j}(\omega t - kx)} - A_R \mathrm{e}^{\mathrm{j}(\omega t + kx + \beta_1)}\right] \tag{8.107}$$

$$u_{\text{exp}} = \frac{1}{\rho c}\left[A_A \mathrm{e}^{\mathrm{j}(\omega t - kx + \beta_2)} - A_B \mathrm{e}^{\mathrm{j}(\omega t + kx + \beta_3)}\right] \tag{8.108}$$

$$u_T = \frac{A_T}{\rho c} \mathrm{e}^{\mathrm{j}(\omega t - kx + \beta_4)} \tag{8.109}$$

Continuity of acoustic pressure and volume velocity at the junction of the inlet pipe and the expansion chamber, where the coordinate system origin, $x = 0$ (see Figure 8.12), will be defined, gives:

$$A_I + A_R \mathrm{e}^{\mathrm{j}\beta_1} = A_A \mathrm{e}^{\mathrm{j}\beta_2} + A_B \mathrm{e}^{\mathrm{j}\beta_3} \tag{8.110}$$

and

$$S_{\text{duct}}(A_I - A_R \mathrm{e}^{\mathrm{j}\beta_1}) = S_{\text{exp}}(A_A \mathrm{e}^{\mathrm{j}\beta_2} - A_B \mathrm{e}^{\mathrm{j}\beta_3}) \tag{8.111}$$

At the junction of the expansion chamber and exit pipe, at $x = L$, continuity of acoustic pressure and volume velocity gives:

$$A_T \mathrm{e}^{\mathrm{j}(-kL + \beta_4)} = A_A \mathrm{e}^{\mathrm{j}(-kL + \beta_2)} + A_B \mathrm{e}^{\mathrm{j}(kL + \beta_3)} \tag{8.112}$$

and

$$S_{\text{duct}} A_T \mathrm{e}^{\mathrm{j}(-kL + \beta_4)} = S_{\text{exp}}[A_A \mathrm{e}^{\mathrm{j}(-kL + \beta_2)} - A_B \mathrm{e}^{\mathrm{j}(kL + \beta_3)}] \tag{8.113}$$

Considerable algebraic manipulation of Equations (8.110) to (8.113) allows the transmitted sound pressure amplitude to be written in terms of the incident sound pressure amplitude as:

$$A_T = \frac{A_I \mathrm{e}^{\mathrm{j}kL}\ \mathrm{e}^{-\mathrm{j}\beta_4}}{\cos kL + \dfrac{\mathrm{j}}{2}\left(\dfrac{S_{\text{duct}}}{S_{\text{exp}}} + \dfrac{S_{\text{exp}}}{S_{\text{duct}}}\right)\sin kL} \tag{8.114}$$

Substituting Equation (8.114) into (8.103) gives:

$$\begin{aligned}\text{TL} =& 10\log_{10}\left|\frac{A_I}{A_T}\right|^2 = 10\log_{10}\left[\cos^2(kL) + \frac{1}{4}\left(\frac{S_{\text{duct}}}{S_{\text{exp}}} + \frac{S_{\text{exp}}}{S_{\text{duct}}}\right)^2 \sin^2 kL\right] \\ =& 10\log_{10}\left[1 + \frac{1}{4}\left(\frac{S_{\text{duct}}}{S_{\text{exp}}} - \frac{S_{\text{exp}}}{S_{\text{duct}}}\right)^2 \sin^2 kL\right]\end{aligned} \tag{8.115}$$

Equation (8.115) was derived using 1-D wave analysis rather than lumped analysis, so it takes into account the effect of acoustic axial modes but it is not valid if acoustic cross-modes exist in the chamber (see Section 8.10.3.3). Equation (8.115) is plotted in Figure 8.13 as a function of expansion ratio, $m = S_{\text{exp}}/S_{\text{duct}}$ versus normalised frequency, calculated as the wavenumber, k, multiplied by the expansion chamber length, L, divided by π. The cross-sectional area, S_{exp}, is always greater than S_{duct}. Note that when $ka > 1.85$, higher order modes begin to propagate and notionally when $ka > 4$, the energy in the higher order modes exceeds the energy in the plane-wave mode and experimental data will generally show smaller attenuation values than predicted in the figure. Also, Equation (8.115) has been derived without any consideration of resistive impedance, so effectively, damping has been excluded from the analysis, which explains why the minima in TL are zero, instead of some positive number, which would occur if damping were included.

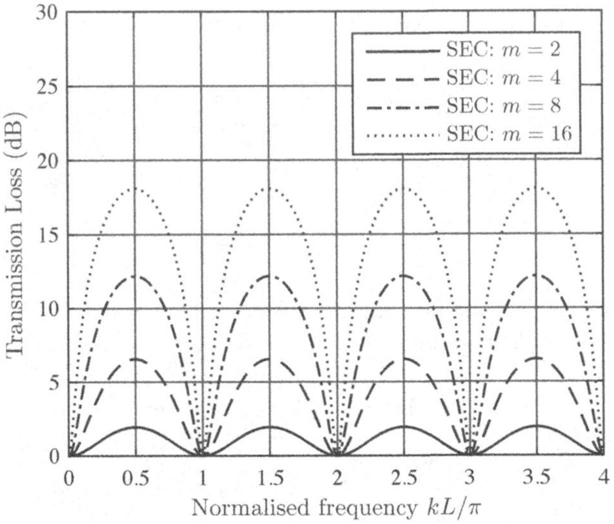

FIGURE 8.13 Transmission loss as a function of normalised frequency, kL/π, for a simple expansion chamber muffler of length, L, and various inlet duct to muffler cross-sectional area ratios, m.

8.8.5 Small Engine Exhaust

Small gasoline engines are commonly muffled using an expansion chamber and tailpipe. An important consideration in the design of such a muffling system is the effect of the back pressure imposed by the muffler on the performance of the engine. This matter has been both experimentally and analytically investigated, with the interesting result that an optimum tailpipe length has been shown to exist for any particular muffler configuration (Watters et al., 1959).

A design procedure based on the cited reference, which takes account of the observed optimum tailpipe length, is:

1. Determine the exhaust volume flow rate, V_0 (m^3/s), the speed of sound in the hot exhaust gas, c (m/s), and the effective ratio of specific heats of the exhaust gas, γ.
2. Assume that the term involving the quality factor, Q, in Equation (8.93) may be neglected, and use the equation to determine the resonance frequency, ω_0 (rad/s), to give the required IL (or NR) at the expected engine speed (rpm). Note that ω_0 must always be smaller than ω (defined in 3. below).

3. Assume a volume, V (m^3), for the expansion chamber. Enter the required quantities and determine a tailpipe cross-sectional area, S (m^2), which satisfies the following:

$$\left[\frac{1}{\gamma} - \frac{18V_0}{\omega V} - \frac{2V_0^2}{S^2c^2}\right] \frac{\omega_0^2 S^{1.5} V}{10\sqrt{\pi} f_m V_0^2} > 1 \tag{8.116}$$

   Note that $\omega = \pi \times \text{rpm} \times (\text{No. of cylinders}) \times \text{stroke}/30$, where stroke $= 2$ for a 2-stroke engine and 4 for a 4-stroke engine and for small steel pipes the factor $f_m \approx 8 \times 10^{-3}$. The quantity, f_m, may be calculated more accurately using Equations (8.240) to (8.242).

4. Calculate the required effective length, ℓ_e, of the tailpipe (including the end correction) using:

$$\ell_e = \frac{S}{V}\left[\frac{c}{\omega_0}\right]^2 \quad \text{(m)} \tag{8.117}$$

   The physical tailpipe length, ℓ_d, may be calculated from the effective length in Equation (8.117) using Equations (8.25) and (8.28) and $\ell_d = \ell_e - 2\ell_0$. If ℓ_d is less than three times the diameter, then set it equal to ten times the diameter and use Equation (8.116) to re-calculate S.

5. Use Equation (8.93) to calculate the IL over the frequency range of interest. The quality factor, Q, which is only neglected in step 2 but not in this step, may be calculated as for an expansion chamber, using Equation (8.57). Note that the length and area terms and the acoustic resistance in Equation (8.57) refer to the tailpipe and V is the expansion chamber volume.

Experimental investigation has shown that the prediction of Equation (8.93) is fairly well confirmed for low frequencies. However, at higher frequencies for which the tailpipe length is an integer multiple of half wavelengths, a series of pass bands are encountered for which the IL is less than predicted. Departure from prediction is dependent on engine speed and can be expected to begin at about half the frequency for which the tailpipe is one-half wavelength long. This latter frequency is called the first tailpipe resonance.

8.8.6 Low-Pass Filter

A device commonly used for the suppression of pressure pulsations in a flowing gas is the low-pass filter. Such a device, sometimes referred to as a Helmholtz filter, may take various physical forms but, whatever the form, all have the same basic elements as the filter illustrated in Figure 8.14.

In Figure 8.14(a) the acoustical system is shown schematically as two expansion chambers, b and d, interconnected by a pipe, g, and in turn connected to a source and load by pipes, a and e respectively. For convenience, it will be assumed that pipe, e, is not part of the low-pass filter but is part of the load and is thus included in the load impedance, Z_L. In Figure 8.14(b) the equivalent acoustical circuit is shown as a system of interconnected impedances, identified by subscripts corresponding to the elements in Figure 8.14(a). Also shown in the figure are the acoustic volume velocities, v (m^3/s), through the various elements and the acoustic pressure, p (Pa), generated by the source. The impedances of pipes a and g are given by Equation (8.13) and the impedances of volumes b and d are given by Equation (8.43).

At low frequencies, where the elements of the acoustic device illustrated in Figure 8.14(a) are all much less than one-half wavelength long, the details of construction are unimportant; in this frequency range, the device may be analysed by reference to the equivalent acoustical circuit shown in Figure 8.14(b). The location of the inlet and discharge tubes in each of the chambers only take on importance above this frequency range.

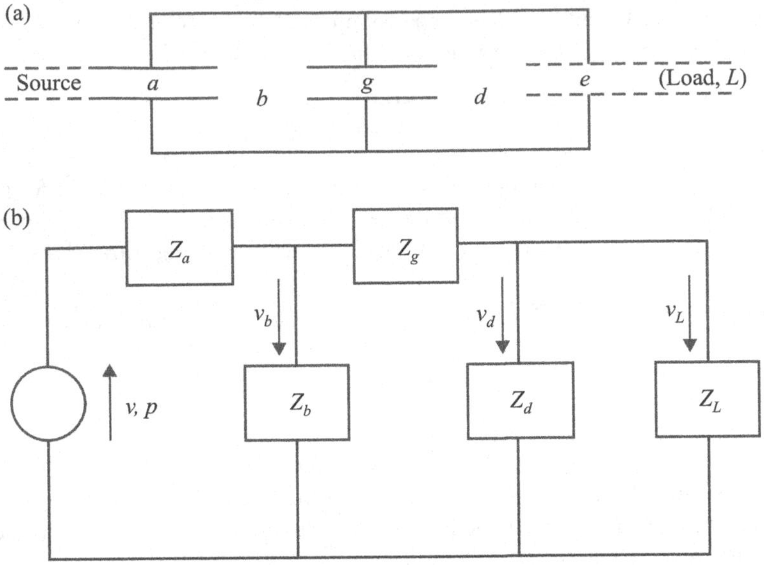

FIGURE 8.14 Low-pass filter: (a) acoustical system; (b) equivalent acoustical circuit.

The impedances of the devices coupled to the filter affect its performance; thus, the low-pass filter cannot be analysed in isolation. However, only the details of the performance and not the basic function will be affected.

To simplify the analysis, certain assumptions will be made about the source and load attached to the filter. The source will be modelled either as a constant-volume-velocity (infinite internal impedance) or a constant-acoustic-pressure (zero internal impedance) source (see Section 8.6 for a discussion of internal source impedance). The further simplifying assumptions will be made that the inlet pipe, (a), is part of the lowpass filter but the discharge pipe, (e), is not. The discharge pipe is assumed to be part of the original system, contributing to the load impedance, Z_L; that is, the discharge pipe length is unchanged by insertion of the filter.

The possibilities for the load impedance are myriad, but discussion will be restricted to just two, which are quite common. It will be assumed that either the filter is connected to a very long piping system, such that any pressure fluctuation entering the load is not reflected back towards the filter, or that the filter abruptly terminates in a short tailpipe (less than one-tenth of a wavelength long) looking into free space. In the former case, the impedance at the entry to pipe, e, of Figure 8.14(a) is simply the characteristic impedance of the pipe, so that where S_L is the cross-sectional area of the tailpipe of effective length, ℓ_e, the load impedance is:

$$Z_L = \rho c / S_L \tag{8.118}$$

For the case of the short tailpipe the following further simplifications are made:

1. the reactive part of the radiation impedance is accounted for as an end correction (see Section 8.7.1.4), which determines the effective length of pipe e; and
2. the real part of the radiation impedance is negligible compared to the pipe mass reactance in series with it. Thus, in this case, Equation (8.10) gives the terminating impedance as:

$$Z_L = \mathrm{j}\rho\omega\ell_e(1 - M)/S_L \tag{8.119}$$

For lengths of pipe between the two extreme cases, the terminating impedance is the sum of the inductive impedance, calculated using Equation (8.18), and the resistive impedance, calculated using Equation (8.34). In all cases, the assumption is implicit that only plane waves propagate in the pipe; that is, the pipe diameter is less than 0.586 of a wavelength of sound. For larger diameter pipes in which cross-modes propagate, the analysis is much more complex and will not be considered here.

The effect of introducing the low-pass filter will be described in terms of an IL. IL provides a measure of the reduction in acoustical power delivered to the load by the source when the filter is interposed between the source and load. Referring to Figure 8.14(b), the IL for the case of the constant-volume-velocity source is:

$$\mathrm{IL} = 20\log_{10}\left|\frac{v}{v_L}\right| \tag{8.120}$$

while for the case of the constant-acoustic-pressure source, for which it is assumed that the pipe (a) of Figure 8.14 is part of the low-pass filter assembly and pipe e is part of the load, the IL is:

$$\mathrm{IL} = 20\log_{10}\left|\frac{p}{v_L Z_L}\right| \tag{8.121}$$

Here, v (m^3/s) is the assumed constant amplitude of the volume-velocity source, and for the constant-acoustic-pressure source, p (Pa) is the constant amplitude of the acoustic pressure of the gas flowing from the source, v_L (m^3/s) is the amplitude of the acoustic volume velocity through the load and Z_L is the load impedance. Note that p is the acoustic pressure incident on the load, Z_L, prior to insertion of the acoustic filter, and this is the same acoustic pressure emitted by the source. The acoustic pressure incident on the load after insertion of the acoustic filter is $v_L Z_L$.

In the following analysis, acoustic resistances within the filter have been neglected for the purpose of simplifying the presentation, and because resistance terms only significantly affect the IL at system resonance frequencies. It is also difficult to estimate acoustic resistances accurately, although Equation (8.34) may be used to obtain an approximate result, as discussed in Section 8.7.1.5.

Reference to Section 8.8.1 and to Figure 8.14 allows the following system of equations to be written for the constant-volume-velocity source:

$$v = v_b + v_d + v_L \tag{8.122}$$

$$0 = -v_b Z_b + v_d(Z_g + Z_d) + v_L Z_g \tag{8.123}$$

$$0 = -v_d Z_d + v_L Z_L \tag{8.124}$$

Alternatively, if the system is driven by a constant-acoustic-pressure source, Equation (8.122) is replaced with:

$$p = Z_a(v_b + v_d + v_L) + v_b Z_b \tag{8.125}$$

In the following equations, V represents the volume, S represents the cross-sectional area and ℓ represents the effective length of an element, which is identified by the associated subscript.

Use of Equations (8.120), (8.122), (8.123) and (8.124) gives the following result for the constant-volume-velocity source:

$$\mathrm{IL} = 20\log_{10}\left|1 + \frac{Z_g}{Z_b} + \frac{Z_L}{Z_d} + \frac{Z_L}{Z_b} + \frac{Z_g Z_L}{Z_b Z_d}\right| \tag{8.126}$$

For the long tailpipe termination, use of Equations (8.13), (8.43) and (8.118) allows Equation (8.126) to be written as:

$$\mathrm{IL} = 10\log_{10}\left[\left[1 - \frac{V_b \ell_g}{S_g}\left(\frac{\omega}{c}\right)^2\right]^2 + \left[\frac{V_b + V_d}{S_L}\left(\frac{\omega}{c}\right) - \frac{V_b V_d \ell_g}{S_L S_g}\left(\frac{\omega}{c}\right)^3\right]^2\right] \tag{8.127}$$

Similarly, for the short tailpipe termination (see Equation (8.119)), Equation (8.126) takes the following alternative form:

$$\mathrm{IL} = 10\log_{10}\left[1 - \left[\frac{V_b\ell_g}{S_g} + \frac{V_d\ell_L}{S_L} + \frac{V_b\ell_L}{S_L}\right]\left(\frac{\omega}{c}\right)^2 + \frac{V_bV_d\ell_L\ell_g}{S_LS_g}\left(\frac{\omega}{c}\right)^4\right]^2 \tag{8.128}$$

Use of Equations (8.121), (8.123), (8.124) and (8.125) gives the following result for the constant-acoustic-pressure source:

$$\mathrm{IL} = 20\log_{10}\left|\frac{Z_a}{Z_d} + \frac{Z_a}{Z_L} + \left(1 + \frac{Z_a}{Z_b}\right)\left(1 + \frac{Z_g}{Z_d} + \frac{Z_g}{Z_L}\right)\right| \tag{8.129}$$

For the long tailpipe termination, use of Equations (8.13), (8.43) and (8.118) allows Equation (8.129) to be written as:

$$\begin{aligned}\mathrm{IL} =& 10\log_{10}\left\{\left[1 - \left(\frac{\ell_aV_d}{S_a} + \frac{\ell_aV_b}{S_a} + \frac{\ell_gV_d}{S_g}\right)\left(\frac{\omega}{c}\right)^2 + \frac{\ell_a\ell_gV_bV_d}{S_aS_g}\left(\frac{\omega}{c}\right)^4\right]^2\right.\\ &\left. + \left[\left(\frac{\ell_aS_L}{S_a} + \frac{\ell_gS_L}{S_g}\right)\left(\frac{\omega}{c}\right) - \frac{\ell_a\ell_gV_bS_L}{S_aS_g}\left(\frac{\omega}{c}\right)^3\right]^2\right\}\end{aligned} \tag{8.130}$$

Similarly, for the short tailpipe termination (see Equation (8.119)), Equation (8.129) takes the following alternative form:

$$\begin{aligned}\mathrm{IL} =& 20\log_{10}\left|1 + \frac{\ell_aS_L}{\ell_LS_a} + \frac{\ell_gS_L}{\ell_LS_g} - \left[\frac{\ell_aV_d}{S_a} + \frac{\ell_aV_b}{S_a} + \frac{\ell_gV_d}{S_g}\right.\right.\\ &\left.\left. + \frac{\ell_a\ell_gV_bS_L}{S_aS_g\ell_L}\right]\left(\frac{\omega}{c}\right)^2 + \frac{\ell_a\ell_gV_bV_d}{S_aS_g}\left(\frac{\omega}{c}\right)^4\right|\end{aligned} \tag{8.131}$$

The IL may be expected to be negative when the entire quantity within the outer brackets of Equations (8.127), (8.128), (8.130) and (8.131) is less than 1 and this may sometimes occur in a narrow frequency band in the low-frequency range, corresponding to the filter fundamental resonance frequency. In this frequency band, the low-pass filter of Figure 8.14 may possibly amplify introduced noise. However, as the frequency increases above this band but is still below the half wavelength resonances (corresponding to the filter dimensions) encountered at high frequencies, the IL increases. As the frequency range is extended upward, above the half wavelength resonance, other resonances will be encountered when the chamber or tube length dimensions approach integer multiples of one-half wavelength. These resonances will introduce higher-frequency pass bands, resulting in additional TL minima.

The possibility of high-frequency pass bands associated with half wavelength resonances suggests that their numbers can be minimised by choice of dimensions so that resonances tend to coincide, thus minimising the number of pass bands and the subsequent frequency range over which the filter will perform poorly. Allowing resonances to have coincident resonance frequencies results in higher sound pressure amplitudes within the filter at passband frequencies, which increases the intensity of nonlinear effects. It is found, in practice, that when the number of pass bands is minimised, the loss in filter performance is minimal; that is, the filter continues to perform as a low-pass filter even in the frequency range of the predicted pass bands, although at reduced effectiveness. Presumably, inherent resistive losses and possibly nonlinear effects, particularly at high sound pressure levels, such as may be encountered in a pumping system, account for the better than predicted performance. As stated earlier, these losses and nonlinear effects were neglected in the preceding analysis.

The following iterative procedure is recommended to obtain an approximate low-pass filter design which can then be modified slightly if necessary to achieve a required IL at any desired frequency, ω, by using Equations (8.126) to (8.131).

1. Select the desired highest fundamental resonance frequency, f_{0h}, for the filter to be approximately 0.6 times the fundamental frequency of the pressure pulsations.
2. Select two equal chamber volumes, V_b and V_d, to be as large as practicable.
3. Choose the length and diameter of the chambers so that the length is of similar magnitude to the diameter.
4. If possible, configure the system so that the choke tube, g, connecting the two volume chambers exists completely inside the chambers, which are then essentially one large chamber divided into two by a baffle between them (see Figure 8.14(a)).
5. Make the choke tube as long as possible, consistent with it being completely contained within the two chambers, such that its ends are at least one tube diameter from the end of each chamber.
6. Choose the choke tube diameter to be as small as possible, but within the range 0.5 to 1× the inlet pipe diameter and consistent with a pressure loss of less than 0.5% of the line pressure. Means for calculating pressure losses are discussed in Section 8.14.
7. Calculate the fundamental resonance frequency of the system using the following approximate expression:

$$f_0 \approx \frac{c}{2\pi}\left[\frac{S_g}{\ell_g}\left[\frac{1}{V_b}+\frac{1}{V_d}\right]\right]^{1/2} \tag{8.132}$$

   This expression is quite accurate for a constant-volume-velocity source if the discharge pipework is much shorter than the choke tube (and is of similar or larger diameter), or if the discharge pipework is sufficiently long that pressure waves reflected from its termination may be ignored. For other situations, Equation (8.132) serves as an estimate only and can overestimate the resonance frequency by up to 50%.

8. If f_0, calculated using Equation (8.132), is less than the frequency, $f_{0,h}$, calculated in (1) above, then modify your design by:
   (a) reducing the chamber volumes, and/or
   (b) increasing the choke tube diameter.
9. If f_0, calculated using Equation (8.132), is greater than the frequency, $f_{0,h}$, calculated in (1) above, then modify your design by doing the reverse of what is described in (8) above.
10. Repeat steps (7), (8) and (9) until the quantity f_0, calculated using Equation (8.132), is between ± 5% of the frequency, $f_{0,h}$, calculated in step (1) or if the required IL at the pressure pulsation frequency is achieved.

It is also possible to solve Equation (8.132) directly by choosing all parameters except for one (e.g. choke tube cross-sectional area, S_g) and solving for the parameter not chosen. If the value of the calculated parameter is not satisfactory (e.g. too much pressure loss), then the other chosen parameters will need to be adjusted within acceptable bounds until a satisfactory solution is reached, following much the same iterative procedure as described in the preceding steps.

8.9 4-Pole Method

In previous sections, an electrical circuit method of analysis was used to determine the acoustic performance of mufflers. Another analysis technique called the 4-pole, transmission matrix or transfer matrix method is an alternative theoretical tool. It is especially useful for evaluating the acoustical performance of more complex muffler systems, which are not amenable to electrical circuit analysis. It is also capable of extending the frequency range covered by electrical circuit analysis, although the frequency range is limited to the region where only plane waves propagate in the duct and muffler sections (see Section 8.10.3.3). The 4-pole analysis technique is derived from a method to model electrical networks, as there exist acoustic to electrical analogies, which are listed in Table 8.4.

Figure 8.15 shows a typical configuration of a muffler, which comprises several acoustic elements. An acoustic source, such as a reciprocating engine, is attached to an upstream duct, which is connected to the inlet of a muffler. The geometry of the muffler could consist of various element types, such as transverse tube resonators, expansion chambers, Helmholtz resonators and perforated tubes. The outlet of the muffler is connected to a downstream tailpipe section, which is terminated by an acoustic impedance. For example, the tailpipe might radiate sound into a free field. Each of these components has an acoustic impedance and can be represented by a $[2 \times 2]$ square matrix, which is referred to as a 4-pole transmission line matrix. More complicated muffler configurations, where the flow branches, can also be modelled using this technique, but the matrices become larger, and the interested reader should consult Munjal (2014). By multiplying these transmission matrices together, and knowing the acoustic excitation, the sound pressure and volume velocity at the exit can be determined. Essentially, the acoustic duct and muffler together act like an acoustic filter. The application of the 4-pole method to a simple case is described in the following paragraphs. Afterward, a table is presented of various acoustic elements that will enable the analysis of more complicated systems.

The 4-pole transmission matrix method relies on the principle of plane-wave propagation inside the duct network. To ensure that the transmission matrix method can be used with validity, it is important to estimate the cut-on frequency, which is defined as the frequency below which only plane waves propagate inside the duct (see Equation (8.228)). Section 8.10.3.3 describes how to estimate this frequency.

Referring to Figure 8.15, the acoustic source has an acoustic impedance, Z_{src}, and the end of the acoustic duct has a termination acoustic impedance, Z_T, which, in the example shown in the figure, is the radiation impedance of an unflanged duct radiating into a free field (see Section 4.6 and Levine and Schwinger (1948)).

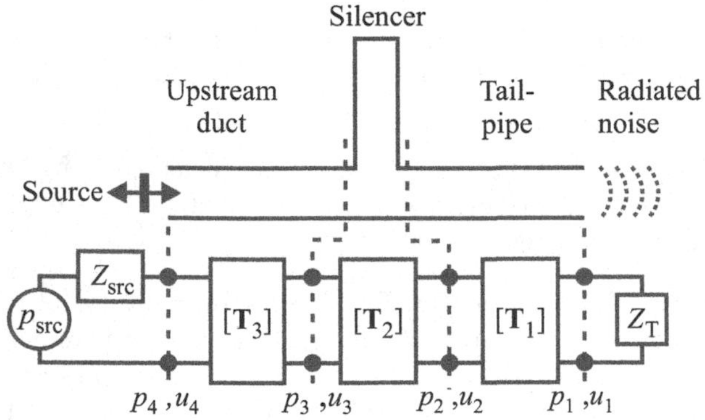

FIGURE 8.15 Schematic of a typical muffler configuration and the equivalent 4-pole transmission matrix representation.

The pressure and volume velocity upstream and downstream of an individual acoustic element are related by a 4-pole transmission matrix as:

$$\begin{bmatrix} p_2 \\ S_2 u_2 \end{bmatrix} = \begin{bmatrix} T_{11} & T_{12} \\ T_{21} & T_{22} \end{bmatrix} \begin{bmatrix} p_1 \\ S_1 u_1 \end{bmatrix} \tag{8.133}$$

where p_i is the acoustic pressure at point, i, along the system, S_i is the cross-sectional area of the duct at point, i, and u_i is the acoustic particle velocity (not the mean flow velocity) at point, i. The elements of the matrix, T_{ab}, are complex numbers.

Note that references such as Munjal et al. (2006) and Munjal (2008, 2013, 2014) calculate the pressure, p, and *mass volume velocity*, $\rho S u$, when using the 4-pole transmission matrix method, rather than using the pressure and *volume velocity*, Su. Munjal also uses alternative definitions for characteristic acoustic impedance, $Y = c/S$ (instead of $Z = \rho c/S$), and acoustic impedance, $Z = p/(\rho S u)$ (instead of $Z = p/(Su)$), which differ from the commonly accepted definitions for these terms, as listed in Table 1.3. In the following sections on 4-pole analysis, some equations have been sourced from Munjal's work and when this has been done they have been modified to be compatible with the impedance definitions used in this book, while still referencing the text from which they originated.

The equations describing the response of the system shown in Figure 8.15 can be written as:

$$\begin{bmatrix} p_{\text{src}} \\ S_{\text{src}} u_{\text{src}} \end{bmatrix} = \begin{bmatrix} 1 & Z_{\text{src}} \\ 0 & 1 \end{bmatrix} \begin{bmatrix} p_4 \\ S_4 u_4 \end{bmatrix} \tag{8.134}$$

$$\begin{bmatrix} p_4 \\ S_4 u_4 \end{bmatrix} = [\mathbf{T}_3] \begin{bmatrix} p_3 \\ S_3 u_3 \end{bmatrix} \tag{8.135}$$

$$\begin{bmatrix} p_3 \\ S_3 u_3 \end{bmatrix} = [\mathbf{T}_2] \begin{bmatrix} p_2 \\ S_2 u_2 \end{bmatrix} \tag{8.136}$$

$$\begin{bmatrix} p_2 \\ S u_2 \end{bmatrix} = [\mathbf{T}_1] \begin{bmatrix} p_1 \\ S_1 u_1 \end{bmatrix} \tag{8.137}$$

$$\begin{bmatrix} p_1 \\ S u_1 \end{bmatrix} = \begin{bmatrix} 1 & Z_T \\ 0 & 1 \end{bmatrix} \begin{bmatrix} 0 \\ S_1 u_1 \end{bmatrix} \tag{8.138}$$

where the $[2 \times 2]$ 4-pole transmission matrices, $[\mathbf{T}_i]$, depend on the configuration of each acoustic element and are covered in Section 8.9.2. Equations (8.134) to (8.138) can be combined to give:

$$\begin{bmatrix} p_{\text{src}} \\ S_{\text{src}} u_{\text{src}} \end{bmatrix} = \begin{bmatrix} 1 & Z_{\text{src}} \\ 0 & 1 \end{bmatrix} \mathbf{T}_3 \mathbf{T}_2 \mathbf{T}_1 \begin{bmatrix} 1 & Z_T \\ 0 & 1 \end{bmatrix} \begin{bmatrix} 0 \\ S_1 u_1 \end{bmatrix} = \begin{bmatrix} T_{11} & T_{12} \\ T_{21} & T_{22} \end{bmatrix} \begin{bmatrix} 0 \\ S_1 u_1 \end{bmatrix} \tag{8.139}$$

Equation (8.139) is an expression that relates the acoustic pressure and volume velocity at the acoustic source to that at the duct outlet. If one wants to evaluate the acoustic performance of the system, such as the TL, IL or NR, then the T_{ij} elements of the overall transmission matrix can be used in the expressions in Section 8.9.1 to evaluate the required performance metric.

8.9.1 Acoustic Performance Metrics

As described in Section 8.2, there are three metrics that can be used to evaluate the acoustic performance of a muffler. These acoustic metrics and their associated equations expressed in terms of the elements of the total 4-pole matrix (see Equation (8.145)) for the entire acoustic network are described as follows:

Insertion Loss (IL) was mathematically defined in Equation (8.1) as the reduction in the radiated sound power level from when the duct system has a straight section of duct installed instead of the proposed muffler, to when the straight section of duct is replaced with the proposed muffler. Calculation of insertion loss requires knowledge of the source and termination acoustic impedances, as well as the 4-pole parameters that describe the straight duct prior to installation of the muffler, and the 4-pole parameters that describe the muffler. The IL is calculated as (Prasad and Crocker (1981), Eq. 13):

$$\mathrm{IL} = 20\log_{10}\left|\frac{Z_R T_{11} + T_{12} + Z_{\mathrm{src}} Z_R T_{21} + Z_{\mathrm{src}} T_{22}}{Z_R T'_{11} + T'_{12} + Z_{\mathrm{src}} Z_R T'_{21} + Z_{\mathrm{src}} T'_{22}}\right| \quad \text{(dB)} \tag{8.140}$$

where T_{ij} are the 4-pole parameters of the muffler system, T'_{ij} are the 4-pole parameters of the system when the muffler has been replaced with a straight section of duct, Z_R is the acoustic radiation impedance from the end of the muffler (or straight pipe) and Z_{src} is the acoustic impedance of the source. It is assumed that the diameter of the tailpipe attached to the muffler is identical to the diameter of the straight duct prior to installation of the muffler, and that the volume velocity and impedance of the source is unaffected by installation of the muffler.

Transmission Loss (TL) was mathematically defined in Equation (8.3). TL is independent of the source impedance, and can be calculated using the 4-pole transmission matrix formulation as (Munjal, 2013, Eq. (5.7), p. 168; Munjal, 2014, Eq. (3.27), p. 105):

$$\mathrm{TL} = 10\log_{10}\left[\left\{\frac{1+M_n}{1+M_1}\right\}^2\left(\frac{1}{4}\right)\left|\frac{Z_{A1}}{Z_{An}}T_{11} + \frac{T_{12}}{Z_{An}} + Z_{A1}T_{21} + \frac{Z_{An}T_{22}}{Z_{A1}}\right|^2\right] \tag{8.141}$$

where M_i is the Mach number of the mean flow in the duct at location i, $Z_{A1} = S_u/\rho c$ is the acoustic impedance of the duct upstream of the muffler and $Z_{An} = S_d/\rho c$ is the acoustic impedance of the duct downstream of the muffler.

Noise Reduction (NR) was mathematically defined in Equation (8.4), and can be calculated using the elements of the total 4-pole transmission matrix as (Prasad and Crocker, 1983, Eq. 2):

$$\mathrm{NR} = 10\log_{10}\left[\left|T_{11} + \frac{T_{12}}{Z_d}\right|^2\right] \tag{8.142}$$

where Z_d is the downstream duct acoustic impedance.

8.9.2 4-Pole Matrices of Various Acoustic Elements

There are various types of acoustic elements used in mufflers, which can be modelled using 4-pole transmission matrices. Examples are shown in Table 8.5 and will be described in more detail in the following sections. The markers on the illustrations, u, for upstream, and d, for downstream, indicate the connection points for each element. The geometry of each element is described by its length, L, cross-sectional area, S, and volume, V. An example is shown in Figure 8.16(b), where the downstream -d- connection point of one 4-pole element is connected to a following 4-pole element at its upstream -u- connection. For more complicated systems, such as mufflers with multiple branches, reverse flows, flows through perforated tubes, etc, the interested reader should consult Munjal (2008) and Munjal (2014).

TABLE 8.5 Examples of acoustic elements that can be described by 4-pole transmission matrices

Acoustic element	Description and section number
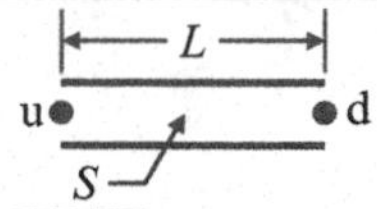	Straight duct. See Section 8.9.3
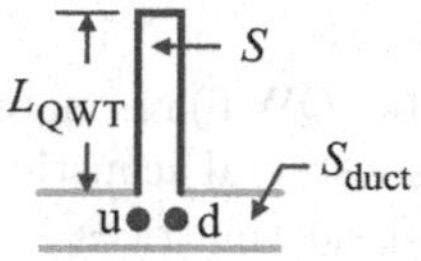	Quarter-wavelength tube (QWT). See Section 8.9.4
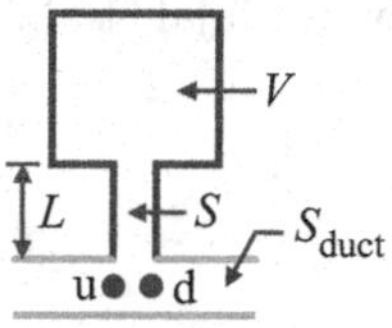	Helmholtz resonator. See Section 8.9.5
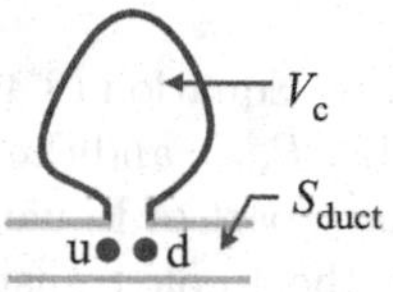	Arbitrary volume. See Section 8.9.5
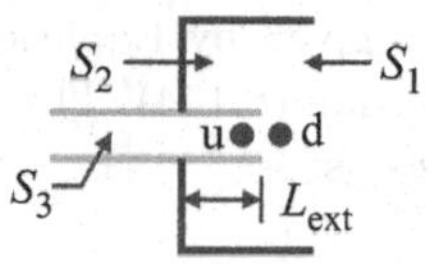	Expansion or contraction segments. See Section 8.9.6

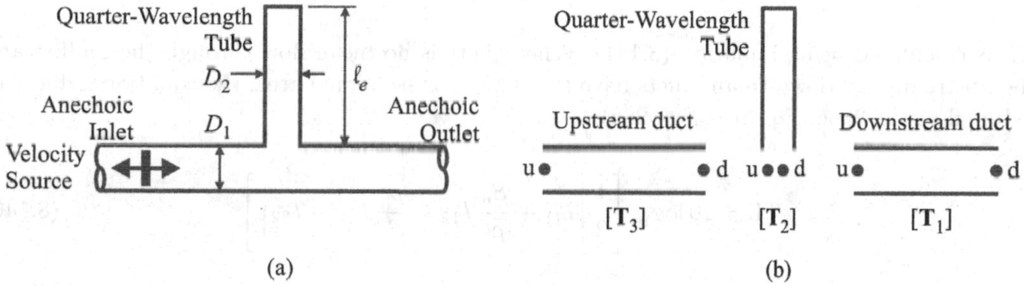

FIGURE 8.16 (a) Schematic of a QWT attached to a circular main exhaust duct and (b) the acoustic elements comprising the system.

8.9.3 Straight Duct

The simplest acoustic element is a straight duct that has a uniform gas temperature. The 4-pole transmission matrix for a straight segment of duct of length, L, is given by (Munjal, 2014, p. 114, Eq. (3.70); Munjal et al., 2006, p. 289, Eq. (9.15)):

$$[\mathbf{T}] = e^{-\mathrm{j}Mk_cL} \begin{bmatrix} \cos(k_cL) & \mathrm{j}\dfrac{\rho c}{S}\sin(k_cL) \\ \mathrm{j}\dfrac{S}{\rho c}\sin(k_cL) & \cos(k_cL) \end{bmatrix} \tag{8.143}$$

where ρ (kg/m^3) is the density of the gas, $k_c = k/(1 - M^2)$, $k = \omega/c$ (m^{-1}) is the wavenumber, $\omega = 2\pi f$ (radians/s) is the circular frequency, f (Hz) is the frequency of excitation, c (m/s) is the speed of sound, U (m/s) is the mean flow velocity through the cross-sectional area S (m^2) and $M = U/c$ is the Mach number.

8.9.4 Quarter-Wavelength Tube (QWT)

Figure 8.16(a) shows a schematic of a quarter-wavelength tube (QWT) resonator muffler attached as a side branch to a circular main duct. The system comprises several acoustic elements as shown in Figure 8.16(b), which can be described by 4-pole transmission matrices.

The 4-pole matrix for a side-branch resonator is given by (Munjal et al., 2006, Eq. (9.21), p. 291):

$$[\mathbf{T}_2] = \begin{bmatrix} 1 & 0 \\ \dfrac{1}{Z_{\text{side}}} & 1 \end{bmatrix} \tag{8.144}$$

where Z_{side} is the acoustic impedance of the resonator given by Equation (8.46), which comprises the sum of the resistive impedance of the entrance to the tube, R_{side}, and the reactive impedance of the resonator cavity, which is represented by the imaginary part of Equation (8.46).

Returning to the QWT muffler shown in Figure 8.16(b), the 4-pole transmission matrices for the upstream, $\mathbf{T}_3$, and downstream, $\mathbf{T}_1$, duct segments are given by Equation (8.143). The 4-pole transmission matrix for the QWT, $\mathbf{T}_2$, is given by Equation (8.144). The total transmission matrix is given by the multiplication of these three matrices as:

$$\begin{bmatrix} T_{11} & T_{12} \\ T_{21} & T_{22} \end{bmatrix} = \mathbf{T}_3\mathbf{T}_2\mathbf{T}_1 \tag{8.145}$$

TL is calculated using Equation (8.141). When there is no mean flow through the muffler and the upstream and downstream ducts have the same cross-sectional area, the equation reduces to (Munjal et al., 2006, Eq. (9.11), p. 286):

$$\text{TL} = 20\log_{10}\left[\frac{1}{2}\left|T_{11} + \frac{S_{\text{u}}}{\rho c}T_{12} + \frac{\rho c}{S_{\text{u}}}T_{21} + T_{22}\right|\right] \tag{8.146}$$

where S_{u} is the cross-sectional area of the inlet duct. If the ducts have different cross-sectional areas, then Equation (8.141) should be used. This example of the QWT and duct system, comprises the QWT with upstream and downstream straight lossless ducts that have anechoic terminations either side of the QWT. These straight ducts provide no acoustic attenuation, and hence their 4-pole transmission matrices $\mathbf{T}_3$ and $\mathbf{T}_1$ can be ignored in the calculations of the TL. In this case, it is only necessary to determine the elements of the 4-pole matrix in Equation (8.144). If the upstream and downstream ducts provide attenuation, or if the complex values of the pressure and particle velocities at the terminations are to be calculated, then it is necessary to include the 4-pole transmission matrices for these ducts.

Figure 8.17 shows the TL of a QWT muffler, predicted using Equation (8.146)), versus normalised frequency $k\ell_e/\pi$, where k is the wavenumber and ℓ_e is the effective length of the QWT, for a range of area ratios of the QWT to main duct $N = S/S_{\text{u}}$. The theoretical model was implemented using MATLAB® and the script, `beranek_ver_fig10_11_quarter_wave_tube_duct_4_pole.m`, is available online (see Appendix F).

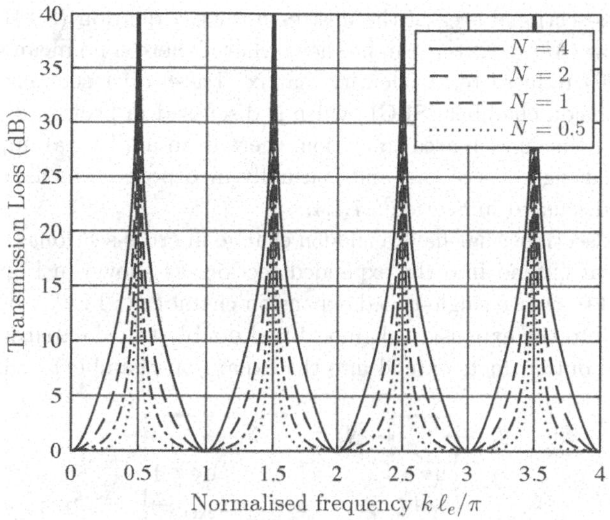

FIGURE 8.17 Transmission loss versus normalised frequency, $k\ell_e/\pi$, of a QWT muffler for a range of area ratios of the QWT to main duct $N = S/S_u = 0.5, 1, 2, 4$.

8.9.5 Helmholtz Resonators

A Helmholtz resonator was discussed in detail in Section 8.8.2 and Singh et al. (2008) provides a tutorial on the use of the 4-pole transmission matrix method applied to it. To apply the 4-pole method to calculate the TL of a Helmholtz resonator, the impedance, Z_{side}, in Equation (8.47) can be used in Equation (8.144).

8.9.6 Sudden Expansion and Contraction

Sudden changes in the cross-sectional area of ducts is common in duct networks. The sudden change in area creates a discontinuity in the acoustic impedance that causes sound waves to be reflected upstream. Figure 8.18 shows (a) a sudden expansion and (b) a sudden contraction in the cross-sectional area of a duct.

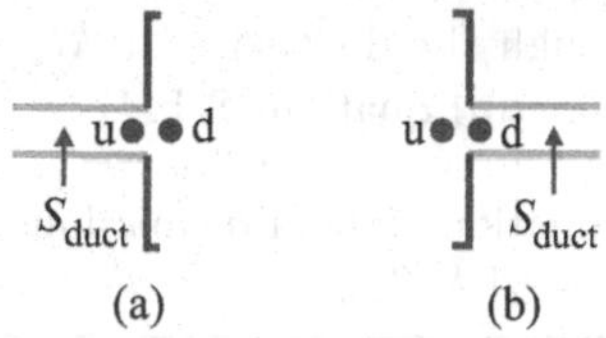

FIGURE 8.18 (a) Sudden area expansion, and (b) sudden area contraction where $S_2 = S_3 - S_1$.

The 4-pole transmission matrix is (Munjal, 2008, p.800):

$$\mathbf{T} = \begin{bmatrix} 1 & K_d M_d Z_{Ad} \\ 0 & 1 \end{bmatrix} \tag{8.147}$$

where M_d is the Mach number of flow in the duct, $Z_{Ad} = \rho c/S_d$ is the characteristic acoustic impedance of the duct, where S_d is the cross-sectional area of the downstream duct, and the term K_d is given by:

$$\text{Sudden expansion: } K_d = [(S_d/S_u) - 1]^2 \tag{8.148}$$

$$\text{Sudden contraction: } K_d = (1 - S_d/S_u)/2 \tag{8.149}$$

where S_u is the cross-sectional area of the upstream duct. Equation (8.147) is applicable when the flow velocity is low ($M_d < 0.2$). It can be shown that if there is no mean gas flow, i.e. $M_d = 0$, then Equation (8.147) reduces to an identity matrix. These duct segments can be be used to create a simple expansion chamber (SEC), which is discussed in Section 8.9.7.

At the location of the sudden area transition, there is an additional impedance that is associated with the oscillating mass of gas, and is usually incorporated as an end correction to the duct length, as was discussed in Section 8.7.1.4.

A variant of duct sections that have a sudden change in cross-sectional area are sections that incorporate tubes that extend into the expanded section, as shown in Figure 8.19. These duct elements can be used to create single-tuned expansion chamber (STEC, where either the inlet or outlet duct extends into the expansion chamber) and double-tuned expansion chamber (DTEC, where both inlet and outlet ducts extend into the expansion chamber) mufflers.

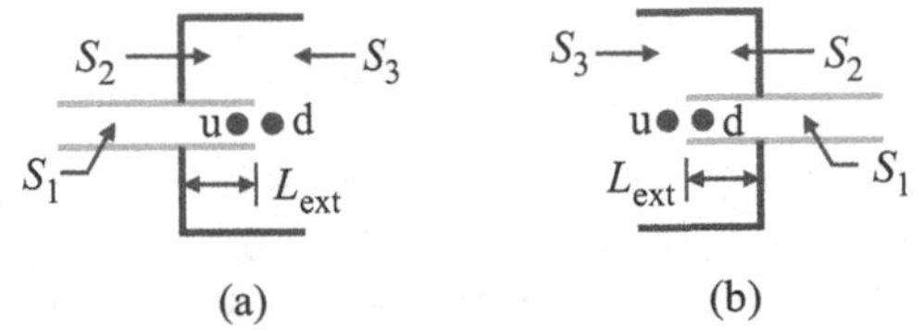

FIGURE 8.19 (a) Sudden area expansion with extension tube, and (b) sudden area contraction with extension tube.

The transmission matrix for these duct transitions is (Munjal, 2014, Eq. (3.95), p. 118):

$$\mathbf{T} = \begin{bmatrix} 1 & KM_1Z_{A1} \\ \dfrac{C_2S_2}{C_2S_2Z_2 + S_3M_3Z_{A3}} & \dfrac{C_2S_2Z_2 - M_1Z_{A1}(C_1S_1 + S_3K)}{C_2S_2Z_2 + S_3M_3Z_{A3}} \end{bmatrix} \tag{8.150}$$

$$Z_2 = -\mathrm{j}\frac{\rho c}{S_2}\cot kL_{\mathrm{ext}} \tag{8.151}$$

$$Z_{Ai} = \rho c/S_i \quad (i = 1, 2, 3) \tag{8.152}$$

where the parameters C_1, C_2 and K are listed in Table 8.6, U_i is the mean flow velocity through the cross-section of area, S_i, $M_i = U_i/c$ is the Mach number of flow through the cross-section of area, S_i, and L_{ext} is the length of the extension tube, as shown in Figure 8.19. It can be shown that if there is no flow through the duct so that $M_1 = M_3 = 0$, then Equation (8.150) will simplify to an expression similar to Equation (8.144).

TABLE 8.6 Parameters for the evaluation of the transmission matrix for ducts with cross-sectional discontinuities (Munjal, 2014, Tables 3.1 & 3.2)

Type	C_1	C_2	K
Expansion	-1	1	$[(S_1/S_3) - 1]^2$
Contraction	-1	-1	$\frac{1}{2}[1 - (S_1/S_3)]$

Examples of the use of these elements for a simple expansion chamber and double-tuned expansion chamber are provided in the following sections.

8.9.7 Simple Expansion Chamber (SEC)

A simple expansion chamber muffler is shown in Figure 8.20 and comprises a sudden expansion, a straight duct and a sudden contraction. The upstream inlet duct has a cross-sectional area of

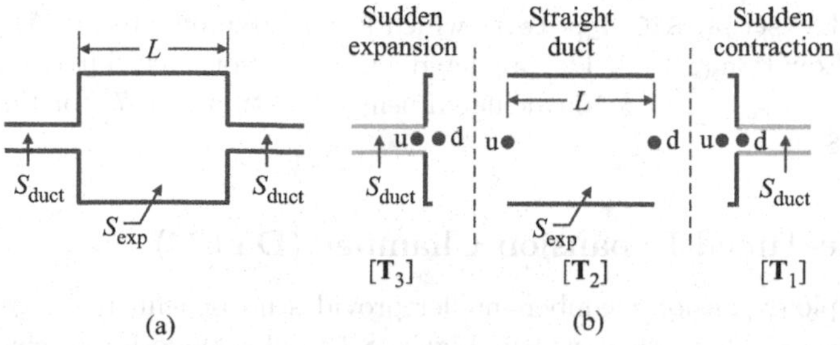

FIGURE 8.20 (a) Schematic of a simple expansion chamber (SEC), and (b) the 4-pole elements of the muffler.

S_{duct} and is connected to a larger diameter tube that has a cross-sectional area of S_{exp}. The interface between the inlet and the expansion chamber is a sudden expansion, and there are no duct extensions within the expansion chamber ($L_{\text{ext}} = 0$). The expansion chamber has a length of L and is terminated with a sudden contraction to the outlet duct, which has a cross-sectional area of S_{duct}. It is assumed that there is no mean flow ($M_i = 0$). The 4-pole transmission matrix approach is used to derive the TL of the muffler.

By using Equation (8.147), when there is no mean gas flow through the sudden expansion and contraction sections ($M = 0$), it can be shown that the transmission matrices are identity matrices. The 4-pole matrix for the straight duct is the same as Equation (8.143). The 4-pole matrices for each of these segments are:

$$\text{contraction: } \mathbf{T}_1 = \begin{bmatrix} 1 & 0 \\ 0 & 1 \end{bmatrix} \tag{8.153}$$

$$\text{straight duct: } \mathbf{T}_2 = \begin{bmatrix} \cos(kL) & \dfrac{\mathrm{j}\rho c}{S_{\text{exp}}}\sin(kL) \\ \dfrac{\mathrm{j}S_{\text{exp}}}{\rho c}\sin(kL) & \cos(kL) \end{bmatrix} \tag{8.154}$$

$$\text{expansion: } \mathbf{T}_3 = \begin{bmatrix} 1 & 0 \\ 0 & 1 \end{bmatrix} \tag{8.155}$$

The combined 4-pole matrix for the expansion chamber muffler is thus:

$$\mathbf{T}_{\text{exp}} = \mathbf{T}_3\mathbf{T}_2\mathbf{T}_1 = \begin{bmatrix} \cos(kL) & \dfrac{\mathrm{j}\rho c}{S_{\text{exp}}}\sin(kL) \\ \dfrac{\mathrm{j}S_{\text{exp}}}{\rho c}\sin(kL) & \cos(kL) \end{bmatrix} \tag{8.156}$$

The RHS of Equation (8.156) is the same as for a straight duct, implying that for the case of no flow, the expansion chamber can be treated as a straight duct. The difference in the overall TL between an expansion chamber and a straight duct connected at each end to ducts of the same cross-sectional area is a result of the downstream and upstream connecting ducts having a different cross-sectional area to the expansion chamber.

The TL is calculated using Equation (8.146), which was derived using the general Equation (8.141) and can be reduced to (see Equation (8.115) and (Munjal, 2014, p. 87)):

$$\begin{aligned} \text{TL} &= 10\log_{10}\left[\cos^2(kL) + \frac{1}{4}\left(\frac{S_{\text{duct}}}{S_{\text{exp}}} + \frac{S_{\text{exp}}}{S_{\text{duct}}}\right)^2 \sin^2(kL)\right] \\ &= 10\log_{10}\left[1 + \frac{1}{4}\left(\frac{S_{\text{duct}}}{S_{\text{exp}}} - \frac{S_{\text{exp}}}{S_{\text{duct}}}\right)^2 \sin^2(kL)\right] \end{aligned} \tag{8.157}$$

which is identical to Equation (8.115), which was derived using 1-D wave analysis.

The MATLAB® script, `SEC_4_pole.m`, which is available online (see Appendix F), can be used to evaluate Equation (8.157). The predicted TL versus normalised frequency, kL/π, for area ratios of $m = S_{\text{exp}}/S_{\text{duct}} = 2, 4, 8, 16$ calculated using Equation (8.157) (or Equation (8.115)), is shown in Figure 8.13.

8.9.8 Double-Tuned Expansion Chamber (DTEC)

The TL of a simple expansion chamber muffler provides no benefit at integer multiples of the normalised frequency, kL/π, as shown in Figure 8.13. This undesirable characteristic can be improved by designing the expansion and contraction parts of the muffler to have tubes that extend into the volume of the expansion chamber, as shown in Figure 8.21(a), where the length of the protruding tubes are not zero. Choosing the length of one of the extension tubes to be equal to $L/2$, where L is the overall length of the expansion chamber, will tend to improve the TL at frequencies of $kL/\pi = 1, 3, 5, \cdots$. Similarly, choosing the length of the other extension tube to be equal to $L/4$, will tend to improve the TL at frequencies of $kL/\pi = 2, 6, 10, \cdots$.

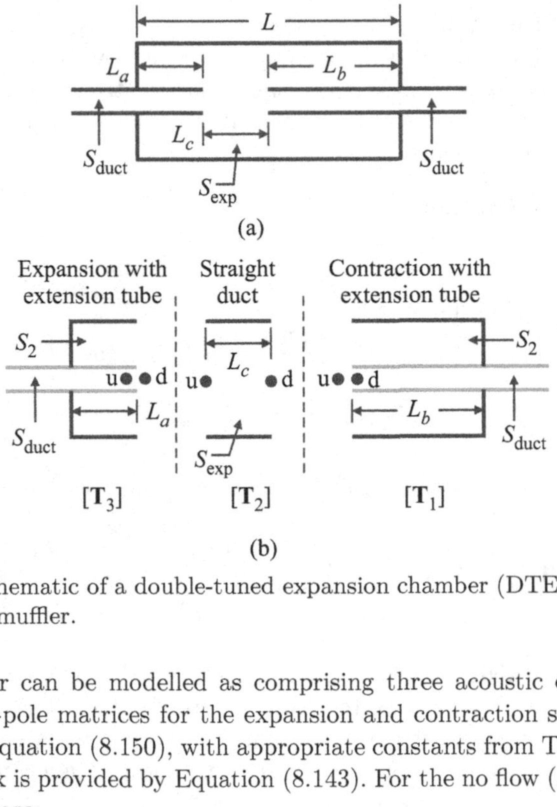

FIGURE 8.21 (a) Schematic of a double-tuned expansion chamber (DTEC) muffler, and (b) the 4-pole elements of the muffler.

The DTEC muffler can be modelled as comprising three acoustic elements, as shown in Figure 8.21(b). The 4-pole matrices for the expansion and contraction sections with extension tubes originate from Equation (8.150), with appropriate constants from Table 8.6. For a straight duct, the 4-pole matrix is provided by Equation (8.143). For the no flow ($M = 0$) case, the three transmission matrices are:

$$\text{contraction: } \mathbf{T}_1 = \begin{bmatrix} 1 & 0 \\ \left(\mathrm{j}\dfrac{S_2}{\rho c}\tan kL_b\right) & 1 \end{bmatrix} \tag{8.158}$$

$$\text{straight duct: } \mathbf{T}_2 = \begin{bmatrix} \cos(kL_c) & \dfrac{\mathrm{j}\rho c}{S_{\text{exp}}}\sin(kL_c) \\ \dfrac{\mathrm{j}S_{\text{exp}}}{\rho c}\sin(kL_c) & \cos(kL_c) \end{bmatrix} \tag{8.159}$$

$$\text{expansion: } \mathbf{T}_3 = \begin{bmatrix} 1 & 0 \\ \left(\mathrm{j}\dfrac{S_2}{\rho c}\tan kL_a\right) & 1 \end{bmatrix} \tag{8.160}$$

where $S_2 = S_{\text{exp}} - S_{\text{duct}}$, L_a and L_b are the lengths of the extension tubes in the expansion and contraction sections, respectively, and L_c is the length of the straight duct. The combined 4-pole transmission matrix for the DTEC muffler is:

$$\mathbf{T}_{\text{DTEC}} = \mathbf{T}_3\mathbf{T}_2\mathbf{T}_1 \tag{8.161}$$

and the TL can be calculated using Equation (8.146), with the elements of the transmission matrix from T_{DTEC}.

Figure 8.22 shows the TL of a double-tuned expansion chamber muffler where the lengths of the extension tubes are $L_a = L/4$ and $L_b = L/2$, where L is the length of the expansion chamber, for area ratios $m = S_{\text{exp}}/S_{\text{duct}} = 2, 4, 8, 16$. The results shown in the figure are for the "no flow" case. In the presence of flow, the peak TL values will decrease and the values between the peaks will increase as a result of the resistive impedance contributed by the flow (see Equation (8.34)).

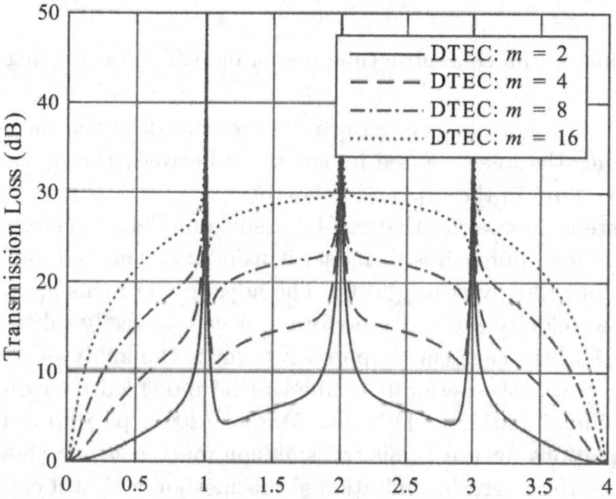

FIGURE 8.22 Transmission loss of double-tuned expansion chamber muffler for no flow, $M = 0$, where the lengths of the extension tubes are $L_a = L/2$ and $L_b = L/4$, and L is the length of the expansion chamber, for area ratios $m = S_{\text{exp}}/S_{\text{duct}} = 2, 4, 8, 16$.

When designing the muffler, the appropriate effective acoustic lengths for the two extension tubes within the expansion chamber are chosen so as to prevent the dips in the TL versus frequency plot. The effective acoustic length of each extension tube is its physical geometric length plus a small end-correction length. Hence, when constructing the muffler, the physical geometric lengths of the extension tubes are made slightly smaller than the required effective acoustic lengths.

Chaitanya and Munjal (2011) presented an empirical relation for the additional end-correction lengths, ℓ_0 as:

$$\ell_0 = d_{\text{ext}}\left(a_0 + a_1\left[\frac{D_{\text{exp}}}{d_{\text{ext}}}\right] + a_2\left[\frac{t_w}{d_{\text{ext}}}\right] + a_3\left[\frac{D_{\text{exp}}}{d_{\text{ext}}}\right]^2 + a_4\left[\frac{D_{\text{exp}}t_w}{d_{\text{ext}}^2}\right] + a_5\left[\frac{t_w}{d_{\text{ext}}}\right]^2\right) \tag{8.162}$$

where t_w is the wall thickness of the extension tubes, d_{ext} (m) is the diameter of the extension tubes, D_{exp} (m) is the diameter of the expansion chamber $a_0 = 0.005177$, $a_1 = 0.0909$, $a_2 = 0.537$, $a_3 = -0.008594$, $a_4 = 0.02616$ and $a_5 = -5.425$.

8.9.9 Concentric Tube Resonator (CTR)

A concentric tube resonator is similar to a double-tuned expansion chamber muffler, except that a tube with perforated holes connects the upstream and downstream extension tubes, as shown in Figure 8.23. A MATLAB® script, `TL_CTR_perforated.m`, for implementing this method can be found on the website listed in Appendix F.

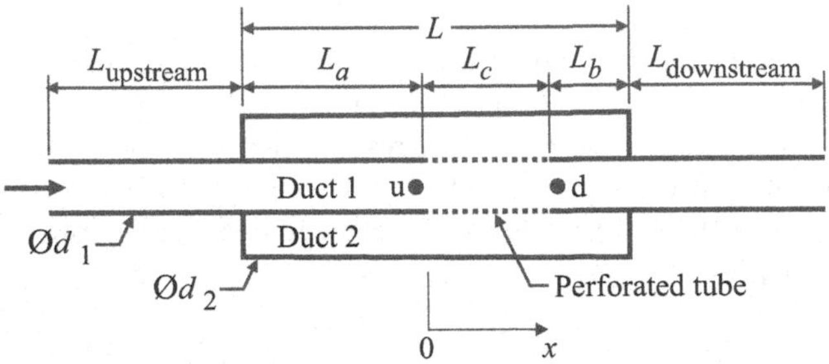

FIGURE 8.23 Geometry of a concentric tube resonator (CTR) comprising a perforated tube.

The effect of the perforated tube is to reduce the pressure drop that occurs from the sudden expansion, and to reduce the losses caused by the shear between the moving fluid in the main duct and the stationary fluid in the expansion chamber.

There are two different approaches that can be used to model a perforated tube using 4-pole transmission matrices. One approach is to model it using a segmented approach, such as done by Sullivan (1979a,b) and Elnady et al. (2010). The advantage of this approach is that if there is a decrease in the flow velocity across the perforate, it can be easily taken into account within each segment. In addition, the segmented approach permits the analysis of complicated muffler configurations. Alternatively, the perforated tube can be modelled as a distributed parameter system, as done by Munjal (2014, p. 119) and Munjal (2008, p. 806). This method has the advantage that the equations for the 4-pole transmission matrix have a closed-form solution, as presented in this section. However, the limitation of this method is that it can be inaccurate when the mean flow decreases along the perforate, as the distributed parameter approach assumes it to be constant. The 4-pole transmission matrix for a concentric tube resonator is more complicated to derive than other muffler components and involves a number of calculation steps described below.

The percentage of open area of a perforated tube is the ratio of the open hole area to the average area between holes. For hole perforations arranged in a square pattern, as shown in Figure 8.24(a), the percentage of open area is calculated as:

$$P_{\text{open}} = \frac{\pi d_h^2}{4d^2} \times 100 = 0.7854\frac{d_h^2}{d^2} \times 100 \quad (\%) \tag{8.163}$$

where d_h is the hole diameter, and d is the grid spacing. For hole perforations arranged in a hexagonal pattern, as shown in Figure 8.24(b), the percentage of open area is calculated as:

$$P_{\text{open}} = \frac{\pi d_h^2}{2\sqrt{3} \times d^2} \times 100 = 0.9069\frac{d_h^2}{d^2} \times 100 \quad (\%) \tag{8.164}$$

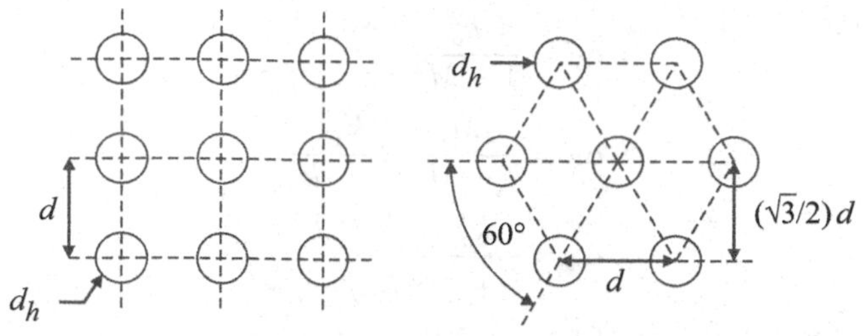

FIGURE 8.24 Hole perforations arranged in a (a) square pattern, and (b) hexagonal pattern.

The holes in the perforated tube couple the acoustic response between the main duct and the outer expansion chamber. If there is no grazing flow in the tube, the perforated tube has an acoustic impedance given by (Sullivan and Crocker, 1978), (Sullivan, 1979b):

$$Z_A = \frac{\rho c}{S_d}\left[\frac{0.006 + \mathrm{j}k(t_w + 0.75d_h)}{(P_{\text{open}}/100)}\right] \tag{8.165}$$

where $k = \omega/c$ (m^{-1}) is the wavenumber, ω (radians/s) is the circular frequency, c (m/s) is the speed of sound of the medium, d_h (m) is the hole diameter, t_w (m) is the thickness of the perforated tube and S_d (m^2) is the cross-sectional area of the tube with the perforated holes.

If there is grazing flow in the tube (i.e. the gas flow passes parallel to the perforated tube and there is no mean flow to the outer expansion chamber), the acoustic impedance is given by the empirical equation:

$$Z_A = \frac{\rho c}{S_d}\left[\frac{7.337\times10^{-3}\,(1+72.23M_1) + \mathrm{j}\,2.2245\times10^{-5}(1+51\,t_w)(1+204\,d_h)\,f}{(P_{\text{open}}/100)}\right] \tag{8.166}$$

where M_1 is the mean flow Mach number in the main duct (duct 1 in Figure 8.23), and f is the frequency in Hz. Equation (8.166) is an empirical expression determined by Rao and Munjal (1986) from numerous experimental measurements of flow through perforated tubes. The expression is valid for $0.05 \le M_1 \le 0.2$, $3\% \le P_{\text{open}} \le 10\%$, $1 \le t_w \le 3$ mm, $1.75 < d_h < 7$ mm.

To determine the transmission matrix, the eigenvectors and eigenvalues (see Section A.7) must be calculated for the following matrix (Munjal, 2008, p. 806):

$$\begin{bmatrix} -\alpha_1 & -\alpha_3 & -\alpha_2 & -\alpha_4 \\ -\alpha_5 & -\alpha_7 & -\alpha_6 & -\alpha_8 \\ 1 & 0 & 0 & 0 \\ 0 & 1 & 0 & 0 \end{bmatrix} \tag{8.167}$$

Note the irregular numbering of the α elements in the matrix, which are defined below. Section A.7 describes how to calculate the eigenvectors and eigenvalues of a matrix. To calculate them using the software MATLAB®, the matrix in Equation (8.167) is defined as (say) `alpha_matrix`, and the eigenvectors and eigenvalues are calculated using the MATLAB® command: "`[psi_eigenvectors,beta_eigenvalues] = eig(alpha_matrix);`".

The matrix of Equation (8.167) has four eigenvectors, ψ_i, each of which has dimensions $[4 \times 1]$, and four eigenvalues, β_i. The α elements of the matrix in Equation (8.167) are given by the following equations, which have been derived from similar equations in (Munjal, 2008, p. 806) by setting $M_2 = 0$.

$$\alpha_1 = \frac{\mathrm{j}M_1}{1-M_1^2} \times \frac{k_a^2 + k^2}{k} \tag{8.168}$$

$$\alpha_2 = \frac{k_a^2}{1 - M_1^2} \tag{8.169}$$

$$\alpha_3 = \frac{\mathrm{j}M_1}{1 - M_1^2} \times \frac{k_a^2 - k^2}{k} \tag{8.170}$$

$$\alpha_4 = -\frac{k_a^2 - k^2}{1 - M_1^2} \tag{8.171}$$

$$\alpha_5 = \alpha_7 = 0 \tag{8.172}$$

$$\alpha_6 = -(k_b^2 - k^2) \tag{8.173}$$

$$\alpha_8 = k_b^2 \tag{8.174}$$

where $M_1 = U_1/c$ is the Mach number of the mean flow in duct 1 (the main duct), U_1 is the mean flow velocity in duct 1 and the modified wavenumbers are given by:

$$k_a^2 = k^2 - \frac{\mathrm{j}4k}{d_1(Z_A S_{d1}/(\rho c))}; \quad k_b^2 = k^2 - \frac{\mathrm{j}4kd_1}{(d_2^2 - d_1^2)(Z_A S_{d1}/(\rho c))} \tag{8.175}$$

where d_1, d_2 are the diameter of the perforated tube and the expansion chamber, respectively, S_{d1} is the cross-sectional area of the perforated tube and Z_A is the impedance of the perforated tube given by Equation (8.165) if there is no flow ($M_1 = 0$), or Equation (8.166) if there is flow ($M_1 \neq 0$). For the configuration shown in Figure 8.23, regardless of whether there is flow in the inner duct or not, the outer chamber is sealed, so there is no mean flow and $M_2 = 0$.

If the four eigenvectors, $\boldsymbol{\psi}_i$, of the matrix in Equation (8.167) have been calculated such that first entry in the eigenvectors is 1 (note that it is not necessary for the eigenvectors to be normalised in this manner; however, the results obtained using the normalised eigenvectors should be the same as those obtained using the non-normalised eigenvectors, which is a good programming error check), then the following relations hold (Munjal, 2014, p. 122):

$$\psi_{1,i} = 1 \tag{8.176}$$

$$\psi_{2,i} = -\left(\frac{\beta_i^2 + \alpha_1\beta_i + \alpha_2}{\alpha_3\beta_i + \alpha_4}\right) \tag{8.177}$$

$$\psi_{3,i} = 1/\beta_i \tag{8.178}$$

$$\psi_{4,i} = \psi_{2,i}/\beta_i = \psi_{2,i}\,\psi_{3,i} \tag{8.179}$$

The eigenvectors and eigenvalues are used to calculate the matrix $[\mathbf{A}(x)]$, whose row elements are defined as:

$$\mathrm{A}_{1,i}(x) = \psi_{3,i}\,e^{\beta_i x} \tag{8.180}$$

$$\mathrm{A}_{2,i}(x) = \psi_{4,i}\,e^{\beta_i x} \tag{8.181}$$

$$\mathrm{A}_{3,i}(x) = -\frac{\psi_{1,i}\,e^{\beta_i x}}{\mathrm{j}k + M_1\beta_i} \tag{8.182}$$

$$\mathrm{A}_{4,i}(x) = -\frac{\psi_{2,i}\,e^{\beta_i x}}{\mathrm{j}k} \tag{8.183}$$

where $i = 1, \cdots, 4$. The matrix $[\mathbf{A}(x)]$ is evaluated at locations $x = 0$ and $x = L_c$, where L_c is the length of the perforated tube (see Figure 8.23). The next step is to evaluate the matrix:

$$\begin{bmatrix} T_{11} & T_{12} & T_{13} & T_{14} \\ T_{21} & T_{22} & T_{23} & T_{24} \\ T_{31} & T_{32} & T_{33} & T_{34} \\ T_{41} & T_{42} & T_{43} & T_{44} \end{bmatrix} = [\mathbf{A}(0)]\,[\mathbf{A}(L_c)]^{-1} \tag{8.184}$$

The 4-pole transmission matrix is given by (Munjal, 2008, p. 808) (Munjal, 2014, p. 124):

$$\begin{bmatrix} p_u \\ \rho c S_u u_u \end{bmatrix} = \begin{bmatrix} T_a & T_b \\ T_c & T_d \end{bmatrix} \begin{bmatrix} p_d \\ \rho c S_d u_d \end{bmatrix} \tag{8.185}$$

where the subscripts u and d refer to the locations upstream and downstream of the perforated tube, respectively, and the elements of the matrix are defined as:

$$T_a = T_{11} + A_1 A_2 \tag{8.186}$$

$$T_b = T_{13} + B_1 A_2 \tag{8.187}$$

$$T_c = T_{31} + A_1 B_2 \tag{8.188}$$

$$T_d = T_{33} + B_1 B_2 \tag{8.189}$$

$$A_1 = (X_1 T_{21} - T_{41})/F_1 \tag{8.190}$$

$$A_2 = T_{12} + X_2 T_{14} \tag{8.191}$$

$$B_1 = (X_1 T_{23} - T_{43})/F_1 \tag{8.192}$$

$$B_2 = T_{32} + X_2 T_{34} \tag{8.193}$$

$$X_1 = -\mathrm{j}\tan(kL_a) \tag{8.194}$$

$$X_2 = +\mathrm{j}\tan(kL_b) \tag{8.195}$$

$$F_1 = T_{42} + X_2 T_{44} - X_1(T_{22} + X_2 T_{24}) \tag{8.196}$$

where T_{ij} are elements of the transmission matrix, $\boldsymbol{T}$, in Equation (8.184).

As was shown for the double-tuned expansion chamber (DTEC) muffler in Section 8.9.8, the lengths of the extension tubes must be selected so that the dips in the TL spectrum are minimised. For the concentric tube resonator muffler described here, selection of the appropriate lengths of the extension tubes is more complicated because of the influence of the impedance of the perforated tube, which varies with the geometry of the tube, flow speed and temperature gradients within the gas medium. Some authors have suggested empirical models can be used to make a first estimate of the length adjustment of the extension tubes (Munjal, 2013, p. 176, Eq. (5.16); Chaitanya and Munjal, 2011; Ramya and Munjal, 2014). However, these length corrections still need to be adjusted by examining the TL spectrum from simulations until a qualitatively satisfactory result is obtained such that the dips in the TL spectrum are minimised.

The resulting lengths of the extension tubes determined from simulations are termed the acoustic lengths, $L_{a,a}, L_{b,a}$. When the muffler is constructed, the physical or geometric lengths, $L_{a,g}, L_{b,g}$, of the extension tubes are slightly shorter than the acoustic lengths, as there is an end-correction length associated with each extension tube as it interacts with the perforated tube.

The required physical or geometric lengths of the inlet, outlet and perforated tubes are related to the physical length, L (m), of the expansion chamber (see Figure 8.23) by:

$$\text{Upstream tube:} \quad L_{a,g} = (L/2) - \ell_0 \tag{8.197}$$

$$\text{Downstream tube:} \quad L_{b,g} = (L/4) - \ell_0 \tag{8.198}$$

$$\text{Perforated tube:} \quad L_{c,g} = (L/4) + 2\ell_0 \tag{8.199}$$

where the first estimate of ℓ_0 for 3-D simulation is (Ramya and Munjal, 2014, Eq. (13)):

$$\begin{aligned} \ell_0\ (\text{mm}) = 0.673 & \left[1 + 0.5864(P_{\text{open}}/100)^{-0.6402}\right] \left[1 + 50.24(t_w)^{0.7902}\right] \\ & \times \left[1 + 79.22(d)^{0.8513}\right] \left[1 + 56.603(d_h)\right] \\ & \times \left[1 - 0.723(d_2/d_1)^{-1.478}\right] \left[1 + 3.44 \times 10^{-3}(L)^{-1.952}\right] \end{aligned} \tag{8.200}$$

where P_{open} (%) is the percentage open area of the perforated tube and is less than 100, t_w is the thickness of the perforated tube (m), d_1 is the diameter of the main duct (m), d_2 is the diameter of the expansion chamber (m) and d_h is the diameter of the holes in the perforated tube (m).

8.9.10 Exhaust Gas Temperature Variations

There are many practical applications where the temperature of a gas varies along the length of a duct, such as an exhaust system. As the temperature of gas changes there is a change in the speed of sound and density of the gas, which affects the acoustic behaviour of the system. This section describes the 4-pole transmission matrix of a straight duct where the gas has a linear temperature gradient. The MATLAB® script, `file:temp_gradient_spl_along_duct_4pole_sujith.m`, for implementing this method can be found on the website listed in Appendix F.

Sujith (1996) derived the 4-pole transmission matrix for a duct with linear and exponential temperature gradients, based upon the work from a previous paper (Sujith et al., 1995). The equations for the 4-pole transmission matrix presented in Sujith (1996) have inconsistencies that have been corrected here and in Howard (2013a). An example application of the theory is shown in Howard (2013b) which shows a QWT muffler that has temperature gradients along the lengths of ducts. Readers who are interested in simulating complicated systems using ANSYS finite element analysis software should refer to Howard and Cazzolato (2014, Section 3.6).

Figure 8.25 shows a schematic of a linear temperature distribution in a circular duct of diameter $2a$ (i.e. radius a) and length L, with zero mean flow. The ends of the duct have acoustic particle velocities u_1 and u_2, and gas temperatures T_1 and T_2 at axial locations $x_1 = L$ and $x_2 = 0$, respectively. The locations of the inlet and outlet have been defined in this way to be consistent with Sujith (1996) and Howard (2013a).

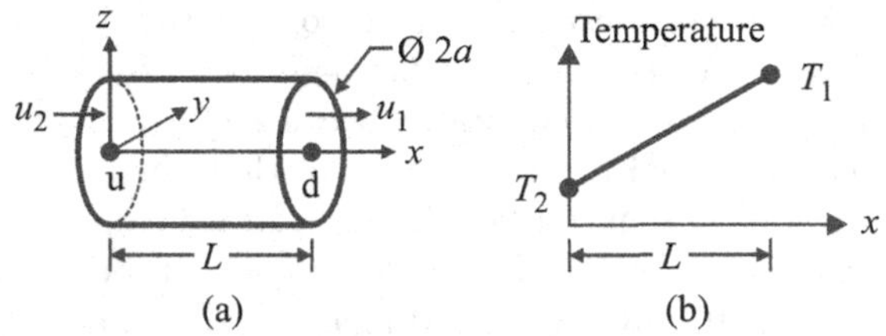

FIGURE 8.25 (a) Schematic of a duct segment with a linear temperature gradient, and (b) linear temperature gradient along the duct axis.

The linear temperature distribution in the duct is given by

$$T(x) = T_2 + gx \tag{8.201}$$

where g is the gradient of the temperature distribution given by

$$g = \frac{T_1 - T_2}{L} \tag{8.202}$$

The temperature-dependent speed of sound can be calculated using Equation (1.8), and the density of the gas can be calculated by rearranging Equation (1.8) as:

$$\rho = \frac{MP_0}{RT} \tag{8.203}$$

where $R = 8.314$ J.mol^{-1} K^{-1} is the universal gas constant, T is the gas temperature in Kelvin, $M = 0.029$ kg.mol^{-1} is the molecular weight of air, and P_0 is the static pressure in the duct, which is equal to atmospheric pressure for an open duct with no flow.

The definition for the 4-pole transmission matrix with a temperature gradient differs from Equation (8.133), as the density and speed of sound of the gas changes with temperature and position along the duct. Also note that Equation (8.205) evaluates the pressure and the acoustic particle velocity, whereas previous descriptions (in this book) of the 4-pole method evaluated the pressure and the acoustic volume velocity. The pressure and acoustic particle velocities at the ends of the duct are related by the 4-pole transmission matrix as:

$$\begin{bmatrix} p_2 \\ u_2 \end{bmatrix} = \begin{bmatrix} T_{11} & T_{12} \\ T_{21} & T_{22} \end{bmatrix} \begin{bmatrix} p_1 \\ u_1 \end{bmatrix} \tag{8.204}$$

$$\begin{bmatrix} p_2 \\ u_2 \end{bmatrix} = \mathbf{T} \begin{bmatrix} p_1 \\ u_1 \end{bmatrix} \tag{8.205}$$

where p_i is the acoustic pressure at ends of the duct, u_i is the acoustic particle velocity at ends of the duct, and the 4-pole transmission matrix is:

$$\mathbf{T} = \begin{bmatrix} T_{11} & T_{12} \\ T_{21} & T_{22} \end{bmatrix} \tag{8.206}$$

where the elements of the transmission matrix are (Howard, 2013a):

$$\begin{aligned} T_{11} = &\left[\frac{\pi\omega\sqrt{T_1}}{2\nu}\right] \\ &\times \left[J_1\left(\frac{\omega\sqrt{T_1}}{\nu}\right) Y_0\left(\frac{\omega\sqrt{T_2}}{\nu}\right) - J_0\left(\frac{\omega\sqrt{T_2}}{\nu}\right) Y_1\left(\frac{\omega\sqrt{T_1}}{\nu}\right)\right] \end{aligned} \tag{8.207}$$

$$\begin{aligned} T_{12} = \mathrm{j} &\left[\frac{\pi\omega\sqrt{T_1}}{2\nu}\right] \times [\mathrm{sgn}(g)] \times \left[\rho_1\sqrt{\gamma R_s T_1}\right] \\ &\times \left[J_0\left(\frac{\omega\sqrt{T_2}}{\nu}\right) Y_0\left(\frac{\omega\sqrt{T_1}}{\nu}\right) - J_0\left(\frac{\omega\sqrt{T_1}}{\nu}\right) Y_0\left(\frac{\omega\sqrt{T_2}}{\nu}\right)\right] \end{aligned} \tag{8.208}$$

$$\begin{aligned} T_{21} = \mathrm{j} &\left[\frac{\pi\omega\sqrt{T_1}}{2\nu}\right] \times [\mathrm{sgn}(g)] \times \left[\frac{1}{\rho_2\sqrt{\gamma R_s T_2}}\right] \\ &\times \left[J_1\left(\frac{\omega\sqrt{T_2}}{\nu}\right) Y_1\left(\frac{\omega\sqrt{T_1}}{\nu}\right) - J_1\left(\frac{\omega\sqrt{T_1}}{\nu}\right) Y_1\left(\frac{\omega\sqrt{T_2}}{\nu}\right)\right] \end{aligned} \tag{8.209}$$

$$\begin{aligned} T_{22} = &\left[\frac{\pi\omega\sqrt{T_1}}{2\nu}\right] \times \left[\frac{\rho_1\sqrt{\gamma R_s T_1}}{\rho_2\sqrt{\gamma R_s T_2}}\right] \\ &\times \left[J_1\left(\frac{\omega\sqrt{T_2}}{\nu}\right) Y_0\left(\frac{\omega\sqrt{T_1}}{\nu}\right) - J_0\left(\frac{\omega\sqrt{T_1}}{\nu}\right) Y_1\left(\frac{\omega\sqrt{T_2}}{\nu}\right)\right] \end{aligned} \tag{8.210}$$

The symbols used in these equations are defined in Table 8.7. The parameter ν is defined as:

$$\nu = \frac{|g|}{2}\sqrt{\gamma R_s} \tag{8.211}$$

and the specific gas constant, R_s, is defined as:

$$R_s = R/M \tag{8.212}$$

The function $\mathrm{sgn}(g) = |g|/g$ is the sign of the value of g.

TABLE 8.7 Symbols used for the 4-pole transmission matrix of a duct with a linear temperature gradient

Symbol	Description
a	Radius of the duct
c	Speed of sound at ambient temperature
g	Linear temperature gradient
j	Unit imaginary number $= \sqrt{-1}$
J_n	Bessel function of the nth order
k	Wavenumber
L	Length of the duct
p_1, p_2	Pressure at the ends of the duct
P_0	Static pressure in the duct
R	Universal gas constant
R_s	Specific gas constant
S	Cross-sectional area of the duct
T	Temperature of the fluid
T_1, T_2	Temperatures of fluid at the ends of the duct
u_1, u_2	Particle velocities at the ends of the duct
Y_n	Neumann function of the nth order
x	Axial coordinate along the duct
ρ_1, ρ_2	Density of fluid at ends of duct
ω	Angular frequency
ν	Coefficient defined in Equation (8.211)
γ	Ratio of specific heats (C_P/C_V)

The 4-pole transmission matrix theory can be implemented readily using the software package MATLAB®. Another software package called DeltaEC (Los Alamos National Laboratories, 2012), which is intended for the analysis of thermoacoustic systems, can be used to predict the sound pressure level inside ducts with temperature gradients.

An example of using ANSYS to evaluate the TL of a duct with a QWT reactive muffler that has linear temperature gradients in each duct segment is provided by Howard (2013b).

8.9.11 Source and Termination Impedances

The performance of a muffler is often quantified by its IL, as described in Section 8.2. Although the measurement of IL is relatively easy, the theoretical prediction of IL requires knowledge of the acoustic impedance of the noise source and the exhaust termination, which is challenging to measure in practice. Bodén (1995, 2007) provides good summaries of measurement techniques to characterise the acoustic impedance of a source.

It is sometimes assumed for convenience, rather than accuracy, that an internal combustion engine is represented as a constant-acoustic-volume-velocity source, such that the source impedance is infinite ($Z_{\text{src}} \to \infty$), and does not vary with time. However, as the crankshaft in an internal combustion engine rotates, valves open and close and pistons alter the volume in the cylinders, and hence the acoustic impedance of the engine changes with time. Therefore, the assumption of a constant-acoustic-volume velocity amplitude is inaccurate.

Several experimental measurement methods are available to measure the acoustic impedance of an operating machine and the attached exhaust system. These methods can be classified either as direct methods that involve the use of an additional sound source, or indirect methods (or multi-load methods) that do not involve the use of an additional sound source. Direct methods are suitable when the additional sound source is capable of generating sound pressure levels greater than generated by the machine being investigated, which can be difficult to achieve when applied to internal combustion engines (Bodén, 1991, 1995). Alternatively, the indirect method involves evaluating the impedance for two exhaust configurations of known impedance, which is called the two-load method, and is the most commonly used technique. Examples of different termination impedances that could be used include an anechoic termination, or radiation into a free field with two different lengths of exhaust duct.

An acoustic source is characterised by its source strength and source impedance. Equivalent circuits of two acoustic systems are shown in Figure 8.26 for a source that is characterised as a (a) constant-acoustic-pressure-amplitude source, $p_{\rm src}$, or a (b) constant-volume-velocity-amplitude source, $u_{\rm src}$, and the source has an impedance, $Z_{\rm src}$.

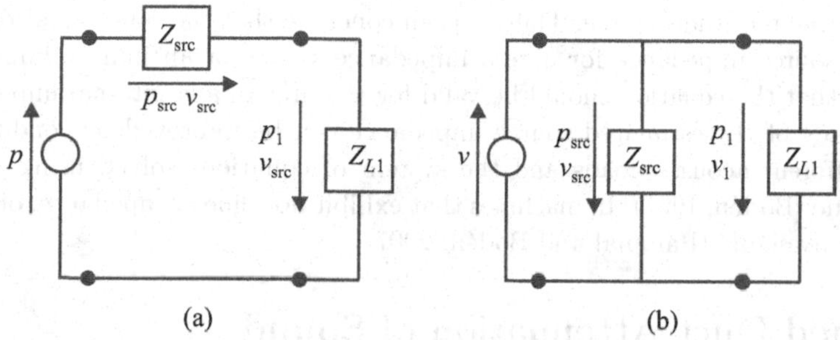

FIGURE 8.26 Equivalent electrical circuit representation for an acoustic system with a (a) constant-acoustic-pressure-amplitude source, or (b) constant-volume-velocity-amplitude source.

To determine the source impedance, which is assumed to be linear and time invariant, the sound pressure, p_1, in the exhaust duct is measured somewhere near the acoustic source. The impedance of the exhaust duct can be represented as a known impedance, Z_{L1}, and includes the termination impedance, which, for example, could be due to the exhaust outlet radiating into a free space. For the circuit shown in Figure 8.26(a), the pressure drop around the circuit can be written as:

$$p = v_{\rm src}Z_{\rm src} + p_1 = v_{\rm src}Z_{\rm src} + v_{\rm src}Z_{L1} = v_{\rm src}(Z_{\rm src} + Z_{L1}) \tag{8.213}$$

Thus:

$$p = \frac{p_1(Z_{\rm src} + Z_{L1})}{Z_{L1}} \tag{8.214}$$

However, Equation (8.214) has two unknowns: the source pressure, $p_{\rm src}$, and the source impedance, $Z_{\rm src}$, where $Z_{\rm src}$ is the desired quantity. To resolve this issue, another test must be conducted using a different acoustic load, Z_{L2} (which has been measured), that causes a measured sound pressure in the exhaust duct, p_2. In the circuit diagram in Figure 8.26(a), this corresponds to the pressure drop, p_2 across load, Z_{L2}. Using the same analysis as above, we obtain:

$$p = \frac{p_2(Z_{\rm src} + Z_{L2})}{Z_{L2}} \tag{8.215}$$

As p is the same for both loads (constant-acoustic-pressure-amplitude assumption), we can equate Equations (8.214) and (8.215) to give:

$$Z_{\rm src} = Z_{L1}Z_{L2}\frac{p_1 - p_2}{p_2 Z_{L1} - p_1 Z_{L2}} \tag{8.216}$$

For the circuit shown in Figure 8.26(b), the total volume velocity, v around the circuit is:

$$v = v_{\rm src} + v_1 = \frac{p_{\rm src1}}{Z_{\rm src}} + \frac{p_1}{Z_{L1}} = \frac{p_1}{Z_{\rm src}} + \frac{p_1}{Z_{L1}} \tag{8.217}$$

If we now replace the load with Z_{L2}, the pressure drop across this new load is p_2 and using the same analysis as above:

$$v = v_{\text{src}} + v_2 = \frac{p_{\text{src2}}}{Z_{\text{src}}} + \frac{p_2}{Z_{L2}} = \frac{p_2}{Z_{\text{src}}} + \frac{p_2}{Z_{L2}} \tag{8.218}$$

As v is the same for both loads (constant-volume-velocity-amplitude assumption), and the circuit In Figure 8.26(b) shows that the pressure drop across the source is equal to the pressure drop across the load in each case, we can equate Equations (8.217) and (8.218) to give:

$$\frac{p_1}{Z_{\text{src}}} + \frac{p_1}{Z_{L1}} = \frac{p_2}{Z_{\text{src}}} + \frac{p_2}{Z_{L2}} \tag{8.219}$$

Rearranging Equation (8.219) results in Equation (8.216), which was obtained for a constant-acoustic-pressure-amplitude source. Thus, we can conclude that the same equation may be used to obtain the source impedance for a zero impedance source or an infinite impedance source, which implies that the equation should be valid for a source of any internal impedance.

The accuracy of the estimated source impedance can be improved by conducting multiple tests using different acoustic loads and the system of equations solved using a least-squares fitting technique (Bodén, 1991). In machines that exhibit non-linear impedance, other estimation techniques are available (Rämmal and Bodén, 2007).

8.10 Lined Duct Attenuation of Sound

Dissipative muffling devices are often used to muffle fans in air conditioning systems and other induced-draft systems. For example, a dissipative-type muffler may be used to successfully control the noise of a fan used in a wood-dust collection system. The dissipation of sound energy is generally accomplished by introducing a porous lining on one or more of the walls of the duct through which the induced draft and unwanted sound travel. In some cases, (for example, the wood-dust collection system), the porous lining material must be protected with a suitable facing for mechanical protection of the liner as well as to control contamination of the liner, which could compromise its acoustic performance.

A liner may also be designed to make use of the nonlinear resistance of a simple hole through which, or across which, a modest flow is induced. This principle has been used in the design of a muffler for noise control of a large subsonic wind tunnel, in which a dissipative liner was used that consists of perforated metal plates of about 4.8% open area, which are cavity backed.

When the dimensions of the lined duct are large compared to the free-field wavelength of the sound that propagates, higher order mode propagation may be expected (see Section 8.10.3.3). Since each mode will be attenuated at its own characteristic rate, which is dependent on the design of the dissipative muffler, the IL of a dissipative muffler will depend on which modes propagate and these will depend on which modes are introduced at the entrance to the muffler. In general, the modes introduced at the entrance and the energy distribution among them are unknown and consequently, a definitive value for muffler IL at high frequencies cannot be provided.

Two approaches have been taken to describe the attenuation provided by a lined duct when higher order mode propagation may be expected. Either the attenuation rate in decibels per unit length for the least attenuated mode may be provided, from which a lower bound for IL for any length duct may be calculated, or the IL for a given length of lined duct is calculated assuming equal energy distribution among all possible cut-on modes at the entrance. In the following discussion, the rate of attenuation of the least attenuated mode is provided. For comparison, in Section 8.10.3 the IL for rectangular lined ducts for the alternative case of equal energy distribution among all possible propagating modes is compared with the predicted IL for the least attenuated mode.

8.10.1 Locally-Reacting and Bulk-Reacting Liners

When analysing sound propagation in a duct with an absorbent liner, one of two possible alternative assumptions is commonly made. The older, simpler and more completely investigated assumption is that the liner may be treated as locally reacting and this assumption results in a great simplification of the analysis (Morse, 1939). In this case, the liner is treated as though it may be characterised by a local impedance, which is independent of whatever occurs at any other part of the liner, and the assumption is implicit that sound propagation does not occur in the material in any other direction than normal to the surface.

Alternatively, when sound propagation in the liner parallel to the surface is not prevented, analysis requires that the locally-reacting assumption be replaced with an alternative assumption that the liner is bulk reacting. In this case, sound propagation in the liner is taken into account.

The assumption that the liner is locally reacting has the practical consequence that for implementation, sound propagation in the liner must be restricted in some way to normal to the surface. This may be done by the placement of solid partitions in the liner to prevent propagation in the liner parallel to the surface. An example of a suitable solid partition is a thin impervious metal or wooden strip, or perhaps a sudden density change in the liner. Somewhat better performance may be achieved by such use of solid partitions but their use is expensive, and it is common practice to omit any such devices in liner construction.

A generalised analysis (Bies et al., 1991), will be used here to discuss the design of dissipative liners. The analysis accounts for both locally-reacting and bulk-reacting liners as limiting cases. It allows for the possible effects of a facing, sometimes employed to protect the liner, and it also allows investigation of the effects of variation of design parameters. An error in the labelling of the ordinates of the figures in the reference has been corrected in the corresponding figures in this text.

8.10.2 Liner Specifications

The dissipative devices listed as numbers 4 to 6 in Table 8.1 make use of wall-mounted porous liners, and thus it will be advantageous to briefly discuss liners and their specifications before discussing their use in the listed devices. Generally, sound-absorbing liners are constructed of some porous material such as fibreglass or rockwool, but any porous material may be used. However, it will be assumed in the following discussion that, whatever the liner material, it may be characterised by a specific acoustic impedance and flow resistivity. In general, for materials of generally homogeneous composition, such as fibrous porous materials, the specific acoustic impedance is directly related to the material flow resistivity and thickness.

Section 1.17 and Appendix D provide discussions of flow resistance and the related quantity, flow resistivity. As shown in Appendix D, the flow resistivity of a uniform porous material is a function of the material density. Consequently, a liner consisting of a layer of porous material may be packed to a specified density, ρ_m, to give any required flow resistance.

Usually, some form of protective facing is provided, but the facing may be omitted where, for example, erosion or contamination due to air flow or mechanical abuse is not expected. The protective facing may consist of a spray-on polyurethane coating, an impervious lightweight plastic sheet (20 µm thick polyester film is suitable), or a perforated heavy-gauge metal facing, or some equivalent construction or combination of constructions. Some possible protective facings are illustrated in Figure 8.27.

Referring to Figure 8.27, it is to be noted that the spacing between elements, **A** and **C**, which is unspecified, is provided by a spacer, **B**, and is essential for good performance. Spacer **B** can be implemented in practice using a coarse wire mesh, with holes of at least 12 mm × 12 mm in size and made of wire between 1 and 2 mm thick. If element **C** is either fibreglass cloth or a fine screen, then the flow resistance of the element is negligible and the element plays no part in the

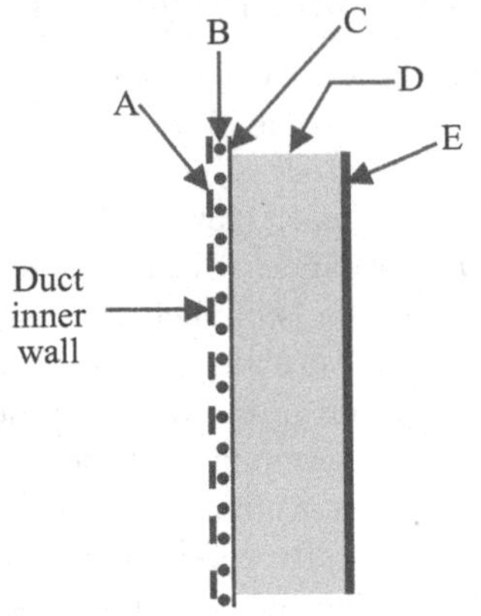

FIGURE 8.27 Protective facings for duct liners. The elements of the liner are: A, 20 gauge perforated facing, minimum 25% open area; B, spacer such as 2 mm thick wire mesh; C, light plastic sheet or fibreglass cloth or fine mesh metal screen; D, fibrous material of specified flow resistance; E, rigid wall or air cavity backing. Maximum flow speeds up to 8 m/s do not require A, B or C. Speeds up to 10 m/s require that the fibrous material of D be coated with a binder to prevent erosion of the fibres. Speeds up to 25 m/s require B and C, while speeds up to 90 m/s require A, B and C. Higher speeds are not recommended. In many installations, A is used regardless of flow speed, to provide mechanical protection for D.

predicted attenuation. On the other hand, if the element **C** is of the form of a limp impervious (plastic) membrane then it may play an important part in the predicted attenuation, as shown by Bies et al. (1991) and as discussed in Appendix D.

Care must be taken in the use of perforated facings for liner protection. Often, the result of flow across the regularly spaced holes in the perforated facing is the generation of "whistling" tones. Alternatively, a small induced flow through the perforations will prevent whistling. The perforated facing should have a minimum open area of 25% to ensure that its effect on the performance of the liner is negligible.

Alternatively, where a perforated facing of percentage open area, P_{open}, is less than 25% (see element **A** of Figure 8.27), the effect of the facing may be taken into account by redefining element **C** of Figure 8.27. The effect is the same as either adding a limp membrane covering, if there is none, or increasing the surface density of a limp membrane covering the porous material. The acoustic impedance, Z_{Ah} of the air in the holes of a perforated plate is given by Equation (8.21) and if the impedance of the solid part of the perforated plate is included the total impedance is given by Equation (8.24), which can be rewritten as:

$$Z_{Ap} = R_{Ap} + \mathrm{j}X_{Ap} = R_{Ap} + \mathrm{j}\omega M_{Ap} \tag{8.220}$$

where R_{Ap} is the acoustic resistance of all the holes in the perforated panel, which is the acoustic resistance, R_A for one hole divided by the number of holes, N. Using Equation (8.22), the overall acoustic resistance may be written as:

$$R_{Ap} = \frac{R_A}{N} = \frac{100 R_A S}{P_{\text{open}} S_p} \tag{8.221}$$

where S is the area of one hole and S_p is the area of the perforated panel. The mass reactance, M_{Ap} for all the holes can be obtained by removing the $R_A S$ terms from Equation (8.24), dividing the result by jω and using the relation, $k = \omega/c$, to give for the mass reactance:

$$M_{Ap} = \frac{\dfrac{100\rho}{P_{\text{open}} k S_p} \tan[k\ell_e(1-M)]}{1 + \dfrac{100\rho}{k\,m P_{\text{open}}} \tan[k\ell_e(1-M)]} \tag{8.222}$$

where m is the mass per unit area (surface mass) of the perforated facing, k is the wavenumber, ρ is the density of air and ℓ_e is the effective length of one hole. The effective mass per unit area (or specific mass reactance) of the perforated plate is equal to M_A multiplied by the total area, S_p of the perforated plate.

If σ' is the surface density of the limp membrane, then the effective surface density, σ, to be used when entering the design charts to be discussed in Section 8.10.3 is obtained by adding to the limp membrane mass, the effective mass per unit area of the perforated plate, which is its specific mass reactance, so that the effective surface density is:

$$\sigma = \sigma' + M_{Ap} S_p \tag{8.223}$$

where ρ is the density of air, P_{open} is the percentage of open area of the perforated facing and ℓ_e is the effective length of the holes in the perforated facing, defined in Equation (8.30). The holes must be a distance apart of at least $2a$ (where a is the radius of the holes) for Equation (8.223) to be valid. For a percentage open area, P_{open}, greater than 25%, the effect of a perforated facing on the lined duct attenuation is generally negligible (Cummings, 1976).

For low frequencies and no flow and considering only the mass reactance of the holes in the perforated panel and not the solid part, $\tan k\ell_e \approx k\ell_e$ and $1/m = 0$. For this case, the effective surface density is:

$$\sigma \simeq \sigma' + 100\rho\ell_e/P_{\text{open}} \tag{8.224}$$

8.10.3 Lined Duct Mufflers

In the following sections, isotropic bulk-reacting and locally-reacting liners are considered and design charts are provided in Figures 8.28 to 8.33, which allow determination of the rate of attenuation of the least attenuated mode of propagation for some special cases, which are not optimal in terms of maximising the attenuation, but which are not very sensitive to the accuracy of the value of flow resistance of the liner. However, the general design problem requires use of a computer program (Bies et al., 1991). Alternative design procedures for determination of the IL of cylindrical and rectangular lined ducts have been described in the literature (Ramakrishnan and Watson, 1992).

Procedures are available for the design of lined ducts for optimum sound attenuation but generally, the higher the attenuation the more sensitive is the liner flow resistance specification and the narrower is the frequency range over which the liner is effective (Cremer, 1953; Cremer et al., 1967).

The design charts shown in Figures 8.28 to 8.33 may be used for estimating the attenuation of the least attenuated propagating mode in lined ducts of both rectangular and circular cross-section. The design charts can be read directly for a rectangular-section duct lined on two opposite sides. For a rectangular-section duct lined on all 4 sides, the total attenuation is the arithmetic sum of the decibel attenuations attributed to the two sets of opposite sides. For a lined circular-section duct, the value of $2h$ is the open width of a square cross-section duct of the same cross-sectional area as the circular section duct, and the values of attenuation given in the charts must be multiplied by two.

In Figure 8.28, attenuation data for a bulk-reacting liner with zero mean flow in the duct are shown for various values of the flow resistivity parameter, $R_1\ell/\rho c$, and ratio, ℓ/h, of liner

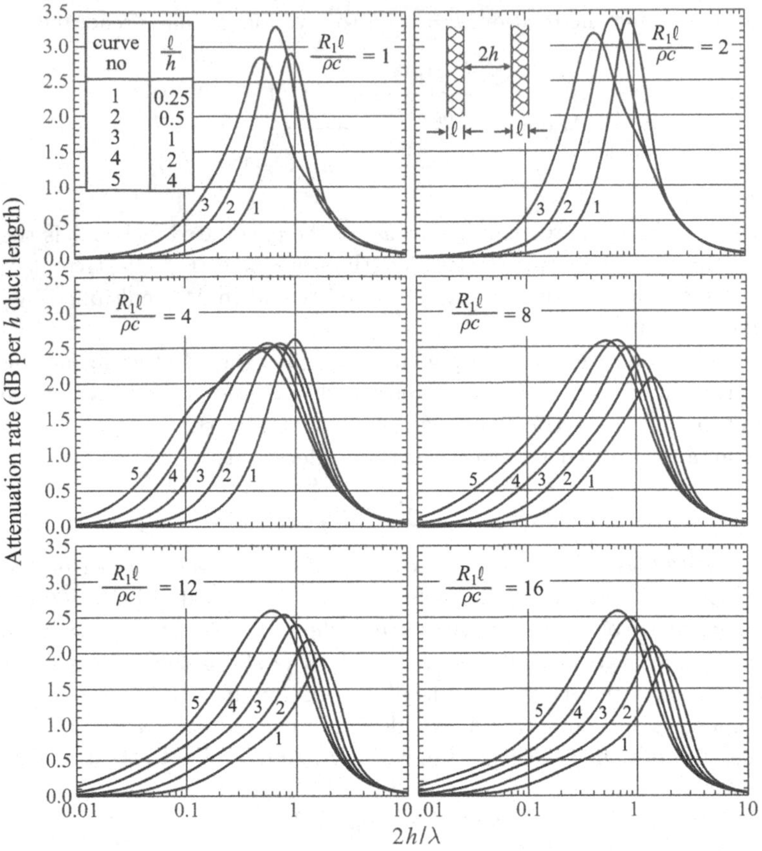

FIGURE 8.28 Predicted octave band attenuations for a rectangular duct lined on two opposite sides. Lined circular ducts or square ducts lined on all four sides give twice the attenuation shown here. The quantity ρ is the density of fluid flowing in the duct, c is the speed of sound in the duct, ℓ is the liner thickness, h is the half width of the airway, σ is the surface density of a limp membrane covering the liner, R_1 is the liner flow resistivity. Bulk-reacting liner with no limp membrane covering (density ratio $\sigma/\rho h = 0$). Zero mean flow ($M = 0$).

thickness to half duct width for a length of duct equal to half of the duct width. In this figure, the density parameter, $\sigma/\rho h$, is equal to zero, implying no plastic covering of the liner and no perforated plate covering unless its open area exceeds 25%. Figures 8.29 and 8.30 are identical to Figure 8.28 except that the Mach number of flow through the lined duct is 0.1 and −0.1, respectively (positive Mach number implies flow in the same direction as sound propagation). In all three figures, 8.28 to 8.30, it is assumed that the flow resistance of the liner is the same in the direction normal to the duct axis as it is in the direction parallel to the duct axis.

In Figure 8.31, data are shown for the same cases as for Figure 8.28, except that the liner is assumed to be locally reacting. In practice, this is realised by placing rigid partitions in the liner normal to the duct axis so that sound propagation in the liner parallel to the duct axis is inhibited.

The data in Figures 8.32 and 8.33 are for various masses of limp membrane (usually plastic and including any perforated liner; see Equation (8.223)) covering of the liner. If a perforated plate is used as well as the limp membrane, a spacer (usually wire mesh) must be placed between the limp membrane and perforated plate; otherwise, the performance will be severely degraded. Densities of typical limp membrane liners are given in Table D.3 in Appendix D. Figure 8.32 is for zero mean flow through the duct while Figure 8.33 is for a mean flow of Mach number = 0.1,

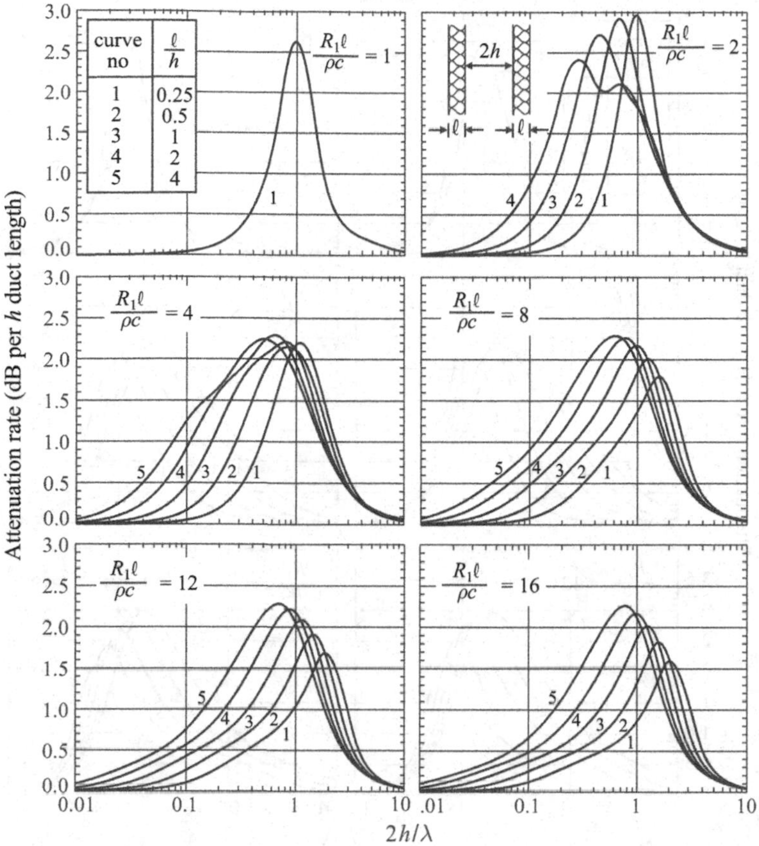

FIGURE 8.29 Predicted octave band attenuations for a rectangular duct lined on two opposite sides. Lined circular ducts or square ducts lined on all four sides give twice the attenuation shown here. The quantity ρ is the density of fluid flowing in the duct, c is the speed of sound in the duct, ℓ is the liner thickness, h is the half width of the airway, σ is the surface density of a limp membrane covering the liner, R_1 is the liner flow resistivity. Bulk-reacting liner with no limp membrane covering (density ratio $\sigma/\rho h = 0$). Mean flow of Mach number, $M = 0.1$ (same direction as sound propagation).

with the figures on the left for flow in the same direction ($M = 0.1$) as sound propagation and the figures on the right representing flow in the opposite direction ($M = -0.1$) to sound propagation.

In Figures 8.28 to 8.33, the duct is assumed to be rectangular in cross-section with two opposite walls lined, as shown in the insets. The open section (airway) of the duct is $2h$ wide, while the liner thickness on either side is ℓ. In determining the thickness, ℓ (see Figure 8.27), elements A and B and the spacing between them are generally neglected, so that ℓ refers to the thickness of element D. If the duct is lined on only one side, then the attenuation would be the same as for a duct lined on two sides, but with the open duct width twice that of the duct lined on one side with the same liner thickness.

The octave band attenuation predictions shown in Figures 8.28 to 8.33 can be used with reasonable reliability for estimating the attenuation of broadband noise in a duct, provided that calculated attenuations greater than 50 dB are set equal to 50 dB (as this seems to be a practical upper limit). On the other hand, the theory on which the predictions are based gives values of attenuation at single frequencies that are difficult to achieve in practice. Use of octave band averaging (see Equation (8.225)) has resulted in a significant smoothing and a reduction in peak

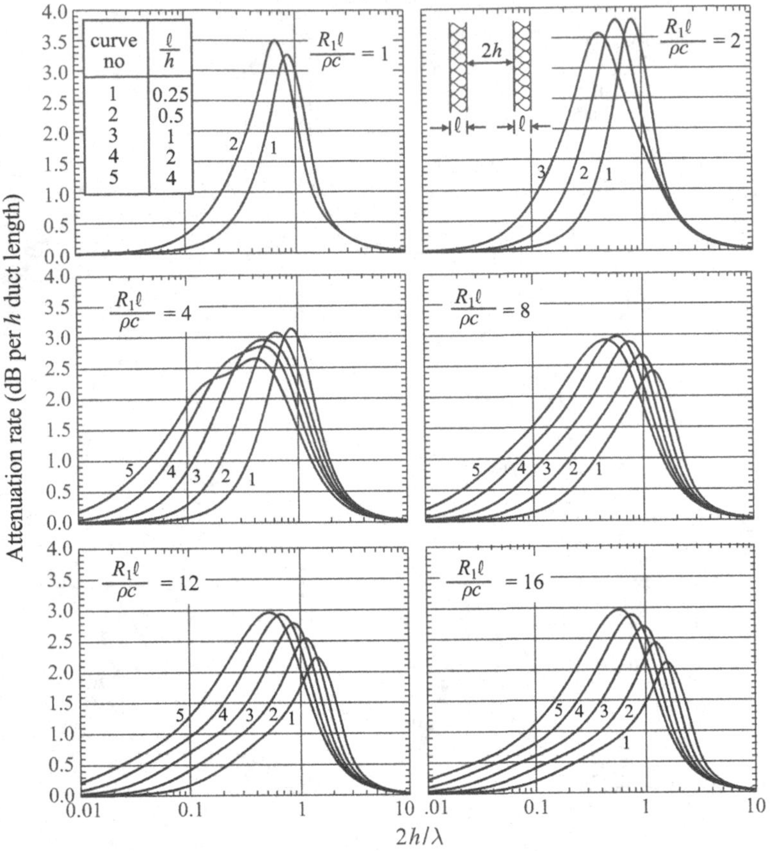

FIGURE 8.30 Predicted octave band attenuations for a rectangular duct lined on two opposite sides. Lined circular ducts or square ducts lined on all four sides give twice the attenuation shown here. The quantity ρ is the density of fluid flowing in the duct, c is the speed of sound in the duct, ℓ is the liner thickness, h is the half width of the airway, σ is the surface density of a limp membrane covering the liner, R_1 is the liner flow resistivity. Bulk-reacting liner with no limp membrane covering (density ratio $\sigma/\rho h = 0$). Mean flow of Mach number, $M = -0.1$ (opposite direction to sound propagation).

values of the single-frequency predictions. Any point on a curve in the figures is the average response of 20 single-frequency predictions, $Att_i, i = 1, 20$ distributed within an octave band encompassing the point:

$$Att_{av} = -10\log_{10}\sum_{i=1}^{20} 10^{-(Att_i/10)} \tag{8.225}$$

As explained above, the observed IL for any lined duct will depend on both the properties of the duct and the propagating sound field that is introduced at the lined duct entrance. In the figures, the assumption is implicit that only the least attenuated mode propagates and in general, the least attenuated mode corresponds to the plane-wave mode in an unlined duct. In Table 8.8, experimentally determined values for IL are listed for comparison with the predicted IL for several lined rectangular ducts based on the following assumptions:

1. equal energy distribution among all propagating modes at the entrance (Ramakrishnan and Watson, 1992); and
2. the least attenuated mode only propagating.

As expected, the least attenuated mode approach underestimates the observed IL at large duct dimension to wavelength ratios. On the other hand, the calculated IL based on the alter-

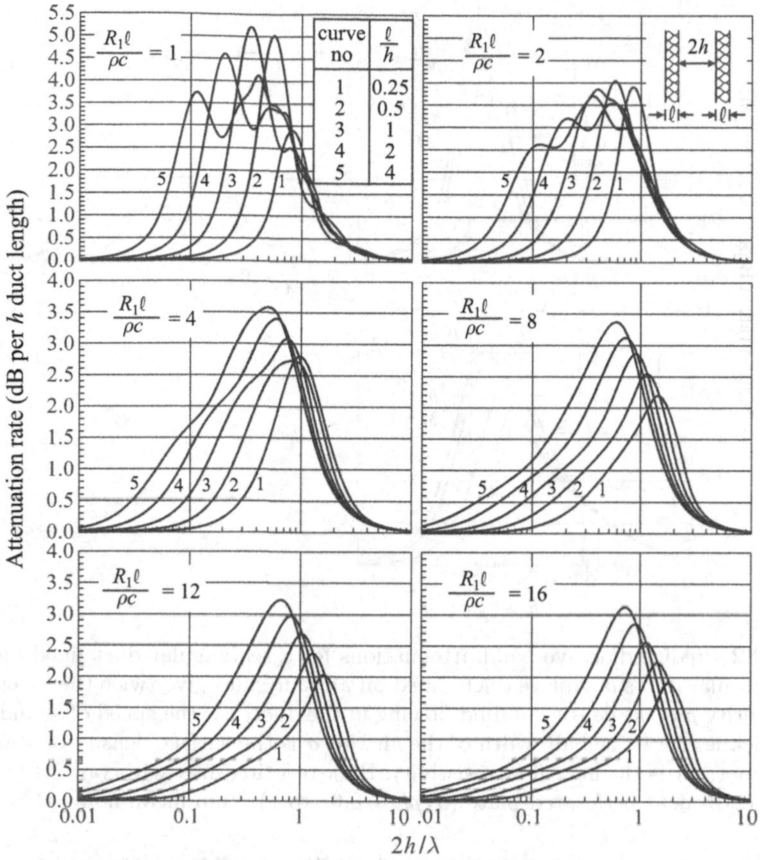

FIGURE 8.31 Predicted octave band attenuations for a rectangular duct lined on two opposite sides. Lined circular ducts or square ducts lined on all four sides give twice the attenuation shown here. The quantity ρ is the density of fluid flowing in the duct, c is the speed of sound in the duct, ℓ is the liner thickness, h is the half width of the airway, σ is the surface density of a limp membrane covering the liner, R_1 is the liner flow resistivity. Locally-reacting liner with no limp membrane covering (density ratio $\sigma/\rho h = 0$). Zero mean flow ($M = 0$).

native assumption of equal energy distribution among all modes does not ensure accuracy of prediction either, because, in practice, the assumption of equal energy distribution is unlikely to be satisfied even approximately.

8.10.3.1 Flow Effects

The assumption is implicit in the presentation of Figures 8.28 to 8.33 that, where flow is present, any velocity gradients are small and unimportant. Essentially, uniform flow in any duct cross-section (commonly referred to as plug flow) is assumed and this assumption generally will be adequate for the range of flow Mach numbers shown. The Mach number is the ratio of the stream speed in the duct to the local speed of sound, and is thus also dependent on the local temperature. In the figures, a negative Mach number indicates sound propagation against or opposite to the flow, while a positive number indicates sound propagation in the direction of flow.

The primary effect of flow is to alter the effective phase speed of a propagating wave by convecting the sound wave with the flow (Bies, 1976). Thus, the effective phase speed is $c(1+M)$, where M, the Mach number, is less than 0.3. Consequently, the effective length of a duct lining

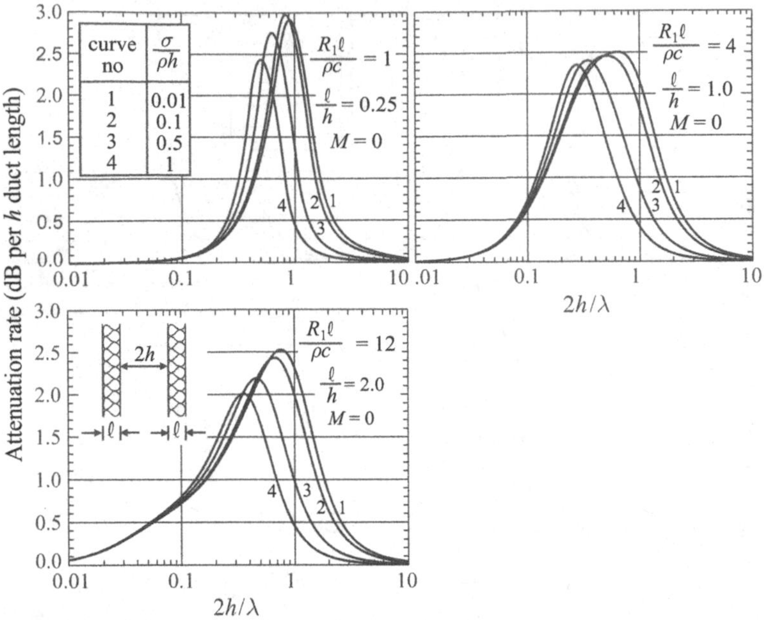

FIGURE 8.32 Predicted octave band attenuations for a rectangular duct lined on two opposite sides. Lined circular ducts or square ducts lined on all four sides give twice the attenuation shown here. The quantity ρ is the density of fluid flowing in the duct, c is the speed of sound in the duct, ℓ is the liner thickness, h is the half width of the airway, σ is the surface density of a limp membrane covering the liner, R_1 is the liner flow resistivity. Bulk-reacting liner with various densities of limp membrane or equivalent perforated plate ($\sigma/\rho h = 0.01$ to 1). Zero mean flow ($M = 0$).

is decreased (as measured in wavelengths) for downstream propagation, resulting in a decrease in attenuation, and increased for upstream propagation, resulting in an increase in attenuation, as may be verified by inspection of the curves in Figures 8.28 to 8.33.

In reality, of course, the matter is not quite so simple as suggested above. The introduction of flow, in general, renders the problem of describing acoustic propagation and attenuation in a lined duct very complicated. For example, shear in the flow has the opposite effect to convection; sound propagating in the direction of flow is refracted into the lining, resulting in increased attenuation. Sound propagating opposite to the flow is refracted away from the lining, resulting in less attenuation than when shear is not present. At Mach numbers higher than those shown in Figures 8.28 to 8.33, where effects such as shear become important, the following empirical relation is suggested as a guide to expected behaviour:

$$D_M = D_0 \left[1 - 1.5M + M^2\right] \tag{8.226}$$

where $-0.3 < M < 0.3$.

In the equation, D_0 is the attenuation (in dB per unit length) predicted for a liner without flow and D_M is the attenuation for the same liner with plug flow of Mach number M.

Flow can strongly affect the performance of a liner, both beneficially and adversely, depending on the liner design. In addition to the refraction effects mentioned earlier, flow may improve or degrade the impedance matching of the wall to the propagating wave, resulting in more or less attenuation (Kurze and Allen, 1971; Mungar and Gladwell, 1968). The introduction of flow may also generate noise; for example, as discussed in Section 8.15.

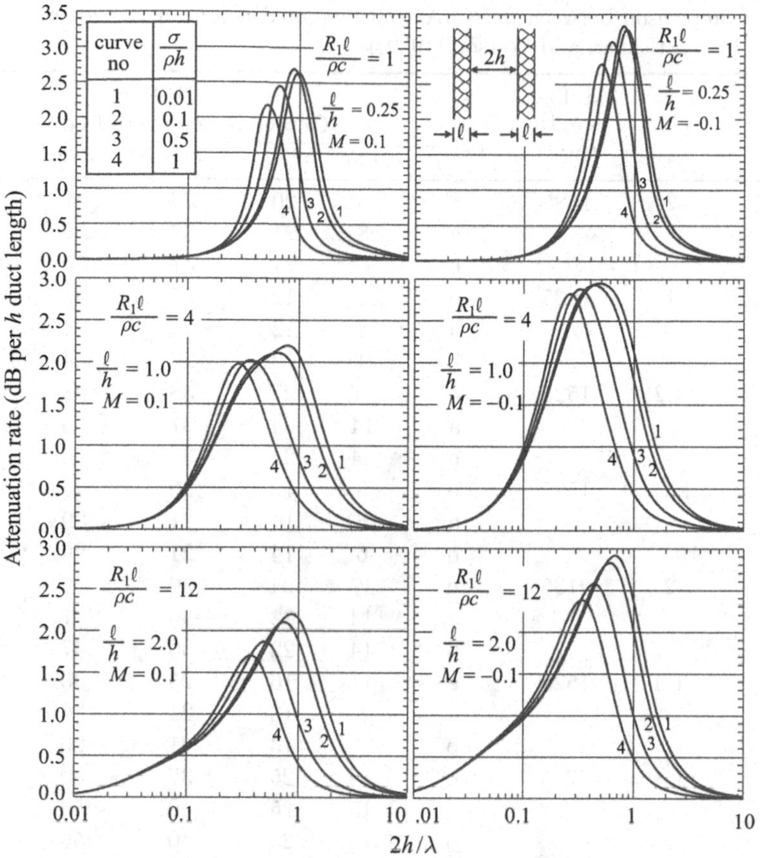

FIGURE 8.33 Predicted octave band attenuations for a rectangular duct lined on two opposite sides. Lined circular ducts or square ducts lined on all four sides give twice the attenuation shown here. The quantity ρ is the density of fluid flowing in the duct, c is the speed of sound in the duct, ℓ is the liner thickness, h is the half width of the airway, σ is the surface density of a limp membrane covering the liner, R_1 is the liner flow resistivity. Bulk-reacting liner with various densities of limp membrane or equivalent perforated plate ($\sigma/\rho h = 0.01$ to 1). Mean flow ($M = 0.1$). For the figures on the left, the flow is in the same direction as sound propagation and for the figures on the right, the flow is in the opposite direction to sound propagation.

8.10.3.2 Temperature Effects

As the gas in a duct increases in temperature, the speed of sound, and hence the wavelength of sound will increase. This will result in sound of a particular frequency being attenuated by a different amount. In most cases, the attenuation will be less than for a room temperature duct but in some instances, it may be greater. For muffler attenuations provided by muffler manufacturers for room temperature, it is best to plot the attenuation as a function of 1/3-octave or octave band centre frequency and then multiply the frequency axis scale by $\sqrt{T_0/T_n}$, where T_n is the temperature in the duct and T_0 is the temperature at which the muffler performance was estimated and where all temperatures are in Kelvin. When undertaking the calculations described in Section 8.10.3, the x-axis scale should be multiplied by $\sqrt{T_0/T_n}$ and the flow resistance of the liner should also be multiplied by the same amount (Munjal et al., 2006, p. 322).

TABLE 8.8 Comparison between measured and predicted insertion loss for rectangular splitter mufflers (after Ramakrishnan and Watson (1992))

Muffler unit size (mm) $= 2\ell + 2h$	ℓ/h	Muffler length (mm)	See note below	Insertion loss (dB) Octave band centre frequency (Hz)					
				125	250	500	1000	2000	4000
305	2	1525	e	8	20	38	47	51	34
			a	8	21	37	50	50	34
			b	8	21	37	50	50	34
305	1	1525	e	4	12	27	41	37	20
			a	4	12	26	44	36	13
			b	4	12	26	44	36	13
408	3.2	1525	e	17	27	38	48	50	31
			a	14	24	37	50	50	26
			b	14	24	37	50	50	26
408	1.13	1525	e	5	12	20	26	16	9
			a	6	14	21	29	17	8
			b	6	14	21	29	17	5
610	2	2135	e	17	24	36	49	33	18
			a	14	22	36	47	29	13
			b	14	22	36	47	27	9
610	1.4	1525	e	11	16	25	30	17	11
			a	8	13	21	26	12	7
			b	8	13	21	26	10	3
610	1.4	2775	e	18	25	37	50	30	16
			a	15	25	40	50	22	14
			b	15	25	40	50	22	9

a = equal energy among all possible significant modes at entrance.
b = least attenuated mode.
e = experimental data.

8.10.3.3 Higher Order Mode Propagation

In the formulation of the curves in Figures 8.28 to 8.33, it has been explicitly assumed that only plane waves propagate and are attenuated. For example, inspection of the curves in the figures shows that they all tend to the same limit at high frequencies; none shows any sensible attenuation for values of the frequency parameter, $2h/\ell$, greater than about three. As has been shown theoretically (Cremer, 1953), high-frequency plane waves tend to beam down the centre of a lined duct; any lining tends to be less and less effective, whatever its properties in attenuating plane waves, as the duct width to wavelength ratio grows large.

Waves that reflect multiple times from the walls of a duct may also propagate. Such waves, called higher order modes or cross-modes, propagate at frequencies above a minimum frequency, called the cut-on frequency, f_{co}, which characterises the particular mode of propagation. For example, in a hard-wall, rectangular cross-section duct, only plane waves may propagate when the largest duct cross-sectional dimension is less than 0.5 wavelengths, while for a circular duct, the required duct diameter is less than 0.586 wavelengths.

Thus, for rectangular section ducts:

$$f_{\text{co}} = \frac{c}{2L_y} \tag{8.227}$$

where L_y is the largest duct cross-sectional dimension. For circular section ducts:

$$f_{co} = 0.586\frac{c}{d} \tag{8.228}$$

where d is the duct diameter.

For ducts of greater dimensions than these, or for any ducts that are lined with acoustically absorptive material (i.e. soft-walled ducts), higher order modes may propagate as well as plane waves but, in general, the plane waves will be least rapidly attenuated at frequencies below the first cut-on frequency of an equivalent unlined duct. As plane waves are least rapidly attenuated, their behaviour will control the performance of a duct in the frequency range in which they are dominant; that is, in the range of wavelength parameter, $2h/\lambda$, generally less than about one. For explanation of the special properties of higher order mode propagation, it will be convenient to restrict attention to ducts of rectangular cross-section and to begin by generalising the discussion of modal response of a rectangular enclosure considered in Chapter 6. Referring to the discussion of Section 6.2.1, it may be concluded that a duct is simply a rectangular room for which one dimension is infinitely long. Letting $k_x = \omega/c_x$, and using Equation (6.16), Equation (6.3) can be rewritten as follows, to give an expression for the phase speed, c_x, along the x-axis in an infinite rectangular section duct for any given frequency, $\omega = 2\pi f$:

$$c_x = \omega\left[\left(\frac{\omega}{c}\right)^2 - \left(\frac{\pi n_y}{L_y}\right)^2 - \left(\frac{\pi n_z}{L_z}\right)^2\right]^{-1/2} \quad \text{(Hz)} \tag{8.229}$$

For a circular section duct of radius, a, Morse and Ingard (1968) give the following for the phase speed along the x-axis.

$$c_x = \omega\left[\left(\frac{\omega}{c}\right)^2 - \left(\frac{\pi \alpha_{m,n}}{a}\right)^2\right]^{-1/2} \quad \text{(Hz)} \tag{8.230}$$

Values of $\pi\alpha_{m,n}$ for the lowest order modes are given in Table 8.9.

TABLE 8.9 Values of $\pi\alpha_{m,n}$ for circular section ducts (after Morse and Ingard (1968))

$m\backslash n$	0	1	2	3	4
0	0	3.83	7.02	10.17	13.32
1	1.84	5.33	8.53	11.71	14.86
2	3.05	6.71	9.97	13.17	16.35
3	4.20	8.02	11.35	14.59	17.79
4	5.32	9.28	12.68	15.96	19.2
5	6.42	10.52	13.99	17.31	20.58

Consideration of Equations (8.229) and (8.230) shows that for the case where at least one of the mode numbers, n_x, n_y, m or n is non-zero for both rectangular and circular section ducts, there will be a frequency, ω, for which the phase speed, c_x, is infinite. This frequency is called either the cut-on or cut-off frequency but "cut-on" is the most commonly used term. Below cut-on, the phase speed is imaginary and no wave propagates; any acoustic wave generated by a source in the duct or transmitted into the duct from outside will decay exponentially as it travels along the duct. Above cut-on, the wave will propagate at a phase speed, c_x, which depends on frequency. With increasing frequency, the phase speed, c_x, measured as the trace along the duct, rapidly diminishes and tends to the free-field wave speed, as illustrated in Figure 8.34. Evidently, the speed of propagation of higher order modes is dispersive (frequency-dependent) and for any given frequency, each mode travels at a speed different from that of all other modes.

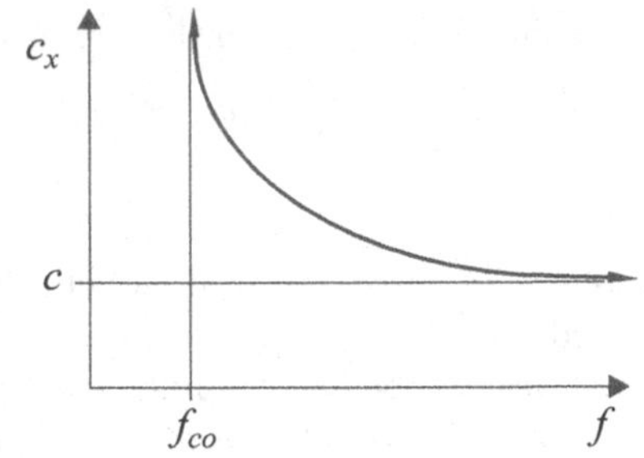

FIGURE 8.34 Phase speed of a higher order mode propagating in a duct as a function of frequency.

Alternatively, letting the length, L_x, in Equation (6.17) tend to infinity gives the following equation for the cut-on frequencies for propagating higher order modes in a rectangular section duct, characterised by mode numbers n_y and n_z:

$$f_{ny,nz} = \frac{c}{2}\sqrt{\left(\frac{n_y}{L_y}\right)^2 + \left(\frac{n_z}{L_z}\right)^2} \tag{8.231}$$

Referring to Equation (8.231), it can be observed that the result is the same as would be obtained by setting $n_x = 0$ in Equation (6.17) and in the latter case, sound propagation is between the opposite walls but not along the x-axis. Similarly, at cut-on, wave propagation is between opposite walls and, consequently, the phase speed along the duct (x-axis) is infinite as the disturbance everywhere is in phase.

Equation (6.19) may be rewritten for the case of the infinitely long room for wave travel in the positive x-axis direction to give an expression for the propagating higher order mode, characterised by mode numbers, n_y and n_z, as:

$$p = \hat{p}\cos\left[\frac{\pi n_y y}{L_y}\right]\cos\left[\frac{\pi n_z z}{L_z}\right]\mathrm{e}^{\mathrm{j}(\omega t - k_x x)} \tag{8.232}$$

Referring to Equation (8.232), it can be observed that a higher order mode is characterised by nodal planes parallel to the axis of the duct and wave fronts at any cross-section of the duct, which are of opposite phase on opposite sides of such nodes.

For a circular section duct of radius, a, the cut-on frequencies are obtained by setting equal the two terms in square brackets in Equation (8.230) to give:

$$f_{m,n} = \frac{c\alpha_{m,n}}{2a} \tag{8.233}$$

At frequencies above the cut-on of the first higher order mode, the number of cut-on modes increases rapidly, being a quadratic function of frequency. Use of Equation (8.231) for representative values for rectangular cross-section ducts of width, L_y, and height, L_z, has allowed counting of cut-on modes and empirical determination of the following equation for the number of cut-on modes, N, in terms of the geometric mean of the duct cross-sectional dimensions, $L = \sqrt{L_y L_z}$, which is valid up to about the first 25 cut-on modes:

$$N = 2.57(fL/c)^2 + 2.46(fL/c) \tag{8.234}$$

A similar procedure has been used to determine the following empirical equation for the number, N, of cut-on modes of a circular cross-section duct of diameter, D, which is valid up to the first 15 cut-on modes:

$$N = (fD/c)^2 + 1.5(fD/c) \tag{8.235}$$

It should be noted that, in practice, one must always expect slight asymmetry in any duct of circular cross-section and consequently, there will always be two modes of slightly different frequency where analysis predicts only one and they will be oriented normal to each other. The prediction of Equation (8.235) is based on the assumption of a perfectly circular cross-section duct and should be multiplied by two to determine the expected number of propagating higher order modes in a practical duct.

The effect of a porous liner adds a further complexity to higher order mode propagation, since the phase speed in the liner may also be dispersive. For example, reference to Appendix D shows that the phase speed in a fibrous material tends to zero as the frequency tends to zero. Consequently, in fibrous, bulk-reacting liners, where propagation of sound waves is not restricted to a direction that is normal to the surface, cut-on of the first few higher order modes may occur at much lower frequencies than predicted based on the dimensions of the airway. This because the wavelength at cut-on is set by the duct cross-sectional geometry, so if the phase speed reduces, the cut-on frequency must also reduce.

The effect of mean flow in a duct is to decrease the frequency of cut-on, and it is the same for either upstream or downstream propagation, since cut-on, in either case, is characterised by wave propagation back and forth between opposite walls of the duct.

For ducts that are many wavelengths across, as are commonly encountered in air conditioning systems, one is concerned with cross-mode as well as plane-wave propagation. Unfortunately, attenuation in this case is very difficult to characterise, as it depends on the energy distribution among the propagating modes, as well as the rates of attenuation of each of the modes. Thus, in general, attenuation in the frequency range for which the frequency parameter, $2h/\lambda$, is greater than about unity, cannot be described in terms of attenuation per unit length as shown in Figures 8.28 to 8.33.

Experimentally determined attenuation will depend on the nature of the source and the manner of the test. Doubling or halving the test duct length will not give twice or half of the previously observed attenuation; that is, no unique attenuation per unit length can be ascribed to a dissipative duct for values of $2h/\lambda$ greater than unity. However, at these higher frequencies, the attenuation achieved in practice will in all cases, be greater than that predicted by Figures 8.28 to 8.33. This is because much of the energy at higher frequencies is contained in higher order modes, which are attenuated much more than indicated in the figures, which are only applicable to the least attenuated mode.

The consequence of the possibility of cross-mode propagation is that the performance of a lined duct is dependent on the characteristics of the sound field introduced at the entrance to the attenuator. Since sound is absorbed by the lining, sound that repeatedly reflects at the wall will be more quickly attenuated than sound that passes at grazing incidence. Thus, sound at all angles of incidence at the entrance to a duct will very rapidly attenuate, until only the axial propagating portion remains. Empirically, it has been determined that such an effect may introduce an additional attenuation, as shown in Figure 8.35. The attenuation shown in the figure is to be treated as a correction to be added to the total expected attenuation of a lined duct. An example of an application of this correction is the use of a lined duct to vent an acoustic enclosure, in which case, the sound entering the duct may be approximated as randomly incident.

8.10.4 Cross-Sectional Discontinuities

Generally, the open cross-section of a lined duct is made continuous with a primary duct, in which the sound to be attenuated is propagating. The result of the generally soft lining is to present to the sound, an effective sudden expansion in the cross-section of the duct. This is because the liner must have the same inside dimensions as the duct section that it is replacing. The liner is much thicker than the original duct wall and the liner is relatively transparent to sound transmission, resulting in an effective expansion of the duct cross-section at the beginning of the

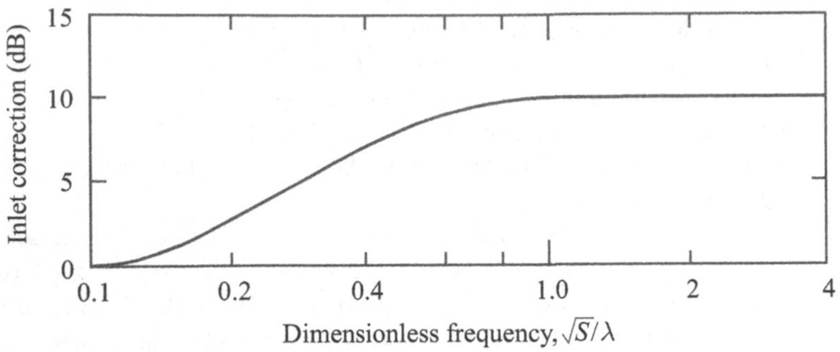

FIGURE 8.35 Duct inlet correction for random-incidence sound. The quantity, λ, is the sound wavelength, and S is the cross-sectional area of the open duct section. The data are empirical. For a silencer with multiple baffles, S is the cross-sectional area of one airway, not the total silencer.

lined section. The expansion affects the sound propagation in a similar way as the expansion chamber considered earlier. The effect of the expansion on the attenuation of a lined duct may be estimated using Figure 8.36. In using this figure the attenuation due to the lining alone is first estimated (using one of Figures 8.28 to 8.33), and the estimate is used to enter Figure 8.36 to find the corrected attenuation (see Section 8.2 for discussion of transmission loss).

8.10.5 Splitter Mufflers

Splitter mufflers consist of baffles that divide the airway into separate passages, as illustrated in Figure 8.37. The splitters are filled with sound-absorbing material, which is selected in the same way as for a lined duct and the IL of each airway is calculated in the same way as for a lined duct. The liner thickness used in the calculations is half the splitter thickness, which is why the liner thickness, ℓ, on the inside of the wall of a rectangular duct containing splitters is half the splitter thickness, 2ℓ. As the IL of each air passage between splitters is the same in most mufflers, the overall IL of the splitter muffler is the same as the IL of a single passage. The IL of a single passage is calculated by assuming it to be a lined duct with a liner thickness equal to half the splitter thickness. It can be shown theoretically as well as numerically (Munjal et al., 2006) that this analysis gives the same IL results for a splitter with a solid partition in its centre to physically separate adjacent passages, as for a splitter with no partition, which is why parallel splitter mufflers do not have a partition in their centre.

In cases where the airways or splitter thicknesses are not the same, or when the flow speeds in the airways differ, the IL is calculated for each airway (using half the splitter thickness as the thickness of a liner in a lined duct) and the insertion losses are combined using Equation (8.236) to give the total insertion loss, $\mathrm{IL_{tot}}$, as:

$$\mathrm{IL_{tot}} = -10\log_{10}\frac{1}{N}\sum_{i=1}^{N}10^{-\mathrm{IL}_i/10} \quad \text{(dB)} \tag{8.236}$$

where IL_i is the insertion loss for airway, i, and N is the number of airways. In gas turbine installations, where a muffler is placed downstream and in the vicinity of a bend, the flow speed can vary considerably from one muffler airway to the next, resulting in different attenuations in each airway.

If a single splitter muffler is replaced with two mufflers in series having the same cross-sectional dimensions and the same total length, the IL will be greater, due to there being additional entrance and exit losses to include. Also, if the splitters in the second muffler are aligned with the airways in the first muffler (staggered arrangement), additional attenuation can

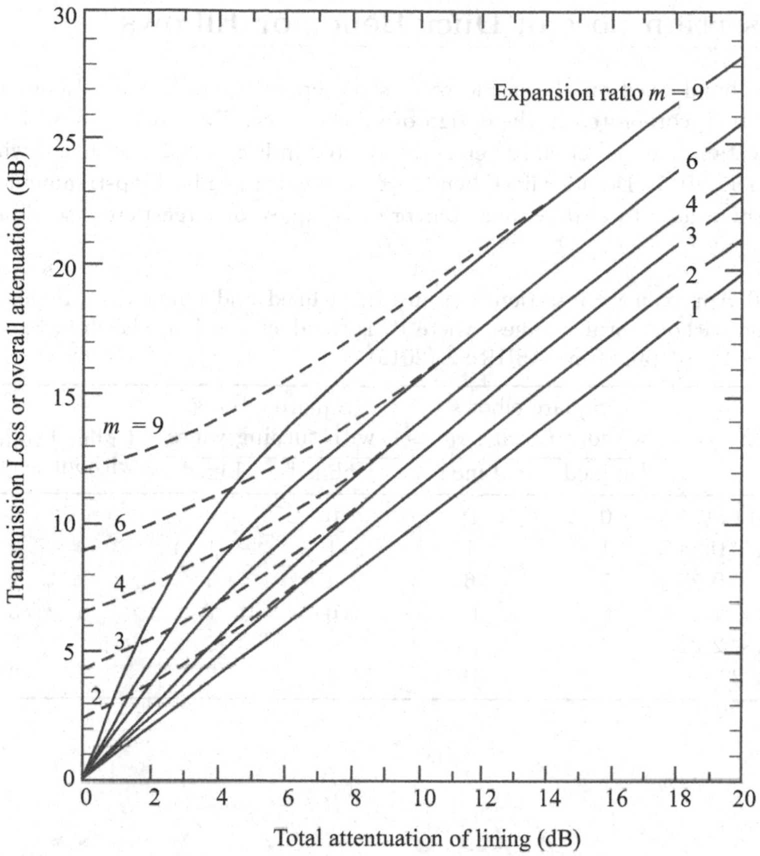

FIGURE 8.36 Transmission loss (TL) of a lined expansion chamber as a function of the area expansion ratio, m, and the total attenuation of the lining. The solid curves show TL for $kL = 0, \pi, \ldots, n\pi$, and the dashed curves show TL for $kL = \pi/2, 3\pi/2, \ldots, (2n+1)\pi/2$. The quantity, k, is the wavenumber and L is the length of the expansion chamber.

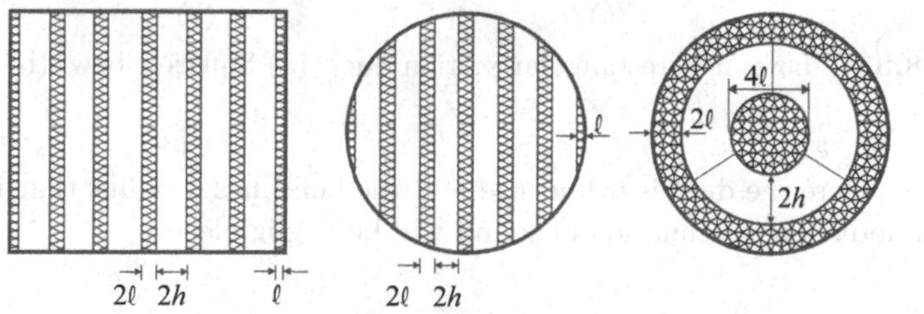

FIGURE 8.37 Cross-section of a duct containing a parallel splitter muffler. (a) Rectangular section duct. (b) Circular section duct with parallel baffles. (c) Circular section duct with centre pod.

be achieved, ranging from 1 dB in the 63 Hz and 125 Hz octave bands, to 3 dB in the 250 Hz and 500 Hz bands, to 4 dB in the 1000 Hz band, to 5 dB in the 2000 Hz band and 6 dB in the 4000 Hz and 8000 Hz octave bands (Mechel, 1994).

8.11 Insertion Loss of Duct Bends or Elbows

The case of a bend lined on the inside was listed separately in Table 8.1, but such a device might readily be incorporated in the design of a lined duct. Table 8.10 shows IL data for lined and unlined rectangular and circular bends (illustrated in Figure 8.38) with and without turning vanes (ASHRAE, 2015). Data for lined bends are for bends lined both upstream and downstream for a distance of at least three duct diameters or three times the largest cross-sectional dimension.

TABLE 8.10 Approximate insertion loss, in dB, of lined and unlined rectangular and circular bends with and without turning vanes, where W is the duct width as shown in Figure 8.38 and λ is the wavelength. Adapted from ASHRAE (2015)

	Square elbows without turning vanes		Square elbows with turning vanes		Unlined round elbows without turning vanes
	Unlined	Lined	Unlined	Lined	
$W/\lambda < 0.14$	0	0	0	0	0
$0.14 \leq W/\lambda < 0.28$	1	1	1	1	1
$0.28 \leq W/\lambda < 0.55$	5	6	4	4	2
$0.55 \leq W/\lambda < 1.11$	8	11	6	7	3
$1.11 \leq W/\lambda < 2.22$	4	10	4	7	3
$W/\lambda > 2.22$	3	10	4	7	3

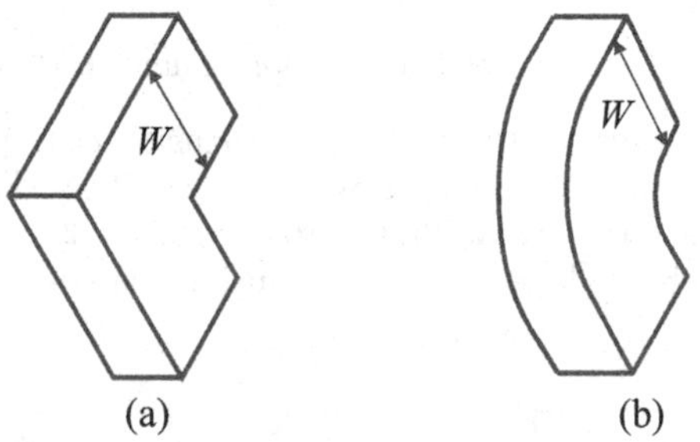

FIGURE 8.38 Elbows in a rectangular section duct. (a) Square elbow. (b) Round elbow.

In situations where the duct is cylindrical and the bend has a radius that is large compared to the duct diameter, the attenuation of sound will be negligible.

8.12 Insertion Loss of Unlined Ducts

An unlined duct will also exhibit sound attenuation properties. The amount of sound attenuation for an unlined rectangular duct may be estimated using Table 8.11 and for an unlined circular duct it may be estimated using Table 8.12. The attenuation for an unlined circular section duct is unaffected by external insulation.

TABLE 8.11 Approximate insertion loss of unlined, square-section, sheet metal ducts in dB/m[a]. Adapted from ASHRAE (2015), which contains an error in the P/S value. Numbers in brackets are from VDI2081-1 (2003)

Section height and width (m)	P/S (m^{-1})[b]	Octave band centre frequency (Hz)				
		63	125	250	500	≥1000
0.15	27	0.98(0.6)	0.66(0.6)	0.33(0.45)	0.33(0.3)	0.33(0.3)
0.305	13	1.15(0.6)	0.66(0.6)	0.33(0.45)	0.20(0.3)	0.20(0.2)
0.61	6.6	0.82(0.6)	0.66(0.6)	0.33(0.3)	0.10(0.15)	0.10(0.15)
1.22	3.3	0.49(0.45)	0.33(0.3)	0.23(0.15)	0.07(0.1)	0.07(0.05)
1.83	2.2	0.33	0.33	0.16	0.07	0.07

[a]If the duct is externally insulated then double these values.

[b]The quantity P is the perimeter and S is the area of the duct cross-section.

TABLE 8.12 Approximate insertion loss of unlined circular sheet metal ducts in dB/m. Data from ASHRAE (2015) and data in brackets from VDI2081-1 (2003)

Duct diameter D, (mm)	Octave band centre frequency (Hz)				
	63	125	250	500	≥1000
$D \leq 200$	0.10(0.10)	0.10(0.10)	0.16(0.15)	0.16(0.15)	0.33(0.30)
$200 < D \leq 400$	0.10(0.05)	0.10(0.10)	0.10(0.10)	0.16(0.15)	0.23(0.20)
$400 < D \leq 800$	0.07(0.00)	0.07(0.05)	0.07(0.05)	0.10(0.01)	0.16(0.15)
$800 < D \leq 1600$	0.03(0.00)	0.03(0.00)	0.03(0.00)	0.07(0.05)	0.07(0.05)

8.13 Effect of Duct End Reflections

The sudden change of cross-section at the end of a duct mounted flush with a wall or ceiling results in additional attenuation due to the sudden change in impedance experienced by the travelling sound wave causing it to be partially reflected at the end of the duct. The additional attenuation has been measured for circular and rectangular ducts, and empirical results are listed in Table 8.13 (ASHRAE, 2007).

Tables 8.12 and 8.13 can also be used for rectangular section ducts, by calculating an equivalent diameter (or hydraulic diameter), D, using $D = \sqrt{4S/\pi}$, where S is the duct cross-sectional area.

8.14 Pressure Loss Calculations for Muffling Devices

The introduction of reactive or dissipative muffling systems in a duct will impose a pressure loss. For example, an engine muffler will impose a back pressure on the engine, which can strongly affect the mechanical power generated. The total pressure drop of a muffling system is a combination of friction and dynamic losses through the system. The former friction losses, which are generally least important, will be proportional to the length of travel along tubes or ducts, while the latter dynamic losses will occur at duct discontinuities; for example, at contractions, expansions and bends. However, the calculation of friction losses can also be done using the same approach and equation as used for dynamic pressure losses, provided that the factor K of Equation (8.237) below can be estimated.

In this section, means will be provided for estimating expected pressure losses for various mufflers and duct configurations. Friction losses will be considered first.

TABLE 8.13 Duct reflection loss (dB). Adapted from ASHRAE (2007) and ASHRAE (2015)

Duct diameter	Octave band centre frequency (Hz)					
D, (mm)	63	125	250	500	1000	2000
150	18(20)	12(14)	7(9)	3(5)	1(2)	0(1)
200	15(18)	10(12)	5(7)	2(3)	1(1)	0(0)
250	14(16)	8(11)	4(6)	1(2)	0(1)	0(0)
300	12(14)	7(9)	3(5)	1(2)	0(1)	0(0)
400	10(12)	5(7)	2(3)	1(1)	0(0)	0(0)
510	8(10)	4(6)	1(2)	0(1)	0(0)	0(0)
610	7(9)	3(5)	1(2)	0(1)	0(0)	0(0)
710	6(8)	2(4)	1(1)	0(0)	0(0)	0(0)
810	5(7)	2(3)	1(1)	0(0)	0(0)	0(0)
910	4(6)	2(3)	0(1)	0(0)	0(0)	0(0)
1220	3(5)	1(2)	0(1)	0(0)	0(0)	0(0)
1830	1(3)	0(1)	0(0)	0(0)	0(0)	0(0)

Applies to ducts terminating flush with a wall or ceiling and located several duct diameters from other room surfaces.

If closer to other surfaces use the entry for next larger duct.

Numbers in brackets are for ducts terminated in free space or at an acoustic suspended ceiling.

8.14.1 Pressure Losses Due to Friction

For the case of laminar flow, friction losses depend on the Reynolds number and are small. However, when the Reynolds number is greater than 2000, the flow will be turbulent, and the pressure loss will be independent of Reynolds number, Re (see Equation (8.242)). Only the latter case is considered here, as it provides a useful upper bound on friction losses.

The following equation may be used to estimate the expected pressure loss for flow through a duct:

$$\Delta P = \frac{1}{2}\rho U^2 K \tag{8.237}$$

where, for friction losses

$$K = f_m \left(\frac{LP_D}{4S}\right) \tag{8.238}$$

and where f_m is the friction factor, ΔP (Pa) is the pressure loss, U (m/s) is the mean flow speed through the duct, S (m^2) is the duct cross-sectional area, P_D (m) is the duct cross-sectional perimeter, L (m) is the length of the duct and ρ is the density of the gas in the duct. For splitters in mufflers, the length, L, does not include the nose or tail sections. The friction factor, f_m is equal to 0.05 for a duct lined with perforated sheet metal (such as used for a dissipative muffler). For other duct liner types or unlined ducts, f_m can be calculated using an iterative equation as (ASHRAE, 2013):

$$f_m = \left[2\log_{10}\left(\frac{\epsilon P_D}{14.8S} + \frac{2.51}{Re\sqrt{f_m}}\right)\right]^{-2} \tag{8.239}$$

where a first estimate for f_m can be obtained using (ASHRAE, 2001):

$$f_m = \begin{cases} f'_m; & \text{if } f'_m \geq 0.018 \\ 0.85f'_m + 0.0028; & \text{if } f'_m < 0.018 \end{cases} \tag{8.240}$$

$$\text{where } f'_m = 0.11\left(\frac{\epsilon P_D}{4S} + \frac{68}{Re}\right)^{0.25}$$
$$= 0.11\left(\frac{\epsilon P_D}{4S} + \frac{2.56\times10^{-4}P_D}{SU}\right)^{0.25} \quad \text{for standard air} \tag{8.241}$$

As can be seen, the friction loss depends on the pipe or duct roughness, ϵ (m), which is usually taken as 1.5×10^{-4} m for galvanised steel ducts, pipes and tubes, such as considered in connection with engine mufflers, expansion chambers and low-pass filters, 9×10^{-4} m for fibreglass-lined ducts, and 0.003 m for spray coated fibreglass liners (ASHRAE, 2005). This latter value may be used to calculate the pressure loss due to friction losses as a result of gas flowing through a dissipative muffler. The quantity, Re, is the Reynolds number, given by:

$$Re = \frac{4SU\rho}{P_D\mu} \tag{8.242}$$

where μ is the dynamic viscosity of the gas flowing through the muffler (1.84×10^{-5} Pa s for air at 20°C).

8.14.2 Dynamic Pressure Losses

Dynamic pressure losses are also calculated using Equation (8.237), but in this case, the value of the coefficient, K depends on the geometry of the discontinuity and the mean flow speed, U. Values of K for unlined duct expansion, contraction and bend sections may be determined by making reference to Figure 8.39, where various geometries and analytical expressions are summarised (ASHRAE, 2005). The expressions provided in Figure 8.39 were derived by curve fitting empirical data in ASHRAE (2005). Data for other fittings and duct sizes are available from ASHRAE (2016).

8.14.3 Splitter Muffler Pressure Loss

The maximum allowed pressure loss associated with splitter mufflers is usually an important design criterion. Thus, splitter mufflers are usually designed so that the total cross-sectional area of the airways in the muffler is equal to the cross-sectional area of the duct to which the muffler is attached. Splitters are usually orientated vertically so that their thickness is in the horizontal direction to minimise the build up of contamination on the splitter faces. Note that the liner thickness on the duct wall should be half the thickness of the liner between airways, as illustrated in Figure 8.37. This is because the liner that separates two airways acts acoustically as if there were a solid partition through its centre, which is parallel to the duct axis and which extends the full height of the liner.

The total pressure loss due to flow through splitters is the sum of the friction loss caused by flow along the liner that makes up the splitter and the dynamic losses due to the change in cross-sectional shape at the entrance and exit of the splitters.

The calculation of the friction pressure loss pressure loss for splitter mufflers uses Equations (8.237) and (8.238), with $f_m = 0.05$, corresponding to a perforated panel liner. In this case, S is the open duct area between splitters and P_D is the perimeter of the open area between two splitters minus the top and bottom lengths. That is, P_D is twice the splitter height. The flow speed, U is the flow speed through one splitter airway (flow speed in the duct upstream of the muffler divided by the number of airways, if all airways are the same) and L is the splitter length in the axial duct direction minus the length of the nose and tail sections. As the pressure is uniform across the upstream muffler face and also across the downstream muffler face, the

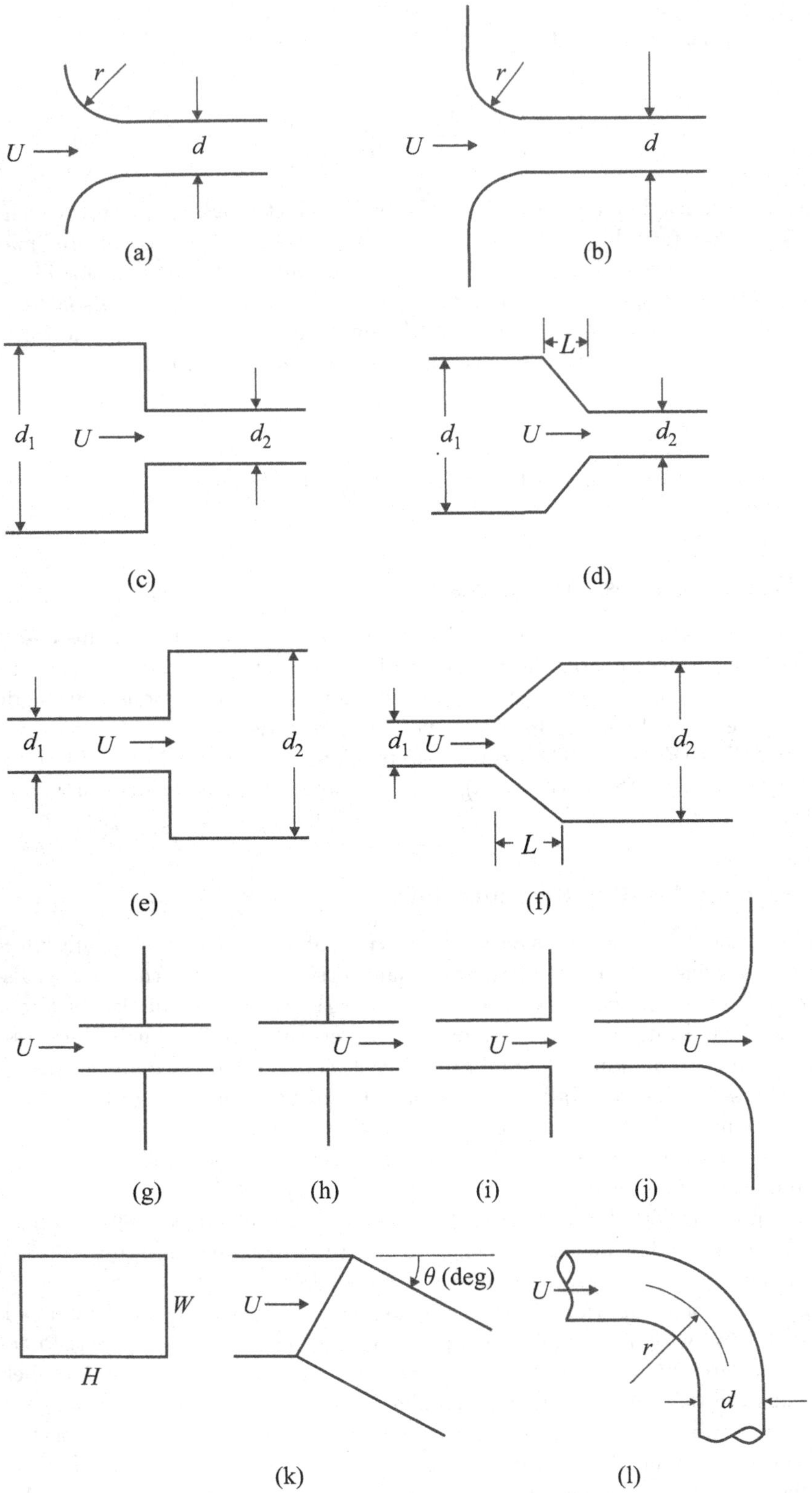

FIGURE 8.39 (Caption on next page.)

FIGURE 8.39 Dynamic pressure loss coefficients. U = speed of flow through the element (m/s).

(a) Contracting bellmouth:

$$K = 0.03 + \frac{0.97}{\exp(y/0.655)}; \qquad y = 10r/d$$

(b) Contracting bellmouth with wall:

$$K = 0.03 + \frac{0.47}{\exp(y/0.655)}; \qquad y = 10r/d$$

(c) Step contraction:

$$K = -4.412 + 4.845[1 - \exp(-y/0.414)]$$
$$y = d_2/d_1 \leq 1$$

(d) Gradual limited contraction:

$$K = \{0.358 - 0.311/[1 + (z/2.426)^{2.929}]\}\{0.223 - 0.0135\exp(y/0.357)\}$$
$$y = d_2/d_1 \leq 1; \qquad z = (d_1 - d_2)/L \geq 0$$

(e) Step expansion:

$$K = 1.0$$

(f) Gradual limited expansion:

$$K = [0.88 - 1.014\exp(-z/0.422)](1.37 - 1.38y)$$
$$y = d_1/d_2 \leq 1; \qquad z = (d_2 - d_1)/L \geq 0$$

(g), (h), (i), (j) Sharp edge, inward contraction, outward expansions:

$$K = 1.0$$

(k) Mitred rectangular duct bends:

$$K = K_{MB}K_{RE}$$
$$K_{MB} = 0.34\left[\frac{\theta}{45}\right]^{1.82}[0.663 + 0.458\exp(-y/2.888)]$$
$$K_{RE} = 1.0 + \frac{0.613}{\mu} - \frac{0.213}{\mu^2}$$

where Reynolds number $Re = 6.63 \times 10^4 Ud$; $\mu = Re \times 10^{-4}$; $y = H/W$, $d = 2HW/(W + H)$.

(l) Rounded duct bends:

$$K = 0.108 + 1.414\exp(-d/37.55)$$

d = duct diameter, (mm) and $r/d = 1.5$, where r is the radius of the duct bend.

For duct diameters, d, greater than 200 mm, set $K = 0.11$.

pressure drop through each airway of the muffler must be the same. Thus, if the airways have different liners, the flow speed through each one will self-adjust so that the total pressure loss through each is the same.

The dynamic pressure loss due to the entrance and exit to the splitters must be added to the friction pressure loss due to flow in the splitter airway. Equation (8.237) may also be used to calculate the entrance and exit losses, but the corresponding K values are found using Figure 8.40.

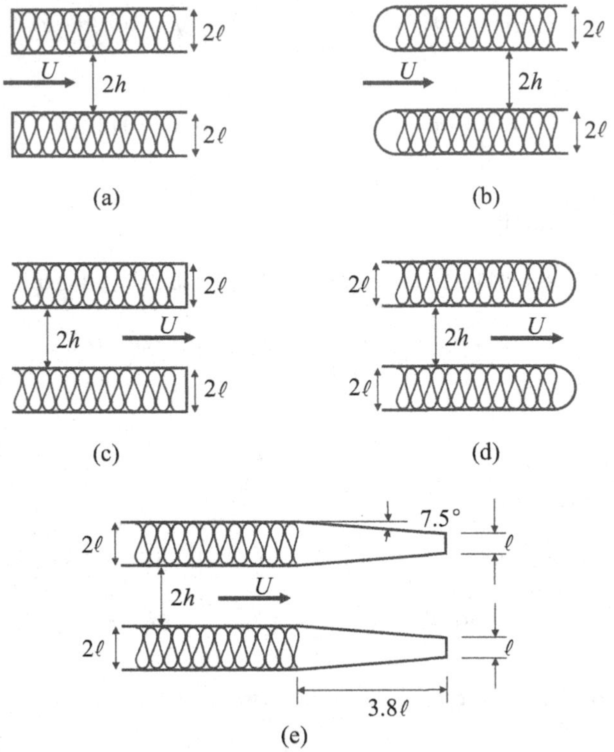

FIGURE 8.40 Dynamic pressure loss factors for splitter mufflers (Munjal et al., 2006; ISO 14163, 1998): (a) square edge inlet, $K = 0.5(1 + h/\ell)^{-1}$; (b) rounded inlet, $K = 0.05(1 + h/\ell)^{-1}$; (c) square edge exit, $K = (1 + h/\ell)^{-2}$; (d) rounded exit, $K = 0.7(1 + h/\ell)^{-2}$; (e) tapered exit, 7.5°, $K = 0.6(1 + h/\ell)^{-2}$. Only one airway is shown here. The full muffler cross-section is illustrated in Figure 8.37.

An alternative procedure for calculating the overall (inlet + airway + exit) pressure loss of a splitter muffler of length, L, is provided in VDI2081-1 (2003). This standard uses Equation (8.237), with K given by:

$$K = a_1 \left[\frac{h}{h+\ell}\right]^{b_1} + a_2 \left[\frac{h}{h+\ell}\right]^{b_2} \frac{L}{d_h} \tag{8.243}$$

where L is the length of the splitter in the direction of flow, h is half the width of the air gap between adjacent splitters, ℓ is half the splitter thickness (and equal to the liner thickness on the duct wall), d_h is defined in Equation (8.253) and the coefficients, a_1, a_2, b_1 and b_2 are defined in Table 8.14. The calculated value applies to baffles with square inlets and exits. However, the estimated pressure losses calculated using Equation (8.237) with K calculated using Equation (8.243) are very conservative and much greater than those calculated using Equation (8.237) with K calculated using Equation (8.238) and Figure 8.40.

TABLE 8.14 Values of a_1, a_2, b_1 and b_2 for use in Equation (8.243) (adapted from VDI2081-1 (2003))

Coefficient	Thickness of splitter 2ℓ (mm)		
	100	200	300
a_1	0.235	0.255	0.294
a_2	0.017	0.015	0.0167
b_1	−2.78	−2.82	−2.83
b_2	−2.70	−2.91	−2.95

The pressure loss coefficients for splitter mufflers assume smooth uniform flow into the muffler and no obstruction (such as a bend) close to the muffler entrance or exit. Where these conditions do not occur, the pressure loss factor must be multiplied by a K_{pf} factor, values of which are listed in ASHRAE (2015), Table 27. Some of these values (which apply to all types of mufflers) are listed in Table 8.15. Where a condition listed in Table 8.15 exists on both the inlet and outlet, the dynamic pressure loss factor, K must be multiplied by the value of K_{pf} for the inlet as well as K_{pf} for the outlet.

TABLE 8.15 Values of K_{pf} for mufflers for various entrance and exit conditions that occur within 3 to 4 duct diameters of the inlet or exit (adapted from ASHRAE (2005))

Condition	K_{pf} for inlet (entrance)	K_{pf} for outlet (exit)
Radius elbow with turning vanes	1.05	1.5
Radius elbow without turning vanes	1.1	1.9
Mitre elbow	1.3	2.0
Fan	1.1–1.3	1.2–1.4
Plenum chamber	1.1–1.3	2.0
Doubling size of duct	—	1.4

Another pressure loss contributor that must be accounted for is the contribution from the transition sections that connect the splitter muffler ends to the upstream and downstream ducts. As the open area of the splitter muffler should be the same as the upstream and downstream duct cross-sectional areas, the total face area of the splitter muffler will be larger than the duct cross-sectional area. This is because some of the face area is taken up with the duct liner. The expansion and contraction pressure losses due to the transitions that connect the duct to the splitter muffler can be estimated using Figure 8.39(f) and (d), respectively, and Equation (8.237). The expansion pressure loss at the inlet and the contraction loss at the outlet are added to the inlet loss and outlet loss, respectively, calculated using Figure 8.40.

8.14.4 Circular Muffler Pressure Loss

A procedure for calculating the pressure loss due to flow through a lined circular muffler with a cylindrical centre pod, as shown in Figure 8.37, is provided in VDI2081-1 (2003). Equation (8.237) is used with the value of K given by:

$$K = 0.981 + \frac{0.0346 L_{\text{eff}} \left[D^2/(D^2 - d^2)\right]^2}{(D - d)} \tag{8.244}$$

where d is the diameter of the centre pod, $D = d + 4h$ is the outer diameter of the airway in the muffler, $2h$ is the width of the airway between the centre pod and the lining on the wall of the muffler, and L_{eff} is the effective length of the muffler. The effective length, L_{eff}, is the actual muffler length minus the entrance and exit rounded sections.

8.14.5 Staggered Splitter Pressure Loss

Staggered splitters are sometimes used when two splitter mufflers are placed in series in a duct. As mentioned in Section 8.10.5, this results in a small amount of additional attenuation over what would be achieved if the splitters in one muffler lined up with the splitters in the following muffler. When the splitters are staggered, a splitter in the upstream muffler is lined up with an airway in the downstream muffler. This results in additional pressure loss in the system. There exists no published data on the additional pressure loss associated with the staggered configuration, probably because it is dependent on how close the exit of the upstream muffler is to the entrance of the downstream muffler. For circular section ducts, a separation of 3 duct diameters should be sufficient to result in negligible additional pressure loss with the staggered configuration over that experienced with the non-staggered configuration. For rectangular section ducts, the separation should exceed a distance of $12S/P$ where S is the duct cross-sectional area and P is the duct cross-sectional perimeter.

8.15 Flow-Generated Noise

Muffling devices depend for their success on the introduction of discontinuities in the conduits of an air handling system. Some simple examples have been considered in previous sections. The introduction of discontinuities at the boundaries of a fluid-conducting passage will produce disturbances in the fluid flow, which will result in noise generation. Regularly spaced holes in the facing of a perforated liner can result in fairly efficient "whistling", with the generation of associated tones. Such "whistling" can be avoided by choice of the shape or formation of the hole edge. For example, those holes that provide parallel edges crosswise to the mean flow will be more inclined to whistle than those that do not. Alternatively, lightly pressurising the liner with an air supply to cause some small flow to pass through the holes will inhibit "whistling".

Aside from the problem of "whistling", noise will be generated at bends and discontinuities in duct cross-sections. Fortunately, the associated noise-generating mechanisms are remarkably inefficient at low flow speeds and generally can be ignored. However, the efficiencies of the mechanisms commonly encountered increase with either the cube or the fifth power of the free stream local Mach number. An upper bound on flow speed for noise reduction for any muffling system is thus implied. At higher flow speeds "self-noise" generated in the device will override the noise reduction that it provides.

The problem of self-noise has long been recognised in the air-conditioning industry. The discussion of this section will depend heavily on what information is available from the latter source. The expressions provided in this section were derived by curve fitting of empirical data. The importance of self-noise generation in automotive muffling systems is well-known. In the latter case, the high sound pressure levels commonly encountered result in oscillating fluid motion of sufficiently large amplitude to generate enough noise to reduce the noise reduction achievable.

8.15.1 Straight, Unlined Air Duct Noise Generation

The sound power generated in an octave band of centre frequency, f, in an unlined duct of cross-sectional area, S, containing air flow travelling with a speed, U (m/s), is given by (VDI2081-1, 2003) as:

$$L_{WB} = 7 + 50\log_{10} U + 10\log_{10} S - 2 - 26\log_{10}\left(1.14 + 0.02\frac{f}{U}\right) \quad (\text{dB re } 10^{-12}\text{W}) \quad (8.245)$$

8.15.2 Mitred Bend Noise Generation

In the following discussion of flow noise generated at a mitred bend, reference should be made to Figure 8.41. At the inner (convex), sharp corner, flow separation occurs. Further downstream, flow reattachment occurs. The point of reattachment, however, is unsteady, resulting in an effective fluctuating drag force on the fluid. As shown in Section 4.3.2, such a fluctuating force acting on the stream can be interpreted as a dipole noise source. For the case considered here, the axis of the dipole is oriented parallel to the stream, and all frequencies propagate. The sound power from this source increases with the sixth power of the stream speed and, as shown in Section 4.3.4, the inner corner noise source will thus increase in efficiency with the cube of the local Mach number, where the efficiency is the sound power divided by the stream power, $S\rho U^3/2$ and where S is the stream cross-sectional area.

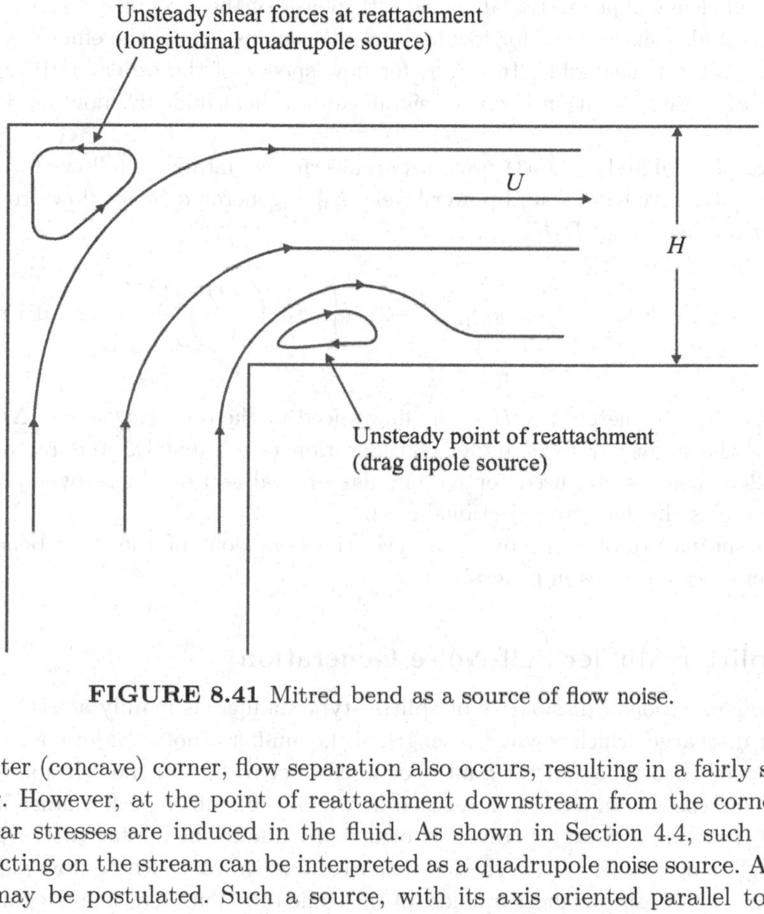

FIGURE 8.41 Mitred bend as a source of flow noise.

At the outer (concave) corner, flow separation also occurs, resulting in a fairly stable bubble in the corner. However, at the point of reattachment downstream from the corner, very high unsteady shear stresses are induced in the fluid. As shown in Section 4.4, such a fluctuating shear stress acting on the stream can be interpreted as a quadrupole noise source. A longitudinal quadrupole may be postulated. Such a source, with its axis oriented parallel to the stream, radiates sound at all frequencies. The sound power produced by this type of source increases with the eighth power of the free-stream speed and, as shown in Section 4.4, this results in the outer corner source efficiency increasing with the fifth power of the local Mach number.

Let the density of the fluid be ρ (kg/m^3), the cross-sectional area of the duct be S (m^2) and the free-stream speed be U (m/s); then the mechanical stream power level, L_{Ws} referenced to 10^{-12} W, is:

$$L_{Ws} = 30\log_{10} U + 10\log_{10} S + 10\log_{10}\rho + 117 \qquad (\text{dB re } 10^{-12}\text{W}) \tag{8.246}$$

A dimensionless number, called the Strouhal number, is defined in terms of the octave band centre frequency, f (Hz), the free-stream speed, U (m/s), and the height of the elbow, H (see

Figure 8.41), as:

$$N_s = f\,H/U \tag{8.247}$$

Experimental data for the sound power, L_{WB}, generated by a mitred bend without turning vanes is described by the following empirical equation:

$$L_{WB} - L_{Ws} = -10\log_{10}\left[1 + 0.165N_s^2\right] + 30\log_{10}U - 103 \qquad (\text{dB re } 10^{-12}\text{W}) \tag{8.248}$$

The quantity, L_{WB}, is the octave band sound power level, and N_s is the Strouhal number corresponding to the octave band centre frequency.

The right-hand side of Equation (8.248) may be interpreted as a measure of the efficiency of conversion of stream power into acoustic power or noise. Note that for small Strouhal numbers, the efficiency is proportional to the cube of the free-stream speed, but for large Strouhal numbers, the efficiency is proportional to the fifth power of the stream speed. Consideration of Equation (8.248) also shows that, for low flow speeds such as 1 m/s, the efficiency is very small and noise generation is negligible. However, for flow speeds of the order of 10 m/s or greater, the efficiency of noise generation becomes significant; rather suddenly, flow noise assumes importance.

The standard, VDI2081-1 (2003) provides an alternative means as follows for calculation of the octave or 1/3-octave band sound power levels, L_{WB}, generated by air flow around a bend in a circular section duct for $(f\,D/U) > 1$:

$$L_{WB} = 50\log_{10}U + 30\log_{10}D + 10\log_{10}\Delta f - 21.5\left[\log_{10}\left(\frac{f\,D}{U}\right)\right]^{1.268} + 12 \quad (\text{dB re } 10^{-12}\text{W}) \tag{8.249}$$

where D is the duct diameter (m), U is the flow speed in the duct (m/s) and Δf is the bandwidth (Hz) of the frequency band under consideration (see Table 1.2 and Equation (1.115)). Equation (8.249) can also be used for rectangular or oval section ducts by replacing D with $\sqrt{4S/\pi}$, where S is the duct cross-sectional area.

Owing to the nature of "self-noise" sources, the behaviour of the duct bend is probably typical of duct discontinuities in general.

8.15.3 Splitter Muffler Self-Noise Generation

The self-noise generation of dissipative or splitter-type mufflers is mainly a result of turbulence at the muffler discharge, which is why the length of the muffler is not a major contributing factor. As the flow speed through a splitter muffler or lined duct increases, the self-noise generated by this flow also increases until a point is reached at which the muffler performance is significantly compromised. As the self-noise generation is principally due to the muffler discharge, it is better to have a single long muffler rather than two shorter ones in cases where flow noise could be a limiting factor. Of course, two short mufflers of the same total length as a single long muffler will result in greater attenuation (although more flow noise) due to more instances of entry and exit attenuation. In some cases, an installation may appear noisier than desired or expected and a suggestion may be made to increase the length of the muffler. If the noise is already dominated by muffler self-noise, then increasing its length will have no significant benefit.

The self-noise of a range of commercial air conditioning dissipative-type mufflers may be shown to obey the following relation (Iqbal et al., 1977):

$$L_{WB} - L_{Ws} = 50\log_{10}U - 155 + C \qquad (\text{dB re } 10^{-12}\text{W}) \tag{8.250}$$

Here L_{WB} is the octave band sound power level of self-noise at the downstream end of the mufflers, L_{Ws} is again the sound power level of the free stream (Equation (8.246)) and U is the stream speed. Dependence on Strouhal number is implied by the frequency band corrections, C, which were empirically determined by measurement. Values of C are given in Table 8.16.

TABLE 8.16 Correction number, C, for Equation (8.250)

	Octave band centre frequency (Hz)							
	63	125	250	500	1000	2000	4000	8000
Correction, C	0	0	0	−4	−13	−13	−19	−22

A more recent approach to the calculation of sound generation in parallel splitter dissipative mufflers is outlined in ISO 14163 (1998). In that standard, the following expression is given for the estimation of the octave band sound power, L_{WB}, produced at the discharge of a parallel splitter muffler with a perforated metal liner, where the last term was provided by Munjal et al. (2006).

$$L_{WB} = B + 10\log_{10}(nP_s cS) + 60\log_{10} M + 10\log_{10}\left[1 + \left(\frac{c}{4fh}\right)^2\right] - 10\log_{10}\left[1 + \left(\frac{f\delta}{U}\right)^2\right] - 25\log_{10}\left(\frac{T}{T_0}\right) \quad \text{(dB re } 10^{-12}\text{W)} \tag{8.251}$$

and the symbols are defined in Table 8.17. For both calculation methods discussed here, the self-noise level is only for a single airway. To obtain the total self-noise level it is necessary to add logarithmically the contributions from each airway (see Equation (1.98)).

The term "static pressure" is the pressure over and above atmospheric pressure, and it is measured using a pressure tapping in the side of a duct containing a flowing fluid. The maximum static pressure referred to by fan manufacturers is the static pressure corresponding to zero flow. In a muffler, the total pressure is the static pressure plus the pressure associated with the movement of the exhaust gas in the muffler, $\rho U^2/2$.

As can be seen in Equation (8.251), the generated sound power is inversely proportional to temperature raised to the power of 2.5 and inversely proportional to the speed of sound raised to the power of 3. As the speed of sound is proportional to the square root of temperature, it follows that in cases where the exhaust gas temperature is above ambient, the sound power generated by the splitters will be reduced compared to the ambient case.

An alternative approach to calculating the self-noise generation of splitter mufflers in an octave band of centre frequency, f, is provided by VDI2081-1 (2003):

$$L_{WB} = 43.6\log_{10} U - 0.5\log_{10}\Delta P_t + 10\log_{10} S - 14.9\left(\log_{10}\frac{f\,d_h}{U}\right) - 1.4\left(\log_{10}\frac{f\,d_h}{U}\right)^2 + 2.2\left(\log_{10}\frac{f\,d_h}{U}\right)^3 - 0.5\left(\log_{10}\frac{f\,d_h}{U}\right)^4 + 12.2 \quad \text{(dB re } 10^{-12}\text{W)} \tag{8.252}$$

where ΔP_t is the pressure loss from the inlet to the discharge of the muffler. The hydraulic diameter, d_h, of a single airway of cross-sectional area, S_a and perimeter, P_a, is given by:

$$d_h = \frac{4S_a}{P_a} \quad \text{(m)} \tag{8.253}$$

As for the other methods, the VDI2081-1 (2003) method is for a single airway and for mufflers with multiple airways (corresponding to multiple splitters), the contributions from each airway must be added logarithmically (see Equation (1.98)) to obtain the total self-noise level.

TABLE 8.17 Definition of symbols used in Equations (8.251) and (8.252)

Symbol	Description
B	is dependent on the type of muffler and frequency and a value of 58 is normally chosen for HVAC equipment. However, for gas turbine mufflers, a value of 68 may be more realistic.
P_s	is the static pressure in the muffler airway (Pa) (see explanation below)
n	is the number of airway passages (typically number of splitters+1)
c	is the speed of sound in the gas in the duct (m/s)
U	is the average flow speed in the narrowest cross-section of the muffler airway (m/s)
f	is the octave band centre frequency (Hz)
$M = U/c$	is the Mach number of the flow in the muffler airway
S	is the area of the narrowest cross-section of the muffler airway (m^2)
$2h$	is the maximum transverse dimension of the muffler airway (m)
T	is the temperature of the gas flowing through the muffler in K
T_0	is the reference temperature = 293K
δ	is a length scale characterising the spectral content of the noise. A value of 0.02 has been found appropriate for HVAC equipment (ISO 14163, 1998). For a gas turbine exhaust, a higher value of approximately 0.04 may be more appropriate

The standard, VDI2081-1 (2003), states that flow noise in circular mufflers with lining on all interior walls and **no** centre pod is negligible. However, the same standard provides means for calculating flow noise in circular mufflers with a lining on the interior walls and a centre sound-absorbing pod of circular cross-section. The self-noise generated in an octave band of centre frequency, f, for a circular duct of internal diameter, D, containing a central pod, is given by (VDI2081-1, 2003):

$$L_{WB} = 80.3\log_{10}U - 20.4\log_{10}\Delta P_t + 15.5\log_{10}D - 71.4\left(\log_{10}\frac{fD}{U}\right) + 64.7\left(\log_{10}\frac{fD}{U}\right)^2 - 26.7\left(\log_{10}\frac{fD}{U}\right)^3 + 3.4\left(\log_{10}\frac{fD}{U}\right)^4 + 32.8 \quad \text{(dB re } 10^{-12}\text{W)} \tag{8.254}$$

8.15.4 Grille Noise

For an air conditioning duct terminated by a grille, the noise is directly proportional to the pressure drop across the grille, which is characterised by the coefficient, K, defined by Equation (8.237). Values of K are usually available from the grille manufacturer, but range from 3 for parallel louvres, 5 for louvres with an open damper, 8 for louvres with a partially closed damper and approximately 6 for a stagnation disk outlet. The overall sound power level, L_W, generated by the grille is (Baumann and Coney, 2006, p. 636):

$$L_W = 10 + 10\log_{10}(SK^3U^6) \qquad (\text{dB re } 10^{-12}\text{ W}) \tag{8.255}$$

where S is the grille area and U is the speed of flow in the upstream duct. The 1/3-octave sound power spectrum levels may be calculated using:

$$L_W(1/3\text{-oct}) = \begin{cases} 10\log_{10}(SK^3U^6) - 12.3\left[\log_{10}(f/f_p)\right]^2; & f \le f_p \\ 10\log_{10}(SK^3U^6) - 11.5\left[\log_{10}(f/f_p)\right]^{1.6}; & f > f_p \end{cases} \qquad (\text{dB re } 10^{-12}\text{ W}) \tag{8.256}$$

where $f_p = 140U$ is the frequency of maximum noise level and f is the 1/3-octave band centre frequency.

The octave band levels are obtained by logarithmically summing the three 1/3-octave band levels making up the octave band. Octave band levels should add up logarithmically to the overall level calculated using Equation (8.255). If they do not, then the octave band dB levels should be reduced by the difference if the octave band sum is greater than the overall level calculated using Equation (8.255) and vice versa.

8.15.5 Exhaust Stack Pin Noise

In many gas turbine installations, the interior of the exhaust stacks is lined with heat insulating material, and this is fixed to the stack walls using steel pins. These pins protrude from the surface of the insulating material into the open duct cross-section, and the high-speed gas flow interacts with the pins to produce unwanted noise. The level of this noise can be calculated based on laboratory measurements on a scale model, as described by Peters et al. (2010). The equation for calculating the sound power level, L_{Wp} (dB), generated by the pins in a full-scale exhaust is calculated from the measured sound power level, L_{Wm} (dB), for the scale model using the following relation, where the subscript, p, refers to the full-scale system and the subscript, m, refers to the model.

$$L_{Wp} = L_{Wm} + 10\log_{10}\left[\frac{(M^5\rho cS_pS_{st}T)_p}{(M^5\rho cS_pS_{st}T)_m}\right] \qquad (\text{dB re } 10^{-12}\text{W}) \tag{8.257}$$

where $M = U/c$ is the Mach number of the flow, ρ is the density of gas in the stack, c is the speed of sound in the gas in the stack, T is the temperature of the gas in the stack in Kelvin, S_p is the total surface area of all of the pins protruding into the stack (pin circumference × length of the pin protruding into the stack × number of pins) and S_{st} is the cross-sectional area of the stack.

Values of sound power level, L_{Wm}, obtained from model tests, which can be used in Equation (8.257), are provided in Table 8.18. The other required model values are $M = 0.05$, $T = 293$ Kelvin, $\rho c = 400$ MKS rayls, $S_p = 0.075$ m^2 and $S_{st} = 0.071$ m^2.

8.15.6 Self-Noise Generation of Air Conditioning System Elements

Air conditioning system elements such as air openings, damper valves and duct junctions all contribute to self-noise generation in an air handling system. These elements are treated in detail in ASHRAE (2015) and VDI2081-1 (2003).

TABLE 8.18 Sound power levels generated by pins in internally insulated exhausts (model results from Peters et al. (2010))

1/3-octave band centre frequency (Hz)	Pin sound power from model tests L_W (dB)	1/3-octave band centre frequency (Hz)	Pin sound power from model tests L_W (dB)
100	51.5	630	50.5
125	51.3	800	50.8
160	49.8	1000	47.2
200	48.3	1250	47.8
250	48.6	1600	44.7
315	46.0	2000	41.2
400	44.0	2500	37.1
500	44.4		

8.16 Duct Break-Out Noise

8.16.1 Break-Out Sound Transmission

In most modern office buildings, air conditioning ductwork takes much of the space between suspended ceilings and the floor above. Noise (particularly low-frequency rumble noise) radiated out of the ductwork walls of rectangular cross-section is, in many cases, sufficient to cause annoyance to the occupants of the spaces below. In some cases, noise radiated into the ductwork from one space, propagated through the duct, and radiated out through the duct walls into another space, may cause speech privacy problems. Noise transmitted out through a duct wall is referred to as breakout transmission. Breakout transmission is not usually considered to be a problem for ducts with a circular cross-section.

To predict in advance the extent of likely problems arising from noise "breaking out" of the duct walls, it is useful to calculate the sound pressure level outside of the duct from a knowledge of the sound power introduced into the duct by the fan or by other external sources further along the duct. A prediction scheme, which is applicable in the frequency range between 1.5 times the fundamental duct wall resonance frequency and half the critical frequency of a flat panel equal in thickness to the duct wall, will be described. In most cases, the fundamental duct wall resonance frequency, f_{0s}, is well below the frequency range of interest and can be ignored. If this is not the case, f_{0s} may be calculated (Cummings, 1980) and the TL for the 1/3-octave frequency bands adjacent to and including f_{0s} should be reduced by 5 dB from that calculated using the following prediction scheme. Also, in most cases, the duct wall critical frequency is well above the frequency range of interest. If this is in doubt, the critical frequency may be calculated using Equation (7.3) and then the TL predictions for a flat panel (see Section 7.2.6) may be used at frequencies above half the duct wall critical frequency.

The sound power level entering a room from an air-conditioning duct is approximately equal to the sound power level radiated from the duct walls minus the transmission loss of the ceiling below the duct. A more accurate calculation, which should be used, involves using the duct radiated sound power level to calculate the reverberant sound pressure level in the space containing the duct (see Equation (6.43)). Equation (6.35) is then used to calculate the sound intensity incident on the ceiling as a result of the reverberant field, and this is then multiplied by the ceiling area to obtain the sound power incident on the ceiling due to the reverberant field. Equation (1.89) is then used to convert this to an incident sound power level and the field incidence ceiling TL is arithmetically subtracted from the incident sound power level to obtain the sound power level entering the room due to the reverberant field in the ceiling cavity. The sound power level incident on the ceiling due to the direct field radiated by the bottom duct wall is equal to the total sound power level radiated from all 4 duct walls (see Equation (8.258) below) plus $10\log_{10}$(area of bottom wall/area of all 4 walls). After subtracting the normal incidence TL (see Equation (7.60) with $\theta = 0$ in Section 7.2.6), this sound power level is combined logarithmically (see Equation (1.98)) with the reverberant field sound power level contribution. The sound pressure level at any location in the room below the ceiling is then calculated using Equation (6.44), with the ceiling as the noise source.

The total sound power level, L_{W02}, radiated out of all 4 walls of a rectangular section duct is given by (Vér, 1983, p. 704):

$$L_{W02} = L_{Wi} - \text{TL}_{\text{out}} + 10\log_{10}\left[\frac{P_D L}{S}\right] + C \quad (\text{dB re } 10^{-12}\text{W}) \tag{8.258}$$

where L_{Wi} is the sound power level of the sound field propagating down the duct at the beginning of the duct section of concern (usually the fan output sound power level (dB) less any propagation losses from the fan to the beginning of the noise-radiating duct section), TL_{out} is the field incidence TL of the duct wall, S is the duct cross-sectional area, P_D is the cross-sectional perimeter, L is the duct length radiating the power and C is a correction factor to account for gradually decreasing values of L_{Wi} as the distance from the noise source increases. For short, unlined ducts, C is usually small enough to ignore. For unlined ducts longer than 2 m or for any length of lined duct, C is calculated using:

$$C = 10\log_{10}\left[\frac{1 - \text{e}^{-(\tau_a + \Delta/4.34)L}}{(\tau_a + \Delta/4.34)L}\right] \tag{8.259}$$

where Δ is the sound attenuation (dB/m) due to internal ductwork losses, which is 0.1 dB/m for unlined ducts (do not use tabulated values in ASHRAE (2015) as these include losses due to breakout) and may be calculate for lined ducts as described in Section 8.10.3. The parameter, τ_a can be calculated as:

$$\tau_a = (P_D/S)10^{-\text{TL}_{\text{out}}/10} \tag{8.260}$$

The quantity, TL_{out}, may be calculated (ASHRAE, 2015) using the following procedure. First of all, the cross-over frequency from plane-wave response to multi-modal response is calculated using:

$$f_{cr} = 612/\sqrt{ab}\ ; \quad (\text{Hz}) \tag{8.261}$$

where a is the larger and b the smaller duct cross-sectional dimension in metres.

At frequencies below f_{cr}, the quantity TL_{out} may be calculated using:

$$\text{TL}_{\text{out}} = 10\log_{10}\left[\frac{fm^2}{(a+b)}\right] - 13 \ (\text{dB}); \qquad f < f_{cr} \tag{8.262}$$

and at frequencies above f_{cr} and below $0.5f_c$ (see equation (7.4)):

$$\text{TL}_{\text{out}} = 20\log_{10}(fm) - 45 \ (\text{dB}); \qquad f_{cr} < f < f_c/2 \tag{8.263}$$

In the preceding equations, m (kg/m^2) is the mass/unit area of the duct walls and f (Hz) is the octave band centre frequency of the sound being considered.

The minimum allowed value for TL_{out} is given by:

$$\text{TL}_{\text{out}} = 10\log_{10}\left[\frac{P_D L}{S}\right] \quad (\text{dB}) \tag{8.264}$$

The maximum allowed value for TL_{out} is 45 dB. For frequencies above half the critical frequency of a flat panel (see Section 7.2.1), TL predictions for a flat panel are used.

If it is desired to calculate the breakout sound power for circular and oval ducts, Equations (8.258) to (8.260) may be used, but the value of $\mathrm{TL}_{o}ut$ is difficult to predict accurately with an analytical model, although it is generally much higher than that for rectangular section ducts of the same cross-sectional area. It is recommended that the guidelines outlined by ASHRAE (2015) for the estimation of $\mathrm{TL}_{o}ut$ be followed closely.

8.16.2 Break-In Sound Transmission

Let L_{W01} be the sound power that is incident on the exterior of an entire length of ductwork and assume that the incoming sound power is divided equally into each of the two opposing axial directions. Then the sound power entering into a rectangular section duct of cross-sectional dimensions a and b, and length, L, from a noisy area and propagating in one axial direction in the duct is:

$$L_{Wi} = L_{W01} - \mathrm{TL}_{\mathrm{in}} - 3 \quad (\text{dB re } 10^{-12}\text{W}) \tag{8.265}$$

For $a \geq b$, and $f < f_0$, where $f_0 = c/2a$ and c is the speed of sound in free space, the duct TL, $\mathrm{TL}_{\mathrm{in}}$, for sound radiated into the duct is the larger of the following two quantities (ASHRAE, 1987):

$$\mathrm{TL}_{\mathrm{in}} = \begin{cases} \mathrm{TL}_{\mathrm{out}} - 4 - 10\log_{10}(a/b) + 20\log_{10}(f/f_0) \\ 10\log_{10}(L/a + L/b) \end{cases} \tag{8.266}$$

For $f > f_0$:

$$\mathrm{TL}_{\mathrm{in}} = \mathrm{TL}_{\mathrm{out}} - 3 \quad (\text{dB}) \tag{8.267}$$

8.17 Lined Plenum Attenuator

A lined plenum chamber is often used in air conditioning systems as a device to smooth fluctuations in the air flow. It may also serve as a sound attenuation device. As shown in Table 8.1, such a device has dimensions that are large compared to a wavelength. The plenum thus acts like a small room and, as such, an absorptive liner, which provides a high random incidence sound-absorption coefficient, is of great benefit. In general, the liner construction does not appear to be critical to the performance of a plenum, although some form of liner is essential. The geometry of such a device is shown in Figure 8.42(a) and, for later reference, Figure 8.42(b) shows the essential elements used in the following discussion.

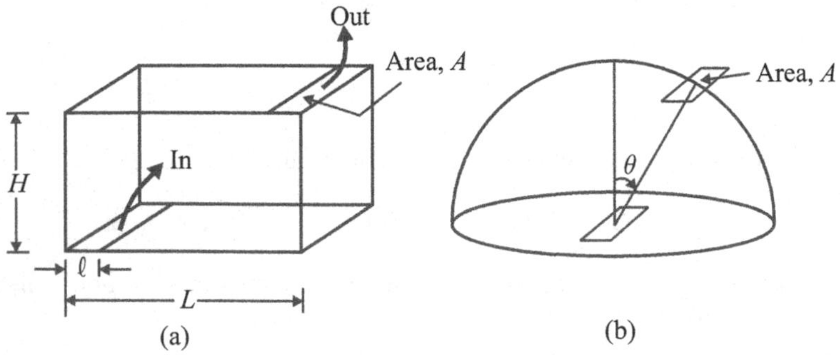

FIGURE 8.42 Lined plenum chamber: (a) physical acoustic system; (b) essentials of the source field.

In all cases discussed in this section, the values obtained for TL will be the same as those for IL, provided that the plenum chamber is lined with sound-absorbing material and providing that the TL is greater than 5 dB.

8.17.1 Wells' Method

There are a number of analytical models that have been developed in the past by various authors for estimating the sound attenuation performance of a plenum chamber. The oldest known model is the Wells model (Wells, 1958) and this will be discussed first of all.

The acoustic power, W_o, which leaves the exit consists of two parts for the purpose of this analysis, a direct field, W_D, and a reverberant field, W_R. The acoustic power in the reverberant field is related to the input power W_i as:

$$W_R = W_i A_{\text{exit}} / R_c \tag{8.268}$$

where

$$R_c = S\bar{\alpha}/(1-\bar{\alpha}) \tag{8.269}$$

In the preceding equations, A_{exit} (m^2) is the cross-sectional area of the plenum exit hole, R_c (m^2) is the plenum room constant, S (m^2) is the total wall area of the plenum and $\bar{\alpha}$ is the mean Sabine wall absorption coefficient.

Referring to Figure 8.42(b), the power transmission in the direct field is (Wells, 1958):

$$W_D = \frac{W_i A_{\text{exit}}}{2\pi r^2}\cos\theta \tag{8.270}$$

where θ is the angular direction, and r is the line of sight distance from the plenum chamber entrance to the exit. It is recommended (ASHRAE, 1987) that for an inlet opening nearer to the edge than to the centre of the plenum chamber wall, the factor of 2 in the preceding equation be deleted. If this is done, the TL (see Section 8.2) of the plenum is:

$$\text{TL} = -10\log_{10}\frac{W_o}{W_i} = -10\log_{10}\left[\frac{A_{\text{exit}}}{R_c} + \frac{A_{\text{exit}}\cos\theta}{\pi r^2}\right] \tag{8.271}$$

Equation (8.271) is only valid at frequencies for which the plenum chamber dimensions are large compared to a wavelength of sound and also only for frequencies above the cut-on frequency for higher order mode propagation in the inlet duct (see Section 8.10.3.3).

If the room constant, R_c, is made large, then the effectiveness of the plenum may be further increased by preventing direct line of sight with the use of suitable internal baffles. When internal baffles are used, the second term in Equation (8.271), which represents the direct field contribution, should be discarded. However, a better alternative is to estimate the IL of the baffle by using the procedure outlined in Chapter 7, Section 7.5 and adding the result arithmetically to the TL of the plenum. The assumption implied in adding the IL of the baffle to the TL of the chamber is that the baffle does not affect the sound power generated by the upstream noise source, which means that the IL will be equal to the TL, and this is a reasonable approximation in this case.

Equation (8.271) agrees with measurement for high frequencies and for small values of TL, but predicts values lower than the observed TL by 5 to 10 dB at low frequencies, which is attributed to neglect of reflection at the plenum exit and the modal behaviour of the sound field in the plenum.

8.17.2 ASHRAE (2015) Method

Mouratidis and Becker (2004) published a modified version of Wells' method and showed that their version approximated their measured data more accurately. Their analysis also includes equations describing the high-frequency performance. The Mouratidis and Becker approach has been adopted by ASHRAE (2015). Their expression for the TL at frequencies above the first mode cut-on frequency, f_{co} (see Equation (8.227)), in the inlet duct is:

$$\text{TL} = b\left[\frac{A_{\text{exit}}}{\pi r^2} + \frac{A_{\text{exit}}}{R_c}\right]^n \tag{8.272}$$

which is a similar form to Equation (8.271). If the inlet is closer to the centre of the wall than the corner, then a factor of 2 is included in the denominator of the first term in the preceding equation. The values of the constants, b and n, are 3.505 and -0.359, respectively. The preceding equation only applies to the case where the plenum inlet is directly in line with the outlet. When the value of θ in Figure 8.42(b) is non-zero, corrections must be added to the TL calculated using the preceding equation, and these are listed in Table 8.19 as the numbers in brackets.

TABLE 8.19 Corrections (dB) to be added to the TL calculated using Equation (8.272) or Equation (8.273) for various angles θ defined by Figure 8.42(b). The numbers not in brackets correspond to frequencies below the inlet duct cut-on frequency and the numbers in brackets correspond to frequencies above the duct cut-on frequency. The absence of numbers for some frequencies indicates that no data are available for these cases. Adapted from Mouratidis and Becker (2003)

1/3-octave band centre frequency (Hz)	Angle, θ (degrees)				
	15	22.5	30	37.5	45
80	0	−1	−3	−4	−6
100	1	0	−2	−3	−6
125	1	0	−2	−4	−6
160	0	−1	−2	−3	−4
200	0(1)	−1(4)	−2(9)	−3(14)	−5(20)
250	1(2)	2(4)	3(8)	5(13)	7(19)
315	4(1)	6(2)	8(3)	10(4)	14(5)
400	2(1)	4(2)	6(3)	9(4)	13(6)
500	1(0)	3(1)	6(2)	10(4)	15(5)
630	(1)	(2)	(3)	(5)	(7)
800	(1)	(2)	(2)	(3)	(3)
1000	(1)	(2)	(4)	(6)	(9)
1250	(0)	(2)	(4)	(6)	(9)
1600	(0)	(1)	(1)	(2)	(3)
2000	(1)	(2)	(4)	(7)	(10)
2500	(1)	(2)	(3)	(5)	(8)
3150	(0)	(2)	(4)	(6)	(9)
4000	(0)	(2)	(5)	(8)	(12)
5000	(0)	(3)	(6)	(10)	(15)

For frequencies below the duct cut-on frequency, f_{co}, Mouratidis and Becker (2004) give the following expression for estimating the plenum TL as:

$$\text{TL} = A_f S + W_e \tag{8.273}$$

where the coefficients, A_f and W_e, are given in Table 8.20. Again, when the value of θ in Figure 8.42(b) is non-zero, corrections must be added to the TL calculated using the preceding equation, and these are listed in Table 8.19 as the numbers not in brackets.

TABLE 8.20 Values of the coefficients in Equation (8.273). The numerical values given in the headings for W_e represent the thickness (ranging from 25 mm to 200 mm) of sound-absorbing material attached to the plenum wall. The material normally used is fibreglass or rockwool with an approximate density of 40 kg/m^3. Adapted from ASHRAE (2015) and Mouratidis and Becker (2003)

1/3-octave band centre frequency (Hz)	A_f		W_e				
	Plenum volume		25 mm fabric facing	50 mm fabric facing	100 mm perforated facing	200 mm perforated facing	25 mm solid metal facing
	< 1.4m^2	> 1.4m^2					
50	1.4	0.3	1	1	0	1	0
63	1.0	0.3	1	2	3	7	3
80	1.1	0.3	2	2	3	9	7
100	2.3	0.3	2	2	4	12	6
125	2.4	0.4	2	3	6	12	4
160	2.0	0.4	3	4	11	11	2
200	1.0	0.3	4	10	16	15	3
250	2.2	0.4	5	9	13	12	1
315	0.7	0.3	6	12	14	14	2
400	0.7	0.2	8	13	13	14	1
500	1.1	0.2	9	13	12	13	0

For an end in, side out plenum configuration (ASHRAE, 2015) corrections listed in Table 8.21 must be added to Equations (8.272) and (8.273) in addition to the corrections in Table 8.19.

TABLE 8.21 Corrections to be added to the TL calculated using Equations (8.272) and (8.273) for plenum configurations with an inlet on one end and an outlet on one side. The numbers not in brackets correspond to frequencies below the inlet duct cut-on frequency and the numbers in brackets correspond to frequencies above the duct cut-on frequency. The absence of numbers for some frequencies indicates that no data are available for these cases. Adapted from Mouratidis and Becker (2003)

1/3-octave band centre frequency (Hz)	Elbow effect correction	1/3-octave band centre frequency (Hz)	Elbow effect correction
50	2	630	(3)
63	3	800	(3)
80	6	1000	(2)
100	5	1250	(2)
125	3	1600	(2)
160	0	2000	(2)
200	−2(3)	2500	(2)
250	−3(6)	3150	(2)
315	−1(3)	4000	(2)
400	0(3)	5000	(1)
500	0(2)		

8.17.3 More Complex Methods

Two other methods have been published for predicting the TL of plenum chambers (Cummings, 1978; Ih, 1992). Cummings looked at high- and low-frequency range models for lined plenum chambers and Ih investigated the TL for unlined chambers. For the low-frequency model, it was assumed that only plane waves existed in both the inlet and outlet ducts and higher order modes existed in the plenum chamber. This low-frequency model is complicated to evaluate and the reader is referred to Cummings' original paper or the summary paper of Li and Hansen (2005).

For Cummings' high-frequency model, it was assumed that higher order modes existed in the inlet and outlet ducts as well as the plenum chamber. After a rather complicated analysis, the end result is that given by Equation (8.271), except that the chamber room constant, R_c, is calculated using Equation (8.269) but with the Sabine absorption coefficient, $\bar{\alpha}$, replaced by the statistical absorption coefficient, α_{st}.

Ih (1992) presented a model for calculating the TL of unlined plenum chambers. As Ih's model assumes a piston-driven, rigid-wall chamber, it is not valid above the inlet and outlet duct cut-on frequencies. However, it is the only model available for unlined plenum chambers.

8.18 Water Injection

Water injection has been investigated, both for the control of the noise of large rocket engines and for the control of steam venting noise. In both cases, the injection of large amounts of water has been found to be quite effective in decreasing high-frequency noise, at the expense of a large increase in low-frequency noise. This effect is illustrated in Figure 8.43 for the case of water injection to reduce steam venting noise.

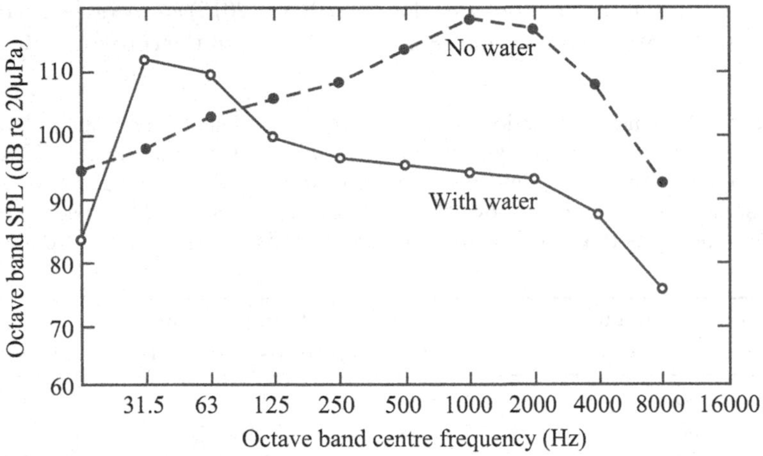

FIGURE 8.43 Effect of water injection on the reduction of steam venting noise.

In both cases, a mass flow rate of water equal to the mass flow rate of exhaust gas was injected directly into the flow to significantly cool the hot gases. Even more water would possibly prevent the low-frequency build-up, but the quantity of water required would be much greater.

Watters et al. (1955) investigated the effectiveness of water sprays used to cool hot exhaust gases in gas turbine exhausts. They provided the data shown here in Figure 8.44 for varying amounts of water, but cautioned against its universal use.

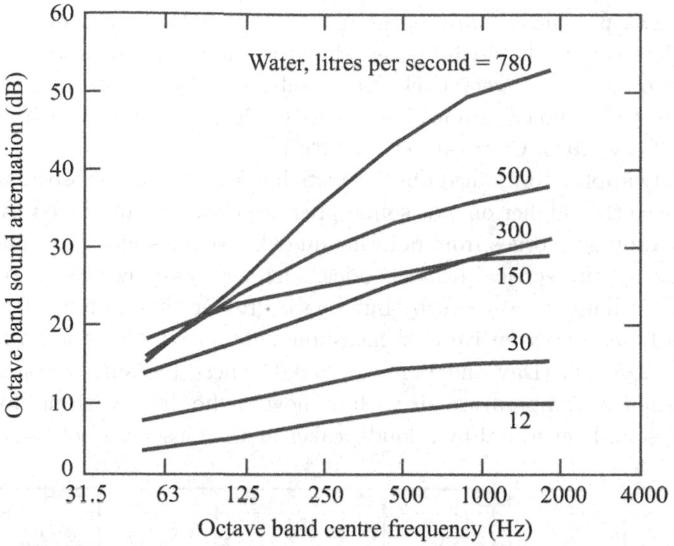

FIGURE 8.44 Octave band sound attenuation measured through several water spray systems (after Watters et al. (1955)).

8.19 Directivity of Exhaust Ducts

The sound radiation from the exit of a duct may be quite directional, as shown in Figures 8.45 and 8.46. Figure 8.45 is based on model studies (Sutton, 1990) in the laboratory and includes ducts of round, square and rectangular cross-section. To get the rectangular section data to collapse on to the ka-axis for circular ducts, where $2a$ is the duct diameter, it was necessary, during the model studies, to multiply the rectangular duct dimension ($2d$) in the direction of the observer by $\sqrt{4/\pi}$ to get $2a$, and this is the value of a used to calculate ka prior to reading the DI from the figure. Also, as k is the wavenumber of sound at the duct exit, the duct exhaust

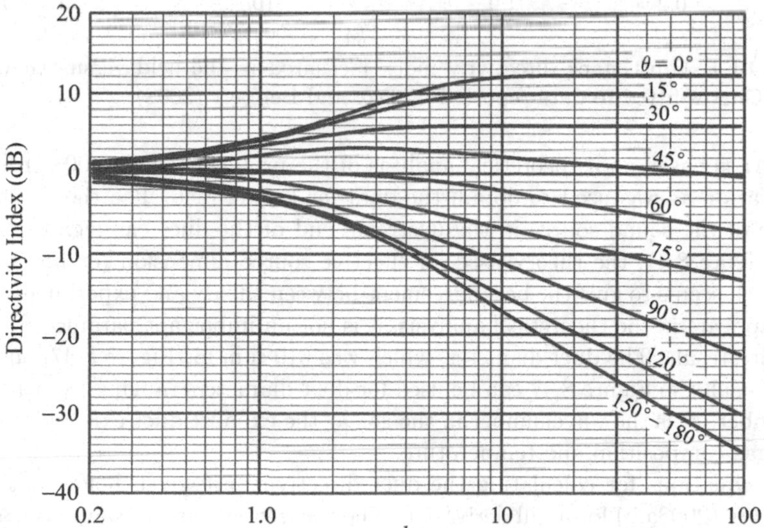

FIGURE 8.45 Exhaust stack directivity index measured in the laboratory versus ka where a is the stack inside radius. Curves fitted to data reported by Sutton (1990) and Dewhirst (2002).

temperature and gas properties must be used to calculate the speed of sound which is used to calculate k. However, it should be noted that the directivity of a vertical hot exhaust in the presence of a cross flow is very likely to be substantially different to that of an ambient temperature exhaust with no cross flow due in part to the strong temperature gradient near the hot exhaust exit (Leav, 2020; Cazzolato et al., 2021).

Figure 8.45 only applies to unlined ducts. Ducts lined with sound-absorbing material radiate more directionally so that higher on-axis sound pressure levels are produced. As shown in Figure 8.45, a distinct advantage accrues from pointing an exhaust duct upwards. The figures show the importance of placing any vertical exhaust duct, such as a cooling tower, so that line of sight from the exhaust discharge to any nearby building is greater than about 30°.

Figure 8.46 is based on extensive field measurements on circular ducts ranging in diameter from 305 mm to 1215 mm (Day and Bennett, 2008). These measurements are for an exhaust duct containing ambient temperature air with no flow in the duct, with no flow across the duct exit and with the sound generated by a loudspeaker at the closed end of the duct.

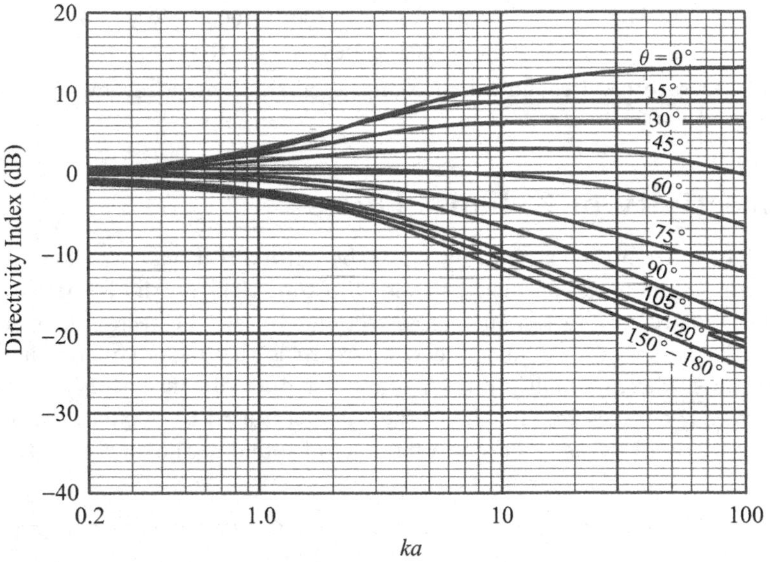

FIGURE 8.46 Exhaust stack directivity index measured in the field versus ka where a is the stack radius. Curves fitted to data reported by Day and Bennett (2008).

Figure 8.47 is based on a theoretical analysis of the problem (Davy, 2008a,b), for the same duct configuration as was used to obtain the data for Figure 8.46. The theory shows that the type and size of the sound source in the upstream end of the duct can significantly affect the results. For Figure 8.47, the theory used an effective source dimension (or line source length), $s_d = \pi/k$, which seems to give the best agreement between theory and experiment. Although the duct length appears in the theoretical analysis, it is not an important parameter in terms of the final directivity result. The duct diameter, which was 610 mm for Figure 8.47, only has a slight effect on the results so Figure 8.47 can be used for duct diameters ranging from 100 mm to 2 m. The wavenumber, k, is the wavenumber in the gas at the exhaust duct exit and is dependent on the exhaust gas composition and temperature.

The final equations for calculating the duct directivity are given below but the reader is referred to Davy (2008a,b) for a full derivation. The theoretical curves (Davy, 2008a,b) of Figure 8.47 and the laboratory measured curves of Figure 8.45 show much greater attenuations at large angles than the field measured curves of Figure 8.46. Reasons for this are discussed later in this section.

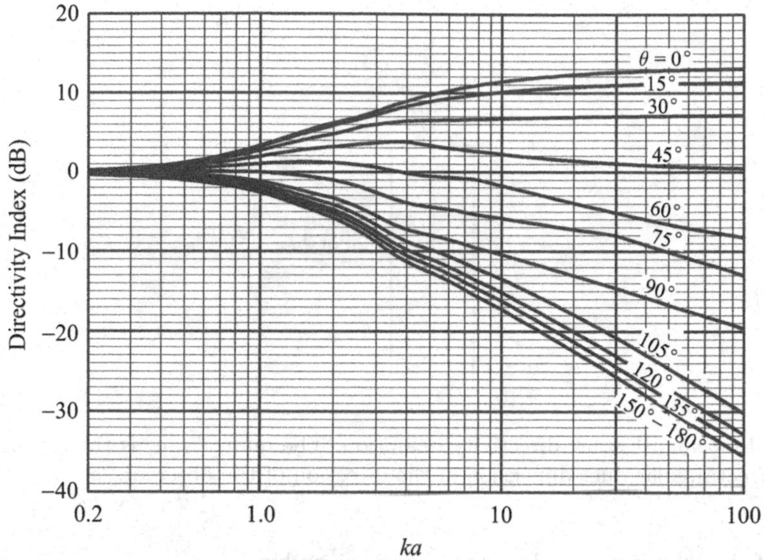

FIGURE 8.47 Exhaust stack directivity calculated using Davy's theory (Davy, 2008a,b) for a $2a$ = 605 mm diameter duct. The curves do not change significantly with duct diameter for diameters between 100 mm and 2 m.

The directivity index is given by the intensity in any one direction divided by the intensity averaged over all possible directions which in this case, is a sphere. Thus:

$$\mathrm{DI}(\theta) = 10\log_{10}(I(\theta)/I_{av}) \tag{8.274}$$

As Equation (8.274) is a ratio, absolute values of the above quantities are not needed. To calculate the average intensity, I_{av}, a radial distance of unity is assumed and the intensity, $I(\theta)$, will be evaluated at the same location for an arbitrary source strength. Thus, the normalised average intensity is given by:

$$I_{av} = \frac{1}{4\pi}\int_0^{2\pi} \mathrm{d}\phi \int_0^{\pi} I(\theta)\sin(\theta)\mathrm{d}\theta = \frac{1}{2}\int_0^{\pi} I(\theta)\sin(\theta)\mathrm{d}\theta \tag{8.275}$$

where ϕ is the azimuthal angle around the perimeter of the stack and for a circular section stack, it is assumed that the sound intensity does not vary with the value of this angle. Even though the directivity index, $\mathrm{DI}(\theta)$, and the intensity, $I(\theta)$, will vary with azimuthal angle for non-circular cross-section ducts, this variation will be ignored when calculating the directivity index in this section. The normalised sound intensity, $I(\theta)$, at angular location, θ, from the axis of an exhaust stack (see Figure 8.48) is given by (Davy, 2008a,b), for $|\theta| \le \pi/2$ as:

$$I(\theta) = \frac{1}{\rho c} p_1^2(\theta) p_2^2(\theta) \tag{8.276}$$

For $\pi/2 < |\theta| \le \pi$, the normalised sound intensity is given by:

$$I(\theta) = \frac{I(\theta = \pi/2)}{1 - kz\cos(\theta)} \quad \text{where } z = \frac{bd}{b+d} \tag{8.277}$$

The first term, $p_1^2(\theta)$, in Equation (8.276) accounts for the expected sound pressure level at a particular location due to a point source and the second term, $p_2^2(\theta)$, effectively compensates

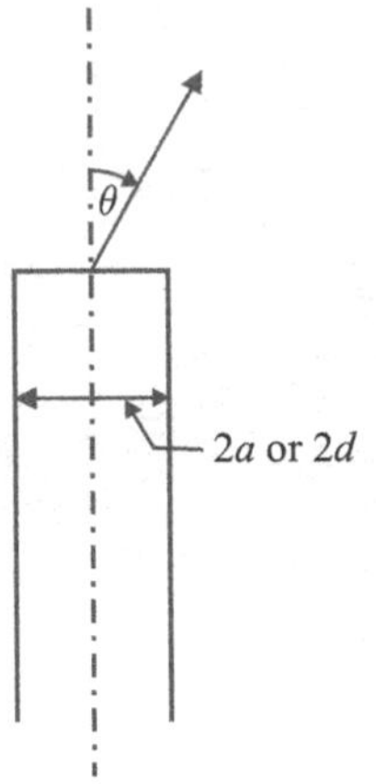

FIGURE 8.48 Exhaust stack directivity definition. The quantity, a, is the circular duct radius and $2d$ is the rectangular duct dimension in the direction of the observer.

for the size of the duct opening. The first term is given by:

$$p_1^2(\theta) = \int_{-\pi/2}^{\pi/2} \frac{w(\phi)}{[2\rho c \sigma(\phi)]^2} \left(\frac{\sin[kd(\sin\theta - \sin\phi)]}{kd(\sin\theta - \sin\phi)} \right)^2 \mathrm{d}\phi \tag{8.278}$$

which must be evaluated using numerical integration. The radiation efficiency, $\sigma(\phi)$, of the duct exit is given by Davy (2008a,b) as:

$$\sigma(\phi) = \begin{cases} \dfrac{1}{\pi/(2k^2bd) + \cos(\phi)}; & \text{if } |\phi| \le \phi_\ell \\ \dfrac{1}{\pi/(2k^2bd) + 1.5\cos(\phi_\ell) - 0.5\cos(\phi)}; & \text{if } \phi_\ell < |\phi| \le (\pi/2) \end{cases} \tag{8.279}$$

where $2d$ is the duct cross-sectional dimension in the direction of the observer and $2b$ is the dimension at 90° to the direction of the observer, for a rectangular section duct. For a circular duct of radius, a, the relationship, $d = b = \pi a/4$, in the preceding equations has been shown by Davy (2008a,b) to be appropriate.

As the graphs in Figures 8.45 to 8.47 are in terms of ka for a circular-section duct of radius, a, then for rectangular-section ducts, the scale on the x-axis (which is kd) must be multiplied by $4/\pi$. The limiting angle, ϕ_ℓ, is defined as:

$$\phi_\ell = \begin{cases} 0; & \text{if } \sqrt{\pi/(2kd)} \ge 1.0 \\ \arccos\left[\sqrt{\pi/(2kd)}\right]; & \text{if } \sqrt{\pi/(2kd)} < 1.0 \end{cases} \tag{8.280}$$

The quantity, $w(\phi)$, in Equation (8.278) is defined by (Davy, 2008a,b):

$$w(\phi) = \left(\frac{\sin[ks_d \sin(\phi)]}{ks_d \sin(\phi)} \right)^2 (1 - \alpha_{st})^{(L/2d)\tan|\phi|} \tag{8.281}$$

where L is the length of the exhaust stack (from the noise source to the stack opening) and α_{st} is the statistical absorption coefficient of the duct walls, usually set equal to 0.05. The quantity, ks_d, is a function of the source size and setting it equal to π seems to give results that agree with experimental measurements made using loudspeakers (Davy, 2008a,b). The quantity, $w(\phi)$, is made up of two physical quantities. The first term in large brackets represents the directivity of the sound source at the end of the duct, which, for the purposes of Equation (8.281), corresponds to a line source of length, $2s_d$, where s_d is the radius of the loudspeaker sound source. Although this model works well for the loudspeaker sound sources used to obtain the experimental data in Figures 8.45 and 8.46, it may not be the best model for an industrial noise source such as a fan with the result that the directivity pattern of sound radiation from a stack driven by an industrial noise source could be slightly different to the directivities presented here.

The second term in Equation (8.281) accounts for the effect of reflections from the duct walls on the angular distribution of sound propagation in the duct.

The quantity, $p_2(\theta)$, in Equation (8.276) is defined as:

$$p_2(\theta) = \begin{cases} p_2(0); & \text{if } \cos(\phi_\ell) \leq \cos(\theta) \\ \dfrac{p_2(0)\cos(\theta) + \cos(\phi_\ell) - \cos(\theta)}{\cos(\phi_\ell)}; & \text{if } 0 \leq \cos(\theta) < \cos(\phi_\ell) \end{cases} \tag{8.282}$$

where $p_2(0) = 1 + p_b p_d$, where

$$p_b = \begin{cases} \sin(kb); & \text{if } kb \leq \pi/2 \\ 1; & \text{if } kb > \pi/2 \end{cases} \quad \text{and where} \quad p_d = \begin{cases} \sin(kd); & \text{if } kd \leq \pi/2 \\ 1; & \text{if } kd > \pi/2 \end{cases} \tag{8.283}$$

In comparing Figures 8.45, 8.46 and 8.47, it can be seen that the field measurements seem not to have as large a directivity as measured in the laboratory or predicted by the theory. One explanation may be attributed to noise breaking out through the walls of the duct used in the field measurements affecting the smaller measured sound pressure levels radiated from the duct exit. The ducts used for the laboratory measurements were double-walled and smaller and did not suffer breakout noise to anywhere near the same extent as the ducts used for the field measurements. Another possible explanation for the smaller directivity levels at angles greater than 90° for the field measurements may be scattering, which adds to the sound pressure levels calculated due to diffraction alone. For noise barriers, this limit has been set to $L_s = 24$ dB and there is no reason to assume anything different in practice for directivity from exhaust stacks. So at long distances, if we assume that the maximum sound pressure level is at $\theta = 0$, then the sound pressure level at angle, θ, relative to that at $\theta = 0$ is:

$$L_p(\theta) - L_p(0) = 10\log_{10}\left(\frac{I(\theta)}{I(0)} + 10^{-L_s/10}\right) \tag{8.284}$$

Note that the curves in Figures 8.45 to 8.47 all show greater differences between the zero degree direction and the large angles than predicted by Equation (8.284) with L_s=24 dB. This is for different reasons in each case: the curves in Figure 8.45 were measured in still air in an anechoic room, where scattering was far less than it would be outdoors; the curves in Figure 8.46 were measured within 1 or 2 metres of the duct exhaust where scattering is not as important as at larger distances; and the curves in Figure 8.47 were derived using a theoretical analysis that does not include scattering.

Curves showing the measured data at 0, 30, 45, 60, 90 and 120 degrees are provided in Figure 8.49 for the field measured data, which gives an idea of the variation that may be expected in practice. The data measured in the laboratory exhibited a similar amount of scatter about the mean value.

The results in Figures 8.45 to 8.47 only apply for unlined ducts. Ducts lined with sound-absorbing material radiate more directionally so that higher on-axis sound pressure levels are produced. As shown in Figures 8.45 to 8.47, a distinct advantage accrues from pointing an exhaust duct upwards. Alternatively, the figures show the importance of placing cooling towers such that line of sight to any nearby building is greater than about 30°.

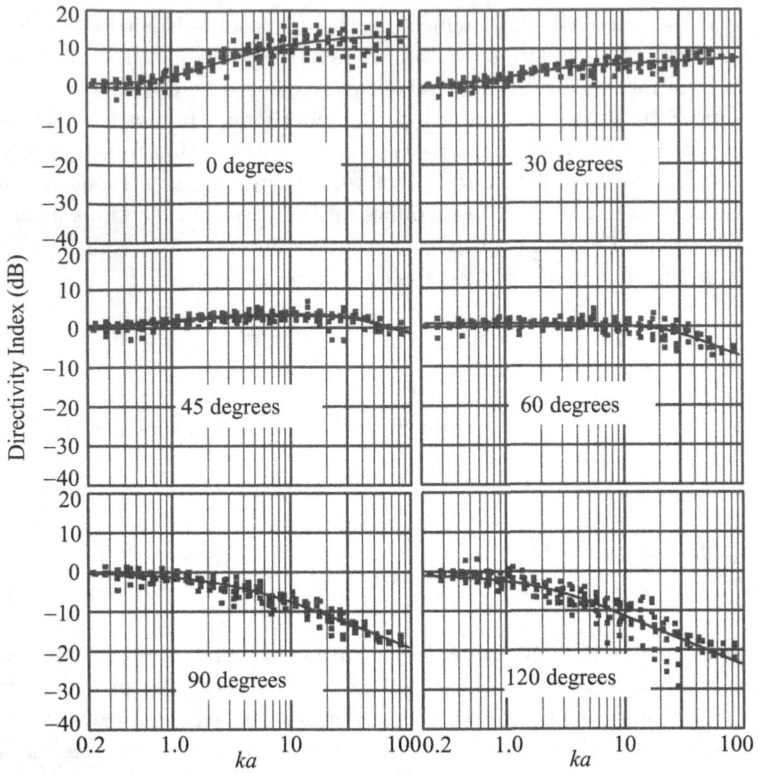

FIGURE 8.49 Scatter in the directivity index data measured in the field and reported by Day and Bennett (2008).

Equation (8.281) is the best description that seems to be available to describe the energy propagation in a duct as a function of angle and this expression implies that cross-modes will contribute to the exit sound field, even for small values of ka. Thus, it is difficult to obtain an accurate estimate of the radiated sound power from sound pressure level measurements at the duct exit as not all of the energy propagation is normal to the plane of the duct cross section at the exit.

The sound power can either be determined using the methods outlined in Chapter 4 or, alternatively, if the sound pressure level, L_p, is measured at some distance, r, and some angle, θ, from the duct outlet, the sound power level, L_W, radiated by the duct outlet may be calculated with the help of Figure 8.47 and Equation (8.285) as:

$$L_W = L_p - \mathrm{DI}_\theta + 10\log_{10} 4\pi r^2 + 10\log_{10}\frac{400}{\rho c} + A_E \tag{8.285}$$

where the directivity index, DI_θ, may be obtained from Figure 8.46 or 8.47 and the propagation attenuation, A_E, may be calculated as described in Section 5.3.

If it is desired to add an exhaust stack to the duct outlet, the resulting noise reduction may be calculated from a knowledge of the insertion loss, IL_s, of the stack due to sound propagation through it, and both the angular direction and distance from the stack axis to the receiver location (if different from the values for the original duct outlet). The attenuation, A_{Es}, must also be taken into account if it is different for propagation from the duct outlet without the stack. Thus:

$$\mathrm{NR} = \mathrm{IL}_s + A_{Es} - A_E + 20\log_{10}(r_s/r) + \mathrm{DI}_\theta - \mathrm{DI}_s \tag{8.286}$$

where the subscript, s, refers to quantities with the stack in place.

The sound pressure level, L_{ps}, at location r_s with the stack in place is given by:

$$L_{ps} = L_W - \mathrm{IL}_s + \mathrm{DI}_s - 10\log_{10} 4\pi r_s^2 - 10\log_{10}\frac{400}{\rho c} - A_{Es} \qquad (8.287)$$

where the directivity, DI_s, of the exhaust stack in the direction of the receiver is obtained using Figure 8.46 or 8.47. Without the stack in place, the sound pressure level may be determined using Equation (8.285).

8.19.1 Hot Exhausts Subject to Cross-Flow

The directivity of sound radiation from a vertical hot exhaust stack subjected to cross flow can increase measured sound levels significantly at distant community locations over what would be expected for a cold exhaust with no cross flow (Leav, 2020; Leav et al., 2021; Cazzolato et al., 2021). Increases of up to 10 dB are caused by wind blowing across a hot vertical exhaust stack, resulting in the plume being bent in the downwind direction causing substantially more severe downward refraction of sound rays than would occur with cross flow over a cold exhaust stack. This problem can be significantly reduced by around 6 dB by using a perforated cylinder extension of the exhaust stack (Cazzolato et al., 2021).

9

Vibration Control

LEARNING OBJECTIVES

In this chapter the reader is introduced to:

- vibration isolation for single- and multi-degree-of-freedom systems;
- damping, stiffness and mass relationships;
- types of vibration isolators;
- vibration absorbers;
- vibration measurement;
- when damping of vibrating surfaces is and is not effective for noise control;
- damping of vibrating surfaces;
- measurement of damping.

9.1 Introduction

Vibration is oscillatory motion of a body or surface about a mean position and occurs to some degree in all industrial machinery. It may be characterised in terms of acceleration, velocity, displacement, surface stress or surface strain amplitude, and associated frequency. On a particular structure, the vibration and relative phase will usually vary with location. Although high levels of vibration are sometimes useful (for example, vibrating conveyors and sieves), vibration is generally undesirable, as it often results in excessive noise, mechanical wear, structural fatigue and possible failure of the vibrating structure or structures connected to it.

Many noise sources commonly encountered in practice are associated with vibrating surfaces, and with the exception of aerodynamic noise sources, the control of vibration is an important part of any noise control program. Vibration can also be a source of annoyance or equipment damage. The minimisation of vibration at a sensitive receiver can be treated as a source–path–receiver problem, in much the same way as a noise problem was treated in Chapter 1. That is, there is a source of unwanted vibration, a path that transmits the vibration and a receiver that is sensitive to it. Vibration control may take the form of modification of the source, the transmission path or the receiver. These control approaches are discussed later in this section.

Any structure can vibrate and will generally do so when excited mechanically (for example, by forces generated by some mechanical equipment) or when excited acoustically (for example, by the acoustic field of noisy machinery). Any vibrating structure will have preferred modes (or vibration amplitude patterns) in which it will vibrate and each mode of vibration will respond most strongly at its resonance frequency. A mode will be characterised by a particular spatial

DOI: 10.1201/9780367814908-9

amplitude of response distribution, having nodes and antinodes. Nodes are lines of nil or minimal response across which there will be abrupt phase changes from in-phase to opposite phase relative to a reference and antinodes are regions of maximal response between nodes.

If an incident force field is coincident with a structural mode, both in spatial distribution and frequency, it will strongly drive that mode. The response will become stronger with better matching of the force field shape to the modal response shape of the structure. When driven at resonance, the structural mode response will only be limited by the damping of the mode. As will be discussed in Section 9.7, it is also possible to drive structural modes at frequencies other than their resonance frequencies. To avoid excessive vibration and associated problems, it is important in any mechanical system to ensure that coincidence of excitation frequency and structural resonance frequencies is avoided as much as possible.

With currently available analytical tools (e.g. statistical energy analysis, finite element analysis – see Chapter 11), it is often possible to predict at the design stage the dynamic behaviour of a machine and any possible vibration problems. However, vibration problems do appear regularly in new as well as old installations, and vibration control then becomes a remedial exercise instead of the more economic design exercise.

With the principal aim of noise control, a number of alternative forms of vibration control are listed below. Receiver control is not considered here as it is not appropriate in cases where people are receiving the vibration. However, it can be appropriate in cases where the receiver is a sensitive item of equipment.

Most situations encountered in practice, which involve excessive vibration, are complex, involving excitation forces at many different frequencies acting in many different directions. There may also be many different receivers with varying degrees of sensitivity to vibration and the noise that results from it. Additionally, there are almost always practical issues and financial considerations that limit the extent of control that can be implemented. However, the principles discussed in this chapter are a good starting point for understanding and solving solving most vibration problems and the associated excessive noise.

1. **Modification of the vibration generating mechanism (source control)**. This may be accomplished most effectively at the design stage by choosing the process that minimises jerk, or the time rate of change of force. In a punch press, this may be done by reducing the peak level of tension in the press frame and releasing it over a longer period of time as, for example, by surface grinding the punch on a slight incline relative to the face of the die. Another way of achieving this in practice is to design tools that apply the load to the part being processed over as long a time period as possible, while at the same time, minimising the peak load. Another example may be to change a reciprocating engine source with a rotating engine such as an electric motor. In some cases, this form of source control may include dynamic balancing, changing cam profiles or reducing reciprocating masses (Ungar, 2019b). Source control can also include detuning the vibration source so that its forcing frequencies do not match frequencies that are transmitted well by the path from the source to receiver (such as structural resonance frequencies). Source control is case specific and not amenable to generalisation; however, it is often the most cost-effective approach and frequently leads to an inherently better process.
2. **Modification of the dynamic characteristics (or mechanical input impedance) of a supporting structure (path control)** to reduce its ability to respond to the input energy. This form of control essentially suppresses the transfer of vibrational energy from the source to the noise-radiating structure. This may be achieved by stiffness or mass changes to the structure, especially at the point of attachment to the vibration source, changing the orientation of the source so that the forces it generates do not align with the directions in which the structure most easily

responds to those forces, or using a vibration absorber or a vibration neutraliser. Alternatively, the radiating surface may be modified to minimise the radiation of sound to the environment. This may sometimes be done by choice of an open structure; for example, a perforated surface instead of a solid surface. Finally the motion of the transmission structure at the sensitive receiver location may be reduced by adding impedance discontinuities along the transmission path to block the energy transmission (Ungar, 2019b). Detuning the transmission path so that it does not respond well to the frequencies generated by the source is another form of structural modification that can be considered in some cases (Ungar, 2019b).

3. **Isolation of the source of vibration from the noise-radiating structure (path control)** by means of flexible couplings or mounts that serve to reduce the vibratory forces transmitted to the supporting structure.
4. **Dissipation of vibrational energy in the structure by means of vibration damping** (path control), which converts mechanical energy into heat. This is usually achieved by use of some form of damping material.
5. **Active control (path control)**, which may be used either to modify the dynamic characteristics of a structure or to enhance the effectiveness of vibration isolators. Active control is discussed in Hansen (2001), Coleman and Remington (2006) and Hansen et al. (2013).

As has been mentioned, the first approach will not be discussed further and the fifth approach will be discussed elsewhere; aspects of the remaining three approaches, isolation, damping and alteration of the mechanical input impedance, will now be discussed with emphasis on noise control.

9.2 Vibration Isolation

Vibration isolation is considered on the basis that structure-borne vibration from a source to some structure, which then radiates noise, may indeed be as important as, or perhaps more important than, direct radiation from the vibration source itself. Almost any stringed musical instrument provides a good example of this point. In every case, the vibrating string is the obvious energy source but the sound that is heard seldom originates at the string, which is a very poor radiator; rather, a sounding board, cavity or electrical system is used as a secondary and very much more efficient sound radiator.

When one approaches a noise control problem, the source of the unwanted noise may be obvious, but the path by which it radiates sound may be obscure. Indeed, determining the propagation path may be the primary problem to be solved. Unfortunately, no general specification of simple steps to be taken to accomplish this task can be given. On the other hand, if an enclosure for a noisy machine is contemplated, then good vibration isolation between the machine and enclosure, between the machine and any pipework or other mechanical connections to the enclosure, and between the enclosure and any protrusions through it should always be considered as a matter of course. Stated another way, the best enclosure can be rendered ineffective by structure-borne vibration. Thus, it is important to control all possible structural paths of vibration, as well as airborne sound, for the purpose of noise control.

The transmission of vibratory motions or forces from one structure to another may be reduced by interposing a relatively flexible isolating element between the two structures. This is called vibration isolation, and when properly designed, the vibration amplitude of the driven structure is largely controlled by its inertia. An important design consideration is the resonance frequency of the isolated structure on its vibration isolation mount. At this frequency, the isolating element will amplify by a large amount the force transmission between the structure and its mount. Only at frequencies greater than 1.4 times the resonance frequency will the force transmission be reduced. Thus, the resonance frequency must be arranged to be well below the range of frequencies to be isolated. Furthermore, adding damping to the vibrating system, for the purpose of reducing the vibratory response at the resonance frequency, has the effect of decreasing the isolation that otherwise would be achieved at higher frequencies. An alternative way of reducing excessive vibration at resonance (for example, when a machine is run up to speed and its rotational frequency passes through the isolator resonance frequency) is to use snubbers as discussed in Section 9.2.1 or vibration absorbers as discussed in Section 9.4.1.

Two types of vibration-isolating applications will be considered: (1) those where the intention is to prevent transmission of vibratory forces from a machine to its foundation, and (2) those where the intention is to reduce the transmission of motion of a foundation to a device mounted on it. Rotating equipment such as motors, fans, turbines, etc., mounted on vibration isolators are examples of the first type. An electron microscope mounted resiliently in the basement of a hospital is an example of the second type.

9.2.1 Single-Degree-of-Freedom Systems

To understand vibration isolation, it is useful to gain familiarity with the behaviour of single-degree-of-freedom systems, such as illustrated in Figure 9.1 (Tse et al., 1979; Rao, 2016; Inman, 2014). In the figure, the two cases considered here are illustrated with a spring, mass and dashpot. In the first case, the mass is driven by an externally applied, single-frequency, periodic force represented for convenience (see Figure 9.4 and the associated discussion) as $\hat{F}\mathrm{e}^{\mathrm{j}\omega t}$, while in the second case, the base is assumed to move with some specified vibration displacement, $\hat{y}_1\mathrm{e}^{\mathrm{j}\omega t}$ (Tse et al., 1979). Vibration displacement, y at a single frequency, ω radians/sec, is related to vibration velocity, v, by $y = v/\omega$.

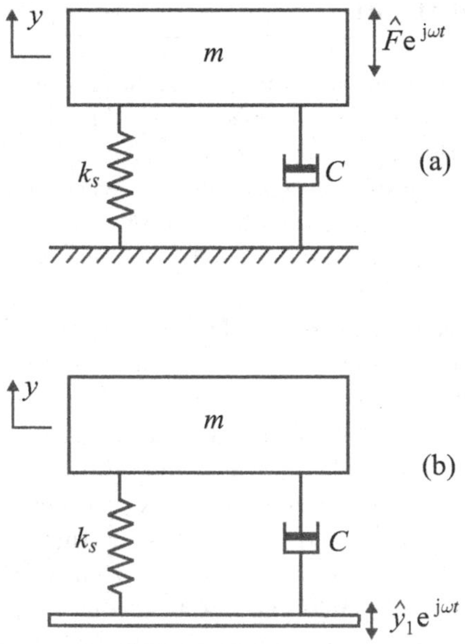

FIGURE 9.1 Single-degree-of-freedom system: (a) forced mass, rigid base; (b) vibrating base.

The equation of motion for the single-degree-of-freedom oscillator of mass, m (kg), damping coefficient, C (N-s/m), stiffness, k_s (N/m), displacement, y (m), and cyclic forcing function, $\hat{F}\mathrm{e}^{\mathrm{j}\omega t}$ (N) at radian frequency, $\omega = 2\pi f$, with f the frequency in Hz, shown in Figure 9.1(a) is:

$$m\ddot{y} + C\dot{y} + k_s y = F = \hat{F}\mathrm{e}^{\mathrm{j}\omega t} \tag{9.1}$$

For sinusoidal motion, acceleration, a can be written as $a = \ddot{y} = -\omega^2 y$ and velocity, v, can be written as $v = \dot{y} = \mathrm{j}\omega y$. In this case, Equation (9.1) can be written as:

$$-m\omega^2 y + \mathrm{j}C\omega y + k_s y = F = \hat{F}\mathrm{e}^{\mathrm{j}\omega t} \tag{9.2a,b}$$

and

$$\frac{y}{F} = \frac{1}{k_s - m\omega^2 + \mathrm{j}C\omega} = \frac{(k_s - m\omega^2) - \mathrm{j}C\omega}{(k_s - m\omega^2)^2 + (C\omega)^2} \tag{9.3a,b}$$

where y and F are complex numbers, indicating that the displacement is not in-phase with the force, as a result of the presence of damping. The amplitude of the complex ratio, $|y/F|$, may be written as:

$$\left|\frac{y}{F}\right| = \sqrt{\frac{(k_s - m\omega^2)^2 + (C\omega)^2}{[(k_s - m\omega^2)^2 + (C\omega)^2]^2}} = \left[(k_s - m\omega^2)^2 + (C\omega)^2\right]^{-1/2} \tag{9.4a,b}$$

In the absence of any excitation force, F, or damping, C, the system, once disturbed, will vibrate sinusoidally at a constant amplitude (dependent on the amplitude of the original disturbance) at its undamped resonance frequency, f_0, which is sometimes called the natural frequency. Solution of Equation (9.2) with $F = C = 0$ gives for the undamped resonance frequency:

$$f_0 = \frac{1}{2\pi}\sqrt{\frac{k_s}{m}} \quad \text{(Hz)} \tag{9.5}$$

The static deflection, d, of the mass supported by the spring is given by $d = mg/k_s$ where g is the acceleration of gravity, so that Equation (9.5) may be written in the following alternative form:

$$f_0 = \frac{1}{2\pi}\sqrt{\frac{g}{d}} \quad \text{(Hz)} \tag{9.6}$$

Substitution of the value of g equal to 9.81 m/s gives the following useful equation (where d is in metres):

$$f_0 = 0.50/\sqrt{d} \quad \text{(Hz)} \tag{9.7}$$

The preceding analysis is for an ideal system in which the spring has no mass, which does not reflect the actual situation. If the mass of the spring is denoted m_s, and it is uniformly distributed along its length, it is possible to get a first order approximation of its effect on the resonance frequency of the mass-spring system by using Rayleigh's method and setting the maximum kinetic energy of the mass, m, plus the spring mass, m_s, equal to the maximum potential energy of the spring. The velocity of the spring is zero at one end and a maximum of v_m at the other end. Thus, the kinetic energy in the spring may be written as:

$$KE_s = \frac{1}{2}\int_0^L v_m^2 \mathrm{d}m_s \tag{9.8}$$

where v_m is the velocity of the segment of spring of mass, $\mathrm{d}m_s$, and L is the length of the spring. The quantities, v_m and $\mathrm{d}m_s$, may be written as:

$$v_m = \frac{xv}{L} = \frac{x\dot{y}}{L} \text{ and } \mathrm{d}m_s = \frac{m_s}{L}\mathrm{d}x \tag{9.9a,b}$$

where x is the distance from the spring support to segment, $\mathrm{d}m_s$. Thus, the KE in the spring may be written as:

$$KE_s = \frac{1}{2}\int_0^L \left|\frac{x\dot{y}}{L}\right|^2 \frac{m_s}{L}\mathrm{d}x = \frac{m_s|\dot{y}^2|}{2L^3}\int_0^L x^2 \mathrm{d}x = \frac{1}{2}\frac{m_s|\dot{y}^2|}{3} \tag{9.10a,b,c}$$

Equating the maximum KE in the mass, m, and spring, with the maximum PE in the spring gives:

$$\frac{1}{2}\frac{m_s}{3}|\dot{y}|^2 + \frac{1}{2}m|\dot{y}|^2 = \frac{1}{2}k_s y^2 \tag{9.11}$$

Substituting $\dot{y} = \mathrm{j}\omega y$ in the above equation gives the resonance frequency as:

$$f_0 = \frac{1}{2\pi}\sqrt{\frac{k_s}{m + (m_s/3)}} \quad \text{(Hz)} \tag{9.12}$$

Thus, according to Equation (9.12), more accurate results for the resonance frequency will be obtained if the suspended mass is increased by one-third of the spring mass in Equation (9.5).

The mass, m_s, of the spring is the mass of the active coils, which, for a spring with flattened ends, is two less than the total number of coils. Alternatively, the number of active coils is equal to the number that are free to move plus 0.5. For a coil spring of overall diameter, D, wire diameter, d, and with n_C active coils of material density ρ_m, the mass is:

$$m_s = n_C \frac{\pi d^2}{4} \pi D \rho_m \tag{9.13}$$

For a coil spring with a helix angle, α (usually just a few degrees), the stiffness (N/m), or the number of Newtons required to stretch it by 1 metre, is given by:

$$k_s = \frac{d^4 \cos\alpha}{8 n_C D^3}\left[\frac{\cos^2\alpha}{G} + \frac{2\sin^2\alpha}{E}\right]^{-1} = \frac{G d^4}{8 n_C D^3}\left[\frac{(1+\nu)\cos\alpha}{1+\nu\cos^2\alpha}\right] \quad \text{(N/m)} \tag{9.14}$$

where E is the modulus of elasticity (Young's modulus) of the spring material, $G = E/[2(1+\nu)]$ is the modulus of rigidity (or shear modulus) and ν is Poisson's ratio (= 0.3 for spring steel). For steel coil springs, the term in square brackets is very close to 1 so this term is usually omitted from equations in textbooks.

Of critical importance to the response of the systems shown in Figure 9.1 is the damping ratio, $\zeta = C/C_c$, where C_c is the critical damping coefficient defined as:

$$C_c = 2\sqrt{k_s m} \quad \text{(kg/s)} \tag{9.15}$$

When the damping ratio is less than unity, the transient response is cyclic, but when the damping ratio is unity or greater, the system transient response ceases to be cyclic.

In the absence of any excitation force, F, but including damping, $\zeta < 1$, the system of Figure 9.1, once disturbed, will oscillate approximately sinusoidally at its damped resonance frequency, f_d. Solution of Equation (9.1) with $F = 0$ and $C \neq 0$ gives for the damped resonance frequency:

$$f_d = f_0\sqrt{1-\zeta^2} \quad \text{(Hz)} \tag{9.16}$$

With a constant-amplitude periodic excitation force, $F = \hat{F}\mathrm{e}^{\mathrm{j}\omega t}$, the system of Figure 9.1 will respond sinusoidally at the driving frequency $\omega = 2\pi f$. Let $f/f_0 = X$, then, using Equations (9.5) and (9.15), Equation (9.4b) can be rewritten as:

$$\frac{|y|}{|F|} = \frac{1}{k_s}\left[\left(1-X^2\right)^2 + 4\zeta^2 X^2\right]^{-1/2} \tag{9.17}$$

Inspection of Equation (9.17) shows that maximum displacement occurs when $X = \sqrt{1-2\zeta^2}$ which corresponds to a frequency of:

$$f_{\text{maxdis}} = f_0\sqrt{1-2\zeta^2} \tag{9.18}$$

The amplitude of velocity is:

$$|u| = |\dot{y}| = 2\pi f|y| = 2\pi f_0 X|y| = X|y|\sqrt{k_s/m} \tag{9.19a,b,c,d}$$

Substituting Equation (9.17) for $|y|$ into Equation (9.19d), results in:

$$\frac{|\dot{y}|}{|F|} = \frac{1}{\sqrt{k_s m}}\left[\left(\frac{1}{X} - X\right)^2 + 4\zeta^2\right]^{-1/2} \tag{9.20}$$

Inspection of Equation (9.20) shows that the maximum velocity amplitude occurs when $X = 1$, which corresponds to the undamped resonance frequency. Thus:

$$f_{\text{maxvel}} = f_0 \tag{9.21}$$

Similarly, it may be shown that the frequency of maximum acceleration amplitude is:

$$f_{\text{maxacc}} = \frac{f_0}{\sqrt{1 - 2\zeta^2}} \tag{9.22}$$

If the system of Figure 9.1 is excited by a transient force that produces a deflection of y_T, the ratio of the transient to the steady state component of deflection is given by (Ungar, 2020):

$$N = \frac{y_T}{y} = \frac{\mathrm{e}^{-2\pi\zeta f_d t}}{\sqrt{1 - \zeta^2}} \tag{9.23}$$

The time, T_N, taken for the transient response to decay to a fraction, $N = N_C$, is obtained by taking natural logarithms of both sides of Equation (9.23) and setting $t = T_N$, to obtain:

$$T_N = \frac{-\log_{\mathrm{e}}\left(N_C\sqrt{1 - \zeta^2}\right)}{2\pi\zeta f_d} \tag{9.24}$$

and the number of cycles, n_N, taken to reach N_C is $n_N = f_d N_C$.

If the structure represented by Figure 9.1 is hysteretically damped, which in practice is the more usual case for a structure, then the viscous damping model is inappropriate. This case may be investigated by setting $C = 0$ and replacing k_s in Equation (9.1) with complex $k_s(1 + \mathrm{j}\eta)$, where η is the structural loss factor. Solution of Equation (9.1) with these modifications gives for the displacement amplitude, $|y'|$, of the hysteretically damped system:

$$\frac{|y'|}{|F|} = \frac{1}{k_s}\left[\left(1 - X^2\right)^2 + \eta^2\right]^{-1/2} \tag{9.25}$$

For the case of hysteretic (or structural) damping, the frequency of maximum displacement occurs when $X = 1$, which corresponds to the undamped resonance frequency of the system, as shown by inspection of Equation (9.25):

$$f'_{\text{maxdis}} = f_0 \tag{9.26}$$

Similarly the frequencies of maximum velocity and maximum acceleration for the case of hysteretic damping may be determined.

The preceding analysis shows clearly that maximum response depends on what is measured and the nature of the damping in the system under investigation. Where the nature of the damping is known, the undamped resonance frequency and the damping coefficient may be determined using appropriate equations; however, in general where damping is significant, resonance frequencies can only be determined by curve fitting frequency response data (Ewins, 2000).

Alternatively, for small damping, the various frequencies of maximum response are essentially all equal to the undamped resonance frequencies.

Referring to Figure 9.1(a) the fraction of the exciting force, F, acting on the mass, m, which is transmitted through the spring to the support, is of interest. Alternatively, referring to Figure 9.1(b), the fraction of the displacement of the base, which is transmitted to the mass, is often of greater interest. Either may be expressed in terms of the transmissibility, T_F, which in Figure 9.1(a) is the ratio of the amplitude of the force transmitted to the foundation ($|F_f| = |k_s y + \mathrm{j}\omega C y|$) to the amplitude of the exciting force, $|F|$, acting on the machine, and in Figure 9.1(b) it is the ratio of the displacement of the machine to the displacement of the foundation. Using Equations (9.3) and (9.17), we can write:

$$T_F = \frac{|F_f|}{|F|} = \frac{|k_s y + \mathrm{j}\omega C y|}{|y(k_s - m\omega^2 + \mathrm{j}C\omega)|} = \sqrt{\frac{1 + (2\zeta X)^2}{(1 - X^2)^2 + (2\zeta X)^2}} \qquad (9.27\text{a,b})$$

Figure 9.2 shows the fraction, expressed in terms of the transmissibility, T_F, of the exciting force (system (a) of Figure 9.1) transmitted from the vibrating body through the isolating spring to the support structure. The transmissibility is shown for various values of the damping ratio, ζ, as a function of the ratio of the frequency of the vibratory force to the resonance frequency of the system.

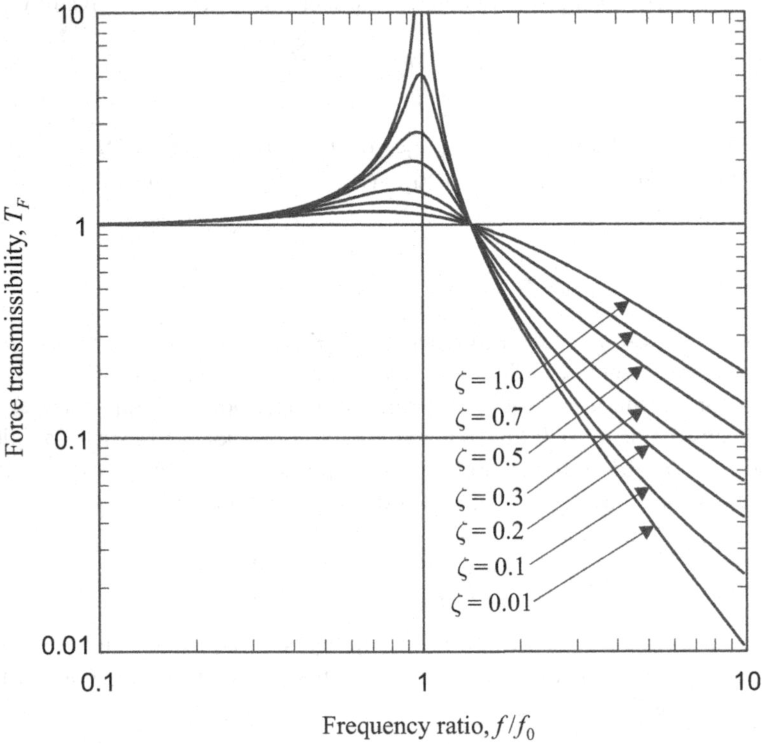

FIGURE 9.2 Force or displacement transmissibility of a viscously damped mass-spring system. The quantities, f and f_0, are the excitation and undamped mass-spring resonance frequencies, respectively, ζ is the system damping ratio and T_F is the fraction of excitation force transmitted by the spring to the foundation. Note that for values of frequency ratio greater than 1.4, the force transmissibility increases with increasing damping ratio.

When the transmissibility is identified with the *displacement* of the mass, m, of the system illustrated in Figure 9.1(b), then Figure 9.2 shows the fraction, T_D, of the exciting displacement amplitude transmitted from the base through the isolating spring to the supported mass, m. The figure allows determination of the effectiveness of the isolation system for a single-degree-of-freedom system.

The vibration amplitude of a single-degree-of-freedom system is dependent on its mass, stiffness and damping characteristics, as well as the amplitude of the exciting force. This conclusion can be extended to apply to multi-degree-of-freedom systems, such as machines and structures. Consideration of Equation (9.27) shows that as X tends to zero, the force transmissibility, T_F, tends to one; in this case, the response is controlled by the stiffness, k_s. When X is approximately one, Equation (9.27) shows that the force transmissibility is approximately inversely proportional to the damping ratio; in this case, the response is controlled by the damping, C. As X tends to large values, Equation (9.27) shows that the force transmissibility tends to zero, as it is proportional to the reciprocal of the square of X; in this case, the response is controlled by the mass, m.

The energy transmissibility, T_E, is related to the force transmissibility, T_F, and displacement transmissibility, T_D, by $T_E = T_F T_D$. As $T_F = T_D$, then $T_E = T_F^2$. The energy transmissibility, T_E, can be related to the expected increase or decrease, ΔL_W, in sound power radiated by the supported structure over that radiated when the vibrating mass is rigidly attached to the support structure as:

$$\Delta L_W = 10 \log_{10} T_E = 20 \log_{10} T_F \qquad (9.28\text{a,b})$$

Differentiation of Equation (9.27) or use of Equation (9.18) gives for the frequency of maximum force transmissibility for a viscously damped system:

$$f_F = f_0 \sqrt{1 - 2\zeta^2} \quad \text{(Hz)} \qquad (9.29)$$

The preceding equations and figures refer to viscous damping (where the damping force is proportional to the vibration velocity), as opposed to hysteretic or structural damping (where the damping force is proportional to the vibration displacement). Generally, the effects of hysteretic damping are similar to those of viscous damping up to frequencies of $f = 10 f_0$. Above this frequency, hysteretic damping results in larger transmission factors than shown in Figure 9.2.

The information contained in Figure 9.2 for the undamped case can be represented in a useful alternative way, as shown in Figure 9.3. However, it must be remembered that this figure only applies to undamped single-degree-of-freedom systems in which the exciting force acts in the direction of motion of the body.

Referring again to Figure 9.2, it can be seen that below resonance (ratio of unity on the horizontal axis) the force transmission is greater than unity and no isolation is achieved. In practice, the amplification obtained below a frequency ratio of 0.5 is rarely of significance so that, although no benefit is obtained from the isolation at these low frequencies, no significant detrimental effect is experienced either. However, in the frequency ratio range 0.5–1.4, the presence of isolators significantly increases the transmitted force and the amplitude of motion of the mounted body. In operation, this range is to be avoided. Above a frequency ratio of 1.4 the force transmitted by the isolators is less than that transmitted with no isolators, resulting in the isolation of vibration; the higher the frequency the greater the isolation. Thus, for an isolator to be successful, its stiffness must be such that the mounted resonance frequency is less than 0.7 times the minimum forcing frequency.

All practical isolators have some damping, and Figure 9.2 shows the effect of damping; increasing the damping decreases the isolation achieved. For best isolation, no damping would be desirable. On the other hand, damping is necessary for installations involving rotating equipment because the equipment rotational speed (and hence forcing frequency) will pass through the mounted resonance frequency on shutdown and start-up. In these cases, the amplitude of the transmitted force will exceed the exciting force and indeed could build up to an alarming level.

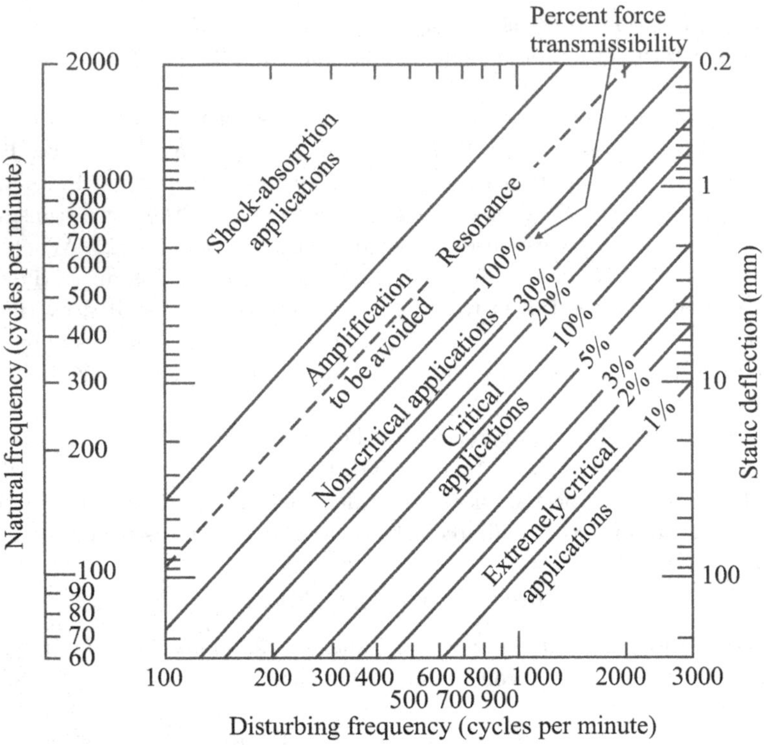

FIGURE 9.3 Force transmissibility as a function of frequency and static deflection for an undamped single-degree-of-freedom vibratory system.

Sometimes the rotational speed of the equipment can be accelerated or decelerated rapidly enough to pass through the region of resonance so quickly that the amplitude of the transmitted force does not have time to build up to the steady-state levels indicated by Figure 9.2. However, in some cases, the rotational speed can only be accelerated slowly through the resonant range, resulting in a potentially disastrous situation, unless a high level of isolator damping (up to $\zeta = 0.5$) or a vibration absorber (see Section 9.4.1) is used.

An external damper can be installed to accomplish the necessary damping, but always at the expense of reduced isolation at higher frequencies. An alternative to using highly damped isolators or vibration absorbers is to use rubber snubbers to limit excessive motion of the machine at resonance. Snubbers can also be used to limit excessive motion. These have the advantage of not limiting high-frequency isolation. Active dampers, which are only effective below a preset speed, are also used in some cases. These also have no detrimental effect on high-frequency isolation and are only effective during machine shutdown and start-up. Air dampers can also be designed so that they are only effective at low frequencies (see Section 9.3.2).

9.2.1.1 Surging in Coil Springs

Surging in coil springs is a phenomenon where high-frequency transmission occurs at frequencies corresponding to the resonance frequencies of wave motion in the coils. This limits the high-frequency performance of such springs and in practical applications, rubber inserts above or below the spring are used to minimise the effect. However, it is of interest to derive an expression for these resonance frequencies so that in isolator design, one can make sure that any machine resonance frequencies do not correspond to surge frequencies.

The analysis proceeds by deriving an expression for the effective Young's modulus for the spring, which is then used to find an expression for the longitudinal wave speed in the spring. Finally, as in Chapter 6 for rooms, the lowest order resonance is the one where the length of the spring is equal to half of a wavelength. Higher order resonances are at multiples of half a wavelength.

Young's modulus is defined as stress over strain so that for a spring of length, L, and extension, x, due to an applied force:

$$E = \frac{\sigma}{\epsilon} = \frac{k_s x/A}{x/L} = \frac{k_s L}{A} \quad \text{(Pa)} \tag{9.30a–c}$$

The longitudinal wave speed in the spring is then:

$$c_{LII} = \sqrt{\frac{E}{\rho}} = \sqrt{\frac{Lk_s/A}{m_s/LA}} = L\sqrt{\frac{k_s}{m_s}} \tag{9.31a–c}$$

The surge frequency, f_s, occurs when the spring length, L, is equal to integer multiples of $\lambda/2$, so that:

$$L = n\frac{\lambda}{2} = n\frac{c_{LII}}{2f_s} = \frac{nL}{2f_s}\sqrt{\frac{k_s}{m_s}}, \qquad n = 1,\ 2,\ 3,\ \ldots\ldots \tag{9.32a–c}$$

Rearranging gives for the surge frequencies:

$$f_s = \frac{n}{2}\sqrt{\frac{k_s}{m_s}} \quad \text{(Hz)} \tag{9.33}$$

9.2.2 Four-Isolator Systems

In most practical situations, more than one isolator is used to isolate a particular machine. This immediately introduces the problem of more than one system resonance frequency at which the force transmission will be large. If possible, it is desirable to design the isolators so that none of the resonance frequencies of the isolated system correspond to any of the forcing frequencies.

The most common example of a multi-degree-of-freedom system is a machine mounted symmetrically on four isolators (Crede, 1965). In general, a machine, or body mounted on springs, has six degrees of freedom. There will be one vertical translational mode of resonance frequency, f_0, one rotational mode about the vertical axis, and in each vertical plane, there will be a rocking mode (see Figure 9.4) and a horizontal motion mode. The rocking and horizontal motions are usually coupled, with resonance frequencies (two for each of the two vertical planes) representing both rocking and horizontal motion occurring simultaneously. The calculation of the corresponding resonance frequencies (four in total) in terms of the resonance frequency, f_0, for vertical motion, will now be considered. The resonance frequency, f_0, may be calculated using either Equation (9.5) or (9.6), as for a single-degree-of-freedom system, with one spring having the combined stiffness of the four shown in Figure 9.4. Note that stiffnesses of springs in parallel add linearly (that is, $k_s = k_{s1} + k_{s2}$, etc.), although when springs are in series, reciprocal stiffnesses add (that is, $1/k_s = 1/k_{s1} + 1/k_{s2}$, etc).

To begin, we will consider coupled horizontal and rocking motion in the $x - y$ plane and define the following parameters.

$$W = (\delta_z/b)\sqrt{(k_{sx}/k_{sy})} \tag{9.34}$$

$$M = a/\delta_z \tag{9.35}$$

$$\Omega = (\delta_z/b)(f_i/f_0) \tag{9.36}$$

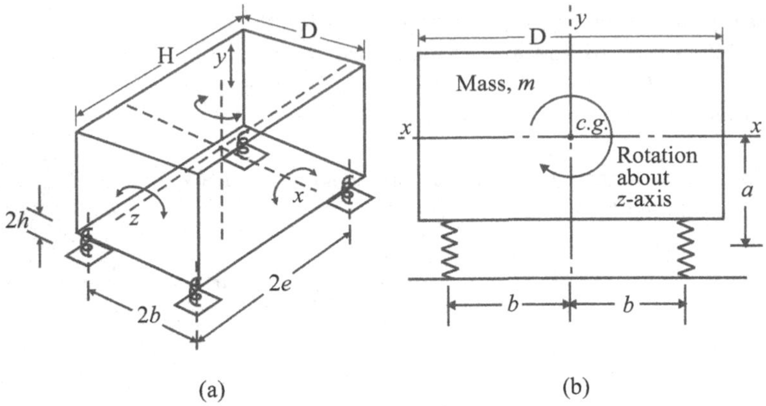

FIGURE 9.4 Vibration modes for a machine mounted on four isolators. The origin of the coordinates is coincident with the assumed centre of gravity (c.g.) at height $a+h$ above the mounting plane.

where k_{sx} and k_{sy} are the stiffnesses of each isolator in the horizontal x and vertical y-directions, respectively. The radius of gyration for rotation about the horizontal z-axis through the centre of gravity is (see Figure 9.4(b), where the dimensions, a and b are also defined):

$$\delta_z = \sqrt{(D^2 + 4(a-h)^2)/12} \tag{9.37}$$

The characteristic equation defining the coupled rocking and horizontal motion in the $x-y$ plane is (Crede, 1965, p.114):

$$\Omega^4 - \Omega^2(1 + W^2 + M^2W^2) + W^2 = 0 \tag{9.38}$$

Solutions of this equation define the two resonance frequencies, $f_i = f_a$ and $f_i = f_b$, which represent coupled rocking and horizontal motion. The solution to this characteristic equation is easily obtained by writing it as a quadratic in terms of the variable, Ω^2. Thus, the solution is:

$$\Omega^2 = 0.5\left[W^2(1+M^2) + 1 \pm \sqrt{[W^2(1+M^2)+1]^2 - 4W^2}\right] \tag{9.39}$$

The resonance frequencies, $f_i = f_a$ and $f_i = f_b$, can be obtained from Equation (9.36), using the two solutions for Ω^2 obtained from Equation (9.39).

For motion in the vertical $y-z$ plane, Equations (9.34) to (9.36) are modified by replacing b with e (see Figure 9.4), k_{sx} with k_{sz} and δ_z with δ_x, which is the radius of gyration for rotation about the x-axis (with D replaced by H in Equation (9.37)).

The resonance frequency of the rotational vibration mode about the vertical y-axis is:

$$f_y = \frac{1}{\pi}\sqrt{(b^2 k_{sz} + e^2 k_{sx})/I_y} \quad \text{(Hz)} \tag{9.40}$$

The quantities, $2b$ and $2e$, are the distances between centrelines of the support springs, k_{sz} is the isolator stiffness in the z-direction, usually equal to k_{sx}, and I_y is the moment of inertia of the body about the y-axis. For a rectangular parallelepiped of uniform density and total mass, m, shown in Figure 9.4, $I_y = m(H^2 + D^2)/12$.

Values for the stiffnesses, k_{sx}, k_{sy} and k_{sz}, are usually available from the isolator manufacturer. Note that for rubber products, static and dynamic stiffnesses are often different. It is the dynamic stiffnesses that are required here, and these are obtained by exciting the supported mass with a shaker and measuring the input force amplitude and spring deflection amplitude as a function of excitation frequency, with the base of the spring mounted on a rigid support.

When placing vibration isolators beneath a machine, it is good practice to use identical isolators and to place them symmetrically with respect to the centre of gravity of the machine. This results in equal loading and deflection of the isolators and is essential if the preceding analysis is to be used to estimate resonance frequencies.

The calculation of the force transmission for a multi-degree-of-freedom system is complex and not usually contemplated in conventional isolator design. However, the analysis of various multi-degree-of-freedom systems has been discussed in the literature (Mustin, 1968; Smollen, 1966). Generally, for a multi-degree-of-freedom system, good isolation is achieved if the frequencies of all the resonant modes are less than about two-fifths of the frequency of the exciting force. However, a force or torque may not excite all the normal modes, and then the natural frequencies (undamped resonance frequencies) of the modes that are not excited do not need to be considered, except to ensure that they do not actually coincide with the forcing frequency.

9.2.3 Two-Stage Vibration Isolation

Two-stage vibration isolation is used when the performance of single stage isolation is inadequate, and it is not practical to use a single stage system with a lower resonance frequency. As an example, two-stage isolation systems have been used to isolate diesel engines from the hull of large submarines.

A two-stage isolator is illustrated in Figure 9.5, where the machine to be isolated is represented as mass, m_2, and the intermediate mass is represented as mass, m_1. The intermediate mass should be as large as possible, but should be at least 70% of the machine mass being supported.

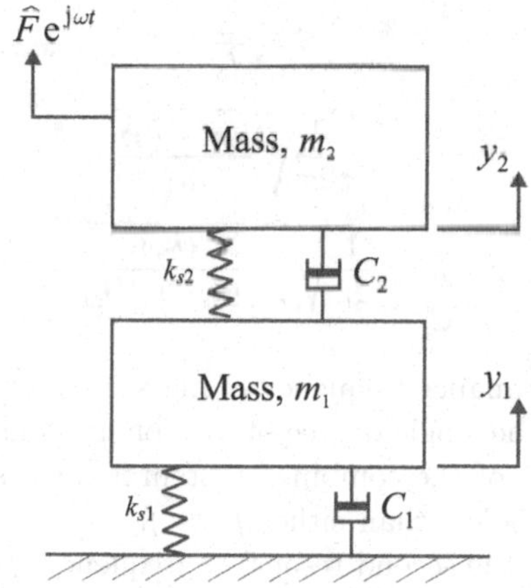

FIGURE 9.5 Two-stage vibration isolator.

The equations of motion of the masses, m_1 and m_2, in Figure 9.5 may be written as:

$$m_1\ddot{y}_1 + C_1\dot{y}_1 + k_{s1}y_1 - C_2(\dot{y}_2 - \dot{y}_1) - k_{s2}(y_2 - y_1) = 0 \tag{9.41}$$

and

$$m_2\ddot{y}_2 + C_2(\dot{y}_2 - \dot{y}_1) + k_{s2}(y_2 - y_1) = F = \hat{F}e^{j\omega t} \tag{9.42a,b}$$

These equations can be solved to give the complex displacements of each mass as:

$$\frac{y_1}{F} = \frac{k_{s2} + j\omega C_2}{(k_{s1} + k_{s2} - \omega^2 m_1 + j\omega C_1 + j\omega C_2)(k_{s2} - \omega^2 m_2 + j\omega C_2) - (k_{s2} + j\omega C_2)^2} \tag{9.43}$$

and

$$\frac{y_2}{F} = \frac{k_{s1} + k_{s2} - \omega^2 m_1 + \mathrm{j}\omega C_1 + \mathrm{j}\omega C_2}{(k_{s1} + k_{s2} - \omega^2 m_1 + \mathrm{j}\omega C_1 + \mathrm{j}\omega C_2)(k_{s2} - \omega^2 m_2 + \mathrm{j}\omega C_2) - (k_{s2} + \mathrm{j}\omega C_2)^2} \tag{9.44}$$

The complex force transmitted to the foundation is:

$$F_T = y_1(k_{s1} + \mathrm{j}\omega C_1) \tag{9.45}$$

and thus the transmissibility, $T_F = |F_T/F|$, is given by:

$$T_F = \left| \frac{(k_{s1} + \mathrm{j}\omega C_1)(k_{s2} + \mathrm{j}\omega C_2)}{(k_{s1} + k_{s2} - \omega^2 m_1 + \mathrm{j}\omega C_1 + \mathrm{j}\omega C_2)(k_{s2} - \omega^2 m_2 + \mathrm{j}\omega C_2) - (k_{s2} + \mathrm{j}\omega C_2)^2} \right| \tag{9.46}$$

The damping coefficients, C_1 and C_2, are found by multiplying the damping ratios, ζ_1 and ζ_2, by the critical damping, C_{c1} and C_{c2}, given by Equation (9.15), using stiffnesses, k_{s1} and k_{s2}, and masses, m_1 and m_2, respectively.

As a two-stage isolation system has two degrees of freedom, it will have two resonance frequencies corresponding to high force transmissibility. The undamped resonance frequencies of the two-stage isolator may be calculated using (Muster and Plunkett, 1988; Ungar and Zapfe, 2006b):

$$\left(\frac{f_a}{f_0}\right)^2 = Q - \sqrt{Q^2 - B^2} \quad \text{and} \quad \left(\frac{f_b}{f_0}\right)^2 = Q + \sqrt{Q^2 - B^2} \tag{9.47a,b}$$

where

$$Q = 0.5\left(B^2 + 1 + \frac{k_{s1}}{k_{s2}}\right) \tag{9.48}$$

and

$$B = \frac{f_1}{f_0} \tag{9.49}$$

$$f_1 = \frac{1}{2\pi}\sqrt{\frac{k_{s1} + k_{s2}}{m_1}} \tag{9.50}$$

$$f_0 = \frac{1}{2\pi}\sqrt{\frac{k_{s1}k_{s2}}{m_2(k_{s1} + k_{s2})}} \tag{9.51}$$

The quantity, f_1, is the resonance frequency of mass, m_1, with mass, m_2, held fixed and f_0 is the resonance frequency of the single-degree-of-freedom system with mass, m_1, removed. The upper resonance frequency, f_b, of the combined system is always greater than either f_1 or f_0, while the lower frequency, f_a, is less than either f_1 or f_0.

At frequencies above twice the second resonance frequency, f_b, the force transmissibility for an undamped system will be approximately equal to $[f^2/(f_1 f_0)]^2$, which is proportional to the fourth power of the excitation frequency, compared to a single-stage isolator, for which it is approximately equal to $(f/f_0)^2$ at frequencies above twice the resonance frequency, f_0.

In Figure 9.6, the force transmissibility for a two-stage isolator for a range of ratios of masses and stiffnesses is plotted for the special case where $\zeta_1 = \zeta_2$.

9.2.4 Practical Considerations for Isolators

The analysis discussed thus far gives satisfactory results for force transmission at relatively low frequencies, if account is taken of the three-dimensional nature of the machine and the fact that several mounts are used. For large machines or structures this frequency range is generally infrasonic, where the concern is for prevention of physical damage or fatigue failure.

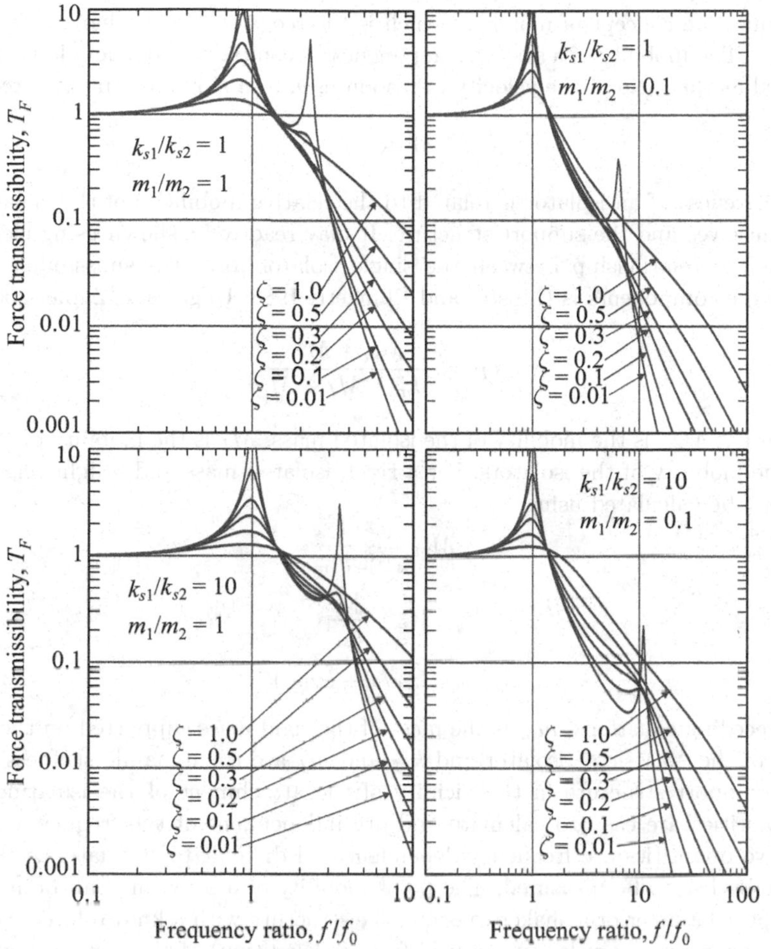

FIGURE 9.6 Force transmissibility for a two-stage vibration isolation system for various values of stiffness and mass ratio. In all figures, $\zeta_1 = \zeta_2 = \zeta$. The higher lines on the graphs to the left of the abscissa location, $f/f_0 = 1.4$, correspond to the lower lines to the right of $f/f_0 = 1.4$, with the crossover point varying between 1.4 and 2.0, depending on the graph.

Unfortunately, the analysis cannot be directly extrapolated into the audio-frequency range, where it is apt to predict attenuations very much higher than those achieved in practice. This is because the assumptions of a rigid machine and a rigid foundation are generally not true. In practice, almost any foundation and almost any machine will have resonances in the audio-frequency range.

Results of both analytical and experimental studies of the high-frequency performance of vibration isolators have been published (Ungar and Dietrich, 1966; Snowdon, 1965). This work shows that the effect of appreciable isolator mass and damping is to significantly increase, over simple classical theory predictions, the transmission of high-frequency forces or displacements. The effects begin to occur at forcing frequencies as low as 10 to 30 times the natural frequency (or undamped resonance frequency) of the mounted mass.

To minimise these effects, the ratio of isolated mass to isolator mass should be as large as possible (1000 : 1 is desirable) and the damping in the isolated structure should be large. The effect of damping in the isolators is not as important, but nevertheless the isolator damping should be minimised.

To continue, the concept of mobility, which is the reciprocal of mechanical impedance, will be introduced. The mobility of a system component is a complex, frequency-dependent quantity, and is defined as the ratio of the velocity of response, v, to a sinusoidal input force, F:

$$M = v/F \tag{9.52}$$

The effectiveness of an isolator is related to the relative mobilities of the isolated mass, the isolators themselves and the support structure. It may readily be shown using electrical circuit analysis that the relationship between the single isolator force transmissibility, T_F, and the mobilities of the components is (Muster and Plunkett, 1988; Ungar and Zapfe, 2006b):

$$T_F = \frac{M_m + M_f}{M_m + M_f + M_i} \tag{9.53}$$

The quantity, M_m, is the mobility of the isolated mass, M_f is the mobility of the foundation and M_i is the mobility of the isolators. For a rigid, isolated mass and a lightweight spring, the mobilities may be calculated using:

$$M_m = \frac{1}{\mathrm{j}\omega m_m} \tag{9.54}$$

$$M_i = \frac{\mathrm{j}\omega}{k_{si}} \tag{9.55}$$

$$M_f = \mathrm{j}(k_{sf}/\omega - \omega m_f)^{-1} \tag{9.56}$$

In the preceding equations, m_m is the mass of the rigid mass supported on the spring, k_{si} is the stiffness of the "massless" isolator and k_{sf} and m_f are the dynamic stiffness and dynamic mass of the support structure in the vicinity of the attachment of the isolating spring. The first two quantities are easy to calculate and are independent of the frequency of excitation. The latter two quantities are frequency-dependent and difficult to estimate, so the foundation mobility usually has to be measured. The point mobility of a structure can be measured using either an impact hammer or a shaker to excite the structure with a known force. For the impact hammer measurement, the velocity (derived from acceleration) of the structure at the point of impact can be measured with an piezoelectric accelerometer attached to the back of the hammer or embedded in it and the force applied to the structure can be measured by a piezo-electric force transducer located on the end of the hammer that strikes the structure. Alternatively, the mobility can be measured using a shaker attached to the structure and excited by a sine sweep signal that covers the frequency range of interest. The force input and acceleration response at the point of force input can be measured using an impedance head between the shaker and the structure. The impedance head contains two piezoelectric crystals, one for measuring the force input and one for measuring the acceleration, which can be converted to velocity. The frequency spectrum of the mobility, which is the ratio of structural velocity to applied force at a particular point on the structure, is obtained by dividing each line on the frequency response spectrum of acceleration versus force, by the frequency, ω radians/s, corresponding to that line. More details of this measurement procedure are provided in Hansen 2018, pp. 243–247.

Equation (9.53) shows that an isolator is ineffective unless its mobility is large when compared with the sum of the mobilities of the machine and foundation. The mobility of a simple structure may be calculated, and that of any structure may be measured (Plunkett, 1954, 1958). Some measured values of mobility for various structures have been published in the literature (Harris and Crede, 1976; Peterson and Plunt, 1982).

Attenuation of more than 20 dB ($T_F < 0.1$) is rare at acoustic frequencies with isolation mounts of reasonable stiffness, and no attenuation at all is common. For this reason, very soft mounts ($f_0 = 5$ to 6 Hz) are generally used where possible. As suggested by Equation (9.53), if a mount is effective at all, a softer mount (M_i larger) will be even more effective.

For a two-stage isolator, Equation (9.53) may be written as (Ungar and Zapfe, 2006b; Muster and Plunkett, 1988):

$$\frac{1}{T_F} = \frac{M_{m2} + M_f + M_i}{M_{m2} + M_f} + \frac{(M_{i2} + M_{m2})(M_{i1} + M_f)}{M_{m1}(M_{m2} + M_f)} \tag{9.57}$$

In Equation (9.57), the first term corresponds to a single isolator system, where the isolator mobility, $M_i = M_{i1} + M_{i2}$, is the sum of the mobilities of the two partial isolators in the two-stage system (as they are acting in series). The subscript, m_2, corresponds to the machine being isolated, the subscript, m_1, corresponds to the intermediate mass, the subscript, $i1$, corresponds to the isolator between the intermediate mass and the support structure and the subscript, $i2$, corresponds to the isolator between the intermediate mass and the machine being isolated.

Note that the second term in Equation (9.57) represents the improvement in performance as a result of using a two-stage isolator and that this improvement is inversely proportional to the mobility of the intermediate mass and thus directly proportional to its mass.

Once the total mobility, M_i, of the isolators has been selected, the optimum distribution of the mobility between isolators 1 and 2 may be calculated using:

$$M_{i1} = r_i M_i \quad \text{and} \quad M_{i2} = (1 - r_i)M_i \tag{9.58a,b}$$

and

$$\text{optimum } r_i = 0.5\,[1 + (M_f - M_{m2})/M_i] \tag{9.59}$$

9.2.5 Moving a Machine to a Different Location on a Floor

Sometimes a machine in a production facility has to be moved, and it is useful to be able to estimate the resulting floor vibration levels under the machine at the new location, especially if the machine is sensitive to floor vibration. If there are several possible choices for the new location, being able to estimate the resulting floor vibration levels at each location enables optimal location of the machine in terms of minimising floor vibration levels. The following analysis shows how to determine the expected vibration velocity levels at any possible new location, based on point impedance (reciprocal of mobility) measurements at the existing and new locations and vibration velocity measurements at the existing location when the machine is operating (Ungar, 2021). To a first approximation, the relationship between force and velocity on the machine at its mounting point at any particular frequency can be represented by a straight line with a negative slope, as shown in Figure 9.7.

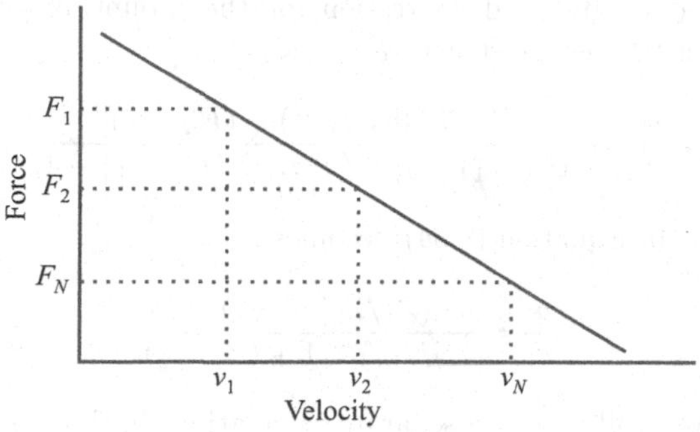

FIGURE 9.7 Relation between force and velocity at a machine/structure interface.

The slope of this straight line and its intercept on the y-axis will depend on the frequency of excitation and can be determined by measuring the force and velocity on the machine at the existing structural attachment point for two different stiffness conditions of the machine attachment to the structure. One condition could be with the machine operating and attached to the structure in the normal way via a mounting foot. The velocity, v_1, at the point where the mounting foot is attached to the structure can be measured. The force, F_1, is calculated using $F_1 = v_1 Z_1$, where Z_1 is the point impedance of the floor at the mount location with the mount removed and the machine not in contact with the floor. The point impedance,Z_1, is the reciprocal of the point mobility, measured using an impedance head and shaker or an impact hammer, as described in Section 9.2.4. The second condition could be the force and velocity (F_N and v_N) on the machine with a flexible mount (of known point impedance, Z_N), placed between the same machine mount and support structure as used to obtain v_1. If the mount is sufficiently soft, its point impedance, Z_N will not be affected by the impedance of the support structure. If it is possible to to completely disconnect the machine attachment point from the structure and operate the machine, then $F_N = Z_N = 0$ and v_N can be measured using an accelerometer on the machine where it would normally be attached to the structure.

The measured values, F_1, F_N, v_1 and v_N can then be used to find the slope and y-axis intercept of the line on the graph representing the relationship between the machine mounting foot's force and velocity (see Figure 9.7). This relationship can then be used together with the point impedance data, Z_2 for the structure at the proposed new attachment point to calculate the expected velocity at that point.

Using the data points corresponding to F_N and F_1, the equation of the line which relates the mounting foot's force to its velocity, v is:

$$F = \frac{F_1 - F_N}{v_1 - v_N} u + F_1 - v_1 \frac{F_1 - F_N}{v_1 - v_N} \tag{9.60}$$

From which we obtain, for a structural attachment point denoted "2", so that $F = F_2$ and $v = v_2$:

$$\frac{F_2}{F_1} = 1 + \left(\frac{v_2 - v_1}{v_1 - v_N}\right)\left(1 - \frac{F_N}{F_1}\right) \tag{9.61}$$

As $v = F/Z$, we can also write:

$$\frac{v_2}{v_1} = \frac{F_2 Z_1}{F_1 Z_2} \tag{9.62}$$

which can be used together with Equation (9.61) to obtain:

$$\frac{v_2}{v_1} = \left(\frac{Z_1}{Z_2}\right)\left[1 + \left(\frac{v_2 - v_1}{v_1 - v_N}\right)\left(1 - \frac{F_N}{F_1}\right)\right] \tag{9.63}$$

which can be rearranged to give and expression for the mounting point velocity, v_2, at the proposed mounting point "2" on the structure:

$$\frac{v_2}{v_1} = \frac{(Z_1/Z_2)[(v_N/v_1) - (F_N/F_1)]}{[(v_N/v_1) - 1] - (Z_1/Z_2)[(F_N/F_1) - 1]} \tag{9.64}$$

For the case where $F_N = 0$, Equation (9.64) becomes:

$$\frac{v_2}{v_1} = \frac{(Z_1/Z_2)/(v_N/v_1)}{(v_N/v_1) - 1 + (Z_1/Z_2)} \tag{9.65}$$

Each frequency considered will require a separate calculation. In this case, the velocity, v_f, of the structure at frequency, f Hz is calculated from the spectral line, a_f, in the acceleration spectrum using $v_f = a_f/(2\pi f)$.

9.2.5.1 Effect of Stiffness of Equipment Mounted on Isolators

If equipment is mounted on a non-rigid frame, which in turn is mounted on isolators, the mounted natural frequency (or undamped resonance frequency) of the assembly will be reduced as shown in Figure 9.8. In this case, the mobility, M_m, of the isolated mass is large because of the non-rigid frame. According to Equation (9.53), the effectiveness of a large value of isolator mobility M_i in reducing the force transmissibility is thus reduced. Clearly, a rigid frame is desirable.

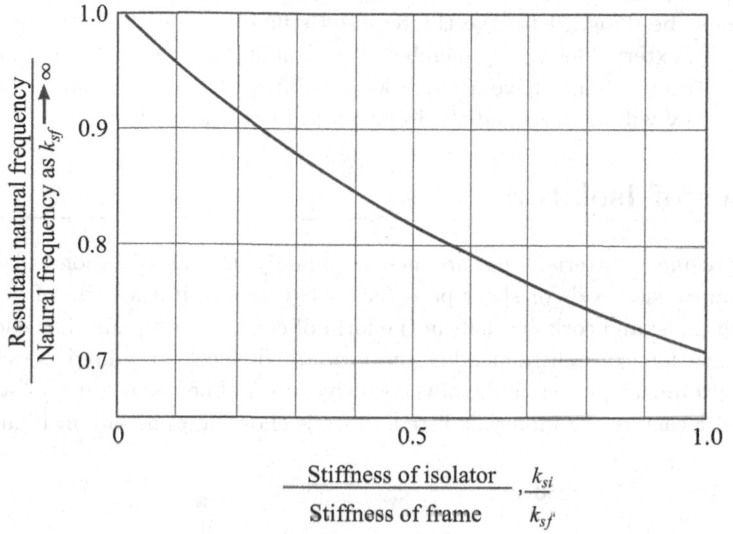

FIGURE 9.8 Natural frequency of equipment having a flexible frame and supported by an isolator, expressed as a fraction of the natural frequency (or undamped resonance frequency) obtained when the frame is infinitely rigid.

9.2.5.2 Effect of Stiffness of Foundations

Excessive flexibility of the foundation is of significance when an oscillatory force generator is to be mounted on it. The force generator could be a fan, or an air conditioning unit, and the foundation could be the roof slab of a building. As a general requirement, if it is desired to isolate equipment from its support structure, the mobility, M_i, of the equipment mounts must be large relative to the foundation mobility, M_f, according to Equation (9.53). A useful criterion is that the mounted resonance frequency should be much lower than the lowest resonance frequency of the support structure.

The equipment rotational frequency must be chosen so that it or its harmonics do not coincide with the resonance frequencies, which correspond to large values of the foundation (or support structure) mobility. If the support structure is flexible, the force generator should be placed on as stiff an area as possible, or supported on stiff beams, which can transfer the force to a stiff part of the foundation. If the vibration mountings cannot be made sufficiently soft, their stiffnesses should be chosen so that the mounted resonance frequency does not coincide with support structure resonances, but lies in a frequency range in which the support structure has a small mobility (see Equation (9.56)).

When the support structure is non-rigid, substantially lower than normal mounting stiffnesses are required. As an example, for a machine speed of 1500 rpm, one manufacturer recommends the following static deflections of the isolators for 95% vibration force isolation (see Equations (9.6) and (9.27)):

Installed in basement: 8.6 mm
On a 10 m floor span: 9.9 mm
On a 12 m floor span: 10.7 mm
On a 15 m floor span: 11.2 mm

9.2.5.3 Superimposed Loads on Isolators

If an external force (such as tension in a drive belt) is applied to a machine mounted on isolators, the isolators must be designed to give the required stiffness under the combined action of the machine mass and external load. Any members transmitting an external force from the machine to the support structure must have a much lower stiffness than the isolators, or the mounted resonance frequency will increase and the isolators will become ineffective.

9.3 Types of Isolators

There are four resilient materials that are most commonly used as vibration isolators: rubber, in the form of compression pads or shear pads (or cones); metal, in the form of various shapes of springs or mesh pads; and cork and felt, in the form of compression pads. The choice of material for a given application is usually dependent on the static deflection required as well as the type of anticipated environment (for example, oily, corrosive, etc.). The usual range of static deflections in general use for each of the materials listed above is shown graphically in Figure 9.9.

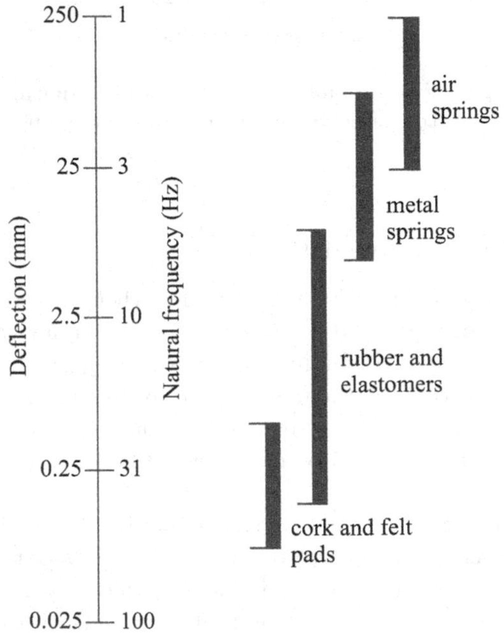

FIGURE 9.9 Ranges of application of different types of isolator.

9.3.1 Rubber

Isolators come in a variety of forms that use rubber in shear or compression, but rarely in tension, due to the short fatigue life experienced by rubber in tension. Isolator manufacturers normally provide the stiffness and damping characteristics of their products. As the dynamic stiffness of rubber is generally greater (by 1.3 to 1.8) than the static stiffness, dynamic data should be obtained whenever possible. Rubber can be used in compression or shear, but the latter use results in greater service life.

The amount of damping can be regulated by the rubber constituents, but the maximum energy that can be dissipated by damping tends to be limited by heat build-up in the rubber, which causes deterioration. Damping in rubber is usually vibration amplitude, frequency and temperature dependent.

Rubber in the form of compression pads is generally used for the support of large loads and for higher frequency applications (above 10 Hz resonance frequency of the load on the isolator system). The stiffness of a compressed rubber pad is generally dependent on its size, and the end restraints against lateral bulging. Pads with raised ribs are usually used, resulting in a combination of shear and compression distortion of the rubber, and a static deflection virtually independent of pad size. However, the maximum loading on pads of this type is generally less than 550 kPa.

The most common use for rubber mounts is for the isolation of medium to lightweight machinery, where the rubber in the mounts acts in shear. Resonance frequencies of machinery on these mounts vary from about 5 Hz upwards, corresponding to mid-frequency range isolation.

Unfortunately natural rubber exhibits poor resistance to sunlight, solvents and oils. Instead, the most commonly used isolators are made from synthetic rubber, such as neoprene, nitriles, acrylates and others, which are oil resistant.

9.3.2 Metal Springs

Next to rubber, metal springs are the most commonly used materials in the construction of vibration isolators. The load-carrying capacity of spring isolators is variable, from the lightest of instruments to the heaviest of buildings. Springs can be produced industrially in large quantities, with only small variations in their individual characteristics. They can be used for low-frequency isolation (resonance frequencies from 1.3 Hz upwards), as it is possible to have large static deflections by suitable choice of material and dimensions.

Metal springs can be designed to provide isolation at virtually any frequency. However, when designed for low-frequency isolation, they have the practical disadvantage of readily transmitting high frequencies. Higher-frequency transmission can be minimised by inserting rubber or felt pads between the ends of the spring and the mounting points, and ensuring that there is no metal-to-metal contact between the spring and the support structure.

Coil springs must be designed carefully to avoid lateral instability. For stable operation, the required ratio of unloaded spring length ℓ_0 to diameter D_0 for a given spring compression ratio, ξ (ratio of change in length when loaded to length unloaded), is shown in Figure 9.10.

Metal springs have little useful internal damping. However, this can be introduced in the form of viscous fluid damping, friction damping or, more cunningly, by viscous air damping. As an example of an air damper, at low frequencies in the region of the mounted resonance, air is pumped in and out of a dashpot by the motion of the spring, hence generating a damping force, but at higher frequencies the air movement and damping force are much reduced and the dashpot becomes an air spring in parallel with the steel spring. This configuration results in good damping at the mounted resonance frequency, and reduced damping at frequencies above resonance, thus giving better overall effectiveness. The accompanying increase in stiffness at higher frequencies normally has little effect on the isolation achieved.

Steel spring types include coil, torsion, coiled wire rope and mesh springs, as well as cantilever and beam springs of single or multi-leaf construction. Coiled wire rope springs are very nonlinear and must be carefully designed and tested, using similar vibration levels to those that will be experienced in practice, if they are to provide adequate isolation. In multi-leaf construction, interface friction between the leaves can provide friction damping, thus reducing higher-frequency

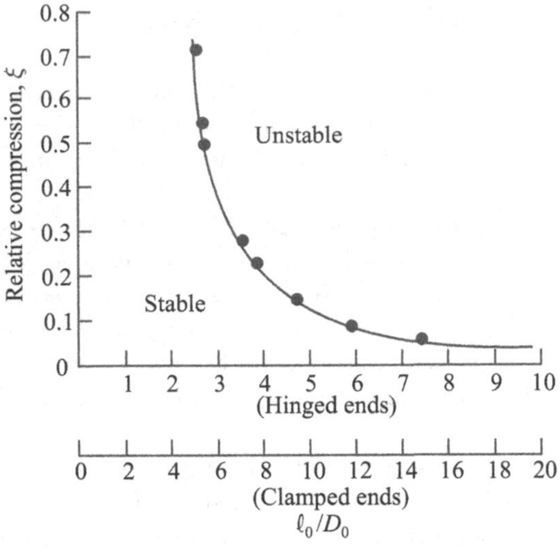

FIGURE 9.10 Stable and unstable values of relative compression for coil springs.

transmission. By putting a suitable lubricant between the leaves, viscous or near viscous damping characteristics can be obtained.

Wire mesh springs consist of a pre-compressed block of wire mesh, which acts as a combined nonlinear spring and damper. Sometimes wire mesh springs are used in conjunction with a steel spring to carry part of the load. Damping is provided by friction within the mesh and between the mesh and the steel spring. This results in damping of the high-frequency vibration transmitted by the coil spring.

9.3.3 Cork

Cork is one of the oldest materials used for vibration isolation. It is generally used in compression and sometimes in a combination of compression and shear. The dynamic stiffness and damping of cork are very much dependent on frequency. Also, the stiffness decreases with increasing load. Generally, the machine or structure to be isolated is mounted on large concrete blocks, which are separated from the surrounding foundation by several layers of cork slabs, 2 to 15 cm thick. For optimum performance, the cork should be loaded to between 50 and 150 kPa. Increasing the cork thickness will lower the frequency above which isolation will be effective. However, large thicknesses, with associated stability problems, are required to achieve isolation at low frequencies. Although oil, water and moderate temperature have little effect on its operating characteristics, cork does tend to compress with age under an applied load. At room temperature its effective life extends to decades; at 90°C it is reduced to less than a year.

9.3.4 Felt

To optimise the vibration isolation effectiveness of felt, the smallest possible area of the softest felt should be used, but in such a way that there is no loss of structural stability or excessive compression under static loading conditions. The felt thickness should be as great as possible. For general purposes, felt mountings of 1 to 2.5 cm thick are recommended, with an area of 5% of the total area of the machine base. In installations where vibration is not excessive, it is not necessary to bond the felt to the machine. Felt has high internal damping ($\zeta \approx 0.13$), which is almost independent of load; thus, it is particularly suitable for reducing vibration at the mounted machine resonance frequency. In most cases, felt is an effective vibration isolator only at frequencies above 40 Hz. Felt is particularly useful in reducing vibration transmission in the audio-frequency range, as its mechanical impedance is poorly matched with most engineering materials. Curves showing the resonance frequency of different grades of felt as a function of static load are illustrated in Figure 9.11.

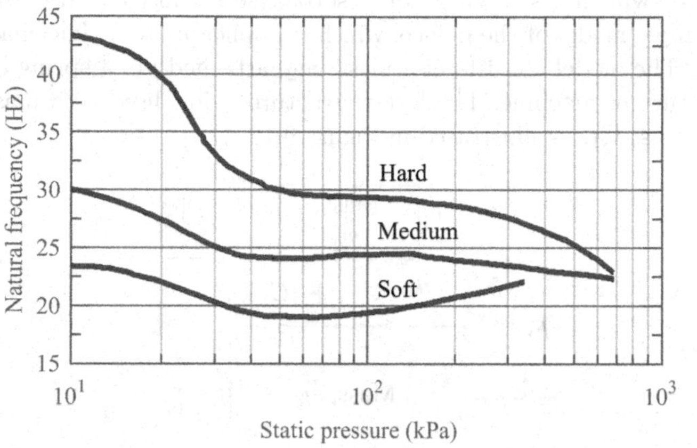

FIGURE 9.11 Natural frequency of 25 mm-thick felt as a function of static load expressed in units of pressure (after Tyzzer and Hardy (1947)).

9.3.5 Air Springs

Although air springs can be used at very low frequencies, they become increasingly difficult and expensive to manufacture as the required resonance frequency is decreased; 0.7 Hz seems to be a practical lower limit and one that is achieved with difficulty. However, resonance frequencies of 1 Hz are relatively common.

Air springs consist of an enclosed volume of air, which is compressed behind a piston or diaphragm. Diaphragms are generally preferred to avoid the friction problems associated with pistons. The static stiffness of air springs is usually less than the dynamic stiffness, as a result of the thermodynamic properties of air. Machine height variations due to air volume changes, which are caused by ambient temperature changes, can be maintained by adding or removing air using a servo-controller. One simple example of an air spring, which is very effective and capable of supporting high loads in the hundreds or thousands of kg range, is the inner tube from a car tyre, supported in a cutaway tire casing.

9.4 Vibration Absorbers, Tuned Mass Dampers and Vibration Neutralisers

A vibration absorber, also known as a tuned mass damper is intended to minimise the vibration of a machine or structure at a resonance frequency, whereas a vibration neutraliser is intended to reduce structural or machine vibrations at a particular excitation (or forcing) frequency.

9.4.1 Vibration Absorbers

One application of a vibration absorber is to reduce the vibrations of a rigid machine mounted on vibration isolators as it passes through a particular speed on startup. This speed corresponds to the resonance frequency of the machine/isolator combination and could result in vibration levels that are sufficiently large to damage the machine and its bearings or damage the vibration isolation system. Other applications of vibration absorbers include the reduction of the swaying of high-rise buildings during earthquakes and high winds, the reduction of bridge vibrations, the reduction of power line vibrations or the vibrations of any other structures at one or more of their troublesome resonance frequencies.

In this section, we will analyse a vibration absorber (sometimes called a tuned mass damper) by considering a simple model of the device, which is applicable to a rigid machine mounted on vibration isolators. The model consists of a mass, m_2, attached via a spring of stiffness, k_{s2}, to the vibrating structure or machine. The latter structure is idealised as a mass, m_1, suspended via a spring of stiffness, k_{s1}, as illustrated in Figure 9.12.

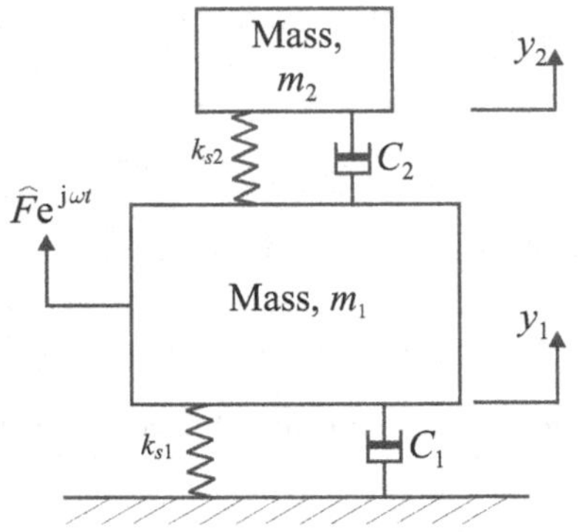

FIGURE 9.12 Vibration absorber system.

Although Figure 9.12 appears similar to Figure 9.5, as they both show two masses above a rigid foundation, with two mass-spring-damper sets to reduce the vibration transmission to the foundation, they are in reality quite different. In Figure 9.5, the top mass m_2, represents a vibrating machine whose vibrations are to be isolated from the foundation. The degree of isolation is enhanced by the presence of an intermediate mass, m_2 and the additional mass-spring-damper system and the intention is to provide better isolation at frequencies above the system resonance with no intermediate mass. On the other hand, in Figure 9.12, the bottom mass, m_1, represents the machine and the purpose of the top mass, m_2, is to reduce the vibration level of the machine at the resonance frequency associated with the machine mass, m_1, and its isolation system. The excitation force, $\hat{F}e^{j\omega t}$, is shown in the figures as acting on the mass that represents the machine, which is in a different location in Figure 9.5 than in Figure 9.12.

The two equations to follow, which describe the motion of the two-degree-of-freedom system shown in Figure 9.12, excited at frequency, ω (radians/sec), are similar to Equations (9.41) and (9.42) for a two-stage isolator, except for differences resulting from the different location of the exciting force.

$$m_1\ddot{y}_1 + C_1\dot{y}_1 + k_{s1}y_1 - C_2(\dot{y}_2 - \dot{y}_1) - k_{s2}(y_2 - y_1) = F = \hat{F}e^{j\omega t} \tag{9.66a,b}$$

and

$$m_2\ddot{y}_2 + C_2(\dot{y}_2 - \dot{y}_1) + k_{s2}(y_2 - y_1) = 0 \tag{9.67}$$

The steady-state solutions for the motion of the two masses are (Soom and Lee, 1983):

$$y_1 = |y_1|\cos(\Omega t + \theta_1) \tag{9.68}$$

and

$$y_2 = |y_2|\cos(\Omega t + \theta_2) \tag{9.69}$$

where θ_1 and θ_2 are phase angles of the motion of the masses relative to the excitation force. The amplitudes, $|y_1|$ and $|y_2|$, are given by Den Hartog (1956, p. 96) and Kelly (1993, p. 583) for a system with no damping on the primary mass ($C_1 = 0$) as:

$$\frac{|y_1|}{|F|} = \frac{1}{k_{s1}}\sqrt{\frac{\left(\frac{2\zeta_2\Omega f_2}{f_1}\right)^2 + \left(\Omega^2 - \left(\frac{f_2}{f_1}\right)^2\right)^2}{\left(\frac{2\zeta_2\Omega f_2}{f_1}\right)^2\left(\Omega^2 - 1 + \frac{m_2\Omega^2}{m_1}\right)^2 + \left[\frac{k_{s2}}{k_{s1}}\Omega^2 - (\Omega^2 - 1)\left(\Omega^2 - \left(\frac{f_2}{f_1}\right)^2\right)\right]^2}} \tag{9.70}$$

where f_1, is the resonance frequency of the mass, m_1, with no absorber, f_2 is the resonance frequency of the absorber mass if mass m_1 is replaced with a rigid foundation, k_{s1}, m_1, k_{s2} and m_2 are defined in Figure 9.12 and ζ_2 is the critical damping ratio of the absorber mass, m_2 and its suspension, defined as:

$$\zeta_2 = \frac{C_2}{C_{c2}} = \frac{C_2}{2\sqrt{k_{s2}m_2}} \tag{9.71a,b}$$

The resonance frequencies of each of the two masses without the other are given by:

$$2\pi f_j = \sqrt{k_{sj}/m_i}\,; \quad j = 1,\ 2 \tag{9.72}$$

The forcing frequency ratio, Ω, is:

$$\Omega = \omega\sqrt{m_1/k_{s1}} = f/f_1 \tag{9.73a,b}$$

where ω is the excitation frequency (radians/sec).

For a vibration absorber, the optimum resonance frequency, f_2, of the added mass-spring system to ensure maximum vibration reduction at the troublesome resonance frequency of the main mass, while minimising the increase in vibration levels at frequencies above and below the frequency to be controlled is (Den Hartog, 1956, p. 100):

$$f_2 = \frac{f_1}{1 + m_2/m_1} \tag{9.74}$$

The frequency, f_2 is always slightly less than f_1, as this ensures that the twin peaks seen on the curves in Figure 9.13 are at the same level. This also causes the mass to vibrate almost out of phase with the structure, resulting in an inertial force that opposes the excitation force.

The form of Equation (9.70) has been changed slightly from that provided in Den Hartog (1956, p. 96) to allow for the different (and unusual) definition of the critical damping used in that book, which is ($C_c = 2m_2\sqrt{k_1/m_1}$). This is different to the more commonly used definition, which is used in this book and provided in Equation (9.71b).

The two undamped resonance frequencies, f_a and f_b, which result from the combination of absorber and machine, may be determined using (de Silva, 2005, p. 49):

$$f_a^2,\ f_b^2 = \frac{1}{2}\left[\left(\frac{f_2}{f_1}\right)^2\left(\frac{m_1 + m_2}{m_1}\right) + 1\right] \mp \frac{1}{2}\sqrt{\left[\left(\frac{f_2}{f_1}\right)^2\left(\frac{m_1 + m_2}{m_1}\right) + 1\right]^2 - \left(\frac{2f_2}{f_1}\right)^2} \tag{9.75}$$

As illustrated in Figure 9.13, the larger the mass ratio, m_2/m_1, the greater will be the frequency separation of the natural frequencies, f_a and f_b, of the system with the absorber from the natural frequency, f_1, of the system without the absorber. The displacement amplitude of mass, m_2, is proportional to m_1/m_2. Thus, m_2 should be as large as possible to minimise its displacement.

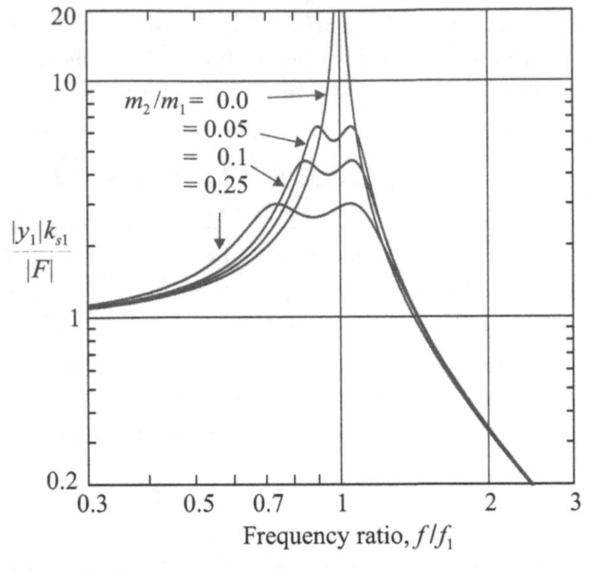

FIGURE 9.13 Response of absorber main mass with varying absorber mass ratios and optimum values for the stiffness and damping ratio. The main mass, m_1, has no damping. The optimum stiffness ratios, k_{s2}/k_{s1}, corresponding to mass ratios, 0.05, 0.1 and 0.25, are 0.0454, 0.0826 and 0.16, respectively. The corresponding optimum damping ratios, ζ_2, are 0.00636, 0.0168 and 0.0548, respectively.

For damping, C_2, added in parallel with the absorber supporting spring of stiffness, k_{s2}, and for $C_1 = 0$, the optimum tuning (for minimising the maximum displacement of the main mass, m_1) requires the following stiffness and damping ratios (Den Hartog (1956) and Kelly (1993, p. 586)):

$$\frac{k_{s2,\mathrm{opt}}}{k_{s1}} = \frac{m_1 m_2}{(m_1 + m_2)^2} = \frac{(m_2/m_1)}{(1 + m_2/m_1)^2} \tag{9.76a,b}$$

$$\zeta^2_{2,\mathrm{opt}} = \left(\frac{C_2}{C_{c2}}\right)^2_{\mathrm{opt}} = \frac{3(m_2/m_1)}{8(1 + m_2/m_1)} \tag{9.77a,b}$$

Note that Equation (9.77) does not show the mass ratio term cubed, as found in Den Hartog (1956), due to the latter reference defining $\zeta_2 = C_2/C_{c1}$ instead of the more usual $\zeta_2 = C_2/C_{c2}$, such as found in (Kelly, 1993, p. 583).

The predicted maximum displacement amplitude, $|y_1|$, of mass, m_1, occurs at one of the system resonance frequencies, f_a or f_b. For optimum damping, $\zeta_{2,\mathrm{opt}}$, the displacement amplitudes at f_a and f_b are equal and are given by:

$$\frac{|y_1|}{|F|} = \frac{1}{k_{s1}}\sqrt{1 + 2(m_1/m_2)} \tag{9.78}$$

In the preceding equation, F is the excitation force and y_1 is the displacement of mass, m_1. A plot showing the effectiveness of an optimum absorber of varying mass is provided in Figure 9.13, where f_1 is the resonance frequency of mass, m_1, with no absorber.

Soom and Lee (1983) repeated the preceding analysis and included damping on the primary mass in addition to damping on the absorber mass. Equation (9.70) then becomes:

$$\frac{|y_1|}{|F|} = \frac{1}{k_{s1}}\left[\left(1 - \Omega^2 - r/q\right)^2 + \left(2\zeta_1\Omega + s/q\right)^2\right]^{-1/2} \tag{9.79}$$

where the critical damping ratio for the main mass, m_1 on its isolation system is:

$$\zeta_1 = \frac{C_1}{C_{c1}} = \frac{C_1}{2\sqrt{k_{s1}m_1}} \tag{9.80a,b}$$

Independent of whether or not the main mass damping, $C_1 = 0$, the displacement of the absorber mass is given by (Soom and Lee, 1983):

$$|y_2| = |y_1| \left[(a/q)^2 + (b/q)^2\right]^{1/2} \tag{9.81}$$

where $|F|$ is the amplitude of the excitation force and:

$$a = (k_{s2}/k_{s1})^2 + 4\zeta_2^2\Omega^2 - (m_2/m_1)(k_{s2}/k_{s1})\Omega^2 \tag{9.82}$$

$$b = -2\zeta_2(m_2/m_1)\Omega^3 \tag{9.83}$$

$$q = \left(\frac{k_{s2}}{k_{s1}} - \frac{m_2\Omega^2}{m_1}\right)^2 + 4\zeta_2^2\Omega^2 \tag{9.84}$$

$$r = (m_2/m_1)(k_{s2}/k_{s1})^2\Omega^2 - (m_2/m_1)^2(k_{s2}/k_{s1})\Omega^4 + 4\zeta_2^2(m_2/m_1)\Omega^4 \tag{9.85}$$

$$s = 2\zeta_2(m_2/m_1)^2\Omega^5 \tag{9.86}$$

Note that Equations (9.70) to (9.78) have been derived on the basis that there was no existing damping in the vibrating machine support system (that is, ζ_1 is assumed to be zero). Soom and Lee (1983) showed that as ζ_1 increases from zero to 0.1, the true optimum for ζ_2 increases from that given by Equation (9.77) by between 2% (for a mass ratio of 0.5) and 7% (for a mass ratio of 0.1) and the true optimum stiffness ratio (k_{s2}/k_{s1}) decreases from that given by Equation (9.76) by between 6% (for a mass ratio of 0.1) and 10% (for a mass ratio of 0.5). Soom and Lee (1983) also showed that there would be little benefit in adding a vibration absorber to a system that already had a damping ratio, ζ_1, greater than 0.2.

9.4.2 Vibration Neutralisers

A vibration neutraliser is similar in construction to a vibration absorber, but differs from it in that a vibration neutraliser targets non-resonant vibration, whereas a vibration absorber targets resonant vibration. The non-resonant vibration targeted by the vibration neutraliser occurs at a forcing frequency that is causing a structure to vibrate at a non-resonance frequency. Thus, the vibration neutraliser resonance frequency is made equal to the forcing frequency that is causing the undesirable structural vibration. The same equations as used for the vibration absorber may be used to calculate the reduction in vibration that will occur at the forcing frequency. First, the vibration amplitude is calculated without the vibration neutraliser installed using the idealised SDOF system characterised by Equation (9.17). Then the vibration amplitude is calculated with the neutraliser attached using Equations (9.70) and (9.79).

9.5 Vibration Measurement

Transducers are available for the direct measurement of instantaneous acceleration, velocity, displacement and surface strain. In noise control applications, the most commonly measured quantity is acceleration, as this is often the most convenient to measure. However, the quantity that is most useful is vibration velocity, as its square is related directly to the structural vibration energy, which, in turn, is often related directly to the radiated sound power (see Section 4.15). Also, most machines and radiating surfaces have a flatter velocity spectrum than acceleration spectrum, which means that the use of velocity signals is an advantage in frequency analysis, as it allows the maximum amount of information to be obtained using an octave or third-octave filter, or spectrum analyser with a limited dynamic range.

For single frequencies or narrow frequency bands with a centre frequency, ω radians/sec, the displacement, d, velocity, v, and acceleration, a, are related as $d\omega^2 = v\omega = a$. In terms of phase angle, velocity leads displacement by 90° and acceleration leads velocity by 90°. For narrow band or broadband signals, velocity can also be derived from acceleration measurements using electronic integrating circuits. Unfortunately, integration amplifies electronic noise at low frequencies and this can be a problem. On the other hand, deriving velocity and acceleration signals by differentiating displacement signals is generally not practical, due primarily to the limited dynamic range of displacement transducers and secondarily to the cost of the differentiating electronics.

One alternative, which is rarely used in noise control, is to bond strain gauges to a surface to measure its vibration displacement levels (National Instruments, 2022). However, this technique will not be discussed further here.

9.5.1 Acceleration Transducers

Vibratory motion for noise control purposes is most commonly measured with an accelerometer attached to the vibrating surface. The accelerometer most generally used consists of a small piezoelectric crystal, loaded with a small weight and designed to have a natural resonance frequency (or undamped resonance frequency) well above the anticipated excitation frequency range. Where this condition may not be satisfied and consequently a problem may exist involving excitation of the accelerometer resonance, mechanical filters are available which, when placed between the accelerometer base and the measurement surface, minimise the effect of the accelerometer resonance, at the expense of the high-frequency response. This results in loss of accuracy at lower frequencies, effectively shifting the ±3 dB error point down in frequency by a factor of five (see Figure 9.14). However, the transverse sensitivity (see below) at higher frequencies is also much reduced by use of a mechanical filter, which, in some cases, is a significant benefit. Sometimes it may also be possible to filter out the accelerometer resonance response using an electrical filter on the output of the amplifier, but this could effectively reduce the dynamic range of the measurements, due to the limited dynamic range of the amplifier.

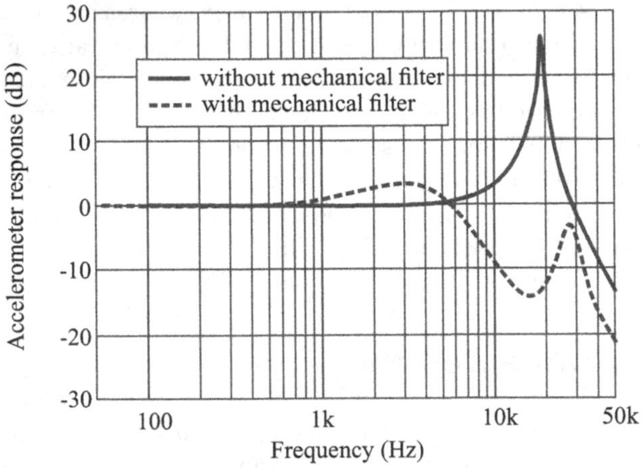

FIGURE 9.14 Typical accelerometer main axis response with and without a mechanical filter.

The mass-loaded piezoelectric crystal accelerometer may be thought of as a one-degree-of-freedom system driven at the base, such as that of case (b) of Figure 9.1. The crystal, which is loaded by a small rigid mass in compression or shear, provides the stiffness and system damping as well as a small contribution to the inertial mass. As may readily be shown (Tse et al., 1979), the response of such a system driven well below resonance is controlled by the system (crystal) stiffness. Within the frequency design range, the difference $(y - y_1)$ (see Figure 9.1(b)) between the displacement, y, of the mass mounted on the crystal and the displacement, y_1, of the base of the accelerometer , results in small stresses in the crystal. The latter stresses are detected, as induced charge on the crystal, by means of some very high-impedance voltage detection circuit, such as that provided by an ordinary sound level meter or a charge amplifier. Although acceleration is the measured quantity, integrating circuitry is commercially available so that velocity and even displacement may also be measured.

Referring to Figure 9.1(b), the difference in displacement, $y - y_1$, is (Tse et al., 1979):

$$y - y_1 = y_1 X^2/|Z|; \quad \text{where } |Z| = \left[\left(1 - X^2\right)^2 + (2X\zeta)^2\right]^{1/2} \quad \text{and } X = f/f_0 \qquad (9.87\text{a,b})$$

Equation (9.87) may also be written in terms of acceleration, a, by replacing y with a/ω^2 in all three places, where ω is the frequency of vibration in radians/s.

In the above equations, X is the ratio of the driving frequency to the resonance frequency of the accelerometer, ζ is the damping ratio of the accelerometer and $|Z|$ is the modulus of the impedance seen by the accelerometer mass, which represents the reciprocal of a magnification factor. The voltage generated by the accelerometer will be proportional to $(y - y_1)$ and thus proportional to the acceleration $y_1 X^2$ divided by the modulus of the impedance, $|Z|$, as shown in Equation (9.87).

If a vibratory motion is periodic it will generally have overtones characterised by frequencies corresponding to multiples of the fundamental frequency of the motion. Alternatively, if the motion is not periodic, the response may be thought of as a continuum of overtones. In any case, if distortion in the measured acceleration is to be minimal, then it is necessary that the magnification factor be essentially constant over the frequency range of interest. In this case, the difference in displacement of the mass mounted on the crystal and the displacement of the base of the accelerometer generates a voltage that is proportional to this displacement difference. According to Equation (9.87), where $y = a/\omega^2$ this voltage is also proportional to the acceleration of the accelerometer base. However, as the magnification factor, $1/|Z|$, in Equation (9.87) is a function of frequency ratio, X, it can only be approximately constant by design over some prescribed range and some distortion will always be present. The percent amplitude distortion is defined as:

$$\text{Amplitude distortion} = [(1/|Z|) - 1)] \times 100\% \qquad (9.88)$$

Distortion occurs when the displacement of the mass inside the accelerometer is so large that the displacement is no longer linearly related to the inertial force that it experiences. This can result from the excitation frequency being too close to the accelerometer resonance frequency or the acceleration amplitude of the accelerometer base being too great. Increasing the damping ratio reduces the displacement of the vibrating mass as the frequency approaches the resonance frequency of the accelerometer and to minimise distortion, the accelerometer should have a damping ratio of between 0.6 and 0.7, giving a useful frequency range of $0 < X < 0.6$.

Where voltage amplification is used, the sensitivity of an accelerometer is dependent on the length of cable between the accelerometer and its amplifier. Any motion of the connecting cable can result in spurious signals. The voltage amplifier must have a very high input impedance to measure low-frequency vibration and not significantly load the accelerometer electrically because the amplifier decreases the electrical time constant of the accelerometer and effectively reduces its sensitivity. Commercially available high impedance voltage amplifiers allow accurate measurement down to about 20 Hz, but are rarely used due to the above-mentioned problems.

Alternatively, charge amplifiers are usually preferred, as they have a very high input impedance and thus do not load the accelerometer output; they allow measurement of acceleration down to frequencies of 0.2 Hz; they are insensitive to cable lengths up to 500 m and they are relatively insensitive to cable movement. Many charge amplifiers also have the capability of integrating acceleration signals to produce signals proportional to velocity or displacement. This facility should be used with care, particularly at low frequencies, as phase errors and high levels of electronic noise may be present, especially if double integration is used to obtain a displacement signal.

Most recently manufactured accelerometers have inbuilt charge amplifiers and thus have a low impedance voltage output that is consistent with the IEPE standard (see Section 3.1.9.1) to ensure that it is compatible with other instruments. Some accelerometers are also TEDS sensors (see Section 3.1.9.2).

The minimum vibration level that can be measured by an accelerometer is dependent on its sensitivity and can be as low as 10^{-4} m/s^2. The maximum level is dependent on size and can be as high as 10^6 m/s^2 for small shock accelerometers. Most commercially available accelerometers at least cover the range 10^{-2} to 5×10^4 m/s^2. This range is then extended at one end or the other, depending on accelerometer type.

The transverse sensitivity of an accelerometer is its maximum sensitivity to motion in a direction at right-angles to its main axis. The maximum value is usually quoted on calibration charts and should be less than 5% of the axial sensitivity. Clearly, readings can be significantly affected if the transverse vibration amplitude at the measurement location is an order of magnitude larger than the axial amplitude.

The frequency response of an accelerometer is regarded as essentially flat over the frequency range for which its electrical output is proportional to within ±5% of its mechanical input. The lower limit has been previously discussed. The upper limit is generally just less than one-third of the resonance frequency. The resonance frequency is dependent on accelerometer size and may be as low as 2500 Hz or as high as 180 kHz. In general, accelerometers with higher resonance frequencies are smaller in size and less sensitive.

When choosing an accelerometer, some compromise must always be made between its weight and sensitivity. Small accelerometers are more convenient to use; they can measure higher frequencies and are less likely to mass load a test structure and affect its vibration characteristics. However, they have low sensitivity, which puts a lower limit on the acceleration amplitude that can be measured. Accelerometers range in weight from miniature 0.65 grams for high-level vibration amplitude measurement (up to a frequency of 18 kHz) on lightweight structures, to 500 grams for low-level ground vibration measurement (up to a frequency of 700 Hz). Thus, prior to choosing an accelerometer, it is necessary to know approximately the range of vibration amplitudes and frequencies to be expected as well as detailed accelerometer characteristics, including the effect of various types of amplifier (see manufacturer's data).

9.5.1.1 Sources of Measurement Error

Temperatures above 100°C can result in small reversible changes in accelerometer sensitivity of up to 12% at 200°C. If the accelerometer base temperature is kept low using a heat sink and mica washer with forced air cooling, then the sensitivity will change by less than 12% when mounted on surfaces having temperatures up to 400°C. Accelerometers cannot generally be used on surfaces characterised by temperatures in excess of 400°C.

Strain variation in the base structure on which an accelerometer is mounted may generate spurious signals. Such effects are reduced using a shear-type accelerometer and are virtually negligible for piezoresistive accelerometers.

Magnetic fields have a negligible effect on an accelerometer output.

Intense electric fields can have a strong effect on the accelerometer output signal. The effect can be minimised by using a differential pre-amplifier with two outputs from the same accelerometer (one from each side of the piezoelectric crystal with the accelerometer casing as a common earth) in such a way that voltages common to the two outputs are cancelled. This arrangement is generally necessary when using accelerometers near large generators or alternators.

Earth loops can occur if the test object is connected to ground and the accelerometer is not electrically isolated from it. This results in a high level 50 Hz tonal peak in the resulting acceleration signal.

Acoustic sensitivity should be accounted for, especially when measuring vibration in high sound pressure level environments, or when using highly sensitive accelerometers.

9.5.1.2 Sources of Error in the Measurement of Transients

If the accelerometer charge amplifier lower limiting frequency is insufficiently low for a particular transient or very low-frequency acceleration waveform, then the phenomenon of **leakage** will occur. This results in the waveform output by the charge amplifier not being the same as the acceleration waveform and errors in the peak measurement of the waveform will occur. To avoid this problem, the lower limiting frequency of the pre-amplifier should be less than $0.008/T_p$ for a square wave transient and less than $0.05/T_p$ for a half-sine transient, where T_p is the period of the transient in seconds. Thus, for a square wave type of pulse of duration 100 ms, the lower limiting frequency set on the charge amplifier should be 0.1 Hz.

Another phenomenon, called **zero shift**, which can occur when any type of pulse is measured, is that the charge amplifier output at the end of the pulse could be negative or positive, but not zero and can take a considerable time longer (up to 1000 times longer than the pulse duration) to decay to zero. Thus, large errors can occur if integration networks are used in these cases. The problem is worst when the accelerometers are being used to measure transient acceleration levels close to their maximum capability. A mechanical filter placed between the accelerometer and the structure on which it is mounted can reduce the effects of zero shift.

The phenomenon of **ringing** can occur when the transient acceleration that is being measured contains frequencies above the useful measurement range of the accelerometer and its mounting configuration. The accelerometer mounted resonance frequency should not be less than $10/T_p$, where T_p is the length of the transient in seconds. The effect of ringing is to distort the charge amplifier output waveform and cause errors in the measurement. The effects of ringing can be minimised by using a mechanical filter between the accelerometer and the structure on which it is mounted.

9.5.1.3 Accelerometer Calibration

In normal use, accelerometers may be subjected to violent treatment, such as dropping, which can alter their characteristics. Thus, the sensitivity should be periodically checked by mounting the accelerometer on a vibration calibrator, which either produces a known value of acceleration at some reference frequency, or on which a reference accelerometer of known calibration may be mounted for comparison.

9.5.1.4 Accelerometer Mounting

Generally, the measurement of acceleration at low to middle frequencies poses few mechanical attachment problems. For example, for measurements below 5 kHz and temperatures below 95°C, an accelerometer may be attached to the test surface simply by using thin, double-sided adhesive tape. For the measurement of higher frequencies, an accelerometer may be attached with a hard epoxy, cyanoacrylate adhesive or by using a stud or bolt. Use of a magnetic base usually limits the upper frequency bound to about 2 kHz. Beeswax may be used on surfaces that are cooler than 40°C, for frequencies below 10 kHz. Thus, for the successful measurement of acceleration at high frequencies, some care is required to ensure the following:

1. that the accelerometer attachment is firm; and
2. that the mass loading provided by the accelerometer is negligible.

With respect to the former, it is suggested that the manufacturer's recommendation for attachment be carefully followed. With respect to the latter, the following is offered as a guide.

Let the mass of the accelerometer be m_a grams. When the mass, m_a, satisfies the appropriate equation that follows, the measured vibration level will be at most 3 dB below the unloaded level due to the mass loading by the accelerometer. For thin plates:

$$m_a \leq 3.7 \times 10^{-4}(\rho c_{LI} h^2/f) \qquad \text{(grams)} \tag{9.89}$$

and for massive structures:

$$m_a \leq 0.013(\rho c_{LI}^2 D_a/f^2) \qquad \text{(grams)} \tag{9.90}$$

where ρ is the plate material density (kg/m^3), h is the plate thickness (mm), D_a is the accelerometer diameter (mm), f is the frequency (Hz) and c_{LI} is the longitudinal wave speed (m/s) in the plate.

As a general guide, the accelerometer mass should be less than 10% of the dynamic mass (or modal mass) of the vibrating structure to which it is attached. The effect of the accelerometer mass on any resonance frequency, f_s, of a structure is given by:

$$f_m = f_s\sqrt{\frac{m_s}{m_s + m_a}} \tag{9.91}$$

and where f_m is the resonance frequency with the accelerometer attached, m_a is the mass of the accelerometer and m_s is the dynamic mass of the structure (often approximated as the mass in the vicinity of the accelerometer). One possible means of accurately determining a structural resonance frequency of a lightweight structure would be to measure it with a number of different weights placed between the accelerometer and the structure, plot measured resonance frequency versus added mass, and extrapolate linearly to zero added mass. If mass loading is a problem, an alternative to an accelerometer is to use a laser Doppler velocimeter, (see Section 9.5.3).

Accelerometer mounting is also discussed in ANSI/ASA S2.61 (2020), where it is stated that when phase measurements are important, such as when structural intensity measurements are taken, then the loss factor, η, of the mounting system should be less than 0.1 and the stiffness of the mounting should be greater than $2m_a(2\pi f)^2$ (N/m). This usually means that, except for very low frequencies, it is necessary to fix the accelerometer to the structure using either a stud or hard cement such as cyanoacrylate. In situations where the sensor must be electrically isolated from the structure to prevent ground loops, but must have a very stiff attachment to the structure, the accelerometer can be mounted using a thin mixture of cyanoacrylate and garnet, which is used in water-jet cutting or sand-blasting. This provides a very high-stiffness connection, but electrically isolates the sensor from the structure (Larizza et al., 2019).

9.5.1.5 Piezoresistive Accelerometers

An alternative type of accelerometer is the piezoresistive type, which relies on the measurement of resistance change in a piezoresistive element (such as a strain gauge) subjected to stress. Piezoresistive accelerometers are less common than piezoelectric accelerometers and generally are less sensitive by an order of magnitude for the same size and frequency response. Piezoresistive accelerometers are capable of measuring down to DC (or zero frequency), are easily calibrated (by turning upside down) and can be used effectively with low impedance voltage amplifiers. However, they require a stable DC power supply to excite the piezoresistive element (or elements).

9.5.2 Velocity Transducers

Measurement of velocity provides an estimate of the energy associated with structural vibration; thus, a velocity measurement is often a useful parameter to quantify sound radiation.

Velocity transducers are generally of three types. The least common is the non-contacting magnetic type consisting of a cylindrical permanent magnet on which is wound an insulated coil. As this type of transducer is only suitable for relative velocity measurements between two surfaces or structures, its applicability to noise control is limited; thus, it will not be discussed further.

The most common type of velocity transducer consists of a moving coil surrounding a permanent magnet. Inductive electromotive force (EMF) is induced in the coil when it is vibrated. This EMF (or voltage signal) is proportional to the velocity of the coil with respect to the permanent magnet. In the 10 Hz to 1 kHz frequency range, for which the transducers are suitable, the permanent magnet remains virtually stationary and the resulting voltage is directly proportional to the velocity of the surface on which it is mounted. Outside this frequency range, the electrical output of the velocity transducer is not proportional to velocity. This type of velocity transducer is designed to have a low natural frequency (below its useful frequency range); thus it is generally quite heavy and can significantly mass-load light structures. Some care is needed in mounting but this is not as critical as for accelerometers, due to the relatively low upper-frequency limit characterising the basic transducer.

The moving coil type of velocity transducer generally covers the dynamic range of 1 to 100 mm/s. Some extend down to 0.1 mm/s while others extend up to 250 mm/s. Sensitivities are generally high, of the order of 20 mV/(mm s^{-1}). Low impedance, inexpensive voltage amplifiers are suitable for amplifying the signal. Temperatures during operation or storage should not exceed 120°C.

A third type of velocity transducer is the laser vibrometer (sometimes referred to as the laser Doppler velocimeter), which is discussed in Section 9.5.3.

Note that velocity signals can also be obtained by integrating accelerometer signals, although this often causes low-frequency electronic noise problems and signal phase errors.

9.5.3 Laser Vibrometers

The laser vibrometer is a specialised and expensive item of instrumentation that uses one or more laser beams to measure the vibration of a surface without any hardware having to contact the surface. They are much more expensive than other transducers but their application is much wider. They can be used to investigate the vibration of very hot surfaces on which it is not possible to mount any hardware and very lightweight structures for which the vibration is affected by any attached hardware.

Laser vibrometers operate on the principle of the detection of the Doppler shift in frequency of laser light that is scattered from a vibrating test object. The object scatters or reflects light from the laser beam, and the Doppler frequency shift of this scattered light is used to measure the component of velocity that lies along the axis of the laser beam. As the laser light has a very high frequency, its direct demodulation is not possible. An optical interferometer is therefore used to mix the scattered light with a reference beam of the same original frequency as the scattered beam before it encountered the vibrating object. A photo-detector is used to measure the intensity of the combined light, which has a beat frequency equal to the difference in frequency between the reference beam and the beam that has been reflected from the vibrating object. For a surface vibrating at many frequencies simultaneously, the beat frequency will contain all of these frequency components in the correct proportions, thus allowing broadband measurements to be made and then analysed in very narrow frequency bands.

Due to the non-contact nature of the laser vibrometer, it can be set up to scan surfaces, and three laser heads may be used simultaneously to scan a surface and evaluate the instantaneous vibration along three orthogonal axes over a wide frequency range, all within a matter of minutes. Sophisticated software provides maps of the surface vibration at any frequency specified by the user.

Currently available laser vibrometer instrumentation has a dynamic range typically of 80 dB or more. Instruments can usually be adjusted using different processing modules so that the minimum and maximum measurable levels can be varied while maintaining the same dynamic range. Instruments are available that can measure velocities up to 20 m/s and down to 1 µm/sec (although not with the same processing electronics) over a frequency range from DC up to 20 MHz.

Laser vibrometers are also available for measuring torsional vibration and consist of a dual beam which is shone onto a rotating shaft. Each back-scattered laser beam is Doppler shifted in frequency by the shaft surface velocity vector in the beam direction. The velocity vector consists of both rotational and lateral vibration components. The processing software separates out the rotational component by taking the difference of the velocity components calculated by the Doppler frequency shift of each of the two beams. The DC part of the signal is the shaft rpm and the AC part is the torsional vibration.

9.5.4 Instrumentation Systems

The instrumentation system that is chosen to be used in conjunction with the transducers just described depends on the level of sophistication desired. Overall or octave band vibration levels can be recorded in the field using a simple vibration meter. If more detailed analysis is required, a portable spectrum analyser can be used. Alternatively, if it is preferable to do the data analysis in the laboratory, samples of the data can be recorded using a high-quality data acquisition system (see Section 3.8) and replayed through a spectrum analyser. This latter method has the advantage of enabling one to re-analyse data in different ways and with different frequency resolutions, which is useful when diagnosing a particular vibration problem.

9.5.5 Units of Vibration

It is often convenient to express vibration amplitudes in decibels. The International Standards Organisation has recommended that the following units and reference levels be used for acceleration and velocity (ISO 1683, 2015). Velocity is measured as a root mean square (RMS) quantity in metres per second and the level reference is one nanometre per second (10^{-9} m/s). The velocity level, L_v, is:

$$L_v = 20\log_{10}(v/v_{\text{ref}}); \qquad v_{\text{ref}} = 10^{-9} \text{ m/s} \tag{9.92}$$

Acceleration is measured as an RMS quantity in metres per second^2 (m/s^2) and the level reference is one micrometre per second squared (10^{-6} m/s^2). The acceleration level, L_a, is:

$$L_a = 20\log_{10}(a/a_{\text{ref}}); \quad a_{\text{ref}} = 10^{-6} \text{ m/s}^2 \tag{9.93}$$

Although there is no standard for displacement, it is customary to measure it as a peak to peak quantity, d, in micrometres (μm) and use a level reference of one picometre ($10^{-6}\mu$m, peak to peak). The displacement level, L_d, is:

$$L_d = 20\log_{10}(d/d_{\text{ref}}); \qquad d_{\text{ref}} = 10^{-6} \ \mu\text{m} \tag{9.94}$$

Vibratory force is measured as an RMS quantity in Newtons (N) and the standard reference quantity is 1 µN. The force level is:

$$L_f = 20\log_{10}(F/F_{\rm ref}); \quad F_{\rm ref} = 10^{-6}{\rm N} \tag{9.95}$$

9.6 Vibration Criteria

Criteria for acceptable vibration levels produced by various items of machinery are detailed in International standards (ISO 7919-3, 2017; ISO 20816-1, 2016; ISO 20816-2, 2017; ISO 20816-4, 2018; ISO 20816-5, 2018; ISO 20816-8, 2018; ISO 20816-9, 2020). These are recommended maximum acceleration or velocity RMS vibration levels measured on various places on the machine or equipment to ensure reliable operation, and the appropriate levels can be obtained from the relevant standard.

Criteria for acceptable levels of building vibration for both people and sensitive equipment, which are in general use, are provided in Ungar (2019a). These criteria are vibration velocities that should not be exceeded in any 1/3-octave band. A different approach, which is much more recent, is used in the international standards, ISO 2631-1 (2010) and ISO 2631-2 (2003), where a single number RMS acceleration value is determined by weighting the overall measured acceleration with a frequency-dependent weighting curve in the frequency range 1 Hz to 80 Hz. If 1/3-octave band measurements, a_i, of the unweighted RMS acceleration are available, the multiplier (or weighting), W_i, to determine the corresponding weighted value for each 1/3-octave band, i, is provided in Table 9.1, which was derived from ISO 2631-2 (2003). The overall weighted RMS acceleration, a_w, is obtained by taking the square root of the sum of the squared weighted RMS acceleration values for all 1/3-octave bands between and including 1 Hz and 80 Hz. Thus:

$$a_w = \left[\sum_i (W_i a_i)^2\right]^{1/2} \tag{9.96}$$

TABLE 9.1 1/3-octave band values of the weighting function, W_i for 1/3-octave band, i

1/3-octave band centre frequency	W_i	1/3-octave band centre frequency	W_i	1/3-octave band centre frequency	W_i	1/3-octave band centre frequency	W_i
1.0	0.833	3.15	0.872	10	0.494	31.5	0.176
1.25	0.907	4	0.818	12.5	0.411	40	0.14
1.6	0.934	5	0.75	16	0.337	50	0.109
2	0.932	6.3	0.669	20	0.274	63	0.0834
2.5	0.91	8	0.582	25	0.22	80	0.0604

ISO 2631-1 (2010) states that the perception threshold of the overall weighted RMS acceleration value for the 50 percentile of the population is 0.015 m/s^2 and the value for the 25 percentile is 0.01 m/s^2. It is thus desirable that the overall weighted RMS acceleration in the orthogonal direction of maximum acceleration inside a residential building should not exceed 0.01 m/s^2, as it has been pointed out (ISO 2631-1, 2010) that vibration becomes annoying for many people at levels that are only slightly higher than the perception threshold. This value can be used, together with the acceptable 1/3-octave band levels at various locations, shown Figure 1 of Ungar (2019a), to obtain overall acceptable weighted acceleration levels as shown in Table 9.2. However, it should be pointed out that these values are suggestions only and have not been ratified in any standard. Also, ISO 2631-1 (2010) should be consulted for additional required measurements in cases where the vibration signal is characterised by a high crest factor (ratio of peak to RMS value).

TABLE 9.2 Suggested maximum overall RMS acceleration levels in the range 1 Hz to 80 Hz in the orthogonal direction of maximum acceleration (usually the vertical direction), where the VC criteria are for the operation of sensitive equipment and are discussed by Ungar (2019a)

Location	Maximum overall weighted RMS acceleration, a_w, (m/s^2)
Workshop	0.04
Office	0.02
Residential building	0.01
Operating theatre	0.005
VC-A	0.0025
VC-B	0.00125
VC-C	0.00062
VC-D	0.00031
VC-E	0.00015

9.7 Damping of Vibrating Surfaces

9.7.1 Damping Methods

Damping of sheet metal structures can be accomplished by the application of a damping material to the metal sheet, such as is used on car bodies. Many types of damping are available from various manufacturers for this purpose. They may take the form of tapes, sheets or sprays, which may be applied like paint. They all make use of some non-hardening, viscoelastic material. For optimum results, the weight of the layer of damping material should be at least equal to that of the base panel.

Damping materials can be applied more efficiently and effectively using a laminated construction (see Figure 9.15) of one or more thin sheet metal layers, each separated by a viscoelastic layer, the whole being bonded together. Very thin layers (approximately 0.4 mm) of viscoelastic material are satisfactory in these constrained-layer systems (see Cremer et al. (1988), pp. 246–255). For a single constrained layer, the greatest vibration reduction of the base structure occurs when the added sheet metal layer is equal in thickness to the base structure. For damping high-frequency vibration, the viscoelastic damping layer should be stiffer than for damping low-frequency vibration. A detailed design procedure for constrained layer damping is provided by Mead (1998).

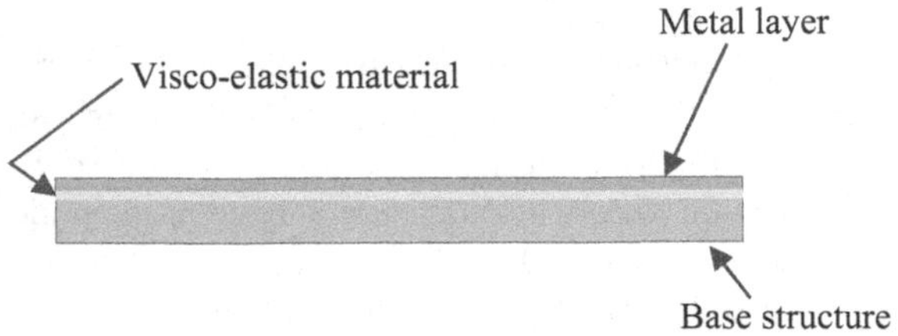

FIGURE 9.15 Constrained layer damping construction.

Riveted metal constructions provide more damping than welded constructions. The damping mechanism is a combination of interfacial friction and air pumping through narrow gaps, although work on structures in a vacuum (Ungar, 1988; Ungar and Zapfe, 2006a) indicates that the latter mechanism dominates. Thus, damped panels can be formed of layered materials that are riveted, bolted or spot-glued together.

9.7.2 When Damping is Effective and Ineffective

In this section, the question of whether or not to apply some form of damping to a vibrating surface for the purpose of noise control is considered. Commercially available damping materials take many forms but, generally, they are expensive and they may be completely ineffective if used improperly. Provided that the structure to be damped is vibrating resonantly, these materials generally will be very effective in damping relatively lightweight structures, and progressively less effective as the structure becomes heavier.

If the structure is driven mechanically by attachment to some other vibrating structure, or by the impact of solid materials, or by turbulent impingement of a fluid, then the response will be dominated by resonant modes and the contribution due to forced modes, as will be explained, will be negligible. Damping will be effective in this case and the noise reduction will be equal to the reduction in surface vibration level (see Equation (4.179)). Damping will be essentially ineffective in all other cases where the structure is vibrating in forced (or non-resonant) response, such as when a structure is excited by a force at a frequency that does not match a structural resonance frequency.

Structures of any kind have preferred patterns known as modes of vibration to which their vibration conforms. A modal mass, stiffness and damping may be associated with each such mode, which has a corresponding resonance frequency at which only a small excitation force is required to make the structure vibrate strongly. Each such mode may conveniently be thought of as similar to the simple one-degree-of-freedom oscillator of Figure 9.1(a), in which the impedance of the base is infinite and its motion is nil. In general, many modes will be excited at once, in which case, the response of a structure may be thought of as the collective responses of as many simple one-degree-of-freedom oscillators (Pinnington and White, 1981).

The load on the structure caused by the acoustic field interacting with it is like an additional small force applied to the structure, which for a single mode can be represented by the mass shown in Figure 9.1(a). However, for a structure vibrating in air, the force due to the acoustic load is generally so small compared to the driving force that the surface displacement, to a very good approximation, is independent of the acoustic load. Also, at frequencies well above a modal resonance frequency, the modal displacement is independent of damping. In this high-frequency range, the system response is said to be mass controlled.

In the consideration of the response of an extended system, such as a panel or structural surface subjected to distributed forces, a complication arises; it is possible to drive structural modes, when the forcing distribution matches the modal displacement distribution, at frequencies other than their resonance frequencies. The latter phenomenon is referred to as forced response. For example, in the mass-controlled frequency range of a panel (see Section 7.2.6), the modes of the panel are driven in forced response well above their resonance frequencies by an incident acoustic wave; their responses are controlled by their modal masses and are essentially independent of the damping.

If the acoustic radiation of a surface or structure is dominated by modes driven well above resonance in forced response, then the addition of damping will have very little or no effect on the sound produced. For example, if a panel is excited by an incident sound field, forced modes will be strongly driven and will contribute most to the radiated sound, although resonant modes may dominate the apparent vibration response. This is because at frequencies below the surface-critical frequency (see Section 7.2.1) the sound-radiating efficiency of the forced vibration modes will be unity, and thus much greater than the efficiency of the large-amplitude resonant vibration modes. In this case, the addition of damping material to the structure may well reduce the overall structural vibration amplitude without reducing the sound radiation amplitude. One example would be the addition of damping material to a panel for the purpose of increasing its transmission loss. In this case, the damping material will generally have only a small effect on the sound transmission loss, except in certain narrow ranges of resonant response where the effect will be large (see Section 7.2.6). These considerations suggest the exercise of caution in the use of surface damping as a means for noise control.

9.8 Measurement of Damping

Damping is associated with the modal response of structures or acoustic spaces; thus, the discussion of Section 6.3.2, in connection with modal damping in rooms, applies equally well to damping of modes in structures and need not be repeated here. In particular, the relationships between loss factor, η, quality factor, Q, and damping ratio, ζ, are the same for modes of rooms and structures. However, whereas the modal density of rooms increases rapidly with increasing frequency and investigation of individual modes is only possible at low frequencies, the modal density of some structures such as panels is constant, independent of frequency, so that in these cases, investigation of individual modes is possible at all frequencies.

Damping takes many forms but viscous and hysteretic damping, described in Section 9.2.1, are the most common. As shown, they can be described relatively simply analytically, and consequently, they have been well investigated. Viscous damping is proportional to the velocity of the structural motion and has the simplest analytical form. Viscous damping is implicit in the definition of the damping ratio, ζ, and is explicitly indicated in Figure 9.1 by the introduction of the dashpot. Damping of modes in rooms is well described by this type of damping (see Section 6.3.2).

Hysteretic (or structural) damping has also been recognised and investigated in the analysis of structures with the introduction of a complex elastic modulus (see Section 9.2.1). Hysteretic damping is represented as the imaginary part of a complex elastic modulus of elasticity of the material, introduced as a loss factor, η, such that the elastic modulus, E, is replaced with $E(1+\mathrm{j}\eta)$. Hysteretic damping is thus proportional to displacement and is well suited to describe the damping of many, though not all, mechanical structures.

For the purpose of loss factor measurement, the excitation of modes in structures may be accomplished either by the direct attachment of a mechanical shaker or by shock excitation using a hammer. When the direct attachment of a shaker is used, the coupling between the shaker and the driven structure is strong. In the case of strong coupling, the mass of the shaker armature and shaker damping become part of the oscillatory system and must be taken into account in the analysis. Alternatively, instrumented hammers are available, which allow direct measurement and recording of the hammer impulse applied to the structure. This information allows direct determination of the structural response and loss factor.

Hysteretic damping can be determined by a curve fitting technique, using the experimentally determined frequency response function (Ewins, 2000) (see Section 12.3.15). For lightweight or lightly damped structures, this method is best suited to the use of instrumented hammer excitation, which avoids the shaker coupling problem mentioned above.

If a simpler, though less accurate, test method is sufficient, then one of the methods described in Section 6.3.2 may suffice. With reference to the latter section, if the reverberation decay method (making use of Equations (6.23) and (6.24)) is used, whereby the vibration decay time (T_{60}) of the structure is measured following the switching-off of the vibration source, then it is important to avoid the problems associated with strong coupling mentioned above to ensure that measured damping is controlled by the structural damping, which is to be measured, and not by the damping of the excitation device. The problems of strong coupling may be avoided by arranging to disconnect the driver from the structure when the excitation is shut off, by using a fuse arrangement. Alternatively, either a hammer or a non-contacting electromagnetic coil, which contains no permanent magnet, may be used to excite the structure. In the latter case, the structure will be excited at twice the frequency of the driving source.

When frequency-band filters are used to process the output from the transducer used to monitor the structural vibration, it is important to ensure that the filter decay rate is much faster than the decay rate that is to be measured, so that the filter decay rate does not control the measured structural vibration decay rate. Typically this means that the following relation must be satisfied, where B is the filter bandwidth (Hz):

$$BT_{60} \geq 16 \tag{9.97}$$

If the steady-state determination of the modal bandwidth is used to determine the damping by using Equation (6.23), then it is necessary that the excitation force is constant over the frequency range of the modal bandwidth, Δf. The measurement requires that the frequency of resonance is determined and that the modal response at the 3 dB down points below and above resonance may be identified (using a sinusoidal excitation signal), as shown in Figure 9.16. In some cases, better results are obtained by using the bandwidth at the 7, 10 or 12.3 dB down points. In these cases, the value of Δf used in Equation (6.23) is one-half, one-third or one-quarter, respectively, of the measured bandwidth, as illustrated for the 7 dB down point in Figure 9.16.

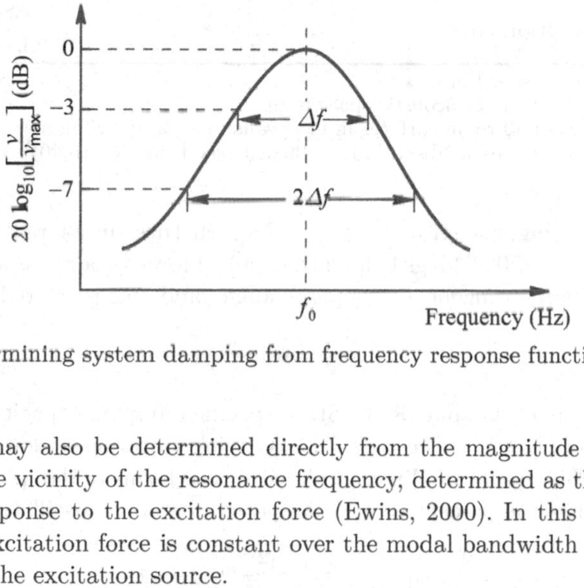

FIGURE 9.16 Determining system damping from frequency response function (FRF) bandwidth measurements.

The bandwidth may also be determined directly from the magnitude of the frequency response function in the vicinity of the resonance frequency, determined as the ratio of the structural acceleration response to the excitation force (Ewins, 2000). In this latter case, it is not important that the excitation force is constant over the modal bandwidth and an impact hammer is often used as the excitation source.

The relationships of the various measures of damping to the loss factor are best summarised in a table such as Table 9.3. The logarithmic decrement, δ (see Equation (6.23)), is one of the oldest methods of determining damping and it depends on the determination of successive amplitudes of a vibrating system as the vibration decays after switching off the excitation source. If A_i is the amplitude of the ith cycle and A_{i+n} is the amplitude n cycles later, then the logarithmic decrement, δ, is:

$$\delta = \frac{1}{n}\log_e\left(\frac{A_i}{A_{i+n}}\right) = \frac{2\pi\zeta}{\sqrt{1-\zeta^2}} \approx \left(\frac{\pi\Delta f}{f_0} = \pi\eta = \frac{\pi}{Q}\right) \tag{9.98a–e}$$

where use has been made of Equation (6.23), Δf and f_0 are defined in Figure 9.16, ζ is the viscous damping ratio, η is the loss factor and Q is the quality factor.

TABLE 9.3 Relationship of loss factor to various other measures of damping

Damping measure	Symbol	Units	$\eta =$
Loss factor	η	—	η
Quality factor	Q	—	$1/Q$
Critical damping ratio	ζ	—	$\dfrac{2\zeta}{\sqrt{1-\zeta^2}}$
Reverberation time (60 dB)	T_{60}	sec	$\dfrac{2.2}{fT_{60}}$
Decay rate	DR	dB/sec	$\dfrac{\text{DR}}{27.3f}$
Wave attenuation	γ	nepers	$\dfrac{c_g\gamma}{\pi f}$
Logarithmic decrement	δ	—	δ/π
Phase angle by which strain lags force	ϵ	radians	$\tan\epsilon$
Modal bandwidth	Δf	Hz	$\dfrac{\Delta f}{f}$
Imaginary part of modulus of elasticity, $(E_r + \mathrm{j}E_i)$	E_i	Pa	E_i/E_r
Sabine absorption coefficient	$\bar{\alpha}$	—	$\dfrac{cS\bar{\alpha}}{25.1fV}$

c_g is the group speed in m/s
V is the volume of the acoustic space in m^3
S is the area of all room surfaces in m^2 (walls, floor and ceiling as a minimum)
$\bar{\alpha}$ is the average Sabine absorption coefficient (see Equation (6.40))

As a result of damping, the strain response of a structure (measured using a strain gauge – see National Instruments (2022)) lags behind the applied force by a phase angle, ϵ. Thus, another measure of damping is the tangent of this phase angle, and this is related to the loss factor by:

$$\eta = \tan\epsilon \tag{9.99}$$

Yet another measure of damping is the SDC (specific damping capacity), which is calculated from the amplitudes of two successive cycles of a decaying vibration (following switching off of the excitation source). This cannot be directly related to the forms of damping measure mentioned above, and is defined as the percent energy dissipated in a single oscillation cycle as:

$$\text{SDC} = \frac{(A_n^2 - A_{n+1}^2)}{A_n^2} \times 100\% \tag{9.100}$$

10

Sound Power and Sound Pressure Level Estimation Procedures

LEARNING OBJECTIVES

In this chapter, procedures are outlined for estimating the noise radiated by:

- fans;
- compressors;
- cooling towers;
- pumps;
- fluid jets;
- control valves;
- fluid flow in pipes;
- boilers;
- turbines;
- internal combustion engines;
- furnaces;
- electric motors;
- generators;
- transformers;
- gears; and
- transportation vehicles.

10.1 Introduction

In this chapter, means are provided for estimating the expected radiated sound power or sound pressure at 1 m, for a variety of types of mechanical equipment. Such information makes consideration of noise control at the design stage possible, a practice which experience has shown to be by far the most cost-effective. The sound power levels provided in the calculations in this chapter can also be used to calculate environmental sound pressure levels (see Chapter 5) and sound pressure levels inside rooms and enclosures (see Chapter 6). The sound pressure levels at 1 m can be used to estimate the noise exposure of the operator of the machine (see Chapter 2). Equations for overall sound power or sound pressure level are for un-weighted levels unless the word, "A-weighted" is used.

The prediction of the sound power level generated by equipment or machinery is generally very difficult, primarily because the possible noise-generating mechanisms are so many and varied in any but the simplest of devices and the magnitude of the noise generated depends on the environment presented to the noise source. It is possible to estimate the noise radiated by some aerodynamic noise sources in terms of an acoustic efficiency factor, as a fraction of the total stream power. However, this method cannot be applied in general to other noise-producing processes and mechanisms. Furthermore, it is not possible to make any sweeping simplifying assumptions on thermodynamic principles, because the power radiated as sound is generally

DOI: 10.1201/9780367814908-10

only a very small part of the power balance of an operating machine. In short, greater efficiency does not necessarily mean less noise.

At the present time, in most instances, quantitative laws for the generation of noise cannot be formulated. However, from the work that has been done and reported thus far, there does seem to emerge a general principle that may be stated as follows. The noise produced by any process seems to be proportional to the rate of change of acceleration of the parts taking part in that process. This has the consequence that of two or more methods of achieving a given end, one can expect the least noise to be produced by the method in which the rates of change of forces are least. Noise produced by the punch press has been shown to follow this rule (Richards, 1979, 1980; Semple and Hall, 1981). The reason is that high rates of change of force result in more energy in the higher frequency part of the spectrum. As the radiation efficiency of structures is greater at higher frequencies, the overall radiated sound power will be greater for the same total vibratory energy generated in the structure. Other examples include helical gears and helical mill cutters, which maintain continuous rather than interrupted or discontinuous contact characteristic of spur gears and non-helical cutters. The continuous contact results in lower rates of change of excitation forces (Richards, 1981).

Current noise prediction schemes are largely empirical and, in a few cases, are augmented with established theoretical considerations. In this chapter, a number of these schemes are reviewed. The noise sources considered are common industrial machines. The schemes presented should be useful for initial estimates of expected noise. However, measured sound power level data are always to be preferred. Thus the procedures described in this chapter should be used as a guide, and only when measured data are not available or when the estimates are not critical.

In the following tables, all sound pressure levels, L_p, are in decibels relative to 20 µPa and all sound power levels, L_W, are in decibels relative to 10^{-12} W. Throughout the rest of this chapter, mechanical power in kilowatts will be represented as kW, or in megawatts as MW. Rotational speed in revolutions per minute (rev min^{-1}) will be represented as rpm.

The types of machines and machinery considered in this chapter are listed in Table 10.1, which provides an indication of what is estimated (sound power level or sound pressure level at 1 m) and means commonly used for noise control. The list of equipment considered here is by no means exhaustive. However, it is representative of much of the equipment commonly found in process and power generation plants. Wherever sound power estimation schemes are provided, sound pressure levels may be estimated using the methods of Chapters 6, 7 and 8. Where sound pressure levels at 1 m are given, then sound pressure levels at other distances may be estimated using the methods of Chapters 4 and 5. In general, the latter information is most useful for estimating sound pressure levels at the machinery operator's location, but sound power levels are more useful for estimating sound pressure levels at greater distances.

In many cases in the following sections, equations are provided to calculate overall sound power levels or sound pressure levels at 1 m distance from the sound source, and these are sometimes followed by further equations that provide octave band levels or tables that provide corrections to the overall level to obtain the octave band levels. When the corrections are listed in tables, the values are arithmetically subtracted from the overall level to obtain the octave band level. Regardless of how the octave band levels are determined, their logarithmic sum (see Equation (1.98)) must equal the overall level. In cases where the two are not equal, the octave band levels must all be further corrected by the same amount so that their logarithmic sum equals the overall level. To find this correction level, which is to be added to all octave band levels, the following steps should be taken. Note that the correction level may be positive or negative.

1. Calculate an overall level by logarithmically summing the octave band levels.
2. Subtract the level calculated in 1. above from the level determined using the relevant overall level equation.
3. Add the result of 2. above to all individual octave band levels.

TABLE 10.1 Noise sources, prediction schemes and controls

Noise source	Quantity predicted L_p at 1 m or L_W or SEL (see Section 2.4.5)	Recommended control
Fans	L_W	Inlet and outlet duct fitted with commercial mufflers; blade passing frequency attenuated with one or more quarter wave tubes.
Air compressors, small and large	L_p or L_W	As for fans, plus an enclosure built around the compressor
Reciprocating		
Rotary screw		
Centrifugal		
Compressors for refrigeration units	L_p	As for fans, plus enclosure
Reciprocating		
Centrifugal		
Axial		
Cooling towers	L_W	Muffler
Propeller		
Centrifugal		
Pumps	L_p	Enclosure
Jet noise	L_p or L_W	Muffler
Gas vents	L_p	Muffler
Steam vents	L_p	Muffler
Control valves	L_p or L_W	Lagging, staging
Gases (including steam)	L_p or L_W	
Liquids	L_p or L_W	
Pipe flow (gases)	L_p or L_W	Mufflers, lagging
Boilers	L_W	Enclosure
Steam and gas turbines	L_W	Enclosure
Reciprocating diesel and gas engines	L_W	Enclosure, muffler, barrier
Furnaces	L_W	Smaller fuel/air jet diameter (even if more jets are needed); fuel oil instead of gas
Electric motors	L_p	Enclosure
Generators	L_W	Enclosure
Transformers	L_W	Enclosure
Gears	L_p	Enclosure
Large wind turbines	L_W	Design improvements
Transportation		
Road traffic	L_p or L_W	
Trains	SEL or L_W	
Aircraft	SEL	

10.2 Fan Noise

Fans are used to move air from one place to another, usually through a duct system. It is a common industrial noise source, which has been well documented (ASHRAE, 1987, 2015). The equations for sound power estimation that have been used in the past (ASHRAE, 1987) and reported in early editions of this text are no longer considered reliable (ASHRAE, 2015). The sound power of fans is dependent on the fan type and can vary considerably, depending on the operating point compared to the point of maximum efficiency. Higher airflow and lower static pressure than optimum, increase noise levels. On the other hand, higher static pressure and lower air flow increase low-frequency rumble noise. Static pressure is the pressure over and above atmospheric pressure that is measured using a pressure tapping in the side of a duct containing a flowing fluid.

There are two major classifications of fan: axial and centrifugal, and within each of these classes, there are several types. **Axial fans** are located in cylindrical tubes in which the airflow is parallel with the fan drive shaft. Noise levels are strongly influenced by the inlet flow conditions and the blade tip speed. There are two main types of axial fan.

- **Tube axial or propeller**. These fans can also have a variable pitch that can be adjusted manually with the fan stationary or automatically with the fan rotating, thus allowing optimum efficiency to be maintained for varying inlet pressures, discharge pressures and flow rates. The noise from this type of fan is characterised by a low-frequency-dominated spectrum shape.

- **Vane axial**. These fans are similar to tube axial fans but incorporate downstream guide vanes. The guide vanes result in higher efficiency and reduced levels of low-frequency noise, although mid- to high-frequency noise levels may be increased.

Centrifugal fans are those for which the fan drive shaft is parallel to the inlet flow and perpendicular to the discharge flow. Noise levels of centrifugal fans are generally lower than they are for axial fans. Centrifugal fans may have blades of any of the following types.

- **Airfoil blades**. These fans have hollow blades with an airfoil shape for maximum efficiency and minimum noise.

- **Backward curved blades**. These fans are suited to high-pressure applications. Unlike other centrifugal fan types, most of the pressure build-up takes place in the impeller, so there is often no need for a housing.

- **Forward curved blades**. These fans have many more blades than backward curved and generate a higher flow rate for the same tip speed. They are also the next quietest fan after the airfoil blade fan.

- **Radial blade**. These fans are used where noise is not a problem and there are high-volume-flow-rate and high-pressure requirements.

All fans generate a tone at the blade pass frequency (BPF) given by:

$$\text{BPF} = N_b \times \text{rpm}/60 \tag{10.1}$$

where N_b is the number of blades on the fan. The extent to which the tone is noticeable depends on the fan type and how close it is to its optimum efficiency. Fan types and some of their noise characteristics are listed in Table 10.2.

TABLE 10.2 Some characteristics of various fan types (based on information in ASHRAE (2015))

Fan type	Description	Noise characteristics
Centrifugal (housed)	Fan with housing (in common use).	BPF less prominent and higher frequency than other fan types.
Forward curved (FC)	Blades curve in same direction as rotation.	Low-frequency rumble from turbulence generated by blade tips if there are not 5 duct diameters of straight discharge duct or if the fan is operating at a lower volume flow than optimal efficiency.
Backward curved (BC)	Blades curve in opposite direction to rotation.	Louder at the BPF than FC fans for the same duty. More energy efficient than FC fans. The BPF becomes more prominent with increasing fan speed. Quieter than FC fans at higher frequencies. Quieter below the BPF as well.
Airfoil	Blades have an airfoil shape to increase efficiency.	Louder at the BPF than FC fans for the same duty. More energy efficient than FC fans. The BPF becomes more prominent with increasing fan speed. Quieter than FC fans at higher frequencies. Quieter below the BPF as well.
Centrifugal (no housing)	Discharges directly into the plenum chamber. An inlet bell is located on the chamber wall adjacent to the fan inlet.	Substantially lower discharge noise but the fan plenum must be the correct size and acoustically treated. Discharge should not be in line with the ductwork or the BPF sound will be magnified.
Vaneaxial	Used in applications where the higher frequency noise can be managed with mufflers.	Lowest low-frequency noise levels. Noise is a function of blade tip speed and inlet airflow symmetry.
Propeller	Most commonly used on condensers and for power exhausts.	Noise is dominated by low frequencies. BPF is usually low-frequency and the level at the BPF is sensitive to inlet obstructions. The shape of the fan inlet also affects noise levels.

Manufacturer's data should always be used for fan sound power levels. These data are typically sound power levels of inlet, discharge and casing. The inlet and discharge levels are those that are inside the duct. To use these levels to calculate the sound power radiated externally through the fan casing and adjacent ductwork, the corrections listed in Table 10.3 may be used, or, alternatively, see Section 8.16. These corrections assume standard rectangular ductwork lined on the inside, with the lining beginning a short distance from the fan. To calculate the sound power emerging from the end of the duct, the attenuation due to duct linings, duct end reflections, duct bends and plenum chambers must be taken into account (see Chapter 8).

TABLE 10.3 Octave band adjustments for sound power levels radiated by fan housings and adjacent ductwork

Octave band centre frequency (Hz)	Value to be subtracted from calculated in-duct sound power level, L_W (dB re 10^{-12}W)
63	0
125	0
250	5
500	10
1000	15
2000	20
4000	22
8000	25

In designing air handling systems, fan noise can be minimised by minimising the resistance to air flow (that is, minimising system pressure losses). In some cases, the BPF tone or its harmonics may be amplified over the manufacturer's specifications due to a number of causes listed below:

- the BPF or its harmonics may correspond to an acoustic resonance in the ductwork;
- inlet flow distortions;
- unstable, turbulent or swirling inlet flow; and
- operation of an inlet volume control damper.

Fan noise in duct systems can be minimised by avoiding the conditions listed above by ensuring the following:

- appropriate sizing of ductwork and duct elements;
- avoiding abrupt changes in duct cross-sectional area or direction and providing smooth airflow through all duct elements;
- providing 5 to 10 duct diameters of straight ductwork between duct elements;
- using variable speed fans instead of dampers for flow control; and
- if dampers are used, locating them a minimum of 3 (preferably 5 to 10) duct diameters away from room air devices such as diffusers.

Where fan sound power levels are unavailable from the manufacturer, estimates for a limited range of types can be obtained by following the procedure outlined in VDI2081-1 (2003). This procedure is based on measurements of a range of fans of different types and sizes. The fundamental equation for the sound power level generated in an octave band of centre frequency, f, by an industrial fan with a volume flow rate of $\dot{V}$ ($\mathrm{m^3/s}$) and a pressure difference between the fan outlet and inlet of ΔP_t (Pa) is given by:

$$L_{WB} = L_{Wsm} + 10\log_{10}\dot{V} + 20\log_{10}\Delta P_t + \Delta L_{WB} \qquad (\text{dB re } 10^{-12}\text{ W}) \tag{10.2}$$

where

$$\Delta L_{WB} = -C_1 - C_2\left[\log_{10}\left(\frac{f}{\pi N}\right) + C_3\right] \qquad (\text{dB re } 10^{-12}\text{ W}) \tag{10.3}$$

and

$$L_{Wsm} = \begin{cases} 34\text{ dB} & \text{for centrifugal fans with backward curved blades} \\ 36\text{ dB} & \text{for centrifugal fans with forward curved blades} \\ 42\text{ dB} & \text{for axial (propeller) fans with downstream diffuser} \end{cases} \tag{10.4}$$

and where N is the fan rotational speed in revolutions per second, $C_1 = C_2 = 5$ and $-0.6 \leq C_3 \leq 0.6$. The value of C_3 is at the lower end of the range for slow-speed fans and at the higher end for high-speed fans.

An additional level of 4 dB must be added to the octave band containing the blade pass frequency for axial fans only. Well designed centrifugal fans need no such addition (VDI2081-1, 2003).

10.3 Air Compressors

10.3.1 Small Compressors

Air compressors are a common source of noise. In this section, several estimation procedures are presented for various types of compressors. If the compressors are of small to medium size, then the data presented in Table 10.4 may be used in the power range shown to estimate sound pressure levels at 1 m. In most cases, the values will be conservative; that is, a little too high.

TABLE 10.4 Estimated sound pressure levels of small air compressors at 1 m distance (dB re 20 µPa)

Octave band centre frequency (Hz)	Air compressor power (kW)		
	Up to 1.5	2–6	7–75
31.5	82	87	92
63	81	84	87
125	81	84	87
250	80	83	86
500	83	86	89
1000	86	89	92
2000	86	89	92
4000	84	87	90
8000	81	84	87

Data from Army, Air Force and Navy, USA (2003).

10.3.2 Large Compressors (Sound Pressure Levels within the Inlet and Exit Piping)

The following equations may be used for estimating the sound power levels generated within the exit piping of large centrifugal, axial and reciprocating compressors (Heitner, 1968).

10.3.2.1 Centrifugal Compressors

The overall sound power level measured at the exit piping inside the pipe is given by:

$$L_W = 20 \log_{10} \text{kW} + 50 \log_{10} U - 46 \quad (\text{dB re } 10^{-12} \text{ W}) \tag{10.5}$$

where U is the impeller tip speed (m/s) ($30 < U < 230$), and kW is the power of the driver motor (in kW). The frequency of maximum noise level is:

$$f_p = 4.1U \quad (\text{Hz}) \tag{10.6}$$

The sound power level in the octave band containing f_p is taken as 4.5 dB less than the overall sound power level. The spectrum rolls off at the rate of 3 dB per octave above and below the octave band of maximum sound power level.

10.3.2.2 Rotary (or Axial) Compressors

The following procedure may be used for estimating the overall sound power level at the exit piping within the pipe:

$$L_W = 68.5 + 20\log_{10}\text{kW} \qquad (\text{dB re } 10^{-12}\text{ W}) \tag{10.7}$$

The frequency of maximum noise output is the second harmonic, or:

$$f_p = B(\text{rpm})/30 \qquad (\text{Hz}) \tag{10.8}$$

where B is the number of blades on the compressor. The spectrum is obtained as follows:

For the 63 Hz octave:

$$L_W = 76.5 + 10\log_{10}\text{kW} \qquad (\text{dB re } 10^{-12}\text{ W}) \tag{10.9}$$

For the 500 Hz octave:

$$L_W = 72 + 13.5\log_{10}\text{kW} \qquad (\text{dB re } 10^{-12}\text{ W}) \tag{10.10}$$

For the octave band containing f_p:

$$L_W = 66.5 + 20\log_{10}\text{kW} \qquad (\text{dB re } 10^{-12}\text{ W}) \tag{10.11}$$

For the octave band containing f_h:

$$L_W = 72 + 13.5\log_{10}\text{kW} \qquad (\text{dB re } 10^{-12}\text{ W}) \tag{10.12}$$

where

$$f_h = f_p^2/400 \tag{10.13}$$

To plot the total spectrum, a straight line is drawn between the L_W for the 63 Hz octave and the L_W for the 500 Hz octave band on an L_W versus $\log_{10} f$ plot. A smooth curve is drawn from the L_W value for the 500 Hz octave band through L_W values for the octave band containing f_p and the octave band containing f_h. The slope continues as a straight line beyond the f_h octave.

10.3.2.3 Reciprocating Compressors

The overall sound power level within the exit piping of a reciprocating compressor can be calculated using:

$$L_W = 106.5 + 10\log_{10}\text{kW} \qquad (\text{dB re } 10^{-12}\text{ W}) \tag{10.14}$$

To determine the spectrum values, the octave band that contains the fundamental frequency, $f_p = B(\text{rpm})/60$, where B is the number of cylinders of the machine, is determined. The sound power level in this band is taken as 4.5 dB less than the overall sound power level calculated using Equation (10.14). The levels in higher and lower octave bands decrease by 3 dB per octave. Implicit in the use of Equation (10.14) is the assumption that the radiated sound power is distributed over the first 15 m of downstream pipe.

10.3.3 Large Compressors (Exterior Sound Pressure Levels)

The sound power radiated by the compressor casing and exit pipe walls can be calculated using the equations for L_W inside the piping, which are included in Section 10.3.2, and subtracting the transmission loss of the casing and exit piping, which is calculated using the following equation (Heitner, 1968):

$$\text{TL} = 17\log_{10}(mf) - 48 \qquad (\text{dB}) \tag{10.15}$$

where m is the surface weight (kg/m^2) of the pipe wall, and f is the octave band centre frequency (Hz). This formula represents a simplification of a complex problem and is based on the assumption of adequate structural rigidity. Thus, for large diameter, thin-walled, inadequately supported pipes, the transmission loss may be less than given above.

Alternatively, the following equations may be used to calculate the overall external sound power levels directly (Edison Electric Institute, 1978).

10.3.3.1 Rotary and Reciprocating Compressors (Including Partially Muffled Inlets)

$$L_W = 90 + 10\log_{10}\text{kW} \qquad (\text{dB re } 10^{-12}\text{ W}) \tag{10.16}$$

10.3.3.2 Centrifugal Compressors (Casing Noise Excluding Air Inlet Noise)

$$L_W = 79 + 10\log_{10}\text{kW} \qquad (\text{dB re } 10^{-12}\text{ W}) \tag{10.17}$$

10.3.3.3 Centrifugal Compressors (Unmuffled Air Inlet Noise Excluding Casing Noise)

$$L_W = 80 + 10\log_{10}\text{kW} \qquad (\text{dB re } 10^{-12}\text{ W}) \tag{10.18}$$

Octave band levels may be derived from the overall levels by arithmetically subtracting the corrections listed in Table 10.5. The octave band levels can then be further corrected by arithmetically adding the difference between the overall level calculated using the appropriate one of Equations (10.16) to (10.18) and the overall level calculated by logarithmically summing the corrected octave band levels. The resulting octave band levels, when summed logarithmically, will equal the value calculated using the appropriate one of Equations (10.16) to (10.18). The same approach should be used whenever octave band levels are calculated by arithmetically subtracting tabulated correction values from an overall level.

TABLE 10.5 Octave band corrections for exterior sound pressure levels radiated by compressors

Octave band centre frequency (Hz)	Correction (dB)		
	Rotary and reciprocating	Centrifugal casing	Centrifugal air inlet
31.5	11	10	18
63	15	10	16
125	10	11	14
250	11	13	10
500	13	13	8
1000	10	11	6
2000	5	7	5
4000	8	8	10
8000	15	12	16

Data from Edison Electric Institute (1978).

10.4 Compressors for Chillers and Refrigeration Units

The compressor is usually the dominant noise source in a refrigeration unit, so it is generally sufficient to consider noise generation from this source alone when considering a packaged chiller. Three types of compressor will be considered: centrifugal, rotary screw and reciprocating. Sound pressure levels measured at 1 m are listed in Table 10.6 for these machines; these levels will not be exceeded by 90–95% of commercially available machines. The machines are identified in the table both by type and by cooling capacity. Speed variations between the different commercially available units are insignificant.

TABLE 10.6 Estimated sound pressure levels of packaged chillers at one metre distance (dB re 20 µPa). These levels are generally higher than observed (see manufacturer's data)

Type and capacity of machine	Octave band centre frequency (Hz)								
	31.5	63	125	250	500	1000	2000	4000	8000
Reciprocating compressors									
35–175 kW	79	83	84	85	86	84	82	78	72
176–700 kW	81	86	87	90	91	90	87	83	78
Rotary screw compressors									
350–1050 kW	70	76	80	92	89	85	80	75	73
Centrifugal compressors									
Under 1750 kW	92	93	94	95	91	91	91	87	80
≥1750 kW	92	93	94	95	93	98	98	93	87

Data from Army, Air Force and Navy, USA (2003).

10.5 Cooling Towers

Various types of cooling towers are illustrated in Figure 10.1. Estimated overall sound power levels, corresponding to the cooling towers illustrated in Figure 10.1, are given by the following equations:

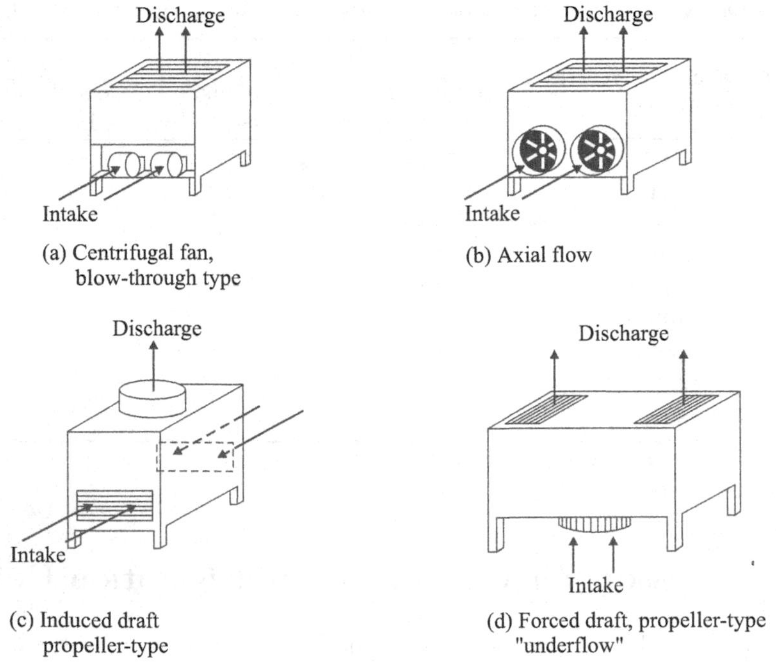

FIGURE 10.1 Principal types of cooling tower (Army, Air Force and Navy, USA, 2003).

1. Propeller-type cooling towers (Army, Air Force and Navy, USA, 2003):

$$L_W = 96.3 + 10\log_{10}\text{kW} \quad (\text{dB re } 10^{-12}\text{ W}) \tag{10.19}$$

2. Centrifugal type cooling towers (Army, Air Force and Navy, USA, 2003):

$$L_W = 86.3 + 10\log_{10}\text{kW} \quad (\text{dB re } 10^{-12}\text{ W}) \tag{10.20}$$

The octave band sound power levels may be calculated by subtracting the corrections listed in Table 10.7 from the appropriate overall values. If the cooling tower fans are operated at half speed, the cooling capacity is reduced by one-third and the noise reductions in octave bands containing frequencies corresponding to fractions and multiples of the half-speed blade passage frequency, BPF, are listed in Table 10.8.

TABLE 10.7 Values (dB) to subtract from overall levels to obtain octave band sound power levels for cooling towers

Octave band centre frequency (Hz)	Propeller type	Centrifugal type
31.5	8	6
63	5	6
125	5	8
250	8	10
500	11	11
1000	15	13
2000	18	12
4000	21	18
8000	29	25

Data from Army, Air Force and Navy, USA (2003).

TABLE 10.8 Values (dB) to subtract from octave band levels for octave bands containing frequencies corresponding to fractions and multiples of the blade pass frequency (BPF) to account for operating cooling tower fans at half speed

BPF fraction or multiple	BPF/8	BPF/4	BPF/2	BPF	2× BPF	4× BPF	8× BPF
Noise reduction (dB)	3	6	9	9	9	6	3

Data from Army, Air Force and Navy, USA (2003).

The average sound pressure levels at various distances from the tower may be calculated using these numbers and the procedures outlined in Chapters 4 and 5. Table 10.9 gives the approximate corrections to add to the calculated logarithmically-averaged sound pressure levels to account for directivity effects at distances greater than 6 m from the tower (see Section 8.19).

Table 10.10 gives estimates for the sound pressure levels close to the intake and discharge openings. Here, and in many places in this chapter, the octave band sound power levels should add up logarithmically to the total sound power level and this should agree with the overall L_W (Equations (10.19) and (10.20) in this case). However, the corrections are just integer numbers and the total of the band levels can sometimes be a little different to the total calculated by the relevant equation. In this case, the amount to be added (which may be negative) to each of the band levels is the overall power level calculated by the relevant equation minus the total calculated by logarithmically summing octave band levels. In some cases, where the overall level exceeds the sum of the octave band levels, the difference may be attributable to sound energy in octave bands not included in the summation, but this is more the exception than the rule.

TABLE 10.9 Approximate corrections (dB) to the average sound pressure level for directional effects of cooling towers

Type of tower and location of measurement	Octave band centre frequency (Hz)								
	31.5	63	125	250	500	1000	2000	4000	8000
Centrifugal fan, blow through type									
Front	+3	+3	+2	+3	+4	+3	+3	+4	+4
Side	0	0	0	−2	−3	−4	−5	−5	−5
Rear	0	0	−1	−2	−3	−4	−5	−6	−6
Top	−3	−3	−2	0	+1	+2	+3	+4	+5
Axial flow fan, blow through type									
Front	+2	+2	+4	+6	+6	+5	+5	+5	+5
Side	+1	+1	+1	−2	−5	−5	−5	−5	−4
Rear	−3	−3	−4	−7	−7	−7	−8	−11	−3
Top	−5	−5	−5	−5	−2	0	0	+2	+4
Induced draft fan, propeller type									
Front	0	0	0	+1	+2	+2	+2	+3	+3
Side	−2	−2	−2	−3	−4	−4	−5	−6	−6
Top	+3	+3	+3	+3	+2	+2	+2	+1	+1
"Underflow" forced draft fan, propeller type.									
Any side	−1	−1	−1	−2	−2	−3	−3	−4	−4
Top	+2	+2	+2	+3	+3	+4	+4	+5	+5

These corrections apply only when there are no reflecting or obstructing surfaces that would modify the normal radiation of sound from the tower. Add these corrections to the average sound pressure level calculated. Do not apply these corrections for close-in positions less than 6 m from the tower.
Data from Army, Air Force and Navy, USA (2003).

TABLE 10.10 Estimated close-in sound pressure levels (dB re 20 µPa) for the intake and discharge openings of various cooling towers

Type of tower and location of measurement	Octave band centre frequency (Hz)								
	31.5	63	125	250	500	1000	2000	4000	8000
Centrifugal fan, blow through type									
Intake	85	85	85	83	81	79	76	73	68
Discharge	80	80	80	79	78	77	76	75	74
Axial flow fan, blow through type									
Intake	97	100	98	95	91	86	81	76	71
Discharge	88	88	88	86	84	82	80	78	76
Induced draft fan, propeller type									
Intake	97	98	97	94	90	85	80	75	70
Discharge	102	107	103	98	93	88	83	78	73

Data from Army, Air Force and Navy, USA (2003).

10.6 Pumps

Estimated sound pressure levels generated by a pump, at a distance of 1 m from its surface, as a function of pump power, are presented in Tables 10.11 and 10.12.

TABLE 10.11 Overall pump sound pressure levels (dB re 20 µPa) at 1 m from the pump

Speed range (rpm)	Drive motor nameplate power	
	Under 75 kW	Above 75 kW
3000–3600	72.3 + 10 $\log_{10}$ kW	85.4 + 3 $\log_{10}$ kW
1600–1800	75.3 + 10 $\log_{10}$ kW	88.4 + 3 $\log_{10}$ kW
1000–1500	70.3 + 10 $\log_{10}$ kW	83.4 + 3 $\log_{10}$ kW
450–900	68.3 + 10 $\log_{10}$ kW	81.4 + 3 $\log_{10}$ kW

Data from Army, Air Force and Navy, USA (2003).

TABLE 10.12 Frequency adjustments for pump sound power levels

Octave band centre frequency (Hz)	Value to be subtracted from overall sound pressure level (dB)
31.5	13
63	12
125	11
250	9
500	9
1000	6
2000	9
4000	13
8000	19

Subtract these values from the overall sound pressure level to obtain octave band sound pressure levels.

Data from Army, Air Force and Navy, USA (2003).

10.7 Jets

10.7.1 General Estimation Procedures

Pneumatic devices quite often eject gas (usually air) in the form of high-pressure jets. Such jets can be very significant generators of noise. The acoustic power generated by a subsonic jet in free space is related to the mechanical stream power by an efficiency factor as:

$$W_a = \eta W_m \qquad \text{(W)} \tag{10.21}$$

The stream mechanical power, W_m, in turn, is equal to the convected kinetic energy of the stream, which for a jet of circular cross-section is:

$$W_m = \rho U^3 \pi d_j^2/8 = U^2 \dot{m}/2 \qquad \text{(W)} \tag{10.22a,b}$$

In the above equations, U is the jet exit velocity (m/s), W_a is the radiated acoustic power (W), W_m is the mechanical stream power (W), $\dot{m}$ is the mass flow rate (kg/s), η is the acoustic efficiency of the jet, d_j is the jet diameter (m) and ρ is the density (kg/m^3) of the flowing gas.

The acoustical efficiency of the jet is approximately (Heitner, 1968):

$$\eta = (T/T_0)^2(\rho/\rho_0)K_a M^5 \tag{10.23}$$

where ρ_0 is the density (kg/m^3) of the ambient gas, K_a is the acoustical power coefficient and is approximately 5×10^{-5}, M is the stream Mach number relative to the ambient gas, T is the jet absolute temperature (K) and T_0 is the absolute temperature (K) of the ambient gas. If

Equation (10.23) gives a value of $\eta > 0.01$, then η is set equal to 0.01. Equation (10.23) is valid for noise radiation due to turbulent flow in both subsonic and supersonic jets and for the range of the quantity $(T/T_0)^2(\rho/\rho_0)$ between 0.1 and 10.

If the jet pressure ratio (ratio of the pressure upstream of the jet to the ambient pressure) is less than 1.89, the jet is subsonic and noise is only due to turbulent flow with the acoustical efficiency estimated using Equation (10.23). The radiated sound power is then found by substituting this acoustical efficiency value into Equation (10.21).

If the pressure ratio is greater than 1.89, the jet will be choked and the gas exit velocity will equal Mach 1. If the nozzle contraction is followed by an expansion, the gas exit velocity will exceed Mach 1. In both cases, there will be shock-wave-generated noise in addition to turbulence noise. Equations (10.21) to (10.23) are used to calculate the turbulence part of the noise. The efficiency of the shock-generated noise is calculated using Figure 10.2 and the associated sound power is calculated using Equation (10.21). The total sound power radiated by the choked jet is the sum of the sound power due to turbulence and that due to shock.

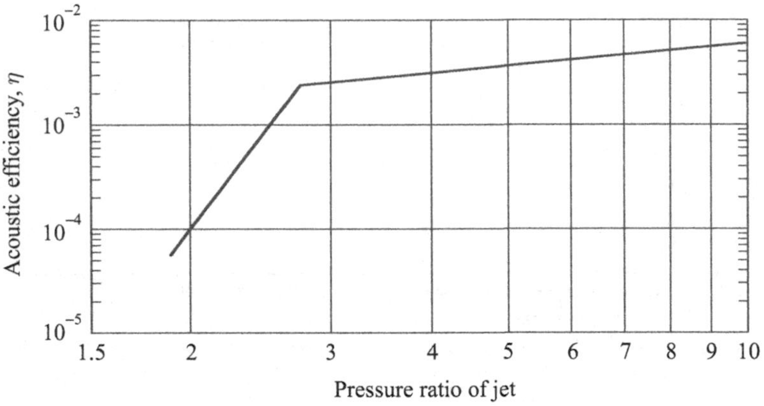

FIGURE 10.2 Acoustic efficiency of shock noise generated by choked jets (adapted from Heitner (1968)).

The overall sound power level of the jet (see Equation (1.89)) is:

$$L_W = 10\log_{10} W_a + 120 \qquad (\text{dB re } 10^{-12}\text{ W}) \tag{10.24}$$

This quantity can be used with the methods of Chapter 6 to estimate the sound pressure level in, for example, a room characterised by a given room constant. On the other hand, in a free field or close to the jet, the overall sound pressure level is:

$$L_p = L_W + \text{DI} - 10\log_{10}(4\pi r^2) \qquad (\text{dB re } 20\,\mu\text{Pa}) \tag{10.25}$$

where DI is the jet directivity index, and r is the distance (m) from the jet orifice to the observation point. Values for the directivity index for a jet are given in Table 10.13 as a function of angle from the jet axis.

The spectrum shape for the jet is illustrated in Figure 10.3 (Ingard, 1959), where the sound pressure level in each of the octave bands is shown relative to the overall sound pressure level calculated using Equation (10.25). The band levels obtained using Figure 10.3 must be adjusted (all by the same number of dB) so that their sum is the same as that obtained using Equation (10.25).

In Figure 10.3, the frequency, f_p (Hz), is defined in terms of a Strouhal number, N_s, as:

$$N_s = f_p\, d_j/U \tag{10.26}$$

TABLE 10.13 Directional correction for jets (data from Heitner (1968))

Angle from jet axis (degrees)	Directivity index, DI (dB)	
	Subsonic	Choked
0	0	−3
20	+1	+1
40	+8	+6
60	+2	+3
80	−4	−1
100	−8	−1
120	−11	−4
140	−13	−6
160	−15	−8
180	−17	−10

Data from Army, Air Force and Navy, USA (2003).

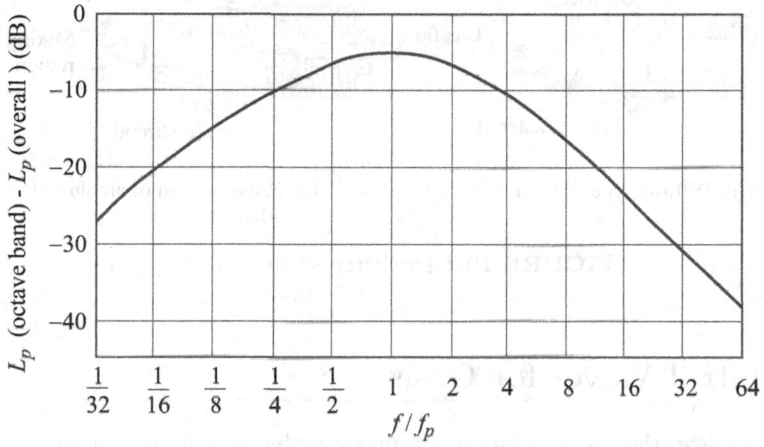

FIGURE 10.3 Noise spectrum for gas jets (adapted from Ingard (1959)).

where N_s is generally about 0.2 for subsonic jets; d_j and U are, respectively, the jet diameter (m) and exit velocity (m/s).

Noise produced by gas and steam vents can be estimated by assuming that they are free jets (unconstricted by pipe walls). The estimation procedure for free jets outlined above is then used to calculate sound power and sound pressure levels. However, for steam vents, if the calculated jet efficiency, η, is less than 0.005, more accurate estimates are obtained by adding 3 dB to the calculated noise levels.

10.7.2 General Jet Noise Control

Means that would be applicable for containment of noise from pneumatic jets were discussed in Chapter 7 and means for attenuation were discussed in Chapter 8. However, these means are generally less efficient than means that alter the noise generating mechanism.

For example, quite large reductions can be obtained with aerodynamic noise sources when the basic noise-producing mechanism is altered. Some methods of reducing aerodynamic noise based on this principle are illustrated in Figure 10.4. The figure refers to silencers that discharge to the atmosphere, but some of these methods can result in appreciable noise reductions, even for discharge in the confinement of a pipe. For example, modified forms of the devices illustrated in the first two parts of Figure 10.4 have been found remarkably effective for the control of valve noise in piping systems. Alternatively, orifice plates (circular plates containing multiple holes) placed downstream from the jet or valve and as close as possible to it also have been found to be very effective. The orifice plate should occupy the entire cross-section of the pipe.

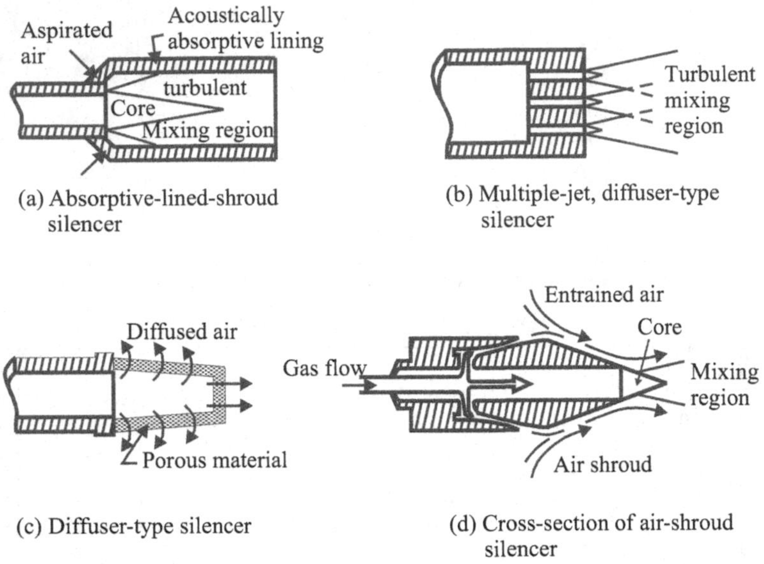

FIGURE 10.4 Discharge silencers for air.

10.8 Control Valves for Gases

A control valve allows the passage of a fluid from one piping system to another while at the same time controlling the pressure ratio between the two systems. For example, a fluid at an absolute pressure, P_1, in one system may be transported to a fluid at a lower absolute pressure, P_2, in the second system, by dissipating potential energy stored in the fluid at the higher pressure during the passage of the fluid through the control valve. The necessary energy dissipation is accomplished by conversion of the pressure head to heat and a very small fraction to sound by intense turbulence and shock formation in the control valve. The absolute pressure is the static pressure, measured with a pressure tapping in the wall of a pipe, plus atmospheric pressure.

Control valves are a common source of noise in many industries, especially when the operation of the valve is characterised by a large pressure drop. The primary noise generating mechanism is the jet of fluid formed between the valve and its seat; thus valve noise is modelled as a confined jet. Consequently, the noise generation mechanisms are turbulent mixing, turbulence-boundary interaction, shock, shock-turbulence interaction and flow separation (Ng, 1980).

10.8.1 Internal Sound Power Generation

The observation that the radiated sound must be some fraction of the potential energy dissipated in the control valve suggests the existence of an energy conversion efficiency factor (or acoustical power coefficient), η, relating the mechanical stream power, W_m, entering the control valve, which is a function of the pressure drop across the valve, to the sound power, W_a, which is generated by the process. The proposed relation for the sound power radiated downstream of the valve is:

$$W_a = W_m \eta r_w \tag{10.27}$$

The mechanical stream power, W_m, is defined in Equation (10.22), and the acoustical power coefficient, η, is defined later in Equations (10.35) to (10.39) for various valve flow conditions. The quantity, r_w, is the ratio of acoustic power propagated downstream of the valve to the total acoustic power generated by the valve and it varies between 0.25 and 1.0, depending on the valve type (see Table 10.14 and IEC 60534-8-3 (2010)). Note that the quantities on the right-hand side of Equation (10.22) can refer to properties in any part of the gas flow. If properties in the valve vena contracta are used, then $d = d_j$ is the diameter of the jet in the vena contracta, which is the region of the flow through the valve where the velocity is a maximum and the pressure is a minimum. If the flow is sonic through the vena contracta, then the stream power is calculated using the speed of sound as the flow speed. Note that the sound power radiated upstream of the valve is not considered significant. It is not clear from IEC 60534-8-3 (2010) what happens to the acoustic power that does not propagate downstream. Presumably, it is dissipated in the valve, as, according to the standard, little escapes through the valve casing.

The control valve noise estimation procedure described here begins with the consideration of possible regimes of operation. Consideration of the regimes of operation, in turn, provides the means for determining the energy conversion efficiency of the stream power entering the control valve.

To facilitate understanding and discussion of the noise generation mechanisms and their efficiencies in a control valve, it will be convenient first to consider the possibility of pressure recovery within the valve, as illustrated in Figure 10.5. In the figure, a valve opening and the

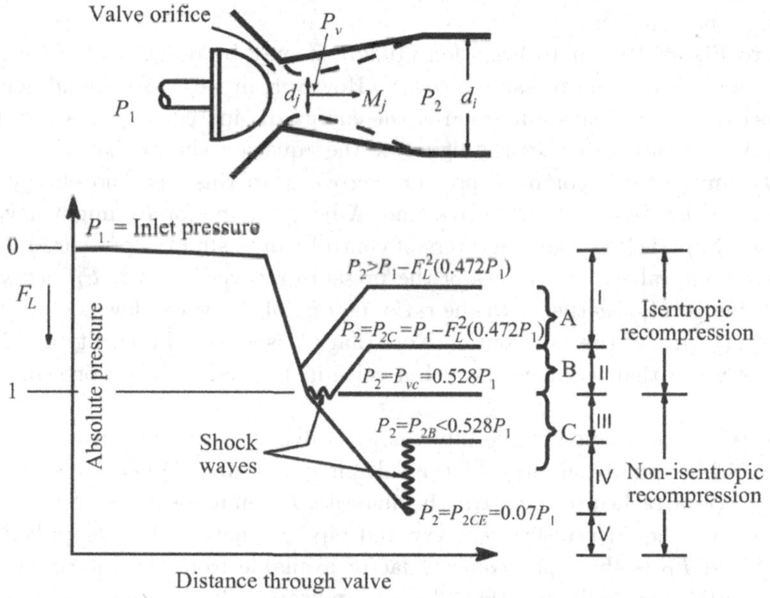

FIGURE 10.5 Schematic representation of the absolute pressure distribution through a control valve with inlet pressure P_1 and outlet pressure P_2. In pressure range A, there is subsonic flow and drag dipole turbulence noise. In pressure range B, there is sonic flow in the vena contracta, and dipole and quadrupole turbulence noise and shock noise. In pressure range C, there is sonic flow and shock noise is dominant. $F_L = 0$ when $P_2 = P_1$ and $F_L = 1$ when $P_2 = 0.528P_1$ (for diatomic gases such as air, after Baumann and Coney (2006)). P_{2C}, P_{vc}, P_{2B}, and P_{2CE} can be calculated using Equations (10.32), (10.30), (10.33a) and (10.33b), respectively.

corresponding pressure distribution through the valve is schematically represented for one inlet pressure, P_1, and five possible outlet pressures, P_2.

Referring to Figure 10.5, when the fluid at higher pressure, P_1, enters the fluid at lower pressure, P_2, through the valve orifice, a confined jet is formed, which is characterised in turn by a vena contracta of diameter, d_j, and minimum absolute pressure, P_v, indicated by the three minima in the figure, after which the absolute pressure rises to P_2. As the figure shows, it is possible that the flow may be sonic ($P_v = P_{vc}$) in the vena contracta, although the pressure ratio, P_1/P_2, across the valve is sub-critical ($P_2 > P_{vc}$). The extent of the pressure recovery is determined by the design of the valve and is determined empirically.

To quantify pressure recovery and the noise generating behaviour of a valve, a quantity, F_L, called the "pressure recovery factor" has been defined. The definition, which forms the basis for its determination, is given in terms of the following relation involving the inlet absolute pressure, P_1, the outlet absolute pressure, P_2, and the absolute pressure, P_v, in the vena contracta:

$$F_L^2 = \frac{P_1 - P_2}{P_1 - P_v} \tag{10.28}$$

The absolute pressure in the vena contracta, which cannot be measured easily and which also cannot be negative, is related to the upstream and downstream pressures by:

$$P_v = P_1 - \left(\frac{P_1 - P_2}{F_L^2}\right) \tag{10.29}$$

This equation is used for calculating the vena contracta pressure for flow regime 1 (sub-sonic flow – see page 628), but for other flow regimes, which are characterised by sonic flow (speed of flow = speed of sound) in the vena contracta, Equation (10.30) to follow is used to calculate the pressure in the vena contracta.

Referring to Figure 10.5 or to Equation (10.28), it may be noted that F_L has the value 1 when $P_2 = P_v$ and there is no pressure recovery. However, in this case, sound generation must be maximal because it is always a fraction of the energy dissipated, which is now maximal. On the other hand, reference to either the figure or the equation shows that F_L has the value 0 when $P_2 = P_1$ and there is complete pressure recovery. In this case, no energy is dissipated and thus no energy has been converted to sound. A better name for F_L might have been "noise recovery factor". Nonetheless, manufacturers of control valves supply "pressure recovery factor" information for their valves, and values of the pressure recovery factor, F_L, for various valves are given in Table 10.14, together with the ratio, C_v/d_o^2, of the valve flow coefficient, C_v, to the inside diameter, d_o (m), of the valve outlet. Note that C_v is a non-SI quantity and is the number of US gallons of water that will flow through a valve in 1 minute when a pressure drop of 1 psi exists.

The quantity, F_d, also listed in the table, is the valve style modifier used later on in Equation (10.51), which is usually supplied by the valve manufacturer. When there are fittings, such as bends or T-pieces attached to the valve, the quantity, F_L, must be replaced by F_{LP}/F_P, where F_{LP} is the combined liquid pressure recovery and pipe geometry factor (supplied by the valve manufacturer) and F_P is the pipe geometry factor available from the International Standard, IEC 60534-2-1 (2011). The column titled "flow to" indicates whether the flow direction is acting to open or close the valve.

When considering noise generation in a control valve using this prediction procedure, three regimes of operation and associated energy dissipation and noise generation mechanisms may be identified. These regimes may be understood by reference to Figure 10.5, where they are identified as A, B and C. In regimes A and B, hydrodynamic dipole and quadrupole sources are responsible for noise generation. Such noise generation is discussed in Chapter 4.

Referring to Figure 10.5, in regime A the flow through the valve is everywhere subsonic and all energy dissipation is accomplished by intense turbulence. In this case, noise generation

TABLE 10.14 Typical values of valve noise prediction factors (after Baumann and Coney (2006)). See IEC 60534-8-3 (2010) for data for other valve types. The column labelled "flow to" indicates whether the flow acts to open or close the valve

Type of valve	Flow to	Percent of capacity	C_v/d_o^2	F_L	F_d	r_w
1. Globe valves						
Single-port parabolic plug	Close	100	25000	0.80	1.00	0.25
	Open	100	20000	0.90	0.46	0.25
	Open	75	15000	0.90	0.36	0.25
	Open	50	10000	0.90	0.28	0.25
	Open	25	5000	0.90	0.16	0.25
	Open	10	2000	0.90	0.10	0.25
v-port plug	Open	100	16000	0.92	0.50	0.25
		50	8000	0.95	0.42	0.25
		30	5000	0.95	0.41	0.25
Four-port cage	Open	100	25000	0.90	0.43	0.25
	Open	50	13000	0.90	0.36	0.25
Six-port cage	Open	100	25000	0.90	0.32	0.25
	Open	50	13000	0.90	0.25	0.25
		Angle of travel (°)				
2. Eccentric rotary plug	Open	50	20000	0.85	0.42	0.25
		30	13000	0.91	0.30	0.25
	Close	50	21000	0.68	0.45	0.25
		30	13000	0.88	0.30	0.25
3. Ball valve segmented	Open	60	18000	0.66	0.75	0.25
		30	5000	0.82	0.63	0.25
4. Butterfly						
Swing-through vane		75	50000	0.56	0.57	0.5
		60	30000	0.67	0.50	0.5
		50	16000	0.74	0.42	0.5
		40	10000	0.78	0.34	0.5
		30	5000	0.80	0.26	0.5
Fluted vane		75	40000	0.70	0.30	0.5
		50	13000	0.76	0.19	0.5
		30	7000	0.82	0.08	0.5

is found to be proportional to the sixth power of the stream speed, which implies that it is dominated by fluctuating forces acting on the fluid and is thus dipole in nature (see Section 4.3). It is suggested here that a fluctuating re-attachment bubble in the region indicated by the dotted lines in Figure 10.5 (see also Figure 8.41) might account for a drag dipole with its axis oriented along the axis of the pipe and which radiates propagating sound at all frequencies. Referring to Figure 10.5, the bound between regimes A and B is seen to be determined by the pressure ratio, P_2/P_1, at which the flow in the vena contracta becomes sonic. The flow speed in the vena contracta will be sonic when the pressure, P_v, in the vena contracta is equal to the critical pressure, P_{vc}, which satisfies the following relation (Landau and Lifshitz, 1959):

$$\frac{P_{vc}}{P_1} = \left(\frac{2}{\gamma+1}\right)^{\gamma/(\gamma-1)} \tag{10.30}$$

For diatomic gases, where $\gamma = 1.4$:

$$P_{vc}/P_1 = 0.528 \tag{10.31}$$

Equations (10.30) and (10.31) are used for Regimes II to V to calculate the vena contracta pressure from the inlet pressure. For Regime I, Equation (10.29) is used. Values of the ratio of specific heats, γ, are given in Table C.2 for various commonly used gases.

Solving Equation (10.28) for the pressure ratio P_2/P_1 and using Equation (10.30) gives the pressure ratio across the valve, $P_2/P_1 = P_{2C}/P_1$, at which the vena contracta flow is sonic, as:

$$\frac{P_{2C}}{P_1} = 1 - F_L^2 \left[1 - \left(\frac{2}{\gamma+1}\right)^{\gamma/(\gamma-1)}\right] \tag{10.32}$$

Equations (10.31) and (10.32) establish the bounds between regimes B and C and between A and B in Figure 10.5.

In regime B, the flow is sonic in the vena contracta and $P_2 > P_{vc}$. In this case, noise generation tends to be proportional to the eighth power of the stream speed. In this regime, the noise generation efficiency of quadrupoles (see Section 4.4) associated with intense turbulence of the jet through the valve orifice will overtake the dipole sources that are still present, so that both are considered to contribute to noise generation. In addition, shocks will be formed, which will be rendered unstable by the passage of turbulent flow through them and these will provide a very efficient noise generation mechanism.

The critical pressure ratio, when $P_2 = P_{vc}$ and flow through the valve is sonic (for diatomic gases), is shown in Figure 10.5 as the bound between regimes B and C. In regime C, the pressure ratio across the valve, P_2/P_1, is less than P_{vc}/P_1 given by Equation (10.31) and the flow is sonic everywhere in the valve. In this regime, energy dissipation is dominated by shock formation and noise generation efficiency is maximal.

The IEC standard (IEC 60534-8-3, 2010) has further divided regimes B and C into regimes II, III and IV, adding a regime V and renaming regime A as regime I (see Figure 10.5). To delineate regions IV and V, the IEC standard defines two additional pressures, P_{2B} and P_{2CE}:

$$P_{2B} = \frac{P_1}{\alpha}\left(\frac{1}{\gamma}\right)^{\gamma/(\gamma-1)}; \quad P_{2CE} = \frac{P_1}{22\alpha} \tag{10.33a,b}$$

where the quantity, α, is defined as:

$$\alpha = \left\{\left(\frac{\gamma+1}{2}\right)^{\gamma/(\gamma-1)} - F_L^2\left[\left(\frac{\gamma+1}{2}\right)^{\gamma/(\gamma-1)} - 1\right]\right\}^{-1} \tag{10.34}$$

Simplifying the equations used for calculating the efficiency in the standard, IEC 60534-8-3 (2010), the following expressions are obtained for the acoustical power coefficients for regimes I to V, defined in Figure 10.5. The acoustical efficiency factors (η_1, η_2, etc.) used in the standard are related to the acoustical power coefficients, η, used here by $\eta = \eta_1 \times F_L^2$, $\eta = \eta_2 \times (P_1 - P_2)/(P_1 - P_{vc})$, $\eta = \eta_3$, η_4, and η_5.

In **regime I**, flow is subsonic everywhere in the valve and $P_2 \geq P_{2C}$:

$$\eta = 10^{-4} F_L^2 M_v^{3.6} = 10^{-4} F_L^2 \left[\left(\frac{2}{\gamma-1}\right)\left(1 - \left(\frac{P_v}{P_1}\right)^{\frac{\gamma-1}{\gamma}}\right)\left(\left(\frac{P_1}{P_v}\right)^{\frac{\gamma-1}{\gamma}}\right)\right]^{1.8} \tag{10.35}$$

where the Mach number, M_v, in the vena contracta is defined in Equation (10.45) and the pressure, P_v, in the vena contracta may be calculated from the upstream and downstream pressures using Equation (10.29). When fittings are attached to the valve, the quantity, F_L, must be replaced with F_{LP}/F_P in Equations (10.34)–(10.39). However, this is generally only a small effect and a good approximation in the absence of the required data is to use F_L.

In **regime II**, flow is sonic in the vena contracta only and $P_{vc} \leq P_2 < P_{2C}$:

$$\eta = 10^{-4}\left(\frac{P_1 - P_2}{P_1 - P_{vc}}\right)\left(\frac{2}{\gamma - 1}\left[\left(\frac{P_1}{\alpha P_2}\right)^{(\gamma-1)/\gamma} - 1\right]\right)^{3.3F_L^2} \tag{10.36}$$

In **regime III**, flow is sonic everywhere in the valve and $P_{2B} \leq P_2 < P_{vc}$:

$$\eta = 10^{-4}\left(\frac{2}{\gamma - 1}\left[\left(\frac{P_1}{\alpha P_2}\right)^{(\gamma-1)/\gamma} - 1\right]\right)^{3.3F_L^2} \tag{10.37}$$

where α is defined in Equation (10.34).

In **regime IV**, flow is sonic everywhere in the valve and $P_{2CE} \leq P_2 < P_{2B}$:

$$\eta = 10^{-4}\frac{1}{\gamma - 1}\left[\left(\frac{P_1}{\alpha P_2}\right)^{(\gamma-1)/\gamma} - 1\right] \times 2^{3.3F_L^2} \tag{10.38}$$

In **regime V**, flow is sonic everywhere in the valve and $P_2 < P_{2CE}$:

$$\eta = 10^{-4}\frac{1}{\gamma - 1}\left[(22)^{(\gamma-1)/\gamma} - 1\right] \times 2^{3.3F_L^2} \tag{10.39}$$

10.8.2 Internal Sound Pressure Level

The mean square sound pressure in the pipe downstream of the valve is approximated as:

$$\langle p_i^2 \rangle = \frac{4W_a \rho_2 c_2}{\pi d_i^2} \tag{10.40}$$

where d_i is the internal diameter of the pipe and W_a is the sound power radiated downstream of the valve, as defined in Equation (10.27).

The density and speed of sound in the pipes upstream and downstream of the valve are related to the pressure and temperature in these sections of pipe (subscript 1 for upstream and subscript 2 for downstream):

$$\rho_i = \frac{P_i M}{RT_i}; \qquad i = 1, 2 \tag{10.41}$$

and

$$c_i = \sqrt{\frac{\gamma R T_i}{M}}; \qquad i = 1, 2 \tag{10.42}$$

If the downstream temperature, T_2, is unknown, it may be assumed to be equal to the upstream temperature.

For regime I (subsonic flow through the vena contracta), the speed of sound, c_v, in the vena contracta is related to the speed of sound, c_1 in the upstream pipe by the temperature and pressure ratios as (IEC 60534-8-3, 2010):

$$\frac{c_v}{c_1} = \sqrt{\frac{T_v}{T_1}} = \sqrt{\left(\frac{P_v}{P_1}\right)^{(\gamma-1)/\gamma}} \tag{10.43a,b}$$

For subsonic flow through the vena contracta (**regime I**), the flow speed, U_v, is calculated using (IEC 60534-8-3, 2010):

$$U_v = \sqrt{\left(\frac{2\gamma}{\gamma-1}\right)\left[1-\left(\frac{P_v}{P_1}\right)^{(\gamma-1)/\gamma}\right]\frac{P_1}{\rho_1}} = c_1\sqrt{\left(\frac{2}{\gamma-1}\right)\left[1-\left(\frac{P_v}{P_1}\right)^{(\gamma-1)/\gamma}\right]} \tag{10.44a,b}$$

The corresponding Mach number in the vena contracta for subsonic flow is:

$$M_v = \frac{U_v}{c_v} = \frac{U_v}{c_1}\frac{c_1}{c_v} = \left[\left(\frac{2}{\gamma-1}\right)\left(1-\left(\frac{P_v}{P_1}\right)^{\frac{\gamma-1}{\gamma}}\right)\left(\left(\frac{P_1}{P_v}\right)^{\frac{\gamma-1}{\gamma}}\right)\right]^{1/2} \tag{10.45a–c}$$

For sonic flow in the vena contracta (**regimes II to V**), the flow speed, U_v, through the vena contracta is sonic and equal to c_{vc}. For these regimes,

$$\frac{c_{vc}}{c_1} = \sqrt{\frac{T_{vc}}{T_1}} = \sqrt{\frac{2}{(\gamma+1)}} \tag{10.46a,b}$$

where T_{vc} is the temperature in the vena contracta when the flow speed is equal to the speed of sound. The preceding and following calculations require as input data the following properties of the gas flowing into the valve: ρ_1, T_1, P_1, γ, M and $\dot{m}$ and the pressure, P_2, at the valve discharge. The mass flow rate, $\dot{m}$, may be calculated from a knowledge of the volume flow rate, $\dot{V}$, at any given pressure, P, and temperature, T, using the universal gas law:

$$\dot{m} = \frac{P\dot{V}M}{RT} \tag{10.47}$$

where M is the molecular mass of the gas (kg/mol – see Table C.2) and R (=8.314 J/mol K) is the universal gas constant in consistent units. The mass flow rate is needed as an input in the valve noise calculations.

The Mach number, M_o at the valve outlet is calculated using:

$$M_o = 4\dot{m}/(\pi d_o^2 \rho_2 c_2) \tag{10.48}$$

where d_o is the valve outlet diameter. If $M_o > 0.3$, the calculation procedures described in this book are no longer valid and the reader should consult IEC 60534-8-3 (2010).

Taking logs of Equation (10.40) gives for the sound pressure level inside the pipe downstream of the valve:

$$L_{pi} = 10\log_{10}\left[\frac{3.18\times10^9 W_a \rho_2 c_2}{d_i^2}\right] \quad \text{(dB re 20 µPa)} \tag{10.49}$$

The spectrum peak frequency, f_p, may be estimated using the following equations (IEC 60534-8-3, 2010):

Regime I: The Mach number through the vena contracta is given by Equation (10.45) and the frequency of the peak noise level is:

$$f_p = \frac{0.2U_v}{d_j} \tag{10.50}$$

where U_v is defined by Equation (10.44) and the jet diameter, d_j is:

$$d_j \approx 4.6\times10^{-3} F_d\sqrt{C_v F_L} \text{ (m)} \tag{10.51}$$

When fittings are attached to the valve, the quantity, F_L, must be replaced with F_{LP}/F_P in Equation (10.51). The flow coefficient, C_v, may be calculated using the procedures outlined in

IEC 60534-2-1 (2011) or it may be supplied by the valve manufacturer. Note that the actual C_v for the valve opening conditions (rather than the valve rated C_v) should be used. Values in Table 10.14 are based on the outlet diameter, d_i (m), of the valve and may be used in the absence of more accurate data. The quantity, F_d, called a valve style modifier, is empirically determined and listed in Table 10.14.

Regimes II and III: The frequency, f_p of the peak noise level is:

$$f_p = \frac{0.2 M_j c_{vc}}{d_j} \tag{10.52}$$

where the freely expanding jet Mach number, M_j, is:

$$M_j = \left[\frac{2}{(\gamma - 1)}\left(\left(\frac{P_1}{\alpha P_2}\right)^{(\gamma-1)/\gamma} - 1\right)\right]^{1/2} \tag{10.53}$$

where α is defined in Equation (10.34).

Regime IV: The frequency, f_p of the peak noise level is:

$$f_p = \frac{0.28 c_{vc}}{d_j \sqrt{M_j^2 - 1}} \tag{10.54}$$

where the freely expanding jet Mach number is given by Equation (10.53).

Regime V: The frequency, f_p, of the peak noise level is given by Equation (10.54), where the freely expanding jet Mach number, M_j, is:

$$M_j = \left[\frac{2}{(\gamma - 1)}\left(\left(22\right)^{(\gamma-1)/\gamma} - 1\right)\right]^{1/2} \tag{10.55}$$

The 1/3-octave band spectrum of the noise inside of the pipe may be estimated as follows, using L_{pi} calculated with Equation (10.49). The level, L_p, of the 1/3-octave band containing the spectrum peak frequency is $L_p = L_{pi} - L_x$. The quantity, L_x, is approximately 8 dB, but must be adjusted up or down in each case considered so that the total sound pressure level, calculated by adding the 1/3-octave band levels as illustrated in Equation (1.98), is equal to L_{pi}. For frequencies greater than the peak frequency, the spectrum rolls off at the rate of 3.5 dB per octave. For frequencies less than the peak frequency, the spectrum rolls off at the rate of 5 dB per octave for the first two octaves and then at the rate of 3 dB per octave at lower frequencies.

10.8.3 External Sound Pressure Level

To calculate the transmission loss of the pipe wall, three frequencies need to be calculated. The first is the ring frequency, given by the following equation:

$$f_r = \frac{c_{LII}}{\pi d_i} \quad \text{(Hz)} \tag{10.56}$$

where d_i is the internal diameter of the pipe downstream of the valve and c_{LII} is the compressional wave speed in the pipe wall. At the ring frequency, the mean circumference is just one compressional wavelength long. Below the ring frequency, the pipe wall bending wave response is controlled by the surface curvature, while above the ring frequency, the pipe wall bending wave response is essentially that of a flat plate and is thus unaffected by the surface curvature.

A second important frequency is the internal coincidence frequency, f_0, of the gas in the pipe, and this is (IEC 60534-8-3, 2010):

$$f_0 = \frac{f_r}{4}\left(\frac{c_2}{343}\right) \quad \text{(Hz)} \tag{10.57}$$

where c_2 is the speed of sound in the downstream fluid in the pipe. At this frequency, the transmission loss of the pipe wall is a minimum.

A third frequency (used in Equation (10.61)) is (IEC 60534-8-3, 2010):

$$f_g = \frac{6.49 \times 10^4}{t c_{LII}} \quad \text{(Hz)} \tag{10.58}$$

where t is the pipe wall thickness (m).

To calculate the sound pressure level on the outside of the downstream pipe and adjacent to the pipe, IEC 60534-8-3 (2010) provides the following expression for the Transmission Loss (TL), which has been inverted here to reflect the definition of TL used in this book.

$$\text{TL} = 10\log_{10}\left[\frac{t^2 f_p^2}{c_2^2}\left(\frac{\rho_2 c_2}{415 G_y} + 1\right)\frac{1}{P_a G_x}\right] + 111.2 \quad \text{(dB)} \tag{10.59}$$

where P_a is atmospheric pressure. Equation (10.59) is applied only to the overall A-weighted internal sound pressure level using Equation (10.62) to get the overall A-weighted external level and is only applicable to control valve noise. The quantities G_x and G_y are defined as:

$$G_x = \begin{cases} \left(\frac{f_0}{f_r}\right)^{2/3}\left(\frac{f_p}{f_0}\right)^4; & f_p < f_0 \\ \left(\frac{f_p}{f_r}\right)^{2/3}; & f_0 \le f_p < f_r \\ 1; & f_p \ge f_0 \text{ and } f_p \ge f_r \end{cases} \tag{10.60a–c}$$

and

$$G_y = \begin{cases} \left(\frac{f_0}{f_g}\right); & f_p < f_0 < f_g \\ \left(\frac{f_p}{f_g}\right); & f_0 \le f_p < f_g \\ 1; & f_0 \ge f_g \text{ and } f_0 > f_p \\ 1; & f_p \ge f_0 \text{ and } f_p \ge f_g \end{cases} \tag{10.61a–c}$$

The overall A-weighted sound pressure level, L_{pAe}, external to the pipe at the outside diameter of the pipe is calculated as (IEC 60534-8-3, 2010):

$$L_{pAe} = L_{pi} - \text{TL} + 5 + L_g \quad \text{(dB re } 20\,\mu\text{Pa)} \tag{10.62}$$

The 5 dB correction term accounts for the many peaks in the internal noise spectrum. The term, L_g, has been introduced in Equation (10.62) and is a correction term to account for the effect of gas flow within the pipe on the sound energy transmitted through the pipe wall. The latter correction term is calculated as:

$$L_g = -16\log_{10}\left(1 - \frac{4\dot{m}}{\pi d_i^2 \rho_2 c_2}\right) \tag{10.63}$$

If the second term in brackets of Equation (10.63) exceeds 0.3, it is set equal to 0.3.

The A-weighted sound pressure level, $L_{pAe,\,1\mathrm{m}}$, external to the pipe at 1 m from the pipe of diameter, d_i, is calculated as (IEC 60534-8-3, 2010):

$$L_{pAe,1m} = L_{pAe} - 10\log_{10}\left[\frac{d_i + 2t + 2}{d_i + 2t}\right] \quad \text{(dB re 20 µPa)} \tag{10.64}$$

The overall, A-weighted sound power level radiated by the pipe is:

$$L_{WA} = L_{pAe} + 10\log_{10}(d_i + 2t) + 10\log_{10}\ell_p + 5 \quad \text{(dB re } 10^{-12}\text{ W)} \tag{10.65}$$

where ℓ_p is the length of downstream pipe radiating sound.

The sound pressure level at any distance from the downstream pipe may be calculated by converting the sound power level to power in watts, then using the analysis of Section 4.5 for a coherent line source and then converting the mean square acoustic pressure to sound pressure level in dB.

Although the international standard, IEC 60534-8-3 (2010), does not address the spectral distribution of the sound pressure level calculated using the preceding procedures, it has been discussed in a paper by Baumann and Hoffmann (1999). In their work, they divide the calculation of the spectrum shape into three regimes as follows, where the measurement band is 1/3-octave, f is the centre frequency of the 1/3-octave band of interest, f_p is the frequency of the peak noise level, f_0 is the internal coincidence frequency of the pipe (Equation (10.57)) and L_B is the 1/3-octave band sound pressure level (unweighted) external to the pipe at 1-metre distance. A similar spectral distribution will apply to the sound power. Note that the final spectrum levels must all be adjusted by adding or subtracting a constant decibel number so that when A-weighted and added together, the result is identical to the A-weighted overall levels from Equations (10.64) and (10.65). For example, if the calculated A-weighted 1/3-octave band levels logarithmically add up to 88.5 dBA and the overall A-weighted level, calculated using Equation (10.65), is 86.9 dBA, then 1.6 dB must be subtracted from each calculated A-weighted 1/3-octave band level.

Range 1, frequency range, $f_p < f_0$

$$\begin{aligned}
L_B &= L_{pAe,1m} - 5 - 40\log_{10}(f_p/f); && f \le f_p \\
L_B &= L_{pAe,1m} - 5; && f_p < f \le f_0 \\
L_B &= L_{pAe,1m} - 5 - 33\log_{10}(f/f_0); && f_0 < f < f_r \\
L_B &= L_{pAe,1m} - 5 - 33\log_{10}(f_r/f_0) - 40\log_{10}(f/f_r); && f_r \le f
\end{aligned} \tag{10.66a–d}$$

Range 2, frequency range, $f_0 \le f_p \le f_r$

$$\begin{aligned}
L_B &= L_{pAe,1m} - 5 - 40\log_{10}(f_0/f) - 7\log_{10}(f_p/f_0); && f \le f_0 \\
L_B &= L_{pAe,1m} - 5 - 7\log_{10}(f_p/f); && f_0 < f \le f_p \\
L_B &= L_{pAe,1m} - 5 - 33\log_{10}(f/f_p); && f_p < f < f_r \\
L_B &= L_{pAe,1m} - 5 - 33\log_{10}(f_r/f_p) - 40\log_{10}(f/f_r); && f_r \le f
\end{aligned} \tag{10.67a–d}$$

Range 3, frequency range, $f_r < f_p$

$$\begin{aligned}
L_B &= L_{pAe,1m} - 5 - 40\log_{10}(f_0/f) - 7\log_{10}(f_r/f_0); && f \le f_0 \\
L_B &= L_{pAe,1m} - 5 - 7\log_{10}(f_0/f) - 7\log_{10}(f_r/f_0); && f_0 < f \le f_r \\
L_B &= L_{pAe,1m} - 5; && f_r < f < f_p \\
L_B &= L_{pAe,1m} - 5 - 40\log_{10}(f/f_p); && f_p \le f
\end{aligned} \tag{10.68a–d}$$

If desired, TL values for each 1/3-octave band can be calculated using the 1/3-octave band values for $L_{pAe,1m}$ and Equations (10.62) and (10.64), with the TL for each 1/3 octave band adjusted by the same number of dB, so that the average is equal to the value calculated in Equation (10.59). For N 1/3-octave band TL values, the average can be calculated using:

$$\text{TL}_{av} = -10\log_{10}\left[\frac{1}{N}\sum_{i=1}^{N} 10^{-\text{TL}_i/10}\right] \tag{10.69}$$

10.8.4 Noise-Reducing Trim

If control valves have noise reducing trim, the radiated noise will be reduced by up to 5 dBA and if the valve has a low noise trim, the reduction is likely to be greater. Details of accounting for this are included in IEC 60534-8-3 (2010). One simple type of noise reducing trim for which the radiated noise calculations are relatively straightforward is the multi-hole cage attached to the valve discharge. The value of F_d for this type of trim on a valve is different to that for a valve with no cage and is usually supplied by the valve manufacturer. The calculation procedure is the same as for a valve with no trim discussed above, except that in Equation (10.51), F_L is replaced with $0.9 - 0.06(\ell/d)$, with a maximum allowed ℓ/d of 4, where ℓ is the length of a single flow passage through the cage and d is the diameter of the passage ($= \sqrt{4A/\pi}$ for any cross-sectional shape), where A is the cross-sectional area of one passage. As for valves with no cage, if fittings are attached F_L is replaced with F_{LP}.

10.8.5 High Exit Velocities

The preceding calculation procedures are only valid for valve exit (downstream of the valve) Mach numbers of 0.3 or less. The Mach number downstream of the valve is calculated using:

$$M_o = \frac{4\dot{m}}{\pi d_o \rho_2 c_2} \tag{10.70}$$

where d_o is the valve exit diameter (use d_i if d_o is unknown). The speed of sound, c_2, in the downstream pipe is given by Equation (10.42) with T_1 replaced with T_2. For higher exit velocities, calculation procedures are available in IEC 60534-8-3 (2010).

10.8.6 Control Valves for Steam

Steam valves are a common source of noise in industrial plants. Sound pressure levels generated by them can be estimated using the procedures outlined previously for gas valves. However, if the calculated jet acoustical power coefficient, η is less than about 0.005, more accurate estimates are obtained by adding 3 dB to the calculated results.

Alternatively, the values listed in Table 10.15 may be used. These are conservative estimates based on measurements of actual steam valves, and include noise radiation from the pipe connected to the valves.

10.8.7 Gas and Steam Control Valve Noise Reduction

As was explained in Section 10.8.1, the control valve necessarily functions as an energy dissipation device and the process of energy dissipation is accompanied by a small amount of energy conversion to noise. Consequently, noise control must take the form of reduction of energy conversion efficiency. In Section 10.8.1 it was shown that energy conversion efficiency increases with the Mach number and becomes even more efficient when shock waves are formed. Evidently, it

TABLE 10.15 Estimated sound pressure levels (dB re 20 µPa at 1 m) for steam valves

Octave band centre frequency (Hz)	Sound pressure level (dB)
31.5	70
63	70
125	70
250	70
500	75
1000	80
2000	85
4000	90
8000	90

This table assumes simple, lightweight thermal wrapping of the pipe but no metal or heavy cover around the thermal wrapping.

Both the valve and the connected piping radiate noise.

Data from Army, Air Force and Navy, USA (2003).

is desirable from a noise reduction point of view to avoid critical flow and shock wave formation. This has been accomplished in practice by effectively providing a series of pressure drop devices across which the pressure drop is less than critical and intense turbulence is induced but shock formation is avoided. There are a number of control valves commercially available built on this principle, but they are generally expensive.

10.9 Control Valves for Liquids

The following prediction procedure follows IEC 60534-8-4 (2015) and considers only single stage valves and noise generated by hydrodynamic processes. Noise resulting from reflections, loose parts or resonances and noise generated by multi-stage valves are not considered here. The procedure outlined in the following paragraphs is different to that described in IEC 60534-8-4 (1994) and previous editions of the textbook. The discussion begins with the definition of a number of variables, and these are then used in following equations to calculate radiated sound power and sound pressure levels.

The pressure ratio, X_{Fz}, at which cavitation can be acoustically detected for an inlet pressure, P_1, of 6×10^5 Pa (the reference pressure used in IEC 60534-8-4 (2015)), is defined as:

$$X_{Fz} = \begin{cases} 0.9\left(1 + 3F_d\sqrt{\dfrac{C_v}{1.17F_L}}\right)^{-0.5} & \text{; all except valves with multihole trims} \\ \left(4.5 + 1650 N_0 d_H^2 / F_L\right)^{-0.5} & \text{; valves with multihole trims} \end{cases} \qquad (10.71\text{a,b})$$

where N_0 is the number of holes in the trim and $d_H = \sqrt{4A_H/\pi}$ is the diameter (or equivalent diameter) of each hole (A_H is the cross-sectional area of each hole). To account for inlet pressures different to 6×10^5 Pa, the quantity, X_{Fz} is replaced in calculations by X_{Fzp1}, defined as:

$$X_{Fzp1} = X_{Fz}\left(\frac{6 \times 10^5}{P_1}\right)^{0.125} \qquad (10.72)$$

The differential pressure, ΔP_c, is:

$$\Delta P_c = \text{smaller of} \begin{cases} (P_1 - P_2) \\ F_L^2(P_1 - P_{va}) \end{cases} \quad \text{(Pa)} \qquad (10.73)$$

where P_1 is the upstream absolute pressure (Pa), P_2 is the downstream absolute pressure (Pa), F_L is the liquid pressure recovery factor provided by the valve manufacturer and P_{va} is the vapour pressure (Pa) for the liquid in the pipe. For water, the vapour pressure may be calculated using the Arden Buck equation (Buck, 1981):

$$P_{va} = 611.21 \exp\left[\frac{(18.678 - T_1/234.5)T_1}{(T_1 + 257.14)}\right] \quad \text{(Pa)} \tag{10.74}$$

where T_1 is the temperature of the water in °C.

The acoustic power, W_a, generated by liquid passing through a control valve is (IEC 60534-8-4, 2015):

$$X_{Fz} = W_a = \eta W_m = \frac{\eta \dot{m} \Delta P_c}{\rho_f} \quad \text{(W)} \tag{10.75a–c}$$

where ρ_f is the density of the liquid and $\dot{m}$ is the mass flow rate. The efficiency factor, η, is different for turbulent flow and cavitating flow and may be calculated using the following equations.

For **turbulent flow**, $X_F \leq X_{Fzp1}$ and $\eta = \eta_{\text{turb}}$, where:

$$X_F = (P_1 - P_2)/(P_1 - P_{va}) \tag{10.76}$$

and

$$\eta_{\text{turb}} = 10^A \left(\frac{U_v}{c_f}\right) \tag{10.77}$$

where $A = -4.6$ for globe valves, $A = -4.3$ for other valve types, c_f is the speed of sound in the liquid (see Equations (1.1) and (1.6)) and U_v is the speed of the flow through the vena contractor, given by:

$$U_v = \frac{1}{F_L}\sqrt{\frac{2\Delta P_c}{\rho_f}} \quad \text{(m/s)} \tag{10.78}$$

For **cavitating flow**, $X_{Fzp1} < X_F < 1$ and $\eta = \eta_{\text{turb}} + \eta_{\text{cav}}$, where

$$\eta_{\text{cav}} = 0.32\eta_{\text{turb}} \left(\frac{P_1 - P_2}{\Delta P_c X_{Fzp1}}\right)^{0.5} \mathrm{e}^{5X_{Fzp1}} \left(\frac{1 - X_{Fzp1}}{1 - X_F}\right)^{0.5} \left(\frac{X_F}{X_{Fzp1}}\right)^5 (X_F - X_{Fzp1})^{1.5} \tag{10.79}$$

The overall internal sound pressure level, L_{pi}, downstream of the control valve is calculated using Equation (10.49), with $\rho_f c_f$ in place of $\rho_2 c_2$.

The internal sound pressure level for the 1/3-octave band with a centre frequency, f, is different for turbulent and cavitating flow.

For **turbulent flow**:

$$L_{pi}(f) = L_{pi} + \Delta(f, \text{turb}) \quad \text{(dB re 20 µPa)} \tag{10.80}$$

where

$$\Delta(f, \text{turb}) = -8 - 10\log_{10}\left[\frac{1}{4}\left(\frac{f}{f_{p,\text{turb}}}\right)^3 + \left(\frac{f_{p,\text{turb}}}{f}\right)\right] \quad \text{(dB)} \tag{10.81}$$

and

$$f_{p,\text{turb}} = \left(\frac{U_v}{d_j}\right)\left(\frac{0.036 F_L^2 C_v F_d^{0.75}}{1.17 X_{Fzp1}^{1.5} D d_v}\right)(P_1 - P_{va})^{-0.57} \quad \text{(Hz)} \tag{10.82}$$

where D is the nominal valve size (m), d_v is the valve orifice diameter (m) U_v is defined in Equation (10.78) and d_j is the jet diameter, defined by:

$$d_j = 4.6 \times 10^{-3} F_d \sqrt{C_v F_L} \quad \text{(m)} \tag{10.83}$$

where F_d and F_L are the valve style modifier and the liquid pressure recovery factor, respectively, provided by the valve manufacturer. For **cavitating flow**:

$$L_{pi}(f) = L_{pi} + 10\log_{10}\left(\frac{\eta_{\text{turb}}}{\eta_{\text{turb}} + \eta_{\text{cav}}}10^{\Delta(f,\text{turb})/10} + \frac{\eta_{\text{cav}}}{\eta_{\text{turb}} + \eta_{\text{cav}}}10^{\Delta(f,\text{cav})/10}\right) \tag{10.84}$$

where

$$\Delta(f, \text{cav}) = -9 - 10\log_{10}\left[\frac{1}{4}\left(\frac{f}{f_{p,\text{cav}}}\right)^{1.5} + \left(\frac{f_{p,\text{cav}}}{f}\right)^{1.5}\right] \quad \text{(dB)} \tag{10.85}$$

$$f_{p,\text{cav}} = 6 f_{p,\text{turb}}\left(\frac{1 - X_F}{1 - X_{Fzp1}}\right)^2\left(\frac{X_{Fzp1}}{X_F}\right)^{2.5} \quad \text{(Hz)} \tag{10.86}$$

The TL for the 1/3-octave band with centre frequency, f, is:

$$\text{TL} = 10 + 10\log_{10}\left(\frac{c_{LI}\rho_m t}{\rho c d_i}\right) + 20\log_{10}\left[\left(\frac{f_r}{f}\right) + \left(\frac{f}{f_r}\right)^{1.5}\right] \quad \text{(dB)} \tag{10.87}$$

The sound pressure level outside of the pipe and adjacent to the pipe for the 1/3-octave band with centre frequency, f, is:

$$L_{pe}(f) = L_{pi}(f) - \text{TL} \quad \text{(dB re 20 µPa)} \tag{10.88}$$

and the sound power level radiated by a pipe of length, ℓ_p, is:

$$L_W(f) = L_{pe}(f) + 10\log_{10}(d_i + 2t) + 10\log_{10}\ell_p + 5 \quad \text{(dB re } 10^{-12}\text{ W)} \tag{10.89}$$

The external sound pressure level at 1 m from the pipe wall is:

$$L_{pe,1\text{m}}(f) = L_{pe}(f) - 10\log_{10}\left(\frac{d_i + 2t + 2}{d_i + 2t}\right) \quad \text{(dB re 20 µPa)} \tag{10.90}$$

Overall sound power and sound pressure levels may be obtained by logarithmically summing the 1/3-octave band levels using Equation (1.98). A-weighted overall levels are obtained by first adding the appropriate A-weighting from Table 2.3 to each 1/3-octave band and logarithmically adding the 1/3-octave A-weighted levels using Equation (1.98). The preceding calculations only apply to single-stage valves. Noise generated by multi-stage valves can be calculated using procedures in IEC 60534-8-4 (2015).

10.9.1 Liquid Control Valve Noise Reduction

Cavitation in liquid valves significantly increases the radiated noise, so it is important to keep $X_F < X_{Fzp1}$ by not having too large a pressure drop across the valve (see Equations (10.71), (10.72) and (10.76)).

10.10 Gas Flow in Pipes

This section has been completely updated since the 5th edition. The calculation of flow noise from pipes remains restricted to flowing gases, as noise due to flowing liquids is generally insignificant. Calculation procedures outlined in previous editions of this book are unreliable.

Estimation of the noise radiated by pipes containing turbulent flow is difficult, especially the estimation of the pipe wall TL. The procedures outlined in previous sections for control valves are restricted to control valve noise sources. Here we outline an approximate method for determining the noise radiated by pipes containing turbulent gas flow, which applies over a restricted frequency range and which is also documented in Baumann and Coney (2006).

The method uses estimates the sound pressure level at a distance of 1 m from the pipe wall due to undisturbed turbulent flow in the pipe and then adds a correction factor, which is a function of the loss coefficient, K, discussed in Section 8.14, to account for local discontinuities that cause a pressure loss and an increase in the turbulence level.

The external sound pressure levels in an octave band with a centre frequency of f, at a distance of 1 m from the pipe wall, may be calculated using:

$$L_p = -3.5 + 40\log_{10} U + 20\log_{10}\rho_1 + 20\log_{10} K - 10\log_{10}\left[\frac{t}{d_i}\left(1+\frac{1.83}{d_i}\right)\right] - 5\log_{10}\left[\frac{f}{f_r}\left(1-\frac{f}{f_r}\right)\right] + \Delta L \tag{10.91}$$

where U is the gas flow speed, ρ_1 is the density of the gas inside the pipe, t is the pipe wall thickness, d_i is the pipe inside diameter, f_r is the pipe ring frequency (Equation (10.56)), K is the friction loss in a length, L, of pipe equal to $10d_i$ (use $K = 0.1$ for a smooth-wall pipe and Equation (8.238) (with $L = 10d_i$) for a pipe with a perforated sheet metal liner) and:

$$X_{Fz} = \Delta L = \begin{cases} 10.4 + 11.4\log_{10}(f/f_p); & f/f_p < 0.5 \\ 7; & 0.5 \le (f/f_p) < 5 \\ 14 - 10\log_{10}(f/f_p); & 5 \le (f/f_p) < 12 \\ 41.9 - 36.1\log_{10}(f/f_p); & (f/f_p) \ge 12 \end{cases} \tag{10.92a–d}$$

where $f_p = 0.2U/d_i$ is the peak noise level frequency corresponding to a Strouhal number of 0.2. If pipe fittings such as elbows are included in the 10 diameter-long pipe section used to calculate K due to friction, then the K factors due to these fittings can be determined using Figure 8.39, and these are added to the value of K for friction losses in the straight section pipe, 10 diameters long (often approximated as $K = 0.1$).

Equation (10.91) is only valid up to a frequency of f_r, where f_r is the pipe ring frequency calculated using Equation (10.56). The sound power radiated by the pipe per metre of length for the pipe length under consideration (10 diameters long) may also be calculated by assuming that it is a coherent line source (see Equation (4.83)), so that:

$$L_W = L_p + 8 + 10\log_{10}(1 + d_i/2) - 10\log_{10}(\rho c/400) \quad \text{(dB re } 10^{-12}\text{ W)} \tag{10.93}$$

10.11 Boilers

For general-purpose boilers, the overall radiated sound power level is given by:

$$L_W = 95 + 4\log_{10}\text{kW} \quad \text{(dB re } 10^{-12}\text{ W)} \tag{10.94}$$

For large power plant boilers, the overall sound power level is given by:

$$L_W = 84 + 15\log_{10}\text{MW} \quad \text{(dB re } 10^{-12}\text{ W)} \tag{10.95}$$

where kW and MW are the rated power of the boiler in kilowatts and megawatts, respectively.

The octave band levels for either type of boiler can be calculated by subtracting the appropriate corrections listed in Table 10.16.

TABLE 10.16 Values to be subtracted from overall unweighted power levels, L_W, to obtain band levels for boiler noise

Octave band centre frequency (Hz)	Octave band corrections (dB)	
	General-purpose boilers	Large power plant boilers
31.5	6	4
63	6	5
125	7	10
250	9	16
500	12	17
1000	15	19
2000	18	21
4000	21	21
8000	24	21

Data from Edison Electric Institute (1978).

10.12 Gas and Steam Turbines

The principal noise sources of gas turbines are the casing, inlet and exhaust. The overall sound power levels contributed by these components of gas turbine noise (with no noise control) may be calculated using the following equations (Army, Air Force and Navy, USA, 2003):

Casing:

$$L_W = 120 + 5\log_{10}\text{MW} \quad (\text{dB re } 10^{-12}\text{ W}) \tag{10.96}$$

Inlet:

$$L_W = 127 + 15\log_{10}\text{MW} \quad (\text{dB re } 10^{-12}\text{ W}) \tag{10.97}$$

Exhaust:

$$L_W = 133 + 10\log_{10}\text{MW} \quad (\text{dB re } 10^{-12}\text{ W}) \tag{10.98}$$

For steam turbines, a good estimate for the overall sound power radiated is given by (Edison Electric Institute, 1978):

$$L_W = 93 + 4\log_{10}\text{kW} \quad (\text{dB re } 10^{-12}\text{ W}) \tag{10.99}$$

The octave band levels for gas and steam turbines may be calculated by subtracting the corrections listed in Table 10.17. The approximate casing noise reductions due to various types of enclosure are listed in Table 10.18.

Noise reductions due to inlet and exhaust mufflers can be calculated using the methods of Chapter 8, or preferably using manufacturer's data. Normally, the inlet and discharge cross-sectional areas of the muffler stacks are very large; thus additional noise reductions will occur due to the directivity of the stacks. This effect can be estimated using Figures 8.45–8.47.

TABLE 10.17 Octave band and overall A-weighted sound power level adjustments (in dB) for gas and steam turbines. Subtract these values from the overall unweighted sound power level, L_W, to obtain octave band and A-weighted sound power levels

Octave band centre frequency (Hz)	Value to be subtracted from overall L_W (dB)			
	Gas turbine			Steam turbine
	Casing	Inlet	Exhaust	
31.5	10	19	12	11
63	7	18	8	7
125	5	17	6	6
250	4	17	6	9
500	4	14	7	10
1000	4	8	9	10
2000	4	3	11	12
4000	4	3	15	13
8000	4	6	21	17
A-weighted (dBA)	2	0	4	5

Data from Army, Air Force and Navy, USA (2003).

TABLE 10.18 Approximate noise reduction of gas turbine casing enclosures

Octave band centre frequency (Hz)	Type 1[a]	Type 2[b]	Type 3[c]	Type 4[d]	Type 5[e]
31.5	2	4	1	3	6
63	2	5	1	4	7
125	2	5	1	4	8
250	3	6	2	5	9
500	3	6	2	6	10
1000	3	7	2	7	11
2000	4	8	2	8	12
4000	5	9	3	8	13
8000	6	10	3	8	14

[a]Glass fibre or mineral wool thermal insulation with a lightweight foil cover over the insulation.

[b]Glass fibre or mineral wool thermal insulation covered with a minimum 20 gauge aluminium or 24 gauge steel.

[c]Enclosing metal cabinet for the entire packaged assembly, with open ventilation holes and with no acoustic absorptive lining inside the cabinet.

[d]Enclosing metal cabinet for the entire packaged assembly, with open ventilation holes and with acoustic absorptive lining inside the cabinet.

[e]Enclosing metal cabinet for the entire packaged assembly with all ventilation holes into the cabinet muffled and with acoustic absorptive lining inside the cabinet.

Data from Army, Air Force and Navy, USA (2003).

10.13 Reciprocating Piston Engines (Diesel or Gas)

The three important noise sources for this type of equipment are the engine exhaust, the engine casing and the air inlet. These sources will be considered in the following subsections.

10.13.1 Exhaust Noise

The overall sound power radiated by an unmuffled reciprocating engine exhaust may be calculated using (Army, Air Force and Navy, USA, 2003):

$$L_W = 120.3 + 10\log_{10}\text{kW} - K - (\ell_{EX}/1.2) \qquad (\text{dB re } 10^{-12}\text{ W}) \tag{10.100}$$

where $K = 0$ for an engine with no turbocharger, $K = 6$ for an engine with a turbocharger and ℓ_{EX} is the length of the exhaust pipe (m). The octave band sound power levels may be calculated by subtracting the corrections listed in Table 10.19 from the overall power level.

The approximate effects of various types of commercially available mufflers are shown in Table 10.20. Exhaust directivity effects may be calculated using Figures 8.45–8.47.

TABLE 10.19 Octave band and overall A-weighted sound power level adjustments for unmuffled engine exhaust noise (Equation (10.100))

Octave band centre frequency (Hz)	Value to be subtracted from the overall sound power level (dB)
31.5	5
63	9
125	3
250	7
500	15
1000	19
2000	25
4000	35
8000	43
A-weighted (dBA)	12

Data from Army, Air Force and Navy, USA (2003).

TABLE 10.20 Approximate insertion loss (dB) of typical reactive mufflers for reciprocating engines

Octave band centre frequency (Hz)	Low-pressure-drop muffler			High-pressure-drop muffler		
	Small	Medium	Large	Small	Medium	Large
63	10	15	20	16	20	25
125	15	20	25	21	25	29
250	13	18	23	21	24	29
500	11	16	21	19	22	27
1000	10	15	20	17	20	25
2000	9	14	19	15	19	24
4000	8	13	18	14	18	23
8000	8	13	18	14	17	23

Refer to manufacturers' literature for more specific data.
Data from Army, Air Force and Navy, USA (2003).

10.13.2 Casing Noise

The overall sound power radiated by the engine casing is given by (Army, Air Force and Navy, USA, 2003):

$$L_W = 94.3 + 10\log_{10}\text{kW} + A + B + C + D \qquad (\text{dB re } 10^{-12}\text{ W}) \tag{10.101}$$

The quantities A, B, C and D are listed in Table 10.21, and the corrections to be subtracted from the overall level to obtain the octave band levels are listed in Table 10.22.

TABLE 10.21 Correction terms to be applied to Equation (10.101) for estimating the overall sound power level (dB re 10^{-12} W) of the casing noise of a reciprocating engine

Speed correction term, A	
Under 600 rpm	−5
600–1500 rpm	−2
Above 1500 rpm	0
Fuel correction term, B	
Diesel only	0
Diesel and natural gas	0
Natural gas only (including a small amount of pilot oil)	−3
Cylinder arrangement term, C	
In-line	0
V-type	−1
Radial	−1
Air intake correction term, D	
Unducted air inlet to unmuffled roots blower	+3
Ducted air from outside the enclosure	0
Muffled roots blower	0
All other inlets (with or without turbocharger)	0

Data from Army, Air Force and Navy, USA (2003).

TABLE 10.22 Frequency adjustments (dB) for casing noise of reciprocating engines: subtract these values from the overall sound power level (Equation (10.101)) to obtain octave band and A-weighted levels

Octave band centre frequency (Hz)	Engine speed under 600 rpm	Engine speed 600–1500 rpm		Engine speed over 600 rpm
		Without roots blower	With roots blower	
31.5	12	14	22	22
63	12	9	16	14
125	6	7	18	7
250	5	8	14	7
500	7	7	3	8
1000	9	7	4	6
2000	12	9	10	7
4000	18	13	15	13
8000	28	19	26	20
A-weighted (dBA)	4	3	1	2

Data from Army, Air Force and Navy, USA (2003).

10.13.3 Inlet Noise

For engines with no turbocharger, inlet noise is negligible in comparison with the casing and exhaust noise. However, for engines with a turbocharger, the following equation may be used to calculate the overall sound power level of the inlet noise (Army, Air Force and Navy, USA, 2003):

$$L_W = 95.3 + 5\log_{10}\text{kW} - \ell/1.8 \qquad (\text{dB re } 10^{-12}\text{ W}) \tag{10.102}$$

where ℓ (m) is the length of the inlet ducting. The octave band levels may be calculated from the overall level by subtracting the corrections listed in Table 10.23.

TABLE 10.23 Frequency adjustments (dB) for the turbocharger air inlet noise calculations. Subtract these values from the overall sound power level (Equation (10.102)) to obtain octave band and A-weighted levels

Octave band centre frequency (Hz)	Value to be subtracted from overall sound power level (dB)
31.5	4
63	11
125	13
250	13
500	12
1000	9
2000	8
4000	9
8000	17
A-weighted (dBA)	3

Data from Army, Air Force and Navy, USA (2003).

10.14 Furnace Noise

Furnace noise is due to a combination of three noise-producing mechanisms:

1. jet noise produced by the entering fuel gas;
2. jet noise produced by the entering air; and
3. noise produced by the combustion process.

Fuel gas flow generates noise and the overall sound power is calculated by using Equations (10.22) and (10.27), together with the procedure for estimating control valve acoustical power coefficients, η (see page 628). However, fuel gas flow noise is dominant for burners having a high fuel gas pressure, whereas for burners using fuel oil, it is negligible. If the pressure drop associated with the fuel gas flowing into the furnace is low (less than 100 kPa), then the noise produced is calculated by assuming a free jet (see Section 10.7).

Air flow (both primary and secondary air flow), generates noise and the overall sound power may be calculated using (Heitner, 1968):

$$L_W = 44\log_{10} U + 17\log_{10}\dot{m} + 55 \quad (\text{dB re } 10^{-12}\text{ W}) \tag{10.103}$$

which can be rewritten as:

$$W = 10^{-12}(U^{4.4}\,\dot{m}^{1.7}10^{5.5}) = U^{4.4}\,\dot{m}^{1.7}10^{-6.5} \quad (W) \tag{10.104}$$

where U is the air velocity (m/s) through the register, $\dot{m}$ is the air mass flow rate (kg/s) and ρ and c are the density and speed of sound, respectively, in the gas surrounding the burner.

To estimate the octave band in which the maximum noise occurs, a Strouhal number, $N_s = 1$, is used (see Equation (10.26)). That is:

$$f_p d/U = 1 \tag{10.105}$$

where d is the smallest dimension of the air opening. The sound pressure level in this octave band is 3 dB below the overall level. Above and below the octave band of maximum noise, the level is reduced at a rate of 5 dB per octave. The total burner noise level is obtained by combining the levels in each octave band for fuel gas flow noise, combustion noise and air-flow noise.

Combustion noise is usually important but sometimes not as significant as that produced by air and gas flow, and may be estimated using (Bragg, 1963):

$$W_a = \eta \dot{m} H \qquad \text{(W)} \tag{10.106}$$

where W_a is the overall acoustical power (W), η is the acoustical efficiency (of the order of 10^{-6}), $\dot{m}$ is the mass flow rate of the fuel (kg/s) and H (Joules/kg) is the heating value of the fuel. Note that 1 calorie = 4.187 Joules. The maximum noise level occurs in the 500 Hz octave band and is 3 dB below the overall level. Above and below the 500 Hz band, the noise level is reduced at the rate of 6 dB per octave.

The total sound power radiated into the combustion chamber is the arithmetic sum of the three component sound powers (in watts) generated by fuel gas flow, air flow and combustion, respectively and the sound pressure level in the furnace may be calculated from the total radiated sound power, W, using the procedures in Chapter 6.

10.15 Electric Motors

10.15.1 Small Electric Motors (below 300 kW)

The overall sound pressure level at 1 m generated by small electric motors can be estimated for totally enclosed, fan cooled (TEFC) motors, using the following equations (Army, Air Force and Navy, USA, 2003):

$$\textbf{Under 40 kW}: \quad L_p = 17.2 + 17\log_{10}\text{kW} + 15\log_{10}\text{rpm} \qquad \text{(dB re 20 µPa)} \tag{10.107}$$

$$\textbf{Over 40 kW}: \quad L_p = 28.2 + 10\log_{10}\text{kW} + 15\log_{10}\text{rpm} \qquad \text{(dB re 20 µPa)} \tag{10.108}$$

where rpm is the rotational speed in revolutions per minute.

Drip-proof (DRPR) motors produce 5 dB less sound pressure level than TEFC motors. The octave band sound pressure levels may be obtained for both types of motor by subtracting the values in Table 10.24 from the corresponding overall levels. A TEFC motor with a quiet fan is likely to be 10 dB quieter than indicated by Equations (10.107) and (10.108).

TABLE 10.24 Octave band level adjustments (dB) for small electric motors

Octave band centre frequency (Hz)	Totally enclosed, fan cooled (TEFC) motor	Drip proof (DRPR) motor
31.5	14	9
63	14	9
125	11	7
250	9	7
500	6	6
1000	6	9
2000	7	12
4000	12	18
8000	20	27

Data from Army, Air Force and Navy, USA (2003).

10.15.2 Large Electric Motors (above 300 kW)

Sound power levels radiated by electric motors with a power rating between 750 kW and 4000 kW are listed in Table 10.25. These levels can be reduced by 5 dBA for slow-speed motors and up to 15 dBA for high-speed motors, if specifically requested by the customer.

For motors rated above 4000 kW, add 3 dB to all levels in Table 10.25. For motors rated between 300 and 750 kW, subtract 3 dB from all levels.

TABLE 10.25 Sound power levels of large electric motors

Octave band centre frequency (Hz)	1800 and 3600 rpm	1200 rpm	900 rpm	720 rpm and lower	250 and 400 rpm vertical
31.5	94	88	88	88	86
63	96	90	90	90	87
125	98	92	92	92	88
250	98	93	93	93	88
500	98	93	93	93	88
1000	98	93	96	98	98
2000	98	98	96	92	88
4000	95	88	88	83	78
8000	88	81	81	75	68

Applies to induction motors rated between 750 and 4000 kW; includes drip-proof and P-1 and WP-2 enclosures (with no acoustical specification by the customer).

Data from Edison Electric Institute (1978).

10.16 Generators

The overall unweighted sound power levels radiated by generators (excluding the driver) can be calculated using the following equation:

$$L_W = 84 + 10\log_{10}\mathrm{MW} + 6.6\log_{10}\mathrm{rpm} \quad (\mathrm{dB\ re\ }10^{-12}\mathrm{\ W}) \tag{10.109}$$

To obtain the octave band levels, the values in Table 10.26 should be subtracted from the overall level calculated using Equation (10.109).

TABLE 10.26 Octave band corrections for generator noise

Octave band centre frequency (Hz)	Value to be subtracted from overall sound power level (dB)
31.5	11
63	8
125	7
250	7
500	7
1000	9
2000	11
4000	14
8000	19

Data from Army, Air Force and Navy, USA (2003).

10.17 Gears

The following equation gives octave band sound pressure levels for gearboxes in all frequency bands at and above 125 Hz, at a distance of 1 m from the gearbox (Army, Air Force and Navy, USA, 2003):

$$L_p = 78.5 + 4\log_{10}\text{kW} + 3\log_{10}\text{rpm} \qquad (\text{dB re } 10^{-12}\text{ W}) \tag{10.110}$$

where kW is the power transmitted by the gearbox, and rpm is the rotational speed of the slowest gear shaft. For the 63 Hz octave band, subtract 3 dB, and for the 31 Hz band subtract 6 dB from the value calculated using Equation (10.110).

These sound pressure levels are applicable to spur gears, and may be reduced somewhat (by up to 30 dB) by replacing the spur gear with a quieter helical or herringbone design. The actual noise reduction (compared to a straight spur gear) is given very approximately by $13+20\log_{10} Q_a$, where Q_a is the number of teeth that would be intersected by a straight line across the gear perimeter parallel to the gear shaft. Q_a can be calculated using:

$$Q_a = \text{int}[(b\sin\alpha)/p_h] \tag{10.111}$$

where "int" means that the calculated value is rounded down, b is the gear width (or thickness), α is the helix angle and p_h is the gear pitch. For double helical or herringbone gears, the number of intersected teeth would only be for one helix, not both.

10.18 Power Transformers

The procedure in this section is for predicting sound radiated from transformers where the core is immersed in either air or oil. Radiated sound is made up of three main components.

1. Core audible sound, which originates in the transformer core and is transmitted acoustically and structurally to the transformer tank, which then radiates the sound energy to the environment.
2. Load audible sound, which is produced by vibrations of the windings and tank walls when the transformer is loaded.
3. Cooling system audible sound produced by operation of the cooling fans and pumps.

Transformers are usually rated for the sound radiation described in item 1 above, which is referred to as the ONAN (oil natural, air natural) rating or NEMA rating, which is an overall A-weighted sound pressure level. The NEMA rating, N_R, is the overall A-weighted sound pressure level arithmetically averaged over specified test locations, with the transformer energised at its rated voltage and frequency, but with no load and no cooling systems operating (NEMA, 2014, p. 70). The measurement locations are at intervals of 1 m, on 2 horizontal contours, at heights of 1/3 and 2/3 of the tank height and located laterally at a distance of 0.3 m from a contour that follows a tight string around the transformer tank. For tanks less than 1.2 m high, only one contour, having a height equal to half the tank height, is used.

With one commonly used estimation procedure, the transformer radiated sound power in octave bands (unweighted) is given by Equation (10.112) in terms of its A-weighted NEMA sound pressure level rating, N_R, measured at a distance of 0.3 m from the tank walls, and the area, S, of the four side walls of the transformer tank in square feet, with the octave band corrections, $C(f)$, provided in Table 10.27 (Army, Air Force and Navy, USA, 2003).

$$L_W = N_R + 10\log_{10} S + C(f) \qquad (\text{dB re } 10^{-12}\text{ W}) \tag{10.112}$$

TABLE 10.27 Values of the correction factor, $C(f)$, of Equation (10.112) for transformer noise

Octave band centre frequency (Hz)	Octave band corrections (dB)		
	Location 1[a]	Location 2[b]	Location 3[c]
31.5	−1	−1	−1
63	5	8	8
125	7	13	13
250	2	8	12
500	2	8	12
1000	−4	−1	6
2000	−9	−9	1
4000	−14	−14	−4
8000	−21	−21	−11

[a]Outdoors, or indoors in a large mechanical room with a large amount of mechanical equipment.

[b]Indoors in small rooms, or large rooms with only a small amount of other equipment.

[c]Any critical location where a problem would result if the transformer should become noisy above its NEMA rating, following installation.

Data from Army, Air Force and Navy, USA (2003).

Bruce and Moritz (1997) made two corrections to Equation (10.112). First they stated that S should be in m^2 and second, they subtracted 2 dB from the octave band corrections, $C(f)$, listed in Table 10.27 so that when A-weighted, the octave band levels add logarithmically to give approximately the same result as Equation (10.112) without the $C(f)$ term. Here, it is recommended that the changes to the definition of S and changes to the values in Table 10.27 suggested by Bruce and Moritz (1997) be implemented with the additional small correction that the surface area, S, of the tank walls should be replaced with the area, S_a of the measurement surface used to obtain N_R, and this is given by (IEEE Std C57.12.90, 2021):

$$S_a = 1.25h\ell_m \quad \text{(m)} \tag{10.113}$$

where the factor of 1.25 accounts for sound radiation by the transformer tank cover, h is the height of the transformer tank and ℓ_m is length of the measurement contour along which sound pressure level measurements are made to determine N_R, given by:

$$\ell_m = 2L + 2W + 2.4 \quad \text{(m)} \tag{10.114}$$

where L is the length of the transformer footprint (including all cooling fins) and W is its width (or depth). Equation (10.114) was derived assuming a rectangular shaped measurement surface. However, IEEE Std C57.12.90 (2021) specifies that the measurement contour should follow the contour of the transformer. This means that the contour could be up to 20% shorter than calculated using Equation (10.114), but this would make a difference of less than 1 dB to the calculated radiated sound power.

Maximum acceptable N_R values (A-weighted overall NEMA audible sound level ratings (NEMA, 2019)) are provided in Table 10.28. However, expected values may be calculated as (Bruce and Moritz, 1997):

$$N_R = 12\log_{10}(\text{MVA}) + K_c \tag{10.115}$$

where $K_c = 55$ for a standard transformer and $K_c = 45$ for a quietened transformer. If required, it is possible to obtain transformers with NEMA ratings 10 to 15 dB below the values specified in column 3 (under "Fan") of Table 10.27, which correspond to $K_c = 45$ in Equation (10.115).

TABLE 10.28 Maximum acceptable overall A-weighted audible sound levels (ONAN), N_R (dBA), for transformers built after 1995 with a BIL (basic insulation level) of 350 kV or less

Transformer power (kVA)	N_R (dBA)		
	Standard		Super quiet core
	No fan	Fan	
10–50[a]	(45)	67	52
50–100[a]	(50)	67	52
100–300[a]	(55)	67	52
301–500[a]	(60)	67	52
501–700[a]	57(62)	67	53
701–1000[a]	58(64)	67	56
1001–1500	60	67	58
1501–2000	61	67	60
2001–2500	62	67	62
2501–3000	63	67	63
3001–4000	64	67	63
4001–5000	65	67	64
5001–6000	66	67	65
6001–7500	67	68	66
7501–10000	68	69	67
10001–12500	69	70	68
12501–15000	70	71	69
15001–20000	71	72	70
20001–25000	72	73	71
25001–30000	73	74	72
30001–40000	74	75	73
40001–50000	75	76	74
50001–60000	76	77	75
60001–80000	77	78	76
90001–100000	78	79	77

See NEMA (2019) for N_R values for transformers with BIL values greater than 350 kV.

[a]Some transformers with a power rating less than 1000 kVA are air filled. NEMA ratings for these with no fan cooling (but with ventilation) are shown in brackets. Ratings with fan cooling are the same as for oil-filled transformers.

As an alternative and likely more accurate approach to estimating transformer radiated sound power, the standard, (NEMA, 2019), states that the audible sound level should be determined according to IEEE Std C57.12.90 (2021), which states that A-weighted sound pressure level measurements should be taken at intervals of 1 m, on 2 horizontal contours, at heights of 1/3 and 2/3 of the tank height and located laterally at a distance of 2 m (not 0.3 m as used for the ONAN rating) from a contour that follows a tight string around the transformer tank. The measurements should be taken for the no load condition with fans (and pumps for the oil when applicable) running and also for the rated load condition with no fans or pumps running. Overall A-weighted measurements are required and 1/3-octave band measurements in the range 50 Hz to 4 kHz may be specified in some cases. For the purposes of environmental sound pressure level predictions, octave band data are usually sufficient. Measured data are averaged logarithmically over all measurement locations (using Equation (1.99)) to produce L_{PN} for the no load condition and L_{PL} for the rated load condition. The overall A-weighted audible sound level, N_a, referred to as the ONAF (oil natural, air forced) or the OFAF (oil forced, air forced) rating is then calculated as:

$$N_a = 10\log_{10}\left(10^{L_{PN}/10} + 10^{L_{PL}/10}\right) \qquad \text{(dB re 20 µPa)} \tag{10.116}$$

The sound power level for each octave band is:

$$L_W = N_a + 10\log_{10} S_a + C_a(f) \tag{10.117}$$

where S_a is defined in Equations (10.113) and (10.114), h is the height of the transformer tank, and for measurements made at a horizontal distance of 2 m from the transformer tank, ℓ_m is defined by Equation (10.114) where the constant, "2.4" is replaced with "16". Values of $C_a(f)$ are 2.4 dB less than the values of $C(f)$ listed in Table 10.27. This ensures that when A-weighted, the octave band sound power levels add up logarithmically to the L_W value of Equation (10.117) without the $C_a(f)$ term.

10.19 Large Wind Turbines (Rated Power Greater than or Equal to 0.2 MW)

The noise output of many modern wind turbines has been reported by Søndergaard (2014) in the form of an overall A-weighted sound power level in dBA re 10^{-12} W for a hub height wind speed of 8 m/s. He showed that the total A-weighted sound power level of turbines with a rated power greater than 200 kW could be described by Equation (10.118) (within ±5 dB).

$$L_{WA} = 8.85\log_{10}(\text{kW}) + 74.9 \qquad (\text{dB re } 10^{-12}\text{W}) \tag{10.118}$$

where kW is the turbine rated power in kilowatts. He also showed that the low-frequency (10 Hz to 160 Hz) A-weighted sound power level of turbines with a rated power greater than 200 kW could be described by Equation (10.119)(within ±5 dB).

$$L_W = 10.28\log_{10}(\text{kW}) + 59.6 \qquad (\text{dBA re } 10^{-12}\text{W}) \tag{10.119}$$

Søndergaard (2014) also provided frequency distribution data shown in Figure 10.6, where the decibel difference between the A-weighted 1/3-octave band sound power level and the total A-weighted sound power level is plotted for turbines with electrical powers greater than 2 MW. The frequency distribution for turbines with powers between 200 kW and 2 MW is not significantly different from that shown in Figure 10.6.

Søndergaard (2014) showed that newer turbines generate slightly lower noise levels than shown in Figure 10.6, between 80 and 400 Hz and slightly higher levels between 630 and 1600 Hz. Equations (10.118) and (10.119) show that as the turbine rated power increases, so too does the noise it produces over the entire frequency spectrum, with low-frequency noise increasing by the same amount as mid- and high-frequency noise.

An interesting result from Sondergard's analysis is that modern pitch-rpm regulated turbines do not produce any more noise as the hub height wind speed increases above about 8 m/s. However, it is quite a different story for stall and active-stall regulated turbines. For these turbines, the noise output increases markedly as the wind speed increases above 8 m/s, by anything from 8 to 12 dBA for a wind speed increase to 12 m/s. The 8 m/s wind speed was chosen as this is the wind speed at which all turbines reach 95% of their rated power.

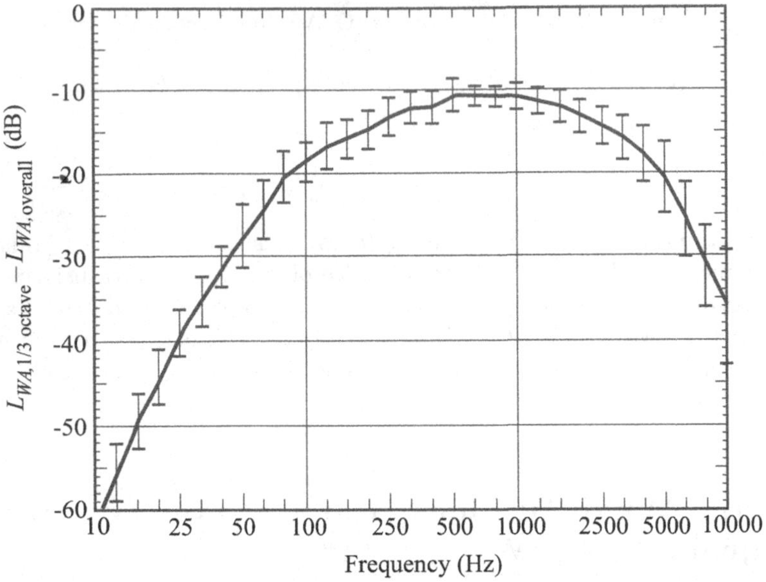

FIGURE 10.6 Wind turbine 1/3-octave band A-weighted sound power level (dB) normalised to the total A-weighted sound power level, for turbines with a power rating greater than 2 MW and for a wind speed of 8 m/s. The error bars show the 95% confidence interval around the mean. Data from Søndergaard (2013).

10.20 Transportation Noise

Environmental noise (including road traffic noise, train noise and aircraft noise) was addressed in Directive 2002/49/EC of the European Parliament and issued in 2002. This directive required that member states use common noise assessment methods to produce strategic noise maps for all major industrial sites, roads, railways and airports on a five-year rolling basis, beginning in 2007. This ambitious requirement has yet to be achieved. However, common noise assessment methods have been developed by the European Commission, Joint Research Centre (Kephalopoulos et al., 2012). These methods are based on the NMPB-2008 noise propagation model, which was discussed in Section 5.7. The processes used to obtain the sound power levels for input to the propagation model to obtain receiver sound pressure levels are outlined here in the relevant sections.

10.20.1 Road Traffic Noise

Traffic noise consists predominantly of engine–exhaust noise and tyre–road interaction noise. Engine–exhaust noise is dependent on the vehicle speed and the gear used, which in turn are dependent on vehicle technology, the grade of the road and driving behaviour. Tyre noise is dependent on the vehicle speed and the quality of the road surface. In automobiles, engine exhaust noise generally predominates in first and second gear, engine exhaust and tyre noise are equally loud in third, while tyre noise predominates in fourth gear. Procedures for measuring traffic noise are discussed in Section 4.14.7.

There are a number of commercial models available for calculating the expected traffic noise for a particular vehicle number and speed and road surface. Two of these will be considered in detail here, the UK DOT (Department of Transport), CoRTN model and the U.S. Department of Transport, Federal Highway Administration (FHWA) Traffic Noise Model, Version 1.1. In addition the model proposed by the European Commission, Joint Research Centre (Kephalopoulos et al., 2012) will also be discussed.

None of the models discussed here account for electric vehicle noise. However, in these cases, it is valid to assume that the noise would be dominated by rolling noise and that, by comparison, propulsion noise is negligible.

10.20.1.1 CNOSSOS Model (European Commission)

The CNOSSOS noise model for transportation noise prediction was described in Kephalopoulos et al. (2012), with the most recent version provided in European Commission (2015). The procedure outlined in this section was derived from European Commission (2015) and involves first dividing the road into uniform sections having the same gradient, vehicle speed and road surface and then calculating the sound power level in octave bands from 63 Hz to 8 kHz for a single vehicle in each vehicle category. This level for each vehicle category is then converted to a sound power per metre of road by taking into account the number of vehicles of that category passing per hour and the vehicle speed. The noise model requires that the road be divided into suitable segments that can each be treated as a point source in a propagation model, with the appropriate length of each segment being longer as the distance from the road increases. A good rule of thumb for a 0.1 dB accuracy is that the receiver should be further away than 10 times the segment length. However, if a section of road can be made up of many identical segments, it may be treated as a continuous, finite, incoherent line source, as discussed in Section 4.5.2.

For the purposes of determining noise emissions, vehicles are classified into 4 categories ($m = 1$ to 4) with a 5th category set aside for hybrid or electric vehicles, although no data are currently available for this latter category. However, electric vehicles could be included by using category 1 data for rolling noise and ignoring the contribution due to propulsion noise. The four categories for which data are available are:

1. light (passenger cars, caravans, trailers, delivery vans ≤ 3.5 metric tons);
2. medium-heavy (buses with 2 axles, mini-buses, motorhomes, delivery vans ≥ 3.5 metric tons);
3. heavy (buses with 3 or more axles, semi-trailers); and
4. powered two-wheelers (part a, mopeds, tricycles < 50 cc and part b, motorcycles, tricycles > 50 cc).

The **effective noise source location** is 0.05 m above the ground and beneath the centre of the vehicle and in the centre of the traffic lane. The sound power level, $L_{W,i,m,\text{line}}$, per metre of road length for octave band, i, and vehicle category, m, radiated by a stream of traffic consisting of Q_m vehicles per hour, travelling at an average speed of v_m, is:

$$L_{W,i,m,\text{line}} = L_{W,i,m} + 10\log_{10}\left(\frac{Q_m}{1000v_m}\right) \quad (\text{dB/m re } 10^{-12}\text{W}) \qquad (10.120)$$

where $L_{W,i,m}$ is the sound power radiated by a single vehicle in octave band, i, ranging from 125 Hz to 8 kHz. The speed, v_m (km/hr), is usually the maximum allowed speed for the vehicle category, although other speeds may be used. To obtain the sound power radiated from a road segment of length, L_road (m), add $10\log_{10}(L_\text{road})$ to Equation (10.120).

The single vehicle sound power, $L_{W,i,m}$, is the logarithmic sum of the sound power of the vehicle power train (propulsion noise), $L_{WP,i,m}$, and the sound power produced by the tyre–road interaction (rolling noise), $L_{WR,i,m}$. Thus:

$$L_{W,i,m} = 10\log_{10}\left(10^{L_{WP,i,m}/10} + 10^{L_{WR,i,m}/10}\right) \quad (\text{dB re } 10^{-12}\text{W}) \qquad (10.121)$$

Propulsion Noise

The propulsion noise sound power level, $L_{WP,i,m}$, may be calculated using:

$$L_{WP,i,m} = A_{P,i,m} + B_{P,i,m}\left(\frac{v_m - v_{\text{ref}}}{v_{\text{ref}}}\right) + \Delta L_{WP,i,m,\text{road}} + \Delta L_{WP,m,\text{acc}} + \Delta L_{WP,m,\text{grad}}$$

$$(\text{dB re } 10^{-12}\text{W}) \quad (10.122)$$

where $v_{\text{ref}} = 70\,(\text{m/s})$, $A_{p,i,m}$ and $B_{P,i,m}$ are defined for each octave band in Table 10.29 for $m = 1$ to 4 and the three correction terms on the right-hand side of the equation are defined in the following paragraphs.

TABLE 10.29 Coefficients, $A_{P,i,m}$ and $B_{P,i,m}$, for propulsion noise

Octave band, i centre frequency (Hz)	$A_{P,i,1}$	$B_{P,i,1}$	$A_{P,i,2}$	$B_{P,i,2}$	$A_{P,i,3}$	$B_{P,i,3}$	$A_{P,i,4a}$	$B_{P,i,4a}$	$A_{P,i,4b}$	$B_{P,i,4b}$
63	94.5	−1.3	101.0	−1.9	104.4	0.0	88.0	4.2	95.0	3.2
125	89.2	7.2	96.5	4.7	100.6	3.0	87.5	7.4	97.2	5.9
250	88.0	7.7	98.8	6.4	101.7	4.6	89.5	9.8	92.7	11.9
500	85.9	8.0	96.8	6.5	101.0	5.0	93.7	11.6	92.9	11.6
1000	84.2	8.0	98.6	6.5	100.1	5.0	96.6	15.7	94.7	11.5
2000	86.9	8.0	95.2	6.5	95.9	5.0	98.8	18.9	93.2	12.6
4000	83.3	8.0	88.8	6.5	91.3	5.0	93.9	20.3	90.1	11.1
8000	76.1	8.0	82.7	6.5	85.3	5.0	88.7	20.6	86.5	12.0

Gradient Contribution to Propulsion Noise

The correction to propulsion noise for vehicles travelling up or down a gradient of $s\%$ (positive for uphill, negative for downhill) is dependent on the vehicle class and is given by the following equations, where $\Delta L_{WP,m,\text{grad}}$ implicitly includes the effect of the slope on vehicle speed. The correction for class, $m = 4$, is 0 dB for all gradients and all correction terms are applied equally to all octave bands.

For $s < -6\%$

$$\Delta L_{WP,m,\text{grad}} = \begin{cases} \min\{12\%; -s\} - 6; & \text{for } m = 1 \\ \dfrac{\min\{12\%; -s\} - 4}{0.7} \times \dfrac{v_m - 20}{100}; & \text{for } m = 2 \\ \dfrac{\min\{12\%; -s\} - 4}{0.5} \times \dfrac{v_m - 10}{100}; & \text{for } m = 3 \end{cases} \quad (10.123\text{a–c})$$

For $-6\% \le s < -4\%$

$$\Delta L_{WP,m,\text{grad}} = \begin{cases} 0; & \text{for } m = 1 \\ \dfrac{(-s - 4)}{0.7} \times \dfrac{v_m - 20}{100}; & \text{for } m = 2 \\ \dfrac{(-s - 4)}{0.5} \times \dfrac{v_m - 10}{100}; & \text{for } m = 3 \end{cases} \quad (10.124\text{a–c})$$

For $-4\% \le s \le 0\%$, $\Delta L_{W,i,m,\text{road}} = 0$ for all vehicle classes.

For $0\% < s \leq 2\%$

$$\Delta L_{WP,m,\mathrm{grad}} = \begin{cases} 0; & \text{for } m=1 \\ s \times \dfrac{v_m}{100}; & \text{for } m=2 \\ \dfrac{s}{0.8} \times \dfrac{v_m}{100}; & \text{for } m=3 \end{cases} \tag{10.125a–c}$$

For $s > 2\%$

$$\Delta L_{WP,m,\mathrm{grad}} = \begin{cases} \dfrac{\min\{12\%; s\} - 2}{1.5} \times \dfrac{v_m}{100}; & \text{for } m=1 \\ \min\{12\%; \mathrm{s}\} \times \dfrac{v_m}{100}; & \text{for } m=2 \\ \dfrac{\min\{12\%; s\}}{0.8} \times \dfrac{v_m}{100}; & \text{for } m=3 \end{cases} \tag{10.126a–c}$$

Acceleration Contribution to Propulsion Noise

The correction to propulsion noise for vehicles accelerating or decelerating is applied where vehicles are crossing traffic lights ($k = 1$) or roundabouts ($k = 2$). The correction is a function of the distance, x (m), from the vehicle to the crossing and for $x > 100$ m, the correction is 0 dB. The correction is the same for all octave bands and is given by:

$$\Delta L_{WP,m,\mathrm{acc}} = C_{P,m,k} \times \max\{(1 - |x|/100);\ 0\} \tag{10.127}$$

where values of $C_{P,m,k}$ are given in Table 10.30. This table also includes the coefficients, $C_{R,m,k}$, for rolling noise which are discussed below. The coefficients are all zero for vehicle category 4 and for other vehicle categories the coefficients are the same for all octave bands.

TABLE 10.30 Coefficients for acceleration and deceleration for both propulsion and rolling noise

m	$C_{P,m,1}$	$C_{P,m,2}$	$C_{R,m,1}$	$C_{R,m,2}$
1	5.5	3.1	−4.5	−4.4
2	9.0	6.7	−4.0	−2.3
3	9.0	6.7	−4.0	−2.3

The acceleration contribution to propulsion noise is often omitted in traffic noise models as it requires the road to be divided into much shorter segments (length of 1 m is acceptable) in the vicinity of the intersection. The $\Delta L_{WP,m,\mathrm{acc}}$ correction can then be calculated to a good approximation for each small segment.

Road Surface Contribution to Propulsion Noise

The correction to propulsion noise to account for the road surface is given by:

$$\Delta L_{WP,i,m,\mathrm{road}} = \min\{\alpha_{i,m};\ 0\} \tag{10.128}$$

where values of $\alpha_{i,m}$ are listed in Table 10.31.

Rolling Noise

The rolling noise sound power level, $L_{WR,i,m}$, may be calculated using:

$$L_{WR,i,m} = A_{R,i,m} + B_{R,i,m}\left(\frac{v_m - v_{\mathrm{ref}}}{v_{\mathrm{ref}}}\right) + \Delta L_{WR,i,m,\mathrm{road}} + \Delta L_{WR,m,\mathrm{acc}} + \Delta L_{WR,i,m,\mathrm{studtyres}} + \Delta L_{WR,m,\mathrm{temp}} \quad (\mathrm{dB\ re\ } 10^{-12}\mathrm{W}) \tag{10.129}$$

TABLE 10.31 Coefficients, $\alpha_{i,m}$ and β_m, for various road surfaces. The minimum and maximum speed values (km/hr) refer to the speed range over which the coefficients are valid

Description	Min. speed	Max. speed	Vehicle cat.	$\alpha_{i,m}$ values: Octave band centre frequency (Hz) 63	125	250	500	1000	2000	4000	8000	β_m
1-layer ZOAB[a]	50	130	1	0.5	3.3	2.4	3.2	−1.3	−3.5	−2.6	0.5	−6.5
			2	0.9	1.4	1.8	−0.4	−5.2	−4.6	−3.0	−1.4	0.2
			3	0.9	1.4	1.8	−0.4	−5.2	−4.6	−3.0	−1.4	0.2
2-layer ZOAB[a]	50	130	1	0.4	2.4	0.2	−3.1	−4.2	−6.3	−4.8	−2.0	−3.0
			2	0.4	0.2	−0.7	−5.4	−6.3	−6.3	−4.7	−3.7	4.7
			3	0.4	0.2	−0.7	−5.4	−6.3	−6.3	−4.7	−3.7	4.7
2-layer ZOAB[a] (fine)	80	130	1	−1.0	1.7	−1.5	−5.3	−6.3	−8.5	−5.3	−2.4	−0.1
			2	1.0	0.1	−1.8	−5.9	−6.1	−6.7	−4.8	−3.8	−0.8
			3	1.0	0.1	−1.8	−5.9	−6.1	−6.7	−4.8	−3.8	−0.8
SMA-NL5[b]	40	80	1	1.1	−1.0	0.2	1.3	−1.9	−2.8	−2.1	−1.4	−1.0
			2	0.0	0.0	0.0	0.0	0.0	0.0	0.0	0.0	0.0
			3	0.0	0.0	0.0	0.0	0.0	0.0	0.0	0.0	0.0
SMA-NL8[b]	40	80	1	0.3	0.0	0.0	−0.1	−0.7	−1.3	−0.8	−0.8	−1.0
			2	0.0	0.0	0.0	0.0	0.0	0.0	0.0	0.0	0.0
			3	0.0	0.0	0.0	0.0	0.0	0.0	0.0	0.0	0.0
Brushed-down concrete	70	120	1	1.1	−0.4	1.3	2.2	2.5	0.8	−0.2	−0.1	1.4
			2	0.0	1.1	0.4	−0.3	−0.2	−0.7	−1.1	−1.0	4.4
			3	0.0	1.1	0.4	−0.3	−0.2	−0.7	−1.1	−1.0	4.4
Optimised brushed-down concrete	70	80	1	−0.2	−0.7	0.6	1.0	1.0	−1.5	−2.0	−1.8	1.0
			2	−0.3	1.0	−1.7	−1.2	−1.6	−2.4	−1.7	−1.7	−6.6
			3	−0.3	1.0	−1.7	−1.2	−1.6	−2.4	−1.7	−1.7	−6.6
Fine-broomed concrete	70	120	1	1.1	−0.5	2.7	2.1	1.6	2.7	1.3	−0.4	7.7
			2	0.0	3.3	2.4	1.9	2.0	1.2	0.1	0.0	3.7
			3	0.0	3.3	2.4	1.9	2.0	1.2	0.1	0.0	3.7
Worked surface	50	130	1	1.1	1.0	2.6	4.0	4.0	0.1	−1.0	−0.8	−0.2
			2	0.0	2.0	1.8	1.0	−0.7	−2.1	−1.9	−1.7	1.7
			3	0.0	2.0	1.8	1.0	−0.7	−2.1	−1.9	−1.7	1.7
Hard elements in herringbone	30	60	1	8.3	8.7	7.8	5.0	3.0	−0.7	0.8	1.8	2.5
			2	8.3	8.7	7.8	5.0	3.0	−0.7	0.8	1.8	2.5
			3	8.3	8.7	7.8	5.0	3.0	−0.7	0.8	1.8	2.5
Hard elements not in herringbone	30	60	1	12.3	11.9	9.7	7.1	7.1	2.8	4.7	4.5	2.9
			2	12.3	11.9	9.7	7.1	7.1	2.8	4.7	4.5	2.9
			3	12.3	11.9	9.7	7.1	7.1	2.8	4.7	4.5	2.9
Quiet hard elements	30	60	1	7.8	6.3	5.2	2.8	−1.9	−6.0	−3.0	−0.1	−1.7
			2	0.2	0.7	0.7	1.1	1.8	1.2	1.1	0.2	0.0
			3	0.2	0.7	0.7	1.1	1.8	1.2	1.1	0.2	0.0
Thin Layer A	30	60	1	1.1	0.1	−0.7	−1.3	−3.1	−4.9	−3.5	−1.5	−2.5
			2	1.6	1.3	0.9	−0.4	−1.8	−2.1	−0.7	−0.2	0.5
			3	1.6	1.3	0.9	−0.4	−1.8	−2.1	−0.7	−0.2	0.5
Thin Layer B	40	130	1	0.4	−1.3	−1.3	−0.4	−5.0	−7.1	−4.9	−3.3	−1.5
			2	1.6	1.3	0.9	−0.4	−1.8	−2.1	−0.7	−0.2	0.5
			3	1.6	1.3	0.9	−0.4	−1.8	−2.1	−0.7	−0.2	0.5

[a]Zeer Open Asfalt Beton which is Dutch for very porous asphalt.

[b]Noise reducing asphalt using two layers of porous asphalt.

where $v_{\text{ref}} = 70\,(\text{m/s})$, $A_{p,i,m}$ and $B_{P,i,m}$ are defined for each octave band in Table 10.32 for $m = 1$ to 3. For $m = 4$, $A_{p,i,4}$ and $B_{P,i,4} = 0$ and the four correction terms on the right-hand side of the equation are defined in the following paragraphs. For category 4 vehicles, the rolling noise sound power is considered to be negligible and the corresponding term is excluded from Equation (10.121).

TABLE 10.32 Coefficients, $A_{R,i,m}$, $B_{R,i,m}$, for rolling noise

Octave band, i centre frequency (Hz)	$A_{R,i,1}$	$B_{R,i,1}$	$A_{R,i,2}$	$B_{R,i,2}$	$A_{R,i,3}$	$B_{R,i,3}$
63	79.7	30.0	84.0	30.0	87.0	30.0
125	85.7	41.5	88.7	35.8	91.7	33.5
250	84.5	38.9	91.5	32.6	94.1	31.3
500	90.2	25.7	96.7	23.8	100.7	25.4
1000	97.3	32.5	97.4	30.1	100.8	31.8
2000	93.9	37.2	90.9	36.2	94.3	37.1
4000	84.1	39.0	83.8	38.3	87.1	38.6
8000	74.3	40.0	80.5	40.1	82.5	40.6

Gradient Contribution to Rolling Noise
The effect of a gradient on rolling noise is an indirect effect resulting from the vehicle speed change. The adjusted speed should be included in all terms in Equation (10.129). No additional term is included explicitly for the gradient effect on rolling noise.

Acceleration Contribution to Rolling Noise
The correction to propulsion noise for vehicles accelerating or decelerating is applied where vehicles are crossing traffic lights ($k = 1$) or roundabouts ($k = 2$). The correction is a function of the distance, x (m), from the vehicle to the crossing and for $x > 100$ m, the correction is 0 dB. The correction is the same for all octave bands and is given by:

$$\Delta L_{WR,m,\text{acc}} = C_{R,m,k} \times \max(1 - |x|/100;\, 0) \tag{10.130}$$

where values of $C_{R,m,k}$ are given in Table 10.30. See the discussion under this heading in the propulsion noise section above for more detail on implementing the acceleration correction in a traffic noise propagation model.

Temperature Effect on Rolling Noise
The correction to rolling noise, to account for the atmospheric annual temperature averages being different to 20°C, is the same for all octave bands and is given by:

$$\Delta L_{WR,m,\text{temp}} = K_m(T_{\text{ref}} - T) \tag{10.131}$$

where $T_{\text{ref}} = 20$°C, T is the annual average air temperature in °C. In the absence of better information, we set $K_m = 0.08$ for category 1 vehicles, $K_m = 0.04$ for category 2 and 3 vehicles and $K_m = 0.0$ for category 4 vehicles. However, it is known that K_m depends on tyre type and road surface type but no data on these effects are currently available.

Effect of Studded Tyres on Rolling Noise
The correction to rolling noise to account for vehicles in category 1 having studded tyres (tyres with embedded retractable tungsten carbide pins for greater traction on ice) is given by:

$$\Delta L_{WR,i,1,\text{studtyres}} = 10\log_{10}\left[\left(1 - \frac{Q_s t_s}{12}\right) + \frac{Q_s t_s}{12} 10^{\Delta_{\text{stud},i}/10}\right] \tag{10.132}$$

where Q_s is the average ratio of vehicles per hour with studded tyres during a period of t_s months, and:

$$\Delta_{\text{stud},i} = \begin{cases} a_i + b_i \log_{10}(50/70); & v < 50 \text{ (km/hr)} \\ a_i + b_i \log_{10}(v/70); & 50 \leq v \leq 90 \text{ (km/hr)} \\ a_i + b_i \log_{10}(90/70); & v > 90 \text{ (km/hr)} \end{cases} \tag{10.133a–c}$$

where values of the coefficients, a_i and b_i, only apply to category 1 vehicles and are listed in Table 10.33.

TABLE 10.33 Coefficients, a_i and b_i, for studded tyres for vehicle category 1

Coefficient	Octave band centre frequency (Hz)					
	≤250	500	1000	2000	4000	8000
a_i	0.0	2.6	2.9	1.5	2.3	9.2
b_i	0.0	−3.1	−6.4	−14.0	−22.4	−11.4

Road Surface Contribution to Rolling Noise
The correction to rolling noise to account for the road surface is given by:

$$\Delta L_{WR,i,m,\text{road}} = \alpha_{i,m} + \beta_m \log_{10}\left(\frac{v_m}{v_{\text{ref}}}\right) \tag{10.134}$$

where $v_{\text{ref}} = 70$ km/hr and $\alpha_{i,m}$ and β_m are given in Table 10.31. A reference road surface is defined for which values of $\alpha_{i,m}$ and β_m are all zero. This road surface consists of average density asphalt concrete 0/11 and stone mastic asphalt 0/11 (SMA) between 2 and 7 years old in a typical maintenance condition. All values of $\alpha_{i,4}$ and β_4 are zero for vehicle categories 4a and 4b. Additional data on the relative noise-producing characteristics of various road surfaces has been provided by Buret et al. (2016).

10.20.1.2 UK DoT model (CoRTN)

A relatively simple procedure to estimate the noise impact of a particular traffic flow, which is based on the model developed by the UK Dept. of Environment (referred to as CoRTN or Calculation of Road Traffic Noise) yields reasonable results, although it can result in significant errors in some cases (UK. DOT, 1988; Delaney et al., 1976), and it is quite dated.

For normal roads, the traffic flow in both directions is combined together to give the total traffic flow used for the sound pressure level calculations. However, if the two carriageways are separated by more than 5 metres, then the sound pressure level contribution at the receiver location due to each carriageway must be calculated separately and the results combined logarithmically using Equation (1.98) to give the total sound pressure level due to both carriageways. The nearside carriageway is treated as for a normal road. However, for the far-side carriageway, the source line is assumed to be 3.5 m in from the far kerb and the effective edge of the carriageway is considered to be 7 m in from the far kerb.

The CoRTN model calculates the A-weighted L_{10} (denoted here as L_{A10}) over 1-hour or 18-hour intervals. The A-weighted L_{10} (18hr) quantity is simply the arithmetic mean of the 18 separate one-hourly values of L_{10} covering the period 6:00 am to 12:00 am on a normal working day. The CoRTN model allows the sound pressure level to be estimated at a distance, d, from the vehicle source using:

$$\begin{aligned} L_{A10}(18\text{hr}) = {} & 29.1 + 10\log_{10} Q + C_{\text{dist}} + C_{\text{use}} + C_{\text{grad}} + C_{\text{cond}} \\ & + C_{\text{ground}} + C_{\text{barrier}} + C_{\text{view}} \quad \text{(dBA)} \end{aligned} \tag{10.135}$$

Most recent regulations are expressed in terms of L_{Aeq} so there has been considerable interest in converting L_{10} estimates to L_{Aeq} estimates. The Transport Research Limited (TRL) (Abbott and Nelson, 2002) developed two conversion equations but these are not reproduced here due to some considerable disagreement with measured data (Kean, 2008). He showed that at 13.5 m from the roadway, the difference between L_{Aeq} and L_{A10} is very close to 3 dBA with L_{A10} being greater. This difference decreases with distance from the roadway and the reader is referred to Kean (2008) for more details.

A more accurate way of estimating L_{A10} (18hr) is to use estimates of L_{A10} (1hr) for each single hour in the 18-hour period:

$$\begin{aligned} L_{A10}(1\text{hr}) = 42.2 + 10\log_{10} q + C_{\text{dist}} + C_{\text{use}} + C_{\text{grad}} + C_{\text{cond}} \\ + C_{\text{ground}} + C_{\text{barrier}} + C_{\text{view}} \quad \text{(dBA)} \end{aligned} \tag{10.136}$$

In Equations (10.135) and (10.136), Q is the total number of vehicles in the 18-hour period between 6:00 am and 12:00 am, q is the number of vehicles per hour, C_{dist} is a correction factor to account for the distance of the observer from the road, C_{use} is a correction factor to account for the percentage of heavy vehicles, C_{grad} is a correction factor to account for the gradient of the road surface, C_{cond} is a correction factor to account for the type and condition of the road surface, C_{ground} is a correction factor to account for the effect of the ground surface and C_{barrier} is a correction factor to account for the presence or otherwise of barriers.

Distance Correction
This correction is given by:

$$C_{\text{dist}} = -10\log_{10}(r/13.5) \tag{10.137}$$

where r is the straight line distance from the source to the observer (dependent on source and observer height). The source line is assumed to be 3.5 m in from the near edge of the road and both carriageways are treated together, except if they are more than 5 m apart (see page 656).

Use Correction
This correction is given by:

$$C_{\text{use}} = 33\log_{10}(v + 40 + (500/v)) + 10\log_{10}(1 + (5P/v)) - 68.8 \quad \text{(dBA)} \tag{10.138}$$

where P is the percentage (0–100%) of heavy vehicles (weighing more than 1525 kg) and v is the average speed (km/h). Information on the average speed on most roads in the metropolitan area is available from the relevant government department responsible for road construction and maintenance. If it cannot be found or determined, the default values in Table 10.34 can be used.

Road Gradient Correction
This correction is $C_{\text{grad}} = 0.3G$ if the measured average speed is used and $C_{\text{grad}} = 0.2G$ if the design speed of the road is used, where G is the percentage gradient (altitude change × 100 / horizontal distance) of the road. Note that no correction is used for vehicles travelling downhill.

TABLE 10.34 Suggested average vehicle speeds for various road types and speed limits

Type of road	Speed limit	Value for v
Rural roads	110 km/h	108
Urban freeway	90 km/h	92
Urban highway	70 km/h	65
Urban street dual carriageway		60
Urban street single carriageway		55
Urban street single congested		50

Road Condition Correction
The correction, C_{cond}, for the road surface is taken as zero for either sealed roads at speeds above 75 km/hr or gravel roads. For speeds below 75 km/hr on impervious sealed roads, the correction is −1 dBA. For porous road surfaces, the correction is −3.5 dBA.

Ground Type Correction
The correction, C_{ground}, for $1.0 < h_r < (d/3 - 1.2)$ is:

$$C_{\text{ground}} = 5.2 P_d \log_{10}\left(\frac{3h_r}{d+3.5}\right) \quad \text{(dBA)} \tag{10.139}$$

where d is the horizontal distance from the edge of the road to the observer (independent of source or receiver height and assumed greater than 4 m), P_d is the proportion (1.0 or less) of absorbent ground (flow resistivity less than 1000 MKS rayls/m) between the edge of the road and the observer and h_r is the height of the observer above the ground. For sound propagation over grass, P_d is set equal to 1.0 for L_{A10} calculations and 0.75 for L_{Aeq} calculations (Kean, 2008). Kean (2008) also points out that the ground correction factor was derived empirically and includes air absorption. Thus, the value of P_d should never be set less than 0.3 so that air absorption is included.

If the observer height, h_r, is greater than $(d/3 - 1.2)$, then $C_{\text{ground}} = 0$. If the observer height is less than 1 m, then:

$$C_{\text{ground}} = 5.2 P_d \log_{10}\left(\frac{3}{d+3.5}\right) \quad \text{(dBA)} \tag{10.140}$$

The relation between r (of Equation (10.137)) and d (as the vehicle source is considered to be 0.5 m above the road and 3.5 m from the edge of the road) is:

$$r = \left[(d+3.5)^2 + (h_r - 0.5)^2\right]^{1/2} \tag{10.141}$$

Barrier Correction
If highway noise barriers exist, then their effect on the sound pressure level at the observer may be calculated using:

$$C_{\text{barrier}} = \sum_{i=0}^{n} A_i X^i \quad \text{(dBA)} \tag{10.142}$$

where $X = 10\log_{10}\delta$ and δ is the difference (in metres) in the following two paths from the source line (3.5 m in from the edge of the road and at a height of 0.5 m) to the observer:

1. shortest path over the top of the barrier; and
2. shortest direct path in the absence of the barrier.

The coefficients, A_i, are listed in Table 10.35 and X^i means the quantity, X, raised to the ith power and $n = 5$ or 7 (see Table 10.35).

When multiple barriers of different heights screen the observer from the road, they should be evaluated separately and only the correction resulting in the lowest sound pressure level should be used.

Low barriers such as twin-beam metal crash barriers can have less effect than soft ground. So if these are used with any proportion, P_d, of soft ground, the combined barrier and ground effect should be calculated by using the lower value of $C_{\text{ground}} + C_{\text{barrier}}$, resulting from the following two calculations.

1. Soft ground correction ($0 < P_d < 1.0$), excluding the barrier correction, C_{barrier}.
2. Hard ground correction, C_{ground}, (with $P_d = 0$ in Equations (10.139) and (10.140)) plus the barrier correction, C_{barrier}.

TABLE 10.35 Coefficients for barrier effect calculations for traffic noise

Coefficient	Shadow zone	Bright zone
A_0	−15.4	0.0
A_1	−8.26	+0.109
A_2	−2.787	−0.815
A_3	−0.831	+0.479
A_4	−0.198	+0.3284
A_5	+0.1539	+0.04385
A_6	+0.12248	—
A_7	+0.02175	—
Above valid for	$-3 \leq X \leq 1.2$	$-4 \leq X \leq 0$

For X outside the limits in the table, the following applies:
Shadow zone; for $X < -3$, $C_{\text{barrier}} = -5.0$
and for $X > 1.2$, $C_{\text{barrier}} = -30.0$
Bright zone; for $X < -4$, $C_{\text{barrier}} = -5.0$
and for $X > 0$, $C_{\text{barrier}} = 0.0$

View to Road Correction

In some cases, the angle of view of the road will include a range of different configurations such as bends in the road, intersections and short barriers. To accommodate this, the overall field of view must be divided into a number of segments, each of which is characterised by uniform propagation conditions. The overall sound pressure level can be found by calculating the sound pressure level due to each segment separately using Equations (10.135) and (10.136) and then adding the contributions (in dB) from each segment together logarithmically as for incoherent sources (see Section 1.11.3). In this case, the following correction is then applied to each segment.

$$C_{\text{view}} = 10\log_{10}(\beta/180) \tag{10.143}$$

where β is the actual field of view in degrees. Note that in such segments, the road is always projected along the field of view and the distance from the segment is measured perpendicular to the extended road, as illustrated in Figure 10.7.

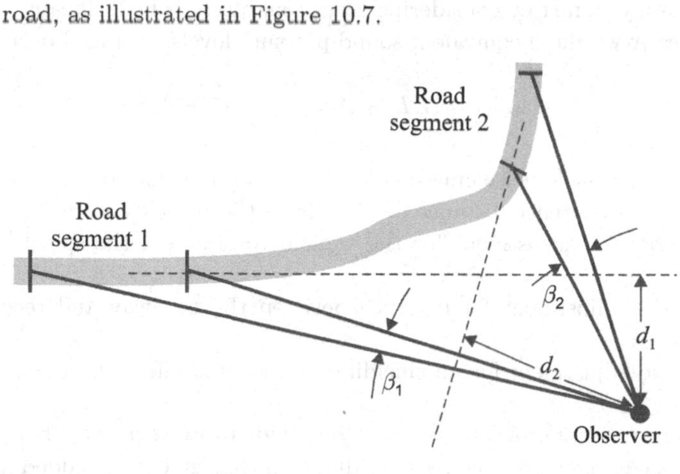

FIGURE 10.7 Arrangement defining β and d for two different road segments.

The segment method just outlined can be simplified if there are two propagation conditions that repeat. For example, if there existed a barrier (or set of buildings) with regular or irregular gaps in it, then all the gaps could be added together to determine an effective β for the gaps and the same could be done for the barrier segments so that only two segments need be considered.

If there are dual carriageways (separated by more than 5 m) or multiple roads, then each road or carriageway is treated separately and the results added together logarithmically using Equation (1.98) However, for best accuracy (not proposed in CoRTN), it is best to treat each lane of traffic separately (as done in the FHWA model in Section 10.20.1.3) and use the actual distance of the centre of the lane of traffic from the edge of the roadway, instead of the arbitrary 3.5 m as specified in CoRTN.

Reflections from buildings and other hard surfaces increase the sound pressure levels at the observer. The following general empirical corrections may be used to estimate the increase:

1. If the observer is within 1 m of a building façade, then the sound pressure level is increased by 2.5 dBA.
2. Sound pressure levels down side streets perpendicular to the road in question are 2.5 dBA higher due to reflections from adjacent houses.
3. Reflective surfaces on the far side of the road increase the level by 1.5 dBA.

Sometimes highway noise barriers are tilted slightly so that reflected waves are directed upwards to minimise reflections between barriers on opposite sides of the highway, which can increase noise levels on the receiver side of each barrier. The preceding analysis also applies for these cases.

10.20.1.3 United States FHWA Traffic Noise Model (TNM)

The United States FHWA Traffic Noise Model (TNM) is intended to be used for predicting noise impacts in the vicinity of highways. This is a reasonably complex model and is available as a software implementation from the FHWA. The guide (FHWA, 1998, 2004) and software package (FHWA, 2015) are available on the FHWA website for no charge. The calculation proceeds by estimating the sound pressure level due to a single lane of a single type of traffic at an observer location. The process is repeated for each lane and each traffic type, and the total sound pressure level for all lanes combined is calculated by logarithmically adding the individual sound pressure levels for each lane using Equation (1.98). In the following description, the term "roadway" may apply to a lane of traffic or to more than 1 lane if the receiver is sufficiently far away from the road that the accuracy gained by considering lanes separately is insignificant.

TNM calculates A-weighted equivalent sound pressure levels, averaged over 1-hour using the following relation:

$$L_{Aeq,1h} = EL_i + A_{\text{traff}(i)} + A_d + A_s \tag{10.144}$$

where

EL_i represents the vehicle noise emission level (maximum sound pressure level emitted by a vehicle pass-by at a reference distance of 15 m from the vehicle centre).

$A_{\text{traff}(i)}$ represents the adjustment for the vehicle volume and the speed for the vehicle of type, i.

A_d represents the adjustment for distance between the roadway and receiver and for the length of roadway.

A_s represents the adjustment for all shielding and ground effects between the roadway and receiver.

Note that the roadway section of interest must be divided into segments that subtend no more than 10° at the receiver. Levels at the receiver due to each segment are added together logarithmically, using Equation (1.98).

When the hourly sound pressure levels are combined together in the appropriate way, as discussed in Chapter 2, the average day-night sound level, L_{dn}, and the community noise equivalent level, L_{den}, can be calculated easily. After the sound pressure levels corresponding to all of the different vehicle types and roadway segments have been calculated for a particular receiver location, they are added together logarithmically to give the total level.

The TNM database for vehicle emission levels includes data for a number of different pavement conditions and vehicle types as well as for vehicles cruising, accelerating, idling and on grades. The database includes 1/3-octave band spectra for cars (2 axles and 4 wheels), medium trucks (2 axles and 6 wheels), heavy trucks (3 or more axles), buses (2 or 3 axles and 6 or more wheels) and motorcycles (2 or 3 wheels). The data are further divided into two source locations: one at pavement height and one at 1.5 m in height (except for heavy trucks for which the upper height is 3.66 m). The database is available in the FHWA Traffic Noise Model technical manual (Menge et al., 1998). However, it is recommended that the FHWA Traffic Noise Model software be obtained from the FHWA (USA) if accurate estimates of traffic noise impact are required. Alternatively, it is possible to take one's own measurements of particular vehicles and use those.

The TNM correction, $A_{\text{traff}(i)}$, for vehicle volume and speed is the same for all vehicle types and is given by the following equation.

$$A_{\text{traff}(i)} = 10\log_{10}\left(\frac{V_i}{v_i}\right) - 13.2 \text{ (dB)} \tag{10.145}$$

where V_i is the vehicle volume in vehicles per hour of vehicle type, i and v_i is the vehicle speed in km/hr.

The adjustment, A_d, for distance from the elemental roadway segment to the receiver and for the length of the roadway segment for all vehicle types and source heights is given by:

$$A_d = 10\log_{10}\left[\left(\frac{15}{d}\right)\left(\frac{\beta}{180}\right)\right] \text{ (dB)} \tag{10.146}$$

where d is the perpendicular distance in metres from the receiver to the line representing the centre of the roadway segment (or its extension) and β is the angle subtended at the receiver (in degrees) by the elemental roadway segment (that is, the field of view – see Figure 10.7). If $d < 0.3$ m and $\beta < 20°$, the following equation should be used:

$$A_d = 10\log_{10}\left[\frac{|d_2 - d_1|}{d_2 d_1}\right] + 12 \quad \text{(dB)} \tag{10.147}$$

where d_1 and d_2 are the distances from the receiver to each end of the roadway segment.

The calculation of the correction factor, A_s, for all shielding and ground effects between the roadway and receiver is quite complicated and is explained in detail in the FHWA Traffic Noise Model technical manual (Menge et al., 1998). Alternatively, the procedures outlined in Chapter 5 may be used.

The FHWA model is regarded as very accurate and more up to date than the CoRTN model, which was developed in 1988.

10.20.1.4 Other Models

There are a number of other traffic noise models that are considerably more complex including the German Road Administration model (RLS-90, "guidelines for noise protection on streets"), the Acoustical Society of Japan model, which was later updated by Takagi and Yamamoto (1994), and the revised version of the joint Nordic prediction method for road traffic noise, published in 1989 and used mainly in Scandinavia. All of these models are similar in that they contain a source model for predicting the noise at the roadside (or close to it) and a propagation model that takes into account ground, barrier and atmospheric effects. The models have all been implemented in specialised software, which in most cases, is available for a reasonable price. Useful reviews of the various models are available in the literature (Saunders et al., 1983; Steele, 2001).

A comprehensive review of the effect of vehicle noise regulations on road traffic noise, changes in vehicle emissions over the past 30 years and recommendations for consideration in the drafting of future traffic noise regulations has been provided by Sandberg (2001).

10.20.1.5 Accuracy of Traffic Noise Models

The accuracy of some of the various prediction models discussed in Section 10.20.1 was investigated by de Lisle (2016). For a large number of measurement locations, he reported that the CoRTN model had an overprediction standard deviation over all 12 locations of 2.6 dB for a rural freeway and a maximum overprediction error of 10 dB. On the other hand, the TNM model had a prediction standard deviation of 3.0 dB (some were higher and some lower than measured data) and a maximum overprediction error of 8 dB. For an urban freeway, the maximum overprediction errors were similar and the standard deviations over all 27 locations were 2.7 dB and 3.3 dB, respectively.

10.20.2 Rail Traffic Noise

Train noise is usually dominated by wheel/rail interaction noise. As the train speed increases, the wheel/rail noise increases, but the locomotive engine noise decreases. Thus, it is often necessary to calculate the contribution from the two types of noise separately.

There are a number of models available for estimating train noise. Perhaps the most well-known are the Nordic model (Nielsen, 1996), the German model (SCHALL 03, 2006), the Dutch model (Nederlands Ministerie volkshuisvesting, ruimtelijke ordening en milieubeheer (in Dutch), 1996), the European Commission model (European Commission, 2015) and the UK Department of Transport Model (UK. DOT, 1995a,b).

The Nordic model (called "NMT") (Nielsen, 1996) is discussed in Section 10.20.2.1. A computer implementation of NMT is available from the Norwegian State Railways. The model calculates octave band sound pressure levels, with band centre frequencies ranging from 63 Hz to 4 kHz, at a specified community location, using train source sound power data with corrections for the state of the track, together with a basic propagation model. The model documentation includes a description of a simplified model as well as means to measure data for new trains.

The German model (SCHALL 03, 2006) includes sound power level estimates for train noise sources for various trains and uses the ISO 9613-2 (1996) propagation model, described in Section 5.6. However, no English translation is currently available which is why the model is not described in detail here.

The Dutch model (Nederlands Ministerie volkshuisvesting, ruimtelijke ordening en milieubeheer (in Dutch), 1996) is very comprehensive and includes sound power level data for train noise sources as well as a useful propagation model. However, no English translation is currently available.

The EU model (European Commission, 2015) is a sound power level estimation scheme and provides data to allow sound power level calculations for rolling noise, aerodynamic noise and impact noise but no data for other noise sources on the train (such as engine noise, inverter noise and fan noise). The model is described in Section 10.20.2.2. The propagation model recommended for use with the sound power data is based on the NMPB-2008 model which is described in Section 5.7.

The UK model (UK. DOT, 1995a) calculates A-weighted sound exposure levels (SEL or L_{AE}) (see Section 2.4.5 and ISO 1996-1 (2016)) and then L_{Aeq} levels at any specified distance from the train track. Lack of sound power information makes it difficult to implement more sophisticated propagation models such as ISO 9613-2 (1996) or NMPB-2008. Another disadvantage of this model is that it only calculates A-weighted overall levels and no calculations or corrections are done in octave or 1/3-octave bands, which makes the model only applicable to the trains emitting noise spectra with a similar shape to those used to gather the data. This model is outlined in Section 10.20.2.3.

A comparison of the various models has been done briefly by van Leeuwen (2000) and a detailed description of train noise estimation and control is provided in Wettschureck et al. (2013).

10.20.2.1 Nordic Prediction Model (1996)

The Nordic prediction model is well described in Nielsen (1996) and allows calculation of both $L_{Aeq,24}$, $L_{A,\mathrm{maxM}}$ and $L_{A,\mathrm{maxF}}$ at any receiver location.

$L_{Aeq,24}$ is the A-weighted, energy-averaged (see Equation (4.142)) sound pressure level at a receiver location averaged over a 24-hour period. Each segment of track is assumed to produce the same amount of energy, and contributions from all track segments that contribute significantly at the receiver under consideration are logarithmically added together to obtain the total $L_{Aeq,24}$.

$L_{A,\mathrm{maxM}}$ is the overall A-weighted, sound power level (energy-averaged over the train length), as the length of train passes the nearest track location to the receiver, and $L_{A,\mathrm{maxF}}$ is the maximum value of $L_{A,\mathrm{maxM}}$ that would be measured by a sound level meter set on "fast" response. The position of the train on a track segment that results in the highest sound pressure level, $L_{A,\mathrm{maxF}}$, at a receiver location will vary from one receiver to the next. The octave band level, $L_{\mathrm{maxF},i}$ for octave band, i, is related to the $L_{\mathrm{maxM},i}$ level in the same octave band by:

$$L_{\mathrm{maxF},i} = L_{\mathrm{maxM},i} - 3r/100 + \begin{cases} 3 & \text{for electric trains} \\ 6 & \text{for diesel trains} \end{cases} \tag{10.148}$$

where r is the distance from the source to the receiver. If $r > 100$ m for electric trains or $r > 200$ m for diesel trains, then $L_{\mathrm{maxF},i} = L_{\mathrm{maxM},i}$. Equation (10.148) applies to all octave bands and the overall A-weighted sound power levels, $L_{A,\mathrm{maxF}}$ and $L_{A,\mathrm{maxM}}$ are found by A-weighting each corresponding octave band level and logarithmically summing the results for each octave band using Equation (1.98).

Train Sound Power Levels

The calculation procedure consists of first dividing the track into segments, the length of which should be less than 50% of the distance from the track to the furthest receiver under consideration. Each track segment should be characterised by a uniform radiated sound power per metre of its length. Next the sound power level per metre of track length is calculated for each train type and for each train speed that characterises the track. The sound pressure levels for each track segment, train type, train speed and each octave band are then determined at each receiver location using train sound power levels per metre of track, the length of the track segment, modifications due to the state of the track and a propagation model. The results at each receiver location are added together logarithmically (see Equation (1.98)) to obtain the total sound pressure level, $L_{eq,24}$, in each octave band. In addition, values of $L_{\mathrm{maxM},i}$ are determined in each octave band for the train position that results in the highest sound pressure level for the particular receiver location. Finally, the octave band results are converted to A-weighted values using Table 2.3 and the A-weighted octave band values are summed logarithmically using Equation (1.98) to find the total A-weighted level, $L_{Aeq,24}$ and $L_{A,\mathrm{maxM}}$.

Train sound power levels/metre of track length, used for calculating $L_{eq,24}$, are estimated for each train type and each train speed using:

$$L_{W,i} = a_i \log_{10}(v/100) + 10 \log_{10} L_{24} + b_i \qquad ((\mathrm{dB/m})\ \mathrm{re}\ 10^{-12}\mathrm{W}) \tag{10.149}$$

where v is the train speed in km/hr, $L_{W,i}$ is the unweighted sound power level for octave band, i and L_{24} is the total passing train length in a 24-hour period. For calculating the value of $L_{Aeq,24}$

at the receiver, the track is divided into segments and all segments that contribute significantly to the sound pressure level at the receiver are taken into account. Each track segment is considered as a point source and the total sound power of that segment is used with a propagation model to determine the contribution of that segment to the octave band levels, $L_{\mathrm{maxM},i}$, at the receiver. Values of a_i and b_i for octave band, i, are given in Table 10.36 for Norwegian, Swedish and Finnish trains that were operating in 1996 when the model was published (Nielsen, 1996). Data for more modern trains are not available, but the model documentation provides procedures for determining sound power levels for any new trains.

The propagation model is used to convert these octave band sound power levels to unweighted octave band sound pressure levels, $L_{eq,24,i}$, at the particular receiver under consideration. The octave band sound pressure levels are converted to A-weighted octave band levels and then added together logarithmically, as indicated in Equation (1.98), to obtain $L_{Aeq,24}$ for a particular train type, train speed and track segment. Finally, the contributions from all different trains, different speeds and different track segments are added together logarithmically, as indicated in Equation (1.98), to obtain the total $L_{Aeq,24}$. Note that in the final $L_{\mathrm{maxM},i}$ calculations, the track length involved in each track segment must be accounted for (see Equation (10.151)).

Train sound power levels/metre of train length, used for calculating $L_{\mathrm{maxM},i}$, are estimated for each train type and each train speed using:

$$L_{W,i,M} = a_i \log_{10}(v/100) + 10\log_{10} v + b_i + 43.8 \qquad ((\mathrm{dB/m})\ \mathrm{re}\ 10^{-12}\mathrm{W}) \tag{10.150}$$

where v is the train speed in km/hr. Values of a_i and b_i for octave band, i, are given in Table 10.36 for Norwegian, Swedish and Finnish trains that were operating in 1996 when the model was published (Nielsen, 1996).

The calculation requires that the train length be divided into small segments and the sound pressure level at the receiver is then calculated for each segment (taking into account the segment length – see Equation (10.152)). The total sound pressure level at the receiver in each octave band, i, is then calculated by logarithmically summing the contributions from all train segments that contribute significantly to the total $L_{\mathrm{maxM,i}}$. Also, the train position when undertaking the calculations for a particular receiver should be the one that results in the maximum overall $L_{A,\mathrm{maxM}}$ level at the receiver. Each receiver will have a different train position that results in the maximum $L_{A,\mathrm{maxM}}$ level.

For calculating the value of $L_{\mathrm{maxM},i}$ at the receiver, the train is placed in the position that produces the highest sound pressure level at the receiver under consideration. The train length is then divided into segments, each of which is considered as a point source and the total sound power of that segment is used with a propagation model to determine the contribution of that segment to the $L_{\mathrm{maxM},i}$ at the receiver.

The propagation model is used to convert these octave band sound power levels to unweighted octave band maximum sound pressure levels, $L_{\mathrm{maxM},i}$ at the particular receiver under consideration. The octave band sound pressure levels are converted to A-weighted octave band levels and then added together logarithmically, as indicated in Equation (1.98), to obtain $L_{A,\mathrm{maxM}}$ for a particular train type and train speed. $L_{A,\mathrm{maxF}}$ is found from $L_{A,\mathrm{maxM}}$ using Equation (10.148).

Correction to Sound Power Levels For Track Conditions

A quantity, ΔL_c (dB), is arithmetically added to the sound power levels ($L_{W,i}$ and $L_{W,i,M}$) in each octave band. Values of ΔL_c for various track conditions are listed in Table 10.37.

Propagation Model

The propagation model assumes downwind and/or atmospheric temperature inversion conditions and generally represents a worst-case estimate for summer weather conditions. The sound pressure level, $L_{eq,24,i}$, at a receiver location in octave band, i, due to one train type, speed and

TABLE 10.36 Coefficients, a and b, for various train types in Scandinavia

Train type	Train class	Train power	Sleeper type	Coefficient	Octave band centre frequency (Hz)						
					63	125	250	500	1000	2000	4000
Norwegian trains											
N-Pass	Passenger	Electric	—	a	10	14	9	14	28	25	24
				b	33	31	35	44	44	41	37
N-Goods	Goods	Electric	—	a	10	14	9	14	28	25	24
				b	36	34	38	45	42	38	36
N-B65	Passenger	Electric	—	a	10	14	9	14	28	25	24
				b	32	32	36	44	42	36	27
N-B70	Passenger	Electric	—	a	10	14	9	14	28	25	24
				b	30	29	33	42	41	36	31
N-B69	Passenger	Electric	—	a	0	0	−6	14	28	25	24
				b	34	33	36	43	40	34	29
Swedish trains											
S-X2	Passenger	Electric	Concrete	a	22	25	20	12	16	29	30
				b	29	28	33	35	36	33	27
S-Pass	Passenger	Electric	Concrete	a	8	0	0	−10	5	15	5
				b	31	32	37	40	42	40	35
S-Pass/W	Passenger	Electric	Wood	a	10	0	0	−5	20	35	35
				b	30	31	40	45	42	38	32
S-X10	Passenger	Electric	Concrete	a	10	6	0	0	20	25	20
				b	33	33	35	37	37	35	28
S-GoodsDi	Goods	Diesel	Concrete	a	−12	−12	−12	12	12	20	18
				b	36	39	41	45	40	39	33
S-Goods	Goods	Electric	Concrete	a	0	0	0	5	5	5	5
				b	32	34	40	44	42	40	34
Finnish trains											
F-Sm	Passenger	Electric	Concrete	a	29	14	−3	15	26	20	18
				b	39	30	27	32	34	33	27
F-Sr1	Passenger	Electric	Concrete	a	24	36	9	36	39	31	24
				b	28	25	32	34	35	35	30
F-Goods	Goods	Electric	Wood	a	−13	3	1	23	27	17	14
				b	31	33	39	45	42	36	33
R-Goods	Goods	Electric	Concrete	a	−1	22	14	31	30	32	26
				b	37	44	46	49	46	45	39

track segment (of length L_{track} metres) is calculated using:

$$L_{eq,24,i} = L_{W,i} + \Delta L_c + \Delta L_d + \Delta L_a + \Delta L_g + \Delta L_s + \Delta L_v + \Delta L_r + 10\log_{10} L_{\text{track}} \quad \text{(dB re 20 µPa)} \quad (10.151)$$

TABLE 10.37 Values of the correction, ΔL_c (dB), for various track conditions

Value of ΔL_c	Track condition
0	Ballasted[a] track with continuously welded rails
Between 1 and 3	Rail or wheel surface rougher than normal
Between −1 and −3	Particularly well maintained tracks
3	Rails with joints
6	10 m track length for each unit of switches or crossings
6	Partial track length on a bridge without ballast
3	Partial track length on a bridge with ballast

[a]Ballast is crushed rock (nominal size 50 to 63 mm) used to support the rail track.

The maximum mean sound pressure level, $L_{\text{maxM},i}$, due to each train type, train speed and train segment (of length L_{train} metres) in each octave band, i, is calculated using:

$$L_{\text{maxM},i} = L_{W,i,M} + \Delta L_c + \Delta L_d + \Delta L_a + \Delta L_g + \Delta L_s + \Delta L_v + \Delta L_r + 10\log_{10} L_{\text{train}} \quad \text{(dB re 20 µPa)} \qquad (10.152)$$

where ΔL_d, ΔL_a, ΔL_g, ΔL_s, ΔL_v and ΔL_r are, respectively, the propagation corrections due to spherical divergence, air absorption, ground effects, screening, vegetation and reflection from surfaces other than the ground. Calculation of these quantities is outlined in the following paragraphs.

Geometrical Spreading, ΔL_d
The geometrical spreading term is defined as $\Delta L_d = -A_{\text{div}}$, where A_{div} is defined in Equation (5.56).

Atmospheric Absorption, ΔL_a
The atmospheric absorption term is defined as $\Delta L_a = -A_a$, where A_a is defined in Equation (5.61).

Ground Effect, ΔL_g
The ground effect term is calculated using a similar method to ISO 9613-2 (1996), so it is recommended that if the Nordic prediction model is used then the ISO 9613-2 (1996) model should be used to calculate the ground effect, so that $\Delta L_g = -A_g$, where A_g is defined in Equation (5.220).

Screening Effect, ΔL_s
The screening effect is calculated somewhat differently than used in current propagation models and is thus summarised here. When multiple screens exist between a train noise source and the receiver, the screen that gives the largest difference between the simplified direct and simplified diffracted paths is used. The diffracted path is the SPR path shown in Figure 10.8. Note that if the screen does not interrupt the line of sight, the difference between direct and diffracted paths is negative. The analysis assumes that the screen is more than three times longer than it is high.

Referring to Figure 10.8 (Nielsen, 1996):

$$\Delta h = \frac{d_1 d_2}{16(d_1 + d_2)} \qquad (10.153)$$

$$h_e = \begin{cases} \text{KP} - \Delta h; & \text{if K is below P} \\ -\text{KP} - \Delta h; & \text{if K is above P} \end{cases} \qquad (10.154\text{a,b})$$

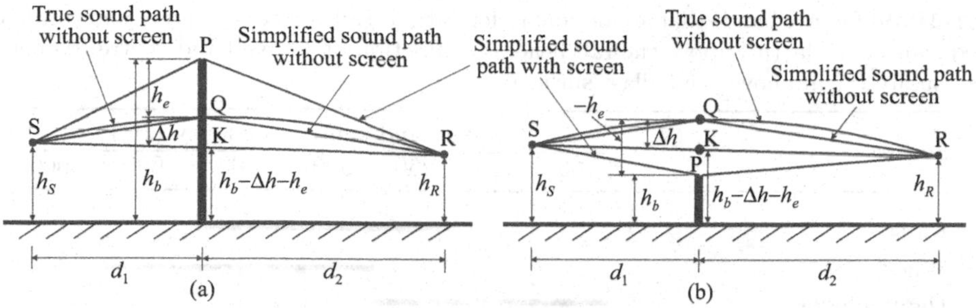

FIGURE 10.8 Definition of variables used for calculating the screen effect, ΔL_s, where S represents the source, P the top of the screen and R the receiver.

The difference between direct and diffracted paths is then:

$$\delta = \begin{cases} \mathrm{SP + PR - SQ - QR}; & \text{if K is below P} \\ \mathrm{2SR - SQ - QR - SP - PR}; & \text{if K is above P} \end{cases} \tag{10.155a,b}$$

The screen effect, ΔL_s, is then:

$$\Delta L_s = -10 C_h \log_{10}(0.094\delta f + 3) \quad \text{(dB)} \tag{10.156}$$

$$C_h = f\, h_b / 250 \tag{10.157}$$

where f is the octave band centre frequency and h_b is the screen height above the ground where the receiver is located.
If $\Delta L_s > 0$, then ΔL_s is set equal to 0.
If $\Delta L_s < -20$, then ΔL_s is set equal to -20.
If $C_h > 1$, then C_h is set equal to 1.

Vegetation Effect, ΔL_v
Including the effects of vegetation is optional in the Nordic prediction model. However, if it is desirable to include vegetation effects, then Equations (5.216) or (5.217) may be used, where $\Delta L_v = -K_v$. Alternatively, Table 5.14 may be used where $\Delta L_v = -A_f$.

Reflection Effect, ΔL_r
The Nordic prediction model only considers reflections from building façades. In this case, ΔL_r, in dB, is given by:

$$\Delta L_r = \begin{cases} 3; & \text{if } 0.5 < d_f \le 2 \text{ (m)} \\ 3(1 - d_f/20); & \text{if } 2 < d_f \le 20 \text{ (m)} \\ 0; & \text{if } d_f > 20 \text{ (m)} \end{cases} \tag{10.158a–c}$$

where d_f is the distance (m) from the façade to the receiver.

Source heights used for the propagation model are dependent on the source type, which for the purposes of the model have been allocated as a function of octave band centre frequency and listed in Table 10.38, which also shows the dominant frequency ranges for the important noise sources.

10.20.2.2 European Commission Model

The European Commission model (European Commission, 2015) for train noise provides methods for calculating the sound power in dB per metre of track length for a particular track section, railway vehicle type and vehicle speed. A track section is a length of track along which properties

TABLE 10.38 Octave bands that are dominant for various railway noise sources and corresponding assumed source height (m) above the rail head as a function of octave band centre frequency (rail head is usually 0.2 m above the ballast surface)

Source	Octave band centre frequency (Hz)						
	63	125	250	500	1000	2000	4000
Rails			—	—	—		
Wheels				—	—	—	
Diesel engine	—	—	—				
Carriages on goods train	—	—	—	—			
Curve screech					—	—	—
Braking					—	—	—
Assumed source height (m)	2.0	1.5	0.8	0.3	0.4	0.5	0.6

such as roughness and radius of curvature are essentially constant. If the track section is too long to be treated as a point source when viewed from the receiver, it is divided into a number of segments, each of which is sufficiently short that it can be treated as a point source. The sound power calculated using this model can then be used with one of the propagation models described in Chapter 5 to calculate the L_{Aeq} (averaged over a reference period of 4, 8, 12 or 24 hours) at a receiver location for that segment of track and that railway vehicle type. In the propagation model, the sound power calculated by a track segment must first be adjusted by adding $10 \log_{10} L_{\text{track}}$ to the sound power level where L_{track} is the length of the track segment. The sound power levels radiated by each railway vehicle type and speed are then added logarithmically (see Equation (1.98)) to obtain the total sound power for a particular track segment. In some cases, it may be desirable to combine 1/3-octave band sound power levels into octave band levels prior to undertaking the propagation analysis to determine the sound pressure levels at a receiver location. The sound pressure level contributions from all track segments that are significant contributors to the sound pressure level at a particular receiver location are then added together logarithmically (see Equation (1.98)) for each 1/3-octave or octave band to obtain the total sound pressure level at the receiver in 1/3-octave or octave bands. The band sound pressure levels are then weighted and combined logarithmically (see Equation (1.98)) to obtain the average L_{Aeq} for the reference time period of 4, 8, 12 or 24 hours.

In the European Commission model (European Commission, 2015), a railway vehicle is defined as any single sub-unit of a train such as a locomotive, self-propelled coach, hauled coach or freight wagon. All noise sources are deemed to be located in the centre of the vehicle or axle and at a height of either 0.5 m (source A) or 4 m (source B) above the top of the rail track. The types and locations of the various sources are listed in Table 10.39. The number of vehicles of each type, travelling per hour on each of the track sections to be included in the noise analysis, is determined first. All vehicle types travelling on each track section are included. A track section is defined as a length of track along which the track properties are essentially constant (such as track roughness and radius of curvature if in a bend).

When calculating sound pressure levels at a receiver location using a propagation model, the *source lines (which are considered to be incoherent) are segmented into equivalent uncorrelated point sources.* The size of individual segments making up a track section is not specified in the model, but they should be sufficiently small that the accuracy of the calculated sound pressure levels is not compromised when the sources are treated as point sources in a propagation model. The further away the receiver is located, the larger will be the acceptable track segment for consideration as a point source. A good rule of thumb for a 0.1 dB accuracy is that the receiver should be further away than 10 times the segment length.

TABLE 10.39 Locations of various noise sources on a rail vehicle

Generic noise type	Noise source	Location
Rolling noise	Wheel and rail roughness generating vibration of and noise radiation from rails, wheels and superstructure	Noise source A
Traction noise	Gears and electric motors	Noise source A
	Diesel engine exhausts	Noise source B
	Louvres, cooling outlets, fans, diesel engine blocks	Noise sources between A and B sound energy distributed proportionally between A and B
Aerodynamic noise	Shrouds and screens	Noise source A
	Roof apparatus and pantograph	Noise source B
Impact noise	Crossings, switches and junctions	Noise source A
Squeal	Wheel rubbing on rail around bends	Noise source A
Bridge noise	Bridge set into vibration by passage of a railway vehicle resulting in bridge noise radiation	Noise source A

The noise emission of railway vehicle traffic on a specified track section is represented by 2 source lines at 0.5 m and 4 m above the top of the rail track. For each track section considered, each source line is characterised by a sound power level per metre per frequency band. This is done by using separate sources for each generic noise type, each vehicle type, and each average speed of the vehicle type (including idling) that is considered. The directivity of the noise is also taken into account as described under the relevant subheading below.

For a *moving railway vehicle*, the sound power level contribution, $L_{W,\text{line},i}$, in 1/3-octave band, i, of a noise source to a source line is given by:

$$L_{W,\text{line},i} = L_{W,0,\text{dir},i} + 10\log 10\left(\frac{Q}{1000\,v}\right) \qquad ((\text{dB/m}) \text{ re } 10^{-12}\text{W}) \tag{10.159}$$

where Q is the average number of vehicles of this type and speed on the track section in 1 hour and v is the vehicle speed in km/hr.

For an *idling vehicle*:

$$L_{W,\text{line},i} = L_{W,0,\text{dir},i} + 10\log 10\left(\frac{T_\text{idle}}{T_\text{ref} L_\text{track}}\right) \qquad (\text{dB/m re } 10^{-12}\text{W}) \tag{10.160}$$

where L_track is the length of the track segment occupied by the idling vehicle, T_idle is the number of hours that the railway vehicle is stationary and idling and T_ref is the reference time over which the noise assessment is being averaged (for example, 4, 8, 12 or 24 hours). Also:

$$L_{W,0,\text{dir},i} = L_{W,0,i} + \Delta L_{W\text{dir,vert},i} + \Delta L_{W\text{dir,horiz},i} \qquad (\text{dB re } 10^{-12}\text{W}) \tag{10.161}$$

where $\Delta L_{W\text{dir,vert},i}$ is defined in Equations (10.176) and (10.177), and $\Delta L_{W\text{dir,horiz},i}$ is defined in Equation (10.175). The quantity, $L_{W,0,i}$, is the source sound power level in 1/3-octave band, i, for the source type under consideration.

Each noise source type is characterised by a separate pair of source lines. The total sound power, $L_{W,\text{lineTOT},i}$, radiated by each source line in 1/3-octave band, i, is found by logarithmically summing the contributions from each source in line, i, as:

$$L_{W,\mathrm{lineTOT},i} = 10 \log 10 \sum_{m=1}^{n} 10^{L_{W,\mathrm{line_m},i}/10} \quad (\mathrm{dB/m\ re\ } 10^{-12}\mathrm{W}) \tag{10.162}$$

where $L_{W,\mathrm{line_m},i}$ is the sound power radiated by source, m and n is the total number of sources. If a track section is too long to be considered as a single point source, then prior to implementing the propagation model, it is necessary to divide the track section into a number of segments (or equivalent number of point sources), each of sound power level, $L_{w,i}$, such that for each track section, $L_{W,\mathrm{lineTOT},i} = L_{w,i} + 10 \log_{10} N_p$, where N_p is the number of equivalent point sources used to characterise the track section. As a guide, the maximum length that should be used for a track segment is approximately 1/10 of the distance to the nearest receiver.

When the 1/3-octave band calculations have been completed, it is necessary to combine each set of three 1/3-octave bands into a single octave band prior to implementing the propagation model. This is done using logarithmic summation in a similar way as was done in Equation (10.162).

The European Commission train noise model provides procedures for calculating the sound power level, $L_{W,0,i}$, radiated by all noise sources, except for traction noise sources on a moving railway vehicle. The calculation of these sound power levels is discussed in the following relevant sections, where the subscript, "0", in the sound power variable, $L_{W,0,i}$, is replaced by the source name. For example, the sound power in 1/3-octave band, i, due to aerodynamic noise is denoted $L_{W,\mathrm{aero},i}$ and the corresponding source line sound power level is denoted $L_{W,\mathrm{line_{aero}},i}$. Thus, in Equation (10.162), values of m have the labels listed in Table 10.40. Note that impact noise, squeal noise and bridge noise are combined with the rolling noise sound power prior to using Equation (10.162).

TABLE 10.40 Labels for the various values of m in Equation (10.162)

Value of m	Label	Description
1	trac	Traction noise due to all traction sources.
2	roll	Rolling noise caused by track and wheel roughness and radiated by the track, wheels and train superstructure. Includes impact noise due to joints, switches and crossings. Also includes bridge noise.
3	aero	Aerodynamic noise.

Traction Noise

Traction noise, $L_{W,\mathrm{traction}}$, includes all of the sources listed in the traction row of Table 10.39. The EU model only includes traction noise data ($L_{W,\mathrm{idling},i}$) for idling railway vehicles and these may be found in Table 10.41. Traction noise databases for railway vehicles moving at a constant speed, accelerating or decelerating do exist but they are not easily accessible. In fact the EU model manual suggests that traction noise data should be measured according to ISO 3095 (2013) for each of the traction noise sources on a stationary railway vehicle with the required engine rotational speed. These sources include the following:

1. diesel engine power train (including inlet, exhaust and engine block), dependent on engine rotational speed;
2. transmission gearboxes, dependent on engine rotational speed;
3. generators, dependent on engine rotational speed;
4. electrical converters and motors, dependent on load;

TABLE 10.41 Coefficients, $L_{W,\text{idling},i}$, for traction noise, expressed as sound power level per vehicle, where source A is 0.5 m above the track and source B is 4 m above the track, and both sources are in the horizontal centre of the railway vehicle (see Table 10.39 for the constituents of source A and source B)

Frequency (Hz)	Diesel locomotive 800 kW		Diesel locomotive 2200 kW		Diesel multiple unit		Electric locomotive		Electric multiple unit	
	Source A	Source B	Source A	Source B	Source A	Source B	Source A	Source B	Source A	Source B
50	98.9	103.2	99.4	103.7	82.6	86.9	87.9	92.2	80.5	84.8
63	94.8	100.0	107.3	112.5	82.5	87.7	90.8	96.0	81.4	86.6
80	92.6	95.5	103.1	106.0	89.3	92.2	91.6	94.5	80.5	83.4
100	94.6	94.0	102.1	101.5	90.3	89.7	94.6	94.0	82.2	81.6
125	92.8	93.3	99.3	99.8	93.5	94.0	94.8	95.3	80.0	80.5
160	92.8	93.6	99.3	100.1	99.5	100.3	96.8	97.6	79.7	80.5
200	93.0	92.9	99.5	99.4	98.7	98.6	104.0	103.9	79.6	79.5
250	94.8	92.7	101.3	99.2	95.5	93.4	100.8	98.7	96.4	94.3
316	94.6	92.4	101.1	98.9	90.3	88.1	99.6	97.4	80.5	78.3
400	95.7	92.8	102.2	99.3	91.4	88.5	101.7	98.8	81.3	78.4
500	95.6	92.8	102.1	99.3	91.3	88.5	98.6	95.8	97.2	94.4
630	98.6	96.8	101.1	99.3	90.3	88.5	95.6	93.8	79.5	77.7
800	95.2	92.7	101.7	99.2	90.9	88.4	95.2	92.7	79.8	77.3
1000	95.1	93.0	101.6	99.5	91.8	89.7	96.1	94.0	86.7	84.6
1250	95.1	92.9	99.3	97.1	92.8	90.6	92.1	89.9	81.7	79.5
1600	94.1	93.1	96.0	95.0	92.8	91.8	89.1	88.1	82.7	81.7
2000	94.1	93.2	93.7	92.8	90.8	89.9	87.1	86.2	80.7	79.8
2500	99.4	98.3	101.9	100.8	88.1	87.0	85.4	84.3	78.0	76.9
3160	92.5	91.5	89.5	88.5	85.2	84.2	83.5	82.5	75.1	74.1
4000	89.5	88.7	87.1	86.3	83.2	82.4	81.5	80.7	72.1	71.3
5000	87.0	86.0	90.5	89.5	81.7	80.7	80.0	79.0	69.6	68.6
6350	84.1	83.4	31.4	30.7	78.8	78.1	78.1	77.4	66.7	66.0
8000	81.5	80.9	81.2	80.6	76.2	75.6	76.5	75.9	64.1	63.5
10000	79.2	78.7	79.6	79.1	73.9	73.4	75.2	74.7	61.8	61.3

5. fans and cooling systems, dependent on fan rotational speed; and
6. intermittent sources such as compressors and valves with a corresponding duty cycle correction for the noise emission.

The level of locomotive noise is also dependent on the load (or the number of vehicles hauled). However, it is sufficient to follow the procedures in ISO 3095 (2013) to obtain the required sound power levels of the above equipment for the various operating conditions.

Aerodynamic Noise

Aerodynamic noise is only relevant for vehicles travelling in excess of 200 km/hr and in this case, the areodynamic noise contribution to 1/3-octave band, i, is given by:

$$L_{W,\text{aero},i} = \begin{cases} L_{W,0,1,i}(v_0) + \alpha_{i,1} \log_{10}\left(\dfrac{v}{v_0}\right) \quad \text{(dB);} \quad \text{for all sources A (height = 0.5 m)} \\ L_{W,0,2,i}(v_0) + \alpha_{i,2} \log_{10}\left(\dfrac{v}{v_0}\right) \quad \text{(dB);} \quad \text{for all sources B (height = 4.0 m)} \end{cases} \tag{10.163a,b}$$

where v_0 = 300 km/hr, values of $L_{W,0,1,i}(v_0)$ and $L_{W,0,2,i}(v_0)$ for each 1/3-octave band, i, are given in Table 10.42 (derived from measurements at two or more measurement locations, with locations for source A at 0.5 m height different to those for source B at 4 m height) and $\alpha_{i,1} = \alpha_{i,2} = 50$ for all 1/3-octave bands.

TABLE 10.42 Coefficients, $L_{W,0,1}$ and $L_{W,0,2}$, for aerodynamic noise at 300 km/hr, expressed as sound power level per vehicle (of length 20 m)(European Commission, 2015)

Frequency (Hz)	$L_{W,0,1}$	$L_{W,0,2}$	Frequency (Hz)	$L_{W,0,1}$	$L_{W,0,2}$	Frequency (Hz)	$L_{W,0,1}$	$L_{W,0,2}$
50	112.6	36.7	316	115.9	35.9	2000	114.7	110.3
63	113.2	38.5	400	116.3	36.3	2500	115.0	110.4
80	115.7	39.0	500	116.2	36.3	3160	114.5	105.6
100	117.4	37.5	630	115.2	36.3	4000	113.1	37.2
125	115.3	36.8	800	115.8	36.2	5000	112.1	37.5
160	115.0	37.1	1000	115.7	36.5	6350	110.6	37.9
200	114.9	36.4	1250	115.7	36.4	8000	109.6	38.4
250	116.4	36.2	1600	114.7	105.2	10 000	108.8	39.2

Rolling Noise

Rolling noise in mainly a result of wheel roughness in the wavelength range from 5 to 500 mm. Roughness is quantified in terms of a roughness level, L_r, defined as:

$$L_r = 20\log_{10}(r) \tag{10.164}$$

where r is the rms of (vertical displacement minus mean displacement) in μm. The roughness level is obtained as a spectrum of roughness wavelength.

The sound power level per vehicle due to rolling noise, $L_{W,\text{roll},i}$, for 1/3-octave band, i, is:

$$L_{W,\text{roll},i} = 10\log_{10}\left(10^{L_{W,\text{TR},i}/10} + 10^{L_{W,\text{VEH},i}/10} + 10^{L_{W,\text{VEHSUP},i}/10}\right) \quad (\text{dB/m re } 10^{-12}\text{W}) \tag{10.165}$$

where:

$$L_{W,\text{TR},i} = L_{R,\text{TOT},i} + L_{H,\text{TR},i} + 10\log_{10} N_a \quad (\text{dB/m re } 10^{-12}\text{W}) \tag{10.166}$$

$$L_{W,\text{VEH},i} = L_{R,\text{TOT},i} + L_{H,\text{VEH},i} + 10\log_{10} N_a \quad (\text{dB/m re } 10^{-12}\text{W}) \tag{10.167}$$

$$L_{W,\text{VEHSUP},i} = L_{R,\text{TOT},i} + L_{H,\text{VEHSUP},i} + 10\log_{10} N_a \quad (\text{dB/m re } 10^{-12}\text{W}) \tag{10.168}$$

where N_a is the number of axles, other variables are defined in the footnotes of Tables 10.43 and 10.44, and:

$$L_{R,\text{TOT},i} = 10\log_{10}\left(10^{L_{R,\text{TR},i}/10} + 10^{L_{R,\text{VEH},i}/10}\right) + A_{3,i} \tag{10.169}$$

The wheel roughness coefficient, $L_{R,\text{VEH},i}$, the track roughness coefficient, $L_{R,\text{TR},i}$ and the filtering coefficient, $A_{3,i}$, are given in Table 10.43. The roughness level to sound power transfer functions, $L_{H,\text{VEH},i}$ and $L_{H,\text{TR},i}$ are given in Table 10.44. The superstructure transfer functions, $L_{H,\text{VEHSUP},i}$, can vary from zero to 140 but the EU standard is that they are all zero (European Commission, 2015). The coefficient, $A_{3,i}$, accounts for the filtering effect of the contact patch between the wheel and rail.

The values in Table 10.43 are in terms of roughness wavelength, λ, and for each railway vehicle speed, they must be converted to frequency using $f = (v/\lambda) \times (10^6/3600)$ where f is in Hz, λ is in mm (see Table 10.43) and v is in km/hr. Linear interpolation can be used to obtain the values of $L_{R,\text{VEH},i}$, $L_{R,\text{TR},i}$ and $A_{3,i}$ corresponding to each 1/3-octave band centre frequency.

TABLE 10.43 Coefficients, $L_{R,\mathrm{VEH},i}$, $L_{R,\mathrm{TR},i}$ and $A_{3,i}$, for wheel roughness, rail roughness and contact filter, respectively

Wave-length (mm)	$L_{R,\mathrm{VEH},i}$			$L_{R,\mathrm{TR},i}$		$A_{3,i}$				
	Brake type 1	Brake type 2	Brake type 3	Track type 1	Track type 2	Axle type 1	Axle type 2	Axle type 3	Axle type 4	Axle type 5
≥400	2.2	−4.0	−5.9	17.1	11.0	0.0	0.0	0.0	0.0	0.0
315	2.2	−4.0	5.9	15.0	10.0	0.0	0.0	0.0	0.0	0.0
250	2.2	−4.0	2.3	13.0	9.0	0.0	0.0	0.0	0.0	0.0
200	2.2	−4.0	2.8	11.0	8.0	0.0	0.0	0.0	0.0	0.0
160	2.4	−4.0	2.6	9.0	7.0	0.0	0.0	0.0	0.0	0.0
120	0.6	−4.0	1.2	7.0	6.0	0.0	0.0	0.0	0.0	0.0
100	2.6	−4.0	2.1	4.9	5.0	0.0	0.0	0.0	0.0	0.0
80	5.8	−4.3	0.9	2.9	4.0	0.0	0.0	0.0	−0.2	−0.2
63	8.8	−4.6	−0.3	0.9	3.0	0.0	−0.2	−0.2	−0.5	−0.6
50	11.1	−4.9	−1.6	−1.1	2.0	−0.2	−0.4	−0.5	−0.9	−1.3
40	11.0	−5.2	−2.9	−3.2	1.0	−0.5	−0.7	−0.9	−1.6	−2.2
31.5	9.8	−6.3	−4.9	−5.0	0.0	−1.2	−1.5	−1.6	−2.5	−3.7
25	7.5	−6.8	−7.0	−5.6	−1.0	−2.0	−2.8	−2.5	−3.8	−5.8
20	5.1	−7.2	−8.6	−6.2	−2.0	−3.0	−4.5	−3.8	−5.8	−9.0
16	3.0	−7.3	−9.3	−6.8	−3.0	−4.3	−7.0	−5.8	−8.5	−11.5
12	1.3	−7.3	−9.5	−7.4	−4.0	−6.0	−10.3	−8.5	−11.4	−12.5
10	0.2	−7.1	−10.1	−8.0	−5.0	−8.4	−12.0	−12.0	−12.0	−12.0
8	−0.7	−6.9	−10.3	−8.6	−6.0	−12.0	−12.5	−12.6	−13.5	−14.0
6.3	−1.2	−6.7	−10.3	−9.2	−7.0	−11.5	−13.5	−13.5	−14.5	−15.0
5	−1.0	−6.0	−10.8	−9.8	−8.0	−12.5	−16.0	−14.5	−16.0	−17.0
4	0.3	−3.7	−10.9	−10.4	−9.0	−13.9	−16.0	−16.0	−16.5	−18.4
3.2	0.2	−2.4	−9.5	−11.0	−10.0	−14.7	−16.5	−16.5	−17.7	−19.5
2.5	1.3	−2.6	−9.5	−11.6	−11.0	−15.6	−17.0	−17.7	−18.6	−20.5
2	3.1	−2.5	−9.5	−12.2	−12.0	−16.6	−18.0	−18.6	−19.6	−21.5
1.6	3.1	−2.5	−9.5	−12.8	−13.0	−17.6	−19.0	−19.6	−20.6	−22.4
1.2	3.1	−2.5	−9.5	−13.4	−14.0	−18.6	−20.2	−20.6	−21.6	−23.5
1	3.1	−2.5	−9.5	−14.0	−15.0	−19.6	−21.2	−21.6	−22.6	−24.5
0.8	3.1	−2.5	−9.5	−14.0	−15.0	−20.6	−22.2	−22.6	−23.6	−25.4

Brake type 1 = Cast iron tread brake ($L_{R,\mathrm{VEH},i}$)
Brake type 2 = Composite brake ($L_{R,\mathrm{VEH},i}$)
Brake type 3 = Disk brake ($L_{R,\mathrm{VEH},i}$)
Track type 1 = EN ISO 3095:2013 (well maintained and very smooth - $L_{R,\mathrm{TR},i}$)
Track type 2 = Average network (normally maintained, smooth - $L_{R,\mathrm{TR},i}$)
Axle type 1 = 50 kN axle load with 360 mm wheel diameter ($A_{3,i}$)
Axle type 2 = 50 kN axle load with 680 mm wheel diameter ($A_{3,i}$)
Axle type 3 = 25 kN axle load with 920 mm wheel diameter ($A_{3,i}$)
Axle type 4 = 50 kN axle load with 920 mm wheel diameter ($A_{3,i}$)
Axle type 5 = 100 kN axle load with 920 mm wheel diameter ($A_{3,i}$)

Squeal Noise

Squeal noise is localised to curved tracks only. Ideally, squeal noise should be measured for a particular track and vehicle, and it should be applied to at least a 50 m length of track. In the absence of measured data, it can be approximately accounted for by arithmetically adding 8 dB to the rolling noise, $L_{W,\mathrm{roll},i}$, in all 1/3-octave frequency bands for a curve with a radius less than 300 m or adding 5 dB for curves with a radius between 300 m and 500 m.

TABLE 10.44 Coefficients, $L_{H,\text{VEH},i}$ and $L_{H,\text{TR},i}$, for the transfer functions of vehicle and track, respectively. (All values are sound power level per axle and no noise reduction measures have been applied)

Freq-ency (Hz)	$L_{H,\text{VEH},i}$				$L_{H,\text{TR},i}$						
	Wheel type 1	Wheel type 2	Wheel type 3	Wheel type 4	Pad type 1	Pad type 2	Pad type 3	Pad type 4	Pad type 5	Pad type 6	Pad type 7
50	75.4	75.4	75.4	75.4	53.3	50.9	50.1	50.9	50.0	49.8	44.0
63	77.3	77.3	77.3	77.3	59.3	57.8	57.2	56.6	56.1	55.9	51.0
80	81.1	81.1	81.1	81.1	67.2	66.5	66.3	64.3	64.1	64.0	59.9
100	84.1	84.1	84.1	84.1	75.9	76.8	77.2	72.3	72.5	72.5	70.8
125	83.3	82.8	82.8	82.8	79.2	80.9	81.6	75.4	75.8	75.9	75.1
160	84.3	83.3	83.3	83.3	81.8	83.3	84.0	78.5	79.1	79.4	76.9
200	86.0	84.1	83.9	84.5	84.2	85.8	86.5	81.8	83.6	84.4	77.2
250	90.1	86.9	86.3	90.4	88.6	90.0	90.7	86.6	88.7	89.7	80.9
316	89.8	87.9	88.0	90.4	91.0	91.6	92.1	89.1	89.6	90.2	85.3
400	89.0	89.9	92.2	89.9	94.5	93.9	94.3	91.9	89.7	90.2	92.5
500	88.8	90.9	93.9	90.1	97.0	95.6	95.8	94.5	90.6	90.8	97.0
630	90.4	91.5	92.5	91.3	99.2	97.4	97.0	97.5	93.8	93.1	98.7
800	92.4	91.5	90.9	91.5	104.0	101.7	100.3	104.0	100.6	97.9	102.8
1000	94.9	93.0	90.4	93.6	107.1	104.4	102.5	107.9	104.7	101.1	105.4
1250	100.4	98.7	93.2	100.5	108.3	106.0	104.2	108.9	106.3	103.4	106.5
1600	104.6	101.6	93.5	104.6	108.5	106.8	105.4	108.8	107.1	105.4	106.4
2000	109.6	107.6	99.6	115.6	109.7	108.3	107.1	109.8	108.8	107.7	107.5
2500	114.9	111.9	104.9	115.9	110.0	108.9	107.9	110.2	109.3	108.5	108.1
3160	115.0	114.5	108.0	116.0	110.0	109.1	108.2	110.1	109.4	108.7	108.4
4000	115.0	114.5	111.0	116.0	110.0	109.4	108.7	110.1	109.7	109.1	108.7
5000	115.5	115.0	111.5	116.5	110.3	109.9	109.4	110.3	110.0	109.6	109.1
6350	115.6	115.1	111.6	116.6	110.0	109.9	109.7	109.9	109.8	109.6	109.1
8000	116.0	115.5	112.0	117.0	110.1	110.3	110.4	110.0	110.0	109.9	109.5
10000	116.7	116.2	112.7	117.7	110.6	111.0	111.4	110.4	110.5	110.6	110.2

Wheel type 1 = Wheel diameter, 920 mm ($L_{H,\text{VEH},i}$)
Wheel type 2 = Wheel diameter, 840 mm ($L_{H,\text{VEH},i}$)
Wheel type 3 = Wheel diameter, 680 mm ($L_{H,\text{VEH},i}$)
Wheel type 4 = Wheel diameter, 1200 mm ($L_{H,\text{VEH},i}$)
Pad type 1 = Mono-block sleeper on soft rail pad ($L_{H,\text{TR},i}$)
Pad type 2 = Mono-block sleeper on medium-stiffness rail pad ($L_{H,\text{TR},i}$)
Pad type 3 = Mono-block sleeper on hard rail pad ($L_{H,\text{TR},i}$)
Pad type 4 = Bi-block sleeper on soft rail pad ($L_{H,\text{TR},i}$)
Pad type 5 = Bi-block sleeper on medium-stiffness rail pad ($L_{H,\text{TR},i}$)
Pad type 6 = Bi-block sleeper on hard rail pad ($L_{H,\text{TR},i}$)
Pad type 7 = Wooden sleepers ($L_{H,\text{TR},i}$)

Impact Noise

Impact noise is a result of the railway vehicle travelling over crossings, switches and rail joints. It is taken into account by combining the effect with rolling noise using Equation (10.170) to obtain a combined impact and rolling noise level. The quantity, $L_{R,(\text{TOT+IMPACT}),i}$, is used in place of $L_{R,(\text{TOT}),i}$ in Equations (10.166), (10.167) and (10.168) to calculate the combined impact and rolling noise resulting from the vibration of the track, wheels and vehicle superstructure.

$$L_{R,(\text{TOT+IMPACT}),i} = 10\log_{10}\left(10^{L_{R,\text{TOT},i}/10} + 10^{L_{R,\text{IMPACT},i}/10}\right) \tag{10.170}$$

Values of $L_{R,\text{IMPACT},i}$ for a single switch, crossing or joint are given in Table 10.45. The values in Table 10.45 are in terms of wavelength, λ, and for each railway vehicle speed, they must be converted to frequency using $f = (v/\lambda) \times (10^6/3600)$ where f is in Hz, λ is in mm (see Table 10.45) and v is in km/hr. Linear interpolation can be used to obtain the value of $L_{R,\text{IMPACT},i}$ corresponding to each 1/3-octave band centre frequency.

TABLE 10.45 Coefficients, $L_{R,\text{IMPACT},i}$, for a single switch, joint or crossing

Wave-length (mm)	Coeff-icient	Wave-length (mm)	Coeff-icient	Wave-length (mm)	Coeff-icient	Wave-length (mm)	Coeff-icient	Wave-length (mm)	Coeff-icient
≥630	22.4	160	20.2	40	16.0	10	−11.0	2.5	−26.6
500	23.8	120	20.4	31.5	13.0	8	−16.5	2	−28.6
400	24.7	100	20.8	25	10.0	6.3	−18.5	1.6	−30.6
315	24.7	80	20.9	20	6.0	5	−21.0	1.2	−32.6
250	23.4	63	19.8	16	1.0	4	−22.5	1	−34.0
200	21.7	50	18.0	12	−4.0	3.2	−24.7	0.8	−34.0

For calculating the noise due to rail track joints, it is necessary to adjust the coefficient given in Table 10.45, which is for only a single joint and will be denoted here as $L_{R,\text{IMPACT,single},i}$. Thus:

$$L_{R,\text{IMPACT},i} = L_{R,\text{IMPACT,single},i} + 10\log_{10}\left(\frac{n_d}{0.01}\right) \qquad (10.171)$$

where n_d is the joint density (number of track joints per metre of track length).

Bridge Noise

The bridge noise contribution is combined with the rolling noise contribution as:

$$L_{R,(\text{TOT}+\text{BRIDGE}),i} = L_{R,\text{TOT},i} + C_{\text{BRIDGE}} \qquad (10.172)$$

where

$$C_{\text{bridge}} = \begin{cases} 1; & \text{predominantly concrete or masonry bridges (any track)} \\ 4; & \text{steel bridges with ballasted track} \end{cases} \qquad (10.173\text{a,b})$$

When a bridge is involved in the section of track being analysed, the quantity, $L_{R,\text{TOT},i}$, in Equations (10.166), (10.167) and (10.168) is replaced with $L_{R,(\text{TOT}+\text{BRIDGE}),i}$ to calculate the combined bridge effect and rolling noise components resulting from the vibration of the track, wheels and vehicle superstructure. If a bridge and impact noise are both involved, then $L_{R,\text{TOT},i}$ in Equations (10.166), (10.167) and (10.168) is replaced with $L_{R,(\text{TOT}+\text{IMPACT}+\text{BRIDGE}),i}$, where

$$L_{R,(\text{TOT}+\text{IMPACT}+\text{BRIDGE}),i} = L_{R,(\text{TOT}+\text{IMPACT}),i} + C_{\text{BRIDGE}} \qquad (10.174)$$

and $L_{R,\text{TOT}+\text{IMPACT},i}$ is defined in Equation (10.170).

Directivity Effect

The directivity of the railway vehicle sources towards the receiver location (whose direction from the railway vehicle noise source is defined by the vertical angle, $\psi > 0$ and the horizontal angle, φ, as shown in Figure 10.9), is taken into account using the following procedure (European Commission, 2010a).

The *directivity in the horizontal plane* is assumed to be dipole in nature and the same for all 1/3-octave bands. For rolling noise, impact noise, squeal noise, braking noise, fan noise and aerodynamic noise, it is given by:

$$\Delta L_{W,\text{dir,hor}} = 10\log_{10}(0.01 + 0.99\sin^2\varphi) \quad \text{(dB)} \qquad (10.175)$$

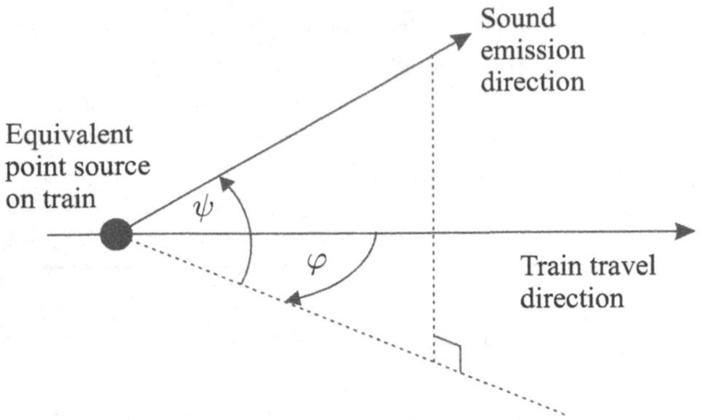

FIGURE 10.9 Definition of the angles used to calculate the directivity corrections.

The *vertical directivity* only applies to sources A (at 0.5 m height) for all sources except the aerodynamic sources. For sources A, it is a function of the 1/3-octave band centre frequency, f_i, and angle, ψ, and is given by:

$$\Delta L_{W,\mathrm{dir,vert},i} = \left| \frac{40}{3}\left[\frac{2}{3}\sin(2\psi) - \sin\psi\right] \times \log_{10}\left[\frac{f_i + 600}{200}\right]\right| \quad \text{(dB) for } -\pi/2 < \psi < \pi/2 \tag{10.176}$$

The *vertical directivity* for aerodynamic source B at height, 4 m, is the same for all 1/3-octave bands and is given by:

$$\Delta L_{W,\mathrm{dir,vert},i} = \begin{cases} 10\log_{10}(\cos^2\psi) \quad \text{(dB)} & \psi < 0 \\ 0 \quad \text{(dB)} & \psi \geq 0 \end{cases} \tag{10.177}$$

10.20.2.3 UK Department of Transport Model

This model involves four stages for estimating noise from moving trains (UK. DOT, 1995a). These are:

1. **Divide the rail line into a number of segments such that the variation of noise produced along any segment is less than 2 dBA.** Greater segmentation will be necessary for situations where there are bends in the track or if there are significant gradient changes or if the screening or ground cover changes, resulting in a greater than 2 dBA difference. A long, straight track with constant gradient and noise propagation properties can usually be considered as a single segment. Clearly, crossings and train stations would also require segmentation of the track for the purpose of noise level calculations.

2. **For each segment determine the reference SEL ($\mathrm{SEL_{ref}}$) at a given speed and at a distance of 25 m from the nearside of the track segment.** SEL is defined in Chapter 2, page 84 as the A-weighted sound exposure level. First, the single vehicle SEL value (SEL_v) is obtained by by measuring the SEL at a distance of 25 m from the track for a train over a range of passing speeds. If a locomotive is involved, the SEL for that should be measured separately and then subtracted from the overall SEL to obtain the SEL for just the rolling stock. The SEL for a single vehicle is then calculated using:
$$\mathrm{SEL}_v = \mathrm{SEL}_T - 10\log_{10} N \tag{10.178}$$

where SEL_T is the overall SEL measured for N identical vehicles in the train. Second, a linear regression of SEL versus $\log_{10}$(speed) is undertaken. Results provided by the Department of Transport - UK (UK. DOT, 1995a) for rolling railway vehicles are:

$$\mathrm{SEL}_v = 31.2 + 20\log_{10} v + C_1 \qquad \text{(dBA)} \tag{10.179}$$

and for locomotives under power or EuroStar fan noise:

$$\mathrm{SEL}_v = 112.6 - 10\log_{10} v + C_1 \qquad \text{(dBA)} \tag{10.180}$$

where v is the vehicle speed in km/h and C_1 is a coefficient dependent on train type and listed in Table 10.46 for a few British trains (UK. DOT, 1995a,b).

TABLE 10.46 Example SEL corrections, C_1, for various single railway vehicles, except for Eurostar for which the correction is for the entire train

Vehicle type	Correction, C_1 (dBA)
Passenger coaches – tread braked	
Class 421 EMU or 422 EMU	10.8
British rail MK I or II	14.8
Passenger coaches – disc braked, 4 axles	
Class 319 EMU	11.3
Class 465 EMU and 466 EMU	8.4
Class 165 EMU and 166 EMU	7.0
British rail MK III or IV	6.0
Passenger coaches – disk braked, 6 axles	15.8
Passenger coaches – disk braked, 8 axles	14.9
Freight vehicles, tread braked, 2 axles	12.0
Freight vehicles, tread braked, 4 axles	15.0
Freight vehicles, disc braked, 2 axles	8.0
Freight vehicles, disc braked, 4 axles	7.5
Diesel locomotives (steady speed)	
Classes 20, 33	14.8
Classes 31, 37, 47, 56, 59, 60	16.6
Class 43	18.0
Diesel locomotives under full power	
Classes 20, 31, 33, 37, 43, 47, 56, 59	0.0
Class 60	−5.0
Electric locomotives	14.8
Eurostar rolling noise	
(2 powered cars separated by 14 or 18 coaches)	17.2
Eurostar fan noise	
(2 powered cars separated by 14 or 18 coaches)	−7.4

Different vehicle types must be considered as separate trains and the L_{Aeq} for each vehicle type (or train) is combined using Equation (1.98) to give the total L_{Aeq} for the entire train. For any specific train type consisting of N identical units, the quantity, $\mathrm{SEL}_{\mathrm{ref}}$, is calculated by adding $10\log_{10} N$ to SEL_v. In addition, the track correction, C_2, from Table 10.47 must also be added so that:

$$\mathrm{SEL}_{\mathrm{ref}} = \mathrm{SEL}_v + 10\log_{10} N + C_2 \qquad \text{(dBA)} \tag{10.181}$$

The values of SEL_{ref} calculated using the procedures just described apply to continuously welded track with timber or concrete sleepers and ballast. For other track types, the corrections in Table 10.47 should be added to the SEL_{ref} values.

TABLE 10.47 Corrections C_2 to be added to SEL_{ref} to account for different track types

Description of rail	Correction, C_2 (dBA)
Jointed track	2.5
Points and crossings	2.5
Slab track	2.0
Concrete bridges and viaducts (excludes shielding by parapet)	1.0
Steel bridges (excludes shielding by parapet)	4.0
Box girder with rails fitted directly to it	9.0

3. **Determine the corrections for distance, air absorption, ground effects, barrier diffraction and angle of view of the observer to the track segment and reflections from buildings and barriers.** The SEL value at the observer is:

$$\text{SEL} = \text{SEL}_{\text{ref}} + C_{\text{dist}} + C_{abs} + C_{\text{ground}} + C_{\text{barrier}} + C_{\text{view}} + C_{\text{refl}} \tag{10.182}$$

For this calculation, the source height used for rolling stock is the rail height. The source height used for locomotive or EuroStar fan noise is 4 m above the track. In both cases, the distance to the observer is the shortest distance from the near side rail.

The corrections listed in Table 10.47 are meant to be applied to overall A-weighted numbers, such as SEL_{ref}. Many of the corrections have been calculated based on typical noise spectra and would not apply to other types of sound source.

The **distance correction** is based on treating the noise source as a line source and for distances greater than 10 m, the correction to be added to SEL_{ref} is:

$$C_{\text{dist}} = -10\log_{10}(r/25) \tag{10.183}$$

where r is the straight line distance from the source to the observer and is defined for a diesel locomotive as:

$$r = \sqrt{d^2 + (h - 4.0)^2} \tag{10.184}$$

and for everything else as:

$$r = \sqrt{d^2 + h^2} \tag{10.185}$$

where h is the difference in height between the track and observer (observer height – track height) and d is the straight line normal distance from the track segment (or in many cases, its extension – see Figure 10.7) to the observer.

The **air absorption correction** is:

$$C_{abs} = 0.2 - 0.008r \tag{10.186}$$

The **ground correction** is:

$$C_{\text{ground}} = \begin{cases} 0; & \text{for } 10 < d \le 25 \ \text{ or } \ H > 6 \text{ (m)} \\ -0.6P_d(6 - H)\log_{10}(d/25); & \text{for } 1.0 < H < 6 \text{ (m)} \\ -3P_d \log_{10}(d/25); & \text{for} H \le 1.0 \text{ (m)} \end{cases} \tag{10.187a–c}$$

where P_d is the fraction of absorbing ground between the source and receiver, d is the horizontal normal distance from the track segment (or in many cases, its extension – see Figure 10.7) to the observer and H is the mean propagation height, = 0.5 × (source height + receiver height) for propagation over flat ground.
When **ballast** is used to support the railway sleepers, the SEL is reduced by a further 1.5 dB for all segments of track except for the one closest to the observer.

The **barrier correction** for the shadow zone (see Figure 10.10a) is:

$$C_{\text{barrier}} = \begin{cases} -21 \text{ (dBA)}; & \text{for } \delta > 2.5 \text{ (m)} \\ -7.75\log_{10}(5.2 + 203\delta) \text{ (dBA)}; & \text{for } 0 < \delta < 2.5 \text{ (m)} \end{cases} \tag{10.188a,b}$$

and for the bright zone (see Figure 10.10b) it is:

$$C_{\text{barrier}} = \begin{cases} 0; & \text{for } \delta > 0.4 \text{ (m)} \\ 0.88 + 2.14\log_{10}(\delta + 10^{-3}) \text{ (dBA)}; & \text{for } 0 < \delta < 0.4 \text{ (m)} \end{cases} \tag{10.189a,b}$$

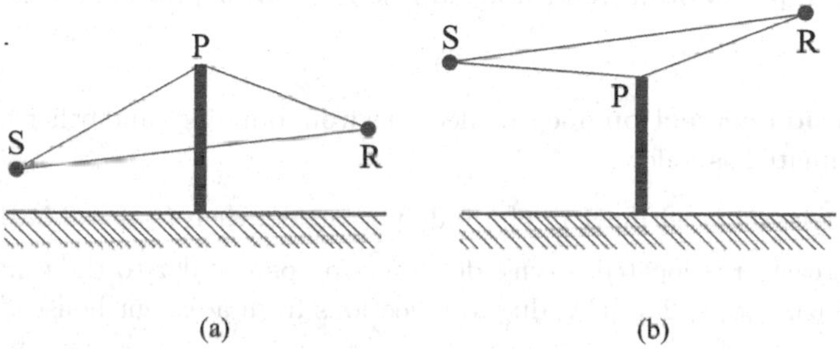

FIGURE 10.10 Geometry for calculating barrier correction: (a) shadow zone; (b) bright zone.

The quantity, δ, used in Equations (11.119) and (11.120), is the difference in the length of the two paths shown in each of the two parts of Figure 10.10. That is:

$$\delta = \text{SP} + \text{PR} - \text{SR} \tag{10.190}$$

where S indicates the source location, R the receiver location and P the barrier edge location adjacent to where the line, SR, pierces or passes over the barrier. If several edges are involved, the reductions can be combined using Equation (1.105) and the procedures in Section 5.3.5.

When multiple barriers of different heights screen the observer from the road, they should be evaluated separately and only the correction resulting in the lowest sound pressure level should be used. *Note that the barrier correction and ground correction are never used at the same time in Equation (10.182).* If the barrier correction is less than the ground correction, then the ground correction is used; otherwise, the barrier correction is used.

The **view correction** (for $\alpha > \beta/2$) for all trains except diesel locomotives under full power is:

$$C_{\text{view}} = 10\log_{10}[\beta - \cos(2\alpha)\sin\beta] - 5 \qquad (10.191)$$

and for diesel locomotives under full power it is:

$$C_{\text{view}} = -10\log_{10}[\sin\alpha\sin(\beta/2)] \qquad (10.192)$$

where β is in radians and is defined along with α in Figure 10.11. Note that α is always less than 90° and is the acute angle between a line drawn through the observer, R, parallel to the track segment and the line bisecting the angle of view, β.

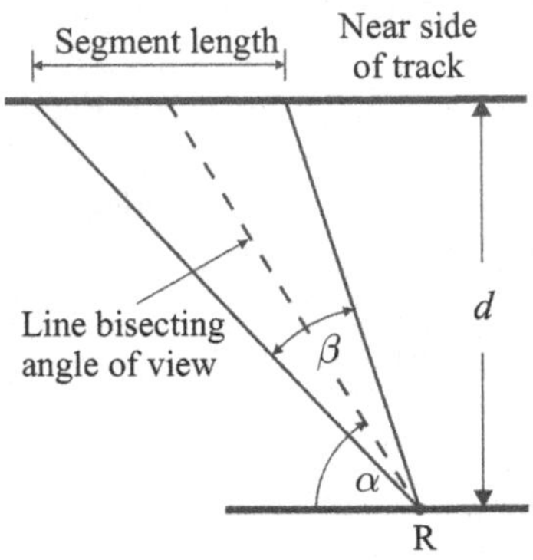

FIGURE 10.11 Top view of the rail track and receiver, showing the angles used to calculate the view correction.

The **Reflection correction** due to reflections from buildings and other hard surfaces may be estimated as follows:

- If the receiver is within 1 m of a building façade, then $C_{\text{refl}} = 2.5$ dBA.
- If the receiver is located down side streets perpendicular to the railway in question, then $C_{\text{refl}} = 2.5$ dBA, due to reflections from adjacent houses.
- If there are reflective surfaces on the far side of the railway from the receiver, $C_{\text{refl}} = 1.5$ dBA.

4. **Convert values of SEL to L_{Aeq}.** This is done for each track segment and for each train using the following relationships:

$$L_{Aeq,6h} = \text{SEL} - 43.3 + 10\log_{10} Q_{\text{NIGHT}} \qquad (\text{dB re } 20\,\mu\text{Pa}) \qquad (10.193)$$

$$L_{Aeq,18h} = \text{SEL} - 48.1 + 10\log_{10} Q_{\text{DAY}} \qquad (\text{dB re } 20\,\mu\text{Pa}) \qquad (10.194)$$

where Q_{NIGHT} is the number of each train type passing the observer in the period, midnight to 6 am and Q_{DAY} is the number of each train type passing the observer during the period, 6 am to midnight.

Overall L_{Aeq} values for each period are obtained by logarithmically summing the component L_{Aeq} values for each train and each track segment using Equation (1.98). The levels calculated using the preceding procedure can be up to 15 dBA higher if the track top surface is corrugated.

10.20.3 Aircraft Noise

The prediction of aircraft noise in the vicinity of airports is traditionally done using complex noise contour generation software. The calculation procedures are complex, as most airports are characterised by many flight paths, many aircraft types and many different engine power settings. Two noise contour programmes that are widely used and are in the public domain are "NOISEMAP" (Wasmer and Maunsell, 2003), which was developed over a number of years under the sponsorship of the US Air Force and "INM" (Integrated Noise Model) (Boeker et al., 2008), which was developed under the sponsorship of the US Department of Transportation (DOT) and the US Federal Aviation Administration (FAA). In May, 2015, the INM was superseded by the Aviation Environmental Design Tool (AEDT) (Koopmann et al., 2016) and INM is no longer in use. The old helicopter noise model (HNM) developed by the FAA has now been included as part of the AEDT. AEDT is available from the FAA for a very modest fee. However, users require a BADA license (EUROCONTROL, 2016), which provides aircraft noise data, to be able to implement AEDT. The US Air Force uses NOISEMAP to predict exposure from all flight activity, including helicopters and fixed wing aircraft, and it uses BOOMMAP and PCBOOM4 for modelling noise exposure due to supersonic aircraft. NOISEMAP software is freely available to the public (Department of Defense, 2016). For this reason it is widely used in the consulting industry for generating noise contours around airports. There is another software package called NoiseMap (NoiseMap Ltd, 2016), which is available for a fee from a consulting company in the UK. It is not related to NOISEMAP and is used to generate noise contours around industrial facilities, but not airports.

In the UK, the model used for civil aviation aircraft is the UK civil aircraft noise contour model (ANCON) (Ollerhead et al., 1999), developed and maintained by the Environmental Research and Consultancy Department of the Civil Aviation Authority (ERCD). However, the AEDT model from the FAA in the USA is also used for civil aircraft in the UK and the NOISEMAP model is used for military aircraft.

In Switzerland, the model used for civil aviation noise prediction is FLULA (Pietrzko and Bütikofer, 2002). In Germany, the models used are SIMUL and AzB (Isermann, 2007) and in France, the model used is IESTA (Malbéqui et al., 2009).

In 2015, the EU Commission published a detailed directive outlining acceptable procedures for calculating the noise impact around airports in Europe (European Commission, 2015). These procedures were required to be implemented by member states prior to December 2018. The aircraft noise part of the directive is over 700 pages long and includes a database for aircraft that were operating at the time of issue. Sound pressure levels at community locations are calculated in terms of sound exposure level, L_{AE}, referenced to 1 second, as defined in ISO 1996-1 (2016).

The outputs from the noise modelling software can be in the form of contours of Sound Exposure Levels (SEL or L_{AE}), Effective Perceived Noise Level (EPNL or L_{EPN}) or Day/Night Equivalent Levels (DNL or L_{dn}). These quantities are defined in Sections 2.4.5 and 2.4.6, respectively. In addition, alternative methods for calculating EPNL are discussed by Raney and Cawthorn (1998) and Zaporozhets and Tokarev (1998).

The noise models all contain extensive databases of values of SEL as a function of distance of the closest approach of an aircraft to an observer for various engine power or thrust settings. Aircraft performance data are used to determine the height above ground and the engine thrust as a function of aircraft load and distance from when the aircraft brakes are released prior to beginning takeoff or the beginning of the runway (landing threshold) (Chapkis et al., 1981).

For a specified flight path and track of a particular aircraft, the noise level at a specified ground location is determined by first calculating the distance to the ground point of the closest part of the flight path. Next, the noise database is used to find the SEL or EPNL corresponding to the distance and thrust setting. Next, adjustments to the level are applied to account for ground attenuation (mainly for shallow angles subtended from the horizontal at the aircraft by a line from the ground point), for fuselage shielding (again for shallow angles only) and for changes to the duration correction in SEL or EPNL as a result of curved flight paths and differences between actual and reference aircraft speeds. The noise contributions to the level at each ground point from all flights into and out of the airport in a given time period are summed on an energy basis to obtain the total noise exposure. This calculation is performed at a large number of ground points to allow contours of equal noise exposure to be generated.

As the calculation procedures are complex, it is recommended that one of the software packages mentioned above is used whenever a community noise exposure assessment as a result of aircraft operations is to be made.

11

Practical Numerical Acoustics

LEARNING OBJECTIVES

In this chapter the reader is introduced to:

- the basic theory underpinning various numerical analysis techniques;
- the use of commercially available software and free software for solving complex sound radiation and transmission problems;
- difference in analysis approaches between low- and high-frequency problems;
- Rayleigh integral analysis;
- Boundary element method (BEM);
- Finite element analysis (FEA);
- Modal coupling analysis; and
- Statistical energy analysis (SEA).

11.1 Introduction

The determination of the sound power radiated by a machine or structure of complex shape at the design stage or the prediction of the distribution of sound in an enclosed space requires the application of numerical techniques. At low frequencies these include the boundary element method (BEM), finite element analysis (FEA) and modal coupling analysis using MATLAB®. At higher frequencies, statistical energy analysis (SEA) is used. There is insufficient space here to provide a complete description of the underlying theory for all of these techniques and a summary is all that will be presented with practical implementation examples available on www.causalsystems.com.

If the expected forcing function can be determined (for example, by suitable measurements on a model in which the load impedance presented to the source is properly represented), then the sound power that will be radiated by the structure can be determined by using one of the above-mentioned numerical or analytical techniques. Such an approach is particularly useful when the effects of modifications to existing structures, for the purpose of noise reduction, are to be investigated. For implementation of these analytical techniques three fundamental steps are necessary.

The first step is the determination and quantification of the force exciting the structure. A given exciting force is generally separated into a sum of sinusoidal components using Fourier analysis. The second step is the determination of the vibrational velocity distribution over the surface of the machine or structure in response to the excitation force. The final step is the calculation of the sound field, and hence the sound power generated by the vibrational response

DOI: 10.1201/9780367814908-11

of the structure or machine surface. A similar approach is needed for the calculation of the distribution of a sound field in a room.

The most fundamental point to understand is that the response of structures and machines can be described in terms of their normal modes of vibration. These normal modes can be excited at resonance or driven at frequencies different from resonance. The response of a machine or structure is always a combination of various vibration modes, most of which are driven off-resonance. In general, the heavier the machine or structure, the higher will be the frequency of the first modal resonance and the fewer will be the number of modes with resonance frequencies in any particular octave or 1/3-octave band. In the analysis of a structure, the first step is to divide the frequency range of interest into octave or 1/3-octave bands which, in turn, will lie either in a low-frequency or a high-frequency region. The low-frequency region is characterised by a paucity of modes in every frequency band, whereas the high-frequency region is defined as that region where there are consistently three or more vibration modes with resonance frequencies in the frequency bands used for the analysis.

11.2 Low-Frequency Region

In the low-frequency region, the surface velocity distribution (or mode shape) is calculated for each vibration mode. For this purpose, a standard numerical analysis procedure such as finite element analysis may be used. For the analysis, the structure is divided into a finite number of surface elements. The element equilibrium and inter-connectivity requirements are satisfied using a system of differential equations. Many commercially available software packages exist, making this method relatively quick and straightforward to apply, even for a three-dimensional structure, once some basic fundamentals have been understood. However, it is only practical to use finite element analysis for the first few (up to 50) vibration modes of a structure. Beyond this, the required element size for accurate prediction becomes too small, the computational process becomes time consuming and prohibitively expensive, and the uncertainty in the accuracy of the model means that the results have a range of possible values. For these reasons, statistical energy analysis can be used for analyses at higher frequencies and is described later in this chapter.

If the overall velocity response of a structure is to be calculated using the finite element method, then knowledge of the damping of each contributing vibration mode (or alternatively a global damping value) is needed. Values of damping cannot be calculated and are generally estimated from measurements on, and experience gained with, similar structures or machines.

The vibrational velocity, $v(\omega, \boldsymbol{r}_0, t)$, at time, t, a point, $\boldsymbol{r}_0$, on a structural surface of area, S_S, due to a sinusoidal excitation force of $F(\omega, \boldsymbol{r}_F, t)$, at forcing frequency, ω, applied at point $\boldsymbol{r}_F$, is (Ewins, 2000):

$$v(\omega, \boldsymbol{r}_0, t) = \mathrm{j}\omega F(\omega, \boldsymbol{r}_F, t) \sum_{\ell=1}^{N_s} \frac{\psi_\ell(\boldsymbol{r}_0)\psi_\ell(\boldsymbol{r}_F)}{\Lambda_\ell Z_\ell} \tag{11.1}$$

where $\psi_\ell(\boldsymbol{r}_0)$ and $\psi_\ell(\boldsymbol{r}_F)$ are the normalised modal responses for mode, ℓ, at locations given by the vectors, $\boldsymbol{r}_0$ and $\boldsymbol{r}_F$, respectively, and N_s (found by iteration) is the number of modes that make a significant contribution to the response at frequency, ω.

The modal mass, Λ, is:

$$\Lambda_\ell = \iint\limits_{S_S} m(\boldsymbol{r}_0)\psi_\ell^2(\boldsymbol{r}_0)\mathrm{d}S_S \tag{11.2}$$

where $m(\boldsymbol{r}_0)$ is the surface mass density (kg/m^2) at location, $\boldsymbol{r}_0$. The integration over a surface has been represented here, and in the remainder of this chapter, by a double integral, which may not reflect the convention used in some texts where the surface is assumed to be represented by a single surface variable and a single integral representation is used. Either representation is acceptable and both are commonly used.

Formulas for the modal mass of simply supported and clamped beams and plates are listed in Howard (2007a). The modal impedance, if hysteretic damping, characterised by loss factor, η_ℓ (the usual case for structures), is assumed, is given by:

$$Z_\ell = \omega_\ell^2 - \omega^2 + \mathrm{j}\eta_\ell\omega_\ell^2 \tag{11.3}$$

Alternatively, if viscous damping, characterised by the critical damping ratio, ζ_ℓ, is assumed, then the modal impedance is given by:

$$Z_\ell = \omega_\ell^2 - \omega^2 + 2\mathrm{j}\zeta_\ell\omega_\ell\omega \tag{11.4}$$

The modal impedance is complex; thus the quantity, $v(\omega,\, \boldsymbol{r}_0, t)$, will be complex, having a magnitude and a phase relative to the forcing function, $F(\omega, t)$. In the preceding equations, ω_ℓ is the resonance frequency, η_ℓ is the structural damping (loss factor) and ζ_ℓ is the viscous damping coefficient (critical damping ratio) for mode, ℓ.

The space- and time-averaged mean square velocity over the structure is given by:

$$\langle v^2\rangle_{S,t} = \langle F^2(\omega, \boldsymbol{r_F})\rangle_t \sum_{\ell=1}^{N_s} \left[\left(\frac{1}{S_S} \iint\limits_{S_S} \psi_\ell^2(\boldsymbol{r}_0)\mathrm{d}S_S \right) \frac{\psi_\ell^2(\boldsymbol{r}_F)}{\Lambda_\ell^2 |Z_\ell|^2} \right] \tag{11.5}$$

As the difference between ω and ω_ℓ becomes large, the contribution due to mode ℓ rapidly becomes small.

If a prototype machine or structure is available, mode shapes and modal damping can be determined from measured data using an experimental procedure known as modal analysis (Ewins, 2000). Modal analysis requires the measurement of the input force to a structure (generated by a shaker or instrumented hammer) and the structural response at a number of locations (Hansen, 2018, pp. 241–268). Software packages, available from manufacturers of most spectrum analysers, will automatically calculate mode shapes, resonance frequencies and damping from these measurements. Again, this method is restricted to the first 10 or so structural resonances. However, the first 20 resonances can be identified if they are sufficiently well separated and some researchers have claimed to have identified up to 100 resonances, but this would be very unusual. From a knowledge of the surface velocity distribution for a given excitation force, the sound pressure field around the structure (and hence the radiated sound power) can be calculated by a number of methods that are described in the following sections. Each method has its advantages and disadvantages. The first method is generally referred to as the Helmholtz integral equation method (Section 11.2.1), which is implemented in the boundary element method using computational software (Section 11.2.2). The second method is referred to as the Rayleigh integral method (Section 11.2.3). The third method involves the use of finite element analysis software (Section 11.2.4). The fourth method involves the use of the calculated mode shapes and resonance frequencies for the acoustic domain (Sections 11.2.5–11.2.6).

11.2.1 Helmholtz Method

In the Helmholtz method (Hodgson and Sadek, 1977, 1983; Koopmann and Benner, 1982), the acoustic pressure field generated by a closed vibrating body is described by the Helmholtz equation, which is just a re-organisation of the wave equation (Equation (1.16) in Chapter 1) such that the RHS is zero.

The acoustic pressure at a point, $\boldsymbol{r}$, outside the vibrating surface at a frequency, ω, is given by Koopmann and Benner (1982) and Brod (1984) as:

$$p(\omega, \boldsymbol{r}) = -\iint\limits_{S_S} \left[p(\omega, \boldsymbol{r}_0)\frac{\partial G(\boldsymbol{r}, \boldsymbol{r}_0)}{\partial \boldsymbol{n}} + \mathrm{j}\rho\omega v(\omega, \boldsymbol{r}_0) G(\boldsymbol{r}, \boldsymbol{r}_0) \right] \mathrm{d}S_S \tag{11.6}$$

where again the time dependence is implicit and the Green's function, G, is defined by:

$$G(\boldsymbol{r}, \boldsymbol{r}_0) = \frac{\mathrm{e}^{-\mathrm{j}k|\boldsymbol{r}-\boldsymbol{r}_0|}}{4\pi|\boldsymbol{r}-\boldsymbol{r}_0|} \tag{11.7}$$

This form of the Green's function applies to sound radiation by a vibrating surface into free space as well as sound radiation into a space enclosed by a vibrating surface.

In the preceding equations, $\boldsymbol{r}$ is the vector distance from the origin of the coordinate system to the point in free space at which the acoustic pressure, $p(\omega, \boldsymbol{r})$, is to be calculated, $\boldsymbol{r}_0$ is the vector distance from the element, $\mathrm{d}S_S$, on the surface of the vibrating structure, to the origin of the coordinate system and $v(\omega, \boldsymbol{r}_0)$ is the normal surface velocity for the surface element, $\mathrm{d}S_S$, at frequency, ω.

The sound power radiated by the structure is found by integrating the product of the acoustic pressure and complex conjugate of the surface velocity over the surface.

11.2.2 Boundary Element Method (BEM)

The boundary element method (BEM) is a numerical implementation of the Helmholtz analysis method discussed in the previous section. The BEM, as its name implies, only involves discretising the boundary of an enclosed space or the boundary of a noise-radiating structure. The method can be used to analyse acoustic problems such as the noise inside an enclosed volume, the noise radiated from a vibrating structure and the acoustic field generated by the scattering of noise by objects in a free-field. On the other hand, finite element analysis (FEA) involves discretising the enclosed volume for interior noise problems and a large space around a noise-radiating structure for exterior noise problems. As a result, the boundary element formulation results in smaller computational models requiring less computer memory than FEA, but the downside is that more computer time may be needed to solve the matrix equations and produce the final result. The difference between the BEM and FEA formulation is illustrated in Figure 11.1, where a finite element model comprises nodes (the dots) and elements (the rectangles formed by the nodes for the FEM) and the BEM model comprises nodes (the dots) and lines.

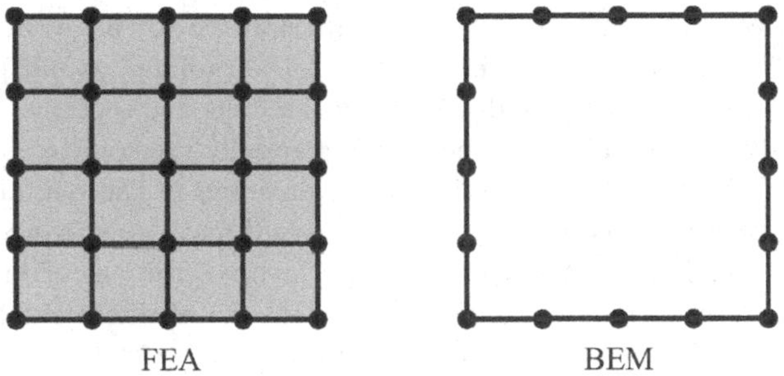

FIGURE 11.1 Node locations for FEA and BEM.

Here the application of the BEM will be illustrated by showing how publicly available BEM software may be used to solve a sound radiation problem inside a rigid-walled room. There are two different boundary element methods that can be used to evaluate an acoustic field generated by a defined forcing function: the direct method and the indirect method. Both of these will be discussed in the following sections.

11.2.2.1 Direct Method

The direct method essentially involves solving equations based on the analysis of Section 11.2.1 with the additional consideration of the case where the pressure is evaluated on the surface of area, S_S. The direct method can be used to calculate the interior acoustic pressure field generated by vibration of an enclosing surface, or the acoustic field external to a vibrating surface, but not both at the same time. In addition, the direct method can be used for solving problems involving scattering of an incident sound field by an object in its path. To determine the pressure in both the interior and exterior field at the same time, other methods are available, such as the variational indirect BEM (Wu, 2000b, p. 83), but will not be discussed here.

Sound field inside or outside of an enclosing surface

The acoustic pressure at point, $\boldsymbol{r}$, and time, t, internal or external to an enclosing surface, with vibration velocity, $v(\omega, \boldsymbol{r}_0, t)$, at point, $\boldsymbol{r}_0$, on the surface is given by (Wu, 2000b) as:

$$p(\omega, \boldsymbol{r}, t) = -\frac{1}{C(\boldsymbol{r})} \iint\limits_{S_S} \left[p(\omega, \boldsymbol{r}_0, t) \frac{\partial G(\boldsymbol{r}, \boldsymbol{r}_0)}{\partial \boldsymbol{n}} + \mathrm{j}\rho\omega v(\omega, \boldsymbol{r}_0, t) G(\boldsymbol{r}, \boldsymbol{r}_0) \right] \mathrm{d}S_S \tag{11.8}$$

where $\mathrm{d}S_S$ is a differential area of the surface, and $C(\boldsymbol{r})$ is a coefficient depending on the location of the point $\boldsymbol{r}$. The quantity, $C(\boldsymbol{r})$, is equal to one when the point $\boldsymbol{r}$ is within the acoustic domain and is equal to $\frac{1}{2}$ when the point $\boldsymbol{r}$ is on the surface (i.e. $\boldsymbol{r} = \boldsymbol{r}_0$). For mathematical completeness $C(\boldsymbol{r}) = 0$ when the point $\boldsymbol{r}$ is not within the acoustic domain or on the enclosing surface; however, this is rarely used as it applies to locations not in the acoustic medium, but inside the material of an enclosing structure or a structure that is within the sound field. These definitions for $C(\boldsymbol{r})$ apply for interior, exterior and scattering acoustic problems. As shown in Figure 11.1, for boundary element models, as the name suggests, only elements on the boundary are defined, and no elements are defined for the acoustic domain. In contrast, for finite element models, the acoustic domain is defined using finite elements. The formulation for the boundary element method requires that the vector, $\boldsymbol{n}$, which is normal to the surface, points away from the acoustic domain. As will be described in the next section, an interior acoustic domain is enclosed by boundary elements with normal vectors, $\boldsymbol{n}$, pointing away from the acoustic domain and enclosed volume. In contrast, for an exterior acoustic domain, which is implicitly infinite in extent, the normal vectors on the boundary elements, $\boldsymbol{n}$, point away from the infinite acoustic domain and inwards to the enclosed volume that is surrounded by the boundary elements. If both interior and exterior acoustic domains are of interest, then the indirect BEM, described in Section 11.2.2.2 must be used.

In Equation (11.8), the time dependence may be ignored as it is the same on both sides of the equation, so the form of the acoustic pressure on the left-hand side will depend on the form of the acoustic pressure and particle velocity used on the right-hand side. So if amplitudes are used on the right-hand side, then the result on the left-hand side will be an acoustic pressure amplitude.

Sound field produced by scattering off an object

For acoustic scattering problems, where an incident sound wave impinges on an object in an infinite acoustic domain, Equation (11.8) is modified slightly so that the total pressure, p, which is the sum of the incident sound pressure when no object is present in the infinite domain, $p_i(\omega, \boldsymbol{r})$, and the scattered sound pressure when the object is present in the infinite domain, p_s, and is given by:

$$p(\omega, \boldsymbol{r}) = -\frac{1}{C(\boldsymbol{r})} \iint\limits_{S_S} \left[p(\omega, \boldsymbol{r}_0) \frac{\partial G(\boldsymbol{r}, \boldsymbol{r}_0)}{\partial \boldsymbol{n}} + \mathrm{j}\rho\omega v(\omega, \boldsymbol{r}_0) G(\boldsymbol{r}, \boldsymbol{r}_0) \right] \mathrm{d}S_S + \frac{p_i(\omega, \boldsymbol{r})}{C(\boldsymbol{r})} \tag{11.9}$$

where the time dependence term has been omitted, as it is the same for both sides of the equation.

The complex acoustic pressure amplitude (ignoring the time variation term, $\mathrm{e}^{\mathrm{j}\omega t}$) associated with an incident plane wave is:

$$p_i(\omega, \boldsymbol{r}) = A\mathrm{e}^{-\mathrm{j}(k_x x + k_y y + k_z z)} \tag{11.10}$$

where k_x, k_y and k_z are the wavenumber components in the x-, y- and z-directions, respectively, such that $k_x^2 + k_y^2 + k_z^2 = k^2 = (\omega/c)^2$ and A is the modulus of the acoustic pressure amplitude. The complex amplitude of the incident sound pressure associated with a spherical wave (or monopole) is (see Equation (4.90)):

$$p_i(\omega, \boldsymbol{r}) = \frac{\mathrm{j}\omega\rho Q_0}{4\pi r}\mathrm{e}^{-\mathrm{j}kr} \tag{11.11}$$

where Q_0 is the monopole source strength amplitude, r is the distance from the monopole to the surface on which the pressure is incident and ρ is the density of the fluid.

One problem with the BEM for the analysis of exterior acoustic problems is that at certain frequencies the Helmholtz integral cannot be solved. This problem is overcome by using the combined Helmholtz integral equation formulation (CHIEF) method proposed by Schenck (1968), and is implemented in many BEM software packages. The details of this method are explained in Wu (2000b) and von Estorff (2000).

11.2.2.2 Indirect Method

The indirect boundary integral formulation of the Helmholtz integral (Equation (11.6)) relies on boundary conditions involving the difference in the acoustic pressure and the difference in the pressure gradient. The indirect method can be used to calculate both interior and exterior acoustic fields as a result of a vibrating surface or acoustic sources, and can include openings that connect an enclosed region to a free-field region, or where free edges occur on a surface such as a stiffening rib attached perpendicularly to a panel. For calculating the acoustic field inside and outside of an enclosing surface, two surfaces are used: the surface that is in contact with the interior acoustic domain has surface normal vectors, $\boldsymbol{n}$, pointing outwards towards the exterior acoustic domain, and the surface that is in contact with the exterior acoustic domain has $\boldsymbol{n}$ pointing inwards towards the interior acoustic domain. The matrices resulting from this indirect method are fully populated and symmetric, which can result in faster solution times compared to solving unsymmetric matrices, such as those associated with the direct method.

The formulation for the indirect boundary element method is given by the expression Wu (2000b):

$$\frac{\partial p(\boldsymbol{r})}{\partial x_i} = -\mathrm{j}\rho\omega u_{x_i}(\boldsymbol{r}) = \iint\limits_{S_S} \left[\delta \mathrm{d}p(\boldsymbol{r}_{S_S})\frac{\partial G(\boldsymbol{r}, \boldsymbol{r}_{S_S})}{\partial x_i} - \delta p(\boldsymbol{r})\frac{\partial^2 G(\boldsymbol{r}, \boldsymbol{r}_{S_S})}{\partial x_i \partial \boldsymbol{n}}\right] \mathrm{d}S_S \tag{11.12a,b}$$

where the time and frequency dependence of all acoustic pressure, $p(\boldsymbol{r})$, and particle velocity, $u_{xi}(\boldsymbol{r})$, terms is implicit, and the subscript, i, refers to the ith direction. The left-hand side of the expression is the gradient of the pressure at any point, $\boldsymbol{r}$, in the acoustic domain in the direction x_i, u_{xi} is the component of acoustic particle velocity in the direction x_i, and the right-hand side is an integral expression over the boundary surface. The term:

$$\delta p(\boldsymbol{r}_{S_S}) = p(\boldsymbol{r}_{S_1}) - p(\boldsymbol{r}_{S_2}) \tag{11.13}$$

is the difference in pressure across the infinitesimally thin surface of the boundary element model and is called the pressure jump or double layer potential, and $\boldsymbol{r}_S$ is a point on the boundary surface. The term:

$$\delta \mathrm{d}p(\boldsymbol{r}_S) = \frac{\partial p(\boldsymbol{r}_{S1})}{\partial \boldsymbol{r}_{S1}} - \frac{\partial p(\boldsymbol{r}_{S2})}{\partial \boldsymbol{r}_{S2}} \tag{11.14}$$

is the difference in gradient of the pressure normal to the surface of the boundary element model at points, $\boldsymbol{r}_{S1}$ and $\boldsymbol{r}_{S2}$, on opposite sides of the boundary surface and is called the single layer potential.

11.2.2.3 Meshing

Equation (11.8) must be evaluated numerically, and this is achieved by discretising the surface with elements, similar to the process used in finite element analysis. If a system with a three-dimensional acoustic volume is to be analysed, then surface patch elements are used, such as rectangular elements with four nodes, triangular elements with three nodes, or other elements with a greater number of nodes along the edges, which permits greater accuracy for modelling the acoustic pressure. For an acoustic system, which can be modelled in two dimensions, where the acoustic domain is an area, the elements for the boundary element model will be line segments with two or more nodes. See Juhl (1993) and Wu (2000b) for descriptions of these element types.

Before solving Equation (11.12), it is necessary to define boundary conditions in terms of the boundary variables. For the direct BEM, the acoustic pressure and acoustic particle velocity are used as the boundary variables. For the indirect BEM, the boundary variables are the difference in acoustic pressure and the difference in the pressure gradient across the boundary. Figure 11.2 shows a schematic of an infinite acoustic domain that is exterior to an enclosed volume, where the normal vector, $\boldsymbol{n}$, points inwards to the volume and away from the acoustic domain. The figure shows the difference in boundary conditions between the direct and indirect boundary element methods.

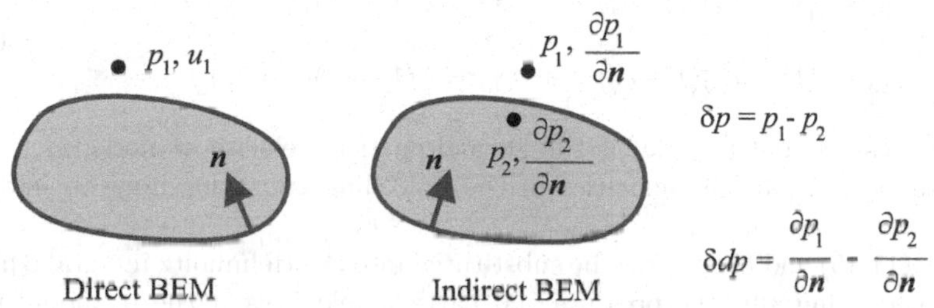

FIGURE 11.2 Direct versus Indirect BEM.

The direct BEM can be used to analyse either an interior problem or an exterior problem, while the indirect BEM can be used to solve an interior and exterior problem at the same time.

The boundary elements have nodes at the vertices, as shown in Figure 11.3, or mid-span between two vertices. The pressure (or velocity) can be calculated anywhere inside the element using a mathematical expression of a weighted sum of the pressures (or velocities) at the nodes, which is the same method as used in FEA.

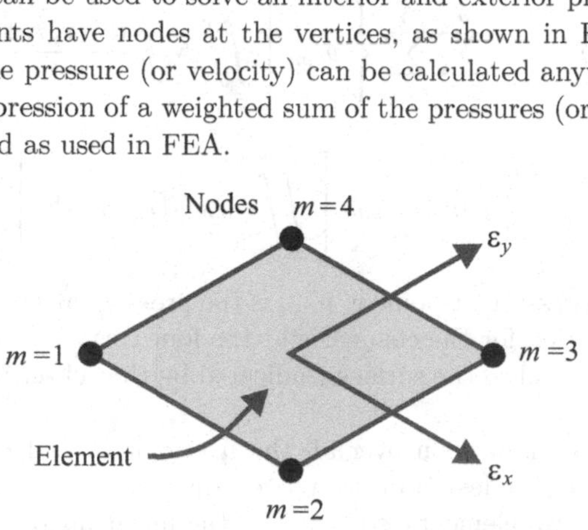

FIGURE 11.3 Coordinate system of a linear quadrilateral boundary element.

11.2.2.4 Problem Formulation

To solve a direct boundary element method problem, it is necessary to describe the problem using matrix equations, which can be solved using computer software. The first step is to evaluate all the nodal acoustic pressures and velocities at all the boundary elements by using the known boundary conditions. Once the nodal values have been determined, the acoustic pressure can be calculated at any point in the acoustic domain.

The surface surrounding the volume, S_S, must be meshed with n_e elements. For this discussion, linear quadrilateral elements will be used, with 4 nodes per element as shown in Figure 11.3, and with the entire mesh comprising n_n nodes. A point within a quadrilateral element is described in terms of normalised coordinates (ϵ_x, ϵ_y), where ϵ_x and ϵ_y vary between ± 1. The acoustic pressure and particle velocity within the boundary element can be written as (Wu, 2000b, p. 35, p. 55) (von Estorff, 2000, p. 18):

$$p(\omega, \epsilon_x, \epsilon_y) = \sum_{m=1}^{4} N_m(\epsilon_x, \epsilon_y) p_m(\omega) \tag{11.15}$$

$$u_n(\omega, \epsilon_x, \epsilon_y) = \sum_{m=1}^{4} N_m(\epsilon_x, \epsilon_y) u_{n,m}(\omega) \tag{11.16}$$

where N_m is the shape function evaluated at the coordinates, ϵ_x, ϵ_y given by

$$\begin{aligned} N_1 &= \tfrac{1}{4}(1-\epsilon_x)(1-\epsilon_y) \qquad & N_2 &= \tfrac{1}{4}(1+\epsilon_x)(1-\epsilon_y) \\ N_3 &= \tfrac{1}{4}(1+\epsilon_x)(1+\epsilon_y) \qquad & N_4 &= \tfrac{1}{4}(1-\epsilon_x)(1+\epsilon_y) \end{aligned} \tag{11.17a,b,c,d}$$

$p_m(\omega)$ is the pressure and $u_{n,m}(\omega)$ is the normal particle velocity at node, m, and frequency, ω. The pressures and particle velocities in the preceding equations may be instantaneous or amplitudes.

Equations (11.15) and (11.16) can be substituted into the Helmholtz integral equation (Equation (11.8)), where initially, the pressure and particle velocities at the n_n nodal points on the boundary elements are evaluated, and hence this procedure is called "collocation". The integration over the surface S_S is done by a summation of each element, so that for each collocation node the Helmholtz equation can be written as (Fahy and Gardonio, 2007, p. 509) (von Estorff, 2000, p. 20):

$$\begin{aligned} C(\boldsymbol{r})\, p(\boldsymbol{r}) = -\sum_{e=1}^{n_e} \sum_{m=1}^{4} \Bigg\{ & p_{e,m} \left[\iint_{S_S} N_{e,m} \frac{\partial G(\boldsymbol{r}, \boldsymbol{r}_0)}{\partial \boldsymbol{n}} \, \mathrm{d}S_S \right] \\ & + \mathrm{j}\rho\omega\, u_{n,e,m} \left[\iint_{S_S} N_{e,m} G(\boldsymbol{r}, \boldsymbol{r}_0) \, \mathrm{d}S_S \right] \Bigg\} \end{aligned} \tag{11.18}$$

where $p(\boldsymbol{r})$ is the sound pressure at point $\boldsymbol{r}$, $p_{e,m}$ is the pressure at node m belonging to element e, $N_{e,m}$ is the shape function for the element e at the four corner nodes $m = 1 \cdots 4$, and $u_{n,e,m}$ is the particle velocity normal to the surface (indicated by the subscript n) at node m belonging to element e.

Equation (11.18) is a summation over all the n_e element, and can be re-arranged into a summation over all the n_n nodes. Note that two adjacent elements will share two nodes on the common edge of the two elements, so that (for the linear quadrilateral elements) the total number of nodes in the mesh will be less than four times the number of elements ($n_n < 4n_e$). In the following analysis, matrices and vectors will be denoted in bold font. This equation can be rearranged into a "global" matrix equation as:

$$[\mathbf{C}]\{\mathbf{p}\} = [\hat{\mathbf{H}}]\{\mathbf{p}\} + [\mathbf{G}]\{\mathbf{u_n}\} \tag{11.19}$$

where $\{\mathbf{p}\}$ and $\{\mathbf{u_n}\}$ are $(n_n \times 1)$ column vectors containing the nodal pressures and particle velocities normal to the boundary surface, respectively. The acoustic pressures and particle velocities are frequency-dependent. The matrix, $\mathbf{C}$, is an $(n_n \times n_n)$ diagonal matrix, with entries given by Chien et al. (1990) and Wu (2000b, p. 32):

$$c_i = \begin{cases} 1 & \text{when the point is in the acoustic domain} \\ 1/2 & \text{when the point is on a smooth surface} \end{cases} \tag{11.20}$$

The matrices $\hat{\mathbf{H}}$ and $\mathbf{G}$ are $(n_n \times n_n)$, and the terms in the matrices are, in general:

$$h_{e,m} = -\int_{-1}^{1}\int_{-1}^{1} \frac{\partial G}{\partial n} N_{e,m}\, \mathrm{d}\epsilon_x \mathrm{d}\epsilon_y \tag{11.21}$$

$$g_{e,m} = -\mathrm{j}\rho\omega \int_{-1}^{1}\int_{-1}^{1} G N_{e,m}\, \mathrm{d}\epsilon_x \mathrm{d}\epsilon_y \tag{11.22}$$

and these integral terms can be evaluated using standard numerical integration methods such as Gaussian quadrature over each boundary element. However, the positions of these terms in the matrices are dependent on the mesh arrangement and the numbering scheme for the nodes.

The matrices $\mathbf{C}$ and $\hat{\mathbf{H}}$ can be combined into a single matrix $\mathbf{H}$ to form the equation (Wu, 2000b, p. 36) (Fahy and Gardonio, 2007, p. 510):

$$[\mathbf{H}]\{\mathbf{p}\} = [\mathbf{G}]\{\mathbf{u_n}\} \tag{11.23}$$

This matrix equation is a set of n_n equations, comprising n_n nodal pressures and n_n nodal normal velocities, so there are a total of $2n_n$ variables. A typical analysis will define (boundary conditions) n_n nodal normal velocities, or n_n nodal acoustic pressures, or a mix thereof. Therefore, Equation (11.23) can be reformulated into a set of n_n equations, with n_n unknown nodal variables that could be a mix of nodal pressures or nodal normal velocities.

Equation (11.23) is then re-arranged into the following form:

$$[\mathbf{A}]\{\mathbf{x}\} = \{\mathbf{b}\} \tag{11.24}$$

where the unknown boundary acoustic pressures or normal particle velocities are in the vector $\{\mathbf{x}\}$ and can be solved for by inverting the matrix $[\mathbf{A}]$. Once all the boundary pressures and velocities are solved for, the pressure at any point $p(\mathbf{r})$ within the acoustic domain can be calculated using the Helmholtz integral equation.

Analysis of acoustic problems using the boundary element method involves the use of computational software. There are commercial and non-commercial (free) software packages that are available. For demonstration purposes, the free software, FastBEM (Advanced CAE Research, 2016), is used in an example available on www.causalsystems.com to calculate the sound pressure level distribution in a three-dimensional enclosure with rigid walls. A MATLAB® script, `file:rect_cav_3D.m`, for the same calculation is available on the website listed in Appendix F.

The pre-processor in the FastBEM software does not include capabilities to construct geometry or mesh the geometry. The standard workflow is to create the mesh for the boundary element model using either Ansys or Nastran, which are both commercial finite element analysis software packages. The example mentioned above involves the creation of the mesh for the boundary element model using Ansys Mechanical APDL. Alternatively, if Ansys is not available, a geometry and mesh can also be created as an STL file using freely available software packages, such as BRL-CAD, FreeCAD, OpenSCAD, and then converted into a format for use with FastBEM, using a free translator program available from the FastBEM website.

11.2.3 Rayleigh Integral Method

The Rayleigh integral method involves evaluating the integral of the surface velocity of the object to calculate the radiated sound power. The Rayleigh method (Hayes and Quantz, 1982; Hansen and Mathews, 1983; Takatsubo et al., 1983) is most effective when the machine or structure can be divided into a number of panels, which are approximately flat, but of any shape. Two broad assumptions must be made. First, it is assumed that the sound fields from adjacent panels do not interact to produce constructive and destructive interference. This assumption is satisfied if the analysis bandwidth is sufficiently large (1/3-octave or octave), or if the ratio of the wavelength of the radiated sound to the panel dimensions is small (less than 1/3). The second assumption is that the sound fields from the two sides of the panel do not interact. This assumption is satisfied if the panel is part of a closed surface or mounted in a rigid baffle, or if the ratio of the wavelength of the radiated sound to the panel dimensions is small (less than 1/3).

The Rayleigh integral for the calculation of the sound pressure, p, at location, $\boldsymbol{r} = (r, \theta, \psi)$, (in the far field) from a radiating panel of area, S_p, at frequency, ω, is:

$$p(\omega, \boldsymbol{r}) = \frac{\mathrm{j}\omega\rho}{2\pi} \iint_{S_p} v(\omega) \left[\frac{\mathrm{e}^{-\mathrm{j}kr}}{r}\right] \mathrm{d}S_p \tag{11.25}$$

where the time dependence is implicit as explained previously, $v(\omega)$ is the normal velocity on the surface element, $\mathrm{d}S_p$, at frequency, ω, and r is the distance from the surface element to the vector location, $\boldsymbol{r}$, in space. In practice, if a number velocity measurements are made on a panel, the integral may be replaced with a sum over the number of measurements with a finite area, $\mathrm{d}S_p$ assigned to each measurement. Note that the areas assigned to each measurement do not need to be identical provided the total of the assigned areas adds up to the total area of the panel.

The advantage of the Rayleigh Integral method is that it can be used to calculate the panel radiated sound power, which can be used for the calculation of interior sound pressure levels (for example, in a vehicle) as well as exterior sound pressure levels.

The sound power radiated by each panel can be calculated by integrating the pressure amplitude over a hemispherical surface in the far field of the panel, using the plane of the panel as the base of the hemisphere. The total sound power radiated by the structure is calculated by adding logarithmically the power due to each panel making up the structure. Best results are obtained if power data in watts are averaged over 1/3-octave or octave frequency bands prior to taking logs to obtain sound power level.

Figure 11.4(a) shows a picture of a flat panel, installed in a baffle, which is vibrating and radiating sound into an infinite half-space. The panel has been discretised into four elements and five nodes. The Rayleigh integral of Equation (11.25) can be used to estimate the far–field sound pressure, which can then be used to estimate the sound power radiated by the structure. Although this system is represented in 2-D, it also applies to 3-D systems. The implicit assumption in the Rayleigh integral is that each element is a small independent source that produces a far–field pressure given by Equation (11.11). The assumption of independent sources is a good engineering approximation only and should be used with caution when analysing vibrating structures that have surfaces that will cause the radiated sound to reflect or diffract, as shown in Figure 11.4(b). Another complication occurs with structures that have a convex

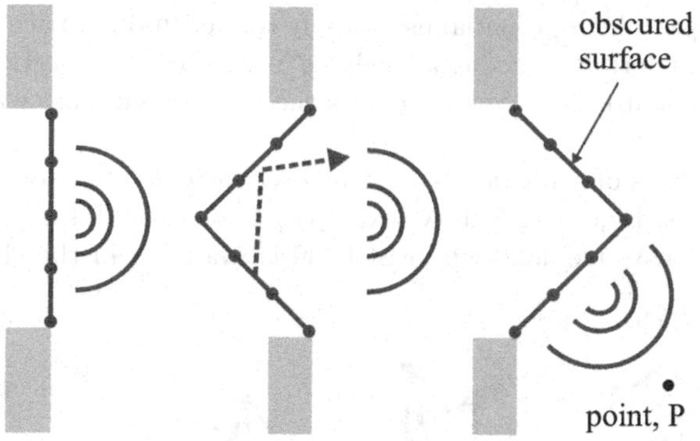

FIGURE 11.4 (a) Unobstructed surface; (b) concave surface with reflections; (c) convex surface with an obscured surface.

shape, as shown in Figure 11.4(c), where only part of the surface might be "visible" to a point in the far field, and other vibrating surfaces will be obscured by the structure. Herrin et al. (2003) describe a method that can be used to determine which surfaces are "visible" to a point in the acoustic field based on evaluating the product of the vector normal to the surface and the vector from a point on the surface to the point in the acoustic field. For structures with characteristics shown in Figures 11.4(b) and (c) the use of the Rayleigh integral could give results that have large errors. In these cases, a numerical method such as the BEM or FEA should be used to estimate the radiated sound field.

11.2.4 Finite Element Analysis (FEA)

Finite element analysis is a numerical method that can be used to calculate the response of a complex structure due to the application of forcing functions, which could be an acoustic field or a distribution of mechanical forces. FEA can also be used to estimate the sound power radiated by a structure or the distribution of the sound field in an enclosed space. Estimating the sound power radiated by a structure generally requires a large numerical model and the associated computer memory requirements can be large. Rather, it is better to use FEA to calculate the response of the noise-radiating structure and then use a numerical evaluation of the Rayleigh integral to calculate the radiated sound power. Alternatively, the sound power radiated by a structure can be determined using a combination of FEA and BEM software and the following steps:

1. the velocity distribution over the surface of the structure is calculated using FEA;
2. the velocity results calculated using FEA are imported into the BEM software and applied as boundary conditions to a model of the structure;
3. the BEM software is used to calculate the far-field sound pressure;
4. Equation (1.84) is used to calculate the sound power by integrating the sound intensity over an area surrounding the vibrating structure, where the intensity in the far field is given by Equation (1.78).

If the structure is excited by an external sound field, then FEA can be used to determine the structural response and also the resonance frequencies and mode shapes of the enclosed sound field. Then the actual sound pressure distribution in the enclosed space can be calculated using modal coupling analysis implemented with a programming tool such as MATLAB®.

The underlying theory for FEA is covered in many textbooks and will not be repeated here. However, its practical implementation using a commercially available FEA package will be discussed in an attempt to help potential users apply the technique to acoustic analysis. Readers who are interested in undertaking acoustic analyses using the Ansys software should consult the book Howard and Cazzolato (2014), which covers the topic extensively with numerous practical examples.

Finite element analysis of acoustic systems involves the discretisation of the acoustic volume into elements and nodes. Figure 11.5 shows examples of acoustic finite elements available in the Ansys software. The Ansys element type FLUID30 is available in the shapes of a hexahedron

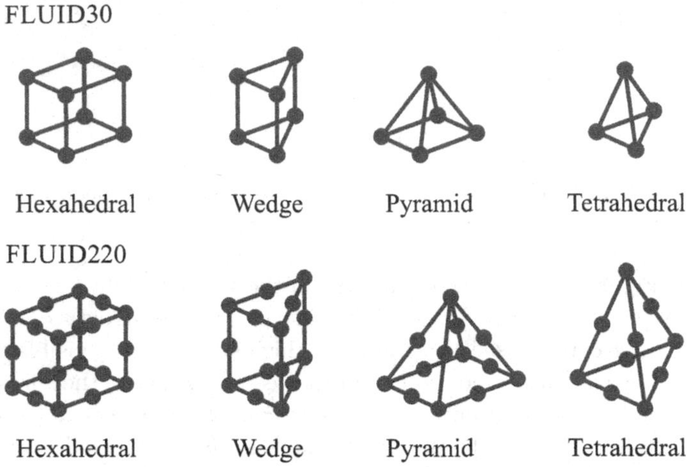

FIGURE 11.5 Shapes of the FLUID30 and FLUID220 acoustic finite elements available in Ansys.

(brick), wedge, pyramid and tetrahedron. It can be seen that the hexahedral element has 8 nodes, where each node has one pressure degree of freedom, and optionally, an additional three translational degrees of freedom when a node is attached to a structure. The element type FLUID220 has the same shape options as the FLUID30 element, but the FLUID220 element has an additional node along each edge. It can be seen that the FLUID220 element in a hexahedral shape has 20 nodes per element. The additional mid-side nodes in the FLUID220 element enable estimation of pressure gradients in an acoustic domain with greater accuracy compared with using FLUID30 elements, but comes at the expense of using a greater number of nodes, and hence larger memory usage.

An enclosed acoustic volume might be surrounded by rigid walls, a flexible structure or walls that provide acoustic damping. Alternatively, the acoustic radiation of a structure into an anechoic field can also be examined.

The analysis of acoustic and structural vibration can be achieved using simple theoretical models for rectangular shaped objects. Any geometry more complex than a rectangular shaped object is onerous to analyse and vibro-acoustic practitioners opt for a numerical method such as the BEM or FEA to solve their particular problem. Later in this chapter, statistical energy analysis (SEA) will be discussed for the analysis of high-frequency problems, which are characterised by a sufficiently high modal density in the radiating structure or enclosed acoustic space. For low-frequency problems, the BEM and FEA can be used.

Finite element analysis of acoustic systems has numerous applications including the acoustic analysis of interior sound fields, sound radiation from structures, the transmission loss of panels, the design of resonator type silences and diffraction around objects. The finite element method takes account of the bi-directional coupling between a structure and a fluid such as air or water.

In acoustic fluid-structure interaction problems, the structural dynamics equation needs to be considered along with the mathematical description of the acoustics of the system, given by the Navier–Stokes equations of fluid momentum and the flow continuity equation. The discretised structural dynamics equation can be formulated using structural finite elements. The fluid momentum (Navier–Stokes) and continuity equations are simplified to get the acoustic wave equation using the following assumptions:

- the acoustic pressure in the fluid medium is determined by the linear wave equation;
- the fluid is compressible where density changes are due to pressure variations;
- the fluid is inviscid with no dissipative effect due to viscosity;
- there is no mean flow of the fluid;
- the mean density and pressure are uniform throughout the fluid and the acoustic pressure is defined as the pressure in excess of the mean pressure; and
- finite element analyses are limited to relatively small acoustic pressures so that the changes in density are small compared with the mean density.

The acoustic wave equation (Equation (1.16)) is used to describe the acoustic response of the fluid. Because viscous dissipation in the fluid is neglected, the equation is referred to as the lossless wave equation. Suitable acoustic finite elements can be derived by discretising the lossless wave equation using the Galerkin method. For a derivation of the acoustic finite element the reader is referred to Craggs (1971).

There are two formulations of finite elements that are used to analyse acoustic problems: pressure and displacement. The most commonly used finite element to analyse acoustic problems is the pressure formulated element. Displacement formulated acoustic elements are less frequently used and will not be discussed here. Instead, the interested reader can consult Howard and Cazzolato (2014), Section 2.5.

11.2.4.1 Pressure Formulated Acoustic Elements

The acoustic pressure p within a single finite element can be written as:

$$p = \sum_{i=1}^{m} N_i p_i \tag{11.26}$$

where N_i is a set of linear shape functions, p_i are acoustic nodal pressures and m is the number of nodes defining the element. For pressure formulated acoustic elements the finite element equation for the fluid in matrix form, including damping, is:

$$[\mathbf{M}_f]\{\ddot{\mathbf{p}}\} + [\mathbf{C}_f]\{\dot{\mathbf{p}}\} + [\mathbf{K}_f]\{\mathbf{p}\} = \{\mathbf{F}_f\} \tag{11.27}$$

where $[\mathbf{M}_f]$ is the equivalent fluid "mass" matrix, $[\mathbf{C}_f]$ is the equivalent fluid "damping" matrix, $[\mathbf{K}_f]$ is the equivalent fluid "stiffness" matrix, $\{\mathbf{F}_f\}$ is a vector of applied "fluid loads", $\{\mathbf{p}\}$ is a vector of unknown nodal acoustic pressures and $\{\ddot{\mathbf{p}}\}$ is a vector of the second derivative of acoustic pressure with respect to time.

The equations of motion for the structure, including damping, where parts of the structure can interact with the fluid are:

$$[\mathbf{M}_s]\{\ddot{\mathbf{y}}\} + [\mathbf{C}_s]\{\dot{\mathbf{y}}\} + [\mathbf{K}_s]\{\mathbf{y}\} = \{\mathbf{F}_s\} \tag{11.28}$$

where $[\mathbf{K}_s]$ is the structural stiffness matrix, $[\boldsymbol{M}_s]$ is the structural mass matrix, $\{\mathbf{F}_s\}$ is a vector of applied "structural loads", $\{\mathbf{y}\}$ is a vector of unknown nodal displacements, and hence $\{\ddot{\mathbf{y}}\}$ is a vector of the second derivative of displacement with respect to time, equivalent to the acceleration of the node. The interaction of the fluid and structure occurs at the interface between the structure and the acoustic elements, where the acoustic pressure exerts a force on the structure and the motion of the structure produces a pressure. To account for the coupling between the structure and the acoustic fluid, additional terms are added to the equations of motion for the structure and fluid (of density, ρ), respectively, as:

$$[\mathbf{M}_s]\{\ddot{\mathbf{y}}\} + [\mathbf{C}_s]\{\dot{\mathbf{y}}\} + [\mathbf{K}_s]\{\mathbf{y}\} = \{\mathbf{F}_s\} + [\mathbf{R}]\{\mathbf{p}\} \tag{11.29}$$

$$[\mathbf{M}_f]\{\ddot{\mathbf{p}}\} + [\mathbf{C}_f]\{\dot{\mathbf{p}}\} + [\mathbf{K}_f]\{\mathbf{p}\} = \{\mathbf{F}_f\} - \rho[\mathbf{R}]^{\mathrm{T}}\{\ddot{\mathbf{y}}\} \tag{11.30}$$

where $[\mathbf{R}]$ is the coupling matrix that accounts for the effective surface area associated with each node on the fluid–structure interface. Equations (11.29) and (11.30) can be formed into a matrix equation including the effects of damping as:

$$\begin{bmatrix} \mathbf{M}_s & 0 \\ \rho\mathbf{R}^{\mathrm{T}} & \mathbf{M}_f \end{bmatrix} \begin{Bmatrix} \ddot{\mathbf{y}} \\ \ddot{\mathbf{p}} \end{Bmatrix} + \begin{bmatrix} \mathbf{C}_s & 0 \\ 0 & \mathbf{C}_f \end{bmatrix} \begin{Bmatrix} \dot{\mathbf{y}} \\ \dot{\mathbf{p}} \end{Bmatrix} + \begin{bmatrix} \mathbf{K}_s & -\mathbf{R} \\ 0 & \mathbf{K}_f \end{bmatrix} \begin{Bmatrix} \mathbf{y} \\ \mathbf{p} \end{Bmatrix} = \begin{Bmatrix} \mathbf{F}_s \\ \mathbf{F}_f \end{Bmatrix} \tag{11.31}$$

where $[\mathbf{C}_s]$ and $[\mathbf{C}_f]$ are the structural and acoustic damping matrices, respectively. Consideration of a single frequency, ω at any one time, allows this equation to be reduced to an expression without differentials as:

$$\begin{bmatrix} -\omega^2\mathbf{M}_s + \mathrm{j}\omega\mathbf{C}_s + \mathbf{K}_s & -\mathbf{R} \\ -\omega_2\rho\mathbf{R}^{\mathrm{T}} & -\omega^2\mathbf{M}_f + \mathrm{j}\omega\mathbf{C}_f + \mathbf{K}_f \end{bmatrix} \begin{Bmatrix} \mathbf{y} \\ \mathbf{p} \end{Bmatrix} = \begin{Bmatrix} \mathbf{F}_s \\ \mathbf{F}_f \end{Bmatrix} \tag{11.32}$$

The important feature to notice about Equation (11.32) is that the matrix on the left-hand side is unsymmetric and solving for the nodal pressures and displacements at frequency, ω, requires the inversion of this unsymmetric matrix, which takes a significant amount of computer resources. The fluid-structure interaction method described above accounts for coupling between structures and fluids, and this is usually only significant if a structure is radiating into a heavier than air medium, such as water or if the structure is very lightweight, such as a car cabin.

Some finite element analysis software packages permit the conversion of the unsymmetric formulation shown in Equation (11.32) to a symmetric formulation for the fluid-structure interaction (Ohayon, 2001). This can be accomplished by defining a transformation variable for the nodal pressures as:

$$\dot{\mathbf{q}} = \mathrm{j}\omega\mathbf{q} = \mathbf{p} \tag{11.33}$$

and substituting this into Equation (11.32) so that the system of equations becomes:

$$\begin{bmatrix} -\omega^2\mathbf{M_s} + j\omega\mathbf{C_s} + \mathbf{K_s} & -j\omega\mathbf{R} \\ -j\omega\mathbf{R}^{\mathrm{T}} & \dfrac{\omega^2\mathbf{M_f}}{\rho} - \dfrac{j\omega\mathbf{C_f}}{\rho} - \dfrac{\mathbf{K_f}}{\rho} \end{bmatrix} \begin{Bmatrix} \mathbf{y} \\ \mathbf{q} \end{Bmatrix} = \begin{Bmatrix} \mathbf{F_s} \\ \dfrac{j}{\omega\rho}\mathbf{F_f} \end{Bmatrix} \tag{11.34}$$

where ρ is the density of the fluid. Equation (11.34) has a symmetric matrix for which the inverse can be found. The resulting equation can then be solved for the vectors of the structural nodal displacements, $\mathbf{y}$, and the transformation variable for nodal pressures, $\mathbf{q}$, faster than the unsymmetric formulation in Equation (11.32). The nodal pressures, $\mathbf{p}$, can then be calculated using Equation (11.33).

A typical structural acoustic finite element model is shown in Figure 11.6, where the structural elements contain displacement degrees of freedom (DOFs), and most of the acoustic volume contains acoustic elements with only pressure degrees of freedom. At the interface between the acoustic fluid and the structure, a thin layer of elements, with pressure and displacement DOFs, one element wide is used to couple the structure to the fluid.

An acoustic domain is modelled by inter-connected acoustic elements as shown in Figure 11.6 in the lower left region of the model. The exterior boundary of the acoustic domain, in the lower left corner of the model, is not connected to another acoustic or structural element, and this arrangement is used to simulate an acoustically rigid wall. Modelling a free surface is achieved by selecting the nodes that lie on the free surface, and if the nodes belong to pure

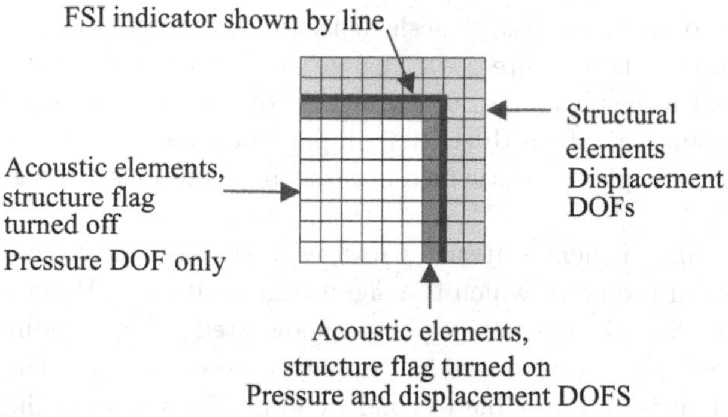

FIGURE 11.6 Structural acoustic finite element model with pressure formulated elements.

acoustic elements (i.e. only pressure DOF), then their pressure is set to zero. Alternatively, if the nodes on the free surface belong to acoustic elements with both pressure and displacement DOFs, a free surface is modelled by not defining any loads, displacement constraints or structure. Either form of acoustic elements can be used, but if the displacement at the nodes is required from the analysis, then acoustic elements with pressure and displacement DOFs should be used. The motion of the surface can then be obtained by examining the response of the nodes on the surface.

The top right corner of Figure 11.6 shows structural elements that are connected to acoustic elements that have both pressure and displacement DOFs. If the (velocity) response of the structure is already known, and one is interested in calculating the resulting acoustic pressure caused by the motion of the structure, then it is possible to conduct a one-way vibro-acoustic simulation, where the structural elements are omitted, and the nodes belonging to the acoustic elements that have both pressure and translational DOFs, which would have been in contact with the structure, are defined with the known (velocity) response. Although this simulation technique of omitting the structural elements reduces the number of elements in the model, it is only appropriate to use for one-way vibro-acoustic simulations, when it can be safely assumed that the reaction force of the fluid back onto the structure does not influence the motion of the structure. For example, this simulation technique could be used to analyse the radiation from an oscillating piston in an infinite plane baffle, where it is assumed that the motion of the piston will be unaffected by the fluid. Similarly, this technique could be used to estimate the sound radiation from a vibrating (heavy) structure where the response of the structure is known, and the structure is in contact with a (light) fluid. However, it would not be appropriate to use this one-way vibro-acoustic simulation technique to estimate the sound field inside a cavity surrounded with a thin-walled structure, as the motion of the fluid will cause the thin-walled structure to vibrate, and, in turn, will influence the sound field inside the cavity. Hence, this problem should be simulated using two-way (bi-directional) fluid-structure interaction, and structural elements should be included in the model.

11.2.4.2 Practical Aspects of Modelling Acoustic Systems with FEA

The following paragraphs describe some practical considerations that should be taken into account when modelling acoustic systems with finite element analysis.

Mesh density. FEA can be useful for low-frequency problems but, as the excitation frequency increases, the number of nodes and elements required in a model increases exponentially, requiring greater computational resources and taking longer to solve. A general rule of thumb is that acoustic models should contain 6 elements per wavelength as a starting point. Accurate models can still be obtained for lower mesh densities; however, caution should be exercised. At regions in a model where there is a change in the acoustic impedance, for example, where the diameter of a duct changes, at a junction of two or more ducts, or at the opening of the throat of a resonator into a duct, it is important that the mesh density is sufficiently high to ensure that the complex acoustic field in these discontinuous areas is modelled accurately.

Mean flow. Many finite element software packages with acoustic finite elements require that there is no mean flow of the fluid, which is a significant limitation. When there is mean flow of fluid, a different formulation of the wave equation is required, which modifies the propagation of the acoustic disturbance (due to "convection"), depending on whether the flow is rotational or irrotational. However, it is still possible to conduct finite element modelling for low speed fluid flow, where the compressibility effects of the flow are negligible, using "no flow" FEA software packages, but some assumptions that underpin the analysis will be violated. When there is mean flow in a duct, aero-acoustic phenomena might be important. For example, consider the situation of mean flow in a duct where the throat of a Helmholtz resonator attaches to the main duct, or over a sharp edge. It is possible that as air flows over the edge of the throat, noise will be generated, similar to blowing air over the top of a glass soda bottle. In some situations, the flow over the structure might cause vortex shedding. Standard finite element models are not able to model these effects.

If the flow speed is significant or it is expected that there will be aero-acoustic phenomena, consider the use of computational fluid dynamics (CFD) software to analyse the problem. However, this software also has limitations for the analysis of acoustic problems. Alternatively, some boundary element analysis software packages are able to model acoustic systems with mean flow, but are not able to model noise generation from shedding type phenomena.

Rigid or flexible boundaries. As described previously, an acoustic domain is formed by a contiguous region of acoustic finite elements. On the exterior boundary of a region of acoustic elements that are not in contact with other acoustic or structural elements, a rigid wall is simulated. It is appropriate to simulate a rigid wall where it is not expected that the motion of the boundary is likely to have any significant affect on the acoustics of the system. However, consider an automobile cabin comprising flexible sheet metal panels. Depending on the stiffness of these panels, acoustic excitation within the enclosure can cause the panels to vibrate, which in turn will affect the acoustic mode shapes and resonance frequencies of the enclosure. As highlighted above, modelling fluid-structure interaction can be computationally complex and can require substantial computer resources to solve. Hence fluid-structure interaction should only be included in the model if it is considered likely to have a significant effect on the results. A second subtle point is the consideration of re-radiation of structures in a different part of the acoustic model. Consider a duct with two Helmholtz resonators attached to reduce sound radiated from the duct exit, as shown in Figure 11.7. A simple acoustic model could be constructed assuming rigid walls. However, if parts of the system are in fact flexible, for example, the wall dividing the two resonators, then high sound pressure levels in the first resonator would vibrate the dividing wall, which would re-radiate sound into the second Helmholtz resonator, and affect the sound field. Alternatively, if the entire system were made from lightweight sheet metal, then vibrations could be transmitted along the duct structure and result in the re-radiation of sound into the main duct. These issues can occur if the sound pressure levels are high, and the structural walls are very flexible. In these circumstances, the walls should be modelled as flexible by using structural elements and a fluid-structure interaction analysis is required.

The results from analyses are usually the acoustic pressure at discrete locations. Sometimes this level of detail is required but often it is not; instead an indicative global sound pressure level may be required for assessment, which will require post-processing of the results from the analysis. The global sound pressure level can be obtained by averaging squared pressures over

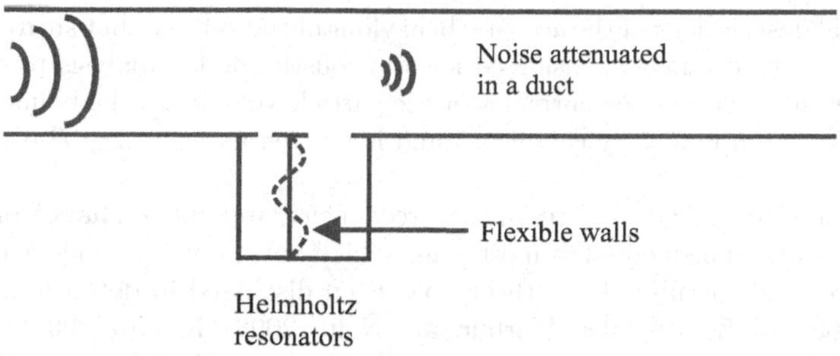

FIGURE 11.7 Re-radiation of sound from a structure into a different part of the acoustic model.

the region of interest and then using Equation (1.87) to convert the result to decibels. This may involve evaluating the sound pressure level at points over a region of interest, and then using Equation (1.96) and (1.86) to calculate the average sound pressure level.

For higher frequency problems, statistical energy analysis methods (see Section 11.3) may be more appropriate and significantly faster in obtaining a solution.

An Ansys script, `file:rigid_cavity_full.inp`, for the same problem of calculating the sound pressures inside a rigid enclosure, as analysed using the BEM method in Section 11.2.2, is available on the website listed in Appendix F.

11.2.5 Numerical Modal Analysis

The resonance frequencies of an acoustic enclosure can be calculated theoretically for simple rectangular or cylindrical shapes. For more complex geometries, a numerical method must be used. Two popular numerical methods that can be used to determine the resonance frequencies and mode shapes for enclosed spaces characterised by complex geometries are finite element and boundary element methods. For systems that involve a structure and an enclosed space, the resonance frequencies and mode shapes for each are usually calculated separately and then the interaction between them is evaluated using modal coupling analysis (see next section). The mode shapes of the enclosed space are calculated assuming rigid boundaries and the mode shapes of the structure are calculated assuming that it is vibrating in a vacuum. Of course it is possible to use BEM and FEA to evaluate the fully coupled system modes and thus avoid the need for modal coupling analysis. However, this uses an enormous amount of computing resources for most practical systems such as transportation vehicles.

The advantage of using BEMs compared to finite element analysis to calculate the resonance frequencies and mode shapes of an acoustic or structural system is that the dimension of the problem can be reduced by one: the resonance frequencies inside a three-dimensional volume can be re-written as a problem involving a surface integral. This results in smaller matrices compared with a finite element formulation. The disadvantage is that these matrices are full (meaning that each entry in the matrix is occupied) and often a non-linear eigenvalue solver is required.

On the other hand, for FEA, although the matrices are larger in size, they are sparse matrices (meaning that there are few entries off the diagonal), so standard linear eigenvalue solvers can be used, and hence the matrices can be solved relatively easily to yield eigenvalues that correspond to system resonance frequencies. These eigenvalues are then used to calculate mode shapes.

For acoustic problems, the resonance frequencies of a volume are given by the eigenvalues of the Helmholtz equation, written in terms of the velocity potential, ϕ, as:

$$\nabla^2\phi + k^2\phi = 0 \tag{11.35}$$

The right-hand side of Equation (11.35) is set to zero, meaning that there is no acoustic source within the volume. To solve Equation (11.35), the boundary conditions must be specified, which are mathematical descriptions of the acoustic behaviour of the surface that surrounds the volume. The typical boundary condition considered for an acoustic modal analysis problem is that the boundaries are rigid such that the normal acoustic particle velocity at the boundary is zero. This is called the "Neumann boundary condition" and is written as $u_n = 0$ or $\partial\phi/\partial\boldsymbol{n} = 0$ along the boundary.

The procedures for calculating resonance frequencies and mode shapes of an acoustic or structural system are implemented in most commercially available finite element and boundary element software, and the underlying theory has been discussed in detail in a number of text books (Wu, 2000b; von Estorff, 2000; Marburg and Nolte, 2008). Kirkup (2007) provides software on a CD-ROM that accompanies his textbook, which can be used for modal analysis of a volume using boundary element analysis.

11.2.6 Modal Coupling Using MATLAB®

Fahy (1985) and (Fahy and Gardonio, 2007, p. 418) describe equations for determining the coupled structural–acoustic displacement response of a system, $w(\boldsymbol{r}_S)$, at some location, $\boldsymbol{r}_S$, on the structure, in terms of the combination and summation of structural and acoustic mode shapes. The structural mode shapes are evaluated by assuming that the structure is vibrating in a vacuum and the acoustic mode shapes of the enclosure surrounded by the structure are evaluated by assuming that the surrounding structure is infinitely rigid. The structural displacement at frequency, ω, is described in terms of a summation over the *in vacuo* normal modes as:

$$w(\boldsymbol{r}_S,\omega) = \sum_{\ell=1}^{N_s} w_\ell(\omega)\varphi_\ell(\boldsymbol{r}_S) \tag{11.36}$$

where the time dependency term, $\mathrm{e}^{\mathrm{j}\omega t}$, has been omitted from both sides of the equation, as explained previously. The quantity, $\varphi_\ell(\boldsymbol{r}_S)$, is the mode shape of the ℓth structural mode at arbitrary location, $\boldsymbol{r}_S$, on the surface of the structure, and $w_\ell(\omega)$ is the modal participation factor (or displacement contribution, expressed as a fraction of the total contributions to the response from all modes) of the ℓth mode at frequency, ω. Theoretically, the value of N_s should be infinity, but this is not possible to implement in practice, so N_s is chosen such that the highest order mode considered has a resonance frequency between twice and four times that of the highest frequency of interest in the analysis, depending on the model being solved and the accuracy required. The N_s structural mode shapes and resonance frequencies can be evaluated using finite element analysis software, and the nodal displacements for a mode, ℓ, are described as a vector, $\boldsymbol{\varphi}_\ell$, and then collated into a matrix, $[\boldsymbol{\varphi}_1,\ \boldsymbol{\varphi}_2, ...\boldsymbol{\varphi}_{Ns}]$, for all the modes from 1 to N_s.

The acoustic pressure in the enclosure at frequency, ω, is described in terms of a summation of the contributions from the acoustic modes of the enclosure space with rigid boundaries as:

$$p(\boldsymbol{r},\omega) = \sum_{n=0}^{N_a} p_n(\omega)\psi_n(\boldsymbol{r}) \tag{11.37}$$

where the time dependency term has been omitted as it is not used in the analysis. The quantity, $\psi_n(\boldsymbol{r})$, is the acoustic mode shape of the nth mode at arbitrary location, $\boldsymbol{r}$, within the enclosure, and p_n is the modal participation factor (or acoustic pressure contribution of mode, n, expressed as a fraction of the total contributions to the response from all modes). Theoretically, the value of N_a should be infinity, but this is not possible to implement in practice so N_a is chosen such that the highest order mode considered has a resonance frequency between twice and four times that of the highest frequency of interest in the analysis, depending on the model being solved and the accuracy required. Note that the $n = 0$ mode is the acoustic bulk compression mode of the fluid in the enclosure and must be included in the summation. When conducting a modal analysis using finite element analysis software, the bulk compression mode of the fluid in the enclosure is the pressure response at 0 Hz. The N_a acoustic mode shapes and resonance frequencies can be evaluated using finite element analysis software, where the nodal pressures for a mode, n, are described as a vector, ψ_n, and then collated into a matrix, $[\psi_0, \psi_1, \psi_2, \ldots \psi_{Na}]$, for all the modes from 0 to N_a.

The equation for the coupled response of the structure for structural mode, ℓ, is:

$$\ddot{w}_\ell + \omega_\ell^2(1 + j\eta_\ell)w_\ell = \frac{S_S}{\Lambda_\ell}\sum_{n=0}^{N_a} p_n C_{n\ell} + \frac{F_\ell}{\Lambda_\ell} \tag{11.38}$$

where the frequency dependence of the pressures, forces and displacements is implicit; that is, these quantities all have a specific and usually different value for each frequency, ω. The quantity, ω_ℓ, is the structural resonance frequency for the ℓth mode, η_ℓ is the structural loss factor for the ℓth mode, which quantifies the hysteric damping of the structure, Λ_ℓ is the modal mass (see Equation (11.2)), F_ℓ is the modal force applied to the structure for the ℓth mode, S_S is the surface area of the structure and $C_{n\ell}$ is the dimensionless coupling coefficient between structural mode, ℓ, and acoustic mode, n, given by the integral of the product of the structural, φ_ℓ, and acoustic, ψ_n, mode shape functions over the surface of the structure, as:

$$C_{n\ell} = \frac{1}{S_S}\iint_{S_S} \psi_n(\boldsymbol{r}_s)\varphi_\ell(\boldsymbol{r}_s)\, \mathrm{d}S_S \tag{11.39}$$

The left-hand side of Equation (11.38) is a standard expression to describe the response of a structure in terms of its modes. The right-hand side of Equation (11.38) describes the forces that are applied to the structure in terms of modal forces. The first term describes the modal force exerted on the structure due to the acoustic pressure in the enclosure acting on the enclosing structure. The second term describes the forces that act directly on the structure. As an example, consider a point force, F_a, acting normal to the structure at nodal location, (x_a, y_a, z_a), for which the mode shapes and resonance frequencies have been evaluated using FEA. As the force acts on the structure at a point, the modal force, F_ℓ, at frequency, ω, for mode, ℓ, is:

$$F_\ell(\omega) = \varphi_\ell(x_a, y_a, z_a)F_a(\omega) \tag{11.40}$$

where $\varphi_\ell(x_a, y_a, z_a)$ is the ℓth structural mode shape at the nodal location, (x_a, y_a, z_a). Tangential forces and moment loadings on the structure can also be included in $F_\ell(\omega)$ and the reader is referred to Soedel (2004) and Howard (2007b) for more information.

The dimensionless coupling coefficient, $C_{n\ell}$, is calculated from finite element model results as:

$$C_{n\ell} = \frac{1}{S_S}\sum_{i=1}^{J_s} \psi_n(\boldsymbol{r}_i)\varphi_\ell(\boldsymbol{r}_i)S_i \tag{11.41}$$

where S_S is the total surface area of the structure in contact with the acoustic fluid, S_i is the area associated with the ith node on the surface (and hence $S_S = \sum_{i=1}^{J_s} S_i$), J_s is the total number of nodes on the surface, $\psi_n(\boldsymbol{r}_i)$ is the acoustic mode shape for the nth mode at node location, $\boldsymbol{r}_i$, and $\varphi_\ell(\boldsymbol{r}_i)$ is the mode shape of the ℓth structural mode at node location, $\boldsymbol{r}_i$. The area associated with each node of a structural finite element is sometimes available and, if so, can be readily extracted from the software. If the area is not available from the FEA model output, it can be calculated by using the nodal coordinates that form the elements.

The equation for the coupled response of the fluid (mode n) is given by:

$$\ddot{p}_n + 2\zeta_n\omega_n\dot{p}_n + \omega_n^2 p_n = -\left(\frac{\rho c^2 S_S}{\Lambda_n}\right)\sum_{\ell=1}^{N_s} \ddot{w}_\ell C_{n\ell} + \left(\frac{\rho c^2}{\Lambda_n}\right)\dot{Q}_n \tag{11.42}$$

where the frequency dependence of p_n, w_ℓ and $\dot{Q}_n$ is implicit. The quantity, ω_n, represents the resonance frequencies of the enclosed volume, ζ_n is the critical damping ratio for mode n, which quantifies the viscous damping of the fluid in the enclosure, ρ is the density of the fluid in the cavity, c is the speed of sound in the fluid, Λ_n is the modal volume (acoustic equivalent to structural modal mass), defined as the integration over the enclosure volume of the square of the mode shape function:

$$\Lambda_n = \iiint\limits_V \psi_n^2(\boldsymbol{r})\,\mathrm{d}V \tag{11.43}$$

and $\dot{Q}_n$ is a modal volume acceleration, which for a single source in the volume at location, $(x_b,\, y_b,\, z_b)$, is defined as:

$$\dot{Q}_n(\omega) = \psi_n(x_b, y_b, z_b)\,\dot{Q}_b(\omega). \tag{11.44}$$

where $\dot{Q}_b$ is the complex amplitude of the volume acceleration at nodal location $(x_b,\, y_b,\, z_b)$, and $\psi_n(x_b,\, y_b,\, z_b)$ is the nth mode shape at the nodal location $(x_b,\, y_b,\, z_b)$. If there is more than one source in the enclosed volume, the modal volume acceleration is obtained as the arithmetic sum of the modal volume accelerations corresponding to each location at which there is a source. A common definition for an acoustic source has units of volume velocity, which in this case, is Q_b, and hence the time derivative of this expression is the source volume acceleration $\dot{Q}_b$.

An important point to note is that because the acoustic mode shapes used in the structural-acoustic modal coupling method are for a rigid-walled enclosure, corresponding to a normal acoustic particle velocity at the wall surface equal to zero, the acoustic velocity at the surface resulting from the modal coupling method is incorrect (Jayachandran et al., 1998). However, the acoustic pressure at the surface is correct, and this is all that is required for correctly coupling the structural vibration and acoustic pressure modal equations of motion.

For simple systems such as rectangular, rigid-walled cavities and simple plates it is possible to write analytical solutions for the mode shapes and resonance frequencies. Anything more complicated than these simple structures nearly always involves the use of a discretised numerical model such as a finite element analysis, in which case, it is necessary to extract parameters from the finite element model to enable the calculation of the coupled response.

Cazzolato (1999) described a method to calculate the acoustic and structural modal masses from a finite element model. When using finite element analysis software to evaluate the acoustic pressure mode shapes, the vectors returned by the software can be normalised to either unity or to the mass matrix. By normalising the mode shapes to the mass matrix, the modal volume of the cavity can be obtained directly; that is:

$$\boldsymbol{\Psi}_n^{\mathrm{T}}[\mathbf{M}_{\mathrm{fea}}]\boldsymbol{\Psi}_n = 1 \tag{11.45}$$

where $\boldsymbol{\Psi}_n$ is the mass normalised mode shape function vector (see explanation below) for the nth mode and $[\mathbf{M}_{\mathrm{fea}}]$ is the fluid element mass matrix defined as:

$$[\mathbf{M}_{\mathrm{fea}}] = \frac{1}{c^2}\iiint\limits_{V_e} [\mathbf{N}][\mathbf{N}]^{\mathrm{T}}\,\mathrm{d}V_e \tag{11.46}$$

where $[\mathbf{N}]$ is the shape function for the acoustic element with a single pressure degree of freedom and V_e is the volume of the element. If the mode shape vectors are normalised to unity; that is, the maximum value in the vector is 1, then:

$$\hat{\boldsymbol{\Psi}}_n^{\mathrm{T}}[\mathbf{M}_{\mathrm{fea}}]\hat{\boldsymbol{\Psi}}_n = \frac{\Lambda_n}{c^2} \tag{11.47}$$

where $\hat{\boldsymbol{\Psi}}_n$ is the mode shape vector normalised to unity for the nth mode and Λ_n is the modal volume of the nth mode. It can be shown that the relationship between the mass normalised mode shape vector, $\boldsymbol{\Psi}_n$, and the unity normalised mode shape vector, $\hat{\boldsymbol{\Psi}}_n$, is (Ewins, 2000):

$$\boldsymbol{\Psi}_n = \frac{c}{\sqrt{\Lambda_n}} \hat{\boldsymbol{\Psi}}_n \tag{11.48}$$

Given that the maximum value of the unity normalised mode shape vector, $\hat{\boldsymbol{\Psi}}_n$, is one, the maximum element of the mass-normalised mode shape vector is equal to the ratio of the speed of sound in air to the square root of the modal volume:

$$\max(\boldsymbol{\Psi}_n) = \frac{c}{\sqrt{\Lambda_n}} \tag{11.49}$$

which can be rewritten as:

$$\Lambda_n = \frac{c^2}{\max(\boldsymbol{\Psi}_n^2)} \tag{11.50}$$

Hence, to extract the acoustic modal volume of a system using finite element analysis software, an acoustic modal analysis is conducted and the results are normalised to the mass matrix. Then Equation (11.50) is used to calculate the acoustic modal volumes for each mode. The unity normalised mode shapes can be calculated as:

$$\hat{\boldsymbol{\Psi}}_n = \frac{\boldsymbol{\Psi}_n}{\max(\boldsymbol{\Psi}_n)} \tag{11.51}$$

Equations (11.38) and (11.42) can form a matrix equation as:

$$\begin{bmatrix} \Lambda_\ell((1+\mathrm{j}\eta_\ell)\omega_\ell^2 - \omega^2) & -S_S[\mathbf{C}_{n\ell}] \\ S_S\omega^2[\mathbf{C}_{n\ell}]^{\mathrm{T}} & \dfrac{\Lambda_n}{\rho c^2}(\omega_n^2 + \mathrm{j}2\zeta_n\omega_n\omega - \omega^2) \end{bmatrix} \begin{bmatrix} \mathbf{w}_\ell \\ \mathbf{p}_n \end{bmatrix} = \begin{bmatrix} \mathbf{F}_\ell \\ \dot{\mathbf{Q}}_n \end{bmatrix} \tag{11.52}$$

where all the ℓ structural and n acoustic modes are included in the matrices, so that the square matrix on the left-hand side of Equation (11.52) has dimensions $(\ell+n) \times (\ell+n)$. The left-hand matrix in Equation (11.52) can be made symmetric by dividing all terms in the second row on both the LHS and RHS of the equation by $-\omega^2$. The structural modal participation factor, w_ℓ, for structural mode, ℓ, is an element of the vector, $\mathbf{w}_\ell$, and the acoustic modal participation factor, p_n, is an element of $\mathbf{p}_n$. $\mathbf{w}_\ell$ and $\mathbf{p}_n$ are frequency-dependent and can be calculated by pre-multiplying each side of Equation (11.52) by the inverse of the square matrix on the left-hand side. Once these factors are calculated, the vibration displacement of the structure can be calculated from Equation (11.36) and the acoustic pressure inside the enclosure can be calculated using Equation (11.37).

This modal coupling method can be implemented using MATLAB®, and scripts are available for download from MATLAB scripts for ENC (2024). An example involving the use of the scripts is described in Howard and Cazzolato (2014).

The method described above can be used to make predictions of the vibro-acoustic response of an enclosed system, but it does have limitations. One mistake that is commonly made is to make numerical calculations with an insufficient number of structural and acoustic modes. This problem affects all numerical methods involving the summation of modes to predict the overall response and has been known since the early 1970s. Cazzolato et al. (2005) demonstrated the errors that can occur with modal truncation and how it can lead to erroneous conclusions. As a start, the analyst should consider including structural and acoustic modes that have resonance frequencies up to two octaves higher than the frequency range of interest. Methods have been proposed to reduce the number of modes required to be included in the analysis by including the effects of the higher-order modes in a residue or pseudo-static correction term (Tournour and Atalla (2000); Gu et al. (2001); Zhao et al. (2002)).

The modal coupling method described above is applicable to vibro-acoustic systems where there is "light" coupling, such as between air and a structure. When the vibroacoustic system is a structure enclosing an air-filled space, applying the modal coupling method with light coupling involves calculating the resonance frequencies and mode shapes of the cavity, assuming that the cavity is enclosed with rigid walls, and the resonance frequencies and mode shapes of the structure, assuming that the structure is in a vacuum. Then the modes of the separated cavity and structure are combined, using Equation (11.52), to determine the coupled response of the system. However, if the vibro-acoustic response of a system is to be calculated where there is "heavy" coupling due to the fluid loading, such as between water and a structure, the modes of the cavity and structural systems are inter-related and cannot be separated. Therefore, this modal coupling method will generate erroneous results because it does not account for the mass loading that the fluid applies on the structure, nor the coupling between fluid modes. In this case, FEA, involving two-way fluid-structure interaction, is an appropriate analysis technique, which is described in Section 11.2.4.

One of the main advantages of using the modal coupling method is that the computational time taken to solve the system of equations is significantly less than conducting a full fluid-structure interaction analysis using finite element analysis. This is important if optimisation studies involving many FEA evaluations are conducted to determine an optimum solution.

11.2.6.1 Acoustic Potential Energy

The acoustic potential energy $E_p(\omega)$ is a useful measure of the acoustic energy contained within a cavity at frequency, ω. This measure can be used to evaluate the effectiveness of noise control in an enclosure and is given by:

$$E_p(\omega) = \frac{1}{4\rho c^2} \iiint\limits_V |p(\boldsymbol{r}, \omega)|^2 \, \mathrm{d}V \tag{11.53}$$

which can be implemented in a finite element formulation as:

$$E_p(\omega) = \frac{1}{4\rho c^2} \sum_{n=1}^{N_a} p_n^2(\omega) V_n \tag{11.54}$$

where p_n is the acoustic pressure at the nth node and V_n is the volume associated with the nth node. This equation can be rearranged so that the acoustic potential energy is calculated in terms of the modal pressure amplitudes as:

$$E_p(\omega) = \sum_{n=1}^{N_a} \Lambda_n |p_n(\omega)|^2 = \mathbf{p}^{\mathrm{H}} \boldsymbol{\Lambda} \mathbf{p} \tag{11.55}$$

where $\mathbf{p}$ is a $(N_a \times 1)$ vector of nodal acoustic pressures, and $\boldsymbol{\Lambda}$ is an $(N_a \times N_a)$ diagonal matrix for which the diagonal terms are:

$$\boldsymbol{\Lambda}(n, n) = \frac{\Lambda_n}{4\rho c^2} \tag{11.56}$$

and Λ_n is defined in Equation (11.43).

11.3 High-Frequency Region: Statistical Energy Analysis

In the high-frequency region, a method generally known as statistical energy analysis (SEA) (Lyon, 1975; Lyon and DeJong, 1995; Sablik, 1985) is used to calculate the flow and storage of vibration and acoustic energy in a complex system, which is made up of two or more subsystems.

An SEA subsystem usually consists of a single wave type (bending, torsional, shear or longitudinal/axial) in a single structural element such as a beam, plate or cylinder or a longitudinal wave type in an acoustic system such as an enclosure. Although SEA has been used extensively in the automotive and aerospace industries, it suffers from inaccuracies due to the statistical nature of the analysis requiring sufficient resonant modes to be excited in each subsystem and also due to the assumptions required to make the analysis tractable, such as weak coupling between subsystems. However, SEA is useful for evaluating the effect of structural design changes on the vibration amplitudes and sound radiation associated with a structural system, even though errors in absolute vibration levels and overall radiated sound power levels are difficult to quantify.

Energy is input to a subsystem by an external vibration excitation source or incident sound field and stored in the vibrational modes of the structural subsystems or acoustic modes of acoustic subsystems. Energy is dissipated by mechanical damping in a structural subsystem or viscous damping in an acoustic subsystem, and is also transferred between the various subsystems, across interconnecting joints. The energy input to the system by one or more excitation forces or sound fields is equal to the total energy in the system (consisting of all of the subsystems) plus that lost due to damping and sound radiation. The energy is assumed to be equally distributed among all vibration modes, so that each mode of a subsystem or system of connected subsystems has equal modal energy. For a given subsystem, the total vibratory power flowing into it is equal to the power dissipated by the subsystem plus the power flowing out of it. Energy is also dissipated at joints between adjacent structural subsystems. By setting up appropriate matrix equations, the modal energy in each part of the structure can be determined and this modal energy is directly related to the area, time and band averaged mean square vibration velocity of a structural subsystem or the area, time and band averaged mean square sound pressure in an acoustic subsystem. The vibration velocity levels of structural subsystems together with their radiation efficiencies can be used to estimate the sound power radiated by a particular subsystem.

For a satisfactory outcome from an SEA analysis, it is necessary to consider frequency-band-averaged data, using at least 1/3-octave bandwidths, and preferably octave bandwidths. It is recommended that there should be at least three modes resonant in the frequency band being considered for each subsystem involved in the analysis, and the modal overlap (see Section 6.3.3) should be at least unity and even higher if possible. A modal overlap factor greater than unity will reduce the variance of the estimated response, but it is not a requirement for the successful application of SEA (Shorter and Langley, 2005; Lyon, 1995). For further discussion on this topic, readers should refer to Mace (2003); Renji (2004); Wang and Lai (2005).

One of the concepts behind SEA is that interconnected systems transfer vibro-acoustic energy between them and the total energy in the system must always be fully accounted for. Consider two generic vibro-acoustic systems as shown in Figure 11.8, which have a mechanism to transfer energy between them. Examples of a vibro-acoustic system could be a vibrating panel, or an enclosure such as a room.

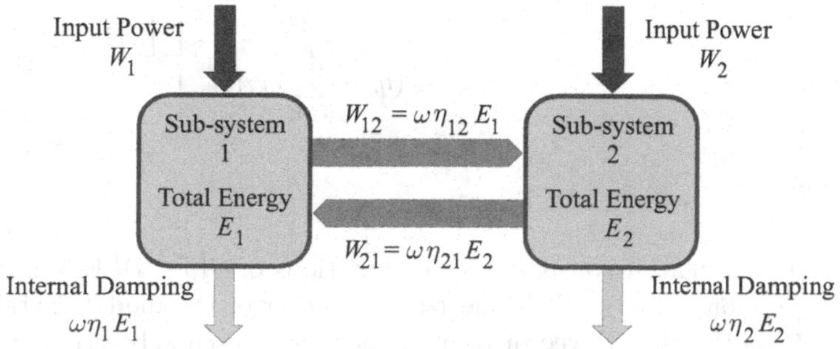

FIGURE 11.8 Two vibro-acoustic systems that are connected and have power flowing into (or out of) them.

Imagine that a system boundary is drawn around subsystem 1 in Figure 11.8. An equation describing the energy balance of subsystem 1 is:

$$W_1 + W_{21} = \omega\eta_1 E_1 + W_{12} \tag{11.57}$$

where W_1 is the external input power entering subsystem 1, W_{21} is the power coming from subsystem 2 into subsystem 1, $\omega\eta_1 E_1$ is the power dissipated in subsystem 1 by damping mechanisms, η_1 is the dissipative or damping loss factor (DLF) in subsystem 1, ω (radians/sec) is the centre frequency of the band, E_1 (joules) is the total energy in subsystem 1, W_{12} is the power (watts) that leaves subsystem 1 and goes into subsystem 2. The left-hand side of Equation (11.57) is the power entering the subsystem, and the right-hand side is the power that is lost by the subsystem by damping and power transferred to a connected subsystem. Equation (11.57) can be re-arranged to give:

$$W_1 = [\omega\eta_1 E_1] + [\omega\eta_{12} E_1] - [\omega\eta_{21} E_2] \tag{11.58}$$

where $W_{12} = [\omega\eta_{12} E_1]$, $W_{21} = [\omega\eta_{21} E_2]$, η_{12} is the coupling loss factor from subsystem 1 to subsystem 2 and η_{21} is the coupling loss factor from subsystem 2 to subsystem 1. The coupling loss factor (CLF) is a ratio of the energy that is transferred from one subsystem to another, which varies with frequency, and can be determined theoretically, numerically using finite element analysis, or experimentally. The determination of CLFs is discussed in detail in Section 11.3.6. The difference between DLFs and CLFs is that the former is a measure of energy lost by a subsystem due to dissipation within the subsystem and the latter is a measure of the energy transferred to another system across a common boundary.

A complex vibro-acoustic system modelled using the SEA framework can be thought of as a network of subsystems, where power flows in and out, and energy is stored within the systems. Altering the (CLFs) or DLFs of the system has the effect of re-routing the power distribution throughout the network.

Consider the general case of k interconnected subsystems, where $\eta_{i,j}$ represents the coupling loss factor for power transmission between subsystem i and subsystem j, η_i represents the internal damping loss for subsystem, i, n_i represents the modal density, in modes/radian/sec for subsystem, i, W_i represents the external power injected into subsystem i and E_i represents the steady state energy in subsystem i. A system of energy balance equations from Equation (11.58) can be formed and put into a matrix equation as:

$$\omega \begin{bmatrix} (\eta_1 + \sum_{i\neq 1}^{k} \eta_{1i})n_1 & (-\eta_{12}n_1) & \cdots & (-\eta_{1k}n_1) \\ (-\eta_{21}n_2) & (\eta_2 + \sum_{i\neq 2}^{k} \eta_{2i})n_2 & \cdots & (-\eta_{2k}n_2) \\ \vdots & \vdots & \ddots & \vdots \\ (-\eta_{k1}n_k) & \cdots & \cdots & (\eta_k + \sum_{i\neq k}^{k} \eta_{ki})n_k \end{bmatrix} \begin{bmatrix} E_1/n_1 \\ E_2/n_2 \\ \vdots \\ E_k/n_k \end{bmatrix} = \begin{bmatrix} W_1 \\ W_2 \\ \vdots \\ W_k \end{bmatrix} \tag{11.59}$$

or in short form as:

$$\omega[\mathbf{C}][\mathbf{E}] = [\mathbf{W}] \tag{11.60}$$

where $[\mathbf{C}]$ is the $(k \times k)$ matrix consisting of combinations of CLFs, DLFs and modal densities as indicated in Equation (11.59), $[\mathbf{E}]$ is the $(k \times 1)$ vector of the modal energies within each subsystem and $[\mathbf{W}]$ is the $(k \times 1)$ vector of input powers to each subsystem. The input power to each subsystem is known (from measurements or from the problem description), and hence the total energy, E_i, in subsystem, i, can be calculated by multiplying the modal energy by the subsystem modal density (modes/radian/sec), where the modal energy, $[\mathbf{E}]$, is calculated using:

$$[\mathbf{E}] = \frac{1}{\omega}[\mathbf{C}]^{-1}[\mathbf{W}] \tag{11.61}$$

Determination of subsystem input powers, DLFs, modal densities and CLFs is discussed in Sections 11.3.3, 11.3.4, 11.3.5 and 11.3.6, to follow. The use of mean square vibration velocities to calculate the sound power radiated by a structural subsystem is discussed in Section 11.3.1.

11.3.1 Subsystem Responses

Following the evaluation of Equation (11.60) and calculation of the energies in each subsystem, the usual practice is to determine the mean square vibration velocity or the mean square acoustic pressure response of all or some subsystems. For structural subsystems the mean square velocity, $\langle v^2 \rangle_{St\Delta}$, of the subsystem is:

$$\langle v^2 \rangle_{St\Delta} = E/M \tag{11.62}$$

where E is the total energy (= modal energy × modal density) in the subsystem, and M is the total mass of the subsystem. For a structural subsystem with a surface (usually plane) of area, S_S, the band-averaged radiated sound power, $W_{\Delta S}$, is calculated using:

$$W_{\Delta S} = S_S \rho c \sigma_{\Delta S} \langle v^2 \rangle_{St\Delta} \tag{11.63}$$

The subscript, Δ, denotes frequency band average, and $\sigma_{\Delta S}$ is the band-averaged radiation efficiency for the structure. Radiation efficiencies and their calculation for various structural elements are discussed in Section 4.15 for resonant response (mechanical excitation) and in Section 6.8.2 for forced response (excitation by an incident acoustic field).

For acoustic subsystems the space, time and band averaged square pressure is:

$$\langle p^2 \rangle_{St\Delta} = E\rho c^2/V \tag{11.64}$$

Analysis of vibro-acoustic problems using SEA can be conducted using commercial or free SEA software or they can be analysed using MATLAB® or spreadsheet software packages such as Microsoft Excel or OpenOffice Calc.

11.3.2 Subsystem Input Impedances

Input Impedances are used to calculate the input power resulting from a point or line force or an incident acoustic field, as will become apparent in Section 11.3.3. They are also used to calculate coupling loss factors (see Section 11.3.6). Point, line and area impedances for various structural and acoustic subsystems are provided in Tables 11.1 to 11.5.

Point moment impedances are reduced by a factor of 2 for each pinned boundary, compared to the case where there is an unpinned boundary; and the beam moment impedance is reduced by a factor of 4 at an end that is free to rotate and translate, compared to the case of an infinitely long beam (Lyon and DeJong, 1995, p. 201). This can be seen in Table 11.1 in the equation for the impedance of a point force applied normal to the axis of a thin, infinite beam, compared with the equation for a point force applied normal to the axis of a thin, semi-infinite beam, which differs by a factor of 4.

To estimate structural input powers arising from externally applied point forces, we need to know the structural point force impedance at the point of application of the force or the point moment impedance at the point of application of the moment. The space and frequency averaged point impedance of a finite system is well approximated by the point impedance of the equivalent infinite system. The infinite system impedance is used to calculate the externally acting input

TABLE 11.1 Summary of point impedance formulae, $Z_{F\infty}$ and $Z_{T\infty}$, for isotropic thin beams excited by a point force or point torque (or moment), respectively (Cremer et al., 2005, p. 273, 288) and (Lyon and DeJong, 1995, p. 201-202), where variables are defined following Table 11.5

Wave type	Structural element	$Z_{F\infty}$	$Z_{T\infty}$
Longitudinal	Infinite thin beam 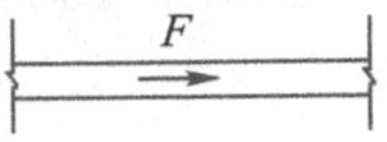	$2S_b\sqrt{E\rho_m}$ $= 2\rho_m S_b c_{LII}$	—
Longitudinal	Semi-infinite thin beam 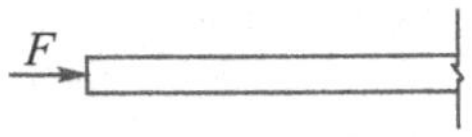	$S_b\sqrt{E\rho_m}$ $= \rho_m S_b c_{LII}$	—
Torsional	Infinite shaft 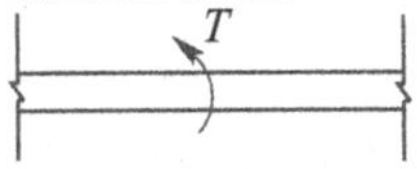	—	$2J_x\sqrt{G\rho_m(J'/J_x)} = 2\rho_m J_x c_T$
Torsional	Semi-infinite shaft 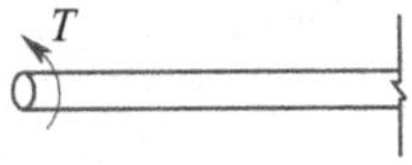	—	$J_x\sqrt{G\rho_m(J'/J_x)} = \rho_m J_x c_T$
Bending	Infinite thin beam 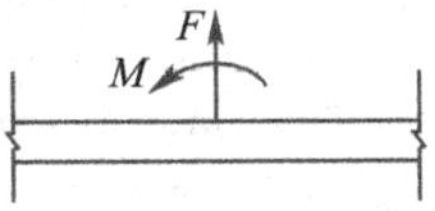	$2\rho_m S_b c_B(1+\mathrm{j})$	$2\rho_m S_b c_B(1-\mathrm{j})/k_B^2$
Bending	Semi-infinite thin beam 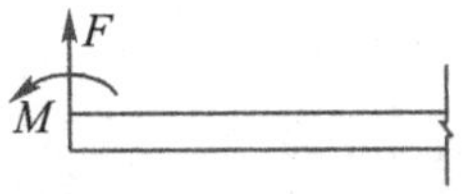	$\frac{1}{2}\rho_m S_b c_B(1+\mathrm{j})$	$\frac{1}{2}\rho_m S_b c_B(1-\mathrm{j})/k_B^2$

force or moment in situations where the input force or moment is applied at distances that are at least 1/4 of a wavelength from a structural boundary. The same applies for junction impedances used to calculate coupling loss factors (Lyon and DeJong, 1995, p. 201). For cases for which the junction or location of an applied external force is less than 1/4 of a wavelength from a structural boundary, the resulting semi-infinite impedance values can be derived from the corresponding infinite impedance values by reducing the infinite impedance value by a factor of 2 for each degree of freedom (DOF) that is unrestrained. Thus, for a free beam end, the factor is 2 for torsional or axial excitation or transmission (as these only involve the one displacement or rotational DOF) and 4 for bending wave excitation and transmission (as this involves a displacement and a rotational DOF).

For structures for which there is no analytical expression for the equivalent infinite system impedance, the frequency and space-averaged, real part of the point force impedance may be found from the average modal density over the bandwidth under consideration, derived here using (Lyon and DeJong, 1995, p. 200):

$$\langle \mathrm{Re}\{Z_F\}\rangle_{S,\Delta} = \frac{2M}{\pi n(\omega)} \tag{11.65}$$

where M is the total mass of the subsystem and $n(\omega)$ is the modal density.

TABLE 11.2 Summary of point impedance formulae, $Z_{F\infty}$ and $Z_{M\infty}$, for thin plates excited by a point force or a point moment (Cremer et al., 2005, p. 275, 288), (Lyon and DeJong, 1995, p. 201, 202) and (Vér, 2006, p. 499), where variables are defined following Table 11.5

Wave type	Structural element	$Z_{F\infty}$	$Z_{M\infty}$
In-plane	Infinite, thin isotropic plate 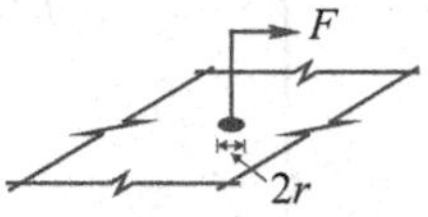	$8\pi m f r^2 \left(1 - \dfrac{\mathrm{j}c_{LI}}{2\pi f r}\right)$	—
In-plane	Semi-infinite, thin isotropic plate 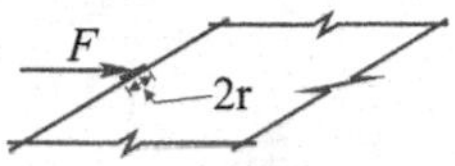	$4\pi m f r^2 \left(1 - \dfrac{\mathrm{j}c_{LI}}{2\pi f r}\right)$	—
Bending	Infinite, thin isotropic plate 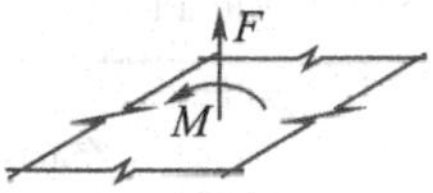	$8\sqrt{Bm}$ $= 8\omega m/k_B^2$ $= 2.3 m c_{LI} h$ $= 8 m \kappa_p c_{LI}$	$\dfrac{16\omega m}{k_B^4}\left[1 - \dfrac{4\mathrm{j}}{\pi}\log_e(0.89 k_B r)\right]^{-1}$
Bending	Semi-infinite, thin isotropic plate 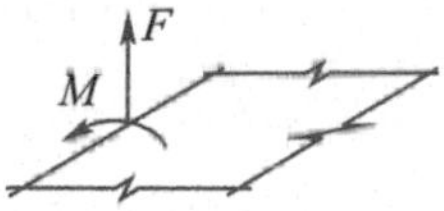	$3.5\sqrt{Bm}$ $= 3.5\omega m/k_B^2$ $= m c_{LI} h$	$\dfrac{5.3\omega m}{k_B^4}\left[1 - 1.46\mathrm{j}\log_e(0.89 k_B r)\right]^{-1}$
Bending	Infinite, thick isotropic plate ($h > 0.8c_s/f$) 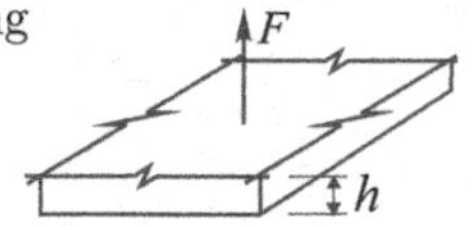	$\dfrac{Gc_s}{\omega^2}\left\{\dfrac{0.063}{H^2} + \dfrac{1}{8}\left(\dfrac{H}{H+1.6}\right)^2 + \mathrm{j}\left(\dfrac{0.001}{H^{1.3}} + \dfrac{0.06\pi c_s}{r\omega}\right)\right\}^{-1}$ Assumes that Poisson's ratio, $\nu = 0.3$.	—
Bending	Infinite, corrugated plate 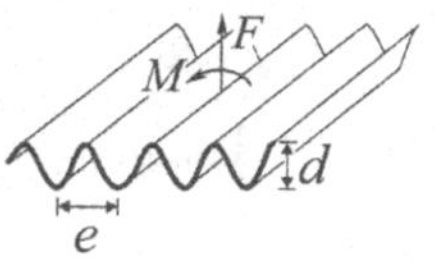	$8[m^2 B_a B_b]^{1/4}$ $\lambda_B \gg e$ (see page 421 for definitions of B_a and B_b)	$\dfrac{16\omega m}{k_B^4}\left[1 - \dfrac{4\mathrm{j}}{\pi}\log_e(0.89 k_B r)\right]^{-1}$ $k_B = \left(\dfrac{\omega^4 m^2}{B_a B_b}\right)^{1/8}$

TABLE 11.3 Summary of point impedance formulae, $Z_{F\infty}$, for isotropic thin cylinders excited by a point force (Vér, 2006; Cremer et al., 2005), where $k_B = \omega/c_B$, with other variables are defined following Table 11.5

Wave type	Structural element	$Z_{F\infty}$ (Cremer et al., 2005, p. 270, 272)	$Z_{F\infty}$ (Vér, 2006, p. 500)
Bending	Infinite, thin tube 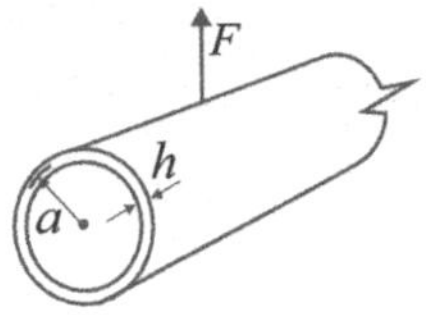	$\begin{cases} 2.3c_{LI}\rho_m h^2; & (f/f_r > 2) \\ \dfrac{2\pi a\rho_m h\sqrt{\omega c_{LI}a/\sqrt{2}}}{(1-\mathrm{j})}; & (f/f_r < 0.77h/a) \\ \dfrac{2.3c_{LI}\rho_m h^2}{0.66}\sqrt{\dfrac{f_r}{f}}; & (0.77h/a < f/f_r < 0.6) \end{cases}$	$\begin{cases} 2.3c_{LI}\rho_m h^2; & (f/f_r > 1.5) \\ 2.3c_{LI}\rho_m h^2 (f_r/f)^{2/3}; & (f/f_r < 0.7) \end{cases}$

TABLE 11.4 Summary of line impedance formulae, $Z_{F\infty}^{\text{line}}$ and $Z_{M\infty}^{\text{line}}$, for beams and isotropic thin plates excited by a line force or moment, respectively (Lyon and DeJong, 1995, p. 201-202), where $k_B = \omega/c_B$ and all variables are those for the plate, defined following Table 11.5

Wave type	Structural element	$Z_{F\infty}^{\text{line}}$	$Z_{M\infty}^{\text{line}}$
Bending	Infinite thin beam 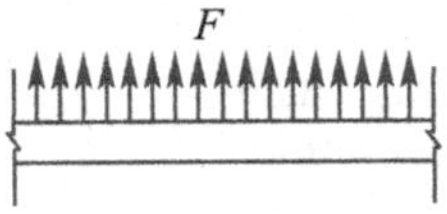	$\mathrm{j}2\pi f\rho_m S_b$	
Longit-udinal	Infinite thin plate 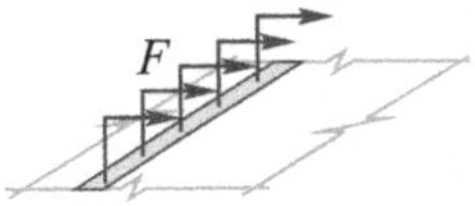	$2h\sqrt{E\rho_m}$ $= 2\rho_m hc_{LI}$	—
Longit-udinal	Semi-infinite thin plate 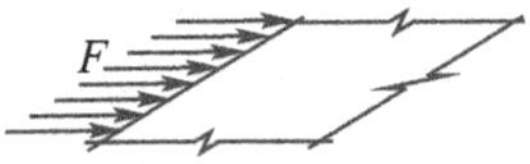	$h\sqrt{E\rho_m}$ $= \rho_m hc_{LI}$	—
Bending	Infinite thin isotropic plate 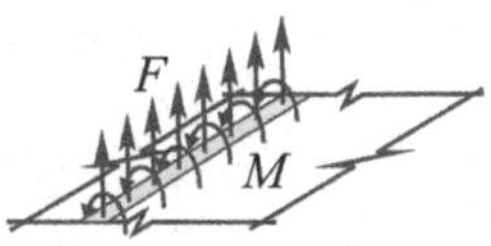	$2\rho_m hc_B(1+\mathrm{j})$	$2\rho_m hc_B(1-\mathrm{j})/k_B^2$
Bending	Semi-infinite, thin isotropic plate 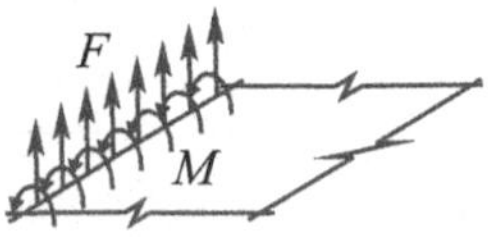	$\frac{1}{2}\rho_m hc_B(1+\mathrm{j})$	$\frac{1}{2}\rho_m hc_B(1-\mathrm{j})/k_B^2$

TABLE 11.5 Summary of point acoustic impedance formulae $Z_A = p/(S_d u_0)$ for a 1-D duct (plane wave propagation) and a 3-D acoustic space (Lyon and DeJong, 1995, p. 201)

Acoustic space type	Illustration	Z_A
1-D duct	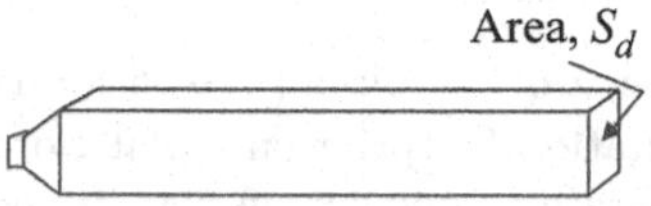	$2\rho c/S_d$
3-D acoustic space. For a source mounted in a wall, the impedance is multiplied by 2 and for a source mounted in a room corner the impedance is multiplied by 8.	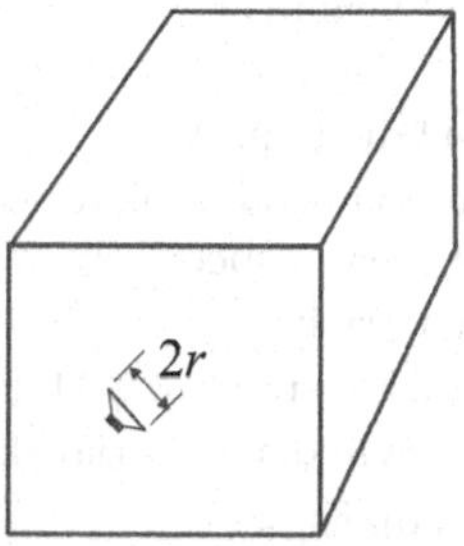	$\frac{\pi \rho f^2}{c}\left(1+\frac{2\mathrm{j}}{2kr}\right)$ Source more than 1/4 wavelength from a room surface.

The variables in Tables 11.1 to 11.5 are defined as follows:

a = cylinder radius (inner radius plus half the wall thickness) or radius of circular beam (m);

B = bending stiffness for a beam = EJ_y (kg m^3 s^{-2}) and bending stiffness/unit width for an isotropic plate = $Eh^3/[12(1-\nu^2)]$ (kg m^2 s^{-2});

c = speed of sound in the gas in a 3-D acoustic space or 1-D duct (m s^{-1});

c_B = bending wave speed (= ω/k_B) (m s^{-1});

$c_{LII} = \sqrt{E/\rho_m}$ = longitudinal wave speed for a beam or "plane-stress" longitudinal wavespeed for a thin-walled cylinder (used for modal density calculations) (m s^{-1});

$c_{LI} = \sqrt{E/[\rho_m(1-\nu^2)]}$ = longitudinal wave speed for a plate or thin-walled cylinder (m s^{-1});

c_s = shear wave speed (= $\sqrt{G/\rho_m}$) (m s^{-1});

c_T = torsional wave speed (m s^{-1});

d = peak to trough height of the corrugation profile for a thin corrugated plate (m);

E = Young's modulus of elasticity for the beam or plate material (kg m^{-1} s^{-2});

f (= $\omega/(2\pi)$) = excitation frequency (Hz);

$f_r = c_{LI}/(2\pi a)$ = ring frequency (Hz);

G = shear modulus of the beam or plate material = $E/[2(1+\nu)]$ (kg m^{-1} s^{-2});

h = plate or thin-walled cylinder thickness (m);

$H = \omega h/(2c_s)$;

J_x = polar second moment of area of the beam cross-section about the longitudinal x-axis (sometimes referred to as the polar area moment of inertia) = $\pi d^4/32$ for a circular-section solid shaft, where d is the shaft diameter; and = $bh(b^2+h^2)/12$ for a rectangular-section shaft of cross-sectional dimensions $b \times h$, (m^4);

J_y = second moment of area of the beam cross-section about the lateral x-axis (sometimes referred to as the area moment of inertia) = $\pi d^4/16$ for a circular-section solid shaft, where d is the shaft diameter; and = $bh^3/12$ for a rectangular section shaft of cross-sectional dimensions $b \times h$, where the b dimension is parallel to the x-axis, (m^4);

J' = torsion constant for the beam cross-section (equal to the polar second moment of area, J_x, for circular section beams) (m^4);

k = wavenumber in the gas in a 3-D acoustic space or 1-D duct (m^{-2});

k_B = bending wavenumber = $[\rho_m S_b \omega^2/(EJ_y)]^{1/4}$ for a beam and
$k_B = (\omega^2 m/B)^{1/4} = \left[\dfrac{12\omega^2 m(1-\nu^2)}{Eh^3}\right]^{1/4}$ for an isotropic plate (m^{-2});

k_s = shear wavenumber = $\sqrt{\dfrac{\rho_m \omega^2}{G}}$ (m^{-2});
m = mass per unit plan area of the plate (= $\rho_m h$ for a flat plate) ($\mathrm{kg\,m^{-2}}$);
$2r$ = diameter of the excitation footprint on a plate, or moment arm, or distance between the two forces used to generate a moment on a plate, or diameter of a circular acoustic source or for a square acoustic source, $2r \approx \sqrt{A}$ (m), where A is the area of the source (m^2);
S_b = cross-sectional area of the 1-D beam (m^2);
S_p = Area of the 2-D plate (m^2);
S_d = cross-sectional area of the 1-D duct (m^2);
$\kappa_p = \sqrt{J_y/S_p} = h/\sqrt{12}$ = radius of gyration of plate section for bending, where J_y and S_p are the second moment of area and area, respectively, of the plate cross-section (m);
ν = Poisson's ratio for the plate material;
ρ = density of the gas in a 3-D acoustic space or 1-D duct ($\mathrm{kg\,m^{-3}}$);
ρ_m = density of the beam, plate or cylinder material ($\mathrm{kg\,m^{-3}}$);
$\omega = 2\pi f$ = excitation frequency (radians/sec)

For a 3D acoustic space the frequency and space-averaged point resistance is:

$$\langle \mathrm{Re}\{Z_A\}\rangle_{S,\Delta} = \frac{\pi n(\omega)\rho c^2}{2V} \approx \frac{\pi \rho f^2}{c} \tag{11.66}$$

where V is the volume of the acoustic space.

For beams excited axially or torsionally or plates excited in bending, the imaginary part of the point impedance is equal to 0. For beams in bending, the imaginary part is equal to the real part. For in plane excitation of plates, the imaginary impedance is a function of the area over which the force acts. For this case, the imaginary impedance is set equal to zero if the area over which the force acts is unknown.

Line impedances are needed to calculate coupling loss factors for plates connected at their edges (see Lyon and DeJong, 1995, p. 202).

Area impedances are needed for calculating coupling loss factors for plates radiating into a room or into free space. They are also needed the calculate the external input power if the external forcing function is an acoustic field. The area impedance of an acoustic space is ρc. The area impedance of a plate is $\mathrm{j}2\pi f \rho_m h$ (Lyon and DeJong, 1995, p. 202), where h is the plate thickness and ρ_m is the plate density.

The area impedance (reciprocal of admittance) for an acoustic space is $Z_{F\infty}^{\mathrm{area}} = \rho c$ (Lyon and DeJong, 1995, p. 202).

The coupling loss factor formulation requires a real impedance for the calculation of the transmission coefficient. This is realised in practice by using the magnitude of the impedance in place of the real impedance when and only when the real impedance is zero.

11.3.3 Subsystem External Input Power

Structural systems. The impedance presented to a point excitation source is not simply the point impedance of the substructure. It also depends on the point impedance of any other substructures connected to the substructure to which the external force is applied. In these cases, it may be preferable to estimate the input power using a measurement (or estimate) of the in situ reverberation time of the substructure to which the force is applied and the space, frequency and time-averaged, squared velocity, $\langle v^2\rangle_{S,\Delta,t}$, over the substructure. Thus, the input power is given by (Hopkins and Robinson, 2014):

$$\Pi_{\text{in}} = \omega \eta M \langle v^2 \rangle_{S,\Delta,t} \tag{11.67}$$

where $\omega = 2\pi f$, M is the subsystem mass and η is the subsystem loss factor, determined using the subsystem reverberation time or one of the other methods described in Section 11.3.4.

The frequency-averaged, subsystem input power can also be obtained from the real part of the equivalent infinite system point input impedance at the point of application of the force or moment using:

$$\Pi_{\text{in}} = \frac{1}{2}\text{Re}\{\hat{F}^* \hat{v}\} = \langle F^2 \rangle_{\Delta,t} \text{Re}\{Y_{F,\infty}\} = \langle v^2 \rangle_{\Delta,t} \text{Re}\{Z_{F,\infty}\} \quad \text{(watts)} \tag{11.68}$$

where $F = \hat{F}\text{e}^{\text{j}\omega t}$, $v = \hat{v}\text{e}^{\text{j}\omega t}$, and $Z_{F,\infty}$ and $Y_{F,\infty}$ are the point impedance and point admittance, respectively, at the point on the structure where the force is applied. The structural input admittance is the reciprocal of the input impedance.

For moment excitation, the moment impedances and mobilities of the structure are used and the force, F, is replaced with the moment, M, and the velocity, v, at the point of application of the force, is replaced with the angular velocity, $\dot{\theta}$.

Acoustic systems. For acoustic systems for which the acoustic pressure, p, has a constant spectral amplitude over the band, the acoustic power input, Π_{in}, can be expressed in terms of the acoustic impedance, Z_A, presented to the sound source as (Lyon and DeJong, 1995, p. 207):

$$\Pi_{\text{in}} = \langle pv \rangle = \frac{\langle p^2 \rangle}{\text{Re}\{Z_A\}} \quad \text{(watts)} \tag{11.69}$$

For acoustic systems for which the acoustic volume velocity, v, has a constant spectral amplitude over the band, the acoustic power input, Π_{in}, is (Lyon and DeJong, 1995, p. 207):

$$\Pi_{\text{in}} = \langle v^2 \rangle \text{Re}\{Z_A\} \quad \text{(watts)} \tag{11.70}$$

For a diffuse pressure field (mean square pressure, $\langle p^2 \rangle$) exciting a plate, and for the case of the modal energy in the plate being much less than that in the acoustic space providing the diffuse pressure field, the input power from the room to the plate is (Lyon and DeJong, 1995, p. 208):

$$\Pi_{\text{in}} = \frac{2\pi^2 \langle p^2 \rangle n_p(\omega)}{\rho_m h k^2} \sigma_{\text{rad}} \quad \text{(watts)} \tag{11.71}$$

where $n_p(\omega)$ is the plate modal density, ρ_m is the plate material density, h is the plate thickness, k is the wavenumber and σ_{rad} is the plate radiation efficiency, given by Equation (11.98). Instead of using Equation (11.71), it is often preferable to model the input power as power applied to the acoustic space by a sound source, such that the reverberant mean square pressure so generated will be incident on the plate. In this case, the power input to the acoustic space can then be calculated directly from the reverberant field sound pressure using (see Equation (6.43)):

$$\Pi_{\text{in}} = \frac{\langle p^2 \rangle S \bar{\alpha}}{4\rho c(1 - \bar{\alpha})} \quad \text{(watts)} \tag{11.72}$$

where S is the total area of the room walls, floor and ceiling, and $\bar{\alpha}$ is the room Sabine absorption coefficient.

For a vibrating plate (of area S_p with a mean square vibration velocity, $\langle v^2 \rangle$) exciting a diffuse sound pressure field in a room, where the modal energy in the plate is much greater than the modal energy in the room, the input power is (Lyon and DeJong, 1995, p. 207):

$$\Pi_{\text{in}} = \langle v^2 \rangle \rho c S_p \sigma_{\text{rad}} \quad \text{(watts)} \tag{11.73}$$

For excitation of a plate by a turbulent boundary layer having a mean square pressure of $\langle p_{\rm TBL}^2 \rangle$, averaged over the frequency band, the input power is (Lyon and DeJong, 1995, p. 208):

$$\Pi_{\rm in} = \frac{0.75 S_p \langle p_{\rm TBL}^2 \rangle U}{\pi^2 f \rho_m h c_B} \quad \text{(watts)} \quad (c_B < 0.75U) \tag{11.74}$$

where c_B is the plate bending wave speed, U is the free stream speed in the fluid, S_p is the plate surface area.

11.3.4 Damping Loss Factors (DLFs)

The damping loss factor (DLF) of a structural subsystem comprises three damping mechanisms that act independently of each other. These are the structural loss factor of the material, $\eta_{j,s}$, the damping that occurs at the junction between one subsystem and another, $\eta_{j,\rm junct}$, and the energy loss caused by acoustic radiation damping, $\eta_{j,\rm rad}$ (Norton and Karczub (2003), p. 407). Hence, the DLF is given by the sum of these three loss factors as:

$$\eta_j = \eta_{j,s} + \eta_{j,\rm junct} + \eta_{j,\rm rad} \tag{11.75}$$

When two subsystems are rigidly connected, it is assumed that the DLF of the junction is less than the structural loss factor of the material, $\eta_{j,\rm junct} < \eta_{j,s}$. When the connection between subsystems is not rigid, the energy loss that occurs at the junction can provide a significant amount of damping (Lyon and DeJong (1995), p. 167), depending on the junction details. Good estimates of the subsystem damping factors are important as these damping factors have a strong influence on the calculated subsystem response values.

Appendix C lists values of internal DLFs for various materials, which are likely values of loss factor for a panel installed in a building, and represents a combination of the material internal loss factor, the support loss factor and the sound radiation loss factor. The lower limits of the values in Appendix C are suitable for the structural loss factor of the material, $\eta_{j,s}$. The higher limit of the internal loss factor, which accounts for the three damping mechanisms, should be used with caution, as the support loss factor, or junction loss factor, $\eta_{j,\rm junct}$, may not be appropriate for the system under investigation. More accurate response estimates using SEA will be obtained if DLF values can be determined from experimental measurements or from published results of the junction configuration under investigation. Section 9.8, as well as Lyon and DeJong (1995, Chapter 9) and Norton and Karczub (2003, p. 412) describe experimental measurement of DLFs.

The damping loss due to sound radiation is usually included in the subsystem DLF except for the case where radiation is into a separate subsystem such as a room, when it is used to calculate the coupling loss factor. In that case, the radiation damping loss factor, for radiation into free space from the structure mounted in a large baffle, should be estimated from the substructure radiation efficiency, and subtracted from the DLF measured for the structure mounted in a large baffle and radiating into free space. The radiation damping loss factor, $\eta_{\rm rad}$, for a structure of mass/unit area, m, excited resonantly and radiating into free space, is related to its radiation efficiency, σ, by (Norton and Karczub, 2003, p. 410):

$$\eta_{\rm rad} = \frac{\rho c \sigma}{\omega m} \tag{11.76}$$

For a plate-like subsystem, $m = \rho_m h$, where ρ_m is the plate density and h is its thickness. For mechanical excitation of a plate-like subsystem (mounted in a large baffle and radiating into free space), the corresponding radiation efficiency (for resonant response) may be calculated using Equations (4.181) to (4.187) in Chapter 4.

For a cylinder excited mechanically (resonant response), the radiation efficiency, σ, is Szechenyi (1971):

$$\sigma = \begin{cases} \dfrac{\sqrt{3}}{7.2}\left(\dfrac{f}{f_r}\right)^{0.5}\left(\dfrac{f}{f_c}\right); & 0.5f_r < f < 0.8f_r \\ \dfrac{\sqrt{3}}{5}\dfrac{f}{f_c}; & f < 0.5f_r \end{cases} \tag{11.77a,b}$$

where $f_r = c_{LI}/(2\pi a)$ is the ring frequency, c_{LI} is the longitudinal wave speed in the cylinder wall (same as in a plate of the same thickness as the wall), a is the radius of curvature of the cylinder and f_c is the critical frequency (same as for a plate of thickness equal to the cylinder wall thickness).

The DLFs for acoustic systems can be derived from 60 dB decay times, T_{60} (see Section 6.6.2). Using Equations (6.23) and (6.79), the DLF for an enclosure is:

$$\eta = \frac{2.2}{fT_{60}} = \frac{2.2cS\bar{\alpha}}{55.26fV} = \frac{2.2 \times 2\pi}{55.26}\frac{cS\bar{\alpha}}{\omega V} = \frac{cS\bar{\alpha}}{4\omega V} = \frac{cS\bar{\alpha}}{8\pi fV} \tag{11.78a–e}$$

Lyon and DeJong (1995, p. 172), provide an alternative formulation for the DLF of an acoustic enclosure as:

$$\eta = \frac{cS}{2\omega V}\frac{\bar{\alpha}}{2-\bar{\alpha}} \tag{11.79}$$

which differs from Equation (11.78d) by a factor of $2/(2-\bar{\alpha})$, as it includes the effect of the direct acoustic field being absorbed, which can be significant for large values of $\bar{\alpha}$.

To find an expression for the DLF of a 1-D diffuse sound field in a duct, we follow the procedure in Section 6.5.1 for a 3-D sound field, resulting in the following expression for the mean square sound pressure, $\langle p^2\rangle$, at any time, t, after the sound source is turned off.

$$\langle p^2\rangle = \langle p_0^2\rangle e^{-c\bar{\alpha}t/L} \tag{11.80}$$

where $\bar{\alpha}$ is the fraction of energy absorbed by each wall, L is the distance between opposite walls (length of a 1-D room) and $\langle p_0^2\rangle$ is the steady state mean square sound pressure before the sound is turned off. Continuing to follow the procedure in Section 6.5.1, the following expression is obtained for the reverberation time in the 1-D duct of length, L.

$$T_{60} = 13.82L/c\bar{\alpha} \tag{11.81}$$

which gives:

$$\eta = \frac{2.2}{fT_{60}} = \frac{2.2c\bar{\alpha}}{13.82Lf} = \frac{c\bar{\alpha}}{2\pi fL} \tag{11.82a–c}$$

11.3.5 Modal Densities

Modal densities, which are needed to calculate coupling loss factors between various 1-D, 2-D and 3-D subsystems, are provided in Tables 11.6, 11.7 and 11.8, respectively. Some of the expressions were derived from Cremer et al. (2005, p. 302).

11.3.5.1 Random Boundary Impedance

When the boundary impedance is random, averaged modal densities for 2-D and 3-D systems are, respectively:

$$n(\omega)_{\text{2-D}} = \frac{kS}{2\pi c_g} \quad \text{and} \quad n(\omega)_{\text{3-D}} = \frac{kV}{2\pi^2 c_g} \tag{11.83a,b}$$

where S is the 2-D system area, V is the 3-D system volume, c_g is the group wave speed and k is the wavenumber of the wave being analysed.

TABLE 11.6 Modal densities, $n(\omega)$, of 1-D subsystems

Beam excited axially

$$n(\omega) = \frac{L_B}{\pi c_g}; \qquad \text{where } c_g = c_{LII} = \sqrt{\frac{E}{\rho_m}}$$

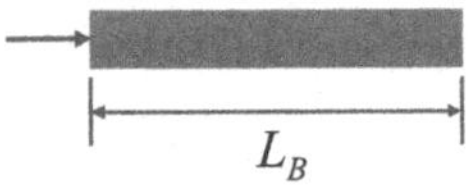

Beam excited torsionally

$$n(\omega) = \frac{L_B}{\pi c_g}; \qquad \text{where } c_g = c_T = \sqrt{\frac{G}{\rho_m}}$$

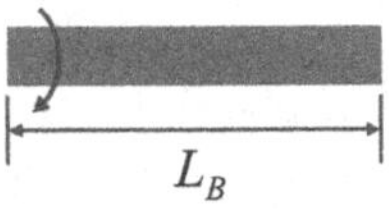

Beam excited in bending

$$n(\omega) = \frac{L_B}{\pi c_g}; \qquad \text{where } c_g = 2c_B = 2\left[\frac{EJ_y\omega^2}{\rho_m S_b}\right]^{1/4}$$

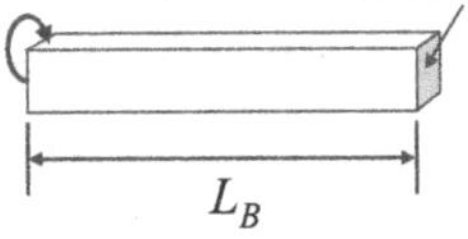

1-D duct excited acoustically

$$n(\omega) = \frac{L}{\pi c}$$

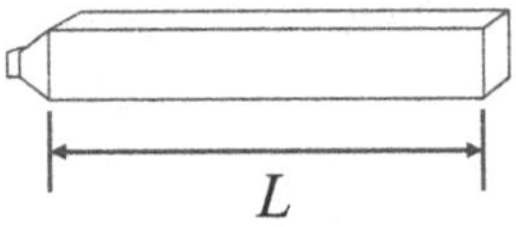

11.3.6 Coupling Loss Factors (CLFs)

When considering subsystems, all wave types could be included, with each wave type represented by a separate subsystem. Thus, a simple beam would have 10 possible coupling loss factors when connected to another beam at right angles at their ends. Five of these can be derived from the other five using modal densities and others may not be important contributors to energy transmission. We could also look at coupling between different waves on the same subsystem, but this form of coupling is usually ignored in practical SEA analysis, as the error in treating different waves as separate and uncoupled is acceptable and within the accuracy expected from an SEA analysis (Lyon and DeJong, 1995, p. 191).

CLF calculations often require the calculation of the transmission coefficient of the junction under consideration, and this is a function of the real part of the infinite impedances of the attached subsystems, provided that the real part is not zero. In cases where the real part is zero but the imaginary part is not, the CLF formulation works if the magnitude of the imaginary part is used instead (Lyon and DeJong, 1995, p. 202).

Note that the admittance, Y, of a stiffness, k, is $Y = \mathrm{j}\omega/k$. If it is a torsional stiffness, the torsional admittance is $Y = \mathrm{j}\omega/k_{\text{tors}}$, where k_{tors} is the torsional stiffness (N-m/rad).

The net power flow, $\Pi_{12} = \omega(\eta_{12}E_1 - \eta_{21}E_2)$ (Lyon and DeJong, 1995, p. 181), from system 1 to subsystem 2 which is a function of the coupling loss factor between subsystems 1 and 2 having modal densities n_1 and n_2, respectively. Note that the coupling loss factor, η_{21}, for power flow from system 2 to system 1 can be calculated from the coupling loss factor, η_{12}, for power flow from system 1 to system 2 using:

$$\eta_{21} = \eta_{12}n_1/n_2 \tag{11.84}$$

TABLE 11.7 Modal densities, $n(\omega)$, of 2-D subsystems

Finite isotropic plate, bending modes
(incl. shear effects) (Lyon and DeJong, 1995, p. 141)

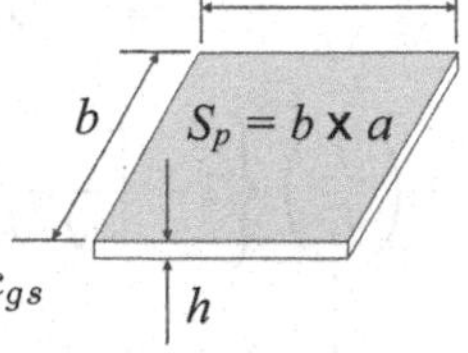

$$n(\omega) = \frac{S\omega}{2\pi}\left[\frac{1}{c_{gB}c_B} + \frac{1}{c_{gs}c_s}\right] = \frac{S\omega}{4\pi}\left[\frac{1}{c_B^2} + \frac{2}{c_s^2}\right]$$

where $c_{gB} = 2c_B = 2\left(B\omega^2/m\right)^{1/4}$ and $c_s = \sqrt{G/\rho} = c_{gs}$

Finite honeycomb-core sandwich plate, bending modes (frequency regions 1 and 3); shear wave modes (frequency region 2)
(Yoerkie et al., 1983, pp. 105–108)

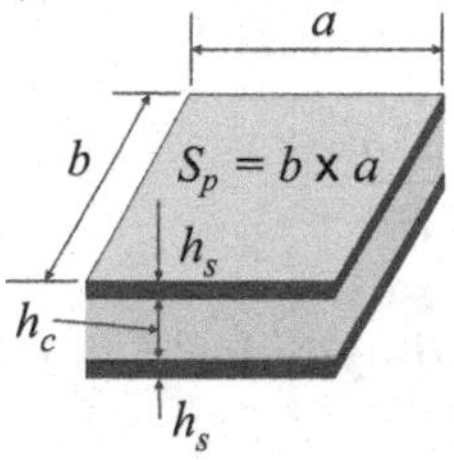

$$n_i(\omega) = \frac{S_p}{4\pi}\sqrt{\frac{m_i}{B_i}};\quad i = 1 \text{ or } 3;\quad n_2(\omega) = \frac{S\omega_2 m_2}{2\pi G h_c}$$

$$m_1 = m_2 = 2\rho_s h_s + \rho_c h_c;\quad m_3 = \rho_s h_s + \rho_c h_c/6$$

$$n(\omega) = n_1(\omega) \text{ for } \omega < \omega_1$$

$$n(\omega) = n_2(\omega) \text{ for } \omega_1 \le \omega \le \omega_2$$

$$n(\omega) = n_3(\omega) \text{ for } \omega > \omega_2$$

$$\omega_1 = \frac{G_c h_c}{\sqrt{m_1 B_1}};\quad B_1 = \frac{E_s h_s (h_s + h_c)^2}{2(1-\nu_s^2)};\quad \omega_2 = \frac{G_c h_c}{\sqrt{m_3 B_3}};\quad B_3 = \frac{E_s h_s^3}{12(1-\nu_s^2)}$$

Finite corrugated plate, bending modes
(Clarkson and Pope, 1983); (Lyon and DeJong, 1995, p. 141)

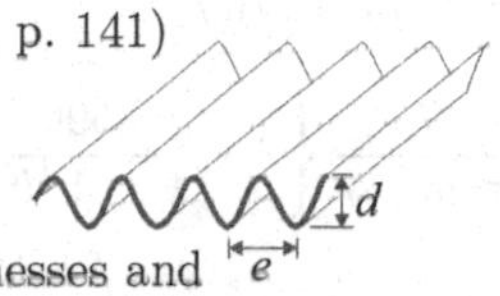

$$n(\omega) = \frac{0.93 S_p}{2\pi c_{LI}}\left[\frac{S_{\hat{y}}}{h^2 I'_y}\right]^{1/4} \quad \text{or} \quad n(\omega) = \frac{S\omega}{4\pi c_{B1} c_{B2}}$$

see pages 400, 403 and 404 to calculate bending stiffnesses and wave speeds, c_{B1} and c_{B2}, across and parallel to the corrugations, respectively.

Finite isotropic plate, in-plane compressional modes

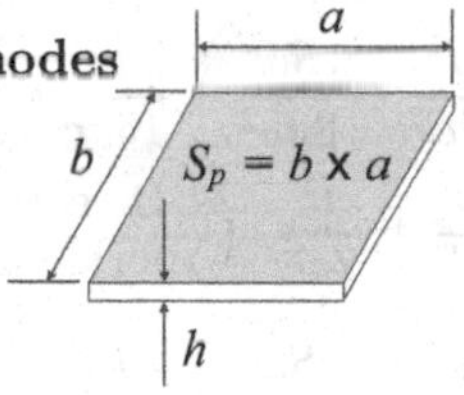

$$n(\omega) = \frac{S_p\omega}{2\pi c_g c_{LI}};\qquad \text{where } c_g = c_{LI} = \sqrt{\frac{E}{\rho_m(1-\nu^2)}}$$

Finite isotropic plate, in-plane shear modes

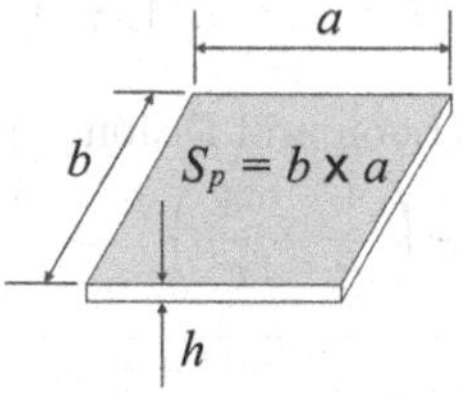

$$n(\omega) = \frac{S_p\omega}{2\pi c_g c_s};\qquad \text{where } c_g = c_s = \sqrt{\frac{G}{\rho_m}} \text{ and}$$

$$G = \frac{E}{2(1+\nu)}$$

2-D thin acoustic cavity (Yoerkie et al., 1983, pp. 65)

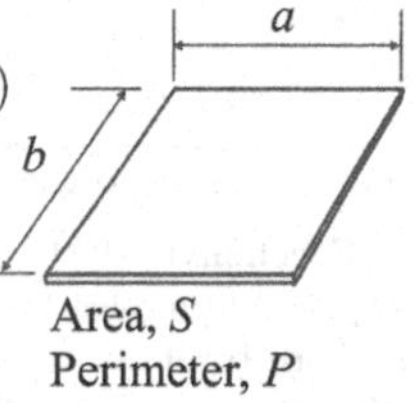

$$n(\omega) = \frac{S\omega}{2\pi c^2} + \frac{P}{2\pi c}$$

TABLE 11.8 Modal densities, $n(\omega)$, of 3-D subsystems

Thin-walled cylinder (Lyon and DeJong, 1995, p. 143)

$$n(\omega) = \frac{N(f_u) - N(f_\ell)}{2\pi\Delta f}; \quad \Delta f = f_u - f_\ell; \quad f_u, f_\ell \text{ are the band upper and lower frequency limits.}$$

$$N(f) = \frac{L\sqrt{3}}{h}\left(\frac{f}{f_r}\right)\left\{1 + \left[\frac{\pi/2}{(f/f_r)^{1/2} + 0.5(f/f_r)^{3.5}}\right]^4\right\}^{-1/4}; \quad f_r = (2\pi a)^{-1}\sqrt{E/(\rho_m(1-\nu^2))}$$

Thin-walled cylinder (Cremer et al., 2005, p. 302)

$$n(\omega) \approx \begin{cases} \dfrac{1.25La}{hc_{LII}}\left(\dfrac{f}{f_1}\right)^{1/2}; & (f/f_1 \le 1); \quad c_{LII} = \sqrt{E/\rho_m} \\ \dfrac{\sqrt{3}La}{c_{LII}h}; & (f/f_1 > 1); \quad f_1 = c_{LII}/(2\pi a) \end{cases}$$

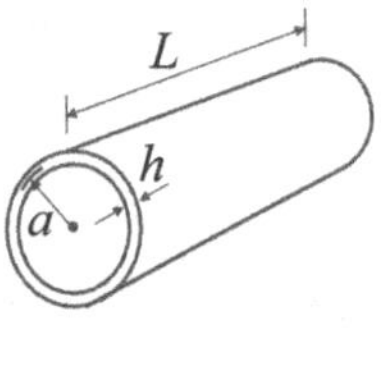

Thin-walled cylinder[1,2] (Szechenyi (1971), with Eq. 10c corrected according to their Fig. 6(a))

$$n(\omega) \approx \begin{cases} \dfrac{2.5S_c}{\pi^2 hc_{LII}}\left(\dfrac{f}{f_1}\right)^{1/2}; \quad f_1 = c_{LII}/(2\pi a) & (f/f_1 \le 0.48) \\ \dfrac{3.6S_c}{\pi^2 hc_{LII}}\left(\dfrac{f}{f_1}\right) & (0.48 < f/f_1 \le 0.83) \\ \dfrac{S_c}{\pi^2 hc_{LII}}\left\{2 + \dfrac{0.596}{F - 1/F}\left[F\cos\left(\dfrac{1.745f_1^2}{F^2 f^2}\right) - \dfrac{1}{F}\cos\left(\dfrac{1.745F^2 f_1^2}{f^2}\right)\right]\right\}; & (f/f_1 > 0.83) \end{cases}$$

2-D and 3-D acoustic cavities

$$n(\omega) = \frac{V\omega^2}{2\pi^2 c^3} + \frac{S\omega}{8\pi c^2} + \frac{P}{16\pi c} \quad \left(\begin{matrix}\text{3-D rectangular room} \\ \text{(see p. 355)}\end{matrix}\right)$$

$$n(\omega) = \frac{S\omega}{4\pi c^2} + \frac{P}{2\pi c} \quad \left(\begin{matrix}\text{2-D rectangular cavity,} \\ \text{Yoerkie et al. (1983, p. 65)}\end{matrix}\right)$$

$$n(\omega) = \frac{V\omega^2}{2\pi^2 c^3} + \frac{S\omega}{8\pi c^2} + \frac{P}{16\pi c} \quad \left(\begin{matrix}\text{cylindrical room} \\ V=\pi a^2 L \quad S=2\pi a(a+L), \\ P=4\pi a+4L \text{ (see p. 356)}\end{matrix}\right)$$

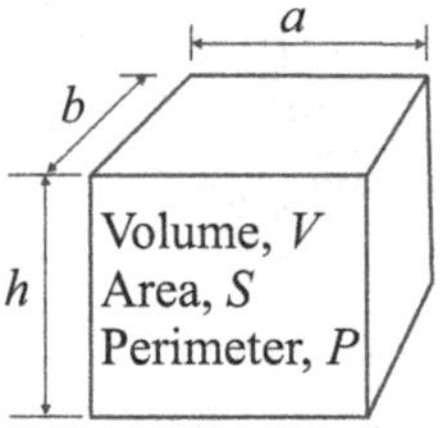

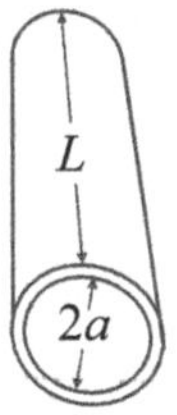

3-D solid (Lyon and DeJong, 1995, p. 146)

$$n(\omega) = \frac{V\omega^2}{2\pi^2}\left(\frac{1}{c_L^3} + \frac{1}{c_s^3}\right)$$

$$c_L = \sqrt{E(1-\nu)/[\rho_m(1+\nu)(1-2\nu)]}$$

$$c_s = \sqrt{G/\rho_m}$$

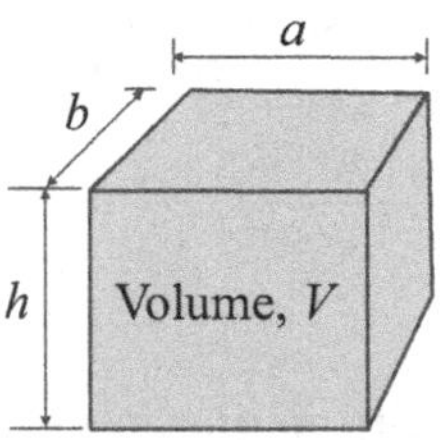

[1]This expression is closest to the exact analysis of Langley (1994)

[2]The results of Szechenyi (1971) are for resonance frequency density (resonance frequencies/Hz) but as there are two possible modes for each resonance frequency, their result has been multiplied by 2 and divided by 2π here to obtain modal density expressed as modes/rad s^{-1} (see Clarkson and Pope (1981), with their definition of c_L requiring a change to $c_{LII} = \sqrt{E/\rho_m}$, as pointed out by Langley (1994)).

Many of the variables in Tables 11.6, 11.7 and 11.8 were defined following Table 11.5. Those that were not defined there are defined below.

c_g is the group wave speed (see Section Section 1.4.5), which is different for different wave types and different for plates and beams;
$c_{LII} = \sqrt{E/\rho_m}$ is the "plane-stress" longitudinal wavespeed for a rod or thin cylinder (m s^{-1});
$c_L = \sqrt{E(1-\nu)/[\rho_m(1+\nu)(1-2\nu)]}$ is the pure longitudinal wavespeed for a 3-D solid;
E_s is Young's modulus of elasticity for the skin material of the sandwich plate (kg m^{-1} s^{-2});
$f_1 = c_{LII}/(2\pi a)$ is the "plane-stress" ring frequency (Hz) (Langley, 1994);
$F = 1.414$ for octave band analysis and 1.122 for 1/3-octave bands;
G_c is the shear modulus for the core of the sandwich plate (kg m^{-1} s^{-2});
$I'_y = B(1-\nu^2)/E$ is the max. 2nd moment of area per unit width about the corrugated plate cross sect. (m^3), where B is the maximum stiffness per unit width;
L is the length of beam, duct or cylinder (m);
P is the perimeter of a plate edge ($= 2(a+b)$) or a 2-D room ($= 2(a+b)$) or a 3-D room ($= 4(a+b+h)$) (m);
S is the interior surface area of a 2-D cavity ($= 2ab$), a 3-D room ($= 2(ab+ah+bh)$) or a cylinder ($= 2\pi a(a+L)$) (m^2);
$S_c = 2\pi aL$ is the surface area of a cylinder (excluding end caps) (m^2);
S_y is the area of a corrugated plate that includes the component due to the corrugations (m^2);
λ_B is the bending wavelength in a plate (m);
ν_s is Poisson's ratio for the skin material of the sandwich plate;
ρ_s is the density of the outside skins of a sandwich plate (kg m^{-3})
ρ_c is the density of the core of a sandwich plate (kg m^{-3})

11.3.6.1 Tunnelling Phenomena

SEA usually only considers resonant transmission from one subsystem to another, but in some cases, non-resonant transmission, or "tunnelling" is important. This phenomenon describes the transmission of energy from one subsystem (1) to another subsystem (3) via non resonant response in a connecting subsystem (2). Sometimes, this non-resonant connection may be the only connection between systems 1 and 3. At other times, there may be additional resonant connections between systems 1 and 2 as well as between systems 2 and 3. An example of the former is three plates connected at right angles with connections that are free to rotate but not translate (pinned connections) and with plate 1 parallel to plate 3 and perpendicular to plate 2, as shown in Figure 11.9(a).

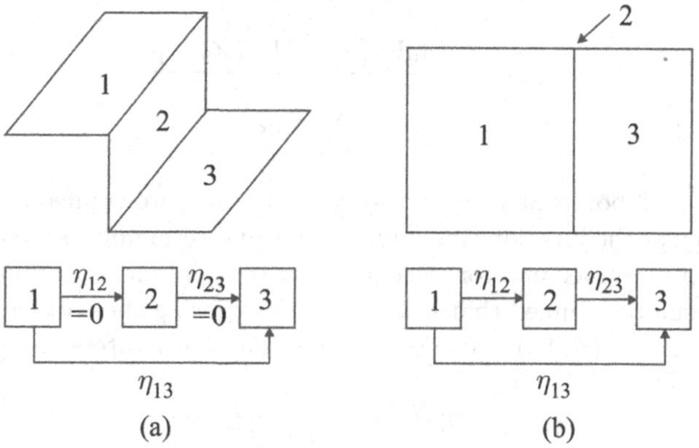

FIGURE 11.9 Illustration of tunnelling phenomena: (a) Two plates connected rigidly by a third plate, transmitting bending waves non-resonantly from plate 1 to 3; (b) Two rooms connected by a common partition (2) that transmits energy non-resonantly from room 1 to room 3 and also resonantly via resonant modes in the partition coupling with resonant modes in the two rooms.

For the case shown in Figure 11.9(a), the resonance frequencies of the in-plane modes of plate 2 are much higher than those of the bending modes in plates 1 and 3. Thus, the low-frequency bending modes of plate 1 couple with the low-frequency modes of plate 3 via forced in-plane motion of plate 2. However, the resonant bending modes of plate 1 cannot couple with the resonant in-plane modes of plate 2, so there is no resonant transmission. An example of the latter, where both resonant and non-resonant transmission exists, is the coupling between two rooms via a partitioning wall, as shown in Figure 11.9(b).

In an SEA analysis, each resonant or non-resonant transmission path is characterised by an individual coupling loss factor and each individual coupling loss factor results in one more row and column in the C-matrix of Equation (11.60).

11.3.6.2 Coupling Loss Factors for Point Connections

Where two substructures are connected at a point junction, the coupling loss factor is given by (Lyon and DeJong, 1995, p. 188):

$$\eta_{12}^{\text{point}} = \frac{\beta_{\text{corr}}}{2\pi f n_1(f)} \frac{2\tau_{12,\infty}}{(2 - \tau_{12,\infty})} \tag{11.85}$$

where $n_1(f) = 2\pi n_1(\omega)$ and:

$$\beta_{\text{corr}} = \left\{1 + \left[\frac{1}{2\pi[\beta_{1,\text{net}} + \beta_{2,\text{net}}]}\right]^8\right\}^{-1/4} = \left\{1 + \left[\frac{1}{2\pi f[\eta_{1,\text{net}} n_1(f) + \eta_{2,\text{net}} n_2(f)]}\right]^8\right\}^{-1/4} \tag{11.86}$$

where $n_i(f)$ is the modal density for subsystem i and $\eta_{i,\text{net}}$ is the damping loss factor for subsystem i, plus some fraction of the loss due to coupling with other systems. The quantity, $\beta_{i,\text{net}}$ is equal to the reciprocal of the (i,i) diagonal term in the inverse of the C-matrix of Equation (11.60) (Lyon and DeJong, 1995, p. 217). Thus, $\beta_{i,\text{net}}$ for $i = 1, N$, where N is the number of subsystems in the analysis, can be found by iteration, beginning with a value of one or two (corresponding to $\beta_{\text{corr}} \approx 1$). The new C-matrix and its inverse are then found and the resulting diagonal values of the new inverse are used to estimate new values of $\beta_{i,\text{net}}$, which are then used to find a new C-matrix.

The correction factor, $2/(2 - \tau_{12,\infty})$, included in Equation (11.85) as proposed by (Lyon and DeJong, 1995, p. 188), to account for the reflected wave having less energy than the incident wave at a junction, was shown by Craik (1997) to be unjustified, so it will not be included in the analysis to follow. The transmission coefficient, $\tau_{12,\infty}$, is:

$$\tau_{12,\infty} = \frac{4q\text{Re}\{Z_{1,\infty}\}\text{Re}\{Z_{2,\infty}\}}{\left|\sum\limits_{i=1}^{N} Z_{i,\infty}\right|^2} \tag{11.87}$$

where q is the number of points at which subsystems 1 and 2 are connected, N is the number of subsystems connected at the junction (including source and receiving subsystems) and $\text{Re}\{Z_\infty\}$ is the real part of the point impedance for the equivalent semi-infinite or infinite system (see Tables 11.1 to 11.3). It is usually assumed that $\beta_{\text{corr}} \approx 1$, and ignoring the correction, $2/(2 - \tau_{12,\infty})$, as explained above, Equation (11.86) becomes the more familiar expression:

$$\eta_{12}^{\text{point}} = \frac{\tau_{12,\infty}}{2\pi f n_1(f)} \tag{11.88}$$

where $f n_1(f) = \omega n_1(\omega)$.

The impedances in Equation (11.87) are moment impedances when bending or torsional waves are being transmitted and/or received. For longitudinal waves transmitted and/or received, point force expressions are used. If beams are connected at right angles at their ends, then semi-infinite moment impedances are used if the bending wave motion is in the plane of the two beams and semi-infinite force impedances are used if the bending wave motion is perpendicular to the plane of the two beams. Note that any end connected element is described by a semi-infinite impedance for the coupling loss factor calculation. If one beam is connected at its end to the centre of another beam, the appropriate (force or moment) semi-infinite impedance is used for the end-connected beam and the appropriate infinite impedance is used for the centre connected beam.

Another consideration is whether the junction connection is rigid or pinned. If it is pinned, no bending waves in the source beam can be transmitted to the receiver beam; bending waves in the source beam can only produce longitudinal waves in the receiver beam.

If a mass, m_a, is part of the junction, its effect can be included by adding its inertance, $\mathrm{j}2\pi f m_a$, to the term inside the modulus lines in the denominator of Equation (11.87). If a subsystem, i, is connected to the junction via a stiffness element of stiffness, k_i, the subsystem impedance, $Z'_{i,\infty}$, to be used in Equation (11.87) in place of $Z_{i,\infty}$ is:

$$Z'_{i,\infty} = \left(\frac{\mathrm{j}\omega}{k_i} + \frac{1}{Z_{i,\infty}}\right)^{-1} \tag{11.89}$$

If bending waves that result in both translational and rotational motion are incident on a junction, it is sufficiently accurate to add the transmission coefficients for translational and rotational motion together to obtain a net transmission coefficient (Lyon and DeJong, 1995, p. 191).

For a 1-D acoustic system, a junction has a common pressure, so that acoustic admittances (reciprocal of acoustic impedances, see table 11.5), Y_∞, add at the junction (Lyon and DeJong, 1995, p. 191). Thus, for a junction consisting of N subsystems:

$$\tau_{12,\infty} = \frac{4\mathrm{Re}\{Y_{1,\infty}\}\mathrm{Re}\{Y_{2,\infty}\}}{\left|\sum\limits_{i=1}^{N} Y_{i,\infty}\right|^2} \tag{11.90}$$

If more than two subsystems are connected, then the impedances (force or moment as appropriate) for the other connected subsystems are also needed to allow use of Equation (11.87).

For all junctions involving point connections, Equations (11.85) and (11.87) are used and the only unknowns are the impedances to be used. For beams connected at their ends or plates connected at their edges, semi-infinite impedances are used and for beams or plates connected more than one quarter of a wavelength from the beam end or plate edge, infinite impedances are used.

11.3.6.3 Coupling Loss Factors for Structural Line Connections

Some examples of coupling loss factors for subsystems connected along a line are illustrated in Table 11.9. The more general approach, outlined by Lyon and DeJong (1995), provides a general expression for the coupling loss factor (averaged over all angles of incidence) for power transmission from subsystem 1 to subsystem 2 as:

$$\eta_{12}^{\mathrm{line}} = \frac{c_{g1} L \beta_{\mathrm{corr}}}{2\omega S_1} \langle \tau_{12,\infty}^{\mathrm{line}}(\theta) \rangle_\theta = \frac{c_{g1} L \beta_{\mathrm{corr}}}{2\omega S_1} \frac{2}{\pi} \int\limits_0^{\pi/2} \tau_{12,\infty}^{\mathrm{line}}(\theta) \cos\theta \, \mathrm{d}\theta \tag{11.91}$$

where c_{g1} is the group wave speed in subsystem 1 for the particular wave type being considered, S_1 is the area of the source structure (usually a plate), $c_{g1}/S_1 = k_1/n_1(f)$, β_{corr} is calculated using iteration as described in Section 11.3.6.2, and the symbols $\langle\ \rangle_\theta$ mean that the transmission

TABLE 11.9 Coupling loss factors for some line connection examples

---

Isotropic plates, edge to edge connection, 180°, bending – bending
(Cremer et al., 1988, p. 351). Valid for plates of the same material.

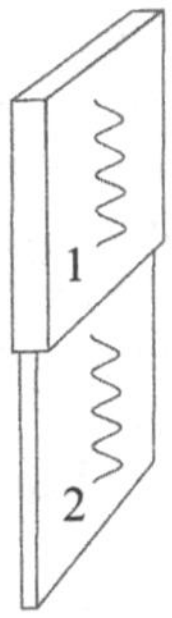

$$\langle \tau_{12}^{\text{line}} \rangle_\theta = \langle \tau_{21}^{\text{line}} \rangle_\theta \approx \tau_{12}^{\text{line}}(0) = \left[\frac{\left(h_a^{-5/4} + h_a^{-3/4} + h_a^{5/4} + h_a^{3/4} \right)}{\left(0.5 h_a^{-2} + h_a^{-1/2} + 1.0 + 0.5 h_a^2 + h_a^{1/2} \right)} \right]^2$$

$h_a = h_2/h_1$, where h_1 and h_2 are the plate thicknesses.

$$\eta_{12} = \frac{c_{B1} L \langle \tau_{12}^{\text{line}} \rangle_\theta}{2\pi f S_1}$$ (Norton and Karczub, 2003, p. 418) (corrected)

where L is the length of the line connection, S_1 is the area of plate 1,
c_{B1} is the bending wave speed in plate 1 and $\langle \tau_{21}^{\text{line}} \rangle_\theta = \frac{2}{\pi} \int_0^{\pi/2} \tau_{21}^{\text{line}}(\theta) \cos\theta \, \mathrm{d}\theta$.

Isotropic plates, T-junction, bending – bending (Peng and Khoo, 2007)

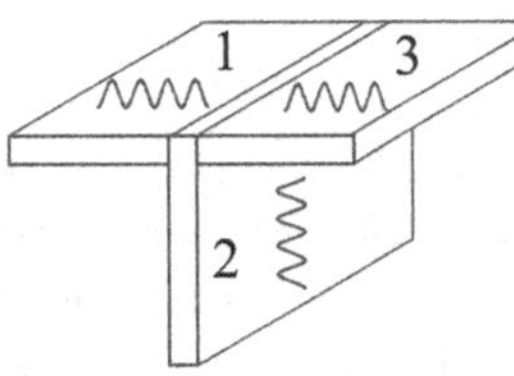

$$\langle \tau_{12}^{\text{line}} \rangle_\theta = \langle \tau_{21}^{\text{line}} \rangle_\theta \approx \frac{2 h_c^{5/2}}{\left[1 + h_c^{5/2} \right] \left[2 + h_a^{5/2} + h_b^{5/2} + \left(h_a^{5/2} + h_b^{5/2} \right)^{-1} \right]}$$

$$\langle \tau_{13}^{\text{line}} \rangle_\theta = \langle \tau_{31}^{\text{line}} \rangle_\theta \approx \frac{2}{\left[1 + h_c^{5/2} \right] \left[2 + h_a^{5/2} + h_b^{5/2} + \left(h_a^{5/2} + h_b^{5/2} \right)^{-1} \right]}$$

$h_a = h_2/h_1$, $h_b = h_3/h_1$ and $h_c = h_2/h_3$ for plates of the same material.

$$\eta_{12} = \frac{c_{B1} L \langle \tau_{12}^{\text{line}} \rangle_\theta}{2\pi f S_1}; \qquad \eta_{13} = \frac{c_{B1} L \langle \tau_{13}^{\text{line}} \rangle_\theta}{2\pi f S_1}$$ (Norton and Karczub, 2003, p. 418) (corrected)

Isotropic plates, edge to edge connection, 90°, bending – bending
(Norton and Karczub (2003, p. 418), Cremer et al. (1988, p. 357))

$$\langle \tau_{12}^{\text{line}} \rangle_\theta = \langle \tau_{21}^{\text{line}} \rangle_\theta = \tau_{12}^{\text{line}}(0) \times \frac{2.754 (h_1/h_2)}{1 + 3.24 (h_1/h_2)} \quad \text{where } \tau_{12}^{\text{line}}(0) = 2 \left(\psi^{1/2} + \psi^{-1/2} \right)^{-2}$$

and $\psi = \left(\rho_{m1} c_{L1}^{3/2} h_1^{5/2} \right) \Big/ \left(\rho_{m2} c_{L2}^{3/2} h_2^{5/2} \right) = \dfrac{h_1^{5/2}}{h_2^{5/2}}$ for plates of the same material.

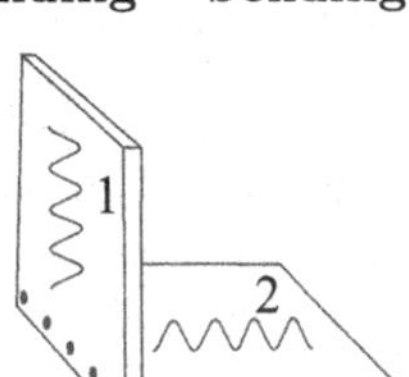

$$\eta_{12} = \frac{c_{B1} L \langle \tau_{12}^{\text{line}} \rangle_\theta}{2\pi f S_1}$$ (Norton and Karczub, 2003, p. 418) (corrected)

Isotropic plates, edge to edge point connection (bolts), 90°, bending – bending
(Panuszka et al., 2005; Norton and Karczub, 2003, p. 418 (corrected))

$$\eta_{12} = \begin{cases} \dfrac{4 N h_1 c_{L1} (\rho_{m1} h_1^3 c_{L1}^2)(\rho_{m2} h_2^3 c_{L2}^2)}{\sqrt{3} \omega S_1 (\rho_{m1} h_1^3 c_{L1}^2 + \rho_{m2} h_2^3 c_{L2}^2)^2}; & \lambda_B < \ell \\[2ex] \dfrac{c_{B1} L \langle \tau_{12}^{\text{line}} \rangle_\theta}{2\pi f S_1}; & \lambda_B \geq \ell \end{cases}$$

where $\omega = 2\pi f$, ℓ = spacing between bolts and N = number of bolts coefficient, τ_θ, is averaged over all angles of the incident waves. For shear waves, $c_g = c_s$, for longitudinal waves, $c_g = c_{LI}$, and for bending waves, $c_g = 2c_B$.

Equation (11.91) can be rewritten in terms of the normal incidence transmission coefficient, $\tau_{12,\infty}^{\text{line}}(0)$, for an infinitely extending structure, as (Lyon and DeJong, 1995, p. 194):

$$\eta_{12}^{\text{line}} = \frac{\beta_{\text{corr}}}{2\pi f n_1(f)} \frac{k_1 L}{\pi} \int_0^{\pi/2} \tau_{12,\infty}^{\text{line}}(\theta)\cos(\theta)\mathrm{d}\theta = \frac{\beta_{\text{corr}} I_{12}^{\text{line}} \tau_{12,\infty}^{\text{line}}(0)}{2\pi f n_1(f)} \tag{11.92a,b}$$

where the correction term, $2/[2 - \tau_{12,\infty}^{\text{line}}(0)]$ has been omitted as advised by Craik (2003), and:

$$\tau_{12,\infty}^{\text{line}}(0) = \frac{4\text{Re}\{Z_{1,\infty}^{\text{line}}\}\text{Re}\{Z_{2,\infty}^{\text{line}}\}}{\left|\sum_{i=1}^{N} Z_{i,\infty}^{\text{line}}\right|^2} \tag{11.93}$$

where there are N plates connected at the junction and $\beta_{\text{corr}} \approx 1$.

The impedances, $Z_{1,\infty}^{\text{line}}$ and $Z_{2,\infty}^{\text{line}}$ are line impedances that depend on the connecting structure type (see Table 11.4). The quantity, I_{12}^{line}, is determined by the integral over angle for the wave type and incident angle range (see Equation 11.92 and Lyon and DeJong (1995, p. 194)).

For two plates with bending wave numbers, k_{B1} and k_{B2}, respectively, connected at right angles along their edges of length, L, an alternative to the equations in Table 11.9 is to use Equations (11.92) and (11.93) together with the following expression for I_{12}^{line} (Lyon and DeJong, 1995, p. 195):

$$I_{12}^{\text{line}} = \frac{L}{4}\left(\frac{k_{B1}^4 k_{B2}^4}{k_{B1}^4 + k_{B2}^4}\right)^{1/4} \tag{11.94}$$

11.3.6.4 Coupling Loss Factors for Area Connections

When two subsystems are connected via an area connection of area, S_p, the coupling loss factor for power transmission from subsystem 1 to subsystem 2 can be rewritten in terms of the normal incidence transmission coefficient, $\tau_{12}^{\text{area}}(0)$, as (Lyon and DeJong, 1995, p. 196):

$$\eta_{12}^{\text{area}} = \frac{\beta_{\text{corr}}}{2\pi f n_1(f)} \frac{k_1^2 S_p}{2\pi} \int_0^{\pi/2} \tau_{12,\infty}^{\text{area}}(\theta)\cos(\theta)\sin(\theta)\mathrm{d}\theta = \frac{\beta_{\text{corr}} I_{12}^{\text{area}}}{2\pi f n_1(f)} \tau_{12,\infty}^{\text{area}}(0) \tag{11.95a,b}$$

where:

$$\tau_{12,\infty}^{\text{area}}(0) = \frac{4\text{Re}\{Z_{1,\infty}^{\text{area}}\}\text{Re}\{Z_{2,\infty}^{\text{area}}\}}{\left|\sum_{i=1}^{N} Z_{i,\infty}^{\text{area}}\right|^2} \tag{11.96}$$

The impedances, $Z_{1,\infty}^{\text{area}}$ and $Z_{2,\infty}^{\text{area}}$ are area impedances that depend on the connecting structure type. The quantity, I_{12}^{area}, is determined by the integral over angle for the wave type and incident angle range (see Equation 11.95 and Lyon and DeJong (1995, p. 196)). The quantity, β_{corr}, can be determined via iteration or set equal to 1.0, as discussed for point connections.

Examples of area connections are illustrated in Figure 11.10. Figure 11.10(a) shows a plate making up all or part of a wall of an enclosure. If plate (2) of area, S_p, perimeter, P, thickness, h, and density, ρ_m is excited mechanically, the resonant response of the plate couples to resonant modes in enclosure (1), with the corresponding coupling loss factor given by (Fahy, 1982, p. 177):

$$\eta_{21} = \frac{\rho c \sigma}{\omega \rho_m h} \tag{11.97}$$

which is identical to Equation (11.76) for radiation damping. The coupling loss factor, η_{12}, is related to the coupling loss factor, η_{21}, by Equation (11.84).

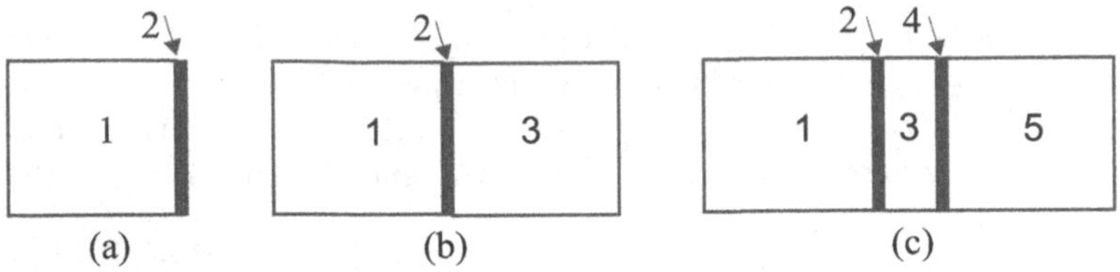

FIGURE 11.10 Three examples of area connected systems: (a) plate coupled with a room; (b) two rooms coupled together via a single wall; and (c) two rooms coupled together via a double wall with an internal cavity.

The plate radiation efficiency, σ, for resonant transmission into an enclosure is (Lyon and DeJong, 1995, p. 199):

$$\begin{aligned}\sigma &= \frac{2Pk^2}{\pi S_p k_{Bp}^3\left(1+\dfrac{\pi k^2}{2k_{Bp}^2}\right)} + \left[\left(\frac{k_{Bp}^2}{k^2}-1\right)^2\left(\frac{\pi k_{Bp}^4}{k^4}+1\right)^2 + \frac{2\pi}{k_{Bp}\sqrt{S_p}}\right]^{-1} \\ &= \frac{2P(f/f_c)^{3/2}}{\pi k S_p\left(1+\dfrac{\pi f}{2f_c}\right)} + \left[\left[(f_c/f)-1\right]^2\left[\pi(f_c/f)^2+1\right]^2 + \frac{2\pi(f/f_c)^{1/2}}{k\sqrt{S_p}}\right]^{-1}\end{aligned} \tag{11.98a,b}$$

where the plate bending wavenumber, $k_{Bp} = \left[\dfrac{12\omega^2\rho_m(1-\nu^2)}{Eh^2}\right]^{1/4} = \left[\dfrac{12\omega^2}{c_{LI}^2h^2}\right]^{1/4} = \omega/c_{Bp}$, P is the plate perimeter and $k = 2\pi/\lambda$ is the wavenumber in air.

The overall coupling loss factor for two acoustic spaces (enclosures) connected together through a solid partition, as shown in Figure 11.10(b), consists of the sum of three coupling loss factors, η_{12} and η_{23} for resonant power transmission and η_{13} for non-resonant power transmission. The coupling loss factors, η_{21} and η_{23} for resonant power transmission are equal and are calculated using Equation (11.97). The coupling loss factor, η_{12} can be calculated from η_{21} using Equation (11.84). The total coupling loss factor between enclosures 1 and 3 is $\eta_{1,3,\text{tot}} = \eta_{12}+\eta_{23}+\eta_{13}$

The coupling loss factor for non-resonant power transmission from enclosures (1) to (3) via partition (2) is:

$$\eta_{13} = \frac{cS_p}{4\omega V_1}\langle\tau_2\rangle \tag{11.99}$$

where S_p is the area of one face of the partition connecting the enclosures, V_1 is the volume of enclosure 1 and $\langle\tau_2\rangle$ is the average transmission coefficient given by (Norton and Karczub, 2003):

$$\langle\tau_2\rangle = \begin{cases} \pi\left\{\dfrac{\pi^9(\rho_m h)^2}{2^{13}\rho^2 S_p}\left[1-\left(\dfrac{10f}{f_c}\right)^2\right] + 1 + \left(\dfrac{\pi f\rho_m h}{\rho c}\right)^2\right\}^{-1}; & f_{1,1} < f < f_c/2 \\ \pi\left[1+\left(\dfrac{\pi f\rho_m h}{\rho c}\right)^2\right]^{-1}; & f \geq f_c/2 \end{cases} \tag{11.100a,b}$$

See Sections 7.2.1 and 7.2.6 for the calculation of f_c and $f_{1,1}$ respectively. Equation (11.100) is slightly different to what is provided in Norton and Karczub (2003). The additional multiplication by π, which accounts for 1/3-octave band field incidence transmission rather than normal incidence, and the "+1" inclusion in both parts of the equation are discussed in Section 7.2.6.1. An alternative and more accurate equation for calculating the non-resonant transmission coefficient for frequencies above $f_c/4$ and below $0.95f_c$ is (Leppington et al., 1987, Equation

(3.12)):

$$\langle\tau_2\rangle = \left(\frac{2\rho}{\rho_m hk(1-\mu^{-2})}\right)^2 \Big\{ \log_e(k\sqrt{S_p}) + 0.16 - U(L_x/L_y) + \frac{1}{4\mu^3}\Big[(2\mu-1)(\mu+1)^2\log_e(\mu-1) + (2\mu+1)(\mu-1)^2\log_e(\mu+1) - 4\mu - 8\mu^3\log_e(\sqrt{\mu})\Big]\Big\} \tag{11.101}$$

where $\mu = f_c/f$ and the function, $U(L_x/L_y)$, which is difficult to evaluate, is approximately zero for plate aspect ratios (long/short side) less than three (Hopkins, 2007, p. 426).

Equations (11.100) and (11.101) agree to within less 1 dB for frequencies less than $f_c/4$ and to within 3 dB for frequencies less than $f_c/2$.

In an SEA model, both resonant and non-resonant transmission paths must be taken into account when calculating the power transmission from room 1 to room 3. This means that the total coupling loss factor is the sum of the three individual loss factors. For frequencies above $0.95f_c$, sound transmission is dominated by resonant transmission. In this case, η_{13} does not contribute significantly to the sound transmission and may be ignored.

If it is of interest to calculate the partition transmission loss, the transmission coefficient that should be used with Equation (7.13) to calculate TL is the sum of the transmission coefficient due to non resonant transmission, τ_2 (Equations (11.100) or (11.101)) and that due to resonant transmission, τ_{res}, given by Equation (11.102) below (Hopkins, 2007, p. 420).

$$\tau_{\text{res}} = \frac{\eta_{12}\eta_{23}V_1S_3\bar{\alpha}_3}{\eta_2\eta_3V_3S_2} \tag{11.102}$$

where $S_3\bar{\alpha}_3$ is the absorption area in room 3 (receiver room). Coupling loss factors are calculated using Equations (11.97) and (11.84), with the radiation efficiency given by Equation (11.98).

Figure 11.10(c) illustrates the configuration for sound transmission between two rooms, (1) and (5), via a double partition with a narrow cavity (3) separating the two partition leaves, (2) and (4). Referring to the figure, the overall coupling loss factor for energy transmission from room 1 to room 5 is the sum of the first element in each of the seven following coupling loss factor calculations. Some researchers suggest that it is necessary to choose either item (7) or items (5) and (6) but not both.

1. η_{12} and η_{21} (resonant transmission, calculated using Equation (11.97)).
2. η_{23} and η_{32} (resonant transmission, calculated using Equation (11.97)).
3. η_{34} and η_{43} (resonant transmission, calculated using Equation (11.97)).
4. η_{45} and η_{54} (resonant transmission, calculated using Equation (11.97)).
5. η_{13} and η_{31} (non-resonant transmission, calculated using Equation (11.103)).
6. η_{35} and η_{53} (non-resonant transmission, calculated using Equation (11.103)).
7. η_{15} and η_{51} (non-resonant transmission, calculated as described below).

The coupling loss factor for non-resonant connection between the two rooms may be found either by evaluating item 7 (η_{15}) or item 5 (η_{13}) and item 6 (η_{35}) in the list above. Using either approach, the resulting equations are valid for frequencies below $0.95f_c$. Above this frequency, transmission is dominated by resonant transmission.

Evaluating η_{15}, requires treating the double wall as a single connection that approximately follows the mass-law, based on the total mass of the partition. Equation (11.99) (with subscript 3 replaced with subscript 5) may be used with τ_2 defined using Equation (11.101) with $\rho_m h$ replaced with $\rho_{m2}h_2 + \rho_{m4}h_4$ and where $\mu = f_c/f$ with f_c the lower critical frequency of panels 2 and 4.

As room (3) in Figure 11.10(c) is a narrow cavity, the coupling loss factors, η_{13} and η_{35}, for non-resonant transmission may be calculated by considering that the sound field in the cavity is mainly incident at grazing incidence, as wave propagation in the cavity is parallel to the face of the partition, resulting in the following expression for the coupling loss factors for non-resonant transmission from the cavity to the rooms (Craik, 2003):

$$\eta_{31} = \frac{\rho^2 c^3 S_2 \sigma_2}{\omega^3 \rho_{m2}^2 h_2^2 V_3} \quad \text{and} \quad \eta_{35} = \frac{\rho^2 c^3 S_4 \sigma_4}{\omega^3 \rho_{m4}^2 h_4^2 V_3} \tag{11.103a,b}$$

where V_3 is the cavity volume, ρ_{m2} and ρ_{m4} are the densities of the material for partition leaves 2 and 4, respectively, h_2 and h_4 are the leaf thicknesses of the partition, $\omega = 2\pi f$, $S_2 = S_4$ is the total area of one face of the partition. The coupling loss factor, η_{13}, for transmission from room 1 to cavity 3 can be obtained from Equation (11.103a) by using Equation (11.84).

The radiation efficiencies, σ_2 and σ_4, of partition leaves 2 and 4 are given by the same expression as for the radiation efficiency at coincidence (due to the same wave speeds in the panel and in air), with the coincidence frequency replaced by the actual frequency, f (Craik, 2003). Thus:

$$\sigma_2 = \sigma_4 = 0.45\sqrt{P_2 f / c} \tag{11.104}$$

where $P_2 = P_4$ is the perimeter of the partition between the cavity wall and the room. That is, it is given by, $P_2 = P_4 = 2(L_x + L_y)$ where L_x and L_y are the dimensions of the face of the cavity. The volume V_3 is the total volume of all cavities between studs and the coupling loss factor represents the entire cavity 3, not just 1 cell between 2 studs.

An alternative equation for the radiation efficiency at the coincidence frequency is (Maidanik (1962) and Cremer et al. (2005, p. 499)):

$$\sigma_2 = \sigma_4 = \sqrt{\frac{L_x f}{c}} + \sqrt{\frac{L_y f}{c}} \tag{11.105}$$

Equation (11.105) gives a result that is approximately a factor of 2 greater than the result calculated using Equation (11.104). Maidanik (1962) has shown that at frequencies below the critical frequency, twice the length of the ribs should be added to the panel dimension that is parallel to the ribs in the panel radiation efficiency calculations. However, this does not apply at the critical frequency and thus neither P_2 in Equation (11.104) nor L_x and L_y in Equation (11.105) should be modified to account for the studs.

Equation (11.105) is the most commonly cited reference for the radiation efficiency of a baffled rectangular panel at the coincidence frequency, and it also appears in (Cremer et al., 2005, p, 499). The alternative Equation (11.104) used by Craik (2003), and also presented in (Cremer et al., 1988, p. 533) is a slight modification of the expression for a baffled beam, which is derived by Lyon and Maidanik (1962). This latter expression agrees numerically with that provided by Leppington et al. (1982) for a baffled rectangular panel, but the numerical values are a factor of 2 less than the values obtained using Equation (11.105). However, Equation (11.105) gives results that agree better with lower frequency equations (see equation (4.181)) at frequencies very close to the coincidence frequency and is recommended here.

If it is of interest to calculate the transmission loss between rooms 1 and 5, the transmission coefficient, τ_{tot} to use with Equation (7.13) to calculate TL is given by $\tau_{\text{tot}} = \tau_{2+4} + \tau_{\text{res}}$ or $\tau_{\text{tot}} = \tau_2\tau_4 + \tau_{\text{res}}$, where τ_2 is given by Equation (11.100) or Equation (11.101). The coefficient, τ_4, may be calculated using the same equations, but using the thickness, density and critical frequency of leaf 4 of the partition rather than leaf 2. The coefficient, τ_{2+4} may be calculated using the same equations but replacing $\rho_m h$ with $\rho_{m2}h_2 + \rho_{m4}h_4$ and using the lowest of the critical frequencies of leaves 2 and 4. The transmission coefficient for resonant transmission, τ_{res}

may be derived from Equation (11.102) to give:

$$\tau_{\mathrm{res}} = \frac{\eta_{12}\eta_{23}V_1 S_3\bar{\alpha_3}}{\eta_2\eta_3 V_3 S_2} \times \frac{\eta_{34}\eta_{45}V_3 S_5\bar{\alpha_5}}{\eta_4\eta_5 V_5 S_4} \tag{11.106}$$

where $S_3\bar{\alpha_3}$ is the absorption area of cavity 3 and $S_5\bar{\alpha_5}$ is the absorption area of room 5. Coupling loss factors are calculated using Equations (11.97) and (11.84), with the radiation efficiency given by Equation (11.104) for coupling loss factors containing a 3 subscript and by Equation (11.98) for coupling loss factors containing a 1 or 5 subscript.

12

Frequency Analysis

LEARNING OBJECTIVES

In this chapter, the reader is introduced to:

- Conversion of a time series signal to the frequency domain;
- Octave and 1/3-octave frequency band analysis;
- Digital filtering;
- Fourier analysis;
- Various frequency analysis tools including frequency response functions and coherence;
- Use of frequency analysis tools in identifying characteristics of a noise source.

12.1 Introduction

Frequency analysis is the process of finding the amplitudes of the component parts that together make up a particular noise. Each component part is defined in terms of a centre frequency and a bandwidth. The reason that frequency analysis is included in this book is that practically all forms of noise level prediction, as well as noise source identification, involve some degree of frequency analysis. With the advent of more sophisticated instrumentation and data acquisition systems coupled with very fast processors, complex frequency analysis (or signal processing) is becoming more commonplace. Thus, it is important that the principles underlying these types of analyses are fully understood by practitioners to ensure the results so obtained are properly interpreted.

Basic frequency analysis was discussed in Chapter 1. In this chapter, more advanced analysis techniques are discussed. These invariably involve transforming a time domain signal to the frequency domain. There are two ways of achieving this. The first requires the use of band-limited digital or analogue filters. The second requires the use of Fourier analysis where the time domain signal is transformed using a Fourier series. This is implemented in practice digitally (referred to as the DFT—discrete Fourier transform) using a very efficient algorithm known as the FFT (fast Fourier transform). Digital filtering is discussed in Section 12.2 and FFT techniques are discussed in Section 12.3.

12.2 Digital Filtering

Spectral analysis is commonly carried out in standardised octave, 1/3-octave, 1/12-octave and 1/24-octave bands, and both analogue and digital filters are available for this purpose. Such

DOI: 10.1201/9780367814908-12

filters are referred to as constant percentage bandwidth filters meaning that the filter bandwidth is a constant percentage of the band centre frequency. For example, the octave bandwidth is always about 70.1% of the band centre frequency, the 1/3-octave bandwidth is 23.2% of the band centre frequency and the 1/12-octave is 5.8% of the band centre frequency, where the band centre frequency is defined as the geometric mean of the upper and lower frequency bounds of the band (see Equation (1.113)).

The stated percentages are approximate, as a compromise has been adopted in defining the bands to simplify and to ensure repetition of the band centre frequencies. The compromise that has been adopted is that 10 times the logarithms to the base ten of the 1/3-octave band centre frequencies are integers or very close to an integer number (see Table 1.2).

Instruments with constant frequency bandwidth filters have been replaced by fast Fourier transform (FFT) analysers, most of which are implemented using a data acquisition card and a computer. When a time-varying signal is filtered using either a constant percentage bandwidth or a constant frequency bandwidth filter, an RMS amplitude signal is obtained, which is proportional to the sum of the total energy content of all frequencies included in the band.

When discussing digital filters and their use, an important consideration is the filter response time, T_R, which is the minimum time required for the filter output to reach steady state. The response time is frequency dependent and can also be expressed in terms of the number of cycles that need to enter the filter before it reaches steady state. The minimum time generally required is the inverse of the filter bandwidth, B (Hz). That is:

$$BT_R = \left(\frac{B}{f}\right)(fT_R) = bn_R \approx 1 \tag{12.1}$$

where f is the centre band frequency, $b = B/f$ is the normalised bandwidth, and $n_R = fT_R$ is the number of cycles required at frequency, f for the filter output to reach steady state. For example, for a 1/3-octave filter, $b = 0.23156$, and the number of cycles, n_R, required for the filter to reach a steady-state output is approximately 4.3. A typical response of a 1/3-octave filter is illustrated in Figure 12.1, where it will be noted that the actual time required to reach steady state is slightly greater than calculated above, perhaps five cycles or more instead of 4.3, depending on the required accuracy.

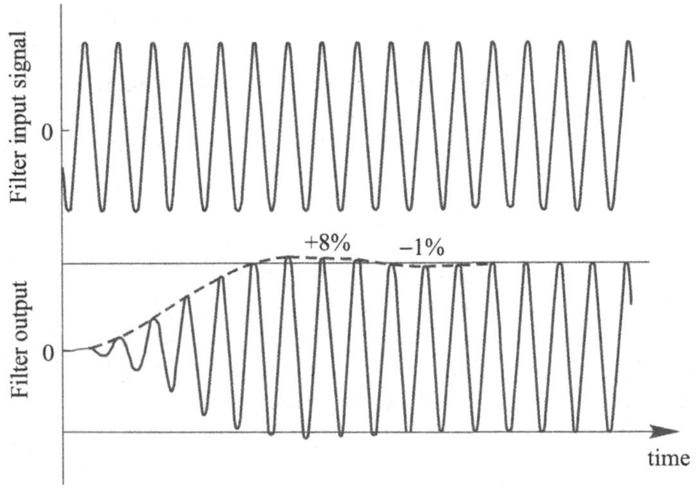

FIGURE 12.1 Typical filter response of a 1/3-octave filter (after Randall (1987)).

Where the RMS value of a filtered signal is required, it is necessary to determine the square root of the average value of the integrated squared output of the filtered signal over some prescribed period of time called the averaging time. The longer the averaging time, the more nearly constant will be the RMS value of the filtered output.

For a sinusoidal input of frequency, f (Hz), or for several sinusoidal frequencies within the band, where f (Hz) is the minimum separation between components, the variation in the average value will be less than 1/4 dB for an averaging time, $T_A \geq 3/f$ seconds.

For many sinusoidal components, or for random noise and $BT_A \geq 1$, the error in the RMS signal may be determined in terms of the statistical error, ϵ, calculated as:

$$\epsilon = 0.5(BT_A)^{-1/2} \tag{12.2}$$

For random noise, the actual error has a 63.3% probability of being within the range $\pm\epsilon$ and a 95.5% probability of being within the range $\pm 2\epsilon$.

The calculated statistical error may be expressed in decibels as:

$$20 \log_{10} \mathrm{e}^{\epsilon} = 4.34(BT_A)^{-1/2} \quad \text{(dB)} \tag{12.3}$$

12.2.1 Octave and 1/3-Octave Filter Rise Times and Settling Times

One important aspect of analysis of very low-frequency sound, using octave or 1/3-octave bands, is the filter rise time, which is the time it takes for the filter to measure the true value of a continuous signal. So if the signal is varying rapidly, especially at low frequencies, it is not possible for the output of the filter to track the rapidly varying input, resulting in a considerable error in the RMS output, especially if the signal has high crest factors (ratio of peak to RMS value of a signal). As discussed in the previous section, the rise time of any digital filter may be estimated approximately using Equation (12.1). More accurate data from Bray and James (2011) for 1/3-octave band filters is assembled in Table 12.1. The table shows the time taken for the filter to achieve 90% accuracy (or less than 1 dB error) of the true magnitude of the signal. It can be seen from the table that for a 1/3-octave band filter centred at 1 Hz, the time taken is approximately 5 seconds or 5 full cycles. It is clear that use of such a 1/3-octave filter will not correctly measure the energy associated with rapidly varying, very low-frequency sound having relatively high crest factors. As the centre frequency of the octave or 1/3-octave band becomes lower, the required sampling time to obtain an accurate measure of the signal becomes larger.

TABLE 12.1 1/3-octave filter rise times for a 1 dB error (6th order filter defined in ANSI/ASA S1.11 (2019)). The rise time decreases by a factor of 10 for each decade increase in frequency. Octave band filters would have rise times of 1/3 of the rise time of a 1/3-octave filter with the same centre frequency

1/3-octave centre frequency (Hz)	Rise time (millisec)	1/3-octave centre frequency (Hz)	Rise time (millisec)	1/3-octave centre frequency (Hz)	Rise time (millisec)
1.00	4989	10.0	499	100	49.9
1.25	3963	12.5	396	125	39.6
1.60	3148	16.0	315	160	31.5
2.00	2500	20.0	250	200	25
2.50	1986	25.0	199	250	19.9
3.15	1578	31.5	158	315	15.8
4.00	1253	40.0	125	400	12.5
5.00	995	50.0	99.5	500	9.95
6.30	791	63.0	79.1	639	7.91
8.00	628	80.0	62.8	800	6.28

One way of increasing the ability of the measurement system to track and measure accurately sound signals with high crest factors, is to increase the bandwidth of the filter used for the measurement and reduce the filter order so it rolls off more slowly. The filter rise time is proportional to the inverse of the filter bandwidth and the filter order. A suggestion by Bray and James (2011) is to use filter bandwidths that are equal to the critical bandwidth of our hearing mechanism, which is approximately 100 Hz at our lowest hearing frequencies. Thus, the filter bandwidth for measuring low-frequency noise and infrasound should be 100 Hz (actually from 0.5 Hz to 100 Hz) if we are to measure crest factors in a similar way to which our hearing mechanism experiences them. Bray and James (2011) recommend using a 4th order Butterworth "Bark 0.5" (see Section 2.1.8.6) bandpass filter centred on 50 Hz with an impulse response of approximately 8.8 milliseconds, which according to Bray and James (2011), simulates the approximately 10 millisecond response of our hearing mechanism. This response is 10 times faster than the "fast" response on a sound level meter, which implies that the "fast" response underestimates the true rise time of our hearing mechanism at these low frequencies. Thus, we hear much higher low-frequency peaks than are measured by the "fast" sound level meter measurements. The rise time of the 4th order filter can be obtained from Tables 1.2 and 12.1 using: rise time $\approx 99.5 \times (4/6) \times (11.6/99.5) = 7.7$, which is close to the 8.8 millisecond value mentioned above. The difference is due to the 8.8 value representing the filter impulse response whereas the 7.7 value represents the time needed for the filter to achieve a 1 dB error.

In addition to filter rise time, it is also important to take into account the settling time of the RMS detector in order to achieve 1/2 dB accuracy, which is in addition to the rise time discussed above. For a single frequency signal of frequency, f, the settling time for 1/2 dB accuracy will be approximately $3/f$ (Randall, 1987). For a random signal, the calculated statistical error, ϵ, is given by Equation (12.3).

There are three types of noise signal used in acoustics to excite systems for the purpose of measuring their acoustical properties.

- **White noise**, which is a signal with uniform spectral energy (that is, equal energy per Hz). White noise has a flat spectral shape when viewed on a narrow band spectrum, but increases at a rate of 3 dB per octave when viewed on an octave band plot.
- **Pink noise**, which is a signal with the same amount of energy in each octave band. Pink noise has a flat spectral shape when viewed on an octave band plot, but has a downwards slope and decreases at 3 dB per octave (doubling of freq) when viewed on a narrow band plot.
- **Pseudo-random noise**, which is discussed in Section 12.3.18.
- **Swept sine**, which is a single frequency signal that gradually increases in frequency during the measurement process.

12.3 Advanced Frequency Analysis

In this section, various aspects of Fourier analysis will be discussed. Fourier analysis provides much more frequency resolution (each component in the frequency spectrum representing smaller frequency spans) than octave or 1/3-octave band analysis. The symbol used here for frequency in a continuous spectrum is f and for a sampled (digital) spectrum it is f_k.

Fourier analysis is the process of transforming a time varying signal into its frequency components to obtain a plot or table of signal amplitude as a function of frequency. A general Fourier representation of a periodic, time varying signal of period, T_p, consisting of a fundamental frequency, $f_1 = 1/T_p$, represented by $x_1(t) = x(t + T_p)$ and various harmonics, n, of frequency f_k, represented by $x_n(t) = x(t + nT_p)$, where $n = 2, 3, \ldots$, takes the form:

$$x(t) = \sum_{n=1}^{\infty} A_n \cos(2\pi n f_1 t) + B_n \sin(2\pi n f_1 t) \tag{12.4}$$

As an example, we can examine the Fourier representation of a square wave shown in Figure 12.2. The first four harmonics in Figure 12.2(b) are described by the first four terms in Equation (12.4), where $B_n = 0$ for all components, $A_n = 4/(\pi n)$ for n odd and $A_n = 0$ for n even. The component characterised by frequency, nf_1, is usually referred to as the nth harmonic of the fundamental frequency, f_1, although some call it the $(n-1)$th harmonic. Use of Euler's well-known equation (Abramowitz and Stegun, 1965) allows Equation (12.4) to be rewritten as:

$$x(t) = \frac{1}{2}\sum_{n=0}^{\infty} \left[(A_n - \mathrm{j}B_n)\,\mathrm{e}^{\mathrm{j}2\pi n f_1 t} + (A_n + \mathrm{j}B_n)\,\mathrm{e}^{-\mathrm{j}2\pi n f_1 t}\right] \tag{12.5}$$

where the $n = 0$ term has been added for mathematical convenience. However, it represents the zero frequency (DC) component of the signal and is usually considered to be zero (see Figure 12.5(a), where the amplitude is constant, which is consistent with $f = 0$ Hz).

A further reduction is possible by defining the complex spectral amplitude components, $X_n = (A_n - \mathrm{j}B_n)/2$ and $X_{-n} = (A_n + \mathrm{j}B_n)/2$. Denoting the complex conjugate by *, the following relation may be written as:

$$X_n = X^*_{-n} \tag{12.6}$$

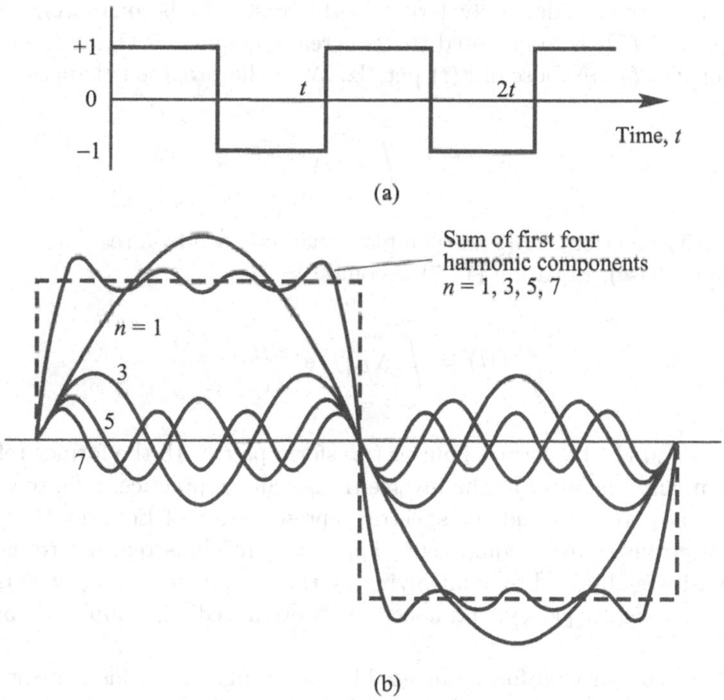

FIGURE 12.2 Example of Fourier analysis of a square wave: (a) periodic square wave in the time domain; (b) the first four harmonic components of the square wave in (a).

The introduction of Equation (12.6) in Equation (12.5) allows the following more compact expression to be written as:

$$x(t) = \sum_{n=-\infty}^{\infty} X_n\,\mathrm{e}^{\mathrm{j}2\pi n f_1 t} \tag{12.7}$$

The spectrum of Equation (12.7) now includes negative as well as positive values of n, giving rise to components $-nf_1$. The spectrum is said to be two sided. The spectral amplitude components, X_n, may be calculated using:

$$X_n = \frac{1}{T_p} \int_{-T_p/2}^{T_p/2} x(t)\, \mathrm{e}^{-\mathrm{j}2\pi n f_1 t}\, \mathrm{d}t \tag{12.8}$$

The spectrum of squared amplitudes is known as the power spectrum. The mean of the instantaneous power of the time-varying signal, $[x(t)]^2$, averaged over the period, T_A, is:

$$W_{\mathrm{mean}} = \frac{1}{T_A} \int_0^{T_A} [x(t)]^2\, \mathrm{d}t \tag{12.9}$$

Substitution of Equation (12.4) in Equation (12.9) and integrating gives:

$$W_{\mathrm{mean}} = \frac{1}{2} \sum_{n=1}^{\infty} \left[A_n^2 + B_n^2\right] \tag{12.10}$$

Equation (12.10) shows that the total power is the sum of the powers of each spectral component.

The previous analysis may be extended to the more general case of random noise by allowing the period, T_p, to become indefinitely large. In this case, X_n becomes $X_D(f)$, a continuous function of frequency, f. It is to be noted that whereas the units of X_n are the same as those of $x(t)$, the units of $X_D(f)$ are those of $x(t)$ per Hz. With the proposed changes, Equation (12.8) takes the form:

$$X_D(f) = \int_{-\infty}^{\infty} x(t)\, \mathrm{e}^{-\mathrm{j}2\pi f t}\, \mathrm{d}t \tag{12.11}$$

The spectral density function, $X_D(f)$, is complex, characterised by a real and an imaginary part (or amplitude and phase). Equation (12.7) becomes:

$$x(t) = \int_{-\infty}^{\infty} X_D(f)\, \mathrm{e}^{\mathrm{j}2\pi f t}\, \mathrm{d}f \tag{12.12}$$

Equations (12.11) and (12.12) form a Fourier transform pair, with the former referred to as the forward transform and the latter as the inverse transform. In practice, a finite sample time, T_s, is always used to acquire data and the spectral representation of Equation (12.7) is the result calculated by spectrum analysis equipment. This latter result is referred to as the spectrum. The spectral density is obtained by multiplying by the sample period, T_s, which is the same as dividing by the sampling frequency, and hence has "normalised" the amplitude by the frequency resolution.

The form of the Fourier transform pair used in spectrum analysis instrumentation is referred to as the discrete Fourier transform (DFT), for which the functions are sampled in both the time and frequency domains. The forward transform (from time to frequency domains), is often written as (Brandt, 2010, p. 180):

$$X(k) = \frac{1}{N} \sum_{n=0}^{N-1} x(n)\, \mathrm{e}^{-\mathrm{j}2\pi n k/N} \qquad k = 0, 1, \;\ldots,\; (N-1) \tag{12.13}$$

where $X(k)$ are complex numbers that can be represented by an amplitude and a phase. The inverse transform, IDFT (from frequency to time domains), is:

$$x(n) = \sum_{k=0}^{N-1} X(k)\,\mathrm{e}^{\mathrm{j}2\pi nk/N} \qquad n = 0, 1, \ \ldots, \ (N-1) \tag{12.14}$$

where n and k represent discrete sample numbers in the time and frequency domains, respectively, and $|X(k)|$ represents the amplitude (requiring scaling to actually be the amplitude) of the kth component in the frequency spectrum. An alternative way of writing Equations (12.13) and (12.14) is to place the $1/N$ term preceding the sum in Equation (12.13) and remove it from Equation (12.14) (Randall, 1987, p. 28). In Equation (12.13), the spacing between frequency components, in Hz, is dependent on the time, T_s, to acquire the N samples of data in the time domain and is equal to $1/T_s$ or f_s/N, where f_s is the sampling frequency in the time domain. The actual frequency, f_k, corresponding to the kth spectral component and the time, t_n, corresponding to the nth time sample are then:

$$f_k = kf_s/N \ \text{ and } \ t_n = n/f_s \tag{12.15}$$

Use of Equation (12.15), allows the $X(k)$ and $x(n)$ in Equations (12.13) and (12.14) to be replaced with $X(f_k)$ and $x(t_n)$, respectively. This terminology will be used for the remainder of this chapter so that spectral terms are written in terms of frequency rather than sample number, as is traditionally done.

The four Fourier transform pairs are shown graphically in Figure 12.3. In Equations (12.13) and (12.14), the functions have not been made symmetrical about the origin, but because of the periodicity of each, the second half of each sum also represents the negative half period to the left of the origin, as can be seen by inspection of Figure 12.3(d). Spectra that include both positive and negative frequencies are referred to as "two-sided spectra". These can be converted to single-sided spectra with only positive frequencies by doubling the amplitudes on the positive frequency side and deleting the negative frequency side.

The frequency components above $f_s/2$ in Figure 12.3(d) can be more easily visualised as negative frequency components and, in practice, the frequency content of the final spectrum must be restricted to less than $f_s/2$. This is explained in Section 12.3.6 where aliasing is discussed.

The discrete Fourier transform is well suited to the digital computations performed by instrumentation or by frequency analysis software on a personal computer. Nevertheless, to obtain N frequency components from N time samples, N^2 complex multiplications are required (Brandt, 2010, p. 181). Fortunately, this is reduced, by the use of the fast Fourier transform (FFT) algorithm, to $N \log_2 N$, which, for a typical case of $N = 1024$, speeds up computations by a factor of 100. This algorithm is discussed in detail by Randall (1987), Kraniauskas (1994) and Brandt (2010).

12.3.1 Relationships Between Various Spectral Quantities

Table 12.2 shows the relationships between the various quantities used in the discussion of spectra in this chapter. The only quantity not defined so far is the % overlap, P, and this is discussed in detail in Section 12.3.7.

12.3.2 Auto Power Spectrum and Power Spectral Density

The auto power spectrum (sometimes called the power spectrum or the autospectrum) is the most common form of spectral representation used in acoustics and vibration. The auto power spectrum is the spectrum of the square of the RMS values of each frequency component, whereas

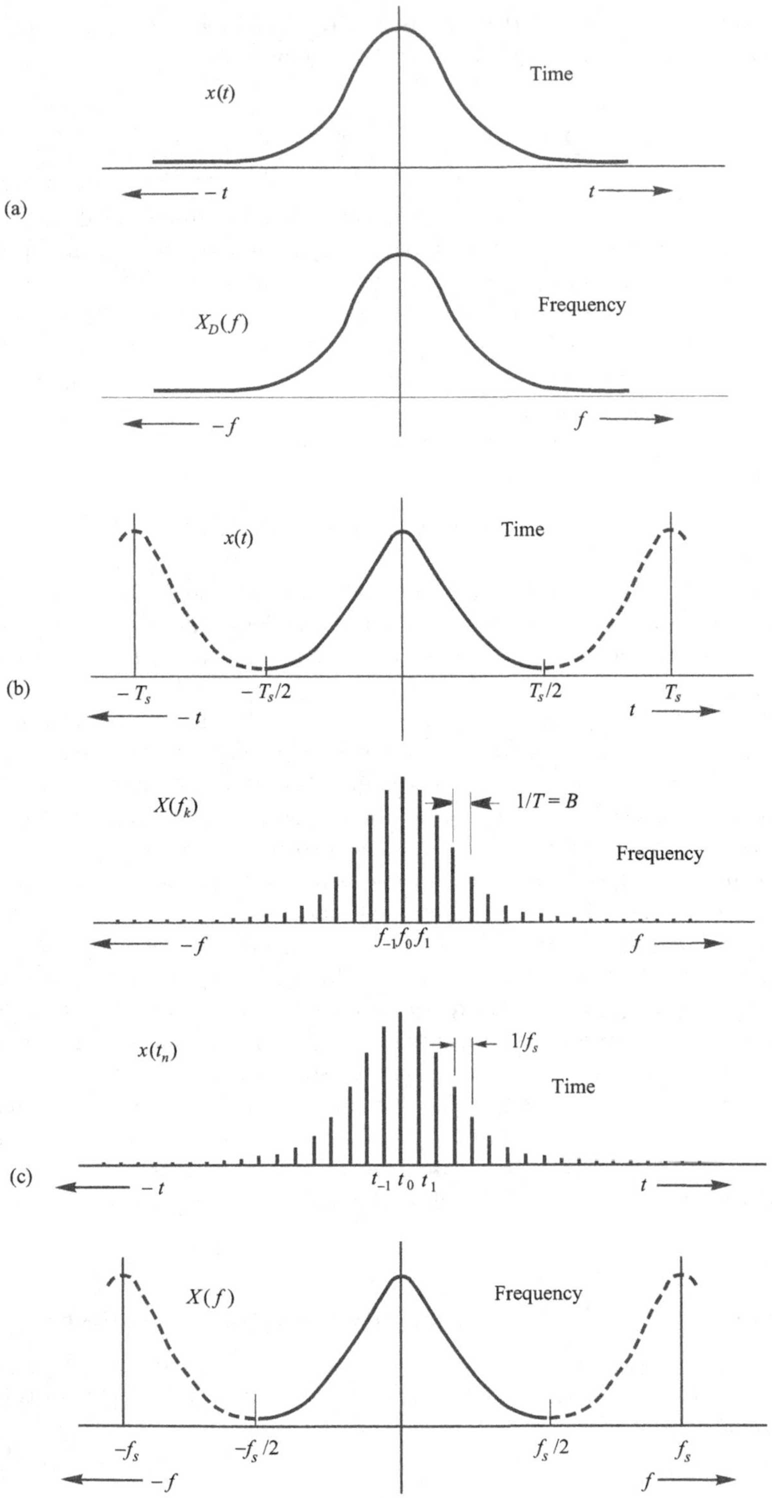

Cont. on next page.

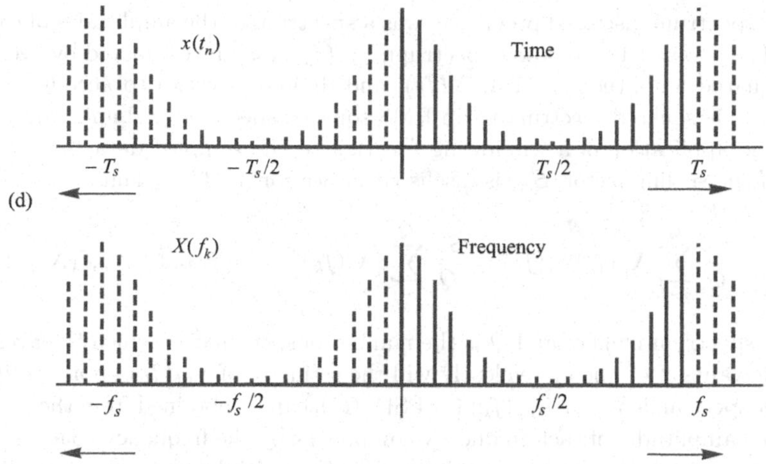

FIGURE 12.3 (Cont.) Various Fourier transform pairs (after Randall (1987)). The dashed lines indicate a periodically repeating sequence:

(a) Integral transform; signal infinitely long and continuous in both the time and frequency domains

$$X_D(f) = \int_{-\infty}^{\infty} x(t)e^{-j2\pi ft}\,dt \qquad \text{and} \qquad x(t) = \int_{-\infty}^{\infty} X_D(f)e^{j2\pi ft}\,df;$$

(b) Fourier series; signal periodic in the time domain and discrete in the frequency domain

$$X(f_k) = \frac{1}{T_s}\int_{-T_s/2}^{T_s/2} x(t)e^{-j2\pi f_k t}\,dt \qquad \text{and} \qquad x(t) = \sum_{k=-\infty}^{\infty} X(f_k)e^{j2\pi f_k t};$$

(c) Sampled function; signal discrete in the time domain and periodic in the frequency domain

$$X(f) = \sum_{n=-\infty}^{\infty} x(t_n)e^{-j2\pi f t_n} \qquad \text{and} \qquad x(t_n) = \frac{1}{f_s}\int_{-f_s/2}^{f_s/2} X(f)e^{j2\pi f t_n}\,df;$$

(d) Discrete Fourier transform; signal discrete and periodic in both the time and frequency domains

$$X(f_k) = \frac{1}{N}\sum_{n=0}^{N-1} x(t_n)e^{-j2\pi n f_k/N} \qquad \text{and} \qquad x(t_n) = \sum_{k=0}^{N-1} X(f_k)e^{j2\pi t_n k/N}.$$

Note that $X(f_k)$ is a spectrum of amplitudes that must be divided by $\sqrt{2}$ to obtain an RMS spectrum.

TABLE 12.2 Relationships between various spectral quantities

Quantity	Relationship to other quantities
Sample rate, f_s (Hz)	$f_s = N\Delta f = N/t$
Time, t, to acquire 1 spectrum (s)	$t = N/f_s = 1/\Delta f$
Number of samples, N, in one spectrum	$N = f_s/\Delta f = f_s t$
Frequency resolution, Δf (Hz)	$\Delta f = f_s/N = 1/t$
Number of spectra to average, n	$n = f_s T/N$
Total sampling time, T (s) with no overlap	$T = nt = nN/f_s = n/\Delta f$
Percent overlap, P, of adjacent spectra	
Total sampling time, T_0 (s), with P percent overlap	$T_0 = nNP/(100 f_s) = nP/(100\Delta f)$

the frequency spectrum discussed previously was a spectrum of the amplitudes of each frequency component. The two-sided auto power spectrum, $S_{xx}(f_k)$, may be estimated by averaging a large number of squared amplitude spectra, $X(f_k)$, and dividing by two to account for conversion from an amplitude squared spectrum to an RMS squared spectrum and an additional scaling to account for the application of a windowing function to the sampled data (see Section 12.3.5). Estimation of the scaling factor, S_A, is discussed in Section 12.3.5.1. Thus:

$$S_{xx}(f_k) \approx \frac{S_A}{Q}\sum_{i=1}^{Q} X_i^*(f_k)X_i(f_k) = \frac{S_A}{Q}\sum_{i=1}^{Q}|X_i(f_k)|^2 \qquad n = 0, 1, \ \ldots, \ (N-1) \tag{12.16}$$

where i is the spectrum number and Q is the number of spectra over which the average is taken. The larger the value of Q, the more closely will the estimate of $S_{xx}(f_k)$ approach its true value.

The power spectral density, $S_{Dxx}(f_k)$ (or PSD(f_k)), can be obtained from the power spectrum by dividing the amplitudes of each frequency component by the frequency spacing, Δf, between adjacent components in the frequency spectrum or by multiplying by the time, T_s, to acquire one record of data. Thus, the two-sided power spectral density is:

$$S_{Dxx}(f_k) \approx \frac{T_s S_A}{Q}\sum_{i=1}^{Q}|X_i(f_k)|^2 = \frac{S_A}{Q\Delta f}\sum_{i=1}^{Q}|X_i(f_k)|^2 \qquad n = 0, 1, \ \ldots, \ (N-1) \tag{12.17}$$

where "two-sided" indicates that the spectrum extends to negative as well as positive frequencies. The time blocks are usually overlapped by up to 50% to decrease the random error in the PSD estimate (Brandt, 2010), as explained in Section 12.3.7.

Although it is often appropriate to express random noise spectra in terms of power spectral density, the same is not true for tonal components. The RMS value of a tone can be read directly from an auto power spectrum, but the value in a PSD must be multiplied by the frequency resolution of the spectrum as well as by the normalised noise bandwidth, which is dependent on which windowing function is used (see Table 12.3).

The auto power spectrum is useful for evaluating tonal components in a spectrum, although for random noise, it is more appropriate to use the power spectral density or PSD function of Equation (12.17). The PSD is a measure of the energy in a 1 Hz-wide frequency band, centred on the spectral line of interest. Thus, a PSD can have any frequency resolution, but the value on the ordinate axis always represents a bandwidth of 1 Hz. Unlike the power spectrum, the value of the PSD is independent of the frequency resolution used to obtain it.

The auto power spectrum is used to evaluate spectra that contain tonal components because, unlike the PSD, it is able to give the true energy content of a tonal component. This is because the bandwidth of a tone is not the same as the frequency spacing in the spectrum and is often much smaller. This results in the spectral amplitude of a tone being independent of the frequency resolution of the FFT analysis, provided that the tonal frequency corresponds to the frequency of one of the spectral lines (see Section 12.3.4 and Section 12.3.5). Thus, dividing a tonal amplitude by the spectral resolution to obtain the PSD will result in a significant error in the tonal amplitude. In real systems, the frequency of a tone may vary slightly during the time it takes to acquire a sufficient number of samples for an FFT and also from one FFT to another during the averaging process to obtain power spectra. In this case, the tone may be spread out in frequency so that its amplitude will depend on the frequency resolution, and a better estimate of the amplitude will be obtained with a coarse frequency resolution. A sufficiently coarse resolution would enable the range of frequency variation to be captured in a single frequency bin in the spectrum.

In cases where the frequency of the tone does not correspond to the centre frequency of one of the spectral lines, there will be an error in its amplitude that will depend on the windowing function used (see Section 12.3.5) and the difference in frequency between the tone and the centre frequency of the nearest spectral line. The maximum possible error is listed in Table 12.3. In this case, the amplitude of the tone will also depend on the frequency resolution of the spectrum.

The problems of errors in tonal amplitudes can be avoided by calculating the PSD from the auto-correlation function, $R_{xx}(\tau)$, which is the covariance of the time series signal, $x(t)$, with itself time shifted by τ seconds. It is defined as:

$$R_{xx}(\tau) = \langle x(t)x(t+\tau)\rangle_t = \mathrm{E}[x(t)\cdot x(t+\tau)] = \lim_{T_s\to\infty}\frac{1}{T_s}\int\limits_{-T_s/2}^{T_s/2} x(t)\cdot x(t+\tau)\,\mathrm{d}t \tag{12.18}$$

where $\mathrm{E}[x]$ is the expected value of x, and $\langle x\rangle_t$ is the time-averaged value of x. The auto-correlation function is discussed in more detail in Section 12.3.17. The PSD, $S_{Dxx}(f_k)$, is obtained from the auto-correlation function by substituting $R_{xx}(\tau)$ for $x(t)$, τ for t, $S_{Dxx}(f_k)$ for $X(f_k)$ and $S_{Dxx}(f)$ for $X_D(f)$ in the equations in the caption of Figure 12.3. For example, the caption for part (c) in the figure is written as:

$$S_{Dxx}(f) = \sum_{k=-\infty}^{\infty} R_{xx}(\tau_k)\mathrm{e}^{-\mathrm{j}2\pi f\tau_k} \tag{12.19}$$

and

$$R_{xx}(\tau_k) = \frac{1}{f_s}\int\limits_{-f_s/2}^{f_s/2} S_{Dxx}(f)\mathrm{e}^{\mathrm{j}2\pi f\tau_k}\,\mathrm{d}f \tag{12.20}$$

However, the auto-correlation function is very computationally intensive to calculate, so this method is not in common use. In fact, the reverse is more often the case: the auto correlation of a dataset is found by taking the inverse FFT of the power spectral density.

For random noise, the frequency resolution affects the spectrum amplitude; the finer the resolution the smaller will be the amplitude. For this reason, we use PSDs for random noise for which the effective frequency resolution is 1 Hz.

In practice, the single-sided power spectrum, $G_{xx}(f_k)$ (positive frequencies only), is the one of interest, and this is expressed in terms of the two-sided auto power spectrum $S_{xx}(f_k)$ as:

$$G_{xx}(f_k) = \begin{cases} 0 & f_k < 0 \\ S_{xx}(f_k); & f_k = 0 \\ 2S_{xx}(f_k); & f_k > 0 \end{cases} \tag{12.21}$$

A similar expression may be written for the single-sided PSD, $G_{Dxx}(f_k)$, as:

$$G_{Dxx}(f_k) = \begin{cases} 0 & f_k < 0 \\ S_{Dxx}(f_k); & f_k = 0 \\ 2S_{Dxx}(f_k); & f_k > 0 \end{cases} \tag{12.22}$$

If sufficient successive spectra, $X_i(f_k)$, are averaged, the result will be zero, as the phases of each spectral component vary randomly from one record to the next. Thus, in practice, auto power spectra are more commonly used, as they can be averaged together to give a more accurate result. This is because auto power spectra are only represented by an amplitude; phase information is lost when the spectra are calculated (see Equation (12.16)). The same reasoning applies to the power spectral density (power per Hz), which is obtained from the auto power spectrum by dividing the amplitude of each frequency component by the frequency spacing, Δf, between adjacent components.

12.3.3 Linear Spectrum

Sometimes the results of a spectral analysis are presented in terms of linear rather than the squared values of an auto power spectrum. Each frequency component in the linear spectrum is calculated by taking the square root of each frequency component in the auto power spectrum.

12.3.4 Leakage

Leakage is the phenomenon that occurs when a DFT uses a finite time window. This results in a spectrum containing discrete frequency components separated by a frequency interval. The number of frequency components and the frequency separation interval, Δf, are set by the sampling frequency, f_s, and the total sampling time (or measurement time), T_s. The frequency separation between adjacent components in the spectrum is given by:

$$\Delta f = \frac{1}{T_s} \tag{12.23}$$

and the number of discrete frequency components, N, in the spectrum is given by:

$$N = T_s f_s \tag{12.24}$$

Each discrete frequency component (or spectral line) in the spectrum is like a bandpass filter with a characteristic response, $W(f_k)$, defined by a sinc function as:

$$W(f_k) = T_s \frac{\sin(\pi f_k T_s)}{\pi f_k T_s} = T_s \operatorname{sinc}(\pi f_k T_s) \qquad \text{where } n = 0, 1, \ \ldots, \ (N-1) \tag{12.25}$$

This means that a sinusoidal signal equal to the exact frequency of a spectral component will be given the correct amplitude value in the frequency spectrum, as the sinc function has a value of unity at frequency, f_k. However, for a sinusoidal signal with a frequency halfway between two spectral lines, the energy of the signal will be split between the two adjacent lines (or frequency bins) and neither line will give the correct result. In fact, for this case, in the absence of any windowing, the error will result in a value that is 36% (or 1.96 dB) too small.

12.3.5 Windowing

When a Fourier analysis is undertaken using the DFT algorithm with a finite sampling period, the resulting frequency spectrum is divided into a number of bands of finite width. Each band may be considered as a filter, the shape of which is dependent on how each sample in a record is weighted. Weighting is necessary to place less importance on the values near the beginning and end of each record. This is because leakage (see Section 12.3.4) occurs when calculating the DFT of a sinusoidal signal with a non-integer number of periods in the time window used for sampling. The error is caused by the truncation of the continuous signal, as a result of using a finite time window, which causes a discontinuity when the two ends of the record are effectively joined in a loop as a result of the DFT. Leakage can be reduced by using a windowing function applied to the time window such that all samples in the time record are not given equal weighting when calculating the DFT. In fact, the window may be configured so that samples near the beginning and end of the time window are weighted much less than samples in the centre of the window. This minimises the effect of the signal being discontinuous at the beginning and end of the time window. The discontinuity without weighting causes side lobes to appear in the spectrum for a single frequency, as shown by the solid curve in Figure 12.4, which is effectively the same as applying a rectangular window weighting function. In this case, all signal samples before sampling begins and after it ends are multiplied by zero, and all values in between are multiplied

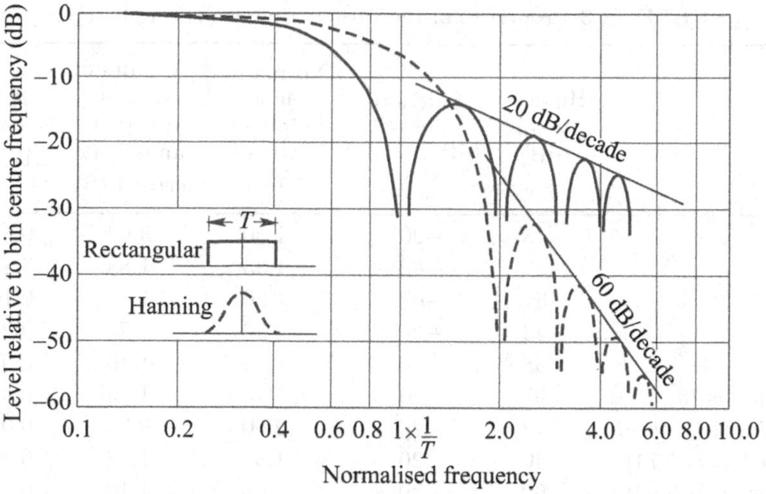

FIGURE 12.4 Comparison of the filter characteristics of the rectangular (solid curve) and Hanning (dashed curve) time weighting functions for a power spectrum (after Randall (1987)).

by one. In the figure, the normalised frequency of $1 \times 1/T$ represents the frequency resolution, or number of Hz between adjacent frequency bins in the spectrum.

A better choice of window is one that places less weight on the signal at either end of the window and maximum weight on the middle of the window. One such weighting, called a Hanning window, is also illustrated in Figure 12.4. The result of weighting the input signal in this way is shown by the dashed curve in the figure. Even though the main lobe is wider, representing poorer frequency resolution, the side lobe amplitudes fall away more rapidly, resulting in less contamination of adjacent frequency bins. The properties of various weighting functions are summarised in Table 12.3. Properties of many more weighting functions are available from Harris (1978).

In the table, the highest side lobe is the number of dB (in the auto power spectrum) that the signal corresponding to the highest side lobe will be attenuated compared to a signal at the filter centre frequency. The "side lobe" fall off is illustrated in Figure 12.4, where the side lobes are the peaks to the right of the normalised frequency of 1.0.

The noise bandwidth in Table 12.3 is an important quantity in spectrum analysis. It is defined as the bandwidth of a rectangular filter that would let pass the same amount of broadband noise as the filter under consideration. It is especially useful in calculating the RMS level of power in a certain bandwidth in a spectrum such as a 1/3-octave band. This is discussed in more detail in Section 12.3.5.3.

The maximum amplitude error is the amount that a signal will be attenuated when it has a frequency that lies exactly midway between the centre frequencies of two adjacent filters (corresponding to a normalised frequency of 0.5 in Figure 12.4). As can be seen in Table 12.3, the maximum amplitude error varies from 0.0 dB for the flat top window to 3.9 dB for the rectangular window. As expected, the 3.92 dB value is twice the value of 1.96 dB mentioned in Section 12.3.4 for the amplitude spectrum, as squaring results in a factor of two in the logarithmic domain. For the flat top window, the frequency bands are 3.77 times wider than for the rectangular window, so the frequency resolution is 3.77 times poorer. As a result of its high-amplitude accuracy in the frequency domain, the flat top window (the name refers to the weighting in the frequency domain whereas the window shape refers to the weighting in the time domain) is the best weighting function for calibration. The high-amplitude accuracy results in the measured amplitude of the spectral component being independent of small variations in signal frequency around the band centre frequency, thus making this window suitable for instrument calibration with a tonal signal.

TABLE 12.3 Properties of the various time weighting functions

Window type	Highest side lobe (dB)	Side lobe fall off (dB/decade)	Normalised noise bandwidth (bins) B_{en}	Maximum power spectrum amplitude error (dB)	$\rho(1)$ Equation (12.57)	Window total energy (dB)
Rectangular	−13	−20	1.00	3.92	0.5	1.0
Triangular	−27	−40	1.33	1.83	0.25	—
Hanning	−32	−60	1.50	1.43	0.167	−4.26
Hamming	−43	−20	1.36	1.75	0.235	−4.01
Blackman	−58	−20	1.73	1.10	0.09	—
Blackman–Harris (3 term)	−67	−20	1.70	1.13	0.096	−5.13
Blackman–Harris (4 term)	−92	−20	2.00	0.83	0.038	−5.88
Kaiser–Bessel ($\alpha = 1.74$)	−40	−20	1.41	1.62	0.208	—
Kaiser–Bessel ($\alpha = 2.59$)	−60	−20	1.68	1.16	0.103	—
Kaiser–Bessel ($\alpha = 3.39$)	−80	−20	1.90	0.9	0.053	—
Kaiser–Bessel ($\alpha = 3.5$)	−82	−20	1.93	0.9	—	—
Gaussian ($\alpha = 3.0$)	−55	−20	1.64	1.25	0.106	−4.51
Poisson ($\alpha = 3.0$)	−24	−20	1.65	1.46	0.151	—
Poisson ($\alpha = 4.0$)	−31	−20	2.08	1.03	0.074	—
Flat top	−93	−20	3.77	0.01	—	−7.56
Welch	−21	−36.5	1.20	2.22	0.345	−2.73

However, the flat top window provides poor frequency resolution. Maximum frequency resolution (and minimum amplitude accuracy) is achieved with the rectangular window, so this is sometimes used to separate two spectral peaks that have a similar amplitude and a small frequency spacing. Good compromises that are commonly used are the Hanning window or the Hamming window, the former used when amplitude accuracy is more important and the latter for separation of closely spaced frequencies.

A signal at a particular frequency will also contribute to the energy in other nearby bands, as can be seen by the shape of the filter curve in Figure 12.4. This spectral leakage can be minimised by having a high negative value for the side lobe fall off in Table 12.3.

When transient signals (that is, signals that occur for a time shorter than the sampling interval) are to be analysed, the best window is a rectangular one. However, if the transients are repetitive and several occur during a data sampling period, then a Hanning weighting function is better.

12.3.5.1 Amplitude Scaling to Compensate for Window Effects

The effect of a non-rectangular window is to remove information and energy from the signal, resulting in an amplitude error. This must be compensated for by using an amplitude correction factor, A_f. This amplitude correction factor is used to calculate a two-sided scaled amplitude spectrum $X_s(f_k)$ from Equation (12.13) as:

$$X_s(f_k) = \frac{A_f}{N} \sum_{n=0}^{N-1} x(t_n) w(n) \mathrm{e}^{-\mathrm{j}2\pi f_k n/N} \qquad \text{where } n = 0, 1, \ \ldots, \ (N-1) \tag{12.26}$$

where A_f/N is a scaling factor, N is the number of discrete frequency components in the spectrum and:

$$A_f = \frac{N}{\sum\limits_{n=0}^{N-1} w(n)} \tag{12.27}$$

where $w(n)$ is the window weighting function (see Section 12.3.5.2) for each sample, n, in the time domain used to calculate the frequency spectrum. If $w(n) = 0$, then the nth sample value is set equal to 0 and if $w(n) = 1$, the sample value is unchanged.

The scaling factor, S_A, for the squared RMS spectrum (single-sided auto power spectrum) in Equation (12.26) is:

$$S_A = \begin{cases} 2A_f^2/N^2; & k > 0 \\ A_f^2/(N^2); & k = 0 \end{cases} \tag{12.28}$$

Taking the RMS spectrum results in dividing Equation (12.26) by $\sqrt{2}$. Scaling results in a two-sided spectrum with components with amplitudes of half the RMS value of the single sided spectrum. Squaring this spectrum then results in amplitude components equal to 1/4 of the RMS value squared and when the two sided are added, the result is half the RMS amplitude squared instead of the RMS amplitude squared. Thus the factor of 2 in Equation (12.28) is needed so that the scaled result is half the RMS amplitude squared for each frequency component in the two-sided spectrum. Also, the assumption implicit in Equation (12.28) is that the spectrum being analysed is tonal so that there is no leakage of energy into adjacent bins.

However, for an auto power spectrum containing energy other than in tones, or for a PSD, an additional term must be included in the scaling factor to account for leakage of energy into adjacent bins as a result of the application of a windowing function. Thus, the scaling factor to be used is given by:

$$S_A = \begin{cases} 2A_f^2/(N^2 B_{en}); & k > 0 \\ A_f^2/(N^2 B_{en}); & k = 0 \end{cases} \tag{12.29}$$

where B_{en} is given in Table 12.3 for various window functions and is defined by:

$$B_{en} = \frac{N \sum\limits_{n=0}^{N-1} w^2(n)}{\left(\sum\limits_{n=0}^{N-1} w(n)\right)^2} \tag{12.30}$$

Equations (12.17), (12.29) and (12.30) are described as a Welch estimate of the PSD and represent the most commonly used method of spectral analysis in instrumentation and computer software, such as the `pwelch` function from MATLAB®. However, this method for obtaining the PSD has associated bias errors that decrease as the frequency resolution in the original frequency spectrum, $X(f_k)$, becomes finer (i.e. smaller). An estimate of the bias error, ϵ_b, for the particular case of a Hanning window with a frequency resolution of Δf is (Schmidt, 1985a,b):

$$\epsilon_b \approx \frac{(\Delta f)^2 G''_{xx}(f_k)}{6\, G_{xx}(f_k)} + \frac{(\Delta f)^4 G''''_{xx}(f_k)}{72\, G_{xx}(f_k)} \tag{12.31}$$

where the prime represents differentiation with respect to frequency and 4 primes represent the fourth derivative. It can be seen from Equation (12.31) that where there are tones (which produce large values of the second derivative in particular), the error will be large.

12.3.5.2 Window Function Coefficients

Each windowing function identified in Table 12.3 requires different equations to calculate the coefficients, $w(k)$. The coefficients represent the quantity that data sample, n, in the time series data is multiplied by, before being included in the dataset used for taking the DFT. These equations are listed below for each of the windows identified in Table 12.3.

Rectangular Window
For a record in the time domain that is a total of N samples in length (producing N discrete frequency components in the frequency domain), the window coefficients corresponding to each sample, k, in the time series record are given by:

$$w(n) = 1; \qquad 1 \leq n \leq N \tag{12.32}$$

Triangular Window
For a record that is a total of N samples in length, the window coefficients corresponding to each sample, n, in the record for N odd are given by:

$$w(n) = \begin{cases} 2n/(N+1); & 1 \leq n \leq (N+1)/2 \\ 2 - 2n/(N+1); & \dfrac{(N+1)}{2} + 1 \leq n \leq N \end{cases} \tag{12.33}$$

and for N even, the coefficients are given by:

$$w(n) = \begin{cases} (2n-1)/N; & 1 \leq n \leq N/2 \\ 2 - (2n-1)/N; & \dfrac{N}{2} + 1 \leq n \leq N \end{cases} \tag{12.34}$$

The coefficients of another version of a triangular window are given by:

$$w(n) = 1 - \left| \frac{2n - N + 1}{L} \right| \tag{12.35}$$

where L can be $N, N+1$, or $N-1$. All alternatives converge for large N.

Hamming Window
For a record that is a total of N samples in length, the window coefficients corresponding to each sample, n, in the record are given by:

$$w(n) = \alpha - \beta \cos\left(\frac{2\pi n}{N-1}\right); \qquad 0 \leq n \leq N-1 \tag{12.36}$$

where the optimum values for α and β are 0.54 and 0.46, respectively.

Hanning Window
The Hanning window is the one most commonly used in spectrum analysis and is the one recommended for PSD analysis.

For a record that is a total of N samples in length, the window coefficients corresponding to each sample, n, in the record are given by:

$$w(n) = 0.5\left[1 - \cos\left(\frac{2\pi n}{N-1}\right)\right]; \qquad 0 \leq n \leq N-1 \tag{12.37}$$

Blackman Window

For a record that is a total of N samples in length, the window coefficients corresponding to each sample, n, in the time series record are given by:

$$w(n) = a_0 - a_1 \cos\left(\frac{2\pi n}{N-1}\right) + a_2 \cos\left(\frac{4\pi n}{N-1}\right); \qquad 0 \le n \le N-1 \tag{12.38}$$

where $a_0 = (1-\alpha)/2$, $a_1 = 1/2$ and $a_2 = \alpha/2$. For an unqualified Blackman window, $\alpha = 0.16$.

Blackman-Harris Window

For a record that is a total of N samples in length, the window coefficients corresponding to each sample, n, in the time series record for a three-term window are given by:

$$w(n) = a_0 - a_1 \cos\left(\frac{2\pi n}{N-1}\right) + a_2 \cos\left(\frac{4\pi n}{N-1}\right); \qquad 0 \le n \le N-1 \tag{12.39}$$

where $a_0 = 0.42323$, $a_1 = 0.49755$ and $a_2 = 0.07922$.

For a four-term window, the equation for the coefficients is given by (Harris, 1979):

$$w(n) = a_0 - a_1 \cos\left(\frac{2\pi n}{N-1}\right) + a_2 \cos\left(\frac{4\pi n}{N-1}\right) - a_3 \cos\left(\frac{6\pi n}{N-1}\right); \qquad 0 \le n \le N-1 \tag{12.40}$$

where $a_0 = 0.35875$, $a_1 = 0.48829$, $a_2 = 0.14128$ and $a_3 = 0.01168$.

Kaiser-Bessel Window

For a record that is a total of N samples in length, the window coefficients corresponding to each sample, n, in the record are given by:

$$w(n) = \frac{I_0\left(\pi\alpha\sqrt{1-\left(\frac{2n}{N-1}-1\right)^2}\right)}{I_0(\pi\alpha)}; \qquad 0 \le n \le N-1 \tag{12.41}$$

where I_0 is the zeroth order modified Bessel function of the first kind. The parameter, α, determines the trade-off between main lobe width and side lobe levels. Increasing α widens the main lobe and increases the attenuation of the side lobes. To obtain a Kaiser-Bessel window that provides an attenuation of β dB for the first side lobe, the following values of $\pi\alpha$ are used:

$$\pi\alpha = \begin{cases} 0.1102(\beta - 8.7); & \beta > 50 \\ 0.5842(\beta - 21)^{0.4} + 0.07886(\beta - 21); & 21 \le \beta \le 50 \\ 0; & \beta < 21 \end{cases} \tag{12.42}$$

A typical value of α is 3 and the main lobe width between the nulls is given by $2\sqrt{1+\alpha^2}$.

Gaussian Window

For a record that is a total of N samples in length, the window coefficients corresponding to each sample, n, in the record are given by:

$$w(n) = \exp\left[-\frac{1}{2}\left(\frac{n-(N-1)/2}{\sigma(N-1)/2}\right)^2\right]; \qquad 0 \le n \le N-1 \quad \text{and} \quad \sigma \le 0.5 \tag{12.43}$$

where σ is the standard deviation of the Gaussian distribution. An alternative formulation is:

$$w(n) = \exp\left[-\frac{1}{2}\left(\frac{\alpha n}{(N-1)/2}\right)^2\right]; \qquad -(N-1)/2 \leq n \leq (N-1)/2 \tag{12.44}$$

where $\alpha = (N-1)/2\sigma$ and $\sigma \leq 0.5$.

Poisson (or Exponential) Window

For a record that is a total of N samples in length, the window coefficients corresponding to each sample, n, in the record are given by:

$$w(n) = \exp\left(-\left|n - \frac{N-1}{2}\right|\frac{1}{\tau}\right) \qquad 0 \leq n \leq N-1 \tag{12.45}$$

For a targeted decay of D dB over half of the window length, the time constant, τ, is given by:

$$\tau = \frac{8.69N}{2D} \tag{12.46}$$

An alternative formulation is given by:

$$w(n) = \exp\left(\frac{-\alpha\,|n|}{(N-1)/2}\right); \qquad -(N-1)/2 \leq n \leq (N-1)/2 \tag{12.47}$$

where $\exp\{x\} = e^x$ and $\alpha = (N-1)/2\tau$.

Flat Top Window

The flat top window is used in frequency analysis mainly for calibration of instrumentation with a tone. The frequency of the calibration tone need not be close to the centre frequency of a bin in the frequency spectrum to obtain an accurate result — in fact, an accurate result will be obtained for any calibration frequency.

For a record that is a total of N samples in length, the window coefficients corresponding to each sample, n, in the record are given by:

$$w(n) = a_0 - a_1\cos\left(\frac{2\pi n}{N-1}\right) + a_2\cos\left(\frac{4\pi n}{N-1}\right) - a_3\cos\left(\frac{6\pi n}{N-1}\right) + a_4\cos\left(\frac{8\pi n}{N-1}\right)$$
$$0 \leq n \leq N-1 \tag{12.48}$$

Coefficients are defined as a_0=0.21557895, a_1=0.41663158, a_2=0.277263158, a_3=0.083578947 and a_4=0.006947368. The coefficients may also be normalised to give a_0=1.0, a_1=1.93, a_2=1.29, a_3=0.388 and a_4=0.0322.

Welch Window

For a record that is a total of N samples in length, the window coefficients corresponding to each sample, n, in the record are given by:

$$w(n) = 1 - \left(\frac{2n - N + 1}{N-1}\right)^2; \qquad 0 \leq n \leq (N-1) \tag{12.49}$$

12.3.5.3 Power Correction and RMS Calculation

It is often of interest to determine the RMS value or dB level of an auto power spectrum over a defined frequency range within the spectrum. For example, one may wish to compute 1/3-octave or octave band decibel levels from an auto power spectrum covering the range from 1 Hz to 10000 Hz. One may think that all one needs to do is add logarithmically (see Section 1.11.3) the various frequency components contained within the band of interest to obtain the required result. Adding logarithmically the frequency component amplitudes in an auto power spectrum is the same as converting the dB amplitude levels to $(\text{RMS})^2$ quantities, adding them together and then converting back to decibels. Unfortunately, finding the RMS value or dB level of a band within a spectrum is not as simple as described above, as energy from each frequency bin leaks into adjacent bins giving a result that is too large. Thus, the calculation needs to be divided by a correction factor, B_{en}, called the "normalised noise bandwidth", as defined in Equation (12.30). As can be seen in Table 12.3, B_{en} is different for each windowing function and for a Hanning window it is 1.5. The RMS value of an auto spectrum between frequency locations, f_{n_1} and f_{n_2}, is thus given by:

$$x_{\text{RMS}}(n_1, n_2) = \sqrt{\frac{\sum\limits_{k=n_1}^{n_2} G_{xx}(f_k)}{B_{en}}} \tag{12.50}$$

where $G_{xx}(f_k)$ is defined in Equation (12.21). The correction factor holds even if the spectrum consists of only a single tone. This is because even for a single tone, energy appears in the two frequency bins adjacent to the one containing the frequency of the tone.

If x represents a sound pressure, then the total sound pressure level between and including spectral lines, n_1 and n_2 is:

$$L_p(n_1, n_2) = 20\log_{10}\left[\frac{x_{\text{RMS}}(n_1, n_2)}{p_{\text{ref}}}\right] = 10\log_{10}\left[\frac{\sum\limits_{k=n_1}^{n_2} G_{xx}(f_k)}{B_{en}\, p_{\text{ref}}^2}\right] \tag{12.51}$$

Calculating the RMS value or dB level of a power spectral density (PSD) requires a slightly different process as the PSD is already scaled such that the area under the PSD versus frequency curve corresponds to the mean square of the signal. In this case, the RMS value of a signal between two spectral lines, n_1 and n_2, is given by:

$$x_{\text{RMS}}(n_1, n_2) = \sqrt{\Delta f \sum_{k=n_1}^{n_2} G_{Dxx}(f_k)} \tag{12.52}$$

where $G_{Dxx}(f_k)$ is the single sided PSD, which is defined in Equation (12.22), and $\Delta f = 1/T_s = f_s/N$ is the frequency resolution of the PSD. However, if the PSD contains prominent tones, the RMS value of the signal at a tonal frequency, f_i, is:

$$x_{\text{RMS}}(n_i) = \sqrt{B_{en} \Delta f G_{Dxx}(f_i)} \tag{12.53}$$

If x_{RMS} represents RMS sound pressure, the total sound pressure level in dB re 20 µPa between and including lines n_1 and n_2 is, $L_p = 20\log_{10}\left[x_{\text{RMS}}(n_1, n_2)/p_{\text{ref}}\right]$.

Even though the PSD represents a power averaged over 1 Hz, it can still be expressed in a finer resolution than 1 Hz. In that case, each spectral line value represents the total energy in a 1 Hz bandwidth centred on the spectral line frequency value and, unlike the power spectrum, the result for a particular frequency is then independent of the frequency resolution used to express the result.

We can use Equations (12.50) and (12.52) to obtain a power spectrum, $G_{xx}(f_k)$, from a PSD, $G_{Dxx}(f_k)$, with the same frequency resolution, Δf, noting that the resulting power spectrum must be divided by B_{en} to obtain $x_{\rm RMS}(n_1, n_2)$. Thus:

$$G_{xx}(f_k) = \Delta f G_{Dxx}(f_k) \tag{12.54}$$

However, if the spectrum contains prominent tones, the power spectrum at the tonal frequencies is related to the PSD as:

$$G_{xx}(f_k) = B_{en} \Delta f G_{Dxx}(f_k) \tag{12.55}$$

12.3.6 Sampling Frequency and Aliasing

The sampling frequency is the frequency at which the input signal is digitally sampled. If the signal contains frequencies greater than half the sampling frequency, then these will be "folded back" and appear as frequencies less than half the sampling frequency. For example, if the sampling frequency is 20000 Hz and the signal contains a component with a frequency of 25000 Hz, then this will appear as 5000 Hz in the resulting spectrum. Similarly, if the signal contains a component that has a frequency of 15000 Hz, this signal also will appear as 5000 Hz in the resulting spectrum. This phenomenon is known as "aliasing" and in a spectrum analyser it is important to have analogue filters that have a sharp roll off for frequencies above about 0.4 times the sampling frequency. Aliasing is illustrated in Figure 12.5.

12.3.7 Overlap Processing

When a limited amount of time is available for collecting data, more accurate results can be obtained by implementing overlap processing, as this allows more spectra to be averaged. For overlap processing, the time series data are divided into a number of records and then a DFT is performed on each segment. For an example case of a 50% overlap, this means that the first segment analysed is the first time record, the second segment is the second half of the first time record appended to the beginning of the first half of the second record, the third segment is the second record, the fourth segment is the second half of the second record, appended to the beginning of the first half of the third record, etc. Even though the same data are used in more than one DFT, the effect of overlap analysis is to provide more spectra to average, which results in a smaller error in the final averaged spectrum. However, the effective number of averages is slightly less than the actual number of averages when overlap processing is used. The effective number of averages is window dependent and can be calculated as:

$$\frac{Q_e}{Q_d} = \frac{Q/Q_d}{1 + 2\sum\limits_{i=1}^{Q} \dfrac{Q-i}{Q}\rho(i)} \tag{12.56}$$

where Q is the number of overlapping segments (or records) used, each of which contains N samples, Q_d is the number of segments that would exist with no overlap and Q_e is the effective or equivalent number of averages, which give the same variance or uncertainty in the averaged DFT as the same number of averages using independent data. The quantity $\rho(i)$ is defined as:

$$\rho(i) = \frac{\left[\sum\limits_{n=0}^{N-1} w(n)w(n+iD)\right]^2}{\left[\sum\limits_{n=0}^{N-1} w^2(n)\right]^2} \tag{12.57}$$

where N is the number of discrete frequency components in the spectrum, $w(n)$ is dependent on the windowing function used (see Section 12.3.5.2), $D = \text{round}[N(1 - P/100)]$ and P is the

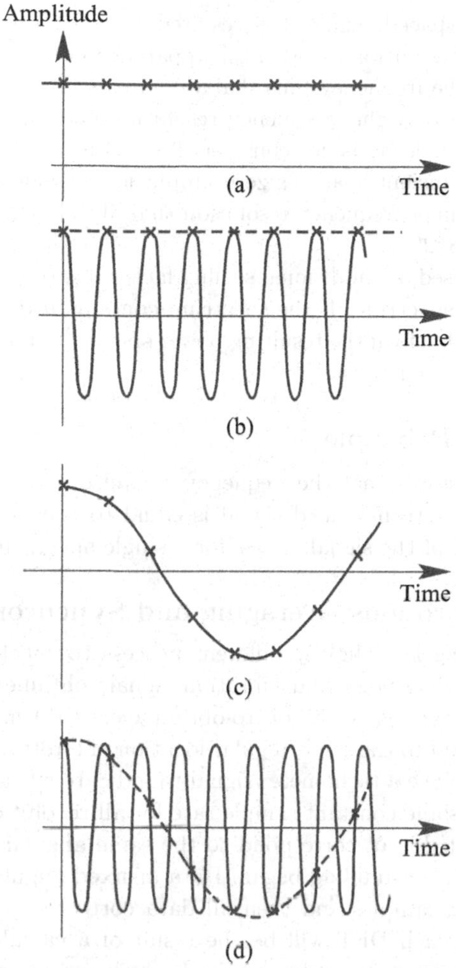

FIGURE 12.5 Illustration of aliasing (after Randall (1987)): (a) zero frequency or DC component; (b) spectrum component at sampling frequency, f_s, interpreted as DC; (c) spectrum component at $(1/N)f_s$; (d) spectrum component at $[(N+1)/N]f_s$ interpreted as $(1/N)f_s$.

percent overlap. For an overlap percentage up to 50% (optimal for Hanning weighting), the only non-zero value of $\rho(i)$ is when $i = 1$. Values of $\rho(1)$ for various windows are included in Table 12.3 for a 50% overlap. Note that "round" means rounding to the nearest integer.

Overlap processing is particularly useful when constructing sonograms (3-D plots of amplitude versus frequency with the third axis being time), as overlap processing results in smaller time intervals between adjacent spectra, resulting in better time resolution. For example, with 50% overlap, three spectra are obtained with the same number of samples, thus representing the same time period as two spectra with non-overlap processing.

Overlap processing can also be used to improve the frequency resolution, Δf, by using more samples, N, in the DFT ($\Delta f = f_s/N$) than is used with non-overlap processing for the same number of effective averages. This is useful when there is a limited length dataset and the maximum possible accuracy and frequency resolution is needed.

12.3.8 Zero Padding

Zero padding is the process of adding zeros to extend the number of samples in a record in the time domain prior to taking the DFT. This results in a frequency spectrum with frequency components more tightly spaced, which has resulted in some users thinking that they have achieved a finer frequency resolution. In fact, the apparent finer frequency resolution is actually an interpolation between the frequency bins that would exist with no zero padding so no more information has been gained. Higher frequency resolution can only be achieved with a longer sampling time. Thus, zero padding is not considered a useful tool in this context. Nevertheless, when one is analysing a transient that has zero amplitude outside the sampling time window, zero padding can result in finer frequency resolution than if only the length of the transient had been used (Brandt, 2010, p. 201).

When zero padding is used, an additional scaling factor has to be introduced to calculate the amplitude of the resulting spectrum. If the spectrum contains n data points and m introduced zeros, the scaling factor with which the resulting power spectral amplitudes have to be multiplied to get the correct values is $(m+n)/n$.

12.3.9 Uncertainty Principle

The uncertainty principle states that the frequency resolution, Δf (equal to the effective filter bandwidth, B), of a Fourier transformed signal is equal to the reciprocal of the time, T_A, to acquire the sampled record of the signal. Thus, for a single spectrum, $\Delta f\, T_A = 1$.

12.3.10 Time Synchronous Averaging and Synchronous Sampling

Time synchronous averaging is a slightly different process to synchronous sampling, although both are intended for use with noise and vibration signals obtained from rotating equipment. With time synchronous averaging, the aim is to obtain averaged time domain data by averaging data samples that correspond to the same angular location of a rotor. This is done prior to taking a DFT. Thus, the idea is to use a tachometer signal to indicate when each revolution begins and then obtain data with the same constant sample rate for all revolutions. If the speed is variable, this results in samples that do not correspond to the same angular locations as they did prior to the first speed change since sampling began. Data at fixed angular intervals are obtained by interpolating between data samples and then all data corresponding to each particular angle are averaged. In this way, each DFT will be the result of a calculation based on the average of a number of synchronised time samples, so the frequency scale will now be replaced with a scale representing multiples of the fundamental frequency. However, the fundamental and its harmonics will be much more clearly visible than if the time samples were separated by fixed time intervals. This is especially true for variable speed rotors. It is still desirable to obtain several records of averaged data to enable the resulting auto power spectra to be averaged. As this method of analysis is for the purpose of identifying tonal signals, it is not suitable for PSD calculations.

Synchronous sampling, on the other hand, is slightly different to the process described above. Rather than taking samples at fixed time intervals and then interpolating and averaging in the time domain, synchronous sampling involves sampling the signal at fixed angular increments of the rotating rotor so that when the rotor speed changes, the interval between time samples changes. A DFT is then taken of each record in the time domain and the resulting frequency spectrum has multiples of the fundamental rotational frequency along its axis. This method is often referred to as "order tracking" (Randall, 1987, p. 219). Again, this method is only used for tonal analysis with the auto power spectrum and as it is intended for tonal analysis, it is not suitable for obtaining a PSD.

12.3.11 Hilbert Transform

The Hilbert transform is often referred to as envelope analysis, as it involves finding a curve that envelopes the peaks in a signal. The signal may be a time domain signal or a frequency domain signal. Any regular harmonic variation in the envelope signal represents an amplitude modulation of the original signal, which can be quantified by taking an FFT of the envelope signal.

The Hilbert transform applied to a time domain (or time series) signal, $x(t)$, can be represented mathematically as:

$$\mathscr{H}\{x(t)\} = \tilde{x}(t) = \frac{1}{\pi}\int_{-\infty}^{\infty} x(\tau)\left(\frac{1}{t-\tau}\right)\,\mathrm{d}\tau = \frac{1}{\pi}x(t) * \left(\frac{1}{t}\right) \tag{12.58}$$

where $*$ represents the convolution operator (see Section 12.3.16).

The Fourier transform of $\tilde{x}(t)$ is given by:

$$\tilde{X}(f) = -\mathrm{j}\,\mathrm{sgn}(f)X(f) = \mathrm{j}\frac{|f|}{f}X(f) \tag{12.59}$$

where sgn(f) is the sign of f. An alternative version of Equation (12.59) is:

$$\tilde{X}(f) = \begin{cases} \mathrm{e}^{-\mathrm{j}\pi/2}X(f); & f > 0 \\ 0; & f = 0 \\ \mathrm{e}^{\mathrm{j}\pi/2}X(f); & f < 0 \end{cases} \tag{12.60}$$

An example of an envelope of an amplitude modulated time domain signal, $m(t)$ (see Section 1.13), is shown in Figure 12.6.

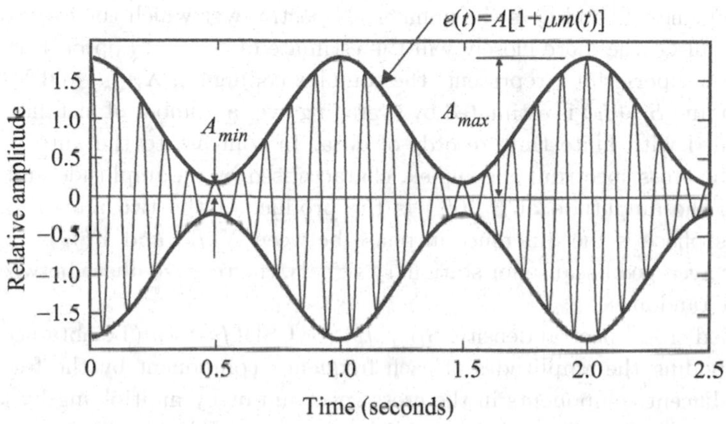

FIGURE 12.6 Envelope analysis with the Hilbert transform.

This type of signal can be analysed using the Hilbert transform of the total time domain signal, $x(t)$, which for our purposes can represent sound pressure as a function of time. To calculate the envelope function, $e(t)$, which is the envelope of the original signal, the following equation is used (Brandt, 2010):

$$e(t) = \sqrt{x^2(t) + \tilde{x}^2(t)} \tag{12.61}$$

or in discrete form at time sample, n, as:

$$e(n) = \sqrt{x^2(n) + \tilde{x}^2(n)} \tag{12.62}$$

The discrete Hilbert transform, $\tilde{x}(n)$, can be calculated from the ordinary discrete Fourier transform, $X(f_k)$, of the original signal as (Brandt, 2010):

$$\tilde{x}(n) = \frac{2}{N}\mathrm{Im}\left\{\sum_{k=0}^{N/2} X(f_k)\mathrm{e}^{\mathrm{j}2\pi n f_k/N}\right\} \tag{12.63}$$

where N is the total number of samples used to calculate the Fourier transform and Im{ } represents the imaginary part of the complex number in brackets.

12.3.12 Cross Spectrum

The cross spectrum is a measure of how much that one signal, represented in the time domain as $x(t)$ and frequency domain as $X(f_k)$, may be related to another signal, represented in the time domain as $y(t)$ and frequency domain as $Y(f_k)$. For example, it may be used to determine the extent to which indoor noise, $y(t)$, in a residence may be caused by exterior noise, $x(t)$, where $y(t)$ is considered to be the system output and $x(t)$ the system input. It is also used to estimate the coherence between two noise or vibration signals, which is another way of quantifying by how much one signal is related to another.

The two-sided cross spectrum and cross-spectral density (CSD) have similar forms to the auto power spectrum and power spectral density of Equations (12.16) and (12.17), respectively, and the single-sided cross spectrum, $G_{yx}(f_k)$, is calculated from the two-sided spectrum in a similar way to the auto power spectrum. The expression for the two-sided spectrum is:

$$S_{yx}(f_k) \approx \frac{S_A}{Q}\sum_{i=1}^{Q} Y_i(f_k)X_i^*(f_k); \qquad n = 0, 1, \ \ldots, \ (N-1) \tag{12.64}$$

where i is the spectrum number, $X_i(f_k)$ and $Y_i(f_k)$ are complex spectral components corresponding to frequency f_k and Q is the number of spectra over which the average is taken. The larger the value of Q, the more closely will the estimate of $S_{yx}(f_k)$ approach its true value. In the equation, the superscript * represents the complex conjugate, $X_i(f_k)$ and $Y_i(f_k)$ are instantaneous spectra and $S_{yx}(f_k)$ is estimated by averaging over a number of instantaneous spectrum products obtained with finite time records of data. In contrast to the auto power spectrum, which is real, the cross spectrum is complex, characterised by an amplitude and a phase.

In practice, the amplitude of $S_{yx}(f_k)$ is the product of the two amplitudes $|X(f_k)|$ and $|Y(f_k)|$, and its phase is the difference in phase between $X(f_k)$ and $Y(f_k)$ $(= \theta_y - \theta_x)$. This function can be averaged because for stationary signals, the relative phase between $x(t)$ and $y(t)$ is fixed and not random.

The two-sided cross-spectral density, $S_{Dyx}(f_k)$ (or CSD(f_k)), can be obtained from the cross spectrum by dividing the amplitudes of each frequency component by the frequency spacing, Δf, between adjacent components in the cross spectrum or by multiplying by the time, T_s, to acquire one record of data. Thus, the two-sided cross-spectral density is:

$$S_{Dyx}(f_k) \approx \frac{T_s S_A}{Q}\sum_{i=1}^{Q} Y_i(f_k)X_i^*(f_k); \qquad n = 0, 1, \ \ldots, \ (N-1) \tag{12.65}$$

The cross-spectral density can also be obtained directly from the cross-correlation function, which is defined as:

$$R_{yx}(\tau) = \langle y(t)x(t-\tau)\rangle_t = \mathrm{E}[y(t)x(t-\tau)] = \lim_{T_s \to \infty} \frac{1}{T_s} \int\limits_{-T_s/2}^{T_s/2} y(t)x(t-\tau)\,\mathrm{d}t \tag{12.66}$$

where E$[X]$ is the expected value of X.

Using the cross-correlation function, the cross-spectral density may be written as:

$$S_{Dyx}(f) = \int\limits_{-\infty}^{\infty} R_{yx}(\tau)\mathrm{e}^{-\mathrm{j}2\pi f\tau}\,\mathrm{d}\tau \tag{12.67}$$

and in the sampled domain (corresponding to part (d) of Figure 12.3),

$$S_{Dyx}(f_k) = \frac{1}{N}\sum_{k=0}^{N-1} R_{yx}(\tau_k)\mathrm{e}^{-\mathrm{j}2\pi nk/N} \tag{12.68}$$

In practice, the one-sided cross spectrum, $G_{yx}(f_k)$, is used instead of the two-sided spectrum of Equation (12.64), where:

$$G_{yx}(f_k) = \begin{cases} 0; & f_k < 0 \\ S_{yx}(f_k); & f_k = 0 \\ 2S_{yx}(f_k); & f_k > 0 \end{cases} \tag{12.69}$$

A similar expression may be written for the single-sided CSD, $G_{Dyx}(f_k)$, as:

$$G_{Dyx}(f_k) = \begin{cases} 0; & f_k < 0 \\ S_{Dyx}(f_k); & f_k = 0 \\ 2S_{Dyx}(f_k); & f_k > 0 \end{cases} \tag{12.70}$$

Note that $G_{yx}(f_k)$ and $G_{Dyx}(f_k)$ are complex, with real and imaginary parts referred to as the co-spectrum and quad-spectrum, respectively. As for auto power spectra and PSDs, the accuracy of the estimate of the cross spectrum improves as the number of records over which the averages are taken increases. The statistical error for a stationary, Gaussian random signal is given as (Randall, 1987):

$$\epsilon = \frac{1}{\sqrt{\gamma^2_{yx}(f_k)Q}}; \qquad n = 0, 1, \ \ldots, \ (N-1) \tag{12.71}$$

where $\gamma^2_{yx}(f_k)$ is the coherence function relating noise signals, $x(t)$ and $y(t)$ (see Section 12.3.13), and Q is the number of averages.

The amplitude of $G_{yx}(f_k)$ gives a measure of how well the two functions $x(t)$ and $y(t)$ correlate as a function of frequency, and the phase angle of $G_{yx}(f_k)$ is a measure of the phase shift between the two signals as a function of frequency.

12.3.13 Coherence

The coherence function is a measure of the degree of linear dependence between two signals, $x(t)$ and $y(t)$, as a function of frequency. Coherence is calculated from the two auto power spectra, $G_{xx}(f_k)$ and $G_{yy}(f_k)$ and the cross spectrum between y and x, $G_{yx}(f_k)$, as (Brandt, 2010, p. 291):

$$\gamma^2_{yx}(f_k) = \frac{|G_{yx}(f_k)|^2}{G_{xx}(f_k)G_{yy}(f_k)}; \qquad n = 0, 1, \ \ldots, \ (N-1) \tag{12.72}$$

By definition, $\gamma^2_{yx}(f_k)$ varies between 0 and 1, with 1 indicating a high degree of linear dependence between the two signals. Thus, in a physical system where $y(t)$ is the output and $x(t)$ is the input signal, the coherence is a measure of the degree to which $y(t)$ is linearly related to $x(t)$. If random noise is present in either $x(t)$ or $y(t)$, then the value of the coherence will diminish. However, if random noise appears in both $x(t)$ and $y(t)$, and the two random noise signals are well correlated, the coherence may not necessarily diminish. Other causes of a diminished coherence are insufficient frequency resolution in the frequency spectrum or poor choice of window function. A further cause of diminished coherence is a time delay, of the same order as the length of the record, between $x(t)$ and $y(t)$. A value of 1 for a coherence measurement represents a very reliable measurement, while a value of less than about 0.7 represents an invalid measurement.

The main application of the coherence function is in checking the validity of frequency response measurements (see Section 12.3.15). The coherence will always be unity by definition if only one spectrum (rather than the average of many spectra) is used to calculate $G_{xx}(f_k)$, $G_{yy}(f_k)$ and $G_{yx}(f_k)$. The greater the number of averages used (minimum of two), the more accurate will be the result. As a guide, the expected random error in a frequency response measurement as a function of the coherence value and number of averages taken is given in Figure 12.7, which is adapted from (Randall, 1987, p. 245).

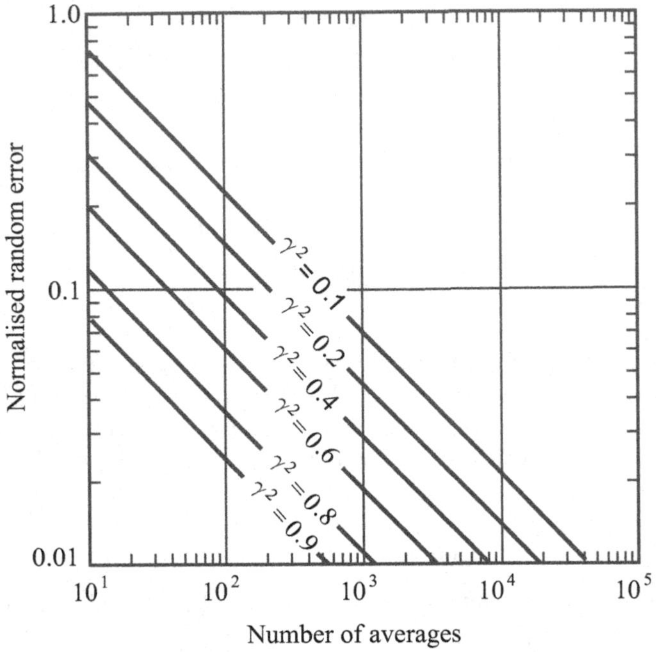

FIGURE 12.7 Accuracy of frequency response function (FRF) estimates versus coherence between the input force to a structure and the resulting structural response.

In general, low coherence values can indicate one or more of five problems:

1. insufficient signal level (turn up gain on the analyser);
2. poor signal-to-noise ratio;
3. presence of other extraneous forcing functions;
4. insufficient averages taken; or
5. leakage.

A more direct application of coherence measurement is the calculation of the signal, S, to noise, N, ratio as a function of frequency, given by:

$$S/N = \frac{\gamma_{yx}^2(f_k)}{1-\gamma_{yx}^2(f_k)}; \qquad k = 0, 1, \ \ldots, \ (N-1) \tag{12.73}$$

In this relatively narrow definition of S/N, the "signal" is that component in the response, $y(t)$, that is caused by $x(t)$ and "noise" refers to anything else in $y(t)$.

Insufficient signal level (case 1 above) is characterised by a rough plot of coherence versus frequency, even though the average may be close to one (see Figure 12.8). To overcome this, the gains of the transducer preamplifiers should be turned up or the attenuator (or gain) setting on the spectrum analyser or data acquisition system adjusted, or both.

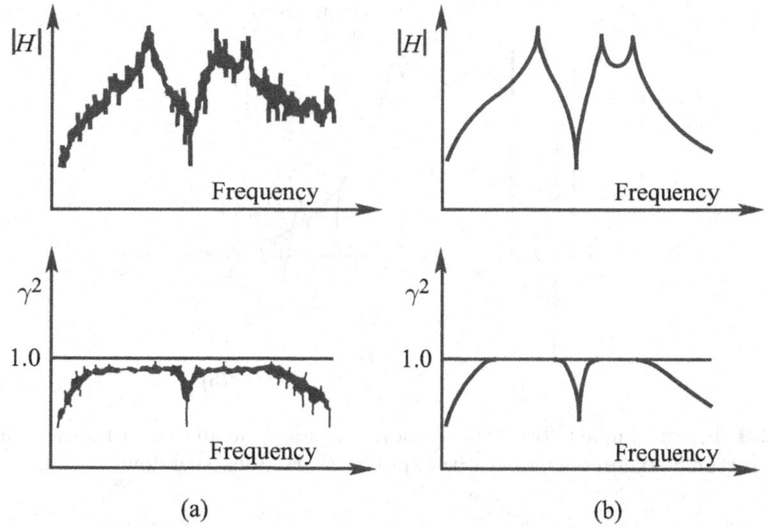

FIGURE 12.8 Effect of insufficient signal level on the frequency response function H, and the coherence, γ^2: (a) measurement with noise due to incorrect attenuator setting (insufficient signal level); (b) same measurement with optimum attenuator setting.

For excitation of a structure by random noise via a shaker (for a structure) or loudspeaker (for an acoustic space), poor coherence between the excitation and response signals is often measured at system resonances. This is partly due to the leakage problem discussed previously and partly because the input force (for a structure) or input sound (for an acoustic space) is very small (close to the instrumentation noise floor) at these frequencies.

Poor signal-to-noise ratio (case 2 above) can have two causes for the case of impact excitation of a structure. The first is due to the bandwidth of the input force being less than the frequency range of interest or the frequency range set on the spectrum analyser. This results in force zeros at higher frequencies, giving the false indication of many high-frequency vibration modes. A force zero causes a false modal indication because any motion associated with zero force means that the impedance is zero, which implies a modal resonance. Conversely, it is not desirable for the bandwidth of the force to extend beyond the frequency range of interest to avoid the problem of exciting vibration modes above the frequency range of interest and thus contaminating the measurements with extraneous signals. A good compromise is for the input force auto power spectrum to be between 10 and 20 dB down from the peak value at the highest frequency of interest.

The second reason for poor signal-to-noise ratio during impact testing is the short duration of the force pulse in relation to the duration of the time domain data block. In many cases, the pulse may be defined by only a few sample points comprising only a small fraction of the total time window, the rest being noise. Thus, when averaged into the measurement, the noise becomes significant. This signal-to-noise ratio problem can be minimised by using an adjustable width rectangular window for the impact force signal and an adjustable length single-sided exponential window for the response signal. This ensures that the force signal and the structural response caused by it are weighted strongly compared with any extraneous noise that may be present. These windows, which are available in many commercial spectrum analysers or computer software, are illustrated in Figure 12.9, which is adapted from the standards, ISO 7626-1 (2011); ISO 7626-2 (2015); ISO 7626-5 (2019).

For cases in which there are many inputs and one or more outputs, it is of interest to estimate the degree of correlation existing between one group of selected inputs, $X(n) = [x_1(n), ..., x_m(n)]$, and one of the outputs, $y[n]$. This is the basis of the concept of multiple coherence, defined as

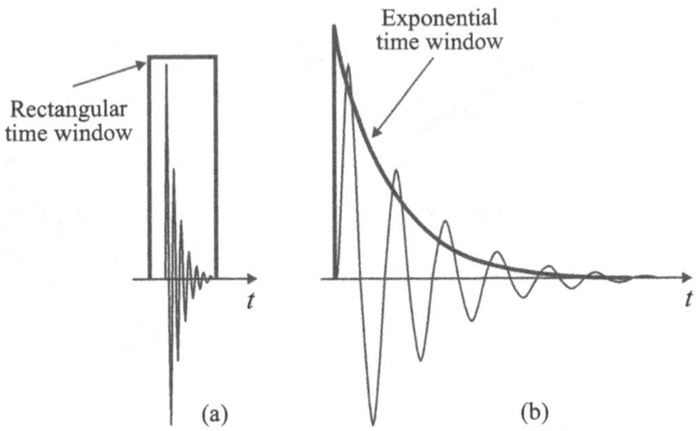

FIGURE 12.9 Typical impact force and response windows available in many commercial DFT analysers: (a) rectangular force window; (b) exponential response window.

(Brandt, 2010, p. 328):

$$\gamma_{yX}^2 = \frac{[G_{yX}(f_k)]\left[G_{XX}^{-1}(f_k)\right]^{\mathrm{H}}[G_{yX}(f_k)]^{\mathrm{H}}}{G_{yy}(f_k)} \tag{12.74}$$

where the superscript H denotes the Hermitian transpose of the matrix, γ_{yX}^2 is the multiple coherence function between the vector of inputs, X, and the output, y, G_{yX} is the m-dimensional vector of the cross spectrum between the inputs, X, and the output, y, G_{XX} is the $(m \times m)$-dimensional matrix of the power spectrum and cross spectrum of the vector of inputs, and G_{yy} is the power spectrum of the output. The power spectrum and cross spectrum quantities in Equation (12.74) can be replaced with their equivalent power spectral-density terms with no change in result.

12.3.14 Coherent Output Power

Coherent output power calculations allow one to determine what contribution a particular sound source may be making to a particular acoustic measurement. The coherent output power process can also be used to eliminate extraneous noise from a signal.

Although coherent power calculations are relatively simple, practical implementation of the procedure requires considerable care. To be able to determine the coherent output power of a sound source, it is necessary to be able to obtain an uncontaminated signal representing just the noise source itself. This usually requires a measurement to be made close to the noise source. If the auto power spectrum of the measurement made close to a noise source is $G_{xx}(f_k), n = 0, 1, \ldots, (N-1)$ and the auto power spectrum of the contaminated measurement at some distant location is $G_{yy}(f_k)$, then the auto power spectrum, $G_{cc}(f_k)$, of the contribution of the noise source at the distant location is given by:

$$G_{cc}(f_k) = |H_1(f_k)|^2 G_{xx}(f_k) = \frac{|G_{yx}(f_k)|^2}{G_{xx}(f_k)}; \qquad k = 0, 1, \ldots, (N-1) \tag{12.75}$$

where $H_1(f_k)$ is defined in Equation (12.77) and N is the number of data points in the auto power spectrum. The term, $G_{cc}(f_k)$, can also be written in terms of the coherence between the two signals, so that:

$$G_{cc}(f_k) = \gamma_{yx}^2(f_k) G_{yy}(f_k) \qquad k = 0, 1, \ldots, (N-1) \tag{12.76}$$

Similar relationships also apply for power spectral densities with the functions, $G(f_k)$, replaced with $G_D(f_k)$ in the above two equations. It is assumed in all relationships that the measurement, $G_{xx}(f_k)$, is uncontaminated by noise not related to the noise source being investigated. In practice, a small amount of contamination leads to the results being slightly less accurate.

It is possible to use Equations (12.75) and (12.76) to calculate the relative contributions of a number of sources to a particular sound pressure measurement, provided the signals $G_{xx}(f_k), k = 0, 1, \ldots, (N-1)$ acquired near each sound source do not contaminate one another significantly.

12.3.15 Frequency Response (or Transfer) Function

The frequency response function (FRF), $H(f_k)$, is defined as:

$$H(f_k) = \frac{Y(f_k)}{X(f_k)} \qquad k = 0, 1, \ldots, (N-1) \tag{12.77}$$

The FRF, $H(f_k)$, is the Fourier transform of the system impulse response function, $h(t_n)$. The FRF is a convenient way of quantifying the relative amplitude of and the phase between two signals as a function of frequency. The impulse response of a system is the system output as a function of time following an impulse input (very short, sudden input)

In practice, it is desirable to average $H(f_k)$ over a number of spectra, but as $Y(f_k)$ and $X(f_k), k = 0, 1, \ldots, (N-1)$ are both instantaneous spectra, it is not possible to average either of these, as they are complex quantities with random phases, so that they eventually average to zero. For this reason, it is convenient to modify Equation (12.77). There are a number of possibilities, one of which is to multiply the numerator and denominator by the complex conjugate of the input spectrum to obtain power spectra, which are only characterised by an amplitude and so can be easily averaged. Thus:

$$H_1(f_k) = \frac{Y(f_k)X^*(f_k)}{X(f_k)X^*(f_k)} = \frac{G_{yx}(f_k)}{G_{xx}(f_k)} \qquad k = 0, 1, \ldots, (N-1) \tag{12.78}$$

A second version is found by multiplying with $Y^*(f)$ instead of $X^*(f)$. Thus:

$$H_2(f_k) = \frac{Y(f_k)Y^*(f_k)}{X(f_k)Y^*(f_k)} = \frac{G_{yy}(f_k)}{G_{xy}(f_k)} \qquad k = 0, 1, \ldots, (N-1) \tag{12.79}$$

Either of the above two forms of frequency response function is amenable to averaging, but $H_1(f_k)$ is the preferred version if the output signal, $y(t)$, is more contaminated by noise than the input signal, $x(t)$, whereas $H_2(f_k)$ is preferred if the input signal, $x(t)$, is more contaminated by noise than the output signal (Randall, 1987).

The frequency response function may also be expressed in spectral-density terms by replacing $G_{xy}(f_k)$, $G_{yx}(f_k), G_{xx}(f_k)$ and $G_{yy}(f_k)$ in Equations (12.78) and (12.79) by their spectral-density equivalents.

12.3.16 Convolution

Convolution is an operation in the time domain that enables us to define the output of a system for any given input, provided that the impulse response of the system is known. The impulse response is the system output when the input is an impulse (or delta function). In practice, due to the difficulty in producing a pure impulse for the input, the impulse response is measured in different ways, as discussed below in this section and also in Section 12.3.18.

12.3.16.1 Continuous Functions

Convolution in the time domain is the operation that is equivalent to multiplication in the frequency domain. Multiplication of two functions, $X(f)$ and $H(f)$, in the frequency domain involves multiplying the spectral amplitude of each frequency component in $X(f)$ with the amplitude of the corresponding frequency component in $H(f)$ and adding the phases of each frequency component in $X(f)$ to the corresponding frequency component in $H(f)$. The equivalent operation of convolution of the two time domain signals, $x(t) * h(t)$, is a little more complex and is given by:

$$y(t) = x(t) * h(t) = \int_{-\infty}^{\infty} x(\tau)h(t-\tau)\,\mathrm{d}\tau \tag{12.80}$$

The Fourier transform of the convolved signal, $y(t) = x(t) * h(t)$, is $\{X(f)H(f)\}$. Taking the inverse Fourier transform (IFT) of $\{X(f)H(f)\}$ returns the orignal signal. That is, $y(t) = \mathrm{IFT}\{X(f)H(f)\}$. In a practical application, $X(f)$ might be the Fourier transform of an input signal to a physical system characterised by a transfer function, $H(f)$, and $Y(f)$ would then be the resulting output from the system.

Deconvolution is the process of determining $h(t)$ of Equation (12.80) from known signals, $y(t)$ and $x(t)$. For example, $h(t)$ may be a system impulse response that is to be determined from input and output signals. Deconvolution is often performed in the frequency domain so that in the absence of any significant noise in the signals:

$$H(f) = Y(f)/X(f) \tag{12.81}$$

The inverse Fourier transform is then taken of $H(f)$ to obtain $h(t)$. If noise is present in the output signal, $y(t)$, then the estimate for $h(t)$ will be in error. The error may be reduced using Weiner deconvolution but this is a complex operation and beyond the scope of this book.

It should be noted that $H(f)$, $Y(f)$ and $X(f)$ are complex numbers at each frequency, f, and can be represented as a magnitude and a phase. Multiplication of two complex numbers to obtain a third complex number requires the magnitudes to be multiplied and the phases to be added, whereas division of one complex number by another requires division of one magnitude by another and subtraction of the denominator phase from the numerator phase.

12.3.16.2 Sampled Data

If the two functions in the time domain are represented by sampled data, $x[n]$, $n = 0, 1, 2, \ldots, (N-1)$ and $h[m]$, $m = 0, 1, 2, \ldots, (M-1)$, such as obtained by a digital data acquisition system, the output of the **convolution**, $y[i], i = 0, 1, 2, \ldots, (N+M-3)$, of the two signals at sample number, i, is given by:

$$y[i] = x[i] * h[i] = h[i] * x[i] = \sum_{n=0}^{N-1} x[n]h[i-n] \tag{12.82}$$

where terms in the sum are ignored if $[i-n]$ lies outside the range from 0 to $(M-1)$. As can be seen from the equation, each sample in the input signal contributes to many samples in the output signal.

Deconvolution is the process of obtaining a system impulse response from the measurement of a system input signal and the output signal that the system generates in the time domain. This discussed in Section 12.3.18 where an MLS signal is used as the input signal and the cross correlation (See Section 12.3.17) of the output and input signals is then the system impulse response function.

Alternatively, the impulse response of a system can be determined using a white noise spectrum as the input and taking the DFT of both the input, $x(n)$, and output, $y(n)$, to obtain $X(f_k)$

and $Y(f_k)$, as well as the transfer function in the frequency domain, $H(f_k) = Y(f_k)/X(f_k)$. However, as $X(f_k)$ and $Y(f_k)$ are obtained using a DFT, the inverse discrete Fourier transform (IDFT) in the frequency domain will no longer represent a linear convolution in the time domain, due to the periodicity of the DFT (see Brandt (2010, Ch. 9)). Thus, $h(n) \neq \text{IDFT}\{H(f_k)\}$. In fact the result for $h(n)$ will be the circular convolution, which is not the same as the linear convolution just discussed. This problem can be avoided by the use of zero padding (see Section 12.3.8) where a number of zeros (equal to the number of samples in the data record) are added to the end of each of the two sampled data records, $x(n)$ and $y(n)$, prior to taking the DFT. If a DFT is then taken of the two sampled time signals, $x(n)$ and $y(n)$, each containing N samples and N zeros following the samples, then $h(n) = \text{IDFT}\{H(f_k)\}$, where $H(f_k) = Y(f_k)/X(f_k)$.

12.3.17 Auto-Correlation and Cross-Correlation Function Estimates

The auto-correlation function of a signal, $x(t)$, in the time domain, is a measure of how similar a signal is to a time-shifted future or past version of itself. If the time shift is $-\tau$ seconds (past version), then the estimated auto-correlation function is (Brandt, 2010, p. 225):

$$\hat{R}_{xx}(\tau) = \frac{1}{T - |\tau|} \int_{-T/2}^{T/2} x(t)x(t-\tau)\,\mathrm{d}t \tag{12.83}$$

where the hat over a variable represents an estimate. The estimated cross correlation between two different time signals, $x(t)$ (the input) and $y(t)$ (the output), where the output is shifted into the past by time $-\tau$) is (Brandt, 2010, p. 225):

$$\hat{R}_{yx}(\tau) = \frac{1}{T - |\tau|} \int_{-T/2}^{T/2} y(t)x(t-\tau)\,\mathrm{d}t \tag{12.84}$$

Auto-correlation and cross-correlation function definitions are applicable to random noise signals only if the average value of one record of samples is not much different to that for subsequent records or the average of many records. Random noise signals typically satisfy this condition sufficiently provided that the signal does not vary too much.

The cross-correlation function can be used to find the acoustic delay between two signals originating from the same source. In this case, the delay is the time difference represented by the maximum value of the cross-correlation function. If the speed of sound is known, then the delay allows one to determine the distance between the two microphones that are providing the two signals.

The auto-correlation function has the property that for real signals, $x(t)$ and $y(t)$:

$$R_{xx}(\tau) = R_{xx}(-\tau) \tag{12.85}$$

and the cross-correlation function has the property:

$$R_{yx}(\tau) = R_{xy}(-\tau) \tag{12.86}$$

As direct calculation of correlation functions is difficult and resource intensive, they are usually estimated from spectra. This is done by first estimating a power spectral density as in Equations (12.17) and (12.22), where the function $X_i(f_k)$ has had a rectangular window applied (which is effectively a multiplication by unity as all $w(k)$ in Equation (12.26) are unity for a rectangular window). The PSD needs to be zero padded with as many zeros as data points to avoid circular convolution (Brandt, 2010, Chapter 9).

We first compute a PSD estimate using a $2N$ record length with the last half of the samples set equal to zero. So we obtain for the two-sided PSD:

$$S_{Dxx}(f_k) = \frac{2S_A}{Q\Delta f}\sum_{i=1}^{Q} X_{iz}(f_k)X_{iz}^*(f_k) = \frac{2S_A}{Q\Delta f}\sum_{i=1}^{Q} |X_{iz}(f_k)|^2 \tag{12.87}$$

where the subscript, z, indicates that the spectrum has been calculated with an equal number of zeros as data samples, added to the dataset and the factor of 2 arises from the use of zero padding. $\Delta f = f_s/N$, f_s is the sample rate, N is the number of samples in each data segment or record, $X_i(f_k)$ is an unscaled spectrum and has been multiplied by 2 to account for the added zeros. As we have N data samples in each record, and another N zeros added to the end, the total length of each record is $2N$. As a rectangular window has been used, the scaling factor, S_A, is simply equal to $1/(4N^2)$, as there are $2N$ samples in the record. The ith unscaled, zero-padded spectrum, $X_{iz}(f_k)$, obtained by passing time samples through a rectangular window, is given by:

$$X_{iz}(f_k) = \frac{1}{2N}\sum_{n=0}^{2N-1} x_i(t_n)\mathrm{e}^{-\mathrm{j}2\pi f_k n/N}; \qquad k = 0, 1, \ \ldots, \ (2N-1) \tag{12.88}$$

where $n = 0, 1, \ldots, 2N-1$, x_i represents the ith complete data record in the time domain, which has been doubled in size using zero padding, t_n is the time that the nth sample was acquired.

The inverse discrete Fourier transform (IDFT) is taken of Equation (12.87) to obtain the quantity $\hat{R}_{xx}(k)$, and this is used to compute an estimate, $\hat{R}^i_{xx}$, for the auto-correlation function for the ith record, using (Brandt, 2010):

$$\hat{R}^i_{xx} = \begin{cases} \dfrac{N-n}{N\Delta t}\hat{R}_{xx}(n); & n = 0, 1, \ \ldots, \ (N-1) \\ \\ \dfrac{n-N}{N\Delta t}\hat{R}_{xx}(n); & n = N, (N+1), \ \ldots, \ (2N-1) \end{cases} \tag{12.89}$$

where Δt is the time interval between samples. As discussed by (Brandt, 2010, p. 225), the resulting $\hat{R}^i_{xx}$ should be FFT shifted and then the outer $N/2$ samples on each side removed to give the final auto-correlation estimate of length, N. As an example of the preceding process, assume a sample vector of $2N = 8$ numbers, $[a, b, c, d, e, f, g, h]$. Applying the FFT shift (moving the second half to the first half location) results in the vector, $[e, f, g, h, a, b, c, d]$ and removing $N/2 = 2$ samples from each side results in $[g, h, a, b]$, which is a vector of $N = 4$ samples.

The cross correlation between two signals, $x(t)$ and $y(t)$, can be computed in a similar way to the auto correlation. In this case, we replace the PSD with the CSD (cross-spectral density) and begin by computing the CSD using 50% zero padding and a $2N$ record length with the last half of the samples set equal to zero. So we obtain for the two-sided CSD:

$$S_{Dyx}(f_k) = \frac{2S_A}{Q\Delta f}\sum_{i=1}^{Q} Y_{iz}(f_k)X_{iz}^*(f_k) \tag{12.90}$$

where $f_k = 0, 1, \ldots, 2N-1$, $Y_{iz}(f_k)$ and $X_{iz}(f_k)$ are unscaled spectra and the two-sided CSD, S_{Dyx}, has been multiplied by 2 to account for the added zeros. As before, the scaling factor, S_A, is simply equal to $1/(4N^2)$, as there are $2N$ samples in the record. The ith unscaled, zero-padded spectrum, $Y_{iz}(f_k)$, obtained by passing time samples through a rectangular window, is similar to that for $X_{iz}(f_k)$, which is given in Equation (12.88), and is given by:

$$Y_{iz}(f_k) = \frac{1}{2N}\sum_{n=0}^{2N-1} y_i(t_n)\mathrm{e}^{-\mathrm{j}2\pi f_k n/N}; \qquad k = 0, 1, \ \ldots, \ (2N-1) \tag{12.91}$$

The inverse discrete Fourier transform (IDFT) is taken of Equation (12.90) to obtain the quantity $\hat{R}_{yx}(k)$, and this is used to compute an estimate, $\hat{R}^i_{yx}$, of the cross-correlation function for the ith record, using (Brandt, 2010, p. 225):

$$\hat{R}^i_{yx} = \begin{cases} \dfrac{N-n}{N\Delta t}\hat{R}_{yx}(n); & n = 0, 1, \; \ldots, \; (N-1) \\[2ex] \dfrac{n-N}{N\Delta t}\hat{R}_{yx}(n); & n = N, (N+1), \; \ldots, \; (2N-1) \end{cases} \tag{12.92}$$

Again, the obtained cross-correlation vector, R^i_{yx}, must be FFT shifted as described following Equation (12.89) and the outer $N/2$ samples removed from each side (Brandt, 2010, p. 225).

12.3.18 Maximum Length Sequence (MLS)

An MLS excitation signal is sometimes used in acoustics to obtain measurements of transfer functions, which include loudspeaker response functions and the noise reduction from the outside to the inside of a house. A schematic of a typical measurement system is shown in Figure 12.10. An MLS signal is a digitally synthesised binary series of samples that have a value of 1 or zero, which is mapped to -1 and 1, respectively, to produce a signal that is symmetric about zero. The clock rate is the number of times a new value is output per second by the shift register generating the binary sequence and this should be at least 2.5 times the maximum frequency of interest in the transfer function to be measured. However, the sample rate is recommended to be 10 times the maximum frequency of interest (Ljung, 1999). For the measurement of transfer functions, MLS analysis has the following advantages.

- It can minimise the effects on the measurement of noise not generated by the test system.
- The spectral content of an MLS signal closely resembles white noise with a flat power spectral shape.
- It is deterministic and this property together with the previous one has attracted the description of pseudo-random noise.
- It has a pre-determined temporal length before repeating. This length, in terms of the total number of samples in a sequence, is $L = 2^M - 1$ samples, where M is the order

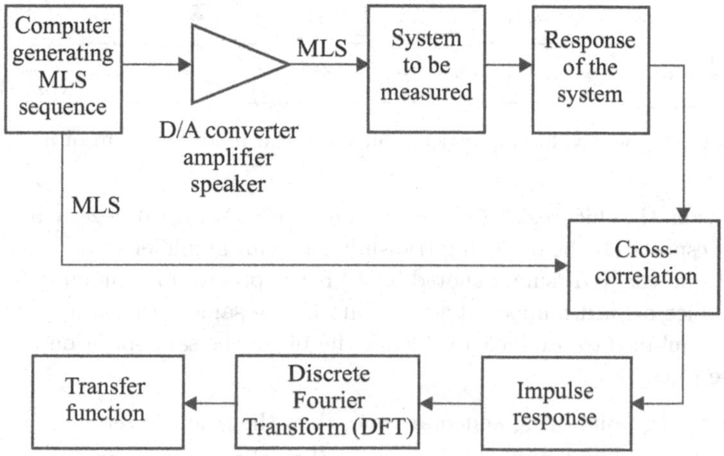

FIGURE 12.10 Schematic arrangement for the measurement of a system transfer function (such as outside to inside sound pressure levels in a house) using an MLS signal.

of the MLS, and this is usually 12 or greater in practical systems. Note that f_s/L, where f_s is the sample rate, must be longer than the impulse response of the system being measured. For the case of measuring noise reductions from the outside to the inside of a house, it would have to be greater than the propagation time of the noise from the outside source to the inside microphone.

- It has a low crest factor, thus transferring a large amount of energy to the system being excited, and thus achieving a very high signal-to-noise ratio (SNR). For every doubling in the number of sequences that are averaged, the SNR is improved by 3 dB.
- The cross-correlation function of the input and output of a system using an MLS is equal to the impulse response of the system between the input and output. As discussed in Section 12.3.12, taking a DFT of the cross-correlation function yields the cross-spectral density of the system and as discussed in Section 12.3.2, taking a DFT of the auto correlation of the system input yields the spectral density of the system input. As discussed in Section 12.3.15, these two quantities can be used to obtain the transfer function spectral density. For the case of the house mentioned above, this transfer function is the noise reduction from outside to inside in dB.
- The use of cross correlation with MLS to obtain transfer functions rejects all noise not correlated with the MLS signal and so MLS is effective in finding transfer functions in noisy environments such as at low frequencies when measuring noise reductions from the outside to the inside of a house.
- MLS suppresses the DC part of the system response.

The MLS sequence is generated digitally using a maximal linear feedback shift register, which cycles through every possible binary value (except all zeros) before repeating. As the signal eventually repeats itself, it is referred to as a periodic signal with a period corresponding to the time between repeats. A shift register of length, M, contains M bits, each of which can have a value of one or zero. The length of an MLS sequence before it repeats itself is related to the number of bits, M, in the shift register used to generate it. If the shift register length is $M = 25$, then the number of samples in the MLS sequence prior to repeating is $2^{25} - 1$ ($N = 33554431$ samples). The -1 is there because the case of all bits in the shift register being zero (which would produce a permanently zero output) is excluded. An MLS generating system with a shift register of length 5 is implemented as illustrated in Figure 12.11.

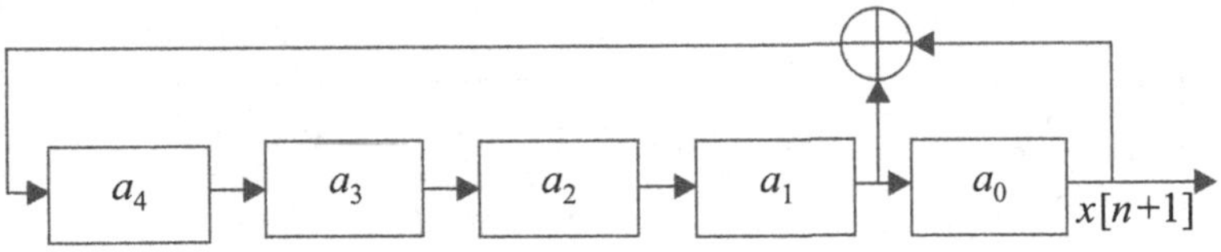

FIGURE 12.11 The next value for register bit a_4 is the modulo 2 sum of a_0 and a_1, indicated by the $\oplus$ symbol.

The output from the shift register chain at time, n, is either a 1 or a zero, and this is mapped to -1 and $+1$, respectively, before being transmitted to an amplifier.

A modulo 2 sum (or XOR sum, denoted by $\oplus$) is the process of combining bits in two binary numbers to produce a third number. The two bits in the same position in the two numbers to be summed are combined to produce a value for the bit in the same position in the new number by following the rules

- If both of the two bits being added are one, then the result is zero.
- If both of the two bits being added are zero, then the result is zero.
- If one of the bits is one and the other is zero, then the result is one.

The values of the bits corresponding to sample number $n+1$ in the MLS being generated with a shift register of length $M=5$ can be calculated recursively from the values corresponding to sample, n, using the relations:

$$\begin{cases} a_4[n+1] = a_0[n] + a_1[n]; \quad \text{(modulo 2 sum)} \\ a_3[n+1] = a_4[n] \\ a_2[n+1] = a_3[n] \\ a_1[n+1] = a_2[n] \\ a_0[n+1] = a_1[n] \end{cases} \tag{12.93}$$

A

Review of Relevant Linear Matrix Algebra

An $(m \times n)$ matrix is a collection of mn numbers (complex or real), a_{ij}, $(i = 1, 2, ..., m,\ j = 1, 2, ..., n)$, written in an array of m rows and n columns:

$$\boldsymbol{A} = \begin{bmatrix} a_{11} & a_{12} & \cdots & a_{1n} \\ a_{21} & a_{22} & \cdots & a_{2n} \\ \vdots & \vdots & \vdots & \vdots \\ a_{m1} & a_{m2} & \cdots & a_{mn} \end{bmatrix} \tag{A.1}$$

The term, a_{ij}, appears in the ith row and jth column of the array. If the number of rows is equal to the number of columns, the matrix is said to be square. An m vector, also referred to as an $(m \times 1)$ vector or column vector, is a matrix with m rows and 1 column:

$$\boldsymbol{a} = \begin{bmatrix} a_1 \\ a_2 \\ \vdots \\ a_m \end{bmatrix} \tag{A.2}$$

A.1 Addition, Subtraction and Multiplication by a Scalar

If two matrices have the same number of rows and columns, they can be added or subtracted. When adding matrices, the individual corresponding terms are added. For example, if:

$$\boldsymbol{A} = \begin{bmatrix} a_{11} & a_{12} & \cdots & a_{1n} \\ a_{21} & a_{22} & \cdots & a_{2n} \\ \vdots & \vdots & \vdots & \vdots \\ a_{m1} & a_{m2} & \cdots & a_{mn} \end{bmatrix} \quad \text{and} \quad \boldsymbol{B} = \begin{bmatrix} b_{11} & b_{12} & \cdots & b_{1n} \\ b_{21} & b_{22} & \cdots & b_{2n} \\ \vdots & \vdots & \vdots & \vdots \\ b_{m1} & b_{m2} & \cdots & b_{mn} \end{bmatrix} \tag{A.3}$$

then:

$$\boldsymbol{A} + \boldsymbol{B} = \begin{bmatrix} a_{11} + b_{11} & a_{12} + b_{12} & \cdots & a_{1n} + b_{1n} \\ a_{21} + b_{21} & a_{22} + b_{22} & \cdots & a_{2n} + b_{2n} \\ \vdots & \vdots & \vdots & \vdots \\ a_{m1} + b_{m1} & a_{m2} + b_{m2} & \cdots & a_{mn} + b_{mn} \end{bmatrix} \tag{A.4}$$

DOI: 10.1201/9780367814908-A

When subtracting matrices, the individual terms are subtracted. Note that matrix addition is commutative, as:

$$\boldsymbol{A} + \boldsymbol{B} = \boldsymbol{B} + \boldsymbol{A} \tag{A.5}$$

It is also associative, as:

$$(\boldsymbol{A} + \boldsymbol{B}) + \boldsymbol{C} = \boldsymbol{A} + (\boldsymbol{B} + \boldsymbol{C}) \tag{A.6}$$

Matrices can also be multiplied by a scalar. Here the individual terms are each multiplied by a scalar. For example, if k is a scalar:

$$k\boldsymbol{A} = \begin{bmatrix} ka_{11} & ka_{12} & \cdots & ka_{1n} \\ ka_{21} & ka_{22} & \cdots & ka_{2n} \\ \vdots & \vdots & \vdots & \vdots \\ ka_{m1} & ka_{m2} & \cdots & ka_{mn} \end{bmatrix} \tag{A.7}$$

A.2 Multiplication of Matrices

Two matrices, $\boldsymbol{A}$ and $\boldsymbol{B}$, can be multiplied together to form the product $\boldsymbol{AB}$ if the number of columns in $\boldsymbol{A}$ is equal to the number of rows in $\boldsymbol{B}$. If, for example, $\boldsymbol{A}$ is an $(m \times p)$ matrix, and $\boldsymbol{B}$ is a $(p \times n)$ matrix, then the product $\boldsymbol{AB}$ is defined by:

$$\boldsymbol{C} = \boldsymbol{AB} \tag{A.8}$$

where $\boldsymbol{C}$ is an $(m \times n)$ matrix, the terms of which are defined by:

$$c_{ij} = \sum_{k=1}^{p} a_{ik} b_{kj} \tag{A.9}$$

Matrix multiplication is associative, with the product of three (or more) matrices defined by:

$$\boldsymbol{ABC} = (\boldsymbol{AB})\boldsymbol{C} = \boldsymbol{A}(\boldsymbol{BC}) \tag{A.10}$$

Matrix multiplication is also distributive, where:

$$\boldsymbol{A}(\boldsymbol{B} + \boldsymbol{C}) = \boldsymbol{AB} + \boldsymbol{AC} \tag{A.11}$$

However, matrix multiplication is not commutative, as, in general:

$$\boldsymbol{AB} \neq \boldsymbol{BA} \tag{A.12}$$

In fact, while the product, $\boldsymbol{AB}$, may be formed, it may not be possible to form the product, $\boldsymbol{BA}$.

The identity matrix, $\boldsymbol{I}$, is defined as the $(p \times p)$ matrix with all principal diagonal elements equal to 1, and all other terms equal to zero:

$$\boldsymbol{I} = \begin{bmatrix} 1 & 0 & \cdots & 0 \\ 0 & 1 & \cdots & 0 \\ & & \ddots & \\ 0 & 0 & \cdots & 1 \end{bmatrix} \tag{A.13}$$

For any $(m \times p)$ matrix, $\boldsymbol{A}$, the identity matrix has the property:

$$\boldsymbol{AI} = \boldsymbol{A} \tag{A.14}$$

Similarly, if the identity matrix, I, is of size $(m \times m)$, then:

$$\boldsymbol{IA} = \boldsymbol{A} \tag{A.15}$$

A.3 Matrix Transposition

If a matrix is transposed, the rows and columns are interchanged. For example, the transpose of the $(m \times n)$ matrix, $\boldsymbol{A}$, denoted by $\boldsymbol{A}^T$, is defined as the $(n \times m)$ matrix, $\boldsymbol{B}$:

$$\boldsymbol{A}^{\mathrm{T}} = \boldsymbol{B} \tag{A.16}$$

where

$$b_{ij} = a_{ji} \tag{A.17}$$

The transpose of a matrix product is defined by:

$$(\boldsymbol{AB})^{\mathrm{T}} = \boldsymbol{B}^{\mathrm{T}}\boldsymbol{A}^{\mathrm{T}} \tag{A.18}$$

This result can be extended to products of more than two matrices, such as:

$$(\boldsymbol{ABC})^{\mathrm{T}} = \boldsymbol{C}^{\mathrm{T}}\boldsymbol{B}^{\mathrm{T}}\boldsymbol{A}^{\mathrm{T}} \tag{A.19}$$

If

$$\boldsymbol{A} = \boldsymbol{A}^{\mathrm{T}} \tag{A.20}$$

then the matrix, $\boldsymbol{A}$, is said to be symmetric.

The Hermitian transpose of a matrix is defined as the complex conjugate of the transposed matrix (when taking the complex conjugate of a matrix, each term in the matrix is conjugated). Therefore, the Hermitian transpose of the $(m \times n)$ matrix, $\boldsymbol{A}$, denoted by $\boldsymbol{A}^H$, is defined as the $(n \times m)$ matrix, $\boldsymbol{B}$:

$$\boldsymbol{A}^{\mathrm{H}} = \boldsymbol{B} \tag{A.21}$$

where:

$$b_{ij} = a^*_{ji} \tag{A.22}$$

If $\boldsymbol{A} = \boldsymbol{A}^H$, then $\boldsymbol{A}$ is said to be a Hermitian matrix.

A.4 Matrix Determinants

The determinant of the (2×2) matrix, $\boldsymbol{A}$, denoted, $|\boldsymbol{A}|$, is defined as:

$$|\boldsymbol{A}| = \begin{vmatrix} a_{11} & a_{12} \\ a_{21} & a_{22} \end{vmatrix} = a_{11}a_{22} - a_{12}a_{21} \tag{A.23}$$

The minor, $\boldsymbol{M}_{ij}$, of the element, a_{ij}, of the square matrix, $\boldsymbol{A}$, is the determinant of the matrix formed by deleting the ith row and jth column from $\boldsymbol{A}$. For example, if $\boldsymbol{A}$ is a (3×3) matrix, then:

$$\boldsymbol{A} = \begin{bmatrix} a_{11} & a_{12} & a_{13} \\ a_{21} & a_{22} & a_{23} \\ a_{31} & a_{32} & a_{33} \end{bmatrix} \tag{A.24}$$

The minor, $\boldsymbol{M}_{11}$, is found by taking the determinant of $\boldsymbol{A}$ with the first column and first row of numbers deleted:

$$\boldsymbol{M}_{11} = \begin{vmatrix} a_{22} & a_{23} \\ a_{32} & a_{33} \end{vmatrix} \tag{A.25}$$

The cofactor, $\boldsymbol{C}_{ij}$, of the element, $\boldsymbol{a}_{ij}$, of the matrix, $\boldsymbol{A}$, is defined by:

$$\boldsymbol{C}_{ij} = (-1)^{i+j}\boldsymbol{M}_{ij} \tag{A.26}$$

The determinant of a square matrix of arbitrary size is equal to the sum of the products of the elements and their cofactors along any column or row. For example, the determinant of the (3×3) matrix, $\boldsymbol{A}$, above can be found by adding the products of the elements and their cofactors along the first row:

$$|\boldsymbol{A}| = a_{11}C_{11} \; + \; a_{12}C_{12} \; + \; a_{13}C_{13} \tag{A.27}$$

Therefore, the determinant of a large square matrix can be broken up into a problem of calculating the determinants of a number of smaller square matrices.

If two matrices, $\boldsymbol{A}$ and $\boldsymbol{B}$, are square, then:

$$|\boldsymbol{AB}| = |\boldsymbol{A}|\,|\boldsymbol{B}| \tag{A.28}$$

A matrix is said to be singular if its determinant is equal to zero.

A.5 Rank of a Matrix

The rank of the $(m \times n)$ matrix, $\boldsymbol{A}$, is the maximum number of linearly independent rows of $\boldsymbol{A}$ and the maximum number of linearly independent columns of $\boldsymbol{A}$. Alternatively, the rank of $\boldsymbol{A}$ is a positive integer, r, such that an $(r \times r)$ submatrix of $\boldsymbol{A}$, formed by deleting $(m - r)$ rows and $(n - r)$ columns, is non-singular, whereas no $((r + 1) \times (r + 1))$ submatrix is non-singular. If the rank of $\boldsymbol{A}$ is equal to the number of columns or the number of rows of $\boldsymbol{A}$, then $\boldsymbol{A}$ is said to have full rank.

A.6 Positive and Nonnegative Definite Matrices

A matrix, $\boldsymbol{A}$, is said to be positive definite if $\boldsymbol{x}^H\boldsymbol{Ax}$ is positive for all non-zero vectors, $\boldsymbol{x}$; if $\boldsymbol{x}^H\boldsymbol{Ax}$ is simply non-negative, then $\boldsymbol{A}$ is said to be non-negative definite. Note that zero is a non-negative number but it is not a positive number.

For $\boldsymbol{A}$ to be positive definite, all of the leading minors must be positive; that is:

$$a_{11} > 0; \qquad \begin{vmatrix} a_{11} & a_{12} \\ a_{21} & a_{22} \end{vmatrix} > 0; \qquad \begin{vmatrix} a_{11} & a_{12} & a_{13} \\ a_{21} & a_{22} & a_{23} \\ a_{31} & a_{32} & a_{33} \end{vmatrix} > 0; \qquad \ldots, \text{etc.} \tag{A.29}$$

For $\boldsymbol{A}$ to be non-negative definite, all of the leading minors must be non-negative.

A.7 Eigenvalues and Eigenvectors

Let $\boldsymbol{A}$ be a (square) $(n \times n)$ matrix. The polynomial, $|\lambda\boldsymbol{I} - \boldsymbol{A}| = 0$, is referred to as the characteristic equation of $\boldsymbol{A}$. The solutions to the characteristic equation are the eigenvalues of $\boldsymbol{A}$. If λ_i is an eigenvalue of $\boldsymbol{A}$, then there exists at least one vector, $\boldsymbol{q}_i$, that satisfies the relationship:

$$\boldsymbol{A}\boldsymbol{q}_i = \lambda_i\boldsymbol{q}_i \tag{A.30}$$

The vector, $\boldsymbol{q}_i$, is an eigenvector of $\boldsymbol{A}$. If the eigenvalue, λ_i, is not repeated, then the eigenvector, $\boldsymbol{q}_i$, is unique. If an eigenvector, λ_i, is real, then the entries in the associated eigenvector, $\boldsymbol{q}_i$, are real; if λ_i is complex, then so too are the entries in $\boldsymbol{q}_i$.

The eigenvalues of a Hermitian matrix are all real, and if the matrix is also positive definite, the eigenvalues are also all positive. If a matrix is symmetric, then the eigenvalues are also all real. Further, it is true that:

$$|\boldsymbol{A}| = \prod_{i=1}^{n} \lambda_i \tag{A.31}$$

If $\boldsymbol{A}$ is singular, then there is at least one eigenvalue equal to zero.

A.8 Orthogonality

If a square matrix, $\boldsymbol{A}$, has the property, $\boldsymbol{A}^H\boldsymbol{A} = \boldsymbol{A}\boldsymbol{A}^H = \boldsymbol{I}$, then the matrix, $\boldsymbol{A}$, is said to be orthogonal. The eigenvalues of $\boldsymbol{A}$ then have a magnitude of unity. If $\boldsymbol{q_i}$ is an eigenvector associated with λ_i, and $\boldsymbol{q_j}$ is an eigenvector associated with λ_j, and if $\lambda_i \neq \lambda_j$ and $\boldsymbol{q_i}^H\boldsymbol{q_j} = 0$, then the vectors, $\boldsymbol{q_i}$ and $\boldsymbol{q_j}$, are said to be orthogonal.

The eigenvectors of a Hermitian matrix are all orthogonal. Further, it is common to normalise the eigenvectors to unit length such that $\boldsymbol{q_i}^H\boldsymbol{q_i} = 1$, in which case the eigenvectors are said to be orthonormal. A set of orthonormal eigenvectors can be expressed as columns of a unitary matrix, $\boldsymbol{Q}$:

$$\boldsymbol{Q} = (q_1,\ q_2, \cdots,\ q_n) \tag{A.32}$$

which means that:

$$\boldsymbol{Q}^{\mathrm{H}}\boldsymbol{Q} = \boldsymbol{Q}\boldsymbol{Q}^{\mathrm{H}} = \boldsymbol{I} \tag{A.33}$$

The set of equations that define the eigenvectors, expressed for a single eigenvector in Equation (A.30), can now be written in matrix form as:

$$\boldsymbol{A}\boldsymbol{Q} = \boldsymbol{Q}\boldsymbol{\Lambda} \tag{A.34}$$

where Λ is the diagonal matrix of eigenvalues:

$$\boldsymbol{\Lambda} = \begin{bmatrix} \lambda_1 & 0 & \cdots & 0 \\ 0 & \lambda_2 & \cdots & 0 \\ & & \ddots & \\ 0 & 0 & \cdots & \lambda_n \end{bmatrix} \tag{A.35}$$

Post-multiplying both sides of Equation (A.34) by $\boldsymbol{Q}^H$ yields:

$$\boldsymbol{A} = \boldsymbol{Q}\boldsymbol{\Lambda}\boldsymbol{Q}^{\mathrm{H}} \tag{A.36}$$

or

$$\boldsymbol{Q}^{\mathrm{H}}\boldsymbol{A}\boldsymbol{Q} = \boldsymbol{\Lambda} \tag{A.37}$$

Equations (A.36) and (A.37) define the orthonormal decomposition of $\boldsymbol{A}$, where $\boldsymbol{A}$ is re-expressed in terms of its eigenvalues and orthonormal eigenvectors.

A.9 Matrix Inverses

The inverse, $\boldsymbol{A}^{-1}$, of the matrix, $\boldsymbol{A}$, is defined by:

$$\boldsymbol{A}\boldsymbol{A}^{-1} = \boldsymbol{A}^{-1}\boldsymbol{A} = \boldsymbol{I} \tag{A.38}$$

The matrix, $\boldsymbol{A}$, must be square and be non-singular for the inverse to be defined.

The inverse of a matrix, $\boldsymbol{A}$, can be derived by first calculating the adjoint, $\hat{\boldsymbol{A}}$, of the matrix. The adjoint, $\hat{\boldsymbol{A}}$, is defined as the transpose of the matrix of cofactors of $\boldsymbol{A}$ (see Section A.4):

$$\hat{\boldsymbol{A}} = \begin{bmatrix} C_{11} & C_{12} & \cdots & C_{1m} \\ C_{21} & C_{22} & \cdots & C_{2m} \\ \vdots & \vdots & \vdots & \vdots \\ C_{m1} & C_{m2} & \cdots & C_{mm} \end{bmatrix} \tag{A.39}$$

The inverse, $\boldsymbol{A}^{-1}$, of the matrix, $\boldsymbol{A}$, is equal to the adjoint of $\boldsymbol{A}$ multiplied by the reciprocal of the determinant of $\boldsymbol{A}$:

$$\boldsymbol{A}^{-1} = \frac{1}{|\boldsymbol{A}|}\hat{\boldsymbol{A}} \tag{A.40}$$

While the definition given in Equation (A.40) is correct, using it to calculate a matrix inverse is inefficient for all but the smallest matrices (as the order of operations increases with the size, m, of the matrix by $m!$). There are a number of algorithms that require of the order of m^3 operations to compute the inverse of an arbitrary square matrix (outlined in many of the standard texts and in numerical methods books such as Press et al. (1986); Anton and Rorres (2013)).

Note that if the matrix, $\boldsymbol{A}$, is not square, or if it is singular, such that the determinant is zero, the inverse is not defined. However, for non-square matrices that are non-singular, a pseudo-inverse can be defined, which provides a least mean squares solution for the vector, $\boldsymbol{x}$, for the problem, $\boldsymbol{Ax} = \boldsymbol{b}$, where the matrix, $\boldsymbol{A}$, has more rows than columns, thus representing an overdetermined system with more equations than there are unknowns. In this case:

$$\boldsymbol{x} = \left(\boldsymbol{A}^{\mathrm{T}}\boldsymbol{A}\right)^{-1}\boldsymbol{A}^{\mathrm{T}}\boldsymbol{b} \tag{A.41}$$

where the pseudo-inverse of $\boldsymbol{A}$ is $\boldsymbol{A}' = \left(\boldsymbol{A}^{\mathrm{T}}\boldsymbol{A}\right)^{-1}\boldsymbol{A}^{\mathrm{T}}$ and $\boldsymbol{A}'\boldsymbol{A} = \boldsymbol{I}$.

If the matrix, $\boldsymbol{A}$, is singular, it is possible to define the Moore–Penrose pseudo-inverse, $\boldsymbol{A}'$, such that $\boldsymbol{A}'\boldsymbol{A}$ acts as the identity matrix on as large a set of vectors as possible. $\boldsymbol{A}'$ has the properties:

$$(\boldsymbol{A}')' = \boldsymbol{A}; \qquad \boldsymbol{A}'\boldsymbol{A}\boldsymbol{A}' = \boldsymbol{A}'; \qquad \boldsymbol{A}\boldsymbol{A}'\boldsymbol{A} = \boldsymbol{A} \tag{A.42}$$

If $\boldsymbol{A}$ is non-singular, then $\boldsymbol{A}^{-1} = \boldsymbol{A}'$.

A.10 Singular Value Decomposition

If a matrix is non-square, it does not have an eigenvalue decomposition, but it can be written in terms of a singular value decomposition, which can then be used to find the pseudo-inverse, thus allowing the best fit solution to be obtained to an overdetermined system of equations. The singular value decomposition of an $m \times n$ matrix, $\boldsymbol{A}$, is:

$$\boldsymbol{A} = \boldsymbol{Q}\boldsymbol{\Lambda}\boldsymbol{V}^{\mathrm{T}} \tag{A.43}$$

where $\boldsymbol{\Lambda}$ has zero elements everywhere except along the diagonal, and $\boldsymbol{Q}$ and $\boldsymbol{V}$ are unitary matrices with orthogonal columns so that:

$$\boldsymbol{Q}^{\mathrm{T}}\boldsymbol{Q} = \boldsymbol{I} \text{ and } \boldsymbol{V}^{\mathrm{T}}\boldsymbol{V} = \boldsymbol{I} \tag{A.44}$$

The columns of the matrix, $\boldsymbol{Q}$, consist of a set of orthonormal eigenvectors of $\boldsymbol{A}\boldsymbol{A}^{\mathrm{T}}$ and the columns of the matrix, $\boldsymbol{V}$, consist of a set of orthonormal eigenvectors of $\boldsymbol{A}^{\mathrm{T}}\boldsymbol{A}$. The diagonal elements of $\boldsymbol{\Lambda}$ are the square roots of the non-zero eigenvalues of both $\boldsymbol{A}\boldsymbol{A}^{\mathrm{T}}$ and $\boldsymbol{A}^{\mathrm{T}}\boldsymbol{A}$.

The pseudo-inverse of $\boldsymbol{A}$ can be calculated from its singular value decomposition using:

$$\boldsymbol{A}' = \boldsymbol{Q}\boldsymbol{\Lambda}'\boldsymbol{V}^{\mathrm{T}} \tag{A.45}$$

where $\boldsymbol{\Lambda}'$ is the pseudo-inverse of $\boldsymbol{\Lambda}$, computed by replacing every non-zero element of $\boldsymbol{\Lambda}$ by its reciprocal and transposing the resulting matrix.

If the matrix, $\boldsymbol{A}$, has complex elements, the transpose operation, T, is replaced with the Hermitian transpose operation, H, in all of the preceding equations in this section.

B

Wave Equation Derivation

The derivation of the acoustical wave equation is based on three fundamental fluid dynamical equations: the continuity (or conservation of mass) equation, Euler's equation (or the equation of motion) and the equation of state. Each of these equations are discussed separately in Sections B.1, B.2 and B.3.

B.1 Conservation of Mass

Consider an arbitrary volume, V, as shown in Figure B.1.

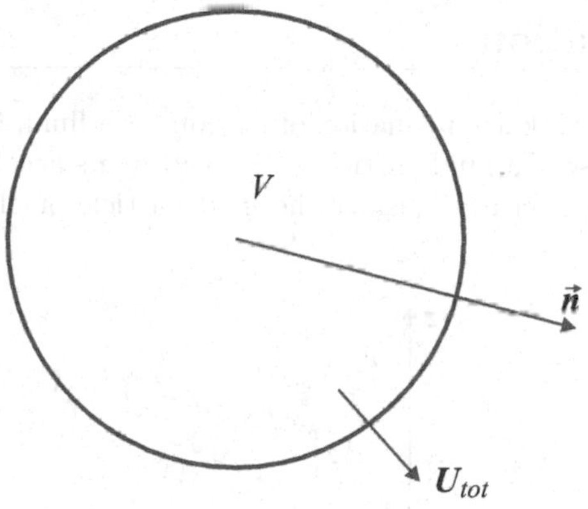

FIGURE B.1 Arbitrary volume for illustrating conservation of mass.

The total mass contained in this volume is $\iiint_V \rho_{\mathrm{tot}}\,\mathrm{d}V$. The law of conservation of mass states that the rate of mass leaving the volume, V, must equal the rate of change of mass in the volume. That is:

$$\iint_S \rho_{\mathrm{tot}} \boldsymbol{U}_{\mathrm{tot}} \cdot \vec{\boldsymbol{n}}\,\mathrm{d}S = -\frac{\mathrm{d}}{\mathrm{d}t}\iiint_V \rho_{\mathrm{tot}}\,\mathrm{d}V \tag{B.1}$$

where ρ_{tot} is the total (mean plus time varying component) density of the fluid contained in the enclosed space of volume, V, at time, t, U_{tot} is the total velocity of fluid in a direction outwards, normal to the enclosing surface, of area, S, at time, t, and $\vec{\boldsymbol{n}}$ is the unit vector normal to the surface, S, at location, $\mathrm{d}S$.

DOI: 10.1201/9780367814908-B

At this stage, it is convenient to transform the area integral on the left-hand side of Equation (B.2) to a volume integral by use of Gauss' integral theorem, which is written as:

$$\iint\limits_S \boldsymbol{\psi} \cdot \vec{\boldsymbol{n}}\, \mathrm{d}S = \iiint\limits_V \nabla \cdot \boldsymbol{\psi}\, \mathrm{d}V \tag{B.2}$$

where $\boldsymbol{\psi}$ is an arbitrary vector and the operator, ∇, is the scalar divergence of the vector, $\boldsymbol{\psi}$. Thus, in Cartesian coordinates:

$$\nabla \cdot \boldsymbol{\psi} = \frac{\partial \boldsymbol{\psi}}{\partial x} + \frac{\partial \boldsymbol{\psi}}{\partial y} + \frac{\partial \boldsymbol{\psi}}{\partial z} \tag{B.3}$$

and Equation (B.1) becomes:

$$\iiint\limits_V \nabla \cdot (\rho_{\mathrm{tot}} \boldsymbol{U}_{\mathrm{tot}})\, \mathrm{d}V = -\frac{\mathrm{d}}{\mathrm{d}t} \iiint\limits_V \rho_{\mathrm{tot}}\, \mathrm{d}V = -\iiint\limits_V \frac{\partial \rho_{\mathrm{tot}}}{\partial t}\, \mathrm{d}V \tag{B.4}$$

Rearranging gives:

$$\iiint\limits_V \left[\nabla \cdot (\rho_{\mathrm{tot}} \boldsymbol{U}_{\mathrm{tot}}) + \frac{\partial \rho_{\mathrm{tot}}}{\partial t} \right] \mathrm{d}V = 0 \tag{B.5}$$

or:

$$\nabla \cdot (\rho_{\mathrm{tot}} \boldsymbol{U}_{\mathrm{tot}}) = -\frac{\partial \rho_{\mathrm{tot}}}{\partial t} \tag{B.6}$$

Equation (B.6) is the continuity equation.

B.2 Euler's Equation

In 1775 Euler derived his well-known equation of motion for a fluid, based on Newton's first law of motion. That is, the mass of a fluid particle multiplied by its acceleration is equal to the sum of the external forces acting on it. Consider the fluid particle of dimensions Δx, Δy and Δz shown in Figure B.2.

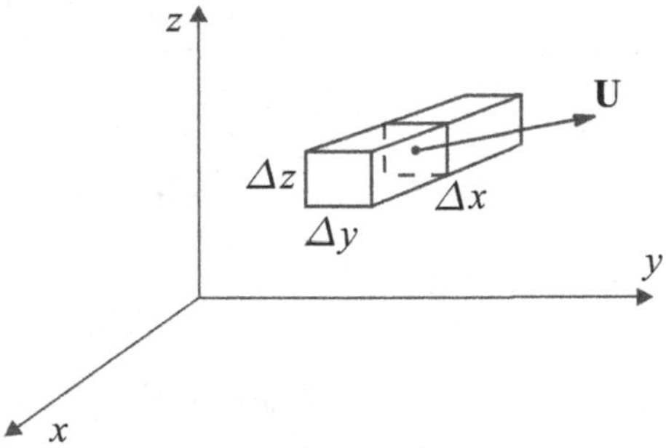

FIGURE B.2 Particle of fluid.

The external forces, F, acting on this particle at time instant, t, are equal to the sum of the pressure differentials across each of the three pairs of parallel forces. Thus:

$$F = \vec{\mathbf{i}} \cdot \frac{\partial P_{\mathrm{tot}}}{\partial x} + \vec{\mathbf{j}} \cdot \frac{\partial P_{\mathrm{tot}}}{\partial y} + \vec{\mathbf{k}} \cdot \frac{\partial P_{\mathrm{tot}}}{\partial z} = \nabla P_{\mathrm{tot}} \tag{B.7}$$

where P_{tot} is the total pressure (mean plus varying component) in the arbitrary volume at time instant, t, $\vec{\mathbf{i}}$, $\vec{\mathbf{j}}$ and $\vec{\mathbf{k}}$ are the unit vectors in the x-, y- and z-directions, and the operator, ∇, is the grad operator, which is the vector gradient of a scalar quantity.

The inertia force of the fluid particle is its mass multiplied by its acceleration and is equal to:

$$m\dot{\boldsymbol{U}}_{\text{tot}} = m\frac{\mathrm{d}\boldsymbol{U}_{\text{tot}}}{\mathrm{d}t} = \rho_{\text{tot}}V\frac{\mathrm{d}\boldsymbol{U}_{\text{tot}}}{\mathrm{d}t} \tag{B.8}$$

Assume that the fluid particle is accelerating in the positive x-, y- and z-directions. Then the pressure across the particle must be decreasing as x, y and z increase, as consistent with the direction of travel from higher to lower pressure, and the external force must be negative. Thus:

$$F = -\nabla P_{\text{tot}}V = \rho_{\text{tot}}V\frac{\mathrm{d}\boldsymbol{U}_{\text{tot}}}{\mathrm{d}t} \tag{B.9}$$

This is the Euler equation of motion for a fluid.

If sound propagation through porous acoustic media were of interest, then it would be necessary to add the term, $A\boldsymbol{U}_{\text{tot}}$, to the right-hand side of Equation (B.9), where A is a constant dependent on the properties of the fluid. The term $\mathrm{d}\boldsymbol{U}_{\text{tot}}/\mathrm{d}t$ on the right side of Equation (B.9) can be expressed in partial derivative form as:

$$\frac{\mathrm{d}\boldsymbol{U}_{\text{tot}}}{\mathrm{d}t} = \frac{\partial\boldsymbol{U}_{\text{tot}}}{\partial t} + (\boldsymbol{U}_{\text{tot}}\cdot\nabla)\boldsymbol{U}_{\text{tot}} \tag{B.10}$$

where:

$$(\boldsymbol{U}_{\text{tot}}\cdot\nabla)\boldsymbol{U}_{\text{tot}} = \frac{\partial\boldsymbol{U}_{\text{tot}}}{\partial x}\cdot\frac{\partial x}{\partial t} + \frac{\partial\boldsymbol{U}_{\text{tot}}}{\partial y}\cdot\frac{\partial y}{\partial t} + \frac{\partial\boldsymbol{U}_{\text{tot}}}{\partial z}\cdot\frac{\partial z}{\partial t} \tag{B.11}$$

B.3 Equation of State

As sound propagation is associated with only very small perturbations to the ambient state of a fluid, it may be regarded as adiabatic. Thus the total pressure $P_{\text{tot}} = P_s + p$ will be functionally related to the total density, $\rho_{\text{tot}} = \rho + \sigma$, as:

$$P_{\text{tot}} = f(\rho_{\text{tot}}) \tag{B.12}$$

Since the acoustic perturbations, p and σ, are small, and P_s and ρ are constant, $\mathrm{d}p = \mathrm{d}P_{\text{tot}}$, $\mathrm{d}\sigma = \mathrm{d}\rho$ and Equation (B.12) can be expanded into a Taylor series as:

$$\mathrm{d}p = \frac{\partial \mathrm{f}}{\partial\rho}\,\mathrm{d}\sigma + \frac{1}{2}\frac{\partial \mathrm{f}}{\partial\rho}(\mathrm{d}\sigma)^2 + \text{ higher order terms} \tag{B.13}$$

The equation of state is derived by using Equation (B.13) and ignoring all of the higher order terms on the right-hand side. This approximation is adequate for moderate sound pressure levels, but becomes less and less satisfactory as the sound pressure level exceeds 130 dB (60 Pa). Thus, for moderate sound pressure levels:

$$\mathrm{d}p = c^2\,\mathrm{d}\sigma \tag{B.14}$$

where $c^2 = \partial\mathrm{f}/\partial\rho$ is assumed to be constant. Integrating Equation (B.14) gives:

$$p = c^2\sigma + \text{const} \tag{B.15}$$

which is the linearised equation of state.

Thus the curve $\mathrm{f}(\rho_{\text{tot}})$ of Equation (B.12) has been replaced by its tangent at P_{tot}, ρ_{tot}. The constant may be eliminated by differentiating Equation (B.15) with respect to time. Thus:

$$\frac{\partial p}{\partial t} = c^2\frac{\partial\sigma}{\partial t} \tag{B.16}$$

Equation (B.16) will be used to eliminate $\partial\sigma/\partial t$ in the wave equation to follow.

B.4 Wave Equation (Linearised)

The wave equation may be derived from Equations (B.6), (B.9) and (B.16) by making the linearising approximations listed below. These assume that the acoustic pressure, p, is small compared with the ambient pressure, P_s, and that P_s is constant over time and space. It is also assumed that the mean velocity, $\boldsymbol{U}$, is equal to 0. Thus:

$$P_{\text{tot}} = P_s + p \approx P_s \tag{B.17}$$

$$\rho_{\text{tot}} = \rho + \sigma \approx \rho \tag{B.18}$$

$$\boldsymbol{U}_{\text{tot}} = \boldsymbol{u} \tag{B.19}$$

$$\frac{\partial P_{\text{tot}}}{\partial t} = \frac{\partial p}{\partial t} \tag{B.20}$$

$$\frac{\partial \rho_{\text{tot}}}{\partial t} = \frac{\partial \sigma}{\partial t} \tag{B.21}$$

$$\nabla P_{\text{tot}} = \nabla p \tag{B.22}$$

Using Equation (B.10), the Euler Equation (B.9) may be written as:

$$-\nabla P_{\text{tot}} = \rho_{\text{tot}} \left[\frac{\partial \boldsymbol{U}_{\text{tot}}}{\partial t} + (\boldsymbol{U}_{\text{tot}} \cdot \nabla)\boldsymbol{U}_{\text{tot}} \right] \tag{B.23}$$

Using Equations (B.17) (B.18) and (B.19), Equation (B.23) may be written as:

$$-\nabla p = \rho \left[\frac{\partial \boldsymbol{u}}{\partial t} + \boldsymbol{u} \cdot \nabla \boldsymbol{u} \right] \tag{B.24}$$

As $\boldsymbol{u}$ is small and $\nabla \boldsymbol{u}$ is approximately the same order of magnitude as $\boldsymbol{u}$, the quantity, $\boldsymbol{u} \cdot \nabla \boldsymbol{u}$, may be neglected and Equation (B.24) written as:

$$-\nabla p = \rho \frac{\partial \boldsymbol{u}}{\partial t} \tag{B.25}$$

Using Equations (B.18), (B.19) and (B.21), the continuity equation, Equation (B.6), may be written as:

$$\nabla \cdot (\rho \boldsymbol{u} + \sigma \boldsymbol{u}) = -\frac{\partial \sigma}{\partial t} \tag{B.26}$$

As $\sigma \boldsymbol{u}$ is so much smaller than ρ, the equality in Equation (B.26) can be approximated as:

$$\nabla \cdot (\rho \boldsymbol{u}) = -\frac{\partial \sigma}{\partial t} \tag{B.27}$$

Using Equation (B.16), Equation (B.27) may be written as:

$$\nabla \cdot (\rho \boldsymbol{u}) = -\frac{1}{c^2} \frac{\partial p}{\partial t} \tag{B.28}$$

Taking the time derivative of Equation (B.28) gives:

$$\nabla \cdot \rho \frac{\partial \boldsymbol{u}}{\partial t} = -\frac{1}{c^2} \frac{\partial^2 p}{\partial t^2} \tag{B.29}$$

Substituting Equation (B.25) into the left side of Equation (B.29) gives:

$$-\nabla \cdot \nabla p = -\frac{1}{c^2} \frac{\partial^2 p}{\partial t^2} \tag{B.30}$$

or:

$$\nabla^2 p = \frac{1}{c^2}\frac{\partial^2 p}{\partial t^2} \tag{B.31}$$

The operator, ∇^2, is the (div grad) or the Laplacian operator, and Equation (B.31) is known as the linearised wave equation or the Helmholtz equation.

The wave equation can be expressed in terms of the particle velocity by taking the gradient of the linearised continuity equation, Equation (B.28). Thus:

$$\nabla(\nabla \cdot \rho \boldsymbol{u}) = -\nabla\left(\frac{1}{c^2}\frac{\partial p}{\partial t}\right) \tag{B.32}$$

Differentiating the Euler Equation (B.25) with respect to time gives:

$$-\nabla\frac{\partial p}{\partial t} = \rho\frac{\partial^2 \boldsymbol{u}}{\partial t^2} \tag{B.33}$$

Substituting Equation (B.33) into (B.32) gives:

$$\nabla(\nabla \cdot \boldsymbol{u}) = \frac{1}{c^2}\frac{\partial^2 \boldsymbol{u}}{\partial t^2} \tag{B.34}$$

However, it may be shown that grad div = div grad + curl curl, or:

$$\nabla(\nabla \cdot \boldsymbol{u}) = \nabla^2\boldsymbol{u} + \nabla \times (\nabla \times \boldsymbol{u}) \tag{B.35}$$

Thus Equation (B.34) may be written as:

$$\nabla^2\boldsymbol{u} + \nabla \times (\nabla \times \boldsymbol{u}) = \frac{1}{c^2}\frac{\partial^2 \boldsymbol{u}}{\partial t^2} \tag{B.36}$$

which is the wave equation for the acoustic particle velocity.

A convenient quantity in which to express the wave equation is the acoustic velocity potential, which is a scalar quantity with no particular physical meaning. The advantage of using velocity potential is that both the acoustic pressure and particle velocity can be derived from it mathematically, simply by using differentiation and no integration (see Equations (1.11) and (1.12)). Thus it is a convenient way to represent the solution of the wave equation for many applications.

It can be shown (Hansen, 2018, page 19) that postulating a velocity potential solution to the wave equation causes some loss of generality and restricts the solutions to those that do not involve fluid rotation. Fortunately, acoustic motion in liquids and gases is nearly always without rotation.

Introducing Equation (1.12) for the velocity potential into Euler's Equation (B.25) gives the following expression:

$$-\nabla p = -\rho\frac{\partial \nabla\phi}{\partial t} = -\rho\nabla\frac{\partial \phi}{\partial t} \tag{B.37}$$

Integrating gives:

$$p = \rho\frac{\partial \phi}{\partial t} + \text{const.} \tag{B.38}$$

Introducing Equation (B.38) into the wave Equation (B.31) for acoustic pressure, integrating with respect to time and dropping the integration constant gives:

$$\nabla^2 \phi = \frac{1}{c^2}\frac{\partial^2 \phi}{\partial t^2} \tag{B.39}$$

This is the preferred form of the Helmholtz equation as both acoustic pressure and particle velocity can be derived from the velocity potential solution by simple differentiation.

C

Properties of Materials

The properties of materials can vary considerably, especially for wood and plastic. The values listed in the table in this appendix have been obtained from a variety of sources, including Simonds and Ellis (1943); Eldridge (1974); Levy (2001); Lyman (1961); Green et al. (1999). The data vary significantly between different sources for plastics and wood and sometimes even for metals; however, the values listed in this appendix reflect those most commonly found.

Where values of Poisson's ratio were unavailable, they were calculated from data for the speed of sound in a 3-D solid using the equation at the end of this table. These data were unavailable for some plastics so for those cases, values for similar materials were used. For wood products, the value for Poisson's ratio has been left blank where no data were available. Poisson's ratio is difficult to report for wood as there are six different ones, depending on the direction of stress and the direction of deformation. Here, only the value corresponding to strain in the longitudinal fibre direction coupled with deformation in the radial direction is listed.

The speed of sound values in column 4 of the table were calculated from the values in columns 2 and 3. Where a range of values occurred in either or both of columns 2 and 3, a median value of the speed of sound was recorded in column 4.

The values in the table should be used with caution and should be considered approximate only, as actual values may vary between samples. The upper end of the range corresponds to values of loss factor for a panel installed in a building and represents a combination of the material internal loss factor, the support loss factor and the sound radiation loss factor. The lower end of the range corresponds to internal loss factors for the material by itself and in the case of metals, it also corresponds to welded structures.

The properties of materials table is followed by a table of properties of gases, which is particularly useful for calculating control valve noise (see Section 10.8).

TABLE C.1 Properties of materials and Gases

Material	Young's modulus, E (10^9 N/m^2)	Density ρ (kg/m^3)	$\sqrt{E/\rho}$ (m/s)	Internal–*in situ*[a] loss factor, η	Poisson's ratio, ν
Fresh water (20°C)	—	998	1497	—	0.5
Sea water (13°C)	—	1025	1530	—	0.5
METALS					
Aluminium sheet	70	2700	5150	0.0001–0.01	0.35
Beryllium	287	1840	12500	0.001–0.01	0.027
Brass	95	8500	3340	0.001–0.01	0.35
Brass (70%Zn 30%Cu)	101	8600	3480	0.001–0.01	0.35

Cont. on next page

Properties of materials (cont.)

Material	Young's modulus, E (10^9 N/m^2)	Density ρ (kg/m^3)	$\sqrt{E/\rho}$ (m/s)	Internal–*in situ*[a] loss factor, η	Poisson's ratio, ν
METALS (Cont.)					
Bronze	96–120	7700–8700	3500	0.001–0.01	0.34
Carbon brick	8.2	1630	2240	0.001–0.01	0.07
Carbon nanotubes	1000	1330–1400	27000	0.001–0.01	0.06
Graphite mouldings	9.0	1700	2300	0.001–0.01	0.07
Chromium	265	7200	6070	0.0001–0.001	0.21
Cobalt	199	8746	4770	0.001–0.01	0.32
Copper (annealed)	128	8900	3790	0.002–0.01	0.34
Copper (rolled)	126	8930	3760	0.001–0.01	0.34
Gold	79	19300	2020	0.001–0.01	0.44
Iron	200	7600	5130	0.0005–0.01	0.30
Iron(white)	180	7700	4830	0.0005–0.01	0.30
Iron (nodular)	150	7600	4440	0.0005–0.01	0.30
Iron (wrought)	195	7900	4970	0.0005–0.01	0.30
Iron (grey (1))	83	7000	3440	0.0005–0.02	0.30
Iron (grey (2))	117	7200	4030	0.0005–0.03	0.30
Iron (malleable)	180	7200	5000	0.0005–0.04	0.30
Lead (annealed)	16.0	11400	1180	0.015–0.03	0.43
Lead (rolled)	16.7	11400	1210	0.015–0.04	0.44
Lead sheet	13.8	11340	1100	0.015–0.05	0.44
Magnesium	44.7	1740	5030	0.01–0.06	0.29
Manganese	187	7260	5075	0.03–0.1	0.35
Molybdenum	315–343	10220	5670	0.05	0.29
Monel metal	180	8850	4510	0.0001–0.02	0.33
Neodymium	390	7000	7460	0.0001–0.03	0.31
Nickel	205	8900	4800	0.001–0.01	0.31
Nickel–iron alloy (Invar)	143	8000	4230	0.0001–0.005	0.33
Platinum	168	21400	2880	0.0001–0.001	0.27
Silver	82.7	10500	2790	0.001–0.03	0.36
Steel (mild)	207	7850	5130	0.001–0.01	0.30
Steel (1% carbon)	210	7840	5170	0.001–0.02	0.29
Stainless steel (302)	200	7910	5030	0.0005–0.01	0.30
Stainless steel (316)	200	7950	5020	0.0005–0.01	0.30
Stainless steel (347)	198	7900	5010	0.0005–0.02	0.30
Stainless steel (430)	230	7710	5460	0.0005–0.03	0.30
Tin	43	7280	2430	0.0004–0.005	0.33
Titanium	116	4500	5080	0.0001–0.01	0.32
Tungsten (drawn)	360	19300	4320	0.0001–0.03	0.34
Tungsten (annealed)	412	19300	4620	0.0001–0.04	0.28
Tungsten carbide	534	13800	6220	0.0001–0.05	0.22
Zinc sheet	96.5	7140	3680	0.0005–0.005	0.33
BUILDING MATERIALS					
Brick	24	2000	3650	0.01–0.05	0.12
Concrete (normal)	18–30	2300	2800	0.005–0.05	0.20

Cont. on next page

Properties of materials (cont.)

Material	Young's modulus, E (10^9 N/m^2)	Density ρ (kg/m^3)	$\sqrt{E/\rho}$ (m/s)	Internal–*in situ*[a] loss factor, η	Poisson's ratio, ν
BUILDING MAT. (Cont.)					
Concrete (aerated)	1.5–2	300–600	2000	0.05	0.20
Concrete (high strength)	30	2400	3530	0.005–0.05	0.20
Masonry block	4.8	900	2310	0.005–0.05	0.12
Cork	0.1	250	500	0.005–0.05	0.15
Fibre board	3.5–7	480–880	2750	0.005–0.05	0.15
Gypsum board	2.1	760	1670	0.006–0.05	0.24
Glass	68	2500	5290	0.0006–0.02	0.23
Glass (Pyrex)	64	2240	5340	0.0006–0.02	0.24
WOOD					
Ash (black)	11.0	450	4940	0.04–0.05	0.46
Ash (white)	12.0	600	4470	0.04–0.05	0.46
Aspen (quaking)	8.1	380	4620	0.04–0.05	0.49
Balsa wood	3.4	160	4610	0.001–0.05	0.23
Baltic whitewood	10.0	400	5000	0.04–0.05	—
Baltic redwood	10.1	480	4590	0.04–0.05	—
Beech	11.9	640	4310	0.04–0.05	0.45
Birch (yellow)	13.9	620	4740	0.04–0.05	0.43
Cedar (white–nthn)	5.5	320	4150	0.04–0.05	0.34
Cedar (red–western)	7.6	320	4870	0.04–0.05	0.38
Compressed					
Hardboard composite	4.0	1000	2000	0.005–0.05	0.25
Douglas fir	9.7–13.2	500	4800	0.04–0.05	0.29
Douglas fir (coastal)	10.8	450	4900	0.04–0.05	0.29
Douglas fir (interior)	8.0	430	4310	0.04–0.05	0.29
Mahogany (African)	9.7	420	4810	0.04–0.05	0.30
Mahogany (Honduras)	10.3	450	4780	0.04–0.05	0.31
Maple	12.0	600	4470	0.04–0.05	0.43
MDF	3.7	770	2190	0.005–0.05	0.25
Meranti (light red)	10.5	340	5560	0.04–0.05	—
Meranti (dark red)	11.5	460	5000	0.04–0.05	—
Oak	12.0	630	4360	0.04–0.05	0.35
Particle board (floor)	2.8	700	1980	0.005–0.05	0.33
Particle board (std)	2.1	625	1830	0.005–0.05	0.33
Pine (radiata)	10.2	420	4930	0.04–0.05	0.26
Pine (other)	8.2–13.7	350–590	4830	0.04–0.06	0.26
Plywood (fir)	8.3	600	4540	0.01–0.05	0.22
Poplar	10.0	350–500	4900	0.04–0.05	0.32
Redwood (old)	9.6	390	4960	0.04–0.05	0.36
Redwood (second growth)	6.6	340	4410	0.04–0.05	0.36
Scots pine	10.1	500	4490	0.04–0.05	0.42
Spruce (Sitka)	9.6	400	4900	0.04–0.05	0.37
Spruce (Engelmann)	8.9	350	5040	0.04–0.05	0.42

Cont. on next page

Properties of materials (cont.)

Material		Young's modulus, E (10^9 N/m^2)	Density ρ (kg/m^3)	$\sqrt{E/\rho}$ (m/s)	Internal–*in situ*[a] loss factor, η	Poisson's ratio, ν
WOOD (Cont.)						
Teak		14.6	550	5150	0.02–0.05	0.2
Walnut (black)		11.6	550	4590	0.04–0.05	0.49
PLASTICS & OTHER						
Lucite		4.0	1200	1830	0.002–0.02	0.35
Melinex		3.0	1390	1470	0.002–0.02	0.4
Plexiglass (acrylic)		3.5	1190	1710	0.002–0.02	0.35
Polycarbonate		2.3	1200	1380	0.003–0.1	0.35
Polyester (thermo)		2.3	1310	1320	0.003–0.1	0.40
Polyethylene						
(high density)		0.7–1.4	940–960	1030	0.003–0.1	0.44
(low density)		0.2–0.5	910–925	600	0.003–0.1	0.44
Polypropylene		1.4–2.1	905	1380	0.003–0.1	0.40
Polystyrene						
(moulded)		3.2	1050	1750	0.003–0.1	0.34
(expanded foam)		0.0012–0.0035	16–32	300	0.0001–0.02	0.30
Polyurethane		1.6	900	1330	0.003–0.1	0.35
PVC		2.8	1400	1410	0.003–0.1	0.40
PVDF		1.5	1760	920	0.003–0.1	0.35
Nylon 6		2.4	1200	1410	0.003–0.1	0.35
Nylon 66		2.7–3	1120–1150	1590	0.003–0.1	0.35
Nylon 12		1.2–1.6	1010	1170	0.003–0.1	0.35
Rubber–neoprene		0.01–0.1	1,100–1,400	95–260	0.05–0.1	0.495
Kevlar 49 cloth		31	1330	4830	0.008	—
Aluminium honeycomb						
Cell size (mm)	Foil thickness (mm)					
6.4	0.05	1.31	72	—	0.0001–0.01	—
6.4	0.08	2.24	96	—	0.0001–0.01	—
9.5	0.05	0.76	48	—	0.0001–0.01	—
9.5	0.13	1.86	101	—	0.0001–0.01	—

[a]Loss factors of materials shown characterised by a very large range are very sensitive to specimen mounting conditions. Use the upper limit for panels used in building construction and the lower limit for metal panels welded together in an enclosure or for unsupported other materials.

Speed of sound for a 1-D solid, $c_{LII} = \sqrt{E/\rho}$; for a 2-D solid (plate), $c_{LI} = \sqrt{E/[\rho(1-\nu^2)]}$; and for a 3-D solid, $c_L = \sqrt{E(1-\nu)/[\rho(1+\nu)(1-2\nu)]}$. For gases, replace E with γP, where γ is the ratio of specific heats (=1.40 for air) and P is the absolute pressure. For liquids, replace E with $V(\partial V/\partial p)^{-1}$, where V is the unit volume and $\partial V/\partial p$ is the compressibility. Note that Poisson's ratio, ν, may be defined in terms of Young's modulus, E, and the material shear modulus, G, as $\nu = E/(2G) - 1$, and it is effectively zero for liquids and gases.

TABLE C.2 Physical properties of some commonly used gases

Gas or vapour	Molecular weight, M kg/mole	Ratio of specific heats, γ	Sutherland's constant at 293.15 K	Viscosity at 293.15 K
Acetylene	0.02604	1.3	–	1.00
Air	0.02897	1.4	120	1.82
Ammonia	0.0173	1.32	370	0.99
Argon	0.03995	1.67	114	2.23
Isobutane	0.05812	1.1	336	0.73
Isobutylene	0.05611	1.11	329	0.79
Carbon dioxide	0.04401	1.3	240	1.47
Carbon monoxide	0.02801	1.4	118	1.74
Chlorine	0.07091	1.31	325	1.34
Ethane	0.03007	1.22	287	0.92
Ethylene	0.02805	1.22	259	1.03
Fluorine	0.019	1.36	129	2.28
Helium	0.004	1.66	79.4	1.96
Hydrogen	0.00202	1.41	72	0.89
Methane	0.01604	1.32	295	1.08
Methyl chloride	0.05049	1.24	380	1.04
Natural gas (representative)	0.01774	1.27	198	1.59
Neon	0.02018	1.64	56	3.13
Nitric oxide	0.06301	1.4	128	1.89
Nitrogen	0.02801	1.4	111	1.76
Oxygen	0.032	1.4	127	2.04
Pentane	0.07215	1.06	128	0.66
Propane	0.0441	1.15	341	0.8
Propylene	0.04208	1.14	322	0.84
Saturated steam	0.01802	1.25–1.32	1064	0.97
Sulphur dioxide	0.06406	1.26	416	1.26

D

Acoustical Properties of Porous Materials

D.1 Flow Resistance and Flow Resistivity

If a constant differential pressure is imposed across a layer of bulk porous material of open cell structure, a steady flow of gas will be induced through the material. Experimental investigation has shown that, for a wide range of materials, the differential pressure, ΔP, and the induced normal velocity, $\boldsymbol{U}$, of the gas at the surface of the material (volume velocity per unit surface area) are linearly related, provided that the normal velocity is small. The ratio of differential pressure in Pascals to normal velocity in m/s is known as the flow resistance R_f (MKS rayls) of the material. It is generally assumed that the gas is air but a flow resistance may be determined for any gas.

If the material is generally of uniform composition, then the flow resistance is proportional to the material thickness. When the measured flow resistance is divided by the test sample thickness in metres, the flow resistivity, R_1, in MKS rayls per metre, is obtained, which is independent of the sample thickness and is characteristic of the material. Flow resistance and flow resistivity of porous materials are discussed in depth in the literature (Bies, 1988; Bies and Hansen, 1979, 1980; Bies, 1981).

The flow resistance of a sample of porous material may be measured using an apparatus that meets the requirements of ASTM C522-03 (2016), such as illustrated in Figure D.1. Flow velocities between 5×10^{-4} and 5×10^{-2} m/s are easily realisable, and yield good results.

As shown in Figure D.1, air under pressure flows through a flow meter, which measures its mass flow rate, and then it enters the bottom of the tube. Higher flow rates should be avoided due to the possible introduction of nonlinear effects. If the flow meter is a volume flow measuring device, valve (6) must be adjusted so that flow through it is choked to ensure that the pressure within the flow meter (7) is the same as that measured by the manometer (8). Air flows through the sample (2) and the pressure drop across the sample is measured by the barocell (11). The sample holder is sealed in the straight, cylindrical part at the top of the conical tube using a tapered sample holder (shown in Figure D.1) or an O-ring mounted in a groove on the inside surface of the cylindrical part into which the sample holder is inserted. The edges of the sample are sealed within the sample holder using silicone sealant or equivalent to ensure that all measured air flow passes through the sample and none flows past the sample edges. The flow resistivity of the specimen shown in Figure D.1 is calculated from the measured quantities as:

$$R_1 = \rho \,\Delta P_s \, S/(\dot{m}\ell) = \Delta P_s \, S/(V_0 \ell) \tag{D.1}$$

DOI: 10.1201/9780367814908-D

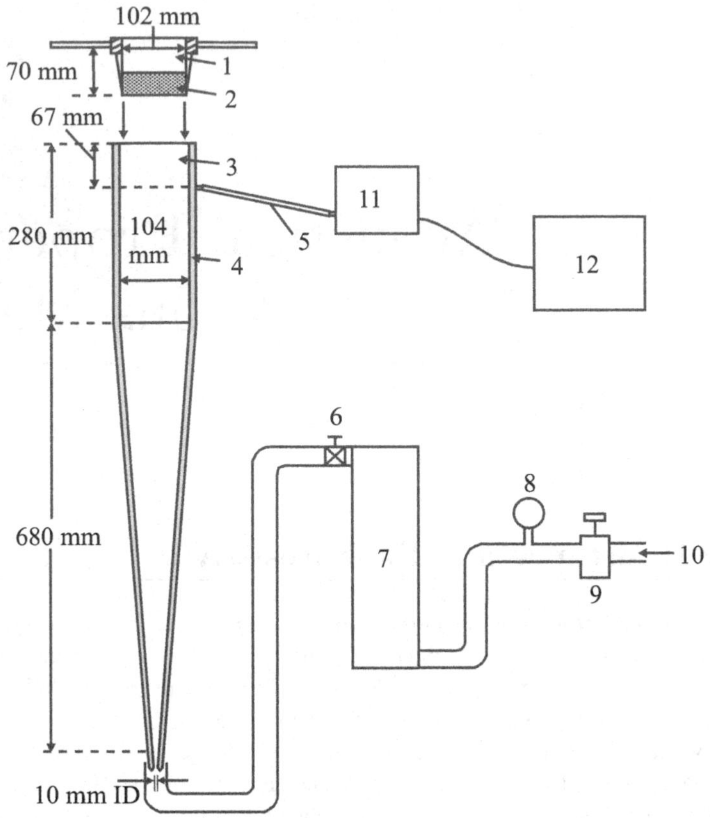

FIGURE D.1 Flow resistance measuring apparatus.

Key

1 sample holder and cutter	5 tube	9 pressure regulator
2 porous material	6 valve	10 air supply
3 O-ring seal	7 flow meter	11 barocell
4 conical tube to ensure uniform air flow through sample	8 manometer	12 electronic manometer

where ρ is the density of the gas (kg/m^3), ΔP_s is the differential static pressure (Pa), S is the specimen cross-sectional area (m^2), $\dot{m}$ is the air mass flow rate (kg/s), V_0 is the volume flow rate through the sample (m^3/s) and ℓ is the specimen thickness (m).

Alternatively, acoustic flow resistance may be measured using a closed end tube, a sound source and any inexpensive microphones arranged as shown in Figure D.2 (Ingard and Dear, 1985). To make a measurement, the sound source is driven with a pure tone signal, preferably below 100 Hz, at a frequency chosen to produce an odd number of quarter wavelengths over the distance $w + \ell$ from the closed end to the sample under test. The first step is to satisfy the latter requirement by adjusting the chosen frequency to achieve a minimum sound pressure level at microphone 1 in the absence of the sample. The sample, which is assumed to be thin compared to a wavelength at the test frequency, is then inserted in the tube and if the sample is flexible, it must be inserted between two rigid, acoustically transparent screens. The sound pressure level is then measured at locations 1 and 3. The normalised flow impedance, $Z_s/\rho c$, is a complex quantity made up of a real term (flow resistance) and an imaginary term (flow reactance) and is given by:

$$\frac{Z_s}{\rho c} = \frac{p_1 - p_2}{\rho c u_1} \tag{D.2}$$

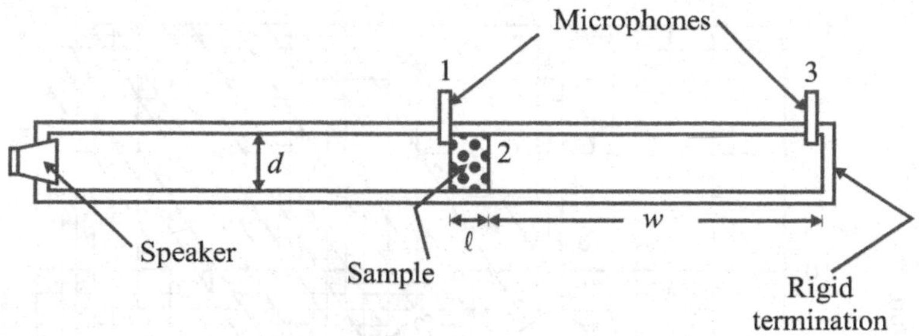

FIGURE D.2 An alternative arrangement for measuring flow resistance.

As the tube is rigidly terminated and the losses along it are assumed small, the amplitude of the reflected wave will be the same as the incident wave and there will be zero phase shift between the incident and reflected waves at the rigid termination. If the coordinate system is chosen so that $x = 0$ corresponds to the rigid end and $x = -L$ (where $L = w + \ell$), at microphone location 1, then the particle velocity (assuming positive time dependence, $e^{j\omega t}$) at location 1 is:

$$u_1 = \frac{jp_3}{\rho c}\sin(kL) \tag{D.3}$$

The acoustic pressure at location 2 is very similar to the pressure at location 1 in the absence of the sample and is given as:

$$p_2 = p_3\cos(kL) \tag{D.4}$$

Thus, Equation (D.2) can be written as:

$$\frac{Z_s}{\rho c} = \frac{-jp_1}{p_3\sin(kL)} + j\cot(kL) \tag{D.5}$$

If L is chosen to be an odd number of quarter wavelengths such that $L = (2n-1)\lambda/4$ where n is an integer (preferably $n = 1$), the normalised flow impedance becomes:

$$\frac{Z_s}{\rho c} = j(-1)^n(p_1/p_3) \tag{D.6}$$

so the normalised flow resistance, $R_1\ell/(\rho c)$, is the absolute value of of the imaginary part of p_1/p_3. Taking into account that the flow reactance is small at low frequencies, the flow resistance is given by the magnitude of the ratio of the acoustic pressure at location 1 to that at location 3. Thus, the flow resistivity (flow resistance divided by sample thickness) is given by:

$$R_1 = \frac{\rho c}{\ell}\left|\frac{p_1}{p_3}\right| = \frac{\rho c}{\ell}10^{(L_{p1}-L_{p3})/20} \tag{D.7}$$

As only the sound pressure level difference between locations 1 and 3 is required, then prior to taking the measurement, microphones 1 and 3 are placed together near the closed end of the tube in the absence of the sample, so that they measure the same sound pressure and the gain of either one is adjusted so that they read the same level. The closed end location is chosen for this as here the sound pressure level varies only slowly with location. The final step is to place the calibrated microphones at positions 1 and 3 as shown in Figure D.1 and measure L_{p1} and L_{p3} and then use Equation (D.7) to calculate the flow resistivity, R_1.

Measured values of flow resistivity for various commercially available sound-absorbing materials are available in the literature (Bies and Hansen, 1980), and can sometimes be obtained from material manufacturers. For fibrous materials with a reasonably uniform fibre diameter

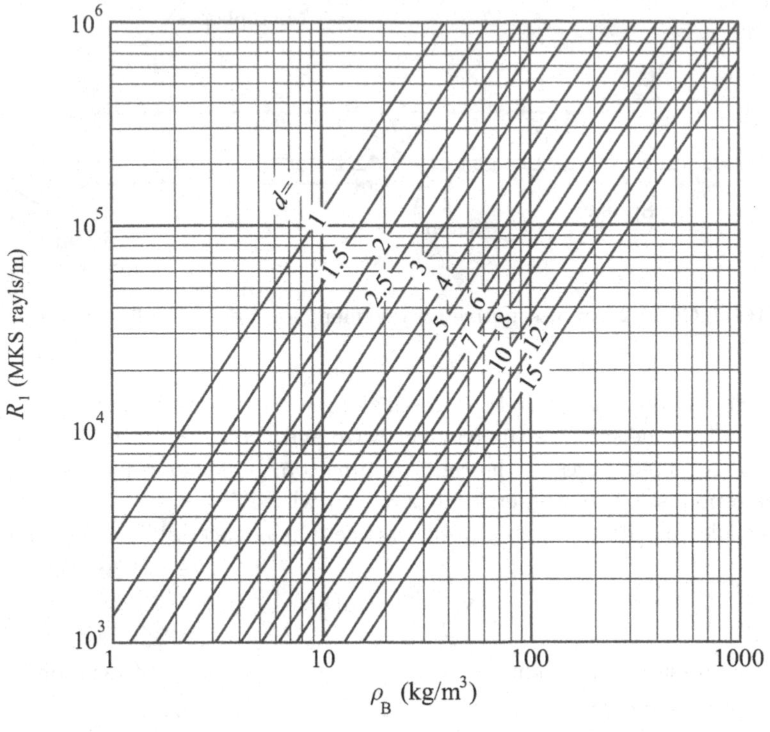

FIGURE D.3 Flow resistivity, R_1, as a function of material bulk density, ρ_B (kg/m^3), and fibre diameter, d (μm).

and with only a small quantity of binder (such that the flow resistance is minimally affected), Figure D.3 may be used to obtain an estimate of flow resistivity.

Figure D.3 is derived from the empirical equation (Bies and Hansen, 1980):

$$R_1 = K_2 d^{-2} \rho_B^{K_1} \tag{D.8}$$

where ρ_B is the bulk density of material, d is the fibre diameter, $K_1 = 1.53$ and $K_2 = 3.18 \times 10^{-9}$.

For fibres with diameters larger than those in Figure D.3, such as polyester, the flow resistivity may be estimated using Equation (D.8), but with different values for K_1 and K_2. In this case, Garai and Pompoli (2005) found that $K_1 = 1.404$ and $K_2 = 25.989d^2$, with d the fibre diameter in metres.

D.2 Parameters for Characterising Sound Propagation in Porous Media

For the purpose of the analysis, the porous, gas-filled medium is replaced by an effective medium, which is characterised in dimensionless variables by a complex density, ρ_m, and complex compressibility, κ. In terms of these quantities, a complex characteristic impedance and propagation coefficient are defined, analogous to the similar quantities for the contained gas in the medium. Implicit in the formulation of the following expressions is the assumption that time dependence has the positive form, $e^{j\omega t}$, consistent with the practice adopted throughout the text.

The characteristic impedance of porous material may be written in terms of the gas density, ρ, the gas speed of sound c, the gas compressibility, κ_0, the porous material complex density,

ρ_m, and the porous material complex compressibility, κ, as:

$$Z_m = \rho c \sqrt{\frac{\rho_m \kappa}{\rho \kappa_0}} \tag{D.9}$$

Similarly, a propagation coefficient, k_m, may be defined as:

$$k_m = \frac{2\pi}{\lambda}(1 - \mathrm{j}\alpha_m) = (\omega/c)\sqrt{\frac{\rho_m \kappa_0}{\rho \kappa}} \tag{D.10}$$

where $\omega = 2\pi f$ is the angular frequency (rad/s) of the sound wave.

The quantities, ρ_m/ρ and κ/κ_0, may be calculated using the following procedure (Bies, 1981). This procedure gives results for fibrous porous materials within 4% of the mean of published data (Delany and Bazley, 1969, 1970), and unlike the Delaney and Bazley model, it tends to the correct limits at both high and low values of the dimensionless frequency, $\rho f/R_1$. However, this model and the Delaney and Bazley model have only been verified for fibreglass and rockwool materials with a small amount of binder and short fibres smaller than 15 µm in diameter, which excludes such materials as polyester and acoustic foam. The normalised compressibility and normalised density of a porous material can be calculated from a knowledge of the material flow resistivity, R_1, using Equations (D.11) to (D.21).

$$\kappa/\kappa_0 = [1 + (1 - \gamma)\tau]^{-1} \tag{D.11}$$

$$\rho_m/\rho = [1 + \sigma]^{-1} \tag{D.12}$$

where γ is the ratio of specific heats for the gas (=1.40 for air), ρ is the density of gas (=1.205 kg/m^3 for air at 20°C), f is the frequency (Hz), R_1 is the flow resistivity of the porous material (MKS rayls/m) and:

$$\tau = 0.592a(X_1) + \mathrm{j}b(X_1) \tag{D.13}$$

$$\sigma = a(X) + \mathrm{j}b(X) \tag{D.14}$$

$$a(X) = \frac{T_3(T_1 - T_3)T_2^2 - T_4^2T_1^2}{T_3^2T_2^2 + T_4^2T_1^2} \tag{D.15}$$

$$b(X) = \frac{T_1^2T_2T_4}{T_3^2T_2^2 + T_4^2T_1^2} \tag{D.16}$$

$$T_1 = 1 + 9.66X \tag{D.17}$$

$$T_2 = X(1 + 0.0966X) \tag{D.18}$$

$$T_3 = 2.537 + 9.66X \tag{D.19}$$

$$T_4 = 0.159(1 + 0.7024X) \tag{D.20}$$

$$X = \rho f/R_1 \tag{D.21}$$

The quantities, $a(X_1)$ and $b(X_1)$, are calculated by substituting $X_1 = 0.856X$ for the quantity, X, in Equations (D.15)–(D.20).

The relationships that have been generally accepted in the past (Delany and Bazley, 1969, 1970), and which are accurate in the flow resistivity range $R_1 = 10^3$ to 5×10^4 MKS rayls/m, are:

$$Z_m = \rho c[1 + C_1X^{-C_2} - \mathrm{j}C_3X^{-C_4}] \tag{D.22}$$

$$k_m = (\omega/c)[1 + C_5X^{-C_6} - \mathrm{j}C_7X^{-C_8}] \tag{D.23}$$

The quantities X, Z_m, k_m, c and ρ have all been defined previously. Values of the coefficients $C_1 - C_8$ are given in Table D.1 for various materials from various references.

TABLE D.1 Values of the coefficients C_1–C_8 for various materials

Material type reference	C_1	C_2	C_3	C_4	C_5	C_6	C_7	C_8
Rockwool/fibreglass Delany and Bazley (1970)	0.0571	0.754	0.087	0.732	0.0978	0.700	0.189	0.595
Polyester Garai and Pompoli (2005)	0.078	0.623	0.074	0.660	0.159	0.571	0.121	0.530
Polyurethane foam of low flow resistivity Dunn and Davern (1986)	0.114	0.369	0.0985	0.758	0.168	0.715	0.136	0.491
Porous plastic foams of medium flow resistivity Wu (1988)	0.212	0.455	0.105	0.607	0.163	0.592	0.188	0.544

D.3 Sound Reduction Due to Propagation through a Porous Material

For the purpose of the calculation, three frequency ranges are defined: low, middle and high, as indicated in Figure D.4. The quantities in the parameters $\rho f/R_1$ and $f\ell/c$ are defined in Sections D.1 and D.2.

In the low-frequency range, the inertia of the porous material is small enough for the material to move with the particle velocity associated with the sound wave passing through it. The transmission loss to be expected in this frequency range can be obtained from Figure D.5. If the material is used as pipe wrapping, the noise reduction will be approximately equal to the transmission loss. In the high-frequency range, the porous material is many wavelengths thick and, in this case, reflection at both surfaces of the layer, as well as propagation losses through the layer, must be taken into account when estimating noise reduction. The reflection loss at an air/porous medium interface may be calculated using Figure D.6 and the transmission loss may be estimated using Figure D.7.

In the middle-frequency range, it is generally sufficient to estimate the transmission loss graphically, with a smooth curve connecting plotted estimates of the low- and high-frequency transmission loss versus log frequency, such that the mid-frequency curve is tangential to the low-frequency curve at the left-hand end and tangential to the high-frequency curve at the right-hand end.

D.4 Measurement of Absorption Coefficients of Porous Materials

D.4.1 Measurement Using the Moving Microphone Method

Statistical absorption coefficients may be determined using impedance tube measurements of the normal incidence specific acoustic impedance, as an alternative to the measurement of the Sabine absorption coefficient using a reverberant test chamber. The statistical absorption coefficient calculated in this way is always less than 1, and always less than the Sabine absorption coefficient, which can exceed 1 (meaning that in principal, there is more energy absorbed than is incident on the test material). This peculiar phenomenon arises from the sample of absorbing material

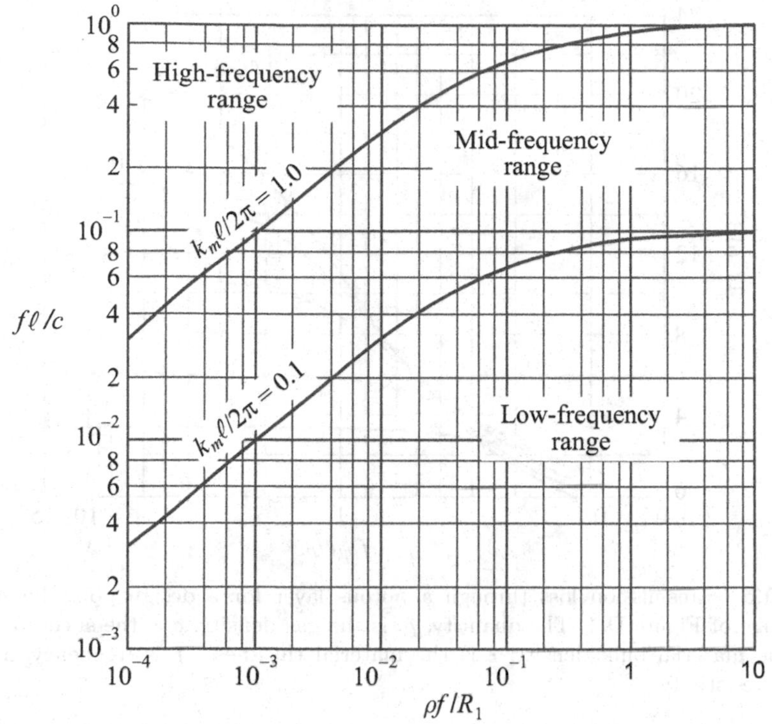

FIGURE D.4 Limits showing when low- and high-frequency models should be used for estimating the transmission loss through a porous layer. The low-frequency model should be used when the design point lies below the $k_m\ell/2\pi = 0.1$ curve, and the high-frequency model should be used when the design point lies above the $k_m\ell/2\pi = 1.0$ curve. The quantity, k_m, is the amplitude of the complex wavenumber of sound in the porous material (see Equations (D.10) and (D.23)), ρ is the gas density, c is the speed of sound in the gas, f is frequency, ℓ is the material thickness and R_1 is the material flow resistivity.

distorting the sound field in its vicinity, so that more energy falls on its surface than fell on the floor in the absence of the material.

When a tonal (single frequency) sound field is set up in a tube terminated with a specific acoustic impedance, Z_s, a pattern of regularly spaced maxima and minima along the tube will result, the locations of which are uniquely determined by the driving frequency and the terminating impedance. The measured absorption coefficients are related to the terminating impedance and the characteristic impedance, ρc, of air.

An impedance tube is relatively easily constructed and therein lies its appeal. Any heavy walled tube may be used for its construction. A source of sound should be placed at one end of the tube and the material to be tested should be mounted at the other end. Means must be provided for probing the standing wave within the tube. An example of a possible configuration is shown in Figure D.8.

The older and simpler method by which the sound field in the impedance tube is explored, using a moveable microphone which traverses the length of the tube, will be described first (see ASTM C384-04 (2016) and ISO 10534-1 (1996)). In this case, the impedance of the test sample is determined from measurements of the sound field in the tube. This method is slow but easily implemented. A much quicker method, which makes use of two fixed microphones and a digital frequency analysis system, is described in Section D.4.2 (see ASTM E1050-19 (2019) and ISO 10534-2 (1998)).

Implicit in the use of the impedance tube is the assumption that only plane waves propagate back and forth in the tube. This assumption puts upper and lower frequency bounds on the

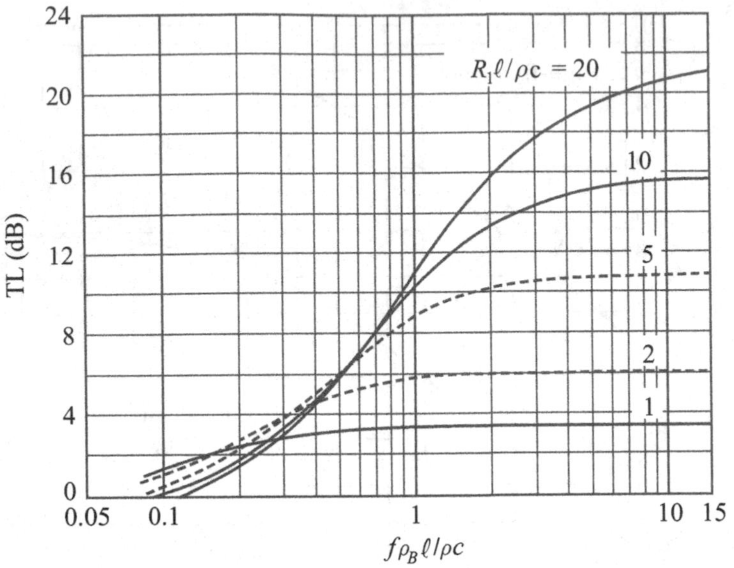

FIGURE D.5 Transmission loss through a porous layer for a design point lying in the low-frequency range of Figure D.4. The quantity, ρ, is the gas density, c is the speed of sound in the gas, ρ_B is the material bulk density, ℓ is the material thickness, f is frequency, and R_1 is the material flow resistivity.

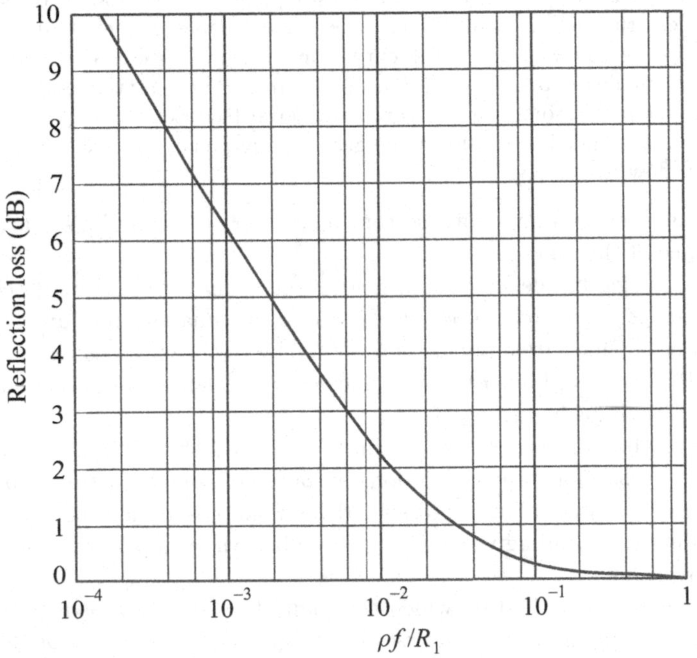

FIGURE D.6 Reflection loss (dB) at a porous material–air interface for a design point in the high-frequency range of Figure D.4. The quantity, ρ, is the gas density, f is the frequency and R_1 is the material flow resistivity.

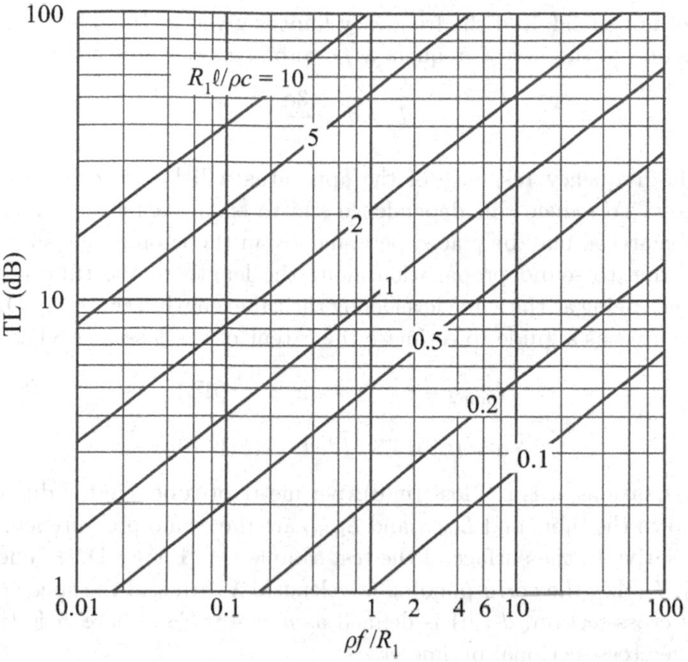

FIGURE D.7 Transmission loss of a porous lining of thickness, ℓ, and normalised flow resistance, $R_1\ell/\rho c$, for a design point lying in the high-frequency region of Figure D.4. The quantity, ρ, is the gas density, c is the speed of sound in the gas and R_1 is the material flow resistivity.

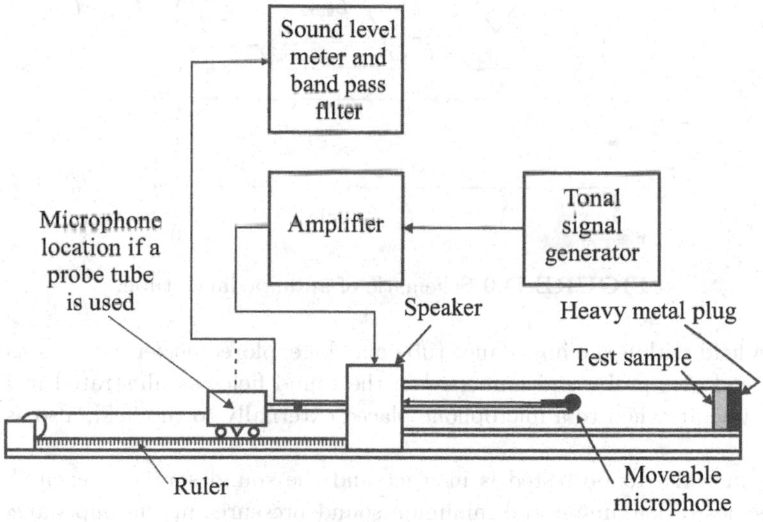

FIGURE D.8 Equipment for impedance tube measurement using the older and simpler method.

use of the impedance tube. Let d be the tube diameter if it is circular in cross-section, or the larger dimension if it is rectangular in cross-section. Then the upper frequency limit (or cut-on frequency), above which measurements are no longer valid, is:

$$f_{\rm co} = \begin{cases} 0.586c/d; & \text{for circular section ducts} \\ 0.5c/d; & \text{for rectangular section ducts} \end{cases} \tag{D.24}$$

Here, c is the speed of sound, and the frequency limit is given in Hz. The required length, L, of the tube is a function of the lowest frequency, f_ℓ, to be tested and is given by:

$$L = d + \frac{3c}{4f_\ell} \tag{D.25}$$

In general, the frequency response of the apparatus will be very much dependent on the material under test. To reduce this dependence and to ensure a more uniform response, some sound-absorptive material may be placed permanently in the tube at the source end.

Energy losses due to sound propagation along the length of the tube result in the sound pressure minima not being at the same level along the tube length (see Figure D.9). The following equations may be used as a guide to estimate the extent of the losses (ASTM C384-04, 2016):

$$L_{\min 2} - L_{\min 1} = a\lambda/2 \qquad \text{(dB)} \tag{D.26}$$

$$a = 0.19137\sqrt{f}/cd \qquad \text{(dB/m)} \tag{D.27}$$

In the above equations, a is the loss (in dB per metre of tube length) due to propagation of the sound wave down the tube, and $L_{\min 1}$ and $L_{\min 2}$ are the sound pressure levels at the first and second minima relative to the surface of the test sample (see Figure D.9). The other quantities are the frequency, f (Hz), the corresponding wavelength, λ (m), and the speed of sound, c (m/s). For tubes of any cross-section, d (m) is defined as $d = 4S/P_D$, where S is the cross-sectional area and P_D is the cross-sectional perimeter.

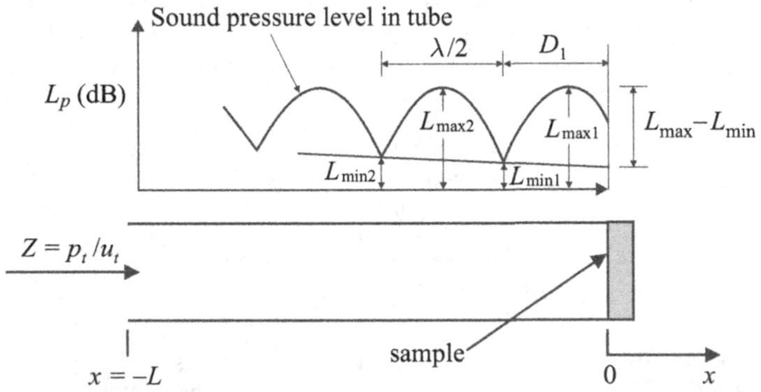

FIGURE D.9 Schematic of an impedance tube.

The sound field within the impedance tube may be explored either with a small microphone placed at the end of a probe and immersed in the sound field, as illustrated in Figure D.8, or with a probe tube attached to a microphone placed externally to the field, also as illustrated in Figure D.8.

When the material to be tested is in place and the sound source is excited with a single frequency, a series of maximum and minimum sound pressures in the impedance tube will be observed. The maxima will be effectively constant in level but the minima will increase in level, according to Equation (D.26), as one moves away from the surface of the test material. For best results, it is recommended that losses in the tube be taken into account by extrapolating the minima back to the surface of the sample by drawing a straight line joining the first two minima to the location corresponding to the surface of the sample on the plot of sound pressure level in dB versus distance along the tube (see Figure D.9). The standing wave ratio, L_0, is then determined as the difference between the maximum level, $L_{\max}$, and the minimum level, $L_{\min}$, that has been extrapolated back to the surface of the sample.

Equation (D.27) may be useful in selecting a suitable tube for exploring the field; the smaller the probe tube diameter, the greater will be the energy loss suffered by a sound wave travelling down it, but the smaller will be the disturbance to the acoustic field being sampled, which could affect the location of the pressure minima in the impedance tube. An external linear scale should be provided for locating the probe. As the sound field will be distorted slightly by the presence of the probe, it is recommended that the actual location of the sampled sound field be determined by replacing the specimen with a heavy solid metal face located at the specimen surface. For such a surface, it is known that he first minimum is $\lambda/4$ away from the solid metal face. Subsequent minima will always be spaced at intervals of $\lambda/2$. Thus, this experiment will allow the determination of how far in front of the end of the probe tube that the sound field is effectively being sampled, as this is the difference between the location of the actual pressure minimum and the theoretical location of the pressure minimum. This difference can then be used to correct the pressure minimum location found using the actual sample. It will be found that the minima will be subject to contamination by acoustic and electronic noise; thus, it is recommended that a narrow band filter, for example, one which is an octave or 1/3-octave wide, be used in the detection system.

It is of interest to derive an expression for the normal incidence absorption coefficient of a sample of acoustic material in an impedance tube, as shown in Figure D.9. For the following analysis, the impedance tube contains the material sample at the right end and the loudspeaker sound source at the left end. For simplicity, it is assumed that there are no losses due to dissipation in the tube; that is, the quantity, a, in Equation (D.27) is assumed to be zero. The loudspeaker is not shown in the figure and the origin is set at the right end of the tube at the face of the test sample. Reference should be made to Section 1.5.4, where it is shown that multiple waves travelling in a single direction may be summed together to give a single wave travelling in the same direction. For the case considered here, the multiple waves travelling in each direction are a result of multiple reflections from each end of the tube. As the origin is at the right end of the tube, the resultant incident wave will be travelling in the positive x-direction. Assuming a phase shift between the incident and reflected waves of β at $x = 0$, the incident wave and reflected wave pressures may be written as:

$$p_i = A\mathrm{e}^{\mathrm{j}(\omega t - kx)} \quad \text{and} \quad p_r = B\mathrm{e}^{\mathrm{j}(\omega t + kx + \beta)} \tag{D.28}$$

where A and B are real amplitudes. The total sound pressure is thus:

$$p_t = A\mathrm{e}^{\mathrm{j}(\omega t - kx)} + B\mathrm{e}^{\mathrm{j}(\omega t + kx + \beta)} \tag{D.29}$$

The first maximum pressure amplitude (closest to the sample) will occur when:

$$\beta = -2kx \tag{D.30}$$

and the first minimum pressure amplitude will occur when:

$$\beta = -2kx + \pi \tag{D.31}$$

Thus:

$$\hat{p}_{\max} = |\mathrm{e}^{-\mathrm{j}kx}|\,(A + B) = A + B \quad \text{and} \quad \hat{p}_{\min} = |\mathrm{e}^{-\mathrm{j}kx}|\,(A - B) = A - B \tag{D.32}$$

and the ratio of maximum to minimum pressure amplitudes is $(A + B)/(A - B)$. The standing wave ratio, L_0, is the difference in decibels between the maximum and minimum sound pressures in the standing wave and is defined as:

$$10^{L_0/20} = \frac{A + B}{A - B} \tag{D.33}$$

Thus, the ratio (B/A) is:

$$\frac{B}{A} = \left[\frac{10^{L_0/20} - 1}{10^{L_0/20} + 1}\right] \tag{D.34}$$

The amplitude of the pressure reflection coefficient squared is defined as $|R_p|^2 = (B/A)^2$, which can be written in terms of L_0 as:

$$|R_p|^2 = \left[\frac{10^{L_0/20} - 1}{10^{L_0/20} + 1}\right]^2 \tag{D.35}$$

The normal incidence absorption coefficient is defined as:

$$\alpha_N = 1 - |R_p|^2 \tag{D.36}$$

and it can also be determined from Table D.2.

TABLE D.2 Normal incidence sound-absorption coefficient, α_N, versus standing wave ratio, L_0 (dB)

L_0	α_N	L_0	α_N	L_0	α_N	L_0	α_N
0	1.000	10	0.730	20	0.331	30	0.119
1	0.997	11	0.686	21	0.301	31	0.107
2	0.987	12	0.642	22	0.273	32	0.096
3	0.971	13	0.598	23	0.247	33	0.086
4	0.949	14	0.555	24	0.223	34	0.077
5	0.922	15	0.513	25	0.202	35	0.069
6	0.890	16	0.472	26	0.182	36	0.061
7	0.854	17	0.434	27	0.164	37	0.055
8	0.815	18	0.397	28	0.147	38	0.049
9	0.773	19	0.363	29	0.132	39	0.04

It is also of interest to continue the analysis to determine the normal specific acoustic impedance of the surface of the sample. This can then be used to determine the statistical absorption coefficient of the sample, which is the absorption coefficient averaged over all possible angles of an incident wave.

The total particle velocity can be calculated using Equations (1.11), (1.12) and (D.28) to give:

$$u_t = \frac{1}{\rho c}(p_i - p_r) \tag{D.37}$$

Thus:

$$u_t = \frac{1}{\rho c}\left(A\mathrm{e}^{\mathrm{j}(\omega t - kx)} - B\mathrm{e}^{\mathrm{j}(\omega t + kx + \beta)}\right) \tag{D.38}$$

The specific acoustic impedance (or characteristic impedance) at any point in the tube may be written as:

$$Z_s = \frac{p_t}{u_t} = \rho c \frac{A\mathrm{e}^{-\mathrm{j}kx} + B\mathrm{e}^{\mathrm{j}kx + \mathrm{j}\beta}}{A\mathrm{e}^{-\mathrm{j}kx} - B\mathrm{e}^{\mathrm{j}kx + \mathrm{j}\beta}} = \rho c \frac{A + B\mathrm{e}^{\mathrm{j}(2kx+\beta)}}{A - B\mathrm{e}^{\mathrm{j}(2kx+\beta)}} \tag{D.39}$$

At $x = 0$, the impedance is the normal specific acoustic impedance, Z_N, of the surface of the sample. Thus:

$$\frac{Z_N}{\rho c} = \frac{p_t}{\rho c u_t} = \frac{A + B\mathrm{e}^{\mathrm{j}\beta}}{A - B\mathrm{e}^{\mathrm{j}\beta}} \tag{D.40}$$

The above impedance equation may be expanded to give:

$$\frac{Z_N}{\rho c} = \frac{A/B + \cos\beta + \mathrm{j}\sin\beta}{A/B - \cos\beta - \mathrm{j}\sin\beta} = \frac{(A/B)^2 - 1 + (2A/B)\mathrm{j}\sin\beta}{(A/B)^2 + 1 - (2A/B)\cos\beta} \tag{D.41}$$

In practice, the phase angle, β, is evaluated by measuring the distance, D_1, of the first sound pressure minimum in the impedance tube from the sample surface. Referring to Equation (D.31) and Figure D.9, the phase angle, β, may be expressed in terms of D_1 (which is a positive number) as:

$$\beta = 2kD_1 + \pi = 2\pi\left(\frac{2D_1}{\lambda} + \frac{1}{2}\right) \tag{D.42}$$

Equation (D.41) may be rewritten in terms of a real and imaginary components as:

$$\frac{Z_N}{\rho c} = R + \mathrm{j}X = \frac{(A/B)^2 - 1}{(A/B)^2 + 1 - (2A/B)\cos\beta} + \mathrm{j}\frac{(2A/B)\sin\beta}{(A/B)^2 + 1 - (2A/B)\cos\beta} \tag{D.43}$$

where β is defined by Equation (D.42) and the ratio, A/B, is defined by the reciprocal of Equation (D.34) where L_0 is the difference in dB between the maximum and minimum sound pressure levels in the tube.

In terms of an amplitude and phase, the normal specific acoustic impedance may also be written as:

$$Z_N/(\rho c) = \xi \mathrm{e}^{\mathrm{j}\psi} \tag{D.44}$$

where

$$\xi = \sqrt{R^2 + X^2} \tag{D.45}$$

and

$$\psi = \tan^{-1}(X/R) \tag{D.46}$$

The statistical absorption coefficient, α_{st}, may be calculated as:

$$\alpha_{st} = \frac{1}{\pi}\int_0^{2\pi} \mathrm{d}\vartheta \int_0^{\pi/2} \alpha(\theta)\cos\theta\sin\theta\,\mathrm{d}\theta \tag{D.47}$$

where the angles, θ and ϑ are defined in Figure 7.8.

Rewriting the absorption coefficient in terms of the pressure amplitude reflection coefficient using:

$$|R_p(\theta)|^2 = 1 - \alpha(\theta) \tag{D.48}$$

Equation (D.47) can then be written as:

$$\alpha_{st} = 1 - 2\int_0^{\pi/2} |R_p(\theta)|^2 \cos\theta\sin\theta\,\mathrm{d}\theta \tag{D.49}$$

For bulk reacting materials, Equations (5.15) and (5.17) may be used to calculate the reflection coefficient, $R_p(\theta)$, and for locally reacting materials, Equation (5.18) may be used. Note that use of these equations requires a knowledge of the normal specific acoustic impedance, Z_N, of the material. This can be determined using Equation (D.43) and an impedance tube in which a sample of the material, in the same configuration as to be used in practice, in terms of how it is backed, is tested. Alternatively, the normal specific acoustic impedance for any configuration may be calculated using the methods described in Section D.5.

Using Equations (D.49) and (5.18), Morse and Bolt (1944) derive the following expression for the statistical absorption coefficient for a locally reactive surface of normal specific acoustic impedance, $Z_N/(\rho c)$, given by Equation (D.44):

$$\alpha_{st} = \left\{\frac{8\cos\psi}{\xi}\right\}\left\{1 - \left[\frac{\cos\psi}{\xi}\right]\log_e(1 + 2\xi\cos\psi + \xi^2) + \left[\frac{\cos(2\psi)}{\xi\sin\psi}\right]\tan^{-1}\left[\frac{\xi\sin\psi}{1+\xi\cos\psi}\right]\right\} \quad \text{(D.50)}$$

Alternatively, Figure D.10 may be used to determine the statistical absorption coefficient. Note that Equation (D.50) and Figure D.10 are based on the explicit assumption that sound propagation within the sample is always normal to the surface. However, calculations indicate that the error in ignoring propagation in the porous material in other directions is negligible.

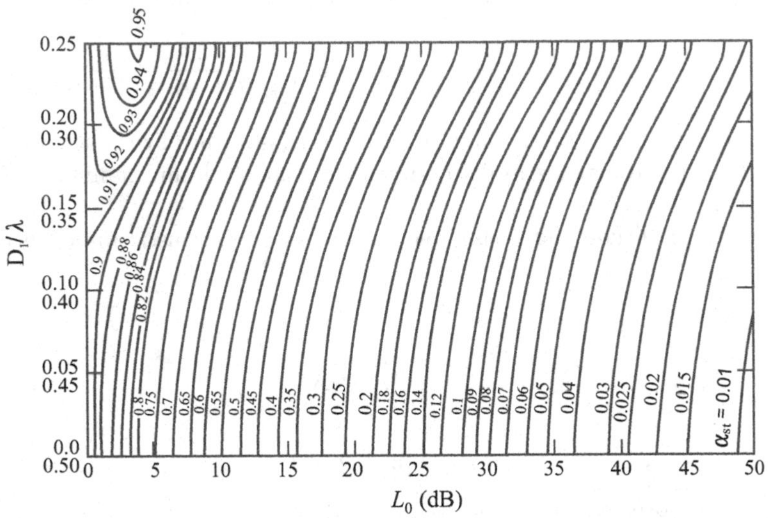

FIGURE D.10 A chart for determining the statistical absorption coefficient, α_{st}, from measurements in an impedance tube of the standing wave ratio, L_0, and position, D_1/λ, of the first minimum sound pressure level. α_{st} is shown parametrically in the chart.

D.4.2 Measurement Using the Two-Microphone Method

The advantage of the two-microphone method over the moving microphone method for determining material absorption coefficients and normal acoustic impedance (discussed in the previous subsection) is that it considerably reduces the time required to determine the normal specific acoustic impedance and normal incidence absorption coefficient of a sample, allowing the sample to be evaluated over the frequency range of interest in one or two measurements. The upper frequency limit for valid data is related to the diameter of the tube used for the measurement in the same way as for the moving microphone method. The lower frequency limit is a function of the spacing, s_1, of the two microphones (which should exceed 1% of the wavelength at the lowest frequency of interest) and the accuracy of the analysis system. However, the microphone spacing must not be larger than 40% of the wavelength at the highest frequency of interest. In some cases, it may be necessary to repeat measurements with two different microphone spacings and two tubes with different diameters.

For this test method, the microphones are mounted through the wall of the tube and flush with its inside surface, as illustrated in Figure D.11. The microphone closest to the sample surface should be located at least one tube diameter from the surface for a rough sample surface and half a tube diameter for a smooth surface.

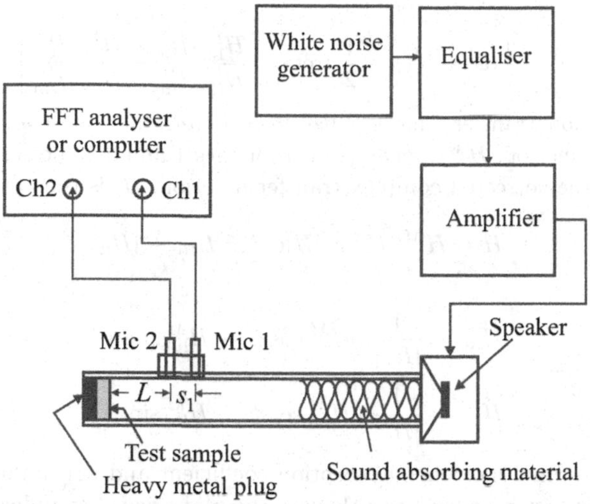

FIGURE D.11 Arrangement for measuring the normal incidence complex reflection coefficient, normal specific acoustic impedance and absorption coefficient of a sample of acoustic material using the two-microphone method.

The sample is mounted in a holder attached to the end of the impedance tube as illustrated in Figure D.11 and described in ASTM E1050-19 (2019) and ISO 10534-2 (1998). An FFT analyser is used to determine the transfer function between the two microphones as a function of frequency. The loudspeaker at the end of the tube is excited with white noise (see Section 12.2.1) and the resulting transfer function (or frequency response function), $H^M(f_n)$, between the two microphones is measured using the FFT analyser (see Section 12.3.15). The measured transfer function is written in terms of its real, $H^M_{\text{Re}}(f_n)$, and imaginary parts, $H^M_{\text{Im}}(f_n)$, as:

$$H^M(f_n) = H^M_{\text{Re}}(f_n) + \text{j}H^M_{\text{Im}}(f_N) = H^M_{\text{Re}} + \text{j}H^M_{\text{Im}} = |H^M|\text{e}^{\text{j}\phi^M} \tag{D.51}$$

It is important that in Equation (D.51) and in the context of Equation (12.78), microphone 1 is the input microphone, X, and microphone 2 is the response microphone, Y.

The measured transfer function, H^M, must be corrected to account for any phase and amplitude mismatch between the microphones and this correction procedure should be undertaken using a test sample that is highly absorptive to maximise the accuracy of the results (ASTM E1050-19 (2019) and ISO 10534-2 (1998)). The correction transfer function is obtained by first measuring the complex transfer function as described above, as:

$$H^{\text{I}} = H^{\text{I}}_{\text{Re}} + \text{j}H^{\text{I}}_{\text{Im}} = |H^{\text{I}}|\text{e}^{\text{j}\phi^{\text{I}}} \tag{D.52}$$

and then repeating the measurement with the positions of the two microphones in the tube wall physically swapped, without changing any cable connections, to obtain the complex transfer function:

$$H^{\text{II}} = H^{\text{II}}_{\text{Re}} + \text{j}H^{\text{II}}_{\text{Im}} = |H^{\text{II}}|\text{e}^{\text{j}\phi^{\text{II}}} \tag{D.53}$$

Care must be taken to ensure the microphone positions are exchanged as accurately as possible, or there will be errors in the calibration factor. The calibration transfer function, H^c, is calculated as:

$$H^c = (H^{\text{I}} \times H^{\text{II}})^{1/2} = |H^c|\text{e}^{\text{j}\phi^c} \tag{D.54}$$

where

$$|H^c| = \left(|H^{\text{I}}| \times |H^{\text{II}}|\right)^{1/2} = \left\{\left[\left(H^{\text{I}}_{\text{Re}}\right)^2 + \left(H^{\text{I}}_{\text{Im}}\right)^2\right] \times \left[\left(H^{\text{II}}_{\text{Re}}\right)^2 + \left(H^{\text{II}}_{\text{Im}}\right)^2\right]\right\}^{1/4} \tag{D.55}$$

and:

$$\phi^c = \frac{1}{2}(\phi^{\mathrm{I}} + \phi^{\mathrm{II}}) = \frac{1}{2}\arctan\left[\frac{H^{\mathrm{I}}_{\mathrm{Im}}H^{\mathrm{II}}_{\mathrm{Re}} + H^{\mathrm{I}}_{\mathrm{Re}}H^{\mathrm{II}}_{\mathrm{Im}}}{H^{\mathrm{I}}_{\mathrm{Re}}H^{\mathrm{II}}_{\mathrm{Re}} - H^{\mathrm{I}}_{\mathrm{Im}}H^{\mathrm{II}}_{\mathrm{Im}}}\right] \quad (D.56)$$

Once the calibration transfer function has been determined with an arbitrary or no test sample, the transfer function, H^M, between microphones 1 and 2 is measured with the required sample in place and the corrected complex transfer function, H, is then calculated as:

$$H = H^M/H^c = |H|\mathrm{e}^{\mathrm{j}\phi} = H_{\mathrm{Re}} + \mathrm{j}H_{\mathrm{Im}} \quad (D.57)$$

where:

$$H_{\mathrm{Re}} = \frac{1}{|H^c|}\left(H^M_{\mathrm{Re}}\cos\phi^c + H^M_{\mathrm{Im}}\sin\phi^c\right) \quad (D.58)$$

$$H_{\mathrm{Im}} = \frac{1}{|H^c|}\left(H^M_{\mathrm{Im}}\cos\phi^c - H^M_{\mathrm{Re}}\sin\phi^c\right) \quad (D.59)$$

To calculate the normal incidence absorption coefficient and normal surface impedance for the test material, it is first necessary to calculate the plane wave, complex amplitude reflection coefficient, R_p (using the corrected transfer function, H). Thus, for each frequency component in the transfer function spectrum:

$$R_p = |R_p|\mathrm{e}^{\mathrm{j}\phi_R} = R_{p,\mathrm{Re}} + \mathrm{j}R_{p,\mathrm{Im}} \quad (D.60)$$

where:

$$R_{p,\mathrm{Re}} = \frac{2H_{\mathrm{Re}}\cos[k(2L+s_1)] - \cos(2kL) - [(H_{\mathrm{Re}})^2 + (H_{\mathrm{Im}})^2]\cos[2k(L+s_1)]}{1 + (H_{\mathrm{Re}})^2 + (H_{\mathrm{Im}})^2 - 2[H_{\mathrm{Re}}\cos ks_1 + H_{\mathrm{Im}}\sin(ks_1)]} \quad (D.61)$$

$$R_{p,\mathrm{Im}} = \frac{2H_{\mathrm{Re}}\cos[k(2L+s_1)] - \sin(2kL) - [(H_{\mathrm{Re}})^2 + (H_{\mathrm{Im}})^2]\sin[2k(L+s_1)]}{1 + (H_{\mathrm{Re}})^2 + (H_{\mathrm{Im}})^2 - 2[H_{\mathrm{Re}}\cos ks_1 + H_{\mathrm{Im}}\sin(ks_1)]} \quad (D.62)$$

where k is defined in Equation (1.30) and s_1 and L are defined in Figure D.11.

The normal incidence absorption coefficient, α_N, is then calculated using:

$$\alpha_N = 1 - |R_p|^2 \quad (D.63)$$

The normal specific acoustic impedance at the face of the sample is then given by

$$Z_N = Z_{NR} + \mathrm{j}Z_{NI} = \rho c\frac{1+R_p}{1-R_p} \quad (D.64)$$

where:

$$Z_{NR} = \rho c\frac{\alpha_N}{2(1-R_{p,\mathrm{Re}}) - \alpha_N} \quad (D.65)$$

and:

$$Z_{NI} = 2\rho c R_{p,\mathrm{Im}}[2(1-R_{p,\mathrm{Re}}) - \alpha_N] \quad (D.66)$$

Measurement errors are discussed in ASTM E1050-19 (2019) and ISO 10534-2 (1998).

D.4.3 Measurement Using the Four-Microphone Method

Absorption coefficients can also be measured using a four-microphone method (Song and Bolton, 2000; ASTM E2611-19, 2019). The advantage of the four-microphone method over the two-microphone method is that the four-microphone method allows evaluation of the porous material complex wavenumber, the complex characteristic impedance, the complex normal incidence reflection coefficient and the normal incidence absorption coefficient. The complex characteristic impedance is only the same as the normal specific acoustic impedance for materials of sufficient thickness that waves reflected upstream (towards the sound source) from the downstream face (furthest from the sound source) are insignificant in amplitude when they arrive back at the upstream face. The four-microphone method also allows the transmission loss of any sample of material or device (such as a muffler – see Chapter 8) to be determined. The measurement method is described in detail in ASTM E2611-19 (2019) and Song and Bolton (2000). The experimental arrangement for the four-microphone method is illustrated in Figure D.12. The analysis of the four-microphone method also serves as an excellent introduction to the transfer matrix method. As for both previous methods involving measurements in a tube, the tube diameter must be sufficiently small so that only plane waves are propagating. That is, the test frequencies must lie below the cut-on frequency of the first higher order mode (see Equation (D.24)). The four microphones are located at axial x-coordinates, x_1, x_2, x_3 and x_4. Note that x_1 and x_2 are negative due to the origin location shown in Figure D.12. The maximum allowed microphone spacings, s_1 and s_2, shown in Figure D.12, are both positive numbers whose values are dependent on $f_{\rm co}$:

$$s_1(\max) = s_2(\max) = 0.8 \times \frac{c}{2f_{\rm co}} \tag{D.67}$$

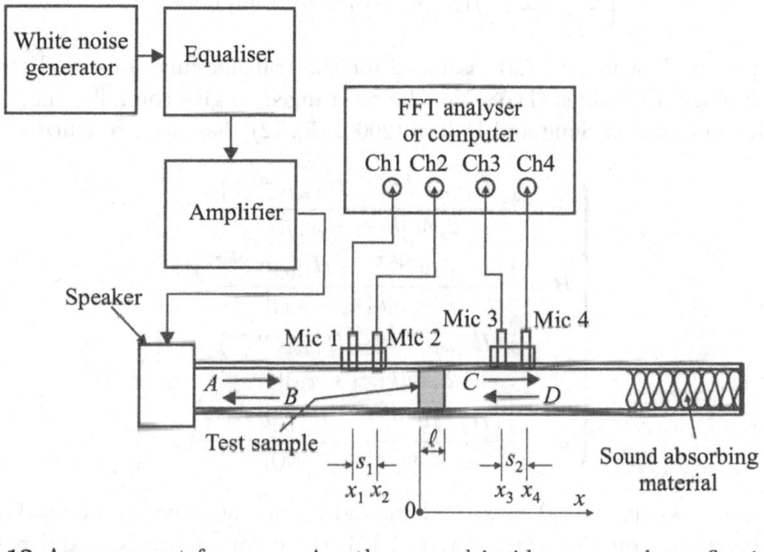

FIGURE D.12 Arrangement for measuring the normal incidence complex reflection coefficient, characteristic impedance, wavenumber and absorption coefficient of a sample of acoustic material using the four-microphone method.

Referring to Figure D.12, the test specimen is inserted at a location between $x = 0$ and $x = \ell$ and the transfer function, $H_{n,\rm ref}$ (see Section 12.3.15) between each microphone, n, and a reference signal, is measured for two different load cases. One load case (case a) should be with a minimally reflecting termination at the end of the right hand side of the tube in Figure D.12. The second load case (case b) could be an open or partially blocked termination, reflecting a portion of the incident wave. Any one of the four microphone signals may be used as the reference signal. In practice, the transfer functions between the acoustic pressures at the microphone locations and the reference signal can be obtained by moving the same microphone to each location or by using different microphones at each location. If the same microphone is used, then it is important that any unused holes in the test tube are plugged when measurements are taken. If four different microphones are used, then they need to be calibrated relative to one another so that any differences between the microphone amplitude and phase responses are accounted for. When a single microphone is used, a convenient reference signal is the input to the loudspeaker that generates the sound in the tube. However, the accuracy is limited by non-linearities (or harmonic distortion) in the response of the loudspeaker so it is better to use a reference microphone located nearer to the loudspeaker than the measurement locations. When more than one microphone is used to obtain measurements at the four microphone locations, it is better to choose one of the measurement microphones to provide the reference signal. Of course, this will result in a transfer function of unity amplitude and zero phase for the microphone position used as the reference.

The transfer functions, denoted $H_{n,\mathrm{ref}}$, always appear in subsequent calculations as ratios and thus can be considered to represent the complex sound pressures at the measurement locations for frequency, ω radians/s (Song and Bolton, 2000). Thus:

$$\begin{cases} \dfrac{\hat{p}_{(x=x_1)}}{\hat{p}_{\mathrm{ref1}}} = H_{1,\mathrm{ref}}(\omega) = A\mathrm{e}^{-\mathrm{j}kx_1} + B\mathrm{e}^{\mathrm{j}kx_1} \\ \dfrac{\hat{p}_{(x=x_2)}}{\hat{p}_{\mathrm{ref1}}} = H_{2,\mathrm{ref}}(\omega) = A\mathrm{e}^{-\mathrm{j}kx_2} + B\mathrm{e}^{\mathrm{j}kx_2} \\ \dfrac{\hat{p}_{(x=x_3)}}{\hat{p}_{\mathrm{ref1}}} = H_{3,\mathrm{ref}}(\omega) = C\mathrm{e}^{-\mathrm{j}kx_3} + D\mathrm{e}^{\mathrm{j}kx_3} \\ \dfrac{\hat{p}_{(x=x_4)}}{\hat{p}_{\mathrm{ref1}}} = H_{4,\mathrm{ref}}(\omega) = C\mathrm{e}^{-\mathrm{j}kx_4} + D\mathrm{e}^{\mathrm{j}kx_4} \end{cases} \tag{D.68}$$

With the frequency dependence, (ω), assumed for the transfer functions, and thus omitted to simplify terminology, Equations (D.68) may be rearranged to give the following, where the sign error in the denominator of Song and Bolton (2000, Eq. (2)) has been rectified:

$$\begin{cases} A = \dfrac{\mathrm{j}\left(H_{1,\mathrm{ref}}\mathrm{e}^{\mathrm{j}kx_2} - H_{2,\mathrm{ref}}\mathrm{e}^{\mathrm{j}kx_1}\right)}{2\sin[k(x_2 - x_1)]} \\ B = \dfrac{\mathrm{j}\left(H_{2,\mathrm{ref}}\mathrm{e}^{-\mathrm{j}kx_1} - H_{1,\mathrm{ref}}\mathrm{e}^{-\mathrm{j}kx_2}\right)}{2\sin[k(x_2 - x_1)]} \\ C = \dfrac{\mathrm{j}\left(H_{3,\mathrm{ref}}\mathrm{e}^{\mathrm{j}kx_4} - H_{4,\mathrm{ref}}\mathrm{e}^{\mathrm{j}kx_3}\right)}{2\sin[k(x_4 - x_3)]} \\ D = \dfrac{\mathrm{j}\left(H_{4,\mathrm{ref}}\mathrm{e}^{-\mathrm{j}kx_3} - H_{3,\mathrm{ref}}\mathrm{e}^{-\mathrm{j}kx_4}\right)}{2\sin[k(x_4 - x_3)]} \end{cases} \tag{D.69}$$

where $(x_2 - x_1) = s_1$, $(x_4 - x_3) = s_2$ and x_1 and x_2 are negative numbers. Using Equation (D.68), together with Equations (1.11) and (1.12), the complex pressure and particle velocity amplitudes (relative to the reference signal) in the acoustic medium adjacent to the two surfaces of the porous material sample may be expressed in terms of the positive and negative going wave amplitudes, A, B, C and D, as:

$$\begin{cases} \dfrac{\hat{p}_{(x=0)}}{\hat{p}_{\mathrm{ref1}}} = A + B; & \dfrac{\hat{u}_{(x=0)}}{\hat{p}_{\mathrm{ref1}}} = \dfrac{A - B}{\rho c} \\ \dfrac{\hat{p}_{(x=\ell)}}{\hat{p}_{\mathrm{ref1}}} = C\mathrm{e}^{-\mathrm{j}k\ell} + D\mathrm{e}^{\mathrm{j}k\ell}; & \dfrac{\hat{u}_{(x=\ell)}}{\hat{p}_{\mathrm{ref1}}} = \dfrac{C\mathrm{e}^{-\mathrm{j}k\ell} - D\mathrm{e}^{\mathrm{j}k\ell}}{\rho c} \end{cases} \tag{D.70}$$

where A, B, C and D are calculated from the transfer function measurements and Equation (D.69).

Equation (D.70) can be rewritten in terms of the normal incidence complex reflection coefficient, $R_p = B/A$, and the normal incidence complex amplitude transmission coefficient,

$\tau_{N,a} = C/A$, to give:

$$\begin{cases} \dfrac{\hat{p}_{(x=0)}}{\hat{p}_{\text{ref1}}} = A[1+R_p]; & \dfrac{\hat{u}_{(x=0)}}{\hat{p}_{\text{ref1}}} = \dfrac{A(1-R_p)}{\rho c} \\ \dfrac{\hat{p}_{(x=\ell)}}{\hat{p}_{\text{ref1}}} = A\tau_{N,a}\mathrm{e}^{-\mathrm{j}k\ell}; & \dfrac{\hat{u}_{(x=\ell)}}{\hat{p}_{\text{ref1}}} = \dfrac{A\tau_{N,a}\mathrm{e}^{-\mathrm{j}k\ell}}{\rho c} \end{cases} \tag{D.71}$$

where the equations containing the transmission coefficient are only valid if the amplitude, D, of the wave reflected from the end of the tube is small compared to the amplitude, C, of the incident wave, implying an almost anechoic termination.

A transfer matrix can be used to relate the acoustic pressure and particle velocity at location $x = 0$ to the acoustic pressure and particle velocity at location $x = \ell$. The appropriate transfer matrix contains elements T_{11}, T_{12}, T_{21} and T_{22}, so that:

$$\begin{bmatrix} \hat{p} \\ \hat{u} \end{bmatrix}_{x=0} = \begin{bmatrix} T_{11} & T_{12} \\ T_{21} & T_{22} \end{bmatrix} \begin{bmatrix} \hat{p} \\ \hat{u} \end{bmatrix}_{x=\ell} \tag{D.72}$$

where the variables on the RHS of Equation (D.72) are calculated using Equation (D.70) for termination conditions, a and b. Note that in Equation (D.72), p_{ref1} cancels out as it appears in all variables on the numerators and denominators.

If the two surfaces of the sample are not the same, it is necessary to measure all of the four transfer functions between the microphone locations and the reference for two different end conditions at the downstream end of the tube. The first condition (condition, a) can be an almost anechoic one (as used for the case where the two sample surfaces are the same), while the second condition (condition, b) can be an open or rigidly closed end. In this case, the elements of the transfer matrix are given by (ASTM E2611-19, 2019):

$$\begin{cases} T_{11} = \dfrac{\hat{p}_{a(x=0)}\hat{u}_{b(x=\ell)} - \hat{p}_{b(x=0)}\hat{u}_{a(x=\ell)}}{\hat{p}_{a(x=\ell)}\hat{u}_{b(x=\ell)} - \hat{p}_{b(x=\ell)}\hat{u}_{a(x=\ell)}} \\ T_{12} = \dfrac{\hat{p}_{b(x=0)}\hat{p}_{a(x=\ell)} - \hat{p}_{a(x=0)}\hat{p}_{b(x=\ell)}}{\hat{p}_{a(x=\ell)}\hat{u}_{b(x=\ell)} - \hat{p}_{b(x=\ell)}\hat{u}_{a(x=\ell)}} \\ T_{21} = \dfrac{\hat{u}_{a(x=0)}\hat{u}_{b(x=\ell)} - \hat{u}_{b(x=0)}\hat{u}_{a(x=\ell)}}{\hat{p}_{a(x=\ell)}\hat{u}_{b(x=\ell)} - \hat{p}_{b(x=\ell)}\hat{u}_{a(x=\ell)}} \\ T_{22} = \dfrac{\hat{p}_{a(x=\ell)}\hat{u}_{b(x=0)} - \hat{p}_{b(x=\ell)}\hat{u}_{a(x=0)}}{\hat{p}_{a(x=\ell)}\hat{u}_{b(x=\ell)} - \hat{p}_{b(x=\ell)}\hat{u}_{a(x=\ell)}} \end{cases} \tag{D.73}$$

If the two surfaces of the sample are the same, $T_{11} = T_{22}$ and reciprocity requires that the determinant of the transfer matrix is unity so that $T_{11}T_{22} - T_{12}T_{21} = 1$. In this case, measurements only need be made for a single load (preferably condition a, which is as close to anechoic as possible). The corresponding transfer matrix becomes:

$$\begin{cases} T_{11} = T_{22} = \dfrac{\hat{p}_{(x=\ell)}\hat{u}_{(x=\ell)} + \hat{p}_{(x=0)}\hat{u}_{(x=0)}}{\hat{p}_{(x=0)}\hat{u}_{(x=\ell)} + \hat{p}_{(x=\ell)}\hat{u}_{(x=0)}} \\ T_{12} = \dfrac{\hat{p}^2_{(x=0)} - \hat{p}^2_{(x=\ell)}}{\hat{p}_{(x=0)}\hat{u}_{(x=\ell)} + \hat{p}_{(x=\ell)}\hat{u}_{(x=0)}} \\ T_{21} = \dfrac{\hat{u}^2_{(x=0)} - \hat{u}^2_{(x=\ell)}}{\hat{p}_{(x=0)}\hat{u}_{(x=\ell)} + \hat{p}_{(x=\ell)}\hat{u}_{(x=0)}} \end{cases} \tag{D.74}$$

D.4.3.1 Amplitude Transmission Coefficient, Anechoic Termination

The normal incidence amplitude transmission coefficient, $\tau_{N,a}$, is related to the energy transmission coefficient, τ_N, of Chapter 7, by $\tau_{N,a} = \sqrt{\tau_N}$. For an anechoic termination, the amplitude transmission coefficient, $\tau_{N,a}$, can be expressed as (Song and Bolton, 2000):

$$\tau_{N,a} = \frac{2\mathrm{e}^{\mathrm{j}k\ell}}{T_{11} + T_{12}/(\rho c) + \rho c T_{21} + T_{22}} \tag{D.75}$$

This transmission coefficient can also be used to determine the transmission loss of a muffler, in which case the sample of porous material is replaced with a muffler. However, the upper limiting frequency for this analysis is the requirement for plane wave propagation in the inlet and discharge ducts servicing the muffler, and this is related to the diameter of these ducts, as indicated by Equation (D.24). The normal incidence transmission loss, TL_N, is given by:

$$\mathrm{TL}_N = -20 \log_{10}(\tau_{N,a}) \tag{D.76}$$

D.4.3.2 Absorption Coefficient, Anechoic Termination

For an almost anechoic termination (achieved using a wedge of porous material inserted into the end of the tube that is opposite to the end with the loudspeaker), the normal incidence absorption coefficient, α_N, can be written terms of the normal incidence, plane wave, complex pressure reflection coefficient, $R_{p,a}$, as:

$$\alpha_N = 1 - |R_{p,a}|^2 \tag{D.77}$$

where $R_{p,a}$ is (Song and Bolton, 2000):

$$R_{p,a} = \frac{T_{11} + T_{12}/(\rho c) - \rho c T_{21} - T_{22}}{T_{11} + T_{12}/(\rho c) + \rho c T_{21} + T_{22}} \tag{D.78}$$

D.4.3.3 Absorption Coefficient, Rigid Termination

If face ($x = \ell$) of the porous material is mounted against a rigid backing, then $\hat{u}_{(x=\ell)} = 0$, so that terms involving T_{22} and T_{12} are zero, in which case, the reflection coefficient to be used in place of $R_{p,a}$ in Equation (D.77) is (Song and Bolton, 2000):

$$R_{p,h} = \frac{T_{11} - \rho c T_{21}}{T_{11} + \rho c T_{21}} \tag{D.79}$$

D.4.3.4 Complex Wavenumber, Impedance and Density of the Test Sample

The transfer matrix for sound normally incident on a finite thickness sample of isotropic, homogeneous, porous, acoustic material is (Allard and Atalla, 2009):

$$\begin{bmatrix} T_{11} & T_{12} \\ T_{21} & T_{22} \end{bmatrix} = \begin{bmatrix} \cos(k_m \ell) & \dfrac{\mathrm{j}\rho_m \omega}{k_m} \sin(k_m \ell) \\ \dfrac{\mathrm{j}k_m}{\omega \rho_m} \sin(k_m \ell) & \cos(k_m \ell) \end{bmatrix} \tag{D.80}$$

The four transfer matrix elements in Equation (D.80) may be directly linked to various properties of such a test sample. Note that as both surfaces of the material are the same, the only end condition needing to be tested is the nearly anechoic one. From Equation (D.80), the wavenumber, k_m, of the test material is (Song and Bolton, 2000):

$$k_m = \frac{1}{\ell} \cos^{-1}(T_{11}) = \frac{1}{\ell} \sin^{-1}\left(\sqrt{-(T_{12}T_{21})}\right) \tag{D.81}$$

the characteristic impedance is (Song and Bolton, 2000):

$$Z_m = \rho_m c_m = \sqrt{\frac{T_{12}}{T_{21}}} \tag{D.82}$$

The complex normal specific acoustic impedance, Z_N, is then calculated using Equation (D.64), with R_p replaced with $R_{p,a}$ or $R_{p,h}$ of Equations (D.78) or (D.79), respectively. The phase speed of sound within the material is:

$$c_m = \omega / k_m \tag{D.83}$$

Thus, the complex density of the material is:

$$\rho_m = Z_m k_m / \omega \tag{D.84}$$

D.4.3.5 Correction of the Measured Transfer Functions Due to Microphone Differences

When more than one test microphone is used, corrections to the measured transfer functions must be made to account for differences in the phase and amplitude responses of the microphones (ASTM E2611-19, 2019). The transfer function, H, used in the calculations is then calculated from the measured transfer function, H^M, using a correction transfer function, H^c, the calculation of which is discussed in the following paragraphs. Thus, for microphone, n, in Figure D.12, where $n = 1$, 2, 3 or 4:

$$H_{n,\mathrm{ref}} = \frac{H^M_{n,\mathrm{ref}}}{H^c_{n,\mathrm{ref}}} \tag{D.85}$$

where one of the four microphones shown in Figure D.12 may be used as the reference one, and $H_{n,\mathrm{ref}}$, $H^c_{n,\mathrm{ref}}$ and $\hat{H}_{n,\mathrm{ref}}$ are all complex numbers, so that, for example:

$$H_{n,\mathrm{ref}} = |H_{n,\mathrm{ref}}| \mathrm{e}^{\mathrm{j}\phi_{n,\mathrm{ref}}} = H_{\mathrm{Re}} + \mathrm{j} H_{\mathrm{Im}} \tag{D.86}$$

The value of the correction transfer function, $H^c_{n,\mathrm{ref}}$, is different for each microphone, but the procedure is the same for all microphones. Of course, the correction transfer function for the microphone chosen as the reference is unity with a phase shift of zero, so it is only necessary to use the following procedure to determine the correction transfer function for the three microphones that are not the reference microphone.

1. The transfer function is first measured between the reference microphone and microphone, n, where n is the microphone number for which the correction transfer function is to be determined. This transfer function is denoted $H^{\mathrm{I}}_{n,\mathrm{ref}}$.
2. Next, the microphone positions are interchanged; that is, the reference microphone is moved to the physical position occupied by microphone, n, in the previous step and microphone, n, is moved to the physical position previously occupied by the reference microphone. However, the connections to the measuring system are not disturbed or changed.
3. The transfer function between the reference microphone and microphone, n, in their new positions is then measured and denoted, $H^{\mathrm{II}}_{n,\mathrm{ref}}$.
4. The correction transfer function, $H^c_{n,\mathrm{ref}}$, for microphone, n, at frequency, ω, is then:

$$H^c_{n,\mathrm{ref}}(\omega) = |H^c_{n,\mathrm{ref}}| \mathrm{e}^{\mathrm{j}\phi^c_{n,\mathrm{ref}}} = \left[H^{\mathrm{I}}_{n,\mathrm{ref}}(\omega) \times H^{\mathrm{II}}_{n,\mathrm{ref}}(\omega) \right]^{1/2} \tag{D.87}$$

The procedure for the case of the speaker driving signal being the reference when only one physical microphone exists (not recommended) is outlined in ASTM E2611-19 (2019).

Measurement errors associated with the four-microphone method are discussed in ASTM E2611-19 (2019).

D.4.4 In-Situ Measurement

In some cases, it is not convenient nor possible to measure absorption coefficients in the laboratory using the aforementioned methods, due to the difficulty of obtaining representative samples of material (such as road surfaces) or the environment surrounding the material to be tested or deformation of the material when installed affecting the absorption coefficient. In such cases, it is desirable to measure the absorption coefficient of the material in-situ. Various methods for undertaking these measurements are explained in Tijs (2013). Since the most common requirement has been the determination of the absorption coefficient of different types of road surfaces (as this affects the level of traffic noise radiated to the surrounding community), two international standards have been produced that describe test methods for this particular application: ISO 13472-1 (2022); ISO 13472-2 (2010). However, the standards also state that the test methods can be used for any surface.

Method one. The test method outlined in ISO 13472-2 (2010), referred to here as "Method one", uses a tube such as an impedance tube (sometimes referred to as a Kundt's tube), open at one end and mounted vertically, such that the open end is in contact with the surface to be measured (Seybert and Han, 2008). A loudspeaker is placed at the end of the tube opposite to the open end. Sound pressure measurements are made inside the tube in the same way as measurements described in Section D.4.2. To minimise sound leakage, the open end of the tube is fitted with a flange, with sealant used between the flange and the sample, as shown in Figure D.13(a). Alternatively, an outer and an inner tube are used with the loudspeaker producing sound in both inner and outer tubes, as shown in Figure D.13(b). The idea of the double tube arrangement is that sound that escapes from the bottom of the inner tube is compensated for by sound leaking into the inner tube from the outer tube. With the tube thus mounted, measurements of surface impedance and sound absorption can be made using the procedures described in Section D.4.2.

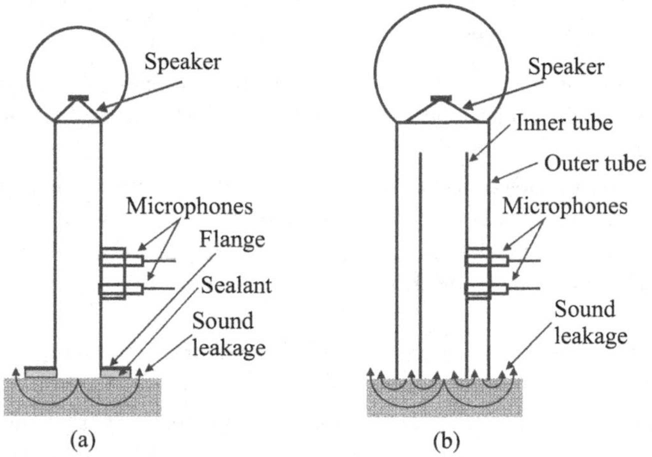

FIGURE D.13 Arrangement for in-situ measurement of absorption coefficient using an impedance tube: (a) single tube; (b) double tube.

Method two. The test method outlined in ISO 13472-1 (2022), referred to here as "Method two", uses a loudspeaker and a single microphone mounted above the surface to be tested, as illustrated in Figure D.14(a).

Although this method is intended mainly for the measurement of the normal incidence absorption coefficient, it can also be used to measure the absorption coefficient for other angles of incidence. The frequency range for reliable measurements is between and including 1/3-octave

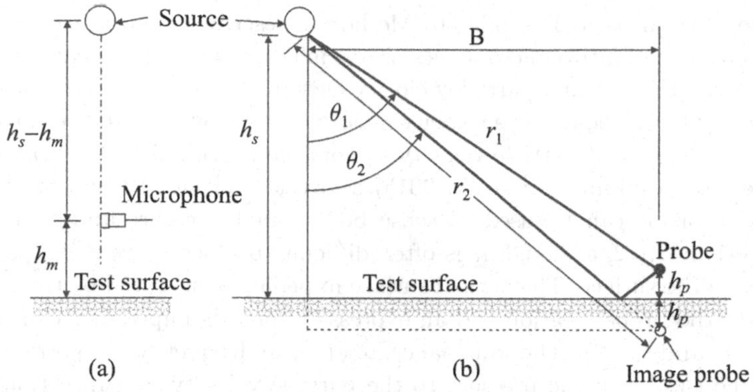

FIGURE D.14 Arrangement for in-situ measurement of absorption coefficient using a single microphone or a sound intensity probe: (a) microphone (method 2) (b) sound intensity probe (method 3)

bands from 315 Hz to 2 kHz, although the method can be extended to cover the low-frequency range down to frequencies included in the 250 Hz 1/3-octave band and the high-frequency range up to frequencies included in the 4 kHz 1/3 octave band, for comparison of different materials, although for frequencies below the 315 Hz 1/3-octave band, results are not necessarily comparable with measurements made using the methods described in Sections D.4.1 and D.4.2. In addition, the accuracy of measurements made using this method is poor when measuring low absorption coefficient values, whereas the accuracy of Method 1 is not considered reliable when measuring absorption coefficient values exceeding 0.15. With Method 2, the incident and reflected signals are separated in time by appropriate separation of the loudspeaker and microphone. The microphone should be as close as possible to the sample surface, but far enough away that the incident and reflected rays can be separated in the impulse response function. ISO 13472-1 (2022) suggests that 250 mm is acceptable. The loudspeaker is usually a point source mounted in a sphere, and it should be sufficiently far above the surface being measured that near field effects are negligible. For measurements above 250 Hz, a distance of 1.25 m is recommended (Londhe et al., 2009; ISO 13472-1, 2022).

The sound absorption coefficient is measured by exciting the loudspeaker with an MLS signal (see Section 12.3.18 and Figure 12.10). The impulse response of the surface under test is obtained by performing a cross correlation between the loudspeaker input, x and the microphone output, y. However, this impulse response is the combined response of the incident signal and reflected signal and needs to be separated by subtracting the impulse response of the incident signal. The incident signal impulse response (or free field response) can be determined when outdoors by pointing the microphone and loudspeaker into the sky. The relative orientation of the loudspeaker and microphone must be the same in both cases, and this is usually achieved by connecting them together with a rigid construction. Once the material surface impulse response has been obtained by subtracting the free field impulse response from the total impulse response, the free field frequency response, H_i, is obtained by taking the Fourier transform of the free field impulse response and the material surface frequency response, H_r, is obtained by taking the Fourier transform of the material surface impulse response. The absorption coefficient, $\alpha(f)$, corresponding to frequency, f, is then (Londhe et al., 2009):

$$\alpha(f) = 1 - |R_p(f)|^2 = 1 - \left[\frac{h_s + h_m}{h_s - h_m}\right] \left|\frac{H_r(f)}{H_i(f)}\right|^2 \tag{D.88}$$

where h_s and h_m are defined in Figure D.14(a).

Method three. This method is similar to Method 2 described above, except that the single microphone is replaced with two microphones or one microphone and one particle velocity sensor to measure acoustic pressure and particle velocity close to the surface to be measured, as illustrated in Figure D.14(b). The particle velocity must be sensed as close to the surface as possible, preferably using a specialised particle velocity sensor (see Section 3.10.1), located preferably at a distance of less than 50 mm (Li et al., 2015) above the surface to be tested. Alternatively, a two-microphone probe can be used to sense both sound pressure and particle velocity, as described in Section 3.10.2, although it is often difficult to place the sensing part of the probe sufficiently close to the surface. The normal surface impedance measured by the probe is actually the impedance at the probe location and an expression for this impedance can be derived from Equations (1.48b) and (1.49). The impedance, $Z_N(f)$, at frequency, f, normal to the surface is the ratio of the total acoustic pressure to the particle velocity in a direction normal to the surface. Using Figure D.14(b), and the previously mentioned equations, we obtain the following expression for the measured normal impedance, $Z_N(f)$, in terms of the signal measured by the microphone and the signal measured by the particle velocity sensor, and also in terms of the spherical wave reflection coefficient, $Q(f)$:

$$Z_N(f) = \frac{p_d(f) + p_r(f)}{u_{n,d}(f) + u_{n,r}(f)} = \rho c \frac{r_2 \mathrm{e}^{-\mathrm{j}kr_1} + Q(f) r_1 \mathrm{e}^{-\mathrm{j}kr_2}}{r_2 \mathrm{e}^{-\mathrm{j}kr_1}\left[\dfrac{(1+\mathrm{j}kr_1)}{\mathrm{j}kr_1}\right] - Q(f) r_1 \mathrm{e}^{-\mathrm{j}kr_2}\left[\dfrac{(1+\mathrm{j}kr_2)}{\mathrm{j}kr_2}\right]} \tag{D.89}$$

where $r_2 = \sqrt{r_1^2 + 4h(r_1 \cos\theta_1 + h)}$ and $\cos\theta_2 = (r_1 \cos\theta_1 + 2h)/r_2$. Rearranging gives an expression for the spherical wave reflection coefficient in terms of the measured normal impedance, Z_N. Thus:

$$Q(f) = \frac{Z_N(f)(1+\mathrm{j}kr_1)/(\mathrm{j}kr_1) - 1}{Z_N(f)(1+\mathrm{j}kr_2)/(\mathrm{j}kr_2) + 1} \tag{D.90}$$

For the purposes of characterising the surface being measured in terms of its absorption coefficient, we need the plane wave reflection coefficient, R_p, which may be considered to be approximately equal to the spherical wave coefficient, Q, if the source is sufficiently far from the test surface and the probe is very close to the test surface. The absorption coefficient at frequency, f, is then given by $\alpha(f) = 1 - |R_p(f)|^2 \approx 1 - |Q(f)|^2$.

D.4.5 Reverberation Room Measurement

The absorption coefficients of a porous material sample, measured in a reverberation room (see Section 6.7.1), are different to those measured using the methods discussed in previous subsections, as in the reverberation room, the sound is incident from many directions simultaneously, whereas for the previously discussed measurements, the sound is incident from a single direction and for the cases where the direction is normal to the surface of the sample, the statistical absorption coefficient is calculated from the normal incidence absorption coefficient. Even the statistical coefficient, so calculated, is different to the absorption coefficient measured in a reverberation room (called the Sabine absorption coefficient) because in a reverberation room, the sound is not incident equally from all angles from grazing to normal and also because the sample being tested diffracts some of the sound into itself, resulting in an effectively larger sample area than is used in the Sabine absorption coefficient calculation. This results in measured Sabine absorption coefficients sometimes exceeding one, whereas the theoretical maximum statistical absorption coefficient is slightly greater than 0.95. In general, Sabine absorption coefficients are always greater than statistical absorption coefficients calculated from measured normal incidence values.

D.5 Calculation of Absorption Coefficients of Porous Materials

D.5.1 Porous Materials with a Backing Cavity

For porous acoustic materials, such as rockwool or fibreglass, the specific normal impedance of Equation (D.44) may also be calculated from the material characteristic impedance and propagation coefficient of Equations (D.9) and (D.10). For a material of infinite depth (or sufficiently thick that waves transmitted through the material from one face and reflected from the opposite face are of insignificant amplitude by the time they arrive back at the first face), the normal specific acoustic impedance is equal to the characteristic impedance of Equation (D.9). For a porous blanket of thickness, ℓ, backed by a cavity of any depth, L (including $L = 0$), with a rigid back, the normal specific acoustic impedance (in the absence of flow past the cavity) may be calculated using an electrical transmission line analogy (Magnusson, 1965) and is given by:

$$Z_N = Z_m \frac{Z_L + \mathrm{j}Z_m \tan(k_m \ell)}{Z_m + \mathrm{j}Z_L \tan(k_m \ell)} \tag{D.91}$$

The quantities, Z_m and k_m, in Equation (D.91) are defined in Equations (D.9) and (D.10). The normal specific acoustic impedance, Z_L, of a rigidly terminated, partitioned backing cavity with no sound absorbing material in it is given by:

$$Z_L = -\mathrm{j}\rho c / \tan(2\pi f L/c) \tag{D.92}$$

and for a rigidly terminated, non-partitioned backing cavity, the normal specific acoustic impedance, Z_L, for a wave incident at angle, θ, is:

$$Z_L = -\mathrm{j}\rho c \cos\theta / \tan(2\pi f L/c) \tag{D.93}$$

where θ is the angle of incidence of the sound wave measured from the normal to the plane of the cavity entrance opening.

A partitioned cavity is one that is divided into compartments by partitions that permit propagation normal to the surface, while inhibiting propagation parallel to the surface of the liner. The depth of each compartment is equal to the overall cavity depth.

If the porous material is rigidly backed so that $L = 0$ or, equivalently, L is an integer multiple of half wavelengths, Equation (D.91) reduces to:

$$Z_N = -\mathrm{j}Z_m / \tan(k_m \ell) \tag{D.94}$$

D.5.2 Multiple Layers of Porous Liner Backed by an Impedance

Equation (D.91) can be easily extended to cover the case of multiple layers of porous material by applying it to each layer successively, beginning with the layer closest to the termination (rigid wall or cavity backed by a rigid wall) with normal specific acoustic impedance Z_L. The normal specific acoustic impedance looking into the ith layer surface that is closest to the termination is:

$$Z_{N,i} = Z_{m,i} \frac{Z_{N,i-1} + \mathrm{j}Z_{m,i} \tan(k_{m,i}\ell_i)}{Z_{m,i} + \mathrm{j}Z_{N,i-1} \tan(k_{m,i}\ell_i)} \tag{D.95}$$

The variables in the above equation have the same definitions as those in Equation (D.91), with the added subscript, i, which refers to the ith layer or the added subscript, $i-1$, which refers to the $(i-1)$th layer.

Equation (D.95) could also be used for materials whose density was smoothly varying, by dividing the material into a number of very thin layers, with each layer having uniform properties.

D.5.3 Porous Liner Covered with a Limp Impervious Layer

If the porous material is protected by covering or enclosing it in an impervious blanket of thickness, h, and mass per unit area, σ', the effective normal specific acoustic impedance, Z_{NB}, at the outer surface of the blanket, which can be used together with Equations (D.44) and (D.50) to find the statistical absorption coefficient of the construction, is:

$$Z_{NB} = Z_N + \mathrm{j}2\pi f\sigma' \tag{D.96}$$

where f is the frequency of the incident tonal sound, or alternatively, the centre frequency of a narrow band of noise. Typical values for σ' and c_L are included in Table D.3 for commonly used covering materials.

TABLE D.3 Properties of commonly used limp impervious wrappings for environmental protection of porous materials

Material	Density (kg/m^3)	Typical thickness (microns = 10^{-6} m)	σ' (kg/m^2)[a]	c_L (2D sheet) (m/s) (approx.)
Polyethylene (LD)	925	6–35	0.0055–0.033	820
Polyurethane	900	6–35	0.005–0.033	1420
Aluminium	2700	2–12	0.0055–0.033	5440
PVC	1400	4–28	0.005–0.033	1540
Melinex (polyester)	1390	15–30	0.021–0.042	1600
Metalised polyester	1400	12	0.017	1600

[a] σ' and c_L are, respectively, the surface density and speed of sound in the wrapping material.

Guidelines for the selection of suitable protective coverings are given by Andersson (1981).

D.5.4 Porous Liner Covered with a Perforated Sheet

If the porous liner were covered with a perforated sheet, the effective specific normal impedance (locally reactive rather than extended reactive) at the outer surface of the perforated sheet is (Bolt, 1947):

$$Z_{NP} = Z_N + \frac{\dfrac{100}{P}\{\mathrm{j}\rho c \tan\left[k\ell_e(1-M)\right] + R_a S_h\}}{1 + \dfrac{100}{\mathrm{j}\omega m P}\{\mathrm{j}\rho c \tan\left[k\ell_e(1-M)\right] + R_a S_h\}} \tag{D.97}$$

where Z_N is the normal specific acoustic impedance of the porous acoustic material with or without a cavity backing (and in the absence of flow), ω is the radian frequency, P is the % open area of the holes, R_a is the acoustic resistance of each hole, S_h is the area of each hole, M is the Mach number of the flow past the holes and m is the mass per unit area of the perforated sheet, all in consistent SI units. The effective length, ℓ_e, of each of the holes in the perforated sheet is:

$$\ell_e = w + \left[\frac{16a}{3\pi}(1 - 0.43a/q)\right]\left[1 - M^2\right] \tag{D.98}$$

where w is the thickness of the perforated sheet, a is the hole radius and $q > 2a$ is the distance between hole centres.

D.5.5 Porous Liner Covered with a Limp Impervious Layer and a Perforated Sheet

In this case, the impedance of the perforated sheet and impervious layer are both added to the normal specific acoustic impedance of the porous acoustic material, so that:

$$Z_{NBP} = Z_N + \frac{\frac{100}{P}\{\mathrm{j}\rho c \tan\left[k\ell_e(1-M)\right] + R_a S_h\}}{1 + \frac{100}{\mathrm{j}\omega m P}\{\mathrm{j}\rho c \tan\left[k\ell_e(1-M)\right] + R_a S_h\}} + \mathrm{j}2\pi f \sigma' \tag{D.99}$$

It is important that the impervious layer and the perforated sheet are separated, using something like a mesh spacer (with a grid size of at least 2 cm); otherwise, the performance of the construction as an absorber will be severely degraded, as the impervious layer will no longer be acting as a limp blanket.

E

Partial Coherence Combination of Sound Pressures

Sound rays arriving at a receiver from a sound source can be combined together to give a total sound pressure level at the receiver using incoherent combination (see Equation (1.98)), coherent combination (see Equation (1.96)) or partially coherent combination. Here we will discuss partially coherent combination, which is applicable to single-frequency sound. However, it can be applied to 1/3-octave or octave bands of noise by dividing the band into a number of single frequencies (at least 20, but more is better). Partial coherence analysis is applied to each single frequency and then the sound pressure level difference between the primary ray and ground reflected ray is logarithmically averaged over the frequencies that were considered for the particular frequency band. Thus, the sound pressure level is calculated at the receiver for both the primary ray by itself (usually the direct ray with no ground reflections) and the total sound pressure level due to partially coherent combination of all rays (including the direct ray) arriving at the receiver. The excess attenuation (which may be negative) is the sound pressure level due to the direct ray by itself minus the total sound pressure level due to all rays combined using partial coherence.

Partial coherence analysis takes into account incoherence effects caused by the following:

- Atmospheric turbulence;
- Frequency band averaging;
- Uncertainty in distance between source and receiver; and
- Uncertainty in source and receiver heights.

The coherence coefficients are taken into account when the complex sound pressures (consisting of a real and imaginary part) due to rays arriving at the receiver from the same source along different paths are combined together. Generally, in sound propagation modelling, it is sufficiently accurate (although not technically rigorous) to consider only the reduction in coherence between the primary sound ray (one with the largest sound pressure level at the receiver) and all other rays but not between two or more non-primary rays. The ray usually considered as the primary ray is the direct ray from the sound source to the receiver location, if it exists. If the line-of-sight is blocked by a screen or barrier, then the direct ray over the top of the barrier with no ground reflections is usually considered to be the primary ray.

The departure from coherence, due to the above-listed causes, between different rays arriving at the receiver from the same source, may be quantified using an overall coefficient of coherence, F_i, for each path, $i, i \geq 2$, between the primary ray ($i = 1$) and the ith ray ($i \geq 2$) (Plovsing,

DOI: 10.1201/9780367814908-E

2006a, p. 74):

$$|p_t|^2 = |p_1|^2 \left[\left| 1 + \sum_{i=2}^{N} F_i \frac{p_i}{p_1} \right|^2 + \sum_{i=2}^{N} (1 - F_i^2) \left| \frac{p_i}{p_1} \right|^2 \right] \tag{E.1}$$

where p_t is the total sound pressure at the receiver due to all rays and p_i, $i = 1, N$, is the complex sound pressure due to ray i, which is:

$$p_i = A_i \, \mathrm{e}^{\mathrm{j}(\omega t + \beta_i)} \tag{E.2}$$

The ratio, p_i/p_1, is given by:

$$\frac{p_i}{p_1} = \frac{A_i}{A_1} \mathrm{e}^{\mathrm{j}(\beta_i - \beta_1)} \tag{E.3}$$

where $A_i/A_1 = r_i/r_1$, $\beta_i - \beta_1 = 360f(\tau_i - \tau_1)$ degrees, r_i is the path length of ray i and τ_i is the propagation time for ray, i (see Sections 5.3.4.6 and 5.3.4.7).

If one or more ground reflections are associated with ray, i, the RHS of Equation (E.3) must be multiplied by the complex reflection coefficient, $R_{i,k} = B_{i,k}\mathrm{e}^{\mathrm{j}\beta_{i,k}}$, corresponding to each ground reflection, where $\beta_{i,k}$ is the phase shift caused by the kth ground reflection of ray, i.

A ratio, p_i/p_1, of pressures is used, as that allows the excess attenuation to be calculated without requiring a knowledge of the absolute value of the sound pressure (which depends on the source sound power).

The excess attenuation due to ground and meteorological effects can be calculated using Equation (E.1) and written in terms of decibels as:

$$\text{Excess attenuation, } L_{p1} - L_{pt} = -10\log_{10} \left\{ \left[\left| 1 + \sum_{i=2}^{N} F_i \frac{p_i}{p_1} \right|^2 + \sum_{i=2}^{N} (1 - F_i^2) \left| \frac{p_i}{p_1} \right|^2 \right] \right\} \tag{E.4}$$

The overall coefficient of coherence, F_i, for the ith ray is made up of the product of the components, F_a, due to the combined effects of frequency averaging in the 1/3-octave band, uncertainty in the distance between source and receiver and uncertainty in the source and receiver heights, and F_t due to turbulence, where each component refers only to the ith path. That is:

$$F_i = F_a F_t \tag{E.5}$$

The component, F_a, of the coefficient of coherence is given by (Salomons and Janssen, 2011):

$$F_a = \mathrm{e}^{-0.5\sigma_\phi^2} \tag{E.6}$$

where σ_ϕ is the standard deviation of the fluctuation of the phase difference, ϕ, between the two rays:

$$\sigma_\phi^2 = \phi^2 \left[\left(\frac{\sigma_f}{f}\right)^2 + \left(\frac{\sigma_d}{d}\right)^2 + \left(\frac{\sigma_{hS}}{h_S}\right)^2 + \left(\frac{\sigma_{hR}}{h_R}\right)^2 \right] \tag{E.7}$$

where

$$\phi = k\Delta d \approx \frac{2\pi f}{\bar{c}} \frac{2h_S h_R}{d} \tag{E.8}$$

and where Δd is the difference in path length between the primary ray and ray, i, k is the wavenumber, $\bar{c}$ is the average speed of sound between the source and receiver, d is the horizontal separation distance between source and receiver, σ_d is the standard deviation of the uncertainty in that distance, h_S is the source height above the local ground with σ_{hS} the standard deviation of its uncertainty and h_R is the receiver height above the local ground with σ_{hR} the standard deviation of its uncertainty. Including uncertainties in the source–receiver distance and source–receiver heights results in a smoother variation of the ground effect with frequency, which is a more realistic result (van Maercke and Defrance, 2007).

The first term on the right of Equation (E.7) accounts for the effect of frequency band integration and is given by:

$$\frac{\sigma_f}{f} = \frac{1}{3}\frac{\Delta f}{f} = \frac{1}{3}\left(2^{B/2} - 2^{-B/2}\right) \tag{E.9}$$

where $B = 1/3$ for 1/3-octave band averaging and $B = 1$ for octave band averaging.

In the absence of better information, the following standard deviation values are used.

$$\begin{aligned} \frac{\sigma_d}{d} &= 0 \\ \frac{\sigma_{h_S}}{h_S} &= \frac{h_S}{10} \\ \frac{\sigma_{h_R}}{h_R} &= \frac{h_R}{10} \end{aligned} \tag{E.10}$$

The coherence coefficient due to turbulence is given by (Salomons et al., 2011):

$$F_t = \mathrm{e}^X \tag{E.11}$$

where

$$X = -5.3888\gamma_T \left(\frac{f}{\bar{c}}\right)^2 q^{5/3} d \tag{E.12}$$

and

$$\gamma_T = \frac{C_T^2}{(\overline{T} + 273.15)^2} + \frac{22C_v^2}{3\bar{c}^2} \tag{E.13}$$

where $\overline{T}$ is the average temperature over the sound propagation path, $\bar{c}$ is the average speed of sound corresponding to the average temperature, d is the horizontal distance between source and receiver, C_T^2 is the turbulence strength due to temperature effects (sometimes called the temperature turbulence structure parameter), and C_v^2 is the turbulence strength due to wind effects (sometimes called the velocity turbulence structure parameter). The quantity, q, is half the mean separation of the direct and reflected ray paths and is defined as:

$$q = \frac{h_S h_R}{h_S + h_R} \tag{E.14}$$

The turbulence strength parameters can be measured using a SODAR system (an acoustic device for determining wind speed, directions and fluctuations at heights between 0 m and 200 m above the ground) or an ultrasonic anemometer, and will usually be found to be altitude dependent. For the purposes of the calculations described here, we only need to know an approximate estimate of the turbulence strength below about 200 m. Typical values of C_T^2 and C_v^2, measured approximately 3 m above the ground, were provided by Daigle (1982). He measured values of C_v^2 ranging from 2.0 in the afternoon to 0.5 in the evening and night, with 1.0 being the most common value (although he did measure a value as low as 0.1 in the early evening). He also measured values of C_T^2 ranging from 8.0 to 10.0 in the afternoon to 0.2 at night. As values of C_T^2 and C_v^2 are difficult to estimate, Salomons et al. (2011) give a value of γ_T for moderate turbulence as 5×10^{-6}, although Bullen (2012) suggests that a value of 10^{-5} gives better agreement with measurements and other models.

F

Files for Use with This Book

F.1 Table of Files for Use with This Book

The following table lists the MATLAB® scripts that are available for use with this book. The online address is www.causalsystems.com. Click on "Textbook - MATLAB® scripts. In addition, many example problems for almost all chapters are available for download, also from www.causalsystems.com. Click on "Textbook - solutions to problems".

Section, Page No.	Filename / Description
Chapter 2 Human Hearing and Noise Criteria	
Section 2.9.8, p.114	`plot_effective_NR_vs_time.m` This MATLAB® script is used to plot the effective Noise Reduction of a hearing protection device in decibels, as a function of the percentage of time that the device is worn during noise exposure.
Chapter 4 Sound Sources and Sound Power	
Section 4.3.1, p.192	`dipole_spl_vs_angle.m` This MATLAB® script is used to plot the directivity of a dipole source versus angle.
Section 4.4.1, p.198	`quadrupole_lateral_spl_vs_angle.m` This MATLAB® script is used to plot the directivity of a lateral quadrupole source versus angle.
Section 4.4.2, p.198	`quadrupole_longitudinal_spl_vs_angle.m` This MATLAB® script is used to plot the directivity of a longitudinal quadrupole source versus angle.
Section 4.6.1, p.205	`radiation_pattern_baffled_piston.m` This MATLAB® script is used to plot the directivity of an oscillating piston in an infinite plane baffle versus angle.
Section 4.7.1, p.212	`single_wall_spl_vs_angle.m` This MATLAB® script is used to plot the sound intensity relative to an equivalent monopole source of an incoherent plane radiator, based on the work by Hohenwarter (1991).

DOI: 10.1201/9780367814908-F

Section, Page No.	Filename / Description
Chapter 5 Outdoor Sound Propagation and Outdoor Barriers	
Section 5.3.4.4, p. 279	`ray_tracing_6thEdition.m` This MATLAB® script includes two methods used for atmospheric acoustic ray tracing in 2D. The code is designed with acoustic rays reflecting at the ground surface (z = 0). However, it can be easily adapted for underwater acoustics with some small tweaks. Method 2 (presented first) uses the Euler numerical discretisation method and Method 1 is based on Lecture 3 of the "Acoustic Remote Sensing and Sea Floor Mapping Course" from TU Delft. More information about the procedure, including all equations needed to implement Methods 1 and 2, is provided in Section 5.3.4.4, p.276 and Section 5.3.4.5, p.285.
Chapter 8 Muffling Devices	
Section 8.9.4, p.510	`beranek_ver_fig10_11_quarter_wave_tube_duct_4_pole.m` This MATLAB® script plots the transmission loss of a quarter-wavelength tube versus normalised frequency.
Section 8.9.7, p.514	`SEC_4_pole.m` This MATLAB® script plots the transmission loss of a simple expansion chamber silencer versus normalised frequency.
Section 8.9.9, p.516	`TL_CTR_perforated.m` This MATLAB® script plots the transmission loss of a concentric tube resonator silencer versus normalised frequency.
Section 8.9.10, p.520	`temp_gradient_spl_along_duct_4pole_sujith.m` This MATLAB® script plots the sound pressure and acoustic particle velocity in a duct with a piston at one end, a rigid termination at the other and where the gas within the duct has a linear temperature gradient along the length of the duct, using the 4-pole transmission matrix method.
Chapter 11 Practical Numerical Acoustics	
Section 11.2.2.4, p.691	`rect_cav_3D.m` This MATLAB® script is used to plot the sound pressure level inside a rigid-walled cavity due to a monopole source, and is used to compare with the predictions using the FastBEM boundary element analysis software.
Section 11.2.4.2, p.699	`rigid_cavity_full.inp` This Ansys Mechanical APDL script is used to calculate the sound pressure inside a rigid-walled cavity due to a monopole source, and used to compare with predictions from the MATLAB® script.

References

Abbott, P. G. and Nelson, P. M. (2002). Converting the UK traffic noise index, $L_{A10,18hr}$, to EU noise indices for noise mapping. Technical Report PE/SE/4502/02, Transport Research Limited, Crowthorne, UK.

Abramowitz, M. and Stegun, I. A. (1965). *Handbook of Mathematical Functions with Formulas, Graphs, and Mathematical Tables.* Dover, New York, NY, USA.

Advanced CAE Research (2016). FastBEM Acoustics. Software. `http://www.fastbem.com/`. Accessed 11 April, 2023.

AIJ-RLB (2004). Recommendations for Loads on Buildings. Architectural Institute of Japan.

Alfredson, R. J. and Scow, B. C. (1976). Performance of three sided enclosures. *Applied Acoustics*, 9:45–55.

Allard, J. and Atalla, N. (2009). *Propagation of Sound in Porous Media: Modelling Sound Absorbing Materials.* Wiley, New York, NY, USA, second edition.

Andersson, P. (1981). Guidelines for film faced absorbers. *Noise and Vibration Control Worldwide*, 12(January–February):16–17.

Anderton, D. and Halliwell, N. A. (1980). Noise from vibration. *Society of Automotive Engineers, Technical paper, 800407.*

ANSI/ASA S1.11 (2019). Electroacoustics: Octave-band and fractional-octave-band filters: Part 1: Specifications. American National Standards Institute.

ANSI/ASA S1.15-1 (2020). Measurement microphones – Part 1: Specifications for laboratory standard microphones. American National Standards Institute.

ANSI/ASA S1.15-2 (2020). Measurement microphones – Part 2: Primary method for pressure calibration of laboratory standard microphones by the reciprocity technique. American National Standards Institute.

ANSI/ASA S1.18 (2018). Method for determining the acoustic impedance of ground surfaces. American National Standards Institute / Acoustical Society of America.

ANSI/ASA S12.10-1 (2020). Acoustics - measurement of airborne noise emitted by information technology and telecommunications equipment – Part 1: Determination of sound power level and emission sound pressure level. American National Standards Institute / Acoustical Society of America.

ANSI/ASA S12.12 (2020). Engineering method for the determination of sound power levels of noise sources using sound intensity. American National Standards Institute / Acoustical Society of America.

ANSI/ASA S12.14 (2020). Methods for the field measurement of the sound output of audible public warning devices installed at fixed locations outdoors. American National Standards Institute / Acoustical Society of America.

ANSI/ASA S12.15 (2016). Portable electric power tools, stationary and fixed electric power tools, and gardening appliances – measurement of sound emitted. American National Standards Institute / Acoustical Society of America.

ANSI/ASA S12.17 (2011). Impulse sound propagation for environmental noise assessment. American National Standards Institute / Acoustical Society of America.

ANSI/ASA S12.18 (2019). Procedures for outdoor measurement of sound pressure level. American National Standards Institute / Acoustical Society of America.

ANSI/ASA S12.19 (2020). Measurement of occupational noise exposure. American National Standards Institute / Acoustical Society of America.

ANSI/ASA S12.2 (1995). Criteria for evaluating room noise. American National Standards Institute / Acoustical Society of America.

ANSI/ASA S12.2 (2019). Criteria for evaluating room noise. American National Standards Institute / Acoustical Society of America.

ANSI/ASA S1.25 (2020). Specification for personal noise dosimeters. American National Standards Institute / Acoustical Society of America.

ANSI/ASA S1.26 (2019). Methods for calculation of the absorption of sound by the atmosphere. American National Standards Institute / Acoustical Society of America.

ANSI/ASA S12.60-1 (2015). Acoustical performance criteria, design requirements, and guidelines for schools, Part 1: permanent schools. American National Standards Institute / Acoustical Society of America.

ANSI/ASA S12.65 (2020). Rating noise with respect to speech interference. American National Standards Institute / Acoustical Society of America.

ANSI/ASA S12.68 (2020). Methods of estimating effective A-weighted sound pressure levels when hearing protectors are worn. American National Standards Institute / Acoustical Society of America.

ANSI/ASA S12.70 (2016). Criteria for evaluating speech privacy in healthcare facilities. American National Standards Institute / Acoustical Society of America.

ANSI/ASA S12.8 (2020). Methods for determination of insertion loss of outdoor noise barriers. American National Standards Institute / Acoustical Society of America.

ANSI/ASA S12.9-1 (2018). Quantities and procedures for description and measurement of environmental sound – Part 1: Basic quantities and definitions. American National Standards Institute / Acoustical Society of America.

ANSI/ASA S12.9-2 (2018). Quantities and procedures for description and measurement of environmental sound – Part 2: Measurement of long-term, wide-area sound. American National Standards Institute / Acoustical Society of America.

ANSI/ASA S12.9-3 (2018). Quantities and procedures for description and measurement of environmental sound – Part 3: Short-term measurements with an observer present. American National Standards Institute / Acoustical Society of America.

ANSI/ASA S12.9-4 (2021). Quantities and procedures for description and measurement of environmental sound – Part 4: Noise assessment and prediction of long-term community response. American National Standards Institute / Acoustical Society of America.

ANSI/ASA S12.9-5 (2020). Quantities and procedures for description and measurement of environmental sound – Part 5: Sound level descriptors for determination of compatible land use. American National Standards Institute / Acoustical Society of America.

ANSI/ASA S12.9-7 (2020). Quantities and procedures for description and measurement of environmental sound – Part 7: Measurement of low-frequency noise and infrasound outdoors and in the presence of wind and indoors in occupied spaces. American National Standards Institute / Acoustical Society of America.

ANSI/ASA S1.42 (2020). Design response of weighting networks for acoustical measurements. American National Standards Institute / Acoustical Society of America.

ANSI/ASA S1.6 (2020). Preferred frequencies and filter band center frequencies for acoustical measurements. American National Standards Institute.

ANSI/ASA S2.20 (2020). Estimating airblast characteristics for single point explosions in air with a guide to evaluation of atmospheric propagation and effects. American National Standards Institute / Acoustical Society of America.

ANSI/ASA S2.61 (2020). Guide to the mechanical mounting of accelerometers. American National Standards Institute / Acoustical Society of America.

ANSI/ASA S3.1 (2018). Maximum permissible ambient noise levels for audiometric test rooms. American National Standards Institute / Acoustical Society of America.

ANSI/ASA S3.2 (2020). Method for measuring the intelligibility of speech over communication systems. American National Standards Institute / Acoustical Society of America.

ANSI/ASA S3.4 (1980). Procedure for the computation of loudness of noise. American National Standards Institute / Acoustical Society of America.

ANSI/ASA S3.4 (2020). Procedure for the computation of loudness of steady sounds. American National Standards Institute / Acoustical Society of America.

ANSI/ASA S3.44-1 (2020). Estimation of noise-induced hearing loss – Part 1: Method for calculating expected noise-induced permanent threshold shift (a modified nationally adopted international standard). American National Standards Institute / Acoustical Society of America.

ANSI/ASA S3.5 (2020). Calculation of the speech intelligibility index. American National Standards Institute / Acoustical Society of America.

ANSI/ASA S3.6 (2018). Specification for audiometers. American National Standards Institute / Acoustical Society of America.

Anton, H. and Rorres, C. (2013). *Elementary Linear Algebra.* Wiley, USA, eleventh edition.

Army, Air Force and Navy, USA (2003). Unified facilities criteria – noise and vibration control. Technical Report UFC 3-450-01, US Department of Defense, Washington, DC, USA.

AS 1055 (1978). Code of practice for noise assessment in residential areas. Standards Australia.

AS 1055 (2018). Acoustics – Description and measurement of environmental noise. Standards Australia.

AS 1055.1 (1997). Acoustics – description and measurement of environmental noise - general procedures. Standards Australia.

AS 1055.3 (1997). Acoustics – Description and measurement of environmental noise - Acquisition of data pertinent to land use. Standards Australia.

AS 2254 (1988). Acoustics – Recommended noise levels for various areas of occupancy in vessels and offshore mobile platforms. Standards Australia.

ASCE/SEI 7-05 (2013). Minimum design loads for buildings and other structures. ASCE, Reston, Virginia.

ASHRAE (1987). *ASHRAE Handbook: Systems and Applications, Chapter 52.* American Society of Heating and Refrigeration Engineers, Atlanta, GA.

ASHRAE (2001). *ASHRAE Handbook: Fundamentals, Chapter 34.* American Society of Heating and Refrigeration Engineers, Atlanta, GA.

ASHRAE (2005). *ASHRAE Handbook: Fundamentals, Chapter 35.* American Society of Heating and Refrigeration Engineers, Atlanta, GA.

ASHRAE (2007). *ASHRAE Handbook: Applications, Chapter 47.* American Society of Heating and Refrigeration Engineers, Atlanta, GA.

ASHRAE (2013). *ASHRAE Handbook: Fundamentals, Chapter 21.* American Society of Heating and Refrigeration Engineers, Atlanta, GA.

ASHRAE (2015). *ASHRAE Handbook: Applications, Chapter 48.* American Society of Heating and Refrigeration Engineers, Atlanta, GA.

ASHRAE (2016). Duct fitting database. `https://www.ashrae.org/resources--publications/bookstore/duct-fitting-database`. Last accessed on July 1, 2022.

AS/NZS 1170-2 (2021). Structural design actions, Part 2: Wind actions. Standards Australia.

AS/NZS 1269-3 (2016). Occupational noise management Part 3: Hearing protector program. Standards Australia.

AS/NZS 1270 (2014). Acoustics - Hearing protectors. Standards Australia.

AS/NZS 2107 (2016). Acoustics – Recommended design sound levels and reverberation times for building interiors. Standards Australia.

ASTM C384-04 (2016). Standard test method for impedance and absorption of acoustical materials by impedance tube method. ASTM International, West Conshohocken, Pennsylvania.

ASTM C423-17 (2017). Standard test method for sound absorption and sound absorption coefficients by the reverberation room method. ASTM International, West Conshohocken, Pennsylvania.

ASTM C522-03 (2016). Standard test method for airflow resistance of acoustical materials. ASTM International, West Conshohocken, Pennsylvania.

ASTM E1007-19 (2019). Standard test method for field measurement of tapping machine impact sound transmission through floor-ceiling assemblies and associated support structures. ASTM International, West Conshohocken, Pennsylvania.

ASTM E1050-19 (2019). Standard test method for impedance and absorption of acoustical materials using a tube, two microphones and a digital frequency analysis system. ASTM International, West Conshohocken, Pennsylvania.

ASTM E1110-06 (2019). Standard classification for determination of articulation class. ASTM International, West Conshohocken, Pennsylvania.

ASTM E1130-16 (2021). Standard test method for objective measurement of speech privacy in open plan spaces using articulation index. ASTM International, West Conshohocken, Pennsylvania.

ASTM E1222-90 (2016). Standard test method for laboratory measurement of the insertion loss of pipe lagging systems. ASTM International, West Conshohocken, Pennsylvania.

ASTM E1265-04 (2021). Standard test method for measuring insertion loss of pneumatic exhaust silencers. ASTM International, West Conshohocken, Pennsylvania.

ASTM E1332-22 (2022). Standard classification for rating outdoor-indoor sound attenuation. ASTM International, West Conshohocken, Pennsylvania.

ASTM E1503-22 (2022). Standard test method for conducting outdoor sound measurements using a digital statistical sound analysis system. ASTM International, West Conshohocken, Pennsylvania.

ASTM E1686-16 (2016). Standard guide for applying environmental noise measurement methods and criteria. ASTM International, West Conshohocken, Pennsylvania.

ASTM E1780-12 (2021). Standard guide for measuring outdoor sound received from a nearby fixed source. ASTM International, West Conshohocken, Pennsylvania.

ASTM E2179-21 (2021). Standard test method for laboratory measurement of the effectiveness of floor coverings in reducing impact sound transmission through concrete floors. ASTM International, West Conshohocken, Pennsylvania.

ASTM E2235-04 (2020). Standard test method for determination of decay rates for use in sound insulation test methods. ASTM International, West Conshohocken, Pennsylvania.

ASTM E2249-19 (2019). Standard test method for laboratory measurement of airborne transmission loss of building partitions and elements using sound intensity. ASTM International, West Conshohocken, Pennsylvania.

ASTM E2611-19 (2019). Standard test method for the measurement of normal incidence sound transmission of acoustical materials based on the transfer matrix method. ASTM International, West Conshohocken, Pennsylvania.

ASTM E2638-10 (2017). Standard test method for objective measurement of the speech privacy provided by a closed room. ASTM International, West Conshohocken, Pennsylvania.

ASTM E3207-21 (2021). Standard classification for determination of low-frequency impact noise ratings. ASTM International, West Conshohocken, Pennsylvania.

ASTM E3222-20a (2020). Standard classification for determination of high-frequency impact sound ratings. ASTM International, West Conshohocken, Pennsylvania.

ASTM E336-19 (2019). Standard test method for measurement of airborne sound attenuation between rooms in buildings. ASTM International, West Conshohocken, Pennsylvania.

ASTM E413-22 (2022). Classification for rating sound insulation. ASTM International, West Conshohocken, Pennsylvania.

ASTM E492-09 (2016). Standard test method for laboratory measurement of impact sound transmission through floor-ceiling assemblies using the tapping machine. ASTM International, West Conshohocken, Pennsylvania.

ASTM E90-09 (2016). Standard test method for laboratory measurement of airborne sound transmission loss of building partitions and elements. ASTM International, West Conshohocken, Pennsylvania.

ASTM E989-21 (2021). Standard classification for determination of impact insulation class (IIC). ASTM International, West Conshohocken, Pennsylvania.

ASTM F1334-18 (2021). Standard test method for determining a-weighted sound power level of vacuum cleaners. ASTM International, West Conshohocken, Pennsylvania.

ASTM F2544-11 (2021). Standard test method for determining A-weighted sound power level of central vacuum power units. ASTM International, West Conshohocken, Pennsylvania.

Astolfi, A., minelli, G., and Puglisi, G. E. (2022). A basic protocol for the acoustic characterization of small and medium-sized classrooms. *Journal of the Acoustical Society of America*, 152:1646–1659.

Attenborough, K., Hayek, S. I., and Lawther, J. M. (1980). Propagation of sound above a porous half-space. *Journal of the Acoustical Society of America*, 68(5):1493–1501.

Attenborough, K., Li, K. M., and Horoshenkov, K. (2007). *Predicting Outdoor Sound.* Taylor & Francis, New York, NY, USA.

Baas, P., Bosveld, F. C., Baltink, H. K., and Holtslag, A. A. M. (2009). A climatology of nocturnal low-level jets at Cabauw. *Journal of Applied Meteorology and Climatology*, 48(8):1627–1642.

Bacon, S. P. (2006). Auditory compression and hearing loss. *Acoustics Today*, 2:30–34.

Bai, M., Ih, J., and Benesty, J. (2013). *Acoustic Array Systems: Theory, Implementation and Application.* John Wiley and Sons, Singapore.

Basner, M. and McGuire, S. (2018). Who environmental noise guidelines for the european region: a systematic review on environmental noise and effects on sleep. *International Journal of Environmental Research and Public Health*, 15(3):519.

Bass, J. (2011). Investigation of the "Den Brook" amplitude modulation methodology for wind turbine noise. *Acoustics Bulletin*, 36(6):18–24.

Baumann, H. D. and Coney, W. B. (2006). Noise of gas flows. In Vér, I. L. and Beranek, L. L., editors, *Noise and Vibration Control Engineering: Principles and Applications*, chapter 15. John Wiley and Sons Inc., Hoboken, NJ, USA, second edition.

Baumann, H. D. and Hoffmann, H. (1999). Method for the estimation of frequency-dependent sound pressures at the pipe exterior of throttling valves. *Noise Control Engineering Journal*, 47:49–55.

Békésy, G. v. and Wever, E. G. (1960). *Experiments in Hearing.* American Institute of Physics by arrangement with McGraw-Hill Book Company, New York, NY, USA.

Bell, A. (2011). How do middle ear muscles protect the cochlea? Reconsideration of the intralabyrinthine pressure theory. *Journal of Hearing Science*, 1:9–23.

Bell, A. (2014). Annoyance from wind turbines: Role of the middle ear muscles. *Acoustics Australia*, 42(1):57.

Beranek, L. L. (1954). *Acoustics*. McGraw-Hill, New York, NY, USA.

Beranek, L. L. (1962). *Music, Acoustics and Architecture*. John Wiley, New York, NY, USA.

Beranek, L. L. (1988). The measurement of power levels and directivity patterns of noise sources. In Beranek, L. L., editor, *Noise and Vibration Control*, chapter 6. Institute of Noise Control Engineering, Washington, DC, USA, revised edition.

Beranek, L. L. and Hidaka, T. (1998). Sound absorption in concert halls by seats, occupied and unoccupied and by the hall's interior surfaces. *Journal of the Acoustical Society of America*, 104:3169–3177.

Beranek, L. L. and Mellow, T. (2012). *Acoustics: Sound Fields and Transducers*. Academic Press, Oxford, UK.

Berglund, B., Lindvall, T., and Schwela, D. H. (1999). *Guidelines for Community Noise*. World Health Organization, Geneva, Switzerland.

Bies, D. A. (1961). Effect of a reflecting plane on an arbitrarily oriented multipole. *Journal of the Acoustical Society of America*, 33:286–288.

Bies, D. A. (1976). Uses of anechoic and reverberant rooms. *Noise Control Engineering Journal*, 7:154–163.

Bies, D. A. (1981). A unified theory of sound absorption in fibrous porous materials. In *Proceedings of the Australian Acoustical Society Annual Conference*, Cowes, Phillip Island, Australia. Australian Acoustical Society.

Bies, D. A. (1988). Acoustical properties of porous materials. In Beranek, L. L., editor, *Noise and Vibration Control*, chapter 10. Institute of Noise Control Engineering, Washington, DC, USA, revised edition.

Bies, D. A. (1992). Circular saw aerodynamic noise. *Journal of Sound and Vibration*, 154:495–513.

Bies, D. A. (1994). An alternative model for combining noise and age-induced hearing loss. *Journal of the Acoustical Society of America*, 95:563–565.

Bies, D. A. (1995). Notes on Sabine rooms. *Acoustics Australia*, 23:97–103.

Bies, D. A. (1996). The half-octave temporary threshold shift. *Journal of the Acoustical Society of America*, 100:2786.

Bies, D. A. (1999). What the inner hair cells of the cochlea sense. *International Journal of Acoustics and Vibration*, 4:1–7.

Bies, D. A. (2000). The hydrodynamic forcing field of the cochlea. In *Proceedings of the 7th International Congress of Sound and Vibration*, pages 2259–2266, Garmisch-Partenkirchen. International Institute of Acoustics and Vibration.

Bies, D. A. and Bridges, G. E. (1993). Sound power determination in the geometric near field of a source by pressure measurements alone. In *Proceedings of the Australian Acoustical Society Annual Conference*, Glenelg, South Australia. Australian Acoustical Society.

Bies, D. A. and Davies, J. M. (1977). An investigation of the measurement of transmission loss. *Journal of Sound and Vibration*, 53:203–221.

Bies, D. A. and Hansen, C. H. (1979). Notes on porous materials for sound absorption. *Bulletin of the Australian Acoustical Society*, 7:17–22.

Bies, D. A. and Hansen, C. H. (1980). Flow resistance information for acoustical design. *Applied Acoustics*, 13:357–391.

Bies, D. A. and Hansen, C. H. (1990). An alternative mathematical description of the relationship between noise exposure and age-induced hearing loss. *Journal of the Acoustical Society of America*, 88:2743–2754.

Bies, D. A., Hansen, C. H., and Bridges, G. E. (1991). Sound attenuation in rectangular and circular cross-section ducts with flow and bulk-reacting liner. *Journal of Sound and Vibration*, 146:47–80.

Blauert, J. (1983). *Spatial Hearing.* MIT Press, Boston, MA, USA.

Bodén, H. (1991). The multiple load method for measuring the source characteristics of time-variant sources. *Journal of Sound and Vibration*, 148(3):437–453.

Bodén, H. (1995). On multi-load methods for determination of the source data of acoustic one-port sources. *Journal of Sound and Vibration*, 180(5):725–743.

Bodén, H. (2007). Recent advances in IC-engine acoustic source characterisation. In Randall, B. and Hahn, E., editors, *ICSV14: Proceedings of 14th International Congress on Sound and Vibration*, Cairns, Australia. International Institute of Acoustics and Vibration, Australian Acoustical Society. Paper 201.

Boeker, E., Dinges, E., He, B., Fleming, G., Roof, C., Gerbi, P., Rapoza, A., and Hemann, J. (2008). Integrated noise model (INM) version 7.0. Technical Report FAA-AEE-08-01, U.S. Department of Transportation, FAA.

Bohne, B., Harding, G. W., and Lee, S. C. (2007). Death pathways in noise-damaged outer hair cells. *Hearing Research*, 223:61–70.

Bolt, R. H. (1947). On the design of perforated facings for acoustic materials. *Journal of the Acoustical Society of America*, 19:917–921.

Bolt, R. H., Labate, S., and Ingard, K. U. (1949). Acoustic reactance of small circular orifices. *Journal of the Acoustical Society of America*, 21:94.

Bork, I. (2000). A comparison of room simulation software: The 2nd Round Robin on room acoustical computer simulation. *Acta Acustica United with Acustica*, 86:943–956.

Bradley, J. S. and Birta, J. A. (2000). Laboratory measurements of the sound insulation of building façade elements. Technical report, Institute for Research in Construction, Canadian National Research Council, Canada.

Bradley, J. S. and Birta, J. A. (2001). On the sound insulation of wood stud exterior walls. *Journal of the Acoustical Society of America*, 110:3086–3096.

Bragg, S. L. (1963). Combustion noise. *Journal of the Institute of Fuel*, 36(January):12–16.

Brandt, A. (2010). *Noise and Vibration Analysis: Signal Analysis and Experimental Procedures.* John Wiley & Sons, New York, NY, USA.

Bray, W. (2012). Relevance and applicability of the soundscape concept to physiological or behavioral effects caused by noise at very low frequencies which may not be audible. *Journal of the Acoustical Society of America*, 132(3):1925.

Bray, W. and James, R. (2011). Dynamic measurements of wind turbine acoustic signals, employing sound quality engineering methods considering the time and frequency sensitivities of human perception. In *Proceedings of Noise-Con 2011*, pages 25–7, Portland, OR, USA.

Brekke, A. (1981). Calculation methods for the transmission loss of single, double and triple partitions. *Applied Acoustics*, 14:225–240.

Brill, L. C., Smith, K., and Wang, L. M. (2018). Building a sound future for students: considering the acoustics in occupied active classrooms. *Acoustics Today*, 14(3):14–22.

Brod, K. (1984). On the uniqueness of solution for all wavenumbers in acoustic radiation. *Journal of the Acoustical Society of America*, 76:1238–1243.

Broner, N. and Leventhall, H. G. (1983). A criterion for predicting the annoyance due to lower level low frequency noise. *Journal of Low Frequency Noise and Vibration*, 2:160–168.

Brooks, T. F. and Humphreys, W. M. (2006). A deconvolution approach for the mapping of acoustic sources (DAMAS) determined from phased microphone arrays. *Journal of Sound and Vibration*, 204:856–879.

Brownell, W. E., Bader, C. R., Bertrand, D., and De Ribaupierre, Y. (1985). Evoked mechanical responses of isolated cochlear hair cells. *Science*, 227:194–196.

Bruce, R. D. (1988). Field measurements: Equipment and techniques. In Beranek, L. L., editor, *Noise and Vibration Control*, chapter 4. Institute of Noise Control Engineering, Washington, DC, USA, revised edition.

Bruce, R. D. and Moritz, C. T. (1997). Sound power level predictions for industrial machinery. In Crocker, M. J., editor, *Encyclopaedia of Acoustics*, chapter 86. John Wiley & Sons, Inc., Hoboken, NJ, USA.

Brüel, P. V. (1977). Do we measure damaging noise correctly? *Noise Control Engineering*, 8(March–April):52–60.

Brüel and Kjær (1973). *Instructions and Applications for 1/2-inch Condenser Microphones. Handbook.* Brüel and Kjær, Copenhagen, Denmark.

Brüel and Kjær (1996). *Microphone Handbook, Volume 1.* Brüel and Kjær, Copenhagen, Denmark.

BS EN 1793-4 (2015). Road traffic noise reducing devices - test method for determining the acoustic performance - Part 4: Intrinsic characteristics - in situ values of sound diffraction. British Standards Institution.

BS EN 352-5 (2020). Hearing protectors. Safety requirements and testing. Active noise reduction ear muffs. British Standards Institution.

BS EN 458 (2016). Hearing protectors. Recommendations for selection, use, care and maintenance. Guidance document. British Standards Institution.

Buck, A. L. (1981). New equations for computing vapor pressure and enhancement factor. *Journal of Applied Meteorology*, 20:1527–1532.

Bullen, R. (2012). The Harmonoise noise prediction algorithm: Validation and use under Australian conditions. In *Acoustics 2012*, Freemantle, WA, Australia. Australian Acoustical Society.

Buret, M., McIntosh, J., and Simpson, C. (2016). Long-term asphalt trial: Results of acoustic tests after three years. *Acoustics Australia*, 44(2):273–281.

Burns, W. and Robinson, D. W. (1970). *Hearing and Noise in Industry.* HMSO, London, UK.

Byrne, K. P., Fischer, H. M., and Fuchs, H. V. (1988). Sealed, close-fitting machine-mounted, acoustic enclosures with predictable performance. *Noise Control Engineering Journal*, 31:7–15.

Cazzolato, B., Leav, O., and Howard, C. Q. (2021). Sound directivity from a 250kw gas turbine exhaust system. In *Proceedings of Acoustics 2021 - Making Waves*, Woolongong, Australia.

Cazzolato, B. S. (1999). *Sensing-Systems for Active Control of Sound Transmission into Cavities.* PhD thesis, University of Adelaide, SA, Australia.

Cazzolato, B. S., Petersen, C. D., Howard, C. Q., and Zander, A. C. (2005). Active control of energy density in a 1D waveguide: A cautionary note. *Journal of the Acoustical Society of America*, 117:3377–3380.

Chaitanya, P. and Munjal, M. (2011). Effect of wall thickness on the end corrections of the extended inlet and outlet of a double-tuned expansion chamber. *Applied Acoustics*, 72(1):65–70.

Chapkis, R. L., Blankenship, G. L., and Marsh, A. H. (1981). Comparison of aircraft noise-contour prediction programs. *Journal of Aircraft*, 18:926–933.

Chapra, S. C. and Canale, R. P. (2011). *Numerical Methods for Engineers*, volume 1221. Mcgraw-Hill, New York.

Chien, C. C., Rajiyah, H., and Atluri, S. N. (1990). An effective method for solving the hypersingular integral equations in 3-D acoustics. *The Journal of the Acoustical Society of America*, 88(2):918–937.

Chien, C. F. and Carroll, M. M. (1980). Sound source above a rough absorbent plane. *Journal of the Acoustical Society of America*, 67:827–829.

Christensen, J. J. and Hald, J. (2004). Beamforming: Technical review. Technical report, Brüel and Kjær, Copenhagen, Denmark.

Clarkson, B. L. and Pope, R. (1981). Experimental determination of modal densities and loss factors of flat plates and cylinders. *Journal of Sound and Vibration*, 77(4):535–549.

Clarkson, B. L. and Pope, R. (1983). Experimental determination of vibration parameters required in the statistical energy analysis method. *Journal of Vibration, Acoustics, Stress, and Reliability in Design*, 105(3):337–344.

Clifford, S. F. and Lataitis, R. J. (1983). Turbulence effects on acoustic wave propagation over a smooth surface. *Journal of the Acoustical Society of America*, 75:1545–1550.

Cobo, P. and Simon, F. (2019). Multiple-layer microperforated panels as sound absorbers in buildings: A review. *Buildings*, 9:53–77.

Cohn, L. F., Harris, R. A., Rolfer, R. L., Duncan, D. L., and Woosley, R. L. (1994). Special noise barrier applications. *Transportation Research Record*, 1416:69–74.

Coleman, J. and Remington, P. J. (2006). Active control of noise and vibration. In Vér, I. L. and Beranek, L. L., editors, *Noise and Vibration Control Engineering: Principles and Applications*, chapter 18. John Wiley and Sons Inc., Hoboken, NJ, USA, second edition.

Craggs, A. (1971). The transient response of a coupled plate-acoustic system using plate and acoustic finite elements. *Journal of Sound and Vibration*, 15:509–528.

Craik, R. J. M. (1997). The relationship between transmission coefficient and coupling loss factor. In *Proceedings of the IUTAM Symposium on Statistical Energy Analysis*, page 349. Springer-Science+Business Media, B.V.

Craik, R. J. M. (2003). Non-resonant sound transmission through double walls using statistical energy analysis. *Applied Acoustics*, 64:325–341.

Crede, C. E. (1965). *Shock and Vibration Concepts in Engineering Design.* Prentice Hall, Englewood Cliffs, NJ, USA.

Cremer, L. (1942). Theorie der schalldämmung dünner wände bei schrägem einfall. *Akustische Zeitschrift*, 7:81–104.

Cremer, L. (1953). Theory regarding the attenuation of sound transmitted by air in a rectangular duct with an absorbing wall, and the maximum attenuation constant during this process. In German. *Acustica*, 3:249–263.

Cremer, L., Heckl, M., and Petersson, B. A. T. (2005). *Structure-borne Sound – Structural Vibrations and Sound Radiation.* Springer-Verlag, Berlin, Germany, third edition.

Cremer, L., Heckl, M., and Ungar, E. E. (1988). *Structure-borne Sound.* Springer-Verlag, New York, NY, USA, second edition.

Cremer, L., Hubert, M., and Kurze, U. J. (1967). Sound absorbing ventilation conduit with side branch chambers. US Patent. US Patent Number 3,353,626.

Cremer, L. and Müller, H. A. (1982). *Principles and Applications of Room Acoustics*, volume 1. Applied Science Publishers, London, UK and New York, NY, USA. English translation by T. J. Schultz.

Crocker, M. and Price, A. (1969). Sound transmission using statistical energy analysis. *Journal of Sound and Vibration*, 9(3):469–486.

Cummer, S. A., Christensen, J., and Alú, A. (2016). Controlling sound with acoustic metamaterials. *Nature Reviews Materials*, 1:16001.

Cummings, A. (1976). Sound attenuation ducts lined on two opposite walls with porous material, with some applications to splitters. *Journal of Sound and Vibration*, 49:9–35.

Cummings, A. (1978). The attenuation of lined plenum chambers in ducts: I. Theoretical models. *Journal of Sound and Vibration*, 61(3):347–373.

Cummings, A. (1980). Low frequency acoustic radiation from duct walls. *Journal of Sound and Vibration*, 71:201–226.

da Silva, G. C. C. and Nunes, M. A. A. (2016). Effect of effective length of the tube on transmission loss of reactive muffler. In Kropp, W., von Estorff, O., and Schulte-Fortkamp, B., editors, *Proceedings of the 45th International Congress on Noise Control Engineering. Internoise 2016*, pages 1588–1597, Hamburg, Germany. International Institute of Noise Control Engineering (I-INCE), German Acoustical Society (DEGA).

Daigle, G. A. (1982). Diffraction of sound by a noise barrier in the presence of atmospheric turbulence. *Journal of the Acoustical Society of America*, 71(4):847–854.

Dang, J., Shadle, C. H., Kawanishi, Y., Honda, K., and Suzuki, H. (1998). An experimental study of the open end correction coefficient for side branch. *Journal of the Acoustical Society of America*, 104:1075–1084.

Daniels, F. B. (1947). Acoustical impedance of enclosures. *Journal of the Acoustical Society of America*, 19:569–571.

Davenport, A. G. (1960). Rationale for determining design wind velocities. *ASCE Journal of the Structural Division*, 86(5):39–68.

Davern, W. A. (1977). Perforated facings backed with porous materials as sound absorbers: An experimental study. *Applied Acoustics*, 10:85–112.

Davies, P. O. A. L. (1992a). Intake and exhaust noise. Technical Report 207, University of Southampton: ISVR, Southampton, UK.

Davies, P. O. A. L. (1992b). Practical flow duct acoustic modelling. Technical Report 213, University of Southampton: ISVR, Southampton, UK.

Davies, P. O. A. L. (1993). Realistic models for predicting sound propagation in flow duct systems. *Noise Control Engineering Journal*, 40:135–141.

Davy, J. L. (1990). Predicting the sound transmission loss of cavity walls. In *Proceedings of Interior Noise Climates, annual conference of the Australian Acoustical Society*, Adelaide, SA, Australia. Australian Acoustical Society.

Davy, J. L. (1991). Predicting the sound insulation of stud walls. In *Proceedings of Internoise 91*, pages 251–254, Sydney, NSW, Australia. Institute of Noise Control Engineering.

Davy, J. L. (1993). The sound transmission of cavity walls due to studs. In *Proceedings of Internoise 93*, pages 975–978, Leuven, Belgium. Institute of Noise Control Engineering.

Davy, J. L. (1998). Problems in the theoretical prediction of sound insulation. In *Proceedings of Internoise 98*, Christchurch, New Zealand. Institute of Noise Control Engineering. Paper No. 44.

Davy, J. L. (2008a). Comparison of theoretical and experimental results for the directivity of panels and openings. In *Proceedings of Acoustics 2008*, Geelong, Australia. Australian Acoustical Society.

Davy, J. L. (2008b). The directivity of the forced radiation of sound from panels and openings including the shadow zone. *Proceedings of Meetings on Acoustics*, 4(1):3833–3838.

Davy, J. L. (2009a). The forced radiation efficiency of finite size flat panels that are excited by incident sound. *Journal of the Acoustical Society of America*, 126:694–702.

Davy, J. L. (2009b). Predicting the sound insulation of walls. *Building Acoustics*, 16:1–20.

Davy, J. L. (2010). The improvement of a simple theoretical model for the prediction of the sound insulation of double leaf walls. *Journal of the Acoustical Society of America*, 127:841–849.

Davy, J. L. (2020). personal communication.

Davy, J. L., Fard, M., Dong, W., and Loverde, J. (2019). Empirical corrections for predicting the sound insulation of double leaf cavity stud building elements with stiffer studs. *Journal of the Acoustical Society of America*, 145:703–713.

Davy, J. L., Guigou-Carter, C., and Villot, M. (2012). An empirical model for the equivalent translational compliance of steel studs. *Journal of the Acoustical Society of America*, 131:4615–4624.

Day, A. and Bennett, B. (2008). Directivity of sound from an open ended duct. In *Proceedings of Acoustics 2008*, Geelong, Australia. Australian Acoustical Society.

de Bree, H. E. and Druyvesteyn, W. F. (2005). A particle velocity sensor to measure the sound from a structure in the presence of background noise. In *Proceedings of Forum Acusticum*, Budapest, Hungary.

de Bree, H. E., Leussink, R., Korthorst, T., Jansen, H., and Lammerink, M. (1996). The microflown: A novel device for measuring acoustical flows. *Sensors and Actuators A*, 54:552–557.

de Jong, R. and Stusnik, E. (1976). Scale model studies of the effects of wind on acoustic barrier performance. *Noise Control Engineering Journal*, 6(3):101–109.

de Lisle, S. (2016). Comparison of road traffic noise prediction models: Cortn, tnm, nmpb, asj rtn. *Acoustics Australia*, 44(3).

de Silva, C. W. (2005). Vibration design and control. In de Silva, C. W., editor, *Vibration and Shock Handbook*, chapter 32. CRC Press, Boca Raton, USA.

de Vries, J. and de Bree, H. E. (2008). Scan & listen: A simple and fast method to find sources. In *SAE Brasil Noise and Vibration Conference*. SAE Technical Paper, SAE Paper 2008-36-0504.

Delaney, M. E., Harland, D. G., Hood, R. A., and Scholes, W. E. (1976). The prediction of noise levels l10 due to road traffic. *Journal of Sound and Vibration*, 48:305–325.

Delany, M. E. and Bazley, E. N. (1969). Acoustical characteristics of fibrous absorbent materials. Technical Report Aero Report AC37, National Physical Laboratory.

Delany, M. E. and Bazley, E. N. (1970). Acoustical properties of fibrous absorbent materials. *Applied Acoustics*, 3:105–106.

D'Elia, M., Humbert, T., and Aur egan, Y. (2020). Effect of flow on an array of helmholtz resonators: Is kevlar a magic layer? *Journal of the Acoustical Society of America*, 148:3392–3396.

Den Hartog, J. P. (1956). *Mechanical Vibrations.* McGraw-Hill, New York, NY, USA, fourth edition.

Department of Defense (2016). Community and environmental noise primer. `http://dodnoise.org/primer_resources`. Last accessed on July 1, 2022.

Dewhirst, M. (2002). Exhaust stack directivity (Honours degree thesis). Technical report, School of Mechanical Engineering, University of Adelaide, SA, Australia.

Dickey, N. S. and Selamet, A. (2001). An experimental study of the impedance of perforated plates with grazing flow. *Journal of the Acoustical Society of America*, 110:2360–2370.

Diehl, G. M. (1977). Sound power measurements on large machinery installed indoors: Two surface measurement. *Journal of the Acoustical Society of America*, 61:449–456.

DIN 45631/A1:2010-03 (2010). Calculation of loudness level and loudness from the sound spectrum - Zwicker method - Amendment 1: calculation of the loudness of time-variant sound; with CD-ROM. In German. German Standard.

DIN 45681 (2005-08) (2005). Acoustics - determination of tonal components of noise and determination of a tone adjustment for the assessment of noise immissions. German Standard.

Dooley, K. A. (2013). Significant infrasound levels a previously unrecognized contaminant in landmark motion sickness studies. *Journal of the Acoustical Society of America*, 134(5):4097–4097.

Dowling, A. and Ffowcs-Williams, J. (1983). *Sound and Sources of Sound.* Ellis Horwood Ltd., Chichester, UK.

Downey, G. and Parnell, J. (2017). Assessing low frequency noise from industry - a practical approach. In *12th ICBEN Congress on Noise as a Public Health Problem*, Zurich, Switzerland.

Driscoll, D. P. and Royster, L. H. (1984). Comparison between the median hearing threshold levels for an unscreened black nonindustrial noise exposed population (NINEP) and four presbycusis data bases. *American Industrial Hygiene Association Journal*, 45:577–593.

Druyvesteyn, W. F. and de Bree, H. E. (2000). A new sound intensity probe; comparison to the Brüel and Kjær p–p probe. *Journal of the Audio Engineering Society*, 48:10–20.

Dunn, I. P. and Davern, W. A. (1986). Calculation of acoustic impedance of multi-layer absorbers. *Applied Acoustics*, 19:321–334.

Dutilleux, G., Defrance, J., Ecotière, D., Gauvreau, B., Bérengier, M., Besnard, F., and Duc, E. L. (2010). NMPB-ROUTES-2008: The revision of the French method for road traffic noise prediction. *Acta Acustica United with Acustica*, 96(3):452–462.

Ebbing, C. (1971). Experimental evaluation of moving sound diffusers in reverberant rooms. *Journal of Sound and Vibration*, 16:99–118.

Edge, Jr., P. M. and Cawthorn, J. M. (1976). Selected methods for quantification of community exposure to aircraft noise. Technical Report TN D-7977, NASA, USA.

Edison Electric Institute (1978). Electric power plant environmental noise guide. Technical report, Edison Electric Institute, New York, NY, USA.

Egan, C. A., Chilekwa, V., and Oldham, D. J. (2006). An investigation of the use of top edge treatments to enhance the performance of a noise barrier using the boundary element method. In *Proceedings of ICSV 2006*. Intenational Institute of Acoustics and Vibration.

Egan, M. D. (1987). *Architectural Acoustics.* McGraw-Hill, New York, NY, USA.

Ekici, I. and Bougdah, H. (2003). A review of research on environmental noise barriers. *Building Acoustics*, 10(4):289–323.

Eldridge, H. J. (1974). *Properties of Building Materials.* Medical and Technical Publishing Co, Lancaster, UK.

Elfert, G. (1988). Vergleich zwischen gemessenen und nach verschiedenen verfahren berechneten schulldruckpegeln in flachräumen. In *Fortschritte der Akustik.* Gmb H. Bad Honnef.

Elnady, T., Åbom, M., and Allam, S. (2010). Modeling perforates in mufflers using two-ports. *Journal of Vibration and Acoustics*, 132(6):061010.

Embleton, T. F. W. (1988). Sound in large rooms. In Beranek, L. L., editor, *Noise and Vibration Control*, chapter 9. Institute of Noise Control Engineering, Washington, DC, USA, revised edition.

Embleton, T. F. W. (1996). Tutorial on sound propagation outdoors. *Journal of the Acoustical Society of America*, 100(1):31–48.

Embleton, T. F. W., Piercy, J. E., and Daigle, G. A. (1983). Effective flow resistivity of ground surfaces determined by acoustical measurements. *Journal of the Acoustical Society of America*, 74(4):1239–1244.

EN1991-1-4 (2010). Eurocode 1: Actions on structures – part 1–4: General actions – wind actions. European Committee for Standardisation.

EPA (2009). Product noise labeling hearing protection devices; proposed rule. Technical report, U.S. Environmental Protection Agency, Washington, DC, USA. 40 CFR Part 211, Docket ID Number EPA-HQ-OAR-2003-0024-0040; FRL-8934-9, Federal Register, Vol. 74, No. 149, p39150-39196.

EPA (2020). US Environmental Protection Agency (EPA), Code of Federal Regulations, Title 40, Chapter I - Environment Protection Agency, Subchapter G - Noise Abatement Programs, Part 211 - Product Noise Labeling. 40 CFR 211.201, `https://www.govregs.com/regulations/expand/title40_chapterI-i26_part211_subpartB_section211.204`, and Electronic Code of Federal Regulations `https://www.ecfr.gov/current/title-40/chapter-I/subchapter-G/part-211`.

EUROCONTROL (2016). Base of Aircraft Data (BADA). `http://www.eurocontrol.int/services/bada`. Last accessed on July 1, 2022.

European Commission (2010a). Common noise assessment methods in Europe (EU Directive 2002/49/EC, Annex II: Assessment Methods for Noise Indicators, Annex II-v9d-ISO9613). Technical report, European Parliament.

European Commission (2010b). Common noise assessment methods in Europe (EU Directive 2002/49/EC, Annex II: Assessment Methods for Noise Indicators, Annex II-v9d-JRC2010). Technical report, European Parliament.

European Commission (2015). Establishing common noise assessment methods according to Directive 2002/49/EC of the European Parliament and of the Council. Technical Report (EU) 2015/996, European Parliament.

European Parliament (2003). Directive 2003/10/EC on the minimum health and safety requirements regarding the exposure of workers to the risks arising from physical agents (noise). `https://osha.europa.eu/en/legislation/directives/82`.

Evans, T. and Cooper, J. (2011). Comparison of predicted and measured wind farm noise levels and implications for assessments of new wind farms. In *Proceedings of Acoustics 2011*, Gold Coast, Australia. Australian Acoustical Society.

Ewins, D. J. (2000). *Modal Testing*. Research Studies Press, Baldock, Hertfordshire, UK, second edition.

Fahy, F. J. (1982). Statistical energy analysis. In White, R. G. and Walker, J. G., editors, *Noise and Vibration*, chapter 7. Ellis Horwood Limited, Chichester, UK.

Fahy, F. J. (1985). *Sound and Structural Vibration: Radiation, Transmission and Response.* Academic Press, Oxford, UK.

Fahy, F. J. (1995). *Sound Intensity.* E&FN Spon, London, UK, second edition.

Fahy, F. J. (2001). *Foundations of Engineering Acoustics.* Academic Press, London, UK.

Fahy, F. J. and Gardonio, P. (2007). *Sound and Structural Vibration: Radiation, Transmission and Response.* Elsevier/Academic Press, London, UK, second edition.

Fahy, F. J. and Thompson, D. (2015). *Fundamentals of Sound and Vibration.* CRC Press, FL, USA, second edition.

Fastl, H. and Zwicker, E. (2007). *Psychoacoustics: Facts and Models.* Springer, New York, NY, USA.

FHWA (1998). FHWA traffic noise model user's guide. Technical report, US Dept of Transportation, Federal Highway Administration, Washington, DC, USA.

FHWA (2004). FHWA traffic noise model user's guide (version 2.5 addendum). Technical report, US Dept of Transportation, Federal Highway Administration, Washington, DC, USA.

FHWA (2015). Traffic noise model, version 2.5 software download. `https://www.fhwa.dot.gov/environment/noise/traffic_noise_model/tnm_v25/`. Last accessed on July 1, 2022.

Fitzroy, D. (1959). Reverberation formula which seems to be more accurate with nonuniform distribution of absorption. *Journal of the Acoustical Society of America*, 31:893–97.

Foss, R. N. (1979). Double barrier noise attenuation and a predictive algorithm. *Noise Control Engineering Journal*, 13:83–91.

Frederiksen, E., Eirby, N., and Mathiasen, H. (1979). Technical review, no. 4. Technical report, Brüel and Kjær, Denmark.

Fujiwara, K., Ando, Y., and Maekawa, Z. (1977a). Noise control by barriers. Part 1: noise reduction by a thick barrier. *Applied Acoustics*, 10:147–159.

Fujiwara, K., Ando, Y., and Maekawa, Z. (1977b). Noise control by barriers. Part 2: noise reduction by an absorptive barrier. *Applied Acoustics*, 10:167–170.

Fuller, C. R. and Mathur, G. (2017). Poro-elastic acoustic metamaterials. In *Proceedings of Internoise 2017*, Hong Kong. Institute of Noise Control Engineering.

Garai, M. and Pompoli, F. (2005). A simple empirical model of polyester fibre materials for acoustical applications. *Applied Acoustics*, 66:1383–1398.

Garifullin, M., Mela, K., Renaux, T., Izabel, D., and Holz, R. (2021). Load-bearing capacity of cold-formed sinusoidal steel sheets. *Journal of Sound and Vibration*, 161:107475.

Garret, S. L. (2020). *Understanding Acoustics: An Experimentalist's View of Sound and Vibration.* Springer/ASA Press, Cham, Switzerland, second edition.

Gill, H. S. (1980). Effect of barriers on the propagation of construction noise. Technical Report 113, ISVR, Southampton, UK.

Glasberg, B. R. and Moore, B. C. J. (2006). Prediction of absolute thresholds and equal loudness contours using a modified loudness model. *Journal of the Acoustical Society of America*, 120:585–588.

Glass, D. C. and Singer, J. E. (1972). Behavioral after effects of unpredictable and uncontrollable aversive events: Although subjects were able to adapt to loud noise and other stressors in laboratory experiments, they clearly demonstrated adverse after effects. *American Scientist*, 60(4):457–465.

Golay, M. J. E. (1947). Theoretical consideration in heat and infra-red detection, with particular reference to the pneumatic detector. *Review of Scientific Instruments*, 18:347–356.

Gomperts, M. C. and Kihlman, T. (1967). The sound transmission loss of circular and slit-shaped apertures in walls. *Acta Acustica United with Acustica*, 18(3):144–150.

Gosele, K. (1965). The damping behavior of reflection-mufflers with air flow. In German. *VDI-Berickre*, 88:123–130.

Gradshteyn, I. S. and Ryshik, I. M. (1965). *Table of Integrals, Series and Products.* Academic Press, New York, NY, USA.

Green, D. W., Winandy, J. E., and Kretschmann, D. E. (1999). Mechanical properties of wood. In *Wood Handbook: Wood as an Engineering Material, Chapter 4 (Gen. Tech. Rep. FPL–GTR–113).* US Department of Agriculture, Forest Service, Forest Products Laboratory, Madison, WI, USA.

Greenwood, D. D. (1990). A cochlear frequency-position function for several species - 29 years later. *Journal of the Acoustical of America*, 87:2592–2605.

Groenewold, M. R., Masterson, E. A., Themann, C. L., and Davis, R. R. (2014). Do hearing protectors protect hearing? *American Journal of Industrial Medicine*, 57(9):1001–1010.

Gu, J., Ma, Z. D., and Hulbert, G. M. (2001). Quasi-static data recovery for dynamic analyses of structural systems. *Finite Elements in Analysis and Design*, 37:825–841.

Hald, J. (2004). Combined NAH and beamforming using the same array. *Sound and Vibration*, pages 18–25.

Hald, J., Morkholt, J., and Gomes, J. (2007). Efficient interior NSI based on various beamforming methods for overview and conformal mapping using SONAH holography for details on selected panels. *SAE Paper 2007-01-2276.*

Hale, M. (1978). A comparison of several theoretical noise abatement prediction methods for pipe lagging systems. In *96th Meeting of the ASA*, HI, USA. Acoustical Society of America.

Hansen, C. H. (1993). Sound transmission loss of corrugated panels. *Noise Control Engineering Journal*, 40:187–197.

Hansen, C. H. (2001). *Understanding Active Noise Cancellation.* E&FN Spon, London, UK.

Hansen, C. H. (2018). *Foundations of Vibroacoustics.* CRC Press, Boca Raton, USA.

Hansen, C. H., Doolan, C. J., and Hansen, K. L. (2017). *Wind Farm Noise: Measurement, Assessment and Control.* Wiley, New York, NY, USA.

Hansen, C. H. and Mathews, M. (1983). Noise reduction through optimum hull design in lightweight tracked vehicles. Technical Report 5074, Bolt, Beranek and Newman Inc.

Hansen, C. H., Snyder, S. D., Qiu, X., Brooks, L. A., and Moreau, D. J. (2013). *Active Control of Sound and Vibration.* E&FN Spon, London, UK, second edition.

Harris, C. M., editor (1979). *Handbook of Noise Control.* McGraw-Hill, New York, NY, USA, second edition.

Harris, C. M. and Crede, C. E. (1976). *Shock and Vibration Handbook.* McGraw-Hill, New York, NY, USA.

Harris, F. J. (1978). On the use of windows for harmonic analysis with the discrete fourier transform. *Proceedings of the IEEE*, 66:51–83.

Hayes, P. A. and Quantz, C. A. (1982). Determining vibration radiation efficiency and noise characteristics of structural designs using analytical techniques. *SAE Technical Paper 820440.*

Hearmon, R. F. S. (1959). The frequency of flexural vibration of rectangular orthotropic plates with clamped or supported edges. *Journal of Applied Mechanics*, 26:537–540.

Heckl, M. (1960). Untersuchungen an orthotropen platted. *Acustica*, 10:109–115.

Heerema, H. and Hodgson, M. R. (1999). Empirical models for predicting noise levels, reverberation times, and fitting densities in industrial workshops. *Applied Acoustics*, 57:51–60.

Heilmann, G., Meyer, A., and Dobler, D. (2008). Beamforming in the time domain using 3D-microphone arrays. In *Proceedings of Acoustics 2008*, Geelong, VIC, Australia. Australian Acoustical Society.

Heitner, I. (1968). How to estimate plant noises. *Hydrocarbon Processing*, 47(12):67–74.

Herrin, D. W., Martinus, F., Wu, T. W., and Seybert, A. F. (2003). A new look at the high frequency boundary element and Rayleigh integral approximations. *SAE Paper 03NVC–114.*

Hessler, G. F., Hessler, D. M., Brandstatte, P., and Bay, K. (2008). Experimental study to determine wind-induced noise and windscreen attenuation effects on microphone response for environmental wind turbine and other applications. *Noise Control Engineering Journal*, 56:300–309.

Hirsh, I. J. and Bilger, R. C. (1955). Auditory-threshold recovery after exposure to pure tones. *Journal of the Acoustical Society of America*, 27:1186–1194.

Hodgson, D. C. and Sadek, M. M. (1977). Sound power as a criterion for forging machine optimization. In *Proceedings of the International Conference on Machine Tool Dynamics and Research*, pages 825–830, London, UK. Imperial College.

Hodgson, D. C. and Sadek, M. M. (1983). A technique for the prediction of the noise field from an arbitrary vibrating machine. *Proceedings of the Institution of Mechanical Engineers*, 179C:189–197.

Hodgson, M. R. (1994a). Are room surface absorption coefficients unique? In *Proceedings of the Wallace Clement Sabine Symposium*, pages 161–164, Cambridge, MA, USA. Acoustical Society of America.

Hodgson, M. R. (1994b). When is diffuse-field theory accurate? In *Proceedings of the Wallace Clement Sabine Symposium*, pages 157–160, Cambridge, MA, USA. Acoustical Society of America.

Hodgson, M. R. (2003). Ray-tracing evaluation of empirical models for prediction noise in industrial workshops. *Applied Acoustics*, 64:1033–1048.

Hohenwarter, D. (1991). Noise radiation of (rectangular) plane sources. *Applied Acoustics*, 33:45–62.

Hopkins, C. (2007). *Sound Insulation.* Butterworth-Heinemann, Oxford, UK.

Hopkins, C. and Robinson, M. (2014). Using transient and steady-state sea to assess potential errors in the measurement of structure-borne sound power input from machinery on coupled reception plates. *Applied Acoustics*, 79:35–41.

Howard, C. Q. (2007a). Modal mass of clamped beams and clamped plates. *Journal of Sound and Vibration*, 301(1-2):410–414.

Howard, C. Q. (2007b). Theoretical and experimental results of the transmission loss of a plate with discrete masses attached. In *Proceedings of the 14th International Congress on Sound and Vibration*, Cairns, QLD, Australia. International Institute of Acoustics and Vibration.

Howard, C. Q. (2013a). The corrected expressions for the four-pole transmission matrix for a duct with a linear temperature gradient and an exponential temperature profile. *Open Journal of Acoustics*, 3(3):62–66.

Howard, C. Q. (2013b). Transmission matrix model of a quarter-wave-tube with gas temperature gradients. In *Proceedings of Acoustics 2013*, Victor Harbor, South Australia, Australia. Australian Acoustical Society. Paper 71.

Howard, C. Q. and Cazzolato, B. S. (2014). *Acoustic Analyses Using MATLAB and ANSYS.* CRC Press, FL, USA.

Howard, C. Q., Cazzolato, B. S., and Hansen, C. H. (2000). Exhaust stack silencer design using finite element analysis. *Noise Control Engineering Journal*, 48:113–120.

Howard, C. Q. and Craig, R. A. (2014a). An adaptive quarter-wave tube that uses the sliding-goertzel algorithm for estimation of phase. *Applied Acoustics*, 78:92–97.

Howard, C. Q. and Craig, R. A. (2014b). Noise reduction using a quarter wave tube with different orifice geometries. *Applied Acoustics*, 76:180–186.

Hurtley, C. (2009). *Night Noise Guidelines for Europe.* WHO Regional Office Europe.

Hutchins, D. A., Jones, H. W., and Russell, L. T. (1984). Model studies of barrier performance in the presence of ground surfaces. *Journal of the Acoustical Society of America*, 75:1807–1826.

IEC 60076-10 (2016). Power transformers – Part 10: Determination of sound levels. International Electrotechnical Commission.

IEC 60076-10-1:2016+AMD1:2020CSV (2016). Power transformers – Part 10-1: Determination of sound levels – application guide. International Electrotechnical Commission.

IEC 60268-16, 4th edn. (2020). Sound system equipment – Part 16: Objective rating of speech intelligibility by speech transmission index. International Electrotechnical Commission.

IEC 60534-2-1 (2011). Industrial-process control valves – Part 2-1: Flow capacity - sizing equations for fluid flow under installed conditions. International Electrotechnical Commission.

IEC 60534-8-3 (2010). Industrial-process control valves - Part 8-3: Noise considerations - control valve aerodynamic noise prediction method. International Electrotechnical Commission.

IEC 60534-8-4 (1994). Industrial-process control valves – Part 8-4: Noise considerations - prediction of noise generated by hydrodynamic flow (superceded). International Electrotechnical Commission.

IEC 60534-8-4 (2015). Industrial-process control valves – Part 8-4: Noise considerations - prediction of noise generated by hydrodynamic flow. International Electrotechnical Commission.

IEC 60645-1 (2017). Electroacoustics - audiometric equipment - Part 1: Equipment for pure-tone audiometry. International Electrotechnical Commission.

IEC 60645-3 (2020). Electroacoustics – audiometric equipment – Part 3: Test signals of short duration. International Electrotechnical Commission.

IEC 60704-2-1 (2020). Household and similar electrical appliances – test code for the determination of airborne acoustical noise – Part 2-1: Particular requirements for vacuum cleaners. International Electrotechnical Commission.

IEC 60704-2-10 (2011). Household and similar electrical appliances – test code for the determination of airborne acoustical noise – Part 2-10: Particular requirements for electric cooking ranges, ovens, grills, microwave ovens and any combination of these. International Electrotechnical Commission.

IEC 60704-2-11 (1998). Household and similar electrical appliances – test code for the determination of airborne acoustical noise – Part 2-11: Particular requirements for electrically-operated food preparation. International Electrotechnical Commission.

IEC 60704-2-13 (2016). Household and similar electrical appliances – test code for the determination of airborne acoustical noise – Part 2-13: Particular requirements for range hoods and other cooking fume extractors. International Electrotechnical Commission.

IEC 60704-2-14 (2019). Household and similar electrical appliances – test code for the determination of airborne acoustical noise – Part 2-14: Particular requirements for refrigerators, frozen-food storage cabinets and food freezers. International Electrotechnical Commission.

IEC 60704-2-15 (2008). Household and similar electrical appliances – test code for the determination of airborne acoustical noise – Part 2-15: Particular requirements for household food waste disposers. International Electrotechnical Commission.

IEC 60704-2-16 (2019). Household and similar electrical appliances – test code for the determination of airborne acoustical noise – Part 2-16: Particular requirements for washer-dryers. International Electrotechnical Commission.

IEC 60704-2-2 (2009). Household and similar electrical appliances – test code for the determination of airborne acoustical noise – Part 2-2: Particular requirements for fan heaters. International Electrotechnical Commission.

IEC 60704-2-3 (2017). Household and similar electrical appliances – test code for the determination of airborne acoustical noise – Part 2-3: Particular requirements for dishwashers. International Electrotechnical Commission.

IEC 60704-2-4 (2011). Household and similar electrical appliances – test code for the determination of airborne acoustical noise – Part 2-4: Particular requirements for washing machines and spin extractors. International Electrotechnical Commission.

IEC 60704-2-5 (2014). Household and similar electrical appliances – test code for the determination of airborne acoustical noise – Part 2-5: Particular requirements for electric thermal storage room heaters. International Electrotechnical Commission.

IEC 60704-2-6 (2012). Household and similar electrical appliances – test code for the determination of airborne acoustical noise – Part 2-6: Particular requirements for tumble dryers. International Electrotechnical Commission.

IEC 60704-2-7 (2020). Household and similar electrical appliances – test code for the determination of airborne acoustical noise – Part 2-7: Particular requirements for fans. International Electrotechnical Commission.

IEC 60704-2-8 (2020). Household and similar electrical appliances – test code for the determination of airborne acoustical noise – Part 2-8: Particular requirements for electric shavers. International Electrotechnical Commission.

IEC 60704-2-9 (2003). Household and similar electrical appliances – test code for the determination of airborne acoustical noise – Part 2-9: Particular requirements for electric hair care appliances. International Electrotechnical Commission.

IEC 61043 (2017). Electroacoustics – instruments for the measurement of sound intensity – measurements with pairs of pressure sensing microphones. International Electrotechnical Commission.

IEC 61252 (2017). Electroacoustics – Specifications for personal sound exposure meters. International Electrotechnical Commission.

IEC 61260-1 (2014). Electroacoustics – Octave-band and fractional-octave-band filters – Part 1: Specifications. International Electrotechnical Commission.

IEC 61400-11 Ed.3.0 (2018). Wind turbines – Part 11: Acoustic noise measurement techniques. International Electrotechnical Commission.

IEC 61672-1 (2013). Electroacoustics - sound level meters – Part 1: Specifications. International Electrotechnical Commission.

IEEE Std 1451.4-2004 (2004). IEEE Standard for a smart transducer interface for sensors and actuators – Mixed-mode communication protocols and Transducer Electronic Data Sheet (TEDS) formats.

IEEE Std 1451.5 (2007). IEEE Standard for a smart transducer interface for sensors and actuators – Wireless communication protocols and and Transducer Electronic Data Sheet (TEDS) formats.

IEEE Std C57.12.90 (2021). Standard test code for liquid-immersed distribution, power, and regulating transformers.

Ih, J. (1992). The reactive attenuation of reactive plenum chambers. *Journal of Sound and Vibration*, 157:93–122.

Ihde, W. M. (1975). Tuning stubs to silence large air handling systems. *Noise Control Engineering*, 5:131–135.

Ingard, K. U. (1959). Attenuation and regeneration of sound in ducts and jet diffusers. *Journal of the Acoustical Society of America*, 31:1206–1212.

Ingard, K. U. (1994). *Notes on Sound Absorption Technology.* Noise Control Foundation, Poughkeepsie, NY, USA, 94-02 edition.

Ingard, K. U. (2010). *Noise Reduction Analysis.* Jones and Bartlett Publishers, Sudbury, MA, USA.

Ingard, K. U. and Dear, T. A. (1985). Measurement of acoustic flow resistance. *Journal of Sound and Vibration*, 103:567–572.

Inman, D. J. (2014). *Engineering Vibration*. Pearson, Chandler, AZ, USA, fourth edition.

Intergovernmental Committee on Surveying and Mapping (2016). `https://elevation.fsdf.org.au/`, Last accessed on 2022-02-06.

Iqbal, M. A., Wilson, T. K., and Thomas, R. J., editors (1977). *The Control of Noise in Ventilation Systems: A Designers Guide*. E&FN Spon, London, UK.

Irwin, J. S. (1979). A theoretical variation of the wind profile power-law exponent as a function of surface roughness and stability. *Atmospheric Environment*, 13(1):191–194.

Isakov, V. and Wu, S. F. (2002). On theory and applications of the hels method in inverse acoustics. *Inverse Problems*, 18:1147–1159.

Isermann, U. (2007). Anleitung zur berechtung von lärmschutzbereichen (azb). Technical report, DLR, Germany.

ISO 10140-2 (2021). Laboratory measurement of sound insulation of building elements – Part 2: Measurement of airborne sound insulation. International Organization for Standardization.

ISO 10140-3 (2021). Acoustics – Laboratory measurement of sound insulation of building elements – Part 3: Measurement of impact sound insulation. International Organization for Standardization.

ISO 10140-4 (2021). Acoustics – Laboratory measurement of sound insulation of building elements – Part 1: Application rules for specific products. International Organization for Standardization.

ISO 10140-5 (2021). Acoustics – Laboratory measurement of sound insulation of building elements – Part 5: Requirements for test facilities and equipment (amended 2014). International Organization for Standardization.

ISO 10302-1 (2011). Gas turbines and gas turbine sets – Measurement of emitted airborne noise – Measurement of airborne noise emitted and structure-borne vibration induced by small air-moving devices – Part 1: Airborne noise measurement. International Organization for Standardization.

ISO 10494 (2018). Gas turbines and gas turbine sets – Measurement of emitted airborne noise – Engineering survey method. International Organization for Standardization.

ISO 10534-1 (1996). Acoustics – Determination of sound absorption coefficient and impedance in impedance tubes – Part 1: Method using standing wave ratio. International Organization for Standardization.

ISO 10534-2 (1998). Acoustics – Determination of sound absorption coefficient and impedance in impedance tubes – Part 2: Transfer-function method. International Organization for Standardization.

ISO 10847 (1997). Acoustics – In-situ determination of insertion loss of outdoor noise barriers of all types. International Organization for Standardization.

ISO 10848-1 (2017). Acoustics – Laboratory measurement of the flanking transmission of airborne and impact sound between adjoining rooms – Part 1: Frame document. International Organization for Standardization.

ISO 11202 (2010). Acoustics – Noise emitted by machinery and equipment – Determination of emission sound pressure levels at a work station and at other specified positions applying approximate environmental corrections. International Organization for Standardization.

ISO 11203 (1995). Acoustics – Noise emitted by machinery and equipment – Determination of emission sound pressure levels at a work station and at other specified positions from the sound power level. International Organization for Standardization.

ISO 11204 (2010). Acoustics – Noise emitted by machinery and equipment – Determination of emission sound pressure levels at a work station and at other specified positions applying accurate environmental corrections. International Organization for Standardization.

ISO 11546-1 (1995). Acoustics – Determination of sound insulation performances of enclosures – Part 1: Measurements under laboratory conditions (for declaration purposes). International Organization for Standardization.

ISO 11546-2 (1995). Acoustics – Determination of sound insulation performances of enclosures – Part 2: Measurements in situ (for acceptance and verification purposes). International Organization for Standardization.

ISO 11654 (1997). Acoustics – Sound absorbers for use in buildings – Rating of sound absorption. International Organization for Standardization.

ISO 11690-1 (2020). Acoustics – Recommended practice for the design of low-noise workplaces containing machinery – Part 1: Noise control strategies. International Organization for Standardization.

ISO 11690-2 (2020). Acoustics – Recommended practice for the design of low-noise workplaces containing machinery – Part 2: Noise control measures. International Organization for Standardization.

ISO 11690-3 (1997). Acoustics – Recommended practice for the design of low-noise workplaces containing machinery – Part 3: Sound propagation and noise prediction in workrooms. International Organization for Standardization.

ISO 12354-1 (2017). Building acoustics – Estimation of acoustic performance of buildings from the performance of elements – Part 1: Airborne sound insulation between rooms. International Organization for Standardization.

ISO 12999-1 (2020). Acoustics – Determination and application of measurement uncertainties in building acoustics – Part 1: sound insulation. International Organization for Standardization.

ISO 13472-1 (2022). Acoustics – Measurement of sound absorption properties of road surfaces in situ – Part 1: Extended surface method. International Organization for Standardization.

ISO 13472-2 (2010). Acoustics – Measurement of sound absorption properties of road surfaces in situ – Part 2: Spot method for reflective surfaces. International Organization for Standardization.

ISO 13475-1 (1999). Acoustics – Stationary audible warning devices used outdoors – Part 1: Field measurements for determination of sound emission quantities. International Organization for Standardization.

ISO 13475-2 (2000). Acoustics – Stationary audible warning devices used outdoors – Part 2: Precision methods for determination of sound emission quantities. International Organization for Standardization.

ISO 14163 (1998). Acoustics – Guidelines for noise control by silencers. International Organization for Standardization.

ISO 15186-1 (2000). Acoustics – Measurement of sound insulation in buildings and of building elements using sound intensity – Part 1: Laboratory measurements. International Organization for Standardization.

ISO 15186-2 (2003). Acoustics – Measurement of sound insulation in buildings and of building elements using sound intensity – Part 2: Field measurements. International Organization for Standardization.

ISO 15186-3 (2002). Acoustics – Measurement of sound insulation in buildings and of building elements using sound intensity – Part 3: Laboratory measurements at low frequencies. International Organization for Standardization.

ISO 16254 (2016). Acoustics – Measurement of sound emitted by road vehicles of category M and N at standstill and low speed operation – Engineering method. International Organization for Standardization.

ISO 16283-1 (2014). Acoustics – Field measurement of sound insulation in buildings and of building elements – Part 1: Airborne sound insulation. International Organization for Standardization.

ISO 16283-2 (2020). Acoustics – Field measurement of sound insulation in buildings and of building elements – Part 2: Impact sound insulation. International Organization for Standardization.

ISO 16283-3 (2016). Acoustics – Field measurement of sound insulation in buildings and of building elements – Part 3: Façade sound insulation. International Organization for Standardization.

ISO 1680 (2013). Acoustics – Test code for the measurement of airborne noise emitted by rotating electrical machines. International Organization for Standardization.

ISO 1683 (2015). Acoustics – Preferred reference values for acoustical and vibratory levels. International Organization for Standardization.

ISO 17201-3 (2019). Acoustics – Noise from shooting ranges – Part 3: Sound propagation calculations. International Organization for Standardization.

ISO 1996-1 (2016). Acoustics – Description, measurement and assessment of environmental noise – Part 1: Basic quantities and assessment procedures (reviewed 2016). International Organization for Standardization.

ISO 1996-2 (2017). Acoustics – Description, measurement and assessment of environmental noise – Part 2: Determination of sound pressure levels. International Organization for Standardization.

ISO 1999 (2013). Acoustics – Estimation of noise-induced hearing loss. International Organization for Standardization.

ISO 20816-1 (2016). Mechanical vibration – Measurement and evaluation of machine vibration – Part 1: General guidelines. International Organization for Standardization.

ISO 20816-2 (2017). Mechanical vibration – Measurement and evaluation of machine vibration – Part 2: Land-based gas turbines, steam turbines and generators in excess of 40 MW, with fluid-film bearings and rated speeds of 1500 r/min, 1800 r/min, 3000 r/min and 3600 r/min. International Organization for Standardization.

ISO 20816-4 (2018). Mechanical vibration – Measurement and evaluation of machine vibration – Part 4: Gas turbines in excess of 3 MW, with fluid-film bearings. International Organization for Standardization.

ISO 20816-5 (2018). Mechanical vibration – Measurement and evaluation of machine vibration – Part 5: Machine sets in hydraulic power generating and pump-storage plants. International Organization for Standardization.

ISO 20816-8 (2018). Mechanical vibration – Measurement and evaluation of machine vibration – Part 8: Reciprocating compressor systems. International Organization for Standardization.

ISO 20816-9 (2020). Mechanical vibration – Measurement and evaluation of machine vibration – Part 9: Gear units. International Organization for Standardization.

ISO 226 (1987). Acoustics – Normal equal-loudness-level contours (withdrawn). International Organization for Standardization.

ISO 226 (2023). Acoustics – Normal equal-loudness-level contours. International Organization for Standardization.

ISO 2631-1 (2010). Mechanical vibration and shock: evaluation of human exposure to whole-body vibration – Part 1: General requirements. International Organization for Standardization.

ISO 2631-2 (2003). Mechanical vibration and shock: evaluation of human exposure to whole-body vibration – Part 2: Vibration in buildings (1 Hz to 80 Hz). International Organization for Standardization.

ISO 266 (1997). Acoustics – Preferred frequencies. International Organization for Standardization.

ISO 2922 (2020). Acoustics – Measurement of airborne sound emitted by vessels on inland waterways and harbours. International Organization for Standardization.

ISO 2923 (1997). Acoustics – Measurement of noise on board vessels. International Organization for Standardization.

ISO 3095 (2013). Acoustics – Railway applications – Measurement of noise emitted by railbound vehicles. International Organization for Standardization.

ISO 3381 (2021). Railway applications – Acoustics – Measurement of noise inside railbound vehicles. International Organization for Standardization.

ISO 3382-2 (2008). Acoustics – measurement of room acoustic parameters – Part 2: Reverberation time in ordinary rooms. International Organization for Standardization.

ISO 354 (2003). Acoustics – Measurement of sound absorption in a reverberation room. International Organization for Standardization.

ISO 362-1 (2015). Measurement of noise emitted by accelerating road vehicles – Engineering method – Part 1: M and N categories. International Organization for Standardization.

ISO 362-2 (2009). Measurement of noise emitted by accelerating road vehicles – Engineering method – Part 2: L category. International Organization for Standardization.

ISO 362-3 (2016). Measurement of noise emitted by accelerating road vehicles – Engineering method – Part 3: Indoor testing M and N categories. International Organization for Standardization.

ISO 3741 (2010). Acoustics – Determination of sound power levels and sound energy levels of noise sources using sound pressure – Precision methods for reverberation test rooms. International Organization for Standardization.

ISO 3743-1 (2010). Acoustics – Determination of sound power levels and sound energy levels of noise sources using sound pressure – Engineering methods for small movable sources in reverberant fields – Part 1: Comparison method for a hard-walled test room. International Organization for Standardization.

ISO 3743-2 (2018). Acoustics – Determination of sound power levels of noise sources using sound pressure – Engineering methods for small, movable sources in reverberant fields – Part 2: Methods for special reverberation test rooms (last reviewed and confirmed in 2011). International Organization for Standardization.

ISO 3744 (2010). Acoustics – Determination of sound power levels and sound energy levels of noise sources using sound pressure – Engineering methods for an essentially free field over a reflecting plane. International Organization for Standardization.

ISO 3745 (2017). Acoustics – Determination of sound power levels and sound energy levels of noise sources using sound pressure – Precision methods for anechoic rooms and hemi-anechoic rooms. International Organization for Standardization.

ISO 3746 (2010). Acoustics – Determination of sound power levels and sound energy levels of noise sources using sound pressure – Survey method using an enveloping measurement surface over a reflecting plane. International Organization for Standardization.

ISO 3747 (2010). Acoustics – Determination of sound power levels and sound energy levels of noise sources using sound pressure – Engineering/survey methods for use in-situ in a reverberant environment. International Organization for Standardization.

ISO 4869-1 (2018). Acoustics – Hearing protectors – Part 1: Subjective method for the measurement of sound attenuation. International Organization for Standardization.

ISO 4869-2 (2018). Acoustics – Hearing protectors – Part 2: Estimation of effective A-weighted sound pressure levels when hearing protectors are worn. International Organization for Standardization.

ISO 4869-2 Spreadsheet (2018). `ISO_4869-2_-_Annex_calculation_(2018-05-02).xlsx`. Microsoft Excel spreadsheet to calculate parameters in ISO 4869-2:2018, `https://standards.iso.org/iso/4869/-2/ed-2/en/`.

ISO 4869-6 (2019). Acoustics – Hearing protectors – Part 6: Determination of sound attenuation of active noise reduction earmuffs. International Organization for Standardization.

ISO 5129 (2013). Acoustics – measurement of sound pressure levels in the interior of aircraft during flight. International Organization for Standardization.

ISO 5130 (2019). Acoustics – measurements of sound pressure level emitted by stationary road vehicles. International Organization for Standardization.

ISO 5135 (2020). Acoustics – determination of sound power levels of noise from air-terminal devices, air-terminal units, dampers and valves by measurement in a reverberation room. International Organization for Standardization.

ISO 5136 (2003). Acoustics – determination of sound power radiated into a duct by fans and other air-moving devices – in-duct method. International Organization for Standardization.

ISO 532 (1975). Acoustics – Method for calculating loudness level. International Organization for Standardization.

ISO 532-1 (2017). Acoustics – Methods for calculating loudness. Part 1: Zwicker method. International Organization for Standardization.

ISO 532-2 (2017). Acoustics – Methods for calculating loudness. Part 2: Moore-Glasberg method. International Organization for Standardization.

ISO 717-1 (2020). Acoustics – Rating of sound insulation in buildings and of building elements – Part 1: Airborne sound insulation. International Organization for Standardization.

ISO 717-2 (2020). Acoustics – Rating of sound insulation in buildings and of building elements – Part 2: Impact sound insulation. International Organization for Standardization.

ISO 7574-1 (1985). Statistical methods for determining and verifying stated noise emission values of machinery and equipment – Part 1: General considerations and definitions. International Organization for Standardization.

ISO 7626-1 (2011). Mechanical vibration and shock – experimental determination of mechanical mobility – Part 1: Basic terms and definitions, and transducer specifications. International Organization for Standardization.

ISO 7626-2 (2015). Mechanical vibration and shock – experimental determination of mechanical mobility – Part 2: Measurements using single-point translation excitation with an attached vibration exciter. International Organization for Standardization.

ISO 7626-5 (2019). Vibration and shock – experimental determination of mechanical mobility – Part 5: Measurements using impact excitation with an exciter which is not attached to the structure. International Organization for Standardization.

ISO 7779 (2018). Acoustics – measurement of airborne noise emitted by information technology and telecommunications equipment. International Organization for Standardization.

ISO 7849-1 (2009). Acoustics – Determination of airborne sound power levels emitted by machinery using vibration measurement – Part 1: Survey method using a fixed radiation factor. International Organization for Standardization.

ISO 7849-2 (2009). Acoustics – Determination of airborne sound power levels emitted by machinery using vibration measurement – Part 2: Engineering method including determination of the adequate radiation factor. International Organization for Standardization.

ISO 7919-3 (2017). Mechanical vibration – Evaluation of machine vibration by measurements on rotating shafts – Part 3: Coupled industrial machines. International Organization for Standardization.

ISO 8253-1 (2010). Acoustics – Audiometric test methods – Part 1: Pure-tone air and bone conduction audiometry. International Organization for Standardization.

ISO 8253-2 (2009). Acoustics – Audiometric test methods – Part 2: Sound field audiometry with pure-tone and narrow-band test signals. International Organization for Standardization.

ISO 8253-3 (2022). Acoustics – Audiometric test methods – Part 3: Speech audiometry. International Organization for Standardization.

ISO 8297 (2021). Acoustics – Determination of sound power levels of multisource industrial plants for evaluation of sound pressure levels in the environment – Engineering method. International Organization for Standardization.

ISO 9612 (2009). Acoustics – Determination of occupational noise exposure – Engineering method. International Organization for Standardization.

ISO 9613-1 (1993). Acoustics – Attenuation of sound during propagation outdoors – Part 1: Calculation of the absorption of sound by the atmosphere. International Organization for Standardization.

ISO 9613-2 (1996). Acoustics – Attenuation of sound during propagation outdoors – Part 2: General method of calculation. International Organization for Standardization.

ISO 9614-1 (1993). Acoustics – Determination of sound power levels of noise sources using sound intensity – Part 1: Measurement at discrete points. International Organization for Standardization.

ISO 9614-2 (1996). Acoustics – Determination of sound power levels of noise sources using sound intensity – Part 2: Measurement by scanning. International Organization for Standardization.

ISO 9614-3 (2002). Acoustics – Determination of sound power levels of noise sources using sound intensity – Part 3: Precision method for measurement by scanning. International Organization for Standardization.

ISO 9645 (1990). Acoustics – Measurement of noise emitted by two-wheeled mopeds in motion – Engineering method. International Organization for Standardization.

ISO/IEC/IEEE 21450 (2010). Information technology – Smart transducer interface for sensors and actuators – Common functions, communication protocols, and Transducer Electronic Data Sheet (TEDS) formats. International Organization for Standardization.

ISO/IEC/IEEE 21451-4 (2010). Information technology – Smart transducer interface for sensors and actuators – Part 4: Mixed-mode communication protocols and Transducer Electronic Data Sheet (TEDS) formats. International Organization for Standardization.

ISO/PAS 20065 (2016). Acoustics – Objective method for assessing the audibility of tones in noise – Engineering method. International Organization for Standardization.

ISO/TR 11688-1 (1995). Acoustics – Recommended practice for the design of low-noise machinery and equipment – Part 1: Planning. International Organization for Standardization.

ISO/TR 11688-2 (1998). Acoustics – Recommended practice for the design of low-noise machinery and equipment – Part 2: Introduction to the physics of low-noise design. International Organization for Standardization.

Jacobsen, F. and de Bree, H.-E. (2005). A comparison of two different sound intensity measurement principles. *Journal of the Acoustical Society of America*, 118:1510–1517.

Jacobsen, F., Figueroa, S. B., and Rasmussen, K. (2004). A note on the concept of acoustic center. *Journal of the Acoustical Society of America*, 115:1468–1473.

Jacobsen, F. and Liu, Y. (2005). Near field acoustic holography with particle velocity transducers. *Journal of the Acoustical Society of America*, 118:3139–3144.

James, R. R. (2012). Wind turbine infra and low-frequency sound. Warning signs that were not heard. *Bulletin of Science, Technology & Society*, 32(2):108–127.

Jansson, E. and Karlsson, K. (1983). Sound levels recorded within the symphony orchestra and risk criteria for hearing loss. *Scandinavian Audiology*, 12:215–221.

Jayachandran, V., Hirsch, S., and Sun, J. (1998). On the numerical modelling of interior sound fields by the modal expansion approach. *Journal of Sound and Vibration*, 210:243–254.

Jensen, F. B., Kuperman, W. A., Porter, M. B., and Schmidt, H. (2011). *Computational Ocean Acoustics*. Springer Science & Business Media.

Jeyapalan, R. K. and Halliwell, N. A. (1981). Machinery noise predictions at the design stage using acoustic modelling. *Applied Acoustics*, 14:361–376.

Jeyapalan, R. K. and Richards, E. J. (1979). Radiation efficiencies of beams in flexural vibration. *Journal of Sound and Vibration*, 67:55–67.

Ji, Z. L. (2005). Acoustic length correction of a closed cylindrical side-branched tube. *Journal of Sound and Vibration*, 283:1180–1186.

Johnson, D. H. and Dudgeon, D. E. (1993). *Array Signal Processing: Concepts and Techniques*. Prentice Hall, Englewood Cliffs, NJ, USA.

Joint Departments of the Army, Air Force, and Navy USA (1983). Power plant acoustics. Technical manual, Washington, DC, USA. Technical Manual TM 5-805-9 / AFM 88-20 / NAVFAC DM-3.14.

Jonasson, H. and Eslon, L. (1981). Determination of sound power levels of external sources. Technical Report SP–RAPP 1981:45, National Testing Institute, Acoustics Laboratory, Borås, Sweden.

Jones, A. D. (1984). Modeling the exhaust noise radiated from reciprocating internal combustion engines: Literature review. *Noise Control Engineering*, 23:12–31.

Joppa, P. D., Sutherland, L. C., and Zuckerwar, A. J. (1996). Representative frequency approach to the effect of bandpass filters on evaluation of sound attenuation by the atmosphere. *Noise Control Engineering Journal*, 44(6):261–273.

Juhl, P. (1993). *The Boundary Element Method for Sound Field Calculations*. PhD thesis, Technical University of Denmark, Denmark. ISSN 0105-3027.

Kalapinski, E. J. and Pellerin, T. (2009). Wind turbine acoustic modeling with the ISO 9613-2 Standard: Methodologies to address constraints. In *Third International Meeting on Wind Turbine Noise*, Aalborg, Denmark.

Kampichler, G., Bülte, H., Koch, F., and Heim, K. (2010). Industrial and marine engines. In Mollenhauer, K. and Tschöke, H., editors, *Handbook of Diesel Engines*, chapter 18, pages 559–608. Springer-Verlag, Berlin, Heidelberg, Germany.

Kanapathipillai, S. and Byrne, K. P. (1991). Calculating the insertion loss produced by a porous blanket pipe lagging. In *Proceedings of Internoise '91*, pages 283–286. Institute of Noise Control Engineering.

Karlsson, K., Lundquist, P. G., and Olausson, T. (1983). The hearing of symphonic musicians. *Scandinavian Audiology*, 12:257–264.

Karlsson, M. and Åbom, M. (2010). Aeroacoustics of t-junctions–an experimental investigation. *Journal of Sound and Vibration*, 329(10):1793–1808.

Kean, S. (2008). Is CoRTN an Leq or L10 procedure? In *Proceedings of Acoustics 2008*, Geelong, Australia. Australian Acoustical Society.

Keen, J. A. (1940). A note on the length of the basilar membrane in man and in various mammals. *Journal of Anatomy*, 74(Pt 4):524–527.

Kelly, S. G. (1993). *Fundamentals of Mechanical Vibrations.* McGraw-Hill, New York, NY, USA.

Kephalopoulos, S., Paviotti, M., and Anfosso-Ledee, F. (2012). Common noise assessment methods in Europe (CNOSSOS-EU). `https://publications.jrc.ec.europa.eu/repository/handle/JRC72550`.

Keränen, J., Airo, E., Olkinuora, P., and Hongisto, V. (2003). Validity of the ray-tracing method for the application of noise control in workplaces. *Acustica - Acta Acustica*, 89:863–874.

Kinsler, L. E., Frey, A. R., Coppens, A. B., and Sanders, J. V. (1999). *Fundamentals of Acoustics.* John Wiley, New York, NY, USA, fourth edition.

Kirkup, S. M. (2007). The boundary element method in acoustics. Integrated sound software. `http://www.boundary-element-method.com`. Last accessed on July 1, 2022.

Knotts, B. and Selamet, A. (2003). Suppression of flow-acoustic coupling in sidebranch ducts by interface modification. *Journal of Sound and Vibration*, 265(5):1025–1045.

Koopmann, G. H. and Benner, H. (1982). Method for computing the sound power of machines based on the Helmholtz integral. *Journal of the Acoustical Society of America*, 71:78–89.

Koopmann, J., Zubrow, A., Hansen, A., Hwang, S., Ahearn, M., and Solman, G. (2016). Aviation Environmental Design Tool (AEDT), Version 2b. User Guide. Technical report, U.S. Department of Transportation, FAA.

Kraak, W. (1981). Investigation on criteria for the risk of hearing loss due to noise. In Tobias, J. V. and Schubert, E. D., editors, *Hearing Research and Theory*, volume 1. Academic Press, San Diego, CA, USA.

Kraak, W., Kracht, L., and Fuder, G. (1977). Die ausbildung von gehörschäden als folge der akkomulation von lärmeinwirkungen. *Acustica*, 38:102–117.

Krahé, D. (2010). Low frequency noise - strain on the brain. In *14th International Meeting on Low Frequency Noise and Vibration and its Control*, Aalborg, Denmark.

Kraniauskas, P. (1994). A plain man's guide to the fft. *IEEE Signal Processing Magazine*, 11(2):24–35.

Krokstad, A., Strom, S., and Sorsdal, S. (1968). Calculation of the acoustical room response by the use of a ray tracing technique. *Journal of Sound and Vibration*, 8:118–125.

Kryter, K. D. (1959). Scaling human reactions to the sound from aircraft. *Journal of the Acoustical Society of America*, 31:1415–1429.

Kryter, K. D. (1970). *The Effects of Noise on Man.* Academic Press, New York, NY, USA.

Kuehler, R., Fedtke, T., and Hensel, J. (2015). Infrasonic and low-frequency insert earphone hearing threshold. *Journal of the Acoustical Society of America*, 137(4):EL347–EL353.

Kurze, U. J. and Allen, C. H. (1971). Influence of flow and high sound level on the attenuation in a lined duct. *Journal of the Acoustical Society of America*, 49:1643.

Kurze, U. J. and Anderson, G. S. (1971). Sound attenuation by barriers. *Applied Acoustics*, 4:35–53.

Kurze, U. J. and Beranek, L. L. (1988). Sound propagation outdoors. In Beranek, L. L., editor, *Noise and Vibration Control*, chapter 7. Institute of Noise Control Engineering, Washington, DC, USA, revised edition.

Kurze, U. J. and Riedel, E. (2013). Silencers. In Müller, G. and Möser, M., editors, *Handbook of Engineering Acoustics*, chapter 11, pages 269–299. Springer-Verlag, Berlin, Heidelberg.

Kuttruff, H. (1985). Stationare schallausbreitung in flachräumen. *Acustica*, 57:62–70.

Kuttruff, H. (1989). Stationare schallausbreitung in langräumen. *Acustica*, 69:53–62.

Kuttruff, H. (1994). Sound decay in enclosures with non-diffuse sound field. In *Proceedings of the Wallace Clement Sabine Symposium*, pages 85–88. Acoustical Society of America.

Kuttruff, H. (2009). *Room Acoustics*. Taylor and Francis, New York, NY, USA, fifth edition.

Lam, Y. W. (1996). A comparison of three diffuse reflection modeling methods used in room acoustics computer models. *Journal of the Acoustical Society of America*, 100:2181–2192.

Lambert, J. (2006). Enhancement of the low frequency sound absorption of porous materials by addition of a thin high flow-resistivity cover layer, Masters thesis. Technical report, State University of New York at Buffalo.

Landau, L. D. and Lifshitz, E. M. (1959). *Fluid Mechanics*. Pergamon Press, Oxford. (translated by J. B. Sykes and W. H. Reid).

Langley, R. S. (1994). The modal density and mode count of thin cylinders and curved panels. *Journal of Sound and Vibration*, 169(1):43–53.

Larizza, F., Howard, C. Q., Grainger, S., and Wang, W. (2019). Detection and location of defects in rolling element bearing using acoustic emission. In *AIAC18 - 18th Australian International Aerospace Congress and HUMS2019 - 11th International Conference on Health and Usage Monitoring Systems*, pages 1000–1005, Melbourne, Victoria, Australia. Institution of Engineers Australia.

Larson, H. (1978). Reverberation at low frequencies. Brüel and Kjær, Copenhagen, Denmark, Technical Review No. 4.

Leav, O., Cazzolato, B., and Howard, C. (2021). Sound propagation through elevated, heated jets in cooler cross-flow: An experimental study. *Journal of the Acoustical Society of America*, 150(1):82–93.

Leav, O. Y. J. (2020). *An investigation into the mechanics of sound propagation through turbulent non-isothermal exhaust jets in cross-flow*. PhD thesis, School of Mechanical Engineering, The University of Adelaide, Adelaide, South Australia, Australia.

Lee, S.-H. and Ih, J.-G. (2003). Empirical model of the acoustic impedance of a circular orifice in grazing mean flow. *The Journal of the Acoustical Society of America*, 114(1):98–113.

Leissa, A. W. (1969). Vibration of plates. Technical Report NASA SP-160, NASA. Reprinted by the Acoustical Society of America, 1993.

Leppington, F. G., Broadbent, E. G., and Heron, K. H. (1982). The acoustic radiation efficiency of rectangular panels. *Proceedings of the Royal Society of London*, A382:245–271.

Leppington, F. G., Heron, K. H., Broadbent, E. G., and Mead, S. M. (1987). Resonant and non-resonant acoustic properties of elastic panels. II. The transmission problem. *Proceedings of the Royal Society of London*, A412:309–337.

Levine, H. and Schwinger, J. (1948). On the radiation of sound from an unflanged circular pipe. *Physics Review*, 73:383–406.

Levy, M., editor (2001). *Handbook of Elastic Properties of Solids, Liquids and Gases, Volumes II and III.* Academic Press, London, UK.

Li, D. (2003). *Vibroacoustic behaviour and noise control studies of advanced composite structures.* PhD thesis, University of Pittsburgh, Pittsburgh, PA, USA.

Li, M., W., v., Tijs, E., van de Ven, M., and Molenaar, A. (2015). Sound absorption measurement of road surface with in situ technology. *Applied Acoustics*, 88:12–21.

Li, X. and Hansen, C. H. (2005). Comparison of models for predicting the transmission loss of plenum chambers. *Applied Acoustics*, 66:810–828.

Liao, G., Luan, C., Wang, Z., Liu, J., Yao, X., and Fu, J. (2021). Acoustic metamaterials: A review of theories, structures, fabrication approaches and applications. *Advanced Materials Technologies*, 6:2000787.

Lighthill, J. M. (1952). On sound generated aerodynamically. *Proceedings of the Royal Society, London*, A211:564.

Lighthill, J. M. (1991). Biomechanics of hearing sensitivity. *Journal of Vibration and Acoustics*, 113:1–13.

Lighthill, J. M. (1996). Recent advances in interpreting hearing sensitivity. In *Proceedings of the Fourth ICSV*, St. Petersburg, Russia. International Institute of Acoustics and Vibration.

Liu, Z., Rumpler, R., and Feng, L. (2018). Broadband locally resonant metamaterial sandwich plate for improved noise insulation in the coincidence region. *Composite Structures*, 200:165–172.

Ljung, L. (1999). *System Identification - Theory for the User.* Prentice Hall, Englewood Cliffs, NJ, USA, second edition.

Ljunggren, S. (1991). Airborne sound insulation of thick walls. *Journal of the Acoustical Society of America*, 89:2238–2345.

Londhe, N., Rao, M. D., and Blough, J. R. (2009). Application of the iso 13472-1 in situ technique for measuring the acoustic absorption coefficient of grass and artificial turf surfaces. *Applied Acoustics*, 70:129–141.

Long, M. (2014). *Architectural Acoustics.* Academic Press, Oxford, UK, second edition.

Los Alamos National Laboratories (2012). DeltaEC: Design environment for low-amplitude thermoacoustic energy conversion. Software, `http://www.lanl.gov/thermoacoustics/DeltaEC.html`. version 6.3b11 (Windows, 18-Feb-12).

Lubman, D. (1969). Fluctuations of sound with position in a reverberation room. *Journal of the Acoustical Society of America*, 44:1491–1502.

Lubman, D. (1974). Precision of reverberant sound power measurement. *Journal of the Acoustical Society of America*, 56:523–533.

Lyman, T., editor (1961). *Metals Handbook.* American Society for Metals, Metals Park, OH, USA.

Lyon, R. H. (1975). *Statistical Energy Analysis of Dynamical Systems: Theory and Application.* MIT Press, Cambridge, MA, USA.

Lyon, R. H. (1995). Statistical energy analysis and structural fuzzy. *The Journal of the Acoustical Society of America*, 97(5):2878–2881.

Lyon, R. H. (2000). *Designing for Product Sound Quality.* Marcel Dekker, New York, NY, USA.

Lyon, R. H. and DeJong, R. G. (1995). *Theory and Application of Statistical Energy Analysis.* Elsevier Inc., London, UK, second edition.

Lyon, R. H. and Maidanik, G. (1962). Power flow between linearly coupled oscillators. *Journal of the Acoustical Society of America*, 34(5):623–639.

Lyons, G. W., Hart, C. R., and Raspet, R. (2021). As the wind blows: turbulent noise on outdoor microphones. *Acoustics Today*, 17(4):20–28.

Maa, D. (1998). Potential of microperforated panel absorber. *Journal of the Acoustical Society of America*, 104:2861–2866.

Mace, B. (2003). Statistical energy analysis, energy distribution models and system modes. *Journal of Sound and Vibration*, 264(2):391–409.

Mackenzie, R. (1979). *Auditorium Acoustics.* Elsevier Applied Science, London, UK.

Macrae, J. H. (1991). Presbycusis and noise-induced permanent threshold shift. *Journal of the Acoustical Society of America*, 90:2513–2516.

Maekawa, Z. (1968). Noise reduction by screens. *Applied Acoustics*, 1(3):157–173.

Maekawa, Z. (1977). Shielding highway noise. *Noise Control Engineering*, 9(1):38–44.

Maekawa, Z. (1985). Simple estimation methods for noise reduction by various shaped barriers. In *Conference on Noise Control Engineering*, Krakow, Poland.

Magnusson, P. C. (1965). *Transmission Lines and Wave Propagation.* Allyn and Bacon, Boston, MA, USA.

Maidanik, G. (1962). Response of ribbed panels to reverberant acoustic fields. *Journal of the Acoustical Society of America*, 34:809–826.

Malbéqui, P., Rozenberg, Y., and Bulté, J. (2009). Aircraft noise prediction in the iesta program. In *Proceedings of the European Conference for Aerospace Sciences, EUCASS*, Versailles, France.

Mammano, F. and Nobili, R. (1993). Biophysics of the cochlea: Linear approximation. *Journal of the Acoustical Society of America*, 93:3320–3332.

Manning, C. J. (1981). The propagation of noise from petroleum and petrochemical complexes to neighbouring communities. Technical Report 4/81, CONCAWE.

Marburg, S. and Nolte, B. (2008). *Computational Acoustics of Noise Propagation in Fluids: Finite and Boundary Element Methods.* Springer-Verlag, Berlin.

Marsh, K. J. (1982). The CONCAWE model for calculating the propagation of noise from open-air industrial plants. *Applied Acoustics*, 15:411–428.

MATLAB scripts for ENC (2024). Software accompanying this book, *Engineering Noise Control.* Available from `https://set.adelaide.edu.au/mechanical-engineering/our-research/acoustics-vibration-and-control#publications-and-software` and `http://www.causalsystems.com`.

May, D. N. and Osman, M. M. (1980). Highway noise barriers: New shapes. *Journal of Sound and Vibration*, 71:73–101.

McGary, M. C. (1988). A new diagnostic method for separating airborne and structure-borne noise radiated by plates with application to propeller aircraft. *Journal of the Acoustical Society of America*, 84:830–840.

McLachlan, N. W. (1941). *Bessel Functions for Engineers.* Oxford University Press, London, UK.

Mead, D. J. (1998). *Passive Vibration Control.* John Wiley & Sons, New York, NY, USA.

Mechel, F. P. (1986). The acoustic sealing of holes in walls. *Journal of Sound and Vibration*, 111:297–336.

Mechel, F. P. (1994). Schalldämpfer. In Heckl, M. and Müller, H., editors, *Taschenbuch der Technischen Akustik*, pages 501–546. Springer, Berlin, Heidelberg.

Mellow, T. and Kärkkäinen, L. (2016). Expansions for the radiation impedance of a rectangular piston in an infinite baffle. *Journal of the Acoustical Society of America*, 140:2867.

Memoli, G., Chisari, L., Eccles, J. P., Caleap, M., Drinkwater, B. W., and Subramanian, S. (2019). Vari–sound: A varifocal lens for sound. In *CHI Conference on Human Factors in Computing Systems*, Glasgow, Scotland.

Menge, C. W., Rossano, C. F., Anderson, G. S., and Bajdek, C. J. (1998). *FHWA Traffic Noise Model, Version 1.0, Technical Manual.* US Dept. Transportation, Washington, DC, USA.

Menounou, P. (2001). A correction to Maekawa's curve for the insertion loss behind barriers. *Journal of the Acoustical Society of America*, 110:1828–1838.

Meyer, E., Mechel, F., and Kurtez, G. (1958). Experiments on the influence of flow on sound attenuation in absorbing ducts. *Journal of the Acoustical Society of America*, 30:165–174.

Meyer, E. and Neumann, E. (1972). *Physical and Applied Acoustics.* Academic Press, New York, NY, USA. Chapter 5. Translated by J. M. Taylor.

Michelsen, R., Fritz, K. R., and Sazenhofan, C. V. (1980). Effectiveness of acoustic pipe wrappings (in German). In *Proceedings of DAGA '80*, pages 301–304, Berlin. VDE-Verlag.

MIL-STD-1474E (2015). Design criteria standard noise limits. Technical report, Department of Defense, USA.

Møller, H. and Pedersen, C. S. (2004). Hearing at low and infrasonic frequencies. *Noise and Health*, 6(23):37.

Moore, B. C. J. (1982). *Introduction to the Psychology of Hearing.* Academic Press, New York, NY, USA, second edition.

Moore, B. C. J. and Glasberg, B. R. (2007). Modeling binaural loudness. *Journal of the Acoustical Society of America*, 121:1604–1612.

Moore, B. C. J., Glasberg, B. R., and Baer, T. (1997). A model for the prediction of thresholds, loudness and partial loudness. *Journal of the Audio Engineering Society*, 45:224–240.

Moreland, J. and Minto, R. (1976). An example of in-plant noise reduction with an acoustical barrier. *Applied Acoustics*, 9:205–214.

Morse, P. M. (1939). The transmission of sound inside pipes. *Journal of the Acoustical Society of America*, 11:205–210.

Morse, P. M. (1948). *Vibration and Sound.* McGraw-Hill, New York, NY, USA, second edition.

Morse, P. M. and Bolt, R. H. (1944). Sound waves in rooms. *Reviews of Modern Physics*, 16:65–150.

Morse, P. M. and Ingard, K. U. (1968). *Theoretical Acoustics.* McGraw-Hill, New York, NY, USA.

Mouratidis, E. and Becker, J. (2003). The aero-acoustic properties of common HVAC plena. Technical Report RP-1026, ASHRAE.

Mouratidis, E. and Becker, J. (2004). The acoustic properties of common HVAC plena. Part 2. *ASHRAE Transactions*, 110:597–606.

MSHA (2021). Subchapter M, Part 62. Occupational noise exposure. In *Title 30 Mineral Resources*, page 445. Mining Safety and Health Admnistration, US Government.

Mungar, P. and Gladwell, G. M. L. (1968). Wave propagation in a sheared fluid contained in a duct. *Journal of Sound and Vibration*, 9:28–48.

Munjal, M. L. (2008). Muffler acoustics. In Mechel, F., editor, *Formulas of Acoustics*, chapter K, pages 793–841. Springer-Verlag, Berlin, Heidelberg, New York, NY, USA, second edition.

Munjal, M. L. (2013). *Noise and Vibration Control*, volume 3 of *IISc Lecture Notes.* World Scientific Publishing Company, IISc Press, Singapore, first edition.

Munjal, M. L. (2014). *Acoustics of Ducts and Mufflers.* John Wiley & Sons, West Sussex, UK, second edition.

Munjal, M. L., Galaitsis, A. G., and Vér, I. L. (2006). Passive silencers. In Vér, I. L. and Beranek, L. L., editors, *Noise and Vibration Control Engineering: Principles and Applications*, chapter 9. John Wiley and Sons Inc., Hoboken, NJ, USA, second edition.

Murphy, W. J., Gong, W., and Karch, S. J. (2022). Personal attenuation ratings versus derated noise reduction ratings for hearing protection devices. *Journal of the Acoustical Society of America*, 152:1074–1089.

Murphy, W. J., Gong, W., Karch, S. J., Federman, J., and Schulz, T. Y. (2023). Erratum: Personal attenuation ratings versus derated noise reduction ratings for hearing protection devices [J. Acoust. Soc. Am. 152(2), 1074–1089 (2022)]. *Journal of the Acoustical Society of America*, 155:50.

Muster, D. and Plunkett, R. (1988). Isolation of vibrations. In Beranek, L. L., editor, *Noise and Vibration Control*, chapter 13. Institute of Noise Control Engineering, Washington, DC, USA, revised edition.

Mustin, G. S. (1968). *Theory and Practice of Cushion Design.* The Shock and Vibration Information Centre, US Department of Defense.

Muzet, A. (2007). Environmental noise, sleep and health. *Sleep Medicine Reviews*, 11(2):135–142.

Nahin, P. J. (1990). Oliver Heaviside. *Scientific American*, 262:122–129.

National Instruments (2022). Measuring strain with strain gages. `https://www.ni.com/en-au/innovations/white-papers/07/measuring-strain-with-strain-gages.html`. Last accessed on July 1, 2022.

Naylor, G. (1993). ODEON: Another hybrid room acoustical model. *Applied Acoustics*, 38:131–143.

Nederlands Ministerie volkshuisvesting, ruimtelijke ordening en milieubeheer (in Dutch) (1996). Reken-en meetvoorschrift railverkeerslawaai. Directie Geluid en Verkeer, The Hague, Netherlands.

NEMA (2014). Dry type transformers for general applications. Technical Report NEMA ST 20-2014, National Electrical Manufacturer's Association.

NEMA (2019). Transformer regulators and reactors. Technical Report NEMA TR 1-2013 (R2019), National Electrical Manufacturer's Association.

Neubauer, R. O. (2000). Estimation of reverberation times in non-rectangular rooms with non-uniformly distributed absorption using a modified Fitzroy equation. In *Proceedings of the 7th International Congress on Sound and Vibration*, pages 1709–1716, Garmisch–Partenkirchen, Germany. International Institute of Acoustics and Vibration.

Neubauer, R. O. (2001). Existing reverberation time formulae – A comparison with computer simulated reverberation times. In *Proceedings of the 8th International Congress on Sound and Vibration*, pages 805–812, Hong Kong. International Institute of Acoustics and Vibration.

Ng, K. W. (1980). Control valve noise generation and prediction. In *Proceedings of Noisexpo, National Noise and Vibration Control Conference*, pages 49–54, Chicago.

Ngo, T. T. and Letchford, C. W. (2009). Experimental study of topographic effects on gust wind speed. *Journal of Wind Engineering and Industrial Aerodynamics*, 97(9–10):426–438.

Nielsen, H. L. (1996). *Rail Traffic Noise: The Nordic Prediction Method.* Nordic Council of Ministers, Copenhagen, Denmark.

Nilsson, A. (2001). Wave propagation and sound transmission in sandwich composite plates. In *Proceedings of the 8th International Congress on Sound and Vibration*, pages 61–70, Hong Kong. International Institute of Acoustics and Vibration.

NIOSH (1998). Subchapter M, Part 62. Occupational noise exposure. In *Criteria for a Recommended Standard: Occupational Noise Exposure, Revised Criteria*, pages 1–3. US Department of Health and Human Services.

NIOSH (2005). Method for calculating and using the Noise Reduction Rating - NRR. Technical report, U.S. Department of Health and Human Services, Public Health Service, Centers for Disease Control and Prevention, National Institute for Occupational Safety and Health, Cincinnati, OH, USA. PDF file.

NIOSH (2018). NIOSH criteria for a recommended standard: Occupational noise exposure. NIOSH web site. `https://www.cdc.gov/niosh/topics/noise/preventhearingloss/hearlosspreventprograms.html` , Accessed February 18, 2022.

NoiseMap Ltd (2016). Noisemap environmental noise mapping software. `http://www.noisemap.ltd.uk/wpress/`. Last accessed on July 1, 2022.

Nordtest (1999). NT-ACOU-104: Ground surfaces: Determination of the acoustic impedance. Nordic Innovation Centre, Espoo, Finland, 1999.

Norton, M. P. and Karczub, D. G. (2003). *Fundamentals of Noise and Vibration Analysis for Engineers.* Cambridge University Press, New York, NY, USA, second edition.

Oerlemans, S. (2013). Wind turbine amplitude modulation: Research to improve understanding as to its cause & effect. Technical report, Renewable UK.

Ohayon, R. (2001). Reduced symmetric models for modal analysis of internal structural-acoustic and hydroelastic-sloshing systems. *Computer Methods in Applied Mechanics and Engineering*, 190(24-25):3009–3019.

Ollerhead, J., Rhodes, D., Viinikainen, M., Monkman, D., and Woodley, A. (1999). The UK civil aircraft noise contour model ANCON: Improvements in version 2. Technical Report 9842, ERCD Dept., Civil Aviation Authority (CAA).

Olsen, E. S. (2005). Acoustical solutions in the design of a measurement microphone for surface mounting. Technical Report Technical Review No. 1, Brüel and Kjær, Copenhagen, Denmark.

OSHA (1995). 1910.95 – Occupational noise exposure. In *Occupational Safety and Health Standards: 1910 Subpart G, Occupational Health and Environmental Control.* US Government.

OSHA (2016). Appendix IV:C. Methods for estimating HPD attenuation. OSHA Instruction TED 01-00-015, Section III: Chapter 5.

Oud, M. (2013). Explanation for suffering from low-frequency sound. *Geluid.* English translation, `http://home.kpn.nl/Oud/publications/2013_03_OudM_journalGeluid.pdf`.

Pan, J. and Bies, D. A. (1988). An experimental investigation into the interaction between a sound field and its boundaries. *Journal of the Acoustical Society of America*, 83:1436–1444.

Pan, J. and Bies, D. A. (1990a). The effect of a semi-circular diffuser on the sound field in a rectangular room. *Journal of the Acoustical Society of America*, 88:1454–1458.

Pan, J. and Bies, D. A. (1990b). The effect of fluid-structural coupling on acoustical decays in a reverberation room in the high-frequency range. *Journal of the Acoustical Society of America*, 87:718–727.

Pan, J. and Bies, D. A. (1990c). The effect of fluid-structural coupling on sound waves in an enclosure: Experimental part. *Journal of the Acoustical Society of America*, 87:708–717.

Pan, J. and Bies, D. A. (1990d). The effect of fluid-structural coupling on sound waves in an enclosure: Theoretical part. *Journal of the Acoustical Society of America*, 87:691–707.

Panton, R. L. and Miller, J. M. (1975). Resonant frequencies of cylindrical Helmholtz resonators. *Journal of the Acoustical Society of America*, 57:1533–1535.

Panuszka, R., Wiciak, J., and Iwaniek, M. (2005). Experimental assessment of coupling loss factors of thin rectangular plates. *Archives of Acoustics*, 30(4):533–551.

Passchier-Vermeer, W. (1968). Hearing loss due to exposure to steady state broadband noise. Technical Report 36, Institute for Public Health Engineering, Netherlands.

Passchier-Vermeer, W. (1977). Hearing levels of non-noise exposed subjects and of subjects exposed to constant noise during working hours. Technical Report B367, Research Institute for Environmental Hygiene, Netherlands.

Pavic, G. (2006). Experimental identification of physical parameters of fluid-filled pipes using acoustical signal processing. *Applied Acoustics*, 67:864–881.

Peat, K. S., Ih, J.-G., and Lee, S.-H. (2003). The acoustic impedance of a circular orifice in grazing mean flow: Comparison with theory. *The Journal of the Acoustical Society of America*, 114(6):3076–3086.

Pedersen, E., van den Berg, F., Bakker, R., and Bouma, J. (2009). Response to noise from modern wind farms in The Netherlands. *Journal of the Acoustical Society of America*, 126(2):634–643.

Peng, S. Z. and Khoo, T. C. P. (2007). Statistical energy analysis (SEA) for energy attenuation factors and coupling loss factors in asymmetric periodic plates. *International Journal of Vehicle Noise and Vibration*, 3(3):263–285.

Peters, H., Kurze, U. J., and Nürnberger, H. (2010). Model tests on flow noise in exhaust stacks of gas turbine power stations. In *36th Annual Meeting of the German Acoustical Society*, Berlin, Germany.

Peterson, B. and Plunt, J. (1982). On effective mobilities in the prediction of structure-borne sound transmission between a source structure and receiving structure: Parts 1 and 2. *Journal of Sound and Vibration*, 82:5171.

Pierce, A. D. (1981). *Acoustics: An Introduction to its Physical Principles and Applications.* McGraw-Hill, New York, NY, USA. Chapter 5.

Piercy, J., Embleton, T., and Sutherland, L. (1977). Review of noise propagation in the atmosphere. *Journal of the Acoustical Society of America*, 61:1403–1418.

Pietrzko, S. and Bütikofer, R. (2002). FLULA - Swiss aircraft noise prediction program. In *Proceedings of Acoustics 2002, Annual Conference of the Australian Acoustical Society*, pages 92–99, Adelaide, SA, Australia. Australian Acoustical Society.

Pinnington, R. J. and White, R. G. (1981). Power flow through machine isolators to resonant and non-resonant beams. *Journal of Sound and Vibration*, 75:179–197.

Piper, B., Barham, R., and Sotirakopoulos, K. (2015). Design for a self-calibrating microphone system using a cluster of mems microphones. In *Proceedings of the 22nd International Congress on Sound and Vibration*, Florence, Italy. International Institute of Acoustics and Vibration.

Plovsing, B. (2006a). Comprehensive outdoor sound propagation model. Part 1: Propagation in an atmosphere without significant refraction. Technical Report AV 1849/00, Delta.

Plovsing, B. (2006b). Comprehensive outdoor sound propagation model. Part 2: Propagation in an atmosphere with refraction. Technical Report AV1851/00, Delta.

Plovsing, B. (2014). Proposal for Nordtest Method: Nord2000–prediction of outdoor sound propagation. Technical report, DELTA Acoustics,.

Plunkett, R. (1954). Experimental measurement of mechanical impedance or mobility. *Journal of Applied Mechanics*, 21:256.

Plunkett, R. (1958). Interaction between a vibrating machine and its foundation. *Noise Control*, 4:18–22.

Pobol, O. (1976). Method of measuring noise characteristics of textile machines. *Measurement Techniques, USSR*, 19:1736–1739.

Porter, M. B. (2011). The bellhop manual and userâĂŹs guide: Preliminary draft. Technical report, Heat, Light, and Sound Research, Inc.

Prasad, M. and Crocker, M. (1983). Studies of acoustical performance of a multi-cylinder engine exhaust muffler system. *Journal of Sound and Vibration*, 90(4):491–508.

Prasad, M. G. and Crocker, M. J. (1981). Insertion loss studies on models of automotive exhaust systems. *The Journal of the Acoustical Society of America*, 70(5):1339–1344.

Press, W. H., Flannery, B. P., Tenkolsky, S. A., and Vettering, W. T. (1986). *Numerical Recipes: The Art of Scientific Computing.* Cambridge University Press, Cambridge, UK, third edition.

Quirt, J. D. (1982). Sound transmission through windows I: Single and double glazing. *Journal of the Acoustical Society of America*, 72:834–844.

Quirt, J. D. and Nightingale, T. (2008). Airborne sound insulation in multi-family buildings. Technical report, Institute for Research in Construction, Canadian National Research Council, Canada.

Quirt, J. D., Nightingale, T., and King, F. (2008). Guide for sound insulation in wood frame construction. Technical Report 66, Institute for Research in Construction, Canadian National Research Council, Canada.

Ramakrishnan, R. and Watson, W. R. (1992). Design curves for rectangular splitter silencers. *Applied Acoustics*, 35:1–24.

Rämmal, H. and Bodén, H. (2007). Modified multi-load method for nonlinear source characterisation. *Journal of Sound and Vibration*, 299(4-5):1094–1113.

Ramya, E. and Munjal, M. (2014). Improved tuning of the extended concentric tube resonator for wide-band transmission loss. *Noise Control Engineering Journal*, 62(4):252–263.

Randall, R. B. (1987). *Frequency Analysis.* Brüel and Kjær, Copenhagen, third edition.

Raney, J. P. and Cawthorn, J. M. (1998). Aircraft noise. In Harris, C. M., editor, *Handbook of Acoustical Measurements and Noise Control.* Acoustical Society of America, New York, NY, USA, third edition.

Rao, K. N. and Munjal, M. (1986). Experimental evaluation of impedance of perforates with grazing flow. *Journal of Sound and Vibration*, 108(2):283–295.

Rao, S. S. (2016). *Mechanical Vibrations.* Pearson, Reading, MA, USA, sixth edition.

Raspet, R., L'Esperance, A., and Daigle, G. A. (1995). The effect of realistic ground impedance on the accuracy of ray tracing. *Journal of the Acoustical Society of America*, 97:683–693.

Raspet, R. and Wu, W. (1995). Calculation of average turbulence effects on sound propagation based on the fast field program formulation. *Journal of the Acoustical Society of America*, 97:147–153.

Rathe, E. J. (1969). Note on two common problems of sound propagation. *Journal of Sound and Vibration*, 10:472–479.

Renji, K. (2004). On the number of modes required for statistical energy analysis-based calculations. *Journal of Sound and Vibration*, 269(3–5):1128–1132.

Reynolds, D. (1981). *Engineering Principles of Acoustics: Noise and Vibration Control.* Allyn & Bacon Inc.

Rice, C. G. (1974). Damage risk criteria for impulse and impact noise. In Crisp, J. D. C., editor, *Noise Shock and Vibration Conference Proceedings*, pages 29–38. Institution of Engineers, Australia, Monash University, Clayton, Victoria, Australia.

Rice, C. G. and Martin, A. M. (1973). Impulse noise damage risk criteria. *Journal of Sound and Vibration*, 28:359–367.

Richards, E. J. (1979). On the prediction of impact noise. Part 2, ringing noise. *Journal of Sound and Vibration*, 65:419–451.

Richards, E. J. (1980). Vibration and noise relationships: Some simple rules for the machinery engineer. *Journal of the Acoustical Society of America*, 68:Supplement 1, S23.

Richards, E. J. (1981). On the prediction of impact noise. Part 3, energy accountancy in industrial machines. *Journal of Sound and Vibration*, 76:187–232.

Rindel, J. H. and Hoffmeyer, D. (1991). Influence of stud distance on sound insulation of gypsum board walls. In *Proceedings of Internoise '91*, pages 279–282. Institute of Noise Control Engineering.

Roe, G. M. (1941). Frequency distribution of normal modes. *Journal of the Acoustical Society of America*, 13(1):1–7.

Royster, L. H., Royster, J. D., and Thomas, W. G. (1980). Representative hearing levels by race and sex in North Carolina industry. *Journal of the Acoustical Society of America*, 68:551–566.

Rudnick, I. (1947). The propagation of an acoustic wave along a boundary. *Journal of the Acoustical Society of America*, 19:348–356.

Rudnik, I. (1957). Propagation of sound in the open air. In Harris, C. M., editor, *Handbook of Noise Control*, chapter 3. McGraw-Hill, New York, NY, USA.

Sabine, W. C. (1993). *Collected Papers on Acoustics.* American Institute of Physics, New York, NY, USA.

Sablik, M. J. (1985). Statistical energy analysis, structural resonances and beam networks. *Journal of the Acoustical Society of America*, 77:1038–1045.

Sakamoto, S., Akiba, Y., Iizuka, R., Watanabe, S., Kominami, R., and Shintani, T. (2022). Experimental study on open-end correction of a helmholtz resonator narrow slit neck. *Noise Control Engineering Journal*, 70(2):128–135.

Salomons, E., van Maercke, D., Defrance, J., and de Roo, F. (2011). The harmonoise sound propagation model. *Acta Acustica United with Acustica*, 97:62–74.

Salomons, E. M. and Janssen, S. A. (2011). Practical ranges of loudness levels of various types of environmental noise, including traffic noise, aircraft noise, and industrial noise. *International Journal of Environmental Research and Public Health*, 8(6):1847–1864.

Salt, A. N. and Hullar, T. E. (2010). Responses of the ear to low frequency sounds, infrasound and wind turbines. *Hearing Research*, 268(1):12–21.

Salt, A. N. and Lichtenhan, J. T. (2014). How does wind turbine noise affect people? *Acoustics Today*, 10:21–28.

Sandberg, U. (2001). Noise emissions of road vehicles: Effect of regulations. Technical Report 01-1, I-INCE Working Party on Noise Emissions of Road Vehicles: International Institute of Noise Control Engineering.

Sanders, C., Wilkins, A., and Walsh, T. (2018). Design of continuously graded elastic acoustic cloaks. *Journal of the Acoustical Society of America*, 43(5):EL31–EL36.

Saunders, R. E., Samuels, S. E., Leach, R., and Hall, A. (1983). An Evaluation of the U.K. DoE Traffic Noise Prediction Method. Technical Report ARR No. 122, Australian Road Research Board, Vermont South, VIC., Australia.

SCHALL 03 (2006). Richtlinie zur Berechnung der Schallimmissionen von Eisenbahnen und Straßenbahnen (Guidelines for the calculation of sound immission near railroads and tram lines), Draft 2006. German Ministry of Transport.

Scharf, B. (1970). Critical bands. In Tobias, J. V., editor, *Foundations of Modern Auditory Theory*. Academic Press, New York, NY, USA.

Schenck, H. (1968). Improved integral formulation for acoustic radiation problem. *Journal of the Acoustical Society of America*, 44:41–58.

Scheuren, J. (2017). Iso 532 – is all well if it ends well? working with alternative loudness standards. In *Internoise 2017*, Hong Kong.

Schmidt, H. (1985a). Resolution bias errors in spectral density, frequency response and coherence function measurements, I: general theory. *Journal of Sound and Vibration*, 101(3):347–362.

Schmidt, H. (1985b). Resolution bias errors in spectral density, frequency response and coherence function measurements, III: application to second-order systems (white noise excitation). *Journal of Sound and Vibration*, 101(3):377–404.

Schomer, P. (2000). Proposed revisions to room noise criteria. *Noise Control Engineering Journal*, 48:85–96.

Schroeder, M. R. (1969). Effect of frequency and space averaging on the transmission responses of multimode media. *Journal of the Acoustical Society of America*, 46:277–283.

Selamet, A. and Ji, Z. L. (2000). Circular asymmetric Helmholtz resonators. *Journal of the Acoustical Society of America*, 107:2360–2369.

Semple, E. C. and Hall, R. E. I. (1981). Mechanisms of noise generation in punch presses and means by which this noise can be reduced. In *Proceedings of the Annual Conference of the Australian Acoustical Society*. Australian Acoustical Society.

Sendra, J. J. (1999). *Computational Acoustics in Architecture*. WIT, Southampton, UK.

Seybert, A. and Han, J. (2008). Measurement of pavement absorption using iso 13472-2. In *Proceedings of Acoustics 08*, pages 7595–7599, Paris, France. European Acoustics Association.

Sharp, B. H. (1973). A study of techniques to increase the sound installation of building elements. Technical Report WR 73-S, Wylie Laboratories.

Sharp, B. H. (1978). Prediction methods for the sound transmission of building elements. *Noise Control Engineering*, 11:53–63.

Shorter, P. and Langley, R. (2005). Vibro-acoustic analysis of complex systems. *Journal of Sound and Vibration*, 288(3):669–699.

Simonds, H. R. and Ellis, C. (1943). *Handbook of Plastics.* D. Van Nostrand Company Inc, New York, NY, USA.

Singh, S., Hansen, C. H., and Howard, C. Q. (2006). The elusive cost function for tuning adaptive Helmholtz resonators. In *Proceedings of the First Australasian Acoustical Societies' Conference*, pages 75–82, Christchurch, New Zealand. New Zealand Acoustical Society.

Singh, S., Hansen, C. H., and Howard, C. Q. (2008). A detailed tutorial for evaluating in-duct net acoustic power transmission in a circular duct with an attached cylindrical Helmholtz resonator using transfer matrix method. In *Acoustics 2008*, pages 1–9, Geelong, Victoria, Australia. Australian Acoustical Society, Australian Acoustical Society. Paper 9.

Singhal, V. (1976). *Acoustics of ducts with flow and its relation to acoustically induced valve-pipe instabilities.* PhD thesis, Deptartment of Aeronautics and Astronautics, Massachusetts Institute of Technology.

Skudrzyk, E. (1971). *Foundations of Acoustics.* Springer-Verlag/Wien, New York, USA.

Smith, B. J. (1971). *Acoustics.* Longman, London, UK.

Smith, S. D., Nixon, C. W., and von Gierke, H. E. (2006). Damage risk criteria for hearing and human body vibration. In Vér, I. L. and Beranek, L. L., editors, *Noise and Vibration Control Engineering: Principles and Applications*, chapter 19. John Wiley and Sons Inc., Hoboken, NJ, USA, second edition.

Smits, J. M. A. and Kosten, C. W. (1951). Sound absorption by slit resonators. *Acta Acustica*, 1(3):114–122.

Smollen, L. E. (1966). Generalized matrix method for the design and analysis of vibration isolation systems. *Journal of the Acoustical Society of America*, 40:195–204.

Snowdon, J. C. (1965). Rubber-like materials, their internal damping and role in vibration isolation. *Journal of Sound and Vibration*, 2:175–193.

Söderqvist, S. (1982). A quick and simple method for estimating the transmission or insertion loss of an acoustic filter. *Applied Acoustics*, 15:347–354.

Soedel, W. (2004). *Vibrations of Plates and Shells.* Marcel Dekker, New York, NY, USA, third edition.

Søndergaard, B. (2013). Low frequency noise from wind turbines: Do the Danish regulations have any impact? In *5th International Meeting on Wind Turbine Noise*, pages 28–30, Denver, Colorado.

Søndergaard, L. S. (2014). Noise and low frequency noise from wind turbines. In *Internoise 2014*, Melbourne, Australia.

Song, B. H. and Bolton, J. S. (2000). A transfer-matrix approach for estimating the characteristic impedance and wave numbers of limp and rigid porous materials. *Journal of the Acoustical Society of America*, 107:1131–1152.

Soom, A. and Lee, M. (1983). Optimal design of linear and nonlinear vibration absorbers for damped systems. *Journal of Vibration, Acoustics, Stress and Reliability in Design*, 105:112–119.

Spoendlin, H. (1975). Neuroanatomical basis of cochlear coding mechanisms. *Audiology*, 14:383–407.

Starr, E. A. (1988). Sound and vibration transducers. In Beranek, L. L., editor, *Noise and Vibration Control*, chapter 3. Institute of Noise Control Engineering, Washington, DC, USA, revised edition.

Steele, C. (2001). A critical review of some traffic noise prediction models. *Applied Acoustics*, 62:271–287.

Stevens, S. S. (1957). On the psychophysical law. *Psychological Review*, 64:153–181.

Stevens, S. S. (1961). Procedure for calculating loudness: Mark VI. *Journal of the Acoustical Society of America*, 33:1577–1585.

Stevens, S. S. (1972). Perceived level of noise by Mark VII and decibels (E). *Journal of the Acoustical Society of America*, 51:575–601.

Stull, R. B. (1988). *An Introduction to Boundary Layer Meteorology*. Kluwer Academic Publishers, Dordrecht, The Netherlands.

Sujith, R. (1996). Transfer matrix of a uniform duct with an axial mean temperature gradient. *The Journal of the Acoustical Society of America*, 100(4):2540–2542.

Sujith, R., Waldherr, G., and Zinn, B. (1995). An exact solution for one-dimensional acoustic fields in ducts with an axial temperature gradient. *Journal of Sound and Vibration*, 184(3):389–402.

Sullivan, J. W. (1979a). A method for modeling perforated tube muffler components. I. Theory. *The Journal of the Acoustical Society of America*, 66(3):772–778.

Sullivan, J. W. (1979b). A method for modeling perforated tube muffler components. II. Applications. *The Journal of the Acoustical Society of America*, 66(3):779–788.

Sullivan, J. W. and Crocker, M. J. (1978). Analysis of concentric-tube resonators having unpartitioned cavities. *The Journal of the Acoustical Society of America*, 64(1):207–215.

Sun, X., Jing, X., Zhang, H., and Shi, Y. (2002). Effect of grazing-bias flow interaction on acoustic impedance of perforated plates. *Journal of Sound and Vibration*, 254(3):557–573.

Sutton, M. (1990). Noise directivity of exhaust stacks (Honours degree thesis). Technical report, School of Mechanical Engineering, University of Adelaide, South Australia.

Swinbanks, M. A. (2015). Supplementary submission to the Senate Wind Turbine Select Committee.

Szechenyi, E. (1971). Modal densities and radiation efficiencies of unstiffened cylinders using statistical methods. *Journal of Sound and Vibration*, 19(1):65–81.

Tadeu, A. J. B. and Mateus, D. M. R. (2001). Sound transmission through single, double and triple glazing. Experimental evaluation. *Applied Acoustics*, 62:307–325.

Takagi, K. and Yamamoto, K. (1994). Calculation methods for road traffic noise propagation proposed by ASJ. In *Proceedings of Internoise 94*, pages 289–294, Yokohama, Japan. Institute of Noise Control Engineering.

Takatsubo, J., Ohno, S., and Suzuki, T. (1983). Calculation of the sound pressure produced by structural vibration using the results of vibration analysis. *Bulletin of the Japanese Society of Mechanical Engineers*, 26:1970–1976.

Tarnow, V. (2002). Measured anisotropic air flow resistivity and sound attenuation of glass wool. *Journal of the Acoustical Society of America*, 111:2735–2739.

Teague, P., Conomos, J., and Alexandrou, V. (2016). Developments in the description and assessment of high intensity impulse noise exposure. In *Acoustics 2016*, Brisbane, Australia.

Tie, T. S., Mo, K. H., Putra, A., Loo, S. C., Alengaram, U. G., and Ling, T.-C. (2020). Sound absorption performance of modified concrete: A review. *Journal of Building Engineering*, 30:1–10.

Tijs, E. H. G. (2013). *Study and Development of an In Situ Acoustic Absorption Measurement Method.* PhD thesis, University of Twente, Enschede, The Netherlands.

Tocci, G. C. and Marcus, E. N. (1978). Prediction method for noise reduction by plant buildings. In *Internoise 78*, San Francisco, USA.

Tonin, R. (1985). Noise control by barriers. Part 2: Noise reduction by an absorptive barrier. *Acoustics Australia*, 13:59–67.

Tournour, M. and Atalla, N. (2000). Pseudostatic corrections for the forced vibroacoustic response of structure-cavity system. *Journal of the Acoustical Society of America*, 107:2379–2386.

Tse, F. S., Morse, I. E., and Hinkle, R. T. (1979). *Mechanical Vibrations: Theory and Applications.* Allyn and Bacon, Boston, second edition.

TU Delft (2020). Seafloor mapping – modelling – underwater propagation – basic principles. `https://ocw.tudelft.nl/wp-content/uploads/Reader_chapter_3.pdf`, Last accessed on 2022-02-06.

Tweed, L. W. and Tree, D. R. (1976). Three methods for predicting the insertion loss of close fitting acoustical enclosures. *Noise Control Engineering*, 10:74–79.

Tyzzer, F. G. and Hardy, H. C. (1947). Properties of felt in the reduction of noise and vibration. *Journal of the Acoustical Society of America*, 19(5):872–878.

UK. DOT (1988). *Calculation of Road Traffic Noise.* HMSO, Department of Transport. London, UK.

UK. DOT (1995a). *Calculation of Railway Noise.* HMSO, Department of Transport. London, UK.

UK. DOT (1995b). *Calculation of Railway Noise. Supplement 1.* HMSO, Department of Transport. London, UK.

Ungar, E. and Zapfe, J. A. (2006a). Structural damping. In Vér, I. L. and Beranek, L. L., editors, *Noise and Vibration Control Engineering: Principles and Applications*, chapter 14. John Wiley and Sons Inc., Hoboken, NJ, USA, second edition.

Ungar, E. and Zapfe, J. A. (2006b). Vibration isolation. In Vér, I. L. and Beranek, L. L., editors, *Noise and Vibration Control Engineering: Principles and Applications*, chapter 13. John Wiley and Sons Inc., Hoboken, NJ, USA, second edition.

Ungar, E. E. (1988). Damping of panels. In Beranek, L. L., editor, *Noise and Vibration Control*, chapter 14. Institute of Noise Control Engineering, Washington, DC, USA, revised edition.

Ungar, E. E. (2019a). Sensitive equipment criteria and footfall-induced vibrations – background and problems. *Noise News International*, 27(2):13–15.

Ungar, E. E. (2019b). Vibration control: an overview and checklist. *Noise News International*, 27(3):18–19.

Ungar, E. E. (2020). Time it takes to approach steady state. *Noise News International*, 28(3):11–12.

Ungar, E. E. (2021). What floor vibration will a machine produce when it is relocated? *Noise News International*, 29(2):7–8.

Ungar, E. E. and Dietrich, C. W. (1966). High-frequency vibration isolation. *Journal of Sound and Vibration*, 4:224–241.

van Leeuwen, H. J. A. (2000). Railway noise prediction models: A comparison. *Journal of Sound and Vibration*, 231(3):975–987.

van Maercke, D. and Defrance, J. (2007). Development of an analytical model for outdoor sound propagation within the harmonoise project. *Acta Acustica United with Acustica*, 93(2):201–212.

VDI 2058 Blatt 3 (2014). Assessment of noise in the working area with regard to specific operations. Standard, The Association of German Engineers (VDI), Düsseldorf, Germany.

VDI2081-1 (2003). Noise generation and noise reduction in air-conditioning systems. Part 1. German Standard.

Vér, I. (1983). Prediction of sound transmission through duct walls: Breakout and pickup. Technical Report 5116, Bolt, Beranek and Newman Inc., Cambridge, MA, USA.

Vér, I. L. (2006). Interaction of sound waves with solid structures. In Vér, I. L. and Beranek, L. L., editors, *Noise and Vibration Control Engineering: Principles and Applications*, chapter 11. John Wiley and Sons Inc., Hoboken, NJ, USA, second edition.

Verheijen, E., Jabben, J., Schreurs, E., and Smith, K. B. (2011). Impact of wind turbine noise in The Netherlands. *Noise and Health*, 13:459–463.

Vinokur, R. Y. (1996). Evaluating sound-transmission effects in multi-layer partitions. *Sound and Vibration Magazine*, July:24–28.

Voldřich, L. (1978). Mechanical properties of basilar membrane. *Acta Otolaryngologica*, 86:331–335.

von Estorff, O., editor (2000). *Boundary Elements in Acoustics: Advances and Applications*. WIT Press, Southampton.

von Gierke, H. E. and Nixon, C. (1976). Effects of intense infrasound on man. In Tempest, W., editor, *Infrasound and Low Frequency Vibration*, pages 197–203. Academic Press, New York, NY, USA.

von Gierke, H. E., Robinson, D., and Karmy, S. J. (1982). Results of the workshop on impulse noise and auditory hazard. *Journal of Sound and Vibration*, 83:579–584.

Wallace, C. E. (1972). Radiation resistance of a rectangular panel. *Journal of the Acoustical Society of America*, 51:946–952.

Wang, C. and Lai, J. (2005). Discussions on "on the number of modes required for statistical energy analysis-based calculations". *Journal of Sound and Vibration*, 281(1–2):475–480.

Wang, C.-N. and Torng, J.-H. (2001). Experimental study of the absorption characteristics of some porous fibrous materials. *Applied Acoustics*, 62:447–459.

Ward, W. D. (1962). Damage risk criteria for line spectra. *Journal of the Acoustical Society of America*, 34:1610–1619.

Ward, W. D. (1974). The safe workday noise dose. In Crisp, J. D. C., editor, *Noise Shock and Vibration Conference Proceedings*, pages 19–28, Monash University, Clayton, Victoria, Australia.

Warnock, A. (2008). Estimation of sound transmission class and impact insulation class rating for steel framed assemblies. Technical Report RP08-7, Canadian National Research Council, Canada.

Wasmer, F. and Maunsell, F. (2003). BaseOps 7.358 user's guide. Technical report, Wasmer Consulting, Gainesville, Florida.

Watanabe, T. and Møller, H. (1990). Hearing thresholds and equal loudness contours in free field at frequencies below 1 kHz. *Journal of Low Frequency Noise Vibration and Active Control*, 9(4):135–148.

Waterhouse, R. V. (1955). Interference patterns in reverberant sound fields. *Journal of the Acoustical Society of America*, 27:247–258.

Watters, B. G., Hoover, R. M., and Franken, P. A. (1959). Designing a muffler for small engines. *Noise Control*, 5:18–22.

Watters, B. G., Labate, S., and Beranek, L. L. (1955). Acoustical behavior of some engine test cell structures. *Journal of the Acoustical Society of America*, 27:449–456.

Waugh, R. (1976). Investigation of sound level conversion as a means of rating ear protector performance. *American Industrial Hygiene Association Journal*, 37(4):239–245.

Waugh, R. (1984). Simplified hearing protector ratings - an international comparison. *Journal of Sound and Vibration*, 93(2):289–305.

Webster, J. C. (1970). Updating and interpreting the speech interference level (SIL). *Journal of the Audio Engineering Society*, 18(2):114–118.

Wells, R. J. (1958). Acoustical plenum chambers. *Noise Control*, 4:9–15.

West, M. and Parkin, P. (1978). Effect of furniture and boundary conditions on the sound attenuation in a landscaped office. *Applied Acoustics*, 11:171–218.

Wettschureck, R. G., Hauck, G., Diehl, R. J., and Willenbrink, L. (2013). Noise and vibration from railroad traffic. In Müller, G. and Möser, M., editors, *Handbook of Engineering Acoustics*, chapter 16. Springer, Berlin, Germany.

Widrow, B. (2001). A microphone array for hearing aids. *Echoes: The newsletter of the Acoustical Society of America*, 11:4–5.

Wiener, F. M. and Keast, D. D. (1959). Experimental study of the propagation of sound over ground. *Journal of the Acoustical Society of America*, 31:724–733.

Wieringa, J. (1980). Representativeness of wind observations at airports. *Bulletin of the American Meteorological Society*, 61(9):962–971.

Wieringa, J. (1992). Updating the Davenport roughness classification. *Journal of Wind Engineering and Industrial Aerodynamics*, 41(1):357–368.

Williams, C. (2015). Digital infrasound sensor user manual. Technical report, Hyperion Technology Group, Tupelo, MS, USA.

Williams, E. G. (1999). *Fourier Acoustics - Sound Radiation and Nearfield Acoustical Holography*. Academic Press, San Diego, CA, USA.

Williams, W. and Dillon, H. (2005). Hearing protector performance and standard deviation. *Noise and Health*, 7(28):51–60.

Windle, R. M. and Lam, Y. W. (1993). Prediction of the sound reduction of profiled metal cladding. In *Proceedings of Internoise '93*, Leuven, Belgium. Institute of Noise Control Engineering.

World Health Organisation (2011). Burden of disease from environmental noise. Quantification of healthy life years lost in Europe. Technical report, WHO Regional Office for Europe, Copenhagen, Denmark.

World Health Organisation (2018). Environmental noise guidelines for the European region. Technical report, WHO Regional Office for Europe, Copenhagen, Denmark.

Wu, Q. (1988). Empirical relations between acoustical properties and flow resistivity of porous plastic open-cell foam. *Applied Acoustics*, 25:141–148.

Wu, S. F. (2000a). On reconstruction of acoustic pressure fields using the Helmholtz equation least squares method. *Journal of the Acoustical Society of America*, 107:2511–2522.

Wu, T. W., editor (2000b). *Boundary Element Acoustics: Fundamentals and Computer Codes*. WIT Press, Boston, MA, USA.

Xiangyang, Z., Kean, C., and Jincai, S. (2003). On the accuracy of the ray-tracing algorithms based on various sound receiver models. *Applied Acoustics*, 64:433–441.

Yang, W., Bai, X., Zhu, W., Kiran, R., An, J., Chua, C. K., and Zhou, K. (2020). 3D printing of polymeric multi-layer micro-perforated panels for tunable wideband sound absorption. *Polymers*, 12:360–376.

Yoerkie, C. A., Moore, J. A., and Manning, J. E. (1983). Development of rotorcraft interior noise control concepts. Technical report, NASA, Hampton, VA, USA.

Yoshioka, H. (2000). Evaluation and prediction of airport noise in Japan. *Journal of the Acoustical Society of Japan (E)*, 21:341–344.

Zaporozhets, O. I. and Tokarev, V. I. (1998). Aircraft noise modelling for environmental assessment around airports. *Applied Acoustics*, 55:99–127.

Zhao, F., Manchaiah, V. K. C., French, D., and Price, S. M. (2010). Music exposure and hearing disorders: An overview. *International Journal of Audiology*, 49:54–64.

Zhao, Y.-Q., Chen, S.-Q., Chai, S., and Qu., Q.-W. (2002). An improved modal truncation method for responses to harmonic excitation. *Computers and Structures*, 80:99–103.

Zhu, P., Mo, F., and Kang, J. (2014). Experimental comparison between STI and SII metrics for the objective rating of speech intelligibility. In *Proceedings of the 21st International Congress on Sound and Vibration*, Beijing, China. International Institute of Acoustics and Vibration.

Zwicker, E. (1958). Über psychologische und methodische grundlagen der lautheirt. *Acustica*, 8:237–258.

Zwicker, E. and Scharf, B. (1965). A model of loudness summation. *Psychological Review*, 72:3–26.

Index